TRIGONOMETRY

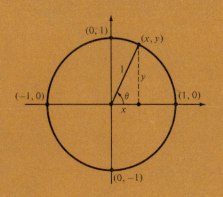

$$\sin \theta = y$$
$$\cos \theta = x$$

$$\tan \theta = \frac{y}{x}$$

$$\cot \theta = \frac{x}{y}$$

$$\sec \theta = \frac{1}{x}$$

$$\csc \theta = \frac{1}{y}$$

$$\tan \theta = \frac{\sin \theta}{\cos \theta}$$
$$\cot \theta = \frac{\cos \theta}{\sin \theta}$$
$$\sec \theta = \frac{1}{\cos \theta}$$
$$\csc \theta = \frac{1}{\sin \theta}$$
$$\cot \theta = \frac{1}{\tan \theta}$$

$$\sin (-\theta) = -\sin \theta$$
$$\cos (-\theta) = \cos \theta$$
$$\tan (-\theta) = -\tan \theta$$

$$\sin^2 \theta + \cos^2 \theta = 1$$
$$\tan^2 \theta + 1 = \sec^2 \theta$$
$$1 + \cot^2 \theta = \csc^2 \theta$$

$$\sin (\theta + \phi) = \sin \theta \cos \phi + \cos \theta \sin \phi$$
$$\cos (\theta + \phi) = \cos \theta \cos \phi - \sin \theta \sin \phi$$
$$\tan (\theta + \phi) = \frac{\tan \theta + \tan \phi}{1 - \tan \theta \tan \phi}$$

$$\sin (\theta - \phi) = \sin \theta \cos \phi - \cos \theta \sin \phi$$
$$\cos (\theta - \phi) = \cos \theta \cos \phi + \sin \theta \sin \phi$$
$$\tan (\theta - \phi) = \frac{\tan \theta - \tan \phi}{1 + \tan \theta \tan \phi}$$

$$\sin 2\theta = 2 \sin \theta \cos \theta$$
$$\cos 2\theta = \cos^2 \theta - \sin^2 \theta$$

$$2 \cos^2 \theta = 1 + \cos 2\theta$$
$$2 \sin^2 \theta = 1 - \cos 2\theta$$

$$\sin 45° = \frac{1}{2}\sqrt{2}, \quad \cos 45° = \frac{1}{2}\sqrt{2}, \quad \tan 45° = 1$$

$$\sin 30° = \frac{1}{2}, \quad \cos 30° = \frac{1}{2}\sqrt{3}, \quad \tan 30° = \frac{1}{3}\sqrt{3}$$

$$\sin 60° = \frac{1}{2}\sqrt{3}, \quad \cos 60° = \frac{1}{2}, \quad \tan 60° = \sqrt{3}$$

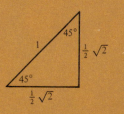

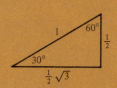

GREEK ALPHABET

Letters		Names	Letters		Names	Letters		Names
A	α	alpha	I	ι	iota	P	ρ	rho
B	β	beta	K	κ	kappa	Σ	σ	sigma
Γ	γ	gamma	Λ	λ	lambda	T	τ	tau
Δ	δ	delta	M	μ	mu	Υ	υ	upsilon
E	ϵ	epsilon	N	ν	nu	Φ	ϕ	phi
Z	ζ	zeta	Ξ	ξ	xi	X	χ	chi
H	η	eta	O	o	omicron	Ψ	ψ	psi
Θ	θ	theta	Π	π	pi	Ω	ω	omega

CALCULUS

WITH ANALYTIC GEOMETRY

CALCULUS
WITH ANALYTIC GEOMETRY

George F. Simmons

Professor of Mathematics, Colorado College

McGRAW-HILL BOOK COMPANY

New York St. Louis San Francisco Auckland Bogotá
Caracas Lisbon London Madrid Mexico Milan
Montreal New Delhi Paris San Juan Singapore
Sydney Tokyo Toronto

This book was set in Times Roman by Progressive Typographers, Inc.
The editors were Peter R. Devine and Jo Satloff;
the designer was Joan E. O'Connor;
the production supervisor was Phil Galea.
The drawings were done by J & R Services, Inc.
Von Hoffmann Press, Inc., was printer and binder.

Cover photograph reproduced from *Scientific Instruments*
by Harriet Wynter and Anthony Turner, London 1975.

CALCULUS WITH ANALYTIC GEOMETRY

    4 5 6 7 8 9 0 VNH VNH 9 4 3 2

ISBN 0-07-057419-7

Library of Congress Cataloging in Publication Data

Simmons, George Finlay, date
    Calculus with analytic geometry.

    Includes bibliographical references and index.
    1. Calculus. 2. Geometry, Analytic. I. Title.
QA303.S5547 1985   515'.15   84-14359
ISBN 0-07-057419-7

For Gertrude Clark,
the great teacher in my life.

Tradition cannot be inherited, and if you want it you must obtain it by great labour. — *T. S. Eliot*

Science and philosophy cast a net of words into the sea of being, happy in the end if they draw anything out besides the net itself, with some holes in it. — *Santayana*

La vraie définition de la science, c'est qu'elle est l'étude de la beauté du monde. (The true definition of science is that it is the study of the beauty of the world.) — *Simone Weil*

To me, logic and learning and all mental activity have always been incomprehensible as a complete and closed picture and have been understandable only as a process by which man puts himself *en rapport* with his environment. It is the battle for learning which is significant, and not the victory. Every victory that is absolute is followed at once by the Twilight of the Gods, in which the very concept of victory is dissolved in the moment of its attainment.

We are swimming upstream against a great torrent of disorganization, which tends to reduce everything to the heat-death of equilibrium and sameness described in the second law of thermodynamics. What Maxwell, Boltzmann, and Gibbs meant by this heat-death in physics has a counterpart in the ethics of Kierkegaard, who pointed out that we live in a chaotic moral universe. In this our main obligation is to establish arbitrary enclaves of order and system. These enclaves will not remain there indefinitely by any momentum of their own after we have once established them. Like the Red Queen, we cannot stay where we are without running as fast as we can.

We are not fighting for a definitive victory in the indefinite future. It is the greatest possible victory to be, to continue to be, and to have been. No defeat can deprive us of the success of having existed for some moment of time in a universe that seems indifferent to us. — *Norbert Wiener*

CONTENTS

Preface   xv
To the Student   xxi

PART I CHAPTER 1 NUMBERS, FUNCTIONS, AND GRAPHS

1.1   Introduction   1
1.2   The Real Line   2
1.3   The Coordinate Plane   7
1.4   Slopes and Equations of Straight Lines   11
1.5   Circles and Parabolas   15
1.6   The Concept of a Function   21
1.7   Types of Functions. Formulas from Geometry   25
1.8   Graphs of Functions   28

CHAPTER 2 THE DERIVATIVE OF A FUNCTION   39

2.1   What Is Calculus? The Problem of Tangents   39
2.2   How to Calculate the Slope of the Tangent   41
2.3   The Definition of the Derivative   46
2.4   Velocity and Rates of Change   50
2.5   Limits and Continuous Functions   55

CHAPTER 3 THE COMPUTATION OF DERIVATIVES   62

3.1   Derivatives of Polynomials   62
3.2   The Product and Quotient Rules   67
3.3   Composite Functions and the Chain Rule   71
3.4   Implicit Functions and Fractional Exponents   75
3.5   Derivatives of Higher Order   80

CHAPTER 4 APPLICATIONS OF DERIVATIVES   87

4.1   Increasing and Decreasing Functions. Maxima and Minima 87
4.2   Concavity and Points of Inflection   92
4.3   Applied Maximum and Minimum Problems   95
4.4   More Maximum-Minimum Problems. Reflection and Refraction 102

vii

4.5   Related Rates   109
4.6   (Optional) Newton's Method for Solving Equations   113
4.7   (Optional) Applications to Economics and Business   116

CHAPTER 5   INDEFINITE INTEGRALS
AND DIFFERENTIAL EQUATIONS   128

5.1   Introduction   128
5.2   The Notation of Differentials   128
5.3   Indefinite Integrals. Integration by Substitution   135
5.4   Differential Equations. Separation of Variables   141
5.5   Motion under Gravity. Escape Velocity and Black Holes   145

CHAPTER 6   DEFINITE INTEGRALS   154

6.1   Introduction   154
6.2   The Problem of Areas   155
6.3   The Sigma Notation and Certain Special Sums   157
6.4   The Area under a Curve. Definite Integrals   159
6.5   The Computation of Areas as Limits   164
6.6   The Fundamental Theorem of Calculus   167
6.7   Properties of Definite Integrals   172

CHAPTER 7   APPLICATIONS OF INTEGRATION   178

7.1   Introduction. The Intuitive Meaning of Integration   178
7.2   The Area between Two Curves   179
7.3   Volumes: The Disc Method   181
7.4   Volumes: The Shell Method   185
7.5   Arc Length   188
7.6   The Area of a Surface of Revolution   192
7.7   Hydrostatic Force   196
7.8   Work and Energy   198

PART II   CHAPTER 8   EXPONENTIAL AND
LOGARITHM FUNCTIONS

8.1   Introduction   208
8.2   Review of Exponents and Logarithms   209
8.3   The Number e and the Function $y = e^x$ 212
8.4   The Natural Logarithm Function $y = \ln x$ 217
8.5   Applications. Population Growth and Radioactive Decay   224
8.6   More Applications. Inhibited Population Growth, etc.   230

CHAPTER 9   TRIGONOMETRIC FUNCTIONS   239

9.1   Review of Trigonometry   239
9.2   The Derivatives of the Sine and Cosine   247
9.3   The Integrals of the Sine and Cosine. The Needle Problem   253
9.4   The Derivatives of the Other Four Functions   257
9.5   The Inverse Trigonometric Functions   259

9.6	Simple Harmonic Motion. The Pendulum	266
9.7	The Hyperbolic Functions	271

CHAPTER 10 METHODS OF INTEGRATION — 276

10.1	Introduction. The Basic Formulas	276
10.2	The Method of Substitution	279
10.3	Certain Trigonometric Integrals	283
10.4	Trigonometric Substitutions	287
10.5	Completing the Square	291
10.6	The Method of Partial Fractions	293
10.7	Integration by Parts	300
10.8	(Optional) Functions That Cannot Be Integrated	305
10.9	(Optional) Numerical Integration	310

CHAPTER 11 FURTHER APPLICATIONS OF INTEGRATION — 318

11.1	The Center of Mass of a Discrete System	318
11.2	Centroids	321
11.3	The Theorems of Pappus	325
11.4	Moment of Inertia	327

CHAPTER 12 INDETERMINATE FORMS AND IMPROPER INTEGRALS — 332

12.1	Introduction. The Mean Value Theorem	332
12.2	The Indeterminate Form 0/0. L'Hospital's Rule	334
12.3	Other Indeterminate Forms	338
12.4	Improper Integrals	343

CHAPTER 13 INTRODUCTION TO INFINITE SERIES — 352

13.1	What Is an Infinite Series?	352
13.2	The Convergence and Divergence of Series	356
13.3	Various Series Related to the Geometric Series	363
13.4	Power Series Considered Informally	369

CHAPTER 14 THE THEORY OF INFINITE SERIES — 376

14.1	Introduction	376
14.2	Convergent Sequences	377
14.3	General Properties of Convergent Series	384
14.4	Series of Nonnegative Terms. Comparison Tests	393
14.5	The Integral Test. Euler's Constant	397
14.6	The Ratio Test and Root Test	403
14.7	The Alternating Series Test. Absolute Convergence	407
14.8	Power Series Revisited. Interval of Convergence	412
14.9	Differentiation and Integration of Power Series	418
14.10	Taylor Series and Taylor's Formula	423
14.11	(Optional) Operations on Power Series	430
14.12	(Optional) Complex Numbers and Euler's Formula	437

PART III

CHAPTER 15 CONIC SECTIONS

15.1	Introduction. Sections of a Cone	448
15.2	Another Look at Circles and Parabolas	450
15.3	Ellipses	454
15.4	Hyperbolas	462
15.5	The Focus-Directrix-Eccentricity Definitions	470
15.6	(Optional) Second Degree Equations. Rotation of Axes	472

CHAPTER 16 POLAR COORDINATES — 479

16.1	The Polar Coordinate System	479
16.2	More Graphs of Polar Equations	483
16.3	Polar Equations of Circles, Conics, and Spirals	488
16.4	Arc Length and Tangent Lines	494
16.5	Areas in Polar Coordinates	499

CHAPTER 17 PARAMETRIC EQUATIONS. VECTORS IN THE PLANE — 506

17.1	Parametric Equations of Curves	506
17.2	(Optional) The Cycloid and Other Similar Curves	512
17.3	Vector Algebra. The Unit Vectors **i** and **j**	520
17.4	Derivatives of Vector Functions. Velocity and Acceleration	525
17.5	Curvature and the Unit Normal Vector	531
17.6	Tangential and Normal Components of Acceleration	536
17.7	(Optional) Kepler's Laws and Newton's Law of Gravitation	540

CHAPTER 18 VECTORS IN THREE-DIMENSIONAL SPACE. SURFACES — 550

18.1	Coordinates and Vectors in Three-Dimensional Space	550
18.2	The Dot Product of Two Vectors	554
18.3	The Cross Product of Two Vectors	559
18.4	Lines and Planes	565
18.5	Cylinders and Surfaces of Revolution	572
18.6	Quadric Surfaces	575
18.7	Cylindrical and Spherical Coordinates	580

CHAPTER 19 PARTIAL DERIVATIVES — 584

19.1	Functions of Several Variables	584
19.2	Partial Derivatives	589
19.3	The Tangent Plane to a Surface	595
19.4	Increments and Differentials. The Fundamental Lemma	598
19.5	Directional Derivatives and the Gradient	601
19.6	The Chain Rule for Partial Derivatives	606
19.7	Maximum and Minimum Problems	613
19.8	(Optional) Constrained Maxima and Minima. Lagrange Multipliers	618
19.9	(Optional) Laplace's Equation, the Heat Equation, and the Wave Equation	624
19.10	(Optional) Implicit Functions	629

CHAPTER 20   MULTIPLE INTEGRALS   635

20.1	Volumes as Iterated Integrals	635
20.2	Double Integrals and Iterated Integrals	639
20.3	Physical Applications of Double Integrals	644
20.4	Double Integrals in Polar Coordinates	648
20.5	Triple Integrals	654
20.6	Cylindrical Coordinates	659
20.7	Spherical Coordinates. Gravitational Attraction	662
20.8	Areas of Curved Surfaces	668
20.9	(Optional) Change of Variables in Multiple Integrals. Jacobians	672

CHAPTER 21   LINE INTEGRALS AND GREEN'S THEOREM   676

21.1	Line Integrals in the Plane	676
21.2	Independence of Path. Conservative Fields	683
21.3	Green's Theorem	689
21.4	What Next?	697

APPENDIXES

A   A VARIETY OF ADDITIONAL TOPICS   699

A.1	More about Numbers: Irrationals, Perfect Numbers, and Mersenne Primes	699
A.2	Archimedes' Quadrature of the Parabola	704
A.3	The Lunes of Hippocrates	706
A.4	Fermat's Calculation of $\int_0^b x^n\, dx$ for Positive Rational n	708
A.5	How Archimedes Discovered Integration	709
A.6a	A Simple Approach to $E = Mc^2$	711
A.6b	Rocket Propulsion in Outer Space	713
A.7	A Proof of Vieta's Formula	714
A.8	An Elementary Proof of Leibniz's Formula $\pi/4 = 1 - \frac{1}{3} + \frac{1}{5} - \frac{1}{7} + \cdots$	715
A.9	The Catenary, or Curve of a Hanging Chain	716
A.10	Wallis's Product	718
A.11	How Leibniz Discovered His Formula $\pi/4 = 1 - \frac{1}{3} + \frac{1}{5} - \frac{1}{7} + \cdots$	720
A.12	Euler's Discovery of the Formula $\Sigma_1^\infty 1/n^2 = \pi^2/6$	722
A.13	A Rigorous Proof of Euler's Formula $\Sigma_1^\infty 1/n^2 = \pi^2/6$	723
A.14	The Sequence of Primes	725
A.15	More about Irrational Numbers. π Is Irrational	732
A.16	Algebraic and Transcendental Numbers. e Is Transcendental	734
A.17	The Series $\Sigma\, 1/p_n$ of the Reciprocals of the Primes	740
A.18	The Bernoulli Numbers and Some Wonderful Discoveries of Euler	742
A.19	Bernoulli's Solution of the Brachistochrone Problem	746
A.20	Evolutes and Involutes	748
A.21	Euler's Formula $\Sigma_{n=1}^\infty 1/n^2 = \pi^2/6$ by Double Integration	751
A.22	Surface Integrals and the Divergence Theorem	753
A.23	Stokes' Theorem	758

B   BIOGRAPHICAL NOTES   763

An Outline of the History of Calculus	764
Pythagoras	765
Euclid	771
Archimedes	776
Pappus	782
Descartes	783
Mersenne	789
Fermat	790
Pascal	797
Huygens	800
Newton	804
Leibniz	810
The Bernoulli Brothers	821
Euler	823
Lagrange	829
Laplace	830
Fourier	831
Gauss	832
Cauchy	838
Abel	839
Dirichlet	841
Liouville	842
Hermite	843
Riemann	844

C   THE THEORY OF CALCULUS   849

C.1	The Real Number System	849
C.2	Theorems about Limits	853
C.3	Some Deeper Properties of Continuous Functions	858
C.4	The Mean Value Theorem	862
C.5	The Integrability of Continuous Functions	866
C.6	Another Proof of the Fundamental Theorem of Calculus	870
C.7	The Existence of $e = \lim_{h \to 0} (1 + h)^{1/h}$	871
C.8	The Validity of Integration by Inverse Substitution	872
C.9	Proof of the Partial Fractions Theorem	873
C.10	The Extended Ratio Tests of Raabe and Gauss	876
C.11	Absolute vs. Conditional Convergence	880
C.12	Dirichlet's Test	886
C.13	Uniform Convergence for Power Series	889
C.14	The Division of Power Series	891
C.15	The Equality of Mixed Partial Derivatives	892
C.16	Differentiation under the Integral Sign	894
C.17	A Proof of the Fundamental Lemma	894
C.18	A Proof of the Implicit Function Theorem	895

D A FEW REVIEW TOPICS 897

D.1 The Binomial Theorem 897
D.2 Mathematical Induction 902

E NUMERICAL TABLES 910

Answers to Odd-Numbered Problems 919
Index 939

PREFACE

It is a curious fact that people who write thousand-page textbooks still seem to find it necessary to write prefaces to explain their purposes. Enough is enough, one would think. However, every textbook—and this one is no exception—is both an expression of dissatisfaction with existing books and a statement by the author of what he thinks such a book ought to contain, and a preface offers one last chance to be heard and understood. Furthermore, anyone who adds to the glut of introductory calculus books should be called upon to justify his action (or perhaps apologize for it) to his colleagues in the mathematics community.

This book is intended to be a mainstream calculus text that is suitable for every kind of course at every level. It is designed particularly for the standard course of three semesters for students of science, engineering, or mathematics. Students are expected to have a background of high school algebra and geometry.

On the other hand, no specialized knowledge of science is assumed, and students of philosophy, history, or economics should be able to read and understand the applications just as easily as anyone else. There is no law of human nature which decrees that people with a strong interest in the humanities or social sciences are automatically barred from understanding and enjoying mathematics. Indeed, mathematics is the stage for many of the highest achievements of the human imagination, and it should attract humanists as irresistibly as a field of wildflowers attracts bees. It has been truly said that mathematics can illuminate the world or delight the mind, and often both. It is therefore clear that a student of philosophy (for example) is just as crippled without a fairly detailed knowledge of this great subject as a student of history would be without a broad understanding of economics and religion. As for students of history, how can they afford to neglect the fact (and it is a fact!) that the rise of mathematics and science in the seventeenth century was the crucial event in the development of the modern world, much more profound in its historical significance than the American, French, and Russian Revolutions combined? We teachers of mathematics have an obligation to help such students with this part of their education, and calculus is an excellent place to start.

The text itself—that is, the 21 chapters without considering the appendixes—is traditional in subject matter and organization. I have placed great emphasis on *motivation* and *intuitive understanding,* and the refinements of theory are downplayed. Most students are impatient with the theory of the subject, and justifiably so, because the essence of calculus does not lie in theorems and how to prove them, but rather in tools and how to use them. My overriding purpose has been to present calculus as a problem-solving art of immense power which is indispensable in all the quantitative sciences. Naturally, I wish to convince the student that the standard tools of calculus are reasonable and legitimate, but not at the expense of turning the subject into a stuffy logical discipline dominated by extra-careful definitions, formal statements of theorems, and meticulous proofs. It is my hope that every mathematical explanation in these chapters will seem to the thoughtful student to be as natural and inevitable as water flowing downhill along a canyon floor. The main theme of our work is what calculus is good for— what it enables us to do and understand—and not what its logical nature is as seen from the specialized (and limited) point of view of the modern pure mathematician.

There are several features of the text itself that it might be useful for me to comment on.

Precalculus Material   Because of the great amount of calculus that must be covered, it is desirable to get off to a fast start and introduce the derivative quickly, and to spend as little time as possible reviewing precalculus material. However, college freshmen constitute a very diverse group, with widely different levels of mathematical preparation. For this reason I have included a first chapter on precalculus material which I urge teachers either to omit altogether or else to skim over as lightly as they think advisable for their particular students. This chapter is written in enough detail so that individual students who need to spend more time on the preliminaries should be able to absorb most of it on their own with a little extra effort.*

Trigonometry   The problem of what to do about trigonometry in calculus courses has no satisfactory solution. Some writers introduce the subject early, partly in order to be able to use trigonometric functions in teaching the chain rule. This approach has the disadvantage of clogging the early chapters of calculus with technical material that is not really essential for the students' primary aims at this stage, which are to grasp the meanings and some of the uses of derivatives and integrals. Another disadvantage of this early introduction of the subject is that many students take only a single semester of calculus, and for these students trigonometry is an unnecessary complication that perhaps they should be spared. The fact is that trigonometry becomes really indispensable only when formal methods of integration must finally be confronted.

---

* A more complete exposition of high school mathematics that is still respectably concise can be found in my little book, *Precalculus Mathematics In a Nutshell* (William Kaufmann, Inc., Los Altos, Calif., 1981), 119 pages.

For these reasons I introduce the calculus of the trigonometric functions in Chapter 9, so that all the ideas will be fresh in the mind when students begin Chapter 10 on methods of integration. A full exposition of trigonometry from scratch is given in Section 9.1. For most students this will be a needed review of material that was learned (and mostly forgotten) in high school. For those who have studied no trigonometry at all, the explanations are complete enough so that they should be able to learn what they need to know from this single section alone.

For teachers who prefer to take up trigonometry early—and there are good reasons for this—I point out that Sections 9.1 and 9.2 can easily be introduced directly after Section 4.5, and Sections 9.3 and 9.4 at any time after Chapter 6. The only necessary adjustments are to warn students away from parts (b), (c), and (d) of Example 2 in Section 9.2, and also to make sure that the following problems are not assigned as homework: in Section 9.2, 15-18; in Section 9.3, 12, 16, 17, 29; and in Section 9.4, 11, 12, and 24.

Problems   For students, the most important parts of their calculus book may well be the problem sets, because this is where they spend most of their time and energy. There are more than 5800 problems in this book, including many old standbys familiar to all calculus teachers and dating back to the time of Euler and even earlier. I have tried to repay my debt to the past by inventing new problems whenever possible. The problem sets are carefully constructed, beginning with routine drill exercises and building up to more complex problems requiring higher levels of thought and skill. The most challenging problems are marked with an asterisk (*). In general, each set contains approximately twice as many problems as most teachers will want to assign for homework, so that a large number will be left over for students to use as review material.

Most of the chapters conclude with long lists of additional problems. Many of these are intended only to provide further scope and variety to the problems sets at the ends of the sections. However, teachers and students alike should treat these additional problems with special care, because some are quite subtle and difficult and should only be attacked by students with ample reserves of drive and tenacity.

I should also mention that there are several sections scattered throughout the book with no corresponding problems at all. Sometimes these sections occur in small groups and are merely convenient subdivisions of what I consider a single topic and intend as a single assignment, as with Sections 6.1, 6.2, 6.3 and 6.4, 6.5. In other cases (Sections 9.7, 14.12, 15.5, 19.4, and 20.9) the absence of problems is a tacit suggestion that the subject matter of these sections should be touched upon only lightly and briefly.

There are a great many so-called "story problems" spread through the entire book. All teachers know that students shudder at these problems, because they usually require nonroutine thinking. However, the usefulness of mathematics in the various sciences demands that we try to teach our students how to penetrate into the meaning of a story problem, how to judge what is relevant to it, and how to translate it from words into sketches and equations. Without these skills—which are equally valuable for students who will become doctors, lawyers, financial analysts, or thinkers of any kind—there is no mathematics education worthy of the name.*

Infinite Series   Any mathematician who glances at Chapter 14 will see at once that infinite series is one of my favorite subjects. In the flush of my enthusiasm, I have developed this topic in greater depth and detail than is usual in calculus books. However, some teachers may not wish to devote this much time and attention to the subject, and for their convenience I have given a shorter treatment in Chapter 13 that may be sufficient for the needs of most students who are not planning to go on to more advanced mathematics courses. Those teachers who consider this subject to be as important as I do will probably use both chapters, the first to give students an overview, and the second to establish a solid foundation and nail down the basic concepts. The spirit of these chapters is quite different, and there is surprisingly little repetition.

Differential Equations and Vector Analysis   Each of these subjects is an important branch of mathematics in its own right. They should be taught in separate courses, after calculus, with ample time to explore their distinctive methods and applications. One of the main responsibilities of a calculus course is to prepare the way for these more advanced subjects and take a few preliminary steps in their direction, but just how far one should go is a debatable question. Some writers on calculus try to include mini-courses on these subjects in large chapters at the ends of their books. I disagree with this practice and believe that few teachers make much use of these chapters. Instead, in the case of differential equations I prefer to introduce the subject as early as possible (Section 5.4) and return to it in a low-key way whenever the opportunity arises (Sections 5.5, 7.8, 8.5, 8.6, 9.6, 17.7, 19.9); and in vector analysis I believe that Green's Theorem is just the right place to stop, with Stokes' Theorem—which after all is one of the most profound and far-reaching theorems in all of mathematics—being left for a later course. For those teachers who wish to include more vector analysis in their calculus course, I give a brief treatment of the divergence theorem and Stokes' Theorem—with problems—in Appendixes A.22 and A.23.

One of the major ways in which this book is unique and quite different from all its competitors can be understood by examining the appendixes, which I will now comment on very briefly. Before doing so, I emphasize that this material is entirely separate from the main text and can be carefully studied, dipped into occasionally, or completely ignored, as each individual student or instructor desires.

---

* I cannot let the opportunity pass without quoting a classic story problem that appeared in the *New Yorker* magazine many years ago. "You know those terrible arithmetic problems about how many peaches some people buy, and so forth? Well, here's one we *like*, made up by a third-grader who was asked to think up a problem similar to the ones in his book: 'My father is forty-four years old. My dog is eight. If my dog was a human being, he would be fifty-six years old. How old would my father plus my dog be if they were both human beings?'"

Appendix A   In teaching calculus over a period of many years, I have collected a considerable number of miscellaneous topics from number theory, geometry, science, etc., which I have used for the purpose of opening doors and forging links with other subjects . . . and also for breaking the routine and lifting the spirits. Many of my students have found these "nuggets" interesting and eye-opening. I have collected most of these topics in this appendix in the hope of making a few more converts to the view that mathematics, while sometimes rather dull and routine, can often be supremely interesting.

Appendix B   This material amounts to a brief biographical history of mathematics, from the earliest times to the mid-nineteenth century. It has two main purposes.

First, I hope in this way to "humanize" the subject, to make it transparently clear that great men created it by great efforts of genius, and thereby to increase the students' interest in what they are studying. The minds of most people turn away from problems — veer off, draw back, avoid contact, change the subject, think of something else at all costs. These people — the great majority of the human race — find solace and comfort in the known and the familiar, and avoid the unknown and unfamiliar as they would deserts and jungles. It is as hard for them to think steadily about a difficult problem as it is to hold together the north poles of two strong magnets. In contrast to this, a tiny minority of men and women are drawn irresistibly to problems: their minds embrace them lovingly and wrestle with them tirelessly until they yield their secrets. It is these who have taught the rest of us most of what we know and can do, from the wheel and the lever to metallurgy and the theory of relativity. I have written about some of these people from our past in the hope of encouraging a few in the next generation.

My second purpose is connected with the fact that many students from the humanities and social sciences are compelled against their will to study calculus as a means of satisfying academic requirements. The profound connections that join mathematics to the history of philosophy, and also to the broader intellectual and social history of Western civilization, are often capable of arousing the passionate interest of these otherwise indifferent students.

Appendix C   In the main text, the level of mathematical rigor rises and falls in accordance with the nature of the subject under discussion. It is rather low in the geometrical chapters, where for the most part I rely on common sense together with intuition aided by illustrations; and it is rather high in the chapters on infinite series, where the substance of the subject cannot really be understood without careful thought. I have constantly kept in mind the fact that most students have very little interest in purely mathematical reasoning for its own sake, and I have tried to prevent this type of material from intruding any more than is absolutely necessary. Some students, however, have a natural taste for theory, and some instructors feel as a matter of principle that all students should be exposed to a certain amount of theory for the good of their souls. This appendix contains virtually all of the theoretical material that by any stretch of the imagination might be considered appropriate for the study of calculus. From the purely mathematical point of view, it is possible for instructors to teach courses at many different levels of sophistication by using — or not using — material selected from this appendix.

In summary, therefore, the main body of this book is straightforward and traditional, while the appendixes make it convenient for teachers with many different interests and opinions to offer a wide variety of courses tailored to the needs of their own classes. I have aimed at the utmost flexibility of use.

Every project of this magnitude obviously depends on the cooperative efforts of many people. On the publisher's staff, I am especially grateful to Peter Devine, who as editor knew very well when to provide gentle guidance and when to let me go my own way; to Jo Satloff, the editorial supervisor, whose sympathy, tact, and highly skilled professionalism mean a great deal to me; and to Joan O'Connor, the designer, whose willingness to listen to an amateur's suggestions is very much appreciated.

Also, I offer my sincere thanks to the publisher's reviewers: Joe Browne, Onondaga Community College; Carol Crawford, United States Naval Academy; Bruce Edwards, University of Florida; Susan L. Friedman, Baruch College; Melvin Hausner, New York University; Louis Hoelzle, Bucks County Community College; Stanley M. Lukawecki, Clemson University; Peter Maserick, Pennsylvania State University; and David Zitarelli, Temple University. These people shared their knowledge and judgment with me in many important ways.

For the flaws and errors that undoubtedly remain, there is no one to blame but myself. I will consider it a kindness if colleagues and student users will take the trouble to inform me of any blemishes they detect, for correction in future editions.

George F. Simmons

TO THE STUDENT

Appearances to the contrary, no writer deliberately sets out to produce an unreadable book; we all do what we can and hope for the best. Naturally, I hope that my language will be clear and helpful to students, and in the end only they are qualified to judge. However, it would be a great advantage to all of us — teachers and students alike — if student users of mathematics textbooks could somehow be given a few hints on the art of reading mathematics, which is a very different thing from reading novels or magazines or newspapers.

In high school mathematics courses most students are accustomed to tackling their homework problems first, out of impatience to have the whole burdensome task over and done with as soon as possible. These students read the explanations in the text only as a last resort, if at all. This is a grotesque reversal of reasonable procedure, and makes about as much sense as trying to put on one's shoes before one's socks. I suggest that students should read the text first, and when this has been thoroughly assimilated, *then and only then* turn to the homework problems. After all, the purpose of these problems is to nail down the ideas and methods described and illustrated in the text.

How should a student read the text in a book like this? Slowly and carefully, and in full awareness that a great many details have been deliberately omitted. If this book contained every detail of every discussion, it would be five times as long, which God forbid! There is an old French proverb: "He who tries to explain everything soon finds himself talking to an empty room." Every writer of a book of this kind tries to walk a narrow path between saying too much and saying too little.

The words "clearly," "it is easy to see," and similar expressions are not intended to be taken literally, and should never be interpreted by any student as a putdown on his or her abilities. These are code-phrases that have been used in mathematical writing for hundreds of years. Their purpose is to give a signal to the careful reader that in this particular place the exposition is somewhat condensed, and perhaps a few details of calculations have been omitted. Any phrase like this amounts to a friendly hint to the student that it might be a good idea to read even more carefully and thoughtfully in order to fill in omissions in the exposition, or perhaps get out a piece of scratch paper to verify omitted details of calculations. Or better yet, make full use of the margins of this book to emphasize points, raise questions, perform little computations, and correct misprints.

CALCULUS

WITH ANALYTIC GEOMETRY

1

NUMBERS, FUNCTIONS, AND GRAPHS

1.1

INTRODUCTION

Everyone knows that the world in which we live is dominated by motion and change. The earth moves in its orbit around the sun; a colony of bacteria grows; a rock thrown upward slows and stops, then falls back to earth with increasing speed; and radioactive elements decay. These are merely a few items in the endless array of phenomena for which mathematics is the most natural medium of communication and understanding. As Galileo said more than 300 years ago, "The Great Book of Nature is written in mathematical symbols."

Calculus is that branch of mathematics whose primary purpose is the study of motion and change. It is an indispensable tool of thought in almost every field of pure and applied science—in physics, chemistry, biology, astronomy, geology, engineering, and even some of the social sciences. It also has many important uses in other parts of mathematics, especially geometry. By any standard, the methods and applications of calculus constitute one of the greatest intellectual achievements of civilization.

The main objects of study in calculus are functions. But what is a function? Roughly speaking, it is a rule or law that tells us how one variable quantity depends upon another. This is the master concept of the exact sciences. It offers us the prospect of understanding and correlating natural phenomena by means of mathematical machinery of great and sometimes mysterious power. The concept of a function is so vitally important for all our work that we must strive to clarify it beyond any possibility of confusion. This purpose is the theme of the present chapter.

The following sections contain a good deal of material that many readers have studied before. Some will welcome the opportunity to review and refresh their ideas. Those who find it irksome to tread the same path over and over may discover some interesting sidelights and stimulating challenges among the additional problems at the end of the chapter. This chapter is intended solely for purposes of review. It can be studied carefully, or lightly, or even skipped altogether, depending on the reader's level of preparation.

The actual subject matter of this course begins in Chapter 2, and it would be very unfortunate if even a single student should come to feel that this preliminary chapter is more of an obstacle than a source of assistance.

1.2

THE REAL LINE

Most of the variable quantities we study—such as length, area, volume, position, time, and velocity—are measured by means of real numbers, and in this sense calculus is based on the real number system. It is true that there are other important and useful number systems—for instance, the complex numbers. It is also true that two- and three-dimensional treatments of position and velocity require the use of vectors. These ideas will be examined in due course, but for a long time to come the only numbers we shall be working with are the real numbers.*

It is assumed in this book that students are familiar with the elementary algebra of the real number system. Nevertheless, in this section we give a brief descriptive survey that may be helpful. For our purposes this is sufficient, but any reader who wishes to probe more deeply into the nature of real numbers will find a more precise discussion in Appendix C.1 at the back of the book.

The real number system contains several types of numbers that deserve special mention: the *positive integers* (or *natural numbers*)

$$1, 2, 3, 4, 5, \ldots ;$$

the *integers*

$$\ldots, -3, -2, -1, 0, 1, 2, 3, \ldots ;$$

and the *rational numbers,* which are those real numbers that can be represented as fractions (or quotients of integers), such as

$$\tfrac{2}{3}, -\tfrac{7}{4}, 4, 0, -5, 3.87, 2\tfrac{1}{4}.$$

A real number that is not rational is said to be *irrational*; for example,

$$\sqrt{2}, \sqrt{3}, \sqrt{2} + \sqrt{3}, \sqrt{5}, \sqrt[3]{5}, \quad \text{and} \quad \pi$$

are irrational numbers.

We take this opportunity to remind the reader that for any positive number a, the symbol $\sqrt{a}$ always means its positive square root. Thus, $\sqrt{4}$ is equal to 2 and not -2, even though $(-2)^2 = 4$. If we wish to designate both square roots of 4, we must write $\pm \sqrt{4}$. Similarly, $\sqrt[n]{a}$ always means the positive nth root of a.

THE REAL LINE

The use of the real numbers for measurement is reflected in the very convenient custom of representing these numbers graphically by points on a horizontal straight line.

---

* The adjective "real" was originally used to distinguish these numbers from numbers like $\sqrt{-1}$, which were once thought to be "unreal" or "imaginary."

Figure 1.1   The real line.

This representation begins with the choice of an arbitrary point as the origin or zero point, and another arbitrary point to the right of it as the point 1. The distance between these two points (the unit distance) then serves as a scale by means of which we can assign a point on the line to every positive and negative integer, as illustrated in Fig. 1.1, and also to every rational number. We call particular attention to the fact that all positive numbers lie to the right of 0, in the "positive direction," and all negative numbers lie to the left. The method of assigning a point to a rational number is shown in Fig. 1.1 for the number $\frac{7}{3} = 2\frac{1}{3}$: the segment between 2 and 3 is subdivided by two points into three equal segments, and the first of these points is labeled $2\frac{1}{3}$. This procedure of using equal subdivisions clearly serves to determine the point on the line which corresponds to any rational number whatever. Furthermore, this correspondence between rational numbers and points can be extended to irrational numbers, for we shall see at the end of this section that the decimal expansion of an irrational number, such as

$$\sqrt{2} = 1.414 \ldots , \qquad \sqrt{3} = 1.732 \ldots , \qquad \pi = 3.14159 \ldots ,$$

can be interpreted as a set of instructions specifying the exact position of the corresponding point.

We have described a one-to-one correspondence between all real numbers and all points on the line which establishes these numbers as a coordinate system for the line. This coordinatized line is called the *real line* (or sometimes the *number line*). It is convenient and customary to merge the logically distinct concepts of the real number system and the real line, and we shall freely speak of points on the line as if they were numbers and of numbers as if they were points on the line. Thus, such mixed expressions as "irrational point" and "the segment between 2 and 3" are quite natural and will be used without further comment.

INEQUALITIES

The left-to-right linear succession of points on the real line corresponds to an important part of the algebra of the real number system, that dealing with inequalities. These ideas play a larger role in calculus than in earlier mathematics courses, so we briefly recall the essential points.

The geometric meaning of the inequality $a < b$ (read "*a* is less than *b*") is simply that *a* lies to the left of *b*; the equivalent inequality $b > a$ ("*b* is greater than *a*") means that *b* lies to the right of *a*. A number *a* is positive or negative according as $a > 0$ or $a < 0$. The main rules used in working with inequalities are the following:

1   If $a > 0$ and $b < c$, then $ab < ac$.
2   If $a < 0$ and $b < c$, then $ab > ac$.
3   If $a < b$, then $a + c < b + c$ for any number *c*.

Rules 1 and 2 are usually expressed by saying that an inequality is preserved on multiplication by a positive number, and reversed on multiplication by a negative number; and rule 3 says that an inequality is preserved when any number (positive or negative) is added to both sides. It is often desirable to replace an inequality $a > b$ by the equivalent inequality $a - b > 0$, with rule 3 being used to establish the equivalence.

If we wish to say that a is positive or equal to 0, we write $a \geq 0$ and read this "a is greater than or equal to zero." Similarly, $a \geq b$ means that $a > b$ or $a = b$. Thus, $3 \geq 2$ and $3 \geq 3$ are both true inequalities.

We also recall that a product of two or more numbers is zero if and only if one of its factors is zero. If none of its factors are zero, it is positive or negative according as it has an even or an odd number of negative factors.

ABSOLUTE VALUES

The absolute value of a number a is denoted by $|a|$ and defined by

$$|a| = \begin{cases} a & \text{if } a \geq 0, \\ -a & \text{if } a < 0. \end{cases}$$

For example, $|3| = 3$, $|-2| = -(-2) = 2$, and $|0| = 0$. It is clear that the operation of forming the absolute value leaves positive numbers unchanged and replaces each negative number by the corresponding positive number. The main properties of this operation are

$$|ab| = |a||b| \qquad \text{and} \qquad |a + b| \leq |a| + |b|.$$

In geometric language, the absolute value of a number a is simply the distance from the point a to the origin. Similarly, the distance from a to b is $|a - b|$.

To solve an equation such as $|x + 2| = 3$, we can write it in the form $|x - (-2)| = 3$ and think of it as saying that "the distance from x to -2 is 3." With Fig. 1.1 in mind, it is evident that the solutions are $x = 1$ and $x = -5$. We can also solve this equation by using the fact that $|x + 2| = 3$ means that $x + 2 = 3$ or $x + 2 = -3$; the solutions are $x = 1$ and $x = -5$, as before.

INTERVALS

The sets of real numbers we shall be dealing with most frequently are intervals. An *interval* is simply a segment on the real line. If its endpoints are the numbers a and b, then the interval consists of all numbers that lie between a and b. However, we may or may not want to include the endpoints themselves as part of the interval.

To be more precise, suppose that a and b are numbers, with $a < b$. The *closed interval* from a to b, denoted by $[a, b]$, includes its endpoints, and therefore consists of all real numbers x such that $a \leq x \leq b$. Parentheses are used to indicate excluded endpoints. The interval (a, b), with both endpoints excluded, is called the *open interval* from a to b; it consists of all x such that $a < x < b$. Sometimes we wish to include only one endpoint in an interval.

Thus, the intervals denoted by $[a, b)$ and $(a, b]$ are defined by the inequalities $a \le x < b$ and $a < x \le b$, respectively. In each of these cases, any number c such that $a < c < b$ is called an *interior point* of the interval (Fig. 1.2).

Strictly speaking, the notations $a \le x \le b$ and $[a, b]$ have different meanings—the first represents a restriction imposed on x, while the second denotes a set—but both designate the same interval. We will therefore consider them to be equivalent and use them interchangeably, and the reader should become familiar with both. However, the geometric meaning of the notation $a \le x \le b$ is more easily grasped by the eye, and for this reason we usually prefer it to the other.

A half-line is often considered to be an interval extending to infinity in one direction. The symbol ∞ (read "infinity") is frequently used in designating such an interval. Thus, for any real number a the intervals defined by the inequalities $a < x$ and $x \le a$ can be written as $a < x < \infty$ and $-\infty < x \le a$, or equivalently as (a, ∞) and $(-\infty, a]$. Remember, however, that the symbols ∞ and $-\infty$ do not denote real numbers; they are used in this manner only as a convenient way of emphasizing that x is allowed to be arbitrarily large (in either the positive or negative direction). As an aid in keeping the notation clear in one's mind, it may be helpful to think of $-\infty$ and ∞ as "fictitious numbers" located at the left and right "ends" of the real line, as suggested in Fig. 1.3. Also, it is sometimes convenient to think of the entire real line itself as an interval, $-\infty < x < \infty$ or $(-\infty, \infty)$.

Sets of numbers described by means of inequalities and absolute values are often intervals. It is clear, for instance, that the set of all x such that $|x| < 2$ is the interval $-2 < x < 2$ or $(-2, 2)$. The following example illustrates some techniques that are useful in a variety of situations.

Example   Solve the inequality $x^3 > x$.

To "solve" an inequality like this means to find all numbers x for which the inequality is true. We write it first as $x^3 - x > 0$, and then in the factored form

$$x(x + 1)(x - 1) > 0. \tag{1}$$

The expression on the left equals zero when $x = 0, -1, 1$. These three points divide the real line into four open intervals, as shown in Fig. 1.4, and inside each of these intervals the expression $x(x + 1)(x - 1)$ has constant sign. For instance, when $x < -1$, we see by inspection that all three factors are negative, and so $x(x + 1)(x - 1)$ is negative; and when $-1 < x < 0$, we see that x and $x - 1$ are negative but $x + 1$ is positive, and so $x(x + 1)(x - 1)$ is positive. We test each interval in this way and record the results in our figure. When this is done, we simply read off the intervals on which (1) is satisfied and write down the solution: $-1 < x < 0$ and $1 < x$, or equivalently $(-1, 0)$ and $(1, \infty)$.

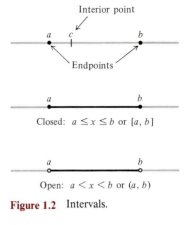

Figure 1.2   Intervals.

Closed: $a \le x \le b$ or $[a, b]$

Open: $a < x < b$ or (a, b)

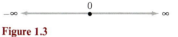

Figure 1.3

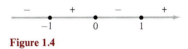

Figure 1.4

We add a few comments on the use of intervals for understanding the geometric significance of the decimal expansion of a real number. In the case of the irrational $\sqrt{2}$, the fact that its decimal expansion is 1.414 . . . means

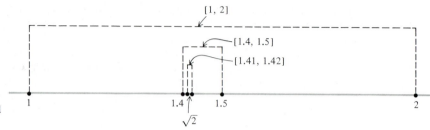

Figure 1.5 $\sqrt{2} = 1.414\ldots$ located geometrically.

that the number $\sqrt{2}$ satisfies each inequality in the following infinite list:

$$1 \le \sqrt{2} \le 2,$$

$$1.4 \le \sqrt{2} \le 1.5,$$

$$1.41 \le \sqrt{2} \le 1.42,$$

$$\ldots$$

This in turn means that the corresponding point lies in each of the following closed intervals with rational endpoints: [1, 2], [1.4, 1.5], [1.41, 1.42], ... This "nested sequence" of intervals is shown in Fig. 1.5. It is geometrically clear that there is one and only one point that lies in all these intervals, and in this sense the decimal expansion of the number $\sqrt{2}$ can be interpreted as a set of instructions specifying the exact position of the point $\sqrt{2}$ on the real line. Since $\sqrt{2}$ is irrational, it is an interior point of every interval in the sequence.

We emphasize that our aims in this book are almost entirely practical. Nevertheless, our discussions often give rise to certain "impractical" questions that some readers may find interesting and appealing. As an example, how do we know that the number $\sqrt{2}$ is irrational? For readers with the time and inclination to pursue such questions — and also because we consider the answers to be worth knowing about for their own sake — we offer food for further thought in occasional appendixes (see Appendix A.1 at the back of the book).

PROBLEMS

1 Find all values of x that satisfy each of the following conditions:
(a) $|x| = 5$;
(b) $|x + 4| = 3$;
(c) $|x - 2| = 4$;
(d) $|x + 1| = |x - 2|$;
(e) $|x + 1| = |2x - 2|$;
(f) $|x^2 - 5| = 4$;
(g) $|x - 3| \le 5$.

2 Solve the following inequalities:
(a) $x(x - 1) > 0$;
(b) $x^4 < x^2$;
(c) $(x - 1)(x + 2) < 0$;
(d) $x^2 - 2 \ge x$;
(e) $x^2(x - 1) \ge 0$;
(f) $(2x + 1)^8(x + 1) \le 0$;
(g) $x^2 + 4x - 21 > 0$;
(h) $2x^2 + x < 3$;
(i) $1 - x \le 2x^2$;
(j) $4x^2 + 10x - 6 < 0$;
(k) $x^3 + 1 < x^2 + x$;
(l) $x^2 + 2x + 4 > 0$.

3 Recall that $\sqrt{a}$ is a real number if and only if $a \ge 0$, and find the values of x for which each of the following is a real number:
(a) $\sqrt{4 - x^2}$;
(b) $\sqrt{x^2 - 9}$;
(c) $\dfrac{1}{\sqrt{4 - 3x}}$;
(d) $\dfrac{1}{\sqrt{x^2 - x - 12}}$.

4 Find the values of x for which each of the following is positive:
(a) $\dfrac{x}{x^2 + 4}$;
(b) $\dfrac{x}{x^2 - 4}$;
(c) $\dfrac{x + 1}{x - 3}$;
(d) $\dfrac{x^2 - 1}{x^2 - 3x}$.

5   Show by a numerical example that the following statement is not true: If $a < b$ and $c < d$, then $ac < bd$. (For this statement to be true, it must be true for *all* numbers a, b, c, d satisfying the stated conditions. A single exception—called a *counterexample*—is therefore sufficient to demonstrate that the statement is not true.)

6   If a, b, c, d are positive numbers such that $a/b < c/d$, show that

$$\frac{a}{b} < \frac{a+c}{b+d} < \frac{c}{d}.$$

7   Show that the number $\frac{1}{2}(a + b)$, called the *arithmetic mean* of a and b, is the midpoint of the interval $a \le x \le b$. (Hint: The midpoint is a plus half the length of the interval.) Find the trisection points of this interval.

8   If $0 < a < b$, show that $a^2 < b^2$ and $\sqrt{a} < \sqrt{b}$.

9   If $0 < a < b$, the number $\sqrt{ab}$ is called the *geometric mean* of a and b. Show that $a < \sqrt{ab} < b$.

10 If a and b are positive numbers, show that $\sqrt{ab} \le \frac{1}{2}(a + b)$.

1.3
THE COORDINATE PLANE

Just as real numbers are used as coordinates for points on a line, pairs of real numbers can be used as coordinates for points in a plane. For this purpose we establish a *rectangular coordinate system* in the plane, as follows.

Draw two perpendicular straight lines in the plane, one horizontal and the other vertical, as shown in Fig. 1.6. These lines are called the *x-axis* and *y-axis*, respectively, and their point of intersection is called the *origin*. Coordinates are assigned to these axes in the manner described earlier, with the origin as the zero point on both and the same unit distance on both. The positive *x*-axis is to the right of the origin and the negative *x*-axis to the left, as before; and the positive *y*-axis is above the origin and the negative *y*-axis below.

Now consider a point P anywhere in the plane. Draw a line through P parallel to the *y*-axis, and let x be the coordinate of the point where this line crosses the *x*-axis. Similarly, draw a line through P parallel to the *x*-axis, and let y be the coordinate of the point where this line crosses the *y*-axis. The numbers x and y determined in this way are called the *x-coordinate* and *y-coordinate* of P. In referring to the coordinates of P, it is customary to write them as an ordered pair (x, y) with the *x*-coordinate written first; we say that

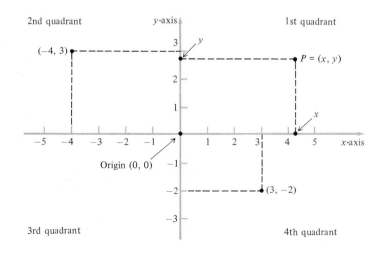

Figure 1.6   The coordinate plane or *xy*-plane.

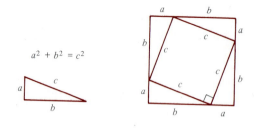

$$a^2 + b^2 = c^2$$

Figure 1.7 The Pythagorean
theorem and a proof.

P has coordinates (x, y).* This correspondence between *P* and its coordinates
establishes a one-to-one correspondence between all points in the plane and
all ordered pairs of real numbers; for *P* determines its coordinates uniquely,
and by reversing the process we see that each ordered pair of real numbers
uniquely determines a point *P* with these numbers as its coordinates. As in
the case of the real line, it is customary to drop the distinction between a
point and its coordinates, and to speak of "the point (x, y)" instead of "the
point with coordinates (x, y)." The coordinates *x* and *y* of the point *P* are
sometimes called the *abscissa* and *ordinate* of *P*. The reader should notice
particularly that points $(x, 0)$ lie on the *x*-axis, that points $(0, y)$ lie on the
y-axis, and that $(0, 0)$ is the origin. Also, the axes divide the plane into four
quadrants, as shown in Fig. 1.6, and these quadrants are characterized as
follows by the signs of *x* and *y*: first quadrant, $x > 0$ and $y > 0$; second
quadrant, $x < 0$ and $y > 0$; third quadrant, $x < 0$ and $y < 0$; fourth quadrant, $x > 0$ and $y < 0$.

When the plane is equipped with the coordinate system described here, it
is usually called the *coordinate plane* or the *xy-plane*.

THE DISTANCE FORMULA

Much of our work involves geometric ideas — right triangles, similar triangles, circles, spheres, cones, etc. — and we assume that students have acquired a reasonable grasp of elementary geometry from earlier mathematics
courses. A major fact of particular importance is the Pythagorean theorem:
In any right triangle, the sum of the squares of the legs equals the square of the
hypotenuse (Fig. 1.7). There are many proofs of this theorem, but the following is probably simpler than most. Let the legs be *a* and *b* and the hypotenuse
c, and arrange four replicas of the triangle in the corners of a square of side
$a + b$, as shown in Fig. 1.7. Then the area of the large square equals 4 times
the area of the triangle plus the area of the small square; that is,

$$(a + b)^2 = 4(\tfrac{1}{2}ab) + c^2.$$

This simplifies at once to $a^2 + b^2 = c^2$, which is the Pythagorean theorem.†

---

* In practice, the use of the same notation for ordered pairs as for open intervals never leads to
confusion, because in any specific context it is always clear which is meant.

† Students who are interested in learning a little about the extraordinary human beings who
created mathematics will find in the back of the book (in Appendix B) a brief account of almost
every person whose contributions are mentioned in the course of our work.

As the first of many applications of this fact, we obtain the formula for the distance d between any two points in the coordinate plane. If the points are $P_1 = (x_1, y_1)$ and $P_2 = (x_2, y_2)$, then the segment joining them is the hypotenuse of a right triangle (Fig. 1.8) with legs $|x_1 - x_2|$ and $|y_1 - y_2|$. By the Pythagorean theorem,

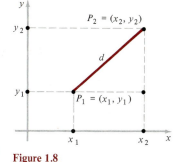

$$d^2 = |x_1 - x_2|^2 + |y_1 - y_2|^2$$
$$= (x_1 - x_2)^2 + (y_1 - y_2)^2,$$

so
$$d = \sqrt{(x_1 - x_2)^2 + (y_1 - y_2)^2}. \tag{1}$$

Figure 1.8

This is the *distance formula*.

Example 1 The distance d between the points $(-4, 3)$ and $(3, -2)$ in Fig. 1.6 is

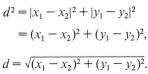

$$d = \sqrt{(-4 - 3)^2 + (3 + 2)^2} = \sqrt{74}.$$

Notice that in applying formula (1) it does not matter in which order the points are taken.

Example 2 Find the lengths of the sides of the triangle whose vertices are $P_1 = (-1, -3)$, $P_2 = (5, -1)$, $P_3 = (-2, 10)$.
  By (1), these lengths are

$$P_1 P_2 = \sqrt{(-1 - 5)^2 + (-3 + 1)^2} = \sqrt{40} = 2\sqrt{10},$$

$$P_1 P_3 = \sqrt{(-1 + 2)^2 + (-3 - 10)^2} = \sqrt{170},$$

$$P_2 P_3 = \sqrt{(5 + 2)^2 + (-1 - 10)^2} = \sqrt{170}.$$

These calculations reveal that the triangle is isosceles, with $P_1 P_3$ and $P_2 P_3$ as the equal sides.

THE MIDPOINT FORMULAS

It is often useful to know the coordinates of the midpoint of the segment joining two given distinct points. If the given points are $P_1 = (x_1, y_1)$ and $P_2 = (x_2, y_2)$, and if $P = (x, y)$ is the midpoint, then it is clear from Fig. 1.9 that x is the midpoint of the projection of the segment on the x-axis, and similarly for y. This tells us (see Problem 7 in Section 1.2) that $x = x_1 + \frac{1}{2}(x_2 - x_1)$ and $y = y_1 + \frac{1}{2}(y_2 - y_1)$, so

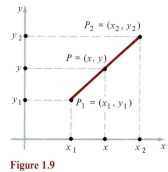

$$x = \tfrac{1}{2}(x_1 + x_2) \quad \text{and} \quad y = \tfrac{1}{2}(y_1 + y_2).$$

Another way of obtaining these formulas is to notice from Fig. 1.9 that $x - x_1 = x_2 - x$, so $2x = x_1 + x_2$ or $x = \frac{1}{2}(x_1 + x_2)$, with the same argument applying to y. Similarly, if P is a trisection point of the segment joining P_1 and P_2, its coordinates can be found from the fact that x and y are trisection points of the corresponding segments on the x-axis and y-axis.

Figure 1.9

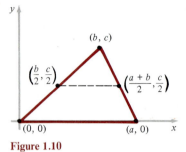

Figure 1.10

Example 3 In any triangle, the segment joining the midpoints of two sides is parallel to the third side and half its length. To prove this by our methods, we begin by noticing that the triangle can always be placed in the position shown in Fig. 1.10, with its third side along the positive x-axis and the left endpoint of this side at the origin. We then insert the midpoints of the other two sides, as shown, and observe that since they have the same y-coordinate, the segment joining them is parallel to the third side lying on the x-axis. The length of this segment is simply the difference between the x-coordinates of its endpoints,

$$\frac{a+b}{2} - \frac{b}{2} = \frac{a}{2},$$

which is half the length of the third side.

This example illustrates the way in which coordinates can be used to give algebraic proofs of many geometric theorems. The device employed here, placing the figure in a convenient position relative to the coordinate system — or equivalently, choosing the coordinate system in a convenient position relative to the figure — has the purpose of simplifying the algebra.

PROBLEMS

1 Draw a sketch indicating the points (x, y) in the plane for which
   (a) $x < 2$;
   (b) $-1 < y \le 2$;
   (c) $0 \le x \le 1$ and $0 \le y \le 1$;
   (d) $x = -1$;
   (e) $y = 3$;
   (f) $x = y$.

2 Use the distance formula to show that the points $(-2, 1)$, $(2, 2)$, and $(10, 4)$ lie on a straight line.

3 Show that the point $(6, 5)$ lies on the perpendicular bisector of the segment joining the points $(-2, 1)$ and $(2, -3)$.

4 Show that the triangle whose vertices are $(3, -3)$, $(-3, 3)$, and $(3\sqrt{3}, 3\sqrt{3})$ is equilateral.

5 The two points $(2, -2)$ and $(-6, 5)$ are the endpoints of a diameter of a circle. Find the center and radius of the circle.

6 Find every point whose distance from each of the two coordinate axes equals its distance from the point $(4, 2)$.

7 Find the point equidistant from the three points $(-9, 0)$, $(6, 3)$, and $(-5, 6)$.

8 If a and b are any two numbers, convince yourself that:
   (a) the points (a, b) and $(a, -b)$ are symmetric with respect to the x-axis;
   (b) (a, b) and $(-a, b)$ are symmetric with respect to the y-axis;

   (c) (a, b) and $(-a, -b)$ are symmetric with respect to the origin.

9 What symmetry statement can be made about the points (a, b) and (b, a)?

10 In each case, place the figure in a convenient position relative to the coordinate system and prove the statement algebraically:
   (a) The diagonals of a parallelogram bisect each other.
   (b) The sum of the squares of the diagonals of a parallelogram equals the sum of the squares of the sides.
   (c) The midpoint of the hypotenuse of a right triangle is equidistant from the three vertices.
   Use the fact stated in (c) to show that when the acute angles of a right triangle are 30° and 60°, the side opposite the 30° angle is half the hypotenuse.

11 In an isosceles right triangle, both acute angles are 45°. If the hypotenuse is h, what is the length of each of the other sides?

12 Let $P_1 = (x_1, y_1)$ and $P_2 = (x_2, y_2)$ be distinct points. If $P = (x, y)$ is on the segment joining P_1 and P_2 and one-third of the way from P_1 to P_2, show that

   $$x = \tfrac{1}{3}(2x_1 + x_2) \quad \text{and} \quad y = \tfrac{1}{3}(2y_1 + y_2).$$

   Find the corresponding formulas if P is two-thirds of the way from P_1 to P_2.

13 Consider an arbitrary triangle with vertices (x_1, y_1), (x_2, y_2), and (x_3, y_3). Find the point on each median which is two-thirds of the way from the vertex to the midpoint of the opposite side. Perform the calculations separately for each median and verify that these three points are all the same, with coordinates

$$\tfrac{1}{3}(x_1 + x_2 + x_3) \qquad \text{and} \qquad \tfrac{1}{3}(y_1 + y_2 + y_3).$$

This proves that the medians of any triangle intersect at a point which is two-thirds of the way from each vertex to the midpoint of the opposite side.

In this section we use the language of algebra to describe the set of all points that lie on a given straight line. This algebraic description is called the *equation of the line*. First, however, it is necessary to discuss an important preliminary concept.

1.4

SLOPES AND EQUATIONS OF STRAIGHT LINES

THE SLOPE OF A LINE

Any nonvertical straight line has a number associated with it that specifies its direction, called its *slope*. This number is defined as follows (Fig. 1.11).

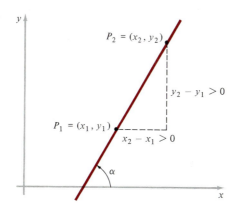

Figure 1.11

Choose any two distinct points on the line, say $P_1 = (x_1, y_1)$ and $P_2 = (x_2, y_2)$. Then the slope is denoted by m and defined to be the ratio

$$m = \frac{y_2 - y_1}{x_2 - x_1}. \tag{1}$$

If we reverse the order of subtraction in both numerator and denominator, then the sign of each is changed, so m is unchanged:

$$m = \frac{y_2 - y_1}{x_2 - x_1} = \frac{y_1 - y_2}{x_1 - x_2}.$$

This shows that the slope can be computed as the difference of the y-coordinates divided by the difference of the x-coordinates — in either order, as long as both differences are formed in the same order. In Fig. 1.11, where P_2 is placed to the right of P_1 and the line rises to the right, it is clear that the slope as defined by (1) is simply the ratio of the height to the base in the indicated right triangle. It is necessary to know that the value of m depends only on the line itself and is the same no matter where the points P_1 and P_2 happen to be located on the line. This is easy to see by visualizing the effect of moving P_1

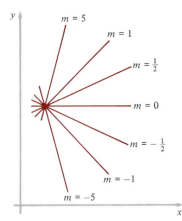

Figure 1.12 A variety of slopes.

and P_2 to different positions on the line; this change gives rise to a similar right triangle, and therefore leaves the ratio in (1) unaltered.

If we choose the position of P_2 so that $x_2 - x_1 = 1$, that is, if we place P_2 1 unit to the right of P_1, then $m = y_2 - y_1$. This tells us that the slope is simply the change in y as a point (x, y) moves along the line in such a way that x increases by 1 unit. This change in y can be positive, negative, or zero, depending on the direction of the line. We therefore have the following important correlations between the sign of m and the indicated directions:

$$m > 0, \quad \text{line rises to the right;}$$
$$m < 0, \quad \text{line falls to the right;}$$
$$m = 0, \quad \text{line horizontal.}$$

Further, the absolute value of m is a measure of the steepness of the line (Fig. 1.12). It is evident from (1) why a vertical line has no slope, for in this case the two points have equal x-coordinates and the denominator in (1) is 0.

If the line under discussion crosses the x-axis, then the angle α from the positive x-direction to the line, measured counterclockwise, is called the *inclination*—or sometimes the *angle of inclination*—of the line. Students who have studied trigonometry will see from Fig. 1.11 that the slope is the tangent of this angle, $m = \tan \alpha$.

EQUATIONS OF A LINE

A vertical line is characterized by the fact that all points on it have the same x-coordinate. If the line crosses the x-axis at the point $(a, 0)$, then a point (x, y) lies on the line if and only if

$$x = a, \tag{2}$$

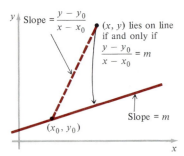

Figure 1.13

as illustrated in Fig. 1.13. The statement that (2) is the equation of the line means precisely this: A point (x, y) lies on the line if and only if condition (2) is satisfied.

Next consider a nonvertical line, and let it be "given" in the sense that we know a point (x_0, y_0) on it and its slope m (Fig. 1.14). If (x, y) is a point in the plane that does not lie on the vertical line through (x_0, y_0), then it is easy to see that this point lies on the given line if and only if the line determined by (x_0, y_0) and (x, y) has the same slope as the given line:

$$\frac{y - y_0}{x - x_0} = m. \tag{3}$$

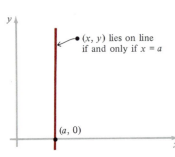

Figure 1.14

This would be the equation of our line except for the minor flaw that the coordinates of the point (x_0, y_0)—which is certainly on the line—do not satisfy the equation (they reduce the left side to the meaningless expression 0/0). This flaw is easily removed by writing equation (3) in the form

$$y - y_0 = m(x - x_0). \tag{4}$$

Nevertheless, we usually prefer the form (3), because its direct connection with the geometric idea illustrated in Fig. 1.14 makes it easy to remember. Either equation (or both) is called the *point-slope equation* of a line, since the line is initially specified by means of a known point on it and its known slope. To grasp more firmly the meaning of equation (4), imagine a point (x, y) moving along the given line. As this point moves, its coordinates x and y change; but even though they change, they are bound together by the fixed relationship expressed by equation (4).

If the known point on the line happens to be the point where the line crosses the y-axis, and if this point is denoted by $(0, b)$, then equation (4) becomes $y - b = mx$ or

$$y = mx + b. \tag{5}$$

The number b is called the *y-intercept* of the line, and (5) is called the *slope-intercept equation* of a line. This form is especially convenient because it tells at a glance the location and direction of a line. For example, if the equation

$$6x - 2y - 4 = 0 \tag{6}$$

is solved for y, we see that

$$y = 3x - 2. \tag{7}$$

Comparing (7) with (5) shows at once that $m = 3$ and $b = -2$, and so (6) and (7) both represent the line that passes through $(0, -2)$ with slope 3. This information makes it very easy to sketch the line. It may seem that (6) and (7) are different equations, so that (6) should be referred to as "an" equation of the line and (7) as "another" equation of the line, but we prefer to regard them as merely different forms of a single equation. Many other forms are possible, for instance,

$$y + 2 = 3x, \qquad x = \tfrac{1}{3}y + \tfrac{2}{3}, \qquad 3x - y = 2.$$

It is reasonable to cut through appearances and speak of any one of these as "the" equation of the line.

More generally, every equation of the form

$$Ax + By + C = 0, \tag{8}$$

where the constants A and B are not both zero, represents a straight line. For if $B = 0$, then $A \neq 0$, and the equation can be written as

$$x = -\frac{C}{A},$$

which is clearly the equation of a vertical line. On the other hand, if $B \neq 0$, then

$$y = -\frac{A}{B}x - \frac{C}{B},$$

and this equation has the form (5) with $m = -A/B$ and $b = -C/B$. Equation (8) is rather inconvenient for most purposes because its constants are not directly related to the geometry of the line. Its main merit is that it is capable of representing all lines, without any need for distinguishing between the vertical and nonvertical cases. For this reason it is called the *general linear equation*.

PARALLEL AND PERPENDICULAR LINES

Two nonvertical straight lines with slopes m_1 and m_2 are evidently parallel if and only if their slopes are equal:

$$m_1 = m_2.$$

The criterion for perpendicularity is the relation

$$m_1 m_2 = -1. \tag{9}$$

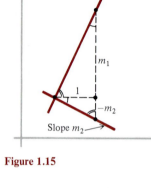

Slope m_1

m_1

1

$-m_2$

Slope m_2

Figure 1.15

This is not obvious, but can be established quite easily by using similar triangles, as follows (Fig. 1.15). Suppose that the lines are perpendicular, as shown in Fig. 1.15. Draw a segment of length 1 to the right from their point of intersection, and from its right endpoint draw vertical segments up and down to the two lines. From the meaning of the slopes, the two right triangles formed in this way have sides of the indicated lengths. Since the lines are perpendicular, the indicated angles are equal and the triangles are similar. This similarity implies that the following ratios of corresponding sides are equal:

$$\frac{m_1}{1} = \frac{1}{-m_2}.$$

This is equivalent to (9), so (9) is true when the lines are perpendicular. The reasoning given here is easily reversed, telling us that if (9) is true, then the lines are perpendicular. Since equation (9) is equivalent to

$$m_1 = -\frac{1}{m_2} \quad \text{and} \quad m_2 = -\frac{1}{m_1},$$

we see that two nonvertical lines are perpendicular if and only if their slopes are negative reciprocals of one another.

The ideas of this section enlarge our supply of tools for proving geometric theorems by algebraic methods.

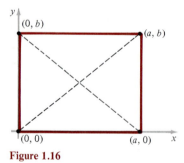

$(0, b)$ (a, b)

$(0, 0)$ $(a, 0)$

Figure 1.16

Example If the diagonals of a rectangle are perpendicular, then the rectangle is a square. To establish this, we place the rectangle in the convenient position shown in Fig. 1.16. The slopes of the diagonals are clearly b/a and $-b/a$. If these diagonals are perpendicular, then

$$\frac{b}{a} = \frac{a}{b}, \quad a^2 = b^2, \quad a^2 - b^2 = 0, \quad \text{and} \quad (a+b)(a-b) = 0.$$

The last equation implies that $a = b$, so the rectangle is a square.

PROBLEMS

1 Plot each pair of points, draw the line they determine, and compute the slope of this line:

(a) $(-3, 1), (4, -1)$;   (b) $(2, 7), (-1, -1)$;
(c) $(-4, 0), (2, 1)$;   (d) $(-4, 3), (5, -6)$;
(e) $(-5, 2), (7, 2)$;   (f) $(0, -4), (1, 6)$.

2 Plot each of the following sets of three points, and use slopes to determine in each case whether all three points lie on a single straight line:

(a) $(5, -1), (2, 2), (-4, 6)$;
(b) $(1, 1), (-5, -2), (5, 3)$;

(c) $(4, 3)$, $(10, 14)$, $(-2, -8)$;

(d) $(-1, 3)$, $(6, -1)$, $(-9, 7)$.

3 Plot each of the following sets of three points, and use slopes to determine in each case whether the points form a right triangle:

(a) $(2, -3)$, $(5, 2)$, $(0, 5)$;

(b) $(10, -5)$, $(5, 4)$, $(-7, -2)$;

(c) $(8, 2)$, $(-1, -1)$, $(2, -7)$;

(d) $(-2, 6)$, $(3, -4)$, $(8, 11)$.

4 Write the equation of each line in Problem 1 using the point-slope form; then rewrite each of these equations in the form $y = mx + b$ and find the y-intercept.

5 Find the equation of the line:

(a) through $(2, -3)$ with slope -4;

(b) through $(-4, 2)$ and $(3, -1)$;

(c) with slope $\frac{2}{3}$ and y-intercept -4;

(d) through $(2, -4)$ and parallel to the x-axis;

(e) through $(1, 6)$ and parallel to the y-axis;

(f) through $(4, -2)$ and parallel to $x + 3y = 7$;

(g) through $(5, 3)$ and perpendicular to $y + 7 = 2x$;

(h) through $(-4, 3)$ and parallel to the line determined by $(-2, -2)$ and $(1, 0)$;

(i) that is the perpendicular bisector of the segment joining $(1, -1)$ and $(5, 7)$;

(j) through $(-2, 3)$ with inclination $135°$.

6 If a line crosses the x-axis at the point $(a, 0)$, the number a is called the x-*intercept* of the line. If a line has x-intercept $a \neq 0$ and y-intercept $b \neq 0$, show that its equation can be written as

$$\frac{x}{a} + \frac{y}{b} = 1.$$

This is called the *intercept form* of the equation of a line. Notice that it is easy to put $y = 0$ and see that the line crosses the x-axis at $x = a$, and to put $x = 0$ and see that the line crosses the y-axis at $y = b$.

7 Put each equation in intercept form and sketch the corresponding line:

(a) $5x + 3y + 15 = 0$;   (b) $3x = 8y - 24$;

(c) $y = 6 - 6x$;   (d) $2x - 3y = 9$.

8 The set of all points (x, y) that are equally distant from the points $P_1 = (-1, -3)$ and $P_2 = (5, -1)$ is the perpendicular bisector of the segment joining these points. Find its equation

(a) by equating the distances from (x, y) to P_1 and P_2, and simplifying the resulting equation;

(b) by finding the midpoint of the given segment and using a suitable slope.

9 Sketch the lines $3x + 4y = 7$ and $x - 2y = 6$, and find their point of intersection. Hint: Their point of intersection is that point (x, y) whose coordinates satisfy both equations simultaneously.

10 Find the point of intersection of each of the following pairs of lines:

(a) $2x + 2y = 2$, $y = x - 1$;

(b) $10x + 7y = 24$, $15x - 4y = 7$;

(c) $3x - 5y = 7$, $15y + 25 = 9x$.

11 Let F and C denote temperature in degrees Fahrenheit and degrees Celsius. Find the equation connecting F and C, given that it is linear and that $F = 32$ when $C = 0$, $F = 212$ when $C = 100$.

12 Find the values of the constant k for which the line $(k - 3)x - (4 - k^2)y + k^2 - 7k + 6 = 0$ is

(a) parallel to the x-axis;

(b) parallel to the y-axis;

(c) through the origin.

13 Show that the segments joining the midpoints of adjacent sides of any quadrilateral form a parallelogram.

14 Show that the lines from any vertex of a parallelogram to the midpoints of the opposite sides trisect a diagonal.

15 Let $(0, 0)$, $(a, 0)$, and (b, c) be the vertices of an arbitrary triangle placed so that one side lies along the positive x-axis with its left endpoint at the origin. If the square of this side equals the sum of the squares of the other two sides, use slopes to show that the triangle is a right triangle. Thus, the converse of the Pythagorean theorem is also true.

The coordinate plane or xy-plane is often called the *Cartesian plane,* and x and y are frequently referred to as the *Cartesian coordinates* of the point $P = (x, y)$. The word "Cartesian" comes from Cartesius, the Latinized name of the French philosopher-mathematician Descartes, who was one of the two principal founders of analytic geometry.* The basic idea of this subject is quite simple: Exploit the correspondence between points and their coordinates to study geometric problems — especially the properties of curves — with the tools of algebra. The reader will see this idea in action throughout

1.5

CIRCLES AND PARABOLAS

---

* The other (also French) was Fermat, a less well known figure than Descartes but a much greater mathematician.

this book. Generally speaking, geometry is visual and intuitive, while algebra is rich in computational machinery, and each can serve the other in many fruitful ways.

Most people who have had a course in algebra are acquainted with the fact that an equation

$$F(x, y) = 0 \qquad (1)$$

usually determines a curve (its *graph*) which consists of all points $P = (x, y)$ whose coordinates satisfy the given equation. Conversely, a curve defined by some geometric condition can usually be described algebraically by an equation of the form (1). It is intuitively clear that straight lines are the simplest curves, and our work in Section 1.4 demonstrated that straight lines in the coordinate plane correspond to linear equations in x and y. We now develop algebraic descriptions of several other curves that will be useful as illustrative examples in the next few chapters.

CIRCLES

The distance formula of Section 1.3 is often useful in finding the equation of a curve whose geometric definition depends on one or more distances.

One of the simplest curves of this kind is a circle, which can be defined as the set of all points at a given distance (the radius) from a given point (the center). If the center is the point (h, k) and the radius is the positive number r (Fig. 1.17), and if (x, y) is an arbitrary point on the circle, then the defining condition says that

$$\sqrt{(x - h)^2 + (y - k)^2} = r.$$

It is convenient to eliminate the radical sign by squaring, which yields

$$(x - h)^2 + (y - k)^2 = r^2. \qquad (2)$$

This is therefore the equation of the circle with center (h, k) and radius r. In particular, if the center happens to be the origin, so that $h = k = 0$, then

$$x^2 + y^2 = r^2$$

is the equation of the circle.

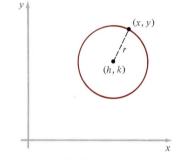

Figure 1.17   Circle.

Example 1   If the radius of a circle is $\sqrt{10}$ and its center is $(-3, 4)$, then its equation is

$$(x + 3)^2 + (y - 4)^2 = 10.$$

Notice that the coordinates of the center are the numbers *subtracted* from x and y in the parentheses.

Example 2   An angle inscribed in a semicircle is necessarily a right angle. To prove this algebraically, let the semicircle have radius r and center at the origin (Fig. 1.18), so that its equation is $x^2 + y^2 = r^2$ with $y \geq 0$. The inscribed angle is a right angle if and only if the product of the slopes of its sides is -1, that is,

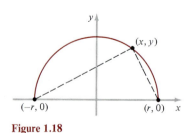

Figure 1.18

$$\frac{y}{x-r} \cdot \frac{y}{x+r} = -1. \tag{3}$$

This is easily seen to be equivalent to $x^2 + y^2 = r^2$, which is certainly true for any point (x, y) on the semicircle, so (3) is true and the angle is a right angle.

It is clear that any equation of the form (2) is easy to interpret geometrically. For instance,

$$(x - 5)^2 + (y + 2)^2 = 16 \tag{4}$$

is immediately recognizable as the equation of the circle with center $(5, -2)$ and radius 4, and this information enables us to sketch the graph without difficulty. However, if the equation has been roughly treated by someone who likes to "simplify" things algebraically, then it might have the form

$$x^2 + y^2 - 10x + 4y + 13 = 0. \tag{5}$$

This is an equivalent but scrambled version of (4), and its constants tell us nothing directly about the nature of the graph. To find out what the graph is, we must "unscramble" by completing the square.* To do this, we begin by rewriting equation (5) as

$$(x^2 - 10x + \quad) + (y^2 + 4y + \quad) = -13,$$

with the constant term moved to the right and blank spaces provided for the insertion of suitable constants. When the square of half the coefficient of x is added in the first blank space and the square of half the coefficient of y in the second, and the same constants are added to the right side to maintain the balance of the equation, we get

$$(x^2 - 10x + 25) + (y^2 + 4y + 4) = -13 + 25 + 4$$

or

$$(x - 5)^2 + (y + 2)^2 = 16. \tag{6}$$

Exactly the same process can be applied to the general equation of the form (5), namely,

$$x^2 + y^2 + Ax + By + C = 0, \tag{7}$$

but there is little to be gained by writing out the details in this general case. However, it is important to notice that if the constant term 13 in (5) is replaced by 29, then (6) becomes

$$(x - 5)^2 + (y + 2)^2 = 0,$$

whose graph is the single point $(5, -2)$. Similarly, if this constant term is replaced by any number greater than 29, then the right-hand side of (6) becomes negative and the graph is empty, in the sense that there are no points (x, y) in the plane whose coordinates satisfy the equation. We therefore see

---

* The form of the equation $(x + a)^2 = x^2 + 2ax + a^2$ is the key to the process of completing the square. Notice that the right side is a perfect square — the square of $x + a$ — precisely because its constant term is the square of half the coefficient of x.

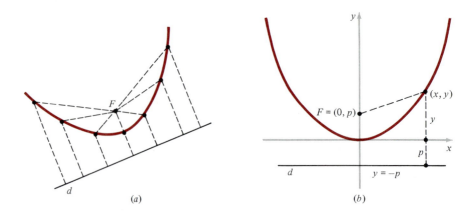

Figure 1.19   Parabola.   (a)   (b)

that the graph of (7) is sometimes a circle, sometimes a single point, and sometimes empty—depending entirely on the constants A, B, and C.

PARABOLAS

The definition we use for a parabola is the following (Fig. 1.19a): It is the curve consisting of all points that are equally distant from a fixed point F (called the *focus*) and a fixed line d (called the *directrix*). The distance from a point to a line is always understood to mean the perpendicular distance.

To find a simple equation for a parabola, we place it in the coordinate system as shown in Fig. 1.19b, with the focus and directrix equally far above and below the x-axis. The line through the focus perpendicular to the directrix is called the *axis* of the parabola; this is the axis of symmetry of the curve, and is the y-axis in the figure. The point on the axis halfway between the focus and the directrix is called the *vertex* of the parabola; in the figure this point is the origin. If (x, y) is an arbitrary point on the parabola, the condition expressed in the definition is stated algebraically by the equation

$$\sqrt{x^2 + (y - p)^2} = y + p. \tag{8}$$

On squaring both sides and simplifying, we obtain

$$x^2 + y^2 - 2py + p^2 = y^2 + 2py + p^2$$

or

$$x^2 = 4py. \tag{9}$$

These steps are reversible, so (8) and (9) are equivalent and (9) is the equation of the parabola whose focus and directrix are located as shown in Fig. 1.19b. Notice particularly that the positive constant p in (9) is the distance from the focus to the vertex, and also from the vertex to the directrix.

If we change the position of the parabola relative to the coordinate axes, we naturally change its equation. Three other positions are shown in Fig. 1.20, each with its corresponding equation and with $p > 0$ in each case. Students should verify the correctness of all three equations. We also point out that each of these four equations can be put in the form

$$y = ax^2 \tag{10}$$

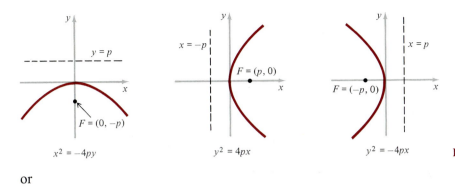

$$x^2 = -4py \qquad\qquad y^2 = 4px \qquad\qquad y^2 = -4px$$

Figure 1.20 Various parabolas.

or

$$x = ay^2.$$

These forms conceal the constant p, with its geometric significance, but as compensation they are more useful in visualizing the overall appearance of the graph. For instance, in (10) the variable x is squared but y is not. This tells us that as a point (x, y) moves out along the curve, y increases much faster than x, and so the curve opens in the y-direction — upward or downward, according as a is positive or negative. It also tells us that the graph is symmetric with respect to the y-axis, because x is squared, and therefore we get the same number y for any number x and its negative.

Example 3 What is the graph of the equation $12x + y^2 = 0$? If this is put in the form $y^2 = -12x$ and compared with the equation on the right in Fig. 1.20, it is clear that the graph is a parabola with vertex at the origin and opening to the left. Since $4p = 12$ and therefore $p = 3$, the point $(-3, 0)$ is the focus and $x = 3$ is the directrix.

Example 4 The graph of $y = 2x^2$ is evidently a parabola with vertex at the origin and opening upward. To find its focus and directrix, the equation must be rewritten as $x^2 = \frac{1}{2}y$ and compared with equation (9). This yields $4p = \frac{1}{2}$, so $p = \frac{1}{8}$. The focus is therefore $(0, \frac{1}{8})$, and the directrix is $y = -\frac{1}{8}$.

We illustrate one last point about parabolas by examining the equation

$$y = x^2 - 4x + 5. \tag{11}$$

If this is written as

$$y - 5 = x^2 - 4x,$$

and if we complete the square on the terms involving x, then the result is

$$y - 1 = (x - 2)^2. \tag{12}$$

If we now introduce the new variables

$$X = x - 2,$$
$$Y = y - 1, \tag{13}$$

then equation (12) becomes

$$Y = X^2.$$

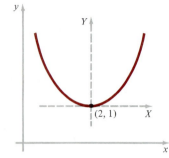

Figure 1.21

The graph of this equation is clearly a parabola opening upward with vertex at the origin of the XY coordinate system. By equations (13), the origin in the XY system is the point (2, 1) in the xy system, as shown in Fig. 1.21. What has happened here is that the coordinate system has been shifted or translated to a new position in the plane, and the axes renamed, and equations (13) express the relation between the coordinates of an arbitrary point with respect to each of the two coordinate systems. In exactly the same way, any equation of the form

$$y = ax^2 + bx + c, \qquad a \neq 0,$$

represents a parabola with vertical axis which opens up or down according as the number a is positive or negative. Similarly, the equation

$$x = ay^2 + by + c, \qquad a \neq 0,$$

represents a parabola with horizontal axis which opens to the right or left according as $a > 0$ or $a < 0$.

In our work up to this stage we have used the static concept of a curve as a certain set of points or geometric figure. It is often possible to adopt the dynamic point of view, in which a curve is thought of as the path of a moving point. For instance, a circle is the path of a point that moves in such a way that it maintains a fixed distance from a given point. When this mode of thought is used — with its advantage of greater intuitive vividness — a curve is often called a *locus*. Thus, a parabola is the locus of a point that moves in such a way that it maintains equal distances from a given point and a given line.

PROBLEMS

1 Find the equation of the circle with the given point as center and the given number as radius:
(a) (4, 6), 3;   (b) $(-3, 7)$, $\sqrt{5}$;
(c) $(-5, -9)$, 7;   (d) $(1, -6)$, $\sqrt{2}$;
(e) $(a, 0)$, a;   (f) $(0, a)$, a.

2 In each case find the equation of the circle determined by the given conditions:
(a) Center (2, 3) and passes through $(-1, -2)$.
(b) The ends of a diameter are $(-3, 2)$ and $(5, -8)$.
(c) Center (4, 5) and tangent to the x-axis.
(d) Center $(-4, 1)$ and tangent to the line $x = 3$.
(e) Center $(-2, 3)$ and tangent to the line $4y - 3x + 2 = 0$.
(f) Center on the line $x + y = 1$, passes through $(-2, 1)$ and $(-4, 3)$.
(g) Center on the line $y = 3x$ and tangent to the line $x = 2y$ at the point (2, 1).

3 In each of the following, determine the nature of the graph of the given equation by completing the square:
(a) $x^2 + y^2 - 4x - 4y = 0$.
(b) $x^2 + y^2 - 18x - 14y + 130 = 0$.

(c) $x^2 + y^2 + 8x + 10y + 40 = 0$.
(d) $4x^2 + 4y^2 + 12x - 32y + 37 = 0$.
(e) $x^2 + y^2 - 8x + 12y + 53 = 0$.
(f) $x^2 + y^2 - \sqrt{2}x + \sqrt{2}y + 1 = 0$.
(g) $x^2 + y^2 - 16x + 6y - 48 = 0$.

4 Find the equation of the locus of a point $P = (x, y)$ that moves in accordance with each of the following conditions, and sketch the graphs:
(a) The sum of the squares of the distances from P to the points $(a, 0)$ and $(-a, 0)$ is $4b^2$, where $b \geq a > 0$.
(b) The distance of P from the point (8, 0) is twice its distance from the point (0, 4).

5 The *quadratic formula* for the roots of the quadratic equation $ax^2 + bx + c = 0$ is

$$x = \frac{-b \pm \sqrt{b^2 - 4ac}}{2a}.$$

Derive this formula from the equation by dividing through by a, moving the constant term to the right side, and completing the square. Under what circumstances does the equation have distinct real roots, equal real roots, and no real roots?

6 At what points does the circle $x^2 + y^2 - 8x - 6y - 11 = 0$ intersect
   (a) the x-axis?   (b) the y-axis?
   (c) the line $x + y = 1$?
   Sketch the figure, and use this picture to judge whether your answers are reasonable or not.

7 Find the equations of all lines that are tangent to the circle $x^2 + y^2 = 2y$ and pass through the point (0, 4). Hint: The line $y = mx + 4$ is tangent to the circle if it intersects the circle at only one point.

8 Find the focus and directrix of each of the following parabolas, and sketch the curves:
   (a) $y^2 = 12x$;   (b) $y = 4x^2$;
   (c) $2x^2 + 5y = 0$;   (d) $4x + 9y^2 = 0$;
   (e) $x = -2y^2$;   (f) $12y = -x^2$;
   (g) $16y^2 = x$;   (h) $24x^2 = y$;
   (i) $y^2 + 8y - 16x = 16$;   (j) $x^2 + 2x + 29 = 7y$.

9 Sketch the parabola and find its equation if it has
   (a) vertex (0, 0) and focus (−3, 0);
   (b) vertex (0, 0) and directrix $y = -1$;
   (c) vertex (0, 0) and directrix $x = -2$;
   (d) vertex (0, 0) and focus $(0, -\frac{1}{3})$;
   (e) directrix $x = 2$ and focus (−4, 0);
   (f) focus (3, 3) and directrix $y = -1$.

10 Find the focus and directrix of each of the following parabolas, and sketch the curves:
   (a) $y = x^2 + 1$;
   (b) $y = (x - 1)^2$;
   (c) $y = (x - 1)^2 + 1$;
   (d) $y = x^2 - x$.

11 Water squirting out of a horizontal nozzle held 4 ft above the ground describes a parabolic curve with the vertex at the nozzle. If the stream of water drops 1 ft in the first 10 ft of horizontal motion, at what horizontal distance from the nozzle will it strike the ground?

12 Show that there is exactly one line with given slope m which is tangent to the parabola $x^2 = 4py$, and find its equation.

13 Prove that the two tangents to a parabola from any point on the directrix are perpendicular.

1.6
THE CONCEPT OF A FUNCTION

The most important concept in all of mathematics is that of a function. No matter what branch of the subject we consider — algebra, geometry, number theory, probability, or any other — it almost always turns out that functions are the primary objects of investigation. This is particularly true of calculus, in which most of our work will be concerned with developing machinery for the study of functions and applying this machinery to problems in science and geometry.

What is a function? Let us begin to answer this question by examining the equation

$$y = x^2$$

and its corresponding graph, which we know is a parabola that opens upward and has its vertex at the origin (Fig. 1.22). In Section 1.5 we thought of this equation as a relation between the variable coordinates of a point (x, y) moving along the curve. We now shift our point of view, and instead think of it as a formula that provides a mechanism for calculating the numerical value of y when the numerical value of x is given. Thus, $y = 1$ when $x = 1$, $y = 4$ when $x = 2$, $y = \frac{1}{4}$ when $x = \frac{1}{2}$, $y = 1$ when $x = -1$, and so on. The value of y is therefore said to *depend on,* or to be *a function of,* the value of x. This dependence can be expressed in functional notation by writing

$$y = f(x) \qquad \text{where} \qquad f(x) = x^2.$$

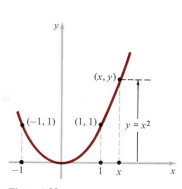

Figure 1.22

The symbol $f(x)$ is read "f of x," and the letter f represents the rule or process — squaring, in this particular case — which is applied to any number x to yield the corresponding number y. The numerical examples just given can therefore be written as $f(1) = 1, f(2) = 4, f(\frac{1}{2}) = \frac{1}{4}$, and $f(-1) = 1$. The meaning of this notation can perhaps be further clarified by observing that

$$f(x + 1) = (x + 1)^2 = x^2 + 2x + 1 \qquad \text{and} \qquad f(x^3) = (x^3)^2 = x^6;$$

that is, the rule f simply produces the square of whatever quantity follows it in parentheses.

We now set aside this special case and formulate the general concept of a function as we shall use it in most of our work.

Let D be a given set of real numbers. A *function* f defined on D is a rule, or law of correspondence, that assigns a single real number y to each number x in D. The set D of allowed values of x is called the *domain* (or *domain of definition*) of the function, and the set of corresponding values of y is called its *range*. The number y that is assigned to x by the function f is usually written $f(x)$ — so that $y = f(x)$ — and is called the *value* of f at x. It is customary to call x the *independent variable* because it is free to assume any value in the domain, and to call y the *dependent variable* because its numerical value depends on the choice of x. Of course, there is nothing essential about the use of the letters x and y here, and any other letters would do just as well.

The reader is doubtless acquainted with the idea of the *graph* of a function f: If we imagine the domain D spread out on the x-axis in the coordinate plane (Fig. 1.23), then to each number x in D there corresponds a number $y = f(x)$, and the set of all the resulting points (x, y) in the plane is the graph. Graphs are visual aids of great value that enable us to see functions in their entirety, and we will examine many in Section 1.8.

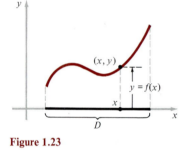

Figure 1.23

Originally, the only functions mathematicians considered were those defined by formulas. This led to the useful intuitive idea that a function f "does something" to each number x in its domain to "produce" the corresponding number $y = f(x)$. Thus, if

$$y = f(x) = (x^3 + 4)^2,$$

then y is the result of applying certain specific operations to x: Cube it, add 4, and square the sum. On the other hand, the following is also a perfectly legitimate function which is given by a verbal prescription instead of a formula:

$$y = f(x) = \begin{cases} 0 & \text{if } x \text{ is a rational number} \\ 1 & \text{if } x \text{ is an irrational number.} \end{cases}$$

All that is really required of a function is that y be uniquely determined — in any manner whatever — when x is specified; beyond this, nothing is said about the nature of the rule f. In discussions that focus on ideas instead of specific problems, such broad generality is often an advantage.

A few additional remarks on usage are perhaps in order. Strictly speaking, the word "function" refers to the rule of correspondence f that assigns a unique number $y = f(x)$ to each number x in the domain. Purists are fond of emphasizing the distinction between the function f and its value $f(x)$ at x. However, once this distinction is clearly understood, most people who work with mathematics prefer to use the word loosely and speak of "the function $y = f(x)$," or even "the function $f(x)$." Further, when the rule is defined by a formula, as in the example $f(x) = x^2$, we also speak of "the function $f(x) = x^2$" or "the function $y = x^2$," or even, if there is no need to refer to either y or $f(x)$, "the function x^2." It is clear that any letter can be used to denote a function. There is nothing sacred about the letter f, but it is the favorite for obvious reasons, and g, h, F, G, H, and many others are also popular. It often happens that we want to discuss functions in general without committing ourselves as to exactly which function we're talking about. The notation $y = f(x)$ is almost invariably used in these discussions.

Note that a function $f(x)$ is not fully known until we know precisely which real numbers are permissible values for the independent variable x. The domain is therefore an indispensable part of the concept of a function. In practice, however, most of the specific functions we deal with are defined only by formulas, such as

$$f(x) = \frac{1}{(x - 1)(x + 2)}, \tag{1}$$

and nothing is said about the domain. Unless we state otherwise, the domain of such a function is understood to be the set of all real numbers x for which the formula makes sense. In this case the only excluded values of x are those that make the denominator zero, since division by zero has no meaning in algebra. The domain of (1) therefore consists of all real numbers except $x = 1$ and $x = -2$.

Example Consider the three functions defined by

$$f(x) = x^2, \tag{2}$$

$$g(x) = \frac{1}{x^2}, \tag{3}$$

$$h(x) = \sqrt{25 - x^2}. \tag{4}$$

The domain of (2) is evidently the set of all real numbers, and its range is the set of all nonnegative real numbers. The domain of (3) is the set of all real numbers except 0, and its range is the set of positive real numbers. In the case of (4), the main thing to keep in mind is that square roots of negative numbers are not real. Thus, the domain here is the set of all x's for which $25 - x^2 \geq 0$, namely the interval $-5 \leq x \leq 5$, and the range is the interval $[0, 5]$.

The functions we work with in calculus are often composite (or compound) functions built up out of simpler ones. As an illustration of this idea, consider the two functions

$$f(x) = x^2 + 3x \quad \text{and} \quad g(x) = x^2 - 1.$$

The single function that results from first applying g to x and then applying f to $g(x)$ is

$$f(g(x)) = f(x^2 - 1) = (x^2 - 1)^2 + 3(x^2 - 1)$$
$$= x^4 + x^2 - 2.$$

Notice that $f(x^2 - 1)$ is obtained by replacing x by $x^2 - 1$ in the formula $f(x) = x^2 + 3x$. The symbol $f(g(x))$ is read "f of g of x." If we apply the functions in the other order (first f, then g), we have

$$g(f(x)) = g(x^2 + 3x) = (x^2 + 3x)^2 - 1$$
$$= x^4 + 6x^3 + 9x^2 - 1,$$

and so $f(g(x))$ and $g(f(x))$ are different. In special cases it can happen that $f(g(x))$ and $g(f(x))$ are the same function of x, for example, if $f(x) = 2x - 3$ and $g(x) = -x + 6$:

$$f(g(x)) = f(-x + 6) = 2(-x + 6) - 3 = -2x + 9,$$

$$g(f(x)) = g(2x - 3) = -(2x - 3) + 6 = -2x + 9.$$

In each of these examples two given functions are combined into a single composite function. In most practical work we proceed in the other direction, and dissect composite functions into their simpler constituents. For example, if

$$y = (x^3 + 1)^7, \tag{5}$$

we can introduce an auxiliary variable u by writing $u = x^3 + 1$ and decompose (5) into

$$y = u^7 \quad \text{and} \quad u = x^3 + 1.$$

We shall see that decompositions of this kind are often useful in the problems of calculus.

PROBLEMS

1 If $f(x) = x^3 - 3x^2 + 4x - 2$, compute $f(1), f(2), f(3)$, $f(0), f(-1)$, and $f(-2)$.

2 If $f(x) = 2^x$, compute $f(1), f(3), f(5), f(0)$, and $f(-2)$.

3 If $f(x) = 4x - 3$, show that $f(2x) = 2f(x) + 3$.

4 What are the domains of $f(x) = 1/(x - 8)$ and $g(x) = x^3$? What is $h(x) = f(g(x))$? What is the domain of $h(x)$?

5 Find the domain of each of the following functions:
  (a) $\sqrt{x}$;
  (b) $\sqrt{-x}$;
  (c) $\sqrt{x^2}$;
  (d) $\sqrt{x^2 - 4}$;

  (e) $\dfrac{1}{x^2 - 4}$;

  (f) $\dfrac{1}{x^2 + 4}$;

  (g) $\sqrt{(x - 1)(x + 2)}$;

  (h) $\dfrac{1}{\sqrt{(x - 1)(x + 2)}}$;

  (i) $\sqrt{3 - 2x - x^2}$;

  (j) $\sqrt{\dfrac{x}{x - 2}}$.

6 If $f(x) = 1 - x$, show that $f(f(x)) = x$.

7 If $f(x) = x/(x - 1)$, compute $f(0), f(1), f(2), f(3)$, and $f(f(3))$. Show that $f(f(x)) = x$.

8 If $f(x) = (ax + b)/(x - a)$, show that $f(f(x)) = x$.

9 If $f(x) = 1/(1 - x)$, compute $f(0), f(1), f(2), f(f(2))$, and $f(f(f(2)))$. Show that $f(f(f(x))) = x$.

10 If $f(x) = ax$, show that $f(x) + f(1 - x) = f(1)$. Also verify that $f(x_1 + x_2) = f(x_1) + f(x_2)$ for all x_1 and x_2.

11 If $f(x) = 2^x$, use functional notation to express the fact that $2^{x_1} \cdot 2^{x_2} = 2^{x_1 + x_2}$.

12 If $f(x) = \log_{10} x$, use functional notation to express the fact that $\log_{10} x_1 x_2 = \log_{10} x_1 + \log_{10} x_2$.

13 A *linear* function is one that has the form $f(x) = ax + b$, where a and b are constants. If $g(x) = cx + d$ is also linear, is it always true that $f(g(x)) = g(f(x))$?

14 If $f(x) = ax + b$ is a linear function with $a \neq 0$, show that there exists a linear function $g(x) = \alpha x + \beta$ such that $f(g(x)) = x$.* Also show that for these two functions it is true that $f(g(x)) = g(f(x))$.

15 A *quadratic* function is one that has the form $f(x) = ax^2 + bx + c$, where a, b, c are constants and $a \neq 0$.
  (a) Find the values of the coefficients a, b, c if $f(0) = 3$, $f(1) = 2, f(2) = 9$.
  (b) Show that, no matter what values may be given to the coefficients a, b, c, the range of a quadratic function cannot be the set of all real numbers.

---

* The symbols α and β are letters of the Greek alphabet whose names are "alpha" and "beta". The letters of this alphabet (see the front endpaper) are used so frequently in mathematics and science that the student should learn them at the earliest opportunity.

In Section 1.6 we discussed the concept of a function at some length. This discussion can be summarized in a few sentences, as follows.

If x and y are two variables that are related in such a way that whenever a numerical value is assigned to x there is determined a single corresponding numerical value for y, then y is called *a function of x* and this is expressed by writing $y = f(x)$. The letter f symbolizes the function itself, which is the operation or rule of correspondence that yields y when applied to x. However, for practical reasons we prefer to speak of "the function $y = f(x)$" instead of "the function f." As a matter of principle, students should clearly understand that a function is not a formula and need not be specified by a formula—even though most of ours are.

In practice, functions often arise from algebraic relations between variables. Thus, an equation involving x and y determines y as a function of x if the equation is equivalent to one that expresses y *uniquely* in terms of x. For example, the equation $4x + 2y = 6$ can be solved for y, $y = 3 - 2x$, and this second equation defines y as a function of x. However, in some cases it happens that the process of solving for y leads to more than one value of y. For example, if the equation is $y^2 = x$, we get $y = \pm \sqrt{x}$. Since this gives two values of y for each positive value of x, the equation $y^2 = x$ does not by itself determine y as a function of x. If we wish, we can split the formula $y = \pm\sqrt{x}$ into two separate formulas, $y = \sqrt{x}$ and $y = -\sqrt{x}$. Each of these formulas defines y as a function of x, so that out of one equation we obtain two functions.

The number of distinct individual functions is clearly unlimited. However, most of those appearing in this book are relatively simple and can be classified into a few convenient categories. It may help students to orient themselves if we give a rough description of these categories in order of increasing complexity.

1.7

TYPES OF FUNCTIONS. FORMULAS FROM GEOMETRY

POLYNOMIALS

The simplest functions of all are the powers of x with nonnegative integral exponents,

$$1, x, x^2, x^3, \ldots, x^n, \ldots.$$

If a finite number of these are multiplied by constants and the results are added, we obtain a general polynomial,

$$p(x) = a_0 + a_1 x + a_2 x^2 + a_3 x^3 + \cdots + a_n x^n.$$

The *degree* of a polynomial is the largest exponent that occurs in it; if $a_n \neq 0$, the degree of $p(x)$ is n. The following are polynomials of degrees 1, 2, and 3:

$$y = 3x - 2, \quad y = 1 - 2x + x^2, \quad y = x - x^3.$$

Polynomials can evidently be multiplied by constants, added, subtracted, and multiplied together, and the results are again polynomials.

RATIONAL FUNCTIONS

If division is also allowed, we pass beyond the polynomials to the rational functions, such as

$$\frac{x}{x^2+1}, \qquad \frac{x+2}{x-2}, \qquad \frac{x^3-4x^2+x+6}{x^2+x+1}, \qquad x+\frac{1}{x}.$$

The general rational function is a quotient of polynomials,

$$\frac{a_0 + a_1x + a_2x^2 + \cdots + a_nx^n}{b_0 + b_1x + b_2x^2 + \cdots + b_mx^m},$$

and a specific function is rational if it is (or can be expressed as) such a quotient. If the denominator here is a nonzero constant, this quotient is itself a polynomial. Thus, the polynomials are included among the rational functions.

ALGEBRAIC FUNCTIONS

If root extractions are also allowed, we pass beyond the rational functions into the larger class of algebraic functions, which will be properly defined in a later chapter. Some simple examples are

$$y=\sqrt{x}, \qquad y=x+\sqrt[3]{x^2+1}, \qquad y=\frac{1}{\sqrt{1-x}}, \qquad y=\sqrt[4]{\frac{x+1}{x-1}}.$$

If the notation of fractional exponents is used, these functions can be written

$$y=x^{1/2}, \qquad y=x+(x^2+1)^{1/3}, \qquad y=(1-x)^{-1/2}, \qquad y=\left(\frac{x+1}{x-1}\right)^{1/4}.$$

TRANSCENDENTAL FUNCTIONS

Any function that is not algebraic is said to be transcendental. The ones studied in calculus are the trigonometric, inverse trigonometric, exponential, and logarithm functions. We do not assume that students have any previous knowledge of these functions. All will be carefully explained in later chapters.

We conclude this section with a brief review of some important functions arising in geometry. A ready grasp of the geometric formulas given in Fig. 1.24 is essential for coping with many examples and problems in the following chapters. These formulas—for the area and circumference of circle, the volume and total surface area of a sphere, and the volume and lateral surface area of a cylinder and a cone—should be understood if possible, but remembered in any event. Each of the first four formulas, those for the circle

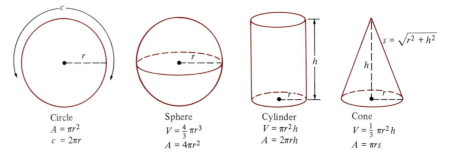

Circle	Sphere	Cylinder	Cone
$A=\pi r^2$	$V=\frac{4}{3}\pi r^3$	$V=\pi r^2 h$	$V=\frac{1}{3}\pi r^2 h$
$c=2\pi r$	$A=4\pi r^2$	$A=2\pi rh$	$A=\pi rs$

Figure 1.24   Geometric formulas.

and the sphere, defines a function of the independent variable r, in which a given positive value of r determines the corresponding value of the dependent variable.

Most of our attention in this book will be directed at functions of a single independent variable, as previously defined and discussed. Nevertheless, we point out that each of the last four formulas in Fig. 1.24 defines a function of the two variables r and h; these variables are called *independent* (of each other) because the value assigned to either need not be related to the value assigned to the other. In special circumstances a function of this kind can be expressed as a function of one variable alone. For example, if the height of a cone is known to be twice the radius of its base, so that $h = 2r$, then the formula for its volume can be written as a function of r or as a function of h:

$$V = \tfrac{1}{3}\pi r^2(2r) = \tfrac{2}{3}\pi r^3 \qquad \text{or} \qquad V = \tfrac{1}{3}\pi \left(\frac{h}{2}\right)^2 h = \tfrac{1}{12}\pi h^3.$$

The formulas in Fig. 1.24 also illustrate the custom of choosing letters for variables that suggest the quantities under discussion, such as A for area, V for volume, r for radius, h for height, and so on.

PROBLEMS

1   In each case, decide whether or not the equation determines y as a function of x, and if it does, find a formula for the function:
   (a) $3x^2 + y^2 = 1$;   (b) $3x^2 + y = 1$;

   (c) $\dfrac{y + 1}{y - 1} = x$;   (d) $x = y - \dfrac{1}{y}$.

2   Split the equation $2x^2 + 2xy + y^2 = 3$ into two equations, each of which determines y as a function of x.

The following problems all involve geometry. In working on such a problem, always draw a sketch and use this sketch as a source of ideas.

3   If an equilateral triangle has side x, express its area as a function of x.

4   The equal sides of an isosceles triangle are 2. If x is the base, express the area as a function of x.

5   If the edge of a cube is x, express its volume, its surface area, and its diagonal as functions of x.

6   A rectangle whose base has length x is inscribed in a fixed circle of radius a. Express the area of the rectangle as a function of x.

7   A string of length L is cut into two pieces, and these pieces are shaped into a circle and a square. If x is the side of the square, express the total enclosed area as a function of x.

8   (a) Is the area of a circle a function of its circumference? If so, what function?
   (b) Is the area of a square a function of its perimeter? If so, what function?

(c) Is the area of a triangle a function of its perimeter? If so, what function?

9   The volume of a sphere is a function of its surface area. Find a formula for this function.

10   A cylinder is inscribed in a sphere with fixed radius a. If h is the height and r is the radius of the base of the cylinder, express its volume and total surface area as functions of r, and also as functions of h.

11   A cylinder is circumscribed about a sphere. If their volumes are denoted by C and S, find C as a function of S.

12   A cylinder has fixed volume V. Express its total surface area as a function of the radius r of its base.

13   A fixed cone has height H and base radius R. If a cylinder with base radius r is inscribed in the cone, express the volume of the cylinder as a function of r.

14   (a) A farmer has 100 ft of fencing with which to build a rectangular chicken pen. If x is the length of one side of the pen, show that the enclosed area is

$$A = 50x - x^2 = 625 - (x - 25)^2.$$

   Use this result to find the largest possible area and the lengths of the sides that yield this largest area.
   (b) Suppose the farmer in part (a) decides to build the pen against a side of the barn, so that he will have to fence only three sides of it. If x is the length of a side perpendicular to the barn wall, find the enclosed area as a function of x. Also find the largest possible area and the lengths of the sides that yield this largest area.

1.8
GRAPHS OF FUNCTIONS

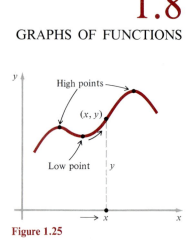

Figure 1.25

The Chinese have a proverb that expresses a fundamental truth about the study of mathematics: One good picture is worth a thousand words. For us, in our study of functions, this means *draw graphs.* Even more, cultivate the habit of thinking graphically, to the point where it becomes almost second nature.

Before getting down to the details of specific functions, we emphasize that it is often possible to think of the graph of a function $y = f(x)$ very concretely, as the path of a moving point (Fig. 1.25). The independent variable x can be visualized as a point moving along the x-axis from left to right; each x determines a value of the dependent variable y, which is the height of the point (x, y) above the x-axis. The graph of the function is simply the path of the point (x, y) as it moves across the coordinate plane, sometimes rising and sometimes falling, and in general varying in height according to the nature of the particular function under consideration. The graph as a whole is intended to provide a clear overall picture of this variation. The graph shown in Fig. 1.25 happens to be a smooth curve with two high points and one low point, but many diverse phenomena are possible.

We now discuss the graphs of a few representative examples of the types of functions described in Section 1.7.

POLYNOMIALS

We have seen that the simplest polynomials are the powers of x with non-negative integral exponents,

$$y = 1, x, x^2, x^3, \ldots, x^n, \ldots.$$

As we know, the graph of $y = 1$ is the horizontal straight line through the point $(0, 1)$, and the graph of $y = x$ is the straight line through the origin with slope 1 (Fig. 1.26*a*). For larger values of the exponent n, the graphs of $y = x^n$ are of two distinct types, depending on whether n is even or odd:

$$y = x^2, x^4, x^6, \ldots$$

and

$$y = x^3, x^5, x^7, \ldots.$$

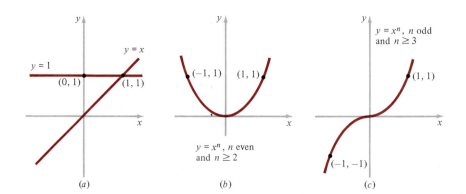

Figure 1.26　Graphs of $y = x^n$.

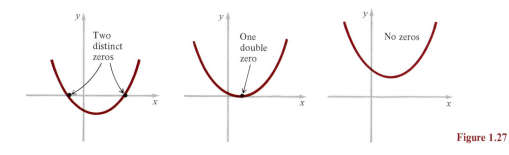

Figure 1.27

These types are shown in parts b and c of Fig. 1.26. As n increases, these curves become flatter near the origin and steeper outside the interval $[-1, 1]$.

We already know that the graphs of all first- and second-degree polynomials, such as

$$y = 2x - 1$$

and

$$y = 3x^2 - 2x + 1,$$

are straight lines and parabolas. These graphs are easy to draw—without plotting points—on the basis of the ideas in Sections 1.4 and 1.5.

For our next remark we need a bit of new terminology. A *zero* of a function $y = f(x)$ is a root of the corresponding equation $f(x) = 0$. Geometrically, the zeros of this function (if it has any) are the values of x at which its graph crosses or touches the x-axis; they are the x-intercepts of this graph.

We now consider the general second-degree polynomial

$$y = ax^2 + bx + c, \qquad a \neq 0. \tag{1}$$

As we know, the graph of this function is a parabola for all values of the coefficients. If we assume that $a > 0$, so that the parabola opens upward, then there are three possibilities for the zeros of (1), and these are shown in Fig. 1.27. Since the roots of the quadratic equation $ax^2 + bx + c = 0$ are given by the quadratic formula

$$x = \frac{-b \pm \sqrt{b^2 - 4ac}}{2a},$$

it is clear that the three possibilities in Fig. 1.27 correspond to the algebraic conditions $b^2 - 4ac > 0$, $b^2 - 4ac = 0$, $b^2 - 4ac < 0$.

The problem of graphing polynomials of degree $n \geq 3$ is not easy. Our discussion of the following example suggests several useful ideas.

Example 1 The graph of

$$y = x^3 - 3x \tag{2}$$

is shown in Fig. 1.28. At present we have no methods available for discovering such important features of this curve as the precise location of the indicated high and low points. This will come later. Nevertheless, a few observations can be made, and these provide at least some details and a good enough

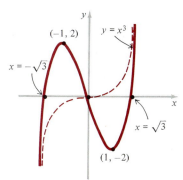

Figure 1.28

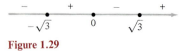

Figure 1.29

impression of the shape of the graph so that students should be able to sketch it for themselves.

We begin by pointing out that if (2) is written in factored form, as

$$y = x(x^2 - 3) = x(x + \sqrt{3})(x - \sqrt{3}), \tag{3}$$

then its zeros are obviously 0, $-\sqrt{3}$, $\sqrt{3}$. These three numbers divide the x-axis into four intervals, as shown in Fig. 1.29, and a quick inspection of (3) tells us that in each interval y has the sign given in this figure. We therefore know, for each interval, whether the graph of (2) lies above or below the x-axis (see Fig. 1.28).

Our second observation relates to the behavior of the graph of (2) when x is numerically large, that is, far to the right and far to the left in Fig. 1.28. If (2) is written in the form

$$y = x^3 \left(1 - \frac{3}{x^2}\right), \qquad x \neq 0,$$

then for large positive or negative values of x the expression in parentheses is nearly 1, and so y is close to x^3. In geometric language, when x is large, the graph of (2) is close to the graph of $y = x^3$, as Fig. 1.28 suggests. In particular, the graph of (2) rises on the far right and falls on the far left.

Students will notice that they can always sketch a graph by laboriously plotting many points and joining these points by a reasonable curve. Nevertheless, this rather clumsy procedure should be adopted only as a last resort, when more imaginative methods fail. The important features of functions and their graphs are much more clearly revealed by the qualitative approach to curve sketching that we have tried to suggest here and will continue to emphasize.

RATIONAL FUNCTIONS

Example 2 The simplest rational function that is not a polynomial is

$$y = \frac{1}{x}. \tag{4}$$

On examining (4), we notice the following facts: y is undefined when $x = 0$; y is positive when x is positive, and is small when x is large and large when x is near 0 on the right; y is negative when x is negative, and is small when x is large and large when x is near 0 on the left. The graph of (4) given in Fig. 1.30 is a direct pictorial version of these statements. In this particular case the graph is also easy to sketch by plotting a few points, as shown in the figure. However, students will profit much more from simply visualizing the behavior of such a function on the various parts of its domain and drawing what they see.

A straight line is called an *asymptote* of a curve if, as a point moves out along an extremity of the curve, the distance from this point to the line approaches 0. It is clear that both the x-axis and the y-axis are asymptotes of the graph shown in Fig. 1.30. The behavior of the function (4) at and near the point $x = 0$, that is, the fact that y is undefined at $x = 0$ and "becomes

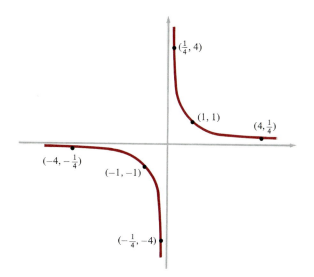

$(\frac{1}{4}, 4)$

$(1, 1)$

$(4, \frac{1}{4})$

$(-4, -\frac{1}{4})$

$(-1, -1)$

$(-\frac{1}{4}, -4)$

Figure 1.30

infinite" near $x = 0$, is described by calling this point an *infinite discontinuity* of the function.

Example 3   In the case of the function

$$y = \frac{x}{x - 1},\qquad(5)$$

it is clear that the point $x = 1$ is particularly interesting, since y is undefined at $x = 1$ and is large when x is near 1 ($x = 1$ is an infinite discontinuity). Also, y is near 1 and slightly greater than 1 when x is large and positive, and is near 1 and slightly less than 1 when x is large and negative.* These observations suggest drawing the vertical and horizontal guidelines shown in Fig. 1.31a. If we notice that $y = 0$ when $x = 0$, and pay attention to the sign of y in each of the intervals $-\infty < x < 0$, $0 < x < 1$, and $1 < x$, then the graph as given in Fig. 1.31a is quite easy to sketch. The lines $x = 1$ and $y = 1$ are both asymptotes.

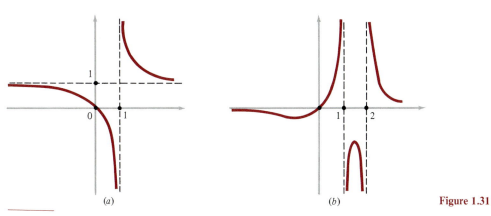

(a) (b) **Figure 1.31**

---

* To see this, test with convenient specific values of x; thus, for example, $y = \frac{10}{9}$ when $x = 10$ and $y = \frac{10}{11}$ when $x = -10$.

Example 4 The function

$$y = \frac{x}{x^2 - 3x + 2} = \frac{x}{(x-1)(x-2)} \qquad (6)$$

is similar to (5) but slightly more complicated. Here the factored form of the denominator reveals two infinite discontinuities, $x = 1$ and $x = 2$. Again, $y = 0$ when $x = 0$, but this time y is small when x is large, since the degree of the denominator is greater than that of the numerator. If we combine these facts with the observable sign of y in each of the intervals $-\infty < x < 0$, $0 < x < 1$, $1 < x < 2$, and $2 < x$, then it is fairly straightforward to sketch the graph as shown in Fig. 1.31b. There is evidently a high point between 1 and 2, and a low point to the left of 0, but at present we are unable to determine the precise location of these points (they occur at $x = \sqrt{2}$ and $x = -\sqrt{2}$).

Example 5 The function

$$y = x + \frac{1}{x} \qquad (7)$$

has an infinite discontinuity at $x = 0$, and is positive or negative according as x is positive or negative. For small positive x's, the first term on the right of (7) is negligible and the second term is large; and for large positive x's, the second term is negligible and y is approximately equal to x. We therefore sketch the part of the graph in the right half-plane as follows: Draw the guideline $y = x$ (Fig. 1.32); insert the two extremities of the curve, approaching this guideline and the positive y-axis, as suggested by the behavior previously stated; and connect these extremities in a reasonable way in the middle, where this part of the graph has an obvious low point. The function behaves similarly—with a corresponding high point—for negative values of x. The y-axis and the line $y = x$ are both asymptotes.

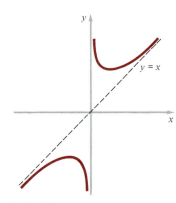

Figure 1.32

Example 6 The denominator of

$$y = \frac{x}{x^2 + 1} \qquad (8)$$

is positive (in fact ≥ 1) for all x, so $y = 0$ when $x = 0$, y is positive when x is positive, and y is negative when x is negative. Also, y is small when x is large, because the degree of the denominator is greater than that of the numerator.* These properties of the function force the graph to have the shape shown in Fig. 1.33.

Figure 1.33

Example 7 In considering the function

$$y = \frac{x^2 - 1}{x - 1}, \qquad (9)$$

---

* Notice that when x is large, $x^2 + 1$ is enormous, and so y is small.

it is natural to factor the numerator, obtaining

$$y = \frac{(x + 1)(x - 1)}{x - 1},$$

and then to cancel the common factor, which yields

$$y = x + 1. \qquad (10)$$

This cancellation is valid *except when* $x = 1$. At this point the value of (10) is 2, but (9) has no value ($y = 0/0$, which is meaningless). To graph (9), we therefore draw the straight line (10) and delete the single point (1, 2), as shown in Fig. 1.34.

Two functions $y = f(x)$ and $y = g(x)$ are said to be *equal* if they have the same domain and if $f(x) = g(x)$ for every x in their common domain. Accordingly, the functions (9) and (10) are not equal, because they have different domains — the point $x = 1$ is in the domain of (10) but is not in the domain of (9). The fact that the graph of (9) has a gap (or hole) corresponding to $x = 1$ is expressed by saying that (9) is *discontinuous* at $x = 1$, or has a *discontinuity* at this point.

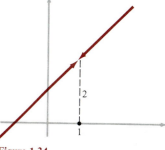

Figure 1.34

ALGEBRAIC FUNCTIONS

Example 8 The functions

$$y = \sqrt{x} \qquad \text{and} \qquad y = \sqrt{25 - x^2} \qquad (11)$$

can be obtained by solving the equations

$$y^2 = x \qquad \text{and} \qquad x^2 + y^2 = 25 \qquad (12)$$

for y and choosing the positive square roots. We know that the graphs of equations (12) are a parabola and a circle, as shown in Fig. 1.35, and so the graphs of (11) are the parts of these curves that lie on or above the x-axis.

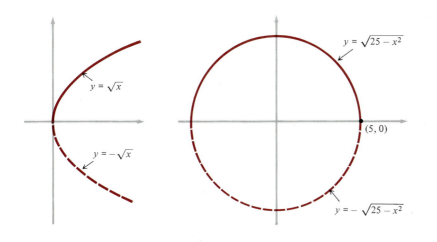

Figure 1.35

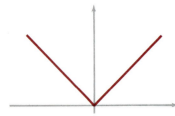

Figure 1.36

Example 9 The graph of the absolute value function,

$$y = |x|,$$

is easy to draw (Fig. 1.36). To see that this function is algebraic, we have only to notice the fact that $|x| = \sqrt{x^2}$ for every value of x.

As these examples show, many of the basic features of a function are made transparently clear by sketching its graph. We are interested less in sketches of high accuracy than in those that display broad general features: where the graph is rising and where falling, the presence of gaps, the presence of high points and low points, and what its approximate shape is. Formulas are obviously important in the study of functions—indeed, they are indispensable whenever our purposes require exact calculations yielding quantitative results. But we should never forget that the primary aim of mathematics is insight, and graphs are invaluable aids for gaining visual insight into the individual characteristics of functions.

PROBLEMS

1 Sketch the graphs of the following polynomials, paying special attention to the location of their zeros and their behavior for large values of x:
(a) $y = x^2 + x - 2$;
(b) $y = x^3 - 3x^2 + 2x$;
(c) $y = (1 - x)(2 - x)(3 - x)$;
(d) $y = x^4 - x^2$;
(e) $y = x^4 - 5x^2 + 4$.

2 Sketch the graphs of the following rational functions:

(a) $y = \dfrac{1}{x^2}$;

(b) $y = \dfrac{1}{x^3}$;

(c) $y = x^2 + \dfrac{1}{x}$;

(d) $y = x^2 + \dfrac{1}{x^2}$;

(e) $y = \dfrac{1}{x^2 + 1}$;

(f) $y = \dfrac{x^2}{x^2 + 1}$;

(g) $y = \dfrac{1}{x^2 - 1}$;

(h) $y = \dfrac{x}{x^2 - 1}$;

(i) $y = \dfrac{x^2}{x^2 - 1}$;

(j) $y = \dfrac{x^2 - 3x + 2}{2 - x}$;

(k) $y = \dfrac{x^3 - x^2}{x - 1}$;

(l) $y = \dfrac{(x + 2)(x - 5)(x^2 + 2x - 8)}{(x - 2)(x^2 - 3x - 10)}$.

3 Sketch the graphs of the following algebraic functions:

(a) $y = \sqrt{(x - 1)(3 - x)}$;

(b) $y = \dfrac{1}{\sqrt{(x - 1)(3 - x)}}$;

(c) $y = \dfrac{1}{\sqrt{x - 1}}$;

(d) $y = \sqrt{\dfrac{x}{3 - x}}$;

(e) $y = \sqrt{\dfrac{4 - x}{x - 2}}$;

(f) $y = \sqrt{\dfrac{x - 4}{x - 2}}$.

4 In each of the following, sketch the graphs of all three functions on a single coordinate system:
(a) $y = |x|$, $y = |x| + 1$, $y = |x| - 1$;
(b) $y = |x|$, $y = |x + 1|$, $y = |x - 1|$;
(c) $y = |x|$, $y = 2|x|$, $y = \frac{1}{2}|x|$.

5 Sketch the graphs of the following functions:

(a) $y = \dfrac{|x|}{x}$;

(b) $y = |2x + 3|$;

(c) $y = x + |x|$;

(d) $y = 2x + |x|$;

(e) $y = x - |x|$;

(f) $y = 1 + x - |x|$;

(g) $y = |x^2 - 1|$.

6 Considering only positive values of x, show that

$$y = \frac{|x + 1| - |x - 1|}{x} = \begin{cases} 2, & 0 < x < 1, \\ \dfrac{2}{x}, & x \geq 1, \end{cases}$$

and sketch the graph.

7 Are any of the following pairs of functions equal?

(a) $f(x) = \dfrac{x}{x}$, $g(x) = 1$.

(b) $f(x) = x^2 - 1$, $g(x) = (x + 1)(x - 1)$.

(c) $f(x) = x$, $g(x) = \sqrt{x^2}$.

(d) $f(x) = x$, $g(x) = (\sqrt{x})^2$.

ADDITIONAL PROBLEMS FOR CHAPTER 1

SECTION 1.2

1 If a and b are positive numbers, prove the inequality $\sqrt{ab} \leq \frac{1}{2}(a + b)$ as Euclid did, by considering a right triangle inscribed in a semicircle (Fig. 1.37).

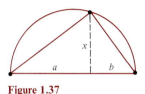

Figure 1.37

2 If a and b are any two numbers, denote the larger by max (a, b) and the smaller by min (a, b). Show that

$$\max (a, b) = \frac{1}{2}(a + b + |a - b|),$$

and find a similar expression for min (a, b).

3 Show that if $a \leq b$ and $c \leq d$, then $a + c \leq b + d$. Use this fact to prove that $|a + b| \leq |a| + |b|$. Hint: Begin by noticing that $-|a| \leq a \leq |a|$ and $-|b| \leq b \leq |b|$.

4 If a is a positive rational number, explain why the following method for calculating the square root of a works. First, choose a rational number which is a reasonable guess at the value of $\sqrt{a}$, and call this initial approximation x_1. Next, divide a by x_1 and average the result with x_1, thereby obtaining a second approximation x_2. Next, divide a by x_2 and average the result with x_2, obtaining a third approximation x_3. This procedure is expressed by the formula

$$x_{n+1} = \frac{1}{2}\left(x_n + \frac{a}{x_n}\right), \qquad n = 1, 2, 3, \dots.$$

Hint: If x_1 is reasonably close to $\sqrt{a}$ but different from it, then $\sqrt{a}$ lies between x_1 and a/x_1 (why?), and so the average of x_1 and a/x_1 is likely to be even closer to $\sqrt{a}$; also note that

$$x_{n+1} - \sqrt{a} = \frac{1}{2}\left(x_n - 2\sqrt{a} + \frac{a}{x_n}\right) = \frac{1}{2x_n}(x_n - \sqrt{a})^2.$$

5 Use the method of Problem 4 to calculate $\sqrt{2}$, first with $x_1 = 1$ and then with $x_1 = \frac{3}{2}$.

6 Use the method of Problem 4 to calculate $\sqrt{3}$, first with $x_1 = 2$ and then with $x_1 = \frac{3}{2}$.

7 If a and b are real numbers with $a < b$, show that there exists at least one rational number c such that $a < c < b$, and hence infinitely many. In particular, between any two irrationals there exist an infinite number of rationals.

8 If a is a nonzero rational number and b is irrational, show that $a + b$, $a - b$, ab, a/b, and b/a are all irrational.

9 If a and b are irrational, is $a + b$ necessarily irrational? Is ab?

10 If a and b are real numbers with $a < b$, show that there exists at least one irrational number c such that $a < c < b$, and hence infinitely many. In particular, between any two rationals there exist an infinite number of irrationals.

SECTION 1.3

11 Give another proof of the Pythagorean theorem by using the equations

$$\frac{a}{c} = \frac{e}{a} \qquad \text{and} \qquad \frac{b}{c} = \frac{d}{b},$$

obtained from similar triangles in Fig. 1.38.

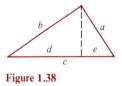

Figure 1.38

12 In each case place the figure in a convenient position relative to the coordinate system and prove the statement algebraically:
(a) The sum of the squares of the distances of any point from two opposite vertices of a rectangle equals the sum of the squares of its distances from the other two vertices.
(b) In any triangle, 4 times the sum of the squares of the medians equals 3 times the sum of the squares of the sides.

13 If $P_1 = (x_1, y_1)$ and $P_2 = (x_2, y_2)$ are distinct points, and if $P = (x, y)$ is located on the segment joining them in such a position that the ratio of its distance from P_1 to its distance from P_2 is q/p, show that

$$x = \frac{px_1 + qx_2}{p + q} \qquad \text{and} \qquad y = \frac{py_1 + qy_2}{p + q}.$$

14 Find the point on the segment joining $(1, 2)$ and $(5, 9)$ that is $\frac{11}{17}$ of the way from the first point to the second.

SECTION 1.4

15 If the line determined by two distinct points (x_1, y_1) and (x_2, y_2) is not vertical, and therefore has slope $(y_2 - y_1)/(x_2 - x_1)$, show that the point-slope form of its equation is the same regardless of which point is used as the given point.

16 Determine what each of the following statements implies about the constants A, B, C in the equation $Ax + By + C = 0$:
(a) The line goes through the origin.
(b) The line is parallel to the y-axis.
(c) The line is perpendicular to the y-axis.
(d) The line goes through $(1, 1)$.
(e) The line is parallel to $5x + 3y = 2$.
(f) The line is perpendicular to $x + 10y = 3$.

17 If the lines $A_1x + B_1y + C_1 = 0$ and $A_2x + B_2y + C_2 = 0$ are not parallel and k is any constant, show that

$$(A_1x + B_1y + C_1) + k(A_2x + B_2y + C_2) = 0$$

is a line through the point of intersection of the given lines. When k is assigned various values, this equation represents various members of the family of all lines through the point of intersection.

18 Given the lines $x + 3y - 2 = 0$ and $2x - y + 4 = 0$, use Problem 17 to find the equation of the line through their point of intersection which
(a) passes through $(-2, 1)$;
(b) is perpendicular to the line $3y + x = 21$;
(c) passes through the origin.

19 The points $(0, 0)$, $(a, 0)$, and (b, c) are the vertices of an arbitrary triangle which is placed in a convenient position relative to the coordinate system.
(a) Find the equation of the line through each vertex perpendicular to the opposite side, and show algebraically that these three lines intersect at a single point.
(b) Find the equation of the perpendicular bisector of each side, and show algebraically that these three lines intersect at a single point. Why is this fact geometrically obvious?
(c) Find the equation of the line through each vertex and the midpoint of the opposite side, and show algebraically that these three lines intersect at a single point. Also, verify that this point is two-thirds of the way from each vertex to the midpoint of the opposite side.

20 Show that each of the following is the equation of a straight line:
(a) $x^3 - x^2y - 2x^2 + 3x - 3y - 6 = 0$.
(b) $3xy^2 + 5y^2 - y^3 - 4y + 12x + 20 = 0$.

21 Show that the distance from a point (x_0, y_0) to a line $Ax + By + C = 0$ is given by

$$\frac{|Ax_0 + By_0 + C|}{\sqrt{A^2 + B^2}}.$$

22 Find the distance between the parallel lines $4x + 3y + 12 = 0$ and $4x + 3y - 38 = 0$.

23 If two intersecting straight lines are given, then it is easy to see that the bisectors of the angles formed by these lines are two other straight lines whose points are equidistant from the given lines. Use this fact to find the equations of the bisectors of the angles formed by the lines
(a) $3x + 4y - 10 = 0$ and $4x - 3y - 5 = 0$;
(b) $y = 0$ and $y = x$.

24 Why is it geometrically obvious (without calculation) that the bisectors of the angles of any triangle intersect at a single point?

SECTION 1.5

25 Find the values of b for which the line $y = 3x + b$ intersects the circle $x^2 + y^2 = 4$.

26 If the line $y = mx + b$ is tangent to the circle $x^2 + y^2 = r^2$, find an equation relating m, b, and r.

27 Find the equation of the locus of a point $P = (x, y)$ that moves in such a way that
(a) its distance from $(0, 0)$ is twice its distance from $(a, 0)$;
(b) the product of its distances from $(a, 0)$ and $(-a, 0)$ is a^2 (this curve is called a *lemniscate*).
In each case, sketch the graph.

28 A line segment of length 6 moves in such a way that its endpoints remain on the x-axis and y-axis. What is the equation of the locus of its midpoint?

29 A point moves in such a way that the ratio of its distances from two fixed points is a constant $k \neq 1$. Show that the locus is a circle.

30 Find the equation of the line which is tangent to the circle $x^2 + y^2 + 8x + 6y + 8 = 0$ at the point $(-8, -2)$.

31 Find the equations of the lines that pass through the point $(1, 3)$ and are tangent to the circle $x^2 + y^2 = 2$.

32 If two circles

$$x^2 + y^2 + A_1x + B_1y + C_1 = 0$$

and

$$x^2 + y^2 + A_2x + B_2y + C_2 = 0$$

intersect in two points, and if k is a constant $\neq -1$, explain why

$$(x^2 + y^2 + A_1x + B_1y + C_1) + k(x^2 + y^2 + A_2x + B_2y + C_2) = 0$$

is the equation of a circle through the points of intersection. If $k = -1$, what does the equation represent?

33 Use Problem 32 to find the equation of the line joining the points of intersection of the circles $x^2 + y^2 = 4x + 4y - 4$ and $x^2 + y^2 = 2y$. Also find these points of intersection.

34 Show that a parabola with focus at the origin, axis the x-axis, and opening to the right has an equation of the form $y^2 = 4p(x + p)$, where $p > 0$.

35 Find the equation of the parabola with focus (1, 1) and directrix $x + y = 0$, and simplify this equation to a form without radicals. Hint: See Problem 21.

36 Let the vertex of the parabola $x^2 = 4py$ be joined to every other point of the parabola. Show that the midpoints of the resulting chords lie on another parabola. Find the focus and directrix of this second parabola.

37 Consider all chords with given slope m that have endpoints on the parabola $x^2 = 4py$. Prove that the locus of the midpoints of these chords is a straight line parallel to the y-axis.

38 A *focal chord* of a parabola is the segment cut by the parabola from a straight line through the focus.
   (a) If A and B are the endpoints of a focal chord, and if the line through A and the vertex intersects the directrix at a point C, show that the line through B and C is parallel to the axis of the parabola.
   (b) Show that the length of a focal chord is twice the distance from its midpoint to the directrix.
   (c) Show that if the two tangents to a parabola are drawn from any point on the directrix, then the points of tangency are the endpoints of a focal chord.

39 Given the two points $A = (4p, 0)$ and $B = (4p, 4p)$, divide the segments OA and AB into equal numbers of equal parts, number the points of division as shown in Fig. 1.39, and join the points of division on AB to the

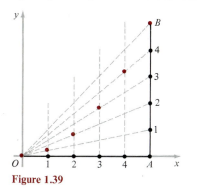

Figure 1.39

origin by straight lines. Show that the points of intersection of each of these lines with the corresponding vertical lines lie on the parabola $x^2 = 4py$.

SECTION 1.6

40 Find the domain of each of the following functions:

(a) $5 - x$;

(b) $\dfrac{x}{2x - 3}$;

(c) $\sqrt{3x - 2}$;

(d) $\sqrt{5 - 3x}$;

(e) $\dfrac{x + 7}{x^2 - 9}$;

(f) $\sqrt[3]{x}$;

(g) $\sqrt{9 - 4x^2}$;

(h) $\dfrac{1}{\sqrt{x + 3}}$;

(i) $\sqrt{7x^2 + 5}$.

41 If $f(x) = ax + b$, show that

$$f\left(\frac{x_1 + x_2}{2}\right) = \frac{f(x_1) + f(x_2)}{2}.$$

Is this true for $f(x) = x^2$?

42 If $f(x) = (1 + x)/(1 - x)$, find

(a) $f(-x)$;

(b) $f\left(\dfrac{1}{x}\right)$;

(c) $f\left(\dfrac{1}{1 - x}\right)$;

(d) $f(f(x))$.

43 If $f(x) = \sqrt[3]{x}$, what function $g(x)$ has the property that $g(f(x)) = x$?

SECTION 1.7

44 The perimeter of a right triangle is 6 and its hypotenuse is x. Express the area as a function of x.

45 A cylinder has fixed total surface area A. Express its volume as a function of the radius r of its base.

46 A cone is inscribed in a sphere with fixed radius a. If r is the radius of the base of the cone, express its volume as a function of r.

47 A cone is circumscribed about a sphere with fixed radius a. If r is the radius of the base of the cone, express its volume as a function of r.

48 If $f(x) = (x - 3)/(x + 1)$, show that $f(f(f(x))) = x$.

49 Let a, b, c, d be given constants with the property that $ad - bc \neq 0$. If $f(x) = (ax + b)/(cx + d)$, show that there exists a function $g(x) = (\alpha x + \beta)/(\gamma x + \delta)$ such that $f(g(x)) = x$. Also show that for these two functions it is true that $f(g(x)) = g(f(x))$.

50 Suppose a function $f(x)$ has the property that $f(x_1 + x_2) = f(x_1) + f(x_2)$ for all x_1 and x_2, from which it follows that

$$f(x_1 + x_2 + \cdots + x_n)$$
$$= f(x_1) + f(x_2) + \cdots + f(x_n).$$

Prove that there is a number a such that $f(x) = ax$ for all rational numbers x. Hint: Decide what a must be, then prove the statement successively for the cases in which x is a positive integer, an integer, the reciprocal of a nonzero integer, and a rational number.*

51 This is a sequel to Problem 50. Suppose a function $f(x)$ has the two properties that $f(x_1 + x_2) = f(x_1) + f(x_2)$ and $f(x_1 x_2) = f(x_1)f(x_2)$ for all x_1 and x_2. If this func-

---

* Without additional hypotheses, it is known to be impossible to prove that $f(x) = ax$ for all real numbers x.

tion has at least one nonzero value, show that $f(x) = x$ for all real numbers x by proving the following statements:

(a) $f(1) = 1$;

(b) $f(x) = x$ if x is rational;

(c) $f(x) > 0$ if $x > 0$ (hint: a positive number is the square of some positive number);

(d) $f(x_1) < f(x_2)$ if $x_1 < x_2$;

(e) $f(x) = x$ for all x (hint: there is a rational number between any two real numbers).

SECTION 1.8

52 Let $p(x) = a_n x^n + a_{n-1} x^{n-1} + \cdots + a_1 x + a_0$ be a polynomial of degree $n \geq 1$, and prove the following statements:

(a) If $p(0) = 0$, then $p(x) = xq(x)$, where $q(x)$ is a polynomial of degree $n - 1$.

(b) If a is any real number, the function $f(x)$ defined by $f(x) = p(x + a)$ is a polynomial of degree n.

(c) If a is a real number for which $p(a) = 0$, that is, if a is a zero of $p(x)$, then $p(x) = (x - a)r(x)$, where $r(x)$ is a polynomial of degree $n - 1$. [Hint: Consider $f(x) = p(x + a)$.]

(d) $p(x)$ has at most n zeros.

53 If n is any integer ≥ 1, show that there exists a polynomial of degree n with n zeros. If n is even, find a polynomial of degree n with no zeros; and if n is odd, find one with only one zero.

54 Let $p(x) = a_n x^n + a_{n-1} x^{n-1} + \cdots + a_1 x + a_0$ be a polynomial of degree $n \geq 1$. If $p(x)$ has n zeros $x_1, x_2, \ldots, x_n$, and is therefore expressible in the form

$$p(x) = a_n(x - x_1)(x - x_2) \cdots (x - x_n),$$

show that

(a) $x_1 x_2 \cdots x_n = (-1)^n \dfrac{a_0}{a_n}$;

(b) $x_1 + x_2 + \cdots + x_n = -\dfrac{a_{n-1}}{a_n}$.

55 A function f is said to be *even* if $f(-x) = f(x)$ for every x in its domain, and it is said to be *odd* if $f(-x) = -f(x)$ for every x in its domain (in each case, it is understood that $-x$ is in the domain of f whenever x is). Determine whether each of the follow-

ing functions is even, odd, or neither:

(a) $f(x) = x^3$;

(b) $f(x) = x(x^3 + x)$;

(c) $f(x) = |x|$;

(d) $f(x) = x + \dfrac{1}{x}$;

(e) $f(x) = x^2 + \dfrac{1}{x}$;

(f) $f(x) = \dfrac{x^3 + x}{x^2 + 1}$;

(g) $f(x) = x^5 + 1$;

(h) $f(x) = x(x + 1)$.

56 What is the distinguishing feature of the graph of an even function? Of an odd function?

57 What can be said about

(a) the product of two even functions?

(b) the product of two odd functions?

(c) the product of an even function and an odd function?

58 If $f(x)$ is an arbitrary function defined on an interval of the form $[-a, a]$, show that $f(x)$ is expressible in one and only one way as the sum of an even function $g(x)$ and an odd function $h(x)$, $f(x) = g(x) + h(x)$. Hint: $f(-x) = g(x) - h(x)$.

59 Write down a second-degree polynomial whose values at 1, 2, and 3 are π, $\sqrt{3}$, and 550.

60 If a and b are positive constants, sketch the graph of

$$y = \frac{b}{2a} (|x + a| + |x - a| - 2|x|).$$

61 The symbol $[x]$ (read "bracket x") is often used to denote the greatest integer $\leq$ a real number x. For example, $[1] = 1$, $[2.1] = 2$, $[\pi] = 3$, and $[-1.7] = -2$. Sketch the graphs of the following functions:

(a) $y = [x]$;

(b) $y = x - [x]$;

(c) $y = \sqrt{x - [x]}$;

(d) $y = [x] + \sqrt{x - [x]}$;

(e) $y = \sqrt{x} - [\sqrt{x}]$, $0 \leq x \leq 9$.

62 Express the number of squares $\leq$ a positive number x in terms of the bracket function defined in Problem 61. Do the same for the number of cubes $\leq x$.

63 If the symbol $\{x\}$ (read "brace x") denotes the distance from a real number x to the nearest integer, graph the following functions:

(a) $y = \{x\}$;

(b) $y = \{2x\}$;

(c) $y = \{4x\}$;

(d) $y = \frac{1}{4}\{4x\}$.

2

THE DERIVATIVE
OF A FUNCTION

We begin our study of calculus with a brief statement of what the subject is about and why it is important. Such a bird's-eye view of the road that lies ahead can help us attain a clarity of purpose and sense of direction that will serve us well among the many technical details that constitute the bulk of our work.

Calculus is usually divided into two main parts, called *differential calculus* and *integral calculus*. Each of these parts has its own unfamiliar terminology, puzzling notation, and specialized computational methods. Getting accustomed to all this takes time and practice, much like the process of learning a new language. Nevertheless, this fact should not prevent us from seeing at the beginning that the central problems of the subject are really quite simple and clear, with nothing strange or mysterious about them.

Almost all the ideas and applications of calculus revolve around two geometric problems that are very easy to understand. Both problems refer to the graph of a function $y = f(x)$. We avoid complications by assuming that this graph lies entirely above the *x*-axis, as shown in Fig. 2.1.

PROBLEM 1 The basic problem of differential calculus is the *problem of tangents:* Calculate the slope of the tangent line to the graph at a given point *P*.

PROBLEM 2 The basic problem of integral calculus is the *problem of areas:* Calculate the area under the graph between the points $x = a$ and $x = b$.

Our work in the rest of this book will be focused on these two problems, on the ideas and techniques that have been developed for solving them, and on the applications that arise from them.*

2.1
WHAT IS CALCULUS? THE PROBLEM OF TANGENTS

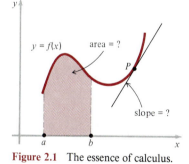

Figure 2.1 The essence of calculus.

---

* For readers who are interested in the origins of words, a *calculus* in ancient Rome was a small stone or pebble used in counting and in gambling, and the Latin verb *calculare* came to mean "to figure out," "to compute," "to calculate." Today a calculus is a method or system of methods for solving quantitative problems of a particular kind, as in calculus of probabilities, calculus of finite differences, tensor calculus, calculus of variations, calculus of residues, etc. Our calculus—the branch of mathematics consisting of differential and integral calculus taken together—is sometimes called *the* calculus to distinguish it from all these other subordinate calculuses.

At first sight these problems appear to be rather limited in scope. We expect them to shed significant light on geometry, and they do. What is very surprising is to find that they also have many profound and far-reaching applications to the various sciences. Calculus pays its way in the great world outside of mathematics through these scientific applications, and one of our major purposes is to introduce the student to as wide a variety of them as possible. At the same time we will continue to emphasize geometry and geometric applications, for this is the context in which the ideas of calculus are most easily understood.

It is sometimes said that calculus was "invented" by those two great geniuses of the late seventeenth century, Newton and Leibniz.* In reality, calculus is the product of a long evolutionary process that began in ancient Greece and continued into the nineteenth century. Newton and Leibniz were indeed remarkable men, and their contributions were of decisive importance, but the subject neither started nor ended with them. The problems stated above were much on the minds of many European scientists of the middle seventeenth century—most notably Fermat—and considerable progress was made on each of them by ingenious special methods. It was the great achievement of Newton and Leibniz to recognize and exploit the close connection between these problems, which no one else had fully understood. Specifically, they were the first to grasp the significance of the *Fundamental Theorem of Calculus,* which says, in effect, that the solution of the tangent problem can be used to solve the area problem. This theorem — certainly the most important in the whole of mathematics — was discovered by each man independently of the other, and they and their successors used it to weld the two halves of the subject together into a problem-solving art of astonishing power and versatility.

As these remarks suggest, we begin our work by undertaking a fairly thorough study of the tangent problem in the next four chapters. Then, in Chapters 6 and 7, we turn to the area problem. From there we push outward in a number of directions, extending our basic concepts and tools to broader classes of functions with a greater variety of significant applications.

Before attempting to calculate the slope of a tangent line, we must first decide what a tangent line *is*, and this is not as easy as it seems.

In the case of a circle there is no difficulty. A tangent to a circle (Fig. 2.2, left) is a line that intersects the circle at only one point, called the point of tangency; lines that are not tangents either intersect the circle at two different points or miss it altogether. This situation reflects the clear intuitive idea

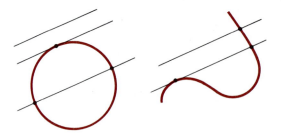

Figure 2.2

---

* The Latin spelling "Leibnitz" is sometimes used in order to suggest the correct pronunciation.

most of us have that a tangent to a curve at a given point is a line that "just touches" the curve at that point.* It also suggests the possibility of defining a tangent to a curve as a line that intersects the curve at only one point. This definition was used successfully by the Greeks in dealing with circles and a few other special curves, but for curves in general it is wholly unsatisfactory. To understand why, consider the curve shown on the right in Fig. 2.2: It has a perfectly acceptable tangent (the lower line) that this definition would reject, and an obvious nontangent (the upper line) that this definition would accept.

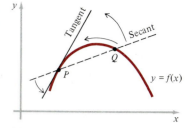

Figure 2.3   Fermat's idea.

The modern concept of a tangent line was originated by Fermat around 1630. As students will come to see, this concept is not only a reasonable statement about the geometric nature of tangents, it is also the key to a practical process for the construction of tangents.

Briefly, the idea is this: Consider a curve $y = f(x)$, and let P be a given fixed point on this curve (Fig. 2.3). Let Q be a second nearby point on the curve, and draw the secant line PQ. The tangent line at P can now be thought of as the limiting position of the variable secant as Q slides along the curve toward P. We shall see in Section 2.2 how this qualitative idea leads at once to a quantitative method for calculating the exact slope of the tangent in terms of the given function $f(x)$.

Let there be no misunderstanding. This way of thinking about tangents is not a minor technical point in the geometry of curves. On the contrary, it is one of the three or four most fruitful ideas that any mathematician has ever had, for without it there would have been no concept of velocity or acceleration or force in physics, no Newtonian dynamics or astronomy, no physical science of any kind except as the mere verbal description of phenomena, and certainly no modern age of engineering and technology.

General discussions have their place, but the time has come to get down to details.

Let $P = (x_0, y_0)$ be an arbitrary fixed point on the parabola $y = x^2$, as shown in Fig. 2.4. As our first illustration of the basic idea of this chapter, we calculate the slope of the tangent to this parabola at the given point P. To begin the process, we choose a second nearby point $Q = (x_1, y_1)$ on the curve.

2.2

HOW TO CALCULATE THE SLOPE OF THE TANGENT

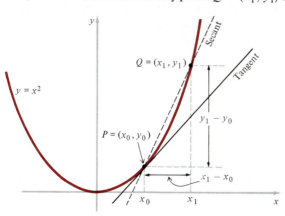

Figure 2.4

_____

* The Latin word *tangere* means "to touch."

Next, we draw the secant line PQ which is determined by these two points. The slope of this secant is evidently

$$m_{\text{sec}} = \text{slope of } PQ = \frac{y_1 - y_0}{x_1 - x_0}. \tag{1}$$

Now for the crucial step: We let x_1 approach x_0, so that the variable point Q approaches the fixed point P by sliding along the curve — much like a bead sliding along a curved wire. As this happens, the secant changes direction and visibly approaches the tangent at P as its limiting position. Also, it is intuitively clear that the slope m of the tangent is the limiting value approached by the slope m_{sec} of the secant. If we use the standard symbol $\to$ to mean "approaches," then the last statement can be expressed in the concise and convenient form

$$m = \lim_{Q \to P} m_{\text{sec}} = \lim_{x_1 \to x_0} \frac{y_1 - y_0}{x_1 - x_0}. \tag{2}$$

The abbreviation "lim," with "$x_1 \to x_0$" written below it, is read "the limit, as x_1 approaches x_0, of . . . ."

We cannot calculate the limiting value m in (2) by simply setting $x_1 = x_0$, because this would give the meaningless result

$$\frac{y_1 - y_0}{x_1 - x_0} = \frac{0}{0}.$$

We must think of x_1 as coming very close to x_0 but remaining distinct from it. However, as this happens, both $y_1 - y_0$ and $x_1 - x_0$ become arbitrarily small, and it isn't at all clear what limiting value their quotient approaches.

The way out of this difficulty is to use the equation of the curve. Since P and Q both lie on the curve, we have $y_0 = x_0^2$ and $y_1 = x_1^2$, so (1) can be written

$$m_{\text{sec}} = \frac{y_1 - y_0}{x_1 - x_0} = \frac{x_1^2 - x_0^2}{x_1 - x_0}. \tag{3}$$

The reason this numerator becomes small is that it contains the denominator $x_1 - x_0$ as a factor. If this common factor is canceled, we obtain

$$m_{\text{sec}} = \frac{y_1 - y_0}{x_1 - x_0} = \frac{x_1^2 - x_0^2}{x_1 - x_0} = \frac{(x_1 - x_0)(x_1 + x_0)}{x_1 - x_0} = x_1 + x_0,$$

and (2) becomes

$$m = \lim_{x_1 \to x_0} \frac{y_1 - y_0}{x_1 - x_0} = \lim_{x_1 \to x_0} (x_1 + x_0).$$

It is now very easy to see what is happening: As x_1 gets closer and closer to x_0, $x_1 + x_0$ becomes more and more nearly equal to $x_0 + x_0 = 2x_0$. Accordingly,

$$m = 2x_0 \tag{4}$$

is the slope of the tangent to the curve $y = x^2$ at the point $(x_0\ y_0)$.

Example 1 The points $(1, 1)$ and $(-\frac{1}{2}, \frac{1}{4})$ lie on the parabola $y = x^2$ (Fig. 2.5). By formula (4), the slopes of the tangents at these points are $m = 2$ and

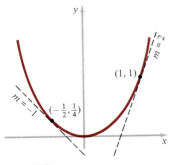

Figure 2.5

$m = -1$. Using the point-slope form of the equation of a line, our two tangent lines clearly have equations

$$\frac{y-1}{x-1} = 2 \quad \text{and} \quad \frac{y-\frac{1}{4}}{x+\frac{1}{2}} = -1.$$

In exactly the same way,

$$\frac{y - x_0^2}{x - x_0} = 2x_0$$

is the equation of the tangent at a general point (x_0, x_0^2) on the curve.

We now introduce the so-called *delta notation*. This is a widely used piece of symbolism that mathematics and science would find it very difficult to do without.

The procedure just described begins by changing the independent variable x from a first value x_0 to a second value x_1. The standard notation for the amount of such a change is Δx (read "delta x"), so that

$$\Delta x = x_1 - x_0 \tag{5}$$

is the change in x in going from the first value to the second. We can also think of the second value as being obtained from the first by adding the change:

$$x_1 = x_0 + \Delta x. \tag{6}$$

It is essential to understand that Δx is not the product of a number Δ and a number x, but a single number called an *increment* of x. An increment Δx can be either positive or negative. Thus, if $x_0 = 1$ and $x_1 = 3$, then $\Delta x = 3 - 1 = 2$; and if $x_0 = 1$ and $x_1 = -2$, then $\Delta x = -2 - 1 = -3$.

The letter Δ is the Greek "*d*"; when it is written in front of a variable, it signifies the difference between two values of that variable. This simple notational device turns out to be extremely convenient, and has spread into almost every part of mathematics and science. We illustrate its role in our present work by using it to reformulate the above calculations.

In view of (5) and (6), formula (3) for the slope of the secant can be written in the form

$$m_{\text{sec}} = \frac{x_1^2 - x_0^2}{x_1 - x_0} = \frac{(x_0 + \Delta x)^2 - x_0^2}{\Delta x}. \tag{7}$$

This time, instead of factoring the numerator, we expand its first term and simplify the result, obtaining

$$(x_0 + \Delta x)^2 - x_0^2 = x_0^2 + 2x_0\,\Delta x + (\Delta x)^2 - x_0^2$$
$$= 2x_0\,\Delta x + (\Delta x)^2$$
$$= \Delta x(2x_0 + \Delta x),$$

so (7) becomes

$$m_{\text{sec}} = 2x_0 + \Delta x.$$

If we insert this in (2) and use the fact that $x_1 \to x_0$ is equivalent to $\Delta x \to 0$,

we find that

$$m = \lim_{\Delta x \to 0} (2x_0 + \Delta x) = 2x_0,$$

as before. Again it is very easy to see what is happening in the indicated limit process: as Δx gets closer and closer to zero, $2x_0 + \Delta x$ becomes more and more nearly equal to $2x_0$.

This second method, using the delta notation, depends on expanding the square $(x_0 + \Delta x)^2$, whereas the first depends on factoring the expression $x_1^2 - x_0^2$. In this particular case neither calculation is noticeably harder than the other. In general, however, expanding is easier than factoring, and for this reason we adopt the method of increments as our standard procedure.

The calculation that we have just carried out for the parabola $y = x^2$ can be described in principle for the graph of any function $y = f(x)$ (Fig. 2.6). We first compute the slope of the secant through the two points P and Q corresponding to x_0 and $x_0 + \Delta x$,

$$m_{\text{sec}} = \frac{f(x_0 + \Delta x) - f(x_0)}{\Delta x}.$$

We then calculate the limit of m_{sec} as Δx approaches zero, obtaining a number m that we interpret geometrically as the slope of the tangent to the curve at the point P:

$$m = \lim_{\Delta x \to 0} \frac{f(x_0 + \Delta x) - f(x_0)}{\Delta x}.$$

The value of this limit is usually denoted by the symbol $f'(x_0)$, read "f prime of x_0," in order to emphasize its dependence on both the point x_0 and the function $f(x)$. Thus, by definition we have

$$f'(x_0) = \lim_{\Delta x \to 0} \frac{f(x_0 + \Delta x) - f(x_0)}{\Delta x}. \tag{8}$$

In this notation, the result of the calculation given above can be expressed as follows: If $f(x) = x^2$, then $f'(x_0) = 2x_0$.

Example 2   Calculate $f'(x_0)$ if $f(x) = 2x^2 - 3x$.

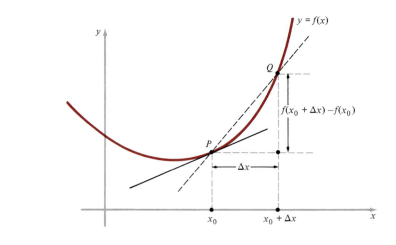

Figure 2.6

Solution   For this function, the numerator of the quotient in (8) is

$$f(x_0 + \Delta x) - f(x_0) = [2(x_0 + \Delta x)^2 - 3(x_0 + \Delta x)] - [2x_0^2 - 3x_0]$$
$$= 2x_0^2 + 4x_0\,\Delta x + 2(\Delta x)^2 - 3x_0 - 3\Delta x - 2x_0^2 + 3x_0$$
$$= 4x_0\,\Delta x + 2(\Delta x)^2 - 3\Delta x$$
$$= \Delta x(4x_0 + 2\Delta x - 3).$$

The quotient in (8) is therefore

$$\frac{f(x_0 + \Delta x) - f(x_0)}{\Delta x} = 4x_0 + 2\Delta x - 3,$$

and

$$f'(x_0) = \lim_{\Delta x \to 0} (4x_0 + 2\Delta x - 3)$$
$$= 4x_0 - 3.$$

We have assumed in the remarks leading to (8) that the curve under discussion actually has a definite tangent at the point P. This is a genuine assumption, because some curves do not have a tangent at every point (Fig. 2.7). However, when a tangent exists, it is clearly necessary for the secant PQ to approach the same limiting position whether Q approaches P from the right or from the left. These two modes of approach correspond, respectively, to Δx approaching zero through only positive or only negative values. It is therefore part of the meaning of (8) that for this limit to exist we must have the same limiting value for both directions of approach.

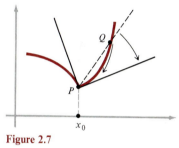

Figure 2.7

PROBLEMS

1   Find the equation of the tangent to the parabola $y = x^2$
   (a) at the point $(-2, 4)$;
   (b) at the point where the slope is 8;
   (c) if the x-intercept of the tangent is 2.

2   Show that the tangent to the parabola $y = x^2$ at a point (x_0, y_0) other than the vertex always has x-intercept $\frac{1}{2}x_0$.

3   A straight line $y = mx + b$ is presumably its own tangent line at any point. Verify this by showing that $f'(x_0) = m$ if $f(x) = mx + b$.

4   Sketch the graph of $y = x - x^2$ on the interval $-2 \le x \le 3$.
   (a) Use the method of increments to compute the slope of the tangent line at an arbitrary point (x_0, y_0) on the curve.
   (b) What are the slopes of the tangent lines at the points $(-1, -2)$, $(0, 0)$, $(1, 0)$, and $(2, -2)$ on the curve? Use these slopes to draw the tangents at these points in your sketch.
   (c) At what point on the curve is the tangent horizontal?

5   Use formula (8) to calculate $f'(x_0)$ if $f(x)$ is equal to
   (a) $x^2 - 4x - 5$;   (b) $x^2 - 2x + 1$;
   (c) $2x^2 + 1$;   (d) $x^2 - 4$.
   The results of these calculations will be needed in the following problems.

6   Sketch the given curve and the tangent line at the given point, and find the equation of this tangent line:
   (a) $y = x^2 - 4x - 5$, $(4, -5)$.
   (b) $y = x^2 - 2x + 1$, $(-1, 4)$.

7   Find the equation of the tangent line to the curve $y = 2x^2 + 1$ that is parallel to the line $8x + y - 2 = 0$.

8   Find the equations of the two lines through the point $(3, 1)$ that are tangent to the curve $y = x^2 - 4$.

9   Prove analytically (that is, without appealing to geometric reasoning) that there is no line through the point $(1, -2)$ that is tangent to the curve $y = x^2 - 4$.

10   Draw the graph of $y = f(x) = |x - 1|$.
   (a) Is there any point on the graph at which there is no tangent line?
   (b) Find $f'(x_0)$ if $x_0 > 1$. If $x_0 < 1$. What can be said about $f'(x_0)$ if $x_0 = 1$?

2.3
THE DEFINITION OF THE DERIVATIVE

If we separate formula (8) in Section 2.2 from its geometric motivation, and also drop the subscript on x_0, then we arrive at our basic definition: Given any function $f(x)$, its *derivative* $f'(x)$ is the new function whose value at a point x is defined by

$$f'(x) = \lim_{\Delta x \to 0} \frac{f(x + \Delta x) - f(x)}{\Delta x}. \tag{1}$$

In calculating this limit, x is held fixed while Δx varies and approaches zero. The indicated limit may exist for some values of x and fail to exist for other values. If the limit exists for $x = a$, then the function is said to be *differentiable at a.* A *differentiable function* is one that is differentiable at each point of its domain. Most of the specific functions considered in this book have this property.

We know that the derivative $f'(x)$ can be visualized in the way suggested by Fig. 2.8, in which $f(x)$ is the variable height of a point P moving along the curve and $f'(x)$ is the variable slope of the tangent at P. Strictly speaking, however, the above definition of the derivative does not depend in any way on geometric ideas. Our thoughts about Fig. 2.8 constitute a *geometric interpretation,* and important as this may be as an aid to understanding, it is not an essential part of the concept of the derivative. In the next section we will meet other equally important interpretations that have nothing to do with geometry. We must therefore be prepared to consider $f'(x)$ purely as a function, and to recognize that it has several interpretations but no necessary connection with any one of them.

The process of actually forming the derivative $f'(x)$ is called the *differentiation* of the given function $f(x)$. This is the fundamental operation of calculus, upon which everything else depends. In principle, we merely follow the computational instructions specified in (1). These instructions can be arranged into a systematic procedure called the *three-step rule.*

STEP 1   Write down the difference $f(x + \Delta x) - f(x)$ for the particular function under consideration, and if possible simplify it to the point where Δx is a factor.

STEP 2   Divide by Δx to form the *difference quotient*

$$\frac{f(x + \Delta x) - f(x)}{\Delta x},$$

and manipulate this to prepare the way for evaluating its limit as $\Delta x \to 0$. In most of the examples and problems of the present chapter, this manipulation involves nothing more than canceling Δx from the numerator and denominator.

STEP 3   Evaluate the limit of the difference quotient as $\Delta x \to 0$. If Step 2 has accomplished its purpose, a simple inspection is usually all that is needed here.

If we remember that the innocent-looking notation $f(x)$ encompasses all conceivable functions, then we will understand that these steps are some-

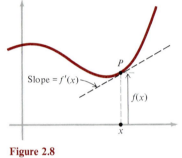

Slope = $f'(x)$

P

$f(x)$

x

Figure 2.8

times easy to carry out and sometimes very difficult. The following examples depend only on elementary algebra, but even this requires a little knowledge and ingenuity.

Example 1   Find $f'(x)$ if $f(x) = x^3$.

STEP 1:

$$f(x + \Delta x) - f(x) = (x + \Delta x)^3 - x^3$$
$$= x^3 + 3x^2\,\Delta x + 3x(\Delta x)^2 + (\Delta x)^3 - x^3$$
$$= 3x^2\,\Delta x + 3x(\Delta x)^2 + (\Delta x)^3$$
$$= \Delta x[3x^2 + 3x\,\Delta x + (\Delta x)^2].$$

STEP 2:

$$\frac{f(x + \Delta x) - f(x)}{\Delta x} = 3x^2 + 3x\,\Delta x + (\Delta x)^2.$$

STEP 3:

$$f'(x) = \lim_{\Delta x \to 0} [3x^2 + 3x\,\Delta x + (\Delta x)^2] = 3x^2.$$

Example 2   Find $f'(x)$ if $f(x) = 1/x$.

STEP 1:

$$f(x + \Delta x) - f(x) = \frac{1}{x + \Delta x} - \frac{1}{x}$$
$$= \frac{x - (x + \Delta x)}{x(x + \Delta x)} = \frac{-\Delta x}{x(x + \Delta x)}.$$

STEP 2:

$$\frac{f(x + \Delta x) - f(x)}{\Delta x} = \frac{-1}{x(x + \Delta x)}.$$

STEP 3:

$$f'(x) = \lim_{\Delta x \to 0} \frac{-1}{x(x + \Delta x)} = -\frac{1}{x^2}.$$

Let us briefly consider what the result of Example 2 can tell us about the graph of the function $y = f(x) = 1/x$. First, $f'(x) = -1/x^2$ is clearly negative for all $x \neq 0$, and since this is the slope of the tangent, all tangent lines point down to the right. Further, when x is near 0, $f'(x)$ is very large, which means that these tangent lines are steep; and when x is large, $f'(x)$ is small, and so these tangent lines are nearly horizontal. It is instructive to verify our observations by examining Fig. 1.30. Generally speaking, derivatives are capable of telling us a great deal about the behavior of functions and the properties of their graphs. We explore this topic more fully in Chapter 4.

Example 3   Find $f'(x)$ if $f(x) = \sqrt{x}$.

STEP 1:

$$f(x + \Delta x) - f(x) = \sqrt{x + \Delta x} - \sqrt{x}.$$

STEP 2:

$$\frac{f(x + \Delta x) - f(x)}{\Delta x} = \frac{\sqrt{x + \Delta x} - \sqrt{x}}{\Delta x}.$$

This is not in a form that is convenient for canceling the Δx's, so we use an algebraic trick to remove the square roots from the numerator. We multiply both numerator and denominator of the last fraction by $\sqrt{x + \Delta x} + \sqrt{x}$, which amounts to multiplying this fraction by 1, and then we use the fact expressed by the algebraic identity $(a - b)(a + b) = a^2 - b^2$:

$$\frac{f(x + \Delta x) - f(x)}{\Delta x} = \frac{\sqrt{x + \Delta x} - \sqrt{x}}{\Delta x} \cdot \frac{\sqrt{x + \Delta x} + \sqrt{x}}{\sqrt{x + \Delta x} + \sqrt{x}}$$

$$= \frac{(x + \Delta x) - x}{\Delta x(\sqrt{x + \Delta x} + \sqrt{x})} = \frac{1}{\sqrt{x + \Delta x} + \sqrt{x}}.$$

Now the next step is easy.

STEP 3:

$$f'(x) = \lim_{\Delta x \to 0} \frac{1}{\sqrt{x + \Delta x} + \sqrt{x}} = \frac{1}{\sqrt{x} + \sqrt{x}} = \frac{1}{2\sqrt{x}}.$$

REMARKS ON NOTATION

There is a slightly disconcerting feature of calculus that we might as well confront here. It is the fact that several different notations are in common use for derivatives, with preference shifting from one to another according to the circumstances in which the symbols are being used. Some may ask, What does it matter which symbols are used? The fact is that it matters a great deal, for good notations can smooth the way and do much of our work for us, while bad ones create a swamp under our feet through which easy movement is almost impossible.

The derivative of a function $f(x)$ has been denoted above by $f'(x)$. This notation has the merit of emphasizing that the derivative of $f(x)$ is another function of x which is associated in a certain way with the given function. If our function is given in the form $y = f(x)$, with the dependent variable displayed, then the shorter symbol y' is often used in place of $f'(x)$.

The main disadvantage of this prime notation for derivatives is that it doesn't suggest the nature of the process by which $f'(x)$ is obtained from $f(x)$.

The notation devised by Leibniz for his version of calculus is better in this respect, and in other ways as well.

To explain Leibniz's notation, we begin with a function $y = f(x)$ and write the difference quotient

$$\frac{f(x + \Delta x) - f(x)}{\Delta x}$$

in the form

$$\frac{\Delta y}{\Delta x},$$

where $\Delta y = f(x + \Delta x) - f(x)$. Here Δy is not just any change in y; it is the specific change that results when the independent variable is changed from x to $x + \Delta x$. As we know, the difference quotient $\Delta y / \Delta x$ can be interpreted as the ratio of the change in y to the change in x along the curve $y = f(x)$, and this is the slope of the secant (Fig. 2.9). Leibniz wrote the limit of this difference quotient, which of course is the derivative $f'(x)$, in the form dy/dx (read "dy over dx"). In this notation, the definition of the derivative becomes

$$\frac{dy}{dx} = \lim_{\Delta x \to 0} \frac{\Delta y}{\Delta x}, \qquad (2)$$

and this is the slope of the tangent in Fig. 2.9. Two slightly different equivalent forms of dy/dx are

$$\frac{df(x)}{dx} \quad \text{and} \quad \frac{d}{dx} f(x).$$

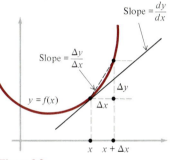

Figure 2.9

In the second of these, the symbol d/dx should be thought of as an operation which can be applied to the function $f(x)$ to yield its derivative $f'(x)$, as suggested by the equation

$$\frac{d}{dx} f(x) = f'(x).$$

The symbol d/dx can be read "the derivative with respect to x of . . ." whatever function of x follows it.

It is important to understand that dy/dx in (2) is a single indivisible symbol. In spite of the way it is written, it is *not* the quotient of two quantities dy and dx, because dy and dx have not been defined and have no independent existence. In Leibniz's notation, the formation of the limit on the right of (2) is symbolically expressed by replacing the letter Δ by the letter d. From this point of view, the symbol dy/dx for the derivative has the psychological advantage that it quickly reminds us of the whole process of forming the difference quotient $\Delta y / \Delta x$ and calculating its limit as $\Delta x \to 0$. There is also a practical advantage, for certain fundamental formulas developed in the next chapter are easier to remember and use when derivatives are written in the Leibniz notation.

But good though it is, this notation is not perfect. For instance, suppose we wish to write down the numerical value of the derivative at a specific point, say $x = 3$. Since dy/dx doesn't display the variable x in the convenient way that $f'(x)$ does, we are forced into using some such clumsy notation as

$$\left(\frac{dy}{dx}\right)_{x=3} \qquad \text{or} \qquad \left.\frac{dy}{dx}\right|_{x=3}.$$

The clear and concise symbol $f'(3)$ is obviously superior to these awkward expressions.

As we have seen, each of the notations described above is good in its own way. All are widely used in the literature of science and mathematics, and to help the student become thoroughly familiar with them, we shall use them freely and interchangeably from now on.

PROBLEMS

In Problems 1 to 12, use the three-step rule to calculate $f'(x)$ if $f(x)$ is equal to the given expression.

1 $ax^2 + bx + c$ (a, b, c constants).
2 $5x - x^3$.
3 $2x^3 - 3x^2 + 6x - 5$.
4 x^4.

5 $x - \dfrac{1}{x}$.
6 $\dfrac{1}{3x + 2}$.

7 $\dfrac{x}{x + 1}$.
8 $\dfrac{1}{x^2}$.

9 $\dfrac{1}{x^3}$.
10 $\dfrac{1}{x^2 + 1}$.

11 $\sqrt{2x}$.
12 $\sqrt{x - 1}$.

13 Consider the part of the curve $y = 1/x$ that lies in the first quadrant, and draw the tangent at an arbitrary point (x_0, y_0) on this curve.

(a) Show that the portion of the tangent which is cut off by the axes is bisected by the point of tangency.
(b) Find the area of the triangle formed by the axes and the tangent, and verify that this area is independent of the location of the point of tangency.

14 Find $f'(x)$ if $f(x) = x^3 - 3x$. Use this result to verify the positions of the high and low points on the curve $y = x^3 - 3x$ that are shown in Fig. 1.28. Hint: At the high and low points the tangent is horizontal.

15 Graph the function $y = f(x) = |x| + x$, and prove that this function is not differentiable at $x = 0$. Hint: In formula (1), first take Δx positive, obtaining one limiting value; then take Δx negative, obtaining a different limiting value. In a situation of this kind we say that the function has a *right derivative* and a *left derivative,* but not a derivative.

2.4
VELOCITY AND RATES OF CHANGE

The concept of the derivative is closely related to the problem of computing the velocity of a moving object. It was this fact that made calculus an essential tool of thought for Newton, in his efforts to uncover the principles of dynamics and understand the motions of the planets. It might appear that only students of physics would find it worthwhile to concern themselves with precise ideas about velocity. However, we shall see that these ideas provide a fairly easy introduction to the general concept of rate of change, and this concept is important in many other fields of study, including the biological and social sciences—especially economics.

In this section we consider a special case of the general velocity problem: that in which the object in question can be thought of as a point moving along a straight line, so that the position of the point is determined by a single coordinate s (Fig. 2.10). The motion is fully known if we know where the moving point is at each moment, that is, if we know the position s as a function of the time t,

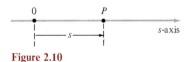

Figure 2.10

$$s = f(t). \tag{1}$$

The time is usually measured from some convenient initial moment $t = 0$.

Example 1   Consider a freely falling object, say a rock dropped from the edge of a cliff 400 ft high (Fig. 2.11). It is known from many experiments that this rock falls

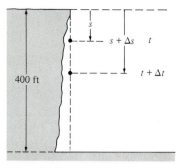

$$s = 16t^2 \qquad\qquad (2)$$

feet in t seconds. We see that when $t = 5$, $s = 400$. The rock therefore hits the ground 5 seconds after it starts to fall, and formula (2) is valid only for $0 \le t \le 5$.

Figure 2.11

Two basic questions can be asked about the motion described in this example. First, what is meant by the velocity of the falling rock at a given instant? And second, how can this velocity be computed from (2)?

We are all familiar with the idea of velocity in its everyday sense, as a number measuring the rate at which distance is being traversed. We speak of walking 3 miles per hour (mi/h), driving 55 mi/h, and so on. We also speak of *average* velocities, and these are the numbers we usually compute. If we drive a distance of 200 mi in 5 hours, then our average velocity is 40 mi/h, because

$$\frac{\text{Distance traveled}}{\text{Elapsed time}} = \frac{200 \text{ mi}}{5 \text{ h}} = 40 \text{ mi/h}.$$

In general,

$$\text{Average velocity} = \frac{\text{distance traveled}}{\text{elapsed time}},$$

and this is a formula that few people would disagree with.

Example 1 (Cont.)   The position function for the falling rock, $f(t) = 16t^2$, tells us that in the first second after the rock is released it falls $f(1) = 16$ ft, in the first 2 seconds $f(2) = 64$ ft, in the first 3 seconds $f(3) = 144$ ft, and so on. The average velocities during each of the first 3 seconds of fall are therefore

$$\frac{16}{1} = 16 \text{ ft/s}, \qquad \frac{64 - 16}{1} = 48 \text{ ft/s}, \qquad \text{and} \qquad \frac{144 - 64}{1} = 80 \text{ ft/s}.$$

The rock is clearly falling faster and faster from moment to moment, but the question of exactly how fast it is falling at any given instant is still unanswered.

To find the velocity v of the rock at a given instant t, we proceed as follows. In the time interval of length Δt between t and a slightly later instant $t + \Delta t$, the rock falls a distance Δs (see Fig. 2.11). The average velocity during this interval is the quotient $\Delta s / \Delta t$. When Δt is small, this average velocity is close to the exact velocity v at the beginning of the interval; that is,

$$v \cong \frac{\Delta s}{\Delta t},$$

where the symbol $\cong$ is read "is approximately equal to." Further, as Δt is made smaller and smaller, this approximation gets better and better, so we have

$$v = \lim_{\Delta t \to 0} \frac{\Delta s}{\Delta t}. \tag{3}$$

Our point of view here is that the velocity v is a direct intuitive concept, and (3) shows us how to compute it. However, it is also possible to regard (3) as the *definition* of the velocity, with the preceding remarks serving as motivation. The limit in (3) is clearly the derivative ds/dt, and carrying out the details we have

$$v = \frac{ds}{dt} = \lim_{\Delta t \to 0} \frac{\Delta s}{\Delta t}$$

$$= \lim_{\Delta t \to 0} \frac{16(t + \Delta t)^2 - 16t^2}{\Delta t}$$

$$= \lim_{\Delta t \to 0} (32t + 16\Delta t) = 32t.$$

This formula tells us that the velocity of the rock after 1, 2, and 3 seconds of fall is 32, 64, and 96 ft/s, and also that the rock hits the ground at 160 ft/s. We notice that the velocity increases by 32 ft/s during each second of fall. This fact is usually expressed by saying that the acceleration of the rock is 32 feet per second per second (ft/s²), or, in the metric system, 9.80 meters per second per second (m/s²).

The reasoning used in this example is valid for any motion along a straight line. For the general motion (1), we therefore calculate the velocity v at time t in exactly the same way; that is, we approximate v more and more closely by the average velocity over a shorter and shorter interval of time beginning at the instant t:

$$v = \lim_{\Delta t \to 0} \frac{\Delta s}{\Delta t} = \lim_{\Delta t \to 0} \frac{f(t + \Delta t) - f(t)}{\Delta t}.$$

We recognize this as the derivative of the function $s = f(t)$, and so the velocity of a point moving on a straight line is simply the derivative of its position function,

$$v = \frac{ds}{dt} = f'(t).$$

Sometimes this is called the "instantaneous" velocity, in order to emphasize that it is calculated at an instant t. However, once this point has been made, it is customary to omit the adjective. The words "velocity" and "speed" are used interchangeably in everyday speech, but in mathematics and physics it is useful to distinguish them from one another. The *speed* of our point is defined to be the absolute value of the velocity,

$$\text{Speed} = |v| = \left| \frac{ds}{dt} \right| = |f'(t)|.$$

The velocity may be positive or negative, depending on whether the point is moving along the line in the positive or negative direction; but the speed, being the magnitude of the velocity, is always positive or zero. The concept of speed is particularly useful in studies of motion along curved paths, for it tells us how fast the point is moving regardless of its direction. In our everyday experience, we learn the speed of a car at any moment by looking at the speedometer.

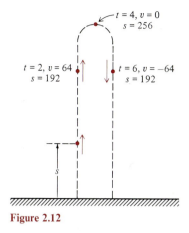

Figure 2.12

Example 2 Consider a projectile fired straight up from the ground with an initial velocity of 128 ft/s. This projectile moves up and down along a straight line. However, the two parts of its path are shown slightly separated in Fig. 2.12, for the sake of visual clarity. Let $s = f(t)$ be the height in feet of the projectile t seconds after firing. If the force of gravity were absent, the projectile would continue moving upward with a constant velocity of 128 ft/s, and we would have $s = f(t) = 128t$. However, the action of gravity causes it to slow down, stop momentarily at the top of its flight, and then fall back to earth with increasing speed. Experimental evidence suggests that the height of the projectile during its flight is given by the formula

$$s = f(t) = 128t - 16t^2. \tag{4}$$

If we write this in the factored form $s = 16t(8 - t)$, we see that $s = 0$ when $t = 0$ and when $t = 8$. The projectile therefore returns to earth 8 seconds after it starts up, and (4) is valid only for $0 \le t \le 8$.

To learn more about the nature of this motion, it is necessary to know the velocity. If the three-step rule for computing derivatives is applied to (4), we find that the velocity at time t is

$$v = \frac{ds}{dt} = 128 - 32t. \tag{5}$$

At the top of its flight the projectile is momentarily at rest, and therefore $v = 0$. By (5), $t = 4$ when $v = 0$; and by (4), $s = 256$ when $t = 4$. In this way we find the maximum height reached by the projectile and the time required to reach this height (see Fig. 1.12). As t increases from 0 to 8, it is clear from (5) that v decreases from 128 ft/s to -128 ft/s; in fact, v decreases by 32 ft/s during each second of flight, and this is expressed by saying that the acceleration is -32 feet per second per second (ft/s²). We notice explicitly that the velocity is positive from $t = 0$ to $t = 4$, when s is increasing; and it is negative from $t = 4$ to $t = 8$, when s is decreasing. In particular, it is easy to see from (5) that $v = 64$ ft/s when $t = 2$ and $v = -64$ ft/s when $t = 6$ (the speed is 64 ft/s at both times).

Velocity is an example of the concept of rate of change, which is basic for all the sciences. For any function $y = f(x)$, the derivative dy/dx is called the *rate of change* of y with respect to x. Intuitively, this is the change in y that would be produced by an increase of one unit in x if the rate of change remained constant (Fig. 2.13). In this terminology, velocity is simply the rate of change of position with respect to time. When time is the independent variable, we often omit the phrase "with respect to time" and speak only of the "rate of change."

Example 3 (a) We know that velocity is important in studying the motion of a point along a straight line, but the way the velocity changes is also impor-

Figure 2.13

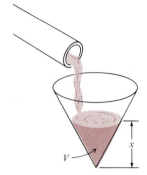

Figure 2.14

tant. By definition, the *acceleration* of a moving point is the rate of change of its velocity v,

$$a = \frac{dv}{dt}.$$

(b) Suppose that water is being pumped into the conical tank shown in Fig. 2.14 at the rate of 5 ft³/min. If V denotes the volume of water in the tank at time t, then

$$\frac{dV}{dt} = 5.$$

The rate of change of the depth x is the derivative dx/dt, and this is not constant. It is intuitively clear that this rate of change is large when the area of the surface of the water is small, and becomes smaller as this area increases.

(c) In economics, the rate of change of a quantity Q with respect to a suitable independent variable is usually called "marginal Q." Thus we have marginal cost, marginal revenue, marginal profit, etc. If $C(x)$ is the cost of manufacturing x pieces of some product, then the marginal cost is dC/dx. In most cases x is a large number, so 1 is small compared with x and dC/dx is approximately $C(x + 1) - C(x)$. For this reason, many economists describe marginal cost as "the cost of producing one more piece."

(d) We know that the area A of a circle in terms of its radius r is given by the formula $A = \pi r^2$, and the derivative of this function is easy to compute by the three-step rule:

$$\frac{dA}{dr} = 2\pi r. \tag{6}$$

This says that the rate of change of the area of a circle with respect to its radius equals its circumference. To understand the geometric reason for this remarkable fact, let Δr be an increment of the radius and ΔA the corresponding increment of the area (Fig. 2.15). It is clear that ΔA is the area of the narrow band around the circle, and this is approximately the product of the circumference $2\pi r$ and the width Δr of the band. The difference quotient $\Delta A/\Delta r$ is therefore close to $2\pi r$, and by letting $\Delta r \to 0$ we obtain (6).

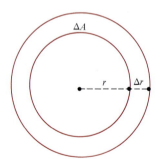

Figure 2.15

We have introduced two topics in this section: velocity, which is the rate of change of the position of a moving object, and rates of change in general. These are themes of major importance for calculus, and we shall return to them repeatedly throughout the rest of this book.

PROBLEMS

According to Problem 1 of Section 2.3, the general quadratic function

$$s = f(t) = at^2 + bt + c$$

has derivative

$$\frac{ds}{dt} = f'(t) = 2at + b.$$

Each of the formulas in Problems 1 to 7 describes the motion of a point along a horizontal line whose positive direction is to the right. In each case use the result stated here to write down the velocity $v = ds/dt$ by inspection. Also, find (a) the times when the velocity is zero, so that the point is momentarily at rest; and (b) the times when the point is moving to the right.

1 $s = 3t^2 - 12t + 7$.

2 $s = 1 - 6t - t^2$.

3 $s = 2t^2 + 28t - 6$.

4 $s = -19 + 10t - 5t^2$.

5 $s = 7t^2 + 2$.

6 $s = 2 + 7t$.

7 $s = (2t - 6)^2$.

8 Two points start from the origin on the s-axis at time $t = 0$ and move along this axis in accordance with the formulas

$$s_1 = t^2 - 6t \quad \text{and} \quad s_2 = 8t - t^2,$$

where s_1 and s_2 are measured in feet and t in seconds.
(a) When will the two points have the same speed?
(b) What are the velocities of the two points at the times when they have the same position?

9 Starting from rest, a certain car moves s feet in t seconds, where $s = 4.4t^2$. How long does it take the car to reach the velocity of 60 mi/h ($= 88$ ft/s)?

10 Assume that a projectile fired straight up from the ground with an initial velocity of v_0 ft/s reaches a height of s feet in t seconds, where

$$s = v_0 t - 16t^2.$$

(a) Find the velocity v at time t.

(b) How much time is required for the projectile to reach its maximum height?
(c) What is the maximum height?
(d) What is the velocity when the projectile hits the ground?
(e) What must the initial velocity be for the projectile to hit the ground 15 seconds after firing?

11 An oil tank is to be drained for cleaning. If there are V gallons of oil left in the tank t minutes after the draining begins, where $V = 40(50 - t)^2$, find
(a) the average rate at which oil drains out of the tank during the first 20 minutes;
(b) the rate at which oil is flowing out of the tank 20 minutes after the draining begins.

12 Consider a square of area A and side s, so that $A = s^2$. If $x = \frac{1}{2}s$, use the idea of Example 3d to make a conjecture about the value of dA/dx. Verify your conjecture by calculation.

13 Suppose a balloon of volume V and radius r is being inflated, so that V and r are both functions of the time t. If dV/dt is constant, what can be said (without calculation) about the behavior of dr/dt as r increases?

It is evident from the preceding sections that the definition of the derivative rests on the concept of the limit of a function, which we have freely used with only the briefest explanation. What is this concept?

Let us consider a function $f(x)$ that is defined for values of x near a point a on the x-axis but not necessarily at a itself. Suppose there exists a number L with the property that $f(x)$ gets closer and closer to L as x gets closer and closer to a (Fig. 2.16). Under these circumstances we say that L is the *limit* of $f(x)$ as x approaches a, and we express this symbolically by writing

$$\lim_{x \to a} f(x) = L. \tag{1}$$

If there is no number L with this property, we say that $f(x)$ *has no limit* as x approaches a, or that $\lim_{x \to a} f(x)$ *does not exist*. Another widely used notation equivalent to (1) is

$$f(x) \to L \quad \text{as} \quad x \to a,$$

which is read "$f(x)$ approaches L as x approaches a." In thinking about the meaning of (1), it is essential to understand that it does not matter what happens to $f(x)$ when x *equals a*; all that matters is the behavior of $f(x)$ for x's that are *near a*.

These informal descriptions of the meaning of (1) are helpful to the intuition and are adequate for most practical purposes. Nevertheless, they are too loose and imprecise to be acceptable as definitions, because of the vagueness of such expressions as "closer and closer" and "approaches." The exact meaning of (1) is too important to be left mainly to the reader's imagination, and at the risk of being overly technical, we will try to give a satisfactory definition as briefly and clearly as possible.

2.5

LIMITS AND CONTINUOUS FUNCTIONS

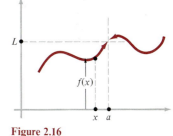

Figure 2.16

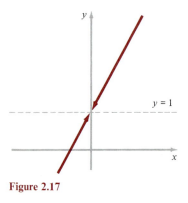

Figure 2.17

We begin by analyzing a specific example:

$$y = f(x) = \frac{2x^2 + x}{x}.$$

This function is not defined for $x = 0$, and for $x \neq 0$ its values are given by the simpler expression

$$f(x) = \frac{x(2x + 1)}{x} = 2x + 1.$$

If we examine the graph (Fig. 2.17), it is clear that $f(x)$ is close to 1 when x is close to 0. In order to give a quantitative description of this qualitative behavior, we need a formula for the difference between $f(x)$ and the limiting value 1:

$$f(x) - 1 = (2x + 1) - 1 = 2x.$$

We see from this formula that $f(x)$ can be made *as close as we please* to 1 by taking x *sufficiently close* to 0. Thus,

$$f(x) - 1 = \tfrac{1}{100} \quad \text{when} \quad x = \tfrac{1}{200},$$

$$f(x) - 1 = \tfrac{1}{1000} \quad \text{when} \quad x = \tfrac{1}{2000},$$

and so on. More generally, let ϵ (epsilon) be any positive number given in advance, no matter how small, and define δ (delta) by $\delta = \tfrac{1}{2}\epsilon$. Then the distance from $f(x)$ to 1 will be smaller than ϵ, provided only that the distance from x to 0 is smaller than δ; that is,

$$\text{if} \quad |x| < \delta = \tfrac{1}{2}\epsilon \quad \text{then} \quad |f(x) - 1| = 2|x| < \epsilon.$$

This assertion is much more precise than the vague statement that $f(x)$ is "close" to 1 when x is "close" to 0. It tells us exactly how close x must be to 0 in order to guarantee that $f(x)$ will attain a previously specified degree of closeness to 1. Of course, x is not permitted to equal 0 here, because $f(x)$ has no meaning for $x = 0$.

The so-called epsilon-delta definition of (1) is now easy to grasp. It is this: For each positive number ϵ there exists a positive number δ with the property that

$$|f(x) - L| < \epsilon$$

for any number x in the domain of the function that satisfies the inequalities

$$0 < |x - a| < \delta.$$

Students should read this definition carefully and be aware of its crucial role in the theory of calculus. However, an intuitive understanding of limits is quite enough for our purposes, and from this point of view the following examples should present no difficulties.

Example 1 First,

$$\lim_{x \to 2} (3x + 4) = 10.$$

Here it is clear that as x approaches 2, $3x$ approaches 6 and $3x + 4$ approaches $6 + 4 = 10$. Next,

$$\lim_{x \to 1} \frac{x^2 - 1}{x - 1} = \lim_{x \to 1} (x + 1) = 2.$$

The first thing we notice here is that the function $(x^2 - 1)/(x - 1)$ is undefined at $x = 1$, since both numerator and denominator equal 0. But this fact is irrelevant, since all that matters is the behavior of the function for x's that are near 1 but different from 1, and for all such x's the function equals $x + 1$ and this is near 2.

Example 2 It is illuminating to consider a few limits that do not exist, for instance

$$\lim_{x \to 0} \frac{x}{|x|}, \qquad \lim_{x \to 0} \frac{1}{x}, \qquad \text{and} \qquad \lim_{x \to 0} \frac{1}{x^2}.$$

The behavior of these limits is most easily understood by looking at the graphs of the functions $x/|x|$, $1/x$, and $1/x^2$ (Fig. 2.18). In the first case the function equals 1 when x is positive and -1 when x is negative (and is undefined for $x = 0$), so there is no single number that the values of the function approach as x approaches 0 from both sides. We can be a bit more specific about the way this limit fails to exist, by writing

$$\lim_{x \to 0+} \frac{x}{|x|} = 1 \qquad \text{and} \qquad \lim_{x \to 0-} \frac{x}{|x|} = -1.$$

The notations $x \to 0+$ and $x \to 0-$ are intended to suggest that the variable x approaches 0 from the positive side (the right) and from the negative side (the left), respectively. The other two limits fail to exist because in each case the values of the function become arbitrarily large in absolute value as x approaches 0. In symbols,

$$\lim_{x \to 0+} \frac{1}{x} = \infty, \qquad \lim_{x \to 0-} \frac{1}{x} = -\infty, \qquad \text{and} \qquad \lim_{x \to 0} \frac{1}{x^2} = \infty.$$

The main rules for calculating with limits are exactly what we would expect. For instance,

$$\lim_{x \to a} x = a;$$

and if c is a constant, then

$$\lim_{x \to a} c = c.$$

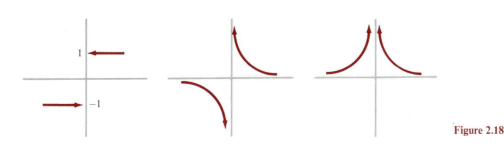

Figure 2.18

Also, if $\lim_{x \to a} f(x) = L$ and $\lim_{x \to a} g(x) = M$, then

$$\lim_{x \to a} [f(x) + g(x)] = L + M,$$

$$\lim_{x \to a} [f(x) - g(x)] = L - M,$$

$$\lim_{x \to a} f(x)g(x) = LM,$$

and

$$\lim_{x \to a} \frac{f(x)}{g(x)} = \frac{L}{M} \qquad (\text{if } M \neq 0).$$

In words, the limit of a sum is the sum of the limits, with similar statements for differences, products, and quotients.

As we penetrate further into our subject, it will often be important for us to know what is meant by a *continuous function.* In everyday speech a "continuous" process is one that proceeds without gaps or interruptions or sudden changes. Roughly speaking, a function $y = f(x)$ is continuous if it displays similar behavior, that is, if a small change in x produces a small change in the corresponding value $f(x)$. The function shown in Fig. 2.19 is continuous at the point a because $f(x)$ is close to $f(a)$ when x is close to a, or, more precisely, because $f(x)$ can be made as close as we please to $f(a)$ by taking x sufficiently close to a. In the language of limits this says that

$$\lim_{x \to a} f(x) = f(a). \qquad (2)$$

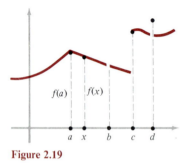

Figure 2.19

Up to this stage our remarks about continuity have been rather loose and intuitive, and intended more to explain than to define. We now adopt equation (2) as the definition of the statement that $f(x)$ is *continuous at a.* The reader should observe that the continuity of $f(x)$ at a requires three things to happen: a must be in the domain of $f(x)$, so that $f(a)$ exists; $f(x)$ must have a limit as x approaches a; and this limit must equal $f(a)$. We can understand these ideas more clearly by examining Fig. 2.19, in which the function is discontinuous in different ways at the points b, c, and d: at the point b, $\lim_{x \to b} f(x)$ exists but $f(b)$ does not; at c, $f(c)$ exists but $\lim_{x \to c} f(x)$ does not; and at d, $f(d)$ and $\lim_{x \to d} f(x)$ both exist but have different values.

The definition given here tells us what it means for a function to be continuous at a particular point in its domain. A function is called a *continuous function* if it is continuous at every point in its domain. In particular, by the properties of limits this is easily seen to be true for all polynomials and rational functions. We will be especially interested in functions that are continuous on intervals. These functions are often described as those whose graphs can be drawn without lifting the pencil from the paper.

With a slight change of notation, we can express the continuity of our function at a point x (instead of a) in either of the equivalent forms

$$\lim_{\Delta x \to 0} f(x + \Delta x) = f(x) \qquad \text{or} \qquad \lim_{\Delta x \to 0} [f(x + \Delta x) - f(x)] = 0;$$

and if we write $\Delta y = f(x + \Delta x) - f(x)$, then this condition becomes

$$\lim_{\Delta x \to 0} \Delta y = 0.$$

The purpose of this is to make it possible to give a very easy proof of a fact that we will need in the next chapter, namely, that *a function which is differentiable at a point is continuous at that point.* The proof occupies only a single line:

$$\lim_{\Delta x \to 0} \Delta y = \lim_{\Delta x \to 0} \frac{\Delta y}{\Delta x} \cdot \Delta x = \left[\lim_{\Delta x \to 0} \frac{\Delta y}{\Delta x} \right] \left[\lim_{\Delta x \to 0} \Delta x \right] = \frac{dy}{dx} \cdot 0 = 0.$$

The converse of this statement is not true, since a function can easily be continuous at a point without being differentiable there (for example, see the point *a* in Fig. 2.19).

We remarked earlier that calculus is a problem-solving art and not a branch of logic. It has more to do with insight nourished by intuitive understanding than it does with careful deductive reasoning. Naturally, we will try to convince the reader of the truth of our statements and the legitimacy of our procedures. However, these efforts will be brief and rather informal, in order to avoid clogging the text with massive indigestible chunks of theoretical material. Those who wish to devote more attention to the purely mathematical side of the subject will find logically rigorous proofs of the major theorems in Appendix C at the back of the book. In particular, the properties of limits stated here are proved in Appendix C.2.

PROBLEMS

Some of the following limits exist, and others do not. Evaluate those that do.

1 $\lim_{x \to 3} (7x - 6)$.

2 $\lim_{x \to 2} \dfrac{10}{3 + x}$.

3 $\lim_{x \to 0} \dfrac{5}{x - 1}$.

4 $\lim_{x \to 2} \dfrac{6}{2x - 4}$.

5 $\lim_{x \to 3} \dfrac{3x - 9}{x - 3}$.

6 $\lim_{x \to 3} \dfrac{x^2 + 3x}{x^2 - x + 3}$.

7 $\lim_{x \to 5} \dfrac{x - 3 - 2x^2}{1 + 3x}$.

8 $\lim_{x \to -3} \dfrac{4x}{x + 3}$.

9 $\lim_{x \to -3} \left(\dfrac{4x}{x + 3} + \dfrac{12}{x + 3} \right)$.

10 $\lim_{x \to 0.001} \dfrac{x}{|x|}$.

11 $\lim_{x \to 7} \dfrac{x^2 + x - 56}{x^2 - 11x + 28}$.

12 $\lim_{x \to -2} \dfrac{(x + 2)(x^2 - x + 3)}{x^2 + x - 2}$.

13 $\lim_{x \to 0} \dfrac{x^2}{|x|}$.

14 $\lim_{x \to 4} \dfrac{x - 4}{\sqrt{x} - 2}$.

15 $\lim_{x \to 4} \dfrac{x - 4}{x - \sqrt{x} - 2}$.

16 $\lim_{x \to 3} \dfrac{\sqrt{x^2 + 16} - 5}{x^2 - 3x}$.

17 If $\lim_{x \to a} f(x) = 4$, $\lim_{x \to a} g(x) = -2$, and $\lim_{x \to a} h(x) = 0$, evaluate the following limits:

(a) $\lim_{x \to a} [f(x) - g(x)]$;

(b) $\lim_{x \to a} [g(x)]^2$;

(c) $\lim_{x \to a} \dfrac{f(x)}{g(x)}$;

(d) $\lim_{x \to a} \dfrac{h(x)}{f(x)}$;

(e) $\lim_{x \to a} \dfrac{f(x)}{h(x)}$;

(f) $\lim_{x \to a} \dfrac{1}{[f(x) + g(x)]^2}$.

18 In many situations we are interested in the behavior of $f(x)$ when x is large and positive. If there exists a number L with the property that $f(x)$ gets closer and closer to L as x gets larger and larger (Fig. 2.20), then we say

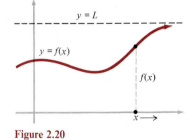

Figure 2.20

that L is the *limit* of $f(x)$ as x approaches infinity, and we symbolize this by writing $\lim_{x \to \infty} f(x) = L$. Evaluate the following limits:

(a) $\lim_{x \to \infty} \dfrac{1}{x}$;

(b) $\lim_{x \to \infty} \left(2 + \dfrac{100}{x} \right)$;

(c) $\lim\limits_{x \to \infty} \dfrac{5x + 3}{2x - 7}$; *

(d) $\lim\limits_{x \to \infty} \dfrac{2x^2 + x - 5}{3x^2 - 7x + 2}$;

(e) $\lim\limits_{x \to \infty} \dfrac{x}{x^2 + 1}$;

(f) $\lim\limits_{x \to \infty} \dfrac{x^2 - 2x + 5}{x^3 + 7x^2 + 2x - 1}$.

---

* Hint: Notice that dividing both numerator and denominator of this quotient by x gives

$$\frac{5x + 3}{2x - 7} = \frac{5 + \dfrac{3}{x}}{2 - \dfrac{7}{x}}.$$

What becomes of the expression on the right as $x \to \infty$?

19 Find the points of discontinuity of the following functions:

(a) $\dfrac{x}{x^2 + 1}$;

(b) $\dfrac{x}{x^2 - 1}$;

(c) $\dfrac{x^2 - 1}{x - 1}$;

(d) $\sqrt{x}$;

(e) $\dfrac{1}{\sqrt{x}}$;

(f) $\sqrt{x^2}$;

(g) $\dfrac{1}{x^2 + x - 12}$;

(h) $\dfrac{1}{x^2 + 4x + 5}$.

ADDITIONAL PROBLEMS FOR CHAPTER 2

SECTION 2.2

1 For what value of b does the graph of $y = x^2 + bx + 1$ have a horizontal tangent at $x = 3$?

2 Find the two points on the curve $y = x - \frac{1}{4}x^2$ at which the tangent passes through the point $(\frac{9}{2}, 0)$.

3 Let $P = (x_0, y_0)$ be a point on the parabola $y = x^2$. Show that a nonvertical line passing through P which does not intersect the curve at any other point is necessarily the tangent at P; that is, show that if the line

$$y - y_0 = m(x - x_0)$$

intersects $y = x^2$ only at (x_0, y_0), then $m = 2x_0$.

4 If (x_1, y_1) and (x_2, y_2) are distinct points on the parabola $y = x^2$, at what point on the curve is the tangent parallel to the chord joining these two given points?

5 The curve $y = x^2$ is a particular parabola, but if a is an unspecified positive constant, $y = f(x) = ax^2$ is a completely general parabola located in a convenient position.

(a) Show that $f'(x_0) = 2ax_0$.

(b) Show that the tangent at a point $P = (x_0, y_0)$ other than the vertex has y-intercept $-y_0$, and use this fact to formulate a geometric method for constructing the tangent at P.

SECTION 2.3

6 Use the three-step rule to calculate $f'(x)$ if $f(x)$ is equal to

(a) $\dfrac{x + 1}{x}$;

(b) $\dfrac{3 - 2x}{x - 2}$;

(c) $\sqrt{3x + 2}$;

(d) $\sqrt{x^2 + 1}$.

7 Sketch the graph of each of the following functions and state where it is not differentiable:

(a) $\sqrt{|x|}$;

(b) $|x^2 - 4|$;

(c) $|2x - 3|$;

(d) $x|x|$.

8 Let $f(x)$ be a function with the property that $f(x_1 + x_2) = f(x_1)f(x_2)$ for all x_1 and x_2. If $f(0) = 1$ and $f'(0) = 1$, show that $f'(x) = f(x)$ for all x.

9 If the derivative $f'(x)$ exists, then it can be calculated from the formula

$$f'(x) = \lim\limits_{\Delta x \to 0} \frac{f(x + \Delta x) - f(x - \Delta x)}{2\Delta x}.$$

Verify this statement for the special case $f(x) = x^2$, and then prove it in general. [To understand the statement, let P, Q, R be the points on the curve $y = f(x)$ that correspond to x, $x + \Delta x$, $x - \Delta x$, and write the slope of the secant through Q and R; and to prove it, notice that $f(x + \Delta x) - f(x - \Delta x) = f(x + \Delta x) - f(x) + f(x) - f(x - \Delta x)$.]

10 Show that the following function is differentiable at $x = 0$:

$$f(x) = \begin{cases} x^2 & \text{if } x \text{ is rational}, \\ 0 & \text{if } x \text{ is irrational}. \end{cases}$$

11 Show that the following function is not differentiable at $x = 0$:

$$f(x) = \begin{cases} x & \text{if } x \text{ is rational}, \\ 0 & \text{if } x \text{ is irrational}. \end{cases}$$

12 If $f(x)$ is a function with the property that $|f(x)| \le x^2$ for all x, prove that $f(x)$ is differentiable at $x = 0$.

13 Consider the function $f(x)$ defined by

$$f(x) = \begin{cases} x^2 & \text{if } x \le a, \\ mx + b & \text{if } x > a, \end{cases}$$

where a, b, m are constants. Find what values m and b

must have (in terms of a) in order for this function to be differentiable at all points.

SECTION 2.4

14 On a certain bicycle trip, the first half of the distance was covered at 30 mi/h and the second half at 20 mi/h. What was the average velocity? (It was not 25 mi/h.)

15 A silver dollar is thrown straight up from the roof of a 200-ft building. After t seconds, it is

$$s = 200 + 24t - 16t^2$$

feet above the ground. When does the dollar begin to fall? What is its speed when it has fallen 1 ft?

16 A capacitor (or condenser) in an electric circuit is a device for storing electric charge. If the amount of charge on a given capacitor at time t is $Q = 3t^2 + 5t + 2$ coulombs, find the current $I = dQ/dt$ in the circuit when $t = 3$.

17 Use the three-step rule to show that the rate of change of the volume of a sphere with respect to its radius equals the surface area.

SECTION 2.5

Evaluate the following limits.

18 $\lim_{x \to 2} \dfrac{2x - x^2}{2 - x}$.

19 $\lim_{x \to 0} \left(x + \dfrac{5}{x} \right)$.

20 $\lim_{x \to 3} \dfrac{x^2 - 6x + 9}{x - 3}$.

21 $\lim_{x \to 0} \dfrac{4x^2 - 5x}{x}$.

22 $\lim_{x \to 0} \dfrac{x^2(1 - x)}{3x}$.

23 $\lim_{x \to 0} \dfrac{x(1 - x)}{3x^2}$.

24 $\lim_{x \to 1} \dfrac{x + 2}{x^2 - 4}$.

25 $\lim_{x \to 2} \dfrac{x + 2}{x^2 - 4}$.

26 $\lim_{x \to 3} \dfrac{2x^2 + 3}{x + 4}$.

27 $\lim_{x \to 0} \dfrac{2 - 3\sqrt{x}}{1 + 9\sqrt{x}}$.

28 $\lim_{x \to 2} \dfrac{x^2 - 6x + 8}{x^2 - 5x + 6}$.

29 $\lim_{x \to -1} \dfrac{x^2 - 2x - 3}{x^2 - 1}$.

30 $\lim_{x \to 1} \dfrac{(x^2 + 3x - 4)^2}{x^2 - 7x + 6}$.

31 $\lim_{x \to 0} \dfrac{2x^2 + x - 6}{x + 2}$.

32 $\lim_{x \to -2} \dfrac{2x^2 + x - 6}{x + 2}$.

33 $\lim_{x \to 3} \dfrac{x - 3}{x^2 + x - 12}$.

34 $\lim_{x \to -4} \dfrac{x - 3}{x^2 + x - 12}$.

35 $\lim_{x \to 3} \dfrac{x^2 - x - 6}{x^2 - 7x + 12}$.

36 $\lim_{x \to 4} \dfrac{x^2 - x - 6}{x^2 - 7x + 12}$.

37 $\lim_{x \to 1} \dfrac{x + \sqrt{x} - 2}{x^3 - 1}$.

38 $\lim_{x \to 1} \dfrac{x^3 - 6x^2 + 3x + 2}{x^3 + x^2 - 3x + 1}$. *

39 $\lim_{x \to 2} \dfrac{x^3 - 4x}{x^3 - 3x^2 + 2x}$.

40 $\lim_{x \to 4} \dfrac{x^3 - 64}{x - 4}$.

41 $\lim_{x \to a} \dfrac{x^3 - a^3}{x^2 - a^2}$.

42 $\lim_{x \to -a} \dfrac{x^4 - a^4}{x^3 + a^3}$.

43 $\lim_{x \to 0} 2^{x^2}$.

44 $\lim_{x \to 0} 2^{-x^2}$.

45 $\lim_{x \to 0} 2^{-1/x^2}$.

46 $\lim_{x \to 0} \dfrac{2^{1/x^2} + 1}{2^{1/x^2} - 1}$.

47 $\lim_{x \to 0} \dfrac{2^{1/x} + 1}{2^{1/x} - 1}$.

48 $\lim_{x \to \infty} \dfrac{x}{\sqrt{x^2 + 1}}$.

49 $\lim_{x \to \infty} \dfrac{x}{\sqrt{x + 1}}$.

50 $\lim_{x \to \infty} \dfrac{2x^3 - x^2 + 7x - 3}{2 - x + 5x^2 - 4x^3}$.

51 $\lim_{x \to \infty} \dfrac{9x^{45} - x^9 + 2}{3x^{45} + x^{29} - 19}$.

52 $\lim_{x \to \infty} 2^x$.

53 $\lim_{x \to \infty} 2^{-x}$.

54 $\lim_{x \to \infty} 2^{1/x}$.

55 $\lim_{x \to \infty} (\sqrt{x + 1} - \sqrt{x})$.

56 $\lim_{x \to \infty} \dfrac{\sqrt{x + 1}}{\sqrt{9x + 1}}$.

57 $\lim_{x \to \infty} \dfrac{2^x - 2^{-x}}{2^x + 2^{-x}}$.

58 Consider the function $f(x)$ defined for $x \neq 0$ by $f(x) = [1/x]$, where $[1/x]$ denotes the greatest integer $\leq 1/x$, as in Additional Problem 61 at the end of Chapter 1. Sketch the graph of this function for $\frac{1}{4} \leq x \leq 2$, and also for $-2 \leq x \leq -\frac{1}{4}$. How does $f(x)$ behave as x approaches 0 from the positive side? From the negative side? Does $\lim_{x \to 0} f(x)$ exist?

59 Follow the directions in Problem 58 for the function $f(x) = (-1)^{[1/x]}$.

60 Follow the directions in Problem 58 for the function $f(x) = |x|(-1)^{[1/x]}$.

61 Consider the function $f(x)$ defined by

$$f(x) = \begin{cases} 0 & \text{if } x \text{ is rational,} \\ 1 & \text{if } x \text{ is irrational.} \end{cases}$$

For every a, $\lim_{x \to a} f(x)$ does not exist. Why?

62 Define a function $f(x)$ by

$$f(x) = \begin{cases} 0 & \text{if } x \text{ is irrational,} \\ 1 & \text{if } x \text{ is a rational number } m/n \\ \dfrac{1}{n} & \text{in lowest terms with } n > 0. \end{cases}$$

Show that $f(x)$ is continuous at irrational points and discontinuous at rational points.

---

* If $x = a$ is a zero of a polynomial $p(x)$, then $x - a$ is a factor of $p(x)$ and the other factor can be found by long division (see Additional Problem 52 at the end of Chapter 1).

3

THE COMPUTATION
OF DERIVATIVES

3.1

DERIVATIVES OF
POLYNOMIALS

Differential calculus — the calculus of derivatives — takes its special flavor and importance from its many applications to the physical, biological, and social sciences. It would be pleasant to plunge into these applications immediately and get to the heart of the matter without any further delay. However, from the point of view of overall efficiency, it is better to postpone this to the next chapter, and instead take a little time now to learn how to calculate derivatives with speed and accuracy.

As we know, the process of finding the derivative of a function is called *differentiation.* In Chapter 2 this process was based directly on the limit definition of the derivative,

$$f'(x) = \lim_{\Delta x \to 0} \frac{f(x + \Delta x) - f(x)}{\Delta x},$$

or equivalently,

$$\frac{dy}{dx} = \frac{d}{dx} f(x) = \lim_{\Delta x \to 0} \frac{\Delta y}{\Delta x}.$$

We have seen that this approach is rather slow and clumsy. Our purpose in the present chapter is to develop a small number of formal rules that will enable us to differentiate large classes of functions quickly, by purely mechanical procedures. In this section we learn how to write down the derivative of any polynomial by inspection, without having to think about limits at all; and by the end of the chapter we will be able to cope quite easily with messy algebraic functions like

$$\frac{x}{\sqrt{x^2 + 1}}, \qquad \left[\frac{x + \sqrt{x + 1}}{x - \sqrt{x + 1}} \right]^{1/3}, \qquad \text{and} \qquad \sqrt{1 + \sqrt{1 + \sqrt{1 + x}}}.$$

Our goal in this phase of our work is computational skill, and, needless to say, such skill comes only with practice.

Students will recall that a polynomial in x is a sum of constant multiples of powers of x in which each exponent is zero or a positive integer:

$$P(x) = a_n x^n + a_{n-1} x^{n-1} + \cdots + a_1 x + a_0.$$

The way a polynomial is put together out of simpler pieces suggests the differentiation rules that we now discuss.

1 *The derivative of a constant is zero,*

$$\frac{d}{dx} c = 0.$$

The geometric meaning of this statement is that a horizontal straight line $y = f(x) = c$ has zero slope. To prove the statement from the definition, we notice that $\Delta y = f(x + \Delta x) - f(x) = c - c = 0$, so

$$\frac{dy}{dx} = \lim_{\Delta x \to 0} \frac{\Delta y}{\Delta x} = \lim_{\Delta x \to 0} \frac{0}{\Delta x} = \lim_{\Delta x \to 0} 0 = 0.$$

2 *If n is a positive integer, then*

$$\frac{d}{dx} x^n = n x^{n-1}.$$

In words, the derivative of x^n is obtained by bringing the exponent n down in front as a coefficient, then subtracting 1 from it to form the new exponent. We already know three special cases of this rule from Chapter 2:

$$\frac{d}{dx} x^2 = 2x, \qquad \frac{d}{dx} x^3 = 3x^2, \qquad \text{and} \qquad \frac{d}{dx} x^4 = 4x^3.$$

To prove this rule in general, we write $y = f(x) = x^n$ and use the binomial theorem* to obtain

$$\Delta y = f(x + \Delta x) - f(x) = (x + \Delta x)^n - x^n$$

$$= \left[x^n + n x^{n-1} \Delta x + \frac{n(n-1)}{2} x^{n-2} (\Delta x)^2 + \cdots + (\Delta x)^n \right] - x^n$$

$$= n x^{n-1} \Delta x + \frac{n(n-1)}{2} x^{n-2} (\Delta x)^2 + \cdots + (\Delta x)^n.$$

---

* For students who have forgotten the details of the binomial theorem, we state it as follows: If n is a positive integer, then

$$(a + b)^n = a^n + n a^{n-1} b + \frac{n(n-1)}{2} a^{n-2} b^2 + \cdots$$

$$+ \frac{n(n-1) \cdots (n-k+1)}{1 \cdot 2 \cdots k} a^{n-k} b^k + \cdots + b^n.$$

The precise form of this expansion can be understood without too much difficulty by simply thinking about the n-factor product

$$(a + b)^n = (a + b)(a + b) \cdots (a + b).$$

To multiply these factors out, we begin by choosing a from each factor, which gives the term a^n. If we next choose b from one factor and a from all the others, this can be done in n ways, so we get $b a^{n-1}$ n times, or $n a^{n-1} b$. Similarly, $n(n-1)/2$ is the number of ways b can be chosen from two factors and a from all the others, etc. The "etc." is explained more fully in Appendix D.1.

This yields

$$\frac{dy}{dx} = \lim_{\Delta x \to 0} \frac{\Delta y}{\Delta x}$$

$$= \lim_{\Delta x \to 0} \left[nx^{n-1} + \frac{n(n-1)}{2} x^{n-2} \Delta x + \cdots + (\Delta x)^{n-1} \right]$$

$$= nx^{n-1},$$

because Δx is a factor of each term in brackets beyond the first.

Our rule remains true when the exponent is a negative integer or a fraction. However, it is convenient to postpone giving a proof of this to a later part of the chapter.

3   *If c is a constant and u = f(x) is a differentiable function of x, then*

$$\frac{d}{dx}(cu) = c\frac{du}{dx}.$$

That is, the derivative of a constant times a function equals the constant times the derivative of the function.* To prove this, we write $y = cu = cf(x)$ and observe that $\Delta y = cf(x + \Delta x) - cf(x) = c[f(x + \Delta x) - f(x)] = c\,\Delta u$, so

$$\frac{dy}{dx} = \lim_{\Delta x \to 0} \frac{\Delta y}{\Delta x} = \lim_{\Delta x \to 0} \frac{c\,\Delta u}{\Delta x} = c \lim_{\Delta x \to 0} \frac{\Delta u}{\Delta x} = c\frac{du}{dx}.$$

Combining rules 2 and 3, we see that

$$\frac{d}{dx} cx^n = cnx^{n-1}$$

for any constant c and any positive integer n.

Example 1   We are now in a position to calculate the following derivatives as fast as we can write:

$$\frac{d}{dx} 3x^7 = 21x^6, \qquad \frac{d}{dx}\left(-\frac{1}{2}x^{12}\right) = -6x^{11}, \qquad \frac{d}{dx} 22x^{101} = 2222x^{100},$$

$$\frac{d}{dx} 55x = 55x^0 = 55, \qquad \frac{d}{dx}\left(\frac{10^{\sqrt{2}} + \log_{10}\pi}{\sqrt{19} + 1024}\right)^{999} = 0.$$

4   *If u = f(x) and v = g(x) are functions of x, then*

$$\frac{d}{dx}(u + v) = \frac{du}{dx} + \frac{dv}{dx}.$$

That is, the derivative of the sum of two functions equals the sum of the individual derivatives. The proof is routine: If we write $y = u + v =$

---

* From now on we assume that every function we deal with is differentiable unless a specific statement is made to the contrary.

$f(x) + g(x)$,   then $\Delta y = [f(x + \Delta x) + g(x + \Delta x)] - [f(x) + g(x)] = [f(x + \Delta x) - f(x)] + [g(x + \Delta x) - g(x)] = \Delta u + \Delta v$, and therefore

$$\frac{dy}{dx} = \lim_{\Delta x \to 0} \frac{\Delta y}{\Delta x} = \lim_{\Delta x \to 0} \frac{\Delta u + \Delta v}{\Delta x} = \lim_{\Delta x \to 0} \left[\frac{\Delta u}{\Delta x} + \frac{\Delta v}{\Delta x} \right]$$

$$= \lim_{\Delta x \to 0} \frac{\Delta u}{\Delta x} + \lim_{\Delta x \to 0} \frac{\Delta v}{\Delta x} = \frac{du}{dx} + \frac{dv}{dx}.$$

In essentially the same way we can show that the derivative of a difference equals the difference of the derivatives,

$$\frac{d}{dx}(u - v) = \frac{du}{dx} - \frac{dv}{dx}.$$

Further, these results can be extended without difficulty to any finite number of terms, as in

$$\frac{d}{dx}(u - v + w) = \frac{du}{dx} - \frac{dv}{dx} + \frac{dw}{dx}.$$

Example 2   It is now easy to differentiate any polynomial. For instance,

$$\frac{d}{dx}(15x^4 + 9x^3 - 7x^2 - 3x + 5) = \frac{d}{dx}15x^4 + \frac{d}{dx}9x^3 - \frac{d}{dx}7x^2 - \frac{d}{dx}3x + \frac{d}{dx}5$$

$$= 60x^3 + 27x^2 - 14x - 3.$$

With a little practice we can omit the middle step and write down the final result immediately by inspection.

Example 3   The function $y = (3x - 2)^4$ is a polynomial but is not in standard polynomial form. None of the rules established so far apply to this function directly, though later we will prove a formula that can be used here. Meanwhile we must first expand by the binomial theorem. This gives

$$y = (3x - 2)^4 = [3x + (-2)]^4$$

$$= (3x)^4 + 4(3x)^3(-2) + \frac{4 \cdot 3}{2}(3x)^2(-2)^2 + \frac{4 \cdot 3 \cdot 2}{1 \cdot 2 \cdot 3}(3x)(-2)^3 + (-2)^4$$

$$= 81x^4 - 216x^3 + 216x^2 - 96x + 16,$$

so

$$\frac{dy}{dx} = 324x^3 - 648x^2 + 432x - 96.$$

Example 4   Even though the letters x and y are often used for the independent and dependent variables, there is obviously nothing to prevent us from using any letters we please, and the calculations work in just the same way. Thus,

$$s = 13t^3 - 11t^2 + 25$$

is a polynomial in t; and by the rules developed in this section, its derivative is clearly

$$\frac{ds}{dt} = 39t^2 - 22t.$$

Example 5 An object moves on a straight line in such a way that its position s at time t is given by

$$s = t^3 + 5t^2 - 8t.$$

What is its acceleration when it is at rest?

The velocity v and acceleration a are

$$v = \frac{ds}{dt} = 3t^2 + 10t - 8 \quad \text{and} \quad a = \frac{dv}{dt} = 6t + 10.$$

The object is at rest when $v = 0$ or

$$3t^2 + 10t - 8 = (3t - 2)(t + 4) = 0,$$

that is, when $t = \frac{2}{3}, -4$. The corresponding values of the acceleration are $a = 14, -14$.

PROBLEMS

1 Find the derivative of each function:
(a) $6x^9$;
(b) 19;
(c) $-15x^4$;
(d) $3x^{500} + 15x^{100}$;
(e) $(x - 3)^2$;
(f) $\frac{1}{5}x^5 + \frac{1}{4}x^4 + \frac{1}{3}x^3 + \frac{1}{2}x^2 + x$;
(g) $x^4 + x^3 + x^2 + x + 1$;
(h) $(x - 2)^5$;
(i) $x^{12} + 2x^6 - 4x^3 - 6x^2$;
(j) $(2x - 1)(3x^2 + 2)$.

2 If s is the position at time t of an object moving on a straight line, find the velocity v and acceleration a:
(a) $s = 12 - 6t + 3t^2$;   (b) $s = 13 - 9t + 6t^3$;
(c) $s = (3t - 2)^2$.

3 Find a function of x whose derivative is the given function:
(a) $3x^2$;   (b) $4x^2$;   (c) $3x^2 + 2x - 5$.

4 Find the line tangent to the curve $y = 3x^2 - 5x + 2$ at the point $(2, 4)$.

5 Find the points on the curve $y = 4x^3 + 6x^2 - 24x + 10$ at which the tangent is horizontal.

6 The line $x = a$ intersects the curve $y = \frac{1}{3}x^3 + 4x + 2$ at a point P and the curve $y = 2x^2 + x$ at a point Q. For what value (or values) of a are the tangents to these curves at P and Q parallel?

7 Find the vertex of the parabola $y = x^2 - 8x + 18$. Hint: The tangent at the vertex is horizontal.

8 Find the vertex of the parabola $y = ax^2 + bx + c$ by the method of Problem 7.

9 What values must the constants a, b, c have if the two curves $y = x^2 + ax + b$ and $y = cx - x^2$ have the same tangent at the point $(3, 3)$?

10 Let p be a positive constant and consider the parabola $x^2 = 4py$ with vertex at the origin and focus at the point $(0, p)$, as shown on the left in Fig. 3.1. Let (x_0, y_0) be a point on this parabola other than the vertex.
(a) Show that the tangent at (x_0, y_0) has y-intercept $-y_0$.
(b) Show that the triangle with vertices (x_0, y_0), $(0, -y_0)$, and $(0, p)$ is isosceles. Hint: Use the distance formula.
(c) Suppose that a source of light is placed at the focus, and assume that each ray of light leaving the focus

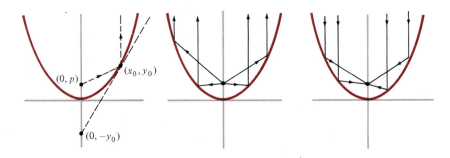

Figure 3.1   A parabolic reflector

is reflected off the parabola in such a way that it makes equal angles with the tangent line at the point of reflection (the angle of incidence equals the angle of reflection). Use (b) to show that after reflection each ray points vertically upward, parallel to the axis (Fig. 3.1, center).*

11   The line through a point on a curve which is perpendic-

ular to the tangent at that point is called the *normal* to the curve at the point. Find the normal to the curve $4y + x^2 = 5$ at the point (1, 1).

12   Consider the normal to the curve $y = x - x^2$ at the point (1, 0). Where does this line intersect the curve a second time?

3.2

THE PRODUCT AND QUOTIENT RULES

In Section 3.1 we learned how to differentiate sums, differences, and constant multiples of functions. We now consider

$$\text{products } uv \quad \text{and} \quad \text{quotients } \frac{u}{v},$$

where u and v are understood to be differentiable functions of x.

Since the derivative of a sum is the sum of the derivatives, it is natural to guess that the derivative of a product might equal the product of the derivatives. However, it is quite easy to construct examples showing that this is not true. For instance, the product of x^3 and x^4 is x^7, so the derivative of the product is $7x^6$, but the product of the individual derivatives is $3x^2 \cdot 4x^3 = 12x^5$. The correct formula for differentiating products is rather surprising.

5†   *The product rule:*

$$\frac{d}{dx}(uv) = u\frac{dv}{dx} + v\frac{du}{dx}. \tag{1}$$

Students may wish to keep in mind the following verbal statement of this rule: The derivative of the product of two functions is the first times the derivative of the second plus the second times the derivative of the first. To prove this, we write $y = uv$ and let the independent variable x be changed by an amount Δx, to $x + \Delta x$. This produces corresponding changes Δu, Δv, Δy in the variables u, v, y, and we have

$$y + \Delta y = (u + \Delta u)(v + \Delta v) = uv + u\,\Delta v + v\,\Delta u + \Delta u\,\Delta v,$$

$$\Delta y = (y + \Delta y) - y = u\,\Delta v + v\,\Delta u + \Delta u\,\Delta v,$$

$$\frac{\Delta y}{\Delta x} = u\frac{\Delta v}{\Delta x} + v\frac{\Delta u}{\Delta x} + \Delta u\frac{\Delta v}{\Delta x}.$$

---

* This is called the *reflection property* of parabolas. To form a three-dimensional idea of the way this property is used in the design of searchlights and automobile headlights, we have only to imagine a mirror constructed by rotating a parabola about its axis and silvering the inside of the resulting surface. Such a parabolic reflector can also be used in reverse (Fig. 3.1, right), to gather faint incoming rays parallel to the axis and concentrate them at the focus. This is the basic principle of radar antennas, radio telescopes, and reflecting optical telescopes. The great telescope on Palomar Mountain in California has a 15-ton glass reflector that is 200 in (almost 17 ft) in diameter, and the accurate grinding of this enormous mirror required 11 years of work.

† We continue the numbering started in Section 3.1.

Taking limits as $\Delta x \to 0$ yields

$$\frac{dy}{dx} = u\frac{dv}{dx} + v\frac{du}{dx} + 0 \cdot \frac{dv}{dx},$$

which is equivalent to (1). We have used the fact that $\Delta u \to 0$ as $\Delta x \to 0$. This expresses the continuity of u, which follows from the differentiability by the argument given in Section 2.5.

Example 1   We first test formula (1) on the factors x^3 and x^4, whose product we already know has derivative $7x^6$. We get

$$\frac{d}{dx}(x^3 \cdot x^4) = x^3\frac{d}{dx}x^4 + x^4\frac{d}{dx}x^3$$

$$= x^3 \cdot 4x^3 + x^4 \cdot 3x^2 = 7x^6,$$

as we should. As a more complicated example, we apply our formula to the function $y = (x^3 - 4x)(3x^4 + 2)$:

$$\frac{dy}{dx} = (x^3 - 4x)\frac{d}{dx}(3x^4 + 2) + (3x^4 + 2)\frac{d}{dx}(x^3 - 4x)$$

$$= (x^3 - 4x)(12x^3) + (3x^4 + 2)(3x^2 - 4)$$

$$= 12x^6 - 48x^4 + 9x^6 - 12x^4 + 6x^2 - 8$$

$$= 21x^6 - 60x^4 + 6x^2 - 8.$$

Notice that we can also proceed by multiplying the two factors at the beginning and then differentiating. This gives

$$y = 3x^7 - 12x^5 + 2x^3 - 8x,$$

so

$$\frac{dy}{dx} = 21x^6 - 60x^4 + 6x^2 - 8,$$

as we expect. Since we can solve this problem without using the product rule, it may appear that this rule is unnecessary. This is indeed true when both factors are polynomials, because the product of two polynomials is also a polynomial. However, in the more complex situations that lie ahead—in which the factors are often different types of functions—it will be clear that the product rule is indispensable.

6   *The quotient rule:*

$$\frac{d}{dx}\left(\frac{u}{v}\right) = \frac{v\,du/dx - u\,dv/dx}{v^2} \tag{2}$$

at all values of x where $v \neq 0$.

Most people find it easier to remember the working instructions given by (2) in words rather than in symbols: The derivative of a quotient is the denominator times the derivative of the numerator minus the numerator times the derivative of the denominator, all divided by the denominator squared. To prove this, we write $y = u/v$ and let x change by an amount Δx. As before, this produces changes $\Delta u, \Delta v, \Delta y$ in the variables $u, v, y,$ and we have

$$y + \Delta y = \frac{u + \Delta u}{v + \Delta v}, \qquad \Delta y = \frac{u + \Delta u}{v + \Delta v} - \frac{u}{v},$$

$$\Delta y = \frac{uv + v\,\Delta u - uv - u\,\Delta v}{v(v + \Delta v)} = \frac{v\,\Delta u - u\,\Delta v}{v(v + \Delta v)},$$

$$\frac{\Delta y}{\Delta x} = \frac{v\,\Delta u/\Delta x - u\,\Delta v/\Delta x}{v(v + \Delta v)}.$$

If we now take limits as $\Delta x \to 0$ we obtain formula (2),

$$\frac{dy}{dx} = \frac{v\,du/dx - u\,dv/dx}{v^2},$$

since $\Delta v \to 0$ as $\Delta x \to 0$.

Example 2　To differentiate the quotient $y = (3x^2 - 2)/(x^2 + 1)$, we follow the verbal prescription,

$$\frac{dy}{dx} = \frac{(x^2 + 1)(d/dx)(3x^2 - 2) - (3x^2 - 2)(d/dx)(x^2 + 1)}{(x^2 + 1)^2}$$

$$= \frac{(x^2 + 1)(6x) - (3x^2 - 2)(2x)}{(x^2 + 1)^2}$$

$$= \frac{6x^3 + 6x - 6x^3 + 4x}{(x^2 + 1)^2} = \frac{10x}{(x^2 + 1)^2}.$$

With practice, calculations like this can be performed very quickly. For instance,

$$\frac{d}{dx}\frac{1}{x^2 + 1} = \frac{(x^2 + 1)(0) - 1(2x)}{(x^2 + 1)^2} = \frac{-2x}{(x^2 + 1)^2},$$

$$\frac{d}{dx}\frac{3x}{x^2 + 1} = \frac{(x^2 + 1)(3) - 3x(2x)}{(x^2 + 1)^2} = \frac{3 - 3x^2}{(x^2 + 1)^2},$$

$$\frac{d}{dx}\frac{2x + 1}{3x - 1} = \frac{(3x - 1)(2) - (2x + 1)(3)}{(3x - 1)^2} = \frac{-5}{(3x - 1)^2}.$$

The quotient rule enables us to extend rule 2 of Section 3.1,

$$\frac{d}{dx}x^n = nx^{n-1}, \tag{3}$$

to the case in which n is a negative integer. To make the negative character of n more visible, we write $n = -m$, where m is a positive integer. Now, using (2) and the fact that (3) is known to be valid for positive integral exponents, we have

$$\frac{d}{dx}x^n = \frac{d}{dx}x^{-m} = \frac{d}{dx}\frac{1}{x^m} = \frac{x^m(0) - 1(mx^{m-1})}{(x^m)^2}$$

$$= \frac{-mx^{m-1}}{x^{2m}} = -mx^{-m-1} = nx^{n-1},$$

which proves our statement. Thus, for example,

$$\frac{d}{dx}x^{-1} = (-1)x^{-2} = -x^{-2}, \qquad \frac{d}{dx}x^{-2} = (-2)x^{-3} = -2x^{-3}, \qquad \text{etc.}$$

Since (3) is clearly true for $n = 0$, we now know that it is valid for all integral exponents.

Example 3 To differentiate

$$y = 3x^2 - \frac{2}{x^3},$$

we write it as

$$y = 3x^2 - 2x^{-3}.$$

Then

$$\frac{dy}{dx} = 6x + 6x^{-4},$$

which can be rewritten as

$$\frac{dy}{dx} = 6x + \frac{6}{x^4}$$

if positive exponents are preferred.

We urge students to memorize the product and quotient rules, and to anchor them in their minds by conscientious practice.

PROBLEMS

1 Differentiate each of the following functions two ways and verify that your answers agree:
  (a) $(x - 1)(x + 1)$;
  (b) $(2x - 6)(3x^2 + 9)$;
  (c) $(3x^2 + 1)(x^3 + 6x)$;
  (d) $(x - 1)(x^4 + x^3 + x^2 + x + 1)$.

2 Differentiate each of the following functions and simplify your answer as much as possible:

  (a) $\dfrac{x + 1}{x - 1}$;

  (b) $\dfrac{1}{x^2 + 2}$;

  (c) $\dfrac{2x^3 + 1}{x + 2}$;

  (d) $\dfrac{3x + 4}{7x + 8}$;

  (e) $\dfrac{3x}{1 + 2x^2}$;

  (f) $\dfrac{4x - x^4}{x^3 + 2}$;

  (g) $\dfrac{1 - x^2}{1 + x^2}$;

  (h) $\dfrac{2x + 1}{1 - x^2}$.

3 Find dy/dx in two ways, first by dividing and then by using the quotient rule, and show that your answers agree:

  (a) $\dfrac{4x + 4}{x}$;

  (b) $\dfrac{2x + 6x^4 - 2x^6}{x^5}$;

  (c) $\dfrac{1 + x^4}{x^2}$.

4 Find all points on the curve $y = 6/x$ where the tangent is parallel to the line $2x + 3y + 1 = 0$.

5 Find the equations of
  (a) the tangent and normal to $y = 6/(x + 2)$ at $(1, 2)$;
  (b) the tangent and normal to $y = 5/(x^2 + 1)$ at $x = 2$;
  (c) the tangent to $y = (x^3 + x)/(x - 1)$ at $(2, 10)$;
  (d) the normal to $y = (1 - 2x + 3x^2)/(1 + x^2)$ at $(0, 1)$.

6 Show that the tangents to the curves $y = (x^2 + 45)/x^2$ and $y = (x^2 - 4)/(x^2 + 1)$ at $x = 3$ are perpendicular to each other.

7 Let P be a point on the first-quadrant part of the curve

$y = 1/x$. Show that the triangle determined by the x-axis, the tangent at P, and the line from P to the origin is isosceles, and find its area.

8 Use the product rule to verify rule 3 of Section 3.1: If c is a constant and u is a function of x, then

$$\frac{d}{dx}(cu) = c\,\frac{du}{dx}.$$

9 Sketch the curve $y = 2/(1 + x^2)$ and find the points on it at which the normal passes through the origin.

10 Verify the location of the high and low points on the graph of

$$y = \frac{x}{x^2 - 3x + 2}$$

as stated in Example 4 of Section 1.8.

Let us consider the problem of differentiating the function

$$y = (x^3 + 2)^5. \tag{1}$$

3.3

COMPOSITE FUNCTIONS AND THE CHAIN RULE

We can do this with the tools we now have by using the binomial theorem to expand the function into the polynomial

$$y = x^{15} + 10x^{12} + 40x^9 + 80x^6 + 80x^3 + 32. \tag{2}$$

It now follows at once that

$$\frac{dy}{dx} = 15x^{14} + 120x^{11} + 360x^8 + 480x^5 + 240x^2. \tag{3}$$

In this case the work of expansion is bothersome but not too difficult. However, few of us would willingly attempt to carry out the same procedure for the function $y = (x^3 + 2)^{100}$. It is much better to develop the chain rule, which enables us to differentiate both functions with equal ease — and a host of others as well.

For this purpose it is important to understand the structure of the function (1). We accomplish this by introducing an auxiliary variable $u = x^3 + 2$, so that (1) can be decomposed into simpler pieces as follows:

$$y = u^5 \quad \text{where} \quad u = x^3 + 2. \tag{4}$$

Working in the other direction, we can reconstruct (1) out of these pieces by substituting the expression for u into $y = u^5$. Such a function is called a *composite function,* or sometimes a *function of a function.* We have already encountered this idea in Section 1.6. In general, suppose that y is a function of u, where u in turn is a function of x, say

$$y = f(u) \quad \text{where} \quad u = g(x). \tag{5}$$

The corresponding composite function is the single function

$$y = f(g(x)), \tag{6}$$

obtained by substituting $u = g(x)$ into $y = f(u)$.

Our position now is this. We assume we are confronted by the composite function (6), and we wish to learn how to differentiate it by decomposing it into the simpler functions (5) and using the presumably simpler derivatives of these functions. This is what the chain rule is all about.

7 *The chain rule: Under the circumstances described above,*

$$\frac{dy}{dx} = \frac{dy}{du} \cdot \frac{du}{dx}. \tag{7}$$

As we see, in this form the chain rule has the appearance of a trivial algebraic identity; it is easily remembered because the Leibniz fractional notation for derivatives suggests that du can be canceled from the two "fractions" on the right. Its intuitive content is easy to grasp if we think of derivatives as rates of change:

> If y changes a times as fast as u
> and u changes b times as fast as x,
> then y changes ab times as fast as x.

Or, in everyday terms, if a car travels twice as fast as a bicycle and the bicycle is four times as fast as a walking man, then the car travels $2 \cdot 4 = 8$ times as fast as the man.

Before looking into the proof of the chain rule, let us see how it applies to the problem just discussed, in which (1) is the given function and (4) is its decomposition. Formula (7) gives

$$\frac{dy}{dx} = \frac{dy}{du} \cdot \frac{du}{dx} = 5u^4 \cdot 3x^2 = 15x^2(x^3 + 2)^4. \tag{8}$$

It is not immediately obvious that this result is the same as (3), but the equivalence is easy to establish.* Further, the derivative of $y = (x^3 + 2)^{100}$ can be computed just as easily in just the same way: We write

$$y = u^{100} \qquad \text{where} \qquad u = x^3 + 2$$

and use (7) to obtain

$$\frac{dy}{dx} = \frac{dy}{du} \cdot \frac{du}{dx} = 100u^{99} \cdot 3x^2 = 300x^2(x^3 + 2)^{99}.$$

As these examples show, the chain rule is a very powerful tool.

We begin the proof of (7) with the usual change Δx in the independent variable x; this produces a change Δu in the variable u, and this in turn produces a change Δy in the variable y. We know that differentiability implies continuity, and so $\Delta u \to 0$ as $\Delta x \to 0$. If we look at the definitions of the three derivatives we are trying to link together,

$$\frac{dy}{dx} = \lim_{\Delta x \to 0} \frac{\Delta y}{\Delta x}, \qquad \frac{dy}{du} = \lim_{\Delta u \to 0} \frac{\Delta y}{\Delta u}, \qquad \frac{du}{dx} = \lim_{\Delta x \to 0} \frac{\Delta u}{\Delta x}, \tag{9}$$

then it is natural to try to complete the proof as follows: By simple algebra we have

$$\frac{\Delta y}{\Delta x} = \frac{\Delta y}{\Delta u} \cdot \frac{\Delta u}{\Delta x}, \tag{10}$$

---

* We hope students did not accept the expansion in (2)—and similarly will not accept the stated equivalence of (8) and (3)—without checking the details for themselves. Total skepticism is the recommended state of mind for studying this (or any similar) book: Take nothing on faith; verify all omitted calculations; believe nothing unless you have seen and understood it for yourself.

so

$$\frac{dy}{dx} = \lim_{\Delta x \to 0} \frac{\Delta y}{\Delta x} = \lim_{\Delta x \to 0} \frac{\Delta y}{\Delta u} \cdot \frac{\Delta u}{\Delta x} = \left[\lim_{\Delta x \to 0} \frac{\Delta y}{\Delta u}\right]\left[\lim_{\Delta x \to 0} \frac{\Delta u}{\Delta x}\right]$$

$$= \left[\lim_{\Delta u \to 0} \frac{\Delta y}{\Delta u}\right]\left[\lim_{\Delta x \to 0} \frac{\Delta u}{\Delta x}\right] = \frac{dy}{du} \cdot \frac{du}{dx}. \tag{11}$$

This reasoning is almost correct, but not quite. The difficulty centers on a possible division by zero. In computing dy/dx from the definition in (9), we know that it is part of the meaning of this formula that the increment Δx is small and approaches zero *but is never equal to zero.* On the other hand, it can happen that Δx induces no actual change in u, so that $\Delta u = 0$, and this possibility invalidates (10) and (11). This flaw can be patched up by an ingenious bit of mathematical trickery. We give the argument in the footnote below for students who wish to examine it.*

It will become clear as we proceed that the chain rule is indispensable for almost all differentiations above the simplest level. An important special case has been illustrated in connection with finding the derivatives of $(x^3 + 2)^5$ and $(x^3 + 2)^{100}$. The general principle here is expressed by the formula

$$\frac{d}{dx}(\ )^n = n(\ )^{n-1}\frac{d}{dx}(\ ),$$

where any differentiable function of x can be inserted in the parentheses. If we denote the function by u, the formula can be written as follows.

8   *The power rule:*

$$\frac{d}{dx}u^n = nu^{n-1}\frac{du}{dx}. \tag{12}$$

At this stage of our work we know that the exponent n is allowed to be any positive or negative integer (or zero). In Section 3.4 we will see that (12) is also valid for all fractional exponents.

---

* We begin with the definition of the derivative dy/du, which is

$$\frac{dy}{du} = \lim_{\Delta u \to 0} \frac{\Delta y}{\Delta u}.$$

This is equivalent to

$$\frac{\Delta y}{\Delta u} = \frac{dy}{du} + \epsilon$$

or

$$\Delta y = \frac{dy}{du}\Delta u + \epsilon\,\Delta u,$$

where $\epsilon \to 0$ as $\Delta u \to 0$. It is assumed in these equations that Δu is a nonzero increment in u, but the last equation is valid even when $\Delta u = 0$. Dividing this by a nonzero increment Δx yields

$$\frac{\Delta y}{\Delta x} = \frac{dy}{du}\frac{\Delta u}{\Delta x} + \epsilon\frac{\Delta u}{\Delta x},$$

and on letting $\Delta x \to 0$ we obtain the chain rule (7), since $\epsilon \to 0$.

Example 1 To differentiate $y = (3x^4 + 1)^7$, we make a routine application of (12):

$$\frac{dy}{dx} = 7(3x^4 + 1)^6 \frac{d}{dx}(3x^4 + 1) = 7(3x^4 + 1)^6 \cdot 12x^3.$$

But to differentiate $y = [(3x^4 + 1)^7 + 1]^5$, we apply (12) twice in succession:

$$\frac{dy}{dx} = 5[(3x^4 + 1)^7 + 1]^4 \frac{d}{dx}[(3x^4 + 1)^7 + 1]$$

$$= 5[(3x^4 + 1)^7 + 1]^4 \cdot 7(3x^4 + 1)^6 \frac{d}{dx}(3x^4 + 1)$$

$$= 5[(3x^4 + 1)^7 + 1]^4 \cdot 7(3x^4 + 1)^6 \cdot 12x^3.$$

After this procedure becomes familiar and more or less automatic, it is often possible to skip the intermediate steps and write down the answer at once.

Example 2 If $y = [(1 - 2x)/(1 + 2x)]^4$, then by (12) and the quotient rule we have

$$\frac{dy}{dx} = 4\left(\frac{1 - 2x}{1 + 2x}\right)^3 \frac{d}{dx}\left(\frac{1 - 2x}{1 + 2x}\right)$$

$$= 4\left(\frac{1 - 2x}{1 + 2x}\right)^3 \cdot \frac{(1 + 2x)(-2) - (1 - 2x)(2)}{(1 + 2x)^2}$$

$$= \frac{-16(1 - 2x)^3}{(1 + 2x)^5}.$$

Example 3 If $y = (x^2 - 1)^3(x^2 + 1)^{-2}$, then by combining (12) with the product rule we have

$$\frac{dy}{dx} = (x^2 - 1)^3 \frac{d}{dx}(x^2 + 1)^{-2} + (x^2 + 1)^{-2} \frac{d}{dx}(x^2 - 1)^3$$

$$= (x^2 - 1)^3 \cdot (-2)(x^2 + 1)^{-3}(2x) + (x^2 + 1)^{-2} \cdot 3(x^2 - 1)^2(2x).$$

To simplify this, we take out the factor $2x(x^2 - 1)^2$, get rid of the negative exponents, and reduce to a common denominator:

$$\frac{dy}{dx} = 2x(x^2 - 1)^2 \left[\frac{-2(x^2 - 1)}{(x^2 + 1)^3} + \frac{3}{(x^2 + 1)^2}\right]$$

$$= 2x(x^2 - 1)^2 \left[\frac{-2(x^2 - 1) + 3(x^2 + 1)}{(x^2 + 1)^3}\right] = \frac{2x(x^2 - 1)^2(x^2 + 5)}{(x^2 + 1)^3}.$$

In Chapter 4 we will be using derivatives as tools in many concrete problems of science and geometry, and it will then be clear that it is worth a little extra effort to put the derivatives we calculate into their simplest possible forms.

There are a few concluding remarks that ought to be made. We have not yet explained why the expression "chain rule" is appropriate. The reason is this. In (7) we are dealing with three variables y, u, and x that are linked together step by step in a chain in such a way that each is dependent on the next. We can suggest this relation by writing

$$y \text{ depends on } u \text{ depends on } x.$$

The formula

$$\frac{dy}{dx} = \frac{dy}{du} \cdot \frac{du}{dx}$$

tells us how to differentiate the first variable with respect to the last by taking into account each individual link in the chain. This formula can easily be extended to more variables. For instance, if x depends in turn on z, then

$$\frac{dy}{dz} = \frac{dy}{du} \cdot \frac{du}{dx} \cdot \frac{dx}{dz};$$

if z depends on w, then

$$\frac{dy}{dw} = \frac{dy}{du} \cdot \frac{du}{dx} \cdot \frac{dx}{dz} \cdot \frac{dz}{dw};$$

and so on. Each new variable adds a new link to the chain and a new derivative to the formula.

PROBLEMS

1   Find dy/dx in each case:
(a) $y = (x^5 - 3x)^4$; (b) $y = (x^2 - 2)^{500}$;
(c) $y = (x + x^2 - 2x^5)^6$; (d) $y = (1 - 3x)^{-1}$;
(e) $y = (12 - x^2)^{-2}$; (f) $y = [1 - (3x - 2)^3]^4$.

2   Find dy/dx in each case:
(a) $y = (5x + 3)^4 (4x - 3)^7$;
(b) $y = (x^2 - 2)^5 (x^2 + 2)^{10}$;
(c) $y = x^2 (9 - x^2)^{-2}$;
(d) $y = (1 - 2x)^{-4} (x^2 - x)^2$.

3   Find ds/dt in each case:

(a) $s = \dfrac{(2t - 1)^3}{(t^2 + 3)^2}$; (b) $s = \dfrac{1}{(2t - 3)^2}$;

(c) $s = \dfrac{6}{(5 - 4t)^3}$; (d) $s = \dfrac{t^4 - 10t^2}{(t^2 - 6)^2}$.

4   In each case find dy/dx by two methods and verify that your answers agree:
(a) $y = (2x - 1)^5 (x + 3)^5 = (2x^2 + 5x - 3)^5$;

(b) $y = \dfrac{1}{(1 - 2x^2)^3} = (1 - 2x^2)^{-3}$;

(c) $y = \dfrac{(3x + 1)^4}{(1 - 2x)^4} = \left(\dfrac{3x + 1}{1 - 2x}\right)^4$.

5   If u is a function of x, express each of the following in terms of u and du/dx:

(a) $\dfrac{d}{dx} u^3$; (b) $\dfrac{d}{dx} (2u - 1)^2$;

(c) $\dfrac{d}{dx} (u^2 - 2)^2$.

6   Find a function $y = f(x)$ for which

(a) $\dfrac{dy}{dx} = 2(x^2 - 1) \cdot 2x$; (b) $\dfrac{dy}{dx} = 4(x^2 - 1)^2 \cdot 2x$;

(c) $\dfrac{dy}{dx} = 2(x^3 - 2) \cdot 3x^2$; (d) $\dfrac{dy}{dx} = 3(x^3 - 2)^2 \cdot 3x^2$.

Most of the functions we have met so far have been of the form $y = f(x)$, in which y is expressed directly — or explicitly — in terms of x. In contrast to this, it often happens that y is defined as a function of x by means of an equation

$$F(x, y) = 0, \tag{1}$$

3.4
IMPLICIT FUNCTIONS AND FRACTIONAL EXPONENTS

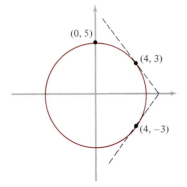

Figure 3.2

which is not solved for y but in which x and y are more or less entangled with each other. When x is given a suitable numerical value, the resulting equation usually determines one or more corresponding values of y. In such a case we say that equation (1) determines y as one or more *implicit functions* of x.

Example 1 (a) The very simple equation $xy = 1$ determines one implicit function of x, which can be written explicitly as

$$y = \frac{1}{x}.$$

(b) The equation $x^2 + y^2 = 25$ determines two implicit functions of x, which can be written explicitly as

$$y = \sqrt{25 - x^2} \quad \text{and} \quad y = -\sqrt{25 - x^2}.$$

As we know, the graphs of these two functions are the upper and lower halves of the circle of radius 5 shown in Fig. 3.2.

(c) The equation $2x^2 - 2xy = 5 - y^2$ also determines two implicit functions. If we use the quadratic formula to solve for y, we find that these functions are

$$y = x + \sqrt{5 - x^2} \quad \text{and} \quad y = x - \sqrt{5 - x^2}.$$

(d) The equation $x^3 + y^3 = 3axy$ $(a > 0)$ determines several implicit functions, but the problem of solving this equation for y is so forbidding that we might as well forget it.

It is rather surprising that we can often calculate the derivative dy/dx of an implicit function without first solving the given equation for y. We start the process by differentiating the given equation through with respect to x, using the chain rule (or power rule) and consciously thinking of y as a function of x wherever it appears. Thus, for example, y^3 is treated as the cube of a function of x and its derivative is

$$\frac{d}{dx} y^3 = 3y^2 \frac{dy}{dx};$$

and $x^3 y^4$ is thought of as the product of two functions of x and its derivative is

$$\frac{d}{dx} (x^3 y^4) = x^3 \cdot 4y^3 \frac{dy}{dx} + y^4 \cdot 3x^2.$$

To complete the process, we solve the resulting equation for dy/dx as the unknown. This method is called *implicit differentiation.* We show how it works by applying it to the equations in Example 1.

Example 2 (a) We can think of the equation $xy = 1$ as stating that two functions of x (namely, xy and 1) are equal. It follows that the derivatives of these functions are equal, and so

$$x \frac{dy}{dx} + y = 0 \quad \text{or} \quad \frac{dy}{dx} = -\frac{y}{x}.$$

In this case it is possible to solve the original equation for y and check our result: Since $y = 1/x$, the formula we have just obtained becomes

$$\frac{dy}{dx} = -\frac{y}{x} = -\frac{1}{x} \cdot y = -\frac{1}{x} \cdot \frac{1}{x} = -\frac{1}{x^2};$$

and differentiating $y = 1/x$ directly also yields

$$\frac{dy}{dx} = -\frac{1}{x^2}.$$

(b) From the equation $x^2 + y^2 = 25$ we get

$$2x + 2y \frac{dy}{dx} = 0 \qquad \text{or} \qquad \frac{dy}{dx} = -\frac{x}{y}.$$

This gives the correct result whichever of the two implicit functions we are thinking about. Thus, at the point $(4, 3)$ on the upper curve in Fig. 3.2, the value of dy/dx is $-\frac{4}{3}$, and at $(4, -3)$ its value is $\frac{4}{3}$.

(c) If we apply this process of implicit differentiation to the equation $2x^2 - 2xy = 5 - y^2$, we obtain

$$4x - 2x \frac{dy}{dx} - 2y = -2y \frac{dy}{dx} \qquad \text{or} \qquad \frac{dy}{dx} = \frac{2x - y}{x - y}.$$

(d) In Example 1(d) the derivative dy/dx is clearly beyond direct calculation. However, it is easily found by our present method: Since $x^3 + y^3 = 3axy$, we have

$$3x^2 + 3y^2 \frac{dy}{dx} = 3ax \frac{dy}{dx} + 3ay \qquad \text{or} \qquad \frac{dy}{dx} = \frac{ay - x^2}{y^2 - ax}.$$

It is apparent that implicit differentiation usually gives an expression for dy/dx in terms of both x and y, instead of in terms of x alone. However, in many cases this is not a real disadvantage. For instance, if we want the slope of the tangent to the graph of the equation at a point (x_0, y_0), all we have to do is substitute x_0 and y_0 for x and y in the formula for dy/dx. This is illustrated in Example 2(b).

We now use implicit differentiation to show that the vital formula

$$\frac{d}{dx} x^n = nx^{n-1} \tag{2}$$

is valid for all fractional exponents $n = p/q$.*

---

* Students who are comfortable with fractional exponents should ignore this footnote. However, for those who have forgotten the meaning of these exponents, we offer a brief review. We begin by recalling that the square root $\sqrt{x}$, the cube root $\sqrt[3]{x}$, and more generally the qth root $\sqrt[q]{x}$, where q is any positive integer, are all defined for $x \geq 0$; if q is odd, $\sqrt[q]{x}$ is also defined for $x < 0$. The definition of fractional exponents now proceeds in two stages: First, $x^{1/q}$ is defined for $q > 0$ by $x^{1/q} = \sqrt[q]{x}$; and second, if p/q is in lowest terms and $q > 0$, $x^{p/q}$ is defined by $x^{p/q} = (x^{1/q})^p$. It is sometimes useful to know (and it is not difficult to prove) that $(x^p)^{1/q} = (x^{1/q})^p$ if $x > 0$. For example $8^{2/3}$ is easy to evaluate both ways, since $8^{2/3} = (8^2)^{1/3} = 64^{1/3} = 4$ and $8^{2/3} = (8^{1/3})^2 = 2^2 = 4$; but $32^{3/5} = (32^3)^{1/5}$ is hard, while $32^{3/5} = (32^{1/5})^3 = 2^3 = 8$ is easy.

For the sake of convenience, we begin the proof of (2) for fractional exponents by introducing y as the dependent variable,

$$y = x^{p/q}.$$

Raising both sides of this to the qth power yields

$$y^q = x^p;$$

and by differentiating implicitly with respect to x and using the known validity of the power rule for integral exponents, we obtain

$$qy^{q-1}\frac{dy}{dx} = px^{p-1}$$

or

$$\frac{dy}{dx} = \frac{p}{q}\frac{x^{p-1}}{y^{q-1}}.$$

But $y^{q-1} = y^q/y = x^p/x^{p/q}$, so

$$\frac{dy}{dx} = \frac{p}{q}\frac{x^{p-1}}{y^{q-1}} = \frac{p}{q}\frac{x^{p-1}}{x^p} \cdot x^{p/q} = \frac{p}{q}x^{p/q-1},$$

and the proof is complete.

Example 3 We immediately have

$$\frac{d}{dx}x^{1/2} = \frac{1}{2}x^{-1/2}, \qquad \frac{d}{dx}x^{-2/3} = -\frac{2}{3}x^{-5/3}, \qquad \frac{d}{dx}x^{5/4} = \frac{5}{4}x^{1/4}.$$

The first of these derivatives is often used in the form

$$\frac{d}{dx}\sqrt{x} = \frac{1}{2\sqrt{x}}.$$

This formula was established directly from the definition in Example 3 of Section 2.3.

Example 4 By the chain rule, the power rule of Section 3.3 is now known to be valid for all fractional exponents. Accordingly,

$$\frac{d}{dx}(4 - x^2)^{-5/2} = -\frac{5}{2}(4 - x^2)^{-7/2}\frac{d}{dx}(4 - x^2)$$

$$= -\frac{5}{2}(4 - x^2)^{-7/2}(-2x) = \frac{5x}{(4 - x^2)^{7/2}}.$$

Example 5 In differentiating expressions containing radicals, it is a good idea to begin by replacing all radical signs by fractional exponents. Thus,

$$\frac{d}{dx}\frac{x}{\sqrt{x^2 - 1}} = \frac{d}{dx}x(x^2 - 1)^{-1/2} = x\left(-\frac{1}{2}\right)(x^2 - 1)^{-3/2}(2x) + (x^2 - 1)^{-1/2}$$

$$= \frac{-x^2}{(x^2 - 1)^{3/2}} + \frac{1}{(x^2 - 1)^{1/2}} = \frac{-x^2 + (x^2 - 1)}{(x^2 - 1)^{3/2}} = \frac{-1}{(x^2 - 1)^{3/2}}.$$

For convenience of reference, we list together all the differentiation rules developed in this chapter.

1 $\dfrac{d}{dx} c = 0.$

2 $\dfrac{d}{dx} x^n = nx^{n-1}$ (n any integer or fraction).

3 $\dfrac{d}{dx} (cu) = c \dfrac{du}{dx}.$

4 $\dfrac{d}{dx} (u + v) = \dfrac{du}{dx} + \dfrac{dv}{dx}.$

5 *The product rule:* $\dfrac{d}{dx} (uv) = u \dfrac{dv}{dx} + v \dfrac{du}{dx}.$

6 *The quotient rule:* $\dfrac{d}{dx} \left(\dfrac{u}{v}\right) = \dfrac{v \, du/dx - u \, dv/dx}{v^2}.$

7 *The chain rule:* $\dfrac{dy}{dx} = \dfrac{dy}{du} \cdot \dfrac{du}{dx}.$

8 *The power rule:* $\dfrac{d}{dx} u^n = nu^{n-1} \dfrac{du}{dx}$ (n any integer or fraction).

These rules will be used in many ways in almost everything we do from this point on. We therefore urge students who have not already done so to commit them to memory and practice them until their use becomes almost automatic. The eminent philosopher A. N. Whitehead might well have had these rules in mind when he said, "Civilization advances by extending the number of important operations which we can perform without thinking about them."

One final remark: Most mistakes in differentiation come from misusing the power rule or the quotient rule. For instance:

Common mistake	Right answer
$\dfrac{d}{dx} (1 + 6x^2)^4 = 4(1 + 6x^2)^3$	$4(1 + 6x^2)^3 \cdot 12x$
$\dfrac{d}{dx} (1 + 2x)^{1/3} = \tfrac{1}{3}(1 + 2x)^{-2/3}$	$\tfrac{1}{3}(1 + 2x)^{-2/3} \cdot 2$

The difficulty with the quotient rule lies in remembering the order of subtraction in the numerator. One way of quickly recalling the correct order is to use the product rule as follows:

$$\frac{d}{dx}\left(\frac{u}{v}\right) = \frac{d}{dx}(uv^{-1}) = u \cdot (-1)v^{-2}\frac{dv}{dx} + v^{-1}\frac{du}{dx}$$

$$= \frac{1}{v}\frac{du}{dx} - \frac{u}{v^2}\frac{dv}{dx} = \frac{v \, du/dx - u \, dv/dx}{v^2}.$$

PROBLEMS

1 Find dy/dx by implicit differentiation:
(a) $3x^3 + 4y^3 + 8 = 0$;   (b) $xy^2 - x^2y + x^2 + 2y^2 = 0$;

(c) $x = y - y^7$;   (d) $x^4y^3 - 3xy = 60$;

(e) $x^3 - y^3 = 4xy$;   (f) $\dfrac{1}{x} + \dfrac{1}{y} = 1$;

(g) $\sqrt{x} + \sqrt{y} = 6$.

2 Find dy/dx by implicit differentiation and also by solving for y and then differentiating, and verify that your two answers are equivalent:
(a) $3xy + 2 = 0$;   (b) $x^2 + y^2 = 9$;
(c) $y^2 = 3x - 1$;   (d) $2x^2 + 3x + y^2 = 12$.

3 Find the derivative in each case:
(a) $x^{4/5}$;   (b) $x^{5/6}$;
(c) $x^{-3/4}$;   (d) $x^{-7/11}$;
(e) $3\sqrt[5]{x^2}$;   (f) $(1 + x^{2/3})^{3/2}$;

(g) $\left(\dfrac{x^3 + 8}{x^2}\right)^{3/4}$;   (h) $\sqrt{1 + \sqrt{1 + x}}$.

4 Find the equation of
(a) the tangent to $y = (5 - 3x)^{1/3}$ at $(-1, 2)$;
(b) the tangent to $x^4 + 16y^4 = 32$ at $(2, 1)$;
(c) the normal to $y = x\sqrt{9 + x^2}$ at the origin;
(d) the normal to $y^2 - 4xy = 12$ at $(1, 6)$.

5 Show that the curves $x^2 + 3y^2 = 12$ and $3x^2 - y^2 = 6$ intersect at right angles at the point $(\sqrt{3}, \sqrt{3})$.

6 Show that for the "curve" $x(x + 6) + y^2 - 4y + 15 = 0$, implicit differentiation gives

$$\frac{dy}{dx} = \frac{x + 3}{2 - y}.$$

Show further that this result is completely meaningless, because there are no points on this "curve."

7 Verify that the normal at any point (x_0, y_0) on the circle $x^2 + y^2 = a^2$ passes through the center.

8 Find a function $y = f(x)$ for which

(a) $\dfrac{dy}{dx} = 3\sqrt{x}$;   (b) $\dfrac{dy}{dx} = 5x\sqrt{x}$.

3.5
DERIVATIVES OF HIGHER ORDER

The derivative of $y = x^4$ is clearly $y' = 4x^3$. But $4x^3$ can also be differentiated, yielding $12x^2$. It is natural to denote this function by y'' and call it the *second derivative* of the original function. By differentiating the second derivative $y'' = 12x^2$ we obtain the *third derivative* $y''' = 24x$, and so on indefinitely. Several notations are in common use for these higher-order derivatives, and students should become familiar with all of them. The successive derivatives of a function $y = f(x)$ can be written as follows:

First derivative	$f'(x)$	y'	$\dfrac{dy}{dx}$ $\dfrac{d}{dx} f(x)$
Second derivative	$f''(x)$	y''	$\dfrac{d^2y}{dx^2}$ $\dfrac{d^2}{dx^2} f(x)$
Third derivative	$f'''(x)$	y'''	$\dfrac{d^3y}{dx^3}$ $\dfrac{d^3}{dx^3} f(x)$
nth derivative	$f^{(n)}(x)$	$y^{(n)}$	$\dfrac{d^ny}{dx^n}$ $\dfrac{d^n}{dx^n} f(x)$

    A few remarks about these notations are perhaps in order. The entries in the first column are read "f prime of x," "f double prime of x," "f triple prime of x," "f upper n of x"; similarly, those in the second column are read "y prime," "y double prime," and so on. The prime notation quickly becomes unwieldy and is not often used beyond the third order. It is sometimes convenient to think of the original function as the zeroth-order derivative and to write $f(x) = f^{(0)}(x)$. The seemingly strange position of the superscripts in the third column can be understood if we remember that the second derivative is the derivative of the first derivative,

$$\frac{d^2y}{dx^2} = \frac{d}{dx}\left(\frac{dy}{dx}\right).$$

On the left side of this, the superscript 2 is attached to the d on top and to the dx on the bottom, and this is consistent with the way these symbols are written on the right.

What are the uses of these higher derivatives? In geometry, as we will see in Chapter 4, the sign of $f''(x)$ tells us whether the curve $y = f(x)$ is concave up or concave down. Also, in a later chapter this qualitative interpretation of the second derivative will be refined into a quantitative formula for the curvature of the curve.

In physics, second derivatives are of very great importance. If $s = f(t)$ gives the position of a moving body at time t, then we know that the first and second derivatives of this position function,

$$v = \frac{ds}{dt} \quad \text{and} \quad a = \frac{dv}{dt} = \frac{d^2s}{dt^2},$$

are the velocity and acceleration of the body at time t. The central role of acceleration arises from Newton's second law of motion, which states that the acceleration of a moving body is proportional to the force acting on it. The basic problem of Newtonian dynamics is to use calculus to deduce the nature of the motion from the given force. We shall begin examining problems of this kind in Chapter 5.

Derivatives of higher order than the second do not have any such fundamental geometric or physical interpretations. However, as we shall see later, these derivatives have their uses too, mainly in connection with expanding functions into infinite series.

All these applications will be discussed in detail at the proper time. Meanwhile, our task is to develop proficiency at performing the calculations.

Example 1   It is easy to find all the derivatives of $y = x^5$:

$$y' = 5x^4, \qquad y'' = 20x^3, \qquad y''' = 60x^2,$$

$$y^{(4)} = 120x, \qquad y^{(5)} = 120, \qquad y^{(n)} = 0 \qquad \text{for } n > 5.$$

The following notation will often be useful. For any positive integer n, the symbol $n!$ (read "n factorial") is defined to be the product of all the positive integers from 1 up to n:

$$n! = 1 \cdot 2 \cdot 3 \ \cdots \ n.$$

Thus, $1! = 1$, $2! = 1 \cdot 2 = 2$, $3! = 1 \cdot 2 \cdot 3 = 6$, $4! = 1 \cdot 2 \cdot 3 \cdot 4 = 24$, etc. If we differentiate $y = x^n$ repeatedly we clearly get

$$y' = nx^{n-1},$$

$$y'' = n(n-1)x^{n-2},$$

$$y''' = n(n-1)(n-2)x^{n-3}, \ \ldots,$$

$$y^{(n)} = n(n-1)(n-2) \ \cdots \ 2 \cdot 1 = n!,$$

$$y^{(k)} = 0 \qquad \text{for } k > n.$$

Example 2 To discover a formula for the nth derivative of $y = 1/x = x^{-1}$, we compute until a pattern emerges:

$$y' = -x^{-2},$$
$$y'' = 2x^{-3},$$
$$y''' = -2 \cdot 3x^{-4} = -3!x^{-4},$$
$$y^{(4)} = 2 \cdot 3 \cdot 4x^{-5} = 4!x^{-5},$$
$$y^{(5)} = -2 \cdot 3 \cdot 4 \cdot 5x^{-6} = -5!x^{-6}.$$

From the evidence so far and the way the process of differentiation works, it is clear that except for sign $y^{(n)}$ is $n!x^{-(n+1)}$. A convenient way of expressing the alternating sign is provided by the number $(-1)^n$, which equals -1 if n is odd and 1 if n is even. We therefore have

$$y^{(n)} = (-1)^n n! x^{-(n+1)}$$

for every positive integer n.

Example 3 Implicit differentiation can be used to find a simple formula for y'' on the circle $x^2 + y^2 = a^2$. To begin the process, we differentiate and obtain

$$2x + 2yy' = 0 \qquad \text{or} \qquad y' = -\frac{x}{y}. \tag{1}$$

Differentiating again by the quotient rule and remembering that y is a function of x, we get

$$y'' = -\frac{y - xy'}{y^2}.$$

When (1) is substituted into this, the formula becomes

$$y'' = -\frac{y - x(-x/y)}{y^2} = -\frac{y^2 + x^2}{y^3} = -\frac{a^2}{y^3},$$

which should be simple enough for anyone.

Example 4 Repeated differentiation enables us to give a relatively easy proof of the binomial theorem. For any positive integer n, we consider the function

$$(1 + x)^n = (1 + x)(1 + x) \; \cdots \; (1 + x).$$

It is obvious that this function is a polynomial of degree n, that is,

$$(1 + x)^n = a_0 + a_1 x + a_2 x^2 + a_3 x^3 + \cdots + a_n x^n, \tag{2}$$

and our problem is to find out what the coefficients are. If we put $x = 0$, we immediately obtain $a_0 = 1$. Next, differentiating both sides of (2) repeatedly yields

$$n(1 + x)^{n-1} = a_1 + 2a_2 x + 3a_3 x^2 + \cdots + na_n x^{n-1},$$
$$n(n - 1)(1 + x)^{n-2} = 2a_2 + 3 \cdot 2a_3 x + \cdots + n(n - 1)a_n x^{n-2},$$
$$n(n - 1)(n - 2)(1 + x)^{n-3} = 3 \cdot 2a_3 + \cdots + n(n - 1)(n - 2)a_n x^{n-3},$$

and so on. These equations hold for all values of x, so we can put $x = 0$ in each of them. This procedure gives the following expressions for the coefficients $a_1, a_2, a_3, \ldots$:

$$a_1 = n, \qquad a_2 = \frac{n(n-1)}{2}, \qquad a_3 = \frac{n(n-1)(n-2)}{2 \cdot 3}, \qquad \ldots \, ,$$

$$a_k = \frac{n(n-1)(n-2) \, \cdots \, (n-k+1)}{1 \cdot 2 \cdot 3 \, \cdots \, k}, \qquad \ldots \, , \qquad a_n = 1.$$

With these coefficients, equation (2) takes the form

$$(1+x)^n = 1 + nx + \frac{n(n-1)}{1 \cdot 2} x^2 + \frac{n(n-1)(n-2)}{1 \cdot 2 \cdot 3} x^3 + \cdots$$

$$+ \frac{n(n-1)(n-2) \, \cdots \, (n-k+1)}{1 \cdot 2 \cdot 3 \, \cdots \, k} x^k + \cdots + x^n, \quad (3)$$

and this is the binomial theorem.*

---

* To obtain the equivalent version given in the footnote of Section 3.1, substitute $x = b/a$ in equation (3) and then multiply by a^n.

PROBLEMS

1 Find the first four derivatives of
  (a) $8x - 3$;
  (b) $8x^2 - 11x + 2$;
  (c) $8x^3 + 7x^2 - x + 9$;
  (d) $x^4 - 13x^3 + 5x^2 + 3x - 2$;
  (e) $x^{5/2}$.

2 Calculate the indicated derivative in each case:

  (a) y'' if $y = \dfrac{x}{1-x}$;   (b) y'' if $y = x^2 - \dfrac{1}{x^2}$;

  (c) $\dfrac{d^2}{dx^2}\left(\dfrac{1-x}{1+x}\right)$;   (d) $\dfrac{d^2}{dx^2}\left(x^3 + \dfrac{1}{x^3}\right)$;

  (e) $\dfrac{d^{500}}{dx^{500}}(x^{131} - 3x^{79} + 4)$.

3 Find a general formula for $y^{(n)}$ in each case:

  (a) $y = \dfrac{1}{1-x}$;   (b) $y = \dfrac{1}{1+3x}$;

  (c) $y = \dfrac{x}{1+x}$.

4 Use implicit differentiation to find a simple formula for y'' in each case:
  (a) $b^2x^2 + a^2y^2 = a^2b^2$;   (b) $y^2 = 4px$;
  (c) $x^{1/2} + y^{1/2} = a^{1/2}$;   (d) $x^3 + y^3 = a^3$;
  (e) $x^4 + y^4 = a^4$.

5 Find a simple formula for y'' on the curve $x^n + y^n = a^n$ and show that your results in parts (c), (d), and (e) of Problem 4 are all special cases of this formula.

6 Find the values of y', y'', and y''' at the point $(4, 3)$ on the circle $x^2 + y^2 = 25$.

7 If s is the position of a moving body at time t, find the time, position, and velocity at each moment when the acceleration is zero:

  (a) $s = 8t^2 - \dfrac{1}{t}$ $(t > 0)$;

  (b) $s = 12t^{1/2} + t^{3/2}$ $(t > 0)$;

  (c) $s = \dfrac{24}{3 + t^2}$ $(t \geq 0)$.

8 (a) What is the 23rd derivative of

$$x^{22} - 501x^{17} + \tfrac{12}{35} x^6 - \pi^3 x^2?$$

  (b) What is the 22nd derivative?

9 If $f(x) = x^3 - 2x^2 - x$, for what values of x is $f'(x) = f''(x)$?

10 Show the following:
  (a) if y' is proportional to x^2, then y'' is proportional to x;
  (b) if y' is proportional to y^2, then y'' is proportional to y^3.

11 It is natural to expect from the chain rule that the formula

$$\frac{d^2y}{dx^2} = \frac{d^2y}{du^2} \cdot \frac{d^2u}{dx^2}$$

might be true. Disprove this guess by considering $y = \sqrt{u}$ where $u = x^2 + 1$.

12 If u and v are functions of x, and $y = uv$, show that

$$y'' = u''v + 2u'v' + uv''.$$

Find a similar formula for y'''.

ADDITIONAL PROBLEMS FOR CHAPTER 3

SECTION 3.1

1 Find the points on the curve $y = x^3 - 3x^2 - 9x + 5$ at which the tangent is horizontal.

2 Find the points on the curve $y = x^3 - x^2$ at which the tangent has slope 1.

3 Find the points on the curve $y = x^3 + x$ at which the tangent has slope 4. What is the smallest value the slope of the tangent to this curve can have, and where on the curve does the slope of the tangent have this smallest value?

4 At what points on the curve $y = x^3 - x^2 + x$ is the tangent parallel to the line $2x - y - 7 = 0$?

5 Find the slope of the tangent to the curve $y = x^4 - 2x^2 + 2$ at any point. For what values of x is the tangent horizontal? For what values of x does the tangent point upward to the right?

6 The curve $y = ax^2 + bx + 2$ is tangent to the line $8x + y = 14$ at the point $(2, -2)$. Find a and b.

7 Find the constants a, b, and c if the curve $y = ax^2 + bx + c$ passes through the point $(-1, 0)$ and is tangent to the line $y = x$ at the origin.

8 If the curve $y = ax^2 + bx + c$ passes through the point $(-1, 0)$ and has the line $3x + y = 5$ as its tangent at the point $(1, 2)$, what values must the constants a, b, and c have?

9 The curves $y = x^2 + ax + b$ and $y = x^3 - c$ have the same tangent at the point $(1, 2)$. What are the values of a, b, and c?

10 Find the equations of the tangents to the curve $y = x^2 - 4x$ that pass through the point $(1, -4)$.

11 If $a \neq 0$, show that the tangent to the curve $y = x^3$ at (a, a^3) intersects the curve a second time at the point where $x = -2a$.

12 Show that the tangents to the curve $y = x^2$ at the points (a, a^2) and $(a + 2, (a + 2)^2)$ intersect on the curve $y = x^2 - 1$.

13 Find the values of a, b, c, and d if the curve $y = ax^3 + bx^2 + cx + d$ is tangent to the line $y = x - 1$ at the point $(1, 0)$ and is tangent to the line $y = 6x - 9$ at the point $(2, 3)$.

14 Use the reflection property of parabolas to show that the two tangents to a parabola at the ends of a chord through the focus are perpendicular to each other.

15 Show that the tangent to the curve $y = x^3 - 2x^2 - 3x + 8$ at the point $(2, 2)$ is one of the normals of $y = x^2 - 3x + 3$.

16 There is only one normal to the parabola $x^2 = 2y$ that passes through the point $(4, 1)$. Find its equation.

17 The point $P = (6, 9)$ lies on the parabola $x^2 = 4y$. Find all points Q on this parabola with the property that the normal at Q passes through P.

SECTION 3.2

18 Differentiate each of the following functions two ways and verify that your answers agree:
(a) $(x^2 - 1)(x^3 - 1)$; (b) $3x^4(x^2 + 2x)$;
(c) $(x^2 - 3)(x - 1)$; (d) $(x + 1)(x^2 - 2x - 3)$.

19 Differentiate each of the following functions and simplify your answer as much as possible:

(a) $\dfrac{x + x^{-1}}{x - x^{-1}}$; (b) $\dfrac{x^2 + 2x + 1}{x^2 - 2x + 1}$;

(c) $\dfrac{x^2}{x^3 + 2}$; (d) $\dfrac{2x + 3}{x^2 + x - 4}$;

(e) $\dfrac{x^3}{1 - x^2}$; (f) $\dfrac{1 - x}{1 + x}$;

(g) $\dfrac{6x^4 + 9}{x - 1}$; (h) $\dfrac{x^2 + 6x + 9}{x^2 - 4x + 4}$.

20 Find dy/dx two ways, first by dividing and then by using the quotient rule, and show that your answers agree:

(a) $\dfrac{9 - x^3}{x^2}$; (b) $\dfrac{5 - 3x}{x^4}$; (c) $\dfrac{x^3 - 6x}{x^4}$.

21 Prove the quotient rule from the product rule as follows: Write $y = u/v$ in the form $yv = u$, differentiate this with respect to x by the product rule, and solve the resulting equation for dy/dx.

22 Extend the product rule to a product of three functions by showing that

$$\frac{d}{dx}(uvw) = vw\frac{du}{dx} + uw\frac{dv}{dx} + uv\frac{dw}{dx}.$$

Hint: Treat uvw as a product $(uv)w$ of two factors. (Notice that the right-hand side of this extended product rule is the sum of all terms in which the derivative of one factor is multiplied by the other factors unchanged. This pattern persists for products of more than three factors.)

23 Use Problem 22 to differentiate
(a) $(x + 1)(x + 2)(x + 3)$;
(b) $(x^2 + 2x)(x^3 + 3x^2)(x^4 + 4)$.

24 Use Problem 22 to show that $(d/dx)\, u^3 = 3u^2\, du/dx$, and apply this formula to calculate

$$\frac{d}{dx}\,(6x^{11} + 9x^5 - 3)^3.$$

25 Sketch the curve $y = 10\sqrt{5}/(1 + x^2)$ and find the points on it at which the normal passes through the origin.

26 Consider the curve $y = a/(1 + x^2)$, where a is a positive constant. For what values of a does there exist a point $P = (x_0, y_0)$ on the first-quadrant part of the curve at which the normal passes through the origin? If the normal at the point for which $x_0 = 2$ passes through the origin, what must be the value of a?

27 There are two points on the curve $y = (x + 4)/(x - 5)$ at which the tangent passes through the origin. Sketch the curve and find these points.

SECTION 3.3

28 Find dy/dx in each case:
(a) $y = (4x^2 - 2)^{12}$;   (b) $y = (x^4 + 1)^{125}$;
(c) $y = (x^4 - x^8)^{16}$;   (d) $y = (x^{-1} - x^{-2})^{-3}$;
(e) $y = (4x^2 + 5)^{-1}$;   (f) $y = (x + x^2 + x^3 + x^4)^5$.

29 Find dy/dx in each case:
(a) $y = (1 + 2x)^3(4 - 5x)^6$;
(b) $y = (x^2 + 1)^{10}(x^2 - 1)^{15}$;
(c) $y = (x^2 - 1)(16 + x^2)^{-3}$;
(d) $y = (4x^3 - 9x^2)^2(3x - 2x^2)^3$.

30 Find dx/dt in each case:

(a) $s = \dfrac{(t + 3t^2)^2}{t + 1}$;   (b) $s = \dfrac{1}{(t^3 - 1)^5}$;

(c) $s = \dfrac{(t^2 + 1)^4}{(t^2 - 1)^3}$;   (d) $s = \dfrac{(1 + 2t^2)^5}{(1 - 3t^3)^4}$.

31 Find a function $y = f(x)$ for which

(a) $\dfrac{dy}{dx} = 12x^3(x^4 + 1)^2$;

(b) $\dfrac{dy}{dx} = 72x^5(x^6 + 1)^5$.

32 Prove the power rule for positive integral exponents n by writing $y = u^n$, expanding $\Delta y = (u + \Delta u)^n - u^n$ by the binomial theorem, and then dividing by Δx. Use the quotient rule to extend this result to negative integral exponents.

SECTION 3.4

33 Find dy/dx by implicit differentiation:

(a) $x^4 + 2xy^3 + 2y^4 = 4$;   (b) $\dfrac{y}{x} - 2x = y$;

(c) $y^2 = \dfrac{x^2 + 2}{x^2 - 2}$;   (d) $x^4y^4 = x^4 + y^4$;

(e) $\sqrt{xy} + 2y = \sqrt{x}$.

34 Find dy/dx by implicit differentiation and also by solving for y and then differentiating, and verify that your two answers are equivalent:
(a) $y^3 = 3x^2 + 5x - 1$;   (b) $y^5 = x^2$;
(c) $4y^2 = 3xy + x^2$;   (d) $x^{3/2} + y^{3/2} = 8$.

35 Find the derivative in each case:
(a) $x^{5/2} - x^{3/2}$;   (b) $(x^2 + 2)^{4/9}$;

(c) $\sqrt[3]{x + \sqrt{x^5}}$;   (d) $\dfrac{x^2}{\sqrt{1 - x^2}}$;

(e) $\sqrt{x} + \dfrac{1}{\sqrt{x}}$;   (f) $\sqrt[4]{2x^2 - 1}$;

(g) $\sqrt{\dfrac{x^2 - 1}{x^2 + 1}}$;   (h) $\sqrt{2 + \sqrt{2 - x}}$.

36 Find the equation of
(a) the tangent to $x^3 + y^3 = 2xy + 5$ at $(2, 1)$;

(b) the tangent to $y = \dfrac{2x}{\sqrt[3]{x^2 - 1}}$ at $(3, 3)$;

(c) the normal to $x^3 + 3xy^3 - xy^2 = xy + 10$ at $(2, 1)$;
(d) the normal to $x^{2/3} + y^{2/3} = 5$ at $(-8, 1)$.

37 Show that the sum of the x- and y-intercepts of any line tangent to the curve $\sqrt{x} + \sqrt{y} = \sqrt{a}$ is equal to a.

38 The curve $x^{2/3} + y^{2/3} = a^{2/3}$ is called a *hypocycloid of four cusps.* Sketch it and show that the tangent at (x_0, y_0) is $x_0^{-1/3}x + y_0^{-1/3}y = a^{2/3}$. Use this equation to show that the segment cut from the tangent by the axes has constant length a, so that a segment of length a with its ends sliding along the axes always touches the curve.

SECTION 3.5

39 Calculate y'' if

(a) $y = (1 + 3x)^{1/3}$;   (b) $y = \dfrac{x}{\sqrt{x + 1}}$;

(c) $y = x^{4/5}$;   (d) $y = x^3\sqrt{x} - 7x$;

(e) $y = \sqrt{x} + \dfrac{1}{\sqrt{x}}$;   (f) $y = (x^2 + 4)^{5/2}$.

40 Find a general formula for $y^{(n)}$ if

(a) $y = \dfrac{1}{1 - 2x}$;   (b) $y = \dfrac{1}{a + bx}$.

41 Show that

$$\frac{d^n}{dx^n}\left[\frac{1}{x(1-x)}\right] = n!\left[\frac{(-1)^n}{x^{n+1}} + \frac{1}{(1-x)^{n+1}}\right].$$

42 Consider the function $f(x)$ defined by

$$f(x) = \begin{cases} x^2 & \text{if } x \geq 0, \\ -x^2 & \text{if } x < 0. \end{cases}$$

Sketch the graph, show that $f'(x) = 2|x|$, and conclude that $f''(0)$ does not exist.

43 For each of the following functions, find $f'''(x)$ and then calculate the limit

$$\lim_{\Delta x \to 0} \frac{f(x + 2\Delta x) - 2f(x + \Delta x) + f(x)}{(\Delta x)^2},$$

and notice that they are equal:
(a) $f(x) = x^3$;　　　(b) $f(x) = 1/x$.

44 Solve Problem 43 after replacing the limit given there by

$$\lim_{\Delta x \to 0} \frac{f(x + \Delta x) - 2f(x) + f(x - \Delta x)}{(\Delta x)^2}.$$

4

APPLICATIONS OF DERIVATIVES

In this chapter we begin to justify the effort we have spent on learning how to calculate derivatives.

Our first applications are based on the interpretation of the derivative as the slope of the tangent line to a curve. The purpose of this work is to enable us to use the derivative as a tool for quickly discovering the most important features of a function and sketching its graph. This art of curve sketching is essential in the physical sciences. It is also one of the most useful skills that calculus can provide for those who need to use mathematics in their study of economics or biology or psychology.

A function $f(x)$ is said to be *increasing* on a certain interval of the x-axis if on this interval $x_1 < x_2$ implies $f(x_1) < f(x_2)$. In geometric language, this means that the graph is rising as the point that traces it moves from left to right. Similarly, the function is said to be *decreasing* (the graph is falling) if $x_1 < x_2$ implies $f(x_1) > f(x_2)$. These concepts are illustrated in Fig. 4.1.

In sketching the graph of a function, it is important to know the intervals on which it is increasing and those on which it is decreasing. The sign of the derivative gives us this information:

A function $f(x)$ is increasing on any interval in which $f'(x) > 0$, and it is decreasing on any interval in which $f'(x) < 0$.

This is geometrically evident if we keep in mind the fact that a straight line

4.1

INCREASING AND DECREASING FUNCTIONS. MAXIMA AND MINIMA

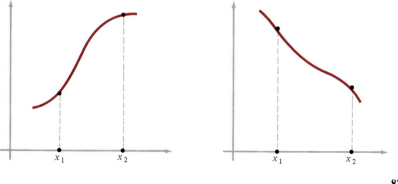

Figure 4.1 Increasing and decreasing functions.

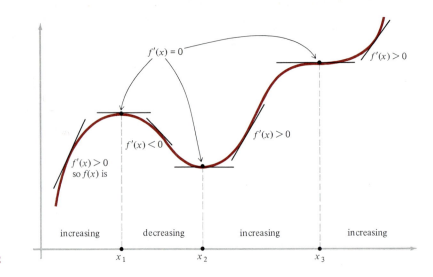

Figure 4.2

points upward to the right if its slope is positive and downward to the right if its slope is negative (Fig. 4.2).

It is clear that a smooth curve can make the transition from rising to falling only by passing over a peak where the slope is zero. Similarly, it can change from falling to rising only by going through a trough where the slope is zero. At such points we have a *maximum* or *minimum value* of the function. We locate these values by finding the *critical points* of the function, which are the solutions of the equation $f'(x) = 0$; that is, we force the tangent to be horizontal by equating the derivative to zero, and we then solve the equation $f'(x) = 0$ to discover where this happens. In Fig. 4.2 the critical points are x_1, x_2, x_3, and the corresponding *critical values* are the values of the function at these points, that is, $f(x_1), f(x_2), f(x_3)$.

It is important to understand that a critical value is not necessarily either a maximum or a minimum. This is shown by $f(x_3)$ in Fig. 4.2; at the critical point x_3 the graph does not pass either over a peak or through a trough, but instead merely flattens out momentarily between two intervals on each of which the derivative is positive.

It should also be pointed out that we are discussing the so-called *relative* (or *local*) maximum or minimum values. These are values that are maximal or minimal compared only with nearby points on the curve. In Fig. 4.2, for instance, $f(x_1)$ is a maximum even though there are many higher points on the curve, off to the right. If we seek the *absolute* maximum of a function, we must compare its relative maxima with one another to determine which (if any) is larger than any other value assumed by the function.

Example 1   To sketch the graph of the polynomial
$$y = f(x) = 2x^3 - 3x^2 - 12x + 12,$$
we begin by computing the derivative and factoring this derivative as completely as possible:
$$f'(x) = 6x^2 - 6x - 12 = 6(x + 1)(x - 2).$$

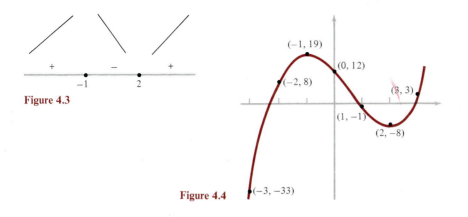

Figure 4.3

Figure 4.4

The critical points are evidently $x = -1$ and $x = 2$, and the corresponding critical values are $y = 19$ and $y = -8$. We now examine the three intervals into which the critical points divide the x-axis, for on each of these intervals $f'(x)$ has constant sign. When $x < -1$, $x + 1$ and $x - 2$ are both negative, so their product is positive and $f'(x) > 0$. When $-1 < x < 2$, $x + 1$ is positive and $x - 2$ is negative, so their product is negative and $f'(x) < 0$. When $x > 2$, $x + 1$ and $x - 2$ are both positive, so their product is positive and $f'(x) > 0$. These results are displayed in Fig. 4.3, where the slanted lines give a schematic suggestion of the direction of the graph in each interval. In Fig. 4.4 we now plot the points $(-1, 19)$ and $(2, -8)$ and sketch a smooth curve through these points, using the information in Fig. 4.3 provided by the sign of the derivative, that is, $f(x)$ is increasing when $x < -1$, decreasing when $-1 < x < 2$, and increasing when $x > 2$. Notice that in Fig. 4.4 we use different units of length on the two axes, as a matter of convenience in drawing a picture of reasonable size.* It is clear that our function has a maximum at $x = -1$ and a minimum at $x = 2$, and also that no absolute maximum or minimum exists. The zeros of a function are always valuable aids in curve sketching when they can be found, but finding them can be quite difficult. We have plotted a few additional points in Fig. 4.4 to suggest that the zeros of this particular function are approximately $-2.2, 0.9$, and 2.9. As a matter of fact, we sometimes sketch the graph of a function to help us guess the approximate location of its zeros, just as we have done here, as a first step toward the numerical calculation of these zeros to any desired degree of accuracy. In Section 4.6 we describe a standard method for carrying out such calculations.

Example 2 The rational function

$$y = \frac{x}{x^2 + 1}$$

---

* The basic idea of a graph as a visual aid displaying the qualitative nature of the function does not require the use of equal units on the two axes. It is only when we work with certain quantitative aspects of the geometry of the plane, such as distances between points, areas of regions, or angles between lines, that it is necessary to use equal units on both axes.

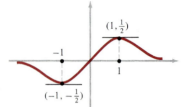

Figure 4.5

was discussed in Example 6 of Section 1.8, and we explained there why the graph has the shape it does (Fig. 4.5). To find the precise location of the indicated maximum and minimum, we calculate the derivative and equate it to zero:

$$y' = \frac{(x^2 + 1)\cdot 1 - x\cdot 2x}{(x^2 + 1)^2} = \frac{1 - x^2}{(x^2 + 1)^2} = 0.$$

The roots of this equation (the critical points) are $x = 1$ and $x = -1$, and so the maximum and minimum occur at $x = 1$ and $x = -1$, respectively. The actual maximum and minimum values are $y = \frac{1}{2}$ and $y = -\frac{1}{2}$. With these facts and our initial awareness of the overall shape of the graph, it is obvious that this function increases on the interval $-1 < x < 1$ and decreases for $x < -1$ and $x > 1$. However, these conclusions can also be drawn directly from the sign of the derivative, which is clearly positive for $-1 < x < 1$ and negative for $x < -1$ and $x > 1$.

These examples, as well as our past experience, suggest a few informal rules that are useful in sketching the graph of a function $f(x)$. If possible and convenient, we should determine

1   The critical points of $f(x)$.
2   The critical values of $f(x)$.
3   The sign of $f'(x)$ between critical points.
4   The zeros of $f(x)$.
5   The behavior of $f(x)$ as $x \to \infty$ and as $x \to -\infty$.
6   The behavior of $f(x)$ near points at which the function is not defined.

However, perhaps the most important rule of all is this: *Don't be a slave to any rule, be flexible, use common sense.* Remember the old Hungarian proverb: "All fixed ideas are wrong, including this one."

Remark 1   Maxima and minima can occur in three ways not covered by the preceding discussion: at *endpoints, cusps,* and *corners.* As examples we consider the three functions

$$y = \sqrt{1 - x^2}, \qquad y = x^{2/3}, \qquad y = 1 - \sqrt{x^2} = 1 - |x|.$$

Their graphs are shown in Fig. 4.6. The first function has the closed interval $-1 \le x \le 1$ as its domain, and at the endpoints it has minima that are not revealed by equating the derivative to zero. The second function has a minimum at $x = 0$ that is a cusp, because its derivative

$$y' = \tfrac{2}{3} x^{-1/3} = \frac{2}{3 \sqrt[3]{x}}$$

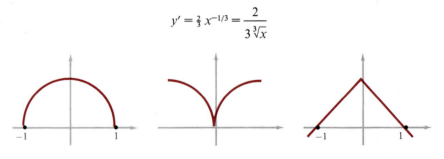

Figure 4.6   Endpoints, cusp, and corner.

is negative to the left of 0 and positive to the right of 0, and becomes infinite near 0. The third function has a maximum at $x = 0$, and this maximum is called a *corner* for obvious reasons. In seeking the maxima and minima of functions, equate the derivative to zero by all means, but do so carefully, keeping these three possibilities in mind as well.

Remark 2 Among other things, mathematicians are professional skeptics. On one side of their nature, they have trained themselves to destroy loose arguments and to believe only those statements which they find it impossible to doubt, in the hope that Ultimate Certainty will reward their efforts. Our statements about increasing and decreasing functions and maxima and minima are supported only by geometric plausibility arguments. The statements themselves are true, but these arguments are a far cry from the proofs that would satisfy a mathematician. However, this book is for students, not mathematicians, so we try to avoid dwelling unnecessarily on theoretical issues. Our main concern is with the use of the tools rather than the tools themselves. Students who are curious about how such things are proved are invited to consult Appendixes C.3 and C.4.

PROBLEMS

Sketch the graphs of the following functions by using the first derivative and the methods of this section; in particular, find the intervals on which each function is increasing and those on which it is decreasing, and locate any maximum or minimum values it may have.

1 $y = x^2 - 2x$.

2 $y = 2 + x - x^2$.

3 $y = x^2 - 6x + 9$.

4 $y = x^2 - 4x + 5$.

5 $y = 2x^3 - 3x^2 + 1$.

6 $y = x^3 - 3x^2 + 3x - 1$.

7 $y = x^3 - x$.

8 $y = x^4 - 2x^2 + 1$.

9 $y = 3x^4 + 4x^3$.

10 $y = 3x^5 - 20x^3$.

11 $y = x + \dfrac{1}{x}$.

12 $y = 2x + \dfrac{1}{x^2}$.

13 $y = \dfrac{1}{x^2 + x}$.

14 $y = \dfrac{x}{(x-1)^2}$.

15 $y = x\sqrt{3 - x}$.

*16 $y = 5x^{2/3} - x^{5/3}$.

17 The function $f(x) = x^3 + x - 1$, being a third-degree polynomial, obviously crosses the x-axis (why?) and therefore has at least one zero. By examining $f'(x)$, show that this function has only one zero. Show similarly that $f(x) = 2x^5 + 5x^3 + 3x - 17$ has one and only one zero.

18 Consider the function $y = x^m(1 - x)^n$, where m and n are positive integers, and show that
(a) if m is even, y has a minimum at $x = 0$;

(b) if n is even, y has a minimum at $x = 1$;
(c) y has a maximum at $x = m/(m + n)$ regardless of whether m and n are even or not.

19 Sketch the graph of a function $f(x)$ defined for $x > 0$ and having the properties

$$f(1) = 0 \quad \text{and} \quad f'(x) = \frac{1}{x} \quad \text{(all } x > 0\text{)}.$$

20 Sketch the graph of a function $f(x)$ with the properties

$$f'(x) < 0 \text{ for } x < 2 \quad \text{and} \quad f'(x) > 0 \text{ for } x > 2$$

(a) if $f'(x)$ is continuous at $x = 2$;
(b) if $f'(x) \to -1$ as $x \to 2-$ and $f'(x) \to 1$ as $x \to 2+$.

21 In each case, sketch the graph of a function with all the stated properties:
(a) $f(1) = 1, f'(x) > 0$ for $x < 1, f'(x) < 0$ for $x > 1$;
(b) $f(-1) = 2$ and $f(2) = -1$, $f'(x) > 0$ for $x < -1$ and $x > 2, f'(x) < 0$ for $-1 < x < 2$;
(c) $f(-1) = 1$ and $f'(-1) = 0, f'(x) < 0$ for $x < -1$ and $-1 < x < 2, f'(x) > 0$ for $x > 2$;
(d) $f'(x) < 0$ for $-2 < x < 0$ and $x > 1, f'(x) > 0$ for $x < -2$ and $0 < x < 1, f'(-2) = f'(0) = 0, f'(1)$ does not exist.

22 Construct a formula for a function $f(x)$ with a maximum at $x = -2$ and a minimum at $x = 1$.

Figure 4.7

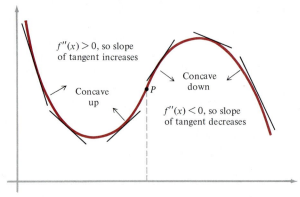

$f''(x) > 0$, so slope of tangent increases

Concave up

Concave down

P

$f''(x) < 0$, so slope of tangent decreases

Figure 4.8

4.2

CONCAVITY AND POINTS OF INFLECTION

One of the most distinctive features of a graph is the direction in which it curves or bends. The graph on the left in Fig. 4.7 curves upward as the point that traces it moves from left to right, and the graph on the right curves downward. The sign of the second derivative gives us this information, as follows.

A positive second derivative, $f''(x) > 0$, tells us that the slope $f'(x)$ is an increasing function of x. This means that the tangent turns counterclockwise as we move along the curve from left to right, as shown on the left-hand side of Fig. 4.8. The curve is said to be *concave up* (the concave side of a curve is its hollow side). Such a curve lies above its tangent except at the point of tangency. Similarly, if the second derivative is negative, $f''(x) < 0$, then the slope $f'(x)$ is a decreasing function, and the tangent turns clockwise as we move to the right (see the right-hand side of Fig. 4.8). Under these circumstances the curve is *concave down*; it lies below its tangent except at the point of tangency.

Most curves are concave up on some intervals and concave down on other intervals. A point like P in Fig. 4.8, across which the direction of concavity changes, is called a *point of inflection*. If $f''(x)$ is continuous and has opposite signs on each side of P, then it must have a zero at P itself. The search for points of inflection is mainly a matter of solving the equation $f''(x) = 0$ and checking the direction of concavity on both sides of each root.

Example 1 To investigate the function

$$y = f(x) = 2x^3 - 12x^2 + 18x - 2$$

for concavity and points of inflection, we calculate

$$f'(x) = 6x^2 - 24x + 18 = 6(x - 1)(x - 3)$$

and

$$f''(x) = 12x - 24 = 12(x - 2).$$

The critical points [the roots of $f'(x) = 0$] are clearly $x = 1$ and $x = 3$, and the corresponding critical values are $y = 6$ and $y = -2$. We have a possible point of inflection at $x = 2$, since this is the only root of $f''(x) = 0$. It is evident that $f''(x)$ is negative for $x < 2$ and positive for $x > 2$, so the graph is concave down on the left of $x = 2$ and concave up on the right. This tells us that we really have a point of inflection at $x = 2$, as indicated in Fig. 4.9.

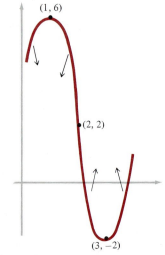

(1, 6)

(2, 2)

(3, −2)

Figure 4.9

Example 2   The rational function

$$y = \frac{1}{x^2 + 1}$$

is very easy to sketch by inspection if we notice the following clues: It is symmetric about the y-axis, its values are all positive, it has a maximum at $x = 0$ because this yields the smallest denominator, and $y \to 0$ as $|x| \to \infty$. It is therefore intuitively clear that the graph has the shape shown in Fig. 4.10. There are evidently two points of inflection, and the only question is, What is their precise location? To discover this, we compute

$$y' = \frac{-2x}{(x^2 + 1)^2}$$

and

$$y'' = \frac{(x^2 + 1)^2 \cdot (-2) + 2x \cdot 2(x^2 + 1) \cdot 2x}{(x^2 + 1)^4}$$
$$= \frac{(x^2 + 1) \cdot (-2) + 8x^2}{(x^2 + 1)^3} = \frac{2(3x^2 - 1)}{(x^2 + 1)^3}.$$

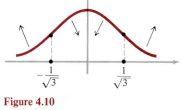

Equating y'' to zero and solving gives $x = \pm 1/\sqrt{3}$, which locates the points of inflection. If we wish, we can verify our first impression about the direction of concavity on various parts of the curve, as shown in Fig. 4.10, by observing that $y'' < 0$ when $x^2 < \frac{1}{3}$ and $y'' > 0$ when $x^2 > \frac{1}{3}$. These facts tell us that the graph is concave down for $-1/\sqrt{3} < x < 1/\sqrt{3}$ and concave up for $x < -1/\sqrt{3}$ or $x > 1/\sqrt{3}$.

Figure 4.10

Remark 1   As we have tried to suggest in these examples, knowing that $f''(x_0) = 0$ is not enough to guarantee that $x = x_0$ furnishes a point of inflection. We must also know that the graph is concave up on one side of x_0 and concave down on the other. The simplest function that shows this difficulty is $y = f(x) = x^4$ (Fig. 4.11). Here $f'(x) = 4x^3$ and $f''(x) = 12x^2$, so $f''(x) = 0$ at $x = 0$. However, $f''(x)$ is clearly positive on both sides of the point $x = 0$, and therefore — as we already know from the graph — this point corresponds to a minimum, not a point of inflection. The function $y = x^5 - 5x^4$ provides a more complicated example of the same phenomenon. Here

$$y' = 5x^4 - 20x^3 \quad \text{and} \quad y'' = 20x^3 - 60x^2 = 20x^2(x - 3).$$

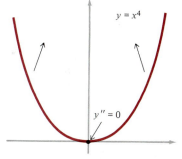

Figure 4.11

The roots of $y'' = 0$ are $x = 0$ and $x = 3$. However, y'' does not change sign at $x = 0$, and so the only point of inflection is at $x = 3$. The graph is concave down on the left of this point and concave up on the right.

Remark 2   The graph of $y = x^{1/3} = \sqrt[3]{x}$ is easy to sketch and has an obvious point of inflection at $x = 0$ (Fig. 4.12). We can also discover this by inspecting the second derivative. We have

$$y' = \frac{1}{3}x^{-2/3}$$

and

$$y'' = -\frac{2}{9}x^{-5/3} = \frac{-2}{9\sqrt[3]{x^5}},$$

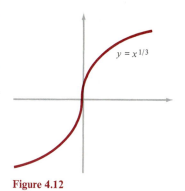

Figure 4.12

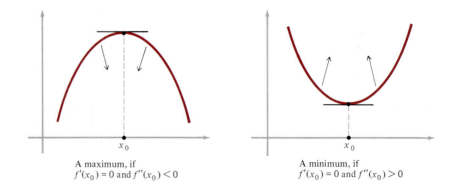

Figure 4.13 The second derivative test.

A maximum, if $f'(x_0) = 0$ and $f''(x_0) < 0$

A minimum, if $f'(x_0) = 0$ and $f''(x_0) > 0$

so y'' is positive if $x < 0$ and negative if $x > 0$. In searching for points of inflection, we must therefore consider not only points at which $y'' = 0$, but also points (if there are any) at which y'' does not exist.

Remark 3 In the so-called *second derivative test*—which we state informally in Fig. 4.13—the sign of the second derivative is used to decide whether a critical point furnishes a maximum or a minimum value. This test is sometimes useful, but its importance is often exaggerated. We will see in the next two sections that in most applied problems it is clear from the context whether we have a maximum or minimum value, and no testing is necessary.

PROBLEMS

For each of the following, locate the points of inflection, find the intervals on which the curve is concave up and those on which it is concave down, and sketch.

1 $y = (x - a)^3 + b$.

2 $y = x^3 - 6x^2$.

3 $y = x^3 + 3x^2 + 4$.

4 $y = 2x^3 + 3x^2 - 12x$.

5 $y = x^4 + 2x^3 + 1$.

6 $y = x^4 - 6x^2$.

7 $y = x^4 - 2x^3$.

8 $y = 3x^5 - 5x^4$.

9 $y = \dfrac{9}{x^2 + 9}$.

10 $y = \dfrac{ax}{x^2 + b^2}$ $(a, b > 0)$.

11 $y = \dfrac{4x^2}{x^2 + 3}$.

12 $y = \dfrac{12}{x^2} - \dfrac{12}{x}$.

13 $y = x - \dfrac{1}{x}$.

14 In each part of this problem, use the given formula for the second derivative of a function to locate the points of inflection, the intervals on which the graph is concave up, and the intervals on which the graph is concave down:
(a) $y'' = 8x^2 + 32x$; (b) $y'' = 15x^3 + 39x$;
(c) $y'' = 3x^4 - 27x^2$; (d) $y'' = (x + 2)(x^2 - 4)$.

15 Sketch the graph of a function $f(x)$ defined for all x such that

(a) $f(x) > 0, f'(x) > 0$, and $f''(x) > 0$;
(b) $f'(x) < 0$ and $f''(x) < 0$.

16 Is it possible for a function $f(x)$ defined for all x to have the three properties $f(x) > 0, f'(x) < 0$, and $f''(x) < 0$? Explain.

17 (a) By sketching, show that $y = x^2 + a/x$ has a minimum but no maximum for every value of the constant a. Also, verify this by calculation.
(b) Find the point of inflection of $y = x^2 - 8/x$.

18 Starting from $x^2 + y^2 = a^2$, calculate d^2y/dx^2 by implicit differentiation and state why its sign should be opposite to the sign of y.

19 Find the value of a that makes $y = x^3 - ax^2 + 1$ have a point of inflection at $x = 1$.

20 Find a and b such that $y = a\sqrt{x} + b/\sqrt{x}$ has $(1, 4)$ as a point of inflection.

21 If k is a positive number $\neq 1$, show that the first quadrant part of the curve $y = x^k$ is
(a) concave up if $k > 1$;
(b) concave down if $k < 1$.

22 If k is a positive number $\neq 1$ and $y = x^k - kx$, show that
(a) if $k < 1$, y has a maximum at $x = 1$;
(b) if $k > 1$, y has a minimum at $x = 1$.

23 Show that the graph of a quadratic function $y = ax^2 +$

$bx + c$ has no points of inflection. Give a condition under which the graph is (a) concave up; (b) concave down.

24 Show that the general cubic curve $y = ax^3 + bx^2 + cx + d$ has a single point of inflection and three possible shapes, depending on whether $b^2 > 3ac$, $b^2 = 3ac$, or $b^2 < 3ac$. Sketch these shapes.

25 In each of the following, sketch the graph of a function with all the stated properties:
(a) $f(0) = 2$, $f(2) = 0$, $f'(0) = f'(2) = 0$, $f'(x) > 0$ for $|x - 1| > 1$, $f'(x) < 0$ for $|x - 1| < 1$, $f''(x) < 0$ for $x < 1$, $f''(x) > 0$ for $x > 1$;

(b) $f(-2) = 6$, $f(1) = 2$, $f(3) = 4$, $f'(1) = f'(3) = 0$, $f'(x) < 0$ for $|x - 2| > 1$, $f'(x) > 0$ for $|x - 2| < 1$, $f''(x) < 0$ for $x > 2$ or $|x + 1| < 1$, $f''(x) > 0$ for $|x - 1| < 1$ or $x < -2$;

(c) $f(0) = 0$, $f(2) = f(-2) = 1$, $f'(0) = 0$, $f'(x) > 0$ for $x > 0$, $f'(x) < 0$ for $x < 0$, $f''(x) > 0$ for $|x| < 2$, $f''(x) < 0$ for $|x| > 2$, $\lim_{x \to \infty} f(x) = 2$, $\lim_{x \to -\infty} f(x) = 2$;

(d) $f(2) = 4$, $f'(x) > 0$ for $x < 2$, $f'(x) < 0$ for $x > 2$, $f''(x) > 0$ for $x \neq 2$, $\lim_{x \to 2} |f'(x)| = \infty$, $\lim_{x \to \infty} f(x) = 2$, $\lim_{x \to -\infty} f(x) = 2$.

4.3
APPLIED MAXIMUM AND MINIMUM PROBLEMS

Among the most striking applications of calculus are those that depend on finding the maximum or minimum values of functions.

Practical everyday life is filled with such problems, and it is natural that mathematicians and others should find them interesting and important. A businessperson seeks to maximize profits and minimize costs. An engineer designing a new automobile wishes to maximize its efficiency. An airline pilot tries to minimize flight times and fuel consumption. In science, we often find that nature acts in a way that maximizes or minimizes a certain quantity. For example, a ray of light traverses a system of lenses along a path that minimizes its total time of travel, and a flexible hanging chain assumes a shape that minimizes its potential energy due to gravity.

Whenever we use such words as *largest*, *smallest*, *most*, *least*, *best*, and so on, it is a reasonable guess that some kind of maximum or minimum problem is lurking nearby. If this problem can be expressed in terms of variables and functions—which is not always possible by any means—then the methods of calculus stand ready to help us understand it and solve it.

Many of our examples and problems deal with geometric ideas, because maximum and minimum values often appear with particular vividness in geometric settings. In order to be ready for this work, students should make sure that they know the formulas for areas and volumes given in Fig. 1.24 of Chapter 1.

We begin with a fairly simple example about numbers.

Example 1 Find two positive numbers whose sum is 16 and whose product is as large as possible.

Solution If x and y are two variable positive numbers whose sum is 16, so that

$$x + y = 16, \tag{1}$$

then we are asked to find the particular values of x and y that maximize their product

$$P = xy. \tag{2}$$

Our initial difficulty is that P depends on two variables, whereas our calculus of derivatives works only for functions of a single independent variable.

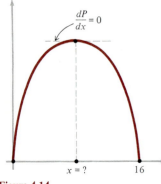

Figure 4.14

Equation (1) gets us over this difficulty. It enables us to express y in terms of x, $y = 16 - x$, and thereby to express P as a function of x alone,

$$P = x(16 - x) = 16x - x^2. \qquad (3)$$

In Fig. 4.14 we give a rough sketch of the graph of (3). Our only purpose here is to provide visual emphasis for the following obvious facts about this function: that $P = 0$ for $x = 0$ and $x = 16$, that $P > 0$ for $0 < x < 16$, and that therefore the highest point (where P has its largest value) is characterized by the condition $dP/dx = 0$, since this condition means that the tangent is horizontal. To solve the problem we compute this derivative from (3),

$$\frac{dP}{dx} = 16 - 2x;$$

we equate this derivative to zero,

$$16 - 2x = 0;$$

and we see that $x = 8$ is the solution of this equation. This is the value of x that maximizes P, and by (1) the corresponding value of y is also 8. It is quite clear from Fig. 4.14 that $x = 8$ actually does maximize P; but if we wish to verify this, we can do so by computing the second derivative,

$$\frac{d^2P}{dx^2} = -2,$$

and by recalling that a negative second derivative implies that the curve is concave down and therefore that we have a maximum — which we already knew from Fig. 4.14. The related problem of making the product P as small as possible within the stated restrictions has no solution, for the restriction that x and y must be positive numbers means that x must belong to the *open* interval $0 < x < 16$, and this part of the graph has no lowest point.

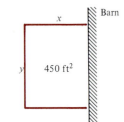

Figure 4.15

Example 2 A rectangular garden 450 ft^2 in area is to be fenced off against rabbits. If one side of the garden is already protected by a barn wall, what dimensions will require the shortest length of fence?

Solution We begin by drawing a sketch and introducing notation that will make it convenient to deal with the area of the garden and the total length of the fence (Fig. 4.15). If L denotes the length of the fence, we are to minimize

$$L = 2x + y \qquad (4)$$

subject to the restriction that

$$xy = 450. \qquad (5)$$

By using (5), L can be written as a function of x alone,

$$L = 2x + \frac{450}{x}. \qquad (6)$$

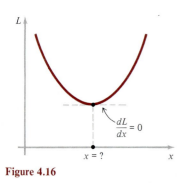

Figure 4.16

A quick sketch (Fig. 4.16) helps us to visualize this function and feel comfortable with its properties, especially the fact that it has a minimum and no maximum (we are only interested in positive values of x). Our next steps are to compute the derivative of (6),

$$\frac{dL}{dx} = 2 - \frac{450}{x^2},$$

and then to equate this derivative to zero and solve the resulting equation,

$$2 - \frac{450}{x^2} = 0, \qquad x^2 = 225, \qquad x = 15.$$

(We ignore the root $x = -15$ for the reason stated.) By (5), the corresponding value of y is $y = 30$, so the garden with the shortest fence is 15 by 30, or twice as long as it is wide.

Example 3 Find the dimensions of the rectangle of greatest area that can be inscribed in a semicircle of radius a.

Solution If we take our semicircle to be the top half of the circle $x^2 + y^2 = a^2$ (Fig. 4.17, left), then our notation is ready and waiting: We must maximize

$$A = 2xy \qquad (7)$$

with the restriction that

$$x^2 + y^2 = a^2. \qquad (8)$$

Since (8) yields $y = \sqrt{a^2 - x^2} = (a^2 - x^2)^{1/2}$, (7) becomes

$$A = 2x(a^2 - x^2)^{1/2}. \qquad (9)$$

It is clear that x lies in the interval $0 < x < a$. On the right in Fig. 4.17 we imagine the extreme cases: When x is close to 0, the rectangle is tall and thin, and when x is close to a, it is short and wide—and in each case the area is small, so somewhere in between we have a maximum area. To locate this maximum, we compute dA/dx from (9), equate it to zero, and solve:

$$2x \cdot \tfrac{1}{2}(a^2 - x^2)^{-1/2} \cdot (-2x) + 2(a^2 - x^2)^{1/2} = 0, \qquad \frac{x^2}{\sqrt{a^2 - x^2}} = \sqrt{a^2 - x^2},$$

$$x^2 = a^2 - x^2, \qquad 2x^2 = a^2, \qquad x = \frac{a}{\sqrt{2}} = \tfrac{1}{2}\sqrt{2}a.$$

Since $y = \sqrt{a^2 - x^2}$, we see that the corresponding value of y is also $\tfrac{1}{2}\sqrt{2}a$, so the dimensions of the largest inscribed rectangle are $2x = \sqrt{2}a$ and $y = \tfrac{1}{2}\sqrt{2}a$, and this rectangle is twice as long as it is wide.

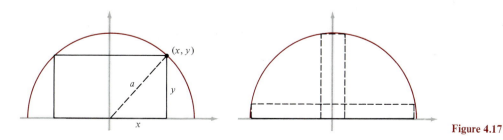

Figure 4.17

There is a more efficient way of solving this problem if we don't care about the actual dimensions of the largest rectangle, but only about its shape. The first step is to notice that (8) determines y as an implicit function of x, so implicit differentiation with respect to x yields

$$2x + 2y\frac{dy}{dx} = 0 \quad \text{or} \quad \frac{dy}{dx} = -\frac{x}{y}. \tag{10}$$

Next, by differentiating (7) with respect to x and using the fact that $dA/dx = 0$ at the maximum, we obtain

$$2x\frac{dy}{dx} + 2y = 0 \quad \text{or} \quad x\frac{dy}{dx} + y = 0. \tag{11}$$

When (10) is inserted in (11), the result is

$$x\left(-\frac{x}{y}\right) + y = 0, \quad -x^2 + y^2 = 0, \quad y^2 = x^2, \quad \text{or} \quad y = x,$$

where the last equation expresses the shape of the rectangle with the largest area. We can also describe this shape by saying that the ratio of the height of the rectangle to its base is

$$\frac{y}{2x} = \frac{x}{2x} = \frac{1}{2}.$$

Example 4   A wire of length L is to be cut into two pieces, one being bent to form a square and the other to form a circle. How should the wire be cut if the sum of the areas enclosed by the two pieces is to be (a) a maximum? (b) a minimum?

Solution   If x denotes the side of the square and r the radius of the circle, as shown on the left in Fig. 4.18, then the sum of the areas is

$$A = x^2 + \pi r^2 \tag{12}$$

where x and r are related by

$$4x + 2\pi r = L. \tag{13}$$

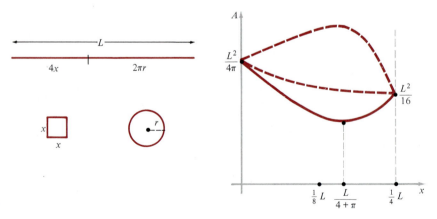

Figure 4.18

We solve (13) for r in terms of x,

$$r = \frac{1}{2\pi}(L - 4x),$$

and use this to express A in terms of x alone,

$$A = x^2 + \pi \cdot \frac{1}{4\pi^2}(L - 4x)^2$$

$$= x^2 + \frac{1}{4\pi}(L - 4x)^2. \tag{14}$$

To solve the problem, we must fully understand the behavior of this function on the interval $0 \le x \le \frac{1}{4}L$. Its values at $x = 0$ and $x = \frac{1}{4}L$ are clearly $L^2/4\pi$ and $L^2/16$, and the first of these values is the larger. This is indicated on the right in Fig. 4.18, along with three possible shapes for the graph. We decide which shape is correct by examining the derivatives of (14). First,

$$\frac{dA}{dx} = 2x + \frac{1}{4\pi} \cdot 2(L - 4x) \cdot (-4)$$

$$= 2x - \frac{2}{\pi}(L - 4x).$$

On setting this equal to zero and solving the resulting equation, we get

$$x - \frac{1}{\pi}(L - 4x) = 0, \qquad \pi x = L - 4x, \qquad x = \frac{L}{4 + \pi}.$$

This number lies between $\frac{1}{8}L$ and $\frac{1}{4}L$. Since the second derivative is positive,

$$\frac{d^2A}{dx^2} = 2 + \frac{8}{\pi} > 0,$$

and therefore the graph is concave up, the two upper possibilities in the figure are eliminated, and we know that the graph has the appearance of the lower curve. These conclusions about the graph enable us to complete the solution of the problem, as follows.

To maximize A, we must choose $x = 0$ and use all the wire for the circle. If we insist that the wire must actually be cut, then (a) has no answer; for no matter how little of the wire is used for the square, we can always increase the total area by using still less.

For (b), the total area is minimized when $x = L/(4 + \pi)$. Accordingly, the length of wire used for the square is $4x = 4L/(4 + \pi)$ and the length used for the circle is

$$L - 4x = L - \frac{4L}{4 + \pi} = \frac{\pi L}{4 + \pi}.$$

We also notice that the minimal area is attained when the diameter of the circle equals the side of the square, since

$$2r = \frac{1}{\pi}(L - 4x) = \frac{1}{\pi} \cdot \frac{\pi L}{4 + \pi} = \frac{L}{4 + \pi}.$$

Example 5 At a price of $1.50, a door-to-door salesperson can sell 500 potato peelers that cost 70 cents each. For every cent that the salesperson lowers the price, the number sold can be increased by 25. What selling price will maximize the total profit?

Solution If x denotes the number of cents the salesperson lowers the price, then the profit on each peeler is $80 - x$ cents and the number sold is $500 + 25x$. The total profit is therefore (in cents)

$$P = (80 - x)(500 + 25x) = 40,000 + 1500x - 25x^2.$$

We maximize this function by setting the derivative equal to zero and solving the resulting equation,

$$\frac{dP}{dx} = 1500 - 50x, \qquad 1500 - 50x = 0, \qquad 50x = 1500, \qquad x = 30.$$

The most advantageous selling price is therefore $1.20.

As these examples show, the mathematical techniques required in most maximum-minimum problems are relatively simple. The hardest part of such a problem is usually "setting it up" in a convenient form. This is the analytical, thinking part of the problem, as opposed to the computational part. We emphasize this because it is clear that calculus is unlikely to be of much value as a tool in the sciences unless one learns how to understand what a problem is about and how to translate its words into appropriate mathematical language. This is what "word problems" or "story problems" are for — to help students learn these critically important skills.

No set of rules for problem solving really works, because the essential ingredients are imagination and intelligence. However, the following general suggestions for coping with maximum-minimum problems may be helpful. They don't guarantee success, but without them progress is unlikely.

1   If geometry is involved — as it often is — make a fairly careful sketch of reasonable size. Show the general configuration. For instance, if a problem is about a general triangle, don't mislead yourself by drawing one that looks like a right triangle or an isosceles triangle. Don't be hasty or sloppy. You hope your sketch will be a source of fruitful ideas, so treat it with respect.

2   Label your figure carefully, making sure you understand which quantities are constant and which are allowed to vary.

3   Be aware of geometric relations among the quantities in your figure, especially those involving right triangles and similar triangles. Write these relations down, and use them when it becomes necessary.

4   If Q is the quantity to be maximized or minimized, try to express Q as a function of a single variable. Draw a quick sketch of this function on a suitable interval, perform little thought experiments in which you visualize the extreme cases, and use derivatives to discover details.

PROBLEMS

1 Find the positive number that exceeds its square by the largest amount. Why would you expect this number to be in the open interval $(0, 1)$?

2 Express the number 18 as the sum of two positive numbers in such a way that the product of the first and the square of the second is as large as possible.

3 Show that the rectangle with maximum area for a given perimeter is a square.*

4 Show that the rectangle with minimum perimeter for a given area is a square.

5 Show that the square has the largest area among all rectangles inscribed in a given fixed circle $x^2 + y^2 = a^2$.

6 If we maximize the perimeter of the rectangle instead of the area in Problem 5, show that the solution is still a square.

7 An east-west and a north-south road intersect at a point O. A diagonal road is to be constructed from a point A east of O to a point B north of O, passing through a town C which is a miles east and b miles north of O. Find the ratio of OA to OB if the triangular area OAB is as small as possible. Show that this minimal area is attained when C bisects the segment AB.

8 A certain poster requires 96 in^2 for the printed message and must have 3-in margins at the top and bottom and a 2-in margin on each side. Find the overall dimensions of the poster if the amount of paper used is a minimum.

9 A university bookstore can get the book *Courtship Rituals of the American College Student* at a cost of $4 a copy from the publisher. The bookstore manager estimates that she can sell 180 copies at a price of $10 and that each 50-cent reduction in the price will increase the sales by 30 copies. What should be the price of the book to maximize the bookstore's total profit?

10 A new branch bank is to have a floor area of 3500 ft^2. It is to be a rectangle with three solid brick walls and a decorative glass front. The glass costs 1.8 times as much as the brick wall per linear foot. What dimensions of the building will minimize the cost of materials for the walls and front?

11 At noon a ship A is 50 mi north of a ship B and is steaming south at 16 mi/h. Ship B is headed west at 12 mi/h. At what time are they closest together, and what is the minimal distance between them?

12 Express the number 8 as the sum of two nonnegative numbers in such a way that the sum of the square of the first and the cube of the second is as small as possible. Also solve the problem if this sum is to be as large as possible.

13 Find two positive numbers whose product is 16 and whose sum is as small as possible.

14 A triangle of base b and height h has acute base angles. A rectangle is inscribed in the triangle with one side on the base of the triangle. Show that the largest such rectangle has base $b/2$ and height $h/2$, so that its area is one-half the area of the triangle.

15 Find the area of the largest rectangle with lower base on the x-axis and upper vertices on the parabola $y = 27 - x^2$.

16 An isosceles triangle has its vertex at the origin, its base parallel to and above the x-axis, and the vertices of its base on the parabola $9y = 27 - x^2$. Find the area of the largest such triangle.

17 If a rectangle has an area of 32 in^2, what are its dimensions if the distance from one corner to the midpoint of a nonadjacent side is as small as possible?

18 If the cost per hour of running a small riverboat is proportional to the cube of its speed through the water, find the speed at which it should be run against a current of a miles per hour to minimize the cost of an upstream journey over a distance of b miles.

19 A Norman window has the shape of a rectangle surmounted by a semicircle. If the total perimeter is fixed, find the proportions of the window (i.e., the ratio of the height of the window to its base) that will admit the most light.

20 Solve the Norman window problem in Problem 19 if the semicircular part is made of stained glass that transmits only half as much light per unit area as does the clear glass in the rectangular part.

21 A trough is to be made from three planks, each 12 in wide. If the cross section has the shape of a trapezoid, how far apart should the tops of the sides be placed to give the trough maximum carrying capacity?

22 Solve Problem 21 if there is one 12-in plank and two 6-in planks.

23 The strength of a rectangular beam is jointly proportional to its width and the cube of its depth.† Find the proportions (ratio of depth to width) of the strongest beam that can be cut from a given cylindrical log.

---

* This was the earliest maximum-minimum problem solved by the methods of calculus (by Fermat, about 1629).

† This means that if x is the width and y is the depth, then the strength S is given by the formula $S = cxy^3$, where c is a constant of proportionality.

24 Among all isosceles triangles with fixed perimeter, show that the triangle of greatest area is equilateral.

25 An isosceles triangle is inscribed in the circle $x^2 + y^2 = a^2$ with its base parallel to the x-axis and one vertex at the point $(0, a)$. Find the height of the triangle with maximum area and show that this triangle is equilateral. (Can you show by geometric reasoning alone that the largest triangle inscribed in the circle is necessarily equilateral?)

26 A wire of length L is to be cut into two pieces, one bent to form a square and the other to form an equilateral triangle. How should the wire be cut if the sum of the areas enclosed by the two pieces is to be (a) a maximum? (b) a minimum? Show that case (b) occurs when the side of the square is $\frac{2}{3}$ the height of the triangle.

*27 A man 6 ft tall wants to construct a greenhouse of length L and width 18 ft against the outer wall of his house by building a sloping glass roof of slant height y from the ground to the wall, as shown in Fig. 4.19. He considers space in the greenhouse to be *usable* if he can stand upright without bumping his head. If the cost of building the roof is proportional to y, find the slope of the roof that minimizes the cost per square foot of usable space. Hint: Notice that this amounts to minimizing y/x.

*28 A fence a feet high is b feet from a wall. Find the length of the shortest ladder that will reach from the ground across the top of the fence to the wall.

*29 A corridor of width a is at right angles to a second corridor of width b. A long, thin, heavy rod is to be pushed along the floor from the first corridor into the second. What is the length of the longest rod that can get around the corner?

*30 A long sheet of paper is a units wide. One corner of the paper is folded over as shown in Fig. 4.20. Find the value of x that minimizes (a) the area of the triangle ABC; (b) the length of the crease AC.

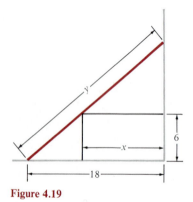

Figure 4.19

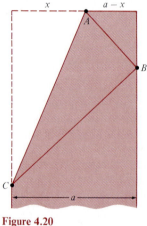

Figure 4.20

4.4

MORE MAXIMUM-MINIMUM PROBLEMS. REFLECTION AND REFRACTION

We continue to develop the basic ideas of Section 4.3 by means of additional examples.

Example 1 A manufacturer of cylindrical soup cans receives a very large order for cans of a specified volume V_0. What dimensions will minimize the total surface area of such a can, and therefore the amount of metal needed to manufacture it?

Solution If r and h are the radius of the base and the height of a cylindrical can (Fig. 4.21, left), then the volume is

$$V_0 = \pi r^2 h \tag{1}$$

and the total surface area is

$$A = 2\pi r^2 + 2\pi r h. \tag{2}$$

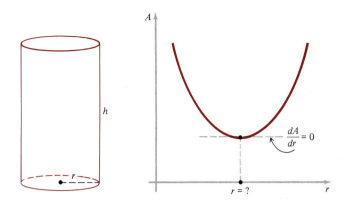

Figure 4.21

We must minimize A, which is a function of two variables, by using the fact that equation (1) relates these variables to one another. We therefore solve (1) for h, $h = V_0/\pi r^2$, and use this to express A as a function of r alone,

$$A = 2\pi r^2 + 2\pi r \cdot \frac{V_0}{\pi r^2}$$

$$= 2\pi r^2 + \frac{2V_0}{r}. \tag{3}$$

The graph of this function (Fig. 4.21, right) shows that A is large when r is small and also when r is large, with a minimum somewhere in between. As usual, to discover the precise location of this minimum, we differentiate (3), equate the derivative to zero, and solve,

$$\frac{dA}{dr} = 4\pi r - \frac{2V_0}{r^2}, \qquad 4\pi r - \frac{2V_0}{r^2} = 0, \qquad 4\pi r^3 = 2V_0,$$

$$2\pi r^3 = V_0. \tag{4}$$

If we want the actual dimensions of the most efficient can, we can solve equation (4) for r and then use this to calculate h,

$$r = \sqrt[3]{\frac{V_0}{2\pi}}, \qquad h = \frac{V_0}{\pi r^2} = \frac{V_0}{\pi}\left(\frac{2\pi}{V_0}\right)^{2/3} = 2\sqrt[3]{\frac{V_0}{2\pi}},$$

from which we observe that $h = 2r$. Or, if we are interested primarily in the shape, we can replace V_0 in (4) by $\pi r^2 h$ and immediately obtain

$$2\pi r^3 = \pi r^2 h \qquad \text{or} \qquad 2r = h.$$

From the point of view of lowering costs for raw materials, this remarkable result tells us that the "best" shape for a cylindrical can is that in which the height equals the diameter of the base.

Example 2   Find the ratio of the height to the diameter of the base for the cylinder of maximum volume that can be inscribed in a sphere of radius R.

Solution   If we sketch a cylinder inscribed in the sphere and label it as shown

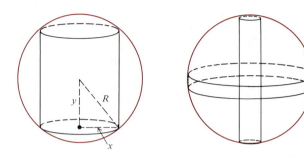

Figure 4.22

on the left in Fig. 4.22, then we see that

$$V = 2\pi x^2 y \tag{5}$$

where

$$x^2 + y^2 = R^2. \tag{6}$$

Visualizing the extreme cases (Fig. 4.22, right) tells us that V is small when x is near zero and also when x is near R, and so between these extremes there is a shape of maximum volume. To find it, we use (6) to write (5) as

$$V = 2\pi y(R^2 - y^2) = 2\pi(R^2 y - y^3),$$

from which we obtain

$$\frac{dV}{dy} = 2\pi(R^2 - 3y^2).$$

Setting this equal to zero to find y and then using (6) to find x gives

$$y = \frac{R}{\sqrt{3}} \quad \text{and} \quad x = \sqrt{R^2 - \tfrac{1}{3}R^2} = \frac{\sqrt{2}}{\sqrt{3}}\, R.$$

The ratio of the height to the diameter of the base for the largest cylinder is therefore

$$\frac{2y}{2x} = \frac{y}{x} = \frac{1}{\sqrt{2}} = \frac{1}{2}\sqrt{2}.$$

This result can be obtained more efficiently by the method of implicit differentiation. If x is taken as the independent variable and y is thought of as a function of x, then (6) yields

$$2x + 2y\frac{dy}{dx} = 0 \quad \text{or} \quad \frac{dy}{dx} = -\frac{x}{y}.$$

From (5) we find that

$$\frac{dV}{dx} = 2\pi\left(x^2\frac{dy}{dx} + 2xy\right) = 2\pi\left[x^2\left(-\frac{x}{y}\right) + 2xy\right]$$

$$= 2\pi\left(\frac{-x^3 + 2xy^2}{y}\right) = \frac{2\pi x}{y}(2y^2 - x^2).$$

It therefore follows that $dV/dx = 0$ when

$$2y^2 = x^2 \quad \text{or} \quad \frac{y}{x} = \frac{1}{\sqrt{2}} = \frac{1}{2}\sqrt{2}.$$

Example 3   If a ray of light travels from a point A to a point P on a flat mirror, and is then reflected to a point B, as shown in Fig. 4.23, then the most careful measurements show that the incident ray and the reflected ray make equal angles with the mirror: $\alpha = \beta$. Assume that the ray of light takes the shortest path from A to B by way of the mirror, and prove this law of reflection by showing that the path APB is shortest when $\alpha = \beta$.

Solution   If we think of the point P as assuming various positions on the mirror, with each position determined by a value of x, then we wish to consider the length L of the path as a function of x. From Fig. 4.23 this function is clearly

$$L = \sqrt{a^2 + x^2} + \sqrt{b^2 + (c - x)^2}$$
$$= (a^2 + x^2)^{1/2} + [b^2 + (c - x)^2]^{1/2},$$

and differentiation yields

$$\frac{dL}{dx} = \tfrac{1}{2}(a^2 + x^2)^{-1/2} \cdot (2x) + \tfrac{1}{2}[b^2 + (c - x)^2]^{-1/2} \cdot 2(c - x) \cdot (-1)$$

$$= \frac{x}{\sqrt{a^2 + x^2}} - \frac{c - x}{\sqrt{b^2 + (c - x)^2}}. \qquad (7)$$

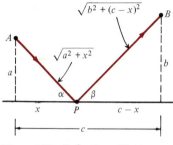

Figure 4.23   Reflection of light.

If we minimize L by equating this derivative to zero, we get

$$\frac{x}{\sqrt{a^2 + x^2}} = \frac{c - x}{\sqrt{b^2 + (c - x)^2}}, \qquad (8)$$

and this equation can be changed in form as follows:

$$\frac{\sqrt{a^2 + x^2}}{x} = \frac{\sqrt{b^2 + (c - x)^2}}{c - x}, \quad \sqrt{\left(\frac{a}{x}\right)^2 + 1} = \sqrt{\left(\frac{b}{c - x}\right)^2 + 1},$$

$$\frac{a}{x} = \frac{b}{c - x}.$$

The equation last written can easily be solved for x. However, there is no need to do this, because the equation as it stands tells us what we want to know: for the angles α and β in the two right triangles shown in the figure, the ratios of the opposite side to the adjacent side are equal, so α and β are equal.

It is fairly clear on intuitive grounds that we have minimized L. If we wish to verify this by the second derivative test, we use (7) to compute

$$\frac{d^2L}{dx^2} = \frac{a^2}{(a^2 + x^2)^{3/2}} + \frac{b^2}{[b^2 + (c - x)^2]^{3/2}}$$

(we omit the details of the computation), and all that remains is to notice that this quantity is positive.

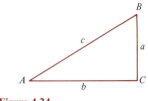

Figure 4.24

Remark 1 The reasoning in Example 3 can be made simpler if we recall from trigonometry the definition of the cosine of a positive acute angle A. If we think of A as one of the acute angles of a right triangle (Fig. 4.24), then by definition

$$\cos A = \frac{b}{c} = \frac{\text{adjacent side}}{\text{hypotenuse}}.$$

Using this, the minimizing condition (8) can be written as

$$\cos \alpha = \cos \beta,$$

so $\alpha = \beta$. For use in the next example, we also recall the definition of the sine of A,

$$\sin A = \frac{a}{c} = \frac{\text{opposite side}}{\text{hypotenuse}}.$$

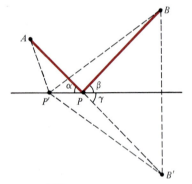

Figure 4.25

Remark 2 The law of reflection discussed in Example 3 was known to the ancient Greeks. However, the fact that a reflected ray of light follows the shortest path was discovered much later, by Heron of Alexandria in the first century A.D. Heron's geometric proof is simple but ingenious. The argument goes as follows. Let A and B be the same points as before (Fig. 4.25), and let B' be the mirror image of B, so that the surface of the mirror is the perpendicular bisector of BB'. The segment AB' intersects the mirror at a point P, and this is the point where a ray of light is reflected in passing from A to B; for $\alpha = \gamma$ and $\gamma = \beta$, so $\alpha = \beta$. The total length of the path is $AP + PB = AP + PB' = AB'$. For any other point P' on the mirror the total length of the path is $AP' + P'B = AP' + P'B'$, and this is greater than AB' because the sum of two sides of a triangle is greater than the third side. This shows that the actual path of our reflected ray of light is the shortest possible path from A to B by way of the mirror.

Example 4 The reflected ray of light previously discussed travels in a single medium at a constant speed. However, in different media (air, water, glass) light travels at different speeds. If a ray of light passes from air into water as shown in Fig. 4.26, it is refracted (bent) toward the perpendicular at the interface. The path APB is clearly no longer the shortest path from A to B. What law determines it? In 1621 the Dutch scientist Snell discovered empirically that the actual path of the ray is that for which

$$\frac{\sin \alpha}{\sin \beta} = \text{a constant}, \tag{9}$$

where this constant is independent of the positions of A and B. This fact is now called *Snell's law of refraction*. Prove Snell's law by assuming that the ray takes the path from A to B that minimizes the total time of travel.

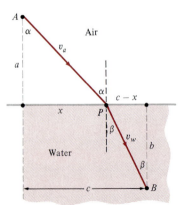

Figure 4.26 Refraction of light.

Solution If the speed of light in air is v_a and in water is v_w, then the total time of travel T is the time in air plus the time in water,

$$T = \frac{\sqrt{a^2 + x^2}}{v_a} + \frac{\sqrt{b^2 + (c - x)^2}}{v_w}$$

$$= \frac{1}{v_a}(a^2 + x^2)^{1/2} + \frac{1}{v_w}[b^2 + (c - x)^2]^{1/2}.$$

If we compute the derivative of this function and notice its meaning in terms of Fig. 4.26, we obtain

$$\frac{dT}{dx} = \frac{1}{v_a}\frac{x}{\sqrt{a^2 + x^2}} - \frac{1}{v_w}\frac{c - x}{\sqrt{b^2 + (c - x)^2}}$$

$$= \frac{\sin \alpha}{v_a} - \frac{\sin \beta}{v_w}. \tag{10}$$

If we now minimize T by equating this to zero, the result is

$$\frac{\sin \alpha}{v_a} = \frac{\sin \beta}{v_w} \qquad \text{or} \qquad \frac{\sin \alpha}{\sin \beta} = \frac{v_a}{v_w}. \tag{11}$$

This is a more revealing form of Snell's law, because it tells us the physical meaning of the constant on the right side of (9): It is the ratio of the speed of light in air to the (smaller) speed of light in water. This constant is called the *index of refraction* of water. If the water in this experiment is replaced by any other translucent medium, such as alcohol, glycerin, or glass, then this constant has a different numerical value, which is the index of refraction of the medium in question.

As in Example 3, we can verify that the configuration (11) actually minimizes T by computing the second derivative and noting that this quantity is positive:

$$\frac{d^2T}{dx^2} = \frac{1}{v_a}\frac{a^2}{(a^2 + x^2)^{3/2}} + \frac{1}{v_w}\frac{b^2}{[b^2 + (c - x)^2]^{3/2}} > 0.$$

But there is another method that is worth mentioning. We begin by observing that dT/dx as given by (10) is a difference of two terms. As x increases from 0 to c, the first term, $(\sin \alpha)/v_a$, increases from 0 to some positive value. The second term, $(\sin \beta)/v_w$, decreases from some positive value to 0. This shows that dT/dx is negative at $x = 0$ and increases to a positive value at $x = c$. The minimum value of T therefore occurs at the only x for which $dT/dx = 0$, and this is precisely the configuration described by (11).

Remark 3   The ideas of Example 4 were discovered in 1657 by the great French mathematician Fermat, and for this reason the statement that a ray of light traverses an optical system along the path that minimizes its total time of travel is called *Fermat's principle of least time*. (It should be noticed that when a ray of light travels in a single uniform medium, "shortest path" is equivalent to "least time," so Example 3 falls under the same principle.) During the next two centuries Fermat's ideas stimulated a broad development of the general theory of maxima and minima, leading first to Euler's creation of the calculus of variations and then to Hamilton's principle of least action, which has turned out to be one of the deepest unifying principles of physical science. Euler expressed his enthusiasm in the following memorable words: "Since the fabric of the world is the most perfect and was established by the wisest Creator, nothing happens in this world in which some reason of maximum or minimum would not come to light."

PROBLEMS

1 A closed rectangular box with a square base is to be made out of plywood. If the volume is given, find the shape (ratio of height to side of base) that minimizes the total number of square feet of plywood that are needed.

2 Solve Problem 1 if the box is open on top.

3 Find the radius of the cylinder of maximum volume that can be inscribed in a cone of height H and radius of base R.

4 Find the height of the cone of maximum volume that can be inscribed in a sphere of radius R.

5 A square piece of tin 24 in on each side is to be made into an open-top box by cutting a small square from each corner and bending up the flaps to form the sides. How large a square should be cut from each corner to make the volume of the box as large as possible?

6 Solve Problem 5 if the given piece of tin is a rectangle 15 by 24 in.

7 A cylindrical can without a top is to be made from a specified weight of sheet metal. Find the ratio of the height to the diameter of the base when the volume of the can is greatest.

8 A cylindrical tank without a top is to have a specified volume. If the cost of the material used for the bottom is three times the cost of that used for the curved lateral part, find the ratio of the height to the diameter of the base for which the total cost is least.

9 Draw a reasonably good sketch of $y = \sqrt{x}$ and mark the point on this graph that seems to be closest to the point $(\frac{3}{2}, 0)$. Then calculate the coordinates of this closest point. Hint: Minimize the square of the distance from the point $(\frac{3}{2}, 0)$ to the point $(x, \sqrt{x})$.

10 Generalize Problem 9 by finding the point on the graph of $y = \sqrt{x}$ that is closest to the point $(a, 0)$ for any $a > 0$.

11 A spy climbs out of a submarine into a rubber boat 2 mi east of a point P on a straight north-south shoreline. He wants to get to a house on the shore 6 mi north of P. He can row 3 mi/h and walk 5 mi/h, and he intends to row directly to a point somewhere north of P and then walk the rest of the way.
  (a) How far north of P should he land in order to get to the house in the shortest possible time?
  (b) How long does the trip take?
  (c) How much longer will it take if he rows directly to P and then walks to the house?

12 Show that the answer to part (a) of Problem 11 does not change if the house is 8 mi north of P.

13 If the rubber boat in Problem 11 has a small outboard motor and can go 5 mi/h, then it is obvious by com-mon sense that the fastest route is entirely by boat. What is the slowest speed for which the fastest route is still entirely by boat?

*14 The intensity of illumination at a point P due to a light source is directly proportional to the strength of the source and inversely proportional to the square of the distance from P to the source. Two light sources of strengths a and b are a distance L apart. What point on the line segment joining these sources receives the least total illumination? If a is 8 times as large as b, where is this point? (Assume that the intensity at any point is the sum of the intensities from the two sources.)

*15 Two towns, A and B, lie on the same side of a straight highway. Their distance apart is c, and their distances from the highway are a and b. Show that the length of the shortest road that goes from A to the highway and then on to B is $\sqrt{c^2 + 4ab}$
  (a) by using calculus;
  (b) without calculus (hint: introduce the "mirror image" of B on the other side of the highway).

16 Find the minimum vertical distance between the curves $y = 16x^2$ and $y = -1/x^2$.

17 An isosceles triangle is circumscribed about a circle of radius R. If x is the height of the triangle, show that its area A is least when $x = 3R$. Hint: Minimize A^2.

18 If the figure in Problem 17 is revolved about the altitude of the triangle, the result is a cone circumscribed about a sphere of radius R. Show that the volume of the cone is least when $x = 4R$, and that this least volume is twice the volume of the sphere.

19 A silo has cylindrical walls, a flat circular floor, and a hemispherical top. For a given volume, find the ratio of the total height to the diameter of the base that minimizes the total surface area.

20 In Problem 19, if the cost of construction per square foot is twice as great for the hemispherical top as for the walls and the floor, find the ratio of the total height to the diameter of the base that minimizes the total cost of construction.

21 What is the smallest value of the constant a for which the inequality $ax + 1/x \geq 2\sqrt{2}$ is valid for all positive numbers x?

*22 There is a refinery at a point A on a straight highway and an oil well at a point B which can be reached by traveling 5 mi along the highway to a point C and then 12 mi across country perpendicular to the highway. If a pipeline is built from A to B, it costs k times as much per mile to build it across country as along the highway, because of the difficult terrain. The line will be built either directly from A to B or along the highway to a point P part of the way toward C and then across country to B, whichever is cheaper. Decide on the cheapest route (a) if $k = 3$; (b) if $k = 2$. (c) What is the largest value of k for which it is cheapest to build the pipeline directly from A to B?

23 A circular ring of radius a is uniformly charged with electricity, the total charge being Q. The force exerted by this charge on a unit charge located at a distance x from the center of the ring, in a direction perpendicular to the plane of the ring, is given by $F = Qx(x^2 + a^2)^{-3/2}$. Sketch the graph of this function and find the value of x that maximizes F.

24 A cylindrical hole of radius x is bored through a sphere of radius R in such a way that the axis of the hole passes through the center of the sphere. Find the value of x that maximizes the complete surface area of the remaining solid. Hint: The area of a segment of height h on a sphere of radius R is $2\pi Rh$.

*25 The sum of the surface areas of a cube and a sphere is given. What is the ratio of the edge of the cube to the diameter of the sphere when (a) the sum of their volumes is a maximum? (b) the sum of their volumes is a minimum?

*26 Consider two spheres of radii 1 and 2 whose centers are 6 units apart. At what point on the line joining their centers will an observer be able to see the most total surface area? (See the hint for Problem 24.)

*27 Find the point on the parabola $y = x^2$ that is closest to the point (6, 3).

4.5
RELATED RATES

If a tank is being filled with water, then the water level rises. To describe how rapidly the water level rises, we speak of the rate of change of the water level or, equivalently, the rate of change of the depth. If the depth is denoted by h, and t is the time measured from some convenient moment, then the derivative dh/dt is the rate of change of the depth. Further, the volume V of water in the tank is also changing, and dV/dt is the rate of change of this volume.

Similarly, any geometric or physical quantity Q that changes with time is *a function of time*, say $Q = Q(t)$, and its derivative dQ/dt is the *rate of change of the quantity*. The problems that we now consider are based on the fact that if two changing quantities are related to one another, then their rates of change are also related.

Example 1 Gas is being pumped into a large spherical rubber balloon at the constant rate of 8 ft³/min. Find how fast the radius r of the balloon is increasing (a) when $r = 2$ ft; (b) when $r = 4$ ft.

Solution The volume of the balloon (Fig. 4.27) is given by the formula

$$V = \tfrac{4}{3}\pi r^3. \qquad (1)$$

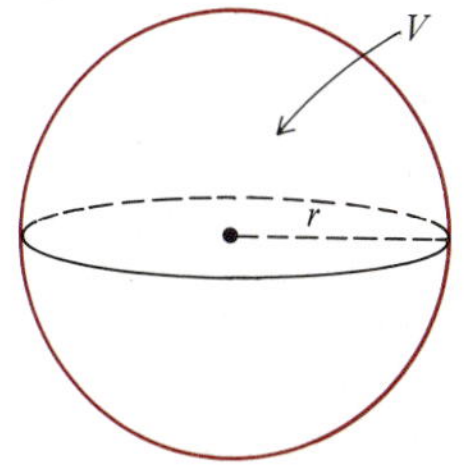

We are told that $dV/dt = 8$, and we are asked to find dr/dt for two specific values of r. It is essential to understand the background of this situation, namely, the fact that V and r are both dependent variables with the time t as the underlying independent variable. With this in mind, it is natural to introduce the rates of change of V and r by differentiating (1) with respect to t,

$$\frac{dV}{dt} = \tfrac{4}{3}\pi \cdot 3r^2 \frac{dr}{dt} = 4\pi r^2 \frac{dr}{dt}, \qquad (2)$$

where the chain rule is needed in the calculation. It follows from (2) that

$$\frac{dr}{dt} = \frac{1}{4\pi r^2} \frac{dV}{dt} = \frac{2}{\pi r^2},$$

since $dV/dt = 8$. In case (a) we therefore have

Figure 4.27

$$\frac{dr}{dt} = \frac{1}{2\pi} \cong 0.16 \text{ ft/min},$$

and in case (b),

$$\frac{dr}{dt} = \frac{1}{8\pi} \cong 0.04 \text{ ft/min}.$$

These conclusions confirm our common-sense awareness that since the volume of the balloon is increasing at a constant rate, the radius increases more and more slowly as the volume grows larger.

Example 2 A ladder 13 ft long is leaning against a wall. The bottom of the ladder is being pulled away from the wall at the constant rate of 6 ft/min. How fast is the top of the ladder moving down the wall when the bottom of the ladder is 5 ft from the wall?

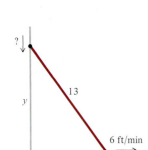

y

6 ft/min

x

Figure 4.28

Solution The first thing we do is draw a diagram of the situation and label it, being careful to use letters to represent quantities that are changing (Fig. 4.28). In terms of this figure, we can clarify our thoughts by stating what is known and what we are trying to find:

$$\frac{dx}{dt} = 6, \qquad -\frac{dy}{dt} = ? \text{ when } x = 5.$$

(The use of the minus sign here can best be understood by thinking of dy/dt as the rate at which y is increasing and $-dy/dt$ as the rate at which y is decreasing. The problem asks for the latter.) Roughly speaking, we know one time derivative and we want to find the other. We therefore seek an equation connecting x and y from which we can obtain a second equation connecting their rates of change. It is clear from the figure that our starting point must be the fact that

$$x^2 + y^2 = 169. \tag{3}$$

When this is differentiated with respect to t, we get

$$2x\frac{dx}{dt} + 2y\frac{dy}{dt} = 0 \quad \text{or} \quad \frac{dy}{dt} = -\frac{x}{y}\frac{dx}{dt} \quad \text{or} \quad -\frac{dy}{dt} = \frac{x}{y}\frac{dx}{dt},$$

and therefore

$$-\frac{dy}{dt} = \frac{6x}{y}, \tag{4}$$

since $dx/dt = 6$. Finally, equation (3) tells us that $y = 12$ when $x = 5$, so (4) yields our conclusion,

$$-\frac{dy}{dt} = \frac{6 \cdot 5}{12} = 2\tfrac{1}{2} \text{ ft/min when } x = 5.$$

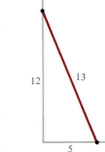

12

13

5

Figure 4.29

Warning: Don't substitute the values $x = 5$, $y = 12$ prematurely. The essence of the problem is the fact that x and y are variables; if we pin them down to specific values too soon, as is done in Fig. 4.29, then this makes it impossible to understand or solve the problem. In other words, preserve the fluidity of the situation until the last possible moment.

Example 3 A conical tank with its vertex down is 12 ft high and 12 ft in diameter at the top. Water is being pumped in at the rate of 4 ft³/min. Find the rate at which the water level is rising (a) when the water is 2 ft deep, and (b) when the water is 8 ft deep.

Solution As before, we begin by drawing and labeling a diagram (Fig. 4.30), with the purpose of visualizing the situation and establishing notation. Our next step is to use this notation to state as follows what is given and what we are trying to find:

$$\frac{dV}{dt} = 4, \qquad \frac{dx}{dt} = ? \text{ when } x = 2 \text{ and } x = 8.$$

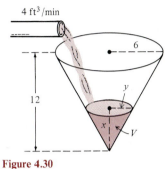

4 ft³/min

Figure 4.30

The changing volume of water in the tank has the shape of a cone, so our starting point is the formula

$$V = \tfrac{1}{3}\pi y^2 x. \tag{5}$$

The only dependent variables we care about are V and x, so we wish to eliminate the superfluous variable y. By examining Fig. 4.30 and using similar triangles, we see that

$$\frac{y}{x} = \frac{6}{12} = \frac{1}{2} \qquad \text{or} \qquad y = \tfrac{1}{2}x, \tag{6}$$

and substituting this in (5) gives

$$V = \frac{\pi}{12} x^3. \tag{7}$$

We are now in a position to introduce the rates of change by differentiating (7) with respect to t, which yields

$$\frac{dV}{dt} = \frac{\pi}{4} x^2 \frac{dx}{dt} \tag{8}$$

or

$$\frac{dx}{dt} = \frac{4}{\pi x^2} \frac{dV}{dt} = \frac{16}{\pi x^2},$$

since $dV/dt = 4$. This formula tells us that when $x = 2$,

$$\frac{dx}{dt} = \frac{4}{\pi} \cong 1.27 \text{ ft/min},$$

and when $x = 8$,

$$\frac{dx}{dt} = \frac{1}{4\pi} \cong 0.08 \text{ ft/min},$$

and the solution is complete.

It is worth noticing that the explicit use of formula (5) can be avoided by using instead the fact that

$$\frac{dV}{dt} = \pi y^2 \frac{dx}{dt}. \tag{9}$$

This says that the rate of change of the volume equals the surface area times the rate of change of the depth, and in this form the statement is true regardless of the shape of the tank. If we now substitute (6) into (9) we get (8), and from here we continue as before.

We summarize the lessons of these examples: In solving a problem involving related rates of change, it is usually a good idea to begin by drawing a careful sketch of the situation being considered. Next, add to the sketch all numerical quantities that remain fixed throughout the problem. Then, denote by letters any quantities — the dependent variables — that change with time, and seek a geometric or physical relation among these variables. Finally, differentiate with respect to the time t to obtain a relation among the various rates of change, and use this to determine the unknown rate the problem asks for.

PROBLEMS

1. A stone dropped into a pond sends out a series of concentric ripples. If the radius r of the outer ripple increases steadily at the rate of 6 ft/s, find the rate at which the area of disturbed water is increasing (a) when $r = 10$ ft, and (b) when $r = 20$ ft.

2. A large spherical snowball is melting at the rate of 2π ft^3/h. At the moment when it is 30 inches in diameter, determine (a) how fast the radius is changing, and (b) how fast the surface area is changing.

3. Sand is being poured onto a conical pile at the constant rate of 50 ft^3/min. Frictional forces in the sand are such that the height of the pile is always equal to the radius of its base. How fast is the height of the pile increasing when the sand is 5 ft deep?

4. A girl 5 ft tall is running at the rate of 12 ft/s and passes under a street light 20 ft above the ground. Find how rapidly the tip of her shadow is moving when she is (a) 20 ft past the street light, and (b) 50 ft past the street light.

5. In Problem 4, find how rapidly the length of the girl's shadow is increasing at each of the stated moments.

6. A light is at the top of a pole 80 ft high. A ball is dropped from the same height from a point 20 ft away from the light. Find how fast the shadow of the ball is moving along the ground (a) 1 second later; (b) 2 seconds later. (Assume that the ball falls $s = 16t^2$ feet in t seconds.)

7. A woman raises a bucket of cement to a platform 40 ft above her head by means of a rope 80 ft long that passes over a pulley on the platform. If she holds her end of the rope firmly at head level and walks away at 5 ft/s, how fast is the bucket rising when she is 30 ft away from the spot directly below the pulley?

8. A boy is flying a kite at a height of 80 ft, and the wind is blowing the kite horizontally away from the boy at the rate of 20 ft/s. How fast is the boy paying out string when the kite is 100 ft away from him?

9. A boat is being pulled in to a dock by means of a rope with one end tied to the bow of the boat and the other end passing through a ring attached to the dock at a point 5 ft higher than the bow of the boat. If the rope is being pulled in at the rate of 4 ft/s, how fast is the boat moving through the water when 13 ft of rope are out?

10. A trough is 10 ft long and has a cross section in the shape of an equilateral triangle 2 ft on each side. If water is being pumped in at the rate of 20 ft^3/min, how fast is the water level rising when the water is 1 ft deep?

11. If a mothball evaporates at a rate proportional to its surface area, show that its radius decreases at a constant rate.

12. A point moves around the circle $x^2 + y^2 = a^2$ in such a way that the x-component of its velocity is given by $dx/dt = -y$. Find dy/dt and decide whether the direction of the motion is clockwise or counterclockwise.

13. A car moving at 60 mi/h along a straight road passes

under a weather balloon rising vertically at 20 mi/h. If the balloon is 1 mi up when the car is directly beneath it, how fast is the distance between the car and the balloon increasing 1 minute later?

14 Most gases obey Boyle's law: If a sample of the gas is held at a constant temperature while being compressed by a piston in a cylinder, then its pressure p and volume V are related by the equation $pV = c$, where c is a constant. Find dp/dt in terms of p and dV/dt.

15 At a certain moment a sample of gas obeying Boyle's law occupies a volume of 1000 in³ at a pressure of 10 lb/in². If this gas is being compressed isothermally at the rate of 12 in³/min, find the rate at which the pressure is increasing at the instant when the volume is 600 in³.

*16 A ladder 20 ft long is leaning against a wall 12 ft high, with its top projecting over the wall. Its bottom is being pulled away from the wall at the constant rate of 5 ft/min. Find how rapidly the top of the ladder is approaching the ground (a) when 5 ft of the ladder projects over the wall; (b) when the top of the ladder reaches the top of the wall.

17 A conical party hat made of cardboard has a radius of 4 in and a height of 12 in. When filled with beer, it leaks at the rate of 4 in³/min. At what rate is the level

of beer falling (a) when the beer is 6 in deep? (b) when the hat is half empty?

18 A hemispherical bowl of radius 8 in is being filled with water at a constant rate. If the water level is rising at the rate of $\frac{1}{3}$ in/s at the instant when the water is 6 in deep, find how fast the water is flowing in

(a) by using the fact that a segment of a sphere has volume

$$V = \pi h^2 \left(a - \frac{h}{3} \right)$$

where a is the radius of the sphere and h is the height of the segment;

(b) by using the fact that if V is the volume of the water at time t, then

$$\frac{dV}{dt} = \pi r^2 \frac{dh}{dt}$$

where r is the radius of the surface and h is the depth.

19 Water is being poured into a hemispherical bowl of radius 3 in at the rate of 1 in³/s. How fast is the water level rising when the water is 1 in deep?

*20 In Problem 19, suppose that the bowl contains a lead ball 2 inches in diameter, and find how fast the water level is rising when the ball is half submerged.

Consider the cubic equation

$$x^3 - 3x - 5 = 0. \qquad (1)$$

4.6

(OPTIONAL) NEWTON'S METHOD FOR SOLVING EQUATIONS

It is possible to solve this equation by exact methods, that is, by formulas yielding a solution in terms of radicals in the same sense that the formula

$$x = \frac{-b \pm \sqrt{b^2 - 4ac}}{2a}$$

provides an exact solution of the quadratic equation $ax^2 + bx + c = 0$. However, if we need a numerical solution of (1) that is accurate only to a few decimal places, then it is more convenient to find this solution by the approximation method to be described here than to try to use the exact solution. Furthermore, while formulas that yield exact solutions in terms of radicals for equations of degree 2, 3, and 4 do exist, it is known to be impossible to solve the general equation of degree 5 or more in terms of radicals. Therefore, in order to solve a fifth-degree equation like $x^5 - 3x^2 + 9x - 11 = 0$, we would be forced to use an approximation method, since no other method is available.

Returning to equation (1), if we denote $x^3 - 3x - 5$ by $f(x)$, then we can easily calculate the following values:

$$f(-2) = -7, \quad f(-1) = -3, \quad f(0) = -5, \quad f(1) = -7, \quad f(2) = -3, \quad f(3) = 13.$$

The pair of values $f(2) = -3$ and $f(3) = 13$ suggests that as x varies continuously from $x = 2$ to $x = 3$, $f(x)$ varies continuously from -3 to 13, and that consequently there is some intermediate value of x where $f(x) = 0$. This is true, but even though it is intuitively obvious, it is quite difficult to give a rigorous proof. We do not attempt such a proof here, but instead take it for granted that if a continuous function $f(x)$ has values $f(a)$ and $f(b)$ with opposite signs, then there is at least one root of the equation $f(x) = 0$ between a and b. This tells us that (1) has a root between $x = 2$ and $x = 3$, and we can take either of these numbers as a first approximation to this root. The approximation $x = 2$ is the better choice, since -3 is closer to 0 than 13 is.

In general, suppose we have a first approximation $x = x_1$ to a root r of an equation $f(x) = 0$. This root is a point where the curve $y = f(x)$ crosses the x-axis, as shown in Fig. 4.31; and the idea of Newton's method is to use the tangent line to the curve at the point where $x = x_1$ as a stepping-stone to a better approximation $x = x_2$. Beginning with the approximation $x = x_1$, we draw the tangent line to the curve at the point $(x_1, f(x_1))$. This line intersects the x-axis at the point $x = x_2$, which in general is a better approximation than x_1. Repeating the process, we use the tangent line at $(x_2, f(x_2))$ to get to the point $x = x_3$, which is a still better approximation. Figure 4.31 illustrates the idea as a geometric procedure, but to apply it in calculations we need a formula. This formula is easily derived as follows.

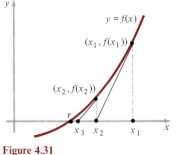

Figure 4.31

The slope of the first tangent line is $f'(x_1)$. If we consider this line to be determined by the points $(x_2, 0)$ and $(x_1, f(x_1))$, then the slope is also

$$\frac{0 - f(x_1)}{x_2 - x_1}, \qquad \text{so} \qquad \frac{0 - f(x_1)}{x_2 - x_1} = f'(x_1).$$

This equation yields

$$-f(x_1) = (x_2 - x_1)f'(x_1) \qquad \text{or} \qquad x_2 - x_1 = -\frac{f(x_1)}{f'(x_1)},$$

so

$$x_2 = x_1 - \frac{f(x_1)}{f'(x_1)}. \tag{2}$$

In this way our first approximation x_1 leads to a second approximation x_2 given by (2); this in turn leads to a third approximation x_3, given by

$$x_3 = x_2 - \frac{f(x_2)}{f'(x_2)};$$

and so on indefinitely.

On applying this method to equation (1), we have

$$f(x) = x^3 - 3x - 5, \qquad f'(x) = 3x^2 - 3, \qquad x_1 = 2,$$

$$f(x_1) = -3, \qquad f'(x_1) = 9, \qquad x_2 = x_1 - \frac{f(x_1)}{f'(x_1)} = 2 - \frac{-3}{9} = 2\tfrac{1}{3}.$$

Using decimals at the next stage, we get

$$x_3 = x_2 - \frac{f(x_2)}{f'(x_2)} \cong 2.33 - \frac{0.66}{13.29} \cong 2.28,$$

rounding off to two decimal places.

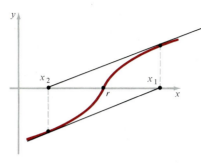

Figure 4.32

In order to keep our computations from becoming too burdensome, we shall be satisfied with two decimal places of accuracy. When two successive approximations are equal in their first two decimal places, we shall consider this as evidence of accuracy. For example, in the case of equation (1) we obtained 2.28 as an approximation to the root after two applications of (2). Another application of (2) leads from 2.28 to the same number, 2.28. We therefore conclude that 2.28 is a root of equation (1) which is accurate to two decimal places.

Newton's method is not restricted to the solution of polynomial equations like (1), but can also be applied to any equation containing functions whose derivatives we can calculate. However, for the sake of simplicity we confine our attention to polynomials in the problems given here.

Remark 1   In some cases, the sequence of approximations produced by Newton's method may fail to converge to the desired root. For example, Fig. 4.32 shows a function for which the approximation x_1 leads to x_2 and x_2 leads back to x_1, so that repetitions of the process do not bring us any closer to the root r than our initial guess. Specific examples of this behavior — and worse — are given in the problems. The mathematical theory providing conditions under which Newton's method is guaranteed to succeed can be found in books on numerical analysis.

Remark 2   The "intuitively obvious" property of continuous functions mentioned in the second paragraph is proved in Appendix C.3.

PROBLEMS

1   By sketching the graph of $y = f(x) = x^3 - 3x - 5$, show that equation (1) has only one real root. Hint: Use the derivative $f'(x) = 3x^2 - 3 = 3(x^2 - 1)$ to locate the maxima and minima of the function and to learn where it is increasing and decreasing.

2   (a) Show that $x^3 + 3x^2 - 6 = 0$ has only one real root, and calculate it to two decimal places of accuracy.
   (b) Show that $x^3 + 3x = 8$ has only one real root, and calculate it to two decimal places of accuracy.

3   Use Newton's method to calculate the positive root of $x^2 + x - 1 = 0$ to two decimal places of accuracy.

4   Calculate $\sqrt{5}$ to two decimal places of accuracy by solving the equation $x^2 - 5 = 0$, and use this result in the quadratic formula to check the answer to Problem 3.

5   Use Newton's method to calculate $\sqrt[3]{10}$ to two decimal places of accuracy.

6   Consider a spherical shell 1 ft thick whose volume equals the volume of the hollow space inside it. Use Newton's method to calculate the shell's outer radius to two decimal places of accuracy.

*7   A hollow spherical buoy of radius 2 ft has specific gravity $\frac{1}{4}$, so that it floats on water in such a way as to displace $\frac{1}{4}$ its own volume. Show that the depth x to which it is submerged is a root of the equation $x^3 - 6x^2 + 8 = 0$, and use Newton's method to calculate this root to two decimal places of accuracy. Hint: The volume of a spherical segment of height h cut from a sphere of radius r is $\pi h^2(r - h/3)$.

8 Suppose that by good luck our first approximation x_1 happens to be the root of $f(x) = 0$ that we are seeking. What does this imply about x_2, x_3, etc.?

9 Show that the function $y = f(x)$ defined by

$$f(x) = \begin{cases} \sqrt{x - r} & x \geq r, \\ -\sqrt{r - x} & x \leq r, \end{cases}$$

has the property illustrated in Fig. 4.32; that is, for any positive number a, if $x_1 = r + a$, then $x_2 = r - a$; and if $x_1 = r - a$, then $x_2 = r + a$.

10 Show that Newton's method applied to the function $y = f(x) = \sqrt[3]{x}$ leads to $x_2 = -2x_1$, and is therefore useless for finding where $f(x) = 0$. Sketch the situation.

11 In Example 1 of Section 4.1 we saw from its graph that the function $y = f(x) = 2x^3 - 3x^2 - 12x + 12$ has positive zeros close to $x = 0.9$ and $x = 2.9$. Use Newton's method to calculate these zeros to two decimal places of accuracy.

4.7

(OPTIONAL) APPLICATIONS TO ECONOMICS AND BUSINESS

Even though calculus is more than three centuries old and its primary uses have always been in the physical sciences, it has continued to find new and fresh applications in other directions. In this section we examine some of the simpler ways in which derivatives can be used in economic theory and business management, where difficult and important decisions must often be made about production levels, costs, prices, inventories, and many other quantities that invite mathematical study.

Perhaps the first function of interest to a manufacturer is the *cost function*, that is, the total cost $C(x)$ of producing x units of a commodity. We might think, for example, of a sugar company spending $C(x)$ dollars to produce x tons of refined sugar out of sugar beets grown by local farmers. Many components make up the total cost. Some, like capital expenditures for buildings and machinery, are fixed and do not depend on x. Others, like wages and the cost of raw materials, are roughly proportional to the amount x produced. If this were all, then the cost function would have the very simple form

$$C(x) = a + bx, \tag{1}$$

where a is the fixed cost and b is the constant running cost per unit.

But this is not all, and most cost functions are not as simple as this. The essential point is the fact that a time restriction is present, and that $C(x)$ is the cost of producing x units of the product *in a given time interval,* say one week. There will then be a fixed cost of a dollars per week, as before, but the variable part of the cost will probably increase more than proportionally to x as the weekly production x increases, because of overtime wages, the need to use older machinery that breaks down more frequently, and other inefficiencies that arise from forcing production to higher and higher levels. The cost function might then have the form

$$C(x) = a + bx + cx^2, \tag{2}$$

or it might be a function even more complicated than this. The general nature of such a cost function is suggested in Fig. 4.33.

A manager who is faced with the decision whether or not to increase production has to know how rapidly the production costs are rising. This is simply the rate of change of C with respect to x, which is the derivative dC/dx. Economists call this derivative the *marginal cost.* This terminology is

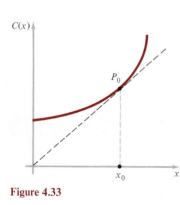

Figure 4.33

(or is thought to be) appropriate for the following reason: The production level x is usually a large number compared with 1, so the change from x to $x + 1$ is a small or "marginal" increase and

$$\frac{dC}{dx} \cong \frac{C(x+1) - C(x)}{1} = C(x+1) - C(x),$$

as shown in Fig. 4.34. Thus, the marginal cost dC/dx can be thought of as "the extra cost of producing one more unit," and this is the definition of marginal cost adopted by those economists who prefer not to use calculus. The significance of this idea can be seen in the upper part of Fig. 4.33, where the increasing steepness of the curve reflects the fact that the marginal cost rises to very high values as the manufacturer strains productive capacity.

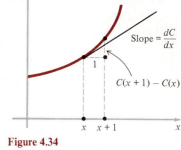

Figure 4.34

The following remarks illustrate the way the concept of marginal cost can be used in economic theory.

It is a reasonable view that the most efficient production level for a manufacturer is that which minimizes the average cost $C(x)/x$. At a point where this average cost has a minimum value its derivative must vanish, so by the quotient rule we have

$$\frac{xC'(x) - C(x)}{x^2} = 0$$

or

$$C'(x) = \frac{C(x)}{x}. \tag{3}$$

This conclusion can be stated in words as follows: *At the peak of operating efficiency, the marginal cost equals the average cost.* Equation (3) has an interesting geometric interpretation for the cost function shown in Fig. 4.33: At the production level $x = x_0$ where condition (3) is satisfied, the line from the origin to the point $P_0 = (x_0, C(x_0))$ is tangent to the graph of $C(x)$, or equivalently, the tangent line at this point passes through the origin. It is worth noticing that not all functions have the property that there is such a point P_0. For example, no such point exists on the graph of the cost function (1). Certain economic theorists believe that the very existence of competitive capitalism depends on the fact that the cost functions for capitalistic production have this property.

It is clearly important for a manufacturer to know all about the cost function, but this is not enough. The manufacturer's overall purpose is to make a profit, and this depends in large measure on how many units x of a product can be sold at a specified price p. Presumably, the higher the price p, the lower the demand x, so the *demand curve* (Fig. 4.35, left) displays x as a decreasing function of p. The nature of this demand curve depends on the product, being relatively flat (or *inelastic*) for bread and motor oil, since people tend to buy what they need without much regard for the cost, and being relatively steep (or *elastic*) for candy, since no one really needs it but more people buy more of it when the price is low. For the sake of convenience in comparing the demand curve with the cost function, economists usually interchange the axes and consider p as a function of x, $p = p(x)$, as shown on the right in Fig. 4.35. This function is called the *demand function*.

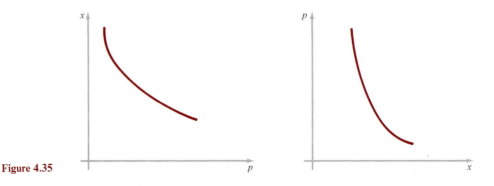

Figure 4.35

A manufacturer's *revenue R(x)* is total income considered as a function of production x, and the marginal revenue $R'(x)$ can be thought of as the extra revenue generated by producing one more unit. Since the *profit P(x)* is income minus expenses, we have

$$P(x) = R(x) - C(x). \qquad (4)$$

A manufacturer will lose money when production is too low, because of fixed costs, and also when production is too high, because of high marginal costs. Unless the manufacturer can operate profitably at some in-between level of production, the business will fail, so we can assume that the profit curve looks like Fig. 4.36. Since differentiating (4) yields

$$P'(x) = R'(x) - C'(x),$$

and since $P'(x) = 0$ at the high point of the profit curve, we obtain a second basic rule of economics:

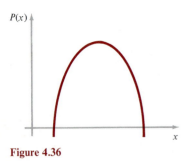

Figure 4.36

> *Profit is maximized when production is adjusted so that marginal revenue equals marginal cost.*

When x units of a commodity are produced and sold at a price of $p(x)$ dollars per unit, then the revenue $R(x)$ is evidently the product of $p(x)$ and x, $R(x) = xp(x)$, and (4) can be written as

$$P(x) = xp(x) - C(x). \qquad (5)$$

If both the demand function $p(x)$ and the cost function $C(x)$ are known, then (5) can be used to calculate the value of x that maximizes profits. It is clear from (5) that this value of x need not be the one that minimizes the average cost, for the latter depends only on the cost function $C(x)$. That is, profit depends on the whims of the marketplace, while efficiency is mainly an internal matter.

This discussion suggests several ways in which derivatives can be used in economics. The most influential contribution to this subject in the twentieth century was perhaps Keynes's *General Theory of Employment, Interest and Money,* which has been characterized as "an endless desert of economics, algebra and abstraction, with trackless wastes of differential calculus, and only an oasis here and there of delightfully refreshing prose."* This may be

---

* Chapter IX of *The Worldly Philosophers,* by Robert L. Heilbroner.

somewhat exaggerated for the sake of its juicy phrases, but nevertheless the general impression is valid — that modern economics makes extensive use of many kinds of mathematics, especially calculus.

We next consider an important and typical problem of business management in the area of inventory control. It is called the *ideal lot problem.*

Example   A large department store sells *N* units of a certain item — perhaps refrigerators — at a constant rate during the year. The units purchased from the distributor in a single order are delivered in one lot. If the store orders all *N* units delivered at the beginning of the year, then it saves on *reorder costs,* such as secretarial time and shipping charges; however, it incurs higher *carrying costs* for warehouse space, insurance, etc., because the average inventory throughout the year is the fairly large number *N/2.* On the other hand, if it places an order every day, then this keeps the average inventory low, but the reorder costs may become quite substantial. By considering both types of cost, determine how many units *x* the store should order in each lot to minimize the total cost *C(x).* (We do not include the wholesale cost of the units themselves, since this is the same per year no matter how they are ordered.)

Solution   With *x* units per lot, there are *N/x* orders per year. We assume that the cost of placing a single order consists of a fixed cost *F* (secretarial time, stationery, postage, receiving charges, etc.) plus a shipping cost *Sx* which is proportional to the size of the order. The total reorder cost for the year is then

$$(F + Sx)\,\frac{N}{x} = \frac{FN}{x} + SN,$$

which clearly increases as the lot size *x* decreases. We further assume that the annual carrying cost for a single unit is a constant *W* that can be determined by the accounting office. Since the average inventory is *x/2* when the lot size is *x,* the total annual carrying cost is

$$W \cdot \frac{x}{2},$$

which increases as *x* increases. The combined cost is therefore

$$C(x) = \frac{W}{2} \cdot x + \frac{FN}{x} + SN,$$

and the graph of this function is easily seen to have the shape indicated in Fig. 4.37. To minimize this total cost, we have only to equate the derivative to zero and solve the resulting equation:

$$\frac{dC}{dx} = \frac{W}{2} - \frac{FN}{x^2} = 0,$$

so the ideal lot size is $x = \sqrt{2FN/W}$.

We should keep in mind that the store might wish to order at least once a year regardless of other considerations. In this case it would not order more than *N* units at a time, and we have the restriction $x \le N$. If *N* is greater than $\sqrt{2FN/W}$, then we see from the figure that this restriction has no effect, but if

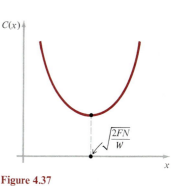

Figure 4.37

N is less than $\sqrt{2FN/W}$ the cost is minimized by ordering N units in each lot, once a year. The ideal lot size will therefore be the smaller of the numbers N, $\sqrt{2FN/W}$.

If N is larger than $\sqrt{2FN/W}$, then the ideal lot size is $\sqrt{2FN/W}$, which is proportional to $\sqrt{N}$. This tells us that if N quadruples, the ideal lot size doubles, and therefore the frequency of reordering should also double to meet the new level of demand. This is more economical than merely quadrupling the order size and keeping the same frequency.

PROBLEMS

1 On separate axes, sketch the graph of the average cost $C(x)/x$ for each of the cost functions (1) and (2). Observe that the first graph has no minimum point, but the second has. Add to the second sketch a graph of the marginal cost $C'(x)$, with due regard for the conclusion (3). In the case of (2), calculate the production level $x = x_0$ that minimizes the average cost, and verify (3) by finding the values of $C'(x_0)$ and $C(x_0)/x_0$.

2 An economist studying a certain appliance business finds that the overhead and wholesale cost involved in handling x electric mixers a week is $56 + 24x$ dollars, and that each week $x = 30 - \frac{1}{2}p$ mixers are sold at a retail price of p dollars apiece. What retail price should she advise the owner to charge in order to earn the greatest profit?

3 (a) Suppose a manufacturer can sell x bicycles per year at a price of $p = 300 - 0.01x$ dollars apiece, and that it costs him $C(x) = 60,000 + 75x$ dollars to produce the x bicycles. For maximum profit, what should his production be and what price should he charge?
   (b) If the government imposes on the manufacturer a tax of \$25 for each bicycle and the other features of the situation are unchanged, how much of the tax should he absorb himself and how much should he pass on to his customers if he wishes to continue making the maximum profit?

4 If the marginal revenue from producing x units of a certain commodity is $40 - \frac{1}{60}x^2$ dollars/unit and the marginal cost is $10 + \frac{1}{60}x^2$ dollars/unit, how many units should be produced to maximize the profit?

5 Consider an arbitrary point (p, x) on the demand curve shown on the left in Fig. 4.35. If p increases by a small amount Δp, and if $-\Delta x$ is the corresponding decrease in x, then the ratio of the percentage decrease in x to the percentage increase in p is

$$\frac{100(-\Delta x/x)}{100(\Delta p/p)} = -\frac{p}{x}\frac{\Delta x}{\Delta p},$$

and the *elasticity of demand $E(p)$* is

$$E(p) = \lim_{\Delta p \to 0}\left(-\frac{p}{x}\frac{\Delta x}{\Delta p}\right) = -\frac{p}{x}\frac{dx}{dp}.$$

This positive function is a useful tool of economic analysis, for it measures the responsiveness of the demand x to changes in the price p.
   (a) If $E(p) < 1$, show that the revenue $R = px$ is increased by increasing the price.
   (b) If $E(p) > 1$, show that the revenue is increased by increasing the demand (or equivalently, decreasing the price).
   (c) Establish the formula

$$\frac{dR}{dp} = x[1 - E(p)],$$

   and use this to deduce that $E(p) = 1$ at a point on the demand curve where the revenue is a maximum.
   (d) Show that $E(p)$ is constant along demand curves of the form $x = ap^{-b}$ where a and b are positive constants.

6 A restaurant owner finds that her customers drink 540 cases per year of a certain California wine. If it costs her \$10 to process each order and her carrying costs are \$3 per case per year, how many cases should she order each time?

7 A printer has agreed to print 135,000 copies of a small advertising card. It costs him \$12 per hour to run his press, which produces 600 impressions per hour. Each impression prints n cards, where n is the number of electrotypes (metal copies of set type) that are used on the press. Each electrotype costs him \$3. How many electrotypes should he use on his press to minimize the cost of the job?

8 A plastics firm receives an order for N units of a

certain item. Many machines are available, each of which can produce n units of the item per hour. The cost of preparing a single machine to produce this particular item is P dollars. Once the machines have been prepared, the production run is fully automated and can be carried out by one skilled supervisor earning W dollars per hour.

(a) How many machines should be used to minimize production costs?
(b) Show that when the production costs are minimal, the cost of preparing the machines equals the supervisor's wages.

ADDITIONAL PROBLEMS FOR CHAPTER 4

SECTION 4.1
Sketch the graphs of the following functions by using the first derivative and the methods of Section 4.1; in particular, find the intervals on which each function is increasing and those on which it is decreasing, and locate any maximum or minimum values it may have.

1 $y = \frac{1}{3}x^3 - \frac{1}{2}x^2 - 2x + \frac{4}{3}$.
2 $y = x^3 + 6x^2 + 12x + 8$.
3 $y = -x^3 + 3x + 2$.
4 $y = x^3 + 3x - 2$.
5 $y = x^4 - 6x^2 + 8x$.
6 $y = (x + 2)^3(x - 4)^3$.
7 $y = x^4 - 4x^3 + 16$.
8 $y = 3x^5 - 10x^3 + 15x + 3$.
9 $y = x^2(x + 1)^2$.
10 $y = x^3(x - 1)^2$.
11 $y = x^2(4 - x^2)$.

*12 $y = \dfrac{x^3}{x + 1}$.

13 $y = \dfrac{x}{(x + 1)^2}$.

14 $y = \frac{16}{3}x^3 + \dfrac{1}{x}$.

15 $y = \dfrac{4(x^2 - 1)}{x^4}$.

16 $y = \dfrac{4(x - 1)}{x^2}$.

17 $y = x^2 + \dfrac{16}{x^2}$.

18 $y = \dfrac{4 - 2x}{1 - x}$.

19 $y = \dfrac{5x^2 + 2}{x^2 + 1}$.

20 $y = \dfrac{5x^2 - 20x + 21}{x^2 - 4x + 5}$.

*21 $y = x^2(x - 4)^{2/3}$.

22 $y = \sqrt{x} + \dfrac{2}{\sqrt{x}} - 2\sqrt{2}$.

SECTION 4.2
For each of the following, locate the points of inflection, find the intervals on which the curve is concave up and those on which it is concave down, and sketch.

23 $y = x^3 + x$.
24 $y = x^3 + 3x^2 + 6x + 7$.
25 $y = x^3 - 12x + 2$.
26 $y = x^4 - 2x^2$.
27 $y = x^4 + 4x^3$.
28 $y = (x + 2)(x - 2)^3$.
29 $y = x^4 - 4x^3 - 2x^2 + 12x - 1$.

30 $y = \dfrac{x}{\sqrt{x^2 + 1}}$.

*31 $y = \dfrac{x^3}{x^2 + 3a^2}$ $(a > 0)$.

*32 $y = \dfrac{1}{x^3 + 1}$.

33 $y = \dfrac{5}{3x^4 + 5}$.

*34 $y = \dfrac{x^3}{(x - 1)^2}$.

35 $y = \dfrac{8}{x^3} - \dfrac{2}{x}$.

36 $y = \dfrac{6}{x} + \dfrac{6}{x^2}$.

37 In each part of this problem, use the given formula for the second derivative of a function to locate the points of inflection, the intervals on which the graph is concave up, and the intervals on which the graph is concave down:
(a) $y'' = x^2(x - 1)(x - 2)^2$;
(b) $y'' = (x^2 + 2)(x + 2)^2(x - 1)(x - 2)$;
(c) $y'' = x(x - 1)(x^2 - 4)(x - 3)$.

38 If $f(x) = (x - a)(x - b)(x - c)$, find the x-coordinate of the point of inflection. Hint: See Additional Problem 22 in Chapter 3.

39 Find the value of a that makes $f(x) = ax^2 + 1/x^2$ have a point of inflection at $x = 1$.

*40 Consider the general cubic curve $y = ax^3 + bx^2 + cx + d$.

(a) Show that the curve has one and only one point of inflection,

$$I = \left(-\frac{b}{3a}, k\right), \quad \text{where} \quad k = \frac{2b^3}{27a^2} - \frac{bc}{3a} + d.$$

(b) Show that the curve has one maximum point and one minimum point if and only if $b^2 - 3ac > 0$.

(c) When the curve has a maximum point P and a minimum point Q, show that the abscissa (x-coordinate) of I is the average of the abscissas of P and Q. Hint: Recall how to find the sum of the roots of a quadratic equation from its coefficients.

(d) Part (c) suggests that our general cubic curve might be symmetric with respect to its point of inflection I. Prove this by (1) introducing a new X-axis and Y-axis by means of

$$X = x + \frac{b}{3a} \quad \text{and} \quad Y = y - k,$$

so that the origin of the XY-system is the point I; (2) showing that the equation of our curve in the XY-system is

$$Y = aX\left(X^2 - \frac{b^2 - 3ac}{3a^2}\right);$$

and (3) observing that this transformed equation is symmetric with respect to the origin of the XY-system.

SECTION 4.3

41 Find the positive number that exceeds its cube by the largest amount.

42 Find two positive numbers x and y such that their sum is 30 and the product xy^4 is a maximum.

43 Find two positive numbers x and y such that their sum is 56 and the product x^3y^5 is a maximum.

44 (Generalization of the preceding problems) Let m and n be given positive integers. If x and y are positive numbers such that $x + y = S$, where S is a constant, show that the maximum value of the product $P = x^m y^n$ is attained when

$$x = \frac{mS}{m + n} \quad \text{and} \quad y = \frac{nS}{m + n}.$$

*45 Express the number 18 as the sum of two positive numbers in such a way that the sum of the square of the first and the fourth power of the second is as small as possible.

46 Find the positive number such that the sum of its cube and 48 times the reciprocal of its square is as small as possible.

47 The sum of three positive numbers is 15. Twice the first plus three times the second plus four times the third is 45. Choose the numbers so that the product of all three is as large as possible.

*48 (A generalization of Problem 6 in Section 4.3) Consider a rectangle with sides $2x$ and $2y$ inscribed in a given fixed circle $x^2 + y^2 = a^2$, and let n be a positive number. We wish to find the rectangle that maximizes the quantity $z = x^n + y^n$. If $n = 2$, it is clear that z has the constant value a^2 for all rectangles. If $n < 2$, show that the square maximizes z; and if $n > 2$, show that z is maximized by a degenerate rectangle in which x or y is zero.

49 Show that of all triangles with given base and given perimeter, the one with the greatest area is isosceles. Hint: Use Heron's formula for the area,

$$A = \sqrt{s(s - a)(s - b)(s - c)},$$

where a, b, c are the sides and s is the semiperimeter (half the perimeter).

50 Show that of all triangles with given base and given area, the one with the least perimeter is isosceles. Hint: If the base lies on the x-axis and is bisected by the origin, and if the third vertex (x, h) has a fixed height above the x-axis, then the triangle is isosceles if $x = 0$.

51 If a and b are positive constants, the region between the parabola $a^2y = a^2b - 4bx^2$ and the x-axis is a parabolic segment of base a and height b. Find the base and height of the largest rectangle with lower base on the x-axis and upper vertices on the parabola.

52 A circle of radius a is divided into two segments by a line L at a distance b from the center. The rectangle of greatest possible area is inscribed in the smaller of these segments. How far from the center is the side of this rectangle that is opposite to the line L?

*53 Two straight fences meet at a point, but not necessarily at right angles. A post stands in the angle between them. If a triangular corral is constructed by building a new straight fence containing this post, show that the fenced-off triangle has minimal area when the old post is in the center of the new fence. (Notice that this generalizes the result of Problem 7 in Section 4.3.)

*54 A line through a fixed point (a, b) in the first quadrant intersects the x-axis at A and the y-axis at B.

Show that the minimum values of AB and $OA + OB$ are

$$(a^{2/3} + b^{2/3})^{3/2} \quad \text{and} \quad (\sqrt{a} + \sqrt{b})^2.$$

*55 (A generalization of Example 4 and Problem 26 in Section 4.3) First, notice that areas of similar figures are proportional to the squares of corresponding lengths, as in Fig. 4.38, where

$$A = c_1 p^2 = c_2 d^2 = c_3 x^2 = c_4 y^2$$

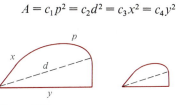

Figure 4.38

for suitable constants c_1, c_2, c_3, c_4. Here p is the perimeter, d is the diameter—the length of the longest chord—and x and y are the indicated lengths. The constants c_1, c_2, c_3, c_4 are evidently the areas when $p = 1$, $d = 1$, $x = 1$, $y = 1$. Now, cut a wire of length L into two pieces and use these pieces as the perimeters p and P of figures of two specified shapes

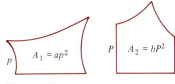

Figure 4.39

(Fig. 4.39), so that $p + P = L$. Then the sum of the areas is

$$A = A_1 + A_2 = ap^2 + bP^2 = ap^2 + b(L - p)^2,$$

where $0 \le p \le L$ (we allow either perimeter to be zero). Show that

(a) the minimum combined area is $abL^2/(a + b)$, corresponding to

$$p = \frac{b}{a + b} L \quad \text{and} \quad P = \frac{a}{a + b} L;$$

(b) the maximum combined area is the larger of the one-figure areas aL^2 and bL^2, corresponding to $p = L$ and $p = 0$.

Also, verify that these conclusions contain as special cases the results of Example 4 and Problem 26 in Section 4.3.

56 A printed page must have A square inches of printed matter and is required to have side margins of width a inches and top and bottom margins of width b inches. Find the length of the printed lines if the page is designed to use the least paper.

57 For a certain printed page, the widths of the four margins (possibly all different) and the area of the printed matter are specified. Show that the least paper is required if the full page is similar in shape to the rectangle of printed matter.

58 A dormer window has the shape of a rectangle surmounted by an equilateral triangle. If the total perimeter is fixed, find the proportions of the window (i.e., the ratio of the height of the window to its base) that will admit the most light.

59 A long strip of sheet metal 8 in wide is to be made into a rain gutter by turning up two sides at right angles to the bottom. If the gutter is to have maximum capacity, how many inches should be turned up on the sides?

60 A playing field is to be built in the shape of a rectangle with a semicircular part at each end, and the perimeter is to be a race track of specified length. Find the proportions of the field that will give the rectangular part as large an area as possible.

61 A farmer wishes to use 5 acres of land along a straight river to construct 6 small pens by means of a fence parallel to the river and 7 fences perpendicular to it. Show that if the total amount of fencing is to be minimized, the parallel fence should be as long as all the others combined.

62 An automobile manufacturer estimates that he can sell 5000 cars a month at \$4000 each, and that he can sell 500 more cars per month for each \$100 decrease in price.
(a) What price per car will bring the largest gross income?
(b) If each car costs \$1600 to make, what price will bring the largest total profit?

63 A manufacturer of electric knives estimates that her weekly production costs are given by the formula $C = 9500 + 8x + 0.00025x^2$, where x is the number of knives manufactured in a week.* The sales department estimates that if the selling price is set at y, then $x = 13{,}000 - 500y$ knives can be sold.† How many knives should be manufactured each week, and what should their selling price be, in order to achieve maximum profit?

*64 The cost for fuel of running a large paddlewheel steamboat at a speed of v miles per hour through still water is $\$v^3/24$ per hour. Other costs—wages, insurance, etc.—are \$108 per hour. What is the most

---

* The overhead is \$9500 per week; the cost of labor and materials is \$8 per knife; and the term $0.00025x^2$, which is small unless x is very large, says—in effect—that the factory has a fixed size and loses efficiency if too much production is attempted.

† This formula says that sales are expected to be 5000 at a selling price of \$16, with a loss of 500 sales for each \$1 increase in price.

economical speed for a certain trip upstream against a current of 2 mi/h?

65 A feed-lot operator has a herd of 200 cows in his pens, each weighing 600 lb. The cost of keeping one cow for a day is 80 cents. The cows are gaining weight at the rate of 8 lb/day. The market price for cows is now $1.25/lb, but it is dropping 1 cent a day. How many days should the operator wait in order to sell his cows for the largest profit?

66 An estimate of the numerical value of a certain quantity is to be determined from n measurements $x_1, x_2, \ldots, x_n$. The *least squares* estimate is the number x that minimizes the sum of the squares,

$$S = (x - x_1)^2 + (x - x_2)^2 + \cdots + (x - x_n)^2.$$

Show that this least squares estimate is the arithmetic mean of the measurements,

$$x = \frac{x_1 + x_2 + \cdots + x_n}{n}.$$

67 As a woman starts jogging across a 300-ft bridge, a man in a canoe passes directly below the center of the bridge. The woman is moving at the rate of 9 ft/s and the man at the rate of 12 ft/s.
  (a) What is the shortest *horizontal distance* between the woman and the man?
  (b) If the bridge is 288 ft high, what is the shortest *distance* between the woman and the man?

SECTION 4.4

68 Find the height of the cylinder of maximum lateral area that can be inscribed in a sphere of radius R. Show that this maximum lateral area is half the surface area of the sphere.

69 A cylinder is generated by revolving a rectangle of given perimeter about one of its sides. What is the shape (ratio of height to diameter of base) of the cylinder of maximum volume?

*70 The cone of smallest possible volume is circumscribed about a given hemisphere. What is the ratio of its height to the diameter of its base?

71 If the volume of a cone is fixed, what shape (ratio of height to diameter of base) minimizes its total surface area?

72 A pyramid has a square base and four equal sloping triangular faces. If the total area of the bottom and faces is given, show that the volume is greatest when the height is $\sqrt{2}$ times the edge of the base.

73 A cylinder is generated by revolving a rectangle about the x-axis, where the base of the rectangle lies on the x-axis and its upper vertices lie on the curve $y = x/(x^2 + 1)$. What is the largest volume such a cylinder can have?

74 (A problem of Kepler) Consider a cylinder with a given fixed distance D from the center of a generator to the farthest point of the cylinder. If this cylinder has the largest possible volume, what is the ratio of its height to the diameter of its base?

75 A solid is formed by cutting hemispherical cavities in the ends of a cylinder. If the total surface area of this solid is given, find the shape of the cylinder (ratio of height to diameter of base) that maximizes the volume of the solid.

76 A given cone has height H and radius of base R. A second cone is inscribed in the first with its vertex at the center of the base of the given cone and its base parallel to the base of the given cone. Find the height of the second cone if its volume is as large as possible.

77 Closed cylindrical cans are to be made with a specified volume. There is no waste involved in cutting the sheet metal that goes into the curved lateral part, but each end is to be cut from a square piece of metal and the scraps discarded. Find the ratio of the height to the diameter of the base that minimizes the cost of sheet metal.

78 A certain tank consists of a cylinder with hemispherical ends. For a given surface area, describe the shape of the tank with maximum volume.

79 A rectangle of tin whose sides are a and b is to be made into an open-top box by cutting a square from each corner and bending up the flaps to form the sides. How large a square should be cut from each corner to make the volume of the box as large as possible?

80 An aquarium is to be 4 ft high and is to have a volume of 88 ft³. The base, ends, and back are to be made of slate, and the front is to be made of special reinforced glass that costs 1.75 times as much as the slate per square foot. What should the dimensions be to minimize the cost of materials?

81 A circular filter paper of radius a is to be formed into a conical filter by folding under a circular sector. Find the ratio of the radius to the depth for the filter of greatest capacity.

82 A frame for a cylindrical lampshade is to be made from a piece of wire 20 ft long. The frame consists of two equal circles, four wires from the upper circle to the lower circle, and two diametral wires in the upper circle. Find the height and radius that will maximize the volume of the cylinder.

83 A box with a lid is to be made from a square sheet of cardboard 18 in on a side by cutting along the dotted lines as shown in Fig. 4.40. The cardboard is then folded up to form the ends and sides, and the flap is folded over to form the lid. What are the dimensions of the box of greatest volume?

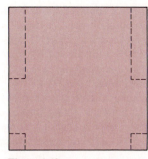

Figure 4.40

84 On a calm day the atmospheric pollution spreading out from a city is directly proportional to the population of the city and inversely proportional to the distance from the city. A retired forester wishes to start a tree nursery somewhere on a straight highway between two cities 60 km apart. If the first city is four times as large as the second, where should the forester locate his nursery to minimize the effect of pollution on his young trees?

85 The x-axis is the southern shore of a lake containing a small island at the point (a, b) in the first quadrant. A woman at the origin can run r meters per second along the shore and can swim s meters per second, where $r > s$. If she wants to reach the island as quickly as possible, how far should she run before she starts to swim?

86 Two towers 30 m apart are 30 and 70 m high, respectively. A taut wire fastened to the top of each tower is anchored to the ground between the towers. How far from the shorter tower will the wire touch the ground if its total length is a minimum? (Can you solve this problem without calculus?)

87 Find the equation of the circle with center at the origin that is internally tangent to the parabola $8y = 48 - x^2$.

88 Sketch the curve $y = \sqrt{x^2 + 16}$ and find the point on it that is closest to the point $(6, 0)$.

*89 Find the point on the parabola $y^2 = 3x$ that is closest to the point $(4, 1)$.

90 What points on the curve $x^2 y = 16$ are closest to the origin?

91 For what points on the circle $x^2 + y^2 = 25$ is the sum of the distances from $(2, 0)$ and $(-2, 0)$ a minimum?

92 Let $P = (x, y)$ be a variable point on the line $ax + by + c = 0$ and let $P_0 = (x_0, y_0)$ be a fixed point not on this line.

(a) If s is the distance from P_0 to P, use the methods of calculus to show that s^2 (and therefore s) is a minimum when PP_0 is perpendicular to the given line.

(b) Show that the minimum distance is

$$\frac{|ax_0 + by_0 + c|}{\sqrt{a^2 + b^2}}.$$

93 A smooth graph not passing through the origin always has a point (x_0, y_0) that is closest to the origin.* If this point is not an endpoint, show that the line from the origin to (x_0, y_0) is perpendicular to the graph.

94 If a, b, c are positive constants, show that $ax + b/x \geq c$ for all positive numbers x if and only if $4ab \geq c^2$.

95 If a, b, c are positive constants, show that $ax^2 + b/x \geq c$ for all positive numbers x if and only if $27ab^2 \geq 4c^3$.

96 Consider the general quadratic function $f(x) = ax^2 + bx + c$ with $a > 0$. By calculating the minimum value of this function, show that $f(x) \geq 0$ for all x if and only if $b^2 - 4ac \leq 0$.

97 By applying the idea of Problem 96 to the function

$$f(x) = (a_1 x + b_1)^2 + (a_2 x + b_2)^2 + \cdots + (a_n x + b_n)^2,$$

establish *Schwarz's inequality*:

$$|a_1 b_1 + \cdots + a_n b_n| \leq (a_1^2 + \cdots + a_n^2)^{1/2}(b_1^2 + \cdots + b_n^2)^{1/2}.$$

Also show that equality holds here if and only if there exists a number x such that $b_i = -a_i x$ for every $i = 1, 2, \ldots, n$.

SECTION 4.5

98 A cubic block of ice is melting at the rate of 6 in³/min. How fast is its surface area changing when its edge is 12 in long?

99 A light is on the ground 50 ft from a building. A man 6 ft tall walks from the light toward the building at 4 ft/s. Find how rapidly the length of his shadow on the building is decreasing (a) when he is 40 ft from the building; (b) when he is 30 ft from the building.

100 Two airplanes are flying westward on parallel courses 9 mi apart. One flies at 425 mi/h and the other at 500 mi/h. How fast is the distance between the planes decreasing when the slower plane is 12 mi farther west than the faster plane?

101 A conical tank with its vertex down is 8 ft high and 4 ft in diameter at the top. It is full of water, but the water is leaking out through a hole in the bottom at the rate of 1 ft³/min. Find the rate at which the water level is falling when the tank is $\frac{7}{8}$ empty.

---

* For the purposes of this problem, interpret the phrase "smooth graph" to mean the graph of a function $y = f(x)$ defined for all x or on a closed interval $a \leq x \leq b$, whose derivative $f'(x)$ exists in the interior of its domain.

102 Assume that water squirts out a hole in the bottom of a tank at a speed proportional to the square root of the depth y of the water. If the tank has the shape of a cone with its vertex down, show that the rate of change of the depth is

$$\frac{dy}{dt} = -\frac{c}{y^{3/2}}$$

where c is a positive constant.

103 Water is being pumped into an open-top cylindrical tank of radius 5 ft at the rate of 6 ft³/min. At the same time, water is squirting out a hole in the bottom of the tank at the rate of $2\sqrt{y}$ ft³/min, where y is the depth of the water. How high must the tank be for the water level to stabilize before it overflows?

104 A long rectangular tank has a sliding panel that divides it into two adjustable tanks of width 4 ft (see Fig. 4.41). Water is pumped into the left compartment at the rate of 12 ft³/min. At the same time the panel is moved steadily to the right at the rate of 1 ft/min. In each of the following situations determine whether the water level is rising or falling, and how fast: (a) the left compartment contains 144 ft³ of water and is 9 ft long; (b) the left compartment contains 144 ft³ of water and is 18 ft long.

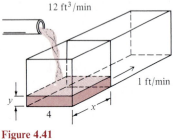

12 ft³/min

1 ft/min

Figure 4.41

105 A large volume V of oil is spilled into a calm sea from a broken tanker. After the initial turbulence has died down, the oil spreads in a circular pattern of radius r and uniform thickness h, where r increases and h decreases in a manner determined by the viscosity and buoyancy of the oil. Laboratory experiments suggest that the thickness is inversely proportional to the square root of the elapsed time, $h = c/\sqrt{t}$. Show that the rate dr/dt at which the oil spreads is inversely proportional to $t^{3/4}$.

106 String of radius $\frac{1}{10}$ in is being wound onto a ball at the rate of 32 in/s. If the ball is assumed to remain spherical and to consist entirely of string with no empty space, find the rate at which its radius is increasing when the radius is 2 in.

107 Thread is being unwound at the rate of a inches per

second from a spool of radius r inches. The unwound part of the thread has length x inches and is stretched taut into a segment PT tangent to the spool at the point T. Find the rate of increase of the distance y from the axis of the spool to the point P at the end of the thread.

108 Meteorologists are interested in the adiabatic expansion or compression of large masses of air, in which temperatures may change but no heat is added or subtracted. The adiabatic gas law for air is $pV^{1.4} = c$, where p is pressure, V is volume, and c is a constant. The volume of a certain insulated chamber of air is decreasing steadily at the rate of 1 ft³/min. Find how rapidly the pressure is increasing at an instant when the pressure is 65 lb/in² and the volume is 13 ft³.

109 If a rocket weighs 1000 lb on the surface of the earth, then it weighs

$$W = \frac{1000}{(1 + r/4000)^2}$$

pounds when it is r miles above the surface of the earth. If the rocket is rising at the rate of 1.25 mi/min, how fast is it losing weight when its altitude is 1000 mi?

110 Wheat is being poured onto a pile at the constant rate of 36 ft³/min. If the pile always has the shape of a cone whose height is half the radius of the base, at what rate is the height increasing when the diameter of the pile is 12 ft?

111 Gravel is being poured onto a pile, forming a cone. If the radius of the base is increasing at the rate of 3 m/min and the height is increasing at the rate of 1 m/min, how rapidly is the volume increasing when the height is 4 m?

112 A chord moves across a circle of radius 5 ft at the rate of 4 ft/min. How fast is the length of the chord decreasing when it is $\frac{4}{5}$ of the way across the circle?

113 A point moves along the parabola $x^2 = 4py$ in such a way that its projection on the x-axis has constant velocity. Show that its projection on the y-axis has constant acceleration.

114 Two points A and B are moving along the x-axis and y-axis, respectively, in such a way that the perpendicular distance k from the origin O to the segment AB remains constant. If A is moving away from O at the rate of $4k$ units per minute, find how fast OB is changing, and whether it is increasing or decreasing, at the moment when $OA = 3k$.

115 One side of a rectangle is increasing at the rate of 7 in/min and the other side is decreasing at the rate of 5 in/min. At a certain moment the lengths of these two sides are 10 and 7 in, respectively. Is the area of the rectangle increasing or decreasing at that moment? How fast?

116 Two concentric circles are expanding. At a certain moment, designated by $t = 0$, the inner radius is 2 ft and the outer radius is 10 ft; and for $t > 0$, these radii are increasing at the steady rates of 4 ft/min and 3 ft/min, respectively. If A is the area between the circles, when will A have its largest value?

*117 Two concentric spheres are expanding. At time $t = 0$, the inner and outer radii r and R have the values r_0 and R_0 feet, respectively. For $t > 0$, these radii are increasing at the steady rates of a and b feet per minute, where $a > b > ar_0^2/R_0^2$. If V is the volume between the spheres, when will V have its largest value?

SECTION 4.6

118 Show that each of the following equations has only one real root, and calculate it to two decimal places:
(a) $x^3 + 5x - 2 = 0$; (b) $x^3 + 2x - 4 = 0$.

119 Calculate each of the following to two decimal places of accuracy:
(a) $\sqrt{11}$; (b) $\sqrt[3]{6.9}$; (c) $\sqrt[4]{19}$.

120 Let a be a given positive number and x_1 a positive number that approximates $\sqrt{a}$.
(a) Show that Newton's method applied to the equation $x^2 - a = 0$ gives $x_2 = \frac{1}{2}(x_1 + a/x_1)$ as the next approximation.
(b) If $x_1 \neq \sqrt{a}$, show that the approximation $\frac{1}{2}(x_1 + a/x_1)$ is greater than $\sqrt{a}$, regardless of whether x_1 is greater than $\sqrt{a}$ or less than $\sqrt{a}$. Hint: Show that the inequality $\frac{1}{2}(x_1 + a/x_1) > \sqrt{a}$ is equivalent to $(\sqrt{x_1} - \sqrt{a/x_1})^2 > 0$.
(c) If the approximation x_1 is too large, i.e., if $x_1 > \sqrt{a}$, show that $\frac{1}{2}(x_1 + a/x_1)$ is a better approximation in the sense that
$$\frac{1}{2}\left(x_1 + \frac{a}{x_1}\right) - \sqrt{a} < x_1 - \sqrt{a}.$$
(d) Assume that the approximation x_1 is too small, i.e., $x_1 < \sqrt{a}$, but is large enough so that $x_1 > \frac{1}{3}\sqrt{a}$. Show that $\frac{1}{2}(x_1 + a/x_1)$ is a better approximation in the sense that
$$\frac{1}{2}\left(x_1 + \frac{a}{x_1}\right) - \sqrt{a} < \sqrt{a} - x_1.$$
Hint: Show that this inequality is equivalent to $x_1 + a/x_1 - 2\sqrt{a} < 2\sqrt{a} - 2x_1$, $3x_1 - 4\sqrt{a} + a/x_1 < 0$, and
$$\frac{(3x_1 - \sqrt{a})(x_1 - \sqrt{a})}{x_1} < 0.$$

121 If a is a given positive number and $\sqrt[3]{a}$ is calculated by applying Newton's method to the equation $x^3 - a = 0$, show that

$$x_2 = \frac{1}{3}\left(2x_1 + \frac{a}{x_1^2}\right).$$

122 Consider a spherical shell 1 ft thick whose volume is twice the volume of the hollow space inside it. Use Newton's method to calculate the shell's outer radius to two decimal places of accuracy.

*123 A conical paper cup is 4 in deep and 4 inches in diameter. Its vertex is pushed up inside, as shown in Fig. 4.42. How far does its tip penetrate the space inside the cup if the new volume is four-fifths of the original volume?

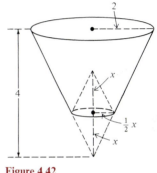

Figure 4.42

SECTION 4.7

124 A foreign car dealer knows that the cost of importing and selling x cars per year is $C(x) = 56,000 + 3500x - 0.01x^2$ dollars. Past experience tells him that he can sell $x = 40,000 - 10p$ cars at p dollars apiece.
(a) How many cars should he import for maximum profit?
(b) What should be his selling price for each car?
(c) What is his maximum profit?

125 If the demand x for a certain commodity is a decreasing linear function of the price p, show that the marginal revenue is also linear and decreasing.

126 A vacuum cleaner manufacturer buys 3000 electric motors a year to install in her machines. If it costs her \$10 to place an order, and if the cost of storing one motor for a year is 24 cents, how many motors should she order at a time, and how often?

*127 The advertising manager for a discount store discovers that by running an ad on any day in the evening newspaper he can boost his daily sales of housewares to \$300 on the next day, after which sales decline by a steady \$5 per day until the next time the ad runs or until sales sink to \$200 per day, whichever comes first. If each insertion of the ad costs \$40, how often should he advertise?

5

INDEFINITE INTEGRALS AND DIFFERENTIAL EQUATIONS

5.1

INTRODUCTION

Our work in the preceding chapters was concerned with the *problem of tangents* as described in Section 2.1—given a curve, find the slope of its tangent; or equivalently, given a function, find its derivative.

In addition to launching the full-scale study of derivatives, Newton and Leibniz also discovered that many problems in geometry and physics depend on "backwards differentiation," or "antidifferentiation." This is sometimes called the *inverse problem of tangents*: Given the derivative of a function, find the function itself.

In this chapter we work with the same derivative rules as in Chapter 3. Here, however, these rules are read backwards, and lead in particular to the "integration" of polynomials. Even these relatively simple procedures have some remarkable applications, which we discuss in Section 5.5.

5.2

THE NOTATION OF DIFFERENTIALS

As we know, the definition of the derivative $f'(x)$ of a function $y = f(x)$ can be stated as follows:

$$f'(x) = \lim_{\Delta x \to 0} \frac{\Delta y}{\Delta x}. \tag{1}$$

It is understood here that Δx is a nonzero change in the independent variable x, and that $\Delta y = f(x + \Delta x) - f(x)$ is the corresponding change in y. In Section 2.3 we introduced the equivalent notation

$$\frac{dy}{dx} \tag{2}$$

for this derivative, and we emphasized there that (2) is a single symbol and not a fraction. However, it is certainly true that (2) looks like a fraction, and in some circumstances it even acts like one. The most important example of this is the chain rule,

$$\frac{dy}{du}\frac{du}{dx} = \frac{dy}{du}\frac{du}{dx} = \frac{dy}{dx},$$

where the correct formula for the derivative of a composite function is obtained by cancelling as if the derivatives were fractions.

128

Our present purpose is to give individual meanings to the pieces of (2), namely, to dy and dx, in such a way that their quotient is indeed the derivative $f'(x)$. Our reasons for doing this are impossible to explain in advance. Suffice it to say that this notational device is a necessary prelude to the powerful computational methods introduced in this chapter — integration by substitution, and the solution of certain differential equations by separating the variables.

We begin by considering the special case in which y is a linear function of x,

$$y = mx + b. \qquad (3)$$

Let $P = (x, y)$ be a point on this line (Fig. 5.1). If x is given an increment Δx and if the corresponding increment in y is Δy, then the slope of the line (3) is

$$m = \frac{\text{change in } y}{\text{change in } x} = \frac{\Delta y}{\Delta x},$$

so

$$\Delta y = m\, \Delta x. \qquad (4)$$

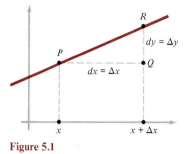

Figure 5.1

When working in this way with increments along a straight line, we denote them by the symbols dx and dy, so that

$$dx = \Delta x \qquad \text{and} \qquad dy = \Delta y,$$

and call them *differentials.* With this notation, (4) becomes

$$dy = m\, dx. \qquad (5)$$

Now consider an arbitrary function

$$y = f(x), \qquad (6)$$

and assume that this function has a derivative at x. If P is the corresponding point on the graph (Fig. 5.2), then the tangent at P is the straight line PR with slope $m = f'(x)$. By the *differentials dx and dy* arising from (6), we mean the increments in the variables x and y that are associated with this tangent line. To state this more precisely, the differential dx of the independent variable x is any increment Δx in x, as shown,

$$dx = \Delta x; \qquad (7)$$

and the differential dy of the dependent variable y is the corresponding increment in y *along the tangent line,* namely,

$$dy = f'(x)\, dx. \qquad (8)$$

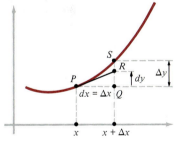

Figure 5.2

Thus, as Fig. 5.2 shows, if $dx = \Delta x = PQ$ is any change in x, then $\Delta y = QS$ and $dy = QR$ are the corresponding changes in y along the curve and along the tangent line, respectively. We observe that (8) reduces to (5) when $f(x) = mx + b$.

If $dx \neq 0$, then we can divide (8) by it and obtain

$$\frac{dy}{dx} = f'(x). \qquad (9)$$

Up to this point equation (9) has been trivially true because its two sides have been merely two different ways of writing the same thing, namely, the derivative of the function $y = f(x)$. The new feature of (9) in our present discussion is that now the Leibniz symbol on the left not only looks like a fraction but *is* a fraction,

$$\frac{dy}{dx} = \frac{\text{differential of } y}{\text{differential of } x}.$$

The Leibniz notation for derivatives makes it particularly easy to produce the differential formula (8) when the function $y = f(x)$ is given, by computing the derivative and multiplying by dx. The calculation in the first column gives the general pattern,

$$y = f(x) \qquad\qquad y = x^2$$
$$\frac{dy}{dx} = f'(x) \qquad\qquad \frac{dy}{dx} = 2x$$
$$dy = f'(x)\, dx \qquad\qquad dy = 2x\, dx,$$

and the calculation in the second column shows how it works for the special case $y = x^2$. A little experience with the use of this notation makes us realize that we can proceed directly from $y = x^2$ to the formula $dy = 2x\, dx$ without bothering to write the intermediate step $dy/dx = 2x$. It is often convenient to write $df(x)$ instead of dy. Thus, for example,

$$d(x^2) = 2x\, dx, \qquad d(5x^4) = 20x^3\, dx, \qquad d\left(\frac{1}{x}\right) = \left(-\frac{1}{x^2}\right) dx = -\frac{dx}{x^2}.$$

Our standard rules for calculating derivatives can now be given useful equivalent formulations in the notation of differentials. For example, if we multiply the product rule

$$\frac{d}{dx}(uv) = u\frac{dv}{dx} + v\frac{du}{dx}$$

by dx, it becomes the differential product rule,

$$d(uv) = u\, dv + v\, du.$$

Here is our basic list, written both ways:

$$\frac{d}{dx}c = 0, \qquad\qquad\qquad dc = 0; \qquad\qquad (10)$$

$$\frac{d}{dx}x^n = nx^{n-1}, \qquad\qquad d(x^n) = nx^{n-1}\, dx; \qquad\qquad (11)$$

$$\frac{d}{dx}(cu) = c\frac{du}{dx}, \qquad\qquad d(cu) = c\, du; \qquad\qquad (12)$$

$$\frac{d}{dx}(u + v) = \frac{du}{dx} + \frac{dv}{dx}, \qquad\qquad d(u + v) = du + dv; \qquad\qquad (13)$$

$$\frac{d}{dx}(uv) = u\frac{dv}{dx} + v\frac{du}{dx}, \qquad\qquad d(uv) = u\, dv + v\, du; \qquad\qquad (14)$$

$$\frac{d}{dx}\left(\frac{u}{v}\right) = \frac{v \, du/dx - u \, dv/dx}{v^2}, \qquad d\left(\frac{u}{v}\right) = \frac{v \, du - u \, dv}{v^2}; \qquad (15)$$

$$\frac{d}{dx} u^n = nu^{n-1} \frac{du}{dx}, \qquad\qquad d(u^n) = nu^{n-1} \, du. \qquad (16)$$

The differential formulas (11) and (16) look the same except for the letters used, but their connotations are quite different; for in (11) we think of x itself as the independent variable, and in (16) we think of u as some unspecified function of x. This point is illustrated by the calculations

$$d(x^4) = 4x^3 \, dx$$

and

$$\begin{aligned} d(x^2 + 1)^4 &= 4(x^2 + 1)^3 \, d(x^2 + 1) \\ &= 4(x^2 + 1)^3 \cdot 2x \, dx \\ &= 8x(x^2 + 1)^3 \, dx. \end{aligned}$$

In order to acquire facility with the differential notation, one must practice using these formulas.

Example 1   If $y = x^4 + 3x^2 + 7$, find dy.

Solution   One way of doing this is to find the derivative,

$$\frac{dy}{dx} = 4x^3 + 6x,$$

and multiply by dx:

$$dy = (4x^3 + 6x) \, dx.$$

We can also use the differential formulas (10) to (13):

$$\begin{aligned} dy = d(x^4 + 3x^2 + 7) &= d(x^4) + 3d(x^2) + d(7) \\ &= 4x^3 \, dx + 3 \cdot 2x \, dx + 0 \\ &= (4x^3 + 6x) \, dx. \end{aligned}$$

We emphasize that a differential on the left side of an equation requires that the right side must also contain a differential. Thus, we never write $dy = 4x^3$, but instead $dy = 4x^3 \, dx$.

Example 2   To find $d(x^2/\sqrt{x^2 + 1})$, we use the differential formula (15):

$$d\left(\frac{x^2}{\sqrt{x^2 + 1}}\right) = \frac{\sqrt{x^2 + 1} \, d(x^2) - x^2 \, d(\sqrt{x^2 + 1})}{x^2 + 1}.$$

But $d(x^2) = 2x \, dx$ and

$$d(\sqrt{x^2 + 1}) = d[(x^2 + 1)^{1/2}] = \frac{1}{2}(x^2 + 1)^{-1/2} \cdot 2x \, dx = \frac{x \, dx}{\sqrt{x^2 + 1}},$$

so

$$d\left(\frac{x^2}{\sqrt{x^2 + 1}}\right) = \frac{\sqrt{x^2 + 1} \cdot 2x \, dx - x^3 \, dx/\sqrt{x^2 + 1}}{x^2 + 1} = \frac{x^3 + 2x}{(x^2 + 1)^{3/2}} \, dx.$$

The method of differentials is particularly useful in implicit differentiation.

Example 3 Assume that y is a differentiable function of x that satisfies the equation $x^2y^3 - 2xy + 5 = 0$, and use differentials to find an expression for dy/dx.

Solution By calculating the differential of each term in this equation, using the rules for products, powers, and constants, we get

$$x^2 \cdot 3y^2\, dy + y^3 \cdot 2x\, dx - 2x\, dy - 2y\, dx = 0.$$

We now collect on the left the terms containing dy and on the right the terms containing dx,

$$(3x^2y^2 - 2x)\, dy = (2y - 2xy^3)\, dx,$$

and this yields our result:

$$\frac{dy}{dx} = \frac{2y - 2xy^3}{3x^2y^2 - 2x}.$$

Most people who use calculus routinely as a tool in their work think of differentials as very small quantities, even though the definitions contain no such requirement. There are several good reasons for this. One such reason can be seen in Fig. 5.2, which shows that the tangent to a curve hugs the curve closely near the point of tangency. This means that when dx is small, the curve is virtually indistinguishable from its tangent, and therefore the differential dy, which is comparatively easy to calculate, provides a very good approximation to the exact increment Δy, which may be harder to calculate.

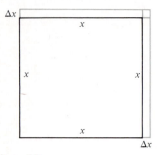

Figure 5.3

Example 4 To see how this idea works in a simple case, let x be the side of a square and $y = x^2$ its area. If each side is increased by a small amount Δx (Fig. 5.3), then the increment in the area is

$$\Delta y = (x + \Delta x)^2 - x^2 = 2x\, \Delta x + \Delta x^2.$$

The term $2x\, \Delta x$ is the sum of the areas of the two bordering rectangles in Fig. 5.3, and the term Δx^2 is the area of the small square in the upper right-hand corner. Since x is the independent variable and therefore $dx = \Delta x$, we have

$$dy = 2x\, dx = 2x\, \Delta x,$$

and it is clear that this is a good approximation to the exact increment Δy when Δx is small compared with x.

Remark 1 In the preceding discussion, differentials were defined in different ways for independent and for dependent variables. It is therefore desirable to verify that the fundamental formula (8) remains valid even if x is not the independent variable, but instead depends on some other variable t. To see this, suppose that

$$y = f(x) \qquad \text{where} \qquad x = g(t).$$

Then y is also a function of t,

$$y = f(g(t)) = F(t),$$

and the chain rule guarantees that

$$\frac{dy}{dt} = \frac{dy}{dx}\frac{dx}{dt},$$

or equivalently,

$$F'(t) = f'(g(t))g'(t). \qquad (17)$$

In this situation, where x and y are both dependent variables and t is the independent variable, the differentials dx and dy are defined by

$$dx = g'(t)\,dt \qquad \text{and} \qquad dy = F'(t)\,dt.$$

But (17) allows us to write

$$dy = F'(t)\,dt = f'(g(t))g'(t)\,dt = f'(x)\,dx,$$

which is (8). Thus, $y = f(x)$ implies $dy = f'(x)\,dx$ in all cases, regardless of whether x is independent or depends on some other variable t.

Remark 2 *The Leibnizian myths about curves and differentials.* The modern concept of limit did not arise until the early nineteenth century, and so no definition of the derivative resembling equation (1) was possible for Leibniz or his immediate successors. What were the early ideas about the nature of derivatives and differentials?

Most of the fruitful mathematical thinking of the period was based on one form or another of the notion of the "infinitely small." Leibniz's attitude toward the equation

$$\frac{dy}{dx} = \lim_{\Delta x \to 0} \frac{\Delta y}{\Delta x}$$

would have been essentially as follows: As Δx approaches zero, both Δy and Δx become "infinitely small" or "infinitesimal" together. It is therefore reasonable to think of the limit dy/dx as the quotient of two infinitesimal quantities denoted by dy and dx and called "differentials." In Leibniz's imagination, an *infinitesimal* was a special kind of number that is not zero and yet is smaller than any other number.

There was also a geometric version of these ideas, in which a curve was thought of as consisting of an infinite number of infinitely small straight line segments (Fig. 5.4). A tangent was a line containing one of these tiny segments. To find the slope of the tangent at a point (x, y), we move an infinitesimal distance along the curve to a point $(x + dx, y + dy)$ and observe that the slope of the infinitesimal segment joining these two points is the quotient dy/dx.

We have suggested that Leibniz introduced his differentials dx and dy to denote corresponding infinitesimal changes in the variables x and y. To get an idea of how these differentials were used, let us suppose that the variables

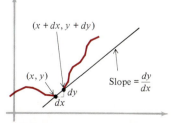

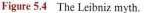

Figure 5.4 The Leibniz myth.

x and y are related by the equation

$$y = x^2. \tag{18}$$

Leibniz would then replace x and y by $x + dx$ and $y + dy$ to obtain

$$y + dy = (x + dx)^2 = x^2 + 2x\,dx + dx^2,$$

which in view of (18) yields

$$dy = 2x\,dx + dx^2. \tag{19}$$

At this stage Leibniz would simply discard the term dx^2 and arrive at our familiar formula

$$dy = 2x\,dx, \tag{20}$$

which after division by dx takes its fractional form

$$\frac{dy}{dx} = 2x. \tag{21}$$

He would justify this step by claiming that any square of an infinitely small number is "infinitely infinitely small," or "an infinitesimal of higher order," and therefore entirely negligible. For Leibniz, the derivative was a genuine quotient, a quotient of infinitesimals as calculated in formula (21) and illustrated in Fig. 5.4, and his form of calculus came to be widely known as "infinitesimal calculus."

It may be instructive to compare this Leibnizian use of infinitesimals with the modern approach based on limits. Thus, with the function $y = x^2$, if Δx is a given nonzero change in x and Δy is the corresponding change in y, then by essentially the same calculation we obtain

$$\Delta y = 2x\,\Delta x + \Delta x^2.$$

Instead of discarding the term Δx^2, in the modern approach we divide through by Δx to obtain the quotient $\Delta y/\Delta x$, then define the derivative to be the limit of this quotient as Δx approaches zero,

$$\frac{dy}{dx} = \lim_{\Delta x \to 0} \frac{\Delta y}{\Delta x} = \lim_{\Delta x \to 0} (2x + \Delta x) = 2x.$$

This produces formula (21) in a way that replaces the use of infinitesimals by a limit calculation.

The ideas of Leibniz worked with almost miraculous ease and effectiveness, and dominated the historical development of calculus and the physical sciences for almost 150 years. However, these ideas were flawed by the fact that infinitesimals in the sense described above clearly do not exist, for there is no such thing as a positive number that is smaller than all other positive numbers. Throughout this period of more than a century the enormous success of calculus as a problem-solving tool was obvious to all, and yet no one was able to give a logically acceptable explanation of what calculus *is*. The fog that obscured its fundamental concepts was at last dispelled in the early nineteenth century by the classical theory of limits. Fortunately the early mathematicians of the modern period—Leibniz himself, the Bernoullis, Euler, Lagrange—had sound intuitive feelings for what was reasonable and correct in the problems they studied. Even though their arguments often lacked rigor from the modern point of view, these pioneers rarely went astray in their conclusions.

If a myth is a veiled, condensed, symbolic expression of a more complicated and perhaps partially hidden truth, then mathematics has its myths just as history and literature do. Leibniz's differentials were banished from "official calculus" by the theory of limits, but nevertheless they remain a living part of the mythology of the subject.*

PROBLEMS

Use the rules for differentials to calculate each of the following.

1 $d(7x^9 - 3x^5 + 34)$.

2 $d(\sqrt{1 - x^2})$.

3 $d(x^2\sqrt{1 - x^2})$.

4 $d\left(\dfrac{x - 2}{x + 3}\right)$.

5 $d(\sqrt{4x - x^2})$.

6 $d\left(\dfrac{x}{\sqrt{a^2 + x^2}}\right)$.

7 $d(3x^{2/3} + 10x^{1/5} - 17x)$.

8 $d\left[\dfrac{(1 - 2x)^3}{3 - 4x}\right]$.

9 $d(x^2\sqrt{3x + 2})$.

10 $d(\sqrt{x + \sqrt{x + 1}})$.

In each of the following cases, assume that y is a differentiable function of x that satisfies the given equation, and use the method of differentials to find an expression for dy/dx. Also find the derivative dx/dy, assuming that x is a differentiable function of y.

11 $16y^3 = 9x^2$.
12 $\sqrt{x} + \sqrt{y} = 4$.
13 $x^3 - 3x^2y + y^3 = 1$.
14 $x^2 + xy - 2y^2 - 3x + 4y + 6 = 0$.
15 Find dy/dx, given that

$$y = \frac{3u - 1}{u^2 - u} \quad \text{and} \quad u = (x^3 + 2)^5.$$

16 Find dy/dx, given that

$$y = \frac{u + 1}{u - 1}, \quad u = \frac{v^3 + 6v - 2}{\sqrt{v - 1}}, \quad v = x^4 + 5x^2 - 3.$$

17 Find dy/dx, given that

$$y^4 - 2u^3 + 3y + 5 = 0, \quad x^3 - u^2 + 2\sqrt{u} - 6 = 0.$$

18 Consider a circle of radius r and area $A = \pi r^2$. If the radius is increased by a small amount Δr, find the increment ΔA and the differential dA. Draw a sketch in the manner of Fig. 5.3 and observe that ΔA is the area of the thin circular ring between two concentric circles. Use the fact that the inner circle has circumference $2\pi r$ to understand geometrically why dA is a good approximation to ΔA.

19 A sphere of radius r has volume $V = \frac{4}{3}\pi r^3$ and surface area $A = 4\pi r^2$. If the radius is increased by a small amount Δr, find ΔV and dV. In the spirit of Problem 18, understand geometrically why dV is a good approximation to ΔV.

20 A coat of paint of thickness 0.02 in is applied to the faces of a cube whose edge is 10 in, thereby producing a slightly larger cube. Use differentials to find approximately the number of cubic inches of paint used. Also find the exact amount used by computing volumes before and after painting.

If $y = F(x)$ is a function whose derivative is known, say, for example,

$$\frac{d}{dx} F(x) = 2x, \tag{1}$$

can we discover what the function $F(x)$ is? It doesn't take much imagination

5.3
INDEFINITE INTEGRALS. INTEGRATION BY SUBSTITUTION

---

*It should be added that a logically acceptable concept of infinitesimals was constructed in the 1960s by the American mathematician Abraham Robinson (see his book *Non-Standard Analysis,* North-Holland Publishing Co., Amsterdam, 1966, especially Sections 1.1 and 10.1). While Robinson's achievement is of great interest to logicians and mathematicians, his ideas depend on mathematical logic and abstract set theory and are not likely to have much influence on the teaching or learning of calculus.

to write down one function with this property, namely $F(x) = x^2$. Moreover, adding a constant term doesn't change the derivative, so each of the functions

$$x^2 + 1, \qquad x^2 - \sqrt{3}, \qquad x^2 + 5\pi,$$

and more generally

$$x^2 + c$$

where c is any constant, also has the property (1). Are there any others? The answer is *no*.

The justification for this answer lies in the following principle:

If $F(x)$ and $G(x)$ are two functions having the same derivative $f(x)$ on a certain interval, then $G(x)$ differs from $F(x)$ by a constant, that is, there exists a constant c with the property that

$$G(x) = F(x) + c$$

for all x in the interval.

To see why this statement is true, we notice that the derivative of the difference $G(x) - F(x)$ is zero on the interval,

$$\frac{d}{dx}[G(x) - F(x)] = \frac{d}{dx}G(x) - \frac{d}{dx}F(x) = f(x) - f(x) = 0.$$

It now follows that this difference itself must have a constant value c, so

$$G(x) - F(x) = c \qquad \text{or} \qquad G(x) = F(x) + c,$$

which is what we wanted to establish.*

This principle allows us to conclude that every solution of equation (1) must have the form $x^2 + c$ for some constant c.

The problem just discussed involved finding an unknown function whose derivative is known. If $f(x)$ is given, then a function $F(x)$ such that

$$\frac{d}{dx}F(x) = f(x) \tag{2}$$

is called an *antiderivative* of $f(x)$, and the process of finding $F(x)$ from $f(x)$ is *antidifferentiation*. We have seen that $f(x)$ does not have a single, uniquely determined antiderivative, but if we can find one antiderivative $F(x)$, then all others have the form

$$F(x) + c$$

for various values of the constant c. For example, $\frac{1}{3}x^3$ is one antiderivative of x^2, and the formula

$$\tfrac{1}{3}x^3 + c$$

comprises all possible antiderivatives of x^2.

---

* The crucial step in this reasoning can be expressed in several different ways: for instance, if the rate of change of a function is zero, then the function cannot change and therefore must be constant; or equivalently, if every tangent line to a graph is horizontal, then the graph can neither rise nor fall and therefore must be a horizontal straight line. The theoretical basis for this inference is examined more closely in Appendix C.4.

For historical reasons, an antiderivative of $f(x)$ is usually called an *integral* of $f(x)$, and antidifferentiation is called *integration.* The standard notation for an integral of $f(x)$ is

$$\int f(x) \, dx, \tag{3}$$

which is read "the integral of $f(x) \, dx$." The equation

$$\int f(x) \, dx = F(x)$$

is therefore completely equivalent to (2). The function $f(x)$ is called the *integrand.* The "elongated S" symbol in (3) is called the *integral sign*; it was introduced by Leibniz in the earliest days of calculus. Its origin will become clear in the next chapter.

To illustrate a point of usage, we remark that the formulas

$$\int x^2 \, dx = \tfrac{1}{3}x^3 \qquad \text{and} \qquad \int x^2 \, dx = \tfrac{1}{3}x^3 + c \tag{4}$$

are both correct, but the first provides one integral while the second provides all possible integrals. For this reason the integral (3) is usually called the *indefinite integral,* in contrast to the definite integrals discussed in the next chapter. The constant c in the second formula of (4) is called the *constant of integration,* and is often referred to as an "arbitrary" constant. Our previous discussion shows that to find all integrals of a given function $f(x)$, it suffices to find one integral by any method that works—calculation, intelligent guessing, or asking a knowledgeable friend—and then to add an arbitrary constant at the end.

Every derivative that we have ever calculated can be reversed and rewritten as an integral. In particular, the power rule

$$\frac{d}{dx} x^n = nx^{n-1} \qquad \text{becomes} \qquad \int nx^{n-1} \, dx = x^n.$$

For our present purposes the formula

$$\frac{d}{dx} \frac{x^{n+1}}{n+1} = x^n$$

is a more convenient version of the power rule. This gives the form of the integral that we shall memorize and use,

$$\int x^n \, dx = \frac{x^{n+1}}{n+1}, \qquad n \neq -1. \tag{5}$$

In words: *To integrate a power, push the exponent up one unit and divide by the new exponent.*

Example 1 The following integrals are all special cases of (5):

$$\int x^3 \, dx = \frac{x^4}{4} = \frac{1}{4}x^4, \quad \int x^{572} \, dx = \frac{x^{573}}{573} = \frac{1}{573}x^{573},$$

$$\int \frac{dx}{x^5} = \int x^{-5}\,dx = \frac{x^{-4}}{-4} = -\frac{1}{4x^4},$$

$$\int \sqrt{x}\,dx = \int x^{1/2}\,dx = \frac{x^{3/2}}{\frac{3}{2}} = \frac{2}{3}x^{3/2}.$$

The reader will notice that when $n = -1$, the right-hand side of (5) has zero denominator and is therefore meaningless. The treatment of this case, that is, the determination of the integral

$$\int \frac{dx}{x},$$

is one of the most important and fascinating parts of calculus. We return to this problem in Chapter 8.

The following additional integration rules are also slightly disguised versions of familiar facts about derivatives:

$$\int cf(x)\,dx = c\int f(x)\,dx \tag{6}$$

and

$$\int [f(x) + g(x)]\,dx = \int f(x)\,dx + \int g(x)\,dx. \tag{7}$$

The first says that a constant factor can be moved from one side of the integral sign to the other. It is important to understand that this does not apply to variable factors, as can be seen from the fact that

$$\int x^2\,dx \neq x\int x\,dx.$$

Formula (7) says that the integral of a sum is the sum of the separate integrals. This applies to any finite number of terms.

To verify (6) and (7), it is enough to notice that they are equivalent to the differentiation formulas

$$\frac{d}{dx}cF(x) = c\frac{d}{dx}F(x)$$

and

$$\frac{d}{dx}[F(x) + G(x)] = \frac{d}{dx}F(x) + \frac{d}{dx}G(x),$$

where $(d/dx)F(x) = f(x)$ and $(d/dx)G(x) = g(x)$.

Example 2 When rules (5), (6), and (7) are combined, they enable us to integrate any polynomial. For instance,

$$\int (3x^4 + 6x^2)\,dx = 3\int x^4\,dx + 6\int x^2\,dx$$

$$= \tfrac{3}{5}x^5 + 2x^3 + c$$

and

$$\int (5 - 2x^5 + 3x^{11})\, dx = 5 \int dx - 2 \int x^5\, dx + 3 \int x^{11}\, dx$$

$$= 5x - \tfrac{1}{3}x^6 + \tfrac{1}{4}x^{12} + c.$$

Observe that $\int dx = \int 1\, dx = x$. In each of these calculations an arbitrary constant is added at the end so that all possible integrals are included.

Example 3   We can also integrate many nonpolynomials that are expressible as linear combinations of powers:

$$\int \sqrt[3]{x^2}\, dx = \int x^{2/3}\, dx = \tfrac{3}{5}x^{5/3} + c;$$

$$\int \frac{2x^3 - x^2 - 2}{x^2}\, dx = \int (2x - 1 - 2x^{-2})\, dx$$

$$= x^2 - x + \frac{2}{x} + c;$$

$$\int \frac{5x^{1/3} - 2x^{-1/3}}{\sqrt{x}}\, dx = \int (5x^{-1/6} - 2x^{-5/6})\, dx$$

$$= 6x^{5/6} - 12x^{1/6} + c.$$

The formula

$$\int u^n\, du = \frac{u^{n+1}}{n+1}, \qquad n \neq -1, \tag{8}$$

appears to be a trivial variation of (5) in which the letter x is replaced by u. However, let us think of u as some function $f(x)$ of x and take du seriously as the differential of u, so that

$$u = f(x)$$

and

$$du = f'(x)\, dx.$$

Then (8) becomes

$$\int [f(x)]^n f'(x)\, dx = \frac{[f(x)]^{n+1}}{n+1}, \qquad n \neq -1, \tag{9}$$

which is a far-reaching generalization of (5).

Example 4   In practice, we usually exploit this idea by explicitly changing the variable in order to reduce a given integral to an integral of the simple form (8). For instance, in the case of

$$\int (3x^2 - 1)^{1/3}4x\, dx,$$

we notice that the differential of the expression in parentheses is $6x\, dx$, which differs from $4x\, dx$ only by a constant factor, so we write

$$u = 3x^2 - 1,$$

$$du = 6x\, dx,$$

$$x\, dx = \tfrac{1}{6} du.$$

This enables us to translate the given integral from the x-notation to the u-notation, as follows:

$$\int (3x^2 - 1)^{1/3} 4x\, dx = \int u^{1/3} \cdot 4 \cdot \tfrac{1}{6}\, du = \tfrac{2}{3} \int u^{1/3}\, du$$

$$= \tfrac{2}{3} \cdot \tfrac{3}{4} u^{4/3} + c = \tfrac{1}{2} u^{4/3} + c;$$

and by returning to the x-notation we obtain our result,

$$\int (3x^2 - 1)^{1/3} 4x\, dx = \tfrac{1}{2}(3x^2 - 1)^{4/3} + c.$$

This method is called *integration by substitution,* because it depends on a substitution or change of variable to simplify the problem. As formula (9) suggests, the success of the method depends on having an integral in which one part of the integrand is essentially the derivative of another part — where "essentially" means "except for a constant factor."

Remark 1 The integral in Example 4 was deliberately constructed so that the method of substitution works. To emphasize this point, we observe that the similar integral

$$\int (3x^2 - 1)^{1/3}\, dx \tag{10}$$

seems to be "simpler" than the one in Example 4, but is actually much more difficult. If we try the substitution that worked before, we get

$$\int (3x^2 - 1)^{1/3}\, dx = \int u^{1/3} \cdot \frac{du}{6x},$$

and there is no practical way to get rid of the x in the denominator. In a later chapter we will study deeper methods that succeed in this type of problem, but just now there is nothing further we can do.

Remark 2 Many students are tempted to try to integrate (10) by writing

$$\int (3x^2 - 1)^{1/3}\, dx = \frac{(3x^2 - 1)^{4/3}}{4/3} = \frac{3}{4}(3x^2 - 1)^{4/3} + c, \tag{11}$$

which is incorrect. To see this, recall that in calculating integrals we can always check our work quite easily, for if we have a suspected integral of a function $f(x)$, we can test it by computing its derivative to see if the result really equals $f(x)$. It is clear that (11) fails this text, for the derivative of the right side is

$$\frac{3}{4} \cdot \frac{4}{3}(3x^2 - 1)^{1/3} \cdot 6x = (3x^2 - 1)^{1/3} 6x,$$

which is not the integrand of (10).

PROBLEMS

Compute the following integrals. Be sure to include the constant of integration in each answer.

1 $\int (x + 1)\, dx.$ **2** $\int (3x - 2)\, dx.$

3 $\int (x^2 + x^3 + x^4)\, dx.$ **4** $\int x^7\, dx.$

5 $\int \dfrac{dx}{\sqrt{x}}.$ **6** $\int (3x^2 + 2x + 1)\, dx.$

7 $\int x^{3/4}\, dx.$ **8** $\int x^2(x^2 - 1)\, dx.$

9 $\int \dfrac{dx}{\sqrt[3]{x}}.$ **10** $\int (600x - 6x^5)\, dx.$

11 $\int \left(\sqrt{x} - \dfrac{1}{\sqrt{x}} \right) dx.$ **12** $\int (2x - 7)\, dx.$

13 $\int \dfrac{3 + 2x}{\sqrt{x}}\, dx.$ **14** $\int \sqrt{3 + 4x}\, dx.$

15 $\int \sqrt{3x^2 + 1}\; x\, dx.$ **16** $\int \dfrac{dx}{(2x - 3)^2}.$

17 $\int x^2(1 - 4x^3)^{1/5}\, dx.$ **18** $\int \dfrac{x\, dx}{\sqrt{5 - 4x^2}}.$

19 $\int x^{2/3}(2 - x^{5/3})^{-5}\, dx.$ **20** $\int \dfrac{(1 + \sqrt{x})^{1/4}}{\sqrt{x}}\, dx.$

21 $\int \dfrac{(2 + 3x)\, dx}{\sqrt{1 + 4x + 3x^2}}.$ **22** $\int \sqrt{x^2 + x^4}\, dx.$

23 Integrate $\int (x^3)^4 \cdot 3x^2\, dx$ as $\int u^4\, du$ and also as $\int 3x^{14}\, dx$, and compare your results.

24 Integrate $\int (x^3 + 1)^2 \cdot 3x^2\, dx$ as $\int u^2\, du$ and also by multiplying out, and compare your results.

25 Find the integral of $3x^2$ that has the value 10 when $x = 2$. Hint: Since every integral of $3x^2$ has the form $y = x^3 + c$, find the value of c that makes $y = 10$ when $x = 2$.

26 Find the integral $F(x)$ of $\sqrt{x}$ with the property that $F(9) = 9$.

We have seen that the equation

$$\int f(x)\, dx = F(x) \tag{1}$$

is equivalent to

$$\frac{d}{dx} F(x) = f(x). \tag{2}$$

This statement can be interpreted in two ways.

  (a) In accordance with the explanation in Section 5.3, we can think of the symbol

$$\int \cdots dx$$

as operating on the function $f(x)$ to produce its integral, or antiderivative, $F(x)$. From this point of view the integral sign and the dx go together as parts of a single symbol; the integral sign specifies the operation of integration, and the only role of the dx is to tell us that x is the "variable of integration."

  (b) A second interpretation is suggested by our treatment of Example 4 in Section 5.3. Let us write (2) in the form

$$dF(x) = f(x)\, dx,$$

so that $f(x)\, dx$ is explicitly seen to be the differential of $F(x)$. If we now take dx in (1) at its face value, as the differential of x, then the integral sign in (1) acts on the differential of a function, namely, on $f(x)\, dx$, and produces the function itself. Thus, the symbol $\int$ for integration (without considering the

5.4

DIFFERENTIAL
EQUATIONS.
SEPARATION OF
VARIABLES

dx as part of the symbol) stands for the operation which is the inverse of the operation denoted by the symbol d.

We shall use both interpretations. However, the second is particularly convenient, not only for the actual procedures used in computing integrals, but also for solving certain simple differential equations.

A *differential equation* is an equation involving an unknown function and one or more of its derivatives. The *order* of such an equation is the order of the highest derivative that occurs in it.

In the process of integration we have been solving first-order differential equations of the form

$$\frac{dy}{dx} = f(x),$$

where $f(x)$ is a given function. Thus, the equation

$$\frac{dy}{dx} = 3x^2 \qquad \text{is equivalent to} \qquad dy = 3x^2\, dx, \tag{3}$$

and we integrate to obtain the solution,

$$\int dy = \int 3x^2\, dx \qquad \text{or} \qquad y = x^3 + c. \tag{4}$$

Notice that a constant of integration arises on both sides here,

$$y + c_1 = x^3 + c_2,$$

but this can be written as $y = x^3 + (c_2 - c_1)$, and no generality is lost by replacing $c_2 - c_1$ by c. Accordingly, it suffices to add a constant of integration to one side only, as we have done in (4).

We can also handle more complicated differential equations. Let us find y if

$$\frac{dy}{dx} = -2xy^2. \tag{5}$$

If we set aside the obvious solution $y = 0$, this can be written as

$$-\frac{dy}{y^2} = 2x\, dx.$$

Integration now yields

$$\frac{1}{y} = x^2 + c$$

or

$$y = \frac{1}{x^2 + c}. \tag{6}$$

This is called the *general solution* of (5), and different choices of c give different *particular solutions*.

We were able to solve equation (5) by the method of *separation of variables,* that is, by isolating the y's from the x's and integrating. In general, if a first-order differential equation can be written in the form

$$g(y)\, dy = f(x)\, dx,$$

with its variables separated, and if we can carry out the integrations, then we have the solution

$$\int g(y)\, dy = \int f(x)\, dx + c. \tag{7}$$

It should be noted that only in very special cases can the variables be separated in this way. For instance, the differential equation

$$\frac{dy}{dx} = \frac{x+y}{x-y} \tag{8}$$

cannot be solved by this method.

Each of the solutions (4) and (6) of equations (3) and (5) consists of a family of curves corresponding to various values of the constant c. These families are shown in Figs. 5.5 and 5.6. The arbitrary constant that appears in the general solution of a first-order equation can be determined by prescribing, as an *initial condition,* the value of the unknown function $y = y(x)$ at a single value of x, say $y = y_0$ when $x = x_0$. In geometric language, an initial condition means that the solution curve is required to pass through a specific point in the plane. Thus, in Fig. 5.6 the upper and lower solid curves correspond to the initial conditions

$$y = 1 \text{ when } x = 0 \qquad \text{and} \qquad y = -1 \text{ when } x = 0,$$

respectively. We shall see in the next section that this terminology is particularly suitable for mechanical problems, where time is the independent variable and the initial positions or initial velocities of moving bodies are specified.

In the problems just discussed, equation (7) was easily solved for y to yield the solution of the given differential equation as a family of *functions.* It is often convenient not to press this point, and to accept a family of *equations* as the general solution, without demanding explicitly displayed functions. We illustrate by finding the most general curve whose normal at each point passes through the origin, and also the particular curve with this property through the point $(2, 3)$. The normal OP has slope y/x (see Fig. 5.7), and the slope of the tangent is the negative reciprocal of this, so our differential equation is

$$\frac{dy}{dx} = -\frac{x}{y}. \tag{9}$$

Separating variables gives $y\, dy = -x\, dx$, and by integrating we get

$$\tfrac{1}{2}y^2 = -\tfrac{1}{2}x^2 + c.$$

If we put $r^2 = 2c$, our general solution of (9) takes the neater form

$$x^2 + y^2 = r^2.$$

This is the family of all circles with center at the origin, as the reader may

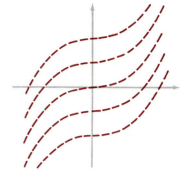

Figure 5.5 $y = x^3 + c$

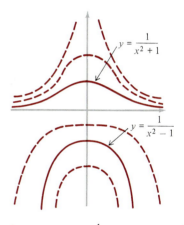

Figure 5.6 $y = \dfrac{1}{x^2 + c}$

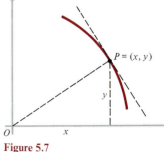

Figure 5.7

have foreseen. By setting $x = 2$ and $y = 3$, we find that $r^2 = 13$, so

$$x^2 + y^2 = 13$$

is the particular solution of (9) passing through the point (2, 3). It is clearly more reasonable to leave this solution as it is than to insist that it be solved for y.

Remark 1 By rights, differential equations should perhaps be called *derivative equations.* However, as we saw in Section 5.2, in the early days of calculus differentials were the primary concepts and derivatives were secondary, and so the term arose in a natural way. In any case, it has been in standard use for hundreds of years and no one dreams of changing it now.

Remark 2 The mathematical description of a physical (or biological or chemical) process is usually given in terms of functions that show how the quantities involved change as time goes on. When we know such a function, we can find its rate of change by calculating the derivative. Often, however, we are faced with the reverse problem of finding an unknown function from given information about its rate of change. This information is usually expressed in the form of an equation involving derivatives of the unknown function. These differential equations arise so frequently in scientific problems that their study constitutes one of the main branches of mathematics. We continue with some important applications of this subject in the next section, and return to it from time to time throughout the rest of our work.

PROBLEMS

Find the general solution of each of the following differential equations.

1 $\dfrac{dy}{dx} = 6x^2 + 4x - 5.$

2 $\dfrac{dy}{dx} = (3x + 1)^3.$

3 $\dfrac{dy}{dx} = 24x^3 + 18x^2 - 8x + 3.$

4 $\dfrac{dy}{dx} = 2\sqrt{y}.$

5 $\dfrac{dy}{dx} = \dfrac{x + \sqrt{x}}{y - \sqrt{y}}.$

6 $\dfrac{dy}{dx} = \sqrt[3]{\dfrac{y}{x}}.$

7 $\dfrac{dy}{dx} = \dfrac{1}{x^2} + x.$

Find the particular solution of each of the following differential equations that satisfies the given initial condition.

8 $\dfrac{dy}{dx} = 10x + 5,\ y = 15$ when $x = 0.$

9 $\dfrac{dy}{dx} = 2xy^2,\ y = 1$ when $x = 2.$

10 $\dfrac{dy}{dx} = \dfrac{x}{y},\ y = 3$ when $x = 2.$

11 $y\dfrac{dy}{dx} = x(y^4 + 2y^2 + 1),\ y = 1$ when $x = 4.$

12 $\dfrac{dy}{dx} = \dfrac{5 + 3x^2}{2 + 2y},\ y = 2$ when $x = -2.$

13 $\dfrac{dy}{dx} = \sqrt{xy},\ y = 64$ when $x = 9.$

In Problems 14 to 19, verify that the given function is a solution of the given differential equation for all choices of the constants A and B.

14 $y = x + Ax^2,\ x\dfrac{dy}{dx} = 2y - x.$

15 $y = Ax + x^3,\ x\dfrac{dy}{dx} = y + 2x^3.$

16 $y = x + A\sqrt{x^2 + 1},\ (x^2 + 1)\dfrac{dy}{dx} = xy + 1.$

17 $y = Ax + \sqrt{x^2 + 1},\ x\dfrac{dy}{dx} = y - \dfrac{1}{\sqrt{x^2 + 1}}.$

18 $y = Ax + \dfrac{B}{x}, \ x^2 \dfrac{d^2y}{dx^2} + x \dfrac{dy}{dx} - y = 0.$

19 $y = Ax + Bx^2, \ x^2 \dfrac{d^2y}{dx^2} - 2x \dfrac{dy}{dx} + 2y = 0.$

20 In a certain barbarous land, two neighboring tribes have hated one another from time immemorial. Being barbarous peoples, their powers of belief are strong, and a solemn curse pronounced by the medicine man of the first tribe deranges the members of the second tribe and drives them to murder and suicide. If the rate of change of the population P of the second tribe is $-\sqrt{P}$ people per week, and if the population is 676 when the curse is uttered, when will they all be dead?

Much of the original inspiration for the development of calculus came from the science of mechanics, and these two subjects have continued to be inseparably connected down to the present day. Mechanics rests on certain basic principles that were first laid down by Newton. The statement of these principles requires the concept of the derivative, and we shall see in this section that their applications depend on integration and the solution of differential equations.

Rectilinear motion is motion along a straight line. In contrast, motion along a curved path is sometimes called *curvilinear motion.* Our present purpose is to study the rectilinear motion of a single *particle,* that is, of a point at which a body of mass m is imagined to be concentrated. In discussing the motion of physical objects, such as cars, bullets, falling rocks, etc., we often ignore the size and shape of the object and think of it as if it were a particle.

The position of our particle is completely determined by its coordinate s with respect to a conveniently chosen coordinate system on the line (Fig. 5.8). Since the particle moves, s is a function of the time t, as measured from some convenient initial instant $t = 0$. We symbolize this by writing $s = s(t)$. As we know from the discussion in Section 2.4, the *velocity* v of the particle is the rate of change of its position,

$$v = \frac{ds}{dt},$$

and the *speed* is the absolute value of the velocity.* In general, the velocity of a moving particle changes with time, and the *acceleration* a is the rate of change of velocity,

$$a = \frac{dv}{dt} = \frac{d}{dt}\left(\frac{ds}{dt}\right) = \frac{d^2s}{dt^2}.$$

This is positive or negative according as v is increasing or decreasing.

The basic assumption of Newtonian mechanics is that force is required in order to change velocity; that is, acceleration is caused by force. The concept of force originates in the subjective feeling of effort that we experience when we change the velocity of a physical object, for instance, when we push a stalled car or throw a rock. In the case of rectilinear motion, we assume that a

5.5
MOTION UNDER GRAVITY. ESCAPE VELOCITY AND BLACK HOLES

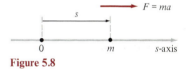

Figure 5.8

___

* We have pointed out before that even though the words "velocity" and "speed" are synonymous in ordinary usage, in physics (and mathematics) they have different meanings. The distinction lies in the fact that the velocity v is sometimes positive and sometimes negative, depending on whether s is increasing or decreasing. On the other hand, the speed is $|v|$, and hence is never negative.

force can be expressed by a number, which is positive or negative according as the force acts in the positive or negative direction.

Newton's second law of motion states that the acceleration of a particle is directly proportional to the force F acting on it and inversely proportional to its mass m,

$$a = \frac{F}{m},\tag{1}$$

or equivalently,

$$F = ma.\tag{2}$$

[The units of measurement for these quantities are always chosen so that the constant of proportionality in equation (1) has the value 1, as shown.] Thus, if the force is doubled, then by (1) the resulting acceleration is also doubled; and if the mass is doubled, the acceleration is cut in half. In this context, the mass of a body can be interpreted as its capacity to resist acceleration.*

From one point of view, equation (2) can be considered as nothing more than a definition of force, for the right-hand side is a quantity that can be calculated by measuring the mass and observing the motion, and this determines the force. On the other hand, the force F is often known in advance from fairly simple physical considerations. The innocent-looking equation $F = ma$ then becomes the second-order differential equation

$$m \frac{d^2s}{dt^2} = F.\tag{3}$$

This equation has profound consequences, for in principle we can find the particle's position at any time t by solving (3) with appropriate initial conditions.†

Example 1 Find the motion of a stone of mass m which is dropped from a point above the surface of the earth.

Solution The most important example of a known force is the familiar "force of gravity." From direct experimental evidence, we know that the force of gravity acting on the stone (this is the *weight* of the stone) is directed downward and has magnitude $F = mg$, where g is the constant acceleration due to gravity near the surface of the earth ($g = 32$ ft/s^2 or 9.80 m/s^2, approximately). If s is the stone's position as measured along a vertical axis, with the positive direction pointing downward and the origin at the initial position of the stone (Fig. 5.9), then equation (3) is

$$m \frac{d^2s}{dt^2} = mg \qquad \text{or} \qquad \frac{d^2s}{dt^2} = g.$$

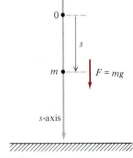

Figure 5.9

---

* *Newton's first law of motion* asserts that if no force acts on a particle, then its velocity does not change, that is, its acceleration is zero. This is clearly a special case of (1).

† The intellectual impact of Newton's $F = ma$ on the seventeenth and eighteenth centuries was even greater than that of Einstein's $E = mc^2$ on the twentieth century.

Integrating this equation twice gives

$$v = \frac{ds}{dt} = gt + c_1, \tag{4}$$

$$s = \tfrac{1}{2}gt^2 + c_1 t + c_2, \tag{5}$$

where c_1 and c_2 are constants of integration. Since the stone is "dropped" (that is, released with no initial velocity) at time $t = 0$ from the point chosen as the origin, the initial conditions are

$$v = 0 \quad \text{and} \quad s = 0 \quad \text{when} \quad t = 0.$$

The condition $v = 0$ when $t = 0$ gives $c_1 = 0$, and $s = 0$ when $t = 0$ gives $c_2 = 0$. We therefore have

$$v = gt, \tag{6}$$

$$s = \tfrac{1}{2}gt^2, \tag{7}$$

at least until the stone hits the ground. If we change the situation and require that the stone be thrown downward with an initial velocity v_0 from the initial position $s = s_0$ at time $t = 0$, then the initial conditions are

$$v = v_0 \quad \text{and} \quad s = s_0 \quad \text{when} \quad t = 0,$$

and (4) and (5) become

$$v = gt + v_0,$$

$$s = \tfrac{1}{2}gt^2 + v_0 t + s_0.$$

It should be pointed out that in this discussion we have ignored the effect of air resistance, and have assumed that the only force acting on the falling stone is the force of gravity. It is possible to take the air resistance into account, but in this case equation (3) is too complicated for us to cope with here. We return to this topic in Chapter 8.

We also remark that if distance is measured in feet and time in seconds, so that g has the numerical value 32, then (6) and (7) become

$$v = 32t \quad \text{and} \quad s = 16t^2.$$

It is clear from the first of these equations that the velocity of the stone increases by 32 ft/s during each second of fall, and of course this is what is meant by the statement that the acceleration due to gravity is 32 feet per second per second (ft/s^2).

Example 2 A stone is thrown upward with an initial velocity of 128 ft/s from the roof of a building 320 ft high. Express its height above the ground as a function of time. Find the maximum height the stone attains. Assuming that the stone misses the building on its way down, how long does it take to hit the ground? What are the velocity and speed of the stone at the moment it hits the ground?

Solution We place the s-axis with its origin on the ground and the positive direction pointing upward (Fig. 5.10). Since the force of gravity is directed

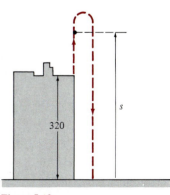

Figure 5.10

downward, and by equation (2) the force and acceleration have the same sign, the acceleration of the stone is given by

$$a = \frac{d^2s}{dt^2} = -32. \tag{8}$$

Integrating this equation yields

$$v = \frac{ds}{dt} = -32t + c_1,$$

and by using the initial condition $v = 128$ when $t = 0$, we get

$$v = \frac{ds}{dt} = -32t + 128. \tag{9}$$

A second integration gives

$$s = -16t^2 + 128t + c_2,$$

and since $s = 320$ when $t = 0$, we obtain

$$s = -16t^2 + 128t + 320 \tag{10}$$

as the height of the stone above the ground at any time t.

To find the maximum height attained by the stone, we write (9) in the form

$$v = -32(t - 4).$$

This tells us that for $t < 4$, the velocity is positive, so the stone is moving upward. When $t = 4$, the velocity is zero and the stone is motionless for an instant. For $t > 4$, the velocity is negative and the stone is falling. We therefore find the maximum height by putting $t = 4$ into equation (10). This gives $s = -16 \cdot 16 + 128 \cdot 4 + 320 = -256 + 512 + 320 = 576$ as the maximum height.

The stone hits the ground when $s = 0$. By using equation (10) we see that this leads us to the sequence of equivalent equations

$$-16t^2 + 128t + 320 = 0,$$

$$-16(t^2 - 8t - 20) = 0,$$

$$(t - 10)(t + 2) = 0.$$

Thus $s = 0$ when $t = 10$ or $t = -2$. The second answer is meaningless in the circumstances, and can be discarded. Therefore the stone hits the ground 10 s after being thrown.

To find the velocity of the stone at the moment it hits the ground, we put $t = 10$ into equation (9): $v = -32 \cdot 10 + 128 = -320 + 128 = -192$. The velocity at that moment is therefore -192 ft/s, and the minus sign tells us that the stone is moving downward. The speed at that moment is $|-192| = 192$ ft/s.

In these examples we have treated the acceleration due to gravity as if it were a constant. This is almost true for moving bodies that stay fairly close to the surface of the earth. However, to study the motion of a body that moves away from the earth into space, we must take account of the fact that the force of gravity varies inversely as the square of the distance from the center of the earth.

Example 3 Suppose a rocket is fired vertically upward with initial velocity v_0, and thereafter coasts with no further expenditure of energy. For larger values of v_0 it rises higher before coming to rest and falling back to earth. What must v_0 be in order for the rocket never to come to rest, and thereby to escape completely from the earth's gravitational attraction?

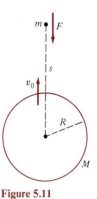

Figure 5.11

Solution According to *Newton's law of gravitation,* any two particles of matter in the universe attract each other with a force that is jointly proportional to their masses and inversely proportional to the square of the distance between them. In the present situation (see Fig. 5.11), this means that the force F attracting the rocket back to earth is given by the inverse square law

$$F = -G \frac{Mm}{s^2},$$

where G is a positive constant, M and m are the masses of the earth and the rocket, and s is the distance from the center of the earth to the rocket.*

We begin our detailed analysis of the problem by observing that in this case Newton's second law of motion $F = ma$ becomes

$$m \frac{d^2s}{dt^2} = -G \frac{Mm}{s^2},$$

and so

$$\frac{d^2s}{dt^2} = -\frac{GM}{s^2}. \tag{11}$$

This tells us at the outset that the motion of the rocket does not depend on the rocket's own mass. We can put the constants here into a more convenient form by noticing that the acceleration d^2s/dt^2 has the value $-g$ when $s = R$, where R is the radius of the earth. This gives

$$-g = -\frac{GM}{R^2} \quad \text{or} \quad GM = gR^2;$$

and since $d^2s/dt^2 = dv/dt$, we can write (11) as

$$\frac{dv}{dt} = -\frac{gR^2}{s^2}. \tag{12}$$

Our next step is to eliminate t from this equation by using the chain rule to write

$$\frac{dv}{dt} = \frac{dv}{ds} \frac{ds}{dt} = \frac{dv}{ds} v.$$

---

* It can be proved—and will be proved in a later chapter—that the gravitational attraction exerted on the rocket by the earth as a whole is the same as that which would be exerted by a particle of mass M located at the center of the earth.

Equation (12) now becomes

$$v\frac{dv}{ds} = -\frac{gR^2}{s^2}.$$

By separating variables and integrating, we obtain

$$\int v\,dv = gR^2 \int -\frac{ds}{s^2}$$

or

$$\tfrac{1}{2}v^2 = \frac{gR^2}{s} + c. \tag{13}$$

To evaluate the constant of integration c, we use the initial condition that $v = v_0$ when $s = R$, so

$$\tfrac{1}{2}v_0{}^2 = gR + c$$

and

$$c = \tfrac{1}{2}v_0{}^2 - gR.$$

With this value of c, equation (13) becomes

$$\tfrac{1}{2}v^2 = \frac{gR^2}{s} + (\tfrac{1}{2}v_0{}^2 - gR). \tag{14}$$

Our final conclusion emerges from (14) as follows: For the rocket to escape from the earth, it must move in such a way that $\tfrac{1}{2}v^2$ is always positive, for if $\tfrac{1}{2}v^2$ vanishes, the rocket stops moving and then falls back to earth. But the first term on the right of (14) evidently approaches zero as s increases. Therefore, in order to guarantee that $\tfrac{1}{2}v^2$ is positive no matter how large s is, we must have $\tfrac{1}{2}v_0{}^2 - gR \geq 0$. This is equivalent to $v_0{}^2 \geq 2gR$ or $v_0 \geq \sqrt{2gR}$. The quantity $\sqrt{2gR}$ is usually called the *escape velocity* for the earth. We can easily estimate its value by using the approximations $g \cong 32$ ft/s^2 and $R \cong$ 4000 mi:

$$\sqrt{2gR} \cong \sqrt{2 \cdot 32 \text{ ft/s}^2 \cdot 4000 \text{ mi}}$$

$$\cong \sqrt{2 \cdot 32 \cdot \tfrac{1}{5280} \text{ mi/s}^2 \cdot 4000 \text{ mi}}$$

$$\cong 7 \text{ mi/s} \cong 25{,}000 \text{ mi/h}.$$

Remark 1   In just the same way as in this example, the quantity $\sqrt{2g'R'}$ is the escape velocity for any planet, satellite, or star, where R' and g' are understood to be the radius and the acceleration due to gravity at the surface. If the radius of such a body is decreased while the mass is unchanged, the escape velocity at the surface increases. Why?

Remark 2   Most normal stars are maintained in their gaseous, puffed-up state by radiation pressure from within, which is generated by the burning of nuclear fuel. When the nuclear fuel gives out, the star undergoes gravitational collapse into a very much smaller sphere of essentially the same mass. The crushed, degenerate matter of these collapsed stars can sustain two types of equilibrium, depending on the mass of the star. *White dwarfs* are those that result when the mass is less than about 1.3 solar masses, and *neutron stars* arise when the mass is between 1.3 and 2 solar masses. For heavier stars no equilibrium is possible, and collapse continues until the escape velocity at the surface reaches the speed of light. Collapsed stars of this type are completely invisible, since no radiation can ever escape. These are the so-called *black holes.*

PROBLEMS

1 In Example 2, how long after the stone is thrown does it pass the roof of the building on its way down? What are the velocity and speed at that moment?

2 In Example 2, if the stone were simply dropped from the roof, what would s be as a function of time? How long would the stone fall?

3 In Example 2, the origin of the s-axis is at ground level. If the origin is placed at the top of the building, what are the formulas for v and s that correspond to (9) and (10)?

4 A ball is thrown upward from the top of a cliff 96 ft high with an initial velocity of 64 ft/s. Find the maximum height of the ball above the ground below. Assuming that the ball misses the cliff on its way down, how long does it take to hit the ground?

5 A bag of ballast is accidentally dropped from a balloon which is stationary at an altitude of 4900 m. How long does it take for the bag to hit the ground?

6 With what velocity must an arrow be shot upward in order to fall back to its starting point 10 seconds later? How high will it rise?

7 A boy at the top of a cliff 299 ft high throws a rock straight down, and it hits the ground $3\frac{1}{4}$ seconds later. With what velocity does the boy throw the rock?

8 A woman standing on a bridge throws a stone straight up. Exactly 5 seconds later the stone passes the woman on the way down, and 1 second after that it hits the water below. Find the initial velocity of the stone and the height of the bridge above the water.

9 A stone is dropped from the roof of a building 256 ft high. Two seconds later a second stone is thrown downward from the roof of the same building with an

initial velocity of v_0 feet per second. If both stones hit the ground at the same time, what is v_0?

10 How much time does a train traveling 144 km/h take to stop if it has a constant negative acceleration of 4 m/s²? How far does the train travel in this time?

11 A man standing on the ground throws a stone straight up. Neglecting the height of the man, find the maximum height of the stone in terms of the initial velocity v_0. What is the smallest value of v_0 that will make it possible for the stone to land on top of a 144-ft building?

12 On the surface of the moon the acceleration due to gravity is approximately $\frac{1}{6}$ that at the surface of the earth, and on the surface of the sun it is approximately 29 times as great as at the surface of the earth. If a person on earth can jump with enough initial velocity to rise 5 ft, how high will the same initial velocity carry that person (a) on the moon? (b) on the sun?

13 Newton's law of gravitation implies that the acceleration due to gravity at the surface of a planet (or the moon or the sun) is directly proportional to the mass of the planet and inversely proportional to the square of the radius.

   (a) If g_m denotes the acceleration due to gravity at the surface of the moon, use the fact that the moon has approximately $\frac{3}{11}$ the radius and $\frac{1}{81}$ the mass of the earth to show that g_m is approximately $g/6$.

   (b) Use part (a) to show that the escape velocity for the moon is approximately 1.5 mi/s.

14 Show that the point between the earth and the moon where the two exert equal but opposite gravitational forces on a particle is $\frac{9}{10}$ of the way from the center of the earth to the center of the moon.

ADDITIONAL PROBLEMS FOR CHAPTER 5

SECTION 5.3

Compute the following integrals. Be sure to include the constant of integration in each answer.

1 $\int (3x^4 - 7x^3 + 10)\,dx.$

2 $\int \dfrac{dx}{\sqrt[3]{x^4}}.$

3 $\int \dfrac{x^3 - 3x^2 + x - 2\sqrt{x}}{x}\,dx.$

4 $\int \left(x + \dfrac{1}{x}\right)^2 dx.$

5 $\int x(x+1)^2\,dx.$
6 $\int (x+3)(x^2-1)\,dx.$
7 $\int (51x^2 - 108x^3)\,dx.$

8 $\int \dfrac{x^3 + 2}{x^2}\,dx.$

9 $\int (2 - \sqrt{x})(3 + \sqrt{x})\,dx.$
10 $\int \sqrt{x}\,(7x^2 - 5x + 3)\,dx.$
11 $\int \sqrt{2 - 3x}\,dx.$
12 $\int (3 + 7x^2)^9 5x\,dx.$
13 $\int (5x + 2)^{164}\,dx.$
14 $\int (3 - 4x)^{3/4}\,dx.$

15 $\int \dfrac{5x\,dx}{\sqrt{1 + x^2}}.$

16 $\int \sqrt{3x^2 - 2}\,x\,dx.$

17 $\int \dfrac{x^2\,dx}{\sqrt{2x^3 - 1}}.$

18 $\int \dfrac{dx}{\sqrt[3]{(7x + 3)^2}}.$

19 $\int \dfrac{(x - 1)\,dx}{\sqrt[3]{x^2 - 2x + 3}}.$

20 $\int \dfrac{dx}{x\sqrt{3x}}.$

21 $\int \dfrac{x\,dx}{\sqrt[3]{(2 - x^2)^2}}.$

22 $\int \dfrac{x\,dx}{\sqrt{(x^2 - 4)^3}}.$

23 $\int \left(1 + \dfrac{1}{x}\right)^2 \dfrac{dx}{x^2}.$

24 $\int \dfrac{x^2\,dx}{(2 + 3x^3)^3}.$

25 $\int (x^2 + 2x + 1)^{2/3}\,dx.$
26 $\int x\sqrt[3]{1 + x^2}\,dx.$
27 $\int x\sqrt[3]{1 + x}\,dx.$

28 $\int \dfrac{\sqrt{2x^6 + x^4}}{x}\,dx.$

29 $\int (x^3 + x + 32)^{9/2}(3x^2 + 1)\,dx.$
30 $\int (x^2 + 1)^7 x^3\,dx.$
31 $\int (x^3 - 1)^{1/3} x^5\,dx.$

SECTION 5.4

32 Find the general solution of each of the following differential equations:

(a) $\dfrac{dy}{dx} = 2y^2(4x^3 + 4x^{-3});$

(b) $\dfrac{dy}{dx} = \sqrt{(x^2 - x^{-2})^2 + 4}.$

33 Find the indicated particular solution of each of the following differential equations:

(a) $\dfrac{dy}{dx} = \dfrac{x(1 + y^2)^2}{y(1 + x^2)^2},\ y = 1$ when $x = 2;$

(b) $\dfrac{dy}{dx} = \sqrt{xy - 4x - y + 4},\ y = 8$ when $x = 5.$

34 The equation $x^2 = 4py$ represents the family of all parabolas with vertex at the origin and axis the y-axis. Find the family of curves that intersect the curves of this given family at right angles. Hint: Show first that the slope of the tangent at every point $(x, y)\,(y \neq 0)$ on each curve of the given family is $2y/x.$

35 Solve Problem 34 if the given family is $xy = c.$
36 Find y as a function of x if $dy/dx + y/x = 0.$
37 Equation (8) in Section 5.4 can be written as

$$\dfrac{dy}{dx} = \dfrac{1 + y/x}{1 - y/x},$$

and this suggests the substitution $z = y/x.$ Use this idea to replace y by z as the dependent variable, and show that the variables can be separated in the resulting differential equation. Notice that the necessary integrations are beyond our capacity at the present stage, so in spite of our progress we have reached a temporary dead end.

SECTION 5.5

38 A ball is thrown vertically upward with an initial velocity of 78 ft/s from the roof of a building 400 ft high.

Find the distance s from the ground up to the ball t seconds later. If the ball misses the building on the way down, how long does it take to hit the ground?

39 (a) A bullet is fired downward with a velocity of 400 ft/s from an airplane 20,000 ft above the ocean. Neglecting air resistance, how long does it take the bullet to reach the water, and what is its velocity at the moment of impact?

(b) If the bullet is merely dropped from the airplane, how long does it take to fall, and what is its velocity on impact?

40 Show that a rock thrown straight up from the ground takes just as long to rise to its highest point as it does to fall back to its initial position. How is the velocity with which it hits the ground related to its initial velocity? Answer the same question for its speed.

41 A ball is dropped out of a window 19.6 m above the ground. At the same time another ball is thrown straight down from a window 79.6 m above the ground. If both balls reach the ground at the same moment, find the initial velocity of the second ball.

42 An automobile is traveling in a straight line at a velocity of v_0 feet per second. The driver suddenly applies the brakes, and the car stops in T seconds after traveling S feet. If the brakes produce a constant negative acceleration of $-a_0$ ft/s², find formulas for T and S in terms of v_0 and a_0.

43 An astronaut stands on the edge of a cliff and drops a stone. She observes that it takes 4 s for the stone to fall to the ground at the bottom. On earth, this would mean that the cliff is 256 ft high. How high is the cliff (a) if the astronaut is on the moon, where the acceleration due to gravity is approximately 5.5 ft/s²? (b) if she is on Jupiter, where the acceleration due to gravity is approximately 85 ft/s²?

44 The results of Problem 13 in Section 5.5 are given in the second column of the following table:

	Earth	Moon	Jupiter	Saturn	Sun
Mass (earth = 1)	1	$\frac{1}{81}$	317	95	332,000
Radius (mi)	4000	1100	43,000	36,000	432,000
Acceleration of gravity	g	$g/6$	$2.6g$	$1.2g$	$29g$
Escape velocity (mi/s)	7	1.5	38	23	400

Verify the rough approximations given in the third and fourth rows for Jupiter, Saturn, and the sun.

45 If the sun could be crushed into a smaller sphere with the same mass, estimate what its new radius would have to be in order to increase the escape velocity at its surface to the speed of light (approximately 186,000 mi/s).

6

DEFINITE INTEGRALS

6.1
INTRODUCTION

At the beginning of Chapter 2 we described calculus as the study of methods for calculating two important quantities associated with curves, namely,

1   Slopes of tangent lines to curves, and
2   Areas of regions bounded by curves.

Of course, this description gives an oversimplified view of the subject, for it emphasizes calculus as a tool in the service of geometry but says nothing about its indispensable role in the study of science. Nevertheless, it explains the traditional division of calculus into two distinct parts: differential calculus, which deals with slopes of tangent lines, and integral calculus, which is concerned with areas.

The problem of areas was of great interest to the ancient Greeks. They knew a good deal about the areas of triangles, circles, and related configurations, but any other figure presented a new and usually insoluble problem. Archimedes was able to apply a technique called the *method of exhaustion* to calculate the area of a segment of a parabola, and also to calculate a few other particular geometric quantities. But for almost two thousand years this handful of calculations by Archimedes stood as the isolated achievement of a great genius, unmatchable by others. However, by the middle of the seventeenth century several European thinkers—most notably Fermat and Pascal—began to push the method of exhaustion beyond the point where Archimedes had left it. The decisive breakthrough was achieved a little later by Newton and Leibniz, who showed that if a quantity can be computed by exhaustion, then it can also be computed much more easily by using antiderivatives. This crucial discovery is called the Fundamental Theorem of Calculus. It binds together the two parts of the subject, and is undoubtedly (as we have said before) the most important single fact in the whole of mathematics.

This is the path we follow in the present chapter. Since calculations will seem to play a prominent part in our work, it is even more necessary than usual for students to keep firmly in mind that the underlying ideas are more important than the calculations.

154

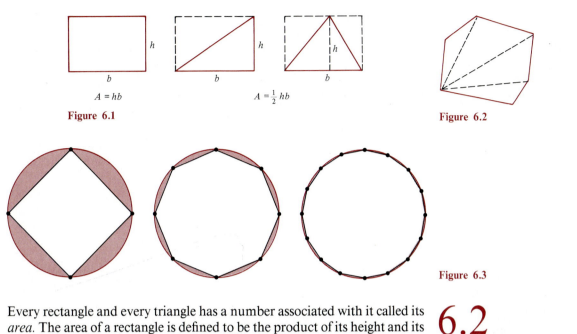

Figure 6.1

$A = hb$ $A = \frac{1}{2}hb$

Figure 6.2

Figure 6.3

Every rectangle and every triangle has a number associated with it called its *area.* The area of a rectangle is defined to be the product of its height and its base, and the area of a triangle is one-half the product of the height and the base (Fig. 6.1). Since a polygon can always be decomposed into triangles (Fig. 6.2), its area is the sum of the areas of these triangles.

The circle is a more difficult figure. The Greeks solved the problem of finding its area in a very natural way. First, they approximated this area by inscribing a square (Fig. 6.3). Then they improved the approximation step by step by doubling and redoubling the number of sides, that is, by inscribing a regular octagon, than a regular 16-gon, and so on. The areas of these inscribed polygons evidently approach the exact area of the circle more and more closely. This idea yields the familiar formula

$$A = \pi r^2 \tag{1}$$

for the area A of a circle in terms of its radius r. The details of the reasoning are as follows. Suppose that the circle has inscribed in it a regular polygon with a large number of sides (Fig. 6.4). Each of the small isosceles triangles shown in the figure has area $\frac{1}{2}hb$, and the sum of these areas equals the area of the polygon, which closely approximates the area of the circle. If p denotes the perimeter of the polygon, then we see that

$$A_{\text{polygon}} = \tfrac{1}{2}hb + \tfrac{1}{2}hb + \cdots + \tfrac{1}{2}hb$$
$$= \tfrac{1}{2}h(b + b + \cdots + b) = \tfrac{1}{2}hp.$$

Now let c be the circumference of the circle, so that $c = 2\pi r$ by the definition of π.* Then, as the number of sides of the polygon increases, h approaches r (in symbols, $h \to r$), $p \to c$, and therefore

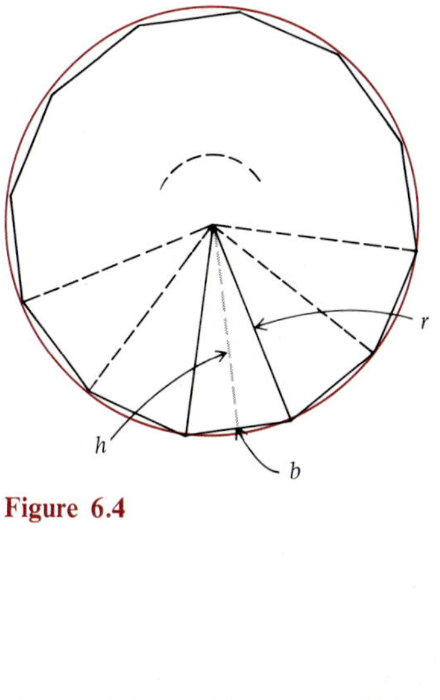

$$A_{\text{polygon}} = \tfrac{1}{2}hp \to \tfrac{1}{2}rc = \tfrac{1}{2}r(2\pi r) = \pi r^2,$$

---

* That is, π is defined to be the ratio of the circumference to the diameter, so $\pi = c/2r$ and therefore $c = 2\pi r$.

6.2

THE PROBLEM OF AREAS

Figure 6.4

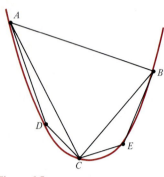

Figure 6.5

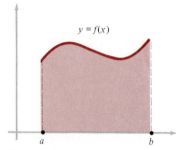

Figure 6.6

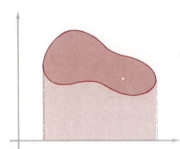

Figure 6.7

which establishes (1). The phrase "method of exhaustion" is clearly a good description of this process, because the area of the circle is "exhausted" by the areas of the inscribed polygons.

We next examine the procedure by which Archimedes calculated the area of a parabolic segment, that is, the area of the part of the parabola in Fig. 6.5 bounded by the arbitrary chord AB and the arc $ADCEB$. There is no convenient way to inscribe regular polygons in this figure, so Archimedes used triangles instead. His first approximation was the triangle ABC, where the vertex C is chosen as that point where the tangent to the parabola is parallel to AB. His second approximation was obtained by adding to the triangle ABC the two triangles ACD and BCE, where the vertex D is the point where the tangent is parallel to AC and the vertex E is the point where the tangent is parallel to BC. To obtain his third approximation, he inscribed triangles in the same way in each of the four regions still not included (one such region is that between the arc CE and the chord CE), so his third approximation was the sum of the areas of the triangles ABC, ACD, BCE, and the four new triangles. By continuing to exhaust the parabolic segment in this way, he was able to show that its area is exactly four-thirds the area of the first triangle ABC. The details of his argument are a bit complicated; and since our interest here is mainly in the idea of the method of exhaustion, we give these details in Appendix A.2 for students who wish to pursue the matter.

The general problem before us is that of finding the area of a region with a curved boundary. However, most of our work will be concentrated on a special case of this general area problem — namely, finding the area under the graph of a function $y = f(x)$ between two vertical lines $x = a$ and $x = b$, as shown in Fig. 6.6. Such a region has a boundary that is curved only along its upper edge, and is therefore much easier to work with. A knowledge of this special case is often enough to enable us to cope with more complicated regions. To understand how this is possible, notice in Fig. 6.7 that the area of a region whose entire boundary is curved can often be obtained by subtracting the area under its lower edge from the area under its upper edge, where each of the latter areas is of the special type shown in Fig. 6.6.

In Section 6.4 and thereafter, we will denote an area of the type shown in Fig. 6.6 by the standard symbol

$$\int_a^b f(x)\,dx, \tag{2}$$

which is read "the definite integral from a to b of $f(x)\,dx$." The reason for this notation will become clear in Section 6.4. For the present, however, we warn students in advance not to confuse the definite integral (2) with the indefinite integral (or antiderivative)

$$\int f(x)\,dx \tag{3}$$

introduced in Chapter 5. In spite of the fact that these two integrals have the same family name and look very much alike, they are totally different entities: The definite integral (2) is a number, and the indefinite integral (3) is a function (or a collection of functions).

At first sight it might appear that the problem of calculating areas is a matter of geometry and nothing more—interesting to mathematicians, perhaps, but with no practical uses in the real world outside of mathematics. This is not the case at all. It will become clear in the next chapter that many important concepts and problems in physics and engineering depend on exactly the same kinds of ideas as those used in calculating areas. As examples we mention the concepts of work and energy in physics, and also the engineering problem of finding the total force acting against the face of a dam due to the pressure of the water in a reservoir. Finding areas is therefore much more than merely a game mathematicians play for their own diversion. Nevertheless, for the sake of clarity we confine our attention in this chapter to the area problem itself, and in Chapter 7 we begin to sample the immense range of applications of the underlying idea.

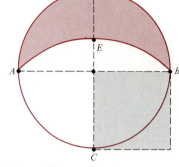

Figure 6.8 Squaring the lune.

Remark As a matter of historical interest, it appears that the first person to find the exact area of a figure bounded by curves was Hippocrates of Chios, the most famous Greek mathematician of the fifth century B.C. To understand what he did, consider the circle shown in Fig. 6.8, with the points A, B, C, D at the ends of the horizontal and vertical diameters. Using C as a center, describe the circular arc AEB connecting A and B. The crescent-shaped figure bounded by the arcs ADB and AEB is called a *lune of Hippocrates,* after the man who made the remarkable discovery that its area is exactly equal to the area of the shaded square whose side is the radius of the circle. Thus Hippocrates "squared the lune," even though he was unable to square the circle itself. The details of his proof are given in Appendix A.3.

In order to clarify our discussion of definite integrals in the next section, we introduce here a standard mathematical notation used for abbreviating long sums. This is called the "sigma notation," because it uses the Greek letter Σ (sigma). In the Greek alphabet the letter Σ corresponds to our letter S, which is the first letter of the word *sum.* This helps us to remember the purpose of the sigma notation, which is to suggest the idea of summation or addition.

Thus, if $a_1, a_2, \ldots, a_n$ are any given numbers, their sum is denoted by

$$\sum_{k=1}^{n} a_k. \tag{1}$$

This symbol is read "the sum from $k = 1$ to n of a_k." The idea compressed in (1) is that we are to write down each of the numbers a_k as the subscript k varies from 1 to n (namely, $a_1, a_2, \ldots, a_n$) and then add all these numbers together:

$$\sum_{k=1}^{n} a_k = a_1 + a_2 + \cdots + a_n.$$

We write $k = 1$ below the Σ in (1), and n above it, to say that the sum starts with the term a_k with k replaced by 1, and stops with the term a_k with k replaced by n. The letter k used as the subscript here is called the *index of summation.* Any other letter (i or j, for instance) would do just as well.

6.3

THE SIGMA NOTATION AND CERTAIN SPECIAL SUMS

We give a few specific examples of the sigma notation:

$$\sum_{k=1}^{3} \frac{k}{k^2+1} = \frac{1}{1^2+1} + \frac{2}{2^2+1} + \frac{3}{3^2+1};$$

$$\sum_{k=1}^{4} (-1)^{k+1} \frac{1}{k^2} = \frac{1}{1^2} - \frac{1}{2^2} + \frac{1}{3^2} - \frac{1}{4^2};$$

$$\sum_{k=1}^{n} k = 1 + 2 + \cdots + n;$$

$$\sum_{k=1}^{n} 2k = 2 + 4 + \cdots + 2n;$$

$$\sum_{k=1}^{n} (2k-1) = 1 + 3 + \cdots + (2n-1).$$

The last three of these sums are evidently the sum of the first n integers, the sum of the first n even numbers, and the sum of the first n odd numbers.

The following are some formulas from elementary algebra that are needed in the next section:

$$\sum_{k=1}^{n} k = 1 + 2 + \cdots + n = \frac{n(n+1)}{2}; \tag{2}$$

$$\sum_{k=1}^{n} k^2 = 1^2 + 2^2 + \cdots + n^2 = \frac{n(n+1)(2n+1)}{6}; \tag{3}$$

$$\sum_{k=1}^{n} k^3 = 1^3 + 2^3 + \cdots + n^3 = \left[\frac{n(n+1)}{2}\right]^2. \tag{4}$$

These formulas are usually proved by the method of mathematical induction. However, an easier way to prove (2) is to write the sum once as shown, and then again in reverse order,

$$s = 1 + \quad 2 \quad + \cdots + n,$$
$$s = n + (n-1) + \cdots + 1.$$

By adding these equations and noticing that each column on the right adds up to $n+1$ and there are n columns, we get $2s = n(n+1)$, from which (2) follows at once.

There is yet another way of proving (2) that is worth knowing about because it can easily be adapted to yield (3) and (4) as well, and further formulas of the same type. It depends on the simple fact that $(k+1)^2 = k^2 + 2k + 1$, or equivalently

$$(k+1)^2 - k^2 = 2k + 1. \tag{5}$$

If we let $k = 1, 2, \ldots, n$ in (5), and write these equations one below the other, we obtain

$$2^2 - 1^2 = 2 \cdot 1 + 1,$$
$$3^2 - 2^2 = 2 \cdot 2 + 1,$$
$$\cdots$$
$$(n+1)^2 - n^2 = 2 \cdot n + 1.$$

When these equations are added with due attention to the cancellations on the left, the result is

$$(n + 1)^2 - 1^2 = 2 \left(\sum_{k=1}^{n} k \right) + n;$$

and solving for the sum in parentheses yields (2):

$$\sum_{k=1}^{n} k = \tfrac{1}{2}[(n + 1)^2 - 1^2 - n] = \tfrac{1}{2}[n^2 + n]$$

$$= \frac{n(n + 1)}{2}.$$

PROBLEMS

1 Find the numerical value of

(a) $\displaystyle\sum_{i=1}^{5} i^2$;   (b) $\displaystyle\sum_{j=1}^{5} 2^j$;   (c) $\displaystyle\sum_{k=50}^{53} k$.

2 Use the sigma notation to write
   (a) $3 + 9 + 27 + 81$;
   (b) $3 - 5 + 7 - 9 + 11 - 13$;
   (c) $\tfrac{1}{5} + \tfrac{1}{10} + \tfrac{1}{15} + \cdots + \tfrac{1}{45}$.

3 Prove formula (3) by using the expansion $(k + 1)^3 = k^3 + 3k^2 + 3k + 1$ and the method suggested in the text.

4 Prove formula (4) similarly, by using the expansion $(k + 1)^4 = k^4 + 4k^3 + 6k^2 + 4k + 1$.

5 Use (2), (3), and (4) to find closed formulas [(2), (3), and (4) are closed] for the sum of the first $n - 1$ (instead of the first n) integers, squares, and cubes:
   (a) $1 + 2 + \cdots + (n - 1) = ?$
   (b) $1^2 + 2^2 + \cdots + (n - 1)^2 = ?$
   (c) $1^3 + 2^3 + \cdots + (n - 1)^3 = ?$

6 Use the method suggested in the text to discover and prove closed formulas for (a) $1^4 + 2^4 + \cdots + n^4$; (b) $1^5 + 2^5 + \cdots + n^5$.

We begin by restating the problem we are trying to solve. Let $y = f(x)$ be a given nonnegative function defined on a closed interval $a \le x \le b$, as shown in Fig. 6.9. How do we calculate the area of the shaded region in the figure, that is, the area of the region under the graph, above the x-axis, and between the vertical lines $x = a$ and $x = b$?

Closed intervals like the one mentioned here will occur quite often in our discussion, so we use the briefer notation [a, b]. Also, most of the functions we study will be continuous. The reader will recall that this means the following: From the intuitive point of view, the graph consists of a single piece, with no gaps or holes; and more precisely, for each point c in [a, b] we must have

$$\lim_{x \to c} f(x) = f(c).$$

Such a function has several basic properties that we wish to recognize explicitly: It is bounded, in the sense that there exists a constant K such that $|f(x)| \le K$ for all x in [a, b]; and it assumes maximum and minimum values, in the sense that the graph has a highest point and a lowest point.*

We return to Fig. 6.9, with the specific assumption that the function $y = f(x)$ is continuous on [a, b]. How do we find the area of the shaded

6.4

THE AREA UNDER A CURVE. DEFINITE INTEGRALS

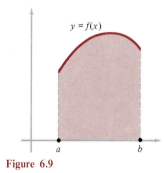

Figure 6.9

---

* Rigorous proofs of these facts are given in Appendix C.3.

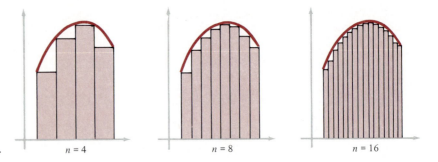

Figure 6.10   Approximating the area.

region? If we take the nature of this region into account — that is, the fact that only the upper edge is curved — then the method of exhaustion suggests the following approximation procedure.

Let n be a positive integer and divide the interval $[a, b]$ into n equal subintervals. Using each subinterval as a base, construct the tallest rectangle that lies entirely under the graph. Write down the sum s_n of the areas of all these rectangles. This sum approximates the area under the graph, and the approximation is improved by taking larger values of n, or equivalently, by dividing $[a, b]$ into a larger number of smaller subintervals. Finally, calculate the exact area under the graph by finding the limiting value approached by the approximating sums s_n as n approaches infinity:

$$\text{Area of region} = \lim_{n \to \infty} s_n. \tag{1}$$

The effect of this procedure is suggested in Fig. 6.10.

We now describe this idea with greater precision by introducing some suitable notation.

Again, let n be a positive integer and divide the interval $[a, b]$ into n equal subintervals by inserting $n - 1$ equally spaced points of division x_1, x_2, . . . , x_{n-1} between a and b. If we denote a by x_0 and b by x_n, then the endpoints of these subintervals are

$$a = x_0 < x_1 < x_2 < \cdots < x_{n-1} < x_n = b, \tag{2}$$

and the subintervals themselves are

$$[x_0, x_1], [x_1, x_2], \quad . . . \quad , [x_{n-1}, x_n]. \tag{3}$$

We denote the length of the kth subinterval by Δx_k, so that

$$\Delta x_k = x_k - x_{k-1}. \tag{4}$$

Since the subintervals are equal in length, it is clear that $\Delta x_k = (b - a)/n$. Let m_k denote the minimum value of $f(x)$ on the kth subinterval $[x_{k-1}, x_k]$. Then this minimum value is assumed at some point $\bar{x}_k$ in the subinterval:

$$f(\bar{x}_k) = m_k, \qquad x_{k-1} \leq \bar{x}_k \leq x_k.$$

For the particular curve shown in Fig. 6.11, $\bar{x}_k$ is easily seen to be the left endpoint of the subinterval when the curve is rising and the right endpoint when it is falling. Since the area of each inscribed rectangle is the product of its height and its base, the approximating sum s_n of the areas of all these

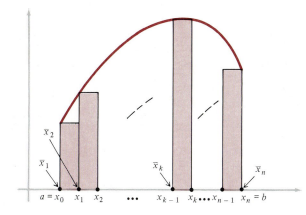

Figure 6.11 Using lower sums.

rectangles is clearly

$$s_n = f(\bar{x}_1)\,\Delta x_1 + f(\bar{x}_2)\,\Delta x_2 + \cdots + f(\bar{x}_k)\,\Delta x_k + \cdots + f(\bar{x}_n)\,\Delta x_n.$$

If we use the sigma notation to abbreviate this sum, we get

$$s_n = \sum_{k=1}^{n} f(\bar{x}_k)\,\Delta x_k, \tag{5}$$

and (1) becomes

$$\text{Area of region} = \lim_{n \to \infty} \sum_{k=1}^{n} f(\bar{x}_k)\,\Delta x_k. \tag{6}$$

This formula is all right as far as it goes, but from several points of view it is inconvenient and unduly restrictive. We broaden its scope and deepen its meaning in a series of remarks.

Remark 1 It is not necessary that the subintervals (3) be equal in length. In fact, the underlying theory is greatly simplified if this restriction is removed. We therefore allow the subintervals (3) to be *equal or unequal* in length, so that the increments (4) may be different from one another. In formula (6), it is now no longer enough to require that n approach infinity; we must also require that the length of the longest subinterval approach zero. Since the latter condition includes the former, we replace (6) by

$$\text{Area of region} = \lim_{\max \Delta x_k \to 0} \sum_{k=1}^{n} f(\bar{x}_k)\,\Delta x_k, \tag{7}$$

where $\max \Delta x_k$ denotes the length of the longest subinterval.

Remark 2 The sum (5) is called a *lower sum* because it uses inscribed rectangles and approximates the area of the region from below. We can also approximate the area from above, as follows. Roughly speaking, we now use each subinterval as a base and construct the shortest rectangle whose top lies entirely above the curve.

To express this in symbols, let M_k denote the maximum value of $f(x)$ on the kth subinterval $[x_{k-1}, x_k]$. As before, this maximum value is assumed at some point $\bar{\bar{x}}_k$ in the subinterval:

$$f(\bar{\bar{x}}_k) = M_k, \qquad x_{k-1} \leq \bar{\bar{x}}_k \leq x_k.$$

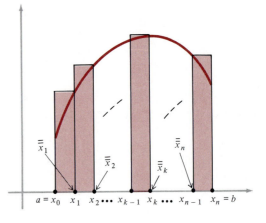

Figure 6.12 Using upper sums.

The sum of the areas of the circumscribed rectangles is therefore

$$S_n = \sum_{k=1}^{n} f(\overline{\overline{x}}_k)\, \Delta x_k. \tag{8}$$

This is called an *upper sum* because it approximates the area of the region from above, as shown in Fig. 6.12. Geometric intuition tells us that the area of our region can just as well be obtained as the limit of upper sums, so we have

$$\text{Area of region} = \lim_{\max \Delta x_k \to 0} \sum_{k=1}^{n} f(\overline{\overline{x}}_k)\, \Delta x_k. \tag{9}$$

However, entirely apart from intuition — which is sometimes misleading — it can be proved as a theorem of pure mathematics that the limits in (7) and (9) both exist and have the same value for any continuous function. The details are given in Appendix C.5.

Further, if x_k^* is taken to be *any* point in the kth subinterval $[x_{k-1}, x_k]$, then we clearly have

$$s_n \le \sum_{k=1}^{n} f(x_k^*)\, \Delta x_k \le S_n.$$

It therefore follows from the theorem just stated that both (7) and (9) can be replaced by the formula

$$\text{Area of region} = \lim_{\max \Delta x_k \to 0} \sum_{k=1}^{n} f(x_k^*)\, \Delta x_k, \tag{10}$$

where the only restriction placed on x_k^* is that $x_{k-1} \le x_k^* \le x_k$.

Remark 3   The limit in (10) is symbolized by the standard Leibniz notation

$$\int_a^b f(x)\, dx, \tag{11}$$

which is read (as we said in Section 6.2) "the *definite integral* from a to b of $f(x)\, dx$." If we write down the definition of (11),

$$\int_a^b f(x)\,dx = \lim_{\max \Delta x_k \to 0} \sum_{k=1}^n f(x_k^*)\,\Delta x_k, \tag{12}$$

then every part of the symbol on the left-hand side is intended to remind us of the corresponding part of the approximating sum on the right-hand side. The *integral sign* $\int$ is an elongated letter S, as in "sum," chosen because of the similarity between a definite integral and a sum of small quantities; the passage to the limit in (12) is suggested by replacing the letter Σ by the symbol $\int$. Also, the usual symbol Δ for an increment is replaced by the letter d to remind us of this limit operation, just as in the Leibniz notation dy/dx for the derivative. The numbers a and b attached to the integral sign are called the *lower* and *upper limits of integration*.* Limits of integration are always present in a definite integral, and help distinguish it from the similar-appearing but very different indefinite integral

$$\int f(x)\,dx.$$

The function $f(x)$ in (11) is called the *integrand*—the thing being integrated—and the variable x is the *variable of integration*. The role of the dx as an important intuitive component of definite integrals will become much clearer in the next chapter.

Remark 4 In this discussion we have adopted the naive but reasonable attitude that the area of the region under the graph clearly exists, and that all we have to do is devise a method for computing it. However, the following example shows that the situation is more complicated than this.

Consider the function $f(x)$ defined on [0, 1] by

$$f(x) = \begin{cases} 0 & \text{if } x \text{ is rational,} \\ 1 & \text{if } x \text{ is irrational.} \end{cases}$$

The graph is suggested in Fig. 6.13, and the very discontinuous nature of this function is determined by the fact that at least one irrational number lies between every pair of rationals and at least one rational number lies between every pair of irrationals. What is the area of the region under this graph? It is quite easy to see that every lower sum is 0 and every upper sum is 1, so the area calculated by (7) is 0 and the area calculated by (9) is 1. Also, the limit on the right of (12) does not exist. Does the concept of area have any meaning in a situation like this?

This bizarre example suggests the following indirect but more logical approach to the problem of area. If we are given a bounded nonnegative function $f(x)$ defined but not necessarily continuous on [a, b], we begin by examining the limit on the right of (12). If this limit exists, then we define its value to be the *area* of the region under the graph, and we say that the function $f(x)$ is *integrable* on [a, b]. And if this limit does not exist, then it is meaningless to speak of the area of the region. Almost all the functions we encounter in practice are continuous, and the theorem stated in Remark 2

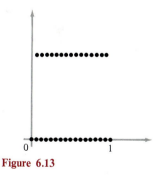

Figure 6.13

---

* Here the word "limit" has nothing to do with the limit concepts that are the basis of calculus. It is used in its loose, everyday sense, meaning "border" or "boundary." The limits of integration tell us where the integration begins and where it ends; they specify the left-hand and right-hand borders of the region.

guarantees that every continuous function is integrable, so these logical fine points will have little practical significance for most of our work. Nevertheless, these issues are interesting and important from the point of view of the theory of calculus, and students should be aware of them even though we choose not to emphasize them.

The definite integral which is defined here is sometimes called the *Riemann integral,* after the great nineteenth-century German mathematician Bernhard Riemann, who was the first to give a careful discussion of integrals of discontinuous functions.

6.5
THE COMPUTATION OF AREAS AS LIMITS

The concepts discussed in Section 6.4 suggest an actual procedure for calculating areas. We now examine how this procedure works in a few specific cases.

Example 1 Consider the function $y = f(x) = x$ on the interval $[0, b]$. The region under this graph (Fig. 6.14) is a triangle with height b and base b, so its area is obviously $b^2/2$. However, it is of some interest to verify that our limit process gives the same result.

Let n be a large positive integer and divide the interval $[0, b]$ into n equal subintervals by means of $n - 1$ equally spaced points

$$x_1 = \frac{b}{n}, \qquad x_2 = \frac{2b}{n}, \qquad \ldots, \qquad x_{n-1} = \frac{(n-1)b}{n}. \tag{1}$$

The bases of the rectangles are $\Delta x_k = b/n$, and if we use upper sums as shown in Fig. 6.14, then the heights of the rectangles are

$$f(x_1) = \frac{b}{n}, \qquad f(x_2) = \frac{2b}{n}, \qquad \ldots, \qquad f(x_n) = \frac{nb}{n},$$

and we have

$$S_n = \left(\frac{b}{n}\right)\left(\frac{b}{n}\right) + \left(\frac{2b}{n}\right)\left(\frac{b}{n}\right) + \cdots + \left(\frac{nb}{n}\right)\left(\frac{b}{n}\right)$$

$$= \frac{b^2}{n^2}(1 + 2 + \cdots + n).$$

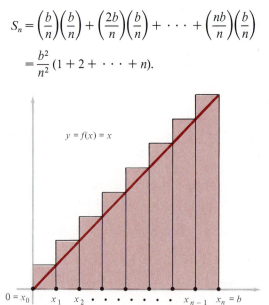

Figure 6.14

By using formula (2) in Section 6.3, we can write this as

$$S_n = \frac{b^2}{n^2} \cdot \frac{n(n+1)}{2} = \frac{b^2}{2} \cdot \frac{n}{n} \cdot \frac{n+1}{n} = \frac{b^2}{2}\left(1 + \frac{1}{n}\right).$$

We therefore conclude that

$$\text{Area of region} = \lim_{n\to\infty} S_n = \lim_{n\to\infty} \frac{b^2}{2}\left(1 + \frac{1}{n}\right) = \frac{b^2}{2},$$

which we knew at the beginning. In the notation of definite integrals, this result is

$$\int_0^b x \, dx = \frac{b^2}{2}. \tag{2}$$

In this example we chose to use equal subintervals and upper sums. There was no compulsion to make these choices; our motive was only to make the calculations as easy as possible.

Example 2 Now consider the function $y = f(x) = x^2$ on the interval $[0, b]$, as shown in Fig. 6.15. Let n be a large positive integer and again divide the interval $[0, b]$ into n equal subintervals of length $\Delta x_k = b/n$ by using the points of division (1). We again use upper sums S_n, so the heights of the successive rectangles are easily seen to be

$$f(x_1) = \left(\frac{b}{n}\right)^2, \qquad f(x_2) = \left(\frac{2b}{n}\right)^2, \qquad \ldots, \qquad f(x_n) = \left(\frac{nb}{n}\right)^2,$$

and we have

$$S_n = \left(\frac{b}{n}\right)^2\left(\frac{b}{n}\right) + \left(\frac{2b}{n}\right)^2\left(\frac{b}{n}\right) + \cdots + \left(\frac{nb}{n}\right)^2\left(\frac{b}{n}\right)$$

$$= \frac{b^3}{n^3}(1^2 + 2^2 + \cdots + n^2).$$

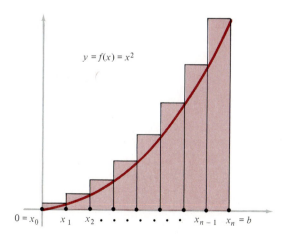

$y = f(x) = x^2$

$0 = x_0 \quad x_1 \quad x_2 \cdot \cdot \cdot \cdot \cdot \cdot \cdot \cdot \cdot \cdot x_{n-1} \quad x_n = b$

Figure 6.15

This time we use formula (3) in Section 6.3 to write

$$S_n = \frac{b^3}{n^3} \cdot \frac{n(n+1)(2n+1)}{6} = \frac{b^3}{6} \cdot \frac{n}{n} \cdot \frac{n+1}{n} \cdot \frac{2n+1}{n}$$

$$= \frac{b^3}{6}\left(1 + \frac{1}{n}\right)\left(2 + \frac{1}{n}\right).$$

As $n \to \infty$ this clearly yields

$$\text{Area of region} = \lim_{n\to\infty} S_n = \frac{b^3}{3},$$

or equivalently,

$$\int_0^b x^2\, dx = \frac{b^3}{3}. \tag{3}$$

This calculation produces a result which we did *not* know at the beginning.

In Problem 1 we ask students to show in the same way that

$$\int_0^b x^3\, dx = \frac{b^4}{4}. \tag{4}$$

It is natural to conjecture from (2), (3), and (4) that the formula

$$\int_0^b x^n\, dx = \frac{b^{n+1}}{n+1} \tag{5}$$

is probably true for *all* positive integers $n = 1, 2, 3, \ldots$. The validity of (5) was established for the cases $n = 3, 4, \ldots, 9$ by the Italian mathematician Cavalieri in 1635 and 1647, but his laborious geometric methods bogged down at $n = 10$. A few years later Fermat discovered a beautiful argument that proves (5) at one stroke for all positive integers. This argument is somewhat aside from our main purpose, so we give the details in Appendix A.4.

PROBLEMS

1   Use upper sums to show that the area under the graph of $y = x^3$ over the interval $[0, b]$ is $b^4/4$.

2   Find the area under the graph of $y = x$ over the interval $[0, b]$ by using lower sums instead of the upper sums of Example 1.

3   Find the area under the graph of $y = x^2$ over the interval $[0, b]$ by using lower sums instead of the upper sums of Example 2.

4   Solve Problem 1 by using lower sums instead of upper sums.

5   As we know, every parabola with vertex at the origin which opens upward has an equation of the form $y = ax^2$. It is easy to see from Example 2 that

$$\int_0^b ax^2\, dx = a\frac{b^3}{3}.$$

Use this to prove the theorem of Archimedes stated in Section 6.2 for the special case in which the chord AB is perpendicular to the axis of the parabola.

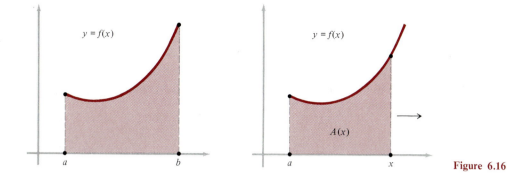

Figure 6.16

As our main achievement so far in this chapter, we have formulated a rather complicated definition of the definite integral of a continuous function as the limit of approximating sums,

$$\int_a^b f(x)\,dx = \lim_{\max \Delta x_k \to 0} \sum_{k=1}^{n} f(x_k^*)\,\Delta x_k. \tag{1}$$

6.6
THE FUNDAMENTAL THEOREM OF CALCULUS

We have also considered several examples of the use of this definition in calculating the values of certain simple integrals, such as

$$\int_0^b x\,dx = \frac{b^2}{2}, \qquad \int_0^b x^2\,dx = \frac{b^3}{3}, \qquad \text{and} \qquad \int_0^b x^3\,dx = \frac{b^4}{4}. \tag{2}$$

These calculations have had two purposes: to emphasize the essential nature of the integral by giving students some direct experience with approximating sums; and also to suggest the severe limitations of this method as a practical tool for evaluating integrals. Thus, for example, how can we possibly use limits of sums to find the numerical values of such complicated integrals as

$$\int_0^1 \frac{x^4\,dx}{\sqrt[3]{7+x^5}} \qquad \text{and} \qquad \int_1^2 \left(1 + \frac{1}{x}\right)^4 \frac{dx}{x^2}? \tag{3}$$

This is clearly out of the question, so where do we go from here? What is evidently needed is a much more efficient and powerful method of computing integrals, and we find this method in the ideas of Newton and Leibniz.

The Newton-Leibniz approach to the problem of calculating the integral (1) depends on an idea that seems paradoxical at first sight. In order to solve this problem, we replace it by an apparently harder problem. Instead of asking for the *fixed* area on the left in Fig. 6.16, we ask for the *variable* area produced when the right-hand border is considered to be moveable, so that the area is a function of x, as suggested on the right in Fig. 6.16. If this area function is denoted by $A(x)$, then clearly $A(a) = 0$ and $A(b)$ is the fixed area on the left in the figure. Our aim is to find an explicit formula for $A(x)$, and then to determine the desired fixed area by setting $x = b$. There are several steps in this process, which we consider separately for the sake of clarity.

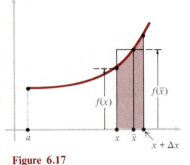

Figure 6.17

STEP 1   We begin by establishing the crucial fact that

$$\frac{dA}{dx} = f(x). \tag{4}$$

This says that *the rate of change of the area A with respect to x is equal to the length of the right edge of the region.* To prove this statement, we must appeal to the definition of the derivative,

$$\frac{dA}{dx} = \lim_{\Delta x \to 0} \frac{A(x + \Delta x) - A(x)}{\Delta x}.$$

Now $A(x)$ is the area under the graph between a and x, and $A(x + \Delta x)$ is the area between a and $x + \Delta x$. Hence $A(x + \Delta x) - A(x)$ is the area between x and $x + \Delta x$ (see the shaded region in Fig. 6.17). It is easy to see that this area is exactly equal to the area of a rectangle with the same base whose height is $f(\bar{x})$, where $\bar{x}$ is a suitably chosen point between x and $x + \Delta x$.* This enables us to complete the proof of (4) as follows:

$$\frac{dA}{dx} = \lim_{\Delta x \to 0} \frac{A(x + \Delta x) - A(x)}{\Delta x} = \lim_{\Delta x \to 0} \frac{f(\bar{x}) \, \Delta x}{\Delta x}$$
$$= \lim_{\Delta x \to 0} f(\bar{x}) = f(x),$$

since $f(x)$ is continuous. To explain the last step here in a bit more detail, we point out that $\Delta x \to 0$ is equivalent to $x + \Delta x \to x$; since $\bar{x}$ is caught between x and $x + \Delta x$, we also have $\bar{x} \to x$, and the continuity of the function now yields the conclusion that $f(\bar{x}) \to f(x)$.

STEP 2   Equation (4) makes it possible for us to achieve our goal of finding a formula for the area function $A(x)$. The reasoning goes this way. By (4), $A(x)$ is one of the antiderivatives of $f(x)$. But if $F(x)$ is *any* antiderivative of $f(x)$, then we know from Chapter 5 that

$$A(x) = F(x) + c \tag{5}$$

for some value of the constant c. To determine c, we put $x = a$ in (5) and obtain $A(a) = F(a) + c$; but since $A(a) = 0$, this yields $c = -F(a)$. Therefore

$$A(x) = F(x) - F(a) \tag{6}$$

is the desired formula.

STEP 3   All that remains is to observe that

$$\int_a^b f(x) \, dx = A(b) = F(b) - F(a),$$

by (6) and the meaning of $A(x)$.

---

* When this statement is expressed in formal language, it is called the *first mean value theorem of integral calculus.*

We summarize our conclusions by formally stating the Fundamental Theorem of Calculus:

If $f(x)$ is continuous on a closed interval $[a, b]$, and if $F(x)$ is any antiderivative of $f(x)$, so that $(d/dx) F(x) = f(x)$ or equivalently

$$\int f(x)\, dx = F(x), \tag{7}$$

then

$$\int_a^b f(x)\, dx = F(b) - F(a). \tag{8}$$

This theorem transforms the difficult problem of evaluating definite integrals by calculating limits of sums into the much easier problem of finding antiderivatives. To find the value of $\int_a^b f(x)\, dx$, we therefore no longer have to think about sums at all; we merely find an antiderivative $F(x)$ in any way we can — by inspection, routine calculation, ingenious calculation, or looking it up in a book — and then compute the number $F(b) - F(a)$.

For instance, in Section 6.5 we used a good deal of algebraic ingenuity to obtain the formulas (2). Now, with the aid of the Fundamental Theorem, we see these formulas as obvious consequences of the following simple facts:

$$\int x\, dx = \frac{x^2}{2}, \quad \int x^2\, dx = \frac{x^3}{3}, \quad \text{and} \quad \int x^3\, dx = \frac{x^4}{4}.$$

More generally, for any exponent $n > 0$ we clearly have

$$\int_a^b x^n\, dx = \frac{b^{n+1}}{n+1} - \frac{a^{n+1}}{n+1}, \quad \text{because} \quad \int x^n\, dx = \frac{x^{n+1}}{n+1}.$$

Remark 1 In the process of working problems, it is often convenient to use the *bracket symbol*,

$$F(x) \Big]_a^b = F(b) - F(a), \tag{9}$$

which is read "$F(x)$ bracket a, b." This symbol means exactly what (9) says it does: To find its value, we write the value of $F(x)$ when x has the upper value b, and subtract the value of $F(x)$ when x has the lower value a. For example, $x^2]_3^4 = 4^2 - 3^2 = 16 - 9 = 7$. By using this notation, (8) can be written in the form

$$\int_a^b f(x)\, dx = F(x) \Big]_a^b.$$

Remark 2 It should be clear from this discussion that *any* antiderivative of $f(x)$ will do in (8). In case students are in doubt about this, they should recall that if $F(x)$ is one antiderivative, then any other can be obtained by adding a suitable constant c to form $F(x) + c$; and since

$$F(x) + c \Big]_a^b = [F(b) + c] - [F(a) + c] = F(b) - F(a),$$

the constant c has no effect on the result. We may therefore ignore constants of integration when finding antiderivatives for the purpose of computing definite integrals.

Example 1 Evaluate each of the following definite integrals:

(a) $\displaystyle\int_{-1}^{2} x^4 \, dx;$ (b) $\displaystyle\int_{1}^{16} \frac{dx}{\sqrt{x}};$ (c) $\displaystyle\int_{8}^{27} \sqrt[3]{x} \, dx;$ (d) $\displaystyle\int_{13}^{14} (x-13)^{10} \, dx.$

Solution In each case an antiderivative is easy to find by inspection:

(a) $\displaystyle\int_{-1}^{2} x^4 \, dx = \frac{1}{5}x^5 \Big]_{-1}^{2} = \frac{1}{5}[32-(-1)] = \frac{33}{5};$

(b) $\displaystyle\int_{1}^{16} \frac{dx}{\sqrt{x}} = 2\sqrt{x} \Big]_{1}^{16} = 2(4-1) = 6;$

(c) $\displaystyle\int_{8}^{27} \sqrt[3]{x} \, dx = \frac{3}{4}x^{4/3} \Big]_{8}^{27} = \frac{3}{4}(81-16) = \frac{195}{4};$

(d) $\displaystyle\int_{13}^{14} (x-13)^{10} \, dx = \frac{1}{11}(x-13)^{11} \Big]_{13}^{14} = \frac{1}{11}(1-0) = \frac{1}{11}.$

The Fundamental Theorem establishes a strong connection between definite integrals and antiderivatives. This connection has made it customary to use the integral sign to denote an antiderivative, as in (7), and to replace the word "antiderivative" by the term "indefinite integral." The reader is familiar with these usages from Chapter 5. From this point on we will often drop the adjective (indefinite, definite) and use the word "integral" alone to refer to either the function (7) or the number (8), relying on the context and the reader's perception of what is going on to avoid confusion. As an infallible aid in keeping track of which is which, we emphasize the fact that a definite integral always has limits of integration attached to it, and that an indefinite integral never has such limits.

From our experience in Chapter 5, we know—or can calculate—many indefinite integrals, so many definite integrals are now within our reach. In particular, the definite integrals (3) are not at all difficult to compute, as we now show.

Example 2 Evaluate

$$\int_{0}^{1} \frac{x^4 \, dx}{\sqrt[3]{7+x^5}}.$$

Solution In the interest of clarity, we consider separately the problem of finding the indefinite integral. The indicated substitution yields

$$\int \frac{x^4 \, dx}{\sqrt[3]{7+x^5}} = \int (7+x^5)^{-1/3}x^4 \, dx = \int u^{-1/3}\left(\frac{1}{5}\, du\right) = \frac{1}{5}\int u^{-1/3}\, du$$

$$u = 7 + x^5 \qquad\qquad = \frac{1}{5}\cdot\frac{3}{2}u^{2/3}$$

$$du = 5x^4 \, dx \qquad\qquad = \frac{3}{10}(7+x^5)^{2/3}.$$

By the Fundamental Theorem we therefore have

$$\int_0^1 \frac{x^4\,dx}{\sqrt[3]{7+x^5}} = \frac{3}{10}\,(7+x^5)^{2/3}\Big]_0^1 = \frac{3}{10}\,(4-7^{2/3}) = \frac{3}{10}\,(4-\sqrt[3]{49}).$$

Example 3　Evaluate

$$\int_1^2 \left(1+\frac{1}{x}\right)^4 \frac{dx}{x^2}.$$

Solution　Here we have

$$\int \left(1+\frac{1}{x}\right)^4 \frac{dx}{x^2} = \int u^4(-du) = -\frac{1}{5}\,u^5$$

$$u = 1+\frac{1}{x} \qquad\qquad\qquad = -\frac{1}{5}\left(1+\frac{1}{x}\right)^5.$$

$$du = -\frac{dx}{x^2}$$

The Fundamental Theorem now yields

$$\int_1^2 \left(1+\frac{1}{x}\right)^4 \frac{dx}{x^2} = -\frac{1}{5}\left(1+\frac{1}{x}\right)^5\Big]_1^2$$

$$= -\frac{1}{5}\left(\frac{243}{32}-32\right) = \frac{781}{160}.$$

Remark 3　Newton and Leibniz are commonly credited with discovering calculus at about the same time but independently of each other. Yet the concepts of the derivative as the slope of the tangent, and the definite integral as the area under a curve, were familiar to many thinkers who preceded them. Under these circumstances, why are Newton and Leibniz given the lion's share of the credit for creating this new branch of mathematics, which played such a central role in the rise of science as the dominant feature of Western civilization? Mostly because they were the principal discoverers of the Fundamental Theorem of Calculus. They, and they alone, understood its importance and began to construct the necessary supporting machinery, and also applied it with spectacular success to problems in science and geometry.

Nevertheless, historians of science have traced the roots of the Fundamental Theorem back to hints in the earlier geometric work of Barrow and Pascal, whose writings are known to have influenced Newton and Leibniz. As Newton said in one of his rare moments of self-deprecation — he was not a modest man — "If I have seen farther, it is by standing on the shoulders of giants." One of these giants was Fermat, who certainly knew the area formula stated in Fig. 6.18. This suggests — as we look back with 20-20 hindsight — that he must therefore also have known the Fundamental Theorem itself, which seems such a short step away. But unfortunately he failed to notice it.

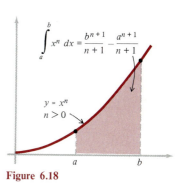

$$\int_a^b x^n\,dx = \frac{b^{n+1}}{n+1} - \frac{a^{n+1}}{n+1}$$

$$y = x^n$$
$$n > 0$$

Figure 6.18

PROBLEMS

A sketch is a necessary part of the solution of almost any problem involving a geometric quantity, and students should form the habit of drawing one as a matter of routine. If drawn with reasonable (but not excessive) care, such a sketch can help us avoid errors by reminding us of what we are doing, and often acts as a valuable source of ideas.

1 Use integration to find the area of the triangle bounded by the line $y = 2x$, the x-axis, and the line $x = 3$. Check your answer by elementary geometry.

2 Use integration to find the area of the triangle bounded by the axes and the line $3x + 2y = 6$. Check your answer by elementary geometry.

3 Find the area between each parabola and the x-axis:
   (a) $x^2 + y = 4$;   (b) $4x^2 + 9y = 36$;
   (c) $4x^2 + 12y = 24x$.

4 Each curve has one arch above the x-axis. Find the area of the region under the arch.
   (a) $y = -x^3 + 4x$.   (b) $y = x^3 - 9x$.
   (c) $y = 2x^2 - x^3$.   (d) $y = x^4 - 6x^2 + 8$.
   (e) $y = x^3 - 5x^2 + 2x + 8$.

5 Find the area bounded by the given curve, the x-axis, and the given vertical lines:
   (a) $y = x^2$, $x = -2$ and $x = 3$;
   (b) $y = x^3$, $x = 0$ and $x = 2$;
   (c) $y = 3x^2 + x + 2$, $x = 1$ and $x = 2$;
   (d) $y = x^2 - 3x$, $x = -3$ and $x = -1$;
   (e) $y = 2x + \dfrac{1}{x^2}$, $x = 1$ and $x = 3$;
   (f) $y = \dfrac{1}{\sqrt{x + 3}}$, $x = 1$ and $x = 6$;
   (g) $y = 3x^2 + 2$, $x = 0$ and $x = 3$;
   (h) $y = 2x + 3$, $x = 0$ and $x = 3$;
   (i) $y = \sqrt{2x + 3}$, $x = -1$ and $x = 3$;
   (j) $y = \dfrac{1}{\sqrt{2x + 3}}$, $x = -1$ and $x = 3$;
   (k) $y = \dfrac{1}{(2x + 3)^2}$, $x = -1$ and $x = 3$.

6 If n is positive, then
$$\int_{-1}^{1} x^n \, dx = \frac{x^{n+1}}{n + 1}\bigg]_{-1}^{1}.$$
Why is this calculation incorrect if n is a negative number $\neq -1$?

7 Find the value of each definite integral:
   (a) $\displaystyle\int_{-1/3}^{2/3} \frac{dx}{\sqrt{3x + 2}}$;
   (b) $\displaystyle\int_{0}^{1} (2x + 3) \, dx$;
   (c) $\displaystyle\int_{-1}^{0} 7x^6 \, dx$;
   (d) $\displaystyle\int_{1}^{4} \sqrt{x} \, dx$;
   (e) $\displaystyle\int_{0}^{2} \sqrt{4x + 1} \, dx$;
   (f) $\displaystyle\int_{-1}^{2} (x + 1)^2 \, dx$;
   (g) $\displaystyle\int_{2a}^{3a} \frac{x \, dx}{(x^2 - a^2)^2}$;
   (h) $\displaystyle\int_{0}^{2b} \frac{x \, dx}{\sqrt{x^2 + b^2}}$;
   (i) $\displaystyle\int_{0}^{1} (x - x^2) \, dx$;
   (j) $\displaystyle\int_{-1}^{2} (1 + x)(2 - x) \, dx$;
   (k) $\displaystyle\int_{0}^{a} (a^2 x - x^3) \, dx$;
   (l) $\displaystyle\int_{0}^{1} (x + 1)^9 \, dx$;
   (m) $\displaystyle\int_{0}^{b} (\sqrt{b} - \sqrt{x})^2 \, dx$;
   (n) $\displaystyle\int_{0}^{1} x^2(1 - x^2) \, dx$;
   (o) $\displaystyle\int_{0}^{1} x^2(1 - x)^2 \, dx$;
   (p) $\displaystyle\int_{1}^{2} \left(x + \frac{1}{x}\right)^2 dx$.

6.7

PROPERTIES OF DEFINITE INTEGRALS

ALGEBRAIC AND GEOMETRIC AREAS

In the previous sections we considered the area of the region under the curve $y = f(x)$ between $x = a$ and $x = b$, and two assumptions were more or less explicit: (1) $f(x) \geq 0$ throughout the interval, and (2) $a < b$. However, the formula defining the definite integral as the limit of approximating sums, namely

$$\int_{a}^{b} f(x) \, dx = \lim_{\max \Delta x_k \to 0} \sum_{k=1}^{n} f(x_k^*) \, \Delta x_k, \tag{1}$$

is independent of these assumptions.

For example, suppose that the curve lies below the x-axis, as shown on the left in Fig. 6.19. In this case we would hesitate to speak of the region "under the curve," but we can certainly describe it as the region "bounded by the

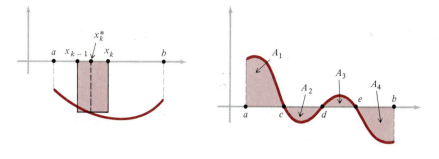

Figure 6.19

curve and the x-axis, between $x = a$ and $x = b$." Each term of the sum (1) is clearly negative because $f(x_k^*) < 0$. Accordingly, $f(x_k^*)\, \Delta x_k$ is the negative of the area of the shaded rectangle, the integral is the negative of the area of the region, and consequently

$$\text{Area of the region} = -\int_a^b f(x)\, dx.$$

Similarly, if the curve lies partly above the x-axis and partly below it, as shown on the right in Fig. 6-19, then the integral (1) can be thought of as a sum of positive and negative terms, corresponding to parts of the region lying above and below the x-axis:

$$\int_a^b f(x)\, dx = A_1 - A_2 + A_3 - A_4, \tag{2}$$

where the areas A_1, A_2, A_3, A_4 are understood to be positive. The integral (2) is often called the *algebraic area* of the region bounded by the curve and the x-axis, because it counts areas of regions above the x-axis with a positive sign and areas of regions below the x-axis with a negative sign.* The actual area of the region bounded by the curve and the x-axis, with each part counted as a positive number, is called the *geometric area:*

$$A_1 + A_2 + A_3 + A_4 = \int_a^c - \int_c^d + \int_d^e - \int_e^b. \tag{3}$$

To find the geometric area, we must sketch the graph, locate the crossing points, and calculate each integral on the right of (3) separately, so that they can be combined with the correct signs.

MISCELLANEOUS PROPERTIES

If we drop the condition $a < b$ and instead assume that $a > b$, we can still retain the purely numerical definition (1) for the definite integral. The only change is that as we traverse the interval from a to b the increments Δx_k are negative. This yields the equation

$$\int_a^b f(x)\, dx = -\int_b^a f(x)\, dx, \tag{4}$$

---

* The discussion in Section 6.6 leading to the Fundamental Theorem of Calculus extends without essential change to integrals of this type. An alternative proof of the Fundamental Theorem, based on entirely different ideas, is given in Appendix C.6.

which is valid for all numbers a and b ($a \neq b$). Also, since (4) says that interchanging the limits of integration changes the sign of the integral, it is natural to take the equation

$$\int_a^a f(x)\, dx = 0 \tag{5}$$

as the definition of the integral on the left.

If $a < b$, and if c is any number between a and b, it is easy to see from (1) that

$$\int_a^b f(x)\, dx = \int_a^c f(x)\, dx + \int_c^b f(x)\, dx. \tag{6}$$

Properties (4) and (5) allow us to conclude that (6) is true for any three numbers $a, b, c,$ regardless of their relation to one another.

We list several further properties of definite integrals that follow in a routine way from the definition (1):

$$\int_a^b cf(x)\, dx = c \int_a^b f(x)\, dx; \tag{7}$$

$$\int_a^b [f(x) + g(x)]\, dx = \int_a^b f(x)\, dx + \int_a^b g(x)\, dx; \tag{8}$$

$$\text{If } f(x) \leq g(x) \text{ on } [a, b], \text{ then } \int_a^b f(x)\, dx \leq \int_a^b g(x)\, dx. \tag{9}$$

In words, property (7) says that a constant factor can be moved across the integral sign, and (8) says that the integral of a sum is the sum of the separate integrals.

VARIABLE LIMITS OF INTEGRATION

We have used x as the "variable of integration" in writing the definite integral

$$\int_a^b f(x)\, dx. \tag{10}$$

However, (10) is a fixed number whose value does not depend on which letter is used for this variable. Instead of (10), we could equally well write

$$\int_a^b f(t)\, dt, \qquad \int_a^b f(u)\, du,$$

or any similar expression, and the meaning would be the same. Letters used in this way are often called *dummy variables*.

In most situations it doesn't matter what letters are used, as long as the ideas are clearly understood. However, sometimes we wish to construct a new function $F(x)$ by integrating a given function $f(x)$ from a fixed lower limit to a *variable* upper limit, as in

$$F(x) = \int_a^x f(x)\, dx. \tag{11}$$

It is evident that this usage can be confusing, because the letter x is used with two different meanings on the right: as the upper limit of integration above the integral sign, and as a dummy variable behind the integral sign. For this reason, it is customary to write (11) in the form

$$F(x) = \int_a^x f(t)\, dt, \qquad (12)$$

with t used as the dummy variable in place of x.

The function $F(x)$ defined by (12) has two properties that make it important. First, it exists whenever the integrand is continuous on the interval between a and x. And second, we proved in Section 6.6 that the derivative of this function is simply the value of the integrand at the upper limit:

$$\frac{d}{dx} F(x) = \frac{d}{dx} \int_a^x f(t)\, dt = f(x). \qquad (13)$$

This provides a satisfactory theoretical solution of the problem of finding an indefinite integral for a given continuous function $f(x)$. As a practical matter, it may be very difficult—or even impossible—to calculate

$$\int f(x)\, dx = F(x)$$

in any recognizable form involving familiar functions. But even if we can't find a formula for $F(x)$, it is at least some consolation to know that in principle an indefinite integral always exists, namely, the function defined by (12).

Example 1 The problem of finding an explicit formula for the indefinite integral

$$\int \frac{dx}{\sqrt[3]{x^{10}+1}} = F(x)$$

is beyond our reach now, and will always be beyond our reach. However, if we don't require an explicit formula, but only a well-defined function, then

$$F(x) = \int_0^x \frac{dt}{\sqrt[3]{t^{10}+1}}$$

will do.

Example 2 Let us try to calculate

$$\frac{d}{dx} \left(\int_0^x \frac{dt}{1+t^2} \right).$$

At this stage of our work we have no way of carrying out the integration to find a formula for the function in parentheses, so that this function can be differentiated. But this doesn't matter. By (13) we immediately have

$$\frac{d}{dx} \left(\int_0^x \frac{dt}{1+t^2} \right) = \frac{1}{1+x^2},$$

so no integration is necessary.

PROBLEMS

1 In each of the following cases, compute the geometric area of the region bounded by the x-axis and the given curves:
(a) $y = 3x - x^2$, $x = 1$, $x = 4$;
(b) $y = x^2 - 2x$, $x = 1$, $x = 4$;
(c) $y = 4 + 4x^3$, $x = -2$, $x = 1$;

(d) $y = x - \dfrac{8}{x^2}$, $x = 1$, $x = 4$.

2 Find the area bounded by the axes and the given curve:
(a) $y = \sqrt{4 - x}$;   (b) $\sqrt{x} + \sqrt{y} = \sqrt{a}$.

3 Find the area bounded by $y^2 = x^3$ and $x = 4$.

4 Find the area enclosed by the loop of $y^2 = x(x - 4)^2$.

5 If $a < c < b$ and $f(x) \geq 0$ on $[a, b]$, draw a suitable picture and explain why equation (6) is an obvious relation among areas.

6 If $f(x) \geq 0$ on $[a, b]$ and $c > 0$, draw a suitable picture and explain why equation (7) is an obvious statement about areas. Do the same for equations (8) and (9) if both $f(x)$ and $g(x)$ are nonnegative on $[a, b]$.

7 If $f(x)$ is an *even* function, that is, if $f(-x) = f(x)$, show geometrically or otherwise that

$$\int_{-a}^{a} f(x)\, dx = 2 \int_{0}^{a} f(x)\, dx.$$

8 Verify the equation in Problem 7 by calculating the following integrals of even functions:

$$\int_{-2}^{2} x^2\, dx \quad \text{and} \quad \int_{-19}^{19} (1 + x^{24})\, dx.$$

9 If $f(x)$ is an *odd* function, that is, if $f(-x) = -f(x)$, show geometrically or otherwise that

$$\int_{-a}^{a} f(x)\, dx = 0.$$

10 Verify the equation in Problem 9 by computing the following integrals of odd functions:

$$\int_{-2}^{2} x^5\, dx \quad \text{and} \quad \int_{-7}^{7} \frac{x\, dx}{\sqrt{x^2 + 11}}.$$

11 The graph of $y = x^2$, $x \geq 0$, can be considered to be the graph of $x = \sqrt{y}$, $y \geq 0$. Show by geometry that this implies the validity of the equation

$$\int_{0}^{a} x^2\, dx + \int_{0}^{a^2} \sqrt{y}\, dy = a^3, \qquad a > 0.$$

Check this by calculating the integrals.

12 Generalize Problem 11 by finding and checking a similar equation for $y = x^n$, where n is any positive number.

13 Use the known area of a circle to find the value of the integral

$$\int_{-a}^{a} \sqrt{a^2 - x^2}\, dx.$$

14 The graph of the equation

$$\frac{x^2}{a^2} + \frac{y^2}{b^2} = 1, \qquad a > b > 0,$$

is called an *ellipse*. Sketch it, and use the result of Problem 13 to find the enclosed area.

15 Show that

(a) $\dfrac{d}{dx} \displaystyle\int_{x}^{b} f(t)\, dt = -f(x)$;

(b) $\dfrac{d}{dx} \displaystyle\int_{a}^{u(x)} f(t)\, dt = f(u(x)) \dfrac{du}{dx}$.

16 In each of the following, compute the indicated derivative:

(a) $\dfrac{d}{dx} \displaystyle\int_{1}^{x+2} \dfrac{dt}{t}$;

(b) $\dfrac{d}{dx} \displaystyle\int_{2x}^{5} t^3\, dt$;

(c) $\dfrac{d}{dx} \displaystyle\int_{1}^{x} \dfrac{dt}{1 + t}$;

(d) $\dfrac{d}{dx} \displaystyle\int_{x}^{1} \dfrac{dt}{1 + t^4}$;

(e) $\dfrac{d}{dx} \displaystyle\int_{1}^{x^2} \dfrac{dt}{\sqrt{t + \sqrt{t + 1}}}$.

ADDITIONAL PROBLEMS FOR CHAPTER 6

SECTION 6.5

1 Show that

$$\int_{0}^{b} \sqrt{x}\, dx = \tfrac{2}{3} b^{3/2}$$

by taking $x_k = k^2 b/n^2$ and $x_k^* = x_k$ in formula (12) of Section 6.4. Notice that this problem illustrates the calculation of an integral as a limit by using subintervals of different lengths.

*2 Show that

$$\int_{1}^{b} \frac{1}{x^2}\, dx = 1 - \frac{1}{b}$$

by using equal subintervals and taking $x_k^* = \sqrt{x_{k-1} x_k}$

in formula (12) of Section 6.4. Note that $x_{k-1} <$ $x_k^* < x_k$ by Problem 9 in Section 1.2. Hint: It will be necessary to use a variation of the idea behind the formula,

$$\frac{1}{1 \cdot 2} + \frac{1}{2 \cdot 3} + \frac{1}{3 \cdot 4} + \cdots + \frac{1}{n(n+1)}$$

$$= \left(\frac{1}{1} - \frac{1}{2}\right) + \left(\frac{1}{2} - \frac{1}{3}\right) + \left(\frac{1}{3} - \frac{1}{4}\right) + \cdots$$

$$+ \left(\frac{1}{n} - \frac{1}{n+1}\right) = 1 - \frac{1}{n+1}.$$

3 Show that

$$\int_1^b \frac{1}{\sqrt{x}}\, dx = 2(\sqrt{b} - 1)$$

by using equal subintervals and taking

$$x_k^* = \frac{x_{k-1} + x_k + 2\sqrt{x_{k-1}x_k}}{4} = \left(\frac{\sqrt{x_{k-1}} + \sqrt{x_k}}{2}\right)^2$$

in formula (12) of Section 6.4.

SECTION 6.6

4 Find the area between each parabola and the x-axis:
(a) $x^2 + 3y = 9$;   (b) $3x^2 + 4y = 48$;
(c) $x^2 + 4x + 2y = 0$.

5 The part of the curve $b^2 y = 4h(bx - x^2)$ that lies above the x-axis forms a parabolic arch with height h and base b. Sketch the graph and check these statements. Use integration to show that the area under this arch is two-thirds the area of the rectangle with the same height and base.

6 Each curve has one arch above the x-axis. Find the area under this arch.
(a) $y = 10 - x - 2x^2$.   (b) $y = -x^3 - 4x^2 - 4x$.
(c) $y = x^3 + 2x^2 - 8x$.   (d) $y = x^4 - 6x^2 + 9$.
(e) $y = x\sqrt{1 - x}$.

7 Find the area bounded by the given curve, the x-axis, and the given vertical lines:
(a) $y = x^2 + 2x + 1$, $x = -1$ and $x = 1$;
(b) $y = \sqrt{x + 2}$, $x = 2$ and $x = 7$;
(c) $y = \sqrt[3]{3 - x}$, $x = -5$ and $x = 3$;
(d) $y = x\sqrt{5 - x^2}$, $x = 0$ and $x = \sqrt{5}$;
(e) $y = \dfrac{x}{(x^2 + 1)^2}$, $x = 0$ and $x = 3$.

8 Find the value of each definite integral:
(a) $\displaystyle\int_0^1 x(x^2 + 2)^3\, dx$;   (b) $\displaystyle\int_{-1}^0 3x^2(3 + x^3)^2\, dx$;

(c) $\displaystyle\int_0^a x\sqrt{a^2 - x^2}\, dx$;   (d) $\displaystyle\int_0^a x\sqrt{a^2 + x^2}\, dx$;

(e) $\displaystyle\int_{-2}^4 (8 - 4x + x^2)\, dx$;

(f) $\displaystyle\int_8^{27} (2x^{-2/3} + 8x^{1/3})\, dx$;

(g) $\displaystyle\int_0^1 \sqrt{9 - 8x}\, dx$;

(h) $\displaystyle\int_2^3 \frac{dx}{(3x - 5)^{5/2}}$;

(i) $\displaystyle\int_0^{\sqrt3} \frac{x\, dx}{\sqrt{4 - x^2}}$;

(j) $\displaystyle\int_0^2 \sqrt{1 + x^3}\, x^2\, dx$;

(k) $\displaystyle\int_0^b (b^{2/3} - x^{2/3})^3\, dx$.

SECTION 6.7

9 In each of the following cases, compute the geometric area of the region bounded by the x-axis and the given curves:
(a) $y = 6 - 3x^2$, $x = 0$, $x = 2$;
(b) $y = x^2 + 2x$, $x = -3$, $x = 0$;
(c) $y = x^2 - x - 2$, $x = 1$, $x = 3$;
(d) $y = x^3 - 3x$, $x = -2$, $x = 3$.

10 In each of the following cases, compute both the algebraic and geometric areas of the region bounded by the x-axis and the given curves:
(a) $y = 3x^5 - x^3$, $x = -1$, $x = 1$;
(b) $y = (x^2 - 4)(9 - x^2)$.

11 Compute

(a) $\dfrac{d}{dx} \displaystyle\int_0^{x^4} \frac{dt}{1 + t}$;   (b) $\dfrac{d}{dx} \displaystyle\int_1^{1+x^2} \frac{dt}{t}$;

(c) $\dfrac{d}{dx} \displaystyle\int_0^{x^3} \frac{dt}{\sqrt{3t + 7}}$;   (d) $\dfrac{d}{dx} \displaystyle\int_0^{x^5} \frac{t\, dt}{\sqrt{1 + t^2}}$.

12 Verify the results obtained in parts (c) and (d) of Problem 11 by actually carrying out the integration and then differentiating.

13 Show that

$$\frac{d}{dx} \int_{u_1(x)}^{u_2(x)} f(t)\, dt = f(u_2(x))\frac{du_2}{dx} - f(u_1(x))\frac{du_1}{dx}.$$

14 Compute

(a) $\dfrac{d}{dx} \displaystyle\int_{x^2}^{x^3} \frac{dt}{t}$;

(b) $\dfrac{d}{dx} \displaystyle\int_{1-x}^{1+x} \frac{1 + t}{t}\, dt$;

(c) $\dfrac{d}{dx} \left[\displaystyle\int_1^{x^3} \sqrt[4]{t^3 + 1}\, dt + \int_{x^3}^5 \sqrt[4]{t^3 + 1}\, dt \right]$.

7

APPLICATIONS
OF INTEGRATION

7.1

INTRODUCTION. THE INTUITIVE MEANING OF INTEGRATION

In Chapter 6 we accomplished two major purposes. First, we approximated the area under a given curve by certain sums and found the exact area by forming the limit of these sums. And second, we learned how to calculate the numerical value of this limit by using the much more powerful method provided by the Fundamental Theorem of Calculus. Almost the whole content of Chapter 6 can be compressed into the following statement: If $f(x)$ is continuous on $[a, b]$, then

$$\lim_{\max \Delta x_k \to 0} \sum_{k=1}^{n} f(x_k^*)\,\Delta x_k = \int_a^b f(x)\,dx$$

$$= F(x)\Big]_a^b = F(b) - F(a), \tag{1}$$

where $F(x)$ is any indefinite integral of $f(x)$.

There are many other quantities in geometry and physics that can be treated in essentially the same way. Among these are volumes, arc lengths, surface areas, and such basic physical quantities as the work done by a variable force acting over a given distance. In each case the process is the same: An interval of the independent variable is divided into small subintervals, the quantity in question is approximated by certain corresponding sums, and the limit of these sums yields the exact value of the quantity in the form of a definite integral — which is then evaluated by means of the Fundamental Theorem.

Once we have seen the details of this limit-of-sums process being carried out for the area under a curve, as was done in Chapter 6, it is unnecessary and boring to think through these details over and over again for each new quantity that we meet. The notation needed for this is complicated and repetitive, and actually impedes the intuitive understanding that we wish to cultivate.

In this spirit, we turn briefly to Fig. 7.1 and consider the easy, intuitive way of constructing the definite integral in (1). We think of the area under the curve as composed of a great many thin vertical rectangular strips. The typical strip shown in the figure has height y and width dx, and therefore area

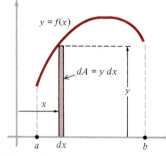

Figure 7.1

$$dA = y\,dx = f(x)\,dx, \tag{2}$$

since $y = f(x)$. This area is called the *differential element of area,* or simply the *element of area*; it is located at an arbitrary position within the region, and this position is specified by a value of x between a and b. We now think of the total area A of the region as the result of adding up these elements of area dA as our typical strip sweeps across the region. This act of addition or summation can be symbolized by writing

$$A = \int dA. \tag{3}$$

Since the element of area sweeps across the region as x increases from a to b, we can express the idea in (3) with greater precision by writing

$$A = \int dA = \int y \, dx = \int_a^b f(x) \, dx. \tag{4}$$

We reach a true definite integral only in the last step in (4), where the variable of integration and the limits of integration become visibly present. In this way we glide smoothly over the messy details and set up the definite integral for the area directly, without having to think about limits of sums at all.

From this point of view, integration is the act of calculating the whole of a quantity by cutting it up into a great many convenient small pieces and then adding up these pieces. It is this intuitive Leibnizian approach to the process of integration that we intend to illustrate and reinforce in the following sections.

7.2

THE AREA BETWEEN TWO CURVES

Suppose we are given two curves $y = f(x)$ and $y = g(x)$, as shown in Fig. 7.2, with points of intersection at $x = a$ and $x = b$ and with the first curve lying above the second on the interval $[a, b]$. In setting up an integral for the area between these curves, it is natural to use thin vertical strips as indicated. The height of such a strip is the distance $f(x) - g(x)$ from the lower curve to the upper, and its base is dx. The element of area is therefore

$$dA = [f(x) - g(x)] \, dx,$$

and the total area is

$$A = \int dA = \int_a^b [f(x) - g(x)] \, dx. \tag{1}$$

We integrate from the smaller limit of integration a to the larger b so that the increment (or differential) dx will be positive. It should also be pointed out that a and b are the values of x for which the two functions yield the same y's; that is, they are the solutions of the equation $f(x) = g(x)$.

We urge students not to be satisfied with merely memorizing formula (1) and applying it mechanically to area problems. Our aim is the mastery of a method, and this aim is better served by thinking geometrically and constructing the needed formula from scratch for each individual problem. The method applies equally well to finding areas by using thin horizontal strips, which are often more convenient. In this case the width of a typical strip will be dy, and the total area will be found by integrating with respect to y.

As an aid to students we give an outline of the steps that should be followed in finding an area by integration.

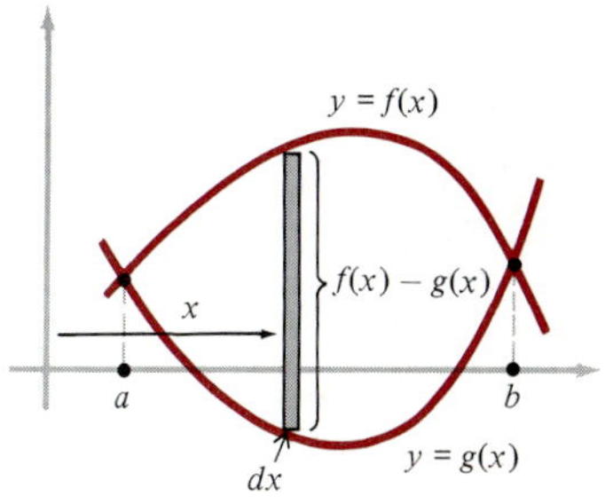

Figure 7.2

STEP 1 Sketch the region whose area is to be found. Write down on the sketch the equations of the bounding curves and find their points of intersection.

STEP 2 Decide whether to use thin vertical strips that have width dx or thin horizontal strips that have width dy, and draw a typical strip on the sketch.

STEP 3 By looking at the sketch and using the equations of the bounding curves, write down the area dA of the typical strip as the product of its length and its width. Express dA entirely in terms of the variable (x or y) appearing in the width.

STEP 4 Integrate dA between appropriate x or y limits, these limits being found by examining the sketch.

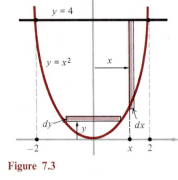

Figure 7.3

Example 1 The region bounded by the curves $y = x^2$ and $y = 4$ is shown in Fig. 7.3. If we use vertical strips, then the length of our typical strip is $4 - x^2$ and its area is $dA = (4 - x^2)\, dx$. The total area of the region is therefore

$$\int_{-2}^{2} (4 - x^2)\, dx = 4x - \tfrac{1}{3}x^3 \Big]_{-2}^{2}$$
$$= (8 - \tfrac{8}{3}) - (-8 + \tfrac{8}{3}) = \tfrac{32}{3}.$$

We urge students to use symmetry whenever possible, in order to simplify the calculations. In this case the left-right symmetry of the figure suggests that we integrate only from $x = 0$ to $x = 2$ to find the right half of the area, and then double the result to obtain the total area:

$$2\int_{0}^{2} (4 - x^2)\, dx = 2(4x - \tfrac{1}{3}x^3)\Big]_{0}^{2} = 2(8 - \tfrac{8}{3}) = \tfrac{32}{3}.$$

As this calculation shows, it is often an advantage (only a slight advantage in this case) to have 0 as one of the limits of integration.

If we decide to use horizontal strips, then the length of the strip is the value of x (in terms of y) at the right end minus the value of x at the left end. This is $\sqrt{y} - (-\sqrt{y})$, so $dA = [\sqrt{y} - (-\sqrt{y})]\, dy = 2\sqrt{y}\, dy$ and the total area is

$$\int_{0}^{4} 2\sqrt{y}\, dy = \tfrac{4}{3}y^{3/2}\Big]_{0}^{4} = \tfrac{32}{3}.$$

The answer is the same as before, which is not surprising but is nevertheless reassuring.

We emphasize once again how important a good sketch is for understanding and carrying out these procedures.

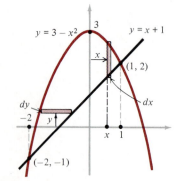

Figure 7.4

Example 2 The region bounded by the curves $y = 3 - x^2$ and $y = x + 1$ is shown in Fig. 7.4. We find where the curves intersect by solving the equations simultaneously. We do this by equating the y's, which gives

$$3 - x^2 = x + 1,$$
$$x^2 + x - 2 = 0,$$

$$(x+2)(x-1) = 0,$$

$$x = -2, 1.$$

The points of intersection are thus $(-2, -1)$ and $(1, 2)$. The length of the indicated vertical strip is $(3 - x^2) - (x + 1) = 2 - x^2 - x$, so the area of the region is found by integrating the element of area as x goes from -2 to 1,

$$\int_{-2}^{1} (2 - x^2 - x) \, dx = (2x - \tfrac{1}{3}x^3 - \tfrac{1}{2}x^2) \Big]_{-2}^{1}$$

$$= (2 - \tfrac{1}{3} - \tfrac{1}{2}) - (-4 + \tfrac{8}{3} - 2) = 4\tfrac{1}{2}.$$

It is inconvenient to use horizontal strips in this problem, because a horizontal strip clearly reaches from the left half of the parabola to the line if $y < 2$, and from the left half of the parabola to the right half if $y > 2$.

PROBLEMS

In Problems 1 to 9, sketch the curves and find the areas of the regions they bound.

1 $y = x^2, y = 2x.$
2 $y = x^2, x = y^2.$
3 $y = x^2 + 2, y = 4 - x^2.$
4 $y = 4x^3 + 3x^2 + 2, y = 2.$
5 $y = x^2 - 2x, y = 3.$
6 $y = x^3 - 3x, y = x \, (x \geq 0).$
7 $y = x^4 - 4x^2, y = -4.$
8 $y = x^3 - 4x, y = 5x \, (x \geq 0).$

*9 $y = 2x + \dfrac{9}{x^2}, y = -2x + 13.$

10 Find the area in Example 2 by integrating with respect to y, first with one integrand from $y = -1$ to $y = 2$, and then with another integrand from $y = 2$ to $y = 3$.

11 Find in two ways the area under $y = x^2$ from $x = 0$ to $x = 4$.

12 Find in two ways the area under $y = x^3$ from $x = 0$ to $x = 2$.

13 Find the area bounded by
(a) the x-axis and $y = x^2 - x^3$;
(b) the y-axis and $x = 2y - y^2$.

14 The area between $x = y^2$ and $x = 4$ is divided into two equal parts by the line $x = a$. Find a.

*15 Find the area between $y = x^3$ and its tangent at $x = 1$.

16 Find the area above the x-axis bounded by $y = 1/x^2$, $x = 1$, and $x = b$, where b is some number greater than 1. The result will depend on b. What happens to this area as $b \to \infty$?

17 Solve Problem 16 with $y = 1/x^2$ replaced by $y = 1/\sqrt{x}$.

18 Solve Problem 16 with $y = 1/x^2$ replaced by $y = 1/x^p$, where p is a fixed positive number greater than 1. What happens if p is a fixed positive number less than 1?

7.3

VOLUMES: THE DISK METHOD

If the region under a curve $y = f(x)$ between $x = a$ and $x = b$ is revolved about the x-axis, it generates a three-dimensional figure called a *solid of revolution*. The symmetrical shape of this solid makes its volume easy to compute.

The situation is illustrated in Fig. 7.5. On the left we show the region itself, together with a typical thin vertical strip of thickness dx whose base lies on the x-axis. When the region is revolved about the x-axis, this strip generates a thin circular disk shaped like a coin, as shown on the right, with radius $y = f(x)$ and thickness dx. The volume of this disk is our *element of volume* dV. Since the disk is a cylinder, its volume is clearly the area of the circular face times the thickness,

$$dV = \pi y^2 \, dx = \pi f(x)^2 \, dx. \tag{1}$$

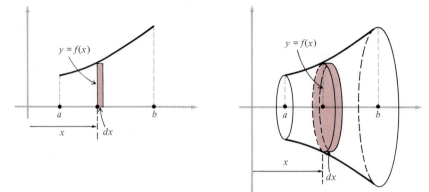

Figure 7.5

We now imagine that the solid of revolution is filled with a very large number of very thin disks like this, so that the total volume is the sum of all the elements of volume as our typical disk sweeps through the solid from left to right, that is, as x increases from a to b:

$$V = \int dV = \int \pi y^2 \, dx = \int_a^b \pi f(x)^2 \, dx. \tag{2}$$

This is another fundamental formula that students should *not* memorize. Instead, it is much better to understand it so clearly in terms of the formula for the volume of a cylinder that memorization is unnecessary.

Students may feel that formula (2) cannot give the *exact* volume of the solid, because it doesn't take into account the volume of the small "peeling" around the outside of the disk in Fig. 7.5. However, just as in the calculation of areas, this slight apparent error visible in the figure — due to using disks instead of actual slices — disappears as a consequence of the limit process that is part of the meaning of the integral sign. We can therefore calculate volumes using formula (2) and have full confidence that our results will be exactly correct, not merely approximations.

Example 1 A sphere can be thought of as the solid of revolution generated by revolving a semicircle about its diameter (Fig. 7.6). If the equation of the semicircle is $x^2 + y^2 = a^2$, $y \geq 0$, then $y = \sqrt{a^2 - x^2}$ and the element of volume is

$$dV = \pi y^2 \, dx = \pi(a^2 - x^2) \, dx.$$

By using the left-right symmetry of the sphere, we can find its total volume by integrating dV from $x = 0$ to $x = a$ and multiplying by 2:

$$V = 2 \int_0^a \pi(a^2 - x^2) \, dx = 2\pi\left(a^2 x - \tfrac{1}{3}x^3\right)\Big]_0^a$$

$$= \tfrac{4}{3}\pi a^3. \tag{3}$$

This result confirms the well-known (but little-understood) formula from elementary geometry. If we integrate dV only from $x = a - h$ to $x = a$, we obtain the formula for the volume of a segment of a sphere of thickness h,

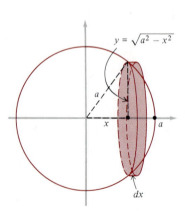

Figure 7.6

$$V = \int_{a-h}^{a} \pi(a^2 - x^2)\, dx = \pi\left(a^2x - \tfrac{1}{3}x^3\right)\Big]_{a-h}^{a}$$
$$= \pi\{\tfrac{2}{3}a^3 - [a^2(a - h) - \tfrac{1}{3}(a - h)^3]\}$$
$$= \pi h^2(a - \tfrac{1}{3}h),$$

after some algebraic simplification. It should be noticed that this formula reduces to (3) when $h = 2a$.

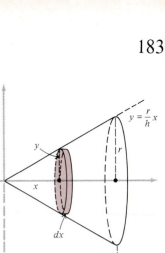

Figure 7.7

Example 2 Another important formula from elementary geometry states that a cone of height h and radius of base r has volume $V = \tfrac{1}{3}\pi r^2 h$; or equivalently, the volume is one-third the volume of the circumscribed cylinder. To obtain this formula by integration, and thereby to understand the origin of the factor $\tfrac{1}{3}$, we think of the cone as the solid of revolution generated by revolving the right triangle shown in the first quadrant of Fig. 7.7 about its base. The hypotenuse of this triangle is clearly part of the straight line $y = (r/h)x$, so the element of volume is

$$dV = \pi y^2\, dx = \frac{\pi r^2}{h^2} x^2\, dx.$$

We now obtain the total volume by integrating dV from $x = 0$ to $x = h$,

$$V = \int_0^h \frac{\pi r^2}{h^2} x^2\, dx = \frac{\pi r^2}{h^2} \cdot \frac{1}{3} x^3\Big]_0^h = \tfrac{1}{3}\pi r^2 h.$$

For obvious reasons, the method of these examples is usually called the *disk method*. The same idea can be applied to solids of other types, in which the element of volume is not necessarily a circular disk. Suppose that each cross section of a solid made by a plane perpendicular to a certain line is a triangle or square or some other geometric figure whose area is easy to find. Then our element of volume dV is the product of this area and the thickness of a thin slice, and we can calculate the total volume of the solid by the *method of moving slices*.

Example 3 A wedge is cut from the base of a cylinder of radius a by a plane passing through a diameter of the base and inclined at an angle of 45° to the base. To find the volume of this wedge, we first draw a careful sketch (Fig. 7.8). A slice perpendicular to the edge of the wedge, as shown, has a triangular face. By using the notation established in the figure, we see that the volume of this slice is

$$dV = \tfrac{1}{2}\sqrt{a^2 - y^2} \cdot \sqrt{a^2 - y^2}\, dy$$
$$= \tfrac{1}{2}(a^2 - y^2)\, dy,$$

so the volume of the wedge is

$$V = 2\int_0^a \tfrac{1}{2}(a^2 - y^2)\, dy = a^2y - \tfrac{1}{3}y^3\Big]_0^a$$
$$= \tfrac{2}{3}a^3.$$

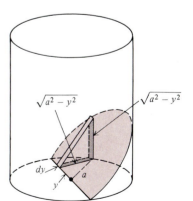

Figure 7.8

A vertical slice parallel to the edge of the wedge evidently has a rectangular face (students should draw their own sketches). If x is the distance from the edge to this slice, then it is easy to see that this time the element of volume is

$$dV = 2x\sqrt{a^2 - x^2}\, dx,$$

and therefore

$$V = \int_0^a 2x\sqrt{a^2 - x^2}\, dx$$

$$= -\tfrac{2}{3}(a^2 - x^2)^{3/2}\Big]_0^a = \tfrac{2}{3}a^3,$$

as before.

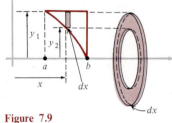

Figure 7.9

Remark 1   The following minor variation of the disk method is often useful, and is necessary for many of the problems at the end of this section. Suppose the strip being revolved about an axis is separated from this axis by a certain distance, as suggested on the left in Fig. 7.9. In this case the element of volume generated by the strip is a disk with a hole in it—what might be described as a *washer* (this washer is moved out to the right in the figure for the sake of clarity). The volume of this washer is the volume of the disk minus the volume of the hole,

$$dV = \pi(y_1{}^2 - y_2{}^2)\, dx.$$

The total volume of the solid of revolution is therefore

$$V = \int dV = \int_a^b \pi(y_1{}^2 - y_2{}^2)\, dx,$$

where y_1 and y_2, the outer and inner radii of the washer, are determined as functions of x from the given conditions of the problem. This procedure for calculating volumes is called—naturally enough—the *washer method*. It applies to solids of revolution that have hollow spaces inside them.

Remark 2   Formula (3) for the volume of a sphere was discovered by Archimedes in the third century B.C. The method he employed was a very beautiful and ingenious early form of integration. The details are given in Appendix A.5.

PROBLEMS

1   Find the volume of the solid of revolution generated when the area bounded by the given curves is revolved about the x-axis:
   (a) $y = \sqrt{x}$, $y = 0$, $x = 4$;
   (b) $y = 2x - x^2$, $y = 0$;
   (c) $y^3 = x$, $y = 0$, $x = 1$;
   (d) $y = x$, $y = 1$, $x = 0$;
   (e) $x = 2y - y^2$, $x = 0$;
   (f) $x^{2/3} + y^{2/3} = a^{2/3}$, first quadrant.
2   Problem 14 in Section 6.7 is concerned with the ellipse

$$\frac{x^2}{a^2} + \frac{y^2}{b^2} = 1, \qquad a > b > 0.$$

If the area inside this ellipse is revolved about the x-axis, the resulting solid (which resembles a football) is called a *prolate spheroid*. Find its volume. [If $a < b$, the solid is called an *oblate spheroid*. Observe that the volume formula is the same regardless of how a and b are related to each other, and also that it reduces to formula (3) when $b = a$.]

3   The horizontal cross section of a certain pyramid at a distance x down from the top is a square of side $(b/h)x$, where h is the height of the pyramid and b is the side of the base. Show that the volume of the pyramid is one-third the area of the base times the height.

4 A horn-shaped solid is generated by a moving circle perpendicular to the y-axis whose diameter lies in the xy-plane and extends from $y = 27x^3$ to $y = x^3$. Find the volume of this solid between $y = 0$ and $y = 8$.

5 The square bounded by the axes and the lines $x = 2$, $y = 2$ is cut into two parts by the curve $y^2 = 2x$. Show that these parts generate equal volumes when revolved about the x-axis.

6 The two areas described in Problem 5 are revolved about the line $x = 2$. Find the volumes generated.

7 A tent consists of canvas stretched from a circular base of radius a to a vertical semicircular rod fastened to the base at the ends of a diameter. Find the volume of this tent.

8 The base of a solid is a quadrant of a circle of radius a. Each cross section perpendicular to one edge of the base is a semicircle whose diameter lies in the base. Find the volume.

9 The base of a certain solid is the circle $x^2 + y^2 = a^2$. Each plane perpendicular to the x-axis intersects the solid in a square cross section with one side in the base of the solid. Find its volume.

10 If the area bounded by the parabola $y = H - (H/R^2)x^2$ and the x-axis is revolved about the y-axis, the resulting bullet-shaped solid is a segment of a paraboloid of revolution with height H and radius of base R. Show that its volume is half the volume of the circumscribing cylinder.

11 If the circle $(x - b)^2 + y^2 = a^2$ $(0 < a < b)$ is revolved about the y-axis, it generates a doughnut-shaped solid called a *torus*. Find the volume of this torus by the washer method. Hint: If necessary, use the result of Problem 13 in Section 6.7. (Notice the remarkable fact that the volume of the torus is the product of the area of the circle and the distance traveled by its center as it revolves about the y-axis.)

12 Find the volume of the solid formed by revolving the area inside the curve $x^2 + y^4 = 1$ about (a) the x-axis; (b) the y-axis.

13 The base of a certain solid is an equilateral triangle of side a, with one vertex at the origin and an altitude along the x-axis. Each plane perpendicular to the x-axis intersects the solid in a square cross section with one side in the base of the solid. Find the volume.

14 Each plane perpendicular to the x-axis intersects a certain solid in a circular cross section whose diameter lies in the xy-plane and extends from $x^2 = 4y$ to $y^2 = 4x$. The solid lies between the points of intersection of these curves. Find its volume.

15 The base of a certain solid is the region bounded by the parabola $x^2 = 4y$ and the line $y = 9$, and each cross section perpendicular to the y-axis is a square with one side in the base. If a plane perpendicular to the y-axis cuts this solid in half, how far from the origin is this plane?

*16 Two great circles lying in planes that are perpendicular to each other are drawn on a wooden sphere of radius a. Part of the sphere is then shaved off in such a way that each cross section of the remaining solid that is perpendicular to the common diameter of the two great circles is a square whose vertices lie on these circles. Find the volume of this solid.

*17 The axes of two cylinders, each of radius a, intersect at right angles. Find the common volume. Hint: Consider cross sections parallel to the plane of the two axes.

18 Consider the area in the first quadrant under the curve $x^2 y^3 = 1$ and to the right of $x = 1$. By integrating from $x = 1$ to $x = b$ and then letting $b \to \infty$, show that this area is infinite, but that on revolving it about the x-axis we obtain a finite volume.

7.4 VOLUMES: THE SHELL METHOD

There is another method of finding volumes which is often more convenient than those described in Section 7.3.

To understand this method, consider the region shown on the left in Fig. 7.10, that is, the region in the first quadrant bounded by the axes and the

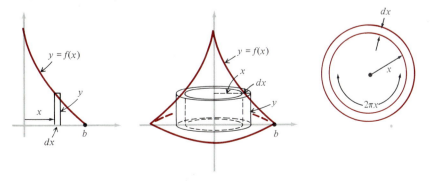

Figure 7.10

indicated curve $y = f(x)$. If this region is revolved about the x-axis, then the thin vertical strip in the figure generates a disk, and we can calculate the total volume of the solid by adding up (or integrating) the volumes of these disks from $x = 0$ to $x = b$. This, of course, is the disk method described in Section 7.3. However, if the region is revolved about the y-axis, as shown in the center of the figure, then we get an entirely different solid of revolution and the vertical strip generates a thin-walled cylindrical shell. This shell can be thought of as resembling a soup can whose top and bottom have been removed, or perhaps a thin-walled cardboard mailing tube. Its volume dV is essentially the area of the inner surface ($2\pi xy$) times the thickness of the wall (dx), so

$$dV = 2\pi xy \, dx. \tag{1}$$

As the radius x of this shell increases from $x = 0$ to $x = b$, we can see from Fig. 7.10 that the resulting series of concentric shells fills the solid of revolution from the axis outward, in much the same way as the concentric layers of an onion fill the onion from the center outward. The total volume of this solid is therefore the sum—or integral—of the elements of volume dV,

$$V = \int dV = \int 2\pi xy \, dx = \int_0^b 2\pi x f(x) \, dx, \tag{2}$$

since $y = f(x)$. In principle, this volume V can also be found by using horizontal disks generated by thin horizontal strips; however, this could turn out to be difficult, since the given equation $y = f(x)$ would have to be solved for x in terms of y.

Just as in our other applications of integration, formulas (1) and (2) give brief expression to a complex process involving limits of sums; and as usual, we omit the details of this process in the interests of clarity.

Also as usual, we suggest that students would be wise not to memorize formula (2). This formula is somewhat similar to the corresponding formula for the disk method, and students who try to memorize them and use them without thinking about their meaning will almost certainly confuse them and come to grief. It is better to sketch a figure and construct (1) directly from the visible evidence of this figure, and then form (2) by integration. Also, this approach has the further advantage that we are not tied to any particular notation, and can easily adapt the basic idea to solids of revolution about various axes.

Example 1 In Example 1 of Section 7.3 we calculated the volume of a sphere by the disk method. We now solve the same problem by the shell method (see Fig. 7.11). The volume of the shell shown in the figure is

$$dV = 2\pi x(2y) \, dx$$
$$= 4\pi x \sqrt{a^2 - x^2} \, dx.$$

The volume of the sphere is therefore

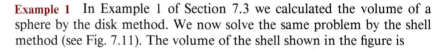

$$V = 4\pi \int_0^a x\sqrt{a^2 - x^2} \, dx = 4\pi(-\tfrac{1}{3})(a^2 - x^2)^{3/2} \Big]_0^a$$

$$= -\frac{4\pi}{3}(a^2 - x^2)^{3/2} \Big]_0^a = \tfrac{4}{3}\pi a^3.$$

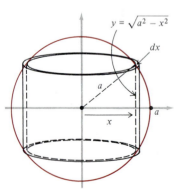

Figure 7.11

In this connection we can profitably consider a related problem: If a vertical hole of diameter a is bored through the center of the sphere, find the remaining volume. For this, it clearly suffices to integrate dV as the radius x of the shell varies from $x = a/2$ to $x = a$, so

$$V = 4\pi \int_{a/2}^{a} x\sqrt{a^2 - x^2}\, dx = -\frac{4\pi}{3}(a^2 - x^2)^{3/2}\Big]_{a/2}^{a}$$

$$= \frac{4\pi}{3}\left(\frac{3}{4}a^2\right)^{3/2} = \frac{4\pi}{3}\left(\frac{3\sqrt{3}}{8}a^3\right) = \frac{\sqrt{3}}{2}\pi a^3.$$

This problem could be solved by the washer method, but the shell method is much more convenient.

Example 2 The region in the first quadrant above $y = x^2$ and below $y = 2 - x^2$ is revolved about the y-axis (Fig. 7.12). To find the volume by the shell method, we observe that the height of our typical shell is $y = (2 - x^2) - x^2 = 2 - 2x^2$, so

$$dV = 2\pi xy\, dx = 2\pi x(2 - 2x^2)\, dx$$

$$= 4\pi(x - x^3)\, dx;$$

and since the curves intersect at $x = \pm 1$, we have

$$V = 4\pi \int_0^1 (x - x^3)\, dx$$

$$= 4\pi(\tfrac{1}{2}x^2 - \tfrac{1}{4}x^4)\Big]_0^1 = \pi.$$

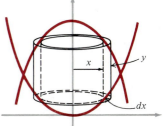

Figure 7.12

Notice that if we attempt to solve this problem by the disk method, then it is necessary to calculate two separate integrals—one referring to the volume below the points of intersection of the two curves, and the other to the volume above.

PROBLEMS

1 Solve the problem of the sphere with the hole bored through it (Example 1) by the washer method.

2 Solve the problem in Example 2 by the disk method.

In Problems 3 to 8, sketch the region bounded by the given curves and use the shell method to find the volume of the solid generated by revolving this region about the given axis.

3 $y = \sqrt{x}$, $x = 4$, $y = 0$; the y-axis.

4 $x^2 = 4y$, $y = 4$; the x-axis.

5 $y = x^3$, $x = 3$, $y = 0$; the y-axis.

6 $x = y^2$, $x^2 = 8y$; the x-axis.

7 $y = \dfrac{1}{x}$, $x = a$, $x = b$ ($0 < a < b$), $y = 0$; the y-axis.

8 $y = x^2$, $y = \tfrac{1}{4}(3x^2 + 1)$; the y-axis.

9 The region bounded by $y = x/\sqrt{x^3 + 8}$, the x-axis, and the line $x = 2$ is revolved about the y-axis. Find the volume of the solid generated in this way. (Observe that the washer method is not practical in this problem.)

10 A hole of radius $\sqrt{3}$ is bored through the center of a sphere of radius 2. Find the volume removed.

11 Consider the region in the first quadrant bounded by $y = 4 - x^2$ and the axes.
   (a) Use both the disk method and the shell method to find the volume of the solid generated when this region is revolved about the y-axis.
   *(b) Use both methods to find the volume of the solid generated when this region is revolved about the x-axis.

12 Let r and h be positive numbers. The region bounded by the line $x/r + y/h = 1$ and the axes is revolved about the y-axis. Use the shell method to obtain the standard formula for the volume of a cone.

13 A spherical ring is the solid that remains after drilling a hole through the center of a solid sphere. If the sphere has radius a and the ring has height h, prove the remarkable fact that the volume of the ring depends on h but not on a.

14 The parabola $a^2 y = bx^2$, $0 \leq y \leq b$, is revolved about the y-axis. Use the shell method to show that the volume of the resulting paraboloid is one-half the volume of the cylinder with the same height and base.

15 The region in the first quadrant above $y = 3x^2$ and below $y = 4 - 6x^2$ is revolved about the y-axis. Find the volume generated in this way.

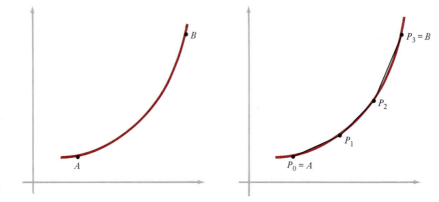

Figure 7.13

7.5
ARC LENGTH

An *arc* is the part of a curve that lies between two specific points A and B, as shown on the left in Fig. 7.13. Physically, the length of an arc is a very simple concept. Mathematically, it is somewhat more complicated. From the physical point of view, we merely bend a piece of string to fit the curve from A to B, mark the points corresponding to A and B, straighten out the string, and measure its length with a ruler.

This process can be carried out by means of an approximation procedure that lends itself to mathematical treatment, as follows. Divide the arc AB into n parts by using points $P_0 = A$, P_1, P_2, . . . , $P_n = B$; place pins at these points; and let the string stretch in short straight-line paths from each pin to the next. We illustrate this idea on the right in Fig. 7.13 with $n = 3$. The part of this string between A and B is evidently shorter than the arc, since a straight line is the shortest distance between two points. However, if we take larger and larger values of n, and at the same time require that the pins be placed closer and closer together, then the length of the string should approach the length of the arc. We now express this idea in mathematical language and derive a practical method of calculating arc length by integration.

Let us assume that the arc under discussion is the graph of a continuous function $y = f(x)$ for $a \leq x \leq b$. We partition the interval $[a, b]$ into n subintervals by using points $x_0 = a, x_1, \ . \ . \ . \ , x_{k-1}, x_k, \ . \ . \ . \ , x_n = b$, as shown in Fig. 7.14. We let P_k be the point (x_k, y_k), where $y_k = f(x_k)$. The total length of the polygonal path $P_0 P_1 \cdot \cdot \cdot P_{k-1} P_k \cdot \cdot \cdot P_n$ is the sum of the lengths of the chords joining each point to the next. If we write

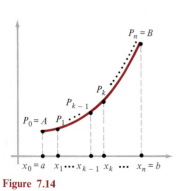

Figure 7.14

$$\Delta x_k = x_k - x_{k-1} \qquad \text{and} \qquad \Delta y_k = y_k - y_{k-1}, \qquad k = 1, 2, \ . \ . \ . \ , n,$$

then it is clear that by the Pythagorean theorem we have

$$\text{Length of } k\text{th chord} = \sqrt{(\Delta x_k)^2 + (\Delta y_k)^2}$$

$$= \sqrt{1 + \left(\frac{\Delta y_k}{\Delta x_k}\right)^2}\, \Delta x_k. \qquad (1)$$

We now assume that $y = f(x)$ is not only continuous but also differentiable. This permits us to replace the ratio inside the radical, which is the slope of the chord joining P_{k-1} to P_k, by the value of the derivative at some point x_k^* between x_{k-1} and x_k:

$$\frac{\Delta y_k}{\Delta x_k} = f'(x_k^*), \qquad x_{k-1} < x_k^* < x_k.$$

The justification for this step lies in the fact that the chord is parallel to the tangent at some point on the curve between P_{k-1} and P_k.* This enables us to write (1) as

$$\text{Length of } k\text{th chord} = \sqrt{1 + [f'(x_k^*)]^2}\, \Delta x_k,$$

so the total length of the polygonal path is

$$\sum_{k=1}^{n} \sqrt{1 + [f'(x_k^*)]^2}\, \Delta x_k. \qquad (2)$$

We now obtain our conclusion by forming the limit of these sums as n approaches infinity and the length of the longest subinterval approaches zero:

$$\text{Length of arc } AB = \lim_{\max \Delta x_k \to 0} \sum_{k=1}^{n} \sqrt{1 + [f'(x_k^*)]^2}\, \Delta x_k$$

$$= \int_a^b \sqrt{1 + [f'(x)]^2}\, dx, \qquad (3)$$

provided $f'(x)$ is continuous so that this integral exists.

At first sight, formula (3) may appear to be rather hard to keep in mind. However, if we use the Leibniz notation dy/dx instead of $f'(x)$, then the following intuitive approach makes this formula much easier to grasp and remember. Let the letter s denote the variable arc length from A to a variable point on the curve, as shown in Fig. 7.15. Let s be allowed to increase by a small amount ds, so that ds is the differential *element of arc length*, and let dx and dy be the corresponding changes in x and y. We think of ds as so small that this part of the curve is virtually straight, and therefore ds is the hypotenuse of a tiny right triangle called the *differential triangle*. For this triangle the Pythagorean theorem says that

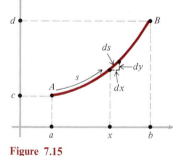

Figure 7.15

$$ds^2 = dx^2 + dy^2, \qquad (4)$$

and this simple equation is the source of all wisdom in the calculation of arc lengths.† If we solve (4) for ds, then factor dx and remove it from the radical,

---

* This highly plausible assertion is called the *mean value theorem*. This theorem is one of the cornerstones of the theory of calculus, and is discussed and proved in Appendix C.4.

† Parentheses are usually omitted in writing squares of differentials. Thus, ds^2 means $(ds)^2$ and not $d(s^2)$, etc.

we clearly get

$$ds = \sqrt{dx^2 + dy^2}$$

$$= \sqrt{\left(1 + \frac{dy^2}{dx^2}\right) dx^2} = \sqrt{1 + \left(\frac{dy}{dx}\right)^2}\, dx. \qquad (5)$$

We now touch again on the basic theme of this chapter and point out that the total length of the arc AB can be thought of as the sum — or integral — of all the elements of arc ds as ds sweeps along the curve from A to B. In view of (5), this yields

$$\text{Length of arc } AB = \int ds = \int_a^b \sqrt{1 + \left(\frac{dy}{dx}\right)^2}\, dx, \qquad (6)$$

which is (3). This formula tells us that x is the variable of integration and that y is to be treated as a function of x. However, it is sometimes more convenient to treat x as a function of y. In this case we replace (5) by

$$ds = \sqrt{dx^2 + dy^2}$$

$$= \sqrt{\left(\frac{dx^2}{dy^2} + 1\right) dy^2} = \sqrt{\left(\frac{dx}{dy}\right)^2 + 1}\, dy, \qquad (7)$$

which is obtained by factoring dy instead of dx out of the radical. With y as the variable of integration, the integral for the length of the arc AB is then

$$\int ds = \int_c^d \sqrt{\left(\frac{dx}{dy}\right)^2 + 1}\, dy, \qquad (8)$$

which is sometimes easier to evaluate than (6).

Most mathematicians remember formulas (6) and (8) not by memorizing them as they stand, but instead by starting with (4) and mentally performing as needed the simple manipulations in (5) and (7). This way, the whole package of ideas is almost impossible to forget.

Example   Find the length of the curve $y^2 = 4x^3$ between the points $(0, 0)$ and $(2, 4\sqrt{2})$.

Solution   This curve is shown in Fig. 7.16, and the arc in question is the indicated piece of the curve in the first quadrant. If we solve for y, then we get

$$y = 2x^{3/2}, \qquad \text{so} \qquad \frac{dy}{dx} = 3x^{1/2}.$$

Formula (6) now yields

$$\text{Length of arc} = \int_0^2 \sqrt{1 + 9x}\, dx = \frac{1}{9} \int_0^2 (1 + 9x)^{1/2}\, 9dx$$

$$= \tfrac{1}{9} \cdot \tfrac{2}{3}(1 + 9x)^{3/2} \Big]_0^2 = \tfrac{2}{27}(19\sqrt{19} - 1).$$

This calculation should serve as a warning, for if we try to find the length of an arc on almost any familiar curve, then the resulting integral will probably be impossible for us to work out. At this stage we must choose our problems

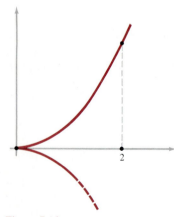

Figure 7.16

very carefully in order for the integrals to be computable. We should also be aware of our urgent need for more integration techniques. Filling this need is the main purpose of the next three chapters.

Remark 1 It is possible to give an example of a continuous curve $y = f(x)$, $a \le x \le b$, that does not have a length.* This very surprising fact suggests that the underlying theory of arc length is more complicated than it seems. In the preceding discussion we found it necessary to assume that the function $y = f(x)$ has a continuous derivative. Such a curve is called a *smooth curve,* and the word "arc" is usually restricted to mean a piece of a curve with this property. A smooth curve is often described geometrically by saying that it has a "continuously turning tangent."

Remark 2 Students may have the impression that equations (4) and (5) — which are equivalent to each other — are only approximately correct, because the differential triangle in Fig. 7.15 is only a "quasi-triangle" whose "hypotenuse" is not even a straight-line segment. But this is not the case. These equations are exactly correct, as the following argument shows. We know that (3) is valid, so the arc length s in Fig. 7.15 can be written as

$$s = \int_a^x \sqrt{1 + [f'(t)]^2} \, dt,$$

using t as the dummy variable of integration. It is clear that s is a function of the upper limit x; and if we calculate the derivative of this function by using formula (13) in Section 6.7, then we get

$$\frac{ds}{dx} = \sqrt{1 + [f'(x)]^2} = \sqrt{1 + \left(\frac{dy}{dx}\right)^2},$$

which is equivalent to (5).

PROBLEMS

In Problems 1 to 8, find the length of the specified arc of the given curve.

1 $y^2 = x^3$ between $(0, 0)$ and $(4, 8)$.

2 $y = \dfrac{1}{4} x^4 + \dfrac{1}{8x^2}$, $1 \le x \le 2$.

3 $y = \dfrac{1}{3} x^3 + \dfrac{1}{4x}$, $1 \le x \le 3$.

4 $y = \dfrac{1}{3}\sqrt{x}(3 - x)$, $0 \le x \le 3$.

5 $x = \dfrac{1}{2} y^3 + \dfrac{1}{6y}$, $1 \le y \le 3$.

6 $y = \dfrac{5}{12}x^{6/5} - \dfrac{5}{8}x^{4/5}$, $1 \le x \le 32$.

7 $y = \dfrac{1}{3}(2 + x^2)^{3/2}$, $0 \le x \le 3$.

8 $y = \dfrac{2}{3}(1 + x^2)^{3/2}$, $0 \le x \le 3$.

9 Let A, B, C be positive constants. Show that the length of an arc of the curve $y = A(B + Cx^2)^{3/2}$ can be calculated by means of an integral not involving a square root if

(a) $A = \frac{2}{3}$ and $B^2C = 1$, in which case the curve is $y = (2/3B^3)(B^3 + x^2)^{3/2}$.

(b) $B = 2$ and $3A\sqrt{C} = 1$, in which case the curve is $y = (1/3\sqrt{C})(2 + Cx^2)^{3/2}$.

Show that each of these curves includes Problems 7 and 8 as special cases.

10 The curve $x^{2/3} + y^{2/3} = a^{2/3}$ is called an *astroid* or a

---

* The curious reader can find descriptions of some of these bizarre curves on pp. 104–107 of N. Ya. Vilenkin, *Stories About Sets,* Academic Press, 1968.

hypocycloid of four cusps. Sketch it and find its total length.

11 If $0 < a < b$ and n is not equal to 1 or -1, show that the length of

$$y = \frac{x^{n+1}}{n+1} + \frac{1}{4(n-1)} \cdot \frac{1}{x^{n-1}}$$

between $x = a$ and $x = b$ can be calculated by means of an integral not involving a square root. Notice that

Problems 2 and 3 are special cases of this result.

12 In each case set up the integral for the arc length, but do not attempt to evaluate it (these integrals are beyond our capacity at the present stage of our work):
(a) $y = \sqrt{x}$, $1 \le x \le 4$;
(b) $y = x^2$, $0 \le x \le 1$;
(c) $y = x^3$, $0 \le x \le 1$;
(d) the part of $y = -x^2 + 4x - 3$ lying above the x-axis.

7.6

THE AREA OF A SURFACE OF REVOLUTION

Let us consider a smooth curve lying above the x-axis, as shown on the left in Fig. 7.19. When this curve is revolved about the x-axis, it generates a *surface of revolution.* We now set ourselves the problem of calculating the area of such a surface.

For reasons that will become clear, we begin by considering a very simple surface of revolution, the curved lateral part of a cone whose base has radius r and whose slant height is L. If this cone is cut down the side from the vertex to the base — that is, along a generator — and laid out flat, as shown in Fig. 7.17, then we get a sector of a circle of radius L whose curved edge has length $2\pi r$, and the lateral area A of the cone equals the area of this sector. It is geometrically clear that the ratio of the area of the sector to the complete area of the circle equals the ratio of the length of the curved edge to the complete circumference of the circle, that is,

$$\frac{A}{\pi L^2} = \frac{2\pi r}{2\pi L}, \qquad \text{so} \qquad A = \pi r L.$$

The lateral surface of the cone can evidently be thought of as the surface of revolution swept out by a generator as it revolves around the axis. If the formula is written as

$$A = L \cdot 2\pi(\tfrac{1}{2}r),$$

then we see that the lateral area of a cone equals the product of the length of a generator and the distance traveled by the midpoint on its journey around the axis.

Next, we generalize slightly and find the area of the surface of revolution generated when a line segment of length L is revolved about an axis at a distance r from its midpoint.* This area is the lateral area of a frustum of a

---

* In the case of a cone, one end of the segment lies on the axis and forms the vertex of the cone.

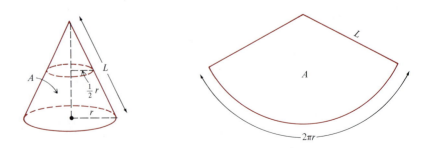

Figure 7.17

cone, as shown in Fig. 7.18. If we denote this area by A, then A is the difference between the lateral areas of the two cones in the figure, so

$$A = \pi r_1 L_1 - \pi r_2 L_2 = \pi(r_1 L_1 - r_2 L_2).$$

By similar triangles, it is clear that

$$\frac{L_2}{r_2} = \frac{L_1}{r_1} \qquad \text{or} \qquad r_1 L_2 = r_2 L_1.$$

This enables us to write A in the form

$$A = \pi(r_1 L_1 - r_1 L_2 + r_2 L_1 - r_2 L_2) = \pi[r_1(L_1 - L_2) + r_2(L_1 - L_2)]$$

$$= \pi(L_1 - L_2)(r_1 + r_2) = (L_1 - L_2) \cdot 2\pi \left(\frac{r_1 + r_2}{2}\right) = L \cdot 2\pi r.$$

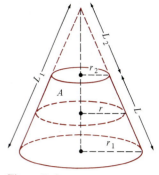

Figure 7.18

We therefore conclude that in this case as well, the area of the surface of revolution equals the product of the length of the segment and the distance traveled by the midpoint on its journey around the axis.

We now apply these ideas to the general area problem stated at the beginning of this section. Our approach will be intuitive and geometric.

We begin by approximating the smooth curve $y = f(x)$ by a polygonal path consisting of many short line segments connecting nearby points on the curve, as shown on the left in Fig. 7.19. The surface generated by revolving the curve about the x-axis will have approximately the same area as the surface generated by revolving this polygonal path about the x-axis (Fig. 7.19, center). The latter surface is evidently made up of a number of pieces, each of which is shaped like a frustum of a cone. This situation suggests the fundamental idea illustrated on the right in the figure. If the element of arc length ds is revolved about the x-axis, then it generates a thin ribbonlike element of area dA; and if the midpoint of ds is at a distance y from the x-axis, then the above discussion tells us that

$$dA = 2\pi y\, ds = 2\pi y\sqrt{1 + \left(\frac{dy}{dx}\right)^2}\, dx.$$

We obtain the total area A of the surface by forming the sum — or integral — of all the elements of area dA as dA sweeps along the complete surface,

$$A = \int dA = \int 2\pi y\, ds = \int_a^b 2\pi y\sqrt{1 + \left(\frac{dy}{dx}\right)^2}\, dx,$$

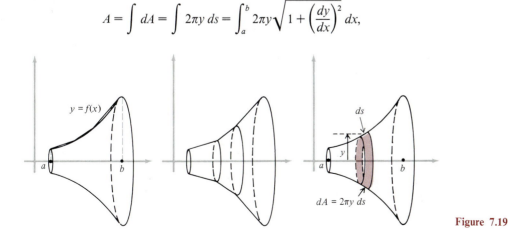

Figure 7.19

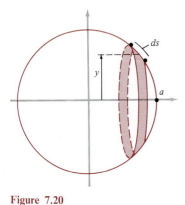

Figure 7.20

where y is assumed to be known as a function of x [$y = f(x)$]. If we choose instead to revolve our curve about the y-axis, and thereby to generate an entirely different surface of revolution, then in the same way its area is given by

$$A = \int 2\pi x \, ds.$$

The underlying idea in both of these formulas can be expressed by writing

$$A = \int 2\pi \, (\text{radius of revolution}) \, ds.$$

In using this formula to perform an actual calculation, the element of arc length ds must be written in terms of a convenient variable of integration and appropriate limits of integration must be provided.

Example Find the surface area of a sphere of radius a.

Solution The surface of this sphere can be considered as the surface of revolution generated by revolving the semicircle $y = \sqrt{a^2 - x^2}$ about the x-axis (Fig. 7.20). Since

$$\frac{dy}{dx} = \frac{d}{dx}(a^2 - x^2)^{1/2} = \frac{-x}{\sqrt{a^2 - x^2}},$$

we can use the left-right symmetry of the figure and write

$$A = \int 2\pi y \, ds = 2\int_0^a 2\pi y \sqrt{1 + \left(\frac{dy}{dx}\right)^2} \, dx$$

$$= 4\pi \int_0^a \sqrt{a^2 - x^2} \sqrt{1 + \frac{x^2}{a^2 - x^2}} \, dx$$

$$= 4\pi \int_0^a a \, dx = 4\pi a^2.$$

It is also possible to use y as the variable of integration. The calculation is a little more complicated, but it may be instructive for students to see how it works. Since $x = \sqrt{a^2 - y^2}$ in the first quadrant, we have

$$\frac{dx}{dy} = \frac{d}{dy}(a^2 - y^2)^{1/2} = \frac{-y}{\sqrt{a^2 - y^2}},$$

and therefore

$$A = \int 2\pi y \, ds = 2\int_0^a 2\pi y \sqrt{\left(\frac{dx}{dy}\right)^2 + 1} \, dy$$

$$= 4\pi \int_0^a y \sqrt{\frac{y^2}{a^2 - y^2} + 1} \, dy = 4\pi a \int_0^a \frac{y \, dy}{\sqrt{a^2 - y^2}}$$

$$= 4\pi a(-\tfrac{1}{2}) \int_0^a (a^2 - y^2)^{-1/2}(-2y \, dy) = 4\pi a(-\tfrac{1}{2})2\sqrt{a^2 - y^2} \Big]_0^a$$

$$= 4\pi a^2,$$

as before.

Remark   In addition to discovering the volume of a sphere, Archimedes also found its surface area by means of a brilliant piece of insight that links these two quantities to each other. His idea was to split up the solid sphere into a large number of small "pyramids," as follows. Imagine that the surface of our sphere of radius a is divided into many tiny "triangles," as suggested in Fig. 7.21. Of course, these little figures are not actually triangles, since there are no straight lines on the surface of a sphere. However, they are so small that each figure is nearly flat and they are nearly triangles. Let each such "triangle" be used as the base of a "pyramid" of height a whose vertex is the center of the sphere. If A_k is the area of the base of our tiny "pyramid" and V_k is its volume, for $k = 1, 2, \ldots, n$, then we know that $V_k = \frac{1}{3}A_k a$. (The fact that the volume of a pyramid is one-third the area of the base times the height was discovered by Democritus two centuries before the time of Archimedes.) By adding these equations for $k = 1, 2, \ldots, n$, we obtain

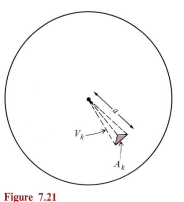

Figure 7.21

$$\sum_{k=1}^{n} V_k = \sum_{k=1}^{n} \tfrac{1}{3}A_k a = \frac{1}{3}\left(\sum_{k=1}^{n} A_k\right)a.$$

Since all our "pyramids" fill the solid sphere, this tells us that the volume V and surface area A of the sphere are related by the equation

$$V = \tfrac{1}{3}Aa.$$

But now Archimedes' discovery that $V = \frac{4}{3}\pi a^3$ enables us to write this equation in the form

$$\tfrac{4}{3}\pi a^3 = \tfrac{1}{3}Aa,$$

so

$$A = 4\pi a^2,$$

just as in the example.

PROBLEMS

In Problems 1 to 6, find the area of the surface of revolution generated by revolving the given arc about the indicated axis.

1 $y = \dfrac{1}{4}x^4 + \dfrac{1}{8x^2}$, $1 \le x \le 2$, the y-axis.

2 $y = \frac{1}{3}\sqrt{x}(3 - x)$, $0 \le x \le 3$, the x-axis.

3 $y = \frac{1}{3}(2 + x^2)^{3/2}$, $0 \le x \le 2$, the y-axis.

4 $y = x^2$, $0 \le x \le 2$, the y-axis.

5 $y = x^3$, $0 \le x \le 1$, the x-axis.

6 $y = 2\sqrt{x}$, $2 \le x \le 8$, the x-axis.

7   The arc of the parabola $x^2 = 4py$ between $(0, 0)$ and $(2p, p)$ is revolved about the y-axis. Find the area of the surface of revolution (a) by integrating with respect to x; (b) by integrating with respect to y.

8   The loop of $9y^2 = x(3 - x)^2$ is revolved about the y-axis. Find the area of the surface generated in this way.

9   Find the area of the surface generated by revolving the astroid (or hypocycloid of four cusps) $x^{2/3} + y^{2/3} = a^{2/3}$ about the y-axis.

10   Consider a cylinder circumscribed about a sphere of radius a. Let two planes perpendicular to the axis of the cylinder intersect the sphere. If these planes are at a distance h apart, show that the area of the zone that lies between them on the sphere is $2\pi ah$. (It is a remarkable fact that this is the same as the area between these planes on the lateral surface of the cylinder. Note also that if $h = 2a$ this result yields the formula for the total surface area of the sphere.)

7.7
HYDROSTATIC FORCE

In the previous sections of this chapter, we have seen how integration can be used to answer many natural questions that arise in geometry. In the next two sections we consider several applications to physics and engineering. Applications of this type usually require, in addition to the pertinent mathematical knowledge, a reasonable grasp of the basic scientific principles involved. In the problems we study, these scientific principles are very simple — so simple, indeed, that students with no previous experience in science should be able to understand them easily. The main theme of our work continues to be the idea that the whole of a quantity can be calculated by dividing it into many convenient small pieces and adding up these pieces by means of integration.

In this section we undertake a brief excursion into the science of *hydrostatics*, which studies the behavior of liquids at rest. In particular, we calculate the force exerted outward against the walls of an open container by water at rest inside the container. The containers we consider can be anything from a small fishbowl to the reservoir behind a gigantic dam.

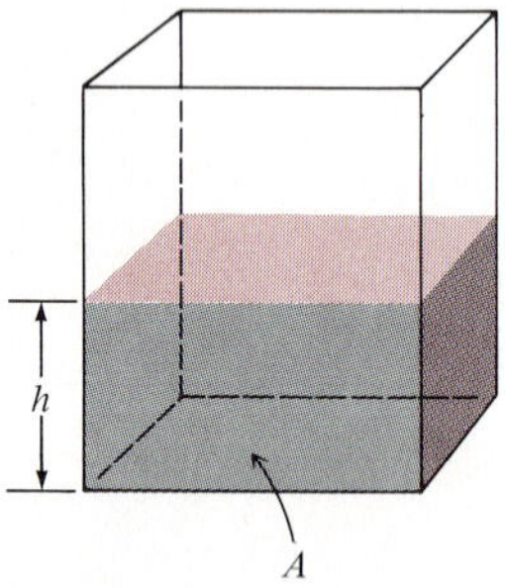

If a tank with a rectangular bottom and vertical sides is filled with water to a depth h (Fig. 7.22), then the force exerted downward on the bottom is equal to the weight of the water contained in the tank. If A is the area of the bottom, then this force is given by the formula

$$F = whA, \tag{1}$$

where w is the weight-density of the water, which is approximately 62.5 lb/ft³ or $\frac{1}{32}$ ton/ft³. It is obviously necessary for the units of measurement in (1) to be compatible. In our work we measure h in feet, A in square feet, and w in pounds or tons per cubic foot. The force F is then expressed in pounds or tons.

If we divide (1) by A, then the resulting quantity

$$p = wh \tag{2}$$

is the *pressure*, or *force per unit area*, exerted by the water on the bottom of the tank. The pressure at a given depth h below the surface can therefore be thought of as the weight of a column of water h units high that rests on a horizontal base whose area is 1 square unit. Formula (2) is quite remarkable, for it states that the pressure is proportional to the depth alone, and that the size and shape of the container are completely irrelevant. For example, at a depth of 4 ft in a swimming pool the pressure is the same as it is at a depth of 4 ft in a nearby lake (namely, 250 lb/ft²) regardless of the size of the lake; and we find the same pressure at the bottom of a vertical glass tube 1 inch in diameter if we plug the bottom with a cork and fill it with 4 ft of water. Furthermore, it can be verified experimentally that at any point in a liquid the pressure is the same in all directions. This means that a flat plate below the surface has the same pressure acting on one face at a given depth whether it is placed horizontally, vertically, or at an angle, and this pressure is normal (perpendicular) to the face of the plate. As skin divers know from personal experience, the water pressure on the eardrums depends only on how deep they are, and not at all on the angle at which the head is tilted.

In order to find the total force exerted by the water against the bottom of

Figure 7.22

the tank in Fig. 7.22, it is enough to multiply the pressure at the bottom by the area of the base,

$$F = pA,$$

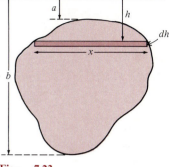

which is merely formula (1). It is more difficult to find the force against one of the sides, because the pressure is not constant there but increases as the depth increases. Instead of pursuing this particular problem, we consider a more general situation.

In Fig. 7.23 we show a flat plate of unspecified shape submerged vertically in a body of water. To find the total force exerted by the water against one face of this plate, we imagine this face to be divided into a large number of thin horizontal strips. The typical strip shown in the figure is at a depth h below the surface. Its width dh is so small compared with h that the pressure is essentially constant over the entire strip, and has the value $p = wh$. The area of the strip is $dA = x\, dh$, so the *element of force dF* acting against the strip is given by

Figure 7.23

$$dF = p\, dA = wh \cdot x\, dh.$$

The total force F acting against the whole face of the plate is now obtained by integrating these elements of force as our typical strip sweeps across the plate from top to bottom,

$$F = \int dF = \int_a^b wh \cdot x\, dh. \tag{3}$$

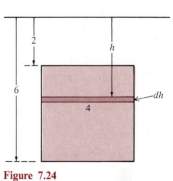

In order to carry out the indicated integration in a specific problem, it is necessary to know x as a function of h, and this is determined geometrically from the shape of the plate. As in the preceding sections of this chapter, it is better to understand and apply the ideas used in constructing formula (3) than to try to memorize this formula and use it without thinking. We repeat the crux of the method: Thin horizontal strips are used because the pressure can be treated as essentially constant over all of such a strip, and the force acting against this strip is then simply the pressure times the area.

Example 1 A vertical gate in a dam has the shape of a square 4 ft on a side, the upper edge being 2 ft below the surface of the water (Fig. 7.24). Find the total force this gate must withstand.

Figure 7.24

Solution In this case $x = 4$ and h increases from 2 to 6, so

$$F = \int_2^6 wh \cdot 4\, dh = 2wh^2 \Big]_2^6 = 2w \cdot 32 = 2 \text{ tons.}$$

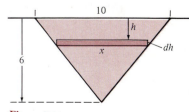

Example 2 A triangular dam in a ditch is 10 ft across the top and 6 ft deep (Fig. 7.25). Find the force of the water against this dam when the water is at the top and ready to spill over.

Figure 7.25

Solution   By similar triangles we see that

$$\frac{x}{10} = \frac{6-h}{6}, \qquad \text{so} \qquad x = \tfrac{5}{3}(6-h).$$

Since h increases from 0 to 6, we have

$$F = \int_0^6 wh \cdot \tfrac{5}{3}(6-h)\,dh = \tfrac{5}{3}w(3h^2 - \tfrac{1}{3}h^3)\Big]_0^6$$

$$= 60w = 1\tfrac{7}{8} \text{ tons} = 3750 \text{ lb}.$$

PROBLEMS

In Problems 1 to 4, it is assumed that the face of a dam adjacent to the water is vertical and has the stated shape. In each case find the total force against the dam.

1 A rectangle 150 ft wide and 12 ft high; water 8 ft deep.

2 An isosceles trapezoid 200 ft wide at the top, 100 ft wide at the bottom, and 20 ft high; reservoir full of water.

3 An isosceles triangle 60 ft wide at the top and 20 ft high in the center; reservoir full of water.

4 An isosceles trapezoid 90 ft across the top, 60 ft across the bottom, and 20 ft high; water 12 ft deep.

In Problems 5 to 8, it is assumed that a vertical gate in the face of a dam has the stated shape. In each case find the total force of the water against the gate.

5 A triangle 4 ft wide at the top and 5 ft high, with upper edge 1 ft below the water surface.

6 An isosceles trapezoid 6 ft wide at the top, 8 ft wide at the bottom, and 6 ft high, with upper edge 4 ft below the water surface.

7 A triangle 4 ft wide at the bottom and 4 ft high, with the upper vertex 2 ft below the water surface.

8 A semicircle 4 ft in diameter with its diameter at the water surface.

9 A cylindrical barrel 4 ft high and 3 ft in diameter stands upright and is half filled with oil that weighs 50 lb/ft³. What is the total force of the oil against the lateral wall of the barrel?

10 If the barrel in Problem 9 lies on its side, what is the force of the oil against one of the circular ends?

11 A rectangular gate in a vertical dam is 5 ft wide and 6 ft high. Find the force against this gate when the water level is 8 ft above its top. How much higher must the water rise to double the force?

12 The vertical ends of a water trough are isosceles triangles with base 3 ft and height 2 ft. Find the force against one end when the trough is full of water.

13 The end of a swimming pool is a rectangle inclined 45° to the horizontal. If the edge at the surface is 12 ft long and the submerged edge 10 ft long, find the force the water exerts against this rectangle.

*14 A rectangular tank is filled with two nonmixing liquids whose densities are w_1 and w_2, where $w_1 < w_2$. In one side of the tank there is a square window $3\sqrt{2}$ ft on a side with one of its diagonals vertical and the upper vertex 1 ft below the surface, and with the other diagonal on the boundary between the liquids. Find the force the liquids exert against the window.

7.8
WORK AND ENERGY

It is a common experience that in moving an object against a force acting on it, as in lifting a heavy stone, we have the sensation of expending effort or doing work. Even before we define the physical concept of work, we are convinced that it takes twice as much work to lift a 20-lb stone a given distance as it does to lift a 10-lb stone, and also that the work done in lifting a stone 3 ft is three times that done in lifting it 1 ft. These ideas point the way to our basic definition: If a constant force F acts through a distance d, then the *work* done during this process is the product of the force and the distance through which it acts,

$$W = F \cdot d. \tag{1}$$

It is understood here that the force acts in the direction of the motion.

As we know, the "weight" of an object is the force with which the object is attracted to the earth by gravity. For a given object moving at or near the surface of the earth, this force remains essentially constant in magnitude and is always directed toward the center of the earth. Thus, if a stone weighing 20 lb is lifted 3 ft, definition (1) tells us that 60 ft-lb of work are done. And if a tractor drags a boulder 18 in by applying a constant force of 2 tons, then the tractor does 36 in-tons (or 3 ft-tons) of work.

This definition is satisfactory as long as the force F is constant. However, many forces do not remain constant during the process of performing work. In a situation like this we divide the process into many small parts and calculate the total work by integrating the elements of work corresponding to these parts.

This idea is illustrated by the operation of stretching a spring, as follows.

Example 1   A certain spring has a natural length of 16 in. When it is stretched x inches beyond its natural length, *Hooke's law* states that the spring pulls back with a restoring force of $F = kx$ pounds, where k is a constant. The constant of proportionality k is called the *spring constant*, and can be thought of as a measure of the stiffness of the spring. For the spring under discussion, 8 lb of force are required to hold it stretched 2 in. How much work is done in stretching this spring from its natural length to a length of 24 in?

Solution   First, the fact that $F = 8$ when $x = 2$ allows us to find k. We have $8 = k \cdot 2$, so $k = 4$ and $F = 4x$. To clarify our ideas, we draw the spring in its unstretched condition, and also after it has been stretched x inches (Fig. 7.26). Now, if we imagine that the spring is stretched a very small additional distance dx, then the force changes very little over this increment of distance and can be treated as essentially constant. The work done against the pull of the spring over this increment of distance is

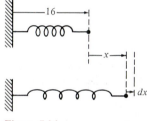

$$dW = F \, dx = 4x \, dx, \qquad (2)$$

Figure 7.26

and the total work done during the complete stretching process is

$$W = \int dW = \int F \, dx = \int_0^8 4x \, dx = 2x^2 \Big]_0^8 = 128 \text{ in-lb},$$

since x increases from 0 to 8 as the length of the spring increases from 16 to 24.

In a similar way, we can consider the work done by any variable force that acts in a given direction as its point of application moves in this direction. If we coordinatize the line of action by introducing an x-axis, and if the point of application of the variable force $F(x)$ moves from $x = a$ to $x = b$, then $dW = F(x) \, dx$ is the *element of work* and

$$W = \int dW = \int_a^b F(x) \, dx \qquad (3)$$

gives the total work done during the process. This formula can be taken either as a definition or as a natural method of computing the work in accordance with the way of thinking described in Example 1. In our next example the same idea is applied to a different situation.

Example 2 According to Newton's law of gravitation, any two particles of matter of masses M and m attract each other with a force F whose magnitude is directly proportional to the product of the masses and inversely proportional to the square of the distance r between them,

$$F = G \frac{Mm}{r^2},$$

where G is the so-called *constant of gravitation.* If M is fixed at the origin, how much work is required to move m from $r = a$ to $r = b$, where $a < b$?

Solution The element of work is

$$dW = F\, dr = GMm\, \frac{dr}{r^2}, \tag{4}$$

so the total work is

$$W = \int dW = GMm \int_a^b \frac{dr}{r^2} = GMm \left(-\frac{1}{r} \right)\Big]_a^b = GMm \left(\frac{1}{a} - \frac{1}{b} \right).$$

If we think of the final position $r = b$ as being chosen farther and farther away, so that $b \to \infty$, then the work W approaches the limiting value GMm/a. This quantity is the work which must be done against the force of attraction to move m from $r = a$ to an infinite distance, that is, to separate the masses completely; it is called the *potential* of the two particles.

Each of the preceding examples is concerned with a variable force acting through a given distance. Our next example is very different. It involves a process in which the parts of a body—in this case, drops of water—are moved different distances against a constant force, and the total work is calculated as the sum of the various bits of work associated with the various parts.

Example 3 Consider a cylindrical tank of radius r and height h, filled with water to a depth D (Fig. 7.27). How much work is done in pumping the water out over the edge of the tank? (As usual, denote the weight-density of water by w.)

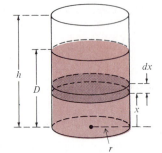

Figure 7.27

Solution The essence of this problem is the fact that each drop of water must be lifted from its initial position up to the edge of the tank and dumped over the side. The work done in this process is the same for all drops which are the same distance below the edge. This suggests that we consider all the water located in a thin horizontal layer of thickness dx at a height x above the bottom of the tank, that we write down the element of work dW needed to lift this entire layer up to the edge of the tank, and that we calculate the total work in our usual way, by adding (or integrating) these elements of work as x increases from 0 to D, so that our typical layer sweeps through all the water in the tank. It is clear from the figure that the volume of the layer is $\pi r^2 \, dx$, so its weight is $w\pi r^2\,dx$, and the work done in lifting this layer through the distance $h - x$ to the top of the tank is

$$dW = w\pi r^2\,dx \cdot (h - x). \tag{5}$$

The total work done in pumping out all the water is therefore

$$W = \int dW = w\pi r^2 \int_0^D (h - x)\,dx$$

$$= w\pi r^2 (hx - \tfrac{1}{2}x^2)\Big]_0^D = w\pi r^2 (hD - \tfrac{1}{2}D^2).$$

We repeat: The crux of the method is the fact that all drops of water in our typical layer are essentially the same distance below the edge of the tank, and can therefore be treated together in calculating the work.

Students should observe that the use of definition (1) in a suitable form is the key to each of these examples. Specifically, formulas (2), (4), and (5) are simply the versions of (1) that are appropriate in each case.

We devote the rest of this section to a brief discussion of the important concept of energy.

Consider a variable force F that acts on a particle of mass m over a given distance along a straight line, which we take to be the x-axis. This force not only does work, but also imparts an acceleration dv/dt to the particle in accordance with Newton's second law of motion,

$$F = m\frac{dv}{dt}, \qquad \text{where } v = dx/dt. \tag{6}$$

This acceleration produced by the force changes the velocity v of m, and therefore also changes its *kinetic energy*—or energy due to motion—which is defined by the formula

$$\text{Kinetic energy} = \tfrac{1}{2}mv^2.$$

We are now in a position to prove the following important theorem of mechanics:

The work done by the force F during the process described above equals the change in the kinetic energy of the particle; and in particular, if the particle starts from rest, then the work done on it equals the kinetic energy it attains.

The proof is easy. We begin by writing (6) in the form

$$F = m\frac{dv}{dt} = m\frac{dv}{dx}\frac{dx}{dt} = mv\frac{dv}{dx}.$$

Formula (3) now yields

$$W = \int_a^b F\,dx = \int_a^b mv\frac{dv}{dx}\,dx = \int_{v_a}^{v_b} mv\,dv$$

$$= \tfrac{1}{2}mv^2\Big]_{v_a}^{v_b} = \tfrac{1}{2}mv_b^2 - \tfrac{1}{2}mv_a^2, \tag{7}$$

so the work W equals the change in the kinetic energy, as stated.

In certain types of physical situations—but not in all—it is possible to introduce the concept of *potential energy*. For instance, if a heavy stone is lifted up from the ground, we say that we have increased its potential energy, because if we then release the stone and let it fall, the force of gravity does work on the stone and increases its kinetic energy. Potential energy is often loosely described as the energy a body possesses due to its position. A better description is to say that the potential energy of a body with respect to a given force is a measure of the capacity of the force to do work on the body. We will give a definition, but first we consider two examples that will help us understand the meaning of this definition.

Example 4 For a particle of mass m moving under the influence of the constant acceleration g due to gravity near the surface of the earth, its weight is mg and its potential energy V with respect to the force of gravity is defined by $V = mgx$, where x is a coordinate giving the vertical position of the particle. The x-axis is assumed to have its positive direction upward, with its origin at any convenient position. Changing the location of the origin will obviously change V, but this does not matter, because the important thing about potential energy is not its own value, but instead the value of dV/dx, which gives its rate of change with respect to x. In this case we have $-dV/dx = -mg$, which is the force of gravity on m in the x-direction.

Example 5 As in Example 2, consider a particle of mass m moving on the positive x-axis under the influence of the inverse square force of attraction exerted by another particle placed at the origin, so that $F(x) = -c/x^2$, where c is a positive constant. Here the potential energy is defined by $V = -c/x$. Notice that this is the negative of the potential defined in Example 2; it is therefore the negative of the work needed to bring this simple physical system to the state of zero potential energy, where m is infinitely far away. For our present purposes it is more important to notice that $-dV/dx = -c/x^2 = F(x)$.

These examples suggest that in suitable circumstances the potential energy V associated with a force $F(x)$ in the direction of the x-axis has the property that $-dV/dx = F(x)$. With this idea in mind, we now continue our examination of equation (7).

In using formula (3) for the calculation in (7), we tacitly assumed that the unspecified force F is a continuous function depending only on the coordinate x over the interval $a \le x \le b$, say $F = F(x)$. (Notice that a frictional force does not have this property; for it depends not only on the location of the particle m, but also on the direction in which it is moving.) By the discussion at the end of Section 6.7, this assumption guarantees that there exists a function $V(x)$ such that $dV/dx = -F(x)$. We can therefore evaluate the work W in (7) as follows:

$$W = \int_a^b F(x)\, dx = \int_b^a -F(x)\, dx = V(x)\Big]_b^a$$

$$= V(a) - V(b). \tag{8}$$

This enables us to write (7) as

$$\tfrac{1}{2}mv_b^2 - \tfrac{1}{2}mv_a^2 = V(a) - V(b)$$

or

$$\tfrac{1}{2}mv_b^2 + V(b) = \tfrac{1}{2}mv_a^2 + V(a). \tag{9}$$

On the left side of (9) we drop the subscript and replace $V(b)$ by $V(x)$, in order to emphasize that v and $V(x)$ are considered to be variables; and on the right side we hold v_a and $V(a)$ fixed. Equation (9) now takes the form

$$\tfrac{1}{2}mv^2 + V(x) = \tfrac{1}{2}mv_a^2 + V(a) = E, \tag{10}$$

where the constant E is called the *total energy* of the particle. The function $V(x)$ is called the *potential energy* of the particle, and (10) states that the sum of the kinetic energy and potential energy is constant. This is the *law of conservation of energy*, which is one of the basic principles of classical physics.

We point out that the definition of $V(x)$ means that this function is determined only to within an additive constant, so in any specific situation the state of zero potential energy can be chosen to suit our convenience. Also, students may wonder about the mild trickery with algebraic signs that takes place in the definition of $V(x)$ and in the calculation (8). The purpose of this is to guarantee the appearance of plus signs instead of minus signs in (10), so that we can speak of the sum of the kinetic and potential energies as being constant instead of their difference.

In Appendix A.6 we indicate a few of the many ways in which these ideas have been altered in twentieth century physics.

PROBLEMS

1 A spring has a natural length of 10 in, and a 12-lb force stretches it $\tfrac{1}{2}$ in. Find the work done in stretching the spring from 10 in to 18 in.

2 A spring has a natural length of 12 in, and a 45-lb force stretches it to 15 in. Find the work done in stretching it from 15 in to 19 in.

3 A spring supporting a railroad car has a natural length of 12 in, and a force of 8000 lb compresses it $\tfrac{1}{2}$ in. Find the work done in compressing it from 12 in to 9 in. (Hooke's law is valid for compressing springs as well as for stretching them.)

4 Find the natural length of a spring if the work done in stretching it from a length of 2 ft to a length of 3 ft is one-fourth the work done in stretching it from 3 ft to 5 ft.

5 A bucket weighing 5 lb when empty is loaded with 60 lb of sand. Unfortunately there is a hole in the bucket, and sand leaks out uniformly at such a rate that a third of the sand is lost when the bucket has been lifted 10 ft. Find the work done in lifting the bucket this distance.

6 A cable 100 ft long that weighs 4 lb/ft is hanging from a windlass. How much work is done in winding it up?

7 Solve Problem 6 if a 300-lb weight is attached to the free end of the cable.

8 A 5-lb monkey is attached to the end of a 30-ft hanging chain that weighs 0.2 lb/ft. It climbs the chain to the top. How much work does it do?

9 Gas in a cylindrical chamber moves a piston by expanding or contracting. Let the cross-sectional area of the cylinder be A, and let its variable volume and length be V and x (Fig. 7.28). If p is the pressure of the gas, then the force the gas exerts on the piston is pA.

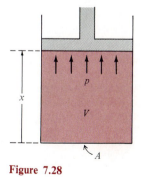

Figure 7.28

(a) If the gas expands from a volume V_1 to a volume V_2, show that the work done by the gas on the piston is

$$W = \int_{V_1}^{V_2} p \, dV.$$

(b) If a force is exerted on the piston to compress the gas from a volume V_1 to a volume V_2, show that the work done on the gas is

$$W = -\int_{V_1}^{V_2} p \, dV.$$

10 If air is compressed or expanded without any loss or gain of heat but with a possible change of temperature, then it obeys the *adiabatic* gas law $pV^{1.4} = c$, where c is a constant. If a cylinder contains 243 in^3 of air at a pressure of 14 lb/in^2, find the work done by the piston on the air in compressing it adiabatically to a volume of 32 in^3. (If this air is compressed slowly so that the heat generated is allowed to escape and the temperature remains constant, the compression is said to be *isothermal.* In this case the pressure and volume are related by *Boyle's law* $pV = c$, and in trying to calculate the work we are led to an integral of the form $\int dV/V$, which is beyond our reach. One of the main purposes of Chapter 8 is to enable us to cope with integrals of this kind, which are important in many applications.)

11 Consider a cylindrical buoy of cross-sectional area 8 ft^2 which is floating upright in water whose weight-density is $w = 62.5$ lb/ft^3. According to Archimedes' principle, a floating body is acted on by an upward buoyant force equal to the weight of the displaced water, and in a state of equilibrium this upward force balances the downward force acting on the body due to gravity.

(a) Show that there is an upward force of $62.5(8x)$ lb acting on the buoy when it is held x feet down from its equilibrium position.

(b) How much work is done in pushing the buoy 1 ft down from its equilibrium position?

*12 A conical buoy that weighs B pounds floats upright in water with its vertex a ft below the surface. A crane on a dock lifts the buoy until its vertex just clears the surface. How much work is done? Hint: When the crane has lifted the buoy x ft, then the force required to hold it in this position is the weight of the buoy minus the upward buoyant force due to the water still displaced, and this can be expressed as a function of x.

13 If an iron ball is attracted to a magnet by a force of $F = 15/x^2$ pounds when the ball is x feet from the magnet, find the work done in pulling the ball away

from the magnet from a point where $x = 2$ to a point where $x = 6$.

14 According to Coulomb's law, two electrons repel each other with a force that is inversely proportional to the square of the distance between them. Suppose one electron is held fixed at the origin on the x-axis. Find the work done in moving a second electron along the x-axis from $x = 2$ to $x = 1$. From $x = a$ to $x = b$, where $0 < b < a$.

15 If two particles of matter of masses M and m are a units apart, how much work must be done to move them twice as far apart?

16 If R is the radius of the earth (about 4000 mi) and g is the acceleration due to gravity at the surface of the earth, then the force of attraction exerted by the earth on a body of mass m is $F = mgR^2/r^2$, where r is the distance from m to the center of the earth. If this body weighs 100 lb at the surface of the earth, what does it weigh at an altitude of 1000 mi? At an altitude of 4000 mi? How much work is required to lift it from the surface to an altitude of 1000 mi?

17 Generalize Problem 16 by finding how much work must be done by a rocket on a satellite of mass m in lifting it to an altitude h above the surface of the earth.

18 Suppose that a hole is drilled straight through the center of the earth, and that a body of mass m is dropped into this hole. As the body falls, the force of attraction exerted on it by the earth is $F = mgr/R$, where r is the distance from m to the center of the earth. (The reason behind this law of force will become clear in a later chapter.) Find the work done by the earth in pulling m from the surface down to the center.

19 A conical tank 10 ft deep and 8 ft across the top is full of water. Find the work done in pumping the water over the top of a nearby 12-ft fence.

20 Find the work done in Problem 19 if the tank is initially filled only to a depth of 5 ft and if the water is pumped just to the top of the tank and over the edge.

21 A spherical tank of radius a is at the top of a tower with its bottom at a distance h above the ground. How much work is needed to fill the tank with water pumped from ground level?

22 A great conical mound of height h is built by the slaves of an oriental monarch, to commemorate a victory over the barbarians. If the slaves simply heap up uniform material found at ground level, and if the total weight of the finished mound is M, how much work do they do?

23 If the same amount of work done on two particles starting from rest causes one to move twice as fast as the other, how are their masses related?

ADDITIONAL PROBLEMS FOR CHAPTER 7

SECTION 7.2

In Problems 1 to 13, sketch the curves and find the areas of the regions they bound.

1 $y = x^2$, $y = x$.

2 $x = 3y + y^2$, $x + y + 3 = 0$.

3 $y = x^4 - 2x^2$, $y = 2x^2$.

4 $y^2 = x^3$, $x = 4$.

5 $y = x^2 - 2x - 3$, $y = 2x + 2$.

*6 $y = \dfrac{2}{\sqrt{x+2}}$, $x + 3y - 5 = 0$.

7 $y = 6x - x^2$, $y = x$.

8 $y^2 = 4x$, $2x - y = 4$.

9 $y^2 = 2x$, $x - y = 4$.

10 $y = 4 - x^2$, $y = 4 - 4x$.

11 $y^2 = -4x + 4$, $y^2 = -2x + 4$.

12 $y = 9 - x^2$, $y = x^2$.

13 $y = 9 - x^2$, $(x + 3)^2 = -4y$.

14   Find the complete area enclosed by $y^2 = 9x^2 - x^4$.

15   Find the area bounded by $y = x^2$, $y = 4$, $y = 2 - x$.

16   Find $c > 0$ so that the area bounded by $y = x^2 - c$ and $y = c - x^2$ equals 9.

*17   Find the area of the region in the second quadrant bounded by the x-axis and the parabolas $y = x^2$, $y = \sqrt{x + 18}$.

*18   Find the area between $4y = x^3$ and its tangent at $x = -2$.

SECTION 7.3

19   Find the volume of the solid of revolution generated when the area bounded by the given curves is revolved about the x-axis:

(a) $y = 2 - x^2$, $y = 1$;

(b) $y = 3x - x^2$, $y = x$;

(c) $y^2 = 4x$, $y = x$;

(d) $y = x^2 + 3$, $y = 4$;

(e) $\sqrt{x} + \sqrt{y} = \sqrt{a}$, $x = 0$, $y = 0$.

20   Find the volume generated by revolving the area bounded by $x = y^2$ and $x = 4$ about

(a) the x-axis;   (b) the y-axis;

(c) the line $y = 2$;   (d) the line $x = 4$;

(e) the line $x = -1$.

21   Find the volume generated by revolving the area bounded by $x = 4y - y^2$ and $x = 0$ about

(a) the y-axis;   (b) the x-axis.

22   Each plane perpendicular to the x-axis intersects a certain solid in a circular cross section whose diameter lies in the xy-plane and extends from $y = x^2$ to $y = 8 - x^2$. The solid lies between the points of intersection of these curves. Find its volume.

23   The base of a certain solid is the circle $x^2 + y^2 = a^2$. Each plane perpendicular to the x-axis intersects the solid in a cross section that is an isosceles right triangle with one leg in the base of the solid. Find the volume.

24   The base of a certain solid is the area bounded by $x^2 = 4ay$ and $y = a$. Each cross section perpendicular to the y-axis is an equilateral triangle with one side lying in the base. Find the volume of the solid.

25   A plane which is perpendicular to the x-axis and contains a circle of radius x^2 moves from $x = a$ to $x = b$. If the center of the circle moves along a curve $y = f(x)$, find the volume of the solid the circle generates.

*26   A solid is generated by revolving about the x-axis the area bounded by a curve $y = f(x)$, the x-axis, and the lines $x = a$ and $x = b$. Its volume is $\pi(b^3 - b^2a)$ for all $b > a$. Find $f(x)$.

27   Find the volume generated by revolving the area bounded by the curves $x^2 = 4ay$, $y = a$, $x = 0$ about

(a) the y-axis;   (b) the x-axis;   (c) the line $y = a$.

28   Let R be a region of area A in a horizontal plane, and suppose that R is bounded by a closed curve C that does not intersect itself. Let P be a point whose height above this plane is h, and form a generalized "cone" by drawing segments connecting P to the points of C. Show that the volume of this cone is $V = \frac{1}{3}Ah$. Hint: If $A(x)$ is the area of the horizontal cross section at a height x above the plane, observe that $A(x) = [(h - x)^2/h^2]A$.

29   A line passes through a vertex of a square of side a and is perpendicular to the plane in which the square lies. As this vertex moves a distance h along the line, the square turns through a complete revolution with the line as the axis. Find the volume of the screw-shaped solid the square generates. What is the volume if the square turns through two complete revolutions while moving the same distance along the line?

30   The square bounded by the axes and the lines $x = 1$, $y = 1$ is cut into two parts by the curve $y = x^n$, where n is a positive constant. Find the value of n for which these two parts generate equal volumes when revolved about the y-axis.

*31   Two oblique circular cylinders of equal height h have a circle of radius a as a common lower base and their upper bases are tangent to each other. Find the common volume.

SECTION 7.4

In Problems 32 to 37, sketch the region bounded by the given curves and use the shell method to find the volume of the solid generated by revolving this region about the given axis.

32 $y = \sqrt[3]{x}$, $x = 8$, $y = 0$; the y-axis.

33 $x = y^2 - 4y$, $x = 0$; the x-axis.

34 $y = 5x - x^2$, $y = 0$; the y-axis.

35 $x = y^3 + 1$, $y + 2x = 2$, $y = 1$; the x-axis.

36 $y = x^2$, $y = x^3$; the y-axis.

37 $2x - y = 12$, $x - 2y = 3$, $x = 4$; the y-axis.

38 The region bounded by the given curves is revolved about the y-axis. Find the volume of the solid of revolution by using both the shell method and the washer method.
(a) $y = 4x - x^2$, $y = 0$.
(b) $y = x^3$, $x = 2$, $y = 0$.

39 The region in the first quadrant between $y = 3x^2$ and $y = \frac{11}{4}x^2 + 1$ is revolved about the y-axis. Find the volume generated in this way.

40 The region bounded by $y^2 = 4x$ and $y = x$ is revolved about the x-axis. Find the volume generated in this way (a) by the shell method; (b) by the washer method.

*41 Consider the torus generated by revolving the circle $(x - b)^2 + y^2 = a^2$ $(0 < a < b)$ about the y-axis. Use the shell method to show that the volume of this torus equals the area of the circle times the distance traveled by its center during the revolution. Hint: At the right moment, change the variable of integration from x to $z = x - b$.

42 Find the volume generated by revolving about the y-axis the region bounded by $y = (x - 1)(x - 2)(x - 3)$ and the x-axis between $x = 1$ and $x = 2$.

SECTION 7.5

In Problems 43 to 49, find the length of the specified arc of the given curve.

43 $9y^2 = 4x^3$ between $(0, 0)$ and $(3, 2\sqrt{3})$.

44 $y = \frac{1}{8}x^4 + \frac{1}{4x^2}$, $1 \le x \le 2$.

45 $y = \frac{1}{6}x^3 + \frac{1}{2x}$, $1 \le x \le 3$.

46 $x = \frac{1}{10}y^5 + \frac{1}{6y^3}$, $1 \le y \le 2$.

47 $y = \frac{1}{24}x^3 + \frac{2}{x}$, $2 \le x \le 4$.

48 $y = \frac{1}{6}\sqrt{x}(4x - 3)$, $1 \le x \le 9$.

49 $y = \frac{5}{48}(1 + 4x^{4/5})^{3/2}$, $1 \le x \le 32$.

50 Let A and B be positive constants. If $0 < a < b$, show

that the problem of finding the length of the arc of the curve

$$y = Ax^3 + \frac{B}{x}$$

for $a \le x \le b$ leads to the integral

$$\int_a^b (3Ax^2 + Bx^{-2})\, dx$$

if $AB = \frac{1}{12}$.

51 Let A and B be positive constants. If $0 < a < b$, find a simple condition relating A and B that makes it possible to calculate the length of the arc of the curve

$$y = Ax^4 + \frac{B}{x^2}$$

between $x = a$ and $x = b$ by means of an integral not involving a square root.

52 Solve Problem 51 for the curve

$$y = Ax^5 + \frac{B}{x^3}.$$

SECTION 7.6

In Problems 53 to 55, find the area of the surface of revolution generated by revolving the given arc about the indicated axis.

53 $y = \frac{2}{3}(1 + x^2)^{3/2}$, $0 \le x \le 3$, the y-axis.

54 $y = \frac{2}{3}x^{3/2} - \frac{1}{2}x^{1/2}$, $0 \le x \le 4$, the y-axis.

55 $y = 2\sqrt{15 - x}$, $0 \le x \le 15$, the x-axis.

56 The loop of $18y^2 = x(6 - x)^2$ is revolved about the x-axis. Find the area of the surface generated in this way.

57 Sketch the graph of $8a^2y^2 = x^2(a^2 - x^2)$ and find the area of the surface generated when this curve is revolved about the x-axis.

SECTION 7.7

58 Find the force due to water pressure against a rectangular floodgate 10 ft wide and 8 ft deep whose upper edge is at the surface of the water.

59 Find the force against the lower half of the floodgate in Problem 58.

In Problems 60 and 61, it is assumed that a vertical gate in the face of a dam has the stated shape. In each case find the total force on the gate.

60 A triangle 6 ft wide and 4 ft high, with upper edge at the water surface.

61 A triangle with base B and height H, with its vertex at the water surface.

62 A rectangular canal lock is 30 ft wide. When the water is 20 ft deep, what is the force of the water against the lock?

63 A rudder has the shape of an isosceles right triangle whose equal legs are 2 ft long. It is submerged vertically in water with one of the equal legs vertical and the other horizontal, and with the horizontal leg 3 ft below the surface and the opposite vertex 1 ft below the surface. Find the force of the water against one face of the rudder.

64 A rectangular gate in a dam has width 10 ft and height 8 ft. Find the force against the gate when the water level is 20 ft above its top.

65 Assume that the gate in Problem 64 cannot withstand a force greater than 100 tons. How high must the water be above the top of the gate in order to break through?

*66 The vertical end of a vat is a segment of a parabola opening upward which is 4 ft across the top and 8 ft deep. What is the force against this end when the vat is full of beer weighing 60 lb/ft^3?

SECTION 7.8

67 A 4-lb force will stretch a spring 6 in. How much work is done in stretching it 3 ft?

68 A spring pulls back with a force of 7 lb when it is stretched from its natural length of 12 in to a length of 13 in. How much work is required to compress it from a length of 11 in to a length of 7 in?

69 Show that the work done in stretching a spring of natural length L from a length a to a length b $(L < a < b)$ is equal to the amount of the stretch $(b - a)$ times the tension in the spring when its length is $\frac{1}{2}(a + b)$.

70 A bag of sand is lifted at the constant rate of 3 ft/s for 10 seconds. At the beginning the bag contains 100 lb of sand, but the sand leaks out at the rate of 4.5 lb/s. How much work is done in lifting this bag?

71 If a certain gas in a cylinder obeys an adiabatic gas law of the form $pV^{5/3} = c$, and if it initially occupies 64 in^3 at a pressure of 128 lb/in^2, find the work it does against the piston in expanding to 8 times its initial volume.

72 Find the work done in compressing 1024 ft^3 of air at a pressure of 27 lb/in^2 down to 243 ft^3 if the air obeys the adiabatic gas law $pV^{1.4} = c$.

73 Generalize Problem 72 by finding the work done in compressing air of initial volume V_1 and pressure p_1

down to a volume V_2, assuming the adiabatic gas law $pV^{1.4} = c$.

*74 A conical buoy that weighs B pounds floats upright in water with its vertex a feet below the surface. If the top of the buoy is $\frac{1}{3}a$ feet out of the water, how much work is done in pushing the buoy down until its top is just at the surface of the water?

*75 A spherical buoy of radius a feet that weighs B pounds has exactly the weight-density w of water, so that it floats with its top just touching the surface. A crane on a dock lifts the buoy until it just clears the water. How much work is done?

76 If two electrons are held fixed at the points $x = 0$ and $x = -1$ on the x-axis, find the work done in moving a third electron along the x-axis from $x = 4$ to $x = 1$.

77 Imagine a *very* deep mine shaft, of depth $D = \frac{1}{2}R$, extending halfway down to the center of the earth (ignore all practical difficulties caused by the internal constitution of the earth). A person whose weight is w at the surface is lifted from the bottom of the shaft to the top. Under the assumption that the weight remains constant during the journey, the work done would be wD. Show that the work done during this process is actually $\frac{3}{4}wD$, by taking into account the fact that the force of gravity below the surface of the earth is proportional to the distance from its center.

78 A tank has the shape of the paraboloid of revolution obtained by revolving $y = x^2$ $(0 \le x \le \sqrt{5})$ about the y-axis. If it is full of water, how much work is required to empty it by pumping all the water out over the edge?

79 Let a cylindrical barrel of diameter 3 ft and height 5 ft be filled to a depth of 2 ft with water and then, above the water, with 2 additional ft of oil that weighs 50 lb/ft^3. Find the work done in pumping the water and oil over the edge of the barrel.

80 A hemispherical tank of radius 8 ft is full of water. If a hole is punched in the bottom, find the work done by gravity in emptying the tank.

81 Two cables are hanging side by side from the ceiling of a gymnasium. The first is an elastic cable of length L and the second is inelastic and has length $2L$. As two gymnasts of equal weight climb down these cables, the weight of the first stretches his cable to a total length of $2L$. Show that when the two gymnasts climb back up to the ceiling, the first does only $\frac{3}{4}$ of the work done by the second.

8

EXPONENTIAL AND LOGARITHM FUNCTIONS

8.1
INTRODUCTION

Our main purpose in this chapter is to learn how to work successfully with the indefinite integral

$$\int \frac{dx}{x}. \tag{1}$$

As we shall see, this purpose compels us to study the special exponential and logarithm functions

$$y = e^x \qquad \text{and} \qquad y = \log_e x. \tag{2}$$

The letter e used in these functions denotes the most important special number in mathematics after π. In decimal form it is an infinite nonrepeating decimal that is known to hundreds of thousands of decimal places; the first few digits are

$$2.7182 \ldots .$$

The ultimate reason for our interest in these matters is that the integral (1) and the functions (2) arise in a great variety of problems involving population growth, radioactive decay, chemical reaction rates, electric circuits, and many other phenomena in physics, chemistry, biology, geology, and virtually every science that uses quantitative methods, including economics, meteorology, oceanography, and even archaeology. This integral and these functions are also indispensable in many branches of pure mathematics.

In order to reach a clear understanding of why the number e and the functions (2) matter so much, it is desirable to broaden the context a bit and consider the more general exponential and logarithm functions

$$y = a^x \qquad \text{and} \qquad y = \log_a x,$$

where a is a positive constant $\neq 1$. This is where we begin, and by adopting this approach we hope to make it perfectly clear that we choose a equal to e for reasons of convenience and simplicity.

208

Students who have managed to get this far in this book certainly have a working grasp of exponents, and perhaps also of logarithms as defined in terms of exponents. Nevertheless, we briefly review the main definitions and facts as they appear in the traditional approach.

We consider expressions of the form a^x where $a > 0$ and x is any real number. It is easy to explain exactly what a^x means if x is an integer n, and we assume students understand this explanation. The following is a brief reminder:

$$\text{If } n > 0, \text{ then } a^n = a \cdot a \cdots a \ (n \text{ factors}), \qquad a^0 = 1, \qquad a^{-n} = \frac{1}{a^n};$$

$$a^m a^n = a^{m+n}, \qquad \text{e.g.,} \qquad a^2 a^3 = (a \cdot a)(a \cdot a \cdot a) = a \cdot a \cdot a \cdot a \cdot a = a^5;$$

$$\frac{a^m}{a^n} = a^{m-n}, \qquad \text{e.g.,} \qquad \frac{a^5}{a^3} = \frac{a \cdot a \cdot a \cdot a \cdot a}{a \cdot a \cdot a} = \frac{a \cdot a \cdot a}{a \cdot a \cdot a} \cdot \frac{a \cdot a}{1} = a \cdot a = a^2;$$

$$(a^m)^n = a^{mn}, \qquad \text{e.g.,} \qquad (a^3)^2 = (a \cdot a \cdot a)(a \cdot a \cdot a) = a \cdot a \cdot a \cdot a \cdot a \cdot a = a^6.$$

Next, in Section 3.4 we summarized the meaning of fractional exponents, and we repeat the essence of this summary here. If $r = p/q$ is a fraction in lowest terms with $q > 0$, then by definition

$$a^r = a^{p/q} = (\sqrt[q]{a})^p, \tag{1}$$

where $\sqrt[q]{a}$ is the unique positive number whose qth power is a.

If the exponent x is an irrational number, then difficulties appear that students might not notice if we didn't mention them. For instance, what is meant by the expression $2^{\sqrt{2}}$? Clearly, it doesn't make sense to multiply 2 by itself $\sqrt{2}$ times. Also, since $\sqrt{2}$ can't be written as a fraction, definition (1) is useless. Is $2^{\sqrt{2}}$ really a definite number with a specific value? The answer is Yes, but this is not at all obvious. A natural way to proceed is to use the fact that any irrational number can be approximated as closely as we please by rational numbers. We can therefore define a^x by

$$a^x = \lim_{r \to x} a^r,$$

where r approaches x through rational values. This way of defining a^x when x is irrational is satisfactory from the logical point of view; however, it is a long and tedious chore to prove rigorously that everything works out as we expect and that the familiar laws of exponents remain valid. We skip over these boring details and merely state the final result, that the laws of exponents continue to hold in the following form:

$$a^{x_1} a^{x_2} = a^{x_1 + x_2}, \qquad \frac{a^{x_1}}{a^{x_2}} = a^{x_1 - x_2}, \qquad (a^{x_1})^{x_2} = a^{x_1 x_2},$$

where x_1 and x_2 are arbitrary real numbers.

The next natural step in this development is to examine the properties of the general *exponential function* $y = a^x$. Here again we simply state the important facts without making any attempt to discuss the logical details of how these facts are established. As above, we assume that a is a positive constant, and also that $a \neq 1$. The case $a = 1$ is of no interest because $1^x = 1$

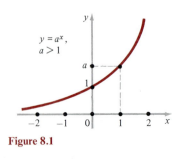

Figure 8.1

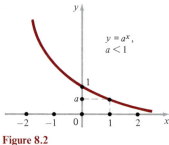

Figure 8.2

for all x. Let us suppose first that $a > 1$. Then $y = a^x$ is a continuous function of x; it is increasing; its values are all positive; and it has the further properties that

$$\lim_{x \to -\infty} a^x = 0 \quad \text{and} \quad \lim_{x \to \infty} a^x = \infty. \tag{2}$$

To sketch the graph, we plot a few points corresponding to several integral values of x, both positive and negative, and then connect these points by a smooth curve, as shown in Fig. 8.1. If $a < 1$, then $y = a^x$ is a decreasing function and its graph has the shape shown in Fig. 8.2.

When this much information about exponents is known or assumed, it is very easy to define logarithms and obtain some of their properties. On the most primitive level, a logarithm is an exponent. Thus, the fact that $100 = 10^2$ says that 2 is the logarithm of 100 to the base 10 (written $2 = \log_{10} 100$); and $4 = 64^{1/3}$ says that $\frac{1}{3}$ is the logarithm of 4 to the base 64 ($\frac{1}{3} = \log_{64} 4$).

More generally, the properties of exponents discussed above show clearly that if a is a positive constant $\neq 1$, then to each positive x there corresponds a unique y such that $x = a^y$. This y is written in the form $y = \log_a x$, and is called the *logarithm* of x to the *base a*. Accordingly,

$$y = \log_a x \quad \text{has the same meaning as} \quad x = a^y, \tag{3}$$

in the sense that each equation expresses the same relation between x and y, with the first written in a form solved for y and the second in a form solved for x. We can state this somewhat differently by saying that the symbol "$\log_a$" is created for the specific purpose of enabling us to solve $x = a^y$ for y in terms of x.

The basic properties of logarithms are direct translations of corresponding properties of exponents. Thus, if $x_1 = a^{y_1}$ and $x_2 = a^{y_2}$, then $x_1 x_2 = a^{y_1} a^{y_2} = a^{y_1 + y_2}$. But $y_1 = \log_a x_1$ and $y_2 = \log_a x_2$, so we have

$$\log_a x_1 x_2 = \log_a x_1 + \log_a x_2.$$

Similarly,

$$\log_a \frac{x_1}{x_2} = \log_a x_1 - \log_a x_2$$

and

$$\log_a x^b = b \log_a x,$$

where b is any real number. Further, (3) tells us that

$$a^{\log_a x} = x \quad \text{and} \quad \log_a a^x = x.$$

We note also that the particular facts

$$\log_a 1 = 0 \quad \text{and} \quad \log_a a = 1$$

are equivalent to $1 = a^0$ and $a = a^1$.

In studying the *logarithm function*

$$y = \log_a x, \tag{4}$$

we consciously think of x and y as variables instead of mere numbers. Our

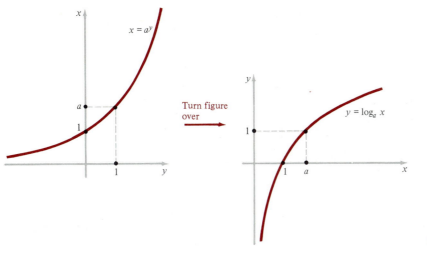

Figure 8.3

starting point is the fact that (4) is equivalent to $x = a^y$. It is clear from this that x must be positive in order for y to exist, so (4) is defined only for $x > 0$. The graph of (4) is easy to obtain from the graph of $x = a^y$ by interchanging the axes, as we show in Fig. 8.3 for the case $a > 1$. In this case $y = \log_a x$ is evidently an increasing continuous function of x. The features of this function that correspond to the properties (2) are

$$\lim_{x \to 0+} \log_a x = -\infty \quad \text{and} \quad \lim_{x \to \infty} \log_a x = \infty.$$

The most convenient logarithm for actual numerical calculations is the logarithm to the base 10, the so-called *common logarithm*. Common logarithms were once widely used by engineers and scientists and students in high school trigonometry courses, but such uses have greatly diminished in these days of computers and hand calculators. However, modern technological changes in the way people do calculations have had no influence whatever on the importance of the logarithm *as a function*; it remains indispensable in the theoretical parts of mathematics and its applications, and these theoretical uses are what concern us in this chapter.

PROBLEMS

1 Express in terms of logarithms:
(a) $4^2 = 16$;
(b) $3^4 = 81$;
(c) $81^{0.5} = 9$;
(d) $32^{4/5} = 16$.

2 Express in terms of exponents:
(a) $\log_{10} 10 = 1$;
(b) $\log_2 8 = 3$;
(c) $\log_5 \frac{1}{25} = -2$;
(d) $\log_6 216 = 3$.

3 Evaluate:
(a) $\log_{10} 10{,}000$;
(b) $\log_2 64$;
(c) $\log_{10} 0.0001$;
(d) $\log_8 4$.

4 Solve for x:
(a) $\log_4 x = 3.5$;
(b) $\log_8 x = \frac{5}{3}$;
(c) $\log_3 x = 5$;
(d) $\log_{32} x = 0.6$.

5 Find the base a:
(a) $\log_a 4 = 0.4$;
(b) $\log_a 8 = -\frac{3}{4}$;
(c) $\log_a 36 = 2$;
(d) $\log_a 7 = \frac{1}{2}$.

6 If $y = \log_a (x + \sqrt{x^2 - 1})$, show that $x = \frac{1}{2}(a^y + a^{-y})$.

7 Show that $\log_a (x + \sqrt{x^2 - 1}) = -\log_a (x - \sqrt{x^2 - 1})$.

8 The magnitude M of an earthquake on the Richter scale is a number that ranges from $M = 0$ for the smallest earthquake that can be detected by instruments to $M = 8.9$ for the greatest known earthquake. M is given by the empirical formula

$$M = \frac{2}{3} \log_{10} \frac{E}{E_0},$$

where E is the energy released by the earthquake in kilowatt-hours and $E_0 = 7 \times 10^{-3}$.

(a) How much energy is released by an earthquake of magnitude 6?

(b) A city whose population is 300,000 uses about 3×10^5 kilowatt-hours (KWh) of electric energy every day. If the energy of an earthquake could somehow be transformed into electric energy, how many days' supply for this city would be provided by the earthquake in part (a)?

(c) The great Alaskan earthquake of 1964 had a Richter magnitude of 8.4. Answer the question in part (b) for this earthquake. Hint: $10^{3/5}$ is approximately 4.

9 In chemistry the pH of a solution is defined by the formula $pH = -\log_{10} [H^+]$, where $[H^+]$ denotes the hydrogen ion concentration as measured in moles per liter.* (One mole—or gram molecular weight—of a substance consists of 6×10^{23} molecules of the substance.) The value of $[H^+]$ for pure water is found by experiment to be 1.00×10^{-7}.

(a) What is the pH of pure water?

(b) A solution is called *acidic* or *basic* (*alkaline*) according as its value of $[H^+]$ is greater or less than that for pure water. What pH's characterize acidic and basic solutions?

_____

* The symbol pH is an abbreviation of the French expression *puissance d'Hydrogène* (power of hydrogen).

8.3

THE NUMBER e AND THE FUNCTION $y = e^x$

The number e is often defined by the limit

$$e = \lim_{n \to \infty} \left(1 + \frac{1}{n}\right)^n. \tag{1}$$

This definition has the advantage of brevity but the serious disadvantage of shedding no light whatever on the significance of this crucial number. We prefer to define e differently, in a manner that reveals as clearly as possible why this number is so important. We then obtain (1) later, as merely one among many explicit formulas for e that can be used in a variety of ways.

Our aim in this section is to study a function $y = f(x)$ that is unchanged by differentiation:

$$\frac{d}{dx} f(x) = f(x). \tag{2}$$

It is far from obvious that any such function exists [we don't count the trivial case $f(x) = 0$]. As we shall see, the desired function turns out to be one of the exponential functions $y = a^x$ for $a > 1$. The central meaning of the number e can now be stated as follows: It is the specific value of the base a that causes the function $f(x) = a^x$ to have the property (2). In this way we understand what purpose e serves. However, we must still give a satisfactory definition and show as simply as possible that this definition accomplishes the stated purpose.

Let us calculate the derivative of $f(x) = a^x$ and see what happens. As usual when differentiating a new type of function, we go back to the definition of the derivative,

$$\frac{d}{dx} f(x) = \lim_{\Delta x \to 0} \frac{f(x + \Delta x) - f(x)}{\Delta x}.$$

It will be convenient here to denote the increment by the single letter h instead of the familiar Δx (Fig. 8.4):

$$\frac{d}{dx} a^x = \lim_{h \to 0} \frac{a^{x+h} - a^x}{h} = \lim_{h \to 0} \frac{a^x a^h - a^x}{h}$$

$$= \lim_{h \to 0} \left(a^x \frac{a^h - 1}{h}\right) = a^x \left(\lim_{h \to 0} \frac{a^h - 1}{h}\right). \tag{3}$$

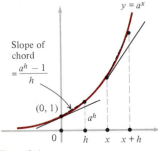

Slope of chord
$= \dfrac{a^h - 1}{h}$

$y = a^x$

$(0, 1)$

a^h

$0 \quad h \quad x \quad x+h$

Figure 8.4

As Fig. 8.4 shows, the quantity in parentheses on the right side of (3) is the slope of the tangent line to the curve $y = a^x$ at the point $(0, 1)$. If this slope equals 1, then the right side of (3) reduces to a^x and this particular function a^x has the property (2). This brings us to our definition: e is the specific value of the base a that produces this result, that is,

$$e \text{ is the number for which } \lim_{h \to 0} \frac{e^h - 1}{h} = 1. \tag{4}$$

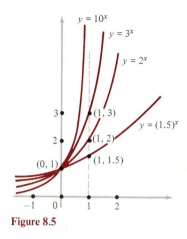

Figure 8.5

We can obtain considerable insight into the nature of the number e by sketching $y = a^x$ for the cases $a = 1.5$, $a = 2$, $a = 3$, and $a = 10$, as shown in Fig. 8.5. These curves tell us that as the base a increases continuously from numbers close to 1 to larger numbers, the slope of the tangent to $y = a^x$ at the point $(0, 1)$ increases continuously from values close to 0 to larger values, and therefore this slope is exactly equal to 1 for some intermediate value of a. This intermediate value is e; and as we hope students will agree, it is geometrically clear from these remarks that e exists. Next, we plot the points on the first three of these curves corresponding to $x = 1$ in order to stress the fact that the slopes of the chords joining these points to $(0, 1)$ are $\frac{1}{2}$, 1, and 2. This is conclusive geometric evidence that the slope of the tangent at $(0, 1)$ is < 1 for the cases $a = 1.5$ and $a = 2$, and plausible evidence that this slope is > 1 for the case $a = 3$; and therefore e is certainly > 2 and probably < 3.

In Fig. 8.6 we show the graph of $y = e^x$ with emphasis placed on its defining characteristic: It is the single member of the family of exponential functions $y = a^x$ ($a > 1$) whose tangent line at the point $(0, 1)$ has slope 1. The function $y = e^x$ is often called *the* exponential function, to distinguish it from its comparatively unimportant relatives.

We can investigate the number e more closely by noting that (4) tells us that

$$\frac{e^h - 1}{h} \text{ is approximately equal to 1,}$$

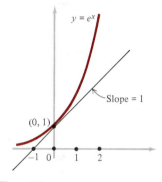

Figure 8.6

and that this approximation gets better and better as h approaches 0. Thus, by simple manipulations, we obtain

$$\frac{e^h - 1}{h} \cong 1, \quad e^h - 1 \cong h, \quad e^h \cong 1 + h, \quad e \cong (1 + h)^{1/h},$$

and finally,

$$e = \lim_{h \to 0} (1 + h)^{1/h}. \tag{5}$$

In words, this says that e is the limit of 1 plus a small number, raised to the power of the reciprocal of the small number, as that small number approaches 0. If we write $h = 1/n$ where n is understood to be a positive integer that $\to \infty$ as $h \to 0$, then (5) yields

$$e = \lim_{n \to \infty} \left(1 + \frac{1}{n}\right)^n,$$

which is (1). This formula enables us to compute rough approximations to e fairly easily, as the following table shows:

n	$\left(1+\dfrac{1}{n}\right)^n$
1	2
2	$\frac{9}{4} = 2\frac{1}{4} = 2.25$
3	$\frac{64}{27} = 2\frac{10}{27} = 2.370$
4	$\frac{625}{256} = 2\frac{113}{256} = 2.441$

However, this is a slow process and the value of e has been computed to great accuracy by other and more efficient methods. To 15 decimal places it is

$$e = 2.718281828459045. \ldots *$$

The number e, like the number π, is woven inseparably into the fabric of both nature and mathematics. Many remarkable properties of e have been discovered over the centuries. For example, e is irrational; indeed, it is not even a root of any polynominal equation with rational coefficients.

However, we must not forget our original purpose in this section, which was to study a function that is unchanged by differentiation. We have now made a good start on this task, in the sense that we have explained the meaning of the following statement and established its validity:

$$\frac{d}{dx} e^x = e^x. \tag{6}$$

An equivalent statement is that $y = e^x$ satisfies the differential equation

$$\frac{dy}{dx} = y.$$

Every function $y = ce^x$ also satisfies this equation, because

$$\frac{dy}{dx} = \frac{d}{dx}(ce^x) = c\frac{d}{dx}e^x = ce^x = y.$$

Further, we assert that these are the *only* functions that are unchanged by differentiation. To prove this, suppose that $y = f(x)$ is any function with this property. Then by the quotient rule,

$$\frac{d}{dx}\left[\frac{f(x)}{e^x}\right] = \frac{e^x f'(x) - f(x)e^x}{e^{2x}} = \frac{e^x f(x) - f(x)e^x}{e^{2x}} = 0.$$

This implies that $f(x)/e^x = c$ for some constant c, so $f(x) = ce^x$, as stated.

By the chain rule, (6) generalizes immediately to

$$\frac{d}{dx} e^u = \frac{de^u}{du}\frac{du}{dx} = e^u \frac{du}{dx}, \tag{7}$$

where $u = u(x)$ is understood to be any differentiable function of x.

---

* Many people remember this much of e by grouping the digits this way,

2.7 1828 1828 45 90 45,

in order to visualize the repeated 1828 followed by 45, then twice 45, then 45 again.

Example 1   In view of (7), the following derivatives are obvious:

$$\frac{d}{dx} e^{4x} = 4e^{4x}, \qquad \frac{d}{dx} e^{x^2} = 2xe^{x^2}, \qquad \frac{d}{dx} e^{1/x} = \left(-\frac{1}{x^2}\right) e^{1/x}.$$

If we write (7) in differential form, as $d(e^u) = e^u \, du$, then by reading this backwards we obtain the integration formula

$$\int e^u \, du = e^u + c. \tag{8}$$

Example 2   To integrate $\int e^{5x} \, dx$, we write

$$\int e^{5x} \, dx = \tfrac{1}{5} \int e^{5x} \, d(5x) = \tfrac{1}{5} e^{5x} + c,$$

where $5x$ plays the role of u in formula (8). This problem is so simple that there is no need to make explicit use of the method of substitution. It suffices to keep in mind what (8) says and make minor adjustments accordingly, as indicated.

Example 3   The integral

$$\int \frac{9x e^{\sqrt{3x^2+2}} \, dx}{\sqrt{3x^2 + 2}}$$

is more complicated. Our only hope is that (8) will see us through, so we write

$$u = \sqrt{3x^2 + 2} = (3x^2 + 2)^{1/2}$$

and

$$du = \tfrac{1}{2}(3x^2 + 2)^{-1/2} 6x \, dx = \frac{3x \, dx}{\sqrt{3x^2 + 2}}.$$

This substitution (or change of variable) enables us to express the given integral in a much simpler form, and thereby to finish the calculation,

$$\int \frac{9x e^{\sqrt{3x^2+2}} \, dx}{\sqrt{3x^2 + 2}} = 3 \int e^u \, du = 3e^u + c = 3e^{\sqrt{3x^2+2}} + c.$$

Students should observe that the complicated appearance of the given integral is only a disguise concealing the relatively simple type displayed in (8). Learning the art of integration is mostly learning to see the type through the disguise.

Example 4   *Continuously compounded interest.* If P dollars is deposited in a bank that pays an interest rate of 8 percent per year, compounded semiannually, then after t years the accumulated amount is

$$A = P(1 + 0.04)^{2t}.$$

More generally, if the interest rate is $100x$ percent ($x = 0.08$ for 8 percent), and if this interest is compounded n times a year, then after t years the accumulated amount is

$$A = P\left(1 + \frac{x}{n}\right)^{nt}.$$

If n is now increased indefinitely, so that the interest is compounded more and more frequently, then we approach the limiting case of continuously compounded interest. To find the formula for A under these circumstances, we observe that (5) yields

$$\left(1 + \frac{x}{n}\right)^{nt} = \left[\left(1 + \frac{x}{n}\right)^{n/x}\right]^{xt} \longrightarrow e^{xt},$$

so

$$A = Pe^{xt}. \tag{9}$$

Ordinary compound interest produces growth in spurts or jumps at the end of each interest period. In contrast to this, we see from (9) that continuously compounded interest produces steady continuous growth of a type called *exponential growth*. In Sections 8.5 and 8.6 we discuss many additional examples of exponential growth as it occurs in the natural sciences.

Remark 1   The function e^x grows very rapidly as x increases; in fact, it grows faster than x^p for any fixed positive exponent p, no matter how large, in the sense that

$$\lim_{x \to \infty} \frac{e^x}{x^p} = \infty.$$

An outline of a proof for the case in which p is a positive integer n is given in Additional Problems 18 to 20.

Remark 2   We have deduced the existence of the limits in (1) and (5) from the definition of e given in (4). However, this definition itself is highly geometric in nature, and some mathematicians might be inclined to dismiss our entire approach to these ideas as "reasoning by wishful thinking." To mollify such critics, and also for the occasional students who might be interested, we provide an independent proof of the existence of these limits in Appendix C.7.

PROBLEMS

In Problems 1 to 10, find the derivative dy/dx of the given function.

1 $y = \frac{1}{2}(e^x + e^{-x})$.

2 $y = \frac{1}{2}(e^x - e^{-x})$.

3 $y = x^2 e^x$.

4 $y = x^2 e^{-x^2}$.

5 $y = e^{e^x}$.

6 $y = x^e + e^x$.

7 $y = \dfrac{ax - 1}{a^2} e^{ax}$.

8 $y = (3x + 1)e^{-3x}$.

9 $y = (2x^2 - 2x + 1)e^{2x}$.

10 $y = e^{1/x^2} + 1/e^{x^2}$.

Evaluate the integrals in Problems 11 to 16.

11 $\int e^{3x}\, dx$.

12 $\int x e^{-x^2}\, dx$.

13 $\int e^{(1/5)x}\, dx$.

14 $\int \dfrac{3\,dx}{e^{2x}}$.

15 $\int 6x^2 e^{x^3}\, dx$.

16 $\int \dfrac{e^{\sqrt{x}}\, dx}{\sqrt{x}}$.

17   Sketch the graph of each of the following functions and find its maximum and minimum points and points of inflection:
(a) $y = e^{-x^2}$;   (b) $y = xe^{x/3}$.

18 Find the base of the largest rectangle that rests on the x-axis and has its upper vertices on the curve $y = e^{-x^2}$.

19 Sketch the curve $y = \frac{1}{2}(e^x + e^{-x})$ and find its length from $x = 0$ to $x = b$ $(b > 0)$.

20 The arc in Problem 19 is revolved about the x-axis. Find the area of the surface of revolution generated in this way.

21 If a particle moves on the x-axis in such a way that its position x at time t is given by $x = Ae^{kt} + Be^{-kt}$, where A, B, and k are constants, show that the particle is repelled from the origin with a force proportional to its distance from the origin. Hint: Use Newton's second law of motion, $F = ma$.

22 If the tangent to $y = e^x$ at the point $x = x_0$ intersects the x-axis at $x = x_1$, show that $x_0 - x_1 = 1$.

23 Graph $y = e^{-x}$, find the area under this curve from $x = 0$ to $x = b$ $(b > 0)$, and find the limit approached by this area as $b \rightarrow \infty$.

24 Verify that $y = e^{-x}$ and $y = e^{2x}$ are both solutions of the differential equation $y'' - y' - 2y = 0$.

25 Evaluate the following limits:

(a) $\lim\limits_{n \to \infty} \left(1 + \dfrac{1}{2n}\right)^{2n}$;

(b) $\lim\limits_{n \to \infty} \left(1 + \dfrac{1}{3n+1}\right)^{3n+1}$;

(c) $\lim\limits_{n \to \infty} \left(1 + \dfrac{1}{n^2}\right)^{n^2}$;

(d) $\lim\limits_{n \to \infty} \left(1 + \dfrac{1}{n}\right)^{2n}$;

(e) $\lim\limits_{n \to \infty} \left(1 + \dfrac{1}{2n}\right)^{n}$.

26 Use the argument in Example 4 to obtain the formula

$$e^x = \lim_{n \to \infty} \left(1 + \frac{x}{n}\right)^n.$$

8.4
THE NATURAL LOGARITHM FUNCTION $y = \ln x$

Logarithms to the base 10—common logarithms—are often taught in high school, starting with the following familiar definition: for any positive number x, $\log_{10} x$ is that number y such that $x = 10^y$. In just the same way, for any positive number x, $\log_e x$ is defined to be that number y such that $x = e^y$. This is illustrated on the left in Fig. 8.7.

The number $\log_e x$ is called the *natural logarithm* of x, for reasons that will become clear in Remark 2. In deference to standard practice at this level, we denote this number by the simpler notation $\ln x$. Thus,

$$y = \ln x \text{ has the same meaning as } x = e^y,$$

in the sense that we are dealing here with a single equation, first written in a form solved for y and then written in a form solved for x. The graph of

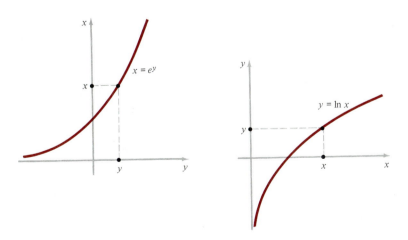

Figure 8.7

$y = \ln x$ is obtained by simply turning over the graph of $x = e^y$ so as to interchange the positions of the axes (Fig. 8.7, right). Just as in Section 8.2, the natural logarithm function $y = \ln x$ is defined only for positive values of x and has the following familiar properties:

$$\ln x_1 x_2 = \ln x_1 + \ln x_2 \quad \text{and} \quad \ln \frac{x_1}{x_2} = \ln x_1 - \ln x_2;$$

$$\ln x^b = b \ln x;$$

$$e^{\ln x} = x \quad \text{and} \quad \ln e^x = x;$$

$$\lim_{x \to 0+} \ln x = -\infty \quad \text{and} \quad \lim_{x \to \infty} \ln x = \infty.$$

Also, $\ln 1 = 0$ and $\ln e = 1$.

We can compute the derivative dy/dx of the function $y = \ln x$ very easily, by differentiating $x = e^y$ implicitly with respect to x:

$$1 = e^y \frac{dy}{dx}, \quad \text{so} \quad \frac{dy}{dx} = \frac{1}{e^y} = \frac{1}{x}.$$

This yields the formula

$$\frac{d}{dx} \ln x = \frac{1}{x},$$

and we immediately have the chain rule extension

$$\frac{d}{dx} \ln u = \frac{1}{u} \frac{du}{dx}, \tag{1}$$

where u is understood to be any function of x.

Example 1 As direct applications of (1) we have

$$\frac{d}{dx} \ln (3x + 1) = \frac{1}{3x + 1} \frac{d(3x + 1)}{dx} = \frac{3}{3x + 1},$$

$$\frac{d}{dx} \ln (1 - x^2) = \frac{1}{1 - x^2} \frac{d(1 - x^2)}{dx} = \frac{-2x}{1 - x^2},$$

$$\frac{d}{dx} \ln \left(\frac{3x}{2x + 1} \right) = \frac{1}{[3x/(2x + 1)]} \frac{(2x + 1) \cdot 3 - 3x \cdot 2}{(2x + 1)^2}$$

$$= \frac{1}{x(2x + 1)}.$$

We point out that the last calculation can be simplified by first writing $\ln [3x/(2x + 1)] = \ln 3 + \ln x - \ln (2x + 1)$, so that

$$\frac{d}{dx} \ln \left(\frac{3x}{2x + 1} \right) = \frac{1}{x} - \frac{2}{2x + 1} = \frac{1}{x(2x + 1)}.$$

The differential version of (1) is $d(\ln u) = du/u$, which leads at once to the main formula of this chapter,

$$\int \frac{du}{u} = \ln u + c. \tag{2}$$

It is understood in (2) that u is positive, because only in this case does $\ln u$ have a meaning. However, it is easy to see that the integrand can always be written with a positive denominator, by juggling the signs. Thus, if $u < 0$ we can write

$$\int \frac{du}{u} = \int \frac{d(-u)}{-u} = \ln(-u) + c. \tag{3}$$

Many writers cover all cases by writing (2) in the form

$$\int \frac{du}{u} = \ln|u| + c.$$

However, we shall not do this, for the reason that most of the applications require a quick transition from logs to exponentials, and the presence of the absolute value sign interferes with the smooth operation of this process. We prefer to use (2) as it is, and to remember as we do this that u must be positive. In situations where u is negative, we easily make the minor adjustments indicated in (3).

Students will recall that the fundamental integration formula

$$\int u^n \, du = \frac{u^{n+1}}{n+1} + c, \qquad n \ne -1,$$

failed to cover one exceptional case, namely, $n = -1$. Formula (2) now fills this gap, since it tells us that

$$\int u^{-1} \, du = \int \frac{du}{u} = \ln u + c.$$

Example 2   The following applications of (2) are easy to carry out by inspection:

$$\int \frac{dx}{x+1} = \ln(x+1) + c,$$

$$\int \frac{dx}{1-2x} = -\frac{1}{2} \int \frac{-2dx}{1-2x} = -\frac{1}{2} \ln(1-2x) + c,$$

$$\int \frac{3x^3 \, dx}{x^4+1} = \frac{3}{4} \int \frac{4x^3 \, dx}{x^4+1} = \frac{3}{4} \ln(x^4+1) + c.$$

In more complicated problems it is desirable to make an explicit substitution or change of variable, in order to diminish the likelihood of accidental error.

In Section 5.4 we discussed the method of separation of variables for solving differential equations. The equation

$$\frac{dy}{dx} = ky \tag{4}$$

is one of the simplest and most important to which this method can be applied. We give the details of this procedure here because the same ideas will be used over and over again in the next two sections, and the sooner students become thoroughly familiar with them, the better:

$$\frac{dy}{y} = k\ dx, \qquad \int \frac{dy}{y} = \int k\ dx, \qquad \ln y = kx + c_1,$$

$$y = e^{kx+c_1} = e^{c_1}e^{kx},$$

and finally,

$$y = ce^{kx},$$

where c is simply a more convenient notation for the constant e^{c_1}. From our point of view, the exponential and logarithm functions find their main reason for being in the fact that they enable us to solve the differential equation (4) in this smooth and straightforward manner. It is also clear from the calculations just given that these functions go together like the two sides of a coin: you can't spend one side without also spending the other.

The next two sections are filled with many far-reaching applications of equation (4) to various fields of science. We hope students will agree that these applications fully justify the attention we have given to this differential equation and to the functions that are necessary for solving it.

Remark 1   We know that $\ln x \to \infty$ as $x \to \infty$. This property of the logarithm is illustrated on the right in Fig. 8.7. However, the graph of $y = \ln x$ rises very slowly, since it is the mirror image of the rapidly rising graph of $x = e^y$. Just how slowly $y = \ln x$ increases can be understood by noticing that it doesn't reach the level $y = 10$ until $x = e^{10} \cong 22{,}000$. The fact that $\ln x$ grows more slowly than x can be expressed by writing

$$\lim_{x \to \infty} \frac{\ln x}{x} = 0. \tag{5}$$

We might try to estimate more accurately how slowly $\ln x$ grows by comparing it with an even smaller function than x, say $\sqrt{x}$ or $\sqrt[3]{x}$. The remarkable fact is that $\ln x$ grows more slowly than *any* positive power of x:

$$\lim_{x \to \infty} \frac{\ln x}{x^p} = 0, \tag{6}$$

where p is any positive constant. Proofs of (5) and (6) are indicated in Problem 13 and Additional Problem 26.

Remark 2   We mention here another way of seeing — with additional clarity — how the number e arises in calculus, where for our present purposes e is defined by the limit

$$e = \lim_{h \to 0} (1 + h)^{1/h}. \tag{7}$$

The idea is to calculate the derivative of $\log_a x$ as if we were doing this for the very first time in history, in an exploratory spirit, without any preconception of what the base a "ought" to be. We begin by applying the definition of the derivative,

$$\frac{d}{dx} \log_a x = \lim_{\Delta x \to 0} \frac{\log_a (x + \Delta x) - \log_a x}{\Delta x}. \qquad (8)$$

Our next step is to manipulate the expression following the limit sign into a more convenient form by using the properties of logarithms discussed in Section 8.2,

$$\frac{\log_a (x + \Delta x) - \log_a x}{\Delta x} = \frac{1}{\Delta x} \log_a \left(\frac{x + \Delta x}{x} \right)$$

$$= \frac{1}{\Delta x} \log_a \left(1 + \frac{\Delta x}{x} \right)$$

$$= \frac{1}{x} \frac{x}{\Delta x} \log_a \left(1 + \frac{\Delta x}{x} \right)$$

$$= \frac{1}{x} \log_a \left(1 + \frac{\Delta x}{x} \right)^{x/\Delta x}.$$

The definition (8) now yields

$$\frac{d}{dx} \log_a x = \lim_{\Delta x \to 0} \left[\frac{1}{x} \log_a \left(1 + \frac{\Delta x}{x} \right)^{x/\Delta x} \right]$$

$$= \frac{1}{x} \lim_{\Delta x \to 0} \left[\log_a \left(1 + \frac{\Delta x}{x} \right)^{x/\Delta x} \right]$$

$$= \frac{1}{x} \log_a \left[\lim_{\Delta x \to 0} \left(1 + \frac{\Delta x}{x} \right)^{x/\Delta x} \right].$$

If we maintain our spirit of research, then the distinctive limit in brackets here attracts our attention. It is natural to simplify its structure a bit by putting $h = \Delta x / x$, and to recognize that $\Delta x \to 0$ is equivalent to $h \to 0$. We now define a new mathematical constant by means of the resulting limit (7), and we at once obtain

$$\frac{d}{dx} \log_a x = \frac{1}{x} \log_a e. \qquad (9)$$

One of our continuing purposes in calculus — though students may find this hard to believe — is to make the formulas we work with as simple as possible. Since $\log_e e = 1$, it is clear that (9) takes its simplest form if the base a is chosen to be the number e:

$$\frac{d}{dx} \log_e x = \frac{1}{x}. \qquad (10)$$

The function $\log_e x$ (or $\ln x$) is called the "natural" logarithm because formula (10) makes it the most convenient logarithm to use in calculus and its applications.

    The ideas described here are those by means of which the great Swiss mathematician Euler (pronounced "oiler") essentially discovered both e and the functions $\ln x$ and e^x in the early eighteenth century.

Remark 3 Students should be informed that some writers define the function $\ln x$ by the formula

$$\ln x = \int_1^x \frac{dt}{t}. \tag{11}$$

These writers are then committed to deriving all the properties of the logarithm from the properties of this integral. Also, it is necessary to define the exponential function in terms of the logarithm instead of the other way around. This approach to the ideas of this chapter has its merits from the theoretical point of view. However, for most students, exponents come before logarithms as naturally as milk comes before cheese; and regardless of the fine points of logic, it is bound to seem perverse and unnatural to begin our subject with (11)—however much it may delight the soul of a mathematician.

PROBLEMS

1 Simplify each of the following:
(a) $e^{\ln 2}$;
(b) $\ln e^3$;
(c) $e^{-\ln x}$;
(d) $\ln e^{1/x}$;
(e) $\ln (1/e^x)$;
(f) $e^{\ln (1/x)}$;
(g) $e^{-\ln (1/x)}$;
(h) $e^{\ln 3 + \ln x}$;
(i) $\ln e^{\ln 1}$;
(j) $\ln e \sqrt[3]{e}$;
(k) $e^{\ln 4 - \ln 3}$;
(l) $\ln (\ln e)$;
(m) $e^{3\ln x + 2\ln y}$;
(n) $e^{3\ln 2}$;
(o) $e^{3 + \ln 2}$;
(p) $e^{x + 2\ln x}$.

2 Find dy/dx in each case:
(a) $y = \ln (3x + 2)$;
(b) $y = \ln (x^2 + 1)$;
(c) $y = \ln (e^x + 1)$;
(d) $y = \ln (e^x)^3$;
(e) $y = x \ln x - x$;
(f) $y = \ln x^2$;
(g) $y = (\ln x)^2$;
(h) $y = \ln (3x^2 - 4x + 5)$;
(i) $y = \dfrac{\ln x}{x}$;
(j) $y = \ln (\ln x)$;
(k) $y = \ln (x + \sqrt{x^2 + 1})$.

3 Find dy/dx in each case:
(a) $\ln xy + 2x - 3y = 4$; (b) $\ln \dfrac{y}{x} - xy = 2$.

4 Find dy/dx in each case. Whenever possible, use properties of logarithms to simplify the function before differentiating. See (a) and (b).
(a) $y = \ln (x\sqrt{x^2 + 1}) = \ln x + \frac{1}{2} \ln (x^2 + 1)$.
(b) $y = \ln \sqrt{\dfrac{x-1}{x+1}} = \dfrac{1}{2} [\ln (x - 1) - \ln (x + 1)]$.
(c) $y = \ln (3x - 2)^4$.
(d) $y = \ln \left(\dfrac{2x + 1}{x + 2}\right)$.
(e) $y = 3 \ln x^4$.
(f) $y = \ln \dfrac{1}{x}$.
(g) $y = 3 \ln 152x$.
(h) $y = 5 \ln 21x + 4 \ln 37x$.
(i) $y = \ln \sqrt[3]{x^6 + 1}$.
(j) $y = \dfrac{1}{3} \ln \dfrac{x^3}{x^3 + 1}$.
(k) $y = \ln [(3x - 7)^4(2x + 5)^3]$.

5 Integrate each of the following:
(a) $\displaystyle\int \frac{dx}{3x + 1}$;
(b) $\displaystyle\int \frac{x\,dx}{3x^2 + 2}$;
(c) $\displaystyle\int \frac{3x^2 + 2}{x}\,dx$;
(d) $\displaystyle\int \frac{x + 1}{x}\,dx$;
(e) $\displaystyle\int \frac{x\,dx}{x + 1}$;
(f) $\displaystyle\int \frac{x\,dx}{x^2 + 1}$;
(g) $\displaystyle\int \frac{x\,dx}{3 - 2x^2}$;
(h) $\displaystyle\int \frac{(2x - 1)\,dx}{x(x - 1)}$;
(i) $\displaystyle\int \frac{\ln x\,dx}{x}$;
(j) $\displaystyle\int \frac{dx}{x \ln x}$;
(k) $\displaystyle\int \frac{dx}{\sqrt{x}(\sqrt{x} + 1)}$;
(l) $\displaystyle\int \frac{e^x - e^{-x}}{e^x + e^{-x}}\,dx$.

6   If c is a positive constant, show that the equation $cx + \ln x = 0$ has exactly one solution. Hint: Sketch the graph of $y = cx + \ln x$ with special attention to the behavior of dy/dx.

7   Show that the equation $x = \ln x$ has no solution
(a) by minimizing $y = x - \ln x$;
(b) geometrically, by considering the graphs of $y = x$ and $y = \ln x$.

8   Find the length of the curve $y = \frac{1}{2}x^2 - \frac{1}{4} \ln x$ between $x = 1$ and $x = 8$.

9   Sketch the graph of $y = x^2 - 18 \ln x$. Locate all maxima, minima, and points of inflection.

10   The area under $y = e^{-x}$ from $x = 0$ to $x = \ln 3$ is revolved about the x-axis. Find the volume generated in this way.

11   The area under $y = 1/\sqrt{x}$ from $x = 1$ to $x = 4$ is revolved about the x-axis. Find the volume generated in this way.

12   Show that the area under $y = 1/x$ from $x = a$ to $x = b$ $(0 < a < b)$ is the same as the area under this curve from $x = ka$ to $x = kb$ for any $k > 0$.

13   Prove that

$$\lim_{x \to \infty} \frac{\ln x}{x} = 0$$

by first showing that for $x > 1$

$$\ln x = \int_1^x \frac{dt}{t} \le \int_1^x \frac{dt}{\sqrt{t}} = 2(\sqrt{x} - 1).$$

Hint: Compare the graphs of $y = 1/t$ and $y = 1/\sqrt{t}$ for $t \ge 1$.

14   Use the result of Problem 13 to show that

$$\lim_{x \to 0+} x \ln x = 0.$$

Hint: Change the variable to $u = 1/x$.

15   Use the result of Problem 14 to sketch the graph of $y = x \ln x$ for all $x > 0$. Locate its minimum and verify that the graph is concave up everywhere.

16   Sketch the graph of $y = (\ln x)/x$ for all $x > 0$, and locate its maximum and point of inflection.

17   The speed at which a signal is transmitted along a cable on the bottom of the ocean is proportional to $x^2 \ln 1/x$, where x is the ratio of the radius of the core of the cable to the radius of the entire cable. What value of x maximizes the speed of transmission?

18   *Logarithmic differentiation* is a technique for computing the derivative of a function like

$$y = \sqrt[3]{(x + 1)(x - 2)(2x + 7)},$$

which is fairly complicated but whose logarithm can be written in a much simpler form:

$$\ln y = \tfrac{1}{3}[\ln (x + 1) + \ln (x - 2) + \ln (2x + 7)].$$

Find dy/dx by differentiating this equation implicitly with respect to x.

19   Use the method of Problem 18 to find dy/dx if

(a) $y = \dfrac{e^x(x^2 - 1)}{\sqrt{6x - 2}}$;   (b) $y = \sqrt[5]{\dfrac{x^2 + 3}{x + 5}}$.

20   The method of logarithmic differentiation (see Problem 18) can also be used to differentiate functions like $y = x^x$, where both the base and the exponent are variable. Thus, we can write

$$\ln y = \ln x^x = x \ln x,$$

or equivalently,

$$y = e^{x \ln x}.$$

Find dy/dx from both equations and use this derivative to find the minimum value of $y = x^x$ for $x > 0$. Sketch the graph.

21   Use the method of Problem 20 to find dy/dx if
(a) $y = x^{x^x}$;   (b) $y = \sqrt[x]{x} = x^{1/x}$.
Sketch the graph of the function in (b) and find its maximum value.

22   In Problem 21(b), the behavior of the function $y = \sqrt[x]{x}$ for large x shows that

$$\lim_{n \to \infty} \sqrt[n]{n} = 1.$$

Find the limits of the following expressions as $n \to \infty$:
(a) $(\ln n)^{1/n}$;   (b) $(n \ln n)^{1/n}$;

(c) $\left(\dfrac{\ln n}{n}\right)^{1/n}$;   (d) $\left(\dfrac{n}{e^n}\right)^{1/n}$.

23   Obtain the limit formula $\lim_{x \to 0}(1 + x)^{1/x} = e$ by using the fact that

$$(1 + x)^{1/x} = e^{\ln(1+x)/x} = e^{[\ln(1+x) - \ln 1]/x}.$$

24   If a is a positive number, show that

$$\lim_{x \to 0} \frac{a^x - 1}{x} = \ln a.$$

Hint: The limit is a value of a certain derivative.

25   Show that

$$\lim_{n \to \infty} n(\sqrt[n]{a} - 1) = \ln a.$$

Hint: Put $x = 1/n$ in Problem 24.

8.5

APPLICATIONS. POPULATION GROWTH AND RADIOACTIVE DECAY

As we emphasized in Section 8.1, our main purpose in this chapter is to develop the mathematical machinery that is necessary for treating a variety of related applications. This machinery is now in place, and the time has come to see what it can do.

Example 1 *Population growth.* Consider a laboratory culture of bacteria with unlimited food and no enemies. If $N = N(t)$ denotes the number of bacteria present at time t, it is natural to assume that the rate of change of N is proportional to N itself.* If the number of bacteria present at the beginning is N_0, and this number doubles after 2 hours (the "doubling time"), how many are there after 6 hours? After t hours?

Solution Even though bacteria come in units and are not continuously divisible, there are so many present, and they are produced at such tiny time intervals, that it is reasonable to treat $N(t)$ as a continuous, even differentiable, function. The assumed law of growth tells us that

$$\frac{dN}{dt} = kN \qquad (k > 0), \tag{1}$$

or, separating variables,

$$\frac{dN}{N} = k \, dt.$$

Integration yields

$$\ln N = kt + c. \tag{2}$$

To determine the value of the constant of integration c, we use the fact that initially (at $t = 0$) we have $N = N_0$. Thus, in equation (2) we have $\ln N_0 = 0 + c$ or $c = \ln N_0$, so (2) becomes

$$\ln N = kt + \ln N_0$$

or

$$\ln N - \ln N_0 = kt, \qquad \ln \frac{N}{N_0} = kt, \qquad \frac{N}{N_0} = e^{kt},$$

and therefore

$$N = N_0 e^{kt}. \tag{3}$$

To find k we use the fact that the population doubles in 2 hours. This gives

$$2N_0 = N_0 e^{2k}, \qquad e^{2k} = 2, \qquad 2k = \ln 2, \qquad k = \tfrac{1}{2} \ln 2,$$

so (3) becomes

$$N = N_0 e^{(t \ln 2)/2}, \tag{4}$$

which gives the population after t hours. Finally, putting $t = 6$ in (4) gives

---

* Briefly, we expect twice as many "births" in a given short interval of time when twice as many bacteria are present.

$N = N_0 e^{3\ln 2} = N_0 e^{\ln 8} = 8N_0$, so the population increases by a factor of 8 in 6 hours.

___

The situation just described is another example of *exponential growth.* This type of growth is characterized by a function of the form (3) where the constant k is positive.

Example 2   *Radioactive decay.* After 3 days, 50 percent of the radioactivity produced by a nuclear explosion has disappeared. How long does it take for 99 percent of this radioactivity to disappear?

Solution   We assume for the sake of simplicity that the radioactivity is entirely due to a single radioactive substance. This substance undergoes *radioactive decay* into nonradioactive substances by means of the spontaneous decomposition of its atoms, at a steady rate that is a characteristic property of the substance itself. Each such decomposition is accompanied by a small burst of radiation, and these bursts are detected and counted by Geiger counters. We are not concerned here with the inner complexities of these remarkable events, but only with the fact that the rate of change of the mass of our substance is negative and is proportional at each moment to the mass of the substance at that moment.* This statement means that if $x = x(t)$ is the mass of the radioactive substance at time t, then

$$\frac{dx}{dt} = -kx \qquad (k > 0), \tag{5}$$

where the minus sign says that x is decreasing. The positive constant k is called the *rate constant*; it clearly measures the speed of the decay process. As before, we separate the variables and integrate,

$$\frac{dx}{x} = -k\,dt, \qquad \ln x = -kt + c. \tag{6}$$

If x_0 is the amount of the substance produced by the explosion, so that $x = x_0$ when $t = 0$, then we see that $c = \ln x_0$, so (6) becomes

$$\ln x = -kt + \ln x_0$$

or

$$\ln x - \ln x_0 = -kt, \qquad \ln \frac{x}{x_0} = -kt, \qquad \frac{x}{x_0} = e^{-kt},$$

and consequently

$$x = x_0 e^{-kt}. \tag{7}$$

In principle at least, x is never zero, because the exponential e^{-kt} never vanishes. It is therefore inappropriate to speak of the "total lifetime" of a

___

* Thus, if the mass of our substance were doubled, we would expect to lose twice as many atoms by decomposition in a given short interval of time.

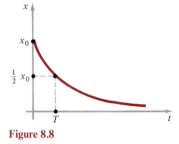

Figure 8.8

radioactive substance. However, it is both convenient and customary to use the concept of half-life: The *half-life* of a radioactive substance is the time required for the substance to decay to half its original amount (Fig. 8.8). If we denote the half-life by T, then (7) yields $\frac{1}{2}x_0 = x_0 e^{-kT}$, so $e^{kT} = 2$ and

$$kT = \ln 2. \tag{8}$$

This equation relates the half-life to the rate constant k, and enables us to find either if the other is known.

In the specific problem we started with, 50 percent of the radioactivity disappears in 3 days. This tells us that the half-life of the substance is 3 days, so by (8) we see that $3k = \ln 2$ or $k = \frac{1}{3}\ln 2$; and in this particular case, (7) becomes

$$x = x_0 e^{-(t\ln 2)/3}.$$

The disappearance of 99 percent of the radioactivity means that 1 percent remains, and therefore $x = \frac{1}{100}x_0$. This happens when t satisfies the equation

$$\tfrac{1}{100}x_0 = x_0 e^{-(t\ln 2)/3},$$

which is equivalent to

$$e^{(t\ln 2)/3} = 100 \qquad \text{or} \qquad \frac{t\ln 2}{3} = \ln 100.$$

Finally, by using tables of natural logarithms (or a calculator) we find that

$$t = \frac{3\ln 100}{\ln 2} = \frac{6\ln 10}{\ln 2} \cong 20 \text{ days}.$$

It should be understood that this example is greatly oversimplified, because an actual nuclear explosion produces many different radioactive by-products with half-lives varying from a fraction of a second to many years. Thus polonium 212 (3 ten-millionths of a second) and krypton 91 (10 seconds) would disappear almost immediately, whereas strontium 90 (28 years) lingers for decades and contributes substantially to the dangers of nuclear fallout.*

The situation just discussed is an example of *exponential decay*. This phrase refers only to the form of the function (7) and the manner in which the quantity x diminishes, and not necessarily to the idea that something or other is disintegrating.

Remark The concepts explained in Example 2 are the basis for a scientific tool of fairly recent development which has been of great significance for geology and archaeology. In essence, radioactive elements occurring in nature (with known half-lives) can be used to assign dates to events that took place from a few thousand to a few billion years ago. For example, the

---

* For students who have not met these ideas before, the number following the name of each of the chemical elements mentioned is the *mass number* (= total number of protons and neutrons in the nucleus) of the particular isotope referred to. For example, strontium as it occurs in nature has four stable isotopes of mass numbers (in the order of their abundance) 88, 86, 87, 84. Several unstable isotopes are produced in nuclear reactions, of which strontium 90 is the best known.

common isotope of uranium (uranium 238) decays through several stages into helium and an isotope of lead (lead 206), with a half-life of 4.5 billion years. When rock containing uranium is in a molten state, as in lava flowing from the mouth of a volcano, the lead created by this decay process is dispersed by currents in the lava; but after the rock solidifies, the lead is locked in place and steadily accumulates alongside the parent uranium. A piece of granite can be analyzed to determine the ratio of lead to uranium, and this ratio permits an estimate of the time that has elapsed since the critical moment when the granite crystallized. Several methods of age determination involving the decay of thorium and the isotopes of uranium into the various isotopes of lead are in current use. Another method depends on the decay of potassium into argon, with a half-life of 1.3 billion years; and yet another, preferred for dating the oldest rocks, is based on the decay of rubidium into strontium, with a half-life of 50 billion years. These studies are complex and susceptible to errors of many kinds; but they can often be checked against one another, and are capable of yielding reliable dates for many events in geological history linked to the formation of igneous rocks. Rocks tens of millions of years old are quite young, ages ranging into hundreds of millions of years are common, and the oldest rocks yet discovered are upwards of 3 billion years old. This of course is a lower limit for the age of the earth's crust, and so for the age of the earth itself. Other investigations, using various types of astronomical data, age determinations for minerals in meteorites, and so on, have suggested a probable age for the earth of about 4.5 billion years.*

These radioactive elements decay so slowly that the methods of age determination based on them are not suitable for dating events that took place relatively recently. This gap was filled by Willard Libby's discovery in the late 1940s of *radiocarbon,* a radioactive isotope of carbon (carbon 14) with a half-life of about 5600 years. By 1950 Libby and his associates had developed the technique of *radiocarbon dating,* which added a second hand to the slow-moving geological clocks just described and made it possible to date events in the later stages of the ice age and some of the movements and activities of prehistoric man. The contributions of this technique to late Pleistocene geology and archaeology have been spectacular.

In brief outline, the facts and principles involved are these. Radiocarbon is produced in the upper atmosphere by the action of cosmic ray neutrons on nitrogen. This radiocarbon is oxidized to carbon dioxide, which in turn is mixed by the winds with the nonradioactive carbon dioxide already present. Since radiocarbon is constantly being formed and constantly decomposing back into nitrogen, its proportion to ordinary carbon in the atmosphere has long since reached an equilibrium state. All air-breathing plants incorporate this proportion of radiocarbon into their tissues, as do the animals that eat these plants. This proportion remains constant as long as a plant or animal lives; but when it dies it ceases to absorb new radiocarbon, while the supply it has at the time of death continues the steady process of decay. Thus, if a piece of old wood has half the radioactivity of a living tree, it lived about 5600 years

---

* For a full discussion of these matters, as well as many other methods and results of the science of geochronology, see F. E. Zeuner, *Dating the Past,* 4th ed., Methuen, London, 1958.

ago, and if it has only one-fourth this radioactivity, it lived about 11,200 years ago. This principle provides a method for dating any ancient object of organic origin, for instance, wood, charcoal, vegetable fiber, flesh, skin, bone, or horn. The reliability of the method has been verified by applying it to the heartwood of giant sequoia trees whose growth rings record 3000 to 4000 years of life, and to furniture from Egyptian tombs whose age is also known independently. There are technical difficulties, but the method is now felt to be capable of reasonable accuracy as long as the periods of time involved are not too great (up to about 50,000 years).

Radiocarbon dating has been applied to thousands of samples, and laboratories for carrying on this work number in the dozens. Among the more interesting age estimates are these: linen wrappings from the Dead Sea scrolls of the Book of Isaiah, recently found in a cave in Palestine and thought to be first or second century B.C., 1917 ± 200 years; charcoal from the Lascaux cave in southern France, site of the remarkable prehistoric paintings, $15,516 \pm 900$ years; charcoal from the prehistoric monument at Stonehenge, in southern England, 3798 ± 275 years; charcoal from a tree burned at the time of the volcanic explosion that formed Crater Lake in Oregon, 6453 ± 250 years. Campsites of ancient people throughout the Western Hemisphere have been dated by using pieces of charcoal, fiber sandals, fragments of burned bison bone, and the like. The results suggest that human beings did not arrive in the New World until about the period of the last Ice Age, some 11,500 years ago, when the level of the water in the oceans was substantially lower than it now is and they could have walked across the Bering Straits from Siberia to Alaska.*

---

* Libby won the 1960 Nobel Prize for chemistry as a consequence of the work described here. His own account of the method, with its pitfalls and conclusions, can be found in his book *Radiocarbon Dating,* 2d ed., Univ. of Chicago Press, 1955.

PROBLEMS

1 The bacteria in a certain culture increase according to the law $dN/dt = kN$. If $N = 2000$ at the beginning and $N = 4000$ when $t = 3$, find (a) the value of N when $t = 1$; and (b) the value of t when $N = 48,000$. Use tables or a calculator.

2 If the rate of increase of the population of a country is 3 percent per year, by what factor does it increase every 10 years? What percentage increase will double the population every 10 years?

3 Sleepyville has 5 times the population of Boomtown. The first is growing at the rate of 2 percent per year, and the second at 10 percent per year. In how many years will they have equal populations?

4 It is often assumed that $\frac{1}{3}$ acre of land is needed to provide food for one person. It is also estimated that there are 10 billion acres of arable land in the world, and therefore a maximum population of 30 billion people can be sustained if no other sources of food are known. The total world population at the beginning of 1970 was 3.6 billion. Assuming that the population continues to increase at the rate of 2 percent per year, when will the maximum population be reached? What will be the population in the year 2000?

5 The half-life of radium is 1620 years. What percentage of a given quantity of radium will remain after 100 years?

6 Cobalt 60, with a half-life of 5.3 years, is extensively used in medical radiology. How long does it take for 90 percent of a given quantity to decay?

7 In a certain chemical reaction a compound C decomposes at a rate proportional to the amount of C that remains. It is found by experiment that 8 g of C diminish to 4 g in 2 hours. At what time will only 1 g be left?

8 "A fool and his money are soon parted." One particular fool loses money in gambling at a rate (in dollars per hour) equal to one-third of the amount he has at any given time. How long will it take him to lose half of his original stake?

9 A cylindrical tank of radius 4 ft and height 10 ft, with its axis vertical, is full of water but has a small hole in the bottom. Assuming that water squirts out of the hole at a speed proportional to the pressure at the bottom of the tank, and that one-fifth of the water leaks out in the first hour, find a formula for the depth of the water left in the tank after t hours.

10 According to *Lambert's law of absorption,* the percentage of incident light absorbed by a thin layer of translucent material is proportional to the thickness of the layer. If sunlight falling vertically on ocean water is reduced to one-half its initial intensity I_0 at a depth of 10 m, show that the formula

$$I = I_0 e^{-(x \ln 2)/10}$$

gives the intensity I at a depth of x meters.

11 According to *Newton's law of cooling,* a body at temperature T cools at a rate proportional to the difference between T and the temperature of the surrounding air. A vat of boiling soup at $100°C$ is brought into a room where the air is $20°C$, and is left to cool. After 1 hour its temperature is $60°C$. How much additional time is required for it to cool to $30°C$?

12 Consider a column of air of cross-sectional area 1 in² extending from sea level up to "infinity." The atmospheric pressure p at an altitude h above sea level is the weight of the air in this column above the altitude h. Assuming that the density of the air is proportional to the pressure (this is a consequence of Boyle's law $pV = k$ at constant temperature), show that p satisfies the differential equation

$$\frac{dp}{dh} = -cp,$$

where c is a positive contant, and deduce that

$$p = p_0 e^{-ch},$$

where p_0 is the atmospheric pressure at sea level. Hint: If h increases by a small amount dh and dp is the corresponding change in p (see Fig. 8.9), then $-dp$ is the

weight of the air in the small portion of the column whose height is dh; and this weight is the density times the volume, so $-dp = (cp)(1 \cdot dh)$.

13 The radiocarbon in living wood decays at the rate of 15.30 disintegrations per minute (dpm) per gram of contained carbon. Using 5600 years as the half-life of radiocarbon, estimate the age of each of the following specimens discovered by archaeologists and tested for radioactivity in 1950:

  (a) a piece of a chair leg from the tomb of King Tutankhamen, 10.14 dpm;

  (b) a piece of a beam of a house built in Babylon during the reign of King Hammurabi, 9.52 dpm;

  (c) dung of a giant sloth found 6 ft 4 in under the surface of the ground inside Gypsum Cave in Nevada, 4.17 dpm;

  (d) a hardwood atlatl (spear-thrower) found in Leonard Rock Shelter in Nevada, 6.42 dpm.

14 Suppose that two chemical substances in solution react together to form a compound. If the reaction occurs by means of the collision and interaction of the molecules of the substances, then we expect the rate of formation of the compound to be proportional to the number of collisions per unit time, which in turn is jointly proportional to the amounts of the substances that are untransformed. A chemical reaction that proceeds in this manner is called a *second-order reaction,* and this law of reaction is often referred to as the *law of mass action.*† Consider a second-order reaction in which x grams of the compound contain ax grams of the first substance and bx grams of the second, where $a + b = 1$. If there are aA grams of the first substance present initially, and bB grams of the second, then the law of mass action says that

$$\frac{dx}{dt} = k(aA - ax)(bB - bx) = kab(A - x)(B - x).$$

If $A \ne B$, show that

$$\frac{B(A - x)}{A(B - x)} = e^{kab(A - B)t} \qquad (*)$$

provides a solution for which $x = 0$ when $t = 0$.‡ Hint: Take the logarithm of both sides and differentiate with respect to t.

15 In Problem 14, find $\lim_{t \to \infty} x(t)$

  (a) by solving equation (*) for x as an explicit function of t and using this function;

  (b) by merely inspecting equation (*).

16 A switch is suddenly closed in an electric circuit, connecting a battery of voltage E to a resistance R and inductance L in series (Fig. 8.10). The battery causes a

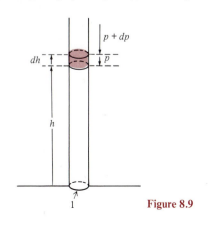

Figure 8.9

---

† For a first-order reaction, see Problem 7.
‡ In Chapter 10 we develop a method for discovering this solution.

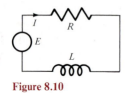

Figure 8.10

variable current $I = I(t)$ to flow in the circuit. By elementary physics, the voltage drop across the resistance is RI and across the inductance is $L\, dI/dt$, and the sum of these two voltage drops must equal the applied voltage E:

$$L\frac{dI}{dt} + RI = E.*$$

---

* Students who are unfamiliar with electric circuits may find it helpful to think of the current I as analogous to the rate of flow of water in a pipe. The battery plays the role of a pump producing pressure (voltage) that causes the water to flow. The resistance is analogous to friction in the pipe, which opposes the flow by producing a drop in the pressure; and the inductance opposes any change in the flow by producing a drop in pressure if the flow is increasing, and an increase in pressure if the flow is decreasing.

By separating the variables and integrating, and using the fact that $I = 0$ when $t = 0$, find the current I as a function of t. Graph this function.

17 Consider a given quantity of gas that undergoes an adiabatic expansion or compression, which means that no heat is gained or lost during the process. The French scientist Poisson showed in 1823 that the pressure and volume of this gas satisfy the differential equation

$$\frac{dp}{p} + \gamma\frac{dV}{V} = 0,$$

where γ is a constant whose value depends on whether the gas is monatomic, diatomic, etc.† Integrate this equation to obtain

$$pV^{\gamma} = c.$$

This is called *Poisson's gas equation* or the *adiabatic gas law,* and is of fundamental importance in meteorology.

---

† For more details on the physical background, see pp. 275–276 of R. A. Millikan, D. Roller, and E. C. Watson, *Mechanics, Molecular Physics, Heat, and Sound,* The M.I.T. Press, Cambridge, Mass., 1965.

8.6
MORE APPLICATIONS. INHIBITED POPULATION GROWTH, etc.

As the reader is certainly aware, the problem of realistically analyzing the growth of a population is not adequately dealt with in Example 1 of Section 8.5. The difficulty with this discussion is that the basic equation,

$$\frac{dN}{dt} = kN \qquad (k > 0),$$

describes only the simplest ideal situation, in which the inner impulse of the population to expand is given a completely free rein; it does not take into account any of the inhibiting factors that put a ceiling on the possible size of a real population. It is obvious, for example, that the human population of the earth can never expand to the stage where there will be only a small fraction of an acre of usable land per person. Long before the point is reached at which the whole surface of the earth becomes a teeming slum, the rate of population growth will be forced down; social, psychological, and economic effects will depress the birth rate, and there will also be an increase in the death rate due to the starvation, disease, and warfare that are the inescapable companions of overpopulation. In our next example we try to recognize some of these factors, and thereby mirror reality a little more closely.

Example 1 *Inhibited population growth.* Consider a small colony of rabbits of population N_0 that is "planted" at time $t = 0$ on a grassy island where they have no enemies. When the population $N = N(t)$ is small, it tends to grow at a rate proportional to itself; but when it becomes larger, there is more and more competition for the limited food and living space, and N grows at a smaller rate. If N_1 is the largest population the island can support, and if the rate of growth of the population N is jointly proportional to N and to $N_1 - N$, so that

$$\frac{dN}{dt} = kN(N_1 - N) \qquad (k > 0), \tag{1}$$

find N as a function of t.

Solution   It should be noticed explicitly at the outset that N increases slowly — that is, dN/dt is small — when N is small, and also when N is large but close to N_1, so that $N_1 - N$ is small. To solve (1), we separate variables and integrate,

$$\int \frac{dN}{N(N_1 - N)} = \int k \, dt. \tag{2}$$

The calculation of the integral on the left side of (2) requires the easily verified algebraic fact that

$$\frac{1}{N(N_1 - N)} = \frac{1}{N_1}\left(\frac{1}{N} + \frac{1}{N_1 - N}\right). \tag{3}$$

With the aid of (3), we can write (2) in the form

$$\frac{1}{N_1}\left(\int \frac{dN}{N} + \int \frac{dN}{N_1 - N}\right) = \int k \, dt,$$

which yields

$$\frac{1}{N_1}[\ln N - \ln (N_1 - N)] = kt + c_1$$

or

$$\frac{1}{N_1} \ln \frac{N}{N_1 - N} = kt + c_1.$$

If we multiply through by N_1, this becomes

$$\ln \frac{N}{N_1 - N} = N_1 kt + c,$$

where $c = N_1 c_1$. Since $N = N_0$ when $t = 0$, we see that $c = \ln [N_0/(N_1 - N_0)]$, so we have

$$\ln \frac{N}{N_1 - N} = N_1 kt + \ln \frac{N_0}{N_1 - N_0},$$

which is equivalent to

$$\frac{N}{N_1 - N} = \frac{N_0}{N_1 - N_0} e^{N_1 kt}.$$

We solve this equation for N by writing

$$N(N_1 - N_0) = N_0 N_1 e^{N_1 kt} - N N_0 e^{N_1 kt},$$

$$N[N_0 e^{N_1 kt} + (N_1 - N_0)] = N_0 N_1 e^{N_1 kt},$$

and

$$N = \frac{N_0 N_1 e^{N_1 kt}}{N_0 e^{N_1 kt} + (N_1 - N_0)}.$$

We can write this in a more convenient form, and thereby obtain our final result, by multiplying the numerator and denominator on the right by $e^{-N_1 kt}$:

$$N = \frac{N_0 N_1}{N_0 + (N_1 - N_0)e^{-N_1 kt}}. \tag{4}$$

It should be observed that (4) gives $N = N_0$ when $t = 0$, and also that $N \to N_1$ as $t \to \infty$, as we expect. The graph of (4) is shown in Fig. 8.11. In ecology and mathematical biology this curve is called the *inhibited growth curve,* or sometimes the *sigmoid growth curve.*

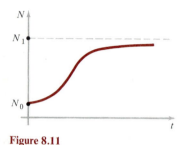

Figure 8.11

In Example 1 of Section 5.5, we discussed the idealized problem of a freely falling body, in which we ignored the effect of air resistance and assumed that the only force acting on the body was the force of gravity. We are now in a position to improve our discussion of this problem by taking air resistance into account.

Example 2 *Falling body with air resistance.* Consider a stone of mass m that is dropped from rest from a great height in the earth's atmosphere. If the only forces acting on the stone are the earth's gravitational attraction mg (where g is the acceleration due to gravity, assumed to be constant) and a retarding force due to air resistance, which is assumed to be proportional to the velocity v, find v as a function of the time t.

Solution Let s be the distance the stone falls in time t, so that the velocity $v = ds/dt$ and the acceleration $a = dv/dt = d^2s/dt^2$. There are two forces acting on the falling stone, a downward force mg due to gravity, and an upward force kv due to air resistance, where k is a positive constant. Newton's second law of motion $F = ma$ says that the total force acting on the stone at any moment equals the product of its mass and its acceleration. With our assumptions, the equation $ma = F$ becomes

$$m\frac{dv}{dt} = mg - kv,$$

or dividing through by m,

$$\frac{dv}{dt} = g - cv, \tag{5}$$

where $c = k/m$. We solve (5) by separating variables and integrating, which gives

$$\int \frac{dv}{g - cv} = \int dt$$

or

$$-\frac{1}{c} \ln (g - cv) = t + c_1;$$

and by changing the notation for constants in a familiar way, we can write this in the form

$$\ln (g - cv) = -ct + c_2$$

or

$$g - cv = c_3 e^{-ct}. \tag{6}$$

The initial condition $v = 0$ when $t = 0$ tells us that $c_3 = g$, so (6) becomes

$$g - cv = ge^{-ct}$$

or

$$v = \frac{g}{c} (1 - e^{-ct}). \tag{7}$$

Since c is positive, this formula tells us that $v \to g/c$ as $t \to \infty$. It is a surprising fact that the velocity of our falling stone does not increase indefinitely, but instead approaches a finite limiting value. This limiting value of v is called the *terminal velocity*. If we differentiate (7), we find that the acceleration is given by the formula $a = ge^{-ct}$, so $a \to 0$ as $t \to \infty$. From the physical point of view, this means that as time goes on the air resistance tends to balance out the force of gravity, so that the total force acting on the stone approaches zero.

Our next example is typical of many problems involving continuously changing mixtures.

Example 3 *Mixing.* Brine containing 2 lb of salt per gallon flows into a tank that initially holds 200 gal of water in which 100 lb of salt are dissolved. If the brine enters the tank at the rate of 10 gal/min, and if the mixture (which is kept uniform by stirring) flows out at the same rate, how much salt is in the tank after 20 minutes? After 100 minutes?

Solution Let x be the number of pounds of salt in the tank after t minutes. The key to thinking about this problem is the following fact:

Rate of change of x = rate at which salt enters tank

$$-\text{rate at which salt leaves tank.} \quad (8)$$

It is clear that salt enters the tank at the rate of $2 \cdot 10 = 20$ lb/min. The concentration of salt at any time is $x/200$ lb/gal, so the rate at which it leaves the tank is $(x/200) \cdot 10 = x/20$ lb/min. Accordingly, (8) becomes

$$\frac{dx}{dt} = 20 - \frac{x}{20} = \frac{400 - x}{20}.$$

By the familiar process of separating variables and integrating, and using the initial condition $x = 100$ when $t = 0$, we obtain

$$x = 400 - 300e^{-t/20}. \tag{9}$$

(As usual when we omit computational details, students should carry these details through for themselves.) By using tables or a calculator, we now find that $x = 289.7$ when $t = 20$, and that $x = 398.0$ when $t = 100$. Also, it is obvious from (9) that $x \to 400$ as $t \to \infty$.

PROBLEMS

1 In Example 1, what is the population when its rate of growth is largest?

2 In a genetics experiment, 50 fruit flies are placed in a glass jar that will support a maximum population of 1000 flies. If 30 days later the population has grown to 200 flies, when will the fly population reach half of the jar's capacity?

3 Let x be the number of people in a community of total population x_1 who have heard a certain rumor t days after the rumor was launched. Common sense suggests that the rate of increase of x, that is, the rate at which this rumor spreads through the community, is proportional to the frequency of contact between those who have heard the rumor and those who have not, and this in turn is jointly proportional to the number of people who have heard the rumor and the number of those who have not. This yields the differential equation

$$\frac{dx}{dt} = cx(x_1 - x),$$

where c is a constant expressing the level of social activity. If the rumor is initially imparted to x_0 individuals ($x = x_0$ when $t = 0$), find x as a function of t. Use this function to show that $x \to x_1$ as $t \to \infty$. Sketch the graph.

4 Rework Example 2 under the more general assumption that the initial velocity is v_0. Show that the terminal velocity is still g/c, and therefore does not depend on v_0. Convince yourself that this is reasonable.

5 A motorboat moving in still water is resisted by the water with a force proportional to its velocity v. Show that the velocity t seconds after the power is shut off is given by the formula $v = v_0 e^{-ct}$, where c is a positive

constant and v_0 is the velocity at the moment the power is shut off. Also, if s is the distance the boat coasts in time t, find s as a function of t and sketch the graph of this function. Hint: Use Newton's second law of motion.

6 Consider the situation described in Problem 5, with the difference that the resisting force is proportional to the square of the velocity v. Find v and s as functions of t, and sketch the graph of the latter function.

7 By the result of Problem 5, the distance s approaches a finite limit as t increases; but in Problem 6 this distance becomes infinite. Because the resisting force seems to be greater in the second case, we would expect the distance traveled to be less than in the first case. Explain this seeming contradiction.

8 A tank initially contains 400 gal of brine in which 100 lb of salt are dissolved. Pure water is run into the tank at the rate of 20 gal/min, and the mixture (which is kept uniform by stirring) is drained off at the same rate. How many pounds of salt remain in the tank after 30 minutes?

9 Rework Problem 8 if instead of pure water, brine containing $\frac{1}{10}$ lb of salt per gallon is run into the tank at 20 gal/min, the mixture being drained off at the same rate.

10 A country has 5 billion dollars of paper money in circulation. Each day 30 million dollars is brought into the banks for deposit and the same amount is paid out. Because of a change of regime, the government decides to issue new paper money displaying pictures of different people, so whenever the old money comes into the banks it is destroyed and replaced by the new money. How long will it take for the paper money in circulation to become 90 percent new?

ADDITIONAL PROBLEMS FOR CHAPTER 8

SECTION 8.3
In Problems 1 to 6, find the derivative dy/dx of the given function.

1 $y = e^{\sqrt{1-x^2}}$.

2 $y = (1 - e^{3x})^2$.

3 $y = e^{x^2 - 2x + 1}$.

4 $y = (e^{4x} - 3)^3$.

5 $y = e^{\sqrt{x}} + \sqrt{e^x}$.

6 $y = \sqrt{e^{2x} + 2x}$.

Evaluate the integrals in Problems 7 to 11.

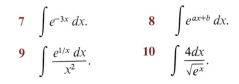

7 $\int e^{-3x}\, dx$.

8 $\int e^{ax+b}\, dx$.

9 $\int \frac{e^{1/x}\, dx}{x^2}$.

10 $\int \frac{4\, dx}{\sqrt{e^x}}$.

11 $\displaystyle\int \frac{e^x \, dx}{\sqrt{e^x + 1}}.$

*12 Find the area between $y = e^x$ and the chord $y = ex - x + 1$.

13 Find the point on the graph of $y = e^{ax}$ at which the tangent line passes through the origin.

14 Evaluate the following limits:

(a) $\displaystyle\lim_{n\to\infty} \left(1 + \frac{1}{4n + 2}\right)^{4n+9};$

(b) $\displaystyle\lim_{n\to\infty} \left(1 + \frac{1}{n}\right)^{n-2};$ (c) $\displaystyle\lim_{n\to\infty} \left(1 + \frac{1}{n}\right)^{3n};$

(d) $\displaystyle\lim_{n\to\infty} \left(1 + \frac{1}{3n}\right)^{n};$ (e) $\displaystyle\lim_{n\to\infty} \left(1 + \frac{1}{2n^2}\right)^{2n}.$

15 Verify that $y = e^{x^2}$ is a solution of the differential equation $y'' - 2xy' - 2y = 0$.

16 Verify that $y = (e^{2x} - 1)/(e^{2x} + 1)$ is a solution of the differential equation $dy/dx = 1 - y^2$.

17 The area under $y = e^x$ from $x = 0$ to $x = 3$ is revolved about the x-axis. Find the volume generated in this way.

*18 Prove that for all $x > 0$ and all positive integers n,

$$e^x > 1 + x + \frac{x^2}{2!} + \frac{x^3}{3!} + \cdots + \frac{x^n}{n!},$$

where the symbol $n!$ (read "n factorial") denotes the product $1 \cdot 2 \cdot 3 \cdots n$. Hint: Since $e^t > 1$ for $t > 0$,

$$e^x = 1 + \int_0^x e^t \, dt > 1 + \int_0^x dt = 1 + x,$$

$$e^x = 1 + \int_0^x e^t \, dt > 1 + \int_0^x (1 + t) \, dt$$

$$= 1 + x + \frac{x^2}{2},$$

and so on.

*19 If n is any given positive integer, prove that $e^x > x^n$ for all sufficiently large values of x. Hint: Use Problem 18 for $n + 1$.

*20 Prove that

$$\lim_{x\to\infty} \frac{e^x}{x^n} = \infty$$

for any positive integer n.

*21 If n is a positive integer, show that $y = x^n e^{-x}$ assumes its maximum value at $x = n$, so that its values at $x = n - 1$ and $x = n + 1$ are less than the maximum. Use this fact to show that

$$\left(\frac{n+1}{n}\right)^n < e < \left(\frac{n}{n-1}\right)^n;$$

and use this in turn to show that

$$\left(1 + \frac{1}{n}\right)^n < e < \left(1 + \frac{1}{n}\right)^{n+1}$$

for every n. When $n = 5$, the second inequality here yields $e < 3$. Verify this.

SECTION 8.4

22 Find dy/dx in each case:
(a) $y = x \ln x^2 - 2x$;   (b) $y = \frac{1}{2} \ln (x^2 + 2x)$;
(c) $y = x^2 \ln x$;   (d) $y = \ln (5x^4 - 7x^3 + 3)$;

(e) $y = \dfrac{\ln x}{x^2}$;   (f) $y = \ln x^5$;

(g) $y = (\ln x)^5$;   (h) $y = \dfrac{1}{\ln x}$;

(i) $y = \sqrt{\ln x}$.

23 Find dy/dx in each case:
(a) $3x - y^2 + \ln xy = 1$;

(b) $x^2 + \ln \dfrac{x}{y} + 3y + 2 = 0$.

24 Find dy/dx in each case:
(a) $y = \ln \sqrt{x}$;

(b) $y = \ln x \sqrt[3]{x}$;

(c) $y = \ln \left(\dfrac{x^2 + 4}{2x + 3}\right)$;

(d) $y = \ln \sqrt{2x^3 - 4x}$;

(e) $y = \ln (x + 1)^5$;

(f) $y = \ln (x^{2} \cdot \sqrt{x^4 + 1})$;

(g) $y = \ln \dfrac{x}{3 - 2x}$;

(h) $y = \ln \sqrt[3]{6x^2 + 3x}$;

(i) $y = \ln \sqrt{\dfrac{4 + x^2}{4 - x^2}}$;

(j) $y = \ln \left(\dfrac{x}{1 + \sqrt{1 + x^2}}\right)$;

(k) $y = x \sqrt{x^2 - 3} - 3 \ln (x + \sqrt{x^2 - 3})$;

(l) $y = -\dfrac{1}{2} \ln \left(\dfrac{2 + \sqrt{x^2 + 4}}{x}\right)$.

25 Integrate each of the following:

(a) $\displaystyle\int \frac{dx}{1 + 2x}$;   (b) $\displaystyle\int \frac{dx}{1 - 3x}$;

(c) $\displaystyle\int_0^1 \frac{x^2 \, dx}{2 - x^3}$;   (d) $\displaystyle\int_0^3 \frac{x \, dx}{x^2 + 1}$;

(e) $\int_0^6 \frac{x \, dx}{x + 3}$;

(f) $\int \frac{dx}{x \sqrt{\ln x}}$;

(g) $\int_0^8 \frac{x^{1/3} \, dx}{1 + 3x^{4/3}}$;

(h) $\int \frac{x \, dx}{1 - x^2}$;

(i) $\int_0^2 \frac{\ln (x + 1) \, dx}{x + 1}$;

(j) $\int \frac{(2x - 1) \, dx}{3x^2 - 3x + 7}$;

(k) $\int \frac{e^x \, dx}{e^x + 1}$;

(l) $\int \frac{(2x + 3) \, dx}{(x + 1)(x + 2)}$;

(m) $\int \frac{(\ln x)^2 \, dx}{x}$;

(n) $\int \frac{\ln \sqrt{x} \, dx}{x}$;

(o) $\int \frac{\ln (\ln x) \, dx}{x \ln x}$;

(p) $\int \frac{1}{x} \ln \left(\frac{1}{x} \right) dx$.

26 If p is a positive constant, show that

$$\lim_{x \to \infty} \frac{\ln x}{x^p} = 0.$$

Hint: Replace x by the variable $y = x^p$.

27 If a and b are positive constants, show that

$$\lim_{x \to \infty} \frac{(\ln x)^a}{x^b} = 0.$$

28 In Problem 27, find the largest value of

$$y = \frac{(\ln x)^a}{x^b} \qquad \text{for } x \geq 1.$$

29 If a is a positive constant, find the length of the curve

$$y = \frac{x^2}{2a} - \frac{a}{4} \ln x$$

between $x = 1$ and $x = 2$. For what value of a is this length a minimum?

30 If a and b are positive constants, find the length of the curve

$$\frac{y}{b} = \left(\frac{x}{a} \right)^2 - \frac{1}{8} \left(\frac{a^2}{b^2} \right) \ln \frac{x}{a}$$

from $x = a$ to $x = 3a$.

31 Use the fact that $a = e^{\log a}$ to find dy/dx in each of the following cases:
(a) $y = 10^x$;
(b) $y = 3^x$;
(c) $y = \pi^x$;
(d) $y = 7^{3x}$;
(e) $y = 6^{x^2 - 2x}$;
(f) $y = 5^{\sqrt{x}}$.

32 Use the idea of Problem 31 to integrate each of the following:

(a) $\int_0^1 2^x \, dx$;

(b) $\int_0^1 10^x \, dx$;

(c) $\int_1^{\sqrt{2}} x3^{-x^2} \, dx$;

(d) $\int_0^1 7^{2x-1} \, dx$;

(e) $\int 3^{-x} \, dx$;

(f) $\int x9^{2x^2} \, dx$;

(g) $\int_0^1 5^{-3x} \, dx$;

(h) $\int \frac{10^{\sqrt{x}} \, dx}{\sqrt{x}}$.

33 Sketch the graph of $y = x^2/5^x$, and locate its maximum and two points of inflection.

34 (a) In changing logarithms from the base a to the base b, one needs the equations $\log_b x = (\log_b a)(\log_a x)$ and $(\log_a b)(\log_b a) = 1$. Prove them.

(b) Compute $\int \frac{dx}{x \log_{10} x}$.

(c) For each choice of the constant $a > 1$, show that $y = (\log_a x)/x$ has a maximum at $x = e$ and a point of inflection at $x = e\sqrt{e}$. Sketch the graph.

35 Find dy/dx if
(a) $y = (\ln x)^x$;
(b) $y = x^{\ln x}$;
(c) $y = (\ln x)^{\ln x}$;
(d) $y = x^{\sqrt{x}}$;
(e) $y = x^{\sqrt[3]{x}}$.

SECTION 8.5

36 The number of bacteria in a culture doubles every hour. How long does it take for a thousand bacteria to produce a billion?

37 The world population at the beginning of 1970 was 3.6 billion. The weight of the earth is 6586×10^{18} tons. If the population of the world continues increasing at a rate of 2 percent per year, and if the average person weighs 120 lb, in what year will the weight of all the people equal the weight of the earth?

38 Cesium 137 is used in medical and industrial radiology. Estimate its half-life if 20 percent decays in 10 years.

39 In a certain chemical reaction a substance S decomposes at a rate proportional to the amount of S not decomposed. If 25 g of this substance is reduced to 10 g in 4 hours, when will 21 g be decomposed?

40 A certain object cools from 120°F to 95°F in half an hour when surrounded by air whose temperature is 70°F. Use Newton's law of cooling to find its temperature at the end of another half hour.

41 A cup of coffee is made with boiling water at 212°F and taken into a room whose air temperature is 72°F. After 20 minutes it has cooled to 100°F. What is its temperature after cooling for a full hour?

42 Assume that the atmospheric pressure p is related to the altitude h above sea level by the differential equation

$$\frac{dp}{dh} = -cp,$$

where c is a positive constant. If p is 15 lb/in^2 at sea level and 10 lb/in^2 at 10,000 ft, find the atmospheric pressure at the top of Mount Everest, where $h \cong$ 30,000 ft.

43 A rocket of total mass m is traveling with velocity v in a distant region of space where the force of gravity is negligible. Its thrust is provided by burning an appropriate fuel and expelling the exhaust products backward at a constant velocity a relative to the rocket. The mass m is therefore variable, and Newton's second law of motion is

$$F = \frac{d}{dt}(mv),$$

which in this case becomes

$$\left(-\frac{dm}{dt}\right)(a - v) = \frac{d}{dt}(mv).$$

(a) Show that $m\dfrac{dv}{dt} = -a\dfrac{dm}{dt}$.

(b) Use part (a) to show that $\dfrac{dm}{dv} = -\dfrac{1}{a}m$.

(c) Use part (b) to show that $m = m_0 e^{-v/a}$ if $v = 0$ and $m = m_0$ when $t = 0$.

(d) The mass m clearly diminishes as the flight progresses, so the velocity v increases. If m_1 is the mass of the initial fuel supply and $\bar{v}$ is the maximum velocity, show that

$$\bar{v} = a \ln \frac{m_0}{m_0 - m_1}.$$

Notice that $m_0 - m_1$ is the so-called *structural mass* of the rocket, i.e., its mass exclusive of fuel.

44 The presence of a certain antibiotic destroys a type of bacteria at a rate jointly proportional to the number N of bacteria and the amount of antibiotic. If there were no antibiotic present, the bacteria would grow at a rate proportional to their number. Assume that the amount of antibiotic is 0 at $t = 0$ and increases at a constant rate. Construct a suitable differential equation for N, solve this equation, and sketch the solution.

45 Assume for the sake of simplicity that uranium 238 decays directly into lead 206 with a half-life of $T = 4.5$ billion years.

(a) If a given quantity of just-solidified volcanic rock contains x_0 atoms of uranium and no lead, show that t years later there are $x = x_0 e^{-kt}$ atoms of uranium and $y = x_0(1 - e^{-kt})$ atoms of lead, where $kT = \ln 2$.

(b) If we can measure the ratio $r = y/x$ in an ancient volcanic rock, and if we have reasonable grounds for believing that all the lead comes from uranium that was locked in the rock when it solidi-

fied, then we can calculate the age of the rock with a fair degree of confidence. Show that this age is given by the formula

$$t = \frac{1}{k}\ln(1 + r) = \frac{T}{\ln 2}\ln(1 + r) \cong \frac{Tr}{\ln 2}$$

when r is small. Hint: Examine the graph of $\ln(1 + r)$ for small values of r.

(c) In a certain rock, r is found to be 0.082. Show that this rock may be about 530 million years old.

46 In the branch of psychology called *psychophysics,* an attempt is made to establish a quantitative connection between the sensation S experienced by a person and the stimulus R that causes this sensation, as in the sensation of heaviness produced by a weight held in the hand. If a small change dR in the stimulus from R to $R + dR$ produces a corresponding change dS in the sensation, then dS is not proportional to dR. Thus, if a weight we hold in our hand is increased from 5 lb to 6 lb, we detect much more of a difference in heaviness than when it is increased from 20 lb to 21 lb. The *Fechner-Weber law* was first formulated by E. H. Weber in 1834 and expounded in detail by G. T. Fechner in 1860, and it played a substantial role in early experimental psychology through the influence of Wilhelm Wundt. This law states that dS is proportional, not to the actual amount dR the stimulus is changed, but to the relative amount it is changed,

$$dS = k\frac{dR}{R}.$$

Find S as a function of R if $S = 0$ when $R = 1$.

SECTION 8.6

47 A flu epidemic hits a city and spreads at a rate jointly proportional to the number of people who are infected and the number of those who are not. If the number of people stricken grows from 10 percent to 20 percent of the population in the first 10 days, how many more days will be required for half the population to be infected?

*48 *Volterra's prey-predator equations* describe an ecological community of the following kind. On an island with plenty of grass, there live x rabbits (the prey) and y foxes (the predator). The number of encounters per unit time between rabbits and foxes is proportional to the product xy of their populations. The rabbits tend to increase at a rate proportional to their number and to decrease at a rate proportional to the product xy. The foxes tend to decrease at a rate proportional to their number and to increase at a rate proportional to xy. This gives the system of differential equations

$$\frac{dx}{dt} = ax - bxy, \qquad \frac{dy}{dt} = -cy + dxy,$$

where a, b, c, d are positive constants.

(a) Show that $x = c/d$ and $y = a/b$ is a solution of the system. These are called the *equilibrium populations*.

(b) Show that any solution $x = x(t)$, $y = y(t)$ satisfies the equation $(x^c e^{-dx})(y^a e^{-by}) = k$, where k is a positive constant. Hint: Eliminate dt from the system by division, separate variables, and integrate.

(c) Use the equation in part (b) to show that neither $x(t)$ nor $y(t)$ can $\to \infty$ as $t \to \infty$.

49 Consider a falling body of mass m and assume that the retarding force due to air resistance is proportional to the square of the velocity. If the body falls from rest, find a formula for the velocity in terms of the distance fallen, and thereby find the terminal velocity in this case. Hint: $dv/dt = (dv/ds)(ds/dt) = v\, dv/ds$.

*50 A torpedo is traveling at a velocity of 60 km/h at the moment it runs out of fuel. If the water resists its motion with a force proportional to the velocity v, and if 1 km of travel reduces v to 40 km/h, find the distance s the torpedo coasts in t hours, and also the total distance it coasts.

51 Brine containing 1 lb of salt per gallon flows at the rate of 10 gal/min into a tank initially filled with 120 gal of pure water. If the concentration is kept uniform by stirring, and the mixture flows out at the same rate, when will the tank contain 40 lb of salt? When will it contain 100 lb of salt?

52 A large tank initially contains 45 lb of salt dissolved in 50 gal of water. Pure water flows in at the rate of 3 gal/min, and the mixture (which is kept uniform by stirring) flows out at the rate of 2 gal/min. When will the tank contain 5 lb of salt? How many gallons of water will be in the tank at that time?

53 An aquarium contains 10 gal of polluted water. A filter is attached to this aquarium which drains off the polluted water at the rate of 5 gal/h and replaces it at the same rate by pure water. How long does it take to reduce the pollution to half its initial level?

TRIGONOMETRIC FUNCTIONS

We continue the program started in Chapter 8 of extending the scope of our work to include broader and broader classes of functions, this time the trigonometric functions. In science, these functions are indispensable tools for the study of periodic phenomena of all kinds, ranging from the back-and-forth movement of the bob of a pendulum clock to the revolution of the planets in their orbits around the sun. And in mathematics — as we shall see in Chapter 10 — almost all of the more advanced methods of integration lean heavily on the trigonometric functions and their properties.

We assume that students have studied trigonometry in high school. Nevertheless, no matter how well the basic facts have been learned, they are easy to forget unless they are needed and used on a day-to-day basis, which they will be through most of the rest of this book. We therefore devote this section to a review of the subject from the beginning. There are a number of fundamental formulas built into this exposition, and these are so important for the purposes of calculus that students should relearn them systematically and thoroughly. Even though our treatment is very brief, it is essentially self-contained; and hard-working students who have no previous experience with trigonometry should be able to get along comfortably with only what they find in these pages.

9.1
REVIEW OF TRIGONOMETRY

RADIAN MEASURE

The most common unit for measuring angles is the degree (1 right angle = 90 degrees = 90°). However, the standard unit for angle measurement in calculus is the *radian*. One radian is the angle which, placed at the center of a circle, subtends an arc whose length equals the radius (Fig. 9.1, left). More generally, the number of radians in an arbitrary central angle (Fig. 9.1, center) is defined to be the ratio of the length of the subtended arc to the radius, $\theta = s/r$; or equivalently, a central angle of θ radians subtends an arc of length θ times the radius, $s = \theta r$. Since the circumference of the circle is $c = 2\pi r$, a complete central angle of $360°$ is equivalent to $2\pi r/r = 2\pi$ radians.

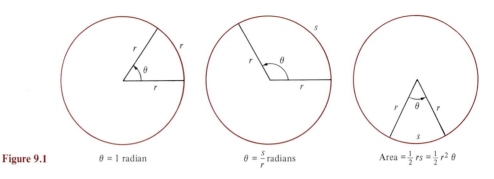

Figure 9.1

$\theta = 1$ radian $\qquad\qquad$ $\theta = \dfrac{s}{r}$ radians $\qquad\qquad$ Area $= \frac{1}{2}\,rs = \frac{1}{2}\,r^2\,\theta$

Thus,

$$2\pi \text{ radians} = 360°, \qquad \pi \text{ radians} = 180°,$$

$$1 \text{ radian} = \frac{180}{\pi} \cong 57.296°, \qquad 1° = \frac{\pi}{180} \cong 0.0175 \text{ radian}.$$

Further, $90° = \pi/2$, $60° = \pi/3$, $45° = \pi/4$, and $30° = \pi/6$, where we here follow the convention of omitting the word "radian" in using radian measure.

Just as the calculus of logarithms is simplified by using the base e, the calculus of the trigonometric functions is simplified by using radian measure. We will point out the specific reason for this in Section 9.2. Throughout our work we will use radian measure routinely and mention degrees only in passing.

It is sometimes useful to know that the area A of the sector whose central angle is θ (Fig. 9.1, right) is given by the formula

$$A = \tfrac{1}{2}rs = \tfrac{1}{2}r^2\theta,$$

since $s = r\theta$. This is easy to prove by using the fact that the area of the sector is to the area of the circle as the arc s is to the circumference:

$$\frac{A}{\pi r^2} = \frac{s}{2\pi r}, \qquad \text{so} \qquad A = \frac{1}{2}\,rs.$$

And this is easy to remember by thinking of the sector as if it were a triangle with height r and base s.

THE TRIGONOMETRIC FUNCTIONS

Consider the unit circle in the xy-plane (Fig. 9.2). If θ is a positive number, let the radius OP start in the position OA and revolve counterclockwise through θ radians. Thus, $\theta = \pi$ produces half a revolution and $\theta = 2\pi$ produces a complete revolution, both counterclockwise. If θ is negative, we let OP revolve clockwise through $-\theta$ radians. See Fig. 9.3. In this way, each real number θ (positive, negative, or zero) determines a unique position of OP in Fig. 9.2, and therefore a unique point $P = (x, y)$ with the property that $x^2 + y^2 = 1$. The sine and cosine of θ are now defined by

$$\sin \theta = y \qquad \text{and} \qquad \cos \theta = x.$$

It is evident from the definition that $-1 \leq \sin \theta \leq 1$, and similarly for $\cos \theta$;

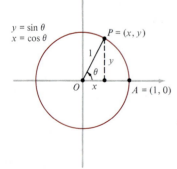

$y = \sin \theta$
$x = \cos \theta$

Figure 9.2

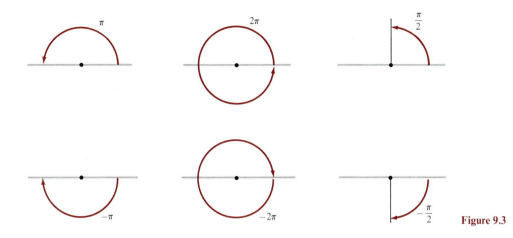

Figure 9.3

and the algebraic signs of these quantities depend on which quadrant the point P happens to lie in. For every θ, the numbers θ and $\theta + 2\pi$ clearly determine the same point P, so

$$\sin(\theta + 2\pi) = \sin \theta \quad \text{and} \quad \cos(\theta + 2\pi) = \cos \theta.$$

Thus the values of $\sin \theta$ and $\cos \theta$ repeat when θ increases by 2π We express this property of $\sin \theta$ and $\cos \theta$ by saying that these functions are *periodic* with *period* 2π.

The remaining four trigonometric functions—the tangent, cotangent, secant, and cosecant—are defined by

$$\tan \theta = \frac{y}{x}, \quad \cot \theta = \frac{x}{y}, \quad \sec \theta = \frac{1}{x}, \quad \csc \theta = \frac{1}{y}.$$

The sine and cosine are the basic functions, and the others can be expressed in terms of these two [see identities (1) to (4) below].

When θ is a positive number $< \pi/2$, the right triangle interpretations of the sine, cosine, and tangent are as follows (see Fig. 9.4):

$$\sin \theta = \frac{\text{opposite side}}{\text{hypotenuse}} = \frac{a}{h},$$

$$\cos \theta = \frac{\text{adjacent side}}{\text{hypotenuse}} = \frac{b}{h},$$

$$\tan \theta = \frac{\text{opposite side}}{\text{adjacent side}} = \frac{a}{b}.$$

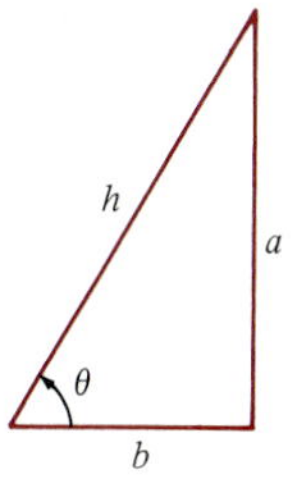

Figure 9.4

We have drawn the right triangle here with base angle equal to the angle θ shown in Fig. 9.2, and the validity of these statements rests on the similarity of the two triangles in the figures (since $\sin \theta = y = y/1$, etc.). In the equivalent forms

$$a = h \sin \theta, \quad b = h \cos \theta, \quad a = b \tan \theta,$$

the right triangle interpretations have many uses in physics and geometry. Nevertheless, the purposes of calculus require that θ be an unrestricted real variable, and for this reason the unit circle definitions are preferable.

IDENTITIES

Several simple relations among our functions are direct consequences of the definitions:

$$\tan \theta = \frac{\sin \theta}{\cos \theta}, \tag{1}$$

$$\cot \theta = \frac{\cos \theta}{\sin \theta}, \tag{2}$$

$$\sec \theta = \frac{1}{\cos \theta}, \tag{3}$$

$$\csc \theta = \frac{1}{\sin \theta}, \tag{4}$$

$$\tan \theta = \frac{1}{\cot \theta}. \tag{5}$$

Altogether, there are 21 fundamental identities that express the main properties of the trigonometric functions and constitute the core of the subject. These identities fall into several natural groups, and are therefore easier to remember than we might expect. We emphasize these groups by enclosing them in boxes.

Our next identities state the effect of replacing θ by $-\theta$. From Fig. 9.5 and the obvious fact that the endpoints of the two radii lie on the same vertical line for all values of θ, we at once have the first two of the identities

$$\sin (-\theta) = -\sin \theta, \tag{6}$$

$$\cos (-\theta) = \cos \theta, \tag{7}$$

$$\tan (-\theta) = -\tan \theta. \tag{8}$$

The third follows easily from (1) combined with (6) and (7).*

Our next group consists of three equivalent versions of the equation $x^2 + y^2 = 1$. Before stating these, we must explain that the symbols $\sin^2 \theta$ and $\cos^2 \theta$ are standard notations for $(\sin \theta)^2$ and $(\cos \theta)^2$. If we write $x^2 + y^2 = 1$ in the form $y^2 + x^2 = 1$, then this yields the first of the identities

$$\sin^2 \theta + \cos^2 \theta = 1, \tag{9}$$

$$\tan^2 \theta + 1 = \sec^2 \theta, \tag{10}$$

$$1 + \cot^2 \theta = \csc^2 \theta. \tag{11}$$

The second and third in this group are obtained by dividing (9) first by $\cos^2 \theta$, and then by $\sin^2 \theta$.

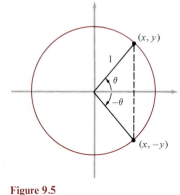

Figure 9.5

---

* It is clear that there are similar identities for the cotangent, secant, and cosecant. However, these are of little significance; and in keeping with our purpose of presenting a stripped-down version of trigonometry, we ignore them.

For obvious reasons, the following are called the *addition formulas*:

$$\sin (\theta + \phi) = \sin \theta \cos \phi + \cos \theta \sin \phi, \tag{12}$$

$$\cos (\theta + \phi) = \cos \theta \cos \phi - \sin \theta \sin \phi, \tag{13}$$

$$\tan (\theta + \phi) = \frac{\tan \theta + \tan \phi}{1 - \tan \theta \tan \phi}. \tag{14}$$

We indicate in Problem 10 a method of proving the first two of these, and the third follows from the first two by a straightforward argument. Write

$$\tan (\theta + \phi) = \frac{\sin (\theta + \phi)}{\cos (\theta + \phi)} = \frac{\sin \theta \cos \phi + \cos \theta \sin \phi}{\cos \theta \cos \phi - \sin \theta \sin \phi}.$$

By dividing both numerator and denominator on the right by $\cos \theta \cos \phi$, we obtain

$$\tan (\theta + \phi) = \frac{\sin \theta/\cos \theta + \sin \phi/\cos \phi}{1 - (\sin \theta/\cos \theta)(\sin \phi/\cos \phi)},$$

which is essentially (14). The corresponding *subtraction formulas* are

$$\sin (\theta - \phi) = \sin \theta \cos \phi - \cos \theta \sin \phi, \tag{15}$$

$$\cos (\theta - \phi) = \cos \theta \cos \phi + \sin \theta \sin \phi, \tag{16}$$

$$\tan (\theta - \phi) = \frac{\tan \theta - \tan \phi}{1 + \tan \theta \tan \phi}. \tag{17}$$

These follow directly from the addition formulas by replacing ϕ by $-\phi$ and using (6), (7), and (8).

The *double-angle formulas* are

$$\sin 2\theta = 2 \sin \theta \cos \theta, \tag{18}$$

$$\cos 2\theta = \cos^2 \theta - \sin^2 \theta. \tag{19}$$

These are the special cases of (12) and (13) obtained by replacing ϕ by θ. (There is also an obvious double-angle formula for the tangent; but this is of minor importance and we omit it.)

The *half-angle formulas* are

$$2 \cos^2 \theta = 1 + \cos 2\theta, \tag{20}$$

$$2 \sin^2 \theta = 1 - \cos 2\theta. \tag{21}$$

These are easy to prove by writing (9) and (19) together, as

$$\cos^2 \theta + \sin^2 \theta = 1,$$

$$\cos^2 \theta - \sin^2 \theta = \cos 2\theta.$$

Adding yields (20), and subtracting yields (21).

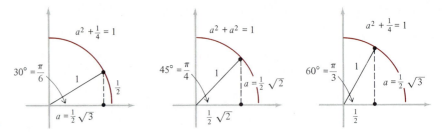

Figure 9.6

VALUES

If we keep firmly in mind the definitions of $\sin\theta$, $\cos\theta$, and $\tan\theta$, then there are several first-quadrant values of θ for which the exact values of these functions are easy to find. All that is necessary is to remember the Pythagorean theorem and look carefully at the three parts of Fig. 9.6:

$$\sin\frac{\pi}{6}=\frac{1}{2} \qquad\qquad \sin\frac{\pi}{4}=\frac{1}{2}\sqrt{2} \qquad \sin\frac{\pi}{3}=\frac{1}{2}\sqrt{3}$$

$$\cos\frac{\pi}{6}=\frac{1}{2}\sqrt{3} \qquad\qquad \cos\frac{\pi}{4}=\frac{1}{2}\sqrt{2} \qquad \cos\frac{\pi}{3}=\frac{1}{2}$$

$$\tan\frac{\pi}{6}=\frac{\frac{1}{2}}{\frac{1}{2}\sqrt{3}}=\frac{1}{3}\sqrt{3} \qquad \tan\frac{\pi}{4}=1 \qquad \tan\frac{\pi}{3}=\frac{\frac{1}{2}\sqrt{3}}{\frac{1}{2}}=\sqrt{3}$$

Also, an inspection of Fig. 9.2 with OP in various positions gives us similar information for the cases $\theta=0$, $\pi/2$, π, $3\pi/2$, and 2π (the entry * means that the quantity is undefined):

$$\sin 0=0 \qquad \sin\frac{\pi}{2}=1 \qquad \sin\pi=0 \qquad \sin\frac{3\pi}{2}=-1 \qquad \sin 2\pi=0$$

$$\cos 0=1 \qquad \cos\frac{\pi}{2}=0 \qquad \cos\pi=-1 \qquad \cos\frac{3\pi}{2}=0 \qquad \cos 2\pi=1$$

$$\tan 0=0 \qquad \tan\frac{\pi}{2}=* \qquad \tan\pi=0 \qquad \tan\frac{3\pi}{2}=* \qquad \tan 2\pi=0$$

In our subsequent work, facts of this kind will often be needed at a moment's notice. They are best learned, not by an effort of memory, but rather by an act of understanding — knowing the definitions of the trigonometric functions and visualizing (or quickly sketching) appropriate pictures. We also emphasize the way the algebraic signs of our functions vary from one quadrant to another. The facts are obvious from the definitions and Fig. 9.2, and are stated in the following table:

Quadrant	1	2	3	4
$\sin\theta$	+	+	−	−
$\cos\theta$	+	−	−	+
$\tan\theta$	+	−	+	−

GRAPHS

The graph of $\sin\theta$ is easy to sketch by looking at Fig. 9.2 and following the way y varies as θ increases from 0 to 2π, that is, as the radius swings around

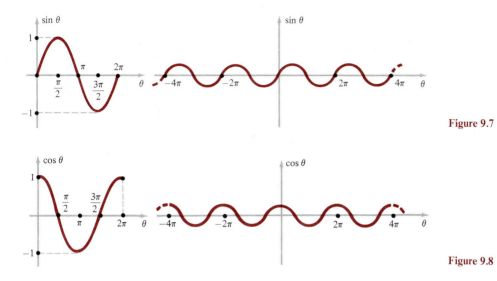

Figure 9.7

Figure 9.8

through one complete counterclockwise revolution. It is clear that sin θ starts at 0, increases to 1, decreases to 0, decreases further to -1, and increases to 0. This gives one complete cycle of sin θ, as shown on the left in Fig. 9.7. The complete graph (on the right in Fig. 9.7) consists of infinitely many repetitions of this cycle, to the right and to the left. The graph of cos θ can be sketched in essentially the same way (Fig. 9.8), the main difference being that cos θ starts at 1, decreases to 0, decreases further to -1, increases to 0, and increases further to 1.

The graph of tan θ is quite different from the graphs of sin θ and cos θ. We point out first that tan θ is periodic with period π:

$$\tan (\theta + \pi) = \frac{\sin (\theta + \pi)}{\cos (\theta + \pi)} = \frac{-\sin \theta}{-\cos \theta} = \tan \theta.$$

This permits us to get the full range of values of tan θ by visualizing the ratio y/x in Fig. 9.2 and allowing θ to increase from $-\pi/2$ to $\pi/2$. The result is the central curve shown in Fig. 9.9, and the complete graph of tan θ consists of infinitely many repetitions of this curve to the right and to the left. The fact that tan $\theta \to \infty$ as $\theta \to \pi/2$ (from the left) is often loosely expressed by writing tan $\pi/2 = \infty$.

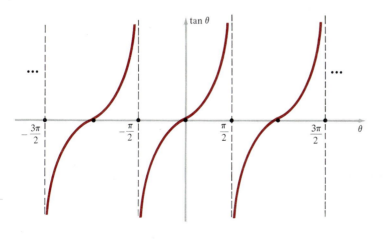

Figure 9.9

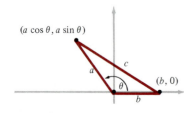

Figure 9.10

LAW OF COSINES

This is a useful tool in a variety of situations in mathematics and physics. It expresses the third side of a triangle (Fig. 9.10) in terms of two given sides a and b and the included angle θ:

$$c^2 = a^2 + b^2 - 2ab \cos \theta.$$

The proof is routine if we place the triangle in the xy-plane as shown in the figure and apply the distance formula to the vertices $(a \cos \theta, a \sin \theta)$ and $(b, 0)$. The square of the side c is clearly

$$c^2 = (a \cos \theta - b)^2 + (a \sin \theta - 0)^2$$
$$= a^2(\cos^2 \theta + \sin^2 \theta) + b^2 - 2ab \cos \theta$$
$$= a^2 + b^2 - 2ab \cos \theta,$$

and the argument is complete. An important application of the law of cosines is made in Problem 10, where it is used to prove identity (16), and thereby identities (12) and (13).

PROBLEMS

1 Convert from degrees to radians:
(a) $15°$;   (b) $105°$;   (c) $120°$;
(d) $75°$;   (e) $150°$;   (f) $135°$;
(g) $225°$;   (h) $210°$;   (i) $630°$;
(j) $900°$.

2 Convert from radians to degrees:
(a) $5\pi/3$;   (b) $7\pi/6$;   (c) $2\pi/9$;
(d) $3\pi/2$;   (e) $4\pi/3$;   (f) 3π;
(g) $7\pi/15$;   (h) $\pi/36$;   (i) $\pi/5$;
(j) $25\pi/3$.

3 A decorative garden is to have the shape of a circular sector of radius r and central angle θ. If the perimeter is fixed in advance, what value of θ will maximize the area of the garden?

4 Find the values of $\sin \theta$, $\cos \theta$, and $\tan \theta$ when θ equals
(a) $-\pi/6$;   (b) $3\pi/4$;   (c) $4\pi/3$;
(d) $-5\pi/4$;   (e) $2\pi/3$;   (f) 17π;
(g) -102π.

5 If the base of an isosceles triangle is 10, express its area A as a function of the vertex angle θ.

6 If the height of an isosceles triangle is h, express its perimeter p as a function of the base angle θ.

7 Express the height H of a flagpole in terms of the length L of its shadow and the angle of elevation θ of the sun.

8 A hunter sits on a platform built in a tree 30 m above the ground. He sees a tiger at an angle of $30°$ below the horizontal. How far is he from the tiger?

9 Sketch the graph of
(a) $\sin 2\theta$ (hint: this curve runs through one complete cycle as 2θ increases from 0 to 2π);
(b) $3 \sin 2\theta$;   (c) $\sin 4\theta$;
(d) $\sin \frac{1}{2}\theta$;   (e) $2 \cos 3\theta$.

10 In this problem we outline a method of proving identities (12) and (13) by first establishing (16). Figure 9.11 shows the unit circle with two arbitrary angles θ and ϕ and their corresponding points $P_\theta = (\cos \theta, \sin \theta)$ and $P_\phi = (\cos \phi, \sin \phi)$.

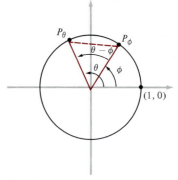

Figure 9.11

(a) Calculate the square of the distance between these points in two ways, by using the distance formula and the law of cosines, and thus prove identity (16),

$$\frac{d}{dx}\sin x = \lim_{\Delta x \to 0} \frac{\sin x \cos \Delta x + \cos x \sin \Delta x - \sin x}{\Delta x}$$

$$= \lim_{\Delta x \to 0}\left[\cos x \left(\frac{\sin \Delta x}{\Delta x}\right) - \sin x \left(\frac{1 - \cos \Delta x}{\Delta x}\right)\right].$$

Using (4) and (5) with θ replaced by Δx now yields

$$\frac{d}{dx}\sin x = (\cos x) \cdot 1 - (\sin x) \cdot 0 = \cos x,$$

which concludes the proof of (1). To prove formula (2), we begin with

$$\frac{d}{dx}\cos x = \lim_{\Delta x \to 0}\frac{\cos(x + \Delta x) - \cos x}{\Delta x}.$$

Since $\cos(x + \Delta x) = \cos x \cos \Delta x - \sin x \sin \Delta x$, we have

$$\frac{d}{dx}\cos x = \lim_{\Delta x \to 0}\frac{\cos x \cos \Delta x - \sin x \sin \Delta x - \cos x}{\Delta x}$$

$$= \lim_{\Delta x \to 0}\left[-\sin x\left(\frac{\sin \Delta x}{\Delta x}\right) - \cos x\left(\frac{1 - \cos \Delta x}{\Delta x}\right)\right]$$

$$= (-\sin x) \cdot 1 - (\cos x) \cdot 0 = -\sin x,$$

and the proof of (2) is complete. The addition formulas for the sine and cosine obviously play essential roles in these arguments, and this is their main use in mathematics.

We now generalize (1) and (2) by means of the chain rule, and obtain the extremely useful formulas

$$\frac{d}{dx}\sin u = \cos u \frac{du}{dx} \tag{6}$$

and

$$\frac{d}{dx}\cos u = -\sin u \frac{du}{dx}. \tag{7}$$

As usual, u is understood to be any differentiable function of x.

Example 1 Find the derivative of each of the following functions:
(a) $y = \sin 5x$; (b) $y = \sin\sqrt{x}$; (c) $y = \cos(2 - 3x^4)$.

Solution For (a), we use (6) with $u = 5x$, so

$$\frac{dy}{dx} = \cos 5x \frac{d}{dx}(5x) = 5\cos 5x.$$

For (b), $u = \sqrt{x} = x^{1/2}$, so

$$\frac{dy}{dx} = \cos\sqrt{x}\frac{d}{dx}(x^{1/2}) = \frac{1}{2\sqrt{x}}\cos\sqrt{x}.$$

For (c), we use (7) with $u = 2 - 3x^4$, so

$$\frac{dy}{dx} = -\sin(2 - 3x^4)\frac{d}{dx}(2 - 3x^4) = 12x^3 \sin(2 - 3x^4).$$

Students must learn to use formulas (6) and (7) in combination with all previous rules of differentiation. In this connection it is necessary to remember the standard notation for powers of trigonometric functions: $\sin^n x$ means $(\sin x)^n$. There is one exception to this usage, for $(\sin x)^{-1}$ is *never* written $\sin^{-1} x$; the latter expression is reserved exclusively for the inverse sine function discussed in Section 9.5.

Example 2 Find the derivative of each of the following functions:
(a) $y = \sin^3 4x$;   (b) $y = e^{\cos x}$;
(c) $y = \ln(\sin x)$;   (d) $y = \sin(\ln x)$.

Solution

(a) $\dfrac{dy}{dx} = 3(\sin 4x)^2 \dfrac{d}{dx}(\sin 4x) = 3(\sin 4x)^2(\cos 4x) \cdot 4$

$$= 12 \sin^2 4x \cos 4x.$$

(b) $\dfrac{dy}{dx} = e^{\cos x} \dfrac{d}{dx}(\cos x) = -\sin x \, e^{\cos x}.$

(c) $\dfrac{dy}{dx} = \dfrac{1}{\sin x} \dfrac{d}{dx}(\sin x) = \dfrac{\cos x}{\sin x} = \cot x.$

(d) $\dfrac{dy}{dx} = \cos(\ln x) \dfrac{d}{dx}(\ln x) = \dfrac{\cos(\ln x)}{x}.$

Example 3 Show that $(d/dx)(\frac{1}{3}\cos^3 x - \cos x) = \sin^3 x.$

Solution

$$\frac{d}{dx}\left(\frac{1}{3}\cos^3 x - \cos x\right) = \frac{1}{3} \cdot 3\cos^2 x(-\sin x) + \sin x$$

$$= \sin x(1 - \cos^2 x) = \sin^3 x.$$

Remark 1 We are now able to explain why radian measure is preferred to degree measure when working with the trigonometric functions in calculus. Let $\sin x°$ and $\cos x°$ denote the sine and cosine of an angle of x degrees. We know that x degrees equals $\pi x/180$ radians, so

$$\sin x° = \sin \frac{\pi x}{180}.$$

By formula (6),

$$\frac{d}{dx}\sin x° = \frac{d}{dx}\sin\frac{\pi x}{180} = \frac{\pi}{180}\cos\frac{\pi x}{180}$$

or

$$\frac{d}{dx}\sin x° = \frac{\pi}{180}\cos x°.$$

We use radian measure routinely in calculus for the sake of simplicity, in order to avoid the repeated occurrence of the nuisance factor $\pi/180$.

Remark 2 If the limit (4) has been solidly established in some way, then the limit (5) can be made to follow from it. We have

$$\lim_{\theta \to 0} \frac{1 - \cos \theta}{\theta} = \lim_{\theta \to 0} \left(\frac{1 - \cos \theta}{\theta} \cdot \frac{1 + \cos \theta}{1 + \cos \theta} \right)$$

$$= \lim_{\theta \to 0} \frac{1 - \cos^2 \theta}{\theta(1 + \cos \theta)}$$

$$= \lim_{\theta \to 0} \frac{\sin^2 \theta}{\theta(1 + \cos \theta)}$$

$$= \lim_{\theta \to 0} \frac{\sin \theta}{\theta} \cdot \frac{\sin \theta}{1 + \cos \theta}$$

$$= \left(\lim_{\theta \to 0} \frac{\sin \theta}{\theta} \right) \left(\lim_{\theta \to 0} \frac{\sin \theta}{1 + \cos \theta} \right)$$

$$= 1 \cdot \frac{0}{1 + 1} = 0,$$

since $\sin \theta \to 0$ and $\cos \theta \to 1$ as $\theta \to 0$.

We return to the problem of proving (4). We first observe that only positive values of θ need to be considered, because if we replace θ by $-\theta$ we have $\sin (-\theta)/(-\theta) = (-\sin \theta)/(-\theta) = (\sin \theta)/\theta$, and the value of the fraction $(\sin \theta)/\theta$ is unchanged.

One method of establishing (4) rests on the following premise. Let P and Q be two nearby points on a unit circle (Fig. 9.13), and let $\overline{PQ}$ and $\overparen{PQ}$ denote the lengths of the chord and the arc connecting these points. Then the ratio of the chord length to the arc length evidently approaches 1 as the two points move together:

$$\frac{\text{Chord length } \overline{PQ}}{\text{Arc length } \overparen{PQ}} \to 1 \quad \text{as} \quad \overparen{PQ} \to 0.$$

With the notation in the figure, this geometric statement is equivalent to

$$\frac{2 \sin \theta}{2\theta} = \frac{\sin \theta}{\theta} \to 1 \quad \text{as} \quad 2\theta \to 0 \quad \text{or} \quad \theta \to 0,$$

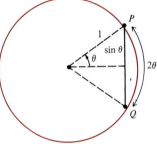

Figure 9.13

and this is (4). There are other ways of proving (4), but this is as direct as any. It amounts to the argument given earlier, but in a little more detail so that the nature of the reasoning is more clearly visible.

Remark 3 Just for fun, we mention another application of the limit (4). This limit can be used to prove *Vieta's formula,* the first (1593) theoretically exact numerical expression involving π:

$$\frac{2}{\pi} = \frac{\sqrt{2}}{2} \cdot \frac{\sqrt{2 + \sqrt{2}}}{2} \cdot \frac{\sqrt{2 + \sqrt{2 + \sqrt{2}}}}{2} \cdots.$$

(The dots signify a so-called "infinite product" whose factors follow the indicated pattern.) For us, this remarkable formula is only a curiosity and has no bearing on our future work, so we give the proof in Appendix A.7.

PROBLEMS

In each of the following problems (1 to 18) find the derivative dy/dx of the given function.

1 $y = \sin (3x - 2)$.
2 $y = \cos (1 - 7x)$.
3 $y = 3 \sin 16x$.
4 $y = \sin^2 x$.
5 $y = \sin x^2$.
6 $y = \sin^2 6x$.
7 $y = 5 \sin 3x + 3 \cos 5x$.
8 $y = \sin^2 x + \cos^2 x$.
9 $y = x \sin x$.
10 $y = x^3 \sin 3x$.
11 $y = \sin^2 3x \cos^2 3x$.
12 $y = \cos^4 x - \sin^4 x$.
13 $y = \frac{1}{5} \sin^5 x - \frac{2}{3} \sin^3 x + \sin x$.
14 $y = \sin (\sin x)$.
15 $y = e^{2x} \sin 3x$.
16 $y = \sin (\ln x^2)$.
17 $y = \ln (\cos x)$.
18 $y = e^{x^2 + \sin x}$.
19 If a is a positive constant, verify that $y = c_1 \sin ax + c_2 \cos ax$ is a solution of the differential equation

$$\frac{d^2y}{dx^2} + a^2y = 0$$

for every choice of the constants c_1 and c_2. (In the Additional Problems we outline a proof of the important fact that every solution of this differential equation has the stated form. For this reason, $y = c_1 \sin ax + c_2 \cos ax$ is called the *general solution* of the differential equation.)

20 Show that $(d/dx) \cos x = -\sin x$ by using the identity $\cos x = \sin (\pi/2 - x)$ and formula (6).

21 Find the angle at which the curve $y = \frac{1}{3} \sin 3x$ crosses the x-axis.

22 Sketch the graph of $y = \sin x + \cos x$ on the interval $0 \le x \le 2\pi$, and find the maximum height of this curve above the x-axis.

23 Find the maximum height of the curve $y = 4 \sin x - 3 \cos x$ above the x-axis.

24 Obtain the second of the following identities by differentiating the first: $\sin 2x = 2 \sin x \cos x$, $\cos 2x = \cos^2 x - \sin^2 x$.

25 Obtain the second of the following identities by differentiating the first: $\sin 3x = 3 \sin x - 4 \sin^3 x$, $\cos 3x = 4 \cos^3 x - 3 \cos x$.

26 Obtain the second of the following identities by differentiating the first with respect to either of the variables, keeping the other fixed:

$$\sin (x + y) = \sin x \cos y + \cos x \sin y,$$
$$\cos (x + y) = \cos x \cos y - \sin x \sin y.$$

27 Show that $y = \sin x$ and $y = \tan x$ have the same tangent at $x = 0$.

28 Show that the function $y = x + \sin x$ $(x \ge 0)$ has no maximum or minimum values even though there are many points where $dy/dx = 0$. Sketch the graph.

29 A regular polygon with n sides is inscribed in a circle of radius r.
   (a) Show that the perimeter of this polygon is $p_n = 2nr \sin (\pi/n)$.
   (b) Find $\lim_{n \to \infty} p_n$, and verify by elementary geometry that your answer is correct.

30 If a, b, c are constants with $ab \ne 0$, show that the graph of $y = a \sin (bx + c)$ is always concave toward the x-axis and that its points of inflection are the points where it crosses the x-axis.

31 Sketch the graphs of $y = \sin x$ and $y = \sin 2x$ together on a single set of axes. These curves have many points of intersection. Find the smallest positive x-coordinate of such a point, and calculate the acute angle at which the curves intersect at this point. Hint: See identity (17) in Section 9.1.

32 The functions $f(x) = \sin x$ and $g(x) = \cos x$ have the following properties: (a) $f'(x) = g(x)$; (b) $g'(x) = -f(x)$; (c) $f(0) = 0$; (d) $g(0) = 1$. If $F(x)$ and $G(x)$ is any pair of functions with the same properties, prove that $F(x) = \sin x$ and $G(x) = \cos x$. Hint: Show that

$$[F(x) - f(x)]^2 + [G(x) - g(x)]^2 = \text{a constant},$$

and find the value of this constant. [This problem has a very remarkable meaning: The functions $\sin x$ and $\cos x$ are *completely described* by properties (a) to (d), and therefore the total nature of these functions — everything that is now known about them or ever will be known — is implicitly contained in these four simple properties.]

In each of the following problems (33 to 43), find the value of the indicated limit.

33 $\lim_{x \to 0} \dfrac{\tan x}{x}$.

34 $\lim_{x \to 0} \dfrac{\sin 3x}{x}$.

35 $\lim\limits_{x\to 0}\dfrac{\tan x}{\sin x}$.

36 $\lim\limits_{x\to 0}\dfrac{\sin 3x}{\sin 5x}$.

41 $\lim\limits_{x\to 0}\dfrac{2x^2 + 2x}{\sin 2x}$.

42 $\lim\limits_{x\to 0}\sin 3x \cot 5x$.

37 $\lim\limits_{x\to 0}\tan 3x \csc 6x$.

38 $\lim\limits_{x\to 0}\dfrac{1 - \cos x}{x^2}$.

43 $\lim\limits_{x\to \pi/2}\dfrac{\cos x}{x - \pi/2}$.

39 $\lim\limits_{x\to \infty} x \sin \dfrac{1}{x}$.

40 $\lim\limits_{x\to \infty} 3x \tan \dfrac{\pi}{x}$.

The differential versions of formulas (7) and (6) in the previous section are

$$d(\cos u) = -\sin u\, du \qquad \text{and} \qquad d(\sin u) = \cos u\, du.$$

These immediately yield the integration formulas

$$\int \sin u\, du = -\cos u + c \tag{1}$$

and

$$\int \cos u\, du = \sin u + c. \tag{2}$$

9.3
THE INTEGRALS OF THE SINE AND COSINE. THE NEEDLE PROBLEM

Example 1   Evaluate $\int \cos 5x\, dx$.

Solution   Let $u = 5x$. Then $du = 5\, dx$, $dx = \tfrac{1}{5}\, du$, and formula (2) gives

$$\int \cos 5x\, dx = \int \cos u \cdot (\tfrac{1}{5}\, du) = \tfrac{1}{5}\int \cos u\, du$$

$$= \tfrac{1}{5}\sin u + c = \tfrac{1}{5}\sin 5x + c.$$

After a little practice, it will be easy for students to make this kind of substitution mentally. In fact, we can dispense with the new variable u altogether, and compress this solution to the following simple steps:

$$\int \cos 5x\, dx = \tfrac{1}{5}\int \cos 5x\, d(5x) = \tfrac{1}{5}\sin 5x + c.$$

Example 2   Evaluate $\int 7x \sin (2 - 9x^2)\, dx$.

Solution   Let $u = 2 - 9x^2$. Then $du = -18x\, dx$, $x\, dx = -\tfrac{1}{18}\, du$, and

$$\int 7x \sin (2 - 9x^2)\, dx = \int 7 \sin u \cdot (-\tfrac{1}{18}\, du)$$

$$= -\tfrac{7}{18}\int \sin u\, du$$

$$= \tfrac{7}{18}\cos u + c = \tfrac{7}{18}\cos (2 - 9x^2) + c.$$

Here the auxiliary variable u plays an important part in our work. It not only emphasizes the need to apply formula (1), but also helps us keep track of the various coefficients and algebraic signs that appear in the calculation — and therefore helps us avoid mistakes.

Example 3 Compute the definite integral

$$\int_{\pi/6}^{\pi/4} \frac{\cos 2x \, dx}{\sin^3 2x}.$$

Solution We begin by finding the indefinite integral. Since $d(\sin 2x) = 2 \cos 2x \, dx$, we put $u = \sin 2x$. This gives $du = 2 \cos 2x \, dx$, so

$$\int \frac{\cos 2x \, dx}{\sin^3 2x} = \int \frac{\frac{1}{2} du}{u^3} = \frac{1}{2} \int u^{-3} \, du = -\frac{1}{4} u^{-2} = -\frac{1}{4 \sin^2 2x}.$$

We remind students that the constant of integration can always be ignored in computing definite integrals, and for this reason we don't bother to write it here. The Fundamental Theorem of Calculus now permits us to complete the solution by writing

$$\int_{\pi/6}^{\pi/4} \frac{\cos 2x \, dx}{\sin^3 2x} = -\frac{1}{4 \sin^2 2x} \Big]_{\pi/6}^{\pi/4} = -\frac{1}{4} - \left(-\frac{1}{3}\right) = \frac{1}{12}.$$

In our next example we discuss an application of these methods to a famous problem about probability that was invented by the French scientist Buffon in the early eighteenth century.

Example 4 *Buffon's needle problem.* A needle 2 in long is tossed at random onto a floor made of boards 2 in wide. What is the probability that the needle falls across one of the cracks?

Solution We begin with a brief digression to explain our use of the word "probability." In mathematics this word means a numerical measure of the likelihood that a certain event will occur. As an example, consider the rectangle shown on the left in Fig. 9.14, in which a portion of the figure is shaded. If a point is chosen at random in this rectangle, for instance by making the rectangle into a target and throwing a dart blindfolded, then the probability of choosing a shaded point is $\frac{1}{4}$. We assume here that each point is just as likely to be chosen as any other, and this number expresses the fact that the proportion of shaded points among all points in the rectangle is $\frac{1}{4}$. In the second rectangle the probability of choosing a shaded point is $\frac{1}{8}$, and in the third rectangle it is $\frac{3}{8}$. We take it as self-evident that the probability of choosing a shaded point equals the ratio of the shaded area to the total area.

Let us now return to the needle problem. We describe the position in which the needle falls on the floor by the two variables x and θ shown in Fig. 9.15; x is the distance OP from the midpoint of the needle to the nearest crack, and θ is the smallest angle between OP and the needle. A toss of the needle amounts to a random choice of the variables x and θ in the intervals

$$0 \le x \le 1 \quad \text{and} \quad 0 \le \theta \le \frac{\pi}{2}, \tag{3}$$

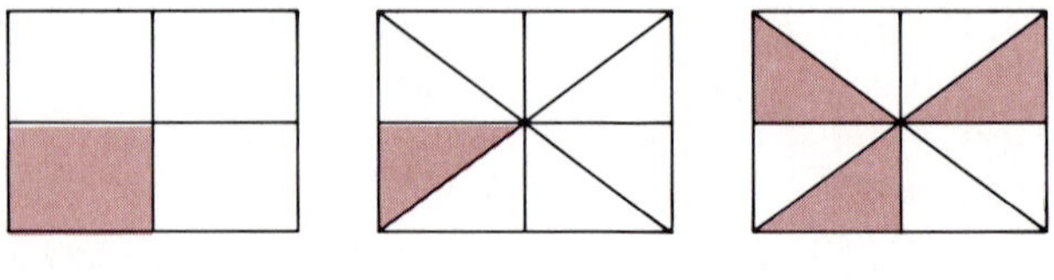

Figure 9.14

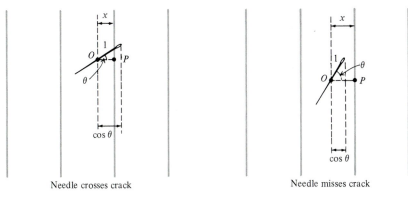

Needle crosses crack   Needle misses crack   **Figure 9.15**

and this in turn amounts to a random choice of a point in the rectangle shown in Fig. 9.16. Furthermore, a close inspection of Fig. 9.15 shows that the event we are interested in, namely, that the needle falls across a crack, corresponds to the inequality

$$x < \cos \theta. \tag{4}$$

This inequality describes the shaded region in Fig. 9.16 under the graph of $x = \cos \theta$. We therefore conclude that the probability of the needle falling across a crack equals the following ratio of areas:

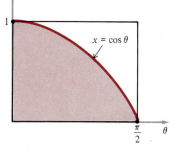

Figure 9.16

$$\frac{\text{Area under curve}}{\text{Area of rectangle}} = \frac{\int_0^{\pi/2} \cos \theta \, d\theta}{\pi/2} = \frac{1}{\pi/2} = \frac{2}{\pi}, \tag{5}$$

which is slightly less than $\frac{2}{3}$. This calculation can be extended at once to the more general situation in which d is the distance between adjacent cracks and the length of the needle is $L \le d$. The inequalities (3) are replaced by

$$0 \le x \le \frac{1}{2}d \quad \text{and} \quad 0 \le \theta \le \frac{\pi}{2},$$

and (4) becomes

$$x < \tfrac{1}{2}L \cos \theta.$$

In this case the probability of the needle falling across a crack is easily seen to be

$$\frac{\text{Area under curve}}{\text{Area of rectangle}} = \frac{\int_0^{\pi/2} \tfrac{1}{2}L \cos \theta \, d\theta}{(\tfrac{1}{2}d)(\pi/2)} = \frac{2L}{\pi d}. \tag{6}$$

(Students should draw their own sketch for this case similar to Fig. 9.16, and in particular should notice the reason for the restriction $L \le d$.)

Remark   We obtained these conclusions about the probability of success in the needle experiment by pure reason alone, without any appeal to experience. However, the "sequence of trials" approach to the concept of probability has some interesting implications for the needle problem. In the case of the 2-in needle and the 2-in floorboards, let us actually perform the experiment of tossing the needle onto the floor a large number of times, say n times, where $n = 100$ or $n = 1000$ depending on our ability to tolerate boredom. Let us also keep careful count of the number k of times the needle falls across a crack. Then the abstract probability that the needle falls across a crack on any one toss should be closely approximated by the ratio k/n, and this approximation should improve as n increases. Roughly speaking, this means that

$$\lim_{n \to \infty} \frac{k}{n} = \frac{2}{\pi},$$

so we should have

$$\frac{k}{n} \cong \frac{2}{\pi};$$

and solving this approximate equation for π yields

$$\pi \cong \frac{2n}{k}$$

for large values of n. In principle, therefore, this provides an experimental method of calculating π. In fact, however, this method is not capable of much accuracy because of the inherent errors that appear in all measurements. We will discuss practical methods of computing π to very great accuracy in Chapter 13.

PROBLEMS

Evaluate the indefinite integrals in Problems 1 to 20.

1 $\int \sin 5x \, dx.$
2 $\int \cos (2x - 5) \, dx.$
3 $\int \sin (1 - 9x) \, dx.$
4 $\int (3 \cos 2x - 2 \sin 3x) \, dx.$
5 $\int 2 \sin x \cos x \, dx.$
6 $\int \cos^2 x \sin x \, dx.$
7 $\int \sin^3 2x \cos 2x \, dx.$
8 $\int \sin x \cos x (\sin x + \cos x) \, dx.$
9 $\int \sin^7 \frac{1}{2} x \cos \frac{1}{2} x \, dx.$
10 $\int 4x \sin x^2 \, dx.$

11 $\int \dfrac{\sin \sqrt{x} \, dx}{\sqrt{x}}.$ 12 $\int \dfrac{\cos (\ln x) \, dx}{x}.$

13 $\int \cos (\sin 2x) \cos 2x \, dx.$

14 $\int \dfrac{\cos x \, dx}{\sin^2 x}.$

15 $\int \dfrac{\sin [(2x - 1)/3] \, dx}{\cos^2 [(2x - 1)/3]}.$

16 $\int \dfrac{\cos x \, dx}{\sin x}.$ 17 $\int \dfrac{\sin x \, dx}{\cos x}.$

18 $\int \dfrac{\cos 3x \, dx}{\sqrt{\sin 3x}}.$

19 $\int (2x + 1) \cos (x^2 + x) \, dx.$
20 $\int (x + \cos x)^4 (1 - \sin x) \, dx.$

Evaluate the definite integrals in Problems 21 to 24.

21 $\displaystyle\int_0^{\pi/5} \sin 5x \, dx.$ 22 $\displaystyle\int_{-\pi/6}^{2\pi/3} \cos 3x \, dx.$

23 $\displaystyle\int_{\pi/4}^{\pi/2} \dfrac{\cos x \, dx}{\sin^2 x}.$ 24 $\displaystyle\int_0^{\sqrt{\pi}} x \cos x^2 \, dx.$

25 Find the area under one arch of $y = \sin 3x$.
26 In the first quadrant, the y-axis and the curves $y = \sin x$ and $y = \cos x$ bound a "triangle-shaped" region. Find its area.
27 Find the area under one arch of $y = 3 \cos 2x$.
28 Find the area under one arch of $y = 6 \sin \frac{1}{2}x$ and above the line $y = 3$.
29 Find the volume generated by revolving about the x-axis the region under $y = \sin x$ and between $x = 0$ and $x = \pi$. Hint: Remember the half-angle formula $2 \sin^2 x = 1 - \cos 2x$.
30 Consider the region between $y = \sin x$ and the x-axis for $0 \le x \le \pi/2$. For what constant c does the line $x = c$ divide this region into two parts of equal areas?

31 Anticipate the results of the next section by deriving the following differentiation formulas:

$$\frac{d}{dx} \tan x = \sec^2 x;$$

$$\frac{d}{dx} \cot x = -\csc^2 x;$$

$$\frac{d}{dx} \sec x = \sec x \tan x;$$

$$\frac{d}{dx} \csc x = -\csc x \cot x.$$

Hint: Express each function in terms of $\sin x$ and $\cos x$.

32 Obtain the following integration formulas from the differentiation formulas in Problem 31:

$$\int \sec^2 x \, dx = \tan x + c;$$

$$\int \csc^2 x \, dx = -\cot x + c;$$

$$\int \sec x \tan x \, dx = \sec x + c;$$

$$\int \csc x \cot x \, dx = -\csc x + c.$$

The results of Problem 31 in Section 9.3 enable us to complete our list of formulas for differentiating the trigonometric functions:

$$\frac{d}{dx} \tan u = \sec^2 u \frac{du}{dx}; \tag{1}$$

$$\frac{d}{dx} \cot u = -\csc^2 u \frac{du}{dx}; \tag{2}$$

$$\frac{d}{dx} \sec u = \sec u \tan u \frac{du}{dx}; \tag{3}$$

$$\frac{d}{dx} \csc u = -\csc u \cot u \frac{du}{dx}. \tag{4}$$

9.4
THE DERIVATIVES OF THE OTHER FOUR FUNCTIONS

These formulas are quite easy to remember if we notice that the derivative of each cofunction (cot, csc) can be obtained from the derivative of the corresponding function (tan, sec) by (a) inserting a minus sign, and (b) replacing each function by its cofunction. Thus, formula (2) is obtained from formula (1) by inserting a minus sign, replacing $\tan u$ by $\cot u$, and replacing $\sec u$ by $\csc u$. In view of this rule, it is only necessary to memorize formulas (1) and (3), because the rule immediately produces the other two.

Example 1 Find dy/dx if $y = \tan^3 4x$.

Solution Since $y = \tan^3 4x = (\tan 4x)^3$, the power rule gives

$$\frac{dy}{dx} = 3(\tan 4x)^2 \cdot \frac{d}{dx} \tan 4x.$$

By formula (1) with $u = 4x$,

$$\frac{d}{dx} \tan 4x = (\sec^2 4x)(4),$$

and by putting the various pieces together we obtain

$$\frac{dy}{dx} = 12 \tan^2 4x \sec^2 4x.$$

Example 2 Find dy/dx if $y = \cot(1 - 3x)$.

Solution By formula (2) with $u = 1 - 3x$,

$$\frac{dy}{dx} = -\csc^2(1 - 3x) \cdot (-3) = 3\csc^2(1 - 3x).$$

The differentiation formulas (1) to (4) immediately produce four new integration formulas:

$$\int \sec^2 u\ du = \tan u + c; \qquad (5)$$

$$\int \csc^2 u\ du = -\cot u + c; \qquad (6)$$

$$\int \sec u \tan u\ du = \sec u + c; \qquad (7)$$

$$\int \csc u \cot u\ du = -\csc u + c. \qquad (8)$$

Example 3 Calculate $\int \sec 3x \tan 3x\ dx$.

Solution This reminds us of (7) with $u = 3x$, so we write

$$\int \sec 3x \tan 3x\ dx = \frac{1}{3}\int \sec 3x \tan 3x\ d(3x) = \frac{1}{3}\sec 3x + c.$$

In this problem the structure of the integral is clear enough so that there is no real need to make an explicit change of variable.

Example 4 Evaluate $\int 3x \sec^2 x^2\ dx$.

Solution This reminds us of (5) with $u = x^2$. Since $du = 2x\ dx$ and $x\ dx = \frac{1}{2}du$, we have

$$\int 3x \sec^2 x^2\ dx = 3\int \sec^2 u \cdot \left(\frac{1}{2}du\right) = \frac{3}{2}\int \sec^2 u\ du$$

$$= \frac{3}{2}\tan u + c = \frac{3}{2}\tan x^2 + c.$$

Here we use the auxiliary variable u as insurance against error. After students have acquired a bit of experience with problems of this type, they will prefer to carry out the integration directly, by inspection.

Example 5 Calculate $\int \tan^2 2x\ dx$.

Solution This integral doesn't resemble any of our types. However, the trigonometric identity $\tan^2 2x + 1 = \sec^2 2x$ connects our problem with formula (5). Once this fact is noticed, we easily write

$$\int \tan^2 2x \, dx = \int (\sec^2 2x - 1) \, dx = \int \sec^2 2x \, dx - \int dx$$

$$= \frac{1}{2} \int \sec^2 2x \, d(2x) - \int dx = \frac{1}{2} \tan 2x - x + c.$$

PROBLEMS

In each of the following problems (1 to 12) calculate dy/dx.

1 $y = \tan 4x^2$.

2 $y = \cot 4x$.

3 $y = \tan^2 (\sin x)$.

4 $y = 3 \cot (1 - x^3)$.

5 $y = \sec^2 x - \tan^2 x$.

6 $y = 2 \sec 3x$.

7 $y = 4 \csc (-6x)$.

8 $y = (\cot x + \csc x)^2$.

9 $y = \sqrt{\csc 2x}$.

10 $y = \cot (\cos x)$.

11 $y = e^{\tan x}$.

12 $y = \ln (\csc x)$.

Evaluate the integral in each of the following problems (13 to 20).

13 $\displaystyle\int \csc^2 6x \, dx$.

14 $\displaystyle\int_0^{\pi/8} \sec^2 2x \, dx$.

15 $\displaystyle\int \frac{dx}{\sin^2 2x}$.

16 $\displaystyle\int \sec^2 \frac{1}{3} x \, dx$.

17 $\displaystyle\int \tan^4 x \sec^2 x \, dx$.

18 $\displaystyle\int_0^{\pi/6} \sec 2x \tan 2x \, dx$.

19 $\displaystyle\int \cot 7x \csc 7x \, dx$.

20 $\displaystyle\int \sec^7 x \tan x \, dx$.

21 Find the area bounded by the curve $y = \tan x \sec^2 x$, the x-axis, and the line $x = \pi/4$.

22 Find the area in the first quadrant bounded by $y = \sec^2 x$, $y = 8 \cos x$, and the y-axis.

23 Find the area in the first quadrant bounded by $y = \sec^2 x$, $y = 2 \tan^2 x$, and the y-axis.

24 The region bounded by the curve $y = \tan x$, the x-axis, and the line $x = \pi/3$ is revolved about the x-axis. Find the volume of the solid of revolution generated in this way.

25 Sketch the graph of the function $y = \tan x + \cot x$ on the interval $0 < x < \pi/2$ and find its minimum value.

26 Solve Problem 25 without calculus, by using the identity

$$\tan x + \cot x = \frac{2}{\sin 2x}.$$

*27 Sketch the graph of the function $y = 8 \csc x - 4 \cot x$ on the interval $0 < x \le \pi/2$ and find its minimum value. Is there a point of inflection?

*28 The classic corridor problem (Problem 29 in Section 4.3) can be expressed as follows. If two corridors of widths a and b meet at right angles (Fig. 9.17), then

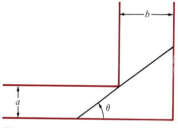

Figure 9.17

the length of the longest thin rod that can be moved in a horizontal position around the corner is the length of the shortest line segment placed like the one in the figure. Find this length by using the angle θ as the independent variable.

29 A revolving light 6 mi offshore from a straight shoreline makes 4 revolutions per minute. How fast is the spot of light moving along the shore at the instant when the beam makes an angle of 30° with the shoreline?

*30 A rope with a ring at one end is looped over two pegs in a horizontal line. The free end is passed through the ring and has a weight suspended from it, so that the rope is held taut. If the rope slips freely through the ring and over the pegs, then the weight will descend as far as possible in order to minimize its potential energy. Find the angle formed at the bottom of the loop.

Our attention in this section is focused on the two integration formulas

$$\int \frac{dx}{\sqrt{1 - x^2}} = \sin^{-1} x \qquad (1)$$

and

$$\int \frac{dx}{1 + x^2} = \tan^{-1} x. \qquad (2)$$

9.5

THE INVERSE TRIGONOMETRIC FUNCTIONS

The unfamiliar functions on the right sides of these equations will be fully explained below. They are called *inverse trigonometric functions,* and are created expressly to enable us to calculate the integrals on the left. These functions have other uses, but this is their primary purpose, the main justification for their existence.

Before we start at the beginning and give a careful and orderly description of these functions, we pause briefly to understand in a rough way how they originate. The difficulty with the integral on the left of (1) is caused by the awkward expression $\sqrt{1 - x^2}$ in the denominator. If we consider this obstacle for a moment, the inside quantity $1 - x^2$ might make us think of the trigonometric expression $1 - \sin^2 \theta$, which of course equals $\cos^2 \theta$. Thus, if we write

$$x = \sin \theta, \tag{3}$$

then we have $\sqrt{1 - x^2} = \sqrt{1 - \sin^2 \theta} = \sqrt{\cos^2 \theta} = \cos \theta$, and the square root sign disappears. But we also have $dx = \cos \theta \, d\theta$, so we can unravel our troublesome integral as follows:

$$\int \frac{dx}{\sqrt{1 - x^2}} = \int \frac{\cos \theta \, d\theta}{\cos \theta} = \int d\theta = \theta. \tag{4}$$

The process of solving (3) for θ in terms of x is symbolized by writing $\theta = \sin^{-1} x$, so (4) yields (1). A similar analysis can be applied to (2), but these remarks are perhaps enough to make our point about the way the inverse trigonometric functions arise—they are forced upon us by the need to calculate certain integrals. Now for the details that make these functions respectable.

THE INVERSE SINE

We know that $\sin \pi/6 = \frac{1}{2}$. Thus, if we are asked to find an angle (in radian measure) whose sine is $\frac{1}{2}$, we can answer at once that $\pi/6$ is such an angle. We are also aware that there are many other angles with this property.

As we have just seen, it is necessary in calculus to have a symbol to denote an angle whose sine is a given number x. There are two such symbols in everyday use,

$$\sin^{-1} x \quad \text{and} \quad \arcsin x.$$

These notations are fully equivalent to each other and can be used interchangeably, though we shall confine ourselves to the first. The first is read "the inverse sine of x," and the second "the arc sine of x," and both mean "an angle whose sine is x." It is essential to understand that in the symbol $\sin^{-1} x$, the -1 is *not* an exponent, and therefore $\sin^{-1} x$ *never* means $1/(\sin x)$. We discuss the reason for this seemingly strange notation in Remark 2.

These ideas can be summarized as follows: The formulas

$$x = \sin y \quad \text{and} \quad y = \sin^{-1} x$$

mean exactly the same thing, in the sense that

$$x = 3y \quad \text{and} \quad y = \frac{1}{3} x$$

mean exactly the same thing. In each case the equation is first written in a form solved for x, and then (the same equation!) in a form solved for y.

In order to sketch the graph of $y = \sin^{-1} x$, it suffices to sketch $x = \sin y$ with y treated as the independent variable—on the horizontal axis—and then to turn the picture over, returning the axes to their customary positions (Fig. 9.18). It is clear that y exists only when x lies in the interval $-1 \leq x \leq 1$. However, for any such x there are infinitely many corresponding y's, and this situation cannot be allowed if $y = \sin^{-1} x$ is to be considered a function. (Recall that a function is single-valued by the very definition of the concept.) We deal with this difficulty by means of a universally understood agreement: The only values of $y = \sin^{-1} x$ we consider are those that lie in the interval $-\pi/2 \leq y \leq \pi/2$, and this restriction is henceforth part of the meaning of the symbol $y = \sin^{-1} x$. The graph of the function $y = \sin^{-1} x$ (it is truly a function now, because of the restriction just described) is the heavy portion of the curve in Fig. 9.18.

THE INVERSE TANGENT

The function $y = \tan^{-1} x$ (the other notation here is $y = \arctan x$) is defined in essentially the same way:

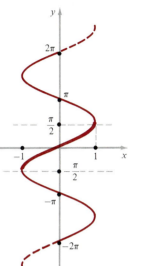

Figure 9.18

$$y = \tan^{-1} x \quad \text{means} \quad x = \tan y \quad \text{and} \quad -\frac{\pi}{2} < y < \frac{\pi}{2}.$$

The symbol $\tan^{-1} x$ is read "the inverse tangent of x," and it means "the angle (in the specified interval) whose tangent is x." The graph of the function $y = \tan^{-1} x$ is the heavy curve in Fig. 9.19.

We now calculate the derivative dy/dx of the function $y = \sin^{-1} x$ by differentiating

$$x = \sin y$$

implicitly with respect to x. The result is

$$1 = \cos y \, \frac{dy}{dx},$$

so

$$\frac{dy}{dx} = \frac{1}{\cos y} = \frac{1}{\sqrt{1 - \sin^2 y}} = \frac{1}{\sqrt{1 - x^2}}.$$

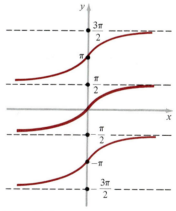

Figure 9.19

We choose the positive square root here because $y = \sin^{-1} x$ is clearly an increasing function (see Fig. 9.18). This result can be written in the form

$$\frac{d}{dx} \sin^{-1} x = \frac{1}{\sqrt{1 - x^2}}, \tag{5}$$

where $-1 < x < 1$. In just the same way we find the derivative of $y = \tan^{-1} x$ by differentiating

$$x = \tan y$$

implicitly with respect to x. This gives

$$1 = \sec^2 y \, \frac{dy}{dx},$$

so

$$\frac{dy}{dx} = \frac{1}{\sec^2 y} = \frac{1}{1 + \tan^2 y} = \frac{1}{1 + x^2}.$$

We therefore have

$$\frac{d}{dx} \tan^{-1} x = \frac{1}{1 + x^2} \tag{6}$$

for all x.

Formulas (5) and (6) are the facts that lead to the main tools of this section. First, we have the chain rule extensions of these formulas, which greatly broaden their scope:

$$\frac{d}{dx} \sin^{-1} u = \frac{1}{\sqrt{1 - u^2}} \frac{du}{dx} \tag{7}$$

and

$$\frac{d}{dx} \tan^{-1} u = \frac{1}{1 + u^2} \frac{du}{dx}. \tag{8}$$

As usual, u is understood to be any differentiable function of x.

Example 1 Find dy/dx for each of the following functions:
(a) $y = \sin^{-1} 4x$;   (b) $y = \sin^{-1} x^3$;   (c) $y = \frac{1}{3} \tan^{-1}(3x - 5)$.

Solution For (a), we use (7) with $u = 4x$, so

$$\frac{dy}{dx} = \frac{1}{\sqrt{1 - (4x)^2}} \frac{d}{dx}(4x) = \frac{4}{\sqrt{1 - 16x^2}}.$$

For (b), we use (7) with $u = x^3$, so

$$\frac{dy}{dx} = \frac{1}{\sqrt{1 - (x^3)^2}} \frac{d}{dx}(x^3) = \frac{3x^2}{\sqrt{1 - x^6}}.$$

For (c), we use (8) with $u = 3x - 5$, so

$$\frac{dy}{dx} = \frac{1}{3} \frac{1}{1 + (3x - 5)^2} \frac{d}{dx}(3x - 5) = \frac{1}{1 + (3x - 5)^2}.$$

Even more important for our future work are the integration formulas equivalent to (7) and (8):

$$\int \frac{du}{\sqrt{1 - u^2}} = \sin^{-1} u + c \tag{9}$$

and

$$\int \frac{du}{1 + u^2} = \tan^{-1} u + c. \tag{10}$$

These formulas are indispensable tools for integral calculus, and all by themselves amply justify the study of trigonometry.

Example 2 Calculate each of the following integrals:

(a) $\displaystyle\int \frac{dx}{\sqrt{1 - 9x^2}}$;   (b) $\displaystyle\int \frac{dx}{1 + 25x^2}$;   (c) $\displaystyle\int \frac{5x^2\,dx}{1 + 4x^6}$.

Solution (a) Put $u = 3x$. Then $du = 3dx$, and by (9)

$$\int \frac{dx}{\sqrt{1 - 9x^2}} = \int \frac{\frac{1}{3}\,du}{\sqrt{1 - u^2}} = \frac{1}{3}\sin^{-1} u + c = \frac{1}{3}\sin^{-1} 3x + c.$$

(b) Put $u = 5x$. Then $du = 5dx$, and by (10)

$$\int \frac{dx}{1 + 25x^2} = \int \frac{\frac{1}{5}\,du}{1 + u^2} = \frac{1}{5}\tan^{-1} u + c = \frac{1}{5}\tan^{-1} 5x + c.$$

(c) To get started here we must notice that $4x^6 = (2x^3)^2$. This suggests putting $u = 2x^3$. Then $du = 6x^2\,dx$, and by (10)

$$\int \frac{5x^2\,dx}{1 + 4x^6} = 5\int \frac{\frac{1}{6}du}{1 + u^2} = \frac{5}{6}\tan^{-1} u + c = \frac{5}{6}\tan^{-1} 2x^3 + c.$$

The crucial feature of this integral is clearly the presence of x^2 in the numerator, for without this factor the method we have used would be unworkable.

Remark 1 As students doubtless suspect, four other inverse trigonometric functions can be defined if we wish to do so. However, these functions are not really needed for the purposes of integration. We can illustrate this point by observing that if $u > 0$ then

$$\int \frac{du}{u\sqrt{u^2 - 1}} = \int \frac{du}{u\sqrt{u^2(1 - 1/u^2)}} = \int \frac{du}{u^2\sqrt{1 - (1/u)^2}}$$

$$= -\int \frac{d(1/u)}{\sqrt{1 - (1/u)^2}} = -\sin^{-1}\frac{1}{u} + c.$$

(If $u < 0$, the factor u^2 under the radical in the second step comes out of the radical as $-u$.) This integral is a standard type that many writers integrate by using the inverse secant—which this calculation clearly shows to be superfluous. The fact of the matter is that the inverse sine and the inverse tangent suffice for all our purposes for calculating integrals, so for the sake of simplicity we ignore the other inverse trigonometric functions. (The notation $\cos^{-1} x$ will occasionally be used, but only for convenience in designating the angle between 0 and π whose cosine is x, where x is a number between 1 and -1.)

Remark 2 Suppose that a variable x is a function of a variable y as shown on the left in Fig. 9.20. In this case, not only does each y (in a certain interval) determine a unique x, but also each x determines a unique y. Thus, y is also a function of x. If the given function is written $x = f(y)$, then the second function is often called the *inverse function* of the first and denoted by the symbol $y = f^{-1}(x)$ [read "f inverse of x"]. The graph of $y = f^{-1}(x)$ is simply the graph of $x = f(y)$ turned over as shown on the right in Fig. 9.20, so that the axes are returned to their normal positions. The essence of this situation is that when two functions are related in this way, then each undoes what the other one does, in the sense that

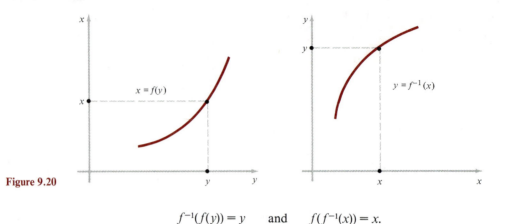

Figure 9.20

$$f^{-1}(f(y)) = y \quad \text{and} \quad f(f^{-1}(x)) = x.$$

It is this reciprocal relation that is suggested by the word "inverse" and the symbol f^{-1}. We have encountered inverse functions in Chapter 8 and also in this section, but we have no special need to develop the subject in detail. We do point out, however, that any increasing or decreasing function $x = f(y)$ obviously has an inverse; and it can be proved that if either function has a nonzero derivative at a point, then so does the other and

$$\frac{dy}{dx} = \frac{1}{dx/dy}.$$

Here again — as in the case of the chain rule — we have a situation in which the Leibniz fractional notation for derivatives strongly suggests a true theorem in the guise of a simple manipulation of differentials.

Remark 3 Formula (10) leads rather quickly (though unrigorously) to the famous Leibniz formula

$$\frac{\pi}{4} = 1 - \frac{1}{3} + \frac{1}{5} - \frac{1}{7} + \cdots, \tag{11}$$

which connects the number π with the odd numbers 1, 3, 5, 7, ... To see this connection, we begin with the formula from elementary algebra for the sum of a geometric series,

$$1 + r + r^2 + r^3 + \cdots = \frac{1}{1-r}. \tag{12}$$

(The reader will perhaps recall that this formula is valid for $|r| < 1$, but here we pay little attention to such cautionary details.) If r in (12) is replaced by $-t^2$ and the resulting equation is reversed, we get

$$\frac{1}{1+t^2} = 1 - t^2 + t^4 - t^6 + \cdots. \tag{13}$$

We now apply (10) to obtain

$$\tan^{-1} x = \int_0^x \frac{dt}{1+t^2} = \int_0^x [1 - t^2 + t^4 - t^6 + \cdots] \, dt$$

$$= x - \frac{x^3}{3} + \frac{x^5}{5} - \frac{x^7}{7} + \cdots,$$

which yields Leibniz's formula (11) when $x = 1$. These ideas and the legitimacy of these procedures will be studied much more carefully in Chapters 13 and 14. Meanwhile, students who wish to satisfy themselves now that (11) is indeed correct will find an elementary but rigorous proof in Appendix A.8.

PROBLEMS

1 Given that $\theta = \sin^{-1}(-\frac{1}{2})$, find $\cos\theta$, $\tan\theta$, $\cot\theta$, $\sec\theta$, $\csc\theta$.

2 Given that $\theta = \tan^{-1}\sqrt{3}$, find $\sin\theta$, $\cos\theta$, $\cot\theta$, $\sec\theta$, $\csc\theta$.

3 Evaluate each of the following:
(a) $\sin^{-1} 1 - \sin^{-1}(-1)$; (b) $\tan^{-1} 1 - \tan^{-1}(-1)$;
(c) $\sin(\sin^{-1} 0.123)$; (d) $\cos(\sin^{-1} 0.6)$;
(e) $\sin(2\sin^{-1} 0.6)$; (f) $\tan^{-1}(\tan \pi/7)$;
(g) $\sin^{-1}(\sin 5\pi/6)$; (h) $\tan^{-1}(\tan[-3\pi/4])$.

Find dy/dx in each of the following problems (4 to 13).

4 $y = \sin^{-1}\dfrac{1}{2}x$.

5 $y = \dfrac{1}{5}\tan^{-1}\dfrac{1}{5}x$.

6 $y = \dfrac{1}{2}\tan^{-1}x^2$.

7 $y = \sin^{-1}\dfrac{x-1}{x+1}$.

8 $y = \tan^{-1}\dfrac{x-1}{x+1}$.

9 $y = x\sin^{-1}x + \sqrt{1-x^2}$.

10 $y = x\tan^{-1}x - \ln\sqrt{1+x^2}$.

11 $y = x(\sin^{-1}x)^2 - 2x + 2\sqrt{1-x^2}\sin^{-1}x$.

12 $y = \dfrac{1}{2}(\sin^{-1}x + x\sqrt{1-x^2})$.

13 $y = \tan^{-1}\dfrac{4\sin x}{3+5\cos x}$.

14 If a is a positive constant, show that

$$\int \frac{du}{\sqrt{a^2 - u^2}} = \sin^{-1}\frac{u}{a} + c$$

and

$$\int \frac{du}{a^2 + u^2} = \frac{1}{a}\tan^{-1}\frac{u}{a} + c.$$

These simple generalizations of formulas (9) and (10) are often more convenient in applications.

Evaluate the integrals in the following problems (15 to 25).

15 $\displaystyle\int_0^{1/2} \frac{dx}{\sqrt{1-x^2}}$.

16 $\displaystyle\int_{-1}^1 \frac{dx}{1+x^2}$.

17 $\displaystyle\int \frac{dx}{\sqrt{1-4x^2}}$.

18 $\displaystyle\int \frac{dx}{1+3x^2}$.

19 $\displaystyle\int_0^{1/2} \frac{dx}{1+4x^2}$.

20 $\displaystyle\int \frac{x\,dx}{1+4x^4}$.

21 $\displaystyle\int \frac{dx}{\sqrt{9-4x^2}}$.

22 $\displaystyle\int \frac{dx}{\sqrt{16-9x^2}}$.

23 $\displaystyle\int \frac{dx}{4+9x^2}$.

24 $\displaystyle\int_{\sqrt{2}}^2 \frac{dx}{x\sqrt{x^2-1}}$.

25 $\displaystyle\int_{-2}^{-\sqrt{2}} \frac{dx}{x\sqrt{x^2-1}}$.

26 A picture hangs on a wall with its base a feet above the level of an observer's eye. If the picture is b feet high and the observer stands x feet from the wall, show that the angle θ subtended by the picture is given by the formula

$$\theta = \tan^{-1}\frac{a+b}{x} - \tan^{-1}\frac{a}{x}.$$

What value of x maximizes this angle?

27 The points $(1, 2)$ and $(2, 1)$ in the first quadrant are joined by two segments to a variable point $(0, y)$ on the y-axis, where $y < 3$. If θ denotes the angle between these segments, what is the largest value θ can have?

28 A balloon is released at eye level and rises at the rate of 5 ft/s. An observer 50 ft away watches the balloon rise. How fast is the angle of elevation increasing 6 seconds after the moment of release?

29 The top of a 15-ft ladder is sliding down a wall. When the base of the ladder is 9 ft from the wall, it is sliding away at the rate of 3 ft/s. (a) What is the angle between the wall and the ladder at that moment? (b) How fast is the angle increasing at that moment?

30 Sketch the curve $y = 1/(1+x^2)$. Find the area of the region under this curve between $x = 0$ and $x = b$, where b is a positive constant. Find the limit of this area as $b \to \infty$.

31 Comment on the legitimacy of the formula

$$\int_0^3 \frac{dx}{\sqrt{1-x^2}} = \sin^{-1} 3.$$

32 Sketch the curve $y = 1/\sqrt{1-x^2}$ on the interval $0 \le x < 1$. Find the area under this curve between $x = 0$ and $x = b$ where $0 < b < 1$. Find the limit of this area as $b \to 1$.

9.6

SIMPLE HARMONIC MOTION. THE PENDULUM

Most people understand that sound is vibration, and for this reason alone the study of vibrations is an important part of science. But vibrations — or oscillations, or waves, or periodic phenomena generally — are much more pervasive than this. They appear in many contexts having little to do with sound, for instance in connection with radio waves, light waves, alternating electric currents, the vibration of atoms in crystals, etc. The study of vibrations in this broader sense is clearly one of the most fundamental themes of physical science, and in any such study sines and cosines play a central role.

One of the simplest types of vibrations occurs when an object or point moves back and forth along a straight line (the x-axis) in such a way that its acceleration is always proportional to its position and is directed in the opposite sense:

$$\frac{d^2x}{dt^2} = -kx, \qquad k > 0. \tag{1}$$

Motion of this kind is called *simple harmonic motion*. To emphasize that the constant k is positive, it is customary to write $k = a^2$ with $a > 0$. The differential equation (1) then takes the form

$$\frac{d^2x}{dt^2} + a^2x = 0. \tag{2}$$

It is easy to see that any function of the form

$$x = A \sin(at + b), \qquad A \neq 0, \tag{3}$$

satisfies equation (2).* We merely calculate

$$\frac{dx}{dt} = Aa \cos(at + b) \qquad \text{and} \qquad \frac{d^2x}{dt^2} = -Aa^2 \sin(at + b) = -a^2x,$$

and observe that

$$\frac{d^2x}{dt^2} + a^2x = 0.$$

It is equally true, though not so easy to see, that every nontrivial solution of (2) can be written in the form (3). We will demonstrate this in Remarks 1 and 2, but meanwhile we take it for granted.

Since the function $\sin(at + b)$ oscillates between -1 and 1, the function (3) oscillates between $-|A|$ and $|A|$. The number $|A|$ is called the *amplitude* of the motion (Fig. 9.21). Also, since the sine is periodic with period 2π, $\sin(at + b)$ is periodic with period $2\pi/a$, because this is the amount t must increase in order to increase $at + b$ by 2π. This number $T = 2\pi/a$ is called the *period* of the motion, and is the time required for one complete cycle. If we measure t in seconds, then the number f of cycles per second satisfies the equation $fT = 1$, and is therefore the reciprocal of the period,

$$f = \frac{1}{T} = \frac{a}{2\pi}.$$

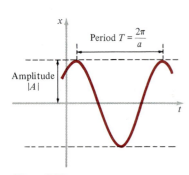

Figure 9.21

---

* We add the condition $A \neq 0$ to avoid the trivial case in which x is identically zero and consequently there is no motion.

This number is called the *frequency* of the motion.

Another equivalent form of the general solution (3) that is often useful is

$$x = A \cos (at + b). \tag{4}$$

This is easily seen from the fact that b in (3) is an arbitrary constant, and can therefore be replaced by the equally arbitrary constant $b + \pi/2$. This gives

$$x = A \sin \left(at + b + \frac{\pi}{2} \right) = A \cos (at + b),$$

since $\sin (\theta + \pi/2) = \cos \theta$.

There are two main interpretations of simple harmonic motion, one geometric and the other physical.

The geometric meaning can be understood by considering a point P that moves with constant angular velocity around a circle of radius A (Fig. 9.22). If this constant angular velocity is denoted by a, then

$$\frac{d\theta}{dt} = a \qquad \text{and therefore} \qquad \theta = at + b,$$

where b is the value of θ when $t = 0$. If Q is the projection of P on the x-axis, then its x-coordinate is

$$x = A \cos \theta = A \cos (at + b).$$

This shows that Q moves back and forth along the x-axis in simple harmonic motion as P moves steadily around the circle in uniform circular motion, and any simple harmonic motion can be visualized in this way.

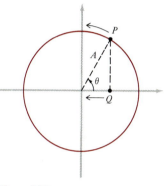

Figure 9.22

The physical meaning appears when we think of equation (1) as describing the motion of a body of mass m rather than merely a point. Newton's second law of motion says that $F = ma$, so equation (1) becomes

$$\frac{1}{m} F = -kx \qquad \text{or} \qquad F = -kmx.$$

A force F of this kind is called a *restoring force*, because its magnitude is proportional to the displacement x and it always acts to pull the body back toward the equilibrium position $x = 0$. We discuss this idea more fully in our first two examples.

Example 1 Consider a cart of mass m attached to a nearby wall by means of a spring (Fig. 9.23). The spring exerts no force when the cart is at its equilibrium position $x = 0$. If the cart is pulled aside to a position x, then the spring exerts a restoring force $F = -kx$, where k is a positive constant whose magnitude is a measure of the stiffness of the spring (see Example 1 in Section 7.8). Suppose that the cart is pulled out to the position $x = x_0$ and released without any initial velocity at time $t = 0$. Discuss its subsequent motion if friction and air resistance are negligible.

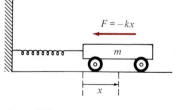

Figure 9.23

Solution We are assuming that the only force acting on the cart is the restoring force $F = -kx$, so by Newton's second law of motion we have

$$m \frac{d^2x}{dt^2} = -kx \qquad \text{or} \qquad \frac{d^2x}{dt^2} + \frac{k}{m} x = 0.$$

It is convenient to write this equation as

$$\frac{d^2x}{dt^2} + a^2 x = 0,$$

where $a = \sqrt{k/m}$. The form of the general solution we prefer here is

$$x = c_1 \sin at + c_2 \cos at, \tag{5}$$

which can be obtained by expanding either (3) or (4). The initial conditions

$$x = x_0 \quad \text{and} \quad v = \frac{dx}{dt} = 0 \quad \text{when} \quad t = 0$$

imply that $c_2 = x_0$ and $c_1 = 0$, so (5) becomes

$$x = x_0 \cos at.$$

It is clear from this that the cart moves in simple harmonic motion with period $T = 2\pi/a = 2\pi\sqrt{m/k}$ and frequency

$$f = \frac{1}{T} = \frac{1}{2\pi} \sqrt{\frac{k}{m}}. \tag{6}$$

We see from (6) that the frequency of this vibration increases if the stiffness k of the spring is increased, and decreases if the mass m of the cart is increased, as our common sense would have led us to expect.

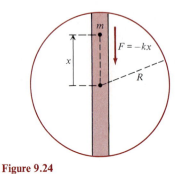

Figure 9.24

Example 2 Suppose that a tunnel is bored straight through the center of the earth from one side to the other, and that a body of mass m is dropped into this tunnel. Assuming as usual that the earth is a perfect sphere of uniform density and radius R of about 4000 mi, the effect of gravity is such that the body is attracted toward the center of the earth by a force F proportional to its distance x from the center (Fig. 9.24).* Show that the body traverses the tunnel from one end to the other and back again with simple harmonic motion, and calculate the period of this motion.

Solution Clearly $F = -kx$ for a suitable constant k. To find the value of this constant we use the fact that $F = -mg$ at the surface of the earth, where $x = R$, so

$$-mg = -kR \quad \text{or} \quad k = \frac{mg}{R}.$$

Newton's second law of motion therefore takes the form

$$m\frac{d^2x}{dt^2} = -\frac{mg}{R} x \quad \text{or} \quad \frac{d^2x}{dt^2} + \frac{g}{R} x = 0.$$

No further discussion is needed in order to conclude that this is simple harmonic motion with period $2\pi\sqrt{R/g}$. A sequence of easy approximate calculations gives

---

* The reason for this law of force will be explained in a later chapter, in connection with triple integrals in spherical coordinates.

$$2\pi \sqrt{\frac{R}{g}} \cong 6.3 \sqrt{\frac{4000 \cdot 5280 \text{ ft}}{32 \text{ ft/s}^2}} \cong 6.3 \sqrt{\frac{500 \cdot 5280}{4}} \text{ s}$$

$$\cong 6.3 \sqrt{\frac{500 \cdot 5280}{4 \cdot 3600}} \text{ min} \cong 6.3 \sqrt{200} \text{ min}$$

$$\cong \tfrac{19}{3} \cdot 14 \text{ min} \cong 89 \text{ min.}$$

The period is of course the total time required for a round trip through the tunnel to the other side of the earth and back again. A one-way trip requires only about 45 minutes, and the journey to the center of the earth only about 22 minutes.

Example 3 A pendulum consists of a bob (a weight) suspended at the end of a light string and allowed to swing back and forth under the action of gravity. As usual, we idealize the situation and consider a particle of mass m at the end of a weightless string of length L (Fig. 9.25). Find the period of this pendulum under the assumption that its oscillations are small.

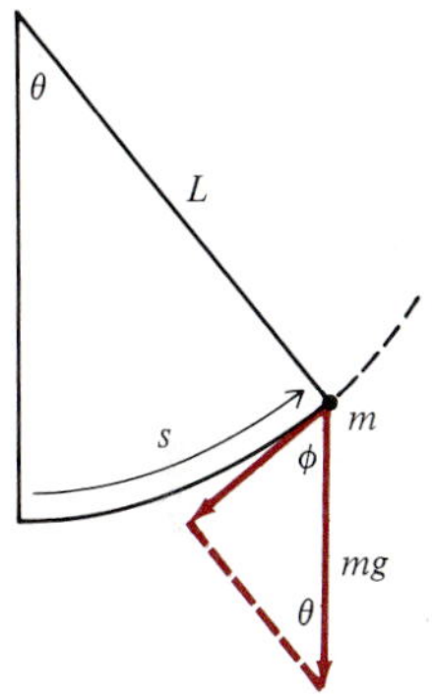

Solution The downward force of gravity on the bob is mg, and this has a component $mg \cos \phi = mg \sin \theta$ tangent to the path. Since $s = L\theta$, the tangential acceleration of the bob is

$$\frac{d^2s}{dt^2} = \frac{d^2(L\theta)}{dt^2} = L \frac{d^2\theta}{dt^2},$$

and Newton's second law applied to the motion of the bob along its circular path is

Figure 9.25

$$mL \frac{d^2\theta}{dt^2} = -mg \sin \theta \qquad \text{or} \qquad \frac{d^2\theta}{dt^2} + \frac{g}{L} \sin \theta = 0. \tag{7}$$

The presence of $\sin \theta$ makes this differential equation impossible to solve, and the motion is not simple harmonic. However, for small oscillations we recall that $\sin \theta$ is approximately equal to θ, so (7) becomes (approximately)

$$\frac{d^2\theta}{dt^2} + \frac{g}{L} \theta = 0.$$

This equation tells us that the angular motion is approximately simple harmonic with period $2\pi\sqrt{L/g}$. When these ideas are analyzed in more detail, it turns out that the period of this oscillation actually depends on the amplitude of the motion, and this is the source of the so-called "circular error" in pendulum clocks.

Remark 1 We return to the matter of proving that (3) is indeed the general solution of (2). By Problem 19 in Section 9.2 we know that every nontrivial solution of (2) has the form

$$x = c_1 \sin at + c_2 \cos at, \tag{8}$$

where the constants c_1 and c_2 are not both zero. To write (8) in the form (3), we begin by setting $A = \sqrt{c_1{}^2 + c_2{}^2}$. Then the point $(c_1/A, c_2/A)$ is a point on the unit circle, and therefore there is an angle b such that

$$\cos b = \frac{c_1}{A} \quad \text{and} \quad \sin b = \frac{c_2}{A}.$$

These equations now enable us to write (8) as

$$x = A(\sin at \cos b + \cos at \sin b)$$
$$= A \sin (at + b),$$

by the addition formula for the sine.

Remark 2 It is also possible to obtain (3) directly from (2), as follows. If we write

$$\frac{d^2x}{dt^2} = \frac{dv}{dt} = \frac{dv}{dx}\frac{dx}{dt} = v\frac{dv}{dx}, \tag{9}$$

then (2) becomes

$$v\frac{dv}{dx} + a^2x = 0 \quad \text{or} \quad v\, dv + a^2x\, dx = 0,$$

and by integrating we get

$$v^2 + a^2x^2 = \text{a constant} \quad \text{or} \quad v^2 + a^2x^2 = a^2A^2,$$

where A is the positive value of x at which $v = 0$. This yields

$$\frac{dx}{dt} = v = \pm a\sqrt{A^2 - x^2} \quad \text{or} \quad \frac{dx}{\sqrt{A^2 - x^2}} = \pm a\, dt,$$

where the choice of sign here depends on whether the velocity v is positive or negative at the moment. We suppose for definiteness that $v > 0$ and integrate again to obtain

$$\sin^{-1}\frac{x}{A} = at + b \quad \text{or} \quad \frac{x}{A} = \sin (at + b),$$

so

$$x = A \sin (at + b),$$

which is (3).

PROBLEMS

1 In each of the following motions calculate the amplitude and period by rewriting in the form $x = A \sin (at + b)$.
   (a) $x = 5 \sin t - 5 \cos t$;
   (b) $x = \sqrt{3} \cos 3t - \sin 3t$;
   (c) $x = \sin t + \cos t$;
   (d) $x = 2\sqrt{3} \sin 2t - 2 \cos 2t$.

2 In any simple harmonic motion of the form (3), show that the velocity v is related to the position x by the equation

$$v^2 = a^2(A^2 - x^2).$$

Deduce that the speed is greatest when the body passes through its equilibrium position, and is zero at the ends of the interval, where the body reverses the direction of its motion.

3 In Example 1, suppose the spring is stretched 3 in by a force of 6 lb. If the cart weighs 12 lb, and if it is pulled out 4 in from its equilibrium position and struck a sudden blow sending it back toward its equilibrium

position at a velocity of 3 ft/s, find the amplitude and period of the resulting simple harmonic motion. Hint: Recall that mass is weight divided by g.

4 A body in simple harmonic motion passes through its equilibrium position at $t = 0, 1, 2, \ldots$. Find a position function of the form (3) if $v = dx/dt = -3$ when $t = 0$.

5 Suppose that a straight tunnel is bored through the earth between any two points on the surface. If tracks are laid, then — neglecting friction — a train placed in the tunnel at one end will roll through the earth under its own weight, stop at the other end, and return. Show that the time required for a complete round trip is the same for all such tunnels, and estimate its value.

*6 A spherical buoy of radius r floats half submerged in water. If it is depressed slightly, Archimedes' principle tells us that a restoring force equal to the weight of the displaced water presses it upward; and if it is released, it will bob up and down. Show that if the friction of the water is negligible, then the motion will be simple harmonic, and find its period.

7 People who manufacture grandfather clocks have a professional interest in pendulums that take 1 second for each swing and thus have a period of 2 seconds. Estimate the length of such a pendulum.

There are certain simple combinations of exponential functions that occur occasionally in applications and to which the name *hyperbolic functions* has been given. The hyperbolic sine and cosine are defined by

$$\sinh x = \tfrac{1}{2}(e^x - e^{-x}) \qquad \text{and} \qquad \cosh x = \tfrac{1}{2}(e^x + e^{-x}).$$

9.7
THE HYPERBOLIC FUNCTIONS

There are also a hyperbolic tangent, cotangent, secant, and cosecant.

These functions satisfy many identities that are very similar to corresponding identities satisfied by the trigonometric functions; for instance,

$$\cosh^2 x - \sinh^2 x = 1.$$

Also, their differentiation and integration properties are quite similar to those of the trigonometric functions, as we see from the formulas

$$\frac{d}{dx} \sinh x = \cosh x \qquad \text{and} \qquad \frac{d}{dx} \cosh x = \sinh x.$$

However, one of the most important properties of the trigonometric functions — that of periodicity — is not possessed by any of the hyperbolic functions.

We mention these functions because students should at least know that they exist, and we mention them here because of their analogies with the trigonometric functions. However, we shall make no use of them in any of our future work, so we now drop the subject without any further ado.

ADDITIONAL PROBLEMS FOR CHAPTER 9

SECTION 9.2

In each of the following problems (1 to 18) find the derivative dy/dx of the given function.

1 $y = \sin(1 - 9x)$.
2 $y = 7\cos(7x - 13)$.
3 $y = \cos^2 x$.
4 $y = \cos x^2$.
5 $y = \cos^2 5x$.
6 $y = 5\sin(1 - 18x)$.
7 $y = \cos^2 3x - \sin^2 3x$.
8 $y = \cos^2 9x + \sin^2 9x$.
9 $y = x^2 \cos x$.
10 $y = \dfrac{\sin x}{x}$.
11 $y = x\sin x + \cos x$.
12 $y = \sqrt{1 + \sin 2x}$.
13 $y = \cos(\cos x)$.
14 $y = e^{\sin^2 x}$.
15 $y = \cos(\sin x)$.
16 $y = \ln(x\sin x)$.
17 $y = \sin(e^{\ln x})$.
18 $y = \ln[\sin(\ln x)]$.

19 Consider the differential equation

$$\frac{d^2y}{dx^2} + a^2y = 0,$$

where a is a positive constant. Use the following steps to prove that every solution of this equation has the form

$$y = c_1 \sin ax + c_2 \cos ax$$

for a suitable choice of the constants c_1 and c_2.
(a) If $y = g(x)$ and $y = h(x)$ are solutions, show that every linear combination $y = c_1\,g(x) + c_2\,h(x)$ is also a solution.
(b) If $y = f(x)$ is a solution, show that

$$a^2[f(x)]^2 + [f'(x)]^2 = a \text{ constant.}$$

Deduce that if $y = f(x)$ is a solution such that $f(0) = f'(0) = 0$, then $f(x) = 0$ for all x.
(c) If $y = f(x)$ is any solution, show that

$$f(x) = c_1 \sin ax + c_2 \cos ax$$

for a suitable choice of the constants c_1 and c_2. Hint: Apply part (b) to

$$f(x) - \frac{1}{a}f'(0) \sin ax - f(0) \cos ax.$$

20 Use Problem 18(b) in Section 9.1 to give another proof of the formula $(d/dx) \sin x = \cos x$.

21 Give another proof of the limit (4) in Section 9.2 by the following steps: If θ is a small positive angle ($0 < \theta < \pi/2$) in the unit circle shown in Fig. 9.26, then
(a) area $\triangle OPQ <$ area sector $OPQ <$ area $\triangle OQR$;
(b) $\frac{1}{2} \sin \theta < \frac{1}{2}\theta < \frac{1}{2} \tan \theta$;

(c) $1 < \dfrac{\theta}{\sin \theta} < \dfrac{1}{\cos \theta}$;   (d) $1 > \dfrac{\sin \theta}{\theta} > \cos \theta$.

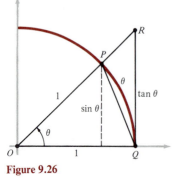

Figure 9.26

*22 Figure 9.27 shows the familiar mechanism of a piston (which moves back and forth in a cylinder) at-

tached at a point P to a connecting rod of length b which in turn is attached to a point Q on a crankshaft

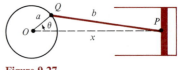

Figure 9.27

that rotates in a circle of radius a with center at O.
(a) Find dx/dt, the velocity of the piston, in terms of $d\theta/dt$, the angular velocity of the crankshaft. Hint: Use the law of cosines.
(b) If the angular velocity of the crankshaft is denoted by the customary symbol ω, show that the speed of the piston is $\omega \cdot OR$, where R is the point in which the line PQ intersects the line through O perpendicular to OP.

*23 A given fixed circle has radius a. A second circle has its center on the given circle, and the arc of the second circle that lies inside the given circle has length s. Show that s has its largest value when a suitable angle θ satisfies the equation $\cot \theta = \theta$.

24 A heavy block of weight W is to be dragged along a flat table by a force F whose line of action is inclined at an angle θ to the line of motion, as shown in Fig. 9.28. The motion is resisted by a frictional force μN

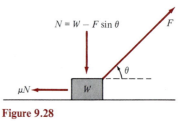

Figure 9.28

which is proportional to the normal force $N = W - F \sin \theta$ with which the block presses perpendicularly against the surface of the table (μ is a constant called the coefficient of friction). The block moves when the forward component of F equals the frictional resistance, i.e., when $F \cos \theta = \mu(W - F \sin \theta)$. Find the direction and magnitude of the smallest force F that will move the block.

In each of the following problems (25 to 36) find the value of the indicated limit.

25 $\displaystyle\lim_{x \to 0} \frac{\tan^3 x}{x^2}$.   **26** $\displaystyle\lim_{x \to 0} \frac{\sin x}{2x}$.

27 $\displaystyle\lim_{x\to\pi}\frac{\sin x}{\pi - x}.$

28 $\displaystyle\lim_{x\to 0}\frac{\sin^2 x}{x}.$

29 $\displaystyle\lim_{x\to 0}\frac{x + \tan x}{\sin x}.$

30 $\displaystyle\lim_{x\to 0}\frac{\tan 3x}{4x}.$

31 $\displaystyle\lim_{x\to 0}\frac{2x}{\sin 3x}.$

32 $\displaystyle\lim_{x\to 0} x\cot 3x.$

33 $\displaystyle\lim_{x\to 0}\frac{\sin 2x}{3x^2 + x}.$

34 $\displaystyle\lim_{x\to 0} x\csc^2\sqrt{3x}.$

35 $\displaystyle\lim_{x\to 2}\frac{\cos \pi/x}{x - 2}.$

36 $\displaystyle\lim_{x\to\pi}\frac{\sin 2x}{\pi - x}.$

SECTION 9.3

Evaluate the indefinite integrals in Problems 37 to 54.

37 $\int \cos 3x\, dx.$

38 $\int \sin (7x + 1)\, dx.$

39 $\int \cos (1 - \tfrac{1}{2}x)\, dx.$

40 $\int \cos^2 7x \sin 7x\, dx.$

41 $\int \sin^5 3x \cos 3x\, dx.$

42 $\int \cos^2 \tfrac{2}{3}x \sin \tfrac{2}{3}x\, dx.$

43 $\int (2 - \cos^2 3x) \sin 3x\, dx.$

44 $\int 3 \sin x \sin 2x\, dx.$

45 $\int x^2 \cos x^3\, dx.$

46 $\int \sqrt{x} \sin x^{3/2}\, dx.$

47 $\int \sin (\cos 2x) \sin 2x\, dx.$

48 $\int \sqrt{\cos 2x} \sin 2x\, dx.$

49 $\displaystyle\int \frac{\cos 4x\, dx}{\sin^2 4x}.$

50 $\displaystyle\int \frac{\sin x\, dx}{\cos^5 x}.$

51 $\displaystyle\int \frac{\sin x\, dx}{(3 + 2 \cos x)^2}.$

52 $\int \sqrt{1 + \sin 2x} \cos 2x\, dx.$

53 $\displaystyle\int \frac{\cos 5x\, dx}{\sqrt{7 - \sin 5x}}.$

54 $\int (1 + 4 \sin 8x)^7 \cos 8x\, dx.$

Evaluate the definite integrals in Problems 55 to 58.

55 $\displaystyle\int_0^{\pi/14} \cos 7x\, dx.$

56 $\displaystyle\int_0^{\pi/18} \sin 6x\, dx.$

57 $\displaystyle\int_0^{\pi/6} \frac{\sin 2x\, dx}{\cos^2 2x}.$

58 $\displaystyle\int_0^{\sqrt[5]{\pi}} 10\, x^4 \sin x^5\, dx.$

59 Find the area bounded by $y = \sin x$ and $y = \cos x$

between the first two positive values of x at which these curves intersect.

60 Find the area bounded by $y = 1 - \cos 2x$ and $y = \cos x - 1$ between $x = 0$ and $x = 2\pi$.

61 Find the area bounded by $y = 4 - 3 \sin 2x$ and $y = 2 \cos 5x - 3$ between $x = 0$ and $x = 3\pi$.

62 Show that if m and n are positive integers, then

$$\int_0^{2\pi} \sin mx \sin nx\, dx = \begin{cases} 0 & \text{if } m \neq n \\ \pi & \text{if } m = n, \end{cases}$$

$$\int_0^{2\pi} \cos mx \cos nx\, dx = \begin{cases} 0 & \text{if } m \neq n \\ \pi & \text{if } m = n, \end{cases}$$

$$\int_0^{2\pi} \sin mx \cos nx\, dx = 0.$$

Hint: See Problem 17 in Section 9.1. (These facts are very important in the theory of Fourier series, which is one of the most useful parts of advanced mathematics from the point of view of applications to science.)

*63 In this problem we ask students to establish the formula

$$\int_a^b \sin x\, dx = \cos a - \cos b \qquad (*)$$

directly from the limit definition of the integral, without making any use of the Fundamental Theorem of Calculus.

(a) Show that

$$\sin x + \sin 2x + \cdots + \sin nx$$
$$= \frac{\cos \tfrac{1}{2} x - \cos (n + \tfrac{1}{2}) x}{2 \sin \tfrac{1}{2}x}.$$

Hint: Write down the identity $2 \sin \theta \sin \phi = \cos (\theta - \phi) - \cos (\theta + \phi)$ for the n cases in which the pair (θ, ϕ) is taken to be $(x, \tfrac{1}{2}x)$, $(2x, \tfrac{1}{2}x)$, . . . , $(nx, \tfrac{1}{2}x)$, and add.

(b) For $b > 0$, the limit definition of the integral gives

$$\int_0^b \sin x\, dx = \lim_{n\to\infty} \sum_{k=1}^{n} \left(\sin \frac{kb}{n} \right) \cdot \frac{b}{n}$$

$$= \lim_{n\to\infty} \frac{b}{n} \sum_{k=1}^{n} \sin \frac{kb}{n}.$$

Use part (a) with $x = b/n$ to show that the value of this limit is $1 - \cos b$.

(c) Use simple area arguments to show that the result in part (b) is also valid for the cases $b = 0$ and $b < 0$.

(d) Use parts (b) and (c) to establish (*).

*64   Establish the formula

$$\int_a^b \cos x \, dx = \sin b - \sin a \qquad \text{(**)}$$

by a line of reasoning similar to that in Problem 63.

65   Give another proof of formula (**) in Problem 64 by using the following reasoning: If the graph of $y = \cos x$ is moved a distance $\pi/2$ to the right, it is translated into the graph of $y = \sin x$; the integral in (**), which represents the area between the curve $y = \cos x$ and the x-axis from $x = a$ to $x = b$, can therefore be written as another integral representing the area between the curve $y = \sin x$ and the x-axis from $x = a + \pi/2$ to $x = b + \pi/2$.

SECTION 9.4

In each of the following problems (66 to 79), calculate dy/dx.

66 $y = \cot (2 - 5x)$.

67 $y = 4 \tan 3x$.

68 $y = \frac{1}{4} \sec^4 x$.

69 $y = \sqrt{\cot 2x}$.

70 $y = \csc (1 - 2x)$.

71 $y = \sec^4 x - \tan^4 x$.

72 $y = 2x + \tan 2x$.

73 $y = \cot^2 5x$.

74 $y = \sec^3 x$.

75 $y = x \tan \dfrac{1}{x}$.

76 $y = \cot (\ln x)$.

77 $y = \sqrt{\sec \sqrt{x}}$.

78 $y = \csc^3 x + \csc x^3$.

79 $y = \tan (\tan x)$.

Evaluate the integral in each of the following problems (80 to 87).

80 $\displaystyle\int \frac{dx}{\cos^2 5x}$.

81 $\int \csc \frac{1}{3}x \cot \frac{1}{3}x \, dx$.

82 $\displaystyle\int_{\pi/6}^{\pi/4} \csc^2 x \cot x \, dx$.

83 $\int \csc^2 3x \, dx$.

84 $\int (2 + 5 \tan x)^7 \sec^2 x \, dx$.

85 $\int \csc^4 x \cot x \, dx$.

86 $\int \sqrt{\cot x} \csc^2 x \, dx$.

87 $\int \cot^3 x \csc^2 x \, dx$.

88   The region under the curve $y = \sec x$ between $x = 0$ and $x = \pi/4$ is revolved about the x-axis. Find the volume of the solid of revolution generated in this way.

89   Solve Problem 88 for the curve $y = \sec^2 x$.

90   Sketch the graph of the function $y = \frac{1}{3} \tan 2x + \cot 2x$ on the interval $0 < x < \pi/4$ and find its minimum value.

91   A racing car is moving around a circular track at a constant speed of 100 km/h. There is a bright light at the center of the track and a straight fence tangent to the track at a point T. How fast is the shadow of the car moving along the fence when the car is $\frac{1}{8}$ lap beyond T?

*92   In Problem 18 of Section 4.4 students were asked to show that the volume of the smallest cone that can be circumscribed about a given sphere of radius a is exactly twice the volume of the sphere. Solve this problem by trigonometric methods, by taking the generating angle of a circumscribed cone (half the vertex angle) as the independent variable.

SECTION 9.5

93   Evaluate each of the following:
   (a) $\tan^{-1} (-\sqrt{3})$;
   (b) $\sin^{-1} \frac{1}{2}\sqrt{3}$;
   (c) $4 \sin^{-1} (-\frac{1}{2}\sqrt{2})$;
   (d) $\sin (\sin^{-1} 0.7)$;
   (e) $\sin^{-1} (\sin 0.7)$;
   (f) $\tan^{-1} (\tan [-1])$;
   (g) $\sin^{-1} (\cos \pi/6)$.

94   If the base b and area A of a triangle are fixed, use geometry alone to find the base angles if the angle opposite the base has its largest value.

Find dy/dx in each of the following problems (95 to 103).

95 $y = \sin^{-1} \frac{1}{3}x$.

96 $y = \frac{1}{2} \tan^{-1} \frac{1}{2}x$.

97 $y = \frac{1}{5} \tan^{-1} x^5$.

98 $y = \sqrt{x} - \tan^{-1} \sqrt{x}$.

99 $y = \tan^{-1} \sqrt{x^2 - 1}$.

100 $y = -\sin^{-1} \dfrac{1}{x}$.

101 $y = \tan^{-1} x + \ln \sqrt{1 + x^2}$.

102 $y = a \sin^{-1} \dfrac{x}{a} + \sqrt{a^2 - x^2}$.

103 $y = \sqrt{x^2 - 1} - \tan^{-1} \sqrt{x^2 - 1}$.

Evaluate the integrals in the following problems (104 to 112).

104 $\displaystyle\int_0^{\sqrt{3}} \frac{dx}{1 + x^2}$.

105 $\displaystyle\int_{-1/2}^{(1/2)\sqrt{3}} \frac{dx}{\sqrt{1 - x^2}}$.

106 $\displaystyle\int \frac{dx}{\sqrt{1 - 16x^2}}$.

107 $\displaystyle\int \frac{dx}{1 + 5x^2}$.

108 $\displaystyle\int_{1/\sqrt{3}}^1 \frac{dx}{x\sqrt{4x^2 - 1}}$.

109 $\displaystyle\int \frac{dx}{\sqrt{25 - 4x^2}}$.

110 $\displaystyle\int \frac{dx}{49 + 36x^2}$.

111 $\displaystyle\int \frac{x^3 \, dx}{1 + x^8}$.

112 $\displaystyle\int \frac{15x^4 \, dx}{\sqrt{1 - x^{10}}}$.

113   A billboard is perpendicular to a straight road and its nearest edge is 18 ft from the road. The billboard is 54 ft wide. As a motorist approaches the billboard along the road, at what point does he see the bill-board in the widest angle?

114 An airplane at an altitude of 7 mi and a speed of 500 mi/h is flying directly away from an observer on the ground. What is the rate of change of the angle of elevation when the airplane is over a point 4 mi away from the observer?

115 A woman is walking along a sidewalk at the rate of 6 ft/s. A police car spotlight 30 ft from the sidewalk follows her as she walks. At what rate is the spotlight turning when the woman is 40 ft past the point on the sidewalk nearest the light?

SECTION 9.6

116 With reference to Example 1, recall the definitions of kinetic and potential energy given in Section 7.8.

(a) Show that the potential energy V of the cart is $\frac{1}{2}kx^2$, where it is understood that $V = 0$ when $x = 0$.

(b) Show directly from Newton's second law of motion

$$m\frac{d^2x}{dt^2} = -kx$$

that the sum of the kinetic and potential energies

of the cart is constant. Hint: Use equation (9) in Section 9.6.

(c) Express the total energy E of the cart in terms of its initial position x_0 and initial velocity v_0.

(d) Express the total energy E of the cart in terms of the amplitude A and frequency f of the vibration.

117 A block of wood 6 in on an edge and weighing 4 lb floats upright in water. If the block is depressed slightly and released, find its period of oscillation assuming that the friction of the water is negligible. Hint: Use $w = 62.5$ lb/ft^3 for the density of water.

118 A body in simple harmonic motion has amplitude A and period T. Find its maximum velocity.

119 Find the amplitude and frequency of the simple harmonic motion $x = 3 \sin 2t + 4 \cos 2t$.

120 If the period of a simple harmonic motion is $2\pi/3$, find a position function of the form (3) that satisfies the conditions $x = 1$ and $v = dx/dt = 3$ when $t = 0$.

*121 Let the pendulum in Example 3 be pulled to one side through an angle α and released. Use the principle of conservation of energy to show that the period T of oscillation is given by the formula

$$T = 4\sqrt{\frac{L}{2g}} \int_0^\alpha \frac{d\theta}{\sqrt{\cos\theta - \cos\alpha}}.$$

10

METHODS OF INTEGRATION

10.1

INTRODUCTION. THE BASIC FORMULAS

If we start with the constants and the seven familiar functions x, e^x, $\ln x$, $\sin x$, $\cos x$, $\sin^{-1} x$, and $\tan^{-1} x$, and go on to build all possible finite combinations of these by applying the algebraic operations and the process of forming a function of a function, then we generate the class of *elementary functions*. Thus,

$$\ln \left[\frac{\tan^{-1} (x^2 + 35x^3)}{e^x + \sin \sqrt{x^3 + 1}} \right]$$

is an elementary function. These functions are often said to have *closed form*, because they can be written down in explicit formulas involving only a finite number of familiar functions.

It is clear that the problem of calculating the derivative of an elementary function can always be solved by a systematic application of the rules developed in the preceding chapters, and this derivative is always an elementary function. However, the inverse problem of integration — which in general is much more important — is very different and has no such clear-cut solution.

As we know, the problem of calculating the indefinite integral of a function $f(x)$,

$$\int f(x) \, dx = F(x), \tag{1}$$

is equivalent to finding a function $F(x)$ such that

$$\frac{d}{dx} F(x) = f(x). \tag{2}$$

It is true that we have succeeded in integrating a good many elementary functions by inverting differentiation formulas. But this doesn't carry us very far, because it amounts to little more than calculating the integral (1) by knowing the answer (2) in advance.

The fact of the matter is this: There does not exist any systematic procedure that can always be applied to any elementary function and leads step by step to a guaranteed answer in closed form. Indeed, there may not even *be* an

276

answer. For example, the function $f(x) = e^{-x^2}$ looks simple enough, but its integral

$$\int e^{-x^2} \, dx \tag{3}$$

cannot be calculated within the class of elementary functions. This assertion is more than merely a report on the present inability of mathematicians to integrate (3); it is a statement of a deep theorem, to the effect that no elementary function exists whose derivative is e^{-x^2}.* We will return to these matters in Section 10.8.

If all this sounds discouraging, it shouldn't be. There is much more that can be done in the way of integration than we have suggested so far, and it is very important for students to acquire a certain amount of technical skill in carrying out integrations whenever they *are* possible. The fact that integration must be considered as more of an art than a systematic process really makes it more interesting than differentiation. It is more like solving puzzles, because there is less certainty and more scope for individual ingenuity. Many students find this an agreeable change from the cut-and-dried routines that make some parts of mathematics rather dull.

Since integration is differentiation read backwards, our starting point must be a short table of standard types of integrals obtained by inverting differentiation formulas as we have done in the previous chapters. Much more extensive tables than the one given below are available in libraries, and with the aid of these tables most of the problems in this chapter can be solved by merely looking them up. However, students should realize that if they follow such a course they will defeat the intended purpose of developing their own skills. For this reason we make no use of integral tables beyond the short list given below. Instead, we urge students to concentrate their efforts on gaining a clear understanding of the various methods of integration and learning how to apply them.

In addition to the method of substitution, which is already familiar to the reader, there are three principal methods of integration to be studied in this chapter: reduction to trigonometric integrals; decomposition into partial fractions; and integration by parts. These methods enable us to transform a given integral in many ways. The object of these transformations is always to break up the given integral into a sum of simpler parts that can be integrated at once by means of familiar formulas. Students should therefore be certain that they have thoroughly memorized all the following basic formulas. These formulas should be so well learned that when one of them is needed it pops into the mind almost involuntarily, like the name of a friend.

---

* Let there be no misunderstanding. The indefinite integral (3) *does* exist, because the function $F(x)$ defined by

$$F(x) = \int_0^x e^{-t^2} \, dt$$

is a perfectly respectable function with the property that

$$\frac{d}{dx} F(x) = e^{-x^2}.$$

[See equations (12) and (13) in Section 6.7.] The difficulty is that it can be proved that there is no way of expressing $F(x)$ as an elementary function.

$$1 \quad \int u^n \, du = \frac{u^{n+1}}{n+1} + c \qquad (n \neq -1)$$

$$2 \quad \int \frac{du}{u} = \ln u + c$$

$$3 \quad \int e^u \, du = e^u + c$$

$$4 \quad \int \cos u \, du = \sin u + c$$

$$5 \quad \int \sin u \, du = -\cos u + c$$

$$6 \quad \int \sec^2 u \, du = \tan u + c$$

$$7 \quad \int \csc^2 u \, du = -\cot u + c$$

$$8 \quad \int \sec u \tan u \, du = \sec u + c$$

$$9 \quad \int \csc u \cot u \, du = -\csc u + c$$

$$10 \quad \int \frac{du}{\sqrt{a^2 - u^2}} = \sin^{-1} \frac{u}{a} + c$$

$$11 \quad \int \frac{du}{a^2 + u^2} = \frac{1}{a} \tan^{-1} \frac{u}{a} + c$$

$$12 \quad \int \tan u \, du = -\ln (\cos u) + c$$

$$13 \quad \int \cot u \, du = \ln (\sin u) + c$$

$$14 \quad \int \sec u \, du = \ln (\sec u + \tan u) + c$$

$$15 \quad \int \csc u \, du = -\ln (\csc u + \cot u) + c$$

The last four formulas are new, and complete our list of the integrals of the six trigonometric functions. Formulas 12 and 13 can be found by a straightforward process:

$$\int \tan u \, du = \int \frac{\sin u \, du}{\cos u} = -\int \frac{d(\cos u)}{\cos u} = -\ln (\cos u) + c$$

and

$$\int \cot u \, du = \int \frac{\cos u \, du}{\sin u} = \int \frac{d(\sin u)}{\sin u} = \ln (\sin u) + c.$$

Many people find that the easiest way to remember these formulas is to think of this process. Formula 14 can be found by an ingenious trick: if we multiply the integrand by $1 = (\sec u + \tan u)/(\sec u + \tan u)$, then we obtain

$$\int \sec u \, du = \int \frac{(\sec u + \tan u) \sec u \, du}{\sec u + \tan u} = \int \frac{(\sec^2 u + \sec u \tan u) \, du}{\sec u + \tan u}$$

$$= \int \frac{d(\sec u + \tan u)}{\sec u + \tan u} = \ln (\sec u + \tan u) + c.$$

A similar trick yields formula 15.

We repeat: These 15 formulas constitute the foundation on which we build throughout the chapter, and they must be at our fingertips.

In the method of substitution we introduce the auxiliary variable u as a new symbol for part of the integrand in the hope that its differential du will account for some other part and thereby reduce the complete integral to an easily recognizable form. Success in the use of this method depends on choosing a fruitful substitution, and this in turn depends on the ability to see at a glance that part of the integrand is the derivative of some other part.

We give several examples to help students review the procedure and make certain that they fully understand it.

10.2
THE METHOD OF SUBSTITUTION

Example 1   Find $\int x \, e^{-x^2} \, dx$.

Solution   If we put $u = -x^2$, then $du = -2x \, dx$, $x \, dx = -\frac{1}{2} \, du$, and

$$\int xe^{-x^2} \, dx = -\frac{1}{2} \int e^u \, du = -\frac{1}{2}e^u = -\frac{1}{2}e^{-x^2} + c.$$

It will be noticed that we insert the constant of integration only in the last step. Strictly speaking, this is incorrect; but we willingly commit this minor error in order to avoid cluttering up the previous steps with repeated c's. We also point out that this integral is easy to calculate even though the similar integral $\int e^{-x^2} \, dx$ is impossible. The reason for this is clearly the presence of the factor x, which is essentially (that is, up to a constant factor) the derivative of the exponent $-x^2$.

Example 2   Find

$$\int \frac{\cos x \, dx}{\sqrt{1 + \sin x}}.$$

Solution   Here we notice that $\cos x \, dx$ is the differential of $\sin x$, and also of $1 + \sin x$. Thus, if we put $u = 1 + \sin x$, then $du = \cos x \, dx$ and

$$\int \frac{\cos x \, dx}{\sqrt{1 + \sin x}} = \int \frac{du}{\sqrt{u}} = \int u^{-1/2} \, du$$

$$= \frac{u^{1/2}}{\frac{1}{2}} = 2\sqrt{u} = 2\sqrt{1 + \sin x} + c.$$

Example 3 Find

$$\int \frac{dx}{x \ln x}.$$

Solution The fact that dx/x is the differential of $\ln x$ suggests the substitution $u = \ln x$, so $du = dx/x$ and

$$\int \frac{dx}{x \ln x} = \int \frac{du}{u} = \ln u = \ln (\ln x) + c.$$

Example 4 Find

$$\int \frac{dx}{\sqrt{9 - 4x^2}}.$$

Solution We put $u = 2x$, so that $du = 2dx$ and

$$\int \frac{dx}{\sqrt{9 - 4x^2}} = \frac{1}{2} \int \frac{du}{\sqrt{9 - u^2}} = \frac{1}{2} \sin^{-1} \frac{u}{3} = \frac{1}{2} \sin^{-1} \frac{2x}{3} + c.$$

Example 5 Find

$$\int \frac{x\,dx}{\sqrt{9 - 4x^2}}.$$

Solution Here the fact that the x in the numerator is essentially the derivative of the expression $9 - 4x^2$ inside the radical suggests the substitution $u = 9 - 4x^2$. Then $du = -8x\,dx$ and

$$\int \frac{x\,dx}{\sqrt{9 - 4x^2}} = -\frac{1}{8} \int \frac{du}{\sqrt{u}} = -\frac{1}{8} \int u^{-1/2}\,du$$

$$= -\frac{1}{8} \frac{u^{1/2}}{\frac{1}{2}} = -\frac{1}{4} \sqrt{u} = -\frac{1}{4} \sqrt{9 - 4x^2} + c.$$

In any particular integration problem the choice of the substitution is a matter of trial and error guided by experience. If our first substitution doesn't work, we should feel no hesitation about discarding it and trying another. Example 5 is similar in appearance to Example 4 and it might be thought that the same substitution will work again, but in fact—as we have seen—it requires an entirely different substitution.

We can establish the validity of the method of substitution as follows, by showing that it is really the chain rule for derivatives read backwards. The essence of the method is this. We start with a complicated integral of the form

$$\int f[g(x)]g'(x)\,dx. \tag{1}$$

If we put $u = g(x)$, then $du = g'(x)\,dx$ and the integral takes the new form

$$\int f(u)\,du.$$

If we can integrate this, so that

$$\int f(u)\,du = F(u) + c,\tag{2}$$

then since $u = g(x)$ we ought to be able to integrate (1) by writing

$$\int f[g(x)]g'(x)\,dx = F[g(x)] + c.\tag{3}$$

All that is needed to justify our procedure is to notice that (3) is a correct result, because

$$\frac{d}{dx}F[g(x)] = F'[g(x)]g'(x) = f[g(x)]g'(x)$$

by the chain rule.

   The method of substitution applies to definite integrals as well as indefinite integrals. The crucial requirement is that the limits of integration must be suitably changed when the substitution is made. This can be expressed as follows:

$$\int_a^b f[g(x)]g'(x)\,dx = \int_c^d f(u)\,du,$$

where $c = g(a)$ and $d = g(b)$. The proof uses (2) and (3) and two applications of the Fundamental Theorem of Calculus,

$$\int_a^b f[g(x)]g'(x)\,dx = F[g(b)] - F[g(a)]$$

$$= F(d) - F(c) = \int_c^d f(u)\,du.$$

Thus, once the original integral is changed into a simpler integral in the variable u, the numerical evaluation can be carried out entirely in terms of u, provided the limits of integration are also correctly changed.

Example 6   Compute

$$\int_0^{\pi/3} \frac{\sin x\,dx}{\cos^2 x}.$$

Solution   We put $u = \cos x$, so that $du = -\sin x\,dx$. Observe that $u = 1$ when $x = 0$ and $u = \frac{1}{2}$ when $x = \pi/3$. By changing both the variable of integration and the limits of integration we obtain

$$\int_0^{\pi/3} \frac{\sin x\,dx}{\cos^2 x} = \int_1^{1/2} \frac{-du}{u^2} = \frac{1}{u}\bigg]_1^{1/2} = 2 - 1 = 1.$$

This technique removes the necessity of returning to the original variable in order to make the final numerical evaluation.

PROBLEMS

Find the following integrals.

1 $\int \sqrt{3 - 2x}\, dx.$

2 $\int \dfrac{2x\, dx}{(4x^2 - 1)^2}.$

31 $\int \dfrac{e^x\, dx}{1 + e^x}.$

32 $\int \dfrac{\cos (\ln x)\, dx}{x}.$

3 $\int \dfrac{\ln x\, dx}{x[1 + (\ln x)^2]}.$

4 $\int \cos x\, e^{\sin x}\, dx.$

33 $\int \tan 3x\, dx.$

34 $\int \dfrac{\sec^2 x\, dx}{\sqrt{1 + \tan x}}.$

5 $\int \sin 2x\, dx.$

6 $\int \dfrac{x\, dx}{\sqrt{16 - x^4}}.$

35 $\int \dfrac{4x\, dx}{\sqrt{x^2 + 1}}.$

36 $\int \dfrac{e^{\sqrt{x}}\, dx}{\sqrt{x}}.$

7 $\int \cot (3x - 1)\, dx.$

8 $\int \sin x \cos x\, dx.$

37 $\int \dfrac{e^x\, dx}{1 + e^{2x}}.$

38 $\int \dfrac{\sin^{-1} x\, dx}{\sqrt{1 - x^2}}.$

9 $\int x\sqrt{x^2 + 1}\, dx.$

10 $\int \dfrac{dx}{x + 2}.$

39 $\int (e^x + 1)^6 e^x\, dx.$

40 $\int 6x^2 e^{-x^3}\, dx.$

11 $\int e^{5x}\, dx.$

12 $\int x \cos x^2\, dx.$

41 $\int \sec^2 5x\, dx.$

42 $\int \cot 4x\, dx.$

13 $\int \csc^2 (3x + 2)\, dx.$

14 $\int \dfrac{dx}{x^2 + 16}.$

43 $\int \csc 2x \cot 2x\, dx.$

44 $\int_2^3 \dfrac{2x\, dx}{x^2 - 3}.$

15 $\int_{-3}^1 \dfrac{dx}{\sqrt{3 - 2x}}.$

16 $\int (x^3 + 1)^2\, dx.$

Compute each of the following definite integrals by making a suitable substitution and changing the limits of integration.

17 $\int \dfrac{\sin x\, dx}{\sqrt{1 - \cos x}}.$

18 $\int \dfrac{(2x + 1)\, dx}{x^2 + x + 2}.$

19 $\int \dfrac{e^{\tan^{-1} x}}{1 + x^2}\, dx.$

20 $\int \dfrac{\sin \sqrt{x}}{\sqrt{x}}\, dx.$

45 $\int_1^2 \dfrac{(2x + 1)\, dx}{\sqrt{x^2 + x + 2}}.$

46 $\int_0^{\pi/4} \tan^2 x \sec^2 x\, dx.$

21 $\int \sec 5x \tan 5x\, dx.$

22 $\int \dfrac{dx}{x\sqrt{\ln x}}.$

47 $\int_1^e \dfrac{\sqrt{\ln x}\, dx}{x}.$

48 $\int_0^{\pi/3} \sec^3 x \tan x\, dx.$

23 $\int \dfrac{\ln x\, dx}{x}.$

24 $\int \dfrac{\sin x\, dx}{\cos^2 x}.$

49 Each of the following integrals is easy to compute for a particular value of n. Find this value and carry out the integration. For example, $\int x^n \sin x^2\, dx$ is easily computed for $n = 1$:

25 $\int_0^{\pi/2} \dfrac{\cos x\, dx}{1 + \sin x}.$

26 $\int \cos 3x\, dx.$

$$\int x \sin x^2\, dx = -\tfrac{1}{2} \cos x^2 + c.$$

27 $\int \dfrac{e^x\, dx}{\sqrt{1 - e^{2x}}}.$

28 $\int \dfrac{dx}{\cos 2x}.$

(a) $\int x^n e^{x^4}\, dx.$
(b) $\int x^n \cos x^3\, dx.$

29 $\int \sin^2 x \cos x\, dx.$

30 $\int_0^3 \tan^2 \tfrac{1}{3}x \sec^2 \tfrac{1}{3}x\, dx.$

(c) $\int x^n \ln x\, dx.$
(d) $\int x^n \sec^2 \sqrt{x}\, dx.$

In the next two sections we discuss several methods for reducing a given integral to one involving trigonometric functions. It will therefore be useful to increase our ability to calculate such trigonometric integrals.

A power of a trigonometric function multiplied by its differential is easy to integrate. Thus,

$$\int \sin^3 x \cos x \, dx = \int \sin^3 x \, d(\sin x) = \tfrac{1}{4} \sin^4 x + c$$

and

$$\int \tan^2 x \sec^2 x \, dx = \int \tan^2 x \, d(\tan x) = \tfrac{1}{3} \tan^3 x + c.$$

Other trigonometric integrals can often be reduced to problems of this type by using appropriate trigonometric identities.

We begin by considering integrals of the form

$$\int \sin^m x \cos^n x \, dx, \tag{1}$$

where one of the exponents is an odd positive integer. If n is odd, we factor out $\cos x \, dx$, which is $d(\sin x)$; and since an even power of $\cos x$ remains, we can use the identity $\cos^2 x = 1 - \sin^2 x$ to express the remaining part of the integrand entirely in terms of $\sin x$. And if m is odd, we factor out $\sin x \, dx$, which is $- d(\cos x)$, and use the identity $\sin^2 x = 1 - \cos^2 x$ in a similar way. The following two examples illustrate the procedure.

Example 1

$$\int \sin^2 x \cos^3 x \, dx = \int \sin^2 x \cos^2 x \cos x \, dx$$

$$= \int \sin^2 x (1 - \sin^2 x) \, d(\sin x)$$

$$= \int (\sin^2 x - \sin^4 x) \, d(\sin x)$$

$$= \tfrac{1}{3} \sin^3 x - \tfrac{1}{5} \sin^5 x + c.$$

Example 2

$$\int \sin^3 x \, dx = \int \sin^2 x \sin x \, dx$$

$$= - \int (1 - \cos^2 x) \, d(\cos x)$$

$$= - \cos x + \tfrac{1}{3} \cos^3 x + c.$$

If one of the exponents in (1) is an odd positive integer that is quite large, it may be necessary to use the binomial theorem, and in such a case an explicit use of the method of substitution may be desirable for the sake of clarity. For instance, every odd positive power of $\cos x$, whether large or small, has the form

$$\cos^{2n+1} x = \cos^{2n} x \cos x = (\cos^2 x)^n \cos x = (1 - \sin^2 x)^n \cos x,$$

where n is a nonnegative integer. If we put $u = \sin x$ and $du = \cos x\, dx$, then

$$\int \cos^{2n+1} x\, dx = \int (1 - \sin^2 x)^n \cos x\, dx$$

$$= \int (1 - u^2)^n\, du.$$

If necessary, the expression $(1 - u^2)^n$ can now be expanded by applying the binomial theorem, and the resulting polynomial in u is easy to integrate term by term.

If both exponents in (1) are nonnegative even integers, then it is necessary to change the form of the integrand by using the half-angle formulas

$$\cos^2 \theta = \tfrac{1}{2}(1 + \cos 2\theta) \qquad \text{and} \qquad \sin^2 \theta = \tfrac{1}{2}(1 - \cos 2\theta). \tag{2}$$

We hope students have thoroughly memorized these important formulas, but if they are forgotten they can easily be recovered by adding and subtracting the identities

$$\cos^2 \theta + \sin^2 \theta = 1,$$

$$\cos^2 \theta - \sin^2 \theta = \cos 2\theta.$$

The uses of (2) are shown in the following examples.

Example 3   The half-angle formula for the cosine enables us to write

$$\int \cos^2 x\, dx = \tfrac{1}{2} \int (1 + \cos 2x)\, dx = \tfrac{1}{2} \int dx + \tfrac{1}{2} \int \cos 2x\, dx$$

$$= \tfrac{1}{2}x + \tfrac{1}{4} \int \cos 2x\, d(2x) = \tfrac{1}{2}x + \tfrac{1}{4} \sin 2x + c.$$

If we wish to express this result in terms of the variable x (instead of $2x$), we use the double-angle formula $\sin 2x = 2 \sin x \cos x$ and write

$$\int \cos^2 x\, dx = \tfrac{1}{2}x + \tfrac{1}{2} \sin x \cos x + c.$$

Example 4   Two successive applications of the half-angle formula for the cosine give

$$\cos^4 x = (\cos^2 x)^2 = \tfrac{1}{4}(1 + \cos 2x)^2 = \tfrac{1}{4}(1 + 2 \cos 2x + \cos^2 2x)$$

$$= \tfrac{1}{4}[1 + 2 \cos 2x + \tfrac{1}{2}(1 + \cos 4x)]$$

$$= \tfrac{3}{8} + \tfrac{1}{2} \cos 2x + \tfrac{1}{8} \cos 4x,$$

so

$$\int \cos^4 x\, dx = \tfrac{3}{8}x + \tfrac{1}{4} \sin 2x + \tfrac{1}{32} \sin 4x + c.$$

As these examples show, the value of the half-angle formulas (2) for this work lies in the fact that they allow us to reduce the exponent by a factor of $\tfrac{1}{2}$ at the expense of multiplying the angle by 2, which is a considerable advantage purchased at very low cost.

Example 5 By using both of the half-angle formulas we get

$$\int \sin^2 x \cos^2 x \, dx = \int \frac{1 - \cos 2x}{2} \cdot \frac{1 + \cos 2x}{2} \, dx$$

$$= \tfrac{1}{4} \int (1 - \cos^2 2x) \, dx = \tfrac{1}{4} \int [1 - \tfrac{1}{2}(1 + \cos 4x)] \, dx$$

$$= \tfrac{1}{8} \int dx - \tfrac{1}{8} \int \cos 4x \, dx = \tfrac{1}{8}x - \tfrac{1}{32} \sin 4x + c.$$

We can also find this integral by combining the results of Examples 3 and 4:

$$\int \sin^2 x \cos^2 x \, dx = \int (1 - \cos^2 x) \cos^2 x \, dx$$

$$= \int \cos^2 x \, dx - \int \cos^4 x \, dx$$

$$= \tfrac{1}{2}x + \tfrac{1}{4} \sin 2x - \tfrac{3}{8}x - \tfrac{1}{4} \sin 2x - \tfrac{1}{32} \sin 4x$$

$$= \tfrac{1}{8}x - \tfrac{1}{32} \sin 4x + c.$$

We next consider integrals of the form

$$\int \tan^m x \sec^n x \, dx,$$

where n is an even positive integer or m is an odd positive integer. Our work is based on the fact that $d(\tan x) = \sec^2 x \, dx$ and $d(\sec x) = \sec x \tan x \, dx$, and we exploit the identity $\tan^2 x + 1 = \sec^2 x$. An example illustrating each case will be enough to show the general method.

Example 6

$$\int \tan^4 x \sec^6 x \, dx = \int \tan^4 x \sec^4 x \sec^2 x \, dx$$

$$= \int \tan^4 x \, (\tan^2 x + 1)^2 \, d(\tan x)$$

$$= \int \tan^4 x \, (\tan^4 x + 2 \tan^2 x + 1) \, d(\tan x)$$

$$= \int (\tan^8 x + 2 \tan^6 x + \tan^4 x) \, d(\tan x)$$

$$= \tfrac{1}{9} \tan^9 x + \tfrac{2}{7} \tan^7 x + \tfrac{1}{5} \tan^5 x + c.$$

Example 7

$$\int \tan^3 x \sec^5 x \, dx = \int \tan^2 x \sec^4 x \sec x \tan x \, dx$$

$$= \int (\sec^2 x - 1) \sec^4 x \, d(\sec x)$$

$$= \int (\sec^6 x - \sec^4 x) \, d(\sec x)$$

$$= \tfrac{1}{7} \sec^7 x - \tfrac{1}{5} \sec^5 x + c.$$

In essentially the same way we can handle integrals of the form

$$\int \cot^m x \csc^n x \, dx,$$

where n is an even positive integer or m is an odd positive integer. Our tools in these cases are the formulas $d(\cot x) = -\csc^2 x \, dx$ and $d(\csc x) = -\csc x \cot x \, dx$, and when necessary we use the identity $1 + \cot^2 x = \csc^2 x$.

Another approach to trigonometric integrals that is sometimes useful is to express each function occurring in the integral in terms of sines and cosines alone.

Example 8 We already know from our work with derivatives that

$$\int \sec x \tan x \, dx = \sec x + c.$$

However, this formula can also be obtained directly, by writing

$$\int \sec x \tan x \, dx = \int \frac{1}{\cos x} \frac{\sin x}{\cos x} \, dx = \int \frac{\sin x \, dx}{\cos^2 x}.$$

If we now put $u = \cos x$ and $du = -\sin x \, dx$, then we get

$$\int \sec x \tan x \, dx = \int \frac{\sin x \, dx}{\cos^2 x}$$

$$= \int \frac{-du}{u^2} = \frac{1}{u} = \frac{1}{\cos x} = \sec x + c.$$

PROBLEMS

Find each of the following integrals.

1 $\int \sin^2 x \, dx.$

2 $\int \sin^4 x \, dx.$

3 $\int \cos^6 x \, dx.$

4 $\int \cos^2 3x \, dx.$

5 $\int \sin^3 x \cos^2 x \, dx.$

6 $\int \sin^2 x \cos^5 x \, dx.$

7 $\displaystyle\int \cos^3 x \, dx.$

8 $\displaystyle\int_0^{\pi/2} \sin^3 x \cos^3 x \, dx.$

9 $\displaystyle\int \sqrt{\sin x} \cos^3 x \, dx.$

10 $\displaystyle\int \sin^3 5x \cos 5x \, dx.$

11 $\displaystyle\int \sin^2 3x \cos^2 3x \, dx.$

12 $\displaystyle\int \frac{dx}{\sin x \cos x}.$

13 $\displaystyle\int_0^{\pi/4} \sec^4 x \, dx.$

14 $\displaystyle\int \frac{dx}{\cos^2 x}.$

15 $\int \tan^5 x \sec^3 x \, dx.$

16 $\int \csc^4 x \, dx.$

17 $\int \cot^2 x \, dx.$

18 $\int \cot^3 x \, dx.$

19 $\displaystyle\int \frac{dx}{\sin^2 4x}.$

20 $\displaystyle\int \cot^2 5x \csc^4 5x \, dx.$

21 $\displaystyle\int \frac{1 + \cos 2x}{\sin^2 2x} \, dx.$

22 $\displaystyle\int \tan^2 x \cos x \, dx.$

23 $\int \sin 3x \cot 3x \, dx.$

24 Find $\int \tan x \, dx$ (which we already know) by the method of Example 7.

25 Use the identity $\tan^2 x = \sec^2 x - 1$ to find
(a) $\int \tan^2 x \, dx$, $\int \tan^4 x \, dx$, $\int \tan^6 x \, dx$;
(b) $\int \tan^3 x \, dx$, $\int \tan^5 x \, dx$, $\int \tan^7 x \, dx$.

26 If n is any positive integer ≥ 2, show that

$$\int \tan^n x \, dx = \frac{\tan^{n-1} x}{n-1} - \int \tan^{n-2} x \, dx.$$

This is called a *reduction formula,* because it reduces the problem of integrating $\tan^n x$ to the problem of integrating $\tan^{n-2} x$.

27 Find the volume of the solid of revolution generated when the indicated region under each of the following curves is revolved about the x-axis:

(a) $y = \sin x, 0 \leq x \leq \pi$;
(b) $y = \sec x, 0 \leq x \leq \pi/4$;
(c) $y = \tan 2x, 0 \leq x \leq \pi/8$;
(d) $y = \cos^2 x, \pi/2 \leq x \leq \pi$.

28 Find the length of the curve $y = \ln(\cos x)$ between $x = 0$ and $x = \pi/4$.

29 Find $\int \sec^3 x \, dx$ by exploiting the observation that $\sec^3 x$ will clearly appear in the derivative of $\sec x$ $\tan x$.

30 Find $\int \csc^3 x \, dx$ by adapting the idea suggested for Problem 29.

10.4
TRIGONOMETRIC SUBSTITUTIONS

An integral involving one of the radical expressions $\sqrt{a^2 - x^2}$, $\sqrt{a^2 + x^2}$, $\sqrt{x^2 - a^2}$ (where a is a positive constant) can often be transformed into a familiar trigonometric integral by using a suitable trigonometric substitution or change of variable.

There are three cases, which depend on the trigonometric identities

$$1 - \sin^2 \theta = \cos^2 \theta, \tag{1}$$

$$1 + \tan^2 \theta = \sec^2 \theta, \tag{2}$$

$$\sec^2 \theta - 1 = \tan^2 \theta. \tag{3}$$

If the given integral involves $\sqrt{a^2 - x^2}$, then changing the variable from x to θ by writing

$$x = a \sin \theta \quad \text{replaces} \quad \sqrt{a^2 - x^2} \quad \text{by} \quad a \cos \theta, \tag{4}$$

because $a^2 - x^2 = a^2 - a^2 \sin^2 \theta = a^2(1 - \sin^2 \theta) = a^2 \cos^2 \theta$. Similarly, if the given integral involves $\sqrt{a^2 + x^2}$, then by identity (2) we see that the substitution

$$x = a \tan \theta \quad \text{replaces} \quad \sqrt{a^2 + x^2} \quad \text{by} \quad a \sec \theta, \tag{5}$$

because $a^2 + x^2 = a^2 + a^2 \tan^2 \theta = a^2(1 + \tan^2 \theta) = a^2 \sec^2 \theta$; and if it involves $\sqrt{x^2 - a^2}$, then by identity (3) the substitution

$$x = a \sec \theta \quad \text{replaces} \quad \sqrt{x^2 - a^2} \quad \text{by} \quad a \tan \theta, \tag{6}$$

because $x^2 - a^2 = a^2 \sec^2 \theta - a^2 = a^2(\sec^2 \theta - 1) = a^2 \tan^2 \theta$. We illustrate these procedures as follows.

Example 1 Find

$$\int \frac{\sqrt{a^2 - x^2}}{x} \, dx.$$

Solution This integral is of the first type, so we write

$$x = a \sin \theta, \quad dx = a \cos \theta \, d\theta, \quad \sqrt{a^2 - x^2} = a \cos \theta.$$

Then

$$\int \frac{\sqrt{a^2 - x^2}}{x} dx = \int \frac{a \cos \theta}{a \sin \theta} a \cos \theta \, d\theta = a \int \frac{\cos^2 \theta}{\sin \theta} d\theta$$

$$= a \int \frac{1 - \sin^2 \theta}{\sin \theta} d\theta = a \int (\csc \theta - \sin \theta) \, d\theta$$

$$= -a \ln (\csc \theta + \cot \theta) + a \cos \theta. \tag{7}$$

This completes the integration, and we now must write the answer in terms of the original variable x. We do this quickly and easily by drawing a right triangle (Fig. 10.1) whose sides are labeled in the simplest way that is consistent with the equation $x = a \sin \theta$ or $\sin \theta = x/a$. This figure tells us at once that

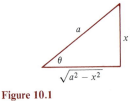

Figure 10.1

$$\csc \theta = \frac{a}{x}, \qquad \cot \theta = \frac{\sqrt{a^2 - x^2}}{x}, \qquad \text{and} \qquad \cos \theta = \frac{\sqrt{a^2 - x^2}}{a},$$

so from (7) we have

$$\int \frac{\sqrt{a^2 - x^2}}{x} dx = \sqrt{a^2 - x^2} - a \ln \left(\frac{a + \sqrt{a^2 - x^2}}{x} \right) + c.$$

Example 2 Find

$$\int \frac{dx}{\sqrt{a^2 + x^2}}.$$

Solution Here we have an integral of the second type, so we write

$$x = a \tan \theta, \qquad dx = a \sec^2 \theta \, d\theta, \qquad \sqrt{a^2 + x^2} = a \sec \theta.$$

This yields

$$\int \frac{dx}{\sqrt{a^2 + x^2}} = \int \frac{a \sec^2 \theta \, d\theta}{a \sec \theta} = \int \sec \theta \, d\theta$$

$$= \ln (\sec \theta + \tan \theta). \tag{8}$$

The substitution equation $x = a \tan \theta$ or $\tan \theta = x/a$ is pictured in Fig. 10.2, and from this figure we obtain

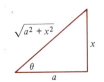

Figure 10.2

$$\sec \theta = \frac{\sqrt{a^2 + x^2}}{a} \qquad \text{and} \qquad \tan \theta = \frac{x}{a}.$$

We therefore continue the calculation in (8) by writing

$$\int \frac{dx}{\sqrt{a^2 + x^2}} = \ln \left(\frac{\sqrt{a^2 + x^2} + x}{a} \right) + c' \tag{9}$$

$$= \ln (\sqrt{a^2 + x^2} + x) + c. \tag{10}$$

Students will notice that since

$$\ln \left(\frac{\sqrt{a^2 + x^2} + x}{a} \right) = \ln (\sqrt{a^2 + x^2} + x) - \ln a,$$

the constant $-\ln a$ has been grouped together with the constant of integration c', and the quantity $-\ln a + c'$ is then rewritten as c. Usually we don't bother to make notational distinctions between one constant of integration and another, because all are completely arbitrary; but we do so here in the hope of clarifying the transition from (9) to (10).

Example 3   Find

$$\int \frac{\sqrt{x^2 - a^2}}{x}\, dx.$$

Solution   This integral is of the third type, so we write

$$x = a \sec \theta, \qquad dx = a \sec \theta \tan \theta\, d\theta, \qquad \sqrt{x^2 - a^2} = a \tan \theta.$$

Then

$$\int \frac{\sqrt{x^2 - a^2}}{x}\, dx = \int \frac{a \tan \theta}{a \sec \theta} a \sec \theta \tan \theta\, d\theta$$

$$= a \int \tan^2 \theta\, d\theta = a \int (\sec^2 \theta - 1)\, d\theta$$

$$= a \tan \theta - a\theta.$$

In this case our substitution equation $\sec \theta = x/a$ is portrayed in Fig. 10.3, which tells us that

$$\tan \theta = \frac{\sqrt{x^2 - a^2}}{a} \qquad \text{and} \qquad \theta = \tan^{-1} \frac{\sqrt{x^2 - a^2}}{a}.$$

The desired integral can therefore be written as

$$\int \frac{\sqrt{x^2 - a^2}}{x}\, dx = \sqrt{x^2 - a^2} - a \tan^{-1} \frac{\sqrt{x^2 - a^2}}{a} + c.$$

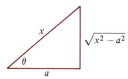

Figure 10.3

There is one feature of these calculations that we have not taken into account. In (4) we tacitly wrote

$$\sqrt{1 - \sin^2 \theta} = \cos \theta$$

without checking the correctness of the algebraic sign. This was careless, because $\cos \theta$ is sometimes negative and sometimes positive. However, the variable θ, which in this case is $\sin^{-1} x/a$, is restricted to the interval $-\pi/2 \le \theta \le \pi/2$, and on this interval $\cos \theta$ is nonnegative, as we assumed. Similar comments apply to the substitutions (5) and (6).

Example 4   As a concrete illustration of the use of these methods, we determine the equation of the *tractrix.* This famous curve can be defined as follows: It is the path of an object dragged along a horizontal plane by a string of constant length when the other end of the string moves along a straight line in the plane. (The word "tractrix" comes from the Latin *tractum,* meaning drag.)

Suppose the plane is the xy-plane and the object starts at the point $(a, 0)$ with the other end of the string at the origin. If this end moves up the y-axis as

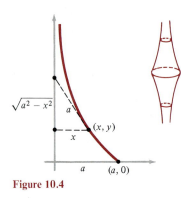

Figure 10.4

shown on the left in Fig. 10.4, then the string is always tangent to the curve, and the length of the tangent between the y-axis and the point of contact is always equal to a. The slope of the tangent is therefore given by the formula

$$\frac{dy}{dx} = -\frac{\sqrt{a^2 - x^2}}{x};$$

and by separating variables and using the result of Example 1, we have

$$y = -\int \frac{\sqrt{a^2 - x^2}}{x}\, dx = a \ln\left(\frac{a + \sqrt{a^2 - x^2}}{x}\right) - \sqrt{a^2 - x^2} + c.$$

Since $y = 0$ when $x = a$, we see that $c = 0$, so

$$y = a \ln\left(\frac{a + \sqrt{a^2 - x^2}}{x}\right) - \sqrt{a^2 - x^2}$$

is the equation of the tractrix, or at least of the part shown in the figure.

If the end of the string moves down the y-axis, then another part of the curve is generated; and if these two parts are revolved about the y-axis, the resulting "double-trumpet" surface shown on the right in Fig. 10.4 is called a *pseudosphere.* In the branch of mathematics concerned with the geometry of curved surfaces, the pseudosphere is a model for Lobachevsky's version of non-Euclidean geometry. It is a surface of constant negative curvature, and the sum of the angles of any triangle on the surface is less than $180°$.

Another famous curve whose equation can be determined by these methods of integration is the *catenary,* which is the curve assumed by a flexible chain or cable hanging between two fixed points. The details are a bit complicated, so we give the discussion in Appendix A.9.

The substitution procedures described in this section can be given a general justification or proof similar to that provided in Section 10.2. Students who are interested in such matters will find the details in Appendix C.8.

PROBLEMS

Find each of the following integrals.

1 $\displaystyle\int \frac{\sqrt{a^2 - x^2}}{x^2}\, dx.$

2 $\displaystyle\int \frac{x^2\, dx}{\sqrt{4 - x^2}}.$

3 $\displaystyle\int \frac{dx}{(a^2 + x^2)^2}.$

4 $\displaystyle\int \frac{dx}{x^2\sqrt{a^2 + x^2}}.$

5 $\displaystyle\int \frac{x^3\, dx}{\sqrt{9 - x^2}}.$

6 $\displaystyle\int \frac{dx}{x\sqrt{a^2 - x^2}}.$

7 $\displaystyle\int \frac{dx}{x\sqrt{a^2 + x^2}}.$

8 $\displaystyle\int \frac{dx}{x + x^3}.$

9 $\displaystyle\int \frac{dx}{\sqrt{x^2 - a^2}}.$

10 $\displaystyle\int \frac{dx}{x^3\sqrt{x^2 - a^2}}.$

11 $\displaystyle\int \sqrt{a^2 + x^2}\, dx.$

12 $\displaystyle\int \frac{x^3\, dx}{a^2 + x^2}.$

13 $\displaystyle\int \frac{dx}{a^2 - x^2}.$

14 $\displaystyle\int \frac{dx}{(a^2 - x^2)^{3/2}}.$

15 $\displaystyle\int \frac{\sqrt{a^2 + x^2}}{x}\, dx.$

16 $\displaystyle\int x^3\sqrt{a^2 + x^2}\, dx.$

17 $\displaystyle\int \frac{\sqrt{x^2 - a^2}}{x^2}\, dx.$

18 $\displaystyle\int \frac{dx}{(x^2 - a^2)^{3/2}}.$

19 $\displaystyle\int x^2\sqrt{a^2 - x^2}\, dx.$

20 $\displaystyle\int (1 - 4x^2)^{3/2}\, dx.$

The following integrals would normally be found in a different way, but this time work them out by using trigonometric substitutions.

21 $\displaystyle\int \frac{x\, dx}{\sqrt{4 - x^2}}.$

22 $\displaystyle\int \frac{x\, dx}{(a^2 - x^2)^{3/2}}.$

23 $\displaystyle\int \frac{dx}{a^2 + x^2}.$ **24** $\displaystyle\int \frac{x\,dx}{4 + x^2}.$

25 $\displaystyle\int x\sqrt{9 - x^2}\,dx.$ **26** $\displaystyle\int \frac{dx}{\sqrt{a^2 - x^2}}.$

27 $\displaystyle\int \frac{x\,dx}{\sqrt{9 + x^2}}.$ **28** $\displaystyle\int \frac{x\,dx}{\sqrt{x^2 - 4}}.$

29 Use integration to show that the area of a circle of radius a is πa^2.

30 In a circle of radius a, a chord b units from the center cuts off a chunk of the circle called a *segment*. Find a formula for the area of this segment.

31 If the circle $(x - b)^2 + y^2 = a^2$ $(0 < a < b)$ is revolved about the y-axis, the resulting solid of revolution is called a *torus* (see Problem 11 in Section 7.3). Use the shell method to find the volume of this torus.

32 Find the length of the parabola $y = x^2$ between $x = 0$ and $x = 1$. Hint: Use the result of Problem 29 in Section 10.3.

33 Find the length of the curve $y = \ln x$ between $x = 1$ and $x = \sqrt{8}$.

34 The given region under each of the following curves is revolved about the x-axis. Find the volume of the solid of revolution.

(a) $y = \dfrac{x^{3/2}}{\sqrt{x^2 + 4}}$ between $x = 0$ and $x = 4$.

(b) $y = \dfrac{1}{x^2 + 1}$ between $x = 0$ and $x = 1$.

(c) $y = \sqrt[4]{4 - x^2}$ between $x = 1$ and $x = 2$.

10.5

COMPLETING THE SQUARE

In Section 10.4 we used trigonometric substitutions to calculate integrals containing $\sqrt{a^2 - x^2}$, $\sqrt{a^2 + x^2}$, and $\sqrt{x^2 - a^2}$. (The case $\sqrt{-a^2 - x^2}$ is clearly of no interest.) By the algebraic device of completing the square, we can extend these methods to integrals involving general quadratic polynomials and their square roots, that is, expressions of the form $ax^2 + bx + c$ and $\sqrt{ax^2 + bx + c}$. We remind students that the process of completing the square is based on the simple fact that

$$(x + A)^2 = x^2 + 2Ax + A^2;$$

this tells us that the right side is a perfect square (the square of $x + A$) because its constant term is the square of half the coefficient of x.

Example 1 Find

$$\int \frac{(x + 2)\,dx}{\sqrt{3 + 2x - x^2}}.$$

Solution Since the coefficient of the term x^2 under the radical is negative, we place the terms containing x in parentheses preceded by a minus sign, leaving space for completing the square,

$$3 + 2x - x^2 = 3 - (x^2 - 2x + \quad) = 4 - (x^2 - 2x + 1)$$
$$= 4 - (x - 1)^2 = a^2 - u^2,$$

where $u = x - 1$ and $a = 2$. Since $x = u + 1$, we have $dx = du$ and $x + 2 = u + 3$, and therefore

$$\int \frac{(x + 2)\,dx}{\sqrt{3 + 2x - x^2}} = \int \frac{(u + 3)\,du}{\sqrt{a^2 - u^2}} = \int \frac{u\,du}{\sqrt{a^2 - u^2}} + 3 \int \frac{du}{\sqrt{a^2 - u^2}}$$

$$= -\sqrt{a^2 - u^2} + 3 \sin^{-1} \frac{u}{a}$$

$$= -\sqrt{3 + 2x - x^2} + 3 \sin^{-1}\left(\frac{x - 1}{2}\right) + c.$$

Example 2   Find

$$\int \frac{dx}{x^2 + 2x + 10}.$$

Solution   We complete the square on the terms containing x, and write

$$x^2 + 2x + 10 = (x^2 + 2x + \quad) + 10 = (x^2 + 2x + 1) + 9$$
$$= (x + 1)^2 + 9 = u^2 + a^2,$$

where $u = x + 1$ and $a = 3$. We now have $du = dx$ or $dx = du$, so

$$\int \frac{dx}{x^2 + 2x + 10} = \int \frac{du}{u^2 + a^2} = \frac{1}{a} \tan^{-1} \frac{u}{a}$$

$$= \frac{1}{3} \tan^{-1} \left(\frac{x + 1}{3} \right) + c.$$

Example 3   Find

$$\int \frac{x \, dx}{\sqrt{x^2 - 2x + 5}}.$$

Solution   We write

$$x^2 - 2x + 5 = (x^2 - 2x + \quad) + 5 = (x^2 - 2x + 1) + 4$$
$$= (x - 1)^2 + 4 = u^2 + a^2,$$

where $u = x - 1$ and $a = 2$. Then $x = u + 1$, $dx = du$, and we have

$$\int \frac{x \, dx}{\sqrt{x^2 - 2x + 5}} = \int \frac{(u + 1) \, du}{\sqrt{u^2 + a^2}} = \int \frac{u \, du}{\sqrt{u^2 + a^2}} + \int \frac{du}{\sqrt{u^2 + a^2}}.$$

The second integral here is the one considered in Example 2 in Section 10.4, so we have

$$\int \frac{du}{\sqrt{u^2 + a^2}} = \ln (u + \sqrt{u^2 + a^2}),$$

and therefore

$$\int \frac{x \, dx}{\sqrt{x^2 - 2x + 5}} = \sqrt{u^2 + a^2} + \ln (u + \sqrt{u^2 + a^2})$$

$$= \sqrt{x^2 - 2x + 5} + \ln (x - 1 + \sqrt{x^2 - 2x + 5}) + c.$$

Example 4   Find

$$\int \frac{dx}{\sqrt{x^2 - 4x - 5}}.$$

Solution   Here we have

$$x^2 - 4x - 5 = (x^2 - 4x + \quad) - 5 = (x^2 - 4x + 4) - 9$$
$$= (x - 2)^2 - 9 = u^2 - a^2,$$

where $u = x - 2$ and $a = 3$. By using the result of Problem 9 in Section 10.4 (or by quickly working out the necessary formula again by putting $u = a \sec \theta$) we complete the calculation as follows:

$$\int \frac{dx}{\sqrt{x^2 - 4x - 5}} = \int \frac{du}{\sqrt{u^2 - a^2}} = \ln (u + \sqrt{u^2 - a^2})$$
$$= \ln (x - 2 + \sqrt{x^2 - 4x - 5}) + c.$$

If an integral involves the square root of a third-, fourth-, or higher-degree polynomial, then it can be proved that there does not exist any general method for carrying out the integration. We shall discuss a few integrals of this kind in Section 10.8.

PROBLEMS

Calculate the following integrals.

1. $\displaystyle\int \frac{dx}{\sqrt{2x - x^2}}.$

2. $\displaystyle\int \frac{dx}{\sqrt{5 + 4x - x^2}}.$

3. $\displaystyle\int \frac{dx}{x^2 + 4x + 5}.$

4. $\displaystyle\int \frac{dx}{x^2 - x + 1}.$

5. $\displaystyle\int \frac{(x + 1) \, dx}{\sqrt{2x - x^2}}.$

6. $\displaystyle\int \frac{(x + 3) \, dx}{\sqrt{5 + 4x - x^2}}.$

7. $\displaystyle\int \frac{x^2 \, dx}{\sqrt{6x - x^2}}.$

8. $\displaystyle\int \frac{(x - 1) \, dx}{\sqrt{x^2 + 4x + 5}}.$

9. $\displaystyle\int \frac{(x + 7) \, dx}{x^2 + 2x + 5}.$

10. $\displaystyle\int \frac{\sqrt{x^2 + 2x - 3}}{x + 1} \, dx.$

11. $\displaystyle\int \frac{dx}{\sqrt{x^2 - 2x - 8}}.$

12. $\displaystyle\int \frac{dx}{\sqrt{5 + 3x - 2x^2}}.$

13. $\displaystyle\int \frac{dx}{\sqrt{4x^2 + 4x + 17}}.$

14. $\displaystyle\int \frac{(4x + 3) \, dx}{(x^2 - 2x + 2)^{3/2}}.$

15. $\displaystyle\int \frac{dx}{(x^2 - 2x - 3)^{3/2}}.$

16. $\displaystyle\int \frac{dx}{(x + 2)\sqrt{x^2 + 4x + 3}}.$

10.6
THE METHOD OF PARTIAL FRACTIONS

We recall that a rational function is a quotient of two polynomials. By taking the denominator of such a quotient to be 1, we see that the polynomials themselves are included among the rational functions. As we know, the simple rational functions

$$2x + 1, \quad \frac{1}{x^2}, \quad \frac{1}{x}, \quad \frac{x}{x^2 + 1}, \quad \text{and} \quad \frac{1}{x^2 + 1}$$

have the following integrals:

$$x^2 + x, \quad -\frac{1}{x}, \quad \ln x, \quad \tfrac{1}{2}\ln (x^2 + 1), \quad \text{and} \quad \tan^{-1} x.$$

Our purpose in this section is to describe a systematic procedure for computing the integral of any rational function, and we shall find that this integral can always be expressed in terms of polynomials, rational functions, logarithms, and inverse tangents. The basic idea is to break up a given rational function into a sum of simpler fractions (called *partial fractions*) which can be integrated by methods discussed earlier.

A rational function is called *proper* if the degree of the numerator is less than the degree of the denominator. Otherwise, it is said to be *improper*. For example,

$$\frac{x}{(x-1)(x+2)^2} \quad \text{and} \quad \frac{x^2+2}{x(x^2-9)}$$

are proper, while

$$\frac{x^4}{x^4-1} \quad \text{and} \quad \frac{2x^3-3x^2+2x-4}{x^2+4}$$

are improper. If we have to integrate an improper rational function, it is essential to begin by performing long division until we reach a remainder whose degree is less than that of the denominator. We illustrate with the second improper function mentioned. Long division yields

$$
\begin{array}{r}
2x - 3 \\
x^2+4\,\overline{\big)\,2x^3 - 3x^2 + 2x - 4} \\
\underline{2x^3 \qquad\quad + 8x} \\
-3x^2 - 6x - 4 \\
\underline{-3x^2 \qquad - 12} \\
-6x + 8
\end{array}
$$

This means that the rational function in question can be written in the form

$$\frac{2x^3-3x^2+2x-4}{x^2+4} = 2x - 3 + \frac{-6x+8}{x^2+4}. \tag{1}$$

By applying this process, any improper rational function $P(x)/Q(x)$ can be expressed as the sum of a polynomial and a proper rational function,

$$\frac{P(x)}{Q(x)} = \text{polynomial} + \frac{R(x)}{Q(x)}, \tag{2}$$

where the degree of $R(x)$ is less than the degree of $Q(x)$. In the particular case of (1), this decomposition by means of long division enables us to carry out the integration quite easily, by writing

$$\int \frac{2x^3-3x^2+2x-4}{x^2+4}\,dx = x^2 - 3x - 6\int \frac{x\,dx}{x^2+4} + 8\int \frac{dx}{x^2+4}$$

$$= x^2 - 3x - 3\ln(x^2+4) + 4\tan^{-1}\frac{x}{2} + c.$$

In the general case (2), these remarks tell us that we can restrict our attention to proper rational functions, since the integration of polynomials is always easy. This restriction is not only convenient, but also necessary, because it is *only* to proper rational functions that the following discussions apply.

In elementary algebra we learned how to combine fractions over a common denominator. We must now learn how to reverse this process and split a given fraction into a sum of fractions having simpler denominators. This procedure is called *decomposition into partial fractions*.

Example 1   It is clear that

$$\frac{3}{x-1} + \frac{2}{x+3} = \frac{3(x+3)+2(x-1)}{(x-1)(x+3)} = \frac{5x+7}{(x-1)(x+3)}. \tag{3}$$

In the reverse process we start with the right side of (3) as our given rational function and seek constants A and B such that

$$\frac{5x + 7}{(x - 1)(x + 3)} = \frac{A}{x - 1} + \frac{B}{x + 3}. \tag{4}$$

(For the sake of understanding the method, let us pretend for a moment that we don't know that $A = 3$ and $B = 2$ will work.) If we clear fractions in (4) by multiplying through by $(x - 1)(x + 3)$, we get

$$5x + 7 = A(x + 3) + B(x - 1) \tag{5}$$

or

$$5x + 7 = (A + B)x + (3A - B). \tag{6}$$

Since (6) is to be an identity in x, we can find A and B by equating coefficients of like powers of x. This gives a system of two equations in the two unknowns A and B,

$$\begin{cases} A + B = 5 \\ 3A - B = 7, \end{cases} \quad \text{whose solution is} \quad A = 3, B = 2.$$

There is another convenient way to find A and B, by using (5) directly. Since (5) must hold for all x, it must hold in particular for $x = 1$ (which removes B) and for $x = -3$ (which removes A). Briefly,

$$x = 1: \qquad 5 + 7 = A(1 + 3) + 0, \qquad 4A = 12, \qquad A = 3;$$

$$x = -3: \quad -15 + 7 = 0 + B(-3 - 1), \qquad -4B = -8, \qquad B = 2.$$

This method is faster than it looks, and can be carried out by inspection. Whichever method we use to find A and B, (4) becomes

$$\frac{5x + 7}{(x - 1)(x + 3)} = \frac{3}{x - 1} + \frac{2}{x + 3},$$

and this is the partial fractions decomposition of the rational function on the left. Of course, the purpose of this decomposition is to enable us to integrate the given function,

$$\int \frac{5x + 7}{(x - 1)(x + 3)}\, dx = \int \left(\frac{3}{x - 1} + \frac{2}{x + 3} \right) dx$$

$$= 3 \ln (x - 1) + 2 \ln (x + 3) + c.$$

The type of expansion used in (4) works in just the same way under more general circumstances, as follows:

Let $P(x)/Q(x)$ be a proper rational function whose denominator $Q(x)$ is an nth-degree polynomial. If $Q(x)$ can be factored completely into *distinct linear factors* $x - r_1, x - r_2, \ldots, x - r_n$, then there exist n constants $A_1, A_2, \ldots, A_n$ such that

$$\frac{P(x)}{Q(x)} = \frac{A_1}{x - r_1} + \frac{A_2}{x - r_2} + \cdots + \frac{A_n}{x - r_n}. \tag{7}$$

The constants in the numerators can be determined by either of the methods suggested in Example 1; and when this is done, the partial fractions decomposition (7) provides an easy way to integrate the given rational function.

Example 2 Find

$$\int \frac{6x^2 + 14x - 20}{x^3 - 4x} \, dx.$$

Solution We factor the denominator by writing $x^3 - 4x = x(x^2 - 4) = x(x + 2)(x - 2)$. Accordingly, we have a decomposition of the form

$$\frac{6x^2 + 14x - 20}{x^3 - 4x} = \frac{6x^2 + 14x - 20}{x(x + 2)(x - 2)} = \frac{A}{x} + \frac{B}{x + 2} + \frac{C}{x - 2} \tag{8}$$

for certain constants A, B, C. To find these constants we clear fractions in (8), which yields

$$6x^2 + 14x - 20 = A(x + 2)(x - 2) + Bx(x - 2) + Cx(x + 2).$$

By setting $x = 0, -2, 2$ (this is the second method in Example 1), we easily see that $A = 5$, $B = -3$, $C = 4$, so (8) becomes

$$\frac{6x^2 + 14x - 20}{x^3 - 4x} = \frac{5}{x} - \frac{3}{x + 2} + \frac{4}{x - 2}.$$

We therefore have

$$\int \frac{6x^2 + 14x - 20}{x^3 - 4x} \, dx = 5 \ln x - 3 \ln (x + 2) + 4 \ln (x - 2) + c.$$

In theory, every polynomial $Q(x)$ with real coefficients can be factored completely into real linear and quadratic factors, some of which may be repeated.* In practice, this factorization is hard to carry out for polynomials of degree 3 or more, except in special cases. Nevertheless, let us assume this has been done, and let us see how the decomposition (7) must be altered to take account of the most general circumstances that can arise.

If a linear factor $x - r$ occurs with multiplicity m, then the corresponding term $A/(x - r)$ in the decomposition (7) must be replaced by a sum of the form

$$\frac{B_1}{x - r} + \frac{B_2}{(x - r)^2} + \cdots + \frac{B_m}{(x - r)^m}.$$

A quadratic factor $x^2 + bx + c$ of multiplicity 1 gives rise to a single term

$$\frac{Ax + B}{x^2 + bx + c},$$

and if this quadratic factor occurs with multiplicity m, then it gives rise to a sum of the form

$$\frac{A_1 x + B_1}{x^2 + bx + c} + \frac{A_2 x + B_2}{(x^2 + bx + c)^2} + \cdots + \frac{A_m x + B_m}{(x^2 + bx + c)^m}.$$

---

* This statement is a consequence of the *Fundamental Theorem of Algebra,* which is discussed in Section 14.12.

This is the whole story, and the theory guarantees that every proper rational function can be expanded into a sum of partial fractions in the manner described above.*

Example 3   Find

$$\int \frac{3x^3 - 4x^2 - 3x + 2}{x^4 - x^2} \, dx.$$

Solution   We have

$$\frac{3x^3 - 4x^2 - 3x + 2}{x^4 - x^2} = \frac{3x^3 - 4x^2 - 3x + 2}{x^2(x + 1)(x - 1)}$$

$$= \frac{A}{x} + \frac{B}{x^2} + \frac{C}{x + 1} + \frac{D}{x - 1}.$$

Clearing fractions gives the identity

$$3x^3 - 4x^2 - 3x + 2 = Ax(x + 1)(x - 1) + B(x + 1)(x - 1)$$
$$+ Cx^2(x - 1) + Dx^2(x + 1).$$

Now put

$$\begin{aligned} x = 0: && 2 = -B, && B = -2; \\ x = 1: && -2 = 2D, && D = -1; \\ x = -1: && -2 = -2C, && C = 1. \end{aligned}$$

Equating coefficients of x^3 gives

$$3 = A + C + D, \quad \text{so} \quad A = 3.$$

Our partial fractions decomposition is therefore

$$\frac{3x^3 - 4x^2 - 3x + 2}{x^4 - x^2} = \frac{3}{x} - \frac{2}{x^2} + \frac{1}{x + 1} - \frac{1}{x - 1},$$

so

$$\int \frac{3x^3 - 4x^2 - 3x + 2}{x^4 - x^2} \, dx = 3 \ln x + \frac{2}{x} + \ln (x + 1) - \ln (x - 1) + c.$$

Example 4   Find

$$\int \frac{2x^3 + x^2 + 2x - 1}{x^4 - 1} \, dx.$$

Solution   We have

$$\frac{2x^3 + x^2 + 2x - 1}{x^4 - 1} = \frac{2x^3 + x^2 + 2x - 1}{(x + 1)(x - 1)(x^2 + 1)}$$

$$= \frac{A}{x + 1} + \frac{B}{x - 1} + \frac{Cx + D}{x^2 + 1},$$

---

* This statement is called the *partial fractions theorem*; it is proved in Appendix C.9. Students will notice that the above description of the partial fractions decomposition assumes that the highest power of x in $Q(x)$ has coefficient 1; this can always be arranged by a minor algebraic adjustment.

so

$$2x^3 + x^2 + 2x - 1 = A(x-1)(x^2+1) + B(x+1)(x^2+1)$$
$$+ Cx(x^2-1) + D(x^2-1).$$

Now put

$$\begin{aligned}
x=1: &\quad 4=4B, &\quad B=1;\\
x=-1: &\quad -4=-4A, &\quad A=1;\\
x=0: &\quad -1=-A+B-D, &\quad D=1.
\end{aligned}$$

Equating coefficients of x^3 gives

$$2=A+B+C, \quad \text{so} \quad C=0.$$

Our partial fractions decomposition is therefore

$$\frac{2x^3+x^2+2x-1}{x^4-1} = \frac{1}{x+1} + \frac{1}{x-1} + \frac{1}{x^2+1},$$

so

$$\int \frac{2x^3+x^2+2x-1}{x^4-1}\,dx = \ln(x+1) + \ln(x-1) + \tan^{-1}x + c.$$

As a final comment, we point out that all the partial fractions that can possibly arise have the form

$$\frac{A}{(x-r)^n} \quad \text{or} \quad \frac{Ax+B}{(x^2+bx+c)^n}, \quad n=1,2,3,\ldots.$$

Functions of the first type can be integrated by using the substitution $u = x - r$, and it is clear that the results are always logarithms or rational functions. A function of the second type in which the quadratic polynomial $x^2 + bx + c$ has no real linear factors, that is, in which the roots of $x^2 + bx + c = 0$ are imaginary, can be integrated by completing the square and making a suitable substitution. When this is done, we get integrals of the form

$$\int \frac{u\,du}{(u^2+k^2)^n}, \quad \int \frac{du}{(u^2+k^2)^n}.$$

The first of these is $\frac{1}{2}\ln(u^2+k^2)$ if $n=1$, and $(u^2+k^2)^{1-n}/2(1-n)$ if $n>1$. When $n=1$, the second integral is calculated by the formula

$$\int \frac{du}{u^2+k^2} = \frac{1}{k}\tan^{-1}\frac{u}{k}.$$

The case $n>1$ can be reduced to the case $n=1$ by repeated application of the *reduction formula*

$$\int \frac{du}{(u^2+k^2)^n} = \frac{1}{2k^2(n-1)} \cdot \frac{u}{(u^2+k^2)^{n-1}} + \frac{2n-3}{2k^2(n-1)}\int \frac{du}{(u^2+k^2)^{n-1}}. \quad (9)$$

We state this complicated formula for the sole purpose of showing that the only functions that arise from the indicated reduction procedure are rational functions and inverse tangents. The formula itself can either be verified by differentiation or obtained from scratch by the methods of the next section.

This discussion shows that the integral of every rational function can be expressed in terms of polynomials, rational functions, logarithms, and inverse tangents. The detailed work can be very laborious, but at least the path that must be followed is clearly visible.

PROBLEMS

1 Express each of the following improper rational functions as the sum of a polynomial and a proper rational function, and integrate:

(a) $\dfrac{x^2}{x-1}$; (b) $\dfrac{x^3}{3x+2}$; (c) $\dfrac{x^3}{x^2+1}$;

(d) $\dfrac{x+3}{x+2}$; (e) $\dfrac{x^2-1}{x^2+1}$.

Find each of the following integrals.

2 $\displaystyle\int \frac{12x-17}{(x-1)(x-2)}\,dx.$ **3** $\displaystyle\int \frac{14x-12}{2x^2-2x-12}\,dx.$

4 $\displaystyle\int \frac{10-2x}{x^2+5x}\,dx.$ **5** $\displaystyle\int \frac{2x+21}{x^2-7x}\,dx.$

6 $\displaystyle\int \frac{9x^2-24x+6}{x^3-5x^2+6x}\,dx.$ **7** $\displaystyle\int \frac{x^2+46x-48}{x^3+5x^2-24x}\,dx.$

8 $\displaystyle\int \frac{16x^2+3x-7}{x^3-x}\,dx.$ **9** $\displaystyle\int \frac{4x^2+11x-117}{x^3+10x^2-39x}\,dx.$

10 $\displaystyle\int \frac{6x^2-9x+9}{x^3-3x^2}\,dx.$ **11** $\displaystyle\int \frac{-4x^2-5x-3}{x^3+2x^2+x}\,dx.$

12 $\displaystyle\int \frac{4x^2+2x+4}{x^3+4x}\,dx.$ **13** $\displaystyle\int \frac{3x^2-x+4}{x^3+2x^2+2x}\,dx.$

14 Use partial fractions to obtain the formula

$$\int \frac{dx}{a^2-x^2} = \frac{1}{2a}\ln\frac{a+x}{a-x}.$$

Also calculate this integral by trigonometric substitution, and verify that the two answers agree.

15 In Section 10.1 the formula

$$\int \sec u\,du = \ln\,(\sec u + \tan u)$$

was obtained by a trick. Derive it by integrating

$$\int \frac{\cos u\,du}{1-\sin^2 u}$$

by partial fractions.

16 Find

(a) $\displaystyle\int \frac{3\sin\theta\,d\theta}{\cos^2\theta-\cos\theta-2}$; (b) $\displaystyle\int \frac{5e^t\,dt}{e^{2t}+e^t-6}.$

17 In Problem 14 of Section 8.5 it is stated that the differential equation

$$\frac{dx}{dt} = kab(A-x)(B-x), \qquad A \neq B,$$

has

$$\frac{B(A-x)}{A(B-x)} = e^{kab(A-B)t}$$

as a solution for which $x=0$ when $t=0$. Derive this solution by using partial fractions.

18 Verify the reduction formula (9) by differentiating the first term on the right.

INTEGRATION BY PARTS

When the formula for the derivative of a product (the product rule) is written in the notation of differentials, it is

$$d(uv) = u\,dv + v\,du \qquad \text{or} \qquad u\,dv = d(uv) - v\,du,$$

and by integrating we obtain

$$\int u\,dv = uv - \int v\,du. \tag{1}$$

This formula provides a method of finding $\int u\,dv$ if the second integral $\int v\,du$ is easier to calculate than the first. The method is called *integration by parts,* and it often works when all other methods fail.

Example 1   Find $\int x \cos x\,dx$.

Solution   If we put

$$u = x, \qquad dv = \cos x\,dx,$$

then

$$du = dx, \qquad v = \sin x,$$

and (1) gives

$$\int x \cos x\,dx = x \sin x - \int \sin x\,dx.$$

This is good luck, because the integral on the right is easy. We therefore have

$$\int x \cos x\,dx = x \sin x + \cos x + c.$$

It is worth noticing that in this example we could have chosen u and dv differently. If we put

$$u = \cos x, \qquad dv = x\,dx,$$

then

$$du = -\sin x\,dx, \qquad v = \tfrac{1}{2}x^2,$$

and (1) gives

$$\int x \cos x\,dx = \tfrac{1}{2}x^2 \cos x + \tfrac{1}{2}\int x^2 \sin x\,dx.$$

This equation is true, but it is completely worthless as a means of solving our problem, because the second integral is harder than the first. We urge students to learn from experience, and to use trial and error as intelligently as possible in choosing u and dv. Also, students should feel free to abandon a choice that doesn't seem to work, and quickly go on to another choice that offers more hope of success.

The method of integration by parts applies particularly well to products of different types of functions, like $x \cos x$ in Example 1, which is a product of a polynomial and a trigonometric function. In using the method, the given differential must be thought of as a product $u \cdot dv$. The part called dv must be something we can integrate, and the part called u should usually be something that is simplified by differentiation, as in our next example.

Example 2 Find $\int \ln x \, dx$.

Solution Here our only choice is

$$u = \ln x, \qquad dv = dx,$$

so

$$du = \frac{dx}{x}, \qquad v = x,$$

and we have

$$\int \ln x \, dx = x \ln x - \int x \frac{dx}{x} = x \ln x - x + c.$$

In some cases it is necessary to carry out two or more integrations by parts in succession.

Example 3 Find $\int x^2 e^x \, dx$.

Solution If we put

$$u = x^2, \qquad dv = e^x \, dx,$$

then

$$du = 2x \, dx, \qquad v = e^x,$$

and (1) gives

$$\int x^2 e^x \, dx = x^2 e^x - 2 \int x e^x \, dx. \tag{2}$$

Here the second integral is easier than the first, so we are encouraged to continue in the same way. When the second integral is integrated by parts with

$$u = x, \qquad dv = e^x \, dx,$$

so that

$$du = dx, \qquad v = e^x,$$

then we get

$$\int x e^x \, dx = x e^x - \int e^x \, dx$$

$$= x e^x - e^x.$$

When this is inserted in (2), our final result is

$$\int x^2 e^x \, dx = x^2 e^x - 2x e^x + 2e^x + c.$$

It sometimes happens that the integral we start with appears a second time during the integration by parts, in which case it is often possible to solve for this integral by elementary algebra.

Example 4 Find $\int e^x \cos x \, dx$.

Solution For convenience we denote this integral by J. If we put

$$u = e^x, \qquad dv = \cos x \, dx,$$

then

$$du = e^x \, dx, \qquad v = \sin x,$$

and (1) yields

$$J = e^x \sin x - \int e^x \sin x \, dx. \tag{3}$$

Now we come to the interesting part of this problem. Even though the new integral is no easier than the old, it turns out to be fruitful to apply the same method again to the new integral. Thus, we put

$$u = e^x, \qquad dv = \sin x \, dx,$$

so that

$$du = e^x \, dx, \qquad v = -\cos x,$$

and obtain

$$\int e^x \sin x \, dx = -e^x \cos x + \int e^x \cos x \, dx. \tag{4}$$

The integral on the right is J again, so (4) can be written

$$\int e^x \sin x \, dx = -e^x \cos x + J. \tag{5}$$

In spite of appearances, we are not going in a circle, because substituting (5) in (3) gives

$$J = e^x \sin x + e^x \cos x - J.$$

It is now easy to solve for J by writing

$$2J = e^x \sin x + e^x \cos x \qquad \text{or} \qquad J = \tfrac{1}{2}(e^x \sin x + e^x \cos x),$$

and all that remains is to insert the constant of integration:

$$\int e^x \cos x \, dx = \tfrac{1}{2} e^x (\sin x + \cos x) + c.$$

The method of this example is often used to make an integral depend on a simpler integral of the same type, and thus to obtain a convenient *reduction formula,* by repeated use of which the given integral can easily be calculated.

Example 5 Find a reduction formula for $J_n = \int \sin^n x \, dx$.

Solution We integrate by parts with

$$u = \sin^{n-1}x, \qquad dv = \sin x \, dx,$$

so that

$$du = (n-1) \sin^{n-2} x \cos x \, dx, \qquad v = -\cos x,$$

and therefore

$$J_n = -\sin^{n-1} x \cos x + (n-1) \int \sin^{n-2} x \cos^2 x \, dx$$

$$= -\sin^{n-1} x \cos x + (n-1) \int \sin^{n-2} x (1 - \sin^2 x) \, dx$$

$$= -\sin^{n-1} x \cos x + (n-1) \int \sin^{n-2} x \, dx - (n-1) \int \sin^n x \, dx$$

$$= -\sin^{n-1} x \cos x + (n-1) J_{n-2} - (n-1) J_n.$$

We now transpose the term involving J_n and obtain

$$nJ_n = -\sin^{n-1} x \cos x + (n-1)J_{n-2},$$

so that

$$J_n = -\frac{1}{n} \sin^{n-1} x \cos x + \frac{n-1}{n} J_{n-2},$$

or equivalently,

$$\int \sin^n x \, dx = -\frac{1}{n} \sin^{n-1} x \cos x + \frac{n-1}{n} \int \sin^{n-2} x \, dx. \qquad (6)$$

The reduction formula (6) allows us to reduce the exponent on $\sin x$ by 2. By repeated application of this formula we can therefore ultimately reduce J_n to J_0 or J_1, according as n is even or odd. But both of these are easy:

$$J_0 = \int \sin^0 x \, dx = \int dx = x \qquad \text{and} \qquad J_1 = \int \sin x \, dx = -\cos x.$$

For example, with $n = 4$ we get

$$\int \sin^4 x \, dx = -\tfrac{1}{4} \sin^3 x \cos x + \tfrac{3}{4} \int \sin^2 x \, dx,$$

and with $n = 2$,

$$\int \sin^2 x \, dx = -\tfrac{1}{2} \sin x \cos x + \tfrac{1}{2} \int dx$$

$$= -\tfrac{1}{2} \sin x \cos x + \tfrac{1}{2}x.$$

Therefore,

$$\int \sin^4 x \, dx = -\tfrac{1}{4} \sin^3 x \cos x + \tfrac{3}{4}(-\tfrac{1}{2} \sin x \cos x + \tfrac{1}{2}x)$$

$$= -\tfrac{1}{4} \sin^3 x \cos x - \tfrac{3}{8} \sin x \cos x + \tfrac{3}{8}x + c.$$

The same result can be achieved by earlier techniques depending on repeated use of the half-angle formulas, but our present methods are more efficient for large exponents. In our next example we illustrate another way in which the reduction formula (6) can be used.

Example 6 Calculate

$$\int_0^{\pi/2} \sin^8 x \, dx.$$

Solution For convenience we write

$$I_n = \int_0^{\pi/2} \sin^n x \, dx.$$

By formula (6) we have

$$I_n = -\frac{1}{n} \sin^{n-1} x \cos x \Big]_0^{\pi/2} + \frac{n-1}{n} \int_0^{\pi/2} \sin^{n-2} x \, dx,$$

so

$$I_n = \frac{n-1}{n} I_{n-2}.$$

We apply this formula with $n = 8$, then repeat with $n = 6$, $n = 4$, $n = 2$:

$$I_8 = \tfrac{7}{8}I_6 = \tfrac{7}{8} \cdot \tfrac{5}{6}I_4 = \tfrac{7}{8} \cdot \tfrac{5}{6} \cdot \tfrac{3}{4}I_2 = \tfrac{7}{8} \cdot \tfrac{5}{6} \cdot \tfrac{3}{4} \cdot \tfrac{1}{2}I_0.$$

Therefore

$$\int_0^{\pi/2} \sin^8 x \, dx = \frac{7}{8} \cdot \frac{5}{6} \cdot \frac{3}{4} \cdot \frac{1}{2} \int_0^{\pi/2} dx = \frac{7}{8} \cdot \frac{5}{6} \cdot \frac{3}{4} \cdot \frac{1}{2} \cdot \frac{\pi}{2} = \frac{35\pi}{256}.$$

Remark 1 The reduction formula (6) can also be used to establish one of the most fascinating formulas of mathematics, *Wallis's infinite product* for $\pi/2$:

$$\frac{\pi}{2} = \frac{2}{1} \cdot \frac{2}{3} \cdot \frac{4}{3} \cdot \frac{4}{5} \cdot \frac{6}{5} \cdot \frac{6}{7} \cdots.$$

For the details of the proof, see Appendix A.10.

Remark 2 In Section 9.5 we stated *Leibniz's formula* for $\pi/4$,

$$\frac{\pi}{4} = 1 - \frac{1}{3} + \frac{1}{5} - \frac{1}{7} + \cdots.$$

For students who are interested in little-known corners of the history of mathematics, we describe in Appendix A.11 how Leibniz himself discovered his formula, by a very ingenious application of integration by parts.

PROBLEMS

Find each of the following integrals by the method of integration by parts.

1 $\int x \ln x \, dx$.
2 $\int \tan^{-1} x \, dx$.
3 $\int x \tan^{-1} x \, dx$.
4 $\int x e^{ax} \, dx$.
5 $\int e^x \sin x \, dx$.
6 $\int e^{ax} \cos bx \, dx$.
7 $\int \sqrt{1 - x^2} \, dx$.
8 $\int \sin^{-1} x \, dx$.

9 $\int x \sin^{-1} x \, dx$.

10 $\int_0^{\pi/2} x \sin x \, dx$.

11 $\int x \cos (3x - 2) \, dx$.

12 $\int \frac{\tan^{-1} x}{x^2} \, dx$.

13 $\int x \sec^2 x \, dx$.
14 $\int \sin (\ln x) \, dx$.
15 $\int \ln (a^2 + x^2) \, dx$.
16 $\int x^2 \ln (x + 1) \, dx$.

17 $\int \frac{\ln x}{x} \, dx$.

18 $\int (\ln x)^2 \, dx$.

19 The region under the curve $y = \cos x$ between $x = 0$ and $x = \pi/2$ is revolved about the y-axis. Find the volume of the resulting solid.

20 Find $\int (\sin^{-1} x)^2 \, dx$. Hint: Make the substitution $y = \sin^{-1} x$.

21 If $P(x)$ is a polynomial, show that

$$\int P(x) e^x \, dx = (P - P' + P'' - P''' + \cdots) e^x.$$

In the next two problems, derive the given reduction formula and apply it to the indicated special case(s).

22 (a) $\int \cos^n x \, dx = \frac{1}{n} \sin x \cos^{n-1} x$

$$+ \frac{n-1}{n} \int \cos^{n-2} x \, dx.$$

(b) $\int_0^{\pi/2} \cos^7 x \, dx$.

(c) $\int_0^{\pi/2} \cos^8 x \, dx$.

23 (a) $\int (\ln x)^n \, dx = x(\ln x)^n - n \int (\ln x)^{n-1} \, dx$.
(b) $\int (\ln x)^5 \, dx$.

24 The region under the curve $y = \sin x$ between $x = 0$ and $x = \pi$ is revolved about the y-axis. Find the volume of the resulting solid (a) by the shell method; and (b) by the washer method.

25 The curve in Problem 24 is revolved about the x-axis. Find the area of the resulting surface of revolution.

10.8

(OPTIONAL) FUNCTIONS THAT CANNOT BE INTEGRATED

At this point we have described all the standard methods of integration that the student is expected to be acquainted with. A few additional techniques of minor importance remain, and two of these are briefly sketched in the problems below; but for most practical purposes we have reached the end of this particular road.

In spite of the many successes achieved by the methods of this chapter, certain integrals have always resisted every attempt to express them in terms of elementary functions, for instance

$$\int e^{-x^2} \, dx, \quad \int \frac{e^x}{x} \, dx, \quad \int \cos x^2 \, dx,$$

$$\int \frac{dx}{\ln x}, \quad \int \sqrt{\sin x} \, dx, \quad \int \frac{\sin x}{x} \, dx.$$

There are also the so-called elliptic integrals, of which

$$\int \sqrt{1 - x^3} \, dx \quad \text{and} \quad \int \frac{dx}{\sqrt{1 - x^4}}$$

are examples.* In the nineteenth century it was finally proved, by the great French mathematician Liouville and his followers, that the problem of working out these integrals in terms of elementary functions is not merely difficult — it is actually impossible.

The full depth of Liouville's ideas cannot be plumbed in a calculus course.† Nevertheless, it is quite possible to gain some impression of how these ideas work without necessarily undertaking a long program of preliminary study.‡

Among other things, Liouville discovered and proved the following theorem:

If $f(x)$ and $g(x)$ are rational functions and $g(x)$ is not a constant, and if $\int f(x)e^{g(x)}\,dx$ is an elementary function, then this integral must have the form

$$\int f(x)e^{g(x)}\,dx = Re^{g(x)}$$

for some rational function R.

We illustrate the value of this theorem by using it to prove that the integral

$$\int \frac{e^x}{x}\,dx \tag{1}$$

is not elementary (that is, cannot be expressed in terms of elementary functions). Suppose, on the contrary, that this integral *is* elementary. Then by Liouville's theorem we know that

$$\int \frac{e^x}{x}\,dx = Re^x$$

for some rational function R. But this means that

$$\frac{e^x}{x} = \frac{d}{dx}(Re^x) \qquad \text{or} \qquad \frac{e^x}{x} = Re^x + R'e^x,$$

so

$$\frac{1}{x} = R + R'. \tag{2}$$

Since R is rational, it can be written in the form $R = P/Q$, where P and Q are polynomials with no common factor. We know that

$$R' = \frac{QP' - PQ'}{Q^2},$$

---

* In general, an elliptic integral is any integral of the form $\int R(x, y)\,dx$, where $R(x, y)$ is a rational function of the two variables x, y and where y is the square root of a polynomial of the third or fourth degree in x. The name "elliptic integral" is used because an integral of this type arises in the problem of finding the circumference of an ellipse.

† Liouville's theory is expounded in full in the monograph by J. F. Ritt, *Integration in Finite Terms,* Columbia University Press, New York, 1948.

‡ See D. G. Mead's article "Integration," in the *American Mathematical Monthly,* vol. 68 (1961), pp. 152–156.

so (2) becomes

$$\frac{1}{x} = \frac{P}{Q} + \frac{QP' - PQ'}{Q^2},$$

which is equivalent to

$$Q^2 = PQx + x(QP' - PQ')$$

or

$$Q(Q - Px - P'x) = -PQ'x. \tag{3}$$

Our purpose is to deduce a contradiction from (3), and we proceed as follows. Let x^n be the highest power of x that can be factored out of the polynomial Q, so that $Q(x) = x^n Q_1(x)$ where $Q_1(x)$ is a polynomial such that $Q_1(0) \neq 0$. We first observe that $n > 0$; for if $n = 0$, so that $Q(0) \neq 0$, then $x = 0$ reduces the right side of (3) to zero but not the left side, which cannot happen because (3) is an identity in x. This implies two facts that we need in order to obtain our final contradiction. First, $P(0) \neq 0$, because P and Q have no common factor and therefore x cannot be a factor of P. Second, we have

$$Q'(x) = x^n Q_1'(x) + nx^{n-1} Q_1(x)$$
$$= x^{n-1}[xQ_1'(x) + nQ_1(x)];$$

and since the polynomial in brackets has a nonzero value when $x = 0$, we know that x^{n-1} is the highest power of x that can be factored out of Q'. These two facts taken together tell us that x^n is the highest power of x that can be factored out of the polynomial on the right side of (3), whereas x^{n+1} can be factored out of the left side. This contradiction brings us to the conclusion that (2) is impossible, so the integral (1) is not elementary.

Remark 1   We know from our work in Section 6.7 that for any continuous integrand the definite integral

$$F(x) = \int_0^x f(t)\, dt \tag{4}$$

exists and has the property that

$$\frac{d}{dx} F(x) = f(x). \tag{5}$$

Since (5) is equivalent to

$$\int f(x)\, dx = F(x),$$

we see that the indefinite integral of every continuous function exists. However, this fact has nothing to do with the issue of whether the integral can be expressed in terms of elementary functions. When such an expression is not possible, formula (4) can be thought of as providing a legitimate and sometimes useful method for creating new functions. For example, the nonelementary function of x defined by

$$\frac{1}{\sqrt{2\pi}} \int_0^x e^{-t^2/2}\, dt$$

has important applications in the theory of probability, and for this reason it has been studied and tabulated and has thereby acquired a certain status as a "known function."

Remark 2 It is easy to see that the integral

$$\int \frac{e^x}{x}\, dx \qquad \text{becomes} \qquad \int \frac{dt}{\ln t}$$

under the substitution $t = e^x$; for $x = \ln t$, $dx = dt/t$, and therefore

$$\int \frac{e^x}{x}\, dx = \int \frac{t}{\ln t}\frac{dt}{t} = \int \frac{dt}{\ln t}.$$

Since we know that the first integral is not elementary, it is clear that the second integral is also not elementary. This is worth noticing because the function of x defined by

$$\int_2^x \frac{dt}{\ln t} \tag{6}$$

is of great importance in the theory of prime numbers, and the behavior of this function for large values of x has been studied exhaustively for more than a century.* [The lower limit of integration in (6) is chosen to be 2 in order to avoid the point $t = 1$, where $\ln t = 0$.]

---

* See pp. 2–4 of H. M. Edwards, *Riemann's Zeta Function,* Academic Press, New York, 1974.

PROBLEMS

1 Consider an integral of the form $\int R(\sin x, \cos x)\, dx$, where the integrand is a rational function of $\sin x$ and $\cos x$. Show that the substitution

$$z = \tan \tfrac{1}{2}x$$

converts this integral into the integral of a rational function of z, which can then be worked out by routine procedures. Hint: Show that

$$\sec^2 \frac{1}{2}x = 1 + z^2, \qquad \cos x = \cos 2\left(\frac{1}{2}x\right) = \frac{1 - z^2}{1 + z^2},$$

$$\sin x = \frac{2z}{1 + z^2}, \qquad \text{and} \qquad dx = \frac{2dz}{1 + z^2}.$$

2 Use the method of Problem 1 to find

(a) $\displaystyle\int \frac{dx}{2 + \cos x}$;   (b) $\displaystyle\int \frac{\sin x\, dx}{2 + \sin x}$.

3 Use the method of Problem 1 to find
(a) $\int \sec x\, dx$;   (b) $\int \tan x\, dx$.
Express your answers in the usual form [i.e., $\ln(\sec x + \tan x)$ and $-\ln(\cos x)$].

4 Use the method of Problem 1 to obtain the following formulas:

(a) $\displaystyle\int \frac{dx}{a + b \sin x} = \int \frac{2dz}{az^2 + 2bz + a}$;

(b) $\displaystyle\int \frac{dx}{a + b \sin x + c \cos x}$

$$= \int \frac{2dz}{(a - c)z^2 + 2bz + (a + c)};$$

(c) $\displaystyle\int \frac{\sin x\, dx}{1 + \sin x} = \int \frac{4z\, dz}{(1 + z)^2(1 + z^2)}$;

(d) $\displaystyle\int \frac{\cos x\, dx}{1 + \cos x} = \int \frac{(1 - z^2)\, dz}{1 + z^2}$.

A *rationalizing substitution* is a change of variable that eliminates radicals or fractional exponents. Find the following integrals by using this idea.

5 $\displaystyle\int \frac{dx}{1 + \sqrt{x}}$.   Hint: Put $u = \sqrt{x}$.

6 $\displaystyle\int \frac{\sqrt{x}+1}{\sqrt{x}-1}\, dx.$

7 $\displaystyle\int \frac{dx}{\sqrt{x}+\sqrt[3]{x}}.$ Hint: Put $u = \sqrt[6]{x}$.

8 $\displaystyle\int \frac{3\sqrt{x}\, dx}{4(1 + x^{3/4})}.$

9 $\displaystyle\int \frac{\sqrt{x}}{1+x}\, dx.$ **10** $\displaystyle\int \frac{x^{2/3}}{1+x}\, dx.$

11 $\displaystyle\int \frac{\sqrt[4]{x}}{1+\sqrt{x}}\, dx.$ **12** $\displaystyle\int \frac{dx}{x(1-\sqrt[4]{x})}.$

13 $\displaystyle\int \frac{\sqrt{x}+2}{x+3}\, dx.$ **14** $\displaystyle\int \frac{\sqrt[3]{x}+1}{x}\, dx.$

15 The special elliptic integral

$$\int \frac{du}{\sqrt{(1-u^2)(1-k^2u^2)}}$$

is called the *elliptic integral of the first kind.* Show that each of the following integrals can be brought into this form by means of the indicated substitution:

(a) $\displaystyle\int \frac{dx}{\sqrt{1-k^2\sin^2 x}} = \int \frac{du}{\sqrt{(1-u^2)(1-k^2u^2)}}$,

    $u = \sin x$;

(b) $\displaystyle\int \frac{dx}{\sqrt{\cos 2x}} = \int \frac{du}{\sqrt{(1-u^2)(1-2u^2)}}$, $u = \sin x$;

(c) $\displaystyle\int \frac{dx}{\sqrt{\cos x}} = 2\int \frac{du}{\sqrt{(1-u^2)(1-2u^2)}}$, $u = \sin \frac{1}{2}x$;

(d) $\displaystyle\int \frac{dx}{\sqrt{\cos x - \cos \alpha}} =$

$$\sqrt{2}\, k \int \frac{du}{\sqrt{(1-u^2)(1-k^2u^2)}},$$

    $u = \sin \frac{1}{2}x$ and $k = \csc \frac{1}{2}\alpha.$

16 Consider the integral in part (b) of Problem 15,

$$\int \frac{dx}{\sqrt{\cos 2x}} = \int \frac{dx}{\sqrt{1-2\sin^2 x}}.$$

Show that the substitution $u = \tan x$ transforms this integral into the special elliptic integral

$$\int \frac{du}{\sqrt{1-u^4}}.$$

17 If p and q are rational numbers, show that the integral

$$\int x^p(1-x)^q\, dx \qquad\qquad (*)$$

is elementary in each of the following cases:
(a) p is an integer (hint: if $q = m/n$ with $n > 0$, put $1 - x = u^n$);
(b) q is an integer;
(c) $p + q$ is an integer $\Big[$ hint:

$$\int x^p(1-x)^q\, dx = \int x^{p+q}\left(\frac{1-x}{x}\right)^q dx\Big].$$

The Russian mathematician Chebyshev proved that these are the only cases for which the integral (*) is elementary.† Accordingly,

$$\int \sqrt{x}\,\sqrt[3]{1-x}\, dx, \quad \int \sqrt[3]{x}\,\sqrt{1-x}\, dx, \quad \int \sqrt[3]{x-x^2}\, dx$$

are not elementary.

18 Use the theorem of Chebyshev stated in Problem 17 to prove that each of the following integrals is not elementary:
(a) $\int \sqrt{1-x^3}\, dx$;
(b) $\int \sqrt{1-x^4}\, dx$;
(c) $\int \sqrt{1-x^n}\, dx$, where n is any integer > 2;

(d) $\displaystyle\int \frac{dx}{\sqrt{1-x^n}}$, where n is any integer > 2.

19 Use Problem 17 to prove that
(a) $\int \sqrt{\sin x}\, dx$ is not elementary (hint: put $u = \sin^2 x$);
(b) $\int \sin^p x\, dx$, where p is a rational number, is elementary if and only if p is an integer;
(c) $\int \sin^p x \cos^q x\, dx$, where p and q are rational numbers, is elementary if and only if p or q is an odd integer or $p + q$ is an even integer.

_____

† See Ritt, p. 37.

10.9
(OPTIONAL) NUMERICAL INTEGRATION

From the point of view of the theorist, the main value of calculus is intellectual; it helps us comprehend the underlying connections among natural phenomena. However, anyone who uses calculus as a practical tool in science or engineering must occasionally face the question of how the theory can be applied to yield useful methods for performing actual numerical calculations.

In this section we consider the problem of computing the numerical value of a definite integral

$$\int_a^b f(x)\,dx \tag{1}$$

in decimal form to any desired degree of accuracy. In order to find the value of (1) by using the formula

$$\int_a^b f(x)\,dx = F(b) - F(a), \tag{2}$$

we must be able to find the indefinite integral $F(x)$ and we must be able to evaluate it at both $x = a$ and $x = b$. When this is not possible, formula (2) is useless. This approach fails even for such simple-looking integrals as

$$\int_0^\pi \sqrt{\sin x}\,dx \qquad \text{and} \qquad \int_1^5 \frac{e^x}{x}\,dx,$$

because there are no elementary functions whose derivatives are $\sqrt{\sin x}$ and e^x/x (see Section 10.8).

Our purpose here is to describe two methods of computing the numerical value of (1) as accurately as we wish by simple procedures that can be applied regardless of whether an indefinite integral can be found or not. The formulas we develop use only simple arithmetic and the values of $f(x)$ at a finite number of points in the interval $[a, b]$. In comparison with the use of the approximating sums that are used in defining the integral (see Section 6.4), the formulas of this section are more efficient in the sense that they give much better accuracy for the same amount of computational labor.

THE TRAPEZOIDAL RULE

Let the interval $[a, b]$ be divided into n equal parts by points $x_0, x_1, \ldots, x_n$ from $x_0 = a$ to $x_n = b$. Let $y_0, y_1, \ldots, y_n$ be the corresponding values of $y = f(x)$. We then approximate the area between $y = f(x)$ and the x-axis, for $x_{k-1} \le x \le x_k$, by the trapezoid whose upper edge is the segment joining the points (x_{k-1}, y_{k-1}) and (x_k, y_k) (see Fig. 10.5). The area of this trapezoid is clearly

$$\tfrac{1}{2}(y_{k-1} + y_k)(x_k - x_{k-1}). \tag{3}$$

If we write

$$\Delta x = x_k - x_{k-1} = \frac{b-a}{n}, \tag{4}$$

then adding the expressions (3) for $k = 1, 2, \ldots, n$ gives the approximation formula

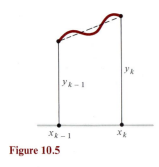

Figure 10.5

$$\int_a^b f(x)\, dx \cong \left(\frac{1}{2} y_0 + y_1 + y_2 + \cdots + y_{n-1} + \frac{1}{2} y_n\right) \Delta x,$$

because each of the y's except the first and the last occurs twice. This formula is called the trapezoidal rule.

Example 1   Use the trapezoidal rule with $n = 4$ to calculate an approximate value for the integral

$$\int_0^1 \sqrt{1 - x^3}\, dx.$$

Here $y = f(x) = \sqrt{1 - x^3}$ and $x_0 = 0$, $x_1 = \frac{1}{4}$, $x_2 = \frac{1}{2}$, $x_3 = \frac{3}{4}$, $x_4 = 1$. We can compute the y's easily by dividing and using a table of square roots:

$$y_0 = 1,$$
$$y_1 = \sqrt{\tfrac{63}{64}} = \sqrt{0.984} = 0.992,$$
$$y_2 = \sqrt{\tfrac{7}{8}} = \sqrt{0.875} = 0.935,$$
$$y_3 = \sqrt{\tfrac{37}{64}} = \sqrt{0.578} = 0.760,$$
$$y_4 = 0.$$

By the trapezoidal rule, we therefore have

$$\int_0^1 \sqrt{1 - x^3}\, dx \cong \frac{1}{4}(0.500 + 0.992 + 0.935 + 0.760 + 0.000) = 0.797.$$

SIMPSON'S RULE

Our second method is based on a more ingenious device than approximating each small piece of the curve by a line segment; this time we approximate each piece by a portion of a parabola that "fits" the curve in a manner to be described.

Again we divide the interval $[a, b]$ into n equal parts, but now we require that n be an *even* integer. Consider the first three points x_0, x_1, x_2 and the corresponding points on the curve $y = f(x)$, as shown in Fig. 10.6. If these points are not collinear, then there is a unique parabola that has vertical axis and that passes through all three points. To see this, recall that the equation of any parabola with vertical axis has the form $y = P(x)$ where $P(x)$ is a quadratic polynomial, and observe that this polynomial can always be written in the form

$$P(x) = a + b(x - x_1) + c(x - x_1)^2. \tag{5}$$

We choose the constants a, b, c to make the parabola pass through the three points under consideration, as indicated in the figure. Three conditions are necessary:

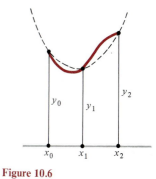

Figure 10.6

$$\text{At } x = x_0, \quad a + b(x_0 - x_1) + c(x_0 - x_1)^2 = y_0; \tag{6}$$
$$\text{At } x = x_1, \quad a = y_1;$$
$$\text{At } x = x_2, \quad a + b(x_2 - x_1) + c(x_2 - x_1)^2 = y_2. \tag{7}$$

Equations (6) and (7) can be solved for the constants b and c. However, it is more convenient to use the definition (4) of Δx and the fact that $a = y_1$ to write these equations in the form

$$-b\,\Delta x + c\,\Delta x^2 = y_0 - y_1,$$
$$b\,\Delta x + c\,\Delta x^2 = y_2 - y_1,$$

from which we obtain

$$2c\,\Delta x^2 = y_0 - 2y_1 + y_2. \tag{8}$$

We now think of the parabola (5) as a close approximation to the curve $y = f(x)$ on the interval $[x_0, x_2]$, and we compute this part of the integral (1) accordingly,

$$\int_{x_0}^{x_2} f(x)\,dx \cong \int_{x_0}^{x_2} [a + b(x - x_1) + c(x - x_1)^2]\,dx$$
$$= \left[ax + \tfrac{1}{2}b(x - x_1)^2 + \tfrac{1}{3}c(x - x_1)^3 \right]_{x_0}^{x_2}.$$

When this is evaluated in terms of Δx, we obtain

$$2a\,\Delta x + \tfrac{2}{3}c\,\Delta x^3.$$

By using (8) and the fact that $a = y_1$, we can write this in the form

$$2y_1\,\Delta x + \tfrac{1}{3}(y_0 - 2y_1 + y_2)\,\Delta x = \tfrac{1}{3}(y_0 + 4y_1 + y_2)\,\Delta x.$$

The same procedure can be applied to each of the intervals $[x_2, x_4]$, $[x_4, x_6]$, ... When the results are all added together, we get the approximation formula

$$\int_a^b f(x)\,dx \cong \frac{1}{3}(y_0 + 4y_1 + 2y_2 + \cdots + 4y_{n-1} + y_n)\,\Delta x,$$

which is called Simpson's rule. We specifically point out the structure of the expression in parentheses: y_0 and y_n occur with coefficient 1, the remaining y's with even subscripts occur with coefficient 2, and the y's with odd subscripts occur with coefficient 4.

Any serious study of a method of approximate calculation must include a detailed estimate of the magnitude of the error committed, so that definite knowledge is available of the level of accuracy attained. We do not pursue this matter here, but merely state that the error in Simpson's rule is known to be at most

$$\frac{M(b - a)}{180}\,\Delta x^4, \tag{9}$$

where M is the maximum value of $f^{(4)}(x)$ on $[a, b]$.

Example 2 Use Simpson's rule with $n = 4$ to calculate an approximate value for the integral

$$\int_0^2 \frac{dx}{1 + x^4}.$$

This time we have $x_0 = 0$, $x_1 = \frac{1}{2}$, $x_2 = 1$, $x_3 = \frac{3}{2}$, $x_4 = 2$. A simple table helps to keep the computations in order:

$$y_0 = 1 \qquad\qquad y_0 = 1.000$$
$$y_1 = \tfrac{16}{17} = 0.941 \qquad 4y_1 = 3.764$$
$$y_2 = \tfrac{1}{2} = 0.500 \qquad 2y_2 = 1.000$$
$$y_3 = \tfrac{16}{97} = 0.165 \qquad 4y_3 = 0.660$$
$$y_4 = \tfrac{1}{17} = 0.059 \qquad y_4 = \underline{0.059}$$
$$6.483$$

Simpson's rule now yields

$$\int_0^2 \frac{dx}{1 + x^4} \cong \frac{1}{6}(6.483) = 1.081.$$

Students who own hand calculators and enjoy working with them have little opportunity to develop their skills in a calculus course, because calculus deals mainly with ideas and very little with numerical computations. However, the methods and problems of this section provide plenty of raw material for these idle calculators.

PROBLEMS

1 Clearly,

$$\int_0^1 \sqrt{x}\, dx = \frac{2}{3} = 0.666 \ldots$$

Calculate the value of this integral approximately with $n = 4$ by using
(a) the trapezoidal rule (recall that $\sqrt{2} = 1.414 \ldots$ and $\sqrt{3} = 1.732 \ldots$);
(b) Simpson's rule.
Since the two rules are almost equally easy to apply, and Simpson's rule is usually more accurate, the trapezoidal rule is rarely used in practical computations.

2 Clearly,

$$\int_0^\pi \sin x\, dx = 2.$$

Calculate the value of this integral approximately by using Simpson's rule with $n = 4$.

3 The exact value of

$$\int_0^\pi \sqrt{\sin x}\, dx$$

is not known. Find its approximate value by using Simpson's rule with $n = 4$.

4 The exact value of

$$\int_1^5 \frac{e^x}{x}\, dx$$

is not known. Use Simpson's rule with $n = 4$ to find its approximate value.

5 The exact value of

$$\int_0^2 e^{-x^2}\, dx$$

is not known, but to 10 decimal places it is 0.8820813908. Calculate this integral approximately by using Simpson's rule with $n = 4$.

6 Find an approximate value for $\ln 2$ by using the fact that

$$\ln 2 = \int_1^2 \frac{dx}{x}$$

and applying Simpson's rule with $n = 4$. (To 10 decimal places, $\ln 2 = 0.6931471806$.)

7 Use the formula

$$\frac{\pi}{4} = \int_0^1 \frac{dx}{1 + x^2}$$

to find an approximate value for π by using Simpson's rule with $n = 4$. (To 10 decimal places, $\pi = 3.1415926536$.)

8 Suppose that the three points on the curve in the derivation of Simpson's rule are collinear. Use (8) to show that in this case $c = 0$, and conclude that under this assumption the curve through the points is a straight line instead of a parabola.

9   Simpson's rule is designed to be exactly correct if $f(x)$ is a quadratic polynomial. It is a remarkable fact that it also gives an exact result for cubic polynomials. Prove this. Hint: Notice that it suffices to establish the statement for $n = 2$; then prove it in this case for the function $f(x) = x^3$; then extend it to any cubic polynomial.

10 Use formula (9) to prove the statement in Problem 9.

ADDITIONAL PROBLEMS FOR CHAPTER 10

SECTION 10.2
Find each of the following integrals.

1   $\int \sqrt{3x + 5}\, dx.$

2   $\int \dfrac{(\ln x)^6\, dx}{x}.$

3   $\int \dfrac{6x\, dx}{1 + 3x^2}.$

4   $\int \dfrac{e^{1/x}\, dx}{x^2}.$

5   $\int \cos (1 - 5x)\, dx.$

6   $\int \sin x \sin (\cos x)\, dx.$

7   $\int \dfrac{\sec \sqrt{x} \tan \sqrt{x}\, dx}{\sqrt{x}}.$

8   $\int \dfrac{x^3\, dx}{\sqrt{1 - x^8}}.$

9   $\int \dfrac{2x\, dx}{1 + x^4}.$

10 $\int \dfrac{x^2 + 5}{x^2 + 4}\, dx.$

11 $\int \cot 4x\, dx.$

12 $\int \dfrac{dx}{\sin 2x}.$

13 $\int \dfrac{dx}{x(\ln x)^2}.$

14 $\int \dfrac{dx}{3 - x}.$

15 $\int \dfrac{\sec^2 x\, dx}{\tan x}.$

16 $\int 10x^4 e^{x^5}\, dx.$

17 $\int \sin \left(\dfrac{3x - 5}{2} \right) dx.$

18 $\int \csc^2 (2 - x)\, dx.$

19 $\int 6x^2 \cot x^3 \csc x^3\, dx.$

20 $\int \dfrac{\sec^2 x\, dx}{\sqrt{1 - \tan^2 x}}.$

21 $\int \dfrac{dx}{x[1 + (\ln x)^2]}.$

22 $\int \cot \pi x\, dx.$

23 $\int \dfrac{dx}{(3x + 5)^2}.$

24 $\int \tan x \sec^4 x\, dx.$

25 $\int \dfrac{dx}{3 - 2x}.$

26 $\int \dfrac{(e^x + 2x)\, dx}{e^x + x^2 - 2}.$

27 $\int x^2 \cos (1 + x^3)\, dx.$

28 $\int \sin (2 - x)\, dx.$

29 $\int x \csc^2 (x^2 + 1)\, dx.$

30 $\int \dfrac{dx}{\sqrt{3 - 4x^2}}.$

31 $\int \dfrac{\cos x\, dx}{1 + \sin^2 x}.$

32 $\int \dfrac{dx}{1 + 4x^2}.$

33 $\int \dfrac{dx}{\tan 2x}.$

34 $\int (\csc x - 1)^2\, dx.$

35 $\int \dfrac{\tan^{-1} x\, dx}{1 + x^2}.$

36 $\int \sqrt[3]{3x - 2}\, dx.$

37 $\int \dfrac{dx}{2x + 1}.$

38 $\int \dfrac{(e^x - e^{-x})\, dx}{e^x + e^{-x}}.$

39 $\int e^{x/3}\, dx.$

40 $\int \dfrac{dx}{\sec 2x}.$

41 $\int \dfrac{\sec^2 (\sin x)\, dx}{\sec x}.$

42 $\int (\csc x - \cot x) \csc x\, dx.$

43 $\int \dfrac{dx}{\sqrt{1 - 25x^2}}.$

44 $\int \dfrac{dx}{16 + 25x^2}.$

45 $\int \dfrac{\sec x \tan x\, dx}{1 + \sec^2 x}.$

46 $\int (1 + \sec x)^2\, dx.$

47 $\int \dfrac{(\ln x)^2\, dx}{x}.$

48 $\int \dfrac{\cos x\, dx}{\sin^2 x}.$

49 $\int \dfrac{\sin x\, dx}{1 + \cos x}.$

50 $\int \dfrac{6 \csc^2 x\, dx}{1 - 3 \cot x}.$

51 $\int \dfrac{dx}{e^{3x}}.$

52 $\int e^x \cos e^x\, dx.$

53 $\int \dfrac{\sin (\ln x)\, dx}{x}.$

54 $\int \dfrac{\csc^2 \sqrt{x}\, dx}{\sqrt{x}}.$

55 $\int \dfrac{\csc 1/x \cot 1/x\, dx}{x^2}.$

56 $\int \dfrac{4\, dx}{3 + 4x^2}.$

57 $\int \dfrac{e^{2x}\, dx}{1 + e^{4x}}.$

58 $\int \dfrac{x\, dx}{\sin x^2}.$

59 $\int x^3 \sqrt{2 + x^4}\, dx.$

60 $\int \dfrac{x\, dx}{\sqrt{2 - x^2}}.$

61 $\int \dfrac{(1 + e^x)\, dx}{e^x + x}.$

62 $\int xe^{x^2}\, dx.$

63 $\int \dfrac{2\, dx}{\sqrt{e^x}}.$

64 $\int x \sin (1 - x^2)\, dx.$

65 $\displaystyle\int \frac{dx}{\sin^2 x}$.

66 $\displaystyle\int \frac{dx}{\sqrt{4 - 9x^2}}$.

67 $\displaystyle\int x \tan x^2 \, dx$.

68 $\displaystyle\int \frac{\sec^2 x \, dx}{\sqrt{\tan x}}$.

69 $\displaystyle\int \frac{x \, dx}{1 + x^2}$.

70 $\displaystyle\int 2e^{2x} \, dx$.

71 $\displaystyle\int xe^{3x^2 - 2} \, dx$.

72 $\displaystyle\int 3x^2 \sin x^3 \, dx$.

73 $\displaystyle\int \sec x \, (\sec x + \tan x) \, dx$.

74 $\displaystyle\int \frac{x^2 \, dx}{9 + x^6}$.

75 $\displaystyle\int x^{2/3}\sqrt{1 + x^{5/3}} \, dx$.

76 $\displaystyle\int \frac{4x^3 \, dx}{1 + x^4}$.

77 $\displaystyle\int \sec^2 x \, e^{\tan x} \, dx$.

78 $\displaystyle\int x \sec^2 x^2 \, dx$.

79 $\displaystyle\int (1 + \cos x)^4 \sin x \, dx$.

80 $\displaystyle\int \frac{(1 + \cos x) \, dx}{x + \sin x}$.

81 $\displaystyle\int \cos (\tan x) \sec^2 x \, dx$.

82 $\displaystyle\int \frac{\csc^2 (\ln x) \, dx}{x}$.

Compute each of the following definite integrals by making a suitable substitution and changing the limits of integration.

83 $\displaystyle\int_0^{\sqrt{2}/2} \frac{2x \, dx}{\sqrt{1 - x^4}}$.

84 $\displaystyle\int_0^{\sqrt{\pi}} x \sin x^2 \, dx$.

85 $\displaystyle\int_{\pi/8}^{\pi/4} \cot 2x \csc^2 2x \, dx$.

86 $\displaystyle\int_0^{\pi/2} \frac{\cos x \, dx}{1 + \sin^2 x}$.

87 $\displaystyle\int_0^4 2x\sqrt{x^2 + 9} \, dx$.

88 $\displaystyle\int_0^3 \frac{x \, dx}{\sqrt{x^2 + 16}}$.

SECTION 10.3

Calculate each of the following integrals.

89 $\displaystyle\int \sin^2 5x \, dx$.

90 $\displaystyle\int \cos^4 3x \, dx$.

91 $\displaystyle\int \cos^2 7x \, dx$.

92 $\displaystyle\int \sin^6 x \, dx$.

93 $\displaystyle\int \sin^5 x \cos^2 x \, dx$.

94 $\displaystyle\int \sin^5 x \, dx$.

95 $\displaystyle\int \cos^3 4x \, dx$.

96 $\displaystyle\int \cos^3 2x \sin 2x \, dx$.

97 $\displaystyle\int \frac{\cos^3 x \, dx}{\sin^4 x}$.

98 $\displaystyle\int \frac{\sin^5 x \, dx}{\sqrt{\cos x}}$.

99 $\displaystyle\int \sin^{3/5} x \cos x \, dx$.

100 $\displaystyle\int \sin^2 x \cos^4 x \, dx$.

101 $\displaystyle\int \sec^6 x \, dx$.

102 $\displaystyle\int \frac{dx}{\cos^4 x}$.

103 $\displaystyle\int \tan^3 x \sec^7 x \, dx$.

104 $\displaystyle\int \cot^4 x \, dx$.

105 $\displaystyle\int \cot^5 x \, dx$.

106 $\displaystyle\int \frac{dx}{\sin^4 3x}$.

107 $\displaystyle\int (\sec 3x + \csc 3x)^2 \, dx$.

108 $\displaystyle\int \frac{dx}{\sec x \tan x}$.

SECTION 10.4

Find each of the following integrals.

109 $\displaystyle\int \sqrt{3 - x^2} \, dx$.

110 $\displaystyle\int \frac{dx}{(a^2 + x^2)^{3/2}}$.

111 $\displaystyle\int \frac{x^2 \, dx}{a^2 + x^2}$.

112 $\displaystyle\int \frac{\sqrt{4 - 9x^2}}{x} \, dx$.

113 $\displaystyle\int x^3\sqrt{a^2 - x^2} \, dx$.

114 $\displaystyle\int \frac{x^3 \, dx}{\sqrt{a^2 + x^2}}$.

115 $\displaystyle\int \frac{\sqrt{a^2 + x^2}}{x^2} \, dx$.

116 $\displaystyle\int \frac{dx}{x^2\sqrt{a^2 - x^2}}$.

117 $\displaystyle\int \frac{dx}{x^4\sqrt{a^2 - x^2}}$.

118 $\displaystyle\int \frac{dx}{x^2 + x^4}$.

119 $\displaystyle\int \frac{x^2 \, dx}{(a^2 + x^2)^2}$.

120 $\displaystyle\int x^3(a^2 - x^2)^{3/2} \, dx$.

121 $\displaystyle\int \frac{dx}{x^2\sqrt{x^2 - 9}}$.

122 $\displaystyle\int \sqrt{x^2 - 1} \, dx$.

123 $\displaystyle\int \frac{dx}{(1 - 9x^2)^{3/2}}$.

124 $\displaystyle\int \frac{x^2 \, dx}{\sqrt{a^2 + x^2}}$.

125 $\displaystyle\int \frac{dx}{x\sqrt{9 + 4x^2}}$.

126 $\displaystyle\int \frac{dx}{\sqrt{9 - (x - 1)^2}}$.

127 $\displaystyle\int \frac{x^2 \, dx}{(a^2 - x^2)^{3/2}}$.

128 $\displaystyle\int \frac{dx}{x^4\sqrt{a^2 + x^2}}$.

129 $\displaystyle\int \frac{x^2 \, dx}{(a^2 + x^2)^{3/2}}$.

130 $\displaystyle\int \frac{\sqrt{a^2 - x^2}}{x^4} \, dx$.

131 $\displaystyle\int \frac{x^2\,dx}{\sqrt{x^2 - a^2}}.$

132 $\displaystyle\int \frac{x^3\,dx}{(x^2 - a^2)^{3/2}}.$

160 $\displaystyle\int \frac{3x^2 - 5x + 4}{x^3 - x^2 + x - 1}\,dx.$

SECTION 10.5
Calculate each of the following integrals.

133 $\displaystyle\int \frac{dx}{\sqrt{65 - 8x - x^2}}.$

134 $\displaystyle\int \frac{dx}{\sqrt{1 + 4x - x^2}}.$

135 $\displaystyle\int \frac{dx}{5x^2 + 10x + 15}.$

136 $\displaystyle\int \frac{(3x - 5)\,dx}{x^2 + 2x + 2}.$

137 $\displaystyle\int \frac{dx}{\sqrt{2 + 2x - 3x^2}}.$

138 $\displaystyle\int \frac{(1 - x)\,dx}{\sqrt{8 + 2x - x^2}}.$

139 $\displaystyle\int \frac{x^2\,dx}{\sqrt{2x - x^2}}.$

140 $\displaystyle\int \frac{x\,dx}{\sqrt{x^2 - 4x + 5}}.$

141 $\displaystyle\int \frac{dx}{3x^2 - 6x + 15}.$

142 $\displaystyle\int \frac{(3x + 4)\,dx}{\sqrt{2x + x^2}}.$

143 $\displaystyle\int \frac{dx}{(x - 1)\sqrt{x^2 - 2x - 3}}.$

144 $\displaystyle\int \frac{(2x - 5)\,dx}{\sqrt{4x - x^2}}.$

145 $\displaystyle\int \frac{(3x + 7)\,dx}{\sqrt{x^2 + 4x + 8}}.$

146 $\displaystyle\int \sqrt{x^2 + 2x + 2}\,dx.$

147 $\displaystyle\int \frac{(2x - 3)\,dx}{(x^2 + 2x - 3)^{3/2}}.$

148 $\displaystyle\int \sqrt{x^2 - 2x}\,dx.$

SECTION 10.6
Find each of the following integrals.

149 $\displaystyle\int \frac{16x + 69}{x^2 - x - 12}\,dx.$

150 $\displaystyle\int \frac{3x - 56}{x^2 + 3x - 28}\,dx.$

151 $\displaystyle\int \frac{-8x - 16}{4x^2 - 1}\,dx.$

152 $\displaystyle\int \frac{12x - 63}{x^2 - 3x}\,dx.$

153 $\displaystyle\int \frac{3x^2 - 10x - 60}{x^3 + x^2 - 12x}\,dx.$

154 $\displaystyle\int \frac{8x^2 + 55x - 25}{x^3 - 25x}\,dx.$

155 $\displaystyle\int \frac{-2x^2 - 18x + 18}{x^3 - 9x}\,dx.$

156 $\displaystyle\int \frac{4x^2 - 2x - 108}{x^3 + 5x^2 - 36x}\,dx.$

157 $\displaystyle\int \frac{-3x^3 + x^2 + 2x + 3}{x^4 + x^3}\,dx.$

158 $\displaystyle\int \frac{9x^2 - 35x + 28}{x^3 - 4x^2 + 4x}\,dx.$

159 $\displaystyle\int \frac{x^2 - 5x - 8}{x^3 + 4x^2 + 8x}\,dx.$

SECTION 10.7
Calculate the integrals in Problems 161 to 176 by the method of integration by parts.

161 $\displaystyle\int x^2 \tan^{-1} x\,dx.$

162 $\displaystyle\int x^2 \cos x\,dx.$

163 $\displaystyle\int \cos(\ln x)\,dx.$

164 $\displaystyle\int x \sin^2 x\,dx.$

165 $\displaystyle\int x^3 \cos x\,dx.$

166 $\displaystyle\int \sqrt{1 + x^2}\,dx.$

167 $\displaystyle\int \frac{\ln x\,dx}{(x + 1)^2}.$

168 $\displaystyle\int \frac{xe^x\,dx}{(x + 1)^2}.$

169 $\displaystyle\int \frac{x^3\,dx}{\sqrt{1 + x^2}}.$

170 $\displaystyle\int x(x + 3)^{10}\,dx.$

171 $\displaystyle\int e^{ax} \sin bx\,dx.$

172 $\displaystyle\int x^n \ln x\,dx\ (n \neq -1).$

173 $\displaystyle\int \frac{x\,dx}{e^x}.$

174 $\displaystyle\int x^2 \sin x\,dx.$

175 $\displaystyle\int x^3 e^{-2x}\,dx.$

176 $\displaystyle\int \ln(x + \sqrt{x^2 + a^2})\,dx.$

177 Find the area under the curve $y = \sin\sqrt{x}$ from $x = 0$ to $x = \pi^2$.

178 Calculate the integral $\displaystyle\int \frac{x^3}{\sqrt{1 + x^2}}\,dx$ by using the identity

$$\frac{x^3}{\sqrt{1 + x^2}} = \frac{x(1 + x^2 - 1)}{\sqrt{1 + x^2}} = x\sqrt{1 + x^2} - \frac{x}{\sqrt{1 + x^2}}.$$

Make sure your answer agrees with the result of Problem 169.

179 Calculate the integral $\displaystyle\int_0^a x^2\sqrt{a - x}\,dx$ (a) by using the substitution $u = \sqrt{a - x}$; (b) by parts.

180 Use integration by parts to show that

$$\int \sqrt{a^2 - x^2}\,dx = x\sqrt{a^2 - x^2} + \int \frac{x^2}{\sqrt{a^2 - x^2}}\,dx.$$

Write $x^2 = -(-x^2) = -(a^2 - x^2 - a^2)$ in the numerator of the second integral and thereby obtain the formula

$$\int \sqrt{a^2 - x^2}\,dx = \tfrac{1}{2}x\sqrt{a^2 - x^2} + \tfrac{1}{2}a^2 \int \frac{dx}{\sqrt{a^2 - x^2}}$$

$$= \tfrac{1}{2}x\sqrt{a^2 - x^2} + \tfrac{1}{2}a^2 \sin^{-1}\frac{x}{a} + c.$$

181 Use the method of Problem 180 to obtain the formula

$$\int (a^2 - x^2)^n \, dx$$

$$= \frac{x(a^2 - x^2)^n}{2n + 1} + \frac{2a^2n}{2n + 1} \int (a^2 - x^2)^{n-1} \, dx.$$

*182 Use the idea of Problem 181 to obtain formula (9) in Section 10.6,

$$\int \frac{dx}{(a^2 + x^2)^n} = \frac{1}{2a^2(n - 1)} \cdot \frac{x}{(a^2 + x^2)^{n-1}}$$

$$+ \frac{2n - 3}{2a^2(n - 1)} \int \frac{dx}{(a^2 + x^2)^{n-1}}.$$

In the next three problems, derive the given reduction formula and apply it to the indicated special case.

183 (a) $\displaystyle\int x^m (\ln x)^n \, dx = \frac{x^{m+1}(\ln x)^n}{m + 1}$

$$- \frac{n}{m + 1} \int x^m (\ln x)^{n-1} \, dx.$$

(b) $\int x^5 (\ln x)^3 \, dx$.

184 (a) $\displaystyle\int x^n e^{ax} \, dx = \frac{1}{a} x^n e^{ax} - \frac{n}{a} \int x^{n-1} e^{ax} \, dx$.

(b) $\int x^3 e^{-2x} \, dx$.

185 (a) $\displaystyle\int \sec^n x \, dx = \frac{1}{n - 1} \sec^{n-2} x \tan x$

$$+ \frac{n - 2}{n - 1} \int \sec^{n-2} x \, dx.$$

(b) $\int \sec^3 x \, dx$ (see Problem 29 in Section 10.3).

11

FURTHER APPLICATIONS OF INTEGRATION

11.1

THE CENTER OF MASS OF A DISCRETE SYSTEM

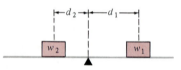

Figure 11.1

Most of the ideas in this chapter are based on the simple physical concept of center of gravity. As we shall see, this concept has implications for geometry, and it turns out to be possible to use it to arrive at a reasonable notion of what ought to be meant by the "center" of a general geometric figure. In this introductory section we confine ourselves to describing the concepts, and make no use of integration.

We begin by considering two children of weights w_1 and w_2 sitting at distances d_1 and d_2 from the fulcrum of a seesaw (Fig. 11.1). As we know, each child can increase the tendency of his or her weight to turn one end down by moving farther out from the fulcrum, and the seesaw balances when

$$w_1 d_1 = w_2 d_2. \tag{1}$$

This principle was discovered by Archimedes, and is known as the *law of the lever*. If we establish a horizontal x-axis with its origin at the fulcrum and the positive direction to the right, then (1) can be written in the form

$$w_1 x_1 + w_2 x_2 = 0 \quad \text{or} \quad \sum_{k=1}^{2} w_k x_k = 0,$$

where $x_1 = d_1$ and $x_2 = -d_2$.

We now extend this discussion by considering the x-axis as a weightless horizontal rod that pivots at the point p, as shown in Fig. 11.2, and we assume that n weights w_k are placed at points x_k, where $k = 1, 2, \ldots, n$. By Archimedes' law, this system of weights will exactly balance, or be in equilibrium about p, if

$$\sum w_k(x_k - p) = 0.$$

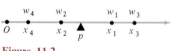

Figure 11.2

More generally, whether this system is in equilibrium or not, the sum $\sum w_k(x_k - p)$ measures the tendency of the system to turn in a clockwise direction about the pivot point p. This sum is called the *moment* of the system about p, and the system is in equilibrium if this moment is zero. If the weights w_k and their positions x_k are given in some arbitrary manner, and if

318

we are free to move the pivot point p, then it is easy to find a point $p = \bar{x}$ at which the system will balance, that is, with the property that the moment of the system about $\bar{x}$ is zero. The required condition is

$$\sum w_k(x_k - \bar{x}) = 0.$$

This is equivalent to

$$\sum w_k x_k - \sum w_k \bar{x} = 0 \quad \text{or} \quad \bar{x} \sum w_k = \sum w_k x_k,$$

so

$$\bar{x} = \frac{\sum w_k x_k}{\sum w_k}. \tag{2}$$

This balancing point $\bar{x}$ is called the *center of gravity* of the given system of weights.

We now recall that the weight of a body at the surface of the earth is simply the force exerted on the body by the gravitational attraction of the earth, and is therefore given by Newton's formula $F = mg$, where m is the mass of the body and g is the acceleration due to gravity (approximately 32 feet per second per second or 9.80 meters per second per second). In the above discussion this means that $w_k = m_k g$, where m_k is the mass of the kth body. Formula (2) can therefore be written as

$$\bar{x} = \frac{\sum m_k g x_k}{\sum m_k g} = \frac{\sum m_k x_k}{\sum m_k}. \tag{3}$$

With the influence of gravity removed from the discussion in this way, and the weights w_k in (2) replaced by the masses m_k in (3), it is customary to call the point $\bar{x}$ the *center of mass* of the system.

It is easy to extend these ideas to a two-dimensional system of masses m_k located at points (x_k, y_k) in a horizontal xy-plane (Fig. 11.3). We define the *moment* of this system about the y-axis by

$$M_y = \sum m_k x_k, \tag{4}$$

which is the sum of each of the masses multiplied by its signed distance from the y-axis. If we think of the xy-plane as a weightless horizontal tray, as suggested by the figure, then in physical language the condition $M_y = 0$ means that this tray — with the given distribution of masses — will balance if

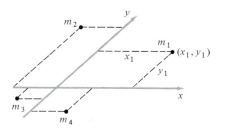

Figure 11.3

it rests on a knife-edge along the y-axis. Similarly, the moment of the system about the x-axis is defined by

$$M_x = \sum m_k y_k. \tag{5}$$

Students should carefully observe the interchange of x's and y's in formulas (4) and (5); to compute M_y we use the x_k's, and to compute M_x we use the y_k's. If we denote the total mass of all the particles in the system by m, so that

$$m = \sum m_k,$$

then the *center of mass* of the system is defined to be the point $(\bar{x}, \bar{y})$, where

$$\bar{x} = \frac{\sum m_k x_k}{\sum m_k} = \frac{M_y}{m} \tag{6}$$

and

$$\bar{y} = \frac{\sum m_k y_k}{\sum m_k} = \frac{M_x}{m}. \tag{7}$$

The center of mass of our system can be interpreted in two ways. First, if equations (6) and (7) are written in the form

$$m\bar{x} = M_y \quad \text{and} \quad m\bar{y} = M_x,$$

then we see that $(\bar{x}, \bar{y})$ is the point at which the entire mass m of the system can be concentrated without changing the total moment about either axis. The second interpretation depends on writing (6) and (7) in the form

$$\sum m_k(x_k - \bar{x}) = 0 \quad \text{and} \quad \sum m_k(y_k - \bar{y}) = 0.$$

If we think of our system in the way described, as a distribution of masses on a weightless horizontal tray, then these equations tell us that the tray will balance if it rests on a knife-edge along the line $x = \bar{x}$ parallel to the y-axis, and also along the line $y = \bar{y}$ parallel to the x-axis. These conditions imply that the tray will balance if it rests on a knife-edge along *any* line through $(\bar{x}, \bar{y})$. It will therefore also balance if supported by a sharp nail *precisely at* the point $(\bar{x}, \bar{y})$.

In the preceding discussion, the xy-coordinate system in Fig. 11.3 provides a frame of reference that is useful for developing the ideas. However, it is clear from the physical meaning of the center of mass that the location of this point is determined by the masses themselves and their individual positions, and does not depend on the particular coordinate system that is used to describe these positions. As a practical consequence, this fact tells us that in any specific situation we are free to choose any coordinate system that seems convenient under the circumstances.

Remark The "center of population" of the United States has been described as the point at which a life-sized flat map of the whole country would balance on a pin if all Americans weighed the same. The location of this point has been calculated from the data in each census. In 1790 it was a few miles east of Baltimore. It has been moving westward ever since, and in 1980 it crossed the Mississippi River and was a few miles south of St. Louis. It is interesting to speculate on what changes in the position of this point would be produced by "weighing" Americans according to age, or wealth, or education, instead of treating them as interchangeable units.

11.2
CENTROIDS

The ideas discussed in Section 11.1 apply to discrete systems of particles located at a finite number of points in a plane. We now consider how integration can be used to generalize these ideas to a continuous distribution of mass throughout a region R in the xy-plane, as shown in Fig. 11.4.

We shall think of R as a thin sheet of homogeneous material — say, a uniform metal plate — whose density δ ($=$ mass per unit area) is constant. To define the moment of this plate about the y-axis, we consider a thin vertical strip of height $f(x)$ and width dx, whose position in the region is specified by the variable x (Fig. 11.4, left). The area of this strip is $f(x)\,dx$ and its mass is $\delta f(x)\,dx$; and since all of its mass is essentially at the same distance x from the y-axis, its moment about this axis is $x\delta f(x)\,dx$. The total moment of the plate about the y-axis is therefore obtained by allowing the strip to sweep across the region, and by integrating — or adding together — all these small contributions to the moment as x increases from a to b,

$$M_y = \int_a^b x\delta f(x)\,dx. \tag{1}$$

This formula can be derived by laboriously constructing approximating sums and then forming the limit of these sums, which leads to (1) by means of the definition of the integral. However, we prefer to continue in the spirit of Chapter 7, and the preceding discussion provides yet another illustration of the power of the Leibnizian approach to integration as we described it in Section 7.1.

Similarly, the moment of the plate about the x-axis is obtained by considering a thin horizontal strip of length $g(y)$ and width dy (Fig. 11.4, right), and is given by the formula

$$M_x = \int_c^d y\delta g(y)\,dy.$$

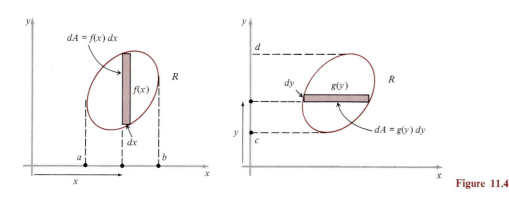

Figure 11.4

The total mass of the plate can evidently be expressed in two ways,

$$m = \int_a^b \delta f(x)\, dx = \int_c^d \delta g(y)\, dy.$$

The *center of mass* $(\bar{x}, \bar{y})$ of the plate is now defined by

$$\bar{x} = \frac{\displaystyle\int_a^b x \delta f(x)\, dx}{\displaystyle\int_a^b \delta f(x)\, dx} = \frac{M_y}{m}$$

and

$$\bar{y} = \frac{\displaystyle\int_c^d y \delta g(y)\, dy}{\displaystyle\int_c^d \delta g(y)\, dy} = \frac{M_x}{m}.$$

These formulas have the following remarkable feature. Since the density δ is assumed to be constant, it can be factored out of the integrals and removed by cancellation, and the formulas for $\bar{x}$ and $\bar{y}$ become

$$\bar{x} = \frac{\displaystyle\int_a^b x f(x)\, dx}{\displaystyle\int_a^b f(x)\, dx} \qquad \text{and} \qquad \bar{y} = \frac{\displaystyle\int_c^d y g(y)\, dy}{\displaystyle\int_c^d g(y)\, dy}. \tag{2}$$

Each denominator here is clearly the total area of the region, and the numerators can be thought of as the moments of this area about the y-axis and x-axis, respectively. The center of mass is therefore determined solely by the geometric configuration of the region R, and does not depend on the density of any mass that this region may contain. For this reason the point $(\bar{x}, \bar{y})$ is called the *centroid* of the region, meaning "point like a center." The examples and problems that follow will make it clear that this terminology is well suited to the geometric concept it is meant to describe.

It will be convenient for our work in the next section if we simplify formulas (2) even further. In the case of $\bar{x}$, the area of the thin vertical strip is an element of area in the sense of Sections 7.1 and 7.2, so we write it as $dA = f(x)\, dx$; and in the case of $\bar{y}$, we similarly have $dA = g(y)\, dy$ for the area of the thin horizontal strip. Formulas (2) can therefore be written in the streamlined form

$$\bar{x} = \frac{\int x\, dA}{\int dA} \qquad \text{and} \qquad \bar{y} = \frac{\int y\, dA}{\int dA}. \tag{3}$$

We emphasize that each dA in these formulas is understood to be the area of a thin strip parallel to the appropriate axis, in order to guarantee that all points in the strip will be essentially at the same distance from this axis. It is also understood here that the process of integration expressed by these symbols is extended over the region under discussion. The limits of integration are omitted deliberately, and don't really need to be written down unless we are performing actual calculations in a specific case.

Example 1   Find the centroid of a rectangle.

Solution   If the rectangle has height h and base b, then we can place the coordinate system so that the origin is at the lower left corner and the point (b, h) is at the upper right corner, as shown in Fig. 11.5. Since the area of this rectangle is hb, we have

$$\bar{x} = \frac{\int_0^b x \cdot h \, dx}{hb} = \frac{1}{hb} \left[\frac{1}{2} hx^2 \right]_0^b$$

$$= \frac{1}{hb} \left[\frac{1}{2} hb^2 \right] = \frac{1}{2} b.$$

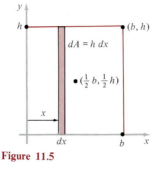

Figure 11.5

In just the same way we find that $\bar{y} = \frac{1}{2} h$, so the centroid is the point $(\frac{1}{2} b, \frac{1}{2} h)$, which is clearly the center of the rectangle.

In general, it appears that the centroid of a region must lie on a line of symmetry of the region, if such a line exists. This is easily seen to be true, as follows. If L is a line of symmetry of a region R, then we can choose this line to be the y-axis (Fig. 11.6), and we wish to convince ourselves that $\bar{x} = 0$. If dA is an arbitrary thin vertical element of area at position x, then by symmetry there is a corresponding element of area at position $-x$; and since $x \, dA + (-x) \, dA = 0$, we have

$$\int x \, dA = 0, \quad \text{and therefore} \quad \bar{x} = \frac{\int x \, dA}{\int dA} = 0.$$

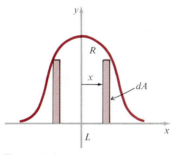

Figure 11.6

(It is clear that this is only a heuristic argument and is far from being a mathematical proof, but it is enough for our present purposes. It is not difficult to convert this argument into a legitimate proof if we wish to take the trouble.) Further, if a region has two distinct lines of symmetry, then the conclusion we have just reached tells us that the centroid must lie on both lines and is therefore the point of intersection of these lines. Accordingly, in every case where a geometric figure has a "center" in the usual sense of the word, this center *is* the centroid. However, as our next example shows, centroids are easily calculated for many regions that are not ordinarily considered to have centers at all. From this point of view, the centroid of a region is a far-reaching generalization of the concept of the center of a geometric figure.

Example 2   Find the centroid of the region in the first quadrant bounded by the axes and the curve $y = 4 - x^2$ (Fig. 11.7).

Solution   By using the vertical strip in the figure, we see that the area of the region is

$$A = \int dA = \int_0^2 (4 - x^2) \, dx = \left[4x - \frac{1}{3} x^3 \right]_0^2 = \frac{16}{3},$$

so

$$\bar{x} = \frac{\int x \, dA}{A} = \frac{3}{16} \int_0^2 x(4 - x^2) \, dx = \frac{3}{16} \left[2x^2 - \frac{1}{4} x^4 \right]_0^2 = \frac{3}{4}.$$

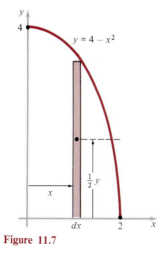

Figure 11.7

Similarly, using a horizontal strip not shown in the figure, we have

$$\bar{y} = \frac{\int y\, dA}{A} = \frac{3}{16} \int_0^4 y\sqrt{4-y}\; dy.$$

To evaluate this integral, we make the substitution $u = 4 - y$, so that $y = 4 - u$ and $dy = -du$, and we also change the limits of integration from $y = 0, 4$ to $u = 4, 0$:

$$\bar{y} = \frac{3}{16} \int_0^4 y\sqrt{4-y}\; dy = \frac{3}{16} \int_4^0 u^{1/2}(4-u)(-du)$$

$$= \frac{3}{16} \int_0^4 (4u^{1/2} - u^{3/2})\; du = \frac{3}{16} \left[\frac{8}{3} u^{3/2} - \frac{2}{5} u^{5/2} \right]_0^4$$

$$= \frac{3}{16} \left(\frac{64}{3} - \frac{64}{5} \right) = \frac{8}{5}.$$

The integration here is a bit complicated because it uses a horizontal strip, and this forces us to solve the equation of the curve for x in terms of y. We therefore describe an alternative method for computing $\bar{y}$ that uses the vertical strip shown in the figure and the result of Example 1. Since the centroid of this rectangular strip is located at its center, the moment of the strip about the x-axis is $\frac{1}{2} y\, dA = \frac{1}{2} y^2\, dx$, and therefore

$$\bar{y} = \frac{\int \frac{1}{2} y^2\, dx}{A} = \frac{3}{32} \int_0^2 (4 - x^2)^2\, dx = \frac{3}{32} \int_0^2 (16 - 8x^2 + x^4)\; dx$$

$$= \frac{3}{32} \left[16x - \frac{8}{3} x^3 + \frac{1}{5} x^5 \right]_0^2 = \frac{3}{32} \left[32 - \frac{64}{3} + \frac{32}{5} \right] = \frac{8}{5},$$

as before.

One more word about centroids. We have discussed centroids of plane regions. We can just as easily speak of the centroid of an arc in the xy-plane or of a region in three-dimensional space. The definitions and formulas are so similar to what we have already done that we won't burden students with detailed explanations. However, we do remark that in finding the centroid of an arc (Fig. 11.8) it may be helpful to think of the arc as a piece of curved wire of constant density 1 ($=$ mass per unit length), so that the mass of a portion of the wire is simply its length. With ds understood to be the element of arc length in the sense of Section 7.5, we therefore have

$$\bar{x} = \frac{\int x\, ds}{\int ds} \quad \text{and} \quad \bar{y} = \frac{\int y\, ds}{\int ds}. \tag{4}$$

Each denominator here is the total length of the arc, and the numerators are the moments of the arc about the y-axis and x-axis, respectively.

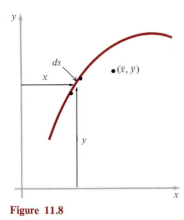

Figure 11.8

PROBLEMS

Find the centroid of the plane region R that is bounded by:

1 $y = x^2$, $y = 0$, $x = 2$.
2 $y = 4x - x^2$ and $y = x$.
3 $y = \sqrt{a^2 - x^2}$ and $y = 0$.
4 $y = \sin x$ and $y = 0$ $(0 \leq x \leq \pi)$.
5 $x^2 = ay$ and $y = a$.
6 $x^2 = ay$ and $y^2 = ax$.
7 $y = \sqrt[3]{x}$, $y = 0$, $x = 8$.
8 $x^2 + y^2 = a^2$ and $x + y = a$ (first quadrant).
9 $x^2 + y^2 = a^2$, $x = a$, $y = a$.
10 $y = 1/x$, $y = 0$, $x = 1$, and $x = 2$.
11 It is known from elementary geometry that the three medians of a triangle intersect at a point that is two-thirds of the way from each vertex to the midpoint of the opposite side. Show that this point is the centroid of the triangle.
12 Find the centroid of the first-quadrant part of the curve $x^{2/3} + y^{2/3} = a^{2/3}$.
13 Find the centroid of the semicircular arc $y = \sqrt{a^2 - x^2}$.
14 The semicircle under $y = \sqrt{b^2 - x^2}$ is removed from the semicircle under $y = \sqrt{a^2 - x^2}$, where $b < a$. Find the centroid of the remaining region. Find the limit of $\bar{y}$ as $b \to a$, and compare with the result of Problem 13.
15 Let $y = f(x)$ be a nonnegative function defined on the interval $a \leq x \leq b$. If the region bounded by this curve, the x-axis, and the lines $x = a$, $x = b$ is revolved about the x-axis, show that the resulting solid of revolution has its centroid on the x-axis with

$$\bar{x} = \frac{\displaystyle\int_a^b xf(x)^2 \, dx}{\displaystyle\int_a^b f(x)^2 \, dx}.$$

16 Use the result of Problem 15 to find the centroid of (a) a cone with height h and radius of base r; and (b) a hemisphere of radius a.

11.3
THE THEOREMS OF PAPPUS

Two beautiful geometric theorems connecting centroids with solids and surfaces of revolution were discovered in the fourth century A.D. by Pappus of Alexandria, the last of the great Greek mathematicians.

First Theorem of Pappus *Consider a plane region that lies completely on one side of a line in its plane. If this region is revolved about the line as an axis, then the volume of the solid generated in this way equals the product of the area of the region and the distance traveled around the axis by its centroid.*

This is easily proved by the following argument. Let the axis of revolution be the x-axis, as shown in Fig. 11.9. Then the distance $\bar{y}$ of the centroid from this axis is defined by

$$\bar{y} = \frac{\int y \, dA}{\int dA} = \frac{\int y \, dA}{A},$$

which is equivalent to

$$A\bar{y} = \int y \, dA$$

or

$$A \cdot 2\pi\bar{y} = \int 2\pi y \, dA.$$

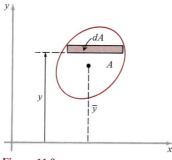

Figure 11.9

All that is needed now is to observe that this equation is precisely the assertion of the theorem, because $2\pi\bar{y}$ is the distance traveled by the centroid and the integral on the right is the volume of the solid as calculated by the shell method.

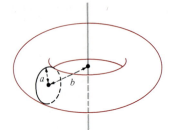

Figure 11.10

Example 1   Find the volume of the torus (doughnut) generated by revolving a circle of radius a about a line in its plane at a distance b from its center, where $b > a$ (Fig. 11.10).

Solution   The centroid of the circle is its center, and this travels a distance $2\pi b$ around the axis. The area of the circle is πa^2, so by the first theorem of Pappus the volume of the torus is

$$V = \pi a^2 \cdot 2\pi b = 2\pi^2 a^2 b.$$

(See Problem 31 in Section 10.4.)

Second Theorem of Pappus   *Consider an arc of a plane curve that lies completely on one side of a line in its plane. If this arc is revolved about the line as an axis, then the area of the surface generated in this way equals the product of the length of the arc and the distance traveled around the axis by its centroid.*

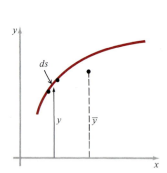

Figure 11.11

The proof is similar to that given above. Again we take the axis to be the x-axis (Fig. 11.11), and we start with the definition of the distance $\bar{y}$ from this axis to the centroid of the arc,

$$\bar{y} = \frac{\int y\,ds}{\int ds} = \frac{\int y\,ds}{s},$$

which is equivalent to

$$s\bar{y} = \int y\,ds$$

or

$$s \cdot 2\pi\bar{y} = \int 2\pi y\,ds.$$

And again this is exactly the assertion of the theorem, because the integral on the right is the area of the surface of revolution.

Example 2   With the aid of this theorem it is easy to see that the surface area of the torus described in Example 1 is

$$A = 2\pi a \cdot 2\pi b = 4\pi^2 ab.$$

Apart from their aesthetic appeal, the theorems of Pappus are useful in two ways. When centroids are known from symmetry considerations—as in the examples—we can use these theorems to find volumes and areas. And also, when volumes and areas are known, we can often use these theorems in reverse to determine the locations of centroids. Both types of applications are illustrated in the following problems.

PROBLEMS

1   Use the known formulas $V = \frac{4}{3}\pi a^3$ and $A = 4\pi a^2$ for the volume and surface area of a sphere of radius a to locate the centroid of (a) the semicircular region under $y = \sqrt{a^2 - x^2}$; (b) the arc $y = \sqrt{a^2 - x^2}$. Compare with Problems 3 and 13 in Section 11.2.

2   Use the centroids found in Problem 1 to find the volume and surface area generated when the semicircular region and the arc are revolved about the line $y = a$.

3   By Problem 10 in Section 7.5, the total length of the curve $x^{2/3} + y^{2/3} = a^{2/3}$ is $6a$. Use this fact and the result of Problem 12 in Section 11.2 to find the area of the surface generated by revolving this curve about (a) the x-axis; (b) the line $x + y = a$.

4   A square with side a is revolved about an axis lying in its plane which intersects it at one of its vertices but at no other points. What should be the position of the axis to yield the largest volume for the resulting solid of revolution? What is this largest volume? What is the corresponding surface area?

5   A regular hexagon with side a is revolved about one of its

sides. What is the volume of the resulting solid of revolution? What is the area of the surface of this solid?

6   The regular hexagon in Problem 5 is revolved about an axis through a vertex which is perpendicular to the line from the center to that vertex. Find the volume and surface area of the resulting solid of revolution.

7   Use Pappus' first theorem to find (a) the volume of a cylinder with height h and radius of base r; (b) the volume of a cone with height h and radius of base r.

8   It is known from elementary geometry that $\pi r L$ is the lateral area of a cone of base radius r and slant height L. Obtain this formula as a consequence of Pappus' second theorem.

11.4
MOMENT OF INERTIA

Consider a rigid body rotating about a fixed axis. For example, the body might be a solid sphere spinning about a diameter, or a solid cube swinging back and forth like a pendulum about a horizontal axis along one of its edges. In order to study motions of this kind, it is necessary to introduce the concept of the *moment of inertia* of the body about the axis. Our purpose in the next few paragraphs is not only to define this concept, but also to explain its intuitive meaning so that students can understand why it matters.

When a rigid body moves in a straight line, all its constituent particles move in the same direction with the same velocity. On the other hand, when a rigid body rotates about an axis, its constituent particles move around circles of different sizes and have different velocities, and for this reason we expect the problem of describing the body's motion to be more difficult. Fortunately, however, this situation is simpler than it seems, and it turns out to be possible to study rotating bodies by using ideas and formulas that are completely analogous to those already familiar for the case of linear motion.

We begin with a brief review of the linear formulas. Consider a particle of mass m moving in a straight line (Fig. 11.12). If its position is given by the variable s, then $v = ds/dt$ and $a = dv/dt$ are its velocity and acceleration. A force F acting on the particle is related to the acceleration by Newton's second law of motion,

$$F = ma \quad \text{or} \quad a = \frac{1}{m} F. \tag{1}$$

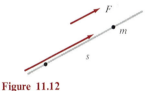

Figure 11.12

The second form of this equation is useful for its clear expression of the idea that the acceleration of the particle is caused by the force and is proportional to this force. This form also helps us think of the mass m of the particle as a measure of its capacity to resist acceleration, because if the force F is the same and m increases, then a decreases.

Now consider a particle of mass m rotating around a fixed axis in a circle of radius r (Fig. 11.13). If its angular position is given by the angle θ as measured from some fixed direction, then $\omega = d\theta/dt$ and $\alpha = d\omega/dt$ are its *angular velocity* and *angular acceleration*. These rotational quantities are related to the corresponding linear quantities s, v, and a, as measured along the circular path, by means of the equations $s = r\theta$, $v = r\omega$, and $a = r\alpha$. The twisting effect of the tangential force F is measured by its *torque* $T = Fr$, which is the

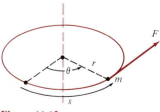

Figure 11.13

product of the force and the distance from its line of action to the axis. We have seen that force produces linear acceleration in accordance with equation (1). In just the same way, torque produces angular acceleration in accordance with the corresponding equation

$$T = I\alpha, \tag{2}$$

where the constant of proportionality I is called the *moment of inertia*. I can be thought of as a measure of the capacity of the system to resist angular acceleration, and in this sense it is the rotational analog of mass.

These remarks describe the conceptual role of the moment of inertia. To discover what its definition must be in order to fit it for this role, we transform (2) by replacing T by Fr and α by a/r, and then we replace F by ma:

$$Fr = I\frac{a}{r}, \qquad mar = I\frac{a}{r}.$$

The last equation tells us that I must be defined by the formula

$$I = mr^2. \tag{3}$$

In this section we are mainly concerned with learning how to use integration to calculate the moment of inertia, about a given axis, of a uniform thin sheet of material of constant density δ ($=$ mass per unit area). It may be helpful to think of such a sheet as a thin plate of homogeneous metal. Our method is to imagine the plate divided into a large number of small pieces in such a way that each piece can be treated as a particle to which formula (3) can be applied. We then find the total moment of inertia by integrating — or adding together — the individual moments of inertia of all these pieces.

Example 1   A uniform thin rectangular plate has sides a and b and density δ. Find its moment of inertia about an axis that bisects the two sides of length a (Fig. 11.14).

Solution   Introduce coordinate axes as indicated in the figure, with the y-axis as the axis of rotation. We concentrate our attention on the thin vertical strip shown in the figure because all of its points are essentially at the same distance x from the axis of rotation. The moment of inertia of this strip about the axis is $x^2 \cdot \delta b \, dx$, so the total moment of inertia of the plate is

$$I = \int_{-a/2}^{a/2} x^2 \cdot \delta b \, dx = \delta b \left[\frac{1}{3} x^3 \right]_{-a/2}^{a/2}$$

$$= \delta b \left[\frac{1}{24} a^3 - \left(-\frac{1}{24} a^3 \right) \right] = \frac{1}{12} \delta a^3 b. \tag{4}$$

It is customary to write the moment of inertia in a form that displays the total mass M. In this case $M = \delta ab$, so

$$I = \tfrac{1}{12} M a^2.$$

Two thin vertical strips that are symmetrically placed with respect to the axis of rotation have the same moment of inertia. In equation (4) we could therefore have written the integral in the form

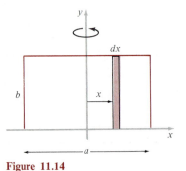

Figure 11.14

$$I = 2 \int_0^{a/2} x^2 \cdot \delta b \, dx = \cdots ,$$

which makes possible a slightly simpler calculation.

Example 2 A uniform thin circular plate has radius a and mass M. Find its moment of inertia about a diameter.

Solution Introduce coordinate axes as shown in Fig. 11.15. If the density of the plate is denoted by δ, then the moment of inertia of the indicated strip about the y-axis is $x^2 \cdot \delta 2y \, dx = x^2 \cdot \delta 2\sqrt{a^2 - x^2} \, dx$, so the total moment of inertia is

$$I = 2 \int_0^a x^2 \cdot \delta 2\sqrt{a^2 - x^2} \, dx = 4\delta \int_0^a x^2 \sqrt{a^2 - x^2} \, dx.$$

To evaluate this integral we make the trigonometric substitution $x = a \sin \theta$, so that $dx = a \cos \theta \, d\theta$ and

$$I = 4\delta a^4 \int_0^{\pi/2} \sin^2 \theta \cos^2 \theta \, d\theta = \tfrac{1}{4}\delta\pi a^4 = \tfrac{1}{4}Ma^2.$$

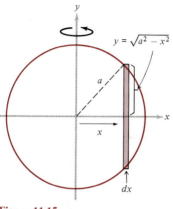

Figure 11.15

(As always, students should verify the omitted details of this calculation for themselves.)

Example 3 Find the moment of inertia of the circular plate in Example 2 about an axis through the center and perpendicular to the plate.

Solution This time the axis is to be imagined as protruding out of the page from the center of the circle (Fig. 11.16), and we divide the area into thin rings with centers at the center of the circle, as shown. The total moment of inertia is therefore

$$I = \int_0^a r^2 \cdot \delta 2\pi r \, dr = 2\pi\delta \left[\frac{1}{4} r^4 \right]_0^a$$

$$= \tfrac{1}{2}\delta\pi a^4 = \tfrac{1}{2}Ma^2.$$

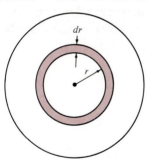

Figure 11.16

Remark 1 We recall that a particle of mass m moving with velocity v has kinetic energy given by the formula

$$\text{K.E.} = \tfrac{1}{2}mv^2,$$

and also that this energy is the amount of work that must be done on the particle to bring it to a stop. On the other hand, if the particle rotates in a circle of radius r, then $v = r\omega$ and we have

$$\text{K.E.} = \tfrac{1}{2}mr^2\omega^2 = \tfrac{1}{2}I\omega^2,$$

and again this is the work required to stop the rotating particle. By comparing these formulas we reinforce the idea that the moment of inertia I plays the same role in rotational motion as is played by the mass m in linear motion.

Remark 2 In addition to its importance in connection with the physics of rotating bodies, the moment of inertia also has significant applications in structural engineering, where it is found that the stiffness of a beam is proportional to the moment of inertia of a cross section of the beam about a horizontal axis through its centroid. This fact is exploited in the design of the familiar steel girders called "I-beams," where flanges at the top and bottom of the beam — as in the letter I — increase the moment of inertia and hence the stiffness of the beam.

PROBLEMS

1 A uniform thin rectangular plate of mass M has sides a and b. Find its moment of inertia about one of the sides of length b.

2 A uniform thin plate of mass M is bounded by the curve $y = \cos x$ and the x-axis between $x = -\pi/2$ and $x = \pi/2$. Find its moment of inertia about the y-axis.

3 Find the moment of inertia of a uniform thin triangular plate of mass M, height h, and base b about its base.

4 Find the moment of inertia of the triangular plate in Problem 3 about an axis parallel to its base and passing through the opposite vertex.

5 A uniform thin circular plate has radius a and mass M. Find its moment of inertia about an axis tangent to the plate.

6 Find the moment of inertia of a uniform straight wire of mass M and length a about an axis perpendicular to the wire at one end.

7 A uniform wire of mass M is bent into a circle of radius a. Find its moment of inertia about a diameter.

8 Find the moment of inertia of a uniform solid cylinder of mass M, height h, and radius a about its axis. Hint: Use the shell method.

9 Find the moment of inertia of a uniform solid cone of mass M, height h, and radius of base a about its axis.

10 Find the moment of inertia of a uniform solid sphere of mass M and radius a about a diameter.

11 If the moment of inertia of a body of mass M about a given axis is $I = Mr^2$, then the number r is called the *radius of gyration* of the body about that axis. This is the distance from the axis at which all the mass of the body could be concentrated at a single point without changing its moment of inertia. Referring to Problems 8 to 10, find the radius of gyration about the indicated axis of (a) the cylinder; (b) the cone; (c) the sphere.

ADDITIONAL PROBLEMS FOR CHAPTER 11

SECTION 11.1

1 Consider the plane distribution of particles whose center of mass $(\bar{x}, \bar{y})$ is defined by equations (6) and (7) in Section 11.1. If $Ax + By + C = 0$ is any line in the plane, then we may suppose (introducing a factor if necessary) that $A^2 + B^2 = 1$; and by Additional Problem 21 in Chapter 1 we see that the signed distance from this line to (x_k, y_k) is

$$d_k = Ax_k + By_k + C,$$

this being positive on one side of the line and negative on the other.

(a) Show that the entire mass $m = \Sigma m_k$ of the system can be concentrated at the center of mass $(\bar{x}, \bar{y})$ without changing the total moment $\Sigma m_k d_k$ about the arbitrary line.

(b) Use part (a) to show that the total moment is zero about every line through $(\bar{x}, \bar{y})$.

2 Consider again the plane distribution of particles discussed in Problem 1.

(a) If the axes are translated as shown in Fig. 11.17, then the old coordinates and the new coordinates

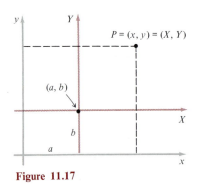

Figure 11.17

of a fixed point P are connected by the transformation equations

$$x = X + a, \qquad y = Y + b.$$

Calculate the center of mass in the new coordinate system, and show that it is the same point as before.

(b) If the axes are rotated through an angle θ as shown in Fig. 11.18, then the old coordinates and the

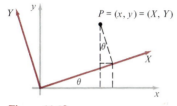

Figure 11.18

new coordinates of a fixed point P are connected by the transformation equations

$$x = X \cos \theta - Y \sin \theta, \qquad y = X \sin \theta + Y \cos \theta.$$

Show that the center of mass as calculated in the new coordinate system is the same point as before.

(c) Deduce that the location of the center of mass is independent of the coordinate system that is used to calculate it.

SECTION 11.2

3 Find the centroid of the plane region R that is bounded by

(a) $y = x^2$ and $y = x$;
(b) $y = 2x^2$ and $y = x^2 + 1$;
(c) $y = 2x - x^2$ and $y = 0$;
(d) $y = x - x^4$ and $y = 0$;
(e) $y^3 = x^2$ and $y = 2$;
(f) $y = x^3$ and $y = 4x$ $(x \geq 0)$;
(g) $y = e^x$, $y = -e^x$, $x = 0$, $x = 1$.

*4 Find the centroid of the part of the curve $y = x^2$ that lies between $x = 0$ and $x = b$.

SECTION 11.3

5 Consider a rectangle with height $2a$ and base $2b$ placed in the xy-plane with its sides parallel to the axes and its center at the point $(0, c)$, where $c \geq \sqrt{a^2 + b^2}$. If this rectangle is rotated counterclockwise through an

angle θ about the point $(0, c)$ and then revolved about the x-axis, show that the volume and surface area of the resulting solid of revolution are the same for all values of θ. What are they?

6 A regular hexagon inscribed in the circle $x^2 + y^2 = 1$ has one of its vertices at the point $(1, 0)$. If this hexagon is revolved about the line $3x + 4y = 25$, find the volume and surface area of the resulting solid of revolution.

SECTION 11.4

7 Show that the moment of inertia of a uniform thin plate in the xy-plane about an axis perpendicular to this plane at the origin is equal to the sum of its moments of inertia about the two coordinate axes. Use this fact to find the moment of inertia of a uniform thin square plate of mass M and side a about an axis through its center and perpendicular to its plane.

8 A uniform thin plate of mass M has the curve

$$\frac{x^2}{a^2} + \frac{y^2}{b^2} = 1$$

as its boundary. Use the method of Problem 7 to find its moment of inertia about an axis through the origin and perpendicular to its plane.

9 Consider a uniform thin plate of mass M in the xy-plane. Let I be its moment of inertia about a line L in this plane, and let I_0 be its moment of inertia about a parallel line L_0 through the centroid. Show that

$$I = I_0 + Md^2,$$

where d is the distance between L and L_0 (this is called the *parallel axis theorem*). Hint: Place the coordinate system so that L_0 is the y-axis and L is the line $x = d$.

*10 Consider a uniform solid body of mass M in three-dimensional space. Let I be its moment of inertia about a line L, and let I_0 be its moment of inertia about a parallel line L_0 through the centroid. Then the *parallel axis theorem* stated in Problem 9 holds in exactly the same form:

$$I = I_0 + Md^2,$$

where d is the distance between L and L_0. Establish this fact, and apply it to find the moment of inertia of (a) a uniform solid sphere of mass M and radius a about a tangent; (b) a uniform solid cube of mass M and edge a about an edge. (Hint: See Problem 7.)

12

INDETERMINATE FORMS AND IMPROPER INTEGRALS

12.1

INTRODUCTION. THE MEAN VALUE THEOREM

In the next few chapters we will need better methods of computing limits than any we have available now. Accordingly, our main purposes in this chapter are to understand the types of limit problems that lie ahead and to acquire the tools that will enable us to solve these problems with maximum efficiency.

In Section 2.5 we saw that the limit of a quotient is the quotient of the limits, in the following sense: If

$$\lim_{x \to a} f(x) = L \qquad \text{and} \qquad \lim_{x \to a} g(x) = M,$$

then

$$\lim_{x \to a} \frac{f(x)}{g(x)} = \frac{L}{M}, \tag{1}$$

provided that $M \neq 0$. Unfortunately, however, it is a fact of life that many of the most important limits are of the form (1) in which both $L = 0$ and $M = 0$. When this happens, formula (1) is useless for calculating the value of the limit and this limit is said to have the *indeterminate form* 0/0 at $x = a$. The expression "indeterminate form" is used because in this case the limit on the left of (1) may very well exist, but nothing can be concluded about its value without further investigation. This is shown by the four examples

$$\frac{x}{x}, \qquad \frac{x^2}{x}, \qquad \frac{x}{x^3}, \qquad \frac{x \sin 1/x}{x},$$

each of which is a quotient of two functions that both approach zero as $x \to 0$. We see from these examples — by cancelling x's from the numerators and denominators — that such a quotient may have the limit 1, or 0, or ∞, or it may have no limit at all, finite or infinite.

Indeterminate forms can sometimes be evaluated by using simple algebraic devices. For example,

$$\lim_{x \to 2} \frac{3x^2 - 7x + 2}{x^2 + 5x - 14} \tag{2}$$

has the indeterminate form 0/0, and this limit is easy to calculate by factoring and cancelling,

$$\lim_{x \to 2} \frac{3x^2 - 7x + 2}{x^2 + 5x - 14} = \lim_{x \to 2} \frac{(x-2)(3x-1)}{(x-2)(x+7)} = \lim_{x \to 2} \frac{3x-1}{x+7} = \frac{5}{9}.$$

In other cases, more complicated methods are required. Thus, the limit

$$\lim_{x \to 0} \frac{\sin x}{x} \qquad\qquad (3)$$

is another indeterminate form of the type 0/0, and in Section 9.2 a geometric argument was used to show that the value of this important limit is 1. In this connection we point out the suggestive fact that the limit (3) can also be evaluated by noticing that it is the derivative of the function sin x at $x = 0$:

$$\lim_{x \to 0} \frac{\sin x}{x} = \lim_{x \to 0} \frac{\sin x - \sin 0}{x - 0}$$

$$= \frac{d}{dx} \sin x \bigg]_{x=0} = \cos x \bigg]_{x=0} = \cos 0 = 1.$$

Indeed, every derivative

$$f'(a) = \lim_{x \to a} \frac{f(x) - f(a)}{x - a} \qquad\qquad (4)$$

is an indeterminate form of the type 0/0, since both numerator and denominator of the fraction on the right approach zero as x approaches a.*

These remarks suggest that there is a close but hidden connection between indeterminate forms and derivatives. And so there is. But to understand this connection, it is first necessary to understand the Mean Value Theorem.

This theorem states that if a function $y = f(x)$ is defined and continuous on a closed interval $a \le x \le b$, and differentiable at each point of the interior $a < x < b$, then there exists at least one number c between a and b for which

$$f'(c) = \frac{f(b) - f(a)}{b - a},$$

or equivalently,

$$f(b) - f(a) = f'(c)(b - a).$$

This assertion is best understood in geometric language (see Fig. 12.1); it says that at some point on the graph between the endpoints, the tangent line is parallel to the chord joining these endpoints. From this point of view the theorem seems obviously true and is difficult to doubt; but in fact, it is a rather deep theorem whose validity depends in a crucial way on the stated hypotheses.

In most of our work we try to avoid dwelling on the theoretical parts of

---

* Students should examine formula (4) together with a suitable sketch, in order to convince themselves that this formula can be taken as the definition of the derivative of an arbitrary function $f(x)$ at a point $x = a$. We have not had occasion to use this version of the definition before, but it will be particularly convenient for our work in the present chapter.

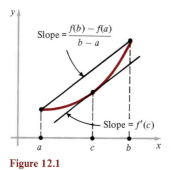

Figure 12.1

calculus. Here, however, we must make an exception, because the central fact of this chapter (L'Hospital's rule, in the next section*) cannot be understood unless we at least know what the Mean Value Theorem says. A complete discussion of this theorem, with all necessary proofs, is provided in Appendix C.4. As usual, this appendix is made available for the few students who wish to delve more deeply into the theoretical foundations of calculus; those whose interests do not lie in this direction may skip it without any serious consequences.

12.2

THE INDETERMINATE FORM 0/0. L'HOSPITAL'S RULE

We remarked earlier that there is a close connection between indeterminate forms and derivatives. We begin to explore this connection with the following simple theorem: *If $f(x)$ and $g(x)$ are both equal to zero at $x = a$ and have derivatives there, then*

$$\lim_{x \to a} \frac{f(x)}{g(x)} = \frac{f'(a)}{g'(a)} = \frac{f'(x)}{g'(x)}\bigg]_{x=a}, \tag{1}$$

provided that $g'(a) \neq 0$. To prove this, it suffices to use $f(a) = 0$ and $g(a) = 0$ to write

$$\frac{f(x)}{g(x)} = \frac{f(x) - f(a)}{g(x) - g(a)} = \frac{[f(x) - f(a)]/(x - a)}{[g(x) - g(a)]/(x - a)} \to \frac{f'(a)}{g'(a)},$$

as stated.

As examples of the use of (1), we easily find the limits (2) and (3) in Section 12.1,

$$\lim_{x \to 2} \frac{3x^2 - 7x + 2}{x^2 + 5x - 14} = \frac{6x - 7}{2x + 5}\bigg]_{x=2} = \frac{5}{9} \tag{2}$$

and

$$\lim_{x \to 0} \frac{\sin x}{x} = \frac{\cos x}{1}\bigg]_{x=0} = \cos 0 = 1. \tag{3}$$

As another example, we have

$$\lim_{x \to 0} \frac{\tan 6x}{e^{2x} - 1} = \frac{6 \sec^2 6x}{2e^{2x}}\bigg]_{x=0} = \frac{6}{2} = 3, \tag{4}$$

a result that would have been hard to find in any other way.

Formula (1) requires the existence of the derivatives of the functions $f(x)$ and $g(x)$ at the single point $x = a$. At other points these functions need not have derivatives, nor indeed even be continuous. However, if the derivatives exist in an interval about a and are continuous at a, then we can obtain formula (1) in another way, by applying the Mean Value Theorem separately to the numerator and denominator,

$$\frac{f(x)}{g(x)} = \frac{f(x) - f(a)}{g(x) - g(a)} = \frac{f'(c_1)(x - a)}{g'(c_2)(x - a)} = \frac{f'(c_1)}{g'(c_2)} \to \frac{f'(a)}{g'(a)}, \tag{5}$$

as $x \to a$. Here c_1 and c_2 lie between x and a, so both approach a as $x \to a$.

---

* L'Hospital is pronounced *low-pee-tal.*

What purpose is served by giving a second alternative proof of formula (1) when the first proof is perfectly satisfactory? The point is this: Formula (1) is a good tool to have, but is still only of limited value, because it often happens in the problems we consider that $f'(a) = g'(a) = 0$, and in this case the right-hand side of (1) is meaningless. However, we can use our second proof to get around this difficulty as follows. Suppose that c_1 and c_2 in (5) can be taken equal to one another, so that the first part of (5) can be written as

$$\frac{f(x)}{g(x)} = \frac{f(x) - f(a)}{g(x) - g(a)} = \frac{f'(c)}{g'(c)}, \tag{6}$$

where c is between x and a. Then in forming the limit as $x \to a$, (6) permits us to replace the quotient

$$\frac{f(x)}{g(x)} \qquad \text{by the quotient} \qquad \frac{f'(x)}{g'(x)}.$$

L'Hospital's rule states that under certain easily satisfied conditions this replacement is legitimate, that is,

$$\lim_{x \to a} \frac{f(x)}{g(x)} = \lim_{x \to a} \frac{f'(x)}{g'(x)}, \tag{7}$$

provided the limit on the right exists. Students should remember that $f(a) = g(a) = 0$ is assumed here, and we also mention that even though ordinary two-sided limits are usually intended in (7), one-sided limits are allowed.*

L'Hospital's rule is named after the French mathematician — a pupil of John Bernoulli — who published it in his book *Analyse des infiniment petits* (182 pp., Paris, 1696), which was the first calculus textbook and enjoyed wide popularity and influence.

Example 1 At the beginning of this section we evaluated the limits (2), (3), and (4) by using formula (1). These limits can also be found by using L'Hospital's rule (7):

$$\lim_{x \to 2} \frac{3x^2 - 7x + 2}{x^2 + 5x - 14} = \lim_{x \to 2} \frac{6x - 7}{2x + 5} = \frac{5}{9},$$

$$\lim_{x \to 0} \frac{\sin x}{x} = \lim_{x \to 0} \frac{\cos x}{1} = 1,$$

$$\lim_{x \to 0} \frac{\tan 6x}{e^{2x} - 1} = \lim_{x \to 0} \frac{6 \sec^2 6x}{2e^{2x}} = 3.$$

The reason (7) works so smoothly in these problems is that in each case the second limit exists and is easy to find by inspection, since the functions

---

* For those students who are interested in the proof of L'Hospital's rule, we here explain as briefly as possible the details of the reasoning that underlies (7). We assume — as stated — that $f(a) = g(a) = 0$, that x approaches a from one side or the other, and that on that side the functions $f(x)$ and $g(x)$ satisfy the following three conditions: (i) both are continuous on some closed interval I having a as an endpoint; (ii) both are differentiable in the interior of I; and (iii) $g'(x) \neq 0$ in the interior of I. With these hypotheses, (6) is an immediate consequence of a technical extension of the Mean Value Theorem known as the Generalized Mean Value Theorem; and if x is now allowed to approach a from the side under consideration, then (7) follows from (6) as indicated above. Those tenacious students who like to nail everything down will find a proof of the Generalized Mean Value Theorem in Appendix C.4.

involved are continuous. The point we wish to make here is that whatever (1) can do, (7) can do just as easily; and as the next example shows, (7) is much more powerful and often works easily when (1) doesn't work at all.

Example 2   L'Hospital's rule (7) proves its value in limit problems like

$$\lim_{x \to 0} \frac{1 - \cos x}{x^2}.$$

Here formula (1) is useless, as we see from the failure of the attempted calculation

$$\lim_{x \to 0} \frac{1 - \cos x}{x^2} = \frac{\sin x}{2x}\bigg]_{x=0} = \frac{0}{0}.$$

The reason for this failure is that (1) assumes that $g'(a) \neq 0$, and this condition is not satisfied here. However, by (7) we have

$$\lim_{x \to 0} \frac{1 - \cos x}{x^2} = \lim_{x \to 0} \frac{\sin x}{2x},$$

if the second limit exists. But this second limit is again of the form $0/0$, so L'Hospital's rule applies a second time and permits us to continue and reach the correct answer,

$$\lim_{x \to 0} \frac{1 - \cos x}{x^2} = \lim_{x \to 0} \frac{\sin x}{2x} = \lim_{x \to 0} \frac{\cos x}{2} = \frac{1}{2}.$$

Another limit that behaves in this way is

$$\lim_{x \to 0} \frac{\sqrt{x + 1} - (1 + \frac{1}{2}x)}{x^2} = \lim_{x \to 0} \frac{\frac{1}{2}(x + 1)^{-1/2} - \frac{1}{2}}{2x}$$

$$= \lim_{x \to 0} \frac{-\frac{1}{4}(x + 1)^{-3/2}}{2} = -\frac{1}{8}.$$

The limits in Example 2 illustrate the great advantage L'Hospital's rule (7) has over formula (1): It is valid whenever the limit on the right exists, regardless of whether $g'(a)$ is zero or not. Thus, as these problems show, if $f'(a) = g'(a) = 0$, then we have another indeterminate form $0/0$ and we can apply L'Hospital's rule a second time,

$$\lim_{x \to a} \frac{f(x)}{g(x)} = \lim_{x \to a} \frac{f'(x)}{g'(x)} = \lim_{x \to a} \frac{f''(x)}{g''(x)},$$

provided the last-written limit exists. As a practical matter, the functions we encounter in this book satisfy the conditions needed for L'Hospital's rule. We therefore apply the rule almost routinely, by continuing to differentiate the numerator and denominator separately as long as we still get the form $0/0$ at $x = a$. As soon as one or the other (or both) of these derivatives is different from zero at $x = a$, we must stop differentiating and hope to evaluate the last limit by some direct method.

Example 3 A careless attempt to apply L'Hospital's rule may yield an incorrect result, as in the calculation

$$\lim_{x \to 0} \frac{\sin 4x}{2x + 3} = \lim_{x \to 0} \frac{4 \cos 4x}{2} = \frac{4}{2} = 2.$$

The correct answer is

$$\lim_{x \to 0} \frac{\sin 4x}{2x + 3} = \frac{0}{3} = 0.$$

In this problem the numerator and denominator of the given quotient are not both equal to zero at $x = 0$, so L'Hospital's rule is not applicable.

Our methods work in just the same way for limits in which $x \to \infty$; that is, if $f(x) \to 0$ and $g(x) \to 0$ as $x \to \infty$, then

$$\lim_{x \to \infty} \frac{f(x)}{g(x)} = \lim_{x \to \infty} \frac{f'(x)}{g'(x)}, \tag{8}$$

if the limit on the right exists. To see this, we put $x = 1/t$ and observe that $t \to 0+$ (recall that this notation means that t approaches zero from the right). Briefly, L'Hospital's rule (7) now gives

$$\lim_{x \to \infty} \frac{f(x)}{g(x)} = \lim_{t \to 0+} \frac{f(1/t)}{g(1/t)} = \lim_{t \to 0+} \frac{f'(x)\, dx/dt}{g'(x)\, dx/dt},$$

which yields (8) after dx/dt is canceled.

Finally, in both forms of L'Hospital's rule, as expressed in formulas (7) and (8), it is easy to see that the procedure remains valid if the value of the limit on the right is ∞ or $-\infty$.

PROBLEMS

Find the following limits.

1 $\displaystyle\lim_{x \to 0} \frac{\sin 3x}{\sin x}.$

2 $\displaystyle\lim_{x \to 1} \frac{\ln x}{x - 1}.$

15 $\displaystyle\lim_{x \to 0} \frac{\sin^2 x + 8x}{e^{2x} - 1}.$

16 $\displaystyle\lim_{x \to 6} \frac{\sqrt{x - 2} - 2}{x^2 - 36}.$

3 $\displaystyle\lim_{x \to 2} \frac{x - 2}{6x^2 - 10x - 4}.$

4 $\displaystyle\lim_{x \to 0} \frac{e^x - e^{-x}}{\sin 5x}.$

17 $\displaystyle\lim_{x \to 0} \frac{x - \sin x}{x - \tan x}.$

18 $\displaystyle\lim_{x \to \pi} \frac{\ln (\cos 2x)}{(x - \pi)^2}.$

5 $\displaystyle\lim_{x \to 0} \frac{\sqrt{x + 9} - 3}{x}.$

6 $\displaystyle\lim_{x \to 1} \frac{4x^3 - 5x + 1}{\ln x}.$

19 $\displaystyle\lim_{x \to \infty} \frac{1/x}{\sin \pi/x}.$

20 $\displaystyle\lim_{x \to 0} \frac{a^x - b^x}{x}.$

7 $\displaystyle\lim_{x \to 0} \frac{\sqrt[3]{x + 1} - (1 + \frac{1}{3}x)}{x^2}.$

8 $\displaystyle\lim_{x \to 0} \frac{\sin^{-1} 3x}{x}.$

21 $\displaystyle\lim_{x \to 0} \frac{\tan 2x - 2x}{x - \sin x}.$

22 $\displaystyle\lim_{x \to 0} \frac{\sin x^3}{\sin^3 x}.$

9 $\displaystyle\lim_{x \to 0} \frac{e^x - 1 - x}{x^2}.$

10 $\displaystyle\lim_{x \to 0} \frac{e^x - 1 - x}{1 - \cos \pi x}.$

23 $\displaystyle\lim_{x \to 0} \frac{\sin 2x - 2 \sin x}{\sin 3x - 3 \sin x}.$

11 $\displaystyle\lim_{x \to 0} \frac{x^3}{\sin x - x}.$

12 $\displaystyle\lim_{x \to 0} \frac{e^{2x} - 1}{\sin 5x}.$

24 $\displaystyle\lim_{x \to 1} \frac{\sqrt[4]{x} - 1}{\sqrt[5]{x} - 1}.$

13 $\displaystyle\lim_{x \to 0} \frac{3x}{\tan x}.$

14 $\displaystyle\lim_{x \to \pi/4} \frac{\ln (\tan x)}{\sin x - \cos x}.$

25 $\displaystyle\lim_{x \to 0} \frac{(e^x - 1)^3}{(x - 2)e^x + x + 2}.$

26 In Fig. 12.2, P is a point on a circle with center O and radius a. The segment AQ equals the arc AP, and the line PQ intersects the line OA at B. Show that OB approaches $2a$ as P approaches A along the circle. Hint: $\triangle QAB$ is similar to $\triangle PRB$.

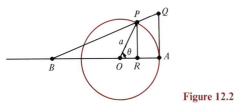

Figure 12.2

27 In Problem 26, let $f(\theta)$ be the area of the triangle ARP and let $g(\theta)$ be the area of the region that remains after the triangle ORP is removed from the sector OAP.

Find formulas for the functions $f(\theta)$ and $g(\theta)$ and evaluate $\lim_{\theta \to 0} f(\theta)/g(\theta)$.

28 L'Hospital's rule (7) works in just the same way if the conditions $f(a) = g(a) = 0$ are replaced by the conditions $\lim_{x \to a} f(x) = \lim_{x \to a} g(x) = 0$. Explain. Use this idea to evaluate

$$\lim_{x \to 0} \frac{(1 + x)^{1/x} - e}{x}.$$

29 The formula

$$f'(x) = \lim_{h \to 0} \frac{f(x + h) - f(x)}{h}$$

is one version of the definition of the derivative. By treating the right-hand side as an indeterminate form, derive this formula from L'Hospital's rule.

12.3

OTHER INDETERMINATE FORMS

For certain applications it is important to know that L'Hospital's rule remains valid for indeterminate forms of the type ∞/∞. That is, if the numerator and denominator of the quotient $f(x)/g(x)$ both become infinite as $x \to a$, then

$$\lim_{x \to a} \frac{f(x)}{g(x)} = \lim_{x \to a} \frac{f'(x)}{g'(x)}, \tag{1}$$

provided the limit on the right exists. The argument is a bit tricky, and is sketched in Remark 2 so that those who wish to skip it can conveniently do so. Just as in Section 12.2, one-sided limits are allowed and (1) extends immediately to the case in which $x \to \infty$; also, it remains valid if the limit on the right is ∞ or $-\infty$.

Example 1 Show that

$$\lim_{x \to \infty} \frac{x^p}{e^x} = 0 \tag{2}$$

for every constant p.

Solution We begin by observing that if $p \leq 0$, then this limit is not an indeterminate form and its value is easily seen to be zero. On the other hand, when $p > 0$, the limit is clearly an indeterminate form of the type ∞/∞. L'Hospital's rule (1) for the case in which $x \to \infty$ therefore gives

$$\lim_{x \to \infty} \frac{x^p}{e^x} = \lim_{x \to \infty} \frac{px^{p-1}}{e^x},$$

if the limit on the right exists; and if this process is continued step by step, we can reduce the exponent for x to zero or a negative number, and the desired conclusion (2) now follows from the above remark about this case. This example gives us important insight into the nature of the exponential function: as $x \to \infty$, e^x increases faster than any positive power of x, however large, and therefore faster than any polynomial.

Example 2 Show that

$$\lim_{x \to \infty} \frac{\ln x}{x^p} = 0 \tag{3}$$

for every constant $p > 0$.

Solution This limit is clearly an indeterminate form of the type ∞/∞, so by L'Hospital's rule we have

$$\lim_{x \to \infty} \frac{\ln x}{x^p} = \lim_{x \to \infty} \frac{1/x}{px^{p-1}} = \lim_{x \to \infty} \frac{1}{px^p} = 0.$$

Expressed in words, (3) tells us that as $x \to \infty$, $\ln x$ increases more slowly than any positive power of x, however small.

We have discussed the limits (2) and (3) before, by clumsier special methods, in Sections 8.3 and 8.4. But our present treatment of these important facts is clearly preferable, because the powerful method of analysis based on L'Hospital's rule extends easily to many similar limits.

The expressions

$$0 \cdot \infty, \qquad \infty - \infty, \qquad 0^0, \qquad \infty^0, \qquad 1^\infty$$

symbolize other types of indeterminate forms that sometimes arise. The product $f(x)g(x)$ where one factor approaches zero and the other becomes infinite ($0 \cdot \infty$) can be reduced to $0/0$ or ∞/∞ by putting the reciprocal of one factor in the denominator. The difference between two functions which are both becoming infinite ($\infty - \infty$) can often be manipulated into a more convenient form. A power $y = f(x)^{g(x)}$ that produces an indeterminate form of one of the other types is best handled by taking logarithms,

$$\ln y = \ln f(x)^{g(x)} = g(x) \ln f(x). \tag{4}$$

This reduces the problem to the more familiar form $0 \cdot \infty$; and since $y = e^{\ln y}$, we then use the continuity of the exponential function to infer that $\lim y = \lim e^{\ln y} = e^{\lim \ln y}$. These generalities are illustrated in the following examples.

Example 3 Evaluate

$$\lim_{x \to 0+} x \ln x. \tag{5}$$

Solution Here x is required to approach zero from the positive side because $\ln x$ is defined only for positive x's. Since $\ln x \to -\infty$ as $x \to 0+$, it is clear that (5) is an indeterminate form of the type $0 \cdot \infty$. The value of this limit is not obvious, because as $x \to 0+$, we cannot tell whether the product $x \ln x$ is influenced more by the smallness of x or by the largeness (in absolute value) of $\ln x$. However, we can easily convert the limit into an indeterminate form of the type ∞/∞ and apply L'Hospital's rule (1), as follows:

$$\lim_{x \to 0+} x \ln x = \lim_{x \to 0+} \frac{\ln x}{1/x} = \lim_{x \to 0+} \frac{1/x}{-1/x^2} = \lim_{x \to 0+} (-x) = 0.$$

Thus, the smallness of x turns out to dominate the behavior of the product $x \ln x$ near $x = 0$.

Example 4   Evaluate

$$\lim_{x \to \pi/2} (\sec x - \tan x). \tag{6}$$

Solution   This is of the type $\infty - \infty$. We convert it into an indeterminate form of the type $0/0$ and apply L'Hospital's rule,

$$\lim_{x \to \pi/2} (\sec x - \tan x) = \lim_{x \to \pi/2} \left(\frac{1}{\cos x} - \frac{\sin x}{\cos x} \right)$$

$$= \lim_{x \to \pi/2} \frac{1 - \sin x}{\cos x} = \lim_{x \to \pi/2} \frac{-\cos x}{-\sin x} = 0.$$

Example 5   Find $\lim_{x \to 0+} x^x$.

Solution   This limit is of the type 0^0, and we reduce it to the simpler type $0 \cdot \infty$ by taking the logarithm. To do this most conveniently, we write $y = x^x$ and observe that

$$\ln y = \ln x^x = x \ln x \to 0 \qquad \text{as} \qquad x \to 0+,$$

by Example 3. This tells us that

$$x^x = y = e^{\ln y} \to e^0 = 1,$$

by the continuity of the exponential function. Therefore we have

$$\lim_{x \to 0+} x^x = 1. \tag{7}$$

Example 6   Find $\lim_{x \to \infty} x^{1/x}$.

Solution   This limit is of the type ∞^0. We write $y = x^{1/x}$ and observe that

$$\ln y = \ln x^{1/x} = \frac{\ln x}{x} \to 0 \qquad \text{as} \qquad x \to \infty,$$

by Example 2. This tells us that

$$x^{1/x} = y = e^{\ln y} \to e^0 = 1,$$

or equivalently,

$$\lim_{x \to \infty} x^{1/x} = 1. \tag{8}$$

Example 7   Show that

$$\lim_{x \to 0} (1 + ax)^{1/x} = e^a \tag{9}$$

for every constant a.

Solution If $a = 0$ this limit is not an indeterminate form, and the statement is clearly true because each side has the value 1. If $a \neq 0$, the limit is an indeterminate form of the type 1^∞. In this case we write $y = (1 + ax)^{1/x}$ and observe that

$$\lim_{x \to 0} \ln y = \lim_{x \to 0} \frac{\ln (1 + ax)}{x} = \lim_{x \to 0} \frac{a/(1 + ax)}{1} = a.$$

This implies that

$$(1 + ax)^{1/x} = y = e^{\ln y} \longrightarrow e^a,$$

which is (9).

Remark 1 *The L'Hospital habit.* Like any mathematical procedure, L'Hospital's rule should be used intelligently, and not purely mechanically. We should try to control the bad habit of automatically applying L'Hospital's rule to every limit problem that comes up. Often there is an easier way, for instance, the use of familiar limits or simple algebraic transformations.

(a) The limit

$$\lim_{x \to \infty} \frac{6x^5 - 2}{2x^5 + 3x^2 + 4}$$

is of the type ∞/∞, and can be found by repeated use of L'Hospital's rule. But it is much simpler to divide both numerator and denominator by x^5 and write

$$\frac{6x^5 - 2}{2x^5 + 3x^2 + 4} = \frac{6 - 2/x^5}{2 + 3/x^3 + 4/x^5} \longrightarrow \frac{6 - 0}{2 + 0 + 0} = 3.$$

(b) The limit

$$\lim_{x \to 0} \frac{\sin^3 x}{x^3}$$

is of the type $0/0$. L'Hospital's rule can be applied, and works, but it is much easier to notice that

$$\frac{\sin^3 x}{x^3} = \left(\frac{\sin x}{x} \right)^3 \longrightarrow 1^3 = 1,$$

because we already know that $(\sin x)/x \to 1$ as $x \to 0$.

(c) The limit

$$\lim_{x \to \infty} \frac{\sqrt{x^2 + 1}}{x}$$

is of the type ∞/∞, and L'Hospital's rule gives

$$\lim_{x \to \infty} \frac{\sqrt{x^2 + 1}}{x} = \lim_{x \to \infty} \frac{x/\sqrt{x^2 + 1}}{1} = \lim_{x \to \infty} \frac{x}{\sqrt{x^2 + 1}} = \frac{1}{\lim\limits_{x \to \infty} \sqrt{x^2 + 1}/x}.$$

This brings us back to the limit we started with, and gets us nowhere. However, it is very easy to insert the denominator into the radical and write

$$\frac{\sqrt{x^2+1}}{x} = \sqrt{\frac{x^2+1}{x^2}} = \sqrt{1+\frac{1}{x^2}} \to \sqrt{1+0} = 1.$$

Remark 2 The argument for L'Hospital's rule (1) in the case ∞/∞ can be briefly sketched as follows. Let $f(x)$ and $g(x)$ both become infinite as $x \to a$ from one side or the other, and suppose that $f'(x)/g'(x) \to L$. We want to show that also $f(x)/g(x) \to L$. For $\bar{x}$ near enough to a on the side under consideration (see Fig. 12.3), $f'(x)/g'(x)$ can be made as close as we please to L between $\bar{x}$ and a. If x is between $\bar{x}$ and a, and if $f(x)$ and $g(x)$ are assumed to satisfy the simple conditions (i) to (iii) stated in the footnote in Section 12.2, then

Figure 12.3

$$\frac{f(x)-f(\bar{x})}{g(x)-g(\bar{x})} = \frac{f'(c)}{g'(c)}$$

for some c between x and $\bar{x}$. Since c is also between $\bar{x}$ and a, we know that $f'(c)/g'(c)$ is close to L. Now hold $\bar{x}$ fixed and let $x \to a$. Then $f(x)$ and $g(x)$ grow very large, $f(\bar{x})/f(x)$ and $g(\bar{x})/g(x)$ become very small, and

$$\frac{f(x)-f(\bar{x})}{g(x)-g(\bar{x})} = \frac{f(x)}{g(x)}\left[\frac{1-f(\bar{x})/f(x)}{1-g(\bar{x})/g(x)}\right]$$

is close to $f(x)/g(x)$. It follows that $f(x)/g(x)$ is close to $f'(c)/g'(c)$, which in turn is close to L, and so $f(x)/g(x)$ is itself close to L when x is close to a, and this is what we wanted to establish.

PROBLEMS

Evaluate the following limits by any method.

1 $\lim\limits_{x \to \infty} \dfrac{18x^3}{3+2x^2-6x^3}$.

2 $\lim\limits_{x \to \infty} \dfrac{\ln(\ln x)}{\ln x}$.

3 $\lim\limits_{x \to \pi/2} \dfrac{\tan x}{1+\sec x}$.

4 $\lim\limits_{x \to \infty} \dfrac{\ln x^2}{\sqrt{x}}$.

5 $\lim\limits_{x \to \pi/2} \dfrac{\tan x}{\tan 3x}$.

6 $\lim\limits_{x \to \infty} \dfrac{x^2}{e^{3x}}$.

7 $\lim\limits_{x \to 0+} \dfrac{\ln x}{\csc x}$.

8 $\lim\limits_{x \to \infty} \dfrac{(\ln x)^{10}}{x}$.

9 $\lim\limits_{x \to 0+} \dfrac{\ln(\sin^2 x)}{\ln x}$.

10 $\lim\limits_{x \to 0} x \cot x$.

11 $\lim\limits_{x \to \infty} x \sin \dfrac{1}{x}$.

12 $\lim\limits_{x \to \pi/2} (\pi - 2x)\tan x$.

13 $\lim\limits_{x \to \infty} (x^2-1)e^{-x^2}$.

14 $\lim\limits_{x \to \infty} x^3 e^{-x}$.

15 $\lim\limits_{x \to \infty} e^{-x} \ln x$.

16 $\lim\limits_{x \to \infty} x\left(\dfrac{\pi}{2} - \tan^{-1} x\right)$.

17 $\lim\limits_{x \to 0+} \sin x \ln x$.

18 $\lim\limits_{x \to 0} x^2 \csc(5\sin^2 x)$.

19 $\lim\limits_{x \to \infty} \left(\dfrac{x^2}{x-1} - \dfrac{x^2}{x+1}\right)$.

20 $\lim\limits_{x \to 0} \left(\dfrac{1}{x} - \dfrac{1}{\sin x}\right)$.

21 $\lim\limits_{x \to 0} \left(\dfrac{1}{x} - \dfrac{1}{e^x-1}\right)$.

22 $\lim\limits_{x \to 1} \left(\dfrac{1}{x-1} - \dfrac{1}{\ln x}\right)$.

23 $\lim\limits_{x \to 0+} (\sin x)^x$.

24 $\lim\limits_{x \to 0+} (\tan x)^{\sin x}$.

25 $\lim\limits_{x \to 0+} x^{\tan x}$.

26 $\lim\limits_{x \to 0+} x^{x^2}$.

27 $\lim\limits_{x \to 0+} (e^x-1)^x$.

28 $\lim\limits_{x \to 0+} x^{\ln(1+x)}$.

29 $\lim\limits_{x \to 0+} (\sin x)^{\sin x}$.

30 $\lim\limits_{x \to 0} (1-\cos x)^{1-\cos x}$.

31 $\lim\limits_{x \to \infty} (\ln x)^{1/x}$.

32 $\lim\limits_{x \to \pi/2-} (\tan x)^{\cos x}$.

33 $\lim\limits_{x \to \infty} (1+e^{ax})^{1/x},\ a>0$.

34 $\lim\limits_{x \to \infty} (x+e^x)^{2/x}$.

35 $\lim\limits_{x \to \infty} (1+ax)^{1/x},\ a>0$.

36 $\lim\limits_{x \to \infty} (1+x^{100})^{1/x}$.

37 $\lim\limits_{x \to 0} (\cos x)^{1/x}$.

38 $\lim\limits_{x \to \pi/2} (\sin x)^{\tan x}$.

39 $\lim\limits_{x \to 1} x^{1/(1-x^2)}$.

40 $\lim\limits_{x \to 0+} (\cos \sqrt{x})^{1/x}$.

41 In spite of the evidence piled up in Problems 23 to 30, indeterminate forms of the type 0^0 do not always have the value 1. Show this by calculating

$$\lim_{x \to 0+} x^{p/\ln x},$$

where p is a nonzero constant.

42 (a) Sketch the graph of the function $y = f(x)$ defined by

$$f(x) = \begin{cases} e^{-1/x^2} & \text{if } x \neq 0, \\ 0 & \text{if } x = 0. \end{cases}$$

(b) Show that $\lim_{x \to 0} x^{-n} e^{-1/x^2} = 0$ for every positive integer n.

(c) Show that $f(x)$ as defined in (a) has an nth derivative $f^{(n)}(x)$ for every positive integer n and every

$x \neq 0$. [We do not ask for a general formula for $f^{(n)}(x)$, but students should carry the calculations far enough to show that $f^{(n)}(x)$ is always given by a formula of a certain form, involving certain constant coefficients.]

(d) Use parts (b) and (c) to show that $f^{(n)}(0) = 0$ for every positive integer n.

*43 As $x \to 0+$, show that

$$\cot x - \frac{1}{x} \to 0 \quad \text{and} \quad \cot x + \frac{1}{x} \to \infty,$$

but that

$$\left(\cot x - \frac{1}{x} \right)\left(\cot x + \frac{1}{x} \right) \to -\frac{2}{3}.$$

44 Use (4) in the text to explain why 1^0, 0^1, and 0^∞ are not indeterminate forms.

When we write down an ordinary definite integral as defined in Chapter 6,

$$\int_a^b f(x)\, dx, \tag{1}$$

12.4
IMPROPER INTEGRALS

we assume that the limits of integration are finite numbers and that the integrand $f(x)$ is continuous on the bounded interval $a \leq x \leq b$. If $f(x) \geq 0$, we are thoroughly familiar with the idea that the integral (1) represents the area of the region shown on the left in Fig. 12.4.

In Chapter 14 it will be necessary to consider so-called *improper integrals* of the form

$$\int_a^\infty f(x)\, dx, \tag{2}$$

in which the upper limit is infinite and the integrand $f(x)$ is assumed to be continuous on the unbounded interval $a \leq x < \infty$.* We define the integral (2)

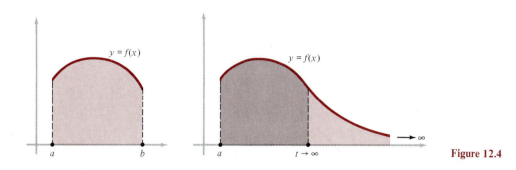

Figure 12.4

---

* The word "improper" is used because of the "impropriety" at the upper limit of integration. If we wish, we can speak of (1) as a *proper integral* because it has no improprieties, but this is neither necessary nor customary.

in the natural way suggested on the right in Fig. 12.4; that is, we integrate from a to a finite but variable upper limit t, and then we allow t to approach ∞ and define (2) by

$$\int_a^\infty f(x)\,dx = \lim_{t \to \infty} \int_a^t f(x)\,dx.$$

If this limit exists and has a finite value, then the improper integral is said to *converge* or to be *convergent*, and this value is assigned to it. Otherwise, the integral is called *divergent*. If $f(x) \geq 0$, then (2) can be thought of as the area of the unbounded region on the right in Fig. 12.4. In this case the area of the region is finite or infinite according as the improper integral (2) converges or diverges.

Example 1

$$\int_0^\infty e^{-x}\,dx = \lim_{t \to \infty} \int_0^t e^{-x}\,dx = \lim_{t \to \infty} [-e^{-x}]_0^t = \lim_{t \to \infty} \left(-\frac{1}{e^t} + 1\right) = 1.$$

This improper integral converges, because the limit exists and is finite.

Students often tend to abbreviate this calculation by writing

$$\int_0^\infty e^{-x}\,dx = [-e^{-x}]_0^\infty = -\frac{1}{e^\infty} + 1 = 1,$$

instead of writing out the limits as we have done in Example 1. This shorthand rarely causes any real difficulties. However, in our work in this section we will always write out the limits for the sake of emphasizing that improper integrals are *defined* as limits.

Example 2

$$\int_1^\infty \frac{dx}{x^2} = \lim_{t \to \infty} \int_1^t \frac{dx}{x^2} = \lim_{t \to \infty} \left[-\frac{1}{x}\right]_1^t = \lim_{t \to \infty} \left(-\frac{1}{t} + 1\right) = 1.$$

This improper integral also converges.

Example 3

$$\int_1^\infty \frac{dx}{x} = \lim_{t \to \infty} \int_1^t \frac{dx}{x} = \lim_{t \to \infty} [\ln x]_1^t = \lim_{t \to \infty} \ln t = \infty.$$

This integral diverges, because the limit is infinite.

Example 4

$$\int_0^\infty \cos x\,dx = \lim_{t \to \infty} \int_0^t \cos x\,dx = \lim_{t \to \infty} \sin t$$

—which does not exist. This integral diverges, because the limit does not exist.

Our next example generalizes Examples 2 and 3 and contains specific information that will be needed in Chapter 14.

Example 5 If p is a positive constant, show that the improper integral

$$\int_1^\infty \frac{dx}{x^p}$$

converges if $p > 1$ and diverges if $p \le 1$.

Solution The case $p = 1$ is settled in Example 3, so we assume that $p \neq 1$. In this case we have

$$\int_1^\infty \frac{dx}{x^p} = \lim_{t \to \infty} \int_1^t \frac{dx}{x^p} = \lim_{t \to \infty} \left[\frac{x^{1-p}}{1-p} \right]_1^t$$

$$= \lim_{t \to \infty} \left[\frac{t^{1-p} - 1}{1-p} \right] = \begin{cases} \dfrac{1}{p-1} & \text{if } p > 1 \\[2mm] \infty & \text{if } p < 1, \end{cases}$$

and this completes the proof.

We consider the geometric meaning of Example 5 by examining Fig. 12.5.

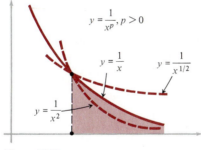

$$y = \frac{1}{x^p},\ p > 0$$

$$y = \frac{1}{x}$$

$$y = \frac{1}{x^{1/2}}$$

$$y = \frac{1}{x^2}$$

Figure 12.5

When the exponent p is allowed to decrease through values greater than 1 (for instance, $p = 4, 3, 2, 1.5$, etc.), then it is easy to see that the graph of $y = 1/x^p$ to the right of $x = 1$ rises; also, the calculation shows that the area of the unbounded region under this graph increases but remains finite. When p reaches 1 this area suddenly becomes infinite, and it remains infinite for all values of $p < 1$. It is indeed remarkable that a region of infinite extent can have a finite area, as happens here when $p > 1$. We will comment further on this phenomenon in Remark 1.

Another type of improper integral arises when the integrand $f(x)$ is continuous on a bounded interval of the form $a \le x < b$ but becomes infinite as x approaches b, as shown in Fig. 12.6. In this case we can integrate from a to a variable upper limit t which is less than b. This integral is a function of t, and we can now ask whether this function approaches a limit as $t \to b$. If so, we use this limit as the definition of the improper integral of $f(x)$ from a to b,

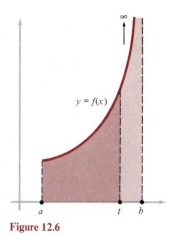

$y = f(x)$

Figure 12.6

$$\int_a^b f(x)\,dx = \lim_{t \to b} \int_a^t f(x)\,dx.$$

As before, this integral is called *convergent* if the limit exists and is finite, and *divergent* otherwise.

Example 6

$$\int_0^1 \frac{dx}{\sqrt{1-x}} = \lim_{t \to 1} \int_0^t \frac{dx}{\sqrt{1-x}} = \lim_{t \to 1}\left[-2\sqrt{1-x}\right]_0^t$$

$$= \lim_{t \to 1}\left[-2\sqrt{1-t}+2\right] = 2.$$

This improper integral clearly converges.

There are several other types of improper integrals which we mention only briefly because the ideas are essentially the same as those already described.

If the impropriety of an integral occurs at the lower limit, we use t as the lower limit and then let $t \to -\infty$ or $t \to a$, as the case may be. If the integrand misbehaves at several points, then the improper integral—if it exists—is obtained by dividing the original interval into subintervals.

Finally, if $f(x)$ is continuous on the entire real line, then we write, *by definition,*

$$\int_{-\infty}^{\infty} f(x)\,dx = \int_{-\infty}^{0} f(x)\,dx + \int_{0}^{\infty} f(x)\,dx,$$

where convergence for the improper integral on the left means that both integrals on the right converge. An integral from $-\infty$ to ∞ can be split at any convenient finite point just as well as at the point $x = 0$.

Example 7

$$\int_{-\infty}^{\infty} \frac{dx}{1+x^2} = \int_{-\infty}^{0} \frac{dx}{1+x^2} + \int_{0}^{\infty} \frac{dx}{1+x^2}$$

$$= \lim_{t \to -\infty} \int_t^0 \frac{dx}{1+x^2} + \lim_{t \to \infty} \int_0^t \frac{dx}{1+x^2}$$

$$= \lim_{t \to -\infty}\left[\tan^{-1} x\right]_t^0 + \lim_{t \to \infty}\left[\tan^{-1} x\right]_0^t$$

$$= \lim_{t \to -\infty}\left(-\tan^{-1} t\right) + \lim_{t \to \infty} \tan^{-1} t = -\left(-\frac{\pi}{2}\right) + \frac{\pi}{2} = \pi.$$

Remark 1 Students may still be skeptical that a region of infinite extent can have a finite area. If so, then the following example may help. Consider the region under the curve $y = 1/2^x$ for $0 \le x < \infty$. This region is shaded in Fig. 12.7, and clearly has a smaller area than the combined area of all the rectangles shown in the figure. But these rectangles have base 1 and heights 1, $\frac{1}{2}$, $\frac{1}{4}$, $\frac{1}{8}$, . . . , so their combined area is exactly

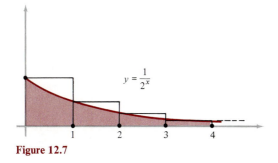

Figure 12.7

$$1 + \tfrac{1}{2} + \tfrac{1}{4} + \tfrac{1}{8} + \cdots = 2.*$$

It follows that the shaded region — of infinite extent! — has finite area less than 2. The area of this region can even be computed exactly; it is

$$\int_0^\infty \frac{dx}{2^x} = \lim_{t \to \infty} \int_0^t \frac{dx}{2^x} = \lim_{t \to \infty} \int_0^t e^{-x\ln 2}\, dx$$

$$= \lim_{t \to \infty} \left[-\frac{1}{\ln 2}\, e^{-x\ln 2} \right]_0^t$$

$$= \lim_{t \to \infty} \left[-\frac{1}{\ln 2} \cdot \frac{1}{2^t} + \frac{1}{\ln 2} \right] = \frac{1}{\ln 2}.$$

Remark 2   Generally speaking, improper integrals play a more substantial role in higher mathematics than they do in calculus. We mention two important examples — which we do not pursue any further in this book — to give students some idea of what we're talking about.

(a) The improper integral

$$\Gamma(p) = \int_0^\infty x^{p-1}\, e^{-x}\, dx$$

(the symbol on the left is capital *gamma* in the Greek alphabet) is called the *gamma function.* This is a very interesting function which is studied in advanced calculus and elsewhere. It has innumerable applications to physics as well as to geometry, number theory, and other parts of pure mathematics.

(b) The improper integral

$$F(p) = \int_0^\infty e^{-px}\, f(x)\, dx$$

has many significant applications to electric circuits, vibrating membranes, and heat conduction, and to the solution of certain types of differential equations. It is a function of p associated with the given function $f(x)$, and is called the *Laplace transform* of $f(x)$.†

---

* This is an infinite geometric series of a kind often studied in high school algebra courses. We shall discuss these series in much more detail in Chapter 13.

† See Chapter 10 in the author's book *Differential Equations with Applications and Historical Notes,* McGraw-Hill, New York, 1972.

PROBLEMS

In each of the following problems, determine whether or not the improper integral converges, and find its value if it does.

1 $\displaystyle\int_3^\infty e^{-2x}\,dx.$

2 $\displaystyle\int_0^\infty \frac{dx}{(1+x)^3}.$

3 $\displaystyle\int_8^\infty \frac{dx}{x^{4/3}}.$

4 $\displaystyle\int_0^\infty \sin x\,dx.$

5 $\displaystyle\int_1^\infty \frac{1}{x^2}\sin\frac{1}{x}\,dx.$

6 $\displaystyle\int_e^\infty \frac{dx}{x\ln x}.$

7 $\displaystyle\int_e^\infty \frac{dx}{x(\ln x)^2}.$

8 $\displaystyle\int_0^\infty e^{-x}\cos x\,dx.$

9 $\displaystyle\int_0^\infty (x-1)e^{-x}\,dx.$

10 $\displaystyle\int_1^\infty \left(\frac{1}{\sqrt{x}}-\frac{1}{\sqrt{x+3}}\right)dx.$

11 $\displaystyle\int_1^\infty \frac{dx}{x(x+2)}.$

12 $\displaystyle\int_3^\infty \frac{dx}{x\sqrt{16+x^2}}.$

13 $\displaystyle\int_0^2 \frac{\ln x\,dx}{\sqrt{x}}.$

14 $\displaystyle\int_0^2 \frac{dx}{4-x^2}.$

15 $\displaystyle\int_{-\infty}^\infty |x|e^{-x^2}\,dx.$

16 $\displaystyle\int_{-\infty}^\infty e^{-x}\cos x\,dx.$

17 Let p be a positive constant. Determine the values of p for which the improper integral

$$\int_0^1 \frac{dx}{x^p}$$

is convergent, and those for which it is divergent.

18 Consider the region under the graph of $y=1/x$ for $x\ge 1$. Even though this region has infinite area, show that the solid of revolution obtained by revolving this region about the x-axis has finite volume, and compute this volume.

19 Consider the region in the first quadrant under the curve $y=1/(x+1)^3$. Find the volume of the solid of revolution generated by revolving this region about (a) the x-axis; (b) the y-axis.

20 The region under the curve $y=4/(3x^{3/4})$ for $x\ge 1$ is revolved about the x-axis. Find the volume of the solid of revolution generated in this way.

21 Show that the surface area of the solid of revolution described in Problem 20 is infinite. As a result of these calculations, we see that a container in the shape of this surface can be filled with paint (it has finite volume), but nevertheless its inner surface cannot be painted (it has infinite surface area). Hint: Use the obvious inequality

$$\frac{1}{x^{3/4}}\sqrt{1+\frac{1}{x^{7/2}}}>\frac{1}{x^{3/4}}$$

to show that

$$\text{Surface area}>\frac{8\pi}{3}\int_1^\infty \frac{dx}{x^{3/4}}=\infty.$$

22 If $a>0$ and the graph of $y=ax^2+bx+c$ lies entirely above the x-axis, show that

$$\int_{-\infty}^\infty \frac{dx}{ax^2+bx+c}=\frac{2\pi}{\sqrt{4ac-b^2}}.$$

23 (*A comparison test*) Let $f(x)$ and $g(x)$ be continuous functions with the property that $0\le f(x)\le g(x)$ for $a\le x<\infty$. Show that

(a) if $\int_a^\infty g(x)\,dx$ converges, then $\int_a^\infty f(x)\,dx$ also converges;

(b) if $\int_a^\infty f(x)\,dx$ diverges, then $\int_a^\infty g(x)\,dx$ also diverges.

24 Use the comparison test in Problem 23 to determine whether each of the following integrals converges or diverges:

(a) $\displaystyle\int_1^\infty \frac{dx}{\sqrt{x^3+5}};$

(b) $\displaystyle\int_2^\infty (x^6-1)^{-1/7}\,dx;$

(c) $\displaystyle\int_2^\infty \frac{\cos^4 5x}{x^3}\,dx;$

*(d) $\displaystyle\int_e^\infty \frac{\ln x}{x^2}\,dx.$

ADDITIONAL PROBLEMS FOR CHAPTER 12

SECTION 12.2
Find the following limits.

1 $\displaystyle\lim_{x\to 0}\frac{\sin 5x}{\sin 2x}.$

2 $\displaystyle\lim_{x\to 2}\frac{\ln(x-1)}{x-2}.$

3 $\displaystyle\lim_{x\to 5}\frac{x^2+x-30}{\sqrt{x-1}-2}.$

4 $\displaystyle\lim_{x\to 1}\frac{\sin \pi x}{1-x^2}.$

5 $\displaystyle\lim_{x\to 4}\frac{x-4}{\sqrt[3]{x+4}-2}.$

6 $\displaystyle\lim_{x\to -3}\frac{x^2+2x-3}{2x^2+3x-9}.$

7 $\displaystyle\lim_{x\to 2}\frac{\tan(2x-4)}{x^3-8}.$

8 $\displaystyle\lim_{x\to 1}\frac{x^3+x^2-2}{\ln x}.$

9 $\displaystyle\lim_{x\to 0}\frac{\sqrt[5]{x+1}-(1+\frac{1}{5}x)}{3x^2}.$

10 $\lim\limits_{x \to 0} \dfrac{\sqrt[4]{x + 16} - (2 + \frac{1}{32}x)}{x^2}$.

11 $\lim\limits_{x \to 3+} \dfrac{\ln(x - 2)}{(x - 3)^2}$.

12 $\lim\limits_{x \to 0} \dfrac{x \sin(\sin x)}{1 - \cos(\sin x)}$.

13 $\lim\limits_{x \to 0} \dfrac{\sin x^3}{x - \sin x}$.

14 $\lim\limits_{x \to 0} \dfrac{e^{x^2} - 1}{x \sin x}$.

15 $\lim\limits_{x \to \infty} \dfrac{e^{3/x} - 1}{\sin 1/x}$.

16 $\lim\limits_{x \to 0+} \dfrac{\tan^{-1} x}{1 - \cos 2x}$.

17 $\lim\limits_{x \to 0} \dfrac{1 - \cos x}{x \sin x}$.

18 $\lim\limits_{x \to 16+} \dfrac{\sqrt[4]{x} - 16}{\sqrt[4]{x} - 2}$.

19 $\lim\limits_{x \to 0+} \dfrac{\sin^{-1} x}{\sin^2 3x}$.

20 $\lim\limits_{x \to 0} \dfrac{2 \cos x - 2 + x^2}{3x^4}$.

21 $\lim\limits_{x \to \pi/2} \dfrac{1 - \sin x}{\cos x}$.

22 $\lim\limits_{x \to 0} \dfrac{2x}{\tan^{-1} x}$.

23 $\lim\limits_{x \to 2} \dfrac{3\sqrt[3]{x - 1} - x - 1}{3(x - 2)^2}$.

24 $\lim\limits_{x \to 1} \dfrac{\ln x}{x^2 - x}$.

25 $\lim\limits_{x \to 0} \dfrac{\sin^2 x + 2 \cos x - 2}{\cos^2 x - x \sin x - 1}$.

26 $\lim\limits_{x \to 0} \dfrac{\sin x - \tan x}{x^3}$.

27 $\lim\limits_{x \to 0} \dfrac{\cos x - 1 + \frac{1}{2}x^2}{x^4}$.

28 $\lim\limits_{x \to 0} \dfrac{\sin x^2 - \sin^2 x}{x^4}$.

29 $\lim\limits_{x \to 0} \dfrac{e^x + e^{-x} - 2}{1 - \cos 4x}$.

30 $\lim\limits_{x \to \pi} \dfrac{1 + \cos x}{(x - \pi)^2}$.

31 $\lim\limits_{x \to 1} \dfrac{x^3 + 3e^{1-x} - 4}{x - \ln x - 1}$.

32 $\lim\limits_{x \to 0} \dfrac{x - \sin x}{x \tan x}$.

33 $\lim\limits_{x \to 0} \dfrac{x^2 \tan x}{\tan x - x}$.

34 $\lim\limits_{x \to 0} \dfrac{1 - \cos^2 x}{x^2}$.

35 $\lim\limits_{x \to 0} \dfrac{\ln(x + 1)}{e^{3x} - 1}$.

36 $\lim\limits_{x \to 0} \dfrac{1 - \cos 2\sqrt{a}\, x}{2x^2}$.

37 $\lim\limits_{x \to 1} \dfrac{x^{10} - 1}{x^9 - 1}$.

38 $\lim\limits_{x \to 0} \dfrac{x - \sin x}{1 - \cos x}$.

39 $\lim\limits_{x \to 0} \dfrac{x - \tan^{-1} x}{x^3}$.

40 $\lim\limits_{x \to \pi} \dfrac{\sin^2 x}{1 + \cos 5x}$.

41 $\lim\limits_{x \to \infty} \dfrac{\tan^2(1/x)}{\ln^2(1 + 4/x)}$.

42 Consider the circular sector of radius 1 shown in Fig. 12.8. The point C is the intersection of the tangent lines at A and B. If $f(\theta)$ is the area of the triangle ABC and $g(\theta)$ is the area of the region that remains when

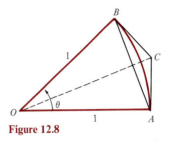

Figure 12.8

the triangle OAB is removed from the sector, evaluate $\lim_{\theta \to 0} f(\theta)/g(\theta)$.

43 Show that

$$\lim_{x \to 0} \dfrac{x^2 \sin(1/x)}{x}$$

is an indeterminate form of the type 0/0 that exists but cannot be evaluated by L'Hospital's rule. What is the value of this limit? Does this example show that L'Hospital's rule is false?

44 Use L'Hospital's rule to establish the following formulas for the direct calculation of the second derivative:

(a) $f''(x) = \lim\limits_{h \to 0} \dfrac{f(x + 2h) - 2f(x + h) + f(x)}{h^2}$;

(b) $f''(x) = \lim\limits_{h \to 0} \dfrac{f(x + h) - 2f(x) + f(x - h)}{h^2}$.

45 If n is a positive integer, show that

$$\lim_{x \to 1} \dfrac{nx^{n+1} - (n + 1)x^n + 1}{(x - 1)^2} = \dfrac{n(n + 1)}{2}.$$

(For the meaning of this rather strange-appearing result, see Problem 46.)

46 If n is a positive integer and $x \neq 1$, the formula

$$1 + x + x^2 + x^3 + \cdots + x^n$$

$$= \dfrac{1 - x^{n+1}}{1 - x} = \dfrac{x^{n+1} - 1}{x - 1}$$

is familiar from high school algebra. Differentiate it to obtain

$$1 + 2x + 3x^2 + \cdots + nx^{n-1}$$

$$= \dfrac{nx^{n+1} - (n + 1)x^n + 1}{(x - 1)^2}, \quad (*)$$

and then take limits as $x \to 1$ and use Problem 45 to derive the formula

$$1 + 2 + 3 + \cdots + n = \dfrac{n(n + 1)}{2}.$$

*47 Multiply equation (*) in Problem 46 by x, differentiate, etc., and thereby derive the formula

$$1^2 + 2^2 + 3^2 + \cdots + n^2 = \frac{n(n+1)(2n+1)}{6}.$$

SECTION 12.3
Evaluate the following limits by any method.

48 $\displaystyle\lim_{x\to\infty} \frac{3x^2 + 9}{x + e^x}.$

49 $\displaystyle\lim_{x\to\frac{1}{2}^-} \frac{\ln(1-2x)}{\tan \pi x}.$

50 $\displaystyle\lim_{x\to 3\pi/2} \frac{2 + \sec x}{\tan x}.$

51 $\displaystyle\lim_{x\to\infty} \frac{\ln x^{100}}{\sqrt[5]{x}}.$

52 $\displaystyle\lim_{x\to\infty} \frac{x + \ln x}{x \ln x}.$

53 $\displaystyle\lim_{x\to 0+} \frac{\ln x}{\cot x}.$

54 $\displaystyle\lim_{x\to 0+} \frac{\ln(\sin x)}{\ln(\sin 2x)}.$

55 $\displaystyle\lim_{x\to\infty} \frac{\ln x}{e^{2x}}.$

56 $\displaystyle\lim_{x\to\infty} \frac{\ln(\ln x)}{\sqrt{x}}.$

57 $\displaystyle\lim_{x\to\infty} \frac{xe^x}{e^{x^2}}.$

58 $\displaystyle\lim_{x\to 0+} x^2 \ln x.$

59 $\displaystyle\lim_{x\to 0+} x^p \ln x, \, p > 0.$

60 $\displaystyle\lim_{x\to 0+} x^2 e^{1/x}.$

61 $\displaystyle\lim_{x\to\infty} x \sin \frac{p}{x}, \, p \neq 0.$

62 $\displaystyle\lim_{x\to 0+} \tan x \ln x.$

63 $\displaystyle\lim_{x\to\pi/2} \left(x - \frac{\pi}{2}\right) \tan 3x.$

64 $\displaystyle\lim_{x\to\pi/2} (2x - \pi) \sec x.$

65 $\displaystyle\lim_{x\to\pi/2} \tan x \ln(\sin x).$

66 $\displaystyle\lim_{x\to\infty} x(e^{1/x} - 1).$

67 $\displaystyle\lim_{x\to 0+} \sin x \ln(\sin x).$

68 $\displaystyle\lim_{x\to 0} \left(\frac{1}{x^2} - \frac{1}{x \sin x}\right).$

69 $\displaystyle\lim_{x\to 0} \left(\frac{1}{1 - \cos x} - \frac{2}{x^2}\right).$

70 $\displaystyle\lim_{x\to 0} \left[\frac{1 + x}{\ln(1 + x)} - \frac{1}{x}\right].$

71 $\displaystyle\lim_{x\to 0} \left[\frac{1}{\ln(1 + x)} - \frac{1}{e^x - 1}\right].$

72 $\displaystyle\lim_{x\to\infty} (x - \sqrt{x^2 + x}).$

73 $\displaystyle\lim_{x\to 0+} x^{\sin x}.$

74 $\displaystyle\lim_{x\to 0+} (\sin x)^{\tan x}.$

75 $\displaystyle\lim_{x\to 0+} (e^x - 1)^{\sin x}.$

76 $\displaystyle\lim_{x\to 1+} (x^2 - 1)^{x-1}.$

77 $\displaystyle\lim_{x\to\pi/2-} (\cos x)^{\cos x}.$

78 $\displaystyle\lim_{x\to\pi/4-} (1 - \tan x)^{1-\tan x}.$

79 $\displaystyle\lim_{x\to 0+} (x + \sin x)^{\tan x}.$

80 $\displaystyle\lim_{x\to 1+} (\ln x)^{\sin(x-1)}.$

81 $\displaystyle\lim_{x\to 0+} [\ln(1 + x)]^x.$

82 $\displaystyle\lim_{x\to 0+} x^{ax^b}, \, b > 0.$

83 $\displaystyle\lim_{x\to 0+} x^{x^x} \, [x^{x^x} = x^{(x^x)}].$

84 $\displaystyle\lim_{x\to\infty} (x + e^{ax})^{b/x}.$

85 $\displaystyle\lim_{x\to\infty} (1 + x^p)^{1/x}, \, p > 0.$

86 $\displaystyle\lim_{x\to\infty} (1 + x)^{e^{-x}}.$

87 $\displaystyle\lim_{x\to 0+} (1 + \csc x)^{\sin^2 x}.$

88 $\displaystyle\lim_{x\to 0} \left(1 + \frac{1}{x}\right)^x.$

89 $\displaystyle\lim_{x\to 0+} (\csc x)^x.$

90 $\displaystyle\lim_{x\to 0+} (\cot x)^x.$

91 $\displaystyle\lim_{x\to\infty} x^{\ln(1+1/x)}.$

92 $\displaystyle\lim_{x\to 0} (1 - 2x)^{3/x}.$

93 $\displaystyle\lim_{x\to\infty} \left(1 + \frac{2}{x}\right)^x.$

94 $\displaystyle\lim_{x\to\infty} \left(1 + \frac{1}{x}\right)^{5x}.$

95 $\displaystyle\lim_{x\to 0} (e^x + 3x)^{1/x}.$

96 $\displaystyle\lim_{x\to 0} (1 + 2x)^{\cot x}.$

97 $\displaystyle\lim_{x\to 0} (1 + 3x)^{\csc x}.$

98 $\displaystyle\lim_{x\to 0} (\cos 2x)^{1/x^2}.$

99 Show that

$$\lim_{x\to\infty} \frac{x + \sin x}{x}$$

is an indeterminate form of the type ∞/∞ that exists but cannot be found by L'Hospital's rule. What is the value of this limit?

100 Find the value a must have if

$$\lim_{x\to\infty} \left(\frac{x + a}{x - a}\right)^x = 4.$$

SECTION 12.4
Determine whether or not each of the following integrals converges, and find its value if it does.

101 $\displaystyle\int_2^\infty e^{-3x} \, dx.$

102 $\displaystyle\int_5^\infty \frac{dx}{x^3}.$

103 $\displaystyle\int_4^\infty \frac{dx}{x\sqrt{x}}.$

104 $\displaystyle\int_0^\infty \frac{x^2 \, dx}{x^3 + 1}.$

105 $\displaystyle\int_0^\infty e^{-x} \sin x \, dx.$

106 $\displaystyle\int_0^\infty xe^{-x} \, dx.$

107 $\displaystyle\int_0^\infty \frac{x \, dx}{x^4 + 1}.$

108 $\displaystyle\int_1^\infty xe^{-x^2} \, dx.$

109 $\displaystyle\int_2^\infty \frac{dx}{4 + x^2}.$

110 $\displaystyle\int_2^\infty \frac{dx}{x^2 - 1}.$

111 $\displaystyle\int_0^\infty \frac{x^2\,dx}{e^{x^3}}.$

112 $\displaystyle\int_e^\infty \frac{\ln x}{x}\,dx.$

113 $\displaystyle\int_e^\infty \frac{dx}{x\ln x\sqrt{\ln x}}.$

114 $\displaystyle\int_0^\infty \frac{dx}{\sqrt[3]{e^x}}.$

115 $\displaystyle\int_0^{\pi/2} \frac{dx}{1-\sin x}.$

116 $\displaystyle\int_0^{\pi/2} \frac{dx}{\sin x}.$

117 $\displaystyle\int_0^2 \frac{\ln x}{x}\,dx.$

118 $\displaystyle\int_0^4 \frac{8dx}{\sqrt{16-x^2}}.$

119 $\displaystyle\int_0^3 \frac{x\,dx}{\sqrt{9-x^2}}.$

120 Let p be a positive constant. Determine the values of p for which the improper integral

$$\int_0^1 \frac{dx}{(1-x)^p}$$

is convergent, and those for which it is divergent.

121 Show that the region in the first quadrant under the curve $y=1/(x+1)^2$ has a finite area but does not have a centroid.

122 If x is a positive constant, show that

$$\int_0^\infty e^{-a^2x^2}\,dx = \frac{1}{a}\int_0^\infty e^{-x^2}\,dx.$$

Without performing any actual integrations, use this fact to show that the centroid of the region between the curve $y=e^{-a^2x^2}$ and the x-axis is $(0,\sqrt{2}/4)$.

13

INTRODUCTION TO INFINITE SERIES

13.1
WHAT IS AN INFINITE SERIES?

We have touched briefly on this subject several times before, but now the time has come to confront it directly.

An *infinite series,* or simply a *series,* is an expression of the form

$$a_1 + a_2 + a_3 + \cdots + a_n + \cdots, \tag{1}$$

where the three dots at the end indicate that the terms continue indefinitely. In other words, there are infinitely many numbers a_n (one for each positive integer n) and (1) is the indicated sum of this infinite collection of terms. The number a_n is called the *nth term* of the series, and is usually some simple function of n. We include it in (1) if we wish to make an explicit statement of the law of formation of the terms. However, if this law of formation is clear from the context, we can write (1) more informally as

$$a_1 + a_2 + a_3 + \cdots \qquad \text{or} \qquad a_1 + a_2 + \cdots.$$

We will often use the sigma notation of Section 6.3 to write the series (1) in the form

$$\sum_{n=1}^{\infty} a_n.$$

This is read "the sum from $n = 1$ to infinity of a_n."

Needless to say, it is quite impossible to perform the operation of addition an infinite number of times—life isn't long enough—and so (1) cannot be interpreted literally and its meaning must be approached in a subtler way. It was one of the great achievements of nineteenth century mathematics to discover that a perfectly reasonable and satisfactory meaning can be given to (1). If we exercise suitable caution, this meaning allows us to work with such expressions just as easily as if they involved only a finite number of terms. In many cases we will actually be able to find the number that is the exact sum of the infinite series, and these sums often turn out to be very surprising indeed.

We will get to all this in Section 13.2, but first we briefly consider a few of the many natural ways in which infinite series arise in mathematics.

A   We discussed the geometric meaning of infinite decimals in Section 1.2. However, this was only an interpretation, and it is often overlooked that an infinite decimal is *defined* in a purely numerical way, as an infinite series,

$$.a_1 a_2 a_3 \ldots a_n \ldots = \frac{a_1}{10} + \frac{a_2}{10^2} + \frac{a_3}{10^3} + \cdots + \frac{a_n}{10^n} + \cdots, \qquad (2)$$

where each of the a's is understood to be one of the ten digits $0, 1, 2, \ldots, 9$.
   Everyone knows that

$$\tfrac{1}{3} = 0.333 \ldots, \qquad (3)$$

but not everyone is sure why, or even what this means. This is not at all surprising, because (3) cannot be fully understood without some acquaintance with infinite series, enough to use (2) to evaluate the right-hand side of (3). We will discuss this and other related issues in Section 13.2.

B   The elementary long division

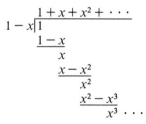

tells us that

$$\frac{1}{1-x} = 1 + x + x^2 + \cdots + x^{n-1} + \frac{x^n}{1-x}. \qquad (4)$$

This process can be carried out to as many steps as we wish, and it is natural to wonder how the function on the left of (4) is related to the infinite series that seems to be forming on the right. That is, is it true that

$$\frac{1}{1-x} = 1 + x + x^2 + \cdots ?$$

Most readers have seen this series before, in connection with geometric progressions in elementary algebra.

C   As we saw in Section 3.5, the binomial theorem can be written in the form

$$(1+x)^n = 1 + nx + \frac{n(n-1)}{1 \cdot 2} x^2 + \frac{n(n-1)(n-2)}{1 \cdot 2 \cdot 3} x^3$$

$$+ \cdots + \frac{n(n-1)(n-2) \cdots (n-k+1)}{1 \cdot 2 \cdot 3 \cdots k} x^k + \cdots, \qquad (5)$$

where n is a positive integer. For example, when $n = 4$, this formula produces the expansion

$$(1 + x)^4 = 1 + 4x + \frac{4 \cdot 3}{1 \cdot 2} x^2 + \frac{4 \cdot 3 \cdot 2}{1 \cdot 2 \cdot 3} x^3 + \frac{4 \cdot 3 \cdot 2 \cdot 1}{1 \cdot 2 \cdot 3 \cdot 4} x^4$$

$$+ \frac{4 \cdot 3 \cdot 2 \cdot 1 \cdot 0}{1 \cdot 2 \cdot 3 \cdot 4 \cdot 5} x^5 + \cdots$$

$$= 1 + 4x + 6x^2 + 4x^3 + x^4.$$

In this case, it is clear that the coefficients of x^5, x^6, etc., are all zero, and the expansion is a finite sum. Formula (5) is proved in elementary algebra only for exponents n that are positive integers, so that only finite sums are involved.

However, in a spirit of curiosity, let us see what happens when we apply (5) to the case $n = \frac{1}{2}$. This gives the expansion

$$\sqrt{1 + x} = (1 + x)^{1/2} = 1 + \frac{1}{2} x + \frac{\frac{1}{2}(-\frac{1}{2})}{1 \cdot 2} x^2 + \frac{\frac{1}{2}(-\frac{1}{2})(-\frac{3}{2})}{1 \cdot 2 \cdot 3} x^3 + \cdots$$

$$= 1 + \frac{x}{2} - \frac{1}{1 \cdot 2} \left(\frac{x}{2}\right)^2 + \frac{1 \cdot 3}{1 \cdot 2 \cdot 3} \left(\frac{x}{2}\right)^3 - \cdots, \quad (6)$$

and this is clearly an infinite series, because none of the coefficients of the various powers of x are zero. Is this formula legitimate? If we test its validity by substituting $x = 0.2$, then we get

$$\sqrt{1.2} = 1 + 0.1000 - 0.0050 + 0.0005 - \cdots.$$

If we use only the first four terms shown on the right, this suggests the approximate value of 1.0955 for $\sqrt{1.2}$, whereas a table gives 1.0954. It seems reasonably clear from this fragment of numerical evidence that (6) may indeed be a valid formula, at least for some nonzero values of x. And this in turn raises the possibility that (5) itself may be valid for certain nonzero x's in all those cases in which the exponent n is not a positive integer, that is, when the right side of (5) is an infinite series.

These remarks suggest many far-reaching questions that are rich in the promise of fruitful applications, but these questions are not easy to answer. One of our purposes in studying infinite series is to replace this kind of speculation and conjecture by definitive answers.

D   Finally, infinite series arise in a very insistent way in the study of differential equations. To see how this happens, let us consider the simple equation

$$\frac{dy}{dx} = y. \quad (7)$$

This equation asks for a function which is unchanged by differentiation, and we know that $y = ce^x$ is such a function for every constant c. But to emphasize the point we wish to make, let us pretend that we don't know any solutions and try to guess one. Since polynomials are the simplest functions of all, we might try one of these first. But we have no idea what degree to choose for this hoped-for polynomial solution. This suggests the use of a bit of creative vagueness, so we leave the degree unspecified and try to find coefficients $a_0, a_1, a_2, a_3, a_4, \ldots$ so that

$$y = a_0 + a_1 x + a_2 x^2 + a_3 x^3 + a_4 x^4 + \cdots \qquad (8)$$

will be a solution of (7). By differentiating (8) term by term we obtain

$$\frac{dy}{dx} = a_1 + 2a_2 x + 3a_3 x^2 + 4a_4 x^3 + \cdots ; \qquad (9)$$

and substituting (8) and (9) in (7) gives

$$a_1 + 2a_2 x + 3a_3 x^2 + 4a_4 x^3 + \cdots = a_0 + a_1 x + a_2 x^2 + a_3 x^3 + \cdots . \quad (10)$$

If we now equate coefficients of equal powers of x in (10), we get

$$a_1 = a_0, \qquad 2a_2 = a_1, \qquad 3a_3 = a_2, \qquad 4a_4 = a_3, \quad \ldots ,$$

so

$$a_1 = a_0, \qquad a_2 = \frac{a_1}{2} = \frac{a_0}{2}, \qquad a_3 = \frac{a_2}{3} = \frac{a_0}{2 \cdot 3}, \qquad a_4 = \frac{a_3}{4} = \frac{a_0}{2 \cdot 3 \cdot 4}, \ldots . \quad (11)$$

At this point we remind students of the factorial notation introduced in Section 3.5. If n is a positive integer, we write

$$n! = 1 \cdot 2 \cdot 3 \cdots n \qquad (12)$$

and call this *n factorial.* Thus, $1! = 1$, $2! = 1 \cdot 2 = 2$, $3! = 1 \cdot 2 \cdot 3 = 6$, $4! = 1 \cdot 2 \cdot 3 \cdot 4 = 24$, $5! = 1 \cdot 2 \cdot 3 \cdot 4 \cdot 5 = 120$, and so on. The definition (12) is meaningless in the case $n = 0$, but for many reasons it is customary to define $0!$ by $0! = 1$. We shall be using factorials often in the next few chapters, so there will be ample opportunity for students to become thoroughly familiar with this notation.

Returning to our problem, we can use the factorial notation to write equations (11) as

$$a_1 = a_0, \qquad a_2 = \frac{a_0}{2!}, \qquad a_3 = \frac{a_0}{3!}, \qquad a_4 = \frac{a_0}{4!}, \quad \ldots .$$

Our tentative solution (8) of the differential equation (7) therefore becomes

$$y = a_0 + a_0 x + \frac{a_0}{2!} x^2 + \frac{a_0}{3!} x^3 + \frac{a_0}{4!} x^4 + \cdots$$

$$= a_0 \left(1 + x + \frac{x^2}{2!} + \frac{x^3}{3!} + \frac{x^4}{4!} + \cdots \right), \qquad (13)$$

where a_0 is an arbitrary constant. On comparing this with the known solutions ce^x, we are led to the natural conjecture that the exponential function e^x equals the infinite series shown in parentheses:

$$e^x = 1 + x + \frac{x^2}{2!} + \frac{x^3}{3!} + \frac{x^4}{4!} + \cdots .$$

It turns out that this formula is indeed true. In fact, for all values of x we can calculate e^x as accurately as we please from the series by taking enough terms, and this is how numerical tables for e^x are constructed. However, students should clearly understand that this discussion is merely suggestive, and is by no means a valid proof. Proofs will come later.

In attempting to solve other differential equations in this way, we are led to other series, some representing familiar functions and some representing previously unknown functions. For example, the differential equation

$$x \frac{d^2y}{dx^2} + \frac{dy}{dx} + xy = 0$$

has so many applications to mathematical physics that it has been given a special name of its own; it is called the *Bessel equation of order zero,* after a famous German astronomer of the nineteenth century. One of its solutions is the infinite series

$$y = 1 - \frac{x^2}{2^2} + \frac{x^4}{2^2 \cdot 4^2} - \frac{x^6}{2^2 \cdot 4^2 \cdot 6^2} + \cdots .$$

This series defines a new function of x which is important enough to have its own standard symbol and its own name; it is denoted by $J_0(x)$ and called the *Bessel function of order zero.*

Our main aim in this chapter is to provide a bird's-eye view of infinite series before we begin a systematic development of the theory of the subject. We will concentrate our attention mostly on specific examples, important formulas, and the uses of such indispensable procedures as the term-by-term differentiation and integration of series of functions. In this way students will be in a better position in Chapter 14 to understand the theory given there, because by then they will have more familiarity with the subject matter and therefore more insight into what the theory is for.

In infinite series, as in almost any part of calculus, there are certain things that we want to be able to do freely with the tools we are studying, and the role of the theory is mostly to justify and legitimize the various procedures that are necessary for carrying out our purposes — such procedures, for example, as the term-by-term differentiation and integration previously mentioned. This situation was well expressed by the famous financier J. P. Morgan in describing the role of lawyers in his business operations. "I don't hire lawyers to tell me what I can't do," he said. "I hire them to find legal ways for me to do what I want to do."

13.2
THE CONVERGENCE AND DIVERGENCE OF SERIES

The idea of "infinite addition" arises in its most transparent form in the famous formula

$$1 + \frac{1}{2} + \frac{1}{4} + \cdots + \frac{1}{2^n} + \cdots = 2. \tag{1}$$

Most people agree that the "sum" on the left is exactly equal to 2, as we have indicated, but not everyone remembers just why. The reason is that we start with 1 (total so far = 1), then add $\frac{1}{2}$ (total so far = $1\frac{1}{2}$), then add $\frac{1}{4}$ (total so far = $1\frac{3}{4}$), and so on, producing the *sequence of partial sums,*

$$1, 1\tfrac{1}{2}, 1\tfrac{3}{4}, \ldots .$$

This sequence visibly approaches 2 as its limit, and this is precisely what we mean by saying that the sum of the series is 2.*

This is an example of a geometric series, a very important type of infinite series that is often studied in high school. To form such a series, we start with a number $a \neq 0$ and a second number r between -1 and 1, and we construct the *geometric progression*

$$a, ar, ar^2, \ldots, ar^n, \ldots .$$

The number r is called the *ratio* because it is the ratio of each term of the progression to its predecessor. Our purpose is to calculate the sum of the *geometric series* formed from this progression,

$$a + ar + ar^2 + \cdots + ar^n + \cdots . \tag{2}$$

We accomplish this as follows. The finite sum of the terms from a to ar^n is called the nth *partial sum* and is denoted by s_n. Thus,

$$s_1 = a + ar,$$
$$s_2 = a + ar + ar^2,$$
$$\cdots$$
$$s_n = a + ar + ar^2 + \cdots + ar^n,$$
$$\cdots .$$

If (2) has an exact sum s, then it certainly seems reasonable that we should approach closer and closer to s as we add more and more terms, that is, as the partial sums s_n get longer and longer. This means that we ought to be able to calculate the sum s as a limit,

$$s = \lim_{n \to \infty} s_n . \tag{3}$$

This is all very well in principle, but what makes an explicit calculation possible is the fact that there exists a simple closed formula for the nth partial sum s_n as a function of n. To find this formula, we multiply s_n by r and write the two sums together, as follows:

$$s_n = a + ar + ar^2 + \cdots + ar^n,$$
$$rs_n = ar + ar^2 + ar^3 + \cdots + ar^{n+1}.$$

We see at once that these sums share many terms in common. This suggests that we should subtract the second equation from the first and take advantage of the indicated cancellations, which yields

$$s_n - rs_n = a - ar^{n+1}$$

---

* The words "sequence" and "series" have essentially the same meaning in ordinary speech, but in mathematics their meanings are quite distinct. A sequence is merely an infinite list of numbers arranged in order, with a first, a second, and so on, whereas a series is an infinite sum of numbers. We will discuss sequences more carefully in the next chapter, but for the present it is enough to understand intuitively what is meant by the statement that a sequence $s_1, s_2, \ldots,$ $s_n, \ldots$ approaches — or converges to — a limit L: It means that s_n can be made as close as we please to L by taking n large enough, that is, by going far enough out in the sequence. This behavior is symbolized by writing

$$\lim_{n \to \infty} s_n = L \qquad \text{or} \qquad s_n \to L.$$

or

$$s_n = \frac{a(1 - r^{n+1})}{1 - r}, \tag{4}$$

since $r \neq 1$. But we know that $r^{n+1} \to 0$ as $n \to \infty$, because $|r| < 1$. It is therefore clear that

$$s_n = \frac{a(1 - r^{n+1})}{1 - r} \to \frac{a}{1 - r} \qquad \text{as } n \to \infty.$$

Thus, assuming that there is an exact sum s for the geometric series (2), we can use formula (4) to calculate it by means of the idea expressed in (3),

$$s = \lim_{n \to \infty} s_n = \frac{a}{1 - r}.$$

This is what we mean when we write

$$a + ar + ar^2 + \cdots + ar^n + \cdots = \frac{a}{1 - r}, \qquad |r| < 1. \tag{5}$$

Examples   It is now easy to understand the full meaning of formula (1): the series on the left is a geometric series with $a = 1$ and $r = \frac{1}{2}$, so by (5) we have

$$1 + \frac{1}{2} + \frac{1}{4} + \cdots + \frac{1}{2^n} + \cdots = \frac{1}{1 - \frac{1}{2}} = 2.$$

Similarly,

$$1 + \frac{2}{5} + \frac{4}{25} + \cdots + \left(\frac{2}{5}\right)^n + \cdots = \frac{1}{1 - \frac{2}{5}} = \frac{5}{3}$$

and

$$2 - \frac{4}{3} + \frac{8}{9} - \cdots + 2\left(-\frac{2}{3}\right)^n + \cdots = \frac{2}{1 - (-\frac{2}{3})} = \frac{6}{5}.$$

Further, if we write the repeating decimal $0.333 \ldots$ as an infinite series and apply (5) at the right stage, then we get the result mentioned in Section 13.1,

$$0.333 \ldots = \frac{3}{10} + \frac{3}{10^2} + \frac{3}{10^3} + \cdots$$

$$= \frac{3}{10} + \frac{3}{10}\left(\frac{1}{10}\right) + \frac{3}{10}\left(\frac{1}{10}\right)^2 + \cdots$$

$$= \frac{\frac{3}{10}}{1 - \frac{1}{10}} = \frac{3}{10}\left(\frac{10}{9}\right) = \frac{1}{3}.$$

We now look back on the ideas discussed above and introduce a small but significant change in our point of view. Further investigation will not uncover any way to calculate the exact sum of the geometric series (2) other than as the limit $s = \lim s_n$. Therefore, instead of simply assuming that the series has a sum and that our task is only to calculate it, it is logically better to reverse our approach and *define* the sum of the series (2) to be the number calculated in this way, which we know exists by this discussion. This altered point of view is extremely important for clarifying our understanding of infinite series in general, because — as we shall see — some series have sums and others do not.

These ideas provide the pattern for our broader study of infinite series, and the sum of any other series

$$a_1 + a_2 + \cdots + a_n + \cdots \qquad (6)$$

is defined and calculated in the same way. Thus, we begin by forming the *sequence of partial sums*

$$s_1 = a_1,$$
$$s_2 = a_1 + a_2,$$
$$s_3 = a_1 + a_2 + a_3,$$
$$\cdots$$
$$s_n = a_1 + a_2 + \cdots + a_n,$$
$$\cdots .$$

If this sequence of longer and longer partial sums approaches a finite limit s, then we say that the series (6) *converges* to s, we write this in the form

$$a_1 + a_2 + \cdots + a_n + \cdots = s,$$

and we call the number s the *sum* of the series. The conclusion we reached about the geometric series can now be expressed as follows: *If $|r| < 1$, then the geometric series $a + ar + ar^2 + \cdots$ converges and its sum is $a/(1 - r)$.*

When an infinite series fails to have a sum, we say that it *diverges*. This means that the partial sums s_n fail to approach a finite limit as $n \to \infty$, and this can happen in several different ways.

For instance, suppose we consider the series

$$1 + 1 + \cdots + 1 + \cdots , \qquad (7)$$

in which $a_n = 1$ for every positive integer n. In this case the nth partial sum s_n is just the sum of n 1s, so $s_n = n$. As $n \to \infty$, it is obvious that s_n grows arbitrarily large and therefore doesn't approach any finite limit. For this reason we say that the series (7) *diverges to ∞*.

A more interesting example of a series that diverges to ∞ is the so-called *harmonic series,*

$$1 + \frac{1}{2} + \frac{1}{3} + \cdots + \frac{1}{n} + \cdots . \qquad (8)$$

The fact that this series diverges isn't at all obvious, mainly because we have no simple formula for s_n. However, if we gather the terms into suitable groups, we see that

$$1 + \tfrac{1}{2} + \overline{\tfrac{1}{3} + \tfrac{1}{4}} + \overline{\tfrac{1}{5} + \tfrac{1}{6} + \tfrac{1}{7} + \tfrac{1}{8}} + \cdots$$
$$> 1 + \tfrac{1}{2} + \underbrace{\tfrac{1}{4} + \tfrac{1}{4}} + \underbrace{\tfrac{1}{8} + \tfrac{1}{8} + \tfrac{1}{8} + \tfrac{1}{8}} + \cdots$$
$$= 1 + \tfrac{1}{2} + \quad \tfrac{1}{2} \quad + \quad\quad \tfrac{1}{2} \quad\quad + \cdots .$$

This shows that by taking n large enough we can arrange that s_n contains as many $\frac{1}{2}$s as we please, and therefore s_n can be made arbitrarily large. In this way we see that the harmonic series (8) diverges to ∞.

Another kind of divergence is illustrated by the series

$$1 + (-1) + 1 + (-1) + \cdots ,$$

in which $a_n = (-1)^{n+1}$ for every positive integer n. (This series is usually written $1 - 1 + 1 - 1 + \cdots$.) Here the sequence of partial sums is easily seen to be $1, 0, 1, 0, \ldots$, so s_n doesn't grow arbitrarily large, but neither does it approach a finite limit. Instead, it oscillates indefinitely between the numbers 1 and 0, and therefore the series clearly has no sum.

A simple test for divergence that is often useful is the *nth term test*:

If the series

$$\sum_{n=1}^{\infty} a_n = a_1 + a_2 + \cdots + a_n + \cdots \tag{9}$$

converges, then $a_n \to 0$ as $n \to \infty$; or equivalently, if a_n does not approach zero as $n \to \infty$, then the series (9) *must necessarily diverge.*

To prove this, let us consider the first of the two equivalent forms of the test and assume that the series (9) converges to the sum s. We must now show that $a_n \to 0$, and the argument can be given as follows. Since both s_n and s_{n-1} are close to s when n is large, they are close to each other, and therefore their difference, which is a_n, must be close to zero.

Remark 1 *Repeating decimals.* The procedure for converting any rational number a/b (in lowest terms) into its decimal expansion is well known: Divide b into a. Let us carry out this procedure in the case of the rational number $\frac{22}{7}$, which is often used as a good approximation to π:

```
                 3.142857142857 . . .
            7 |22.0000000. . . . . . .
               21
                1 0
                 7
                30
                28
                 20
                 14
                  60
                  56
                   40
                   35
                    50
                    49
                    10 · · ·
```

The successive remainders here are 1, 3, 2, 6, 4, 5, 1; as soon as 1 appears a second time, the cycle begins all over again and generates the repeating block of digits 142857. This example illustrates—and almost proves—the fact that *the decimal expansion of any rational number is repeating.*[*] The proof of this general statement consists of little more than noticing the phenomena displayed in the example. When b is divided into a, the remainder at each stage is one of the numbers $0, 1, 2, \ldots, b - 1$. Since there are only a finite number of possible values for these remainders, some remainder necessarily appears a second time, and the division process repeats from that point on to give a repeating decimal. We also note that if the remainder 0 appears, then the decimal terminates, but a terminating decimal can always be thought of as repeating, as in $0.25 = 0.25000. \ldots$

The converse of this statement is also true: *Any repeating decimal is the expansion of a rational number.* To see why this is so, let us examine a typical repeating decimal, say $3.7222. \ldots$ If we split off the nonrepeating part, write the repeating part using powers of 10, and use formula (5) at the proper stage, then we obtain

$$3.7222 \ldots = \frac{37}{10} + 0.0222 \ldots$$

$$= \frac{37}{10} + \frac{2}{10^2} + \frac{2}{10^3} + \frac{2}{10^4} + \cdots$$

$$= \frac{37}{10} + \left[\frac{2}{100} + \frac{2}{100} \left(\frac{1}{10} \right) + \frac{2}{100} \left(\frac{1}{10} \right)^2 + \cdots \right]$$

$$= \frac{37}{10} + \frac{\frac{2}{100}}{1 - \frac{1}{10}} = \frac{37}{10} + \frac{2}{100} \cdot \frac{10}{9}$$

$$= \frac{37}{10} + \frac{2}{90} = \frac{335}{90} = \frac{67}{18},$$

which is a rational number expressed as a fraction in lowest terms. It is evident that a similar procedure works equally well for any repeating decimal, so the statement at the beginning of this paragraph is clearly true.

We can summarize our results by saying that *the rational numbers are exactly those real numbers whose decimal expansions are repeating.* Equivalently, *the irrational numbers are exactly those real numbers whose decimal expansions are nonrepeating.*

Remark 2 It is interesting to observe that both the harmonic series

$$\sum_{n=1}^{\infty} \frac{1}{n} = 1 + \frac{1}{2} + \frac{1}{3} + \frac{1}{4} + \cdots$$

and the geometric series

$$1 + \frac{1}{2} + \frac{1}{4} + \frac{1}{8} + \cdots$$

have positive terms that decrease toward zero, and yet the first diverges while the second converges. This suggests the subtlety we will encounter as we penetrate further into the study of infinite series.

---

[*] Certain rational numbers have two distinct decimal expansions, e.g., $\frac{1}{4} = 0.25000 \ldots = 0.24999. \ldots$ This situation is analyzed in the Additional Problems at the end of the chapter.

One of the attractions of the subject is that it offers so many results that stir the imagination and stimulate curiosity. For instance, it seems reasonably clear from the preceding observation that a series of positive terms will converge if its terms decrease "rapidly enough." This is true. For example, it is not difficult to see that the divergent harmonic series can be made to converge by squaring the positive integers in the denominators; that is, the series

$$\sum_{n=1}^{\infty} \frac{1}{n^2} = 1 + \frac{1}{2^2} + \frac{1}{3^2} + \frac{1}{4^2} + \cdots$$

$$= 1 + \frac{1}{4} + \frac{1}{9} + \frac{1}{16} + \cdots \tag{10}$$

converges.* Further, it is known that the sum of this series is $\pi^2/6$:

$$1 + \frac{1}{4} + \frac{1}{9} + \frac{1}{16} + \cdots = \frac{\pi^2}{6}. \tag{11}$$

The underlying reasons for this very remarkable fact will become accessible to us in Section 13.4.

Another way to change the harmonic series into a convergent series is to change the signs of alternate terms. This produces a series whose sum (another astonishing fact!) is ln 2,

$$1 - \frac{1}{2} + \frac{1}{3} - \frac{1}{4} + \cdots = \ln 2, \tag{12}$$

as we shall see in Section 13.3.

---

* For an easy (but ingenious) proof, we point out that $s_n \to s$ for some $s \le 2$ because the s_n's form an increasing sequence with the property that $s_n < 2$ for every n. The latter fact is not obvious, but follows from the inequalities

$$s_n = 1 + \frac{1}{2 \cdot 2} + \frac{1}{3 \cdot 3} + \cdots + \frac{1}{n \cdot n}$$

$$< 1 + \frac{1}{1 \cdot 2} + \frac{1}{2 \cdot 3} + \cdots + \frac{1}{(n-1)n}$$

$$= 1 + \left(\frac{1}{1} - \frac{1}{2}\right) + \left(\frac{1}{2} - \frac{1}{3}\right) + \cdots + \left(\frac{1}{n-1} - \frac{1}{n}\right) = 2 - \frac{1}{n} < 2.$$

PROBLEMS

1. There is nothing to prevent us from forming the geometric series $a + ar + ar^2 + \cdots$ for any real number r (we still insist that a be different from zero). Show that this series diverges whenever $|r| \ge 1$.

2. Determine whether each of the following geometric series is convergent or divergent, and if convergent find its sum:

(a) $1 + \frac{1}{4} + \frac{1}{16} + \cdots$;   (b) $9 + 3 + 1 + \cdots$;

(c) $2 + \frac{3}{2} + \cdots$;   (d) $\frac{1}{2} + \frac{1}{3} + \cdots$;

(e) $\frac{1}{4} - \frac{1}{20} + \cdots$;   (f) $\sum_{n=0}^{\infty} \left(\frac{2}{3}\right)^n$;

(g) $\sum_{n=0}^{\infty} \left(-\frac{4}{5}\right)^n$;   (h) $\sum_{n=0}^{\infty} \left(\frac{5\sqrt{2}}{7}\right)^n$;

(i) $\sum_{n=0}^{\infty} \left(\frac{8}{5\sqrt{3}}\right)^n$;   (j) $\sum_{n=0}^{\infty} \frac{1}{(3 - \sqrt{5})^n}$;

(k) $\displaystyle\sum_{n=0}^{\infty} 7\left(-\frac{4}{7}\right)^{n}$;

(l) $\displaystyle\sum_{n=0}^{\infty} \frac{2^{n}}{5^{n/2}}$.

3 Show that formula (4) is essentially equivalent to the following factorization formula of elementary algebra:

$$x^{n} - 1 = (x - 1)(x^{n-1} + x^{n-2} + \cdots + x + 1).$$

Give an independent verification of this factorization formula.

4 A certain rubber ball is dropped from a height of 4 ft. Each time it bounces it rises to a height of $\frac{3}{4}h$, where h is the height of the previous bounce. Find the total distance the ball travels.

5 Which of the following series are convergent and which are divergent?

(a) $\sin \pi + \sin 2\pi + \cdots + \sin n\pi + \cdots$.

(b) $\sin \dfrac{\pi}{2} + \sin \dfrac{2\pi}{2} + \cdots + \sin \dfrac{n\pi}{2} + \cdots$.

(c) $\cos \pi + \cos 2\pi + \cdots + \cos n\pi + \cdots$.

(d) $\cos \dfrac{\pi}{2} + \cos \dfrac{3\pi}{2} + \cdots + \cos \dfrac{(2n-1)\pi}{2} + \cdots$.

(e) $\ln \sqrt{3} + \ln \sqrt[3]{3} + \ln \sqrt[4]{3} + \cdots + \ln \sqrt[n]{3} + \cdots$.

(f) $\ln \sqrt{3} + \ln \sqrt[4]{3} + \ln \sqrt[8]{3} + \cdots + \ln \sqrt[2n]{3} + \cdots$.

(g) $\dfrac{1}{10+3} + \dfrac{2}{10+6} + \dfrac{3}{10+9}$

$\quad + \cdots + \dfrac{n}{10+3n} + \cdots$.

(h) $\frac{1}{2} + \frac{1}{4} + \frac{1}{6} + \cdots$.

6 Convert each of the following repeating decimals into a fraction (in lowest terms):
(a) 0.777 . . . ;   (b) 0.151515 . . . ;
(c) 0.639639639 . . . ;   (d) 2.3070707 . . . .

7 If a and b are digits, show that

(a) $0.aaa \ldots = \dfrac{a}{9}$;

(b) $0.ababab \ldots = \dfrac{10a + b}{99}$.

8 The decimal 0.101001000100001 . . . , in which the 1s are followed by successively longer chains of 0s, looks as if it is nonrepeating, and therefore defines an irrational number. Construct an argument that converts this impression into a certainty. Hint: Assume that the decimal is repeating.

9 (The fly problem) Two bicyclists start 20 mi apart and head toward each other, each pedaling at a steady 10 mi/h. At the same time a fly traveling 40 mi/h starts from the front wheel of one bicycle and flies to the front wheel of the other, then turns around and flies back to the front wheel of the first, and continues back and forth in this manner until the bicycles collide and he is crushed between the wheels. How far has the fly flown? The hard way to solve this problem is to express the total distance as an infinite series and find its sum. There is also an easy way. Do it both ways.*

---

* There is a famous anecdote about the first time one of the most brilliant scientists of the twentieth century heard this problem, and how he reacted to it. See P. R. Halmos, "The Legend of John von Neumann," *Amer. Math. Monthly*, **80** (1973), pp. 386–387.

As we shall begin to see in this section, the geometric series

$$1 + x + x^{2} + \cdots + x^{n} + \cdots \tag{1}$$

is the most useful of all infinite series. By using the sigma notation, we can write this series in either of the equivalent forms

$$\sum_{n=1}^{\infty} x^{n-1} \quad \text{or} \quad \sum_{n=0}^{\infty} x^{n}.$$

13.3

VARIOUS SERIES RELATED TO THE GEOMETRIC SERIES

The first form is slightly clumsier than the second, so the second is preferred. Whether we begin a series with the first term ($n = 1$) or the zeroth term ($n = 0$) is purely a matter of notational convenience.

A series of functions like (1) cannot be said to converge or diverge as it stands, since it is not an infinite series of numbers. However, it becomes a series of numbers when we give the variable x a numerical value, and in general we expect that such a series will converge for some values of x and diverge for other values.

We can use the nth term test to conclude that the series (1) diverges if $x \geq 1$

or $x \leq -1$, since in these cases x^n does not approach zero. The fact that this series converges for all other values of x is easily seen from the formula for the nth partial sum,

$$s_n = 1 + x + x^2 + \cdots + x^n = \frac{1 - x^{n+1}}{1 - x}, \tag{2}$$

because for $|x| < 1$ we clearly have $x^{n+1} \to 0$, and therefore $s_n \to 1/(1 - x)$. Our purpose here is to remind students of this formula, which we will need later, and also to restate the central fact of Section 13.2:

If $-1 < x < 1$, the series (1) converges and its sum is

$$1 + x + x^2 + \cdots = \frac{1}{1 - x}; \tag{3}$$

for all other values of x this series diverges.

The interval $-1 < x < 1$ is called the *interval of convergence* of the series.

The geometric series (3) is a rare example of an infinite series whose sum we are able to find by first finding a simple formula for its nth partial sum s_n. Part of the importance of this series lies in the fact that it can be used as a starting point for determining the sums of many other interesting series. One way to do this is to replace x in (3) by various functions, substituting in $-1 < x < 1$ to find the interval of convergence of the new series. For example, replacing x by $-x$ gives

$$1 - x + x^2 - x^3 + \cdots = \frac{1}{1 + x}, \quad -1 < x < 1. \tag{4}$$

Replacing x by x^2 in (4) gives

$$1 - x^2 + x^4 - x^6 + \cdots = \frac{1}{1 + x^2}, \quad -1 < x < 1. \tag{5}$$

Replacing x by x^2 in (3), we obtain

$$1 + x^2 + x^4 + x^6 + \cdots = \frac{1}{1 - x^2}, \quad -1 < x < 1. \tag{6}$$

Notice that this series contains only those terms of (3) with *even* exponents. To find the sum of the series of odd powers alone, we multiply (6) by x and find that

$$x + x^3 + x^5 + x^7 + \cdots = \frac{x}{1 - x^2}, \quad -1 < x < 1. \tag{7}$$

If we replace x by $2x$ in (4), we obtain

$$1 - 2x + 4x^2 - 8x^3 + \cdots = \frac{1}{1 + 2x}, \tag{8}$$

which is valid for $|2x| < 1$, or equivalently, $-\frac{1}{2} < x < \frac{1}{2}$. If we replace x by e^x in (3), so that $e^x < 1$ is required, we find that

$$1 + e^x + e^{2x} + e^{3x} + \cdots = \frac{1}{1 - e^x}, \qquad x < 0.$$

And finally, replacing x by $2x - 1$ in (3), so that $-1 < 2x - 1 < 1$ is required, we obtain

$$1 + (2x - 1) + (2x - 1)^2 + \cdots = \frac{1}{1 - (2x - 1)} = \frac{1}{2 - 2x}, \qquad 0 < x < 1.$$

It is clear that many other series and their corresponding sums can be found in essentially the same way.

The series (3) to (8) all have the special form

$$\sum_{n=0}^{\infty} a_n x^n = a_0 + a_1 x + a_2 x^2 + a_3 x^3 + \cdots , \qquad (9)$$

and are known as *power series*. The numbers $a_0, a_1, a_2, a_3, \ldots$ are called the *coefficients* of the power series (9). The geometric series (3), in which the coefficients are all equal to 1, is evidently the simplest power series and the prototype for all such series.

We learned many chapters ago that to find the derivative or integral of the sum of a finite number of functions, we simply find the derivatives or integrals of the individual terms and add. However, when applied to infinite series of functions, these term-by-term differentiation and integration procedures are filled with danger and can lead to false conclusions. Some sort of careful analysis is needed to determine whether the series with the altered terms converges, and if it does, whether its sum is the desired derivative or integral of the original sum. Nevertheless, if we suspend skepticism for a moment and plunge recklessly ahead just to see what happens, then we find that these procedures lead to many remarkable new formulas, which can be established on a solid foundation afterwards.

For example, differentiating (3) yields

$$1 + 2x + 3x^2 + \cdots = \frac{1}{(1 - x)^2}, \qquad -1 < x < 1. \qquad (10)$$

When we turn to integration, the right sides of (4) and (5) remind us of the familiar formulas

$$\int \frac{dx}{1 + x} = \ln (1 + x) \qquad \text{and} \qquad \int \frac{dx}{1 + x^2} = \tan^{-1} x.$$

By integrating the left side of (4) term by term, we therefore obtain the important series

$$x - \frac{x^2}{2} + \frac{x^3}{3} - \frac{x^4}{4} + \cdots = \ln (1 + x). \qquad (11)$$

It turns out that this equation is valid not only in the open interval $-1 < x < 1$, but also at the right endpoint $x = 1$. In the case $x = 1$, (11) gives the series

$$1 - \frac{1}{2} + \frac{1}{3} - \frac{1}{4} + \cdots = \ln 2, \qquad (12)$$

which was mentioned in Section 13.2. Similarly, integrating (5) yields the inverse tangent series,

$$x - \frac{x^3}{3} + \frac{x^5}{5} - \frac{x^7}{7} + \cdots = \tan^{-1} x. \tag{13}$$

This equation turns out to be valid on the entire closed interval $-1 \le x \le 1$, and for $x = 1$ it gives the famous Leibniz series

$$1 - \frac{1}{3} + \frac{1}{5} - \frac{1}{7} + \cdots = \frac{\pi}{4}. \tag{14}$$

This equation connecting π with the positive odd numbers was one of the most beautiful mathematical discoveries of the seventeenth century, and it made a deep impression on the minds of the earliest workers in the field of calculus.

We emphasize again that these derivations of (10), (11), and (13) have suggestive power only; they *do not* constitute acceptable mathematical proofs. We now provide genuine proofs of (11) and (13), which we hope will help students to understand the type of argument that is necessary. A proof of (10) is outlined in the Additional Problems at the end of the chapter.

THE LOGARITHM SERIES

We begin our rigorous proof of (11) with the following slightly altered form of equation (2):

$$\frac{1}{1 - x} = 1 + x + x^2 + \cdots + x^{n-1} + \frac{x^n}{1 - x}.$$

Replacing x by $-t$ gives

$$\frac{1}{1 + t} = 1 - t + t^2 - \cdots + (-1)^{n-1} t^{n-1} + \frac{(-1)^n t^n}{1 + t}. \tag{15}$$

We now choose any number x such that $-1 < x \le 1$ and integrate both sides of (15) from 0 to x. All the integrals except the last are easy to evaluate, and we obtain

$$\ln(1 + x) = \int_0^x \frac{dt}{1 + t} = x - \frac{x^2}{2} + \frac{x^3}{3} - \cdots + (-1)^{n-1} \frac{x^n}{n} + R_n(x), \tag{16}$$

where

$$R_n(x) = (-1)^n \int_0^x \frac{t^n}{1 + t} dt.$$

Remembering that x is confined to the interval $-1 < x \le 1$, where the endpoint $x = 1$ is included but $x = -1$ is not, we now prove that $R_n(x) \to 0$ as $n \to \infty$. We do this by showing that $|R_n(x)|$ is smaller than some quantity that visibly approaches zero as $n \to \infty$. Since t varies from 0 to x, for the case $0 \le x \le 1$ we clearly have $1 + t \ge 1$, so

$$|R_n(x)| = \int_0^x \frac{t^n}{1 + t} dt \le \int_0^x t^n dt = \frac{x^{n+1}}{n + 1} \to 0.$$

For $-1 < x < 0$ we similarly have $1 + t \geq 1 + x$, so

$$|R_n(x)| = \left| \int_0^x \frac{t^n}{1+t} \, dt \right| \leq \left| \int_0^x \frac{t^n}{1+x} \, dt \right| = \frac{1}{1+x} \cdot \frac{|x|^{n+1}}{n+1} \to 0.$$

Since $R_n(x) \to 0$ for all the values of x considered, we can safely infer from (16) that

$$\ln(1+x) = x - \frac{x^2}{2} + \frac{x^3}{3} - \frac{x^4}{4} + \cdots, \qquad -1 < x \leq 1. \tag{17}$$

We know that $\ln 0$ has no meaning, so there is clearly no hope of extending this result to the other endpoint $x = -1$. We also point out that equation (12) is now a matter of definite knowledge, rather than merely a plausible conjecture.

The series (17) is not a very practical means for computing numerical values of the logarithm, since $1 + x$ is restricted to the interval $0 < 1 + x \leq 2$ and since the series converges so slowly that we must include a great many terms to obtain a reasonably accurate result. We can find a more convenient series as follows. If $-1 < x < 1$, we can replace x by $-x$ in (17) to get

$$\ln(1-x) = -x - \frac{x^2}{2} - \frac{x^3}{3} - \frac{x^4}{4} - \cdots. \tag{18}$$

Subtracting (18) from (17) yields

$$\ln \frac{1+x}{1-x} = 2 \left(x + \frac{x^3}{3} + \frac{x^5}{5} + \cdots \right), \qquad -1 < x < 1.$$

Not only does this series converge much faster than (17), but now the form of the left side enables us to calculate the logarithm of any positive number; for as x varies over the interval $-1 < x < 1$, the expression $y = (1+x)/(1-x)$ ranges over all positive numbers.

THE INVERSE TANGENT SERIES

To give a rigorous proof of (13), we begin by replacing t by t^2 in (15), which yields

$$\frac{1}{1+t^2} = 1 - t^2 + t^4 - \cdots + (-1)^{n-1}t^{2n-2} + \frac{(-1)^n t^{2n}}{1+t^2}.$$

By integrating both sides of this from $t = 0$ to $t = x$, we obtain

$$\tan^{-1} x = \int_0^x \frac{dt}{1+t^2} = x - \frac{x^3}{3} + \frac{x^5}{5} - \cdots + (-1)^{n-1} \frac{x^{2n-1}}{2n-1} + R_n(x), \tag{19}$$

where

$$R_n(x) = (-1)^n \int_0^x \frac{t^{2n}}{1+t^2} \, dt.$$

We now use the same strategy as before to prove that $R_n(x) \to 0$ as $n \to \infty$, provided that $-1 \leq x \leq 1$. But this is easy, because the inequality $1 + t^2 \geq 1$ allows us to write

$$|R_n(x)| = \left| \int_0^x \frac{t^{2n}}{1+t^2} \, dt \right| \le \left| \int_0^x t^{2n} \, dt \right| = \frac{|x|^{2n+1}}{2n+1} \le \frac{1}{2n+1} \to 0.$$

The fact that $R_n(x) \to 0$ for the stated values of x enables us to conclude from (19) that

$$\tan^{-1} x = x - \frac{x^3}{3} + \frac{x^5}{5} - \cdots, \qquad -1 \le x \le 1. \tag{20}$$

Leibniz's formula (14),

$$\frac{\pi}{4} = 1 - \frac{1}{3} + \frac{1}{5} - \frac{1}{7} + \cdots,$$

now follows at once from (20) on putting $x = 1$.

THE COMPUTATION OF π

In principle, Leibniz's formula (14) can be used for computing the numerical value of π, but as a practical matter, the series converges so slowly that this method is of little value. A more efficient procedure is to use the formula

$$\frac{\pi}{4} = \tan^{-1} \frac{1}{2} + \tan^{-1} \frac{1}{3}, \tag{21}$$

then compute the two terms on the right by means of (20). [To establish (21), notice that if $A = \tan^{-1} \frac{1}{2}$ and $B = \tan^{-1} \frac{1}{3}$, then

$$\tan (A + B) = \frac{\tan A + \tan B}{1 - \tan A \tan B} = \frac{\frac{1}{2} + \frac{1}{3}}{1 - \frac{1}{6}} = 1.]$$

However, most of the extended computations of π have been based on the formula

$$\frac{\pi}{4} = 4 \tan^{-1} \frac{1}{5} - \tan^{-1} \frac{1}{239}, \tag{22}$$

which was discovered in 1706 by John Machin, a Scottish astronomer. We shall not pursue the details of these matters any further, but instead simply point out that π has been computed by these methods to more than 500,000 decimal places, of which the first twenty are

$$\pi = 3.14159 \ 26535 \ 89793 \ 23846 \ \ldots.$$

Further information can be found in P. Beckmann, *A History of π* (Golem Press, Boulder, Colo., 1971), especially pp. 140–141 and 180–181.

PROBLEMS

1 Find the region of convergence of each of the following series, and also the sum s in that region:

(a) $1 + (x + 3)^2 + (x + 3)^4 + (x + 3)^6 + \cdots$;

(b) $1 + \dfrac{1}{2x+3} + \dfrac{1}{(2x+3)^2} + \cdots$;

(c) $1 + 2 \sin x + 4 \sin^2 x + 8 \sin^3 x + \cdots$;

(d) $\dfrac{1}{3} \left[1 + \dfrac{4x}{3} + \left(\dfrac{4x}{3} \right)^2 + \left(\dfrac{4x}{3} \right)^3 + \cdots \right]$;

(e) $-\dfrac{1}{4x} \left[1 + \dfrac{3}{4x} + \left(\dfrac{3}{4x} \right)^2 + \left(\dfrac{3}{4x} \right)^3 + \cdots \right]$;

(f) $\frac{1}{4}[1 + (x + \frac{1}{4}) + (x + \frac{1}{4})^2 + \cdots]$.

2 Show that $1 + 2\sin^2\theta + 4\sin^4\theta + 8\sin^6\theta + \cdots = \sec 2\theta$ for suitable values of θ.

3 By interlacing terms properly, add equations (6) and (7) to obtain (3).

4 Find the sum s of the series on the left side of (10),

$$s = 1 + 2x + 3x^2 + 4x^3 + \cdots,$$

by subtracting the series xs and evaluating the result.

5 Use (10) to find the sum of the series

$$\sum_{n=1}^{\infty} \frac{n}{2^n} = \frac{1}{2} + \frac{2}{2^2} + \frac{3}{2^3} + \cdots.$$

6 Find the sum of the series

$$\sum_{n=1}^{\infty} \frac{n^2}{2^n} = \frac{1^2}{2^1} + \frac{2^2}{2^2} + \frac{3^2}{2^3} + \cdots$$

by a method similar to that used in Problem 5. Hint: Begin by multiplying (10) through by x.

7 Obtain the following formulas without attempting to justify the steps. These formulas are all valid at least on the interval $-1 < x < 1$.

(a) $\displaystyle\sum_{n=1}^{\infty} nx^n = \frac{x}{(1-x)^2}$.

(b) $\displaystyle\sum_{n=1}^{\infty} n^2 x^n = \frac{x^2 + x}{(1-x)^3}$.

(c) $\displaystyle\sum_{n=1}^{\infty} n^3 x^n = \frac{x^3 + 4x^2 + x}{(1-x)^4}$.

(d) $\displaystyle\sum_{n=1}^{\infty} n^4 x^n = \frac{x^4 + 11x^3 + 11x^2 + x}{(1-x)^5}$.

(e) $\displaystyle\sum_{n=0}^{\infty} (n+1)x^n = \frac{1}{(1-x)^2}$.

(f) $\displaystyle\sum_{n=0}^{\infty} \frac{(n+1)(n+2)}{2!} x^n = \frac{1}{(1-x)^3}$.

(g) $\displaystyle\sum_{n=0}^{\infty} \frac{(n+1)(n+2)(n+3)}{3!} x^n = \frac{1}{(1-x)^4}$.

8 Prove formula (22) by putting $A = \tan^{-1}\frac{1}{5}$, $B = \tan^{-1}\frac{1}{239}$, and computing successively $\tan 2A$, $\tan 4A$, and $\tan(4A - B)$.

13.4 POWER SERIES CONSIDERED INFORMALLY

We have remarked before that polynomials are the simplest functions of all, and power series can be thought of as polynomials of infinite degree. An expansion of a function $f(x)$ in a power series,

$$f(x) = a_0 + a_1 x + a_2 x^2 + \cdots + a_n x^n + \cdots, \tag{1}$$

is therefore a way of expressing $f(x)$ by means of functions of a particularly simple kind. Our investigations in the preceding sections have led us to the following power series expansions, among others:

$$\frac{1}{1+x} = 1 - x + x^2 - x^3 + \cdots, \qquad \text{valid for } -1 < x < 1; \tag{2}$$

$$\ln(1+x) = x - \frac{x^2}{2} + \frac{x^3}{3} - \cdots, \qquad \text{valid for } -1 < x \le 1; \tag{3}$$

$$\tan^{-1} x = x - \frac{x^3}{3} + \frac{x^5}{5} - \cdots, \qquad \text{valid for } -1 \le x \le 1; \tag{4}$$

$$e^x = 1 + x + \frac{x^2}{2!} + \frac{x^3}{3!} + \cdots, \qquad \text{valid for all } x. \tag{5}$$

We have rigorously proved formulas (2), (3), and (4), but at the present stage of our work (5) is supported only by the plausibility argument given in Section 13.1.

We now add the following important expansions to this list:

$$\sin x = x - \frac{x^3}{3!} + \frac{x^5}{5!} - \cdots, \qquad \text{valid for all } x; \tag{6}$$

$$\cos x = 1 - \frac{x^2}{2!} + \frac{x^4}{4!} - \cdots, \qquad \text{valid for all } x. \qquad (7)$$

The proofs are not difficult, and can be given together. They rest on the formulas

$$\int_0^x \sin t \, dt = 1 - \cos x \qquad \text{and} \qquad \int_0^x \cos t \, dt = \sin x,$$

and also on the following obvious property of definite integrals:

$$\text{If } f(x) \leq g(x) \text{ and } a < b, \text{ then } \int_a^b f(x) \, dx \leq \int_a^b g(x) \, dx.$$

We begin the argument with the inequality

$$\cos x \leq 1.$$

On replacing x by t and integrating both sides of $\cos t \leq 1$ from 0 to a fixed positive number x, we obtain

$$\sin x \leq x.$$

Integrating this in the same way over the same interval yields

$$1 - \cos x \leq \frac{x^2}{2},$$

which is equivalent to

$$\cos x \geq 1 - \frac{x^2}{2}.$$

Another similar integration gives

$$\sin x \geq x - \frac{x^3}{2 \cdot 3} = x - \frac{x^3}{3!}.$$

By continuing this process indefinitely, we generate the two sets of inequalities

$$\sin x \leq x \qquad\qquad \cos x \leq 1$$

$$\sin x \geq x - \frac{x^3}{3!} \qquad\qquad \cos x \geq 1 - \frac{x^2}{2!}$$

$$\sin x \leq x - \frac{x^3}{3!} + \frac{x^5}{5!} \qquad\qquad \cos x \leq 1 - \frac{x^2}{2!} + \frac{x^4}{4!}$$

$$\sin x \geq x - \frac{x^3}{3!} + \frac{x^5}{5!} - \frac{x^7}{7!} \qquad \cos x \geq 1 - \frac{x^2}{2!} + \frac{x^4}{4!} - \frac{x^6}{6!}$$

$$\cdots \qquad\qquad\qquad \cdots .$$

To complete the proofs of (6) and (7) for the positive value of x we are considering, it suffices to show that $x^n/n! \to 0$ as $n \to \infty$. To demonstrate this, we choose a fixed positive integer m so large that $x/m < \frac{1}{2}$, then we put $a = x^m/m!$. For any integer $n > m$ we write $n = m + k$ and observe that

$$0 < \frac{x^n}{n!} = a \cdot \frac{x}{m+1} \cdot \frac{x}{m+2} \cdots \frac{x}{m+k} < a \left(\frac{1}{2}\right)^k.$$

As $n \to \infty$, k also $\to \infty$, so $a\left(\frac{1}{2}\right)^k \to 0$, and we conclude that $x^n/n! \to 0$. The extension of (6) and (7) to negative values of x is left to students, in the problems that follow.

The power series expansions (2) to (7) are among the most important formulas in all of mathematics. In addition to being understood, they should also be memorized.

Formulas (2) to (7) show that many different types of functions can be expanded in power series. These formulas are all special cases of a very general and powerful formula that enables us to find the unique power series expansion (1) of any one of a large class of functions $f(x)$ by finding the values of the coefficients a_n in terms of the function $f(x)$ and its derivatives. It is not possible here to prove the theorems that establish the validity and uniqueness of the expansions formed in this way. This will be done in Chapter 14. Nevertheless, the following informal plausibility considerations will suggest the nature of these theorems; they are useful to students and mathematicians alike as a convenient way of remembering the broad outlines of the subject.

Therefore, let us tentatively assume that $f(x)$ can be expanded in a power series of the form (1). Let us further assume that $f(x)$ has a derivative $f'(x)$, that $f'(x)$ has a derivative $f''(x)$, and so on, so that all of the infinite sequence of derivatives

$$f'(x), f''(x), \ldots, f^{(n)}(x), \ldots$$

actually exist. Finally, let us assume that each of these derivatives can be found by differentiating the series (1) term by term a suitable number of times. We will now see that the values of the coefficients a_n are determined by the values of $f(x)$ and its derivatives at the point $x = 0$, as follows. If we substitute $x = 0$ in (1), then all the terms containing x disappear and we have

$$a_0 = f(0). \tag{8}$$

Next, we differentiate (1) to obtain

$$f'(x) = a_1 + 2a_2 x + 3a_3 x^2 + \cdots + na_n x^{n-1} + \cdots, \tag{9}$$

and putting $x = 0$ yields

$$a_1 = f'(0).$$

Differentiating (9) gives

$$f''(x) = 2a_2 + 2 \cdot 3a_3 x + \cdots + (n-1)na_n x^{n-2} + \cdots, \tag{10}$$

and by substituting $x = 0$ we obtain

$$a_2 = \frac{f''(0)}{2}.$$

Similarly, by differentiating (10) and putting $x = 0$, we find that

$$a_3 = \frac{f'''(0)}{2 \cdot 3} = \frac{f'''(0)}{3!}.$$

By continuing this procedure we clearly obtain the following general formula for all the coefficients:

$$a_n = \frac{f^{(n)}(0)}{n!},$$ (11)

where $f^{(n)}(0)$ is of course the value of the nth derivative of $f(x)$ at $x = 0$. It should be noticed particularly that this formula includes (8) if — as usual — we understand that the zeroth derivative $f^{(0)}(x)$ is just the function $f(x)$ itself, and also that 0! is defined to be 1, as stated in Section 13.1. The expansion (1) now takes the form

$$f(x) = f(0) + f'(0)x + \frac{f''(0)}{2!} x^2 + \frac{f'''(0)}{3!} x^3 + \cdots$$

$$= \sum_{n=0}^{\infty} \frac{f^{(n)}(0)}{n!} x^n.$$ (12)

Example We illustrate this method by using it to give another derivation of formula (5). The calculations are very easy. Our function $f(x)$ in this case is simply e^x, so by computing the successive derivatives we find that

$$f(x) = e^x, \qquad f(0) = e^0 = 1,$$
$$f'(x) = e^x, \qquad f'(0) = e^0 = 1,$$
$$f''(x) = e^x, \qquad f''(0) = e^0 = 1,$$
$$f^{(n)}(x) = e^x, \qquad f^{(n)}(0) = e^0 = 1,$$

and so on. Formula (11) therefore gives $a_n = 1/n!$ for $n = 0, 1, 2, \ldots$, so for this particular function the series (12) is

$$e^x = \sum_{n=0}^{\infty} \frac{x^n}{n!} = 1 + x + \frac{x^2}{2!} + \frac{x^3}{3!} + \cdots + \frac{x^n}{n!} + \cdots,$$

which is (5).

We emphasize once again that the discussion of formulas (11) and (12) given here has suggestive value only, and by no means constitutes a legitimate mathematical proof. These formulas will not become tools of known reliability until *after* the careful theoretical analysis given in Chapter 14.

Remark 1 In Problem 26 of Section 8.3 we asked students to obtain the formula

$$e^x = \lim_{n \to \infty} \left(1 + \frac{x}{n} \right)^n.$$ (13)

This can be used to derive formula (5) in yet another way. By the binomial theorem we have

$$\left(1 + \frac{x}{n} \right)^n = 1 + n \left(\frac{x}{n} \right) + \frac{n(n-1)}{2!} \left(\frac{x}{n} \right)^2 + \frac{n(n-1)(n-2)}{3!} \left(\frac{x}{n} \right)^3$$

$$+ \cdots + \frac{n(n-1)(n-2) \cdots [n - (n-1)]}{n!} \left(\frac{x}{n} \right)^n,$$

or equivalently,

$$\left(1+\frac{x}{n}\right)^n = 1 + x + \left(1-\frac{1}{n}\right)\frac{x^2}{2!} + \left(1-\frac{1}{n}\right)\left(1-\frac{2}{n}\right)\frac{x^3}{3!}$$
$$+ \cdots + \left(1-\frac{1}{n}\right)\left(1-\frac{2}{n}\right)\cdots\left(1-\frac{n-1}{n}\right)\frac{x^n}{n!}.$$

It is plausible to form the limit of this expression as $n \to \infty$ by simultaneously replacing each of the quantities in parentheses by 1 and lengthening the finite sum into an infinite series. When we apply these procedures to (13), we find that

$$e^x = 1 + x + \frac{x^2}{2!} + \frac{x^3}{3!} + \cdots + \frac{x^n}{n!} + \cdots,$$

which is (5) again. We now have three different ways of deriving formula (5), all of which are plausibility arguments rather than genuine proofs. This important formula is of course true, but its truth will not be firmly established until late in Chapter 14.

Remark 2 At the end of Section 13.2, we mentioned the memorable formula

$$\sum_{n=1}^{\infty} \frac{1}{n^2} = 1 + \frac{1}{4} + \frac{1}{9} + \frac{1}{16} + \cdots = \frac{\pi^2}{6}.$$

This formula for the sum of the reciprocals of the squares was discovered by Euler in 1736. His method of discovery was based on the power series (6) for $\sin x$, and is described in Appendix A.12.

PROBLEMS

1 The argument given in the text establishes formulas (6) and (7) for positive values of x, and they are obviously true for the case $x = 0$. Show that these formulas are also valid for negative values of x. Hint: Recall the identities $\sin(-x) = -\sin x$ and $\cos(-x) = \cos x$.

2 Show that differentiating the power series for $\sin x$ gives the power series for $\cos x$.

3 Show that the power series for e^x is unchanged by differentiation.

4 Obtain the expansions (2), (3), (4), (6), and (7) by using formulas (11) and (12). Notice that these calculations produce the expansions but give no information whatever about the x's for which these expansions are valid.

5 Use formulas (11) and (12) to find the power series expansion of the function $f(x) = x^3$.

6 Use formulas (11) and (12) to obtain the first three nonzero terms of the power series expansion of $f(x) = \tan x$. Notice that no simple pattern emerges for the coefficients.

7 Let p be an arbitrary constant and use formulas (11) and (12) to obtain the binomial series

$$(1+x)^p = 1 + px + \frac{p(p-1)}{2!}x^2$$
$$+ \frac{p(p-1)(p-2)}{3!}x^3 + \cdots$$
$$+ \frac{p(p-1)(p-2)\cdots(p-n+1)}{n!}x^n + \cdots.$$

Observe that this power series is a polynomial whenever p is a nonnegative integer, and only in this case.

8 Derive formula (2) from the binomial series in Problem 7.

9 Derive each of the following series from the binomial series in Problem 7:

$$\sqrt{1-x} = 1 - \frac{1}{2}x - \frac{1}{2 \cdot 4}x^2$$
$$- \frac{1 \cdot 3}{2 \cdot 4 \cdot 6}x^3 - \cdots;$$

$$\frac{1}{\sqrt{1-x}} = 1 + \frac{1}{2}x + \frac{1\cdot 3}{2\cdot 4}x^2 + \frac{1\cdot 3\cdot 5}{2\cdot 4\cdot 6}x^3 + \cdots.$$

Obtain the second of these series by differentiating the first.

10 Use the formula

$$\sin^{-1} x = \int_0^x \frac{dt}{\sqrt{1-t^2}}$$

and Problem 9 to obtain the series

$$\sin^{-1} x = x + \frac{1}{2}\frac{x^3}{3} + \frac{1\cdot 3}{2\cdot 4}\frac{x^5}{5} + \frac{1\cdot 3\cdot 5}{2\cdot 4\cdot 6}\frac{x^7}{7} + \cdots.$$

ADDITIONAL PROBLEMS FOR CHAPTER 13

SECTION 13.2

1 Find the sum of each of the following series:

(a) $\dfrac{1+4}{9} + \dfrac{1+8}{27} + \dfrac{1+16}{81} + \cdots$;

(b) $18 - 6 + 2 - \frac{2}{3} + \cdots$;

(c) $\dfrac{\sin\theta}{2} + \dfrac{\sin^2\theta}{4} + \dfrac{\sin^3\theta}{8} + \cdots$;

(d) $\dfrac{1}{2+x^2} + \dfrac{1}{(2+x^2)^2} + \dfrac{1}{(2+x^2)^3} + \cdots.$

2 Describe all convergent series of integers.

3 In the series $\frac{1}{3} - \frac{2}{5} + \frac{3}{7} - \frac{4}{9} + \cdots$, the numerators are the successive positive integers, the denominators are the successive odd numbers starting with 3, and the signs alternate.

(a) Write the series using the sigma notation.

(b) Show that the series diverges.

4 Express each of the following numbers as a repeating decimal:

(a) $\frac{3}{5}$;   (b) $\frac{5}{3}$;   (c) $\frac{27}{25}$;   (d) $\frac{27}{24}$;   (e) $\frac{27}{26}$.

5 Show that a positive rational number a/b (in lowest terms) has a terminating decimal expansion if and only if the positive integer b has the form $b = 2^m 5^n$, where the exponents m and n are nonnegative integers. Check this statement against the results of Problem 4.

6 A terminating decimal such as $\frac{3}{8} = 0.375 = 0.375000\ldots$ can also be written as a repeating decimal ending in an infinite chain of 9s if the last nonzero digit is decreased by one unit, as in $0.375000\ldots = 0.374999\ldots$. Prove this by using formula (5) in Section 13.2.

7 A certain rubber ball is dropped from a height H. Each time it bounces it rises to a height rh, where h is the height of the previous bounce and r is a constant. Show that the total distance the ball travels is $H(1 + r)/(1 - r)$.

8 The Greek philosopher Zeno of Elea (early fifth century B.C.) precipitated a long-lived crisis in Western thought by stating a number of ingenious paradoxes. One of these, often called the *racecourse paradox* and designed to prove that motion is impossible, can be stated as follows:

I can't go from here to there. For to do so, I must first cover half the distance, then half the remaining distance, then half of what still remains, and so on. This process can always be continued, and can never be completed in a finite time.

Refute Zeno.

9 Show that the number

$$0.12345678910111213141516 17\ldots,$$

in which all the positive integers are written down in order after the decimal point, is irrational.

SECTION 13.3

10 To prove equation (10) in Section 13.3, assume that $-1 < x < 1$ and consider the identity

$$\frac{1}{1-x} = 1 + x + x^2 + \cdots + x^n + \frac{x^{n+1}}{1-x}.$$

[We have seen that this identity immediately yields equation (3) because $x^{n+1} \to 0$ as $n \to \infty$. (It is clear that a number numerically less than 1 which is raised to higher and higher powers becomes smaller and smaller and approaches zero.)] Differentiate both sides to obtain

$$\frac{1}{(1-x)^2} = 1 + 2x + 3x^2 + \cdots$$

$$+ nx^{n-1} + \frac{nx^n}{1-x} + \frac{x^n}{(1-x)^2}.$$

Equation (10) will now follow from this if it can be proved that $nx^n \to 0$ as $n \to \infty$. See Problem 11.

11 (a) Prove the following: If a sequence $s_1, s_2, \ldots,$ $s_n, \ldots$ has the property that for all sufficiently large subscripts, say $n \ge n_0$, the ratio $r_n = s_{n+1}/s_n$ of successive terms lies between $-r$ and r $(-r \le r_n \le r)$ for some number r such that $0 < r < 1$, then $s_n \to 0$. Hint: For any fixed $m \ge n_0$, we have $|s_{m+1}/s_m| \le r$, so $|s_{m+1}| \le r|s_m|$; $|s_{m+2}/s_{m+1}| \le r$, so $|s_{m+2}| \le r|s_{m+1}| \le r^2|s_m|$; and in the same way, $|s_{m+n}| \le r^n|s_m|$ for every $n \ge 1$.

(b) If $-1 < x < 1$, use part (a) to prove that $nx^n \to 0$ as $n \to \infty$.

12 Use the ideas in Problems 10 and 11 to prove that if the geometric series $1 + x + x^2 + \cdots$ is differentiated m times term by term, the resulting series converges on the interval $-1 < x < 1$ and has as its sum the mth derivative of $1/(1-x)$. Hint: If $R = (1-x)^{-1}x^{n+1}$ is the last term on the right in the identity of Problem 10, then its mth derivative has the form

$$R^{(m)} = f_0 x^{n+1} + f_1(n+1)x^n + f_2(n+1)nx^{n-1}$$

$$+ \cdots + f_m(n+1)n \cdots (n-m+2)x^{n-m+1},$$

where the f's are functions of x alone and do not depend on n.

13 Consider a series $\sum_{n=0}^{\infty} P(n)x^n$ in which the coefficient of x^n is a polynomial in n of degree k. Show that $P(n)$ can be written in the form

$$P(n) = A_0 + A_1(n+1) + A_2(n+1)(n+2)$$

$$+ \cdots + A_k(n+1)(n+2) \cdots (n+k),$$

and that therefore the sum of the series is

$$s = \frac{A_0}{1-x} + \frac{A_1}{(1-x)^2} + \frac{2!A_2}{(1-x)^3}$$

$$+ \cdots + \frac{k!A_k}{(1-x)^{k+1}}.$$

Hint: See parts (e), (f) and (g) of Problem 7 in Section 13.3.

14 Using only the first term in the series for $\ln(1+x)$, we get the approximation formula

$$\ln(1+x) \cong x \qquad \text{for small } |x|.$$

How can this formula be explained and understood without making any use of infinite series?

15 Using only the first term in the series for $\tan^{-1} x$, we get the approximation formula

$$\tan^{-1} x \cong x \qquad \text{for small } |x|.$$

How can this formula be explained and understood without making any use of infinite series?

SECTION 13.4

16 Find the first two or three nonzero terms of the power series expansion of each of the following functions, first by direct substitution in one of formulas (2) to (7) in Section 13.4, then by using formulas (11) and (12) in that section:

(a) $\dfrac{1}{1+x^3}$;   (b) $\sin x^2$.

17 Use formula (5) in Section 13.4 to find the sums of the following series:

(a) $\sum_{n=2}^{\infty} \dfrac{n-1}{n!}$;   (b) $\sum_{n=2}^{\infty} \dfrac{n+1}{n!}$;

(c) $\sum_{n=2}^{\infty} \dfrac{n^2-1}{n!}$.

14

THE THEORY OF INFINITE SERIES

14.1
INTRODUCTION

In Chapter 13 we presented a roughly drawn sketch of infinite series in which only a few of the high points were mentioned. The proofs of convergence we gave used special methods of limited value that applied only to the particular series under discussion. We briefly described several very powerful methods of working with series, but these remarks were quite informal, and nothing of substance was proved. Our aim in that introductory chapter was mostly to help students enter the subject and grasp its main outlines as quickly as possible, like intelligent travelers who take a preliminary guided tour of a great city for the purpose of general orientation before exploring its streets and squares and markets and museums in more intimate detail. We are now ready for this second phase of our work.

Students of calculus do not always understand that infinite series are primarily tools for the study of functions. For instance, in Section 13.4 we established the power series expansions of the sine and cosine,

$$\sin x = x - \frac{x^3}{3!} + \frac{x^5}{5!} - \cdots \quad \text{and} \quad \cos x = 1 - \frac{x^2}{2!} + \frac{x^4}{4!} - \cdots,$$

but since these functions were presumably well known beforehand, it may not be clear what purpose is served by expressing them in this form. Expansions of known functions have their own importance, especially in the computation of numerical values for these functions. However, in advanced work it often happens that an unknown power series arises from some other source, perhaps as a solution of a differential equation. In such a case the series is used to *define* the otherwise unknown function which is its sum, and so the series itself is the only tool we have for investigating the properties of this function. This situation can best be understood by supposing that the basic facts about $\sin x$ and $\cos x$ — their continuity, the identities they satisfy, their properties with respect to differentiation and integration, etc. — could be discovered only by examining the series given above. For these familiar functions such a tortuous process is of course unnecessary, but for many important functions of higher mathematics there is no practical alternative. Thus, we discuss series of constants in the first part of this chapter as a prelude to studying series of functions — especially power series — in the

376

later sections; and our ultimate motive in studying series of functions is to learn what we can about the sums of such series.

Nevertheless, our interest in series is not confined to their value for these applications, and in the course of our work we will touch on many fascinating topics in pure mathematics which are well worth studying for their own sake. Thus, we will see that the study of series is linked to some of the most interesting parts of the theory of numbers, concerning prime numbers, irrational and transcendental numbers, the nature of the constants e and π, and similar matters. We wish to keep the structure of this chapter as simple as possible and still put the full richness of the subject within easy reach of interested readers. For this reason we place most of this optional material in Appendix A at the back of the book, where it can be examined or not according to the preference of the individual student.

Our intention is to start all over again at the beginning and omit nothing essential. This commits us to a bit of repetition of a few basic ideas already discussed in Chapter 13. However, we will be as brief as possible, and students may find that thinking through a concise summary of these ideas is a useful way to begin building a solid foundation for a broader and deeper understanding of the subject.

14.2 CONVERGENT SEQUENCES

Any reasonably satisfactory study of series must be based on a careful definition of convergence for sequences. However, the behavior of most sequences is easy to understand without elaborate explanations, and a genuine theory of convergent sequences would be an unwelcome obstacle blocking our way to the main concepts of this chapter. We will therefore discuss sequences rather briefly, and try to steer a middle course between excessive informality and tedious detail.

If to each positive integer n there corresponds a definite number x_n, then the x_n's are said to form a *sequence*. We think of the x_n's as arranged in the order of their subscripts,

$$x_1, x_2, \ldots, x_n, \ldots,$$

and we often abbreviate this array to $\{x_n\}$. It is clear that a sequence is nothing but a function defined for all positive integers n, with the emphasis placed on the subscript notation x_n instead of the function notation $x(n)$. The numbers constituting a sequence are called its *terms*. Thus, x_1 and x_2 are the first and second terms of the given sequence, and x_n is the nth term.

Example 1 In each of the following we define a sequence $\{x_n\}$ by giving a formula for its nth term:

(a) $x_n = 1$, that is, $1, 1, 1, \ldots$;
(b) $x_n = [1 - (-1)^n]/2$, that is, $1, 0, 1, 0, \ldots$;
(c) $x_n = 1/n$, that is, $1, \frac{1}{2}, \frac{1}{3}, \frac{1}{4}, \ldots$;
(d) $x_n = (n - 1)/n$, that is, $0, \frac{1}{2}, \frac{2}{3}, \frac{3}{4}, \ldots$;
(e) $x_n = (-1)^{n+1}/n$, that is, $1, -\frac{1}{2}, \frac{1}{3}, -\frac{1}{4}, \ldots$;

(f) $x_n = 1 + \frac{1}{2} + \frac{1}{4} + \cdots + \frac{1}{2^{n-1}}$;

(g) $x_n = 1 + \dfrac{1}{2} + \dfrac{1}{3} + \cdots + \dfrac{1}{n}$;

(h) $x_n = \left(1 + \dfrac{1}{n}\right)^n$.

A sequence like (a), in which all the terms are equal, is called a *constant sequence*. Not every sequence has a simple formula, or even any formula at all. This is shown by the sequence $\{d_n\}$, where d_n is the nth digit after the decimal point in the decimal expansion of π.

It is sometimes convenient to relax the definition and allow a sequence to start with the zeroth term x_0, or even with the second or third term x_2 or x_3, instead of requiring it to begin with the first term x_1. One reason for this is that we want to include sequences like that defined by $x_n = 1/\ln n$, where x_1 is meaningless. In any case, we continue to call the term with subscript n the nth term.

A sequence $\{x_n\}$ is said to be *bounded* if there are two numbers A and B such that $A \le x_n \le B$ for every n, and in this case A is called a *lower bound* and B an *upper bound* for the sequence. A sequence that is not bounded is said to be *unbounded*. In Example 1, it is easy to see that sequences (a) to (f) are bounded, but it is less obvious that (g) is not (hint: $\frac{1}{3} + \frac{1}{4} > \frac{1}{4} + \frac{1}{4} = \frac{1}{2}$, $\frac{1}{5} + \frac{1}{6} + \frac{1}{7} + \frac{1}{8} > \frac{1}{8} + \frac{1}{8} + \frac{1}{8} + \frac{1}{8} = \frac{1}{2}$, and so on). The sequence (h) is also bounded, but this is not evident on inspection and will be established later.

Our main interest is in the concept of the limit of a sequence. Roughly speaking, this refers to the fact that certain sequences $\{x_n\}$ have the property that the numbers x_n get closer and closer to some real number L as n increases. Another way of stating this is to say that $|x_n - L|$ gets smaller as n gets larger. As an illustration, consider the sequence $\{x_n\}$ whose nth term is $x_n = (n-1)/n$:

$$0, \frac{1}{2}, \frac{2}{3}, \frac{3}{4}, \ldots.$$

These numbers seem to "approach" the number 1 as we move farther and farther to the right. As a matter of fact, for each n we have

$$|x_n - 1| = \left|\frac{n-1}{n} - 1\right| = \left|-\frac{1}{n}\right| = \frac{1}{n};$$

and the number $1/n$, and therefore $|x_n - 1|$, *can be made as small as we please by taking n sufficiently large*. We express this behavior by saying that the sequence *has the limit* 1, and we write

$$\lim_{n \to \infty} \frac{n-1}{n} = 1.$$

It is helpful to visualize this behavior in the manner suggested by Fig. 14.1.

The general definition is as follows. A sequence $\{x_n\}$ is said to have a number L as a *limit* if for each positive number ϵ there exists a positive integer n_0 with the property that

$$|x_n - L| < \epsilon \qquad \text{for all } n \ge n_0. \tag{1}$$

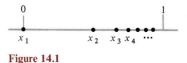

Figure 14.1

When L is related to $\{x_n\}$ in this way, we write

$$\lim_{n \to \infty} x_n = L, \qquad \text{or more briefly,} \qquad \lim x_n = L,$$

and we say that x_n *converges to* L. This is also expressed by saying that x_n *approaches L as n becomes infinite,* which we can write as

$$x_n \to L \qquad \text{as} \qquad n \to \infty.$$

This notation is often abbreviated even further, to $x_n \to L$.

This definition requires that each ϵ, no matter how small, have at least one corresponding n_0 that "works" for it in the sense expressed by (1). In general, we expect that for smaller ϵ's, larger n_0's will be needed; that is, when the required measure of closeness is made smaller, we must go farther out in the sequence to satisfy it.

A sequence is said to *converge* or to be *convergent* if it has a limit. A convergent sequence cannot have two different limits, because it is not possible for x_n to be as close as we please to both of two different numbers for all sufficiently large n's.

A convergent sequence is bounded, but not all bounded sequences are convergent. The sequence 1, 0, 1, 0, . . . of Example 1(b) is a bounded sequence that is not convergent.

It is not always easy to decide whether a given sequence converges, and if it does, what its limit is. The following facts are often useful in problems of this kind: If $x_n \to L$ and $y_n \to M$, then

$$\lim (x_n + y_n) = L + M, \qquad \lim (x_n - y_n) = L - M, \qquad \lim x_n y_n = LM,$$

and, with the additional assumption that $M \neq 0$,

$$\lim \frac{x_n}{y_n} = \frac{L}{M}.$$

These facts can be rigorously proved by carefully using the definition and the properties of inequalities. We omit the details. By using these rules, we can easily perform such feats as calculating

$$\lim \frac{2n^3 + n - 5}{7n^3 - 2n^2 + 4} = \lim \frac{2 + 1/n^2 - 5/n^3}{7 - 2/n + 4/n^3} = \frac{2 + 0 - 0}{7 - 0 + 0} = \frac{2}{7},$$

where the essential first step is to divide both numerator and denominator by the highest power of n occurring in the denominator.

The usual intuitive idea of convergence — that $x_n \to L$ means that x_n can be made "as close as we please" to L by taking n "sufficiently large" — is natural and necessary, and is the way most mathematicians really think about this concept. Accordingly, in most of our work with sequences we shall rely on common sense to tell us how much detail is needed to make an argument convincing.

Example 2 If $|x| < 1$, then $\lim x^n = 0$. Most people are willing to accept this on the grounds that "a number numerically less than 1 which is raised to higher and higher powers gets smaller and smaller." But if a more detailed argument is desired, it can be given as follows. The assertion is clear if $x = 0$, so assume that $0 < |x| < 1$. Then $|x| = 1/(1 + a)$ for some $a > 0$, so by the binomial theorem we have

$$\frac{1}{|x^n|} = \frac{1}{|x|^n} = (1 + a)^n = 1 + na + \text{positive terms} > na.$$

We see from this that $|x^n| < 1/na$; and since $1/na \to 0$, we clearly have $x^n \to 0$.

Example 3   For every x, $\lim x^n/n! = 0$. This is not at all obvious, but the brief argument given in Section 13.4 is both convincing and satisfactory. Students should re-examine this argument.

Example 4   The fact that $\lim (\sqrt{n + 1} - \sqrt{n}) = 0$ will probably seem reasonable after a little thought (it only says that $\sqrt{n}$ is nearly equal to $\sqrt{n + 1}$ for large n), but a definitive argument may not be so easy to find. Such an argument can be constructed by writing the quantity $\sqrt{n + 1} - \sqrt{n}$ as a fraction with denominator 1 and rationalizing the numerator, as follows:

$$\frac{\sqrt{n + 1} - \sqrt{n}}{1} = \frac{\sqrt{n + 1} - \sqrt{n}}{1} \cdot \frac{\sqrt{n + 1} + \sqrt{n}}{\sqrt{n + 1} + \sqrt{n}}$$

$$= \frac{1}{\sqrt{n + 1} + \sqrt{n}} \to 0.$$

In working with sequences in connection with infinite series, we will often need to be able to recognize that a sequence is convergent, even though we know nothing about the numerical value of the limit. In such a case we cannot make any direct use of the definition of a limit. We now discuss a very important method for handling such situations.

A sequence $\{x_n\}$ is said to be *increasing* if

$$x_1 \le x_2 \le x_3 \le \cdots \le x_n \le x_{n+1} \le \cdots,$$

that is, if each term is greater than or equal to the one that precedes it.* Among the sequences listed in Example 1, (a), (d), (f), and (g) are clearly increasing; (h) is also increasing, but this is not obvious on inspection.

Increasing sequences are pleasant to work with because their convergence behavior is particularly easy to determine. We have the following simple criterion: *An increasing sequence converges if and only if it is bounded.* This criterion is not only simple, but also extremely important, because the theory of convergent series given in the rest of this chapter stems directly from it.

This criterion is quite easy to establish. Imagine the terms of the sequence plotted on the real line, as shown in Fig. 14.2, with each term to the right of (or on) its predecessor. If the sequence is unbounded, then its terms simply

---

* Some writers require the terms of an increasing sequence to satisfy the strict inequality $x_n < x_{n+1}$ for all n. However, our definition allows an increasing sequence to be stationary, in the sense that adjacent terms may be equal.

Figure 14.2   A bounded increasing sequence.

march off the page, and the sequence clearly cannot converge. This proves half of the criterion, the "only if" part. To establish the other half, we assume that the sequence is bounded with an upper bound B, as shown in Fig. 14.2, and we must produce a limit for the sequence. Very briefly, we see geometrically that the x_n's, which move steadily to the right and yet cannot penetrate the barrier at B, must "pile up" at some point $L \leq B$, so L is the limit of the sequence and the sequence converges to L.*

This convergence criterion has many important applications, one of which is given in the following example. For this we will need the formula for the sum of a geometric progression,

$$1 + x + x^2 + \cdots + x^{n-1} = \frac{1 - x^n}{1 - x}, \qquad x \neq 1. \tag{2}$$

This formula is familiar from our work in Chapter 13, and can easily be proved in yet another way by dividing $x - 1$ into $x^n - 1$.

Example 5   Our purpose here is to prove that

$$\lim_{n \to \infty} \left(1 + \frac{1}{1!} + \frac{1}{2!} + \cdots + \frac{1}{n!} \right) = e. \tag{3}$$

We accomplish this by discussing together the two closely related sequences $\{x_n\}$ and $\{y_n\}$ defined by

$$x_n = \left(1 + \frac{1}{n} \right)^n \qquad \text{and} \qquad y_n = 1 + \frac{1}{1!} + \frac{1}{2!} + \cdots + \frac{1}{n!}.$$

We will demonstrate that both of these sequences are increasing and bounded, and therefore convergent, and furthermore that they converge to the same limit. Our first step is to show that $\{x_n\}$ is increasing and bounded. By the binomial theorem, x_n can be expressed as the following sum of $n + 1$ terms:

$$x_n = \left(1 + \frac{1}{n} \right)^n = 1 + n \cdot \frac{1}{n} + \frac{n(n-1)}{2!} \cdot \frac{1}{n^2} + \frac{n(n-1)(n-2)}{3!} \cdot \frac{1}{n^3} + \cdots$$
$$+ \frac{n(n-1) \ldots [n - (n-1)]}{n!} \cdot \frac{1}{n^n}$$

---

* A more detailed argument can be given as follows. Any number greater than B is also an upper bound for the sequence, and perhaps there are numbers less than B that are also upper bounds. Let the number L be the *least* upper bound of the sequence. We now show that this number L is the limit approached by the sequence, so that $x_n \to L$. Let $\epsilon > 0$ be given. Because L is an upper bound, we have $x_n \leq L$ for every n. However, there must be an n_0 such that $L - \epsilon < x_{n_0}$; otherwise, if always $x_n \leq L - \epsilon$, then $L - \epsilon$ would be an upper bound and L would not be the *least* upper bound. Since the sequence is increasing, we have $L - \epsilon < x_n \leq L$ for all $n \geq n_0$, so $|x_n - L| < \epsilon$ for all $n \geq n_0$, and we conclude that $x_n \to L$. Even this discussion can be considered a plausibility argument and not a proof, because we assumed that the least upper bound L exists, and this depends on a crucial property of the real number system that is far from obvious. For more about least upper bounds, see Appendix C.1.

$$= 1 + 1 + \frac{1}{2!}\left(1 - \frac{1}{n}\right) + \frac{1}{3!}\left(1 - \frac{1}{n}\right)\left(1 - \frac{2}{n}\right) + \cdots$$

$$+ \frac{1}{n!}\left(1 - \frac{1}{n}\right)\left(1 - \frac{2}{n}\right)\cdots\left(1 - \frac{n-1}{n}\right). \quad (4)$$

As we pass from x_n to x_{n+1} by replacing n by $n + 1$, it is easy to see from this sum that each term after $1 + 1$ increases, and also that another term is added, so $x_n < x_{n+1}$. Further, a term-by-term comparison of (4) with y_n shows that $x_n \leq y_n$. By applying formula (2), we see that the y_n's have 3 as an upper bound,

$$y_n = 1 + 1 + \frac{1}{2} + \frac{1}{2 \cdot 3} + \cdots + \frac{1}{2 \cdot 3 \cdots n}$$

$$\leq 1 + 1 + \frac{1}{2} + \frac{1}{2^2} + \cdots + \frac{1}{2^{n-1}} = 1 + 2\left(1 - \frac{1}{2^n}\right) < 3,$$

and therefore the x_n's also have 3 as an upper bound. Since $\{x_n\}$ is an increasing sequence with 3 as an upper bound, we know that it converges. Its limit is of course the number e, which was introduced in a somewhat different way in Section 8.3,

$$\lim_{n \to \infty}\left(1 + \frac{1}{n}\right)^n = e. \quad (5)$$

Since $x_n \leq y_n < y_{n+1} < 3$, we see that $\{y_n\}$ is also a bounded increasing sequence which approaches a limit $y \geq e$. All that remains is to show that $y \leq e$, for this will yield our main conclusion that $y = e$. If $m < n$ and we consider only the first $m + 1$ terms of (4), then we have

$$1 + 1 + \frac{1}{2!}\left(1 - \frac{1}{n}\right) + \frac{1}{3!}\left(1 - \frac{1}{n}\right)\left(1 - \frac{2}{n}\right) + \cdots$$

$$+ \frac{1}{m!}\left(1 - \frac{1}{n}\right)\left(1 - \frac{2}{n}\right)\cdots\left(1 - \frac{m-1}{n}\right) < x_n < e.$$

If m is held fixed and n is allowed to increase, then we obtain

$$y_m = 1 + 1 + \frac{1}{2!} + \frac{1}{3!} + \cdots + \frac{1}{m!} \leq e,$$

so $y \leq e$. We conclude from this that $y = e$, or

$$\lim_{n \to \infty}\left(1 + 1 + \frac{1}{2!} + \cdots + \frac{1}{n!}\right) = e,$$

which is (3). We also observe that

$$\lim_{n \to \infty}\left(1 - \frac{1}{n}\right)^n = \frac{1}{e}, \quad (6)$$

for this limit can be written as

$$\lim_{n \to \infty}\left(1 - \frac{1}{n+1}\right)^{n+1} = \lim_{n \to \infty}\left(\frac{n}{n+1}\right)^{n+1} = \lim_{n \to \infty}\frac{n/(n+1)}{(1 + 1/n)^n} = \frac{1}{e}.$$

The additional fact that

$$\lim_{n \to \infty} \left(1 - \frac{1}{n}\right)^{-n} = e \tag{7}$$

is an immediate consequence of (6).

Example 6 Most students will recall that a *prime number,* or simply a *prime,* can be defined as an integer $p > 1$ that has no positive factors (or divisors) except 1 and p. These numbers form one of the most interesting of all sequences,

$$2, 3, 5, 7, 11, 13, 17, 19, 23, 29, 31, \ldots . \tag{8}$$

Indeed, the fact that there are infinitely many of them, so that they actually do constitute a sequence, is itself a famous theorem of number theory. The sequence (8) is clearly not convergent, and it may appear that the concept of a convergent sequence has little or no relevance to the primes. However, this impression is quite wrong, for students who wish to pursue the subject will find that the convergence behavior of certain sequences is very close to the heart of the modern theory of prime numbers. We support this remark by stating without proof the following very profound theorem about the approximate size of the nth prime: If p_n denotes the nth prime number, then p_n is "asymptotically equal" to $n \ln n$, in the sense that

$$\lim_{n \to \infty} \frac{p_n}{n \ln n} = 1.$$

We discuss several properties of primes in Appendix A.1, and for readers who are interested in these matters this discussion is continued in a more systematic way in Appendix A.14.

PROBLEMS

1 State whether each of the indicated sequences converges or diverges, and if it converges, find its limit:

(a) $\sqrt[3]{n}$;

(b) $\dfrac{1 + (-1)^n}{n}$;

(c) $\sin \dfrac{\pi}{5n}$;

(d) $\dfrac{10^{10^{10}}\sqrt{n}}{n + 1}$;

(e) $\dfrac{3^n}{2^n + 10^{10}}$;

(f) $\dfrac{\sqrt{n + 2}}{2\sqrt{n}}$;

(g) $\ln(n + 1) - \ln n$;

(h) $\dfrac{n^2}{\sqrt{4n^4 + 5}}$;

(i) $\dfrac{1}{n} - \dfrac{1}{n + 1}$;

(j) $\cos n\pi$;

(k) $\cos \dfrac{(2n + 1)^2 \pi}{2}$;

(l) $\dfrac{5n^3 - 2n}{n^4 + 3n^2 - 10}$;

(m) $n^{(-1)^n}$;

(n) $\dfrac{\sqrt{n} \sin(n!e^n)}{n + 1}$;

(o) $n \sin \dfrac{\pi}{n}$;

(p) $\dfrac{(2 - \sqrt{n})(3 + \sqrt{n})}{4n + 5}$.

2 Show that $n!/n^n \to 0$. Hint: Write it out, and look.

3 The limits of many sequences can be found by replacing the discrete variable n by a continuous variable x and applying L'Hospital's rule for the case $x \to \infty$. Use this method to show that

(a) $\dfrac{\ln n}{n} \to 0$;

(b) $\sqrt[n]{n} \to 1$;

(c) if $|a| < 1$, then $na^n \to 0$;

(d) if k is any positive integer, then $n^k/e^n \to 0$;

(e) if a is any real number, then $(1 + a/n)^n \to e^a$.

4   State whether each of the indicated sequences converges or diverges, and if it converges, find its limit:

(a) $3^{3/n}$;   (b) $e^{-10/n}$;

(c) $n/2^n$;   (d) $\dfrac{\ln(n+1)}{n}$;

(e) $n^2/3^n$;   (f) $n^{1/(n+1)}$;

(g) $(n+10)^{1/(n+10)}$;   (h) $n^2 \sin n\pi$;

(i) $n^2 \cos n\pi$.

5   Find $\lim x_n$ if

(a) $x_n = \sqrt{n}(\sqrt{n+a} - \sqrt{n})$;

(b) $x_n = n\left[\left(a + \dfrac{1}{n}\right)^4 - a^4\right]$.

6   If $0 < a < b$, show that $\lim \sqrt[n]{a^n + b^n} = b$.

7   If $f(x) = \lim_{n\to\infty} (2/\pi) \tan^{-1} nx$, show that $f(x) = x/|x|$ if $x \neq 0$ and $f(0) = 0$. Sketch the graph of this function.

8   If the terms of a sequence $\{x_n\}$ are positive numbers, show that:

(a) the sequence is increasing if $x_{n+1}/x_n \geq 1$ for all n;

(b) the sequence is decreasing if $x_{n+1}/x_n \leq 1$ for all n.*

9   Use Problem 8 to show that $\lim x_n$ exists if

(a) $x_n = \dfrac{1 \cdot 3 \cdot 5 \cdots (2n-1)}{2 \cdot 4 \cdot 6 \cdots (2n)}$;

(b) $x_n = \dfrac{1}{n^2}\left[\dfrac{2 \cdot 4 \cdot 6 \cdots (2n)}{1 \cdot 3 \cdot 5 \cdots (2n-1)}\right]$;

(c) $x_n = \dfrac{1}{n}\left[\dfrac{2 \cdot 4 \cdot 6 \cdots (2n)}{1 \cdot 3 \cdot 5 \cdots (2n-1)}\right]^2$.

10 Find the value of

(a) $\lim \dfrac{(n+1)^n}{n^{n+1}}$;

---

* Naturally, a sequence $\{x_n\}$ is said to be *decreasing* if

$$x_1 \geq x_2 \geq x_3 \geq \cdots \geq x_n \geq x_{n+1} \geq \cdots,$$

i.e., if each term is less than or equal to the one that precedes it.

(b) $\lim \dfrac{(n+1)\ln n - n \ln(n+1)}{\ln n}$.

11 Show that

$$\left[\dfrac{n+1}{n^2} + \dfrac{(n+1)^2}{n^3} + \cdots + \dfrac{(n+1)^n}{n^{n+1}}\right] \to e - 1.$$

12 Show that

(a) $\left(1 + \dfrac{1}{2n+3}\right)^{2n+3} \to e$;

(b) $\left(1 + \dfrac{1}{n^2}\right)^{n^2} \to e$;

(c) $\left(1 + \dfrac{1}{n}\right)^{2n} \to e^2$;

(d) $\left(1 + \dfrac{1}{n^2}\right)^n \to 1$;

(e) $\left(1 + \dfrac{1}{2n}\right)^n \to \sqrt{e}$.

*13 Consider a suitable number of circles of equal size packed in n rows inside an equilateral triangle, as shown in Fig. 14.3. If c_n denotes the number of these circles, then it is clear from the geometry of the situation that $c_1 = 1$, $c_2 = 1 + 2$, $c_3 = 1 + 2 + 3$, and so on. If A is the area of the triangle and A_n is the combined area of the c_n circles, show that

$$\lim_{n\to\infty} \dfrac{A_n}{A} = \dfrac{\pi}{2\sqrt{3}}.$$

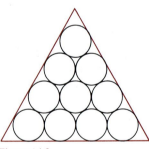

Figure 14.3

14.3

GENERAL PROPERTIES OF CONVERGENT SERIES

We begin with a brief review of the main ideas discussed in Chapter 13. Most people are familiar with the fact that

$$1 + \tfrac{1}{2} + \tfrac{1}{4} + \tfrac{1}{8} + \cdots = 2. \tag{1}$$

However, since we cannot add infinitely many numbers in the same way that we can add finitely many, the meaning of (1) is evidently quite different from the meaning of a statement like

$$1 + 2 + 3 + 4 = 10.$$

What (1) really means is that the sequence of partial sums on the left, that is, the sequence of numbers

$$1,$$

$$1 + \tfrac{1}{2} = 1\tfrac{1}{2},$$

$$1 + \tfrac{1}{2} + \tfrac{1}{4} = 1\tfrac{3}{4},$$

$$1 + \tfrac{1}{2} + \tfrac{1}{4} + \tfrac{1}{8} = 1\tfrac{7}{8},$$

$$\cdots,$$

converges to the number 2 on the right. This suggests the approach we adopt for the general case.

If $a_1, a_2, \ldots, a_n, \ldots$ is a sequence of numbers, then the expression

$$\sum_{n=1}^{\infty} a_n = a_1 + a_2 + \cdots + a_n + \cdots \tag{2}$$

is called an *infinite series,* or simply a *series,* and the a_n's are called its *terms.* We emphasize that until a meaning is assigned to it by a suitable definition, the expression (2) is merely a formal collection of symbols arranged in a certain way, because the indicated operation of adding infinitely many numbers has no meaning in itself. To attach a numerical value to (2) in a natural and useful way, as suggested in the preceding paragraph, we form the sequence of *partial sums*

$$s_1 = a_1,$$

$$s_2 = a_1 + a_2,$$

$$\cdots$$

$$s_n = a_1 + a_2 + \cdots + a_n,$$

$$\cdots.$$

The series (2) is said to *converge,* or to be *convergent,* if the sequence $\{s_n\}$ converges; and if $\lim s_n = s$, then we say that the series *converges to s* or that s is the *sum* of the series, and we express this by writing

$$a_1 + a_2 + \cdots + a_n + \cdots = s \qquad \text{or} \qquad \sum_{n=1}^{\infty} a_n = s.$$

If the series does not converge, then we say that it *diverges* or is *divergent,* and no sum is assigned to it.

At this point a few remarks about notation and usage are in order. As we indicated, the statement that the series $\sum_{n=1}^{\infty} a_n$ converges to the sum s is usually written $\sum_{n=1}^{\infty} a_n = s$. Thus, the notation $\sum_{n=1}^{\infty} a_n$ is used with a dual meaning: to specify a series regardless of convergence or divergence, and also (if the series converges) to denote its sum. Students will find that this ambiguity causes no difficulty in practice.

Another matter concerns the indexing (or numbering) of the terms. It is often more natural to number the terms of a series beginning with $n = 0$; that is, we write some series in the form

$$\sum_{n=0}^{\infty} a_n = a_0 + a_1 + \cdots + a_n + \cdots$$

(and in this case we also write $s_n = a_0 + a_1 + \cdots + a_n$). For example, the series on the left of (1) can be written as

$$\sum_{n=1}^{\infty} \frac{1}{2^{n-1}} \quad \text{or} \quad \sum_{n=0}^{\infty} \frac{1}{2^n},$$

but the latter form is somewhat neater. It is a triviality that any general statement about series written as $\sum_{n=1}^{\infty} a_n$ has an exact analog for series written as $\sum_{n=0}^{\infty} a_n$. For this reason, when no ambiguity is likely or when the distinction is immaterial, we often omit the limits of summation and for the sake of simplicity write $\sum a_n$ instead of $\sum_{n=1}^{\infty} a_n$ or $\sum_{n=0}^{\infty} a_n$. These remarks also apply to series of the form $\sum_{n=k}^{\infty} a_n$ for any integer $k \geq 2$.

We now briefly consider several fundamental examples, some of which we met in Chapter 13.

Example 1 Probably the simplest and most important of all infinite series is the familiar *geometric series*

$$\sum_{n=0}^{\infty} x^n = 1 + x + x^2 + \cdots . \tag{3}$$

By equation (2) in Section 14.2, the nth partial sum of this series is given by the closed formula

$$s_n = 1 + x + x^2 + \cdots + x^n = \frac{1 - x^{n+1}}{1 - x}$$

if $x \neq 1$. If $|x| < 1$, we see from this that $s_n \to 1/(1-x)$, so for these x's we have

$$1 + x + x^2 + \cdots + x^n + \cdots = \frac{1}{1-x}.$$

The series (3) therefore converges to the sum $1/(1-x)$ for $|x| < 1$, and is easily seen to diverge for all other values of x.

Example 2 Another series whose behavior is particularly simple is

$$\sum_{n=1}^{\infty} \frac{1}{n(n+1)} = \frac{1}{1 \cdot 2} + \frac{1}{2 \cdot 3} + \frac{1}{3 \cdot 4} + \cdots = 1.$$

To establish convergence and verify that the sum is 1, we use an ingenious trick due to Leibniz and observe that

$$\frac{1}{n(n+1)} = \frac{1}{n} - \frac{1}{n+1}.$$

This enables us to write the nth partial sum as

$$s_n = \left(\frac{1}{1} - \frac{1}{2}\right) + \left(\frac{1}{2} - \frac{1}{3}\right) + \cdots + \left(\frac{1}{n} - \frac{1}{n+1}\right)$$

$$= 1 - \frac{1}{n+1},$$

which makes it obvious that $s_n \to 1$. Any series whose nth partial sum collapses in this way into a closed formula is called a *telescopic series.*

As these examples suggest, the most direct method for studying the convergence behavior of a series is to find a closed formula for its nth partial sum. The main disadvantage of this approach is that it rarely works, because it is usually impossible to find such a formula. It is this situation that forces us to rely mostly on various indirect methods for establishing the convergence or divergence of series.

The main indirect method rests on the convergence criterion for sequences discussed in Section 14.2, that is, on the fact that an increasing sequence converges if and only if it is bounded. Thus, if the terms of our series are all nonnegative numbers, then we clearly have $s_n \le s_n + a_{n+1} = s_{n+1}$ for every n, and therefore the s_n's form an increasing sequence. It follows in this case that the sequence $\{s_n\}$ of partial sums — *and with it the series* — converges if and only if the s_n's have an upper bound. Our next examples furnish several illustrations of the use of this simple but important idea.

Example 3 The *harmonic series*

$$\sum_{n=1}^{\infty} \frac{1}{n} = 1 + \frac{1}{2} + \frac{1}{3} + \cdots \tag{4}$$

diverges because its partial sums are unbounded, as we saw at the beginning of Section 14.2. To establish this in a bit more detail, let m be a positive integer and choose $n > 2^{m+1}$. Then

$$s_n > 1 + \frac{1}{2} + \frac{1}{3} + \frac{1}{4} + \cdots + \frac{1}{2^{m+1}}$$

$$= \left(1 + \frac{1}{2}\right) + \left(\frac{1}{3} + \frac{1}{4}\right) + \left(\frac{1}{5} + \cdots + \frac{1}{8}\right) + \cdots + \left(\frac{1}{2^m + 1} + \cdots + \frac{1}{2^{m+1}}\right)$$

$$> \frac{1}{2} + 2 \cdot \frac{1}{4} + 4 \cdot \frac{1}{8} + \cdots + 2^m \cdot \frac{1}{2^{m+1}} = (m+1)\frac{1}{2}.$$

This proves that s_n can be made larger than the sum of any number of $\frac{1}{2}$'s, and therefore as large as we please, by taking n large enough, and so the s_n's are unbounded and (4) diverges. A series that behaves in this way is often said to *diverge to infinity,* and we express this behavior by writing

$$\sum_{n=1}^{\infty} \frac{1}{n} = 1 + \frac{1}{2} + \frac{1}{3} + \cdots = \infty.$$

A great many interesting series — some convergent and others divergent — can be obtained from the harmonic series by thinning it out, that is, by deleting terms according to a systematic pattern. For instance, if we remove all terms except reciprocals of powers of 2, what remains is the convergent geometric series

$$\sum_{n=0}^{\infty} \frac{1}{2^n} = 1 + \frac{1}{2} + \frac{1}{4} + \cdots \; ;$$

and if we remove all terms except reciprocals of primes, then—as we shall see in a later section—the resulting series diverges,

$$\sum \frac{1}{p_n} = \frac{1}{2} + \frac{1}{3} + \frac{1}{5} + \frac{1}{7} + \frac{1}{11} + \cdots = \infty.$$

The simplest general principle that is useful in deciding whether a series converges or not is the *nth term test*: *If Σa_n converges, then $a_n \to 0$.* To prove this, we merely observe that $a_n = s_n - s_{n-1} \to s - s = 0$. This result shows that $a_n \to 0$ is a necessary condition for convergence, in the sense that it follows from the convergence of the series Σa_n. Unfortunately, however, it is not a sufficient condition; that is, it does not imply the convergence of the series. This is easy to see by considering the harmonic series $\Sigma 1/n$, which diverges even though $1/n \to 0$. The nth term test is essentially a divergence test, for it is equivalent to the statement that if a_n does not approach zero, then Σa_n must diverge. As examples of its use, we mention the series

$$\sum_{n=1}^{\infty} (-1)^{n+1} = 1 - 1 + 1 - 1 + \cdots$$

and

$$\sum_{n=1}^{\infty} \frac{n}{n+1} = \frac{1}{2} + \frac{2}{3} + \frac{3}{4} + \cdots .$$

The first diverges because the sequence $(-1)^{n+1}$ does not converge at all, and so cannot converge to zero, and the second diverges because $n/(n+1) \to 1 \neq 0$.

Example 4 The series of the reciprocals of the squares,

$$\sum_{n=1}^{\infty} \frac{1}{n^2} = 1 + \frac{1}{4} + \frac{1}{9} + \frac{1}{16} + \cdots , \tag{5}$$

is convergent. As we have already seen in the third footnote of Section 13.2, this follows at once from the fact that the partial sums form an increasing sequence with 2 as an upper bound:

$$s_n = 1 + \frac{1}{2 \cdot 2} + \frac{1}{3 \cdot 3} + \cdots + \frac{1}{n \cdot n}$$

$$< 1 + \frac{1}{1 \cdot 2} + \frac{1}{2 \cdot 3} + \cdots + \frac{1}{(n-1)n}$$

$$= 1 + \left(\frac{1}{1} - \frac{1}{2}\right) + \left(\frac{1}{2} - \frac{1}{3}\right) + \cdots + \left(\frac{1}{n-1} - \frac{1}{n}\right)$$

$$= 2 - \frac{1}{n} < 2.$$

We have stated before that the sum of the series (5) is $\pi^2/6$, and Euler's plausibility argument for this remarkable fact is described in Appendix A.12. A less interesting—but rigorous and elementary—modern proof is given in Appendix A.13.

Example 5 If we recall that $0! = 1$ and $1! = 1$, then it is clear that the series

$$\sum_{n=0}^{\infty} \frac{1}{n!} = 1 + 1 + \frac{1}{2!} + \frac{1}{3!} + \cdots$$

has partial sums $s_0 = 1$, $s_1 = 2$, and, for $n \geq 2$,

$$s_n = 1 + 1 + \frac{1}{2} + \frac{1}{2 \cdot 3} + \cdots + \frac{1}{2 \cdot 3 \cdots n}.$$

If each factor in the denominators is replaced by 2, then we see that

$$s_n \leq 1 + 1 + \frac{1}{2} + \frac{1}{2^2} + \cdots + \frac{1}{2^{n-1}}$$

$$= 1 + 2\left(1 - \frac{1}{2^n}\right) = 3 - \frac{1}{2^{n-1}} < 3,$$

so the series converges with sum ≤ 3. By Example 5 in Section 14.2 we know that the sum of this series is actually e:

$$\sum_{n=0}^{\infty} \frac{1}{n!} = 1 + 1 + \frac{1}{2!} + \frac{1}{3!} + \cdots = e. \qquad (6)$$

We will use this fact below to prove that e is an irrational number.

These examples provide a small supply of specific series of known convergence behavior, where this behavior is decidable by rather elementary means. The value of these familiar series for determining the behavior of new series by various methods of comparison will begin to appear in Section 14.4. First, however, there are several simple properties of convergent series in general that need to be mentioned explicitly.

The effective use of infinite series rests on our freedom to manipulate them by the various processes of algebra. However, we will soon see that carelessness can easily lead to confusion and disaster. It is therefore of prime importance to know exactly which operations are permissible and which are traps for the unwary.

If $\sum_{n=1}^{\infty} a_n$ converges to s, we write

$$a_1 + a_2 + \cdots + a_n + \cdots = s \qquad (7)$$

and call s the "sum" of the series. This well-established terminology is perhaps unfortunate, for it tends to foster the belief that an infinite series can be treated as if it were an ordinary finite sum. In reality, of course, s is not obtained simply by addition, but is the limit of a sequence of finite sums, and the properties of series must be based on this definition and not on any tempting but misleading analogy. As we shall see, many properties of finite sums do carry over to series, but we must always be careful not to assume this without proof.

As an example of the pitfalls that lie around us, consider the familiar fact that rearranging the order of the terms of a finite sum has no effect on the numerical value of that sum. In contrast to this, in Problem 10 we ask students to see for themselves that the sum of a convergent infinite series can be altered by writing its terms—exactly the same terms!—in a different order. This astounding (and fascinating) behavior illustrates the need for caution. It also emphasizes the delicacy of the concepts we are working with, and gives us fair warning that we cannot hope to study infinite series successfully without giving a reasonable level of attention to the underlying theory.

We begin by pointing out that in dealing with finite sums we can freely insert or remove parentheses, as in the expressions

$$1 - 1 + 1 = (1 - 1) + 1 = 1 - (1 - 1) = 1,$$

but this is not true for infinite series. For instance, the series $1 - 1 + 1 - 1 + \cdots$ clearly diverges, but

$$(1 - 1) + (1 - 1) + \cdots = 0 + 0 + \cdots$$

converges to 0, and

$$1 - (1 - 1) - (1 - 1) - \cdots = 1 - 0 - 0 - \cdots$$

converges to 1. These examples show that the insertion or removal of parentheses can change the nature of an infinite series. However, in the case of a convergent series like (7), any series obtained from it by inserting parentheses, such as

$$a_1 + (a_2 + a_3) + (a_4 + a_5 + a_6) + \cdots,$$

still converges and has the same sum. The reason for this is that the partial sums of the new series form a subsequence of the original sequence of partial sums, and therefore necessarily converge to the same limit. In the same way, we see that parentheses can be removed if the resulting series converges.

We next remark that if $a_1 + a_2 + \cdots$ converges to s, then $a_1 + 0 + a_2 + 0 + \cdots$ also converges and has the same sum, because the two sequences of partial sums are $s_1, s_2, \ldots$ and $s_1, s_1, s_2, s_2, \ldots$, and the repetitions in the latter do not interfere with its convergence to s. Similarly, any finite number of 0s can be inserted or removed anywhere in a series without affecting its convergence behavior or (if it converges) its sum.

It is important to observe that when two convergent series are added term by term, the resulting series converges to the expected sum; that is, if $\Sigma_{n=1}^{\infty} a_n = s$ and $\Sigma_{n=1}^{\infty} b_n = t$, then $\Sigma_{n=1}^{\infty}(a_n + b_n) = s + t$. This is easy to prove, for if s_n and t_n are the partial sums, then

$$(a_1 + b_1) + (a_2 + b_2) + \cdots + (a_n + b_n)$$
$$= (a_1 + a_2 + \cdots + a_n) + (b_1 + b_2 + \cdots + b_n)$$
$$= s_n + t_n \rightarrow s + t.$$

Similarly, $\Sigma_{n=1}^{\infty}(a_n - b_n) = s - t$ and $\Sigma_{n=1}^{\infty} ca_n = cs$ for any constant c. It is also convenient to know that if

$$a_1 + a_2 + \cdots = s,$$

then

$$a_0 + a_1 + a_2 + \cdots = a_0 + s \quad \text{and} \quad a_2 + a_3 + \cdots = s - a_1.$$

The first statement is clear from the fact that

$$\lim (a_0 + a_1 + a_2 + \cdots + a_n) = \lim a_0 + \lim (a_1 + a_2 + \cdots + a_n) = a_0 + s,$$

and the second follows in the same way. Thus, any finite number of terms can be added or subtracted at the beginning of a convergent series without disturbing its convergence, and the sums of the various series are related in the expected ways.

We now use several of the properties of series discussed above to prove the following theorem of Euler: *e is irrational.*

Our starting point is equation (6),

$$e = 1 + 1 + \frac{1}{2!} + \cdots + \frac{1}{n!} + \cdots,$$

from which it follows that the number

$$e - 1 - 1 - \frac{1}{2!} - \cdots - \frac{1}{n!} = \frac{1}{(n+1)!} + \frac{1}{(n+2)!} + \cdots \tag{8}$$

is positive for every positive integer n. We assume that e is rational, so that $e = p/q$ for certain positive integers p and q, and we deduce a contradiction from this assumption. Let n in (8) be chosen so large that $n > q$, and define a number a by

$$a = n! \left[e - 1 - 1 - \frac{1}{2!} - \cdots - \frac{1}{n!} \right].$$

Since q divides $n!$, a is a positive integer. However, (8) implies that

$$a = n! \left[\frac{1}{(n+1)!} + \frac{1}{(n+2)!} + \cdots \right]$$

$$= \frac{1}{n+1} + \frac{1}{(n+1)(n+2)} + \cdots$$

$$< \frac{1}{n+1} + \frac{1}{(n+1)^2} + \cdots$$

$$= \frac{1}{n+1} \left[1 + \frac{1}{n+1} + \frac{1}{(n+1)^2} + \cdots \right]$$

$$= \frac{1}{n+1} \cdot \frac{1}{1 - 1/(n+1)} = \frac{1}{n}.$$

This contradiction (there is no positive integer $< 1/n$) completes the argument.

Further information about irrational numbers (π is irrational, etc.) is given in Appendix A.15.

PROBLEMS

1 If Σa_n converges and Σb_n diverges, show that $\Sigma(a_n + b_n)$ diverges. Hint: Assume that it converges and deduce a contradiction.

2 Decide whether each of the following series converges or diverges, and give convincing reasons for your answers:

(a) $\dfrac{1}{500} + \dfrac{1}{505} + \dfrac{1}{510} + \cdots$;

(b) $\displaystyle\sum \left[\dfrac{2}{n} - \left(\dfrac{3}{4} \right)^n \right]$;

(c) $\displaystyle\sum \left(\dfrac{1}{3^n} + \dfrac{1}{4^n} \right)$;

(d) $\displaystyle\sum \left[\dfrac{2}{n(n+1)} - \dfrac{100}{n!} \right]$;

(e) $\displaystyle\sum 2^{-1/n}$;

(f) $\displaystyle\sum \dfrac{1}{\ln 2^n}$;

(g) $\displaystyle\sum \dfrac{1}{\ln 2^{n^2}}$;

(h) $\displaystyle\sum \cos \dfrac{(2n+1)\pi}{2}$;

(i) $\displaystyle\sum \cos \dfrac{n\pi}{4}$.

3 For each of the following series, find the values of x for which the series converges and express the sum as a simple function of x:

(a) $ax + ax^3 + ax^5 + \cdots$, $a \neq 0$;

(b) $\dfrac{1}{x} + \dfrac{1}{x^2} + \dfrac{1}{x^3} + \cdots$;

(c) $x + \dfrac{x}{1+x} + \dfrac{x}{(1+x)^2} + \cdots$;

(d) $\ln x + (\ln x)^2 + (\ln x)^3 + \cdots$.

4 Show that

$$x^2 + \frac{x^2}{1+x^2} + \frac{x^2}{(1+x^2)^2} + \cdots$$

converges for all x and find its sum.

5 Show that $\Sigma 1/e^n$ converges but $\Sigma 1/(e^{\ln n})$ diverges. For what values of x does Σe^{nx} converge?

6 Show that

(a) $\displaystyle\sum_{n=1}^{\infty} [\tan^{-1}(n+1) - \tan^{-1} n] = \pi/4$;

(b) $\displaystyle\sum_{n=1}^{\infty} \ln \left(1 + \dfrac{1}{n} \right) = \infty$.

7 If $f(n) \to L$, show that

$$\sum_{n=1}^{\infty} [f(n) - f(n+1)] = f(1) - L$$

and use this to establish the indicated sums of the following telescopic series:

(a) $\displaystyle\sum_{n=1}^{\infty} \dfrac{1}{4n^2 - 1} = \dfrac{1}{2}$;

(b) $\displaystyle\sum_{n=1}^{\infty} (-1)^{n+1} \cdot \dfrac{2n+1}{n(n+1)} = 1$;

(c) $\displaystyle\sum_{n=1}^{\infty} \dfrac{2n+1}{n^2(n+1)^2} = 1$;

(d) $\displaystyle\sum_{n=1}^{\infty} \dfrac{1}{(4n-1)(4n+3)} = \dfrac{1}{12}$.

8 It follows from Example 4 that the series

$$\frac{1}{2^2} + \frac{1}{3^2} + \frac{1}{5^2} + \frac{1}{7^2} + \frac{1}{11^2} + \cdots ,$$

where the denominators are the squares of the successive primes, converges. Why?

9 A *decimal* $a_0.a_1a_2 \ldots a_n \ldots$ is simply an abbreviated way of writing the infinite series

$$\sum_{n=0}^{\infty} \frac{a_n}{10^n} = a_0 + \frac{a_1}{10} + \frac{a_2}{10^2} + \cdots ,$$

where it is understood that a_0 is an arbitrary integer and each of the a_n's for $n \geq 1$ is one of the digits 0, 1, 2, . . . , 9. Show that every decimal converges.

10 Consider the familiar series (Section 13.3)

$$1 - \tfrac{1}{2} + \tfrac{1}{3} - \tfrac{1}{4} + \tfrac{1}{5} - \tfrac{1}{6} + \tfrac{1}{7} - \tfrac{1}{8} + \cdots = \ln 2, \quad (*)$$

and write under it, as follows, the result of multiplying through by the factor $\tfrac{1}{2}$:

$$\tfrac{1}{2} \quad - \tfrac{1}{4} \quad + \tfrac{1}{6} \quad - \tfrac{1}{8} + \cdots = \tfrac{1}{2} \ln 2.$$

Now add, combining the terms placed in vertical columns, to obtain the series

$$1 + \tfrac{1}{3} - \tfrac{1}{2} + \tfrac{1}{5} + \tfrac{1}{7} - \tfrac{1}{4} + + - \cdots = \tfrac{3}{2} \ln 2. \quad (**)$$

Satisfy yourself (a) that $(**)$ is valid; (b) that this series can be produced by rearranging the terms of the series $(*)$, so that the first two positive terms of $(*)$ are followed by the first negative term, then the next two positive terms by the second negative term, etc.; and (c) that the value of the sum of the series $(*)$ has been mysteriously multiplied in this way by the factor $\tfrac{3}{2}$.†

---

† This phenomenon will be explored from a different point of view, and we hope clarified, in Sections 14.5 and 14.7.

The easiest infinite series to work with are those whose terms are all non-negative numbers. The reason for this — as we saw in Section 14.3 — is that the total theory of these series can be expressed by the following simple statement: *If* $a_n \geq 0$, *then the series* Σa_n *converges if and only if its sequence* $\{s_n\}$ *of partial sums is bounded.*

Thus, in order to establish the convergence of a series of nonnegative terms, it suffices to show that its terms approach zero fast enough to keep the partial sums bounded. But how fast is "fast enough"? One answer to this question can be stated informally as follows: at least as fast as the terms of a known convergent series of nonnegative terms. This idea is contained in a formal statement called the *comparison test*: if $0 \leq a_n \leq b_n$, then

Σa_n converges if Σb_n converges;

Σb_n diverges if Σa_n diverges.

The proof is easy. The first step is to notice that if s_n and t_n are the partial sums of Σa_n and Σb_n, then the assumption yields

$$s_n = a_1 + a_2 + \cdots + a_n \leq b_1 + b_2 + \cdots + b_n = t_n.$$

Our conclusion now follows at once from this inequality and the statement in the preceding paragraph, for if the t_n's are bounded, then so are the s_n's, and if the s_n's are unbounded, then so are the t_n's.

Example 1 The comparison test is easy to apply to the series

$$\sum_{n=1}^{\infty} \frac{1}{3^n + 1} \quad \text{and} \quad \sum_{n=2}^{\infty} \frac{1}{\ln n}.$$

The first series converges, because

$$\frac{1}{3^n + 1} \leq \frac{1}{3^n}$$

and $\Sigma 1/3^n$ converges; and the second diverges, because

$$\frac{1}{n} \leq \frac{1}{\ln n} \tag{1}$$

and $\Sigma 1/n$ diverges. [To verify (1) in the equivalent form $\ln n \leq n$, recall that the graph of $y = \ln x$ lies below the graph of $y = x$.]

It is worth remarking here that we can disregard any finite number of terms at the beginning of a series if we are interested only in deciding whether that series converges or diverges.* This tells us that the condition $0 \leq a_n \leq b_n$ for the comparison test need not hold for all n, but only for all n from some point on. As an illustration, suppose we want to show that $\Sigma(n + 1)/n^n$ converges by comparison with $\Sigma 1/n^2$. The inequality

$$\frac{n + 1}{n^n} \leq \frac{1}{n^2}$$

14.4

SERIES OF
NONNEGATIVE TERMS.
COMPARISON TESTS

---

* On the other hand, if we are interested in the sum of a convergent series, then obviously we must take all of its terms into account.

is not true for all n, but it is true for all $n \geq 4$. The series therefore converges by comparison with the convergent series $\Sigma 1/n^2$.

The comparison test is very simple in principle, but in complicated cases it can be difficult to establish the necessary inequality between the nth terms of the two series being compared. Since limits are often easier to work with than inequalities, the following *limit comparison test* is a more convenient tool for studying many series: *If Σa_n and Σb_n are series with positive terms such that*

$$\lim_{n \to \infty} \frac{a_n}{b_n} = 1, \tag{2}$$

then either both series converge or both series diverge. To establish this, we observe that (2) implies that for all sufficiently large n we have

$$\frac{1}{2} \leq \frac{a_n}{b_n} \leq 2$$

or

$$\frac{1}{2} b_n \leq a_n \leq 2b_n. \tag{3}$$

Since the convergence behavior of a series is not affected by multiplying each of its terms by the same nonzero constant, our conclusion is an easy consequence of the inequalities (3) and the comparison test as extended in the preceding paragraph. Thus, for instance, if Σb_n converges, then $\Sigma 2b_n$ converges and, by the second inequality in (3), Σa_n also converges; etc.

Example 2 The series $\Sigma(n + 2)/(2n^3 - 3)$ converges, because $\Sigma 1/2n^2$ converges and

$$\frac{(n + 2)/(2n^3 - 3)}{1/2n^2} = \frac{2n^3 + 4n^2}{2n^3 - 3} \to 1 \quad \text{as} \quad n \to \infty;$$

and $\Sigma \sin(1/n)$ diverges, because $\Sigma 1/n$ diverges and

$$\frac{\sin 1/n}{1/n} \to 1 \quad \text{as} \quad n \to \infty.$$

[Recall that $\lim_{x \to 0} (\sin x)/x = 1$; see Section 9.2.]

The limit comparison test is slightly more convenient to use if condition (2) is replaced by

$$\lim_{n \to \infty} \frac{a_n}{b_n} = L,$$

where $0 < L < \infty$. The proof is essentially the same and will not be repeated.

Example 2 shows that in using the limit comparison test we must try to guess the probable behavior of Σa_n by estimating the "order of magnitude" of the nth term a_n. That is, we must try to judge whether a_n is approximately equal to a constant multiple of the nth term of some familiar series whose convergence behavior is known to us, such as

$$\sum x^n, \quad \sum \frac{1}{n}, \quad \sum \frac{1}{n^2}, \quad \text{or} \quad \sum \frac{1}{n!}.$$

To apply this method effectively, it is clearly desirable to have at our disposal a "stockpile" of comparison series of known behavior. Our next example provides a family of series that is especially valuable for this purpose. We emphasize once again that the limit comparison test is used *only if the terms of the series being tested are all positive numbers.*

Example 3   If p is a positive constant, then the *p-series*

$$\sum_{n=1}^{\infty} \frac{1}{n^p} = 1 + \frac{1}{2^p} + \frac{1}{3^p} + \frac{1}{4^p} + \cdots \tag{4}$$

diverges if $p \le 1$ and converges if $p > 1$.

To establish this, we notice first that if $p \le 1$, then $n^p \le n$ or $1/n \le 1/n^p$, and so (4) diverges by comparison with the harmonic series $\sum 1/n$. We now prove that (4) converges if $p > 1$ by showing that its partial sums have an upper bound. Let n be given and choose m so that $n < 2^m$. Then

$$s_n \le s_{2^m-1} = 1 + \left(\frac{1}{2^p} + \frac{1}{3^p} \right) + \left(\frac{1}{4^p} + \cdots + \frac{1}{7^p} \right)$$

$$+ \cdots + \left[\frac{1}{(2^{m-1})^p} + \cdots + \frac{1}{(2^m - 1)^p} \right]$$

$$\le 1 + \frac{2}{2^p} + \frac{4}{4^p} + \cdots + \frac{2^{m-1}}{(2^{m-1})^p}.$$

If we put $a = 1/2^{p-1}$, then $a < 1$ since $p > 1$, and

$$s_n \le 1 + a + a^2 + \cdots + a^{m-1} = \frac{1 - a^m}{1 - a} < \frac{1}{1 - a}.$$

This provides an upper bound for the s_n's, and the argument is complete.

As an illustration of the use of this family of series, we see that

$$\sum \frac{1}{\sqrt{n^3 + 3}}$$

converges, because the *p*-series (with $p = \frac{3}{2}$) $\sum 1/n^{3/2}$ converges and

$$\frac{1/\sqrt{n^3 + 3}}{1/n^{3/2}} = \sqrt{\frac{n^3}{n^3 + 3}} \to 1.$$

It is worth noticing that $\sum 1/n^p$ does not necessarily converge if p is a variable > 1. This is shown by the series

$$\sum \frac{1}{n^{1+1/n}},$$

which diverges because $\sum 1/n$ diverges and

$$\frac{1/n^{1+1/n}}{1/n} = \frac{1}{\sqrt[n]{n}} \to 1.$$

(Recall that $\lim_{n \to \infty} n^{1/n} = 1$.)

We conclude this section with some observations on the process of rearranging the terms of a series, which was briefly discussed in Section 14.3. Suppose that Σa_n is a convergent series of nonnegative terms whose sum is s, and form a new series Σb_n by rearranging the a_n's in any way. For instance, Σb_n might be the series

$$a_{10} + a_3 + a_5 + a_1 + a_6 + a_2 + \cdots .$$

Let n be a given positive integer and consider the nth partial sum $t_n = b_1 + b_2 + \cdots + b_n$ of the new series. Since each b is some a, there exists an m with the property that each term in t_n is one of the terms in $s_m = a_1 + a_2 + \cdots + a_m$. This tells us that $t_n \le s_m \le s$, and so Σb_n converges to a sum $t \le s$. On the other hand, the first series is also a rearrangement of the second, and so by the same reasoning we have $s \le t$, and therefore $t = s$. This proves that *if a convergent series of nonnegative terms is rearranged in any manner, then the resulting series also converges and has the same sum.* If this conclusion seems rather obvious and trivial to students, let them recall from Problem 10 in Section 14.3 that it isn't true if we drop the assumption that the terms of the given series are nonnegative numbers.

PROBLEMS

1 Establish the convergence or divergence of the following series by using the comparison test:

(a) $\sum \dfrac{1}{\sqrt{n(n+1)}}$; (b) $\sum \dfrac{1}{\sqrt{n^2(n+1)}}$;

(c) $\sum \dfrac{1}{n^n}$; (d) $\sum \dfrac{1}{(\ln n)^n}$;

(e) $\sum \dfrac{1}{n^{\ln n}}$; (f) $\sum \dfrac{n+1}{n(n-1)}$;

(g) $\sum \dfrac{(2n+3)^n}{n^{2n}}$; (h) $\sum \left(\dfrac{n}{n+1}\right)^{n^2}$.

Determine by any method whether each of the following series converges or diverges.

2 $\sum \dfrac{3}{n^2+1}$.

3 $\sum \dfrac{1+3n^2}{n^3+700}$.

4 $\sum \sin \dfrac{1}{n^2}$.

5 $\sum \cos \dfrac{1}{n^2}$.

6 $\sum \dfrac{1}{3^n+9}$.

7 $\sum \dfrac{1}{(1+1/n)^n}$.

8 $\sum \dfrac{\sqrt{n}}{n^2+5}$.

9 $\sum \dfrac{\ln n}{n}$.

10 $\sum \dfrac{3n+2}{n} \cdot \dfrac{4^n}{5^n+1}$.

11 $\sum \dfrac{1}{n+\sqrt{n}}$.

12 $\sum \dfrac{\ln n}{n^3}$.

13 $\sum \dfrac{1000}{\sqrt[3]{n+1}\,\sqrt[4]{n^3+5}}$.

14 $\sum \dfrac{n^2}{n^2+100}$.

15 $\sum \dfrac{1}{n10^n}$.

16 $\sum \dfrac{1}{5000n}$.

17 $\sum \dfrac{n^2+3n-7}{n^3-2n+5}$.

18 $\sum \dfrac{\sqrt[3]{n+2}}{\sqrt[4]{n^3+3}\,\sqrt[5]{n^3+5}}$.

19 $\sum \dfrac{n^2}{n^5-\pi}$.

20 $\sum \dfrac{3+\cos n}{n^2}$.

21 $\sum \ln(1+1/n^p)$, $p>0$.

22 $\sum \dfrac{\sqrt{n+1}-\sqrt{n}}{n}$.

23 $\sum (1-\cos 1/n)$.

24 $\sum \sqrt{n} \ln \dfrac{n+1}{n}$.

25 $\sum \dfrac{1}{n^p}\left(1+\dfrac{1}{2^p}+\cdots+\dfrac{1}{n^p}\right)$, $p>0$.

26 If Σa_n is a convergent series with nonnegative terms, show that Σa_n^2 also converges. With the same hypotheses, show by examples that $\Sigma \sqrt{a_n}$ is sometimes convergent and sometimes divergent.

27 If p is a positive constant, show that

$$\sum_{n=1}^{\infty} \dfrac{1}{(2n-1)^p} = 1 + \dfrac{1}{3^p} + \dfrac{1}{5^p} + \cdots$$

converges if $p > 1$ and diverges if $p \le 1$.

28 Show that $\Sigma 1/n$ diverges by comparing it with the divergent series $\Sigma \ln(1 + 1/n)$ of Problem 6(b) in Section 14.3. Hint: Compare the graphs of the functions $y = x$ and $y = \ln(1 + x)$ for $x > 0$.

29 Use the idea of Problem 28 to show that

$$\sum_{n=1}^{\infty} \ln \frac{(n+1)^2}{n(n+2)}$$

converges. Also, find the sum of this series.

14.5
THE INTEGRAL TEST. EULER'S CONSTANT

Among the simplest infinite series are those whose terms form a decreasing sequence of positive numbers. In this section we study certain series of this type by means of improper integrals of the form

$$\int_a^{\infty} f(x)\, dx = \lim_{b \to \infty} \int_a^b f(x)\, dx. \tag{1}$$

We recall that the integral on the left is said to be *convergent* if the limit on the right exists (as a finite number), and in this case the value of the integral is by definition the value of the limit. If this limit does not exist, then the integral is called *divergent*. There is an obvious analogy between (1) and the corresponding definition for series,

$$\sum_{n=1}^{\infty} a_n = \lim_{k \to \infty} \sum_{n=1}^{k} a_n.$$

Our purpose is to exploit this analogy by using integrals to obtain information about series.

Consider a series

$$\sum_{n=1}^{\infty} a_n = a_1 + a_2 + \cdots + a_n + \cdots \tag{2}$$

whose terms are positive and decreasing. In most cases the nth term a_n is a function of n given by a simple formula, $a_n = f(n)$. Suppose that the function $y = f(x)$ obtained by substituting the continuous variable x in place of the discrete variable n is a decreasing function of x for $x \ge 1$, as shown in Fig. 14.4. On the left in this figure we see that the rectangles of areas $a_1, a_2, \ldots, a_n$ have a greater combined area than the area under the curve from $x = 1$ to $x = n + 1$, so

$$a_1 + a_2 + \cdots + a_n \ge \int_1^{n+1} f(x)\, dx \ge \int_1^n f(x)\, dx. \tag{3}$$

On the right side of the figure we make the rectangles face to the left, so that

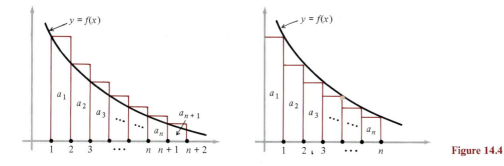

Figure 14.4

they lie under the curve. If we momentarily ignore the first rectangle, with area a_1, then we see that

$$a_2 + a_3 + \cdots + a_n \leq \int_1^n f(x)\,dx;$$

and including a_1 gives

$$a_1 + a_2 + \cdots + a_n \leq a_1 + \int_1^n f(x)\,dx. \tag{4}$$

By combining (3) and (4), we obtain

$$\int_1^n f(x)\,dx \leq a_1 + a_2 + \cdots + a_n \leq a_1 + \int_1^n f(x)\,dx. \tag{5}$$

The point of all this is that the inequalities (5) enable us to establish the *integral test*:

If f(x) is a positive decreasing function for $x \geq 1$ with the property that $f(n) = a_n$ for each positive integer n, then the series and integral

$$\sum_{n=1}^{\infty} a_n \quad \text{and} \quad \int_1^{\infty} f(x)\,dx$$

converge or diverge together.*

The argument is easy, for if the series converges, then the inequality on the left of (5) shows that the integral does also; and if the integral converges, then the inequality on the right shows that the series also converges.

Example 1   (the p-series revisited) If p is a positive constant, then we know from Section 14.4 that the p-series

$$\sum_{n=1}^{\infty} \frac{1}{n^p} = 1 + \frac{1}{2^p} + \frac{1}{3^p} + \cdots \tag{6}$$

converges if $p > 1$ and diverges if $p \leq 1$. It is of some interest to give another proof of this as an illustration of the integral test. Since $a_n = 1/n^p$, we consider the function $f(x) = 1/x^p$ (which clearly satisfies all the stated conditions) and examine the integral

$$\int_1^{\infty} \frac{dx}{x^p}.$$

If $p = 1$, this integral diverges, because

$$\int_1^{\infty} \frac{dx}{x} = \lim_{b \to \infty} \int_1^b \frac{dx}{x} = \lim_{b \to \infty} \ln b = \infty.$$

If $p \neq 1$, then

$$\int_1^{\infty} \frac{dx}{x^p} = \lim_{b \to \infty} \int_1^b x^{-p}\,dx = \lim_{b \to \infty} \left(\frac{b^{1-p} - 1}{1 - p} \right),$$

---

* This test is often called the *Cauchy integral test,* after its discoverer, the eminent nineteenth century French mathematician Augustin Louis Cauchy (pronounced *co-shee*).

and the issue of convergence hangs on the behavior of b^{1-p} as $b \to \infty$. If $p < 1$, so that $1 - p > 0$, then $b^{1-p} \to \infty$ and the integral diverges. If $p > 1$, so that $1 - p < 0$, then $b^{1-p} \to 0$ and the integral converges. By the integral test we now conclude again that the p-series (6) converges if $p > 1$ and diverges if $p \leq 1$.

It is clear that the integral test holds for any interval of the form $x \geq k$, not just for $x \geq 1$. We make use of this remark in our next example, which deals with a class of series whose behavior is not revealed by any of our previous tests.

Example 2 The terms of the series

$$\sum_{n=2}^{\infty} \frac{1}{n \ln n} \tag{7}$$

decrease faster than those of the harmonic series. Nevertheless, it is easy to see by the integral test that (7) diverges, for

$$\int_{2}^{\infty} \frac{dx}{x \ln x} = \lim_{b \to \infty} \int_{2}^{b} \frac{dx}{x \ln x} = \lim_{b \to \infty} [\ln \ln x]_{2}^{b}$$
$$= \lim_{b \to \infty} (\ln \ln b - \ln \ln 2) = \infty.$$

More generally, if p is a positive constant, then

$$\sum_{n=2}^{\infty} \frac{1}{n(\ln n)^p}$$

converges if $p > 1$ and diverges if $p \leq 1$; for if $p \neq 1$, we have

$$\int_{2}^{\infty} \frac{dx}{x(\ln x)^p} = \lim_{b \to \infty} \int_{2}^{b} \frac{dx}{x(\ln x)^p} = \lim_{b \to \infty} \left[\frac{(\ln x)^{1-p}}{1 - p} \right]_{2}^{b}$$
$$= \lim_{b \to \infty} \left[\frac{(\ln b)^{1-p} - (\ln 2)^{1-p}}{1 - p} \right],$$

and this limit exists if and only if $p > 1$.*

We now return to the series (2) and squeeze some additional information out of the inequalities (5). By subtracting the integral that occurs on the left, these inequalities can be written as

$$0 \leq a_1 + a_2 + \cdots + a_n - \int_{1}^{n} f(x) \, dx \leq a_1, \tag{8}$$

and this serves to focus our attention on the quantity in the middle. If we denote this quantity by $F(n)$, so that

$$F(n) = a_1 + a_2 + \cdots + a_n - \int_{1}^{n} f(x) \, dx,$$

then (8) becomes

$$0 \leq F(n) \leq a_1.$$

---

* The series of this example are all called *Abel's series,* after the great Norwegian mathematician Niels Henrik Abel, who first investigated them and determined their convergence behavior.

From our present point of view, the key to this situation is the fact that $\{F(n)\}$ is a decreasing sequence. This follows from the calculation

$$F(n) - F(n+1) = \left[a_1 + a_2 + \cdots + a_n - \int_1^n f(x)\,dx \right]$$
$$- \left[a_1 + a_2 + \cdots + a_{n+1} - \int_1^{n+1} f(x)\,dx \right]$$
$$= \int_n^{n+1} f(x)\,dx - a_{n+1} \geq 0,$$

where the reason for the last-written inequality can be understood by examining the left side of Fig. 14.4. Since any decreasing sequence of nonnegative numbers converges, the limit

$$L = \lim_{n \to \infty} F(n) = \lim_{n \to \infty} \left[a_1 + a_2 + \cdots + a_n - \int_1^n f(x)\,dx \right] \tag{9}$$

exists and satisfies the inequalities $0 \leq L \leq a_1$.

As our main application of these ideas, we deduce the existence of the important limit

$$\lim_{n \to \infty} \left(1 + \frac{1}{2} + \cdots + \frac{1}{n} - \ln n \right). \tag{10}$$

This is easily seen to be the special case of (9) in which $a_n = 1/n$ and $f(x) = 1/x$, because

$$\int_1^n \frac{dx}{x} = \ln x \Big]_1^n = \ln n.$$

The value of the limit (10) is usually denoted by the Greek letter γ *(gamma)*, and is called *Euler's constant*:

$$\gamma = \lim_{n \to \infty} \left(1 + \frac{1}{2} + \cdots + \frac{1}{n} - \ln n \right). \tag{11}$$

This constant occurs quite frequently in several parts of advanced calculus, especially in the theory of the gamma function, and is, along with π and e, one of the most important special numbers of mathematics. Its numerical value, $\gamma = 0.57721\ 56649\ 01532\ 86060\ \ldots$, has been calculated to many hundreds of decimal places. Nevertheless, no one knows whether γ is rational or irrational.

In order to describe some of the uses of Euler's constant, it is convenient to introduce a notation which has been widely accepted in twentieth century mathematics. Let $\{a_n\}$ and $\{b_n\}$ be two sequences, and suppose that $b_n > 0$. We say that "a_n is little-oh of b_n," and symbolize this by writing

$$a_n = o(b_n),$$

if $a_n/b_n \to 0$. In particular, $a_n = o(1)$ means that $a_n \to 0$. An equation of the form $a_n = b_n + o(1)$ means that $a_n - b_n = o(1)$, and so a_n and b_n differ by a quantity that approaches zero as $n \to \infty$. In our work we will use the symbol $o(1)$ to mean any sequence that approaches zero as $n \to \infty$, as in the calculation $[a + o(1)] + 2[b + o(1)] = a + 2b + o(1)$.

With the aid of this notation, (11) can be written in the form

$$1 + \frac{1}{2} + \cdots + \frac{1}{n} = \ln n + \gamma + o(1). \tag{12}$$

Since $\ln n \to \infty$ as $n \to \infty$, this formula displays in a very transparent way the reason for the divergence of the harmonic series. It is also useful for many other purposes, as the following examples show.

Example 3 We can use (12) to give a simple proof of the formula

$$1 - \tfrac{1}{2} + \tfrac{1}{3} - \tfrac{1}{4} + \cdots = \ln 2. \tag{13}$$

Let s_n be the nth partial sum of this series, and observe that

$$s_{2n} = 1 - \frac{1}{2} + \frac{1}{3} - \frac{1}{4} + \cdots + \frac{1}{2n-1} - \frac{1}{2n}$$

$$= \left(1 + \frac{1}{3} + \cdots + \frac{1}{2n-1}\right) - \left(\frac{1}{2} + \frac{1}{4} + \cdots + \frac{1}{2n}\right)$$

$$= \left(1 + \frac{1}{2} + \frac{1}{3} + \cdots + \frac{1}{2n}\right) - 2\left(\frac{1}{2} + \frac{1}{4} + \cdots + \frac{1}{2n}\right)$$

$$= \left(1 + \frac{1}{2} + \frac{1}{3} + \cdots + \frac{1}{2n}\right) - \left(1 + \frac{1}{2} + \cdots + \frac{1}{n}\right)$$

$$= [\ln 2n + \gamma + o(1)] - [\ln n + \gamma + o(1)]$$

$$= \ln 2 + o(1) \to \ln 2.$$

The odd partial sums approach the same limit, because

$$s_{2n+1} = s_{2n} + \frac{1}{2n+1} \to \lim s_{2n} = \ln 2,$$

so the proof of (13) is complete. We emphasize that this method establishes (13) on the basis of (12) alone, without making any use of the power series expansion of $\ln(1 + x)$ as given in Section 13.3.

Example 4 We can also use (12) to obtain the remarkable formula

$$1 + \tfrac{1}{3} - \tfrac{1}{2} + \tfrac{1}{5} + \tfrac{1}{7} - \tfrac{1}{4} + + - \cdots = \tfrac{3}{2}\ln 2, \tag{14}$$

which was the subject of Problem 10 in Section 14.3. This method is similar to that of Example 3. If s_n is the nth partial sum of (14), then, since $2n$ is the nth even number and $2n - 1$ is the nth odd number, we have

$$s_{3n} = \left(1 + \frac{1}{3} - \frac{1}{2}\right) + \left(\frac{1}{5} + \frac{1}{7} - \frac{1}{4}\right) + \cdots + \left(\frac{1}{4n-3} + \frac{1}{4n-1} - \frac{1}{2n}\right)$$

$$= \left(1 + \frac{1}{3} + \frac{1}{5} + \frac{1}{7} + \cdots + \frac{1}{4n-1}\right) - \left(\frac{1}{2} + \frac{1}{4} + \cdots + \frac{1}{2n}\right)$$

$$= \left(1 + \frac{1}{2} + \frac{1}{3} + \cdots + \frac{1}{4n}\right) - \left(\frac{1}{2} + \frac{1}{4} + \cdots + \frac{1}{4n}\right)$$

$$- \left(\frac{1}{2} + \frac{1}{4} + \cdots + \frac{1}{2n}\right)$$

$$= \left(1 + \frac{1}{2} + \frac{1}{3} + \cdots + \frac{1}{4n}\right) - \frac{1}{2}\left(1 + \frac{1}{2} + \cdots + \frac{1}{2n}\right)$$

$$- \frac{1}{2}\left(1 + \frac{1}{2} + \cdots + \frac{1}{n}\right)$$

$$= [\ln 4n + \gamma + o(1)] - \tfrac{1}{2}[\ln 2n + \gamma + o(1)] - \tfrac{1}{2}[\ln n + \gamma + o(1)]$$

$$= \ln 4n - \frac{1}{2}\ln 2n^2 + o(1) = \ln \frac{4n}{\sqrt{2}\,n} + o(1)$$

$$= \ln 2^{3/2} + o(1) \to \ln 2^{3/2} = \tfrac{3}{2}\ln 2.$$

It is easy to see that the partial sums

$$s_{3n+1} = s_{3n} + \frac{1}{4n+1} \quad \text{and} \quad s_{3n+2} = s_{3n} + \frac{1}{4n+1} + \frac{1}{4n+3}$$

approach the same limit, so the proof of (14) is complete.

Remark   The basic idea of the integral test is to compare sums with integrals by looking at their geometric meanings in terms of areas. This idea can also be used to prove the divergence of the series of the reciprocals of the primes, as mentioned in Section 14.3:

$$\sum \frac{1}{p_n} = \frac{1}{2} + \frac{1}{3} + \frac{1}{5} + \frac{1}{7} + \frac{1}{11} + \cdots = \infty.$$

This proof is a bit complicated; and since it is not essential to the main line of thought in this chapter, we give it in Appendix A.17.

PROBLEMS

Use the integral test to determine whether each of the following series converges or diverges.

1   $\displaystyle\sum_{n=1}^{\infty} \frac{n}{e^{n^2}}.$

2   $\displaystyle\sum_{n=1}^{\infty} \frac{1}{n(n+1)}.$

3   $\displaystyle\sum_{n=1}^{\infty} \frac{n}{n^2+1}.$

4   $\displaystyle\sum_{n=1}^{\infty} \frac{1}{n^2+1}.$

5   $\displaystyle\sum_{n=1}^{\infty} \frac{n}{n^4+1}.$

6   $\displaystyle\sum_{n=1}^{\infty} \frac{1}{3n+1}.$

7   $\displaystyle\sum_{n=1}^{\infty} \frac{\tan^{-1} n}{1+n^2}.$

8   $\displaystyle\sum_{n=2}^{\infty} \frac{\ln n}{n^2}.$

9   (a) The series $\sum_{n=3}^{\infty} (\ln n)/n$ diverges by comparison with the harmonic series, since

$$\frac{1}{n} \le \frac{\ln n}{n}$$

for $n \ge 3$. Establish this divergence by means of the integral test.

(b) If p is a positive constant, show that $\sum_{n=3}^{\infty} (\ln n)/n^p$ converges if $p > 1$ and diverges if $p \le 1$.

10   (a) Use the integral test to show that the series $\sum_{n=1}^{\infty} n/e^n$ converges.

(b) What is the sum of the series in (a)? Hint: Assume that the geometric series $\sum_{n=0}^{\infty} x^n = 1 + x + x^2 + \cdots = 1/(1-x)$ can legitimately be differentiated term by term on the interval $-1 < x < 1$.

11   The curve in Fig. 14.5 is the graph of $y = 1/x$. Convince yourself that the combined area of all the infi-

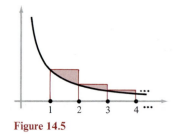

Figure 14.5

nitely many shaded regions is Euler's constant γ. By inspecting the figure, show that the value of γ is between $\frac{1}{2}$ and 1, and is only slightly larger than $\frac{1}{2}$.

12   Use (12) to show that

$$1 + \frac{1}{2} - \frac{2}{3} + \frac{1}{4} + \frac{1}{5} - \frac{2}{6} + + - \cdots = \ln 3.$$

13   If $\{x_n\}$ is the sequence defined by

$$x_n = \frac{1}{n+1} + \frac{1}{n+2} + \cdots + \frac{1}{n+n},$$

then $x_n \to \ln 2$ because

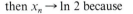

$$x_n = \frac{1}{n} \cdot \frac{1}{1 + 1/n} + \frac{1}{n} \cdot \frac{1}{1 + 2/n} + \cdots + \frac{1}{n} \cdot \frac{1}{1 + n/n}$$

$$\to \int_0^1 \frac{dx}{1 + x} = \ln 2.$$

Establish this fact by using formula (12).

14   The harmonic series

$$1 + \frac{1}{2} + \frac{1}{3} + \cdots + \frac{1}{n} + \cdots$$

diverges very slowly. To grasp how slowly, use (12) to show that in order to get s_n to exceed 10, we must add about 12,000 terms.

14.6
THE RATIO TEST AND ROOT TEST

In the case of the geometric series Σr^n with $r > 0$, the ratio a_{n+1}/a_n of the $(n + 1)$st term to the nth term has the constant value r, since

$$\frac{a_{n+1}}{a_n} = \frac{r^{n+1}}{r^n} = r. \tag{1}$$

We know that this series converges if $r < 1$, essentially because for these r's, condition (1) guarantees that the terms decrease rapidly. Analogy leads us to expect that any series Σa_n of positive terms will also converge if the ratio a_{n+1}/a_n is small for large n, even though this ratio may not have a constant value.

These ideas are made precise in the *ratio test*:

If Σa_n is a series of positive terms such that

$$\lim_{n \to \infty} \frac{a_{n+1}}{a_n} = L, \tag{2}$$

then

(a) *if $L < 1$, the series converges;*
(b) *if $L > 1$, the series diverges;*
(c) *if $L = 1$, the test is inconclusive.*

To establish (a), we assume that $L < 1$ and choose any number r between L and 1, so that $L < r < 1$. Then the meaning of (2) tells us that there exists an n_0 such that $a_{n+1}/a_n \le r$ for all $n \ge n_0$, so

$$\frac{a_{n+1}}{r^{n+1}} \le \frac{a_n}{r^n} \qquad \text{for } n \ge n_0.$$

This says that the sequence $\{a_n/r^n\}$ is decreasing for $n \ge n_0$; in particular, $a_n/r^n \le a_{n_0}/r^{n_0}$ for $n \ge n_0$. Thus, if we put $K = a_{n_0}/r^{n_0}$, then we have

$$a_n \le Kr^n \qquad \text{for } n \ge n_0. \tag{3}$$

But ΣKr^n converges because $r < 1$, and therefore, by the comparison test, (3) implies that Σa_n converges. To prove (b), we simply observe that $L > 1$

implies that $a_{n+1}/a_n \geq 1$, or equivalently, $a_{n+1} \geq a_n$, from some point on, so a_n cannot approach zero, and by the nth term test we know that the series diverges. Part (c) says that if $L = 1$, then no conclusion can be drawn, that is, sometimes the series converges and sometimes it diverges. To demonstrate this, we consider the p-series $\Sigma 1/n^p$. It is clear that for all values of p we have

$$\frac{a_{n+1}}{a_n} = \frac{n^p}{(n+1)^p} = \left(\cdot \frac{n}{n+1}\right)^p \to 1,$$

and yet this series converges if $p > 1$ and diverges if $p \leq 1$.

The ratio test is especially useful for handling series whose nth term a_n is given by a formula that involves various products, since even though a_n itself may be complicated, the ratio a_{n+1}/a_n can often be simplified by cancellations.

Example 1 We know that the series

$$\sum_{n=0}^{\infty} \frac{1}{n!}$$

converges by the argument given in Section 14.3. The ratio test yields the same conclusion much more easily, because

$$L = \lim \frac{a_{n+1}}{a_n} = \lim \frac{1/(n+1)!}{1/n!}$$

$$= \lim \frac{n!}{(n+1)!} = \lim \frac{1}{n+1} = 0.$$

Since $L < 1$, the series converges.

Students should notice our use of the equation $(n+1)! = (n+1)n!$ in this example, because this fact will often be needed in our future work.

Example 2 In the case of the series

$$\sum_{n=0}^{\infty} \frac{3^n}{n!}, \tag{4}$$

it is easy to see that

$$L = \lim \frac{a_{n+1}}{a_n} = \lim \frac{3^{n+1}}{(n+1)!} \cdot \frac{n!}{3^n} = \lim \frac{3}{n+1} = 0.$$

Again we have $L < 1$, so the series converges.

Since the nth term of any convergent series approaches zero, and therefore the convergence of (4) tells us that $3^n/n! \to 0$, we know that $n!$ increases faster than 3^n as $n \to \infty$. Students should try to develop an intuitive feeling for the relative rates of growth of expressions like these as an aid in forming quick but reliable judgments about the probable behavior of series. In this connection we point out here that the numerator 3^n of the nth term of series (4) contributes the 3 to the numerator of the ratio a_{n+1}/a_n after simplification, and that the $n!$ in the denominator contributes the $n+1$ to the denominator of this ratio.

Example 3   For the series

$$\sum_{n=1}^{\infty} \frac{n^{10}}{3^n},$$

we have

$$L = \lim \frac{a_{n+1}}{a_n} = \lim \frac{(n+1)^{10}}{3^{n+1}} \cdot \frac{3^n}{n^{10}}$$

$$= \lim \left(1 + \frac{1}{n}\right)^{10} \cdot \frac{1}{3} = \frac{1}{3}.$$

Again we have $L < 1$, so the series converges by the ratio test.

In this example the convergence of the series tells us that 3^n grows faster than n^{10}, and we see from the calculation of L that the series behaves like the geometric series with $r = \frac{1}{3}$. We also observe that the polynomial factor n^{10} contributes the factor 1 to the calculation of L, and so no such polynomial factor ever has any effect on the outcome of the ratio test.

Example 4   The remark preceding Example 1 is illustrated with special clarity by the series

$$\frac{1}{2} + \frac{1 \cdot 3}{2 \cdot 5} + \frac{1 \cdot 3 \cdot 5}{2 \cdot 5 \cdot 8} + \cdots + \frac{1 \cdot 3 \cdot 5 \cdots (2n-1)}{2 \cdot 5 \cdot 8 \cdots (3n-1)} + \cdots .$$

Here the cancellation of factors yields

$$L = \lim \frac{a_{n+1}}{a_n} = \lim \frac{1 \cdot 3 \cdots (2n-1)(2n+1)}{2 \cdot 5 \cdots (3n-1)(3n+2)} \cdot \frac{2 \cdot 5 \cdots (3n-1)}{1 \cdot 3 \cdots (2n-1)}$$

$$= \lim \frac{2n+1}{3n+2} = \frac{2}{3},$$

and the series converges because $L < 1$.

We now discuss the so-called root test, which is another convenient tool for studying the convergence behavior of series.

Suppose that Σa_n is a series of nonnegative terms with the property that from some point on we have

$$a_n \leq r^n, \qquad \text{where } 0 < r < 1. \tag{5}$$

The geometric series Σr^n clearly converges, so Σa_n also converges by the comparison test. The fact that the inequalities (5) can be written in the form

$$\sqrt[n]{a_n} \leq r < 1 \tag{6}$$

brings us to a convenient statement of the *root test*:

If Σa_n is a series of nonnegative terms such that

$$\lim_{n \to \infty} \sqrt[n]{a_n} = L, \tag{7}$$

then

(a) *if* $L < 1$, *the series converges*;

(b) *if* $L > 1$, *the series diverges*;

(c) *if* $L = 1$, *the test is inconclusive*.

The proof rests on the preceding remarks. For (a), if $L < 1$ and r is any number such that $L < r < 1$, then the meaning of (7) tells us that (6) holds for all sufficiently large n's, and so Σa_n converges. For (b), if $L > 1$, then $\sqrt[n]{a_n} \geq 1$ from some point on, and so $a_n \geq 1$ for all sufficiently large n's, and the series diverges because a_n does not approach zero. Finally, we establish (c) by observing that $L = 1$ for both the divergent series $\Sigma 1/n$ and the convergent series $\Sigma 1/n^2$, since $\sqrt[n]{n} \rightarrow 1$ as $n \rightarrow \infty$.

Example 5   In the case of the series

$$\Sigma \frac{1}{(\ln n)^n},$$

we have

$$L = \lim \sqrt[n]{a_n} = \lim \frac{1}{\ln n} = 0.$$

Since $L < 1$, the series converges.

In general, it is clear that the root test is most likely to be useful for treating series in which a_n is complicated but $\sqrt[n]{a_n}$ is simple, so that $\lim \sqrt[n]{a_n}$ is easy to compute. However, the practical value of the root test is outweighed by its theoretical significance, and this appears mainly in the advanced theory of power series.

Remark 1   The ratio test and the root test were first stated and correctly proved by Cauchy in 1821, as part of the earliest satisfactory exposition of the basic concepts of the theory of series.

Remark 2   We have seen that the ratio test is inconclusive when $\lim a_{n+1}/a_n = 1$, but this is far from the end of the story. If $a_{n+1}/a_n \rightarrow 1$ from above, then we have $a_{n+1}/a_n \geq 1$ or $a_{n+1} \geq a_n$, and Σa_n certainly diverges, because a_n does not approach zero. But if $a_{n+1}/a_n \rightarrow 1$ from below, then there are several more delicate tests that are capable of yielding additional information. The curious reader will find some of these tests discussed in Appendix C.10.

PROBLEMS

Use the ratio test to determine the behavior of the following series.

1 $\displaystyle \Sigma \frac{n}{2^n}$.

2 $\displaystyle \Sigma \frac{n^2}{2^n}$.

3 $\displaystyle \Sigma \frac{n^n}{2^n}$.

4 $\displaystyle \Sigma \frac{n!}{2^n}$.

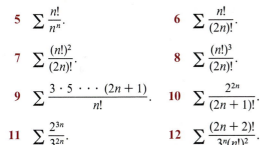

5 $\sum \dfrac{n!}{n^n}$.

6 $\sum \dfrac{n!}{(2n)!}$.

7 $\sum \dfrac{(n!)^2}{(2n)!}$.

8 $\sum \dfrac{(n!)^3}{(2n)!}$.

9 $\sum \dfrac{3 \cdot 5 \, \cdots \, (2n+1)}{n!}$.

10 $\sum \dfrac{2^{2n}}{(2n+1)!}$.

11 $\sum \dfrac{2^{3n}}{3^{2n}}$.

12 $\sum \dfrac{(2n+2)!}{3^n (n!)^2}$.

13 $1 + \dfrac{1 \cdot 3}{2!} + \dfrac{1 \cdot 3 \cdot 5}{3!} + \cdots$

$\qquad + \dfrac{1 \cdot 3 \cdot 5 \, \cdots \, (2n-1)}{n!} + \cdots$.

14 $\sum \dfrac{(n!)^2 3^n}{(2n)!}$.

15 Let Σa_n be a series of positive terms with the following property: There exists a number $r < 1$ and a positive integer n_0 such that $a_{n+1}/a_n \le r$ for all $n \ge n_0$. Show that Σa_n converges even though $\lim a_{n+1}/a_n$ may not exist.

Use the root test to determine the behavior of the following series.

16 $\Sigma(\sqrt[n]{n} - 1)^n$.

17 $\sum \sqrt{n}\left(\dfrac{2n-1}{n+13}\right)^n$.

18 $\sum \dfrac{e^n}{n^n}$.

19 $\sum \left(\dfrac{n+1}{n}\right)^{3n} \cdot \dfrac{1}{3^n}$.

20 $\sum e^{2n}\left(\dfrac{n}{n+1}\right)^{n^2}$.

21 $\sum e^n \left(\dfrac{n}{n+1}\right)^{n^2}$.

22 $\sum \dfrac{n^3}{(\ln 2)^n}$.

23 $\sum \dfrac{n^{10}}{(\ln 3)^n}$.

Most of our attention so far has been directed at series of positive terms. We now wish to consider series with both positive and negative terms. The simplest are those whose terms are alternately positive and negative. These are called *alternating series,* and can be written in the form

$$\sum_{n=1}^{\infty} (-1)^{n+1} a_n = a_1 - a_2 + a_3 - a_4 + \cdots , \tag{1}$$

where the a_n's are all positive numbers. As examples that are already familiar from our previous work, we mention the ln 2 series

$$1 - \frac{1}{2} + \frac{1}{3} - \frac{1}{4} + \cdots = \ln 2, \tag{2}$$

and also the series

$$1 - \frac{1}{3} + \frac{1}{5} - \frac{1}{7} + \cdots = \frac{\pi}{4}, \tag{3}$$

whose sum was discovered by Leibniz in 1673.

It is easy to see that both of the alternating series (2) and (3) have the property that the a_n's form a decreasing sequence that approaches zero:

(i) $a_1 \ge a_2 \ge a_3 \ge \cdots$;

(ii) $a_n \to 0$.

In 1705 Leibniz noticed that these two simple conditions are enough to guarantee that *any* alternating series (1) converges. This fact is called the *alternating series test.*

The essence of the situation lies in the back-and-forth movement of the partial sums of the series (1) under the stated hypotheses, as illustrated in Fig. 14.6. To locate the partial sums $s_1, s_2, s_3, \ldots$, we start at the origin and go to the right a distance a_1 to reach s_1, then go left a smaller distance a_2 to reach s_2, then go right the still smaller distance a_3 to reach s_3, and so on. The behavior of these partial sums is similar to that of a swinging pendulum that

14.7
THE ALTERNATING SERIES TEST. ABSOLUTE CONVERGENCE

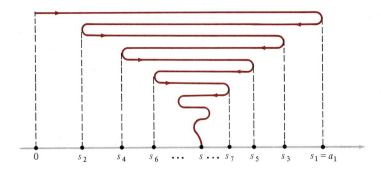

Figure 14.6 The alternating series test.

oscillates back and forth and slowly approaches an equilibrium position which represents the sum s of the series. We suggest that students keep this figure in mind while reading the proof in the following paragraph.

Now for the details of the argument. A typical even partial sum s_{2n} can be written in two ways, as

$$s_{2n} = (a_1 - a_2) + (a_3 - a_4) + \cdots + (a_{2n-1} - a_{2n})$$
$$= a_1 - (a_2 - a_3) - (a_4 - a_5) - \cdots - (a_{2n-2} - a_{2n-1}) - a_{2n},$$

where each expression in parentheses is nonnegative because the a_n's form a decreasing sequence. The first way of writing s_{2n} displays it as the sum of n nonnegative terms, so $s_{2n} \le s_{2n+2}$ and the even partial sums form an increasing sequence, as shown in the figure. The second way of writing s_{2n} shows that $s_{2n} \le a_1$, so the s_{2n}'s have an upper bound. Since every bounded increasing sequence converges, there exists a number s such that

$$\lim_{n \to \infty} s_{2n} = s.$$

But the odd partial sums approach the same limit, because

$$s_{2n+1} = a_1 - a_2 + a_3 - a_4 + \cdots - a_{2n} + a_{2n+1}$$
$$= s_{2n} + a_{2n+1},$$

and therefore

$$\lim_{n \to \infty} s_{2n+1} = \lim_{n \to \infty} s_{2n} + \lim_{n \to \infty} a_{2n+1} = s + 0 = s,$$

since $a_{2n+1} \to 0$ as $n \to \infty$. This tells us that the sequence $\{s_n\}$ of *all* the partial sums converges to the limit s, and therefore the alternating series (1) converges to the sum s under the stated conditions.

Example 1 The alternating series test clearly implies the convergence of series (2) and (3),

$$1 - \frac{1}{2} + \frac{1}{3} - \frac{1}{4} + \cdots = \ln 2 \quad \text{and} \quad 1 - \frac{1}{3} + \frac{1}{5} - \frac{1}{7} + \cdots = \frac{\pi}{4},$$

because $1/n$ and $1/(2n - 1)$ both decrease to zero. However, this test gives us no information at all about the sums of these series, and students will recall that the sums indicated here were established in previous discussions using very different methods of proving convergence.

Example 2 Determine the convergence behavior of the alternating series

(a) $\displaystyle\sum_{n=1}^{\infty} \frac{(-1)^{n+1}n}{1000 + 5n}$;   (b) $\displaystyle\sum_{n=2}^{\infty} (-1)^{n+1} \frac{\ln n}{n}$.

Solution (a) Even though this series is alternating, we nevertheless have $a_n = n/(1000 + 5n) \to \frac{1}{5}$ as $n \to \infty$, so the series diverges by the nth term test.

(b) To prove that this series converges by using the alternating series test, we must show that $a_n = (\ln n)/n$ decreases to zero. We know that $(\ln n)/n \to 0$ by Problem 3(a) in Section 14.2. To demonstrate that the a_n's are decreasing, we note that the function

$$f(x) = \frac{\ln x}{x} \qquad \text{has derivative} \qquad f'(x) = \frac{1 - \ln x}{x^2}.$$

This derivative is negative for $x > e$, so $f(x)$ is a decreasing function for $x > e$, and therefore $a_n \geq a_{n+1}$ for $n \geq 3$. (As usual in considering the matter of convergence, we can disregard the first few terms of the series.) In this case — and in others — we may be convinced that a_n decreases to zero without feeling any need for a detailed verification. However, if there is any doubt at all, we should be prepared to supply such a verification.

ABSOLUTE CONVERGENCE

Why is it that the alternating harmonic series

$$1 - \tfrac{1}{2} + \tfrac{1}{3} - \tfrac{1}{4} + \cdots$$

converges, even though the harmonic series

$$1 + \tfrac{1}{2} + \tfrac{1}{3} + \tfrac{1}{4} + \cdots$$

diverges? The essential reason for the divergence of the harmonic series is that its terms don't decrease quite fast enough, as do the terms of the convergent series $\Sigma 1/2^n$, for example. The partial sums of $\Sigma 1/n$ consist of many small terms that add up to a large total, whereas the terms of $\Sigma 1/2^n$ decrease so fast that no sum of any large number of them can even reach 2.

In contrast to the relatively simple behavior of these series, the alternating harmonic series converges, not only because its terms get small, but also because the well-placed minus signs prevent the partial sums from growing too large and permit them to approach a finite limit.

Some series with terms of mixed signs do not need the assistance of minus signs for convergence, but converge because of the smallness of their terms alone; they would still converge even if all the minus signs were replaced by plus signs. Series of this kind are especially important and are called absolutely convergent; that is, a series Σa_n is said to be *absolutely convergent* if $\Sigma |a_n|$ converges.

These remarks suggest that absolute convergence is a stronger property than ordinary convergence, in the sense that *absolute convergence implies convergence.* This is true and is easy to prove, as follows. Suppose that Σa_n is an absolutely convergent series, so that $\Sigma |a_n|$ converges. The inequalities $0 \leq a_n + |a_n| \leq 2|a_n|$ are clearly valid, because $a_n + |a_n|$ equals 0 or $2|a_n|$

according as $a_n < 0$ or $a_n \geq 0$; and since $\Sigma 2|a_n|$ converges, we know that $\Sigma(a_n + |a_n|)$ also converges by the comparison test. Since $\Sigma(a_n + |a_n|)$ and $\Sigma|a_n|$ both converge, so does their difference, which is Σa_n.

When we try to establish the convergence of a series whose terms have mixed signs, testing for absolute convergence is a good first step, because (as we have just seen) this implies convergence. To do this, we merely change all minus signs to plus signs and test the resulting series of nonnegative terms. We remind students that all our previous tests—the comparison tests, integral test, ratio test, and root test—apply only to series of positive (or nonnegative) terms, and are therefore essentially tests for absolute convergence.

Example 3 Test the following series for absolute convergence, and also for convergence:

(a) $1 - \dfrac{1}{2^2} + \dfrac{1}{3^2} - \dfrac{1}{4^2} + \cdots$;

(b) $1 \pm \dfrac{1}{2^2} \pm \dfrac{1}{3^2} \pm \dfrac{1}{4^2} \pm \cdots$;

(c) $1 - \dfrac{1}{\sqrt{2}} + \dfrac{1}{\sqrt{3}} - \dfrac{1}{\sqrt{4}} + \cdots$.

Solution (a) Here the series $\Sigma|a_n|$ of absolute values is $\Sigma 1/n^2$, which converges. This tells us that series (a) is absolutely convergent, and therefore convergent. This series also converges by the alternating series test.

(b) The intent here is that the plus or minus signs are to be inserted in any manner, either at random or according to some systematic pattern. In either case, it is clear that, just as in part (a), the series is absolutely convergent, and therefore convergent. However, without the concept of absolute convergence we would have no means of determining the convergence behavior of this series.

(c) In this case the series of absolute values is $\Sigma 1/\sqrt{n}$, which is a divergent p-series. Series (c) is therefore not absolutely convergent. Nevertheless, this series is clearly convergent by the alternating series test.

Remark 1 A convergent series that is not absolutely convergent is said to be *conditionally convergent.* As examples we mention series (2) and (3), and also the series in Example 3(c). All series of this kind are capable of startling but fascinating misbehavior, and should be labeled "handle with care." For instance, any such series can be made to converge to any given number as its sum, or even to diverge, by suitably changing the order of its terms. On the other hand, any absolutely convergent series can be rearranged in any manner without changing either its convergence behavior or its sum. In an earlier section we said, "The effective use of infinite series rests on our freedom to manipulate them by the various processes of algebra." Generally speaking, this freedom is available only when we are working with absolutely convergent series. These issues and others that are not normally part of a first course are discussed in detail in Appendix C.11.

Remark 2 The only test we have that establishes convergence rather than absolute convergence is the alternating series test. Several other tests of this kind are presented in Appendix C.12.

Remark 3 Students may have noticed that almost all of our work in the preceding sections of this chapter has been devoted to showing whether a series converges or diverges, and very little to finding the sum of a convergent series. A good reason for this is that the second problem is usually much harder than the first. Thus, the convergence of the series of the reciprocals of the squares is quite easy to establish, but the exact value of its sum, namely, the fact that

$$\sum_{n=1}^{\infty} \frac{1}{n^2} = 1 + \frac{1}{2^2} + \frac{1}{3^2} + \cdots = \frac{\pi^2}{6},$$

is far from obvious and can only be discovered by great ingenuity. Further, the convergence of the series of the reciprocals of the cubes is equally easy to establish, but its sum has *never* been discovered and remains to this day one of the more tantalizing unsolved problems of mathematics:

$$\sum_{n=1}^{\infty} \frac{1}{n^3} = 1 + \frac{1}{2^3} + \frac{1}{3^3} + \cdots = ?$$

A second, and perhaps even more important reason for not pressing harder on the problem of finding the sum of a convergent series is that this problem will be constantly before us in the applications we discuss in the rest of this chapter. That is, we shall be "representing" a given function by a certain kind of series, and we will always give careful attention to the matter of proving that this series actually converges to the given function as its sum. In this way, the sums of many convergent series of constants will be known to us as a minor consequence of our work on convergent series of functions.

PROBLEMS

Classify each of the following series as absolutely convergent, conditionally convergent, or divergent.

1 $\sum (-1)^{n+1} \dfrac{1}{\sqrt{n+10}}.$

2 $\sum (-1)^{n+1} \dfrac{1}{n\sqrt{n}}.$

3 $\sum (-1)^{n+1} \dfrac{n}{n+2}.$

4 $\sum (-1)^{n+1} \dfrac{\sqrt[3]{n}}{\sqrt{n}}.$

5 $\sum (-1)^{n+1} \dfrac{2^n}{n!}.$

6 $\sum (-1)^{n+1} \dfrac{1}{3^{1/n}}.$

7 $\sum (-1)^{n+1} \dfrac{n^3}{1+n^5}.$

8 $\sum (-1)^{n+1} \dfrac{1}{\ln (n+2)}.$

9 $\sum (-1)^{n+1} \dfrac{\sqrt{n}}{\ln n}.$

10 $\sum (-1)^{n+1} \ln \dfrac{1}{n}.$

11 $\sum (-1)^{n+1} \dfrac{\ln n}{\sqrt{n}}.$

12 $\sum (-1)^{2n+1} \dfrac{1}{\sqrt{n}}.$

13 $\sum (-1)^{n+1} \dfrac{\sqrt{n}}{n+3}.$

14 $\sum (-1)^{n+1} \dfrac{1}{5n}.$

15 $\sum (-1)^{n+1} \dfrac{2}{3n^2}.$

16 $\sum (-1)^{n+1} \sin n\pi.$

17 $\sum (-1)^{n+1} \dfrac{\sin^2 n}{n^2}.$

18 $\sum (-1)^{n+1} \dfrac{\sin^2 n}{n^{5/2}}.$

19 $\sum (-1)^{n+1} \dfrac{\sin^2 n}{n^{9/2}}.$

20 $\sum (-1)^{n+1} \ln \sqrt[n]{n}.$

21 $\sum (-1)^{n+1} \dfrac{1}{\sqrt[5]{n^4}}.$

22 $\sum (-1)^{n+1} \dfrac{n^4 3^n}{n!}.$

23 $1 - \dfrac{1}{2} + \dfrac{1}{3!} - \dfrac{1}{4} + \dfrac{1}{5!} - \dfrac{1}{6} + \cdots .$

24 $1 - \dfrac{1}{2} + \dfrac{1}{2} - \dfrac{1}{3} + \dfrac{1}{3} - \dfrac{1}{4} + \dfrac{1}{4} - \cdots .$

25 $1 + \dfrac{1}{2} - \dfrac{1}{3} - \dfrac{1}{4} + \dfrac{1}{5} + \dfrac{1}{6} - \dfrac{1}{7} - \dfrac{1}{8} + \cdots .$

26 We know that if $\{a_n\}$ is a decreasing sequence of positive numbers such that $a_n \to 0$, then the alternating series (1) converges to some number s as its sum. Show that s lies between consecutive partial sums s_n and s_{n+1} and that $|s_n - s| \le a_{n+1}$.

27 State whether each of the following is true or false:
(a) every convergent alternating series is conditionally convergent;
(b) every absolutely convergent series is convergent;
(c) every convergent series is absolutely convergent;
(d) every alternating series converges;
(e) if Σa_n is conditionally convergent, then $\Sigma |a_n|$ diverges;
(f) if $\Sigma |a_n|$ diverges, then Σa_n is conditionally convergent.

28 If the a_n's are all positive numbers, show that the series $-a_1 + a_2 - a_3 + a_4 - \cdots$ converges if and only if the series $a_1 - a_2 + a_3 - a_4 + \cdots$ converges. [This shows that starting an alternating series with a positive term, as in (1), is merely a convenience, not a necessity.]

29 If Σa_n and Σb_n are absolutely convergent, show that
(a) $\Sigma(a_n + b_n)$ is absolutely convergent;
(b) $\Sigma c a_n$ is absolutely convergent for any constant c.

30 If $\Sigma a_n{}^2$ and $\Sigma b_n{}^2$ converge, show that $\Sigma a_n b_n$ is absolutely convergent. Hint: $(a - b)^2 \ge 0$, so $2ab \le a^2 + b^2$.

31 Use Problem 30 to show that if $\Sigma a_n{}^2$ converges, then $\Sigma a_n/n$ is absolutely convergent.

32 In using the alternating series test, it is a common error to check only that $a_n \to 0$; but this is not enough, and convergence cannot be deduced unless both of conditions (i) and (ii) are verified. Demonstrate this fact in the case of the alternating series

$$\tfrac{2}{1} - \tfrac{1}{1} + \tfrac{2}{2} - \tfrac{1}{2} + \tfrac{2}{3} - \tfrac{1}{3} + \cdots$$

by showing that
(a) $a_n \to 0$;
(b) the a_n's do not form a decreasing sequence;
(c) the series diverges (hint: consider s_{2n}).

33 If s is any given number, show that the alternating harmonic series $1 - \tfrac{1}{2} + \tfrac{1}{3} - \tfrac{1}{4} + \cdots$ can be rearranged (that is, its terms can be written in a different order) in such a way that the resulting series converges to s. Hint: Take just enough positive terms to get above s, then just enough negative terms to get below s, etc.

14.8
POWER SERIES REVISITED. INTERVAL OF CONVERGENCE

In the preceding sections of this chapter we concentrated our attention on series whose terms are constants. In the next five sections we turn to the study of power series, whose terms are very simple functions of a variable x. We became somewhat acquainted with power series in Chapter 13, mostly with those that arise in a fairly direct way from the geometric series $1 + x + x^2 + \cdots$. However, we now approach this subject from a different point of view, and we will rarely have occasion to refer to this earlier work.

As we recall, a *power series* is a series of the form

$$\sum_{n=0}^{\infty} a_n x^n = a_0 + a_1 x + a_2 x^2 + \cdots + a_n x^n + \cdots , \tag{1}$$

where the coefficients a_n are constants and x is a variable. Since power series are almost always indexed from $n = 0$ to $n = \infty$, we often simplify the notation by writing (1) as $\Sigma a_n x^n$.

Power series serve many important purposes in both the theory and applications of mathematics. First, they are particularly well suited to computation, and are therefore useful in many numerical problems. Second, they provide an alternative way of expressing many familiar functions, and often add to our understanding of these functions. And finally, in the study of differential equations, we often find that power series are used to define functions that are difficult or impossible to define in any other way. To be more explicit about this, a differential equation can often be used to generate a solution of itself in the form of a power series, as illustrated in Section 13.1;

it is then not only convenient but perfectly reasonable to define a function $f(x)$ by saying that it is the sum of this power series,

$$f(x) = a_0 + a_1 x + a_2 x^2 + \cdots + a_n x^n + \cdots ,$$

provided, of course, that the series converges.

The geometric series

$$\sum x^n = 1 + x + x^2 + \cdots \tag{2}$$

is evidently the simplest power series. We know that this series converges for $|x| < 1$ and diverges for $|x| \geq 1$. In general, we expect a power series to converge for some values of x and to diverge for others.

We will clearly be very interested in knowing the x's for which a given power series $\sum a_n x^n$ converges. For any such x, the sum of the series is a number whose value depends on x and is therefore a function of x. If we denote this function by $f(x)$, then $f(x)$ can be thought of as defined by the equation

$$f(x) = \sum a_n x^n. \tag{3}$$

Sometimes a power series has a known function as its sum. For example, if $|x| < 1$, we know that the series (2) has $1/(1 - x)$ as its sum. However, in general there is no reason to expect that the sum of a convergent power series will turn out to be a function that we can recognize from our previous experience.

We will be concerned with two major groups of questions. First, what properties does the function $f(x)$ defined by (3) have? Is it continuous? Is it differentiable? If it *is* differentiable, can its derivative be calculated by differentiating (3) term by term? And second—turning the whole situation around—if a function $f(x)$ is given beforehand, under what circumstances does it have a power series expansion of the form (3)? How can this expansion be calculated, and what can be said about the x's for which the expansion is valid? These are some of the issues we study in the next few sections.

We begin by determining the nature of the set of points at which an arbitrary power series converges.

First, a few examples. It is clear that every power series converges for $x = 0$. Some series converge *only* for this value of x, for instance,

$$\sum_{n=1}^{\infty} n^n x^n = x + 2^2 x^2 + 3^3 x^3 + 4^4 x^4 + \cdots .^*\tag{4}$$

To see this, it suffices to observe that for any $x \neq 0$ we have $|nx| > 1$ if n is large enough, so that the nth term $(nx)^n$ does not approach zero and the series cannot converge. At the opposite extreme are series like

$$\sum \frac{x^n}{n!} = 1 + x + \frac{x^2}{2!} + \frac{x^3}{3!} + \cdots , \tag{5}$$

which converges for all values of x. We establish this by showing that (5) is

---

* This series is indexed from $n = 1$ to $n = \infty$ because n^n has no meaning when $n = 0$.

absolutely convergent for every x, and this is easy to prove by the ratio test:

$$\frac{|x^{n+1}/(n+1)!|}{|x^n/n!|} = \frac{|x|^{n+1}}{(n+1)!} \cdot \frac{n!}{|x|^n} = \frac{|x|}{n+1} \to 0.$$

(When using the ratio test, it is necessary to test for absolute convergence, because this test applies only to series of positive terms.) As a simple example that lies between these extremes, we have the geometric series (2), which converges on the interval $|x| < 1$ and diverges everywhere else.

Our task is to discover the structure of the set of all x's for which a given power series converges. The examples discussed above show that there are at least three possibilities: This set may consist of the single point $x = 0$, or the entire real line, or a finite interval centered at the origin. We will prove that these are the *only* possibilities.

In order to establish this, we need the following lemma:

If a power series $\Sigma a_n x^n$ converges at x_1, $x_1 \neq 0$, then it converges absolutely at all x with $|x| < |x_1|$; and if it diverges at x_1, then it diverges at all x with $|x| > |x_1|$.

The proof is quite easy. If $\Sigma a_n x_1^n$ converges, then $a_n x_1^n \to 0$. In particular, if n is sufficiently large, we have $|a_n x_1^n| < 1$, and therefore

$$|a_n x^n| = |a_n x_1^n| \left|\frac{x}{x_1}\right|^n < r^n, \tag{6}$$

where $r = |x/x_1|$. Now suppose that $|x| < |x_1|$. Then we have

$$r = \left|\frac{x}{x_1}\right| < 1,$$

and the inequality (6) tells us that $\Sigma |a_n x^n|$ converges by comparison with the convergent geometric series Σr^n. This proves the first statement. To prove the second statement, we assume that $\Sigma a_n x_1^n$ diverges. Then $\Sigma a_n x^n$ cannot converge for any x with $|x| > |x_1|$, for if it does, then what has just been proved implies the absolute convergence—and therefore the convergence—of $\Sigma a_n x_1^n$, and this contradicts our assumption.

We are now in a position to state and prove the main facts about the convergence behavior of an arbitrary power series.

Given a power series $\Sigma a_n x^n$, precisely one of the following is true:

(i)   *The series converges only for $x = 0$.*
(ii) *The series is absolutely convergent for all x.*
(iii) *There exists a positive real number R such that the series is absolutely convergent for $|x| < R$ and divergent for $|x| > R$.*

The argument goes this way. If (i) is not true, then the series converges for some $x_1 \neq 0$, and by the lemma we know that the positive number $r = |x_1|$ has the property that $\Sigma a_n x^n$ is absolutely convergent on the interval $-r < x < r$. Let S be the set of all positive numbers r with this property. If S is unbounded, then (ii) is true. Now suppose that neither (i) nor (ii) is true, so that S is a nonempty set of positive numbers which has an upper bound. By

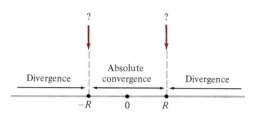

Figure 14.7 The interval of convergence.

the basic completeness property of the real number system, S has a *least* upper bound. This means that there is a smallest number R such that $r \leq R$ for all r's in S.* It is now easy to see that R has the properties stated in (iii).

The positive real number R in case (iii) is called the *radius of convergence* of the power series. We have seen that the series converges absolutely at every point of the open interval $(-R, R)$, and diverges outside the closed interval $[-R, R]$. No general statement can be made about the behavior of the series at the endpoints R and $-R$. There are examples of series that converge at both endpoints, or diverge at both, or converge at one and diverge at the other; to find out what happens for any particular series, we must test each endpoint separately. The set of all x's for which a power series converges is called its *interval of convergence*. These ideas are illustrated in Fig. 14.7.

It is customary to put $R = 0$ when the series converges only for $x = 0$, and to put $R = \infty$ when it converges for all x. This convention allows us to cover all possibilities (except endpoint behavior) in a single statement:

Every power series $\Sigma a_n x^n$ has a radius of convergence R, where $0 \leq R \leq \infty$, with the property that the series converges absolutely if $|x| < R$ and diverges if $|x| > R$.

It should be noticed that if $R = 0$, then no x satisfies the condition $|x| < R$, and if $R = \infty$, then no x satisfies the condition $|x| > R$, so in both of these cases our general statement is true by default.

If a power series $\Sigma a_n x^n$ is given, how do we find its interval of convergence?

The first step is to find the radius of convergence R. There is a simple formula for R that works in many situations:

$$R = \lim \left| \frac{a_n}{a_{n+1}} \right|, \tag{7}$$

provided this limit exists—and has ∞ as an allowed value. This follows directly from the ratio test, because the series converges absolutely, or diverges, according as the number

$$\lim \frac{|a_{n+1} x^{n+1}|}{|a_n x^n|} = \lim \left| \frac{a_{n+1}}{a_n} \right| \cdot |x| = \frac{|x|}{\lim |a_n/a_{n+1}|}$$

is < 1 or > 1, that is, according as

$$|x| < \lim \left| \frac{a_n}{a_{n+1}} \right| \qquad \text{or} \qquad |x| > \lim \left| \frac{a_n}{a_{n+1}} \right|.$$

This establishes (7).

---

* See Appendix C.1 for a general discussion of least upper bounds.

The second step is to test the behavior of the series at the endpoints.

Example 1 Find the interval of convergence of the series

$$\sum \frac{x^n}{n^2} = x + \frac{x^2}{2^2} + \frac{x^3}{3^2} + \cdots.$$

As indicated, this series is understood to be indexed from $n = 1$ to $n = \infty$, because $a_n = 1/n^2$ has no meaning for $n = 0$.

Solution Here we have

$$\left| \frac{a_n}{a_{n+1}} \right| = \frac{1/n^2}{1/(n+1)^2} = \frac{(n+1)^2}{n^2} \to 1,$$

so $R = 1$. In this example the series converges at both endpoints: At $x = 1$ the series is $\Sigma 1/n^2$, which is a convergent p-series, and at $x = -1$ it is $\Sigma(-1)^n/n^2$, which converges by the alternating series test. The interval of convergence is therefore the entire closed interval $[-1, 1]$.

Example 2 Find the interval of convergence of the series

$$\sum \frac{n+2}{3^n} x^n = 2 + \frac{3}{3}x + \frac{4}{3^2}x^2 + \cdots.$$

Solution This time we have

$$\left| \frac{a_n}{a_{n+1}} \right| = \frac{n+2}{3^n} \cdot \frac{3^{n+1}}{n+3} = \frac{n+2}{n+3} \cdot 3 \to 3,$$

so $R = 3$. In this case the series diverges at both endpoints: At $x = 3$ it becomes $2 + 3 + 4 + \cdots$, and at $x = -3$ it becomes $2 - 3 + 4 - \cdots$. The interval of convergence is therefore $(-3, 3)$.

Example 3 Find the interval of convergence of the series

$$\sum \frac{x^n}{n+1} = 1 + \frac{x}{2} + \frac{x^2}{3} + \cdots.$$

Solution Here we have

$$\left| \frac{a_n}{a_{n+1}} \right| = \frac{1/(n+1)}{1/(n+2)} = \frac{n+2}{n+1} \to 1,$$

so $R = 1$. At $x = 1$ the series is the harmonic series $1 + \frac{1}{2} + \frac{1}{3} + \cdots$, which diverges, and at $x = -1$ it is the alternating harmonic series $1 - \frac{1}{2} + \frac{1}{3} - \cdots$, which converges. The interval of convergence is therefore $[-1, 1)$.

Example 4 Find the interval of convergence of the series

$$\sum (-1)^n \frac{x^{2n}}{(2n)!} = 1 - \frac{x^2}{2!} + \frac{x^4}{4!} - \cdots . \tag{8}$$

Solution Formula (7) does not apply directly because half of the coefficients of this power series are zero. Nevertheless, the series can be written in the form

$$1 - \frac{y}{2!} + \frac{y^2}{4!} - \cdots ,$$

where $y = x^2$. Our formula can be applied to this series and yields

$$\left| \frac{a_n}{a_{n+1}} \right| = \frac{1/(2n)!}{1/(2n+2)!} = \frac{(2n+2)!}{(2n)!} = (2n+1)(2n+2) \to \infty,$$

so $R = \infty$ for the y-series. It follows that the x-series also converges for every x, so the desired interval of convergence for (8) is $(-\infty, \infty)$.

If a is a real number, the series

$$\sum_{n=0}^{\infty} a_n(x-a)^n = a_0 + a_1(x-a) + a_2(x-a)^2 + \cdots \tag{9}$$

is called a *power series in $x - a$*. For the sake of emphasis, the special case (1), in which $a = 0$, is often called a *power series in x*. If we put $z = x - a$, then (9) becomes $\sum a_n z^n$, which is a power series in z. If we can determine the interval of convergence of this latter series by the methods previously described, then this information can be used to find the interval of convergence of (9). For instance, if $\sum a_n z^n$ has $[-R, R)$ is its interval of convergence, then it converges for $-R \le z < R$. This means that $-R \le x - a < R$ or $a - R \le x < a + R$. We then say that $[a - R, a + R)$ is the *interval of convergence,* and that R is the *radius of convergence,* of the power series (9).

For the sake of simplicity of notation, we shall confine most of our detailed discussions to power series in x.

PROBLEMS

Find the interval of convergence of each of the following power series.

1. $\sum \frac{n}{4^n} x^n.$

2. $\sum n! x^n.$

3. $\sum \frac{n!}{100^n} x^n.$

4. $\sum \frac{2^n}{n^2} x^n.$

5. $\sum \frac{x^n}{n+4}.$

6. $\sum (-1)^{n+1} \frac{x^n}{\sqrt{n}}.$

7. $\sum \frac{x^n}{n^3+1}.$

8. $\sum \frac{2^n}{(2n)!} x^{2n}.$

9. $\sum \frac{(3n)!}{(2n)!} x^n.$

10. $\sum \frac{x^n}{n(n+1)}.$

11. $\sum \frac{x^{2n+1}}{(-3)^n}.$

12. $\sum \frac{x^n}{n2^n}.$

13. $\sum \frac{(-1)^n}{2n+1} x^{2n+1}.$

14. $\sum n^2 x^n.$

15. $\sum \frac{(-2)^n}{n} x^n.$

16. $\sum \frac{(-1)^{n+1}}{\sqrt[3]{n}\, 2^n} x^n.$

17. $\sum \frac{x^n}{\ln n}.$

18. $\sum \frac{\ln n}{n} x^n.$

19 $\displaystyle\sum \frac{(-1)^n}{n(\ln n)^2} x^n$.

20 $\displaystyle\sum \frac{3^n}{n4^n} x^n$.

21 $\displaystyle\sum \frac{n^2}{2^n} (x - 4)^n$.

22 $\displaystyle\sum \frac{3^n}{n^2 + 1} (x - 1)^n$.

23 $\displaystyle\sum \frac{10^n}{(2n)!} (x - 7)^n$.

24 $\displaystyle\sum \frac{(x - 3)^n}{n^2 2^n}$.

25 $\displaystyle\sum \frac{(2n)!}{n!} (x - 10)^n$.

26 $\displaystyle\sum \frac{(-1)^{n+1}}{n \ln n} (x - 3)^n$.

27 $\displaystyle\sum \frac{\ln n}{e^n} (x - e)^n$.

28 $\displaystyle\sum \frac{n^2}{2^{3n}} (x + 2)^n$.

29 Find the radius of convergence of
(a) the hypergeometric series

$$1 + \frac{ab}{c} x + \frac{a(a + 1)b(b + 1)}{2!c(c + 1)} x^2$$

$$+ \frac{a(a + 1)(a + 2)b(b + 1)(b + 2)}{3!c(c + 1)(c + 2)} x^3 + \cdots \, ;$$

(b) the Bessel function

$$J_0(x) = 1 - \frac{x^2}{2^2} + \frac{x^4}{2^2 \cdot 4^2} - \frac{x^6}{2^2 \cdot 4^2 \cdot 6^2} + \cdots .$$

30 Give an example of a power series with $R = \pi$.

31 If infinitely many coefficients of a power series are nonzero integers, show that $R \le 1$.

14.9

DIFFERENTIATION AND INTEGRATION OF POWER SERIES

Consider a power series $\Sigma a_n x^n$ with a positive radius of convergence R. We saw in Section 14.8 that this series can be used to define a function $f(x)$ whose domain of definition is the interval of convergence of the series. Specifically, for each x in this interval we define $f(x)$ to be the sum of the series,

$$f(x) = a_0 + a_1 x + a_2 x^2 + \cdots + a_n x^n + \cdots . \qquad (1)$$

This relation between the series and the function is often expressed by saying that $\Sigma a_n x^n$ is a *power series expansion* of $f(x)$. For example, we know that if $|x| < 1$, then

$$\frac{1}{1 + x} = 1 - x + x^2 - x^3 + \cdots + (-1)^n x^n + \cdots , \qquad (2)$$

because the geometric series $\Sigma(-1)^n x^n$ converges and has the sum $1/(1 + x)$. Accordingly, the function $f(x) = 1/(1 + x)$ has the series on the right of (2) as a power series expansion.

Polynomials, which are *finite* sums of terms of the form $a_n x^n$, are very simple functions. They are continuous everywhere, and can be differentiated and integrated term by term. As we saw in Chapter 13, the sum of a power series can be a much more complicated function, but it is still simple enough to share these three properties with polynomials inside the interval of convergence.

We give the following formal statement of these very important facts:

(i) *The function $f(x)$ defined by* (1) *is continuous on the open interval* $(-R, R)$.

(ii) *The function $f(x)$ is differentiable on* $(-R, R)$, *and its derivative is given by the formula*

$$f'(x) = a_1 + 2a_2 x + 3a_3 x^2 + \cdots + na_n x^{n-1} + \cdots . \qquad (3)$$

(iii) *If x is any point in* $(-R, R)$, *then*

$$\int_0^x f(t)\, dt = a_0 x + \frac{1}{2} a_1 x^2 + \frac{1}{3} a_2 x^3 + \cdots + \frac{1}{n + 1} a_n x^{n+1} + \cdots . \qquad (4)$$

The proofs of these statements depend on a special kind of convergence called *uniform convergence*. The details can be found in Appendix C.13.

Several comments are in order.

First, we observe that if (ii) is applied to the function $f'(x)$ in (3), then it follows that $f'(x)$ is itself differentiable. This in turn implies that $f''(x)$ is differentiable, and so on. Thus, the original function $f(x)$ has derivatives of all orders. We can summarize the situation this way: *In the interior of its interval of convergence, a power series defines an infinitely differentiable function whose derivatives can be calculated by differentiating the series term by term.* The term-by-term differentiation can be emphasized by writing (3) as

$$\frac{d}{dx}\left(\sum a_n x^n\right) = \sum \frac{d}{dx}(a_n x^n).$$

It is worth knowing that the term-by-term differentiability of a convergent series of functions is usually false; it is true here only because we are dealing with a special kind of series. As a simple example of the failure of this property, we mention the series $\sum_{n=1}^{\infty} (\sin nx)/n^2$, which is absolutely convergent for all x by comparison with the convergent series $\sum 1/n^2$, because $|(\sin nx)/n^2| \le 1/n^2$. The difficulty arises with the term-by-term differentiated series $\sum(\cos nx)/n$, because this series diverges for $x = 0$.

In the case of (iii), if we prefer to avoid using the dummy variable t, then (4) can be written in the form

$$\int f(x)\, dx = a_0 x + \frac{1}{2} a_1 x^2 + \frac{1}{3} a_2 x^3 + \cdots + \frac{1}{n+1} a_n x^{n+1} + \cdots , \quad (5)$$

provided we find an indefinite integral on the left that equals zero when $x = 0$. The term-by-term integration of power series can be emphasized by writing (5) as

$$\int \left(\sum a_n x^n\right) dx = \sum \left(\int a_n x^n\, dx\right).$$

We also point out that it is part of the meaning of (3) and (4) that the differentiated and integrated series on the right sides of these equations converge on the interval $(-R, R)$. We shall prove this at the end of this section. However, before doing so we give several examples of the practical value of the procedures discussed here.

Example 1 Find a power series expansion of $\ln (1 + x)$.

Solution Our starting point is the fact that the derivative of $\ln (1 + x)$ is $1/(1 + x)$, and for $|x| < 1$ this function has the power series expansion given by (2). Now, using (5) and the fact that $\ln (1 + x)$ equals zero when $x = 0$, and integrating (2) term by term, we at once obtain

$$\ln (1 + x) = x - \frac{x^2}{2} + \frac{x^3}{3} - \cdots + (-1)^n \frac{x^{n+1}}{n+1} + \cdots$$

$$= \sum_{n=1}^{\infty} (-1)^{n+1} \frac{x^n}{n}.$$

We know from the preceding discussion that this expansion is valid for $|x| < 1$. As we saw in Section 13.3, it is also valid for $x = 1$, but our present methods give no information on this matter.

Example 2   Find a power series expansion of $\tan^{-1} x$.

Solution   The derivative of this function is $1/(1 + x^2)$, and we see by replacing x by x^2 in (2) that

$$\frac{1}{1 + x^2} = 1 - x^2 + x^4 - \cdots + (-1)^n x^{2n} + \cdots$$

if $|x| < 1$. Using (4) this time, we get

$$\tan^{-1} x = \int_0^x \frac{dt}{1 + t^2} = \int_0^x (1 - t^2 + t^4 - t^6 + \cdots)\, dt$$

$$= x - \frac{x^3}{3} + \frac{x^5}{5} - \frac{x^7}{7} + \cdots = \sum_{n=0}^{\infty} (-1)^n \frac{x^{2n+1}}{2n + 1}$$

for $|x| < 1$. Again, our present methods give no information about what happens at the endpoints $x = \pm 1$.

Example 3   Find a power series expansion of e^x.

Solution   In Section 8.3 we proved that e^x is the only function that equals its own derivative everywhere and has the value 1 at $x = 0$. To construct a power series equal to its own derivative, we use the fact that when such a series is differentiated, the degree of each term drops by 1. We therefore want each term to be the derivative of the one that follows it. Starting with 1 as the constant term, the next should be x, then $\frac{1}{2} x^2$, then $\frac{1}{2 \cdot 3} x^3$, and so on. This produces the series

$$1 + x + \frac{x^2}{2!} + \frac{x^3}{3!} + \cdots + \frac{x^n}{n!} + \cdots, \tag{6}$$

which converges for all x because

$$R = \lim \frac{1/n!}{1/(n + 1)!} = \lim (n + 1) = \infty.$$

We have constructed the series (6) so that its sum is unchanged by differentiation and has the value 1 at $x = 0$. In view of the above remark, this establishes the validity of the expansion

$$e^x = 1 + x + \frac{x^2}{2!} + \frac{x^3}{3!} + \cdots + \frac{x^n}{n!} + \cdots$$

for all x.

Example 4   Find power series expansions of $1/(1 - x)^2$ and $1/(1 - x)^3$.

Solution We begin by noticing that

$$\frac{1}{(1-x)^2} = \frac{d}{dx}\left(\frac{1}{1-x}\right).$$

The next steps are to expand $1/(1-x)$ into the power series Σx^n for $|x| < 1$, and then to differentiate this series term by term:

$$\frac{1}{(1-x)^2} = \frac{d}{dx}(1 + x + x^2 + \cdots + x^n + \cdots)$$

$$= 1 + 2x + 3x^2 + \cdots + nx^{n-1} + \cdots$$

$$= \sum_{n=1}^{\infty} nx^{n-1} = \sum_{n=0}^{\infty} (n+1)x^n.$$

Another differentiation yields

$$\frac{2}{(1-x)^3} = \frac{d}{dx}\left[\frac{1}{(1-x)^2}\right]$$

$$= \frac{d}{dx}(1 + 2x + 3x^2 + \cdots + nx^{n-1} + \cdots)$$

$$= 2 + 3 \cdot 2x + 4 \cdot 3x^2 + \cdots + n(n-1)x^{n-2} + \cdots,$$

so

$$\frac{1}{(1-x)^3} = \frac{1}{2}[2 + 3 \cdot 2x + 4 \cdot 3x^2 + \cdots + n(n-1)x^{n-2} + \cdots]$$

$$= \sum_{n=2}^{\infty} \frac{n(n-1)}{2}x^{n-2} = \sum_{n=0}^{\infty} \frac{(n+2)(n+1)}{2}x^n.$$

Example 5 Find the sum of the series

$$x + \frac{x^2}{2} + \frac{x^3}{3} + \cdots = \sum_{n=1}^{\infty} \frac{x^n}{n}.$$

Solution It is easy to see that $R = 1$, so the series converges to some function $f(x)$ for $|x| < 1$. Differentiating this series clearly simplifies it to

$$f'(x) = 1 + x + x^2 + \cdots + x^n + \cdots = \frac{1}{1-x}.$$

Since $f(0) = 0$, integration now yields

$$f(x) = -\ln(1-x).$$

Example 6 Find the sum of the series

$$x + 4x^2 + 9x^3 + \cdots = \sum_{n=1}^{\infty} n^2 x^n.$$

Solution Again we have $R = 1$, so the series converges to some function $f(x)$ for $|x| < 1$. We can write

$$f(x) = x + 4x^2 + 9x^3 + \cdots + n^2x^n + \cdots = xg(x),$$

where

$$g(x) = 1 + 2^2x + 3^2x^2 + \cdots + n^2x^{n-1} + \cdots .$$

At this point we notice that

$$g(x) = \frac{d}{dx}(x + 2x^2 + 3x^3 + \cdots + nx^n + \cdots)$$

$$= \frac{d}{dx}[x(1 + 2x + 3x^2 + \cdots + nx^{n-1} + \cdots)].$$

By Example 4,

$$1 + 2x + 3x^2 + \cdots + nx^{n-1} + \cdots = \frac{1}{(1-x)^2},$$

so

$$g(x) = \frac{d}{dx}\left[\frac{x}{(1-x)^2}\right] = \frac{1+x}{(1-x)^3}$$

and

$$f(x) = \frac{x+x^2}{(1-x)^3}.$$

In conclusion, we return to the unfinished business of showing that series (3) and (4) converge on the interval $(-R, R)$.

The proof for (4) is easy: Since $\Sigma|a_nx^n|$ converges and

$$\left|\frac{a_nx^n}{n+1}\right| \le |a_nx^n|,$$

the comparison test implies that $\displaystyle\sum\left|\frac{a_nx^n}{n+1}\right|$ converges, and therefore

$$x\sum\frac{a_nx^n}{n+1} = \sum\frac{1}{n+1}a_nx^{n+1}$$

also converges.

The proof for (3) is a bit more complicated. Let x be a point in the interval $(-R, R)$ and choose $\epsilon > 0$ so that $|x| + \epsilon < R$. Since $|x| + \epsilon$ is in the interval, $\Sigma|a_n(|x| + \epsilon)^n|$ converges. In Problem 7 students are asked to show that the inequality

$$|nx^{n-1}| \le (|x| + \epsilon)^n$$

is true for all sufficiently large n's. This implies that

$$|na_nx^{n-1}| \le |a_n(|x| + \epsilon)^n|,$$

so the series $\Sigma|na_nx^{n-1}|$ converges by the comparison test.

PROBLEMS

1 Find power series expansions for the following functions, and determine the values of x for which these expansions are valid:

(a) $\dfrac{1}{(1+x)^2}$; (b) $\dfrac{1}{(1+x)^3}$.

2 Show that

$$\sum_{n=0}^{\infty} \frac{(n+1)(n+2)(n+3)}{6} x^n = \frac{1}{(1-x)^4}.$$

3 Find the sum of each of the following series:

(a) $x + \dfrac{x^3}{3} + \dfrac{x^5}{5} + \cdots + \dfrac{x^{2n+1}}{2n+1} + \cdots$;

(b) $1 + \dfrac{x}{2!} + \dfrac{x^2}{3!} + \cdots + \dfrac{x^{n-1}}{n!} + \cdots$;

(c) $x + 2x^2 + 3x^3 + \cdots + nx^n + \cdots$;

(d) $x + 2x^3 + 3x^5 + \cdots + nx^{2n-1} + \cdots$.

4 Show that

$$\sum_{n=1}^{\infty} \frac{x^n}{n^2} = -\int_0^x \frac{\ln(1-t)}{t}\, dt.$$

5 Show that the Bessel function

$$J_0(x) = 1 - \frac{x^2}{2^2} + \frac{x^4}{2^2 \cdot 4^2} - \frac{x^6}{2^2 \cdot 4^2 \cdot 6^2} + \cdots$$

satisfies the differential equation $xy'' + y' + xy = 0$.

6 Obtain the series

$$\ln(x + \sqrt{1+x^2}) = x - \frac{1}{2}\frac{x^3}{3} + \frac{1\cdot 3}{2\cdot 4}\frac{x^5}{5}$$
$$- \frac{1\cdot 3\cdot 5}{2\cdot 4\cdot 6}\frac{x^7}{7} + \cdots$$

by integrating another series.

7 If $\epsilon > 0$, show that the inequality $|nx^{n-1}| \le (|x| + \epsilon)^n$ is true for all sufficiently large n's. Hint: $n^{1/n}|x|^{1-(1/n)} \to |x|$.

We have solved the problem of determining the general nature of the sum of a power series: Inside the interval of convergence, it is a continuous function with derivatives of all orders. We now investigate the converse problem of starting with a given infinitely differentiable function and expanding it in a power series. In Section 14.9 we established several such expansions for a few special functions with particularly simple derivatives. Our purpose here is to consider a method of much greater generality.

It may seem that the coefficients of a power series are not connected with one another in any necessary way. In fact, however, they are bound together by an invisible chain, which we now make visible.

To this end, let us assume that a function $f(x)$ is the sum of a power series with positive radius of convergence,

$$f(x) = \sum_{n=0}^{\infty} a_n x^n = a_0 + a_1 x + a_2 x^2 + \cdots, \qquad R > 0. \tag{1}$$

By the results of Section 14.9, repeated term-by-term differentiation is legitimate and yields

$$f'(x) = a_1 + 2a_2 x + 3a_3 x^2 + \cdots,$$
$$f''(x) = 1\cdot 2 a_2 + 2\cdot 3 a_3 x + 3\cdot 4 a_4 x^2 + \cdots,$$
$$f'''(x) = 1\cdot 2\cdot 3 a_3 + 2\cdot 3\cdot 4 a_4 x + 3\cdot 4\cdot 5 a_5 x^2 + \cdots,$$

and in general,

$$f^{(n)}(x) = n! a_n + \text{terms containing } x \text{ as a factor.}$$

14.10

TAYLOR SERIES AND TAYLOR'S FORMULA

We know that these series expansions of the derivatives are valid on the open interval $|x| < R$. By putting $x = 0$ in these equations we obtain

$$f(0) = a_0, \qquad f'(0) = a_1, \qquad f''(0) = 1 \cdot 2a_2,$$
$$f'''(0) = 1 \cdot 2 \cdot 3a_3, \qquad \ldots, \qquad f^{(n)}(0) = n!a_n,$$

so

$$a_0 = f(0), \qquad a_1 = f'(0),$$
$$a_2 = \frac{f''(0)}{2!}, \qquad a_3 = \frac{f'''(0)}{3!}, \qquad \ldots, \qquad a_n = \frac{f^{(n)}(0)}{n!}. \qquad (2)$$

These are very remarkable formulas, for they tell us that *if* $f(x)$ has a power series expansion of the form (1), *then* its coefficients must be the numbers given by (2). The series (1) therefore becomes

$$f(x) = f(0) + f'(0)x + \frac{f''(0)}{2!} x^2 + \cdots + \frac{f^{(n)}(0)}{n!} x^n + \cdots . \qquad (3)$$

The power series on the right of (3) is called the *Taylor series* of $f(x)$ [at $x = 0$]. The following conclusion is implicit in this discussion:

> *If a function is represented by a power series with positive radius of convergence, then there is only one such series and it must be the Taylor series of the function.*

Briefly, power series expansions are unique, because (2) tells us that the coefficients are uniquely determined by the function itself. If we use the standard conventions mentioned earlier, that $0! = 1$ and that the zeroth derivative of $f(x)$ is $f(x)$ itself $[f^{(0)}(x) = f(x)]$, then (3) can be written as

$$f(x) = \sum_{n=0}^{\infty} \frac{f^{(n)}(0)}{n!} x^n. \qquad (4)$$

The numbers $a_n = f^{(n)}(0)/n!$ are called the *Taylor coefficients* of $f(x)$.*

Equation (3) was true in the preceding discussion because we started with a convergent series having $f(x)$ as its sum. We now start with a function $f(x)$ that has derivatives of all orders throughout some open interval I containing the point $x = 0$. We can form the Taylor series on the right of (3) and ask the question, Is equation (3) a valid expansion of $f(x)$ on the interval I? Lest there be any misunderstanding, we state as clearly as possible that this equation is *not* always valid, and whether it is or not depends entirely on the individual nature of the function $f(x)$. In Remark 2 we will give an example of an infinitely differentiable function whose Taylor series converges everywhere, *but not to the value of the function,* so in this particular case equation (3) is false.

---

* Brook Taylor (1685–1731) was secretary of the Royal Society and an enthusiastic supporter of Newton in his acrimonious controversy with Leibniz and the Bernoullis about the invention of calculus. Taylor published his power series expansion of a function in 1715, but only as a formula and without any consideration at all of the issue of convergence. This expansion had already been published by John Bernoulli in 1694. Taylor was fully aware of this fact, but chose to ignore it out of partisan malice. In those days science roused passions that today we see only in politics and religion.

In order to put our ideas on a firm foundation, we proceed as follows. Break off the Taylor series on the right side of (3) at the term containing x^n and define the *remainder* $R_n(x)$ by the equation

$$f(x) = f(0) + f'(0)x + \frac{f''(0)}{2!}x^2 + \cdots + \frac{f^{(n)}(0)}{n!}x^n + R_n(x). \tag{5}$$

Then the Taylor series on the right side of (3) converges to the function $f(x)$ precisely when

$$\lim_{n \to \infty} R_n(x) = 0. \tag{6}$$

These equations don't really solve anything, because $R_n(x)$ is defined to be whatever it takes to make (5) true, and (6) is merely the meaning of the statement that the Taylor series converges to the function. This approach is useful only if we can show that $R_n(x)$ can be expressed in a form that makes it feasible to try to prove (6) in the case of particular functions. We emphasize that (6) is not always true, because (3) is not always true. The most convenient general formula for $R_n(x)$ is

$$R_n(x) = \frac{f^{(n+1)}(c)}{(n+1)!}x^{n+1}, \tag{7}$$

where c is some number between 0 and x. When $R_n(x)$ is expressed this way, (5) is called *Taylor's formula with derivative remainder.** The proof of (7) that we give is fairly technical, and is placed in Remark 3 so that students who wish to skip it can easily do so.

We now give several illustrations of the use of (6) and (7). First, however, we observe that

$$\lim_{n \to \infty} \frac{|x|^n}{n!} = 0 \tag{8}$$

for every x, because the series $\Sigma x^n/n!$ is absolutely convergent everywhere. We shall need this fact in our first two examples.

Example 1   Find the Taylor series for the function $f(x) = e^x$, and use (6) and (7) to prove that it converges to e^x for every x.

Solution   We clearly have

$$f(x) = e^x, \qquad f(0) = 1;$$
$$f'(x) = e^x, \qquad f'(0) = 1;$$
$$f''(x) = e^x, \qquad f''(0) = 1;$$

and so on. By substituting in (3) we obtain

$$e^x = 1 + x + \frac{x^2}{2!} + \cdots + \frac{x^n}{n!} + \cdots = \sum_{n=0}^{\infty} \frac{x^n}{n!}; \tag{9}$$

---

* There are other forms of the remainder that we do not discuss.

and to prove the validity of this expansion by our present methods, we examine the remainder $R_n(x)$. For any x, the maximum value of the exponential function on the interval from 0 to x is easily seen to be its value M at the right endpoint. By (7) and (8) we therefore have

$$|R_n(x)| = \left|\frac{f^{(n+1)}(c)}{(n+1)!}x^{n+1}\right| = \left|\frac{e^c}{(n+1)!}x^{n+1}\right| \leq M\frac{|x|^{n+1}}{(n+1)!} \to 0,$$

so (9) is valid. Of course, we established (9) in another way in Section 14.9.

Example 2   Find the Taylor series for $f(x) = \sin x$, and use (6) and (7) to prove that it converges to $\sin x$ for every x.

Solution   We can arrange our work as follows:

$$f(x) = \sin x, \qquad f(0) = 0;$$
$$f'(x) = \cos x, \qquad f'(0) = 1;$$
$$f''(x) = -\sin x, \qquad f''(0) = 0;$$
$$f'''(x) = -\cos x, \qquad f'''(0) = -1;$$

and the subsequent derivatives follow this same pattern. By substituting in (3), we obtain

$$\sin x = x - \frac{x^3}{3!} + \frac{x^5}{5!} - \cdots + (-1)^n\frac{x^{2n+1}}{(2n+1)!} + \cdots$$

$$= \sum_{n=0}^{\infty}(-1)^n\frac{x^{2n+1}}{(2n+1)!}. \tag{10}$$

So far, all we know is that *if* $\sin x$ has a power series expansion, *then* that expansion must be (10). To prove that (10) is actually true for every x, we use (6) and (7). Since either

$$|f^{(n+1)}(x)| = |\sin x| \qquad \text{or} \qquad |f^{(n+1)}(x)| = |\cos x|,$$

it is clear that $|f^{(n+1)}(c)| \leq 1$ for every number c. Therefore by (7) and (8) we have

$$|R_n(x)| = |f^{(n+1)}(c)|\frac{|x|^{n+1}}{(n+1)!} \leq \frac{|x|^{n+1}}{(n+1)!} \to 0,$$

so (10) is true. Students will remember that we proved (10) in Section 13.4 by a very different method, one requiring considerable ingenuity. Our present method has the advantage of being straightforward and systematic.

Example 3   Find the Taylor series for $\cos x$, and prove that it converges to $\cos x$ for every x.

Solution   We could proceed directly, as in Example 2. Instead, this is left for students to carry out in the problems, and we obtain the desired series by differentiating (10),

$$\cos x = 1 - \frac{x^2}{2!} + \frac{x^4}{4!} - \cdots + (-1)^n \frac{x^{2n}}{(2n)!} + \cdots$$

$$= \sum_{n=0}^{\infty} (-1)^n \frac{x^{2n}}{(2n)!}. \tag{11}$$

The validity of this expansion is guaranteed by the results of Section 14.9. Since we have found *a* power series expansion for cos x, we know by the discussion of uniqueness given previously that this series must be the Taylor series.

The three series established in these examples can be used to find power series expansions for many other functions. Thus, since (9) is true for every x, a power series representation for e^{-x^2} can be found by substituting $-x^2$ for x. This yields

$$e^{-x^2} = 1 - x^2 + \frac{x^4}{2!} - \frac{x^6}{3!} + \cdots + (-1)^n \frac{x^{2n}}{n!} + \cdots. \tag{12}$$

We can apply this formula to evaluate the definite integral

$$\int_0^1 e^{-x^2}\, dx,$$

even though the corresponding indefinite integral cannot be calculated in terms of elementary functions.* Term-by-term integration of (12) gives

$$\int_0^1 e^{-x^2}\, dx = \left[x - \frac{x^3}{3} + \frac{x^5}{5 \cdot 2!} - \frac{x^7}{7 \cdot 3!} + \cdots \right]_0^1$$

$$= 1 - \frac{1}{3} + \frac{1}{5 \cdot 2!} - \frac{1}{7 \cdot 3!} + \cdots.$$

This expression for the value of the integral as a series of constants is exact, and can be used to obtain this value in decimal form to any desired degree of accuracy.

Remark 1   Given a function $f(x)$ that is infinitely differentiable in some interval containing the point $x = 0$, we have examined the possibility of expanding this function as a power series in x. More generally, if $f(x)$ is infinitely differentiable in some interval containing the point $x = a$, we can ask about its possible expansion as a power series in $x - a$,

$$f(x) = \sum_{n=0}^{\infty} a_n (x - a)^n = a_0 + a_1(x - a) + a_2(x - a)^2 + \cdots. \tag{13}$$

This problem is equivalent to the first, because (13) is the same as the series $g(w) = a_0 + a_1 w + a_2 w^2 + \cdots$, where $w = x - a$ and $g(w) = f(x)$, and therefore no separate discussion is required. Since $g^{(n)}(0) = f^{(n)}(a)$, the *Taylor*

---

* See Section 10.8.

series of $f(x)$ in powers of $x - a$ (or at $x = a$) is

$$f(x) = \sum_{n=0}^{\infty} \frac{f^{(n)}(a)}{n!} (x - a)^n$$

$$= f(a) + f'(a)(x - a) + \frac{f''(a)}{2!} (x - a)^2 + \cdots . \qquad (14)$$

Some writers refer to (3), which is the special case of (14) corresponding to $a = 0$, as *Maclaurin's series*. However, this custom has no historical justification and is rapidly being abandoned. Whenever we use the phrase "Taylor series" without qualification, we always mean "Taylor series in powers of x," or at $x = 0$.

Remark 2 In Problem 42 of Section 12.3 we asked students to consider the function $f(x)$ defined for all real numbers x by

$$f(x) = \begin{cases} e^{-1/x^2}, & x \neq 0 \\ 0, & x = 0. \end{cases}$$

This function is continuous and has derivatives of all orders for all values of x. Furthermore, every derivative vanishes at $x = 0$, that is, $f^{(n)}(0) = 0$ for every positive integer n. This means that the graph of the function (Fig. 14.8) is extremely flat at the origin — we might even say "infinitely flat." For this function, Taylor's formula (5) becomes

$$f(x) = 0 + 0 + \cdots + 0 + R_n(x).$$

The Taylor series of the function is therefore the series

$$0 + 0 + \cdots + 0 + \cdots ,$$

which converges for every x but converges to $f(x)$ only for $x = 0$. Thus, even though a function has derivatives of all orders everywhere, it still is not necessarily represented by its Taylor series; and if we wish to establish the validity of such a representation, we must invoke solid additional arguments of some kind, as illustrated in the three examples discussed above.

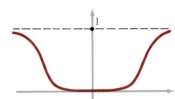

Figure 14.8

Remark 3 We now give a proof of formula (7) for the remainder $R_n(x)$. First, we define a function $S_n(x)$ by writing

$$R_n(x) = S_n(x)x^{n+1}$$

for $x \neq 0$. Next, we hold x fixed and define a function $F(t)$ for $0 \leq t \leq x$ (or $x \leq t \leq 0$) by writing

$$F(t) = f(x) - f(t) - f'(t)(x - t) - \frac{f''(t)}{2!} (x - t)^2 - \cdots$$

$$- \frac{f^{(n)}(t)}{n!} (x - t)^n - S_n(x)(x - t)^{n+1}.$$

Equation (5) shows that $F(0) = 0$. Also, it is obvious that $F(x) = 0$. It follows that $F'(c) = 0$ for some number c between 0 and x.* By differentiating $F(t)$ with respect to t, cancelling, and replacing t by c, we get

$$F'(c) = -\frac{f^{(n+1)}(c)}{n!}(x-c)^n + S_n(x)(n+1)(x-c)^n = 0,$$

so

$$S_n(x) = \frac{f^{(n+1)}(c)}{(n+1)!}$$

and the proof of (7) is complete.

---

* In words, if the graph of our function $F(t)$ touches the t-axis at two points, then it must have a horizontal tangent somewhere in between. This inference rests on the fact that $F(t)$ is continuous on the closed interval $0 \le t \le x$ (or $x \le t \le 0$) and differentiable inside this interval. See Rolle's theorem in Appendix C.4.

PROBLEMS

1 Use (3) to find the Taylor series of each of the following functions, and then use (6) and (7) to prove that each of these expansions is valid for all x:
(a) $\cos x$; (b) e^{-x}; (c) e^{3x}.

2 Obtain the series (11) for $\cos x$ from the series (10) for $\sin x$ by integrating term by term. Hint: Remember that the indefinite integral on the left must equal zero when $x = 0$.

3 Obtain the series (10) for $\sin x$ by differentiating the series (11) for $\cos x$.

4 Find the Taylor series expansion of each of the following functions (hint: any power series that converges to a function in an interval about $x = 0$ must be the Taylor series of that function):
(a) $x^2 e^x$;
(b) xe^{-3x};
(c) $\cos \sqrt{x}$;
(d) $x \sin 5x$;
(e) $\sin x^2$;
(f) $\cos^2 x$ [hint: $\cos^2 x = \frac{1}{2}(1 + \cos 2x)$];
(g) $\sin^2 x$;
(h) $1 - 7 \sin^2 x$;

(i) $f(x) = \dfrac{\sin x}{x}$ for $x \ne 0$, $f(0) = 1$.

In Problems 5 to 9, use any method to obtain each of the given Taylor series expansions as far as indicated, without worrying about convergence.

5 $\tan x = x + \frac{1}{3}x^3 + \frac{2}{15}x^5 + \cdots$.

6 $\sec^2 x = 1 + x^2 + \frac{2}{3}x^4 + \cdots$.

7 $\ln(\cos x) = -\dfrac{x^2}{2} - \dfrac{x^4}{12} - \dfrac{x^6}{45} - \cdots$.

8 $\ln(1 + \sin x) = x - \frac{1}{2}x^2 + \frac{1}{6}x^3 - \frac{1}{12}x^4 + \cdots$.

9 $\ln(1 + e^x) = \ln 2 + \frac{1}{2}x + \frac{1}{8}x^2 - \frac{1}{192}x^4 + \cdots$.

10 Show that a polynomial $P(x) = a_0 + a_1 x + a_2 x^2 + \cdots + a_n x^n$ is its own Taylor series.

11 If $P(x)$ is a polynomial of degree n and a is any number, show that

$$P(x) = P(a) + \frac{P'(a)}{1!}(x-a) + \cdots$$

$$+ \frac{P^{(n)}(a)}{n!}(x-a)^n.$$

12 Find the Taylor series expansion of
(a) $3x^2 - 5x + 7$ in powers of $x - 1$;
(b) x^3 in powers of $x + 2$.
Check the expansions in (a) and (b) by using algebra.

13 (a) Let p be an arbitrary constant and use (3) to obtain the *binomial series*

$$(1 + x)^p = 1 + px + \frac{p(p-1)}{2!}x^2$$

$$+ \frac{p(p-1)(p-2)}{3!}x^3 + \cdots$$

$$+ \frac{p(p-1)(p-2)\cdots(p-n+1)}{n!}x^n + \cdots.$$

(b) Observe that this series terminates and is a polynomial whenever p is a nonnegative integer, and only in this case. In all other cases, show that this series has radius of convergence $R = 1$. The fact that the expansion in (a) is valid for $|x| < 1$ is not easy to establish by the methods of this section; a different type of proof is outlined in Additional Problem 75 at the end of this chapter.

14 Use Problem 13 to write the Taylor series expansion of

$(1 - x^2)^{-1/2}$, and integrate this to get the Taylor series for $\sin^{-1} x$.

15 Find power series representations for

(a) $\displaystyle\int \frac{\sin x}{x}\, dx$; (b) $\displaystyle\int \sqrt{1 + x^3}\, dx$;

(c) $\displaystyle\int \frac{dx}{\sqrt{1 + x^4}}$.

14.11
(OPTIONAL) OPERATIONS ON POWER SERIES

In Section 14.10 we obtained Taylor series for various functions by using the formula $a_n = f^{(n)}(0)/n!$ to find the coefficients. But computing successive derivatives can be difficult and discouraging work if no simple pattern emerges. We can easily appreciate this fact by finding the seventh derivative of such functions as $\tan x$ or $x^5/(1 - x^4)$, because the calculations visibly sink us deeper into the bog with every step. For some functions we were able to establish the validity of their Taylor expansions by proving that $R_n(x) \to 0$ as $n \to \infty$, but this also can be difficult. To avoid such problems, we now discuss several algebraic methods for obtaining valid new Taylor expansions from ones that are already known.

Before we begin, we remind students that power series expansions are unique. This means that if a function $f(x)$ can be expressed as the sum of a power series *by any method,* then this series must be the Taylor series of $f(x)$. For example, we know from Section 14.3 that

$$\frac{1}{1 - x} = 1 + x + x^2 + x^3 + \cdots, \qquad |x| < 1; \tag{1}$$

and by first replacing x by x^4 and then multiplying through by x^5, we find that

$$\frac{1}{1 - x^4} = 1 + x^4 + x^8 + x^{12} + \cdots, \qquad |x| < 1,$$

and

$$\frac{x^5}{1 - x^4} = x^5 + x^9 + x^{13} + x^{17} + \cdots, \qquad |x| < 1.$$

We have deliberately avoided the very laborious task of using the formula $a_n = f^{(n)}(0)/n!$ to verify that the three series on the right are actually the Taylor series of the functions on the left. But these verifications aren't necessary, because this conclusion follows automatically from the uniqueness principle stated above.

MULTIPLICATION

Suppose we are given two power series expansions,

$$f(x) = \Sigma a_n x^n = a_0 + a_1 x + a_2 x^2 + a_3 x^3 + \cdots \tag{2}$$

and

$$g(x) = \Sigma b_n x^n = b_0 + b_1 x + b_2 x^2 + b_3 x^3 + \cdots, \qquad (3)$$

both valid on an interval $|x| < R$. If we ignore the question of convergence for a moment, then we can multiply these series in the same way we multiply two polynomials. That is, we systematically multiply each term of the first series into all the terms of the second series and then collect terms involving the same power of x. First, the term-by-term multiplication—

$$
\begin{array}{ll}
a_0: & a_0 b_0 + a_0 b_1 x + a_0 b_2 x^2 + a_0 b_3 x^3 + \cdots \\
a_1 x: & a_1 b_0 x + a_1 b_1 x^2 + a_1 b_2 x^3 + \cdots \\
a_2 x^2: & a_2 b_0 x^2 + a_2 b_1 x^3 + \cdots \\
a_3 x^3: & a_3 b_0 x^3 + \cdots
\end{array}
$$

$$\cdots.$$

By adding these columns, we obtain the power series

$$a_0 b_0 + (a_0 b_1 + a_1 b_0)x + (a_0 b_2 + a_1 b_1 + a_2 b_0)x^2$$
$$+ (a_0 b_3 + a_1 b_2 + a_2 b_1 + a_3 b_0)x^3 + \cdots. \qquad (4)$$

The form of the coefficient of x^n in (4) is evident: The subscripts of the a's increase as the subscripts of the b's decrease, and their sum remains constant and equals the exponent n on x^n. Briefly, we have multiplied (2) and (3) to obtain

$$f(x)g(x) = \sum_{n=0}^{\infty} \left(\sum_{k=0}^{n} a_k b_{n-k} \right) x^n. \qquad (5)$$

We assert that this *product* of the series (2) and (3) actually converges on the interval $|x| < R$ to the product of the functions $f(x)$ and $g(x)$, as indicated by (5). The proof is not easy, and depends on the absolute convergence of the two series in the given interval. The details can be found in Appendix C.11.

Example 1 Find the Taylor series for $e^x \sin x$.

Solution We know that

$$e^x = 1 + x + \frac{x^2}{2!} + \frac{x^3}{3!} + \cdots \qquad (6)$$

and

$$\sin x = x - \frac{x^3}{3!} + \frac{x^5}{5!} - \cdots. \qquad (7)$$

Our work can be arranged as follows:

$$e^x \sin x = \left(1 + x + \frac{x^2}{2} + \cdots\right)\left(x - \frac{x^3}{6} + \cdots\right)$$

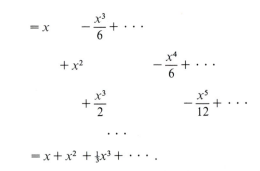

$$= x + x^2 + \tfrac{1}{3}x^3 + \cdots .$$

Since the two given series (6) and (7) converge for all x, the same is true of the product series. It is rarely easy—and usually quite impossible—to recognize the formula for the general term of the product series in this process.

Example 2 Find the Taylor series for $[\ln (1 - x)]/(x - 1)$.

Solution We know that

$$\ln (1 - x) = -\left(x + \frac{x^2}{2} + \frac{x^3}{3} + \cdots\right), \qquad |x| < 1,$$

so

$$\frac{\ln (1 - x)}{x - 1} = \left(\frac{1}{1 - x}\right)[-\ln (1 - x)]$$

$$= (1 + x + x^2 + \cdots)\left(x + \frac{x^2}{2} + \frac{x^3}{3} + \cdots\right)$$

$$= x + (1 + \tfrac{1}{2})x^2 + (1 + \tfrac{1}{2} + \tfrac{1}{3})x^3 + \cdots$$

$$= \sum_{n=1}^{\infty} \left(1 + \frac{1}{2} + \cdots + \frac{1}{n}\right)x^n, \qquad |x| < 1.$$

This is one of the rare cases where the general term of the product series is easy to recognize.

DIVISION

Two power series can be divided by the long division process that is used in elementary algebra for dividing polynomials. Since we are working with power series, the terms are of course arranged in order of increasing exponents, instead of in order of decreasing exponents as is usual with polynomials. In particular cases, however, one or both of the given series may be a polynomial.

Example 3 Find the Taylor series for $\tan x$ by dividing the series for $\sin x$ by the series for $\cos x$.

Solution We have

$$1 - \frac{x^2}{2} + \frac{x^4}{24} - \cdots \,\overline{\left)\,x - \frac{x^3}{6} + \frac{x^5}{120} - \cdots\right.}\quad\overset{\displaystyle x + \frac{1}{3}x^3 + \frac{2}{15}x^5 + \cdots}{}$$

$$x - \frac{x^3}{2} + \frac{x^5}{24} - \cdots$$

$$\tfrac{1}{3}x^3 - \tfrac{1}{30}x^5 + \cdots$$

$$\tfrac{1}{3}x^3 - \tfrac{1}{6}x^5 + \cdots$$

$$\tfrac{2}{15}x^5 + \cdots ,$$

so

$$\tan x = x + \tfrac{1}{3}x^3 + \tfrac{2}{15}x^5 + \cdots . \tag{8}$$

It can be shown that this expansion is valid on the interval $|x| < \pi/2$. Since this is the largest interval with center $x = 0$ on which the denominator $\cos x$ is nonzero, the series (8) has radius of convergence $R = \pi/2$. The problem of discovering a formula for the general term of this series was solved by Euler in 1748. His solution depends on the ideas of Section 14.12 and is outlined in Appendix A.18.

The actual process of dividing one power series by another is clearly not very difficult. The theory that justifies this process is given in Appendix C.14.

SUBSTITUTION

This is a method we have already used: If a power series

$$f(x) = a_0 + a_1 x + a_2 x^2 + \cdots \tag{9}$$

converges for $|x| < R$ and if $|g(x)| < R$, then we can certainly find $f[g(x)]$ by substituting $g(x)$ for x in (9). For example, there is no problem in using the series (1), (6), and (7) to obtain

$$\frac{1}{1 + 2x^2} = \frac{1}{1 - (-2x^2)} = 1 + (-2x^2) + (-2x^2)^2 + \cdots$$

$$= 1 - 2x^2 + 4x^4 - \cdots , \qquad |2x^2| < 1;$$

$$e^{x^4} = 1 + x^4 + \frac{(x^4)^2}{2!} + \frac{(x^4)^3}{3!} + \cdots$$

$$= 1 + x^4 + \frac{x^8}{2!} + \frac{x^{12}}{3!} + \cdots , \qquad \text{all } x;$$

and

$$\sin 3x = 3x - \frac{(3x)^3}{3!} + \frac{(3x)^5}{5!} - \cdots$$

$$= 3x - \frac{27}{3!}x^3 + \frac{243}{5!}x^5 - \cdots , \qquad \text{all } x.$$

In these examples we have substituted the simple functions $g(x) = -2x^2, x^4$, and $3x$ into appropriate power series, but much more is possible. Under suitable conditions, we can actually substitute one power series into another!

Thus, suppose that the function $g(x)$ is given by a power series,

$$g(x) = b_0 + b_1 x + b_2 x^2 + \cdots, \tag{10}$$

and substitute this entire series for x in (9),

$$f[g(x)] = a_0 + a_1 g(x) + a_2 [g(x)]^2 + \cdots$$
$$= a_0 + a_1 [b_0 + b_1 x + \cdots] + a_2 [b_0 + b_1 x + \cdots]^2 + \cdots. \tag{11}$$

Again, this is perfectly legitimate as long as $|g(x)| < R$. However, the series on the right of (11) can now be converted into a power series in x by squaring, cubing, etc., and collecting like powers of x, and it can be proved that the power series formed in this way converges to $f[g(x)]$ whenever (10) is absolutely convergent and $|g(x)| < R$.*

Example 4 Apply the method of substitution to find the Taylor series of $e^{\sin x}$ up to the term containing x^4.

Solution We can use (6) and (7) to write

$$e^{\sin x} = 1 + \left(x - \frac{x^3}{6} + \cdots\right) + \frac{1}{2}\left(x - \frac{x^3}{6} + \cdots\right)^2$$

$$+ \frac{1}{6}\left(x - \frac{x^3}{6} + \cdots\right)^3 + \frac{1}{24}\left(x - \frac{x^3}{6} + \cdots\right)^4 + \cdots$$

$$= 1 + \left(x - \frac{x^3}{6} + \cdots\right) + \frac{1}{2}\left(x^2 - \frac{1}{3}x^4 + \cdots\right) + \frac{1}{6}(x^3 + \cdots)$$

$$+ \frac{1}{24}(x^4 + \cdots) + \cdots$$

$$= 1 + x + \frac{1}{2}x^2 - \frac{1}{8}x^4 + \cdots.$$

If we try to apply this method to $e^{\cos x}$, we find that infinitely many terms contribute to the formation of each coefficient, which is a difficult situation to deal with. For this reason, the method of substitution is not a practical tool unless $b_0 = 0$ in the series (10).

EVEN AND ODD FUNCTIONS

A function $f(x)$ defined on an interval $|x| < R$ is said to be *even* if $f(-x) = f(x)$, and *odd* if $f(-x) = -f(x)$. The Taylor series of the even function $\cos x$ contains only even powers of x, and the Taylor series of the odd function $\sin x$ contains only odd powers of x. These facts are special cases of a general principle: If $f(x)$ is an even function, then its Taylor series has the form

$$a_0 + a_2 x^2 + a_4 x^4 + a_6 x^6 + \cdots;$$

and if $f(x)$ is an odd function, then its Taylor series has the form

$$a_1 x + a_3 x^3 + a_5 x^5 + a_7 x^7 + \cdots.$$

---

* See p. 180 of K. Knopp, *Theory and Application of Infinite Series,* Hafner, 1951.

That is, the Taylor series of an even (odd) function contains only even (odd) powers of x. This is very easy to prove from the uniqueness of power series expansions.* As another example of this phenomenon, we know beforehand that the series on the right side of (8) contains only odd powers of x, because $\tan x$ is an odd function.

Many functions are even and many are odd, but most are neither. However, every function $f(x)$ defined on an interval $|x| < R$ can be expressed as the sum of an even function and an odd function:

$$f(x) = \tfrac{1}{2}[f(x) + f(-x)] + \tfrac{1}{2}[f(x) - f(-x)] = g(x) + h(x),$$

where — as is easily verified —

$$g(x) = \tfrac{1}{2}[f(x) + f(-x)] \qquad \text{is even}$$

and

$$h(x) = \tfrac{1}{2}[f(x) - f(-x)] \qquad \text{is odd.}$$

Further, if $f(x) = \Sigma_{n=0}^{\infty} a_n x^n$, then

$$g(x) = \sum_{n=0}^{\infty} a_{2n} x^{2n} \qquad \text{and} \qquad h(x) = \sum_{n=0}^{\infty} a_{2n+1} x^{2n+1}.$$

Thus, the Taylor series of $f(x)$ splits into two power series, one with even exponents representing the even part of $f(x)$ and one with odd exponents representing the odd part.

Example 5 The even part and the odd part of

$$f(x) = e^x = 1 + x + \frac{x^2}{2!} + \frac{x^3}{3!} + \frac{x^4}{4!} + \frac{x^5}{5!} + \cdots$$

are

$$g(x) = \frac{e^x + e^{-x}}{2} = 1 + \frac{x^2}{2!} + \frac{x^4}{4!} + \cdots$$

and

$$h(x) = \frac{e^x - e^{-x}}{2} = x + \frac{x^3}{3!} + \frac{x^5}{5!} + \cdots.$$

These are the two hyperbolic functions $\cosh x$ and $\sinh x$ that were defined in Section 9.7.

Example 6 Find the sum of the series

$$\frac{x^2}{2} + \frac{x^4}{4} + \cdots.$$

Solution This is the even part of the familiar series

$$f(x) = -\ln(1 - x) = x + \frac{x^2}{2} + \frac{x^3}{3} + \frac{x^4}{4} + \cdots.$$

---

* We have only to point out that if $f(x) = \Sigma a_n x^n$ is even, then $\Sigma a_n x^n = \Sigma(-1)^n a_n x^n$, so by uniqueness we have $a_n = (-1)^n a_n$, and therefore $a_n = -a_n$ or $a_n = 0$ if n is odd. Similar reasoning applies if $f(x)$ is odd.

The sum of the given series is therefore

$$\tfrac{1}{2}[f(x) + f(-x)] = \tfrac{1}{2}[-\ln(1-x) - \ln(1+x)]$$
$$= -\tfrac{1}{2}\ln(1-x^2) = -\ln\sqrt{1-x^2}.$$

PROBLEMS

1 In Example 1, continue the calculation and find the terms of the product series as far as the term containing x^6.

In Problems 2 to 13, use multiplication to show that the given function has the indicated power series expansion.

2 $\dfrac{\sin x}{1-x} = x + x^2 + \dfrac{5}{6}x^3 + \dfrac{5}{6}x^4 + \dfrac{101}{120}x^5 + \cdots$.

3 $e^{x+x^2} = 1 + x + \dfrac{3}{2}x^2 + \dfrac{7}{6}x^3 + \dfrac{25}{24}x^4 + \cdots$.

4 $e^x \cos x = 1 + x - \dfrac{1}{3}x^3 - \dfrac{1}{6}x^4 - \dfrac{1}{30}x^5 + \cdots$.

5 $\dfrac{\tan^{-1} x}{1-x} = x + x^2 + \dfrac{2}{3}x^3 + \dfrac{2}{3}x^4 + \dfrac{13}{15}x^5 + \cdots$.

6 $\ln^2(1-x) = x^2 + x^3 + \dfrac{11}{12}x^4 + \dfrac{5}{6}x^5 + \cdots$.

7 $\dfrac{\cos x}{1-x} = 1 + x + \dfrac{1}{2}x^2 + \dfrac{1}{2}x^3 + \dfrac{13}{24}x^4$

$$+ \dfrac{13}{24}x^5 + \cdots.$$

8 $\tan^2 x = x^2 + \dfrac{2}{3}x^4 + \dfrac{17}{45}x^6 + \cdots$.

9 $\dfrac{e^x}{2+x} = \dfrac{1}{2} + \dfrac{1}{4}x + \dfrac{1}{8}x^2 + \dfrac{1}{48}x^3 + \dfrac{1}{96}x^4 + \cdots$.

10 $\sqrt{1+x}\,\ln(1+x) = x - \dfrac{1}{24}x^3 + \dfrac{1}{24}x^4 - \dfrac{71}{1920}x^5 + \cdots$.

11 $e^{-x}\tan x = x - x^2 + \dfrac{5}{6}x^3 - \dfrac{1}{2}x^4 + \dfrac{41}{120}x^5 + \cdots$.

12 $\dfrac{1-x}{1-x^3} = \dfrac{1}{1+x+x^2} = 1 - x + x^3 - x^4 + x^6 - x^7$

$$+ x^9 - x^{10} + \cdots.$$

13 $\dfrac{\sin x}{1+x} = x - x^2 + \dfrac{5}{6}x^3 - \dfrac{5}{6}x^4 + \dfrac{101}{120}x^5 + \cdots$.

14 By squaring the series for $\sin x$ and $\cos x$, show that $\sin^2 x + \cos^2 x = 1$, at least as far as the x^6 term.

15 If $f(x) = \sum_{n=0}^{\infty} a_n x^n$, use multiplication of series to show that

$$\dfrac{1}{1-x} f(x) = \sum_{n=0}^{\infty} (a_0 + a_1 + \cdots + a_n)x^n.$$

Use this result to write down the series in Problems 2, 5, and 7 by inspection.

16 Use Problem 15 to find the sum of the series $\sum_{n=0}^{\infty}(n+1)x^n$.

17 The binomial series expansion of $1/\sqrt{1-x}$ is

$$\dfrac{1}{\sqrt{1-x}} = 1 + \dfrac{1}{2}x + \dfrac{1\cdot 3}{2^2\cdot 2!}x^2 + \dfrac{1\cdot 3\cdot 5}{2^3\cdot 3!}x^3$$

$$+ \dfrac{1\cdot 3\cdot 5\cdot 7}{2^4\cdot 4!}x^4 + \cdots.$$

Check this by squaring the series and showing that the result is $1 + x + x^2 + x^3 + x^4 + \cdots$, at least as far as the x^4 term.

18 In Example 3, continue the calculation and find the series for $\tan x$ as far as the term containing x^7.

19 Use division to obtain the series expansions given in Problems 2, 5, 7, 9, 12, and 13.

In Problems 20 to 27, use division to obtain the given expansions.

20 $\dfrac{1}{(1-x)^2} = 1 + 2x + 3x^2 + 4x^3 + \cdots$.

21 $\dfrac{1-x}{1-x+x^2} = 1 - x^2 - x^3 + x^5 + x^6$

$$- x^8 - x^9 + \cdots.$$

22 $\dfrac{x^2}{1-x+x^2-x^3} = x^2 + x^3 + x^6 + x^7 + \cdots$.

23 $\dfrac{x}{\sin x} = 1 + \dfrac{1}{6}x^2 + \dfrac{7}{360}x^4 + \cdots$.

24 $\dfrac{1}{1+x+x^2+x^3+\cdots} = 1 - x.$

25 $\dfrac{\sin x}{\ln(1+x)} = 1 + \dfrac{1}{2}x - \dfrac{1}{4}x^2 - \dfrac{1}{24}x^3 + \cdots$.

26 $\sec x = \dfrac{1}{\cos x} = 1 + \dfrac{1}{2}x^2 + \dfrac{5}{24}x^4 + \cdots$.

27 $\dfrac{\sin^{-1} x}{\cos x} = x + \dfrac{2}{3}x^3 + \dfrac{11}{30}x^5 + \cdots$.

In Problems 28 and 29, use the method of substitution to find the given Taylor series.

28 $\ln(1 + \sin x) = x - \frac{1}{2}x^2 + \frac{1}{6}x^3 + \cdots$.

29 $\dfrac{1}{1 - x^2 e^x} = 1 + x^2 + x^3 + \dfrac{3}{2}x^4 + \dfrac{13}{6}x^5$

$$+ \frac{73}{24}x^6 + \cdots.$$

30 Use substitution and the fact that $\sec x = 1/\cos x = 1/[1 - (1 - \cos x)]$ to find the Taylor series for $\sec x$ up to the term containing x^6. What is the radius of convergence?

31 Use multiplication and the result of Problem 30 to find the Taylor series for $\tan x$ up to the term containing x^7.

32 (a) Find the Taylor series for $\sec^2 x$ as far as the term containing x^6 by expanding $\sec^2 x = 1/\cos^2 x = 1/(1 - \sin^2 x)$ as a geometric series in $\sin^2 x$.
  (b) Find the same series by squaring the series found in Problem 30.
  (c) Find the same series by differentiating the series found in Problem 31.

33 Show that a function $f(x)$ defined on an interval $|x| < R$ can be expressed *in only one way* as the sum of an even function $g(x)$ and an odd function $h(x)$.

34 Find the sum of the series $x + \dfrac{x^3}{3} + \dfrac{x^5}{5} + \cdots$.

35 Calculate each of the following limits by first finding the Taylor series of the given function:

(a) $\lim\limits_{x \to 0} \dfrac{1 - \cos x}{x^2}$; (b) $\lim\limits_{x \to 0} \dfrac{x - \sin x}{x^3}$.

In this way, the use of Taylor series often provides a convenient alternative to the use of L'Hospital's rule.

36 Find the sum of each of the following series:
(a) $x + x^2 - x^3 + x^4 + x^5 - x^6 + + - \cdots$;
(b) $x^2 + x^3 + x^4 - x^5 + x^6 + x^7$
$$+ x^8 - x^9 + + + - \cdots;$$
(c) $\dfrac{x}{2!} + \dfrac{x^2}{3!} + \dfrac{x^3}{4!} + \dfrac{x^4}{5!} + \cdots$;
(d) $1 + \dfrac{x^4}{4!} + \dfrac{x^8}{8!} + \dfrac{x^{12}}{12!} + \cdots$.

37 Calculate $f^{(7)}(0)$ if $f(x) = \tan x$ and use this to verify the coefficient of x^7 in the expansion found in Problem 18.

Consider the three familiar power series expansions

$$e^x = 1 + x + \frac{x^2}{2!} + \frac{x^3}{3!} + \frac{x^4}{4!} + \frac{x^5}{5!} + \cdots, \tag{1}$$

$$\cos x = 1 - \frac{x^2}{2!} + \frac{x^4}{4!} - \cdots, \tag{2}$$

and

$$\sin x = x - \frac{x^3}{3!} + \frac{x^5}{5!} - \cdots. \tag{3}$$

14.12

(OPTIONAL) COMPLEX NUMBERS AND EULER'S FORMULA

The second and third of these series seem to be parts of the first series in some nonobvious way that involves changes in the signs of some of the terms. This in turn suggests that the three functions on the left are probably related to one another. There is indeed such a relation, which is known as *Euler's formula*:

$$e^{ix} = \cos x + i \sin x, \tag{4}$$

where $i = \sqrt{-1}$ is the so-called imaginary unit. This formula turns out to be one of the most important facts in the whole of mathematics, with implications that deeply influence mathematics itself and also many of its applications, particularly in the fields of physics and electrical engineering. A full explanation of Euler's formula would require us to develop a fairly complete theory of complex numbers and functions of a complex variable. With apologies, we leave this task to a more advanced course, and instead briefly outline a few of the necessary ideas in a very incomplete way that at least has the merit of lending a little plausibility to formula (4).

Up to this point, all of our work in this book has taken place in the context of the real number system. Nevertheless, the real numbers do have a serious deficiency — not every polynomial equation has a solution. Thus, the quadratic equation $x^2 + 1 = 0$ has no solution in the real number system because there is no real number whose square is -1. This deficiency was so crippling that several hundred years ago mathematicians felt the need to use the seemingly contradictory symbol $\sqrt{-1}$ to signify a solution of $x^2 + 1 = 0$. This symbol was later denoted by the letter i, and was thought of as an imaginary or fictitious number that could be manipulated algebraically just like an ordinary real number, except that $i^2 = -1$. Any qualms these early mathematicians may have felt about the puzzling nature of this "number" were set aside because it was too useful to ignore. Thus, for example, the equation $x^2 + 1 = 0$ was factored by writing it in the equivalent forms $x^2 - i^2 = 0$ or $(x + i)(x - i) = 0$, and its solutions were exhibited as the numbers $x = \pm i$.

Without entering into the details that would be needed to give mathematical respectability to our discussion, we now simply describe the *complex numbers* as all formal expressions $a + bi$, where a and b are real numbers and i is the *imaginary unit* for which $i^2 = -1$. The complex numbers take on the character of a legitimate number system by means of the following general rule for performing calculations: In adding, multiplying, and dividing, follow all the familiar rules of elementary algebra and then simplify wherever possible by using the equation $i^2 = -1$ to remove all powers of i higher than the first, as in

$$i^3 = i^2 \cdot i = (-1) \cdot i = -i, \quad i^4 = i^2 \cdot i^2 = (-1)(-1) = 1, \quad i^5 = i^4 \cdot i = i, \quad (5)$$

and so on.

The complex number $a + bi$ can be identified with the real number a if $b = 0$, so the complex number system constitutes an enlargement of the real number system. Not only does the equation $x^2 + 1 = 0$ acquire the two solutions i and $-i$ in this way, but also every quadratic equation $ax^2 + bx + c = 0$ acquires the two familiar solutions

$$x = \frac{-b \pm \sqrt{b^2 - 4ac}}{2a},$$

which are real and distinct if $b^2 - 4ac > 0$, real and equal if $b^2 - 4ac = 0$, and complex and distinct if $b^2 - 4ac < 0$. For example, the equation $x^2 - 6x + 13 = 0$ has the distinct complex roots

$$x = \frac{6 \pm \sqrt{36 - 52}}{2} = \frac{6 \pm \sqrt{-16}}{2} = \frac{6 \pm 4\sqrt{-1}}{2}$$

$$= \frac{6 \pm 4i}{2} = 3 \pm 2i.$$

Much more than this is true, for every polynomial equation of the form

$$a_n x^n + a_{n-1} x^{n-1} + \cdots + a_1 x + a_0 = 0,$$

where n is a positive integer and the a's are arbitrary real numbers with $a_n \neq 0$, has exactly n roots (some of which may be equal to one another) among the complex numbers. Moreover, this is still true even if the coefficients are complex. This fact is known as the *fundamental theorem of algebra*. It shows that there is no need to construct further enlargements of the complex number system in order to solve all polynomial equations with complex coefficients.*

We now return to our original purpose, which was to gain a little insight into why Euler's formula (4) is true.

A perfectly satisfactory theory of power series can be constructed in which the variable is permitted to be a complex number instead of merely a real number. Within this theory, all of the series (1), (2), and (3) converge for all complex values of the variable. If we replace x in (1) by ix and use (5), then we obtain

$$e^{ix} = 1 + ix + \frac{(ix)^2}{2!} + \frac{(ix)^3}{3!} + \frac{(ix)^4}{4!} + \frac{(ix)^5}{5!} + \cdots$$

$$= 1 + ix - \frac{x^2}{2!} - i\frac{x^3}{3!} + \frac{x^4}{4!} + i\frac{x^5}{5!} - \cdots$$

$$= \left(1 - \frac{x^2}{2!} + \frac{x^4}{4!} - \cdots\right) + i\left(x - \frac{x^3}{3!} + \frac{x^5}{5!} - \cdots\right),$$

which gives Euler's formula

$$e^{ix} = \cos x + i \sin x.$$

If we now replace x by $-x$ and use the fact that $\cos(-x) = \cos x$ and $\sin(-x) = -\sin x$, then this becomes

$$e^{-ix} = \cos x - i \sin x;$$

and by first adding and then subtracting, we at once obtain

$$\cos x = \frac{e^{ix} + e^{-ix}}{2} \quad \text{and} \quad \sin x = \frac{e^{ix} - e^{-ix}}{2i}.$$

These formulas have many uses in advanced mathematics. A particularly interesting application is sketched in Appendix A.18. We also point out that if we put $x = \pi$ in Euler's formula, then we get

$$e^{\pi i} = \cos \pi + i \sin \pi$$

or

$$e^{\pi i} = -1.$$

This beautiful equation, connecting the mysterious and pervasive numbers π, e, and i, is one of the most remarkable facts in mathematics.

The ideas of this section are sketched in such a cursory fashion that they are bound to seem more suggestive than convincing. The eminent British mathematician E. C. Titchmarsh once remarked, "I met a man recently who told me that, so far from believing in the square root of minus one, he did not

---

* There are many proofs of this important theorem, of varying levels of sophistication. See, for example, pp. 269–271 of R. Courant and H. Robbins, *What Is Mathematics?*, Oxford University Press, 1941.

even believe in minus one. This is at any rate a consistent attitude." There is only one way to lift these concepts from the status of reasonable speculations to the realm of certainty, and this is to undertake a careful study of the theory of functions of a complex variable. This subject is one of the richest and most rewarding branches of mathematics, and we heartily recommend it.

ADDITIONAL PROBLEMS FOR CHAPTER 14

SECTION 14.2

1 Find $\lim x_n$ if

(a) $x_n = \left(1 - \frac{1}{2}\right)\left(1 - \frac{1}{3}\right) \cdots \left(1 - \frac{1}{n}\right)$;

(b) $x_n = \left(1 - \frac{1}{2^2}\right)\left(1 - \frac{1}{3^2}\right) \cdots \left(1 - \frac{1}{n^2}\right)$;

(c) $x_n = \frac{1}{n^2 + 1} + \frac{1}{n^2 + 2} + \cdots + \frac{1}{n^2 + n}$;

(d) $x_n = \frac{1}{\sqrt{n^2 + 1}} + \frac{1}{\sqrt{n^2 + 2}} + \cdots + \frac{1}{\sqrt{n^2 + n}}$.

2 If $f(x) = \lim_{n \to \infty} [\lim_{m \to \infty} (2/\pi) \tan^{-1} (m \sin^2 [n! \pi x])]$, show that $f(x) = 0$ when x is rational, and 1 when x is irrational.

3 If $f(x) = \lim_{n \to \infty} [\lim_{m \to \infty} \cos^m(n! \pi x)]$, show that $f(x) = 1$ when x is rational, and 0 when x is irrational.

4 Find the value of each of the following limits:

(a) $\lim_{n \to \infty} \frac{1}{x^n + x^{-n}} (x > 0)$;   (b) $\lim_{n \to \infty} \frac{x^{n+1} + n}{x^n + 2n}$.

5 For each sequence $\{x_n\}$ whose nth term is given, verify that the first three terms are $1, \frac{1}{2}, \frac{1}{3}$ and find the fourth term:

(a) $x_n = \frac{1}{n}$;   (b) $x_n = \frac{1}{2n^3 - 12n^2 + 23n - 12}$;

(c) $x_n = \frac{1}{n2^{(n-1)(n-2)(n-3)}}$.

*6 If a is any given number, define a sequence $\{x_n\}$ (by constructing a suitable formula for x_n in terms of n) which has the property that $x_1 = 1$, $x_2 = \frac{1}{2}$, $x_3 = \frac{1}{3}$, and $x_4 = a$.

*7 The so-called *Fibonacci sequence* 1, 1, 2, 3, 5, 8, 13, . . . is defined recursively by putting $x_1 = 1$, $x_2 = 1$, and $x_n = x_{n-2} + x_{n-1}$ for $n > 2$.* Find a formula for x_n in terms of n. Hint: Make the ingenious guess that x_n has the form $\alpha A^n + \beta B^n$ for suitable values of α, β, A, B; then determine A and B so that the recursion formula is true for all α's and β's; and finally, find α and β so that $x_1 = 1$ and $x_2 = 1$.

8 If $\{x_n\}$ is the Fibonacci sequence defined in Problem 7, show that $\lim x_{n+1}/x_n = (1 + \sqrt{5})/2$.

9 The sequence $\sqrt{2}, \sqrt{2\sqrt{2}}, \sqrt{2\sqrt{2\sqrt{2}}}, \ldots$ can be defined recursively by putting $x_1 = \sqrt{2}$ and $x_{n+1} = \sqrt{2x_n}$ for $n \geq 1$.

(a) Use mathematical induction to prove that $x_n < x_{n+1} < 2$ for every n.* This shows that the sequence is increasing and has 2 as an upper bound, and therefore converges to a limit $x \leq 2$.

(b) Show that $x = 2$ by using the recursion formula.

(c) Show that $x = 2$ by finding an explicit formula for x_n in terms of n.

10 The sequence $\sqrt{2}, \sqrt{2 + \sqrt{2}}, \sqrt{2 + \sqrt{2 + \sqrt{2}}}, \ldots$ can be defined recursively by putting $x_1 = \sqrt{2}$ and $x_{n+1} = \sqrt{2 + x_n}$ for $n \geq 1$. Show that it is increasing with 2 as an upper bound, and find its limit.

*11 If $a > 0$, then the sequence $\sqrt{a}, \sqrt{a + \sqrt{a}}, \sqrt{a + \sqrt{a + \sqrt{a}}}, \ldots$ can be defined recursively as in Problem 10. Show that it converges and find its limit.

12 Let $f(x)$ be an increasing continuous function on the interval $0 \leq x \leq 1$. Define two sequences $\{a_n\}$ and $\{b_n\}$ by

$$a_n = \frac{1}{n} \sum_{k=0}^{n-1} f\left(\frac{k}{n}\right), \qquad b_n = \frac{1}{n} \sum_{k=1}^{n} f\left(\frac{k}{n}\right).$$

(a) Show that

$$a_n \leq \int_0^1 f(x)\, dx \leq b_n \qquad \text{and}$$

$$0 \leq \int_0^1 f(x)\, dx - a_n \leq \frac{f(1) - f(0)}{n}.$$

---

* Fibonacci, or Leonardo of Pisa (ca. 1170–1230), was an Italian businessman who traveled extensively in the Middle East and was chiefly responsible for introducing the Hindu-Arabic numerals (i.e., 1, 2, 3, . . .) into Europe. He encountered his sequence in a problem about the progeny of rabbits. It has since been applied extensively (and eccentrically) to religion, art, the shapes of sea shells, etc., etc.

---

* Recall that the principle of mathematical induction asserts the following: A statement $S(n)$ which is meaningful (in the sense of being either true or false) for each positive integer n is true for all n if (i) $S(1)$ is true; and (ii) $S(n)$ implies $S(n + 1)$. This principle is discussed in detail in Appendix D.2.

(b) Show that both sequences converge to the limit $\int_0^1 f(x)\,dx$.

(c) State a corresponding fact for the interval $a \le x \le b$.

*13 Use Problem 12 to obtain the following limits:

(a) $\dfrac{1}{n^2} + \dfrac{2}{n^2} + \cdots + \dfrac{n}{n^2} \to \dfrac{1}{2}$;

(b) $\dfrac{1^2}{n^3} + \dfrac{2^2}{n^3} + \cdots + \dfrac{n^2}{n^3} \to \dfrac{1}{3}$;

(c) $\dfrac{1}{n+1} + \dfrac{1}{n+2} + \cdots + \dfrac{1}{n+n} \to \ln 2$;

(d) $\dfrac{1}{n+1} + \dfrac{1}{n+2} + \cdots + \dfrac{1}{kn} \to \ln k$;

(e) $\dfrac{n}{n^2+1^2} + \dfrac{n}{n^2+2^2} + \cdots + \dfrac{n}{n^2+n^2} \to \dfrac{\pi}{4}$;

(f) $\dfrac{n}{(n+1)^2} + \dfrac{n}{(n+2)^2} + \cdots + \dfrac{n}{(n+n)^2} \to \dfrac{1}{2}$;

(g) $\dfrac{1}{n}\left(\sin\dfrac{\pi}{n} + \sin\dfrac{2\pi}{n} + \cdots + \sin\dfrac{n\pi}{n}\right) \to \dfrac{2}{\pi}$;

(h) $\dfrac{1}{n}\left(\sin^2\dfrac{\pi}{n} + \sin^2\dfrac{2\pi}{n} + \cdots + \sin^2\dfrac{n\pi}{n}\right) \to \dfrac{1}{2}$;

(i) $\dfrac{1}{n}(\sqrt[n]{e} + \sqrt[n]{e^2} + \cdots + \sqrt[n]{e^n}) \to e - 1$;

(j) $\dfrac{1}{\sqrt{n^2+1^2}} + \dfrac{1}{\sqrt{n^2+2^2}} + \cdots + \dfrac{1}{\sqrt{n^2+n^2}}$
$\to \ln(1+\sqrt{2})$;

(k) $\dfrac{1}{n}\left(\ln\dfrac{1}{n} + \ln\dfrac{2}{n} + \cdots + \ln\dfrac{n}{n}\right) \to -1$.

14 Use part (k) of Problem 13 to show that $\dfrac{\sqrt[n]{n!}}{n} \to \dfrac{1}{e}$.

*15 Use Problem 14 to show that

(a) $\dfrac{1}{n}\sqrt[n]{(n+1)(n+2)\cdots(n+n)} \to \dfrac{4}{e}$;

(b) $\dfrac{1}{n}\sqrt[n]{(2n+1)(2n+2)\cdots(2n+n)} \to \dfrac{27}{4e}$.

*16 If $x_n \to x$, then the sequence of the arithmetic means of the x_n's also converges to x; that is,

$$y_n = \frac{x_1 + x_2 + \cdots + x_n}{n} \to x.$$

Prove this in two steps, as follows.

(a) Begin by assuming that $x = 0$, find a positive integer n_0 such that $|x_n| < \epsilon/2$ for all $n \ge n_0$, and use the fact that for these n's we have

$$|y_n| \le \frac{|x_1 + x_2 + \cdots + x_{n_0-1}|}{n}$$
$$+ \frac{|x_{n_0}| + \cdots + |x_n|}{n}$$
$$< \frac{a}{n} + \frac{\epsilon}{2},$$

where $a = |x_1 + x_2 + \cdots + x_{n_0-1}|$ is a constant.

(b) In the general case, where $x = 0$ is not assumed, use the fact that since $x_n - x \to 0$, we can infer from part (a) that

$$y_n - x$$
$$= \frac{(x_1 - x) + (x_2 - x) + \cdots + (x_n - x)}{n} \to 0.$$

17 Use Problem 16 to show that

(a) $\dfrac{1 + \frac{1}{2} + \cdots + 1/n}{n} \to 0$;

(b) $\dfrac{1 + \sqrt{2} + \sqrt[3]{3} + \cdots + \sqrt[n]{n}}{n} \to 1$.

*18 If $\{x_n\}$ is a sequence of positive numbers such that $x_{n+1}/x_n \to r$, then we also have $\sqrt[n]{x_n} \to r$. Prove this as follows: put $y_n = \ln x_{n+1}/x_n$; show that

$$\frac{y_1 + y_2 + \cdots + y_{n-1}}{n} = \ln\sqrt[n]{x_n} - \ln\sqrt[n]{x_1};$$

and apply Problem 16.

19 Use Problem 18 to show that $\sqrt[n]{n!}/n \to 1/e$.

*20 Wallis's product, which can be expressed in the form

$$\frac{2}{1}\cdot\frac{2}{3}\cdot\frac{4}{3}\cdot\frac{4}{5}\cdot\frac{6}{5}\cdot\frac{6}{7}\cdots\frac{2n}{2n-1}\cdot\frac{2n}{2n+1} \to \frac{\pi}{2},$$

is proved in Appendix A.10. Since $2n/(2n+1) \to 1$, this can also be written as

$$\frac{2^2}{3^2}\cdot\frac{4^2}{5^2}\cdot\frac{6^2}{7^2}\cdots\frac{(2n-2)^2}{(2n-1)^2}\cdot 2n \to \frac{\pi}{2}.$$

By taking square roots and multiplying numerator and denominator by $2\cdot 4\cdot 6\cdots(2n-2)$, establish the formula

$$\lim_{n\to\infty}\frac{(n!)^2 2^{2n}}{(2n)!\sqrt{n}} = \sqrt{\pi},$$

which is needed in Problem 21.

*21 It is a remarkable fact that the function $f(n) = \sqrt{2\pi n}\, n^n e^{-n}$ is a good approximation to $n!$ for large n, in the sense that the relative error approaches zero:

$$\lim_{n\to\infty}\left[\frac{f(n) - n!}{n!}\right] = \lim_{n\to\infty}\left[\frac{f(n)}{n!} - 1\right] = 0.$$

This is equivalent to the statement that

$$\lim_{n \to \infty} \frac{n!}{\sqrt{2\pi n}\ n^n e^{-n}} = 1,$$

which is known as *Stirling's formula.** In addition to its intrinsic interest, this formula is a useful tool (in statistics and the theory of probability) for the approximate numerical calculation of $n!$ when n is large. Prove Stirling's formula by verifying the following statements:

(a) $2/(2n+1) \le \ln(1+1/n)$ (hint: compare the area under the curve $y = 1/x$ from $x = n$ to $x = n + 1$ with the area of the trapezoid whose top is tangent to the curve at $x = n + \frac{1}{2}$);

(b) $e \le \left(1 + \dfrac{1}{n}\right)^{n+1/2}$ $\left[\text{hint: } \left(n + \dfrac{1}{2}\right) \ln\left(1 + \dfrac{1}{n}\right) \le \right.$

$\left. \ln\left(1 + \dfrac{1}{n}\right)^{n+1/2}\right];$

(c) the area A under $y = \ln x$ from $x = 1$ to $x = n$ is

$$\int_1^n \ln x\ dx = n \ln n - n + 1 = \ln\left(\frac{n}{e}\right)^n + 1;$$

(d) the number x_n defined by

$$x_n = \frac{(n/e)^n \sqrt{n}}{n!}$$

is ≤ 1 [hint: compare the area A in part (c) with the area $B = 1 + \ln n! - \ln \sqrt{n}$ of the following figure: Divide the interval from $x = 1$ to $x = n$ into subintervals by the points $\frac{3}{2}, \frac{5}{2}, \ldots, n - \frac{1}{2}$; on the first and last subintervals construct rectangles with heights 2 and $\ln n$; and on the remaining subintervals construct trapezoids whose tops are tangent to the curve $y = \ln x$ at $x = 2, 3, \ldots, n - 1$];

(e) $\{x_n\}$ is an increasing sequence which is bounded by part (d), so $\lim x_n$ exists;

(f) $\lim x_n = \lim \dfrac{x_n{}^2}{x_{2n}} = \dfrac{1}{\sqrt{2\pi}}$

[hint: use the formula established in Problem 20];

(g) part (f) implies Stirling's formula.

---

* James Stirling (1692–1770) began his career by being expelled from Oxford for supporting the defunct Stuart dynasty, and ended it as the successful manager of a mining company. In his salad days he was a friend of Newton, and wrote an essay on infinite series in which he almost discovered the formula that bears his name.

SECTION 14.3

22 If $\sum_{n=1}^{\infty} a_n = s$, what is the sum of the series $\sum_{n=1}^{\infty}(a_n + a_{n+1})$?

23 For what values of x is

$$\frac{1}{x} + \frac{2}{x^3} + \frac{4}{x^5} + \cdots = \frac{x}{x^2 - 2}$$

valid?

24 Find the values of x for which

$$\frac{x}{1+x} - \left(\frac{x}{1+x}\right)^2 + \left(\frac{x}{1+x}\right)^3 - \cdots$$

converges. What is its sum?

*25 By finding a closed formula for the nth partial sum s_n, show that the series $\sum_{n=1}^{\infty} nx^n$ converges to $x/(1-x)^2$ when $|x| < 1$ and diverges otherwise.

26 Find the sum of the series $\frac{1}{3} + \frac{2}{9} + \frac{3}{27} + \cdots$.

*27 Use the fact that $\displaystyle\sum_{n=1}^{\infty} \frac{1}{n^2} = \frac{\pi^2}{6}$ to show that

(a) $\dfrac{1}{1^2} + \dfrac{1}{3^2} + \dfrac{1}{5^2} + \cdots = \dfrac{\pi^2}{8};$

(b) $\dfrac{1}{1^2} - \dfrac{1}{2^2} + \dfrac{1}{3^2} - \dfrac{1}{4^2} + \cdots = \dfrac{\pi^2}{12};$

(c) $\dfrac{1}{1^2} + \dfrac{1}{5^2} + \dfrac{1}{7^2} + \dfrac{1}{11^2} + \dfrac{1}{13^2} + \dfrac{1}{17^2} + \cdots = \dfrac{\pi^2}{9}.$

28 Show that

$$\int_0^\infty \frac{x\ dx}{e^x - 1} = \sum_{n=1}^{\infty} \frac{1}{n^2}$$

by expressing the integrand as a geometric series and integrating term by term.

*29 Show that

(a) $\dfrac{1}{1 \cdot 2} + \dfrac{1}{3 \cdot 4} + \dfrac{1}{5 \cdot 6} + \cdots = \ln 2;$

(b) $\dfrac{1}{2 \cdot 3} + \dfrac{1}{4 \cdot 5} + \dfrac{1}{6 \cdot 7} + \cdots = 1 - \ln 2;$

(c) $\dfrac{1}{1 \cdot 2} - \dfrac{1}{2 \cdot 3} + \dfrac{1}{3 \cdot 4} - \dfrac{1}{4 \cdot 5} + \cdots$

$$= 2 \ln 2 - 1.$$

30 Show that

(a) $\displaystyle\sum_{n=1}^{\infty} \frac{n}{(n+1)!} = \frac{1}{2!} + \frac{2}{3!} + \frac{3}{4!} + \cdots = 1;$

(b) $\displaystyle\sum_{n=1}^{\infty} \frac{1}{n(n+1)(n+2)} = \frac{1}{4};$

(c) $\sum_{n=1}^{\infty} \dfrac{1}{n(n+1)\cdots(n+k)} = \dfrac{1}{k \cdot k!}$.

31 Find the sum of

(a) $\sum_{n=2}^{\infty} \ln\left(1 - \dfrac{1}{n^2}\right)$;

(b) $\sum_{n=1}^{\infty} \dfrac{1}{(n+1)\sqrt{n} + n\sqrt{n+1}}$.

*32 If $f(n) \to L$, show that

$$\sum_{n=1}^{\infty} [f(n) - f(n+2)] = f(1) + f(2) - 2L$$

and use this to establish the following statements:

(a) $\dfrac{1}{1\cdot 3} + \dfrac{1}{2\cdot 4} + \dfrac{1}{3\cdot 5} + \dfrac{1}{4\cdot 6} + \cdots = \dfrac{3}{4}$;

(b) $\dfrac{1}{1\cdot 3} - \dfrac{1}{2\cdot 4} + \dfrac{1}{3\cdot 5} - \dfrac{1}{4\cdot 6} + \cdots = \dfrac{1}{4}$.

*33 If $a_1, a_2, a_3, \ldots$ are the positive integers whose decimal representations do not contain the digit 5, show that $\Sigma 1/a_n$ converges and has sum < 90.

*34 Figure 14.9 shows the region bounded by two circles

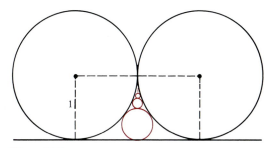

Figure 14.9

of radius 1 that are tangent to each other and by a straight line tangent to both. A sequence of smaller circles, each having the largest possible radius, is inscribed in the region in the manner shown in the figure. It is clear from the geometry of the situation that the lengths of the diameters of these smaller circles are the terms of a convergent series whose sum is 1. Show that this series is

$$\sum_{n=1}^{\infty} \dfrac{1}{n(n+1)} = \dfrac{1}{1\cdot 2} + \dfrac{1}{2\cdot 3} + \dfrac{1}{3\cdot 4} + \cdots.$$

SECTION 14.4

35 Determine whether each of the following series converges or diverges:

(a) $\sum \dfrac{2}{n^2 + n}$;

(b) $\sum \dfrac{n^2 + 31}{10,000n^3}$;

(c) $\sum \dfrac{1}{(n+3)^2}$;

(d) $\sum \dfrac{n}{\sqrt{n^2 + 2}}$;

(e) $\sum \dfrac{1}{[1 + (n-1)/n]^n}$;

(f) $\sum \dfrac{\sqrt{n}}{n+5}$;

(g) $\sum \dfrac{\tan^{-1} n}{n^3}$;

(h) $\sum \dfrac{3}{2 + \sqrt{n}}$;

(i) $\sum \dfrac{(3n+1)^3}{(n^3 + 2)^2}$;

(j) $\sum \dfrac{1}{\sqrt[3]{3n^2 + 1}}$;

(k) $\sum \dfrac{1}{\sqrt{n(n+1)(n+2)}}$;

(l) $\sum \dfrac{2n+3}{n \cdot 3^n}$;

(m) $\sum \dfrac{5n - 7}{(n+5)n!}$;

(n) $\sum \sqrt{\sin^3 \dfrac{1}{n}}$;

(o) $\sum \left(\dfrac{n+1}{2n}\right)^n$;

(p) $\sum \left(\dfrac{n^2 - 1}{n^3 + 3}\right)^{1/3}$;

(q) $\sum \dfrac{\sqrt{n+1}}{n^{n+1/2}}$;

(r) $\sum \dfrac{2^n + 3^n}{3^n + 4^n}$;

(s) $\sum (1 - e^{-1/n})^n$;

(t) $\sum \dfrac{(n+1)^n}{n^{n+1}}$;

(u) $\sum \sin^2 \pi \left(n + \dfrac{1}{n}\right)$;

(v) $\sum \dfrac{[\ln(n+1)]^n}{n^{n+1}}$.

36 If $a > 1$, show that $\Sigma 1/a^{\ln n}$ diverges if $a \le e$ and converges if $a > e$.

37 Prove that any convergent series of positive terms can be rearranged so that its terms form a decreasing sequence.

38 If p is any positive constant, show that $\Sigma 1/(\ln n)^p$ diverges.

39 Show that the series

$$\sum \dfrac{1}{(\ln n)^{\ln n}} \quad \text{and} \quad \sum \dfrac{1}{(\ln \ln n)^{\ln n}}$$

are both convergent. Hint: Express $(\ln n)^{\ln n}$ as a power of n.

40 Show that

$$\sum \dfrac{1}{(\ln n)^{\ln \ln n}}$$

diverges. Hint: $(\ln \ln n)^2 \le \ln n$ for large n (why?).

41 If $a_n \ge 0$ and Σa_n converges, and if $\{b_n\}$ is a bounded sequence of nonnegative numbers, prove that $\Sigma a_n b_n$

also converges. Use the series $1 - \frac{1}{2} + \frac{1}{3} - \frac{1}{4} + \cdots$ to show that this is false if the assumptions $a_n \geq 0$ and $b_n \geq 0$ are dropped.

42 If Σa_n and Σb_n are series of nonnegative terms such that Σa_n^2 and Σb_n^2 both converge, show that $\Sigma a_n b_n$ also converges.

SECTION 14.5

43 Show that

$$\sum_{n=3}^{\infty} \frac{1}{n \ln n \ln \ln n}$$

diverges, and also that if p is a positive constant, then

$$\sum_{n=3}^{\infty} \frac{1}{n \ln n (\ln \ln n)^p}$$

converges if $p > 1$ and diverges if $p \leq 1$.

44 If k is any integer > 1, show that

$$\left(\frac{1}{n+1} + \frac{1}{n+2} + \cdots + \frac{1}{kn} \right) \to \ln k.$$

45 The sum of the convergent series $\Sigma_{n=1}^{\infty} 1/n^3$ is not known. However, if this sum is denoted by s, show that

(a) $\frac{1}{1^3} + \frac{1}{3^3} + \frac{1}{5^3} + \cdots = \frac{7}{8} s$;

(b) $\frac{1}{1^3} - \frac{1}{2^3} + \frac{1}{3^3} - \frac{1}{4^3} + \cdots = \frac{3}{4} s.$

*46 For $p > 1$, the sum of the p-series $\Sigma_{n=1}^{\infty} 1/n^p$ is a function of p called the *zeta function* (the symbol ζ is the Greek letter *zeta*) and denoted by $\zeta(p)$; that is,

$$\zeta(p) = \sum_{n=1}^{\infty} \frac{1}{n^p} = 1 + \frac{1}{2^p} + \frac{1}{3^p} + \cdots.$$

Euler discovered that $\zeta(2) = \pi^2/6$, $\zeta(4) = \pi^4/90$, and $\zeta(6) = \pi^6/945$ (see Appendix A.18), but the value of $\zeta(p)$ is not known when p is odd.

(a) Use the inequalities (5) in Section 14.5 to show that the zeta function satisfies the inequalities

$$\frac{1}{p-1} \leq \zeta(p) \leq \frac{p}{p-1}$$

and

$$1 \leq \zeta(p) \leq \frac{p}{p-1}.$$

(b) Show that $\lim_{p \to 1} \zeta(p) = \infty$ and $\lim_{p \to \infty} \zeta(p) = 1$.

(c) Show that $\lim_{a \to 0+} a \sum_{n=1}^{\infty} \frac{1}{n^{1+a}} = 1.$

*47 Let k be an integer > 1 and show that

$$\sum_{n=1}^{\infty} \frac{a_n(k)}{n} = \ln k,$$

where $a_n(k)$ is defined by

$$a_n(k) = \begin{cases} 1 & \text{if } n \text{ is not a multiple of } k, \\ -(k-1) & \text{if } n \text{ is a multiple of } k. \end{cases}$$

*48 The *Cauchy condensation test* states that if $a_1, a_2, \ldots, a_n, \ldots$ is a decreasing sequence of positive numbers, then the two series

$$\sum_{n=1}^{\infty} a_n = a_1 + a_2 + a_3 + a_4 + \cdots$$

and

$$\sum_{n=0}^{\infty} 2^n a_{2^n} = a_1 + 2a_2 + 4a_4 + 8a_8 + \cdots$$

converge or diverge together. (This statement is called the "condensation test" because it says that a rather small proportion of the terms of the first series determines its convergence behavior.)

(a) Prove the condensation test. Hint: If s_n and t_n are the partial sums, group the terms of the first series into blocks to show that $s_n \leq t_m$ if $n \leq 2^m$, and $t_m \leq 2s_n$ if $2^m \leq n$.

(b) Use the condensation test to show that the series

$$\sum \frac{1}{n} \quad \text{and} \quad \sum \frac{1}{n \ln n}$$

diverge, and that the series

$$\sum \frac{1}{n^p} \quad \text{and} \quad \sum \frac{1}{n(\ln n)^p}$$

converge if $p > 1$.

49 Prove that

$$1 + \frac{1}{3} + \frac{1}{5} + \cdots + \frac{1}{2n-1} = \frac{1}{2} \ln n$$
$$+ \ln 2 + \frac{1}{2} \gamma + o(1).$$

*50 Show that

(a) $\sum_{n=1}^{\infty} \frac{1}{n(2n+1)} = 2 - 2 \ln 2$;

(b) $\sum_{n=1}^{\infty} \frac{1}{n(4n^2-1)} = 2 \ln 2 - 1$;

(c) $\sum_{n=1}^{\infty} \frac{1}{n(9n^2-1)} = \frac{3}{2}(\ln 3 - 1)$;

(d) $\displaystyle\sum_{n=1}^{\infty} \frac{1}{n(16n^2 - 1)} = 3 \ln 2 - 2;$

(e) $\displaystyle\sum_{n=1}^{\infty} \frac{1}{n(36n^2 - 1)} = \frac{3}{2} \ln 3 + 2 \ln 2 - 3;$

(f) $\displaystyle\sum_{n=1}^{\infty} \frac{n}{(4n^2 - 1)^2} = \frac{1}{8};$

(g) $\displaystyle\sum_{n=1}^{\infty} \frac{1}{n(4n^2 - 1)^2} = \frac{3}{2} - 2 \ln 2.$

SECTION 14.6

Use the ratio test to determine the behavior of the following series.

51 $\displaystyle\sum \frac{n}{e^n}.$

52 $\displaystyle\sum n^{1000}(\tfrac{2}{3})^n.$

53 $\displaystyle\sum \frac{1 \cdot 3 \cdot 5 \cdots (2n - 1)}{2 \cdot 4 \cdot 6 \cdots (2n)}.$

54 $\displaystyle\sum \frac{1 \cdot 3 \cdot 5 \cdots (2n - 1)}{1 \cdot 4 \cdot 7 \cdots (3n - 2)}.$

55 $\displaystyle\sum \frac{1 \cdot 6 \cdot 11 \cdots (5n - 4)}{2 \cdot 6 \cdot 10 \cdots (4n - 2)}.$

56 $\displaystyle\sum \frac{1000^n}{n!}.$

57 $\displaystyle\sum \frac{(n + 3)!}{n! 3^n}.$

58 $\displaystyle\sum \frac{2^{2n}}{2 \cdot 4 \cdot 6 \cdots (2n)}.$

59 $\dfrac{1}{2} + \dfrac{1 \cdot 4}{2 \cdot 4} + \dfrac{1 \cdot 4 \cdot 7}{2 \cdot 4 \cdot 6} + \cdots$

$\qquad + \dfrac{1 \cdot 4 \cdot 7 \cdots (3n - 2)}{2 \cdot 4 \cdot 6 \cdots (2n)} + \cdots.$

60 (a) Show that the ratio test fails for the series

$$\sum \frac{1}{2^{n + (-1)^n}}.$$

(b) Show that the root test succeeds for the series in part (a) and tells us that this series converges.

(Thus, the root test works in some cases where the ratio test fails. Even more can be said, for Additional Problem 18 asserts that if $\{a_n\}$ is any sequence of positive numbers, then

$$\frac{a_{n+1}}{a_n} \to L \qquad \text{implies} \qquad \sqrt[n]{a_n} \to L.$$

In principle, therefore, the root test is more powerful than the ratio test.)

61 Consider the series

$$\sum_{n=1}^{\infty} a_n = a + b + a^2 + b^2 + a^3 + b^3 + \cdots,$$

where $0 < a < b < 1$. Show that the ratio test fails, and establish convergence by using the root test.

SECTION 14.7

62 Show that the series

$$\frac{1}{\sqrt{2} - 1} - \frac{1}{\sqrt{2} + 1} + \frac{1}{\sqrt{3} - 1} - \frac{1}{\sqrt{3} + 1} + \cdots$$

diverges. Does this contradict the alternating series test?

63 Use the alternating series test to prove the existence of Euler's constant γ as follows.

(a) Show that the series

$$1 - \int_1^2 \frac{dx}{x} + \frac{1}{2} - \int_2^3 \frac{dx}{x} + \frac{1}{3} - \int_3^4 \frac{dx}{x}$$
$$+ \frac{1}{4} - \int_4^5 \frac{dx}{x} + \cdots$$

converges.

(b) If the sum of the series in (a) is denoted by γ, show that

$$s_{2n-1} = 1 + \frac{1}{2} + \frac{1}{3} + \cdots + \frac{1}{n} - \ln n,$$

so that

$$\lim_{n \to \infty} \left(1 + \frac{1}{2} + \cdots + \frac{1}{n} - \ln n\right) = \gamma.$$

64 Use the alternating series test to show that the improper integral

$$\int_0^{\infty} \frac{\sin x}{x}\, dx$$

converges. Hint: Sketch the graph of $y = (\sin x)/x$ for $x > 0$ and observe that it consists of an infinite number of parts, each covering an interval of length π and lying alternately above and below the x-axis; and then express the integral as an alternating series,

$$\int_0^{\infty} \frac{\sin x}{x}\, dx = \int_0^{\pi} \frac{\sin x}{x}\, dx + \int_{\pi}^{2\pi} \frac{\sin x}{x}\, dx + \cdots$$
$$= a_1 - a_2 + a_3 - a_4 + \cdots.$$

*65 Show that $1 - \frac{1}{4} + \frac{1}{7} - \frac{1}{10} + \cdots = \frac{1}{3} \ln 2 + \pi/3\sqrt{3}.$

*66   Show that

$$\sum_{n=2}^{\infty} (-1)^n \frac{\ln n}{n} = \gamma \ln 2 - \frac{1}{2} (\ln 2)^2.$$

Hint: See equation (9) in Section 14.5.

SECTION 14.8

67   Consider a power series $\Sigma a_n x^n$ and assume that $\lim \sqrt[n]{|a_n|}$ exists, with ∞ as an allowed value. Show that the radius of convergence R of the series is given by the formula

$$R = \frac{1}{\lim \sqrt[n]{|a_n|}}.$$

Use this formula to find the radius of convergence of

(a) $\sum \dfrac{x^n}{n^n}$;   (b) $\sum \dfrac{1}{(\ln n)^n} x^n$;

(c) $\sum \dfrac{n^{n^2}}{(n+1)^{n^2}} x^n.$

68   If the radius of convergence of $\Sigma a_n x^n$ can be calculated from formula (7) in Section 14.8, show that it can also be calculated from the formula in Problem 67. (Hint: See Additional Problem 18.) Show that the latter formula is more powerful than the former by considering the series

$$\sum_{n=1}^{\infty} \frac{x^n}{2^{n+(-1)^{n+1}}} = \frac{x}{2^2} + \frac{x^2}{2^1} + \frac{x^3}{2^4} + \frac{x^4}{2^3} + \cdots .$$

*69   If a power series converges conditionally at a point x_1, or diverges in such a way that its terms are bounded, show that x_1 must be an endpoint of the interval of convergence.

70   Use Problem 69 to find by inspection the radius of convergence of

(a) $1 + \dfrac{x}{3} + \dfrac{x^2}{3^2} + \dfrac{x^3}{3^3} + \cdots$;

(b) $x^2 - \dfrac{x^4}{2} + \dfrac{x^6}{3} - \cdots$;

(c) $\Sigma x^{n!} = x + x + x^2 + x^6 + x^{24} + \cdots$;

(d) $1 + \dfrac{x}{2} + \dfrac{x^2}{3^2} + \dfrac{x^3}{2^3} + \dfrac{x^4}{3^4} + \dfrac{x^5}{2^5} + \dfrac{x^6}{3^6} + \cdots$;

(e) $\sum [2^{(-1)^n} x]^n = 1 + \dfrac{x}{2} + 2^2 x^2 + \dfrac{x^3}{2^3}$

$$+\ 2^4 x^4 + \cdots .$$

71   Find the interval of convergence of $\Sigma a_n x^n$ if its coefficients are chosen in order from among the numbers 2, 3, . . . , 12 by throwing a pair of dice.

SECTION 14.9

72   Consider a power series $\Sigma a_n x^n$ in which the coefficients repeat cyclically, $a_{n+k} = a_n$. Show that
(a) $R = 1$;
(b) the sum is

$$\frac{a_0 + a_1 x + a_2 x^2 + \cdots + a_{k-1} x^{k-1}}{1 - x^k}.$$

73   Find the sum of each of the following series:

(a) $x - \dfrac{x^3}{3^2} + \dfrac{x^5}{5^2} - \dfrac{x^7}{7^2} + \cdots$;

(b) $\dfrac{x^2}{1 \cdot 2} - \dfrac{x^3}{2 \cdot 3} + \dfrac{x^4}{3 \cdot 4} - \dfrac{x^5}{4 \cdot 5} + \cdots$;

(c) $1 + 4x + 9x^2 + 16x^3 + 25x^4 + \cdots$;

(d) $\dfrac{x^4}{4} + \dfrac{x^8}{8} + \dfrac{x^{12}}{12} + \dfrac{x^{16}}{16} + \cdots$;

(e) $x + 2^3 x^2 + 3^3 x^3 + 4^3 x^4 + \cdots$;

(f) $4 + 5x + 6x^2 + 7x^3 + \cdots .$

SECTION 14.10

74   Use any method to obtain each of the following Taylor series expansions as far as indicated:

(a) $e^{\sin x} = 1 + x + \frac{1}{2} x^2 - \frac{1}{8} x^4 + \cdots$;

(b) $\dfrac{1}{1 + e^x} = \dfrac{1}{2} - \dfrac{1}{4} x + \dfrac{1}{48} x^3 + \cdots$;

(c) $e^{x^2 - x} = 1 - x + \frac{3}{2} x^2 - \frac{7}{6} x^3 + \frac{25}{24} x^4 + \cdots .$

75   Consider the binomial series

$$(1 + x)^p = 1 + px + \frac{p(p-1)}{2!} x^2$$

$$+ \frac{p(p-1)(p-2)}{3!} x^3 + \cdots$$

$$+ \frac{p(p-1)(p-2) \cdots (p-n+1)}{n!} x^n + \cdots ,$$

where p is an arbitrary constant. In Problem 13 of Section 14.10 the series on the right was obtained as the Taylor series of the function $(1 + x)^p$, and we also saw that this series converges for $|x| < 1$. We here outline a sequence of steps to prove that the series on the right actually converges *to the function on the left* for these values of x.

(a) Let $f(x)$ denote the sum of the series for $|x| < 1$, calculate $f'(x)$ and $x f'(x)$, and show that

$$(1 + x) f'(x) = p f(x).$$

(b) Define $g(x)$ by

$$g(x) = \frac{f(x)}{(1 + x)^p}.$$

and use part (a) to show that $g'(x) = 0$ for $|x| < 1$, so that $g(x) = c$ for some constant c.

(c) Show that $c = 1$ in part (b), and conclude that

$$(1 + x)^p = f(x).$$

SECTION 14.11

76　If $f_1(x) = \sum_{n=1}^{\infty} nx^n$, calculate $(1 - x)f_1(x)$, and use the result to find a closed formula for $f_1(x)$.

77　Use the idea of Problem 76 to find a closed formula for $f_2(x) = \sum_{n=1}^{\infty} n(n + 1)x^n$.

78　Use the idea of Problems 76 and 77 to find a closed formula for $f_3(x) = \sum_{n=1}^{\infty} n(n + 1)(n + 2)x^n$.

*79　In the notation of Problems 76 to 78, show that

$$\sum_{n=1}^{\infty} n^2 x^n = f_2(x) - f_1(x)$$

and

$$\sum_{n=1}^{\infty} n^3 x^n = f_3(x) - 3f_2(x) + f_1(x).$$

Using these ideas as a starting point, devise a proof of the following theorem: If $p(n)$ is a polynomial in n, then $f(x) = \sum_{n=0}^{\infty} p(n)x^n$ is a rational function.

*80　Show that $1/\sqrt{1 - 2xt + t^2} = \sum_{n=0}^{\infty} P_n(x)t^n$, where $P_n(x)$ is a polynomial of degree n, by substituting $h = 2xt - t^2$ in the binomial series for $1/\sqrt{1 - h}$ [see Problem 17 in Section 14.11]. Find $P_0(x)$, $P_1(x)$, $P_2(x)$, and $P_3(x)$. The polynomials $P_n(x)$ are called the *Legendre polynomials*; they are important in mathematical physics, for instance, in the study of heat flow in solid spheres.

81　Calculate the following limits by using Taylor series:

(a) $\lim\limits_{x \to 0} \dfrac{x \cos x - \sin x}{x^2 \tan x}$;

(b) $\lim\limits_{x \to 0} \dfrac{\sin x - \tan x}{\sin^2 x}$;

(c) $\lim\limits_{x \to 0} \dfrac{\sqrt{1 + x^2} + \cos x - 2}{x^4}$.

15

CONIC SECTIONS

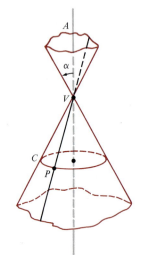

Figure 15.1

In order to understand the central ideas of Chapter 14, it was necessary to pay close attention to the precise wording of definitions and to the details of proofs, so the level of mathematical rigor in that chapter was rather high. However, we now turn to work that is mostly geometric in nature. We shall therefore rely much more heavily on reasoning based on spatial intuition and the kind of insight that can be obtained from carefully drawn figures.

Consider a circle C. Let A be the line through the center of C perpendicular to the plane of C, and let V be a point on A not in the plane of C, as shown in Fig. 15.1. Let P be a point on C, and draw the infinite straight line through P that also passes through V. As P moves around C, the line PV sweeps out a *right circular cone* with *axis A* and *vertex V*. Each of the lines PV is called a *generator* of the cone, and the angle α between the axis and any generator is called the *vertex angle.* The cone shown in Fig. 15.1 has a vertical axis, and the upper and lower portions of the cone that meet at the vertex are called the *nappes* of the cone. In elementary geometry a cone is usually understood to be a solid figure occupying the bounded region of space that lies between V and the plane of C and is inside the surface we have just described. However, in the present context the cone is this surface itself, and is understood to consist of both nappes, extending to infinity in both directions.

The curves obtained by slicing a cone with a plane that does not pass through the vertex are called *conic sections,* or simply *conics.* If the slicing plane is parallel to a generator, the conic is called a *parabola.* Otherwise, the conic is called an *ellipse* or a *hyperbola,* depending on whether the plane cuts just one or both nappes. The hyperbola is to be thought of as a single curve consisting of two "branches," one on each nappe. These three curves are illustrated in Fig. 15.2.

The three curves shown in Fig. 15.2 can be described in another way. Imagine a source of light placed at V and a circular ring placed at C. Then the shadow cast by the ring on a plane will be a parabola, an ellipse, or one branch of a hyperbola, depending on the steepness of the plane. If the plane is parallel to one of the lines joining V to C, we get a parabolic shadow; the shadow will be an ellipse if the plane is less steep than this, and part of a hyperbola if it is more steep.

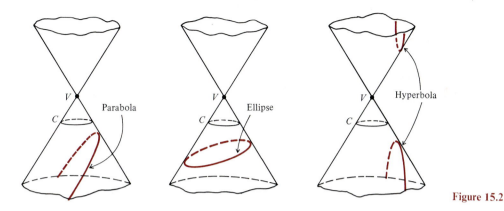

Figure 15.2

It should be noted that if we move each of the slicing planes in Fig. 15.2 parallel to itself until it passes through the vertex, then we get three so-called "degenerate" conic sections, namely, a single straight line, a point, and a pair of intersecting straight lines.

Many important discoveries in both pure mathematics and science have been linked to the conic sections. The classical Greeks — Archimedes, Apollonius and others — studied these beautiful curves for the sheer pleasure of it, as a form of play, without any thought of their possible uses. The first applications appeared almost 2000 years later, at the beginning of the seventeenth century. About the year 1604 Galileo discovered that if a projectile is fired horizontally from the top of a tower and is assumed to be acted on only by the force of gravity — that is, if air resistance and other complicating factors are ignored — then the path of the projectile will be a parabola. One of the great events in the history of astronomy occurred only a few years later, in 1609, when Kepler published his discovery that the orbit of Mars is an ellipse and then went on to suggest that all the planets move in elliptical orbits. And about 60 years after this, Newton was able to prove mathematically that an elliptical planetary orbit implies, and is implied by, an inverse square law of gravitational attraction. This led Newton to formulate and publish (in 1687) his famous theory of universal gravitation as the explanation of the mechanism of the solar system, which has been described as the greatest contribution to science ever made by one man. These developments took place hundreds of years ago, but the study of conic sections is far from outdated even today. Indeed, these curves are important tools for present-day explorations of outer space, and also for research into the behavior of atomic particles. Artificial satellites move around the earth in elliptical orbits, and the path of an alpha particle moving in the electric field of an atomic nucleus is a hyperbola. These examples and many others show that the importance of conic sections, both historically and in modern times, is difficult to exaggerate.

We shall be studying the conic sections as plane curves. For this purpose it is convenient to make use of equivalent definitions that refer only to the plane in which the curves lie and depend on special points in this plane called *foci* (*focus* is the singular). An ellipse can be defined as the set of all points in the plane the sum of whose distances d_1 and d_2 from two fixed points F and F'

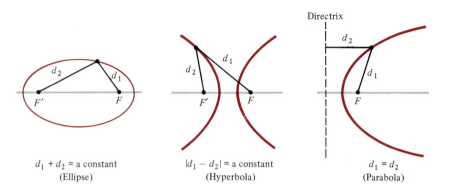

$d_1 + d_2 = $ a constant
(Ellipse)

$|d_1 - d_2| = $ a constant
(Hyperbola)

$d_1 = d_2$
(Parabola)

Figure 15.3

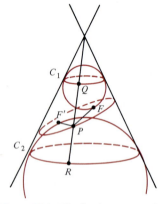

Figure 15.4 The focal property of an ellipse.

(the foci) is constant, as shown on the left in Fig. 15.3. A hyperbola is the set of all points for which the difference $|d_1 - d_2|$ is constant. And a parabola is the set of all points for which the distance to a fixed point F (the focus) equals the distance to a fixed line (called the *directrix*).

There is a simple and elegant argument which shows that the focal property of an ellipse follows from its definition as a section of a cone. This proof uses the two spheres shown in Fig. 15.4, which are internally tangent to the cone along the horizontal circles C_1 and C_2, and are also tangent to the slicing plane at the points F and F'. If P is an arbitrary point on the ellipse, we must show that the sum of the distances $PF + PF'$ is constant in the sense that it does not depend on the particular position of P. To see this, we notice that if Q and R are the points on C_1 and C_2 that lie on the generator through P, then $PF = PQ$ and $PF' = PR$, because any two tangents to a sphere drawn from a common external point have the same length. It follows that $PF + PF' = PQ + PR = QR$; and the argument is completed by observing that QR, as the distance from C_1 to C_2 down a generator, has the same value for every position of P.

With slight modifications this proof also works for the hyperbola and the parabola. In the case of the hyperbola, we use one sphere in each portion of the cone, with both spheres tangent to the slicing plane. And for the parabola we use one sphere tangent to the slicing plane. The focus of the parabola is this point of tangency, and its directrix is the line of intersection of the slicing plane with the plane of the circle along which the sphere is internally tangent to the cone. Students should use these hints to draw suitable pictures and prove for themselves that the focal properties of the hyperbola and the parabola can be derived from their definitions as sections of a cone.

15.2
ANOTHER LOOK AT CIRCLES AND PARABOLAS

Circles and parabolas were discussed fairly thoroughly in Chapter 1. However, that was a long time ago, and it may be helpful to give a very brief review of the main facts in order to assist students in fitting these topics into the context of the present chapter.

CIRCLES

Referring to Fig. 15.4, we see at once that a circle can be thought of as the special case of an ellipse obtained by taking the slicing plane perpendicular to the axis of the cone, so that the foci coincide. Nevertheless, for several reasons it is convenient to discuss circles separately, and to reserve the word "ellipse" for the case in which the foci are two distinct points.

A *circle,* therefore—as we very well know—can be defined as a plane curve consisting of the set of all points at a given fixed distance (called the *radius*) from a given fixed point (called the *center*). If $r > 0$ is the radius and (h, k) is the center, and if (x, y) is an arbitrary point on the circle (see Fig. 15.5), then by using the distance formula we can write the defining condition as

$$\sqrt{(x - h)^2 + (y - k)^2} = r$$

or

$$(x - h)^2 + (y - k)^2 = r^2, \tag{1}$$

which is the equation of the circle in standard form. By squaring the terms on the left of (1) and rearranging, this equation can be written in the form

$$x^2 + y^2 + Ax + By + C = 0. \tag{2}$$

Conversely, by completing the square on the x and y terms, any equation of the form (2) can be written in the form (1), and therefore represents a circle if $r^2 > 0$. As students will remember, there is a slight difficulty with this procedure as a result of the fact that the constant r^2 on the right of (1) may turn out to be zero or a negative number. In these cases, (1) can be thought of as the equation of a single point or the empty set.

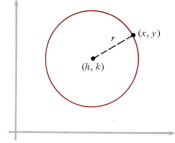

Figure 15.5 A circle.

PARABOLAS

As we saw in Section 15.1, a parabola can be defined as a plane curve consisting of the set of all points P that are equally distant from a given fixed point F and a given fixed line d, as shown on the left in Fig. 15.6. The fixed point is called the *focus,* and the fixed line is called the *directrix.* To find a simple equation for this curve, we introduce the coordinate system shown on the right in the figure, in which the focus is the point $F = (0, p)$, where p is a positive number, and the directrix is the line $y = -p$. If $P = (x, y)$ is an

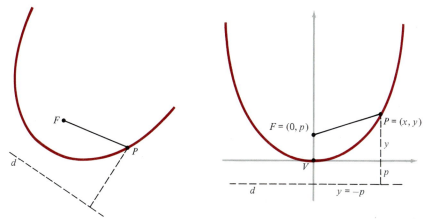

Figure 15.6 A parabola.

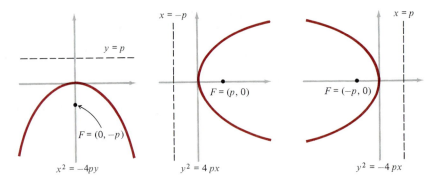

Figure 15.7

$x^2 = -4py$ $\qquad\qquad$ $y^2 = 4px$ $\qquad\qquad$ $y^2 = -4px$

arbitrary point on the parabola, then by using the distance formula the defining condition can be written as

$$\sqrt{x^2 + (y - p)^2} = y + p. \tag{3}$$

By squaring and simplifying we get

$$x^2 + y^2 - 2py + p^2 = y^2 + 2py + p^2$$

or

$$x^2 = 4py. \tag{4}$$

Conversely, by reversing the steps, it can be shown that (3) can be derived from (4). Equation (4) is therefore the equation of this particular parabola in standard form. The line through the focus perpendicular to the directrix is called the *axis* of the parabola, and the point V where the parabola intersects the axis is called the *vertex*. For the parabola (4), the axis is clearly the y-axis, and the vertex is the origin.

If we change the position of the parabola relative to the coordinate axes, we naturally change its equation. Three other simple positions, each with its corresponding equation, are shown in Fig. 15.7. Students should verify the correctness of all three equations. We emphasize that the constant p is always understood to be a positive number; geometrically, it is the distance from the vertex to the focus, and also from the vertex to the directrix.

We illustrate a further point about parabolas by considering the equation

$$x^2 - 8x - y + 19 = 0.$$

If we write this as $x^2 - 8x = y - 19$ and complete the square on x, then the result is

$$(x - 4)^2 = y - 3.$$

If we now introduce new variables x' and y' by writing

$$x' = x - 4,$$

$$y' = y - 3,$$

then our equation becomes

$$x'^2 = y'.$$

The graph of this equation is clearly a parabola with vertical axis whose vertex lies at the origin in the x', y' coordinate system, and this origin is located at the point $(4, 3)$ in the x, y system, as shown in Fig. 15.8. In exactly the same way, any equation of the form

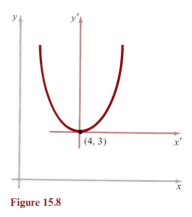

$$x^2 + Ax + By + C = 0, \qquad B \neq 0, \tag{5}$$

represents a parabola with vertical axis. The vertex of this parabola is easily located by completing the square on x, and in this way the equation can be written in the form

$$(x - h)^2 = 4p(y - k) \qquad \text{or} \qquad (x - h)^2 = -4p(y - k),$$

where the point (h, k) is the vertex.* Similarly, any equation of the form

$$y^2 + Ax + By + C = 0, \qquad A \neq 0,$$

Figure 15.8

represents a parabola with horizontal axis, and the geometric features of this parabola can be discovered by completing the square on y and writing the equation as

$$(y - k)^2 = 4p(x - h) \qquad \text{or} \qquad (y - k)^2 = -4p(x - h).$$

We conclude this section by describing the so-called *reflection property* of parabolas. Consider the tangent line at a point $P = (x, y)$ on the parabola $y^2 = 4px$ (Fig. 15.9), where $F = (p, 0)$ is the focus. As shown in the figure, let α be the angle between the tangent and the segment FP, and let β be the angle between the tangent and the horizontal line through P. In Problem 9, students are asked to prove that $\alpha = \beta$.

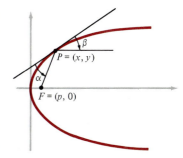

This geometric property of parabolas has many applications. For example, it is used in the design of mirrors for searchlights. To construct such a mirror, revolve the parabola about its axis to form a surface of revolution, then coat the inside with silver paint to make a reflecting surface. If a source of light is placed at F, each ray will be reflected along a line parallel to the axis. The same principle is used in a more important way in the design of mirrors for reflecting telescopes and solar furnaces, where rays of light that are parallel to the axis and come in toward the mirror are reflected in to the focus. This reflection property of parabolas also underlies the design of radar antennas and radio telescopes.

Figure 15.9 The reflection property.

* We point out here that if $B = 0$ is allowed in (5), then the graph of the equation can be one straight line, or two parallel lines, or the empty set. For the particular equation $x^2 - 2x - k = 0$, or equivalently $(x - 1)^2 = k + 1$, these cases correspond to $k = -1$, $k > -1$, and $k < -1$.

PROBLEMS

1 For each of the following equations, determine the nature of the graph by completing the square:
(a) $x^2 + y^2 - 2x - 6y - 15 = 0$;
(b) $x^2 + y^2 + 4x - 18y + 88 = 0$;
(c) $x^2 + y^2 - 10x + 2y + 26 = 0$;
(d) $x^2 + y^2 - 16x + 12y + 96 = 0$;
(e) $x^2 + y^2 + 6x - 14y + 58 = 0$;
(f) $x^2 + y^2 + 14x + 10y + 95 = 0$.

2 If $0 < a < b$, find the radius r and center (h, k) of the circle that passes through the points $(0, a)$ and $(0, b)$ and is tangent to the x-axis at a point to the right of the origin.

3 For each of the following parabolas, find the vertex, focus, and directrix:
(a) $x^2 + 4x - 4y = 0$;
(b) $y^2 - 8x - 2y + 25 = 0$;
(c) $x^2 + 4x + 16y - 76 = 0$;
(d) $y^2 + 12x - 2y + 25 = 0$;
(e) $y = x^2 + 2x + 3$.

4 A searchlight reflector is designed as stated in the text. If it is 2 ft deep and the opening is 5 ft across, find the focus.

5 Find the equation of the parabolic arch with base b and height h that is shown in Fig. 15.10.

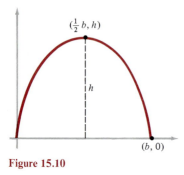

Figure 15.10

6 Show that the area of the parabolic segment in Fig. 15.10 is $\frac{2}{3}hb$. (Notice that this area is four-thirds the area of the triangle with the same base and height, a fact that was discovered and proved by Archimedes. See Appendix A.2.)

7 If the parabolic segment in Fig. 15.10 is revolved about its axis, show that the volume of the resulting solid of revolution is three-halves the volume of the inscribed cone.

8 A parabola with focus F and directrix d that are given and marked on a sheet of paper can be constructed as follows (see Fig. 15.11). On a drafting board, fasten a ruler to the paper with its edge along d, and place the short leg AB of a draftsman's triangle ABC against the edge of the ruler. At the opposite vertex C of the triangle fasten one end of a piece of string whose length is the same as that of the long leg BC of the triangle, and fasten the other end of the string at F. If a pencil point at P keeps the string taut, as shown in the figure, then the point of the pencil draws part of a parabola as the triangle slides along the ruler. Explain why this construction works.

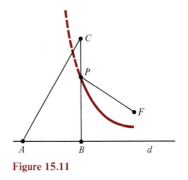

Figure 15.11

9 Prove that $\alpha = \beta$ in Fig. 15.9. Hint: Extend FP through P and use the subtraction formula for the tangent to show that $\tan \alpha = \tan \beta$.

10 Show that the lines tangent to a parabola at the ends of a focal chord (a chord through the focus) intersect at right angles.

*11 Show that the lines tangent to a parabola at the ends of a focal chord intersect on the directrix.

15.3
ELLIPSES

In Section 15.2 we gave "set of all points" definitions for both circles and parabolas. It is also possible to give "locus" definitions, in which each curve is defined—and thought of—as the path of a moving point that satisfies a certain condition as it moves. This language has the advantage of greater pictorial vividness. Thus, a parabola can be defined as the locus of a point that moves in such a way that it maintains equal distances from a given fixed point and a given fixed line.

Similarly, in accordance with Section 15.1, we can define an *ellipse* as the locus of a point P that moves in such a way that the sum of its distances from two fixed points F and F' is constant, as shown on the left in Fig. 15.12. To simplify later equations, we denote this constant by $2a$ and write the defining condition as

$$PF + PF' = 2a. \tag{1}$$

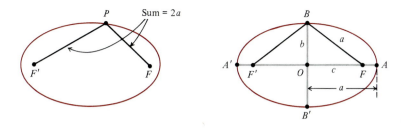

Figure 15.12

The two points F and F' are called the *foci* (plural of *focus*) of the ellipse because of the reflection property discussed in Remark 1. Since circles are not considered to be ellipses in this discussion, F and F' are understood to be two *distinct* points: $F \neq F'$.

The definition provides an easy way to draw an ellipse on a sheet of paper. We begin by fastening the paper to a drawing board with two tacks placed at F and F'. Next, we tie the ends of a piece of string to the tacks and pull the string taut with the point of a pencil. It is clear that if the pencil is moved around while the string is kept taut, then its point draws an ellipse. Because of this construction, the defining condition (1) is often called the *string property* of an ellipse.

We now introduce several standard notations for the dimensions of an ellipse. It is easy to see from the definition that the curve is symmetric with respect to the line through the foci, and also with respect to the perpendicular bisector of the segment FF'. On the right in Fig. 15.12 the segment AA' is called the *major axis* and the segment BB' is called the *minor axis* of the ellipse, and the point O where these axes intersect is called the *center*. The two points A and A' at the ends of the major axis are called the *vertices* of the ellipse. We denote the length of the minor axis by $2b$ and the distance between the foci by $2c$. It is clear that $BF = a$, because $BF + BF' = 2a$ and $BF = BF'$, so

$$a^2 = b^2 + c^2. \tag{2}$$

Since $AF + AF' = 2a$ and $AF' = FA'$, we see that $AA' = 2a$, and so the length of the major axis is $2a$. The numbers a and b are called the *semimajor axis* and the *semiminor axis*.

It is easy to see from equation (2) that $b < a$. If b is very small compared with a, so that the ellipse is long and thin, then (2) shows that c is nearly as large as a, and the foci are near the ends of the major axis; and if b is nearly as large as a, so that the ellipse is nearly circular, then c is small, and the foci are close to the center. The ratio c/a is called the *eccentricity* of the ellipse and is denoted by e:

$$e = \frac{c}{a} = \frac{\sqrt{a^2 - b^2}}{a}. \tag{3}$$

Notice that $0 < e < 1$. Nearly circular ellipses have eccentricity near 0, and long, thin ellipses have eccentricity near 1.

In order to find a simple equation for the ellipse, we take the x-axis along the segment FF' and the y-axis as the perpendicular bisector of this segment.

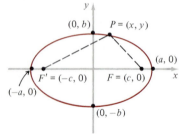

Figure 15.13

Then the foci are $F = (c, 0)$ and $F' = (-c, 0)$, as shown in Fig. 15.13, and the defining condition (1) yields

$$\sqrt{(x - c)^2 + y^2} + \sqrt{(x + c)^2 + y^2} = 2a \qquad (4)$$

as the equation of the curve.

To simplify this equation, we follow the usual procedure for eliminating radicals, namely, solve for one of the radicals and square. If we move the first radical in (4) over to the right-hand side, square both sides, and simplify, then we obtain

$$PF = \sqrt{(x - c)^2 + y^2} = a - \frac{c}{a} x \qquad (5)$$

and

$$PF' = \sqrt{(x + c)^2 + y^2} = a + \frac{c}{a} x, \qquad (6)$$

where (6) follows from (5) because $PF' = 2a - PF$. By squaring again and simplifying, either of these equations gives

$$\left(\frac{a^2 - c^2}{a^2} \right) x^2 + y^2 = a^2 - c^2$$

or

$$\frac{x^2}{a^2} + \frac{y^2}{a^2 - c^2} = 1. \qquad (7)$$

By using (2) to simplify (7) still further, we now put the equation into its final form,

$$\frac{x^2}{a^2} + \frac{y^2}{b^2} = 1. \qquad (8)$$

This argument shows that (8) is satisfied if (4) is. It can be shown, conversely, that (4) is satisfied if (8) is, but we omit the details. Equation (8) is therefore the standard form of the equation of the ellipse shown in Fig. 15.13.

We pause briefly to point out that equation (8) easily yields most of the simpler geometric features of the ellipse that are visible in Fig. 15.13. (i) If $y = 0$, then the equation tells us that $x = \pm a$, and if $x = 0$, then $y = \pm b$, so the curve crosses the x and y axes at the four points $(\pm a, 0)$ and $(0, \pm b)$. (ii) Since both terms x^2/a^2 and y^2/b^2 are nonnegative and their sum is 1, it follows that neither of them can be greater than 1, and so $|x| \leq a$ and $|y| \leq b$. This means that the whole ellipse is contained in the rectangle whose sides are $x = \pm a$ and $y = \pm b$, and is therefore — unlike the parabola — a bounded curve. (iii) If (x, y) satisfies the equation, then so do $(x, -y)$ and $(-x, y)$, so the curve is symmetric with respect to both the x-axis and the y-axis. This tells us that to graph the complete curve it suffices to sketch the graph in the first quadrant and then extend it to the other quadrants by symmetry. The left-right symmetry of the ellipse that is so obvious from equation (8) is really rather remarkable, because most people contemplating Fig. 15.2 for the first time feel quite sure that an ellipse should be an egg-shaped oval which has a "small end" at the part of the ellipse nearest the vertex of the cone and a "big end" at the part farthest from this vertex — but of course this is not true.

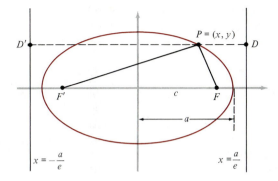

Figure 15.14

We consider again formulas (5) and (6) for the right and left *focal radii PF* and *PF′*, which can be written as

$$PF = a - \frac{c}{a}x = e\left[\frac{a}{e} - x\right] \tag{9}$$

and

$$PF' = a + \frac{c}{a}x = e\left(\frac{a}{e} + x\right) = e\left[x - \left(-\frac{a}{e}\right)\right], \tag{10}$$

where $e = c/a$ is the eccentricity defined earlier. The quantities in brackets can be interpreted (see Fig. 15.14) as the distances *PD* and *PD′* from *P* to the lines $x = a/e$ and $x = -a/e$, respectively. Formulas (9) and (10) can therefore be written in the form

$$\frac{PF}{PD} = e \quad \text{and} \quad \frac{PF'}{PD'} = e. \tag{11}$$

Each of the lines $x = a/e$ and $x = -a/e$ is called a *directrix* of the ellipse. Equations (11) show that *an ellipse can be characterized as the locus of a point that moves in such a way that the ratio of its distance from a fixed point (a focus) to its distance from a fixed line (the corresponding directrix) equals a constant e < 1.* We shall see in Chapter 16 and elsewhere that this way of characterizing ellipses is often very useful.

Example 1 Identify the graph of $16x^2 + 25y^2 = 400$ as an ellipse, and find its vertices, foci, eccentricity, and directrices. Sketch the graph.

Solution First, we divide by 400 to convert the equation into the standard form

$$\frac{x^2}{25} + \frac{y^2}{16} = 1,$$

which on comparison with (8) tells us that the graph is an ellipse. Since $a^2 = 25$ and $b^2 = 16$, we have $a = 5$ and $b = 4$, so the vertices are $(\pm 5, 0)$ and the ends of the minor axis are $(0, \pm 4)$, as shown in Fig. 15.15. Next, $c^2 = a^2 - b^2 = 25 - 16 = 9$, so $c = 3$ and the foci are $(\pm 3, 0)$. Finally, the eccentricity is $e = c/a = \frac{3}{5}$, and the directrices are the vertical lines $x = \pm a/e = \pm \frac{25}{3}$.

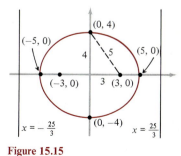

Figure 15.15

In the above discussion it is assumed that the ellipse has its center at the origin and its foci on the x-axis. However, if its center is the origin and its foci lie on the y-axis, then its major axis is vertical and the roles of x and y are interchanged.

Example 2   Show that $9x^2 + 4y^2 = 36$ represents an ellipse, and find its vertices, foci, eccentricity, and directrices. Sketch the graph.

Solution   As before, we divide by 36 to convert the given equation into the recognizable standard form

$$\frac{x^2}{4} + \frac{y^2}{9} = 1.$$

Observe that here *the denominator of the y term is larger,* so we have an ellipse whose major axis is vertical. The semimajor and semiminor axes are clearly $a = 3$ and $b = 2$, so the vertices (see Fig. 15.16) are $(0, \pm 3)$ and the ends of the minor axis are $(\pm 2, 0)$. Since $c^2 = a^2 - b^2 = 9 - 4 = 5$, $c = \sqrt{5}$, and the foci are the points $(0, \sqrt{5})$ and $(0, -\sqrt{5})$ on the y-axis. The eccentricity is $e = c/a = \sqrt{5}/3$, and the directrices are the horizontal lines $y = \pm a/e = \pm 9/\sqrt{5} = \pm \frac{9}{5}\sqrt{5}$.

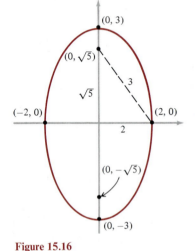

Figure 15.16

Examples 1 and 2 illustrate the fact that if we have an equation of the form

$$\frac{x^2}{(\ )^2} + \frac{y^2}{(\ )^2} = 1$$

with unequal denominators, then the equation represents an ellipse, and the question of whether the foci and major axis lie on the x-axis or the y-axis is determined by which denominator is larger.

In equation (8), x and y are the horizontal and vertical displacements from the axes of the ellipse to the point $P = (x, y)$. If the center is the point (h, k) instead of the origin, then these displacements are $x - h$ and $y - k$, and the equation of the ellipse becomes

$$\frac{(x - h)^2}{a^2} + \frac{(y - k)^2}{b^2} = 1. \tag{12}$$

Example 3   Show that $4x^2 + 16y^2 - 24x - 32y = 12$ is the equation of an ellipse, and find its vertices, foci, eccentricity, and directrices. Sketch the graph.

Solution   The equation can be written as

$$4(x^2 - 6x) + 16(y^2 - 2y) = 12.$$

Completing the squares inside the parentheses, we obtain

$$4(x - 3)^2 + 16(y - 1)^2 = 64$$

or

$$\frac{(x - 3)^2}{16} + \frac{(y - 1)^2}{4} = 1.$$

Comparison with (12) shows that this represents an ellipse with center (3, 1), horizontal major axis, and semiaxes $a = 4$, $b = 2$, and so the vertices (Fig. 15.17) are the points (7, 1), $(-1, 1)$ and the ends of the minor axis are (3, 3), $(3, -1)$. The foci are a distance $c = \sqrt{a^2 - b^2} = \sqrt{12} = 2\sqrt{3}$ to the right and left of the center, and are therefore the points $(3 \pm 2\sqrt{3}, 1)$. The eccentricity is $e = c/a = \frac{1}{2}\sqrt{3}$, and the directrices are vertical lines at a distance $a/e = 8/\sqrt{3} = \frac{8}{3}\sqrt{3}$ to the right and left of the center. Their equations are $x = 3 \pm \frac{8}{3}\sqrt{3}$.

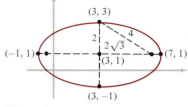

Figure 15.17

Remark 1 Like parabolas, ellipses also have a remarkable reflection property. Let P be a point on an ellipse with foci F and F', and let T be the tangent at P, as shown in Fig. 15.18. If T makes angles α and β with the two focal radii PF and PF', then $\alpha = \beta$. Students are asked to prove this in Problem 9.

This reflection property has no important applications like those we saw in the case of parabolas, but there is at least one mildly amusing consequence. Let the ellipse in the figure be revolved about its major axis to form a surface of revolution, and imagine that a room is built with its walls and ceiling having the shape of the upper part of this surface, with the two foci about shoulder height above the floor. Then a whisper uttered at one focus can be clearly heard a considerable distance away at the other focus even though it is inaudible at intermediate points, because the sound waves bounce off the walls and are reflected to the second focus, and furthermore arrive together because they all travel the same distance. There actually exist several rooms of this kind—known as "whispering galleries"—in certain American museums of science and in the castles of a few eccentric European monarchs.

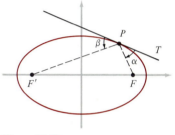

Figure 15.18

Remark 2 Except for small perturbations resulting from the influence of the other planets, each planet in the solar system revolves around the sun in an elliptical orbit with the sun at one focus. As we pointed out in Section 15.1, this phenomenon was discovered empirically by Kepler in the early seventeenth century, and was explained mathematically by Newton in the later decades of the same century. We shall give a detailed treatment of Newton's ideas at the end of Chapter 17.

Most of the planets, including the earth, have orbits that are nearly circular. This can be seen from the eccentricities given in the table in Fig. 15.19. Mercury, however, has a rather eccentric orbit, with $e = 0.21$, as does Pluto, with $e = 0.25$. Other bodies in the solar system have even more eccentric orbits, for instance, the flying mountains known as asteroids. Thus, the asteroid Icarus, which was discovered at Mount Palomar in 1949 and is about 1 mi in diameter, has an orbit so eccentric, with $e = 0.83$, that at its closest approach to the sun (Fig. 15.20) it is halfway between the sun and the orbit of Mercury, and at its farthest it is out beyond the orbit of the earth.*

Mercury	.21	Saturn	.06
Venus	.01	Uranus	.05
Earth	.02	Neptune	.01
Mars	.09	Pluto	.25
Jupiter	.05		

Figure 15.19 Eccentricities of planetary orbits.

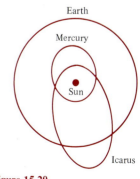

Figure 15.20

---

* The surface temperature of Icarus has been estimated at about 900°F at its closest approach to the sun. Arthur C. Clarke has used this fact as the basis for a fine story, "Summertime on Icarus," in his collection *The Nine Billion Names of God.* See also Chapter 2, "The Little Planets," in Fletcher G. Watson's *Between the Planets* (Doubleday Anchor Books, 1962), especially p. 29.

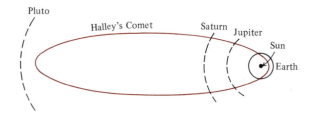

Pluto
Halley's Comet   Saturn   Jupiter
                                        Sun
                                        Earth

Figure 15.21   Orbit of Halley's
Comet, drawn approximately to scale.

One of the most interesting objects in the solar system is Halley's Comet, which has eccentricity $e = 0.98$ and an orbit (Fig. 15.21) about 7 astronomical units wide by 35 astronomical units long. [One astronomical unit (AU) is the semimajor axis of the earth's orbit, approximately 93 million miles or 150 million kilometers.] The period of revolution of this comet around the sun is about 76 years. It appeared in 1910, and again in 1985–1986. It was observed in 1682, and the astronomer Edmund Halley (Newton's friend) successfully predicted its return in 1758, many years after his own death in 1742. This was one of the most convincing successes of Newton's theory of gravitation. At its closest approach, Halley's Comet is only 0.59 AU away from the sun. Its previous visits to the near neighborhood of the sun have been traced back step by step by means of historical records to the year 11 B.C., and perhaps even earlier.*

---

* For a more detailed account of these remarkable events, see P. L. Brown, *Comets, Meteorites and Men* (Taplinger, 1974); or N. Calder, *The Comet Is Coming!* (Viking, 1980).

PROBLEMS

1 Find the equation of the ellipse
   (a) with foci at $(\pm 2, 0)$ and major axis of length 10;
   (b) with foci at $(0, \pm 4)$ and minor axis of length 12;
   (c) with major and minor axes of lengths 4 and 3, respectively, center at the origin, and foci on the y-axis;
   (d) with foci at $(\pm 3, 0)$ and eccentricity $e = \frac{3}{4}$;
   (e) with eccentricity $e = \frac{1}{2}$, center at the origin, and the ends of the major axis at $(0, \pm 6)$;
   (f) with eccentricity $e = \frac{1}{3}$ and the ends of the minor axis at $(0, \pm 10)$.

2 Find the equation of the ellipse
   (a) with vertices $(6, 2)$, $(-4, 2)$ and minor axis of length 6;
   (b) with major axis 8 units long, and foci at $(6, 3)$ and $(2, 3)$;
   (c) with minor axis 6 units long, and foci at $(1, 0)$ and $(1, 6)$;
   (d) with eccentricity $e = \frac{3}{4}$, and ends of the major axis at $(10, 1)$ and $(-6, 1)$.

3 Find the center, vertices, foci, and eccentricity of each of the following ellipses:
   (a) $25x^2 + 9y^2 = 225$;
   (b) $x^2 + 4y^2 = 4$;
   (c) $2(x + 2)^2 + (y - 1)^2 = 2$;
   (d) $x^2 + 4y^2 - 2x = 0$;
   (e) $4x^2 + 9y^2 - 16x + 18y = 11$;
   (f) $x^2 + 2y^2 - 8y = 0$.

4 Consider an equation of the form

$$Ax^2 + By^2 + Cx + Dy + E = 0,$$

where A and B are both positive or both negative and $A \neq B$. Show that the graph is an ellipse, a single point, or the empty set.

5 Write down the integrals that give (a) the first-quadrant area of the circle $x^2 + y^2 = a^2$, and (b) the first-quadrant area of the ellipse $x^2/a^2 + y^2/b^2 = 1$. Show that the second integral is b/a times the first, and in this way obtain the area of the ellipse from the known area of the circle.

6 If a cylindrical drinking glass with some water in it is tilted slightly, as in Fig. 15.22, then the surface of the water is no longer a circle but looks like an ellipse. Prove that it really is an ellipse by modifying the argument used in Fig. 15.4.

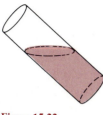

Figure 15.22

7 Find the volume of the solid of revolution obtained by revolving the ellipse $x^2/a^2 + y^2/b^2 = 1$ about
(a) the x-axis;
(b) the y-axis.
If $a > b$, the first solid is called a *prolate spheroid* and the second an *oblate spheroid.*

8 The base of a solid is the region bounded by an ellipse with semiaxes 5 and 3. Find the volume of the solid if each cross section in a plane perpendicular to the major axis is
(a) a square;
(b) an equilateral triangle.

9 Prove that $\alpha = \beta$ in Fig. 15.18. Hint: Extend FP and $F'P$ through P and show that $\tan \alpha = \tan \beta$ by using the subtraction formula for the tangent.

10 Consider two ellipses with the same eccentricity e, both centered at the origin and both with major axis on the x-axis. Suppose that their equations are

$$\frac{x^2}{a_1^2} + \frac{y^2}{b_1^2} = 1 \quad \text{and} \quad \frac{x^2}{a_2^2} + \frac{y^2}{b_2^2} = 1.$$

Show that these ellipses are similar in the sense that
(a) there exists a constant k such that

$$\frac{a_1}{a_2} = \frac{b_1}{b_2} = \frac{c_1}{c_2} = k;$$

(b) if a half-line from the origin O intersects the first ellipse at P_1 and the second at P_2, then

$$\frac{OP_1}{OP_2} = k.$$

11 Show that the line tangent to the ellipse $x^2/a^2 + y^2/b^2 = 1$ at a point $P_1 = (x_1, y_1)$ is

$$\frac{xx_1}{a^2} + \frac{yy_1}{b^2} = 1.$$

12 If tangent lines to the ellipse $x^2/25 + y^2/16 = 1$ intersect the y-axis at $(0, 8)$, find the points of tangency.

13 If tangent lines to the ellipse $x^2/a^2 + y^2/b^2 = 1$ intersect the y-axis at $(0, d)$, where $d > b$, find the points of tangency.

14 Let F be a point which is inside a given circle but is not the center C. Consider a point P that moves in such a way as to be equidistant from F and the circle. Show that the path of P is an ellipse.

15 Show that the point on an ellipse that is closest to a focus is the end of the major axis nearest that focus, and also that the point on the ellipse farthest from this focus is the other end of the major axis.

16 The *apogee* of an earth satellite is its maximum altitude above the surface of the earth during orbit, and its *perigee* is its minimum altitude during orbit.* If R is the radius of the earth, use Problem 15 to show that if a satellite has an elliptical earth orbit with the center of the earth at one focus and semimajor axis a, then

$$2a = 2R + \text{apogee} + \text{perigee}.$$

17 The point of the orbit of a planet nearest the sun is called the *perihelion,* and the point farthest from the sun is called the *aphelion.* If the ratio of the earth's distance from the sun at perihelion to its distance at aphelion is $\frac{29}{30}$, find the eccentricity of the earth's orbit.

18 A line segment moves with one end A on the y-axis and the other end B on the x-axis. A point P fixed on the segment is a units from A and b units from B. Find the equation of the path of P.

19 (a) Show that the length of the part of the ellipse $x^2/a^2 + y^2/b^2 = 1$ $(a > b)$ that lies in the first quadrant is

$$\int_0^a \sqrt{\frac{a^2 - e^2x^2}{a^2 - x^2}} \, dx.$$

(b) Use the change of variable $x = a \sin \theta$ to transform the integral in (a) into

$$a \int_0^{\pi/2} \sqrt{1 - e^2 \sin^2 \theta} \, d\theta.$$

This is called a *complete elliptic integral of the second kind,* and cannot be evaluated by means of elementary functions.

20 Let P be a point on the ellipse $x^2/a^2 + y^2/b^2 = 1$ that does not lie on either axis. If $a > b$, show without using calculus that the distance from P to the origin is greater than b and less than a.

21 There are exactly two lines with given slope m that are tangent to the ellipse $x^2/a^2 + y^2/b^2 = 1$. Find their equations.

---

* These words are also used to mean the corresponding points of the orbit.

22 Consider two circles centered at the origin with radii a and b, where $b < a$. Draw a line through the origin intersecting the smaller circle at Q and the larger circle at R. If the horizontal line through Q and the vertical line through R intersect at $P = (x, y)$, show that P lies on the ellipse $x^2/a^2 + y^2/b^2 = 1$. Hint: Let the coordinates of Q and R be (q, y) and (x, r), and use the fact that Q and R lie on a line through the origin.

15.4
HYPERBOLAS

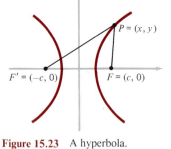

Figure 15.23   A hyperbola.

The ideas of Section 15.1 allow us to define a *hyperbola* as the locus of a point P that moves in such a way that the difference of its distances from two fixed points F and F' (called the *foci*) is constant. If this constant is denoted by $2a$, with $a > 0$, then a little thought will show that the locus consists of two *branches,* as shown in Fig. 15.23, where the right branch is the locus of the equation $PF' - PF = 2a$ and the left branch is the locus of the equation $PF - PF' = 2a$. The defining condition for the complete hyperbola can therefore be written as

$$PF' - PF = \pm 2a. \tag{1}$$

To find a simple equation for the hyperbola, we take the x-axis along the segment FF' and the y-axis as the perpendicular bisector of this segment. If $2c$ denotes the distance between F and F', then $F = (c, 0)$ and $F' = (-c, 0)$, as shown in Fig. 15.23, and (1) becomes

$$\sqrt{(x + c)^2 + y^2} - \sqrt{(x - c)^2 + y^2} = \pm 2a.$$

By moving the second radical to the right-hand side, squaring, and simplifying, we obtain the focal radius formulas

$$PF = \sqrt{(x - c)^2 + y^2} = \pm\left(\frac{c}{a}x - a\right) \tag{2}$$

and

$$PF' = \sqrt{(x + c)^2 + y^2} = \pm\left(\frac{c}{a}x + a\right), \tag{3}$$

where (3) follows from (2) because $PF' = \pm 2a + PF$. As in (1), the plus signs here correspond to the right branch of the curve, and the minus signs to the left branch. By squaring and simplifying, either of these equations gives

$$\left(\frac{c^2 - a^2}{a^2}\right)x^2 - y^2 = c^2 - a^2$$

or

$$\frac{x^2}{a^2} - \frac{y^2}{c^2 - a^2} = 1. \tag{4}$$

To simplify this equation still further, we begin by observing that in the triangle $PF'F$ with P on the right branch we have $PF' < PF + FF'$, because one side of a triangle is less than the sum of the other two sides. Therefore $PF' - PF < FF'$, or $2a < 2c$, and so $a < c$ and $c^2 - a^2$ is a positive number which we denote by b^2,

$$b^2 = c^2 - a^2. \tag{5}$$

This enables us to write (4) as

$$\frac{x^2}{a^2} - \frac{y^2}{b^2} = 1, \tag{6}$$

which is the standard form of the equation of the hyperbola shown in Fig. 15.23.

We now turn to a careful consideration of equation (6) and the light it sheds on the nature of the hyperbola it represents. Our discussion will reveal several additional features of this curve that are not obvious from the definition and that are indicated in greater detail in Fig. 15.24.

Since the equation contains only even powers of x and y, the hyperbola is symmetric with respect to both coordinate axes. They are therefore called the *axes* of the curve, and their intersection is called the *center.* The left-right, up-down symmetry is perhaps the only feature of the hyperbola that is easy to see directly from the definition.

When $y = 0$, the equation gives $x = \pm a$, but when $x = 0$, y is imaginary. Therefore the axis through the foci, called the *principal axis,* intersects the curve at two points called the *vertices,* which are located at a distance a on each side of the center; but the other axis, called the *conjugate axis,* does not intersect the curve at all. The hyperbola thus consists of two separate parts, its symmetrical *branches,* on opposite sides of the conjugate axis.

These facts are easier to see if equation (6) is solved for y,

$$y = \pm \frac{b}{a} \sqrt{x^2 - a^2}. \tag{7}$$

This formula shows that there are no points of the graph in the vertical strip $-a < x < a$, because for these x's the quantity inside the radical is negative. When $x = \pm a$, (7) yields $y = 0$; these two points are the vertices. And now, as x increases from a or decreases from $-a$, we get two distinct values of y that increase numerically as x moves farther to the right or left; this behavior produces the upper and lower arms of each branch of the curve.

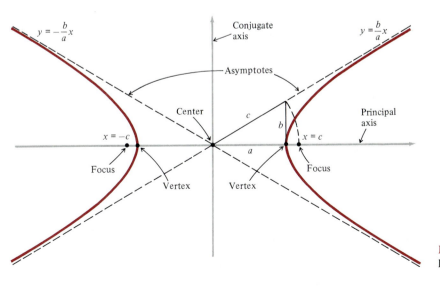

Figure 15.24 Features of a hyperbola.

A very significant feature of the graph can be observed by writing (7) in the form

$$y = \pm \frac{b}{a} x \sqrt{1 - \frac{a^2}{x^2}}. \tag{8}$$

When x is numerically large, the quantity inside the radical in (8) is nearly 1, and for this reason it appears that the hyperbola is very close to the pair of straight lines

$$y = \pm \frac{b}{a} x. \tag{9}$$

We can verify this guess as follows. In the first quadrant, if x is large, then the vertical distance from the hyperbola up to the corresponding line is

$$\frac{b}{a} x - \frac{b}{a} \sqrt{x^2 - a^2} = \frac{b}{a} (x - \sqrt{x^2 - a^2})$$

$$= \frac{b}{a} \frac{(x - \sqrt{x^2 - a^2})(x + \sqrt{x^2 - a^2})}{x + \sqrt{x^2 - a^2}}$$

$$= \frac{ab}{x + \sqrt{x^2 - a^2}}.$$

This clearly approaches zero as $x \to \infty$. The lines (9) are therefore called the *asymptotes* of the hyperbola. The asymptotes provide a convenient guide for sketching a hyperbola whose equation is given: Simply plot the vertices, draw the asymptotes, and fill in the two branches of the curve in a reasonable way, as suggested by the figure.

The triangle shown in the first quadrant of Fig. 15.24 is a convenient mnemonic device for remembering the main geometric features of a hyperbola. Its base a is the distance from the center to the vertex on the right; its height b is the distance from this vertex up to the asymptote in the first quadrant, whose slope is b/a; and since (5) tells us that

$$c^2 = a^2 + b^2,$$

the hypotenuse c of this triangle is also the distance from the center to a focus.

The ratio c/a is called the *eccentricity* of the hyperbola, and is denoted by e:

$$e = \frac{c}{a} = \frac{\sqrt{a^2 + b^2}}{a} = \sqrt{1 + \left(\frac{b}{a}\right)^2}.$$

It is clear that $e > 1$. When e is near 1, then b is small compared with a, and the hyperbola lies in a small angle between the asymptotes. When e is large, then b is large compared with a, the angle between the asymptotes is large, and the hyperbola is rather flat at the vertices.

To understand the significance of the eccentricity, we consider again formulas (2) and (3) for the right and left focal radii PF and PF'. These formulas can be written as

$$PF = \pm(ex - a) = \pm e \left[x - \frac{a}{e}\right] \tag{10}$$

and

$$PF' = \pm(ex + a) = \pm e\left(x + \frac{a}{e}\right) = \pm e\left[x - \left(-\frac{a}{e}\right)\right], \qquad (11)$$

where the plus signs apply to the right branch of the curve (see Fig. 15.25) and the minus signs to the left branch. If P lies on the right branch, as shown in the figure, then the quantities in brackets can be interpreted as the distances PD and PD' from P to the lines $x = a/e$ and $x = -a/e$, respectively. The same statement is true if P lies on the left branch, if the effect of the minus signs is properly taken into account. Therefore, in all cases formulas (10) and (11) can be written in the form

$$\frac{PF}{PD} = e \qquad \text{and} \qquad \frac{PF'}{PD'} = e. \qquad (12)$$

Each of the lines $x = a/e$ and $x = -a/e$ is called a *directrix* of the hyperbola. Equations (12) show that *a hyperbola can be characterized as the locus of a point that moves in such a way that the ratio of its distance from a fixed point (a focus) to its distance from a fixed line (the corresponding directrix) equals a constant $e > 1$.* Just as in the case of ellipses, this way of characterizing hyperbolas will be needed in our future work.

By interchanging the roles of x and y in the preceding discussion, we find that the equation

$$\frac{y^2}{a^2} - \frac{x^2}{b^2} = 1 \qquad (13)$$

represents a hyperbola with vertical principal axis, vertices at $(0, \pm a)$, and foci at $(0, \pm c)$, where $c^2 = a^2 + b^2$. This time the asymptotes are the lines

$$y = \pm \frac{a}{b}x,$$

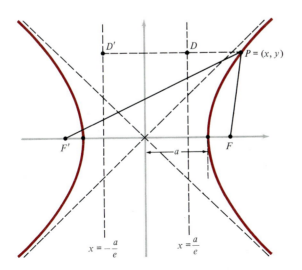

Figure 15.25

as we easily see by writing (13) in a form solved for y,

$$y = \pm\frac{a}{b}\sqrt{x^2 + b^2} = \pm\frac{a}{b}x\sqrt{1 + \frac{b^2}{x^2}}.$$

Notice that the axis containing the foci of a hyperbola is not determined by the relative size of a and b, as it was in the case of an ellipse, but rather by which term is subtracted from which in the standard form of the equation. The numbers a and b can therefore be of any relative size. In particular they can be equal, in which case the asymptotes are perpendicular to each other and the hyperbola is called *rectangular*. The equations

$$x^2 - y^2 = a^2 \qquad \text{and} \qquad y^2 - x^2 = a^2$$

represent rectangular hyperbolas.

Example 1   Find the equation of the hyperbola with foci $(\pm 6, 0)$ and the lines $5y = \pm 2\sqrt{5}\, x$ as asymptotes.

Solution   First, the location of the foci tells us that the principal axis is the x-axis. We see that $c = 6$ and $b/a = \frac{2}{5}\sqrt{5}$, so $a = (\sqrt{5}/2)b$. Since $a^2 + b^2 = c^2$, we have $\frac{5}{4}b^2 + b^2 = 36$, so $b^2 = \frac{4}{9} \cdot 36 = 16$ and $a^2 = \frac{5}{4}b^2 = \frac{5}{4} \cdot 16 = 20$. This shows that

$$\frac{x^2}{20} - \frac{y^2}{16} = 1$$

is the equation of the hyperbola.

Example 2   Determine the principal axis of the hyperbola $6y^2 - 9x^2 = 36$ and find its vertices, foci, and asymptotes.

Solution   The equation can be put in the standard form

$$\frac{y^2}{6} - \frac{x^2}{4} = 1,$$

so the principal axis is the y-axis, $a^2 = 6$, $b^2 = 4$, and $c^2 = a^2 + b^2 = 10$. Hence the vertices are $(0, \pm\sqrt{6})$, the foci are $(0, \pm\sqrt{10})$, and the asymptotes are $y = \pm(\sqrt{6}/2)x$.

Just as in the case of the ellipse, we can easily write the equation of a hyperbola with center (h, k) and principal axis parallel to one of the coordinate axes. The equation is

$$\frac{(x - h)^2}{a^2} - \frac{(y - k)^2}{b^2} = 1 \qquad \text{or} \qquad \frac{(y - k)^2}{a^2} - \frac{(x - h)^2}{b^2} = 1,$$

according as the principal axis is horizontal or vertical. This suggests that we consider equations of the form

$$Ax^2 + By^2 + Cx + Dy + E = 0,$$

where A and B have opposite signs. Such an equation will usually represent a hyperbola, but in certain special cases it may represent a pair of intersecting straight lines. The next example illustrates these possibilities.

Example 3 Identify the graph of

$$16x^2 - 9y^2 - 64x - 18y + E = 0$$

for various values of E.

Solution The procedure is to complete the square on the x and y terms, which yields

$$16(x^2 - 4x) - 9(y^2 + 2y) = -E$$

and

$$16(x - 2)^2 - 9(y + 1)^2 = 55 - E.$$

There are now three cases.

CASE 1 $55 - E > 0$; for example $E = -89$, so that $55 - E = 144$. In this case we have

$$\frac{(x - 2)^2}{9} - \frac{(y + 1)^2}{16} = 1,$$

which is a hyperbola with center $(2, -1)$ and horizontal principal axis.

CASE 2 $55 - E < 0$; for example $E = 199$, so that $55 - E = -144$. Here we have

$$\frac{(y + 1)^2}{16} - \frac{(x - 2)^2}{9} = 1,$$

which is a hyperbola with center $(2, -1)$ and vertical principal axis.

CASE 3 $55 - E = 0$; $E = 55$. This time our equation becomes

$$16(x - 2)^2 - 9(y + 1)^2 = 0$$

or

$$4(x - 2) = \pm 3(y + 1).$$

This represents the two lines

$$y + 1 = \pm \tfrac{4}{3}(x - 2),$$

which are the asymptotes in the first two cases.

Remark 1 Hyperbolas have the following reflection property: The tangent line at any point P on a hyperbola bisects the angle between the focal radii PF and PF'. This means that $\alpha = \beta$ in the notation of Fig. 15.26 (see Problem 21). As a consequence of this, if the hyperbola is revolved about its principal axis to form a surface of revolution, and if the convex sides of each part are silvered to make them reflecting surfaces, then any ray of light that ap-

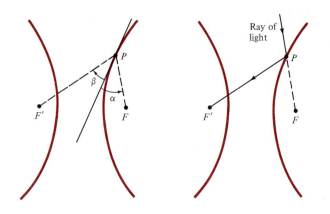

Figure 15.26   The reflection property.

proaches a convex side along a line pointing toward a focus (Fig. 15.26, right) is reflected toward the other focus.

This property of hyperbolas is the essential principle in the design of reflecting telescopes of the Cassegrain type (Fig. 15.27). As the figure shows, one focus of the hyperbolic mirror is at the focus of the parabolic mirror and the other is at the vertex of the parabolic mirror, where an eyepiece or camera is located. Faint parallel rays of starlight are therefore reflected off the parabolic mirror toward its focus, then are intercepted by the hyperbolic mirror and reflected back toward the eyepiece or camera.

Remark 2   There are two kinds of comets. Some are permanent members of the solar system, like Halley's Comet, described in Section 15.3, and travel forever around the sun in elliptical orbits with the sun at one focus. Others enter the solar system at high speeds from outer space, swing around the sun in hyperbolic orbits with the sun at one focus, and then escape into outer space again. The crucial factor is the total energy E of the comet itself, which is the sum of the kinetic energy due to its motion and the potential energy due to the gravitational attraction of the sun. It turns out that if $E < 0$, the orbit is an ellipse, and if $E > 0$, the orbit is a hyperbola. (The case $E = 0$ corresponds to a parabolic orbit, but this is exceedingly unlikely.)

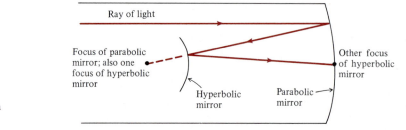

Figure 15.27   Design of Cassegrain telescope.

PROBLEMS

In Problems 1 to 8, sketch the graph of the given hyperbola and find the vertices, foci, asymptotes, eccentricity, and directrices.

1 $\dfrac{x^2}{4} - \dfrac{y^2}{9} = 1.$ 2 $\dfrac{y^2}{36} - \dfrac{x^2}{16} = 1.$

3 $\dfrac{y^2}{4} - \dfrac{x^2}{9} = 1.$ 4 $\dfrac{x^2}{25} - \dfrac{y^2}{16} = 1.$

5 $4y^2 - x^2 = 16.$ 6 $x^2 - 3y^2 = 12.$

7 $y^2 - x^2 = 1.$ 8 $x^2 - 9y^2 = 1.$

In Problems 9 to 16, find the equation of the hyperbola determined by the given conditions.
9 Foci $(0, \pm 5)$, vertex $(0, 3)$.
10 Vertices $(\pm 3, 0)$, focus $(5, 0)$.
11 Vertices $(\pm 3, 0)$, asymptote $y = 2x$.
12 Foci $(-1, 8)$ and $(-1, -2)$, vertex $(-1, 7)$.
13 Foci $(\pm 8, 0)$, $e = \frac{4}{3}$.
14 Vertices $(0, \pm 5)$, $e = 2$.
15 Vertices $(\pm 6, 0)$, directrix $x = 4$.
16 Foci $(1, 1)$ and $(-1, -1)$, difference of focal radii ± 2.

In Problems 17 to 20, identify the graph of the given equation as in the discussion of Example 3.
17 $16x^2 - 3y^2 - 32x - 12y - 44 = 0.$
18 $9y^2 - 7x^2 + 72y - 70x - 94 = 0.$
19 $36x^2 - 25y^2 + 144x - 50y + 119 = 0.$
20 $11y^2 - 12x^2 + 88y + 72x + 300 = 0.$
21 Show that $\alpha = \beta$ in Fig. 15.26.
22 Let F be a point which is outside a given circle. Consider a point P that moves in such a way as to be equidistant from F and the circle. Show that the path of P is one branch of a hyperbola.
23 Suppose that an ellipse and a hyperbola are *confocal*, that is, have the same foci F and F'. Use the reflection properties of the two curves to give a purely geometric proof that they are perpendicular to each other at every point P of intersection.
24 (a) Show that
$$\frac{x^2}{25 - k} + \frac{y^2}{16 - k} = 1$$
represents an ellipse if $k < 16$ and a hyperbola if $16 < k < 25$, and that all these curves are confocal.
(b) Find the first-quadrant point of intersection of the curves given by $k = 0$ and $k = 20$, and find the tangent line to each curve at this point.
(c) Show that the tangent lines in part (b) are perpendicular to each other by showing that the product of their slopes is -1.

25 Consider two hyperbolas with the same eccentricity e, both centered at the origin and both with principal axis on the x-axis. Suppose that their equations are
$$\frac{x^2}{a_1^2} - \frac{y^2}{b_1^2} = 1 \quad \text{and} \quad \frac{x^2}{a_2^2} - \frac{y^2}{b_2^2} = 1.$$
Show that these hyperbolas are similar in the sense that
(a) there exists a constant k such that
$$\frac{a_1}{a_2} = \frac{b_1}{b_2} = \frac{c_1}{c_2} = k;$$
(b) if a half-line from the origin O intersects the first hyperbola at P_1 and the second at P_2, then
$$\frac{OP_1}{OP_2} = k.$$
26 Find the locus of the centers of all circles that are tangent to the y-axis and cut off a segment of length $2a$ on the x-axis.
27 Let F and F' be two points on a sheet of paper whose distance apart is $2c$. Take a piece of string and tie a knot K in it so that the difference between the lengths of the two parts into which K divides the string is $2a$, where $0 < a < c$. Tie the ends of the string to two tacks placed at F and F', and loop the string around the point of a pencil, as shown in Fig. 15.28. If the string is held taut and the knot K is carefully pulled, show that the pencil at P draws one branch of a hyperbola.

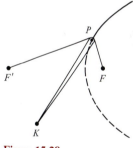

Figure 15.28

28 If two hyperbolas $x^2/a^2 - y^2/b^2 = 1$ and $y^2/A^2 - x^2/B^2 = 1$ have the same asymptotes, show that their eccentricities e and E are related by the equation
$$\frac{1}{e^2} + \frac{1}{E^2} = 1.$$
29 Show that the line tangent to the hyperbola $x^2/a^2 - y^2/b^2 = 1$ at the point $P_1 = (x_1, y_1)$ is
$$\frac{xx_1}{a^2} - \frac{yy_1}{b^2} = 1.$$

30 If tangent lines to the hyperbola $x^2/25 - y^2/16 = 1$ intersect the y-axis at $(0, 8)$, find the points of tangency.

31 If tangent lines to the hyperbola $x^2/a^2 - y^2/b^2 = 1$ intersect the y-axis at $(0, d)$, find the points of tangency.

32 A line through a point P on a hyperbola and parallel to the nearest asymptote intersects the nearest directrix at Q. If F is the corresponding focus, show that $PQ = PF$.

15.5

THE FOCUS-DIRECTRIX-ECCENTRICITY DEFINITIONS

Students have already seen that there are several distinct but equivalent ways of defining the conic sections, each with its own merits. We began with the definition by means of a given cone and a slicing plane that cuts through the cone more or less steeply, yielding our three types of curves by varying the degree of steepness. This three-dimensional approach is vivid and geometric, and provides a clear visual impression of what the curves look like. However, for the purpose of obtaining Cartesian equations for use in precise quantitative studies, we needed two-dimensional characterizations, and for this the focal properties discussed at the end of Section 15.1 turned out to be convenient. The concepts of eccentricity and directrix emerged in the course of our detailed work on ellipses and hyperbolas, and we saw that each of these curves can be given yet another two-dimensional characterization by means of a focus, a directrix, and an eccentricity. Our purpose in this brief section is to show that all three of the conic sections—parabolas, ellipses, and hyperbolas—can in this way be given unified definitions that depend directly on our original concept of these curves as sections of a cone.*

Our discussion is based on Fig. 15.29, which shows a cone with vertex angle α and a slicing plane with tilting angle β. This tilting angle can be defined as the angle between the axis of the cone and a normal line to the plane, but it plays its main role in our argument as the indicated acute angle

---

* For reasons that will soon be clear, circles must be excluded from this discussion, because the necessary geometric constructions are not possible when the slicing plane is perpendicular to the axis of the cone.

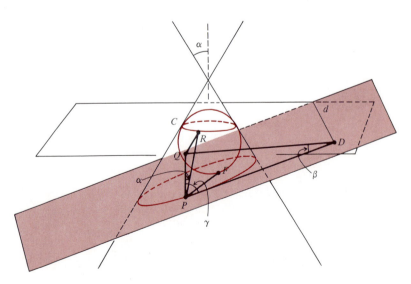

Figure 15.29

of the right triangle PQD. The figure is drawn to illustrate the case of an ellipse, but the argument is valid for the other cases as well.

We begin at the beginning. Let there be inscribed in the cone a sphere which is tangent to the slicing plane at a point F, and tangent to the cone along a circle C. If d is the line in which the slicing plane intersects the plane of the circle C, we shall prove that the conic section has F as its focus and d as its directrix, and the facts about the eccentricity will emerge in the course of our discussion.

To this end, let P be a point on the conic section, let Q be the point where the line through P and parallel to the axis of the cone intersects the plane of C, let R be the point where the generator through P intersects C, and let D be the foot of the perpendicular from P to the line d. Then PR and PF are two segments which are tangent to the sphere from the same point P, and therefore have the same length,

$$PR = PF. \tag{1}$$

Also, from the right triangle PQR we have

$$PQ = PR \cos \alpha;$$

and from the right triangle PQD we have

$$PQ = PD \sin \beta.$$

It follows that

$$PR \cos \alpha = PD \sin \beta,$$

so

$$\frac{PR}{PD} = \frac{\sin \beta}{\cos \alpha}.$$

In view of (1) this means that

$$\frac{PF}{PD} = \frac{\sin \beta}{\cos \alpha}.$$

This can be written in the slightly more convenient form

$$\frac{PF}{PD} = \frac{\cos \gamma}{\cos \alpha}, \tag{2}$$

where γ is the other acute angle in the right triangle PQD. If we now define the eccentricity e by

$$e = \frac{\cos \gamma}{\cos \alpha},$$

then this number is constant for a given cone and a given slicing plane, and (2) becomes

$$\frac{PF}{PD} = e \begin{cases} <1 & \text{for an ellipse} \\ =1 & \text{for a parabola} \\ >1 & \text{for a hyperbola,} \end{cases}$$

where the statements on the right are easily verified by inspecting the figure.

Thus, for a parabola, we see that PD is parallel to a generator of the cone, so $\gamma = \alpha$ and $e = 1$; for an ellipse, we have $\gamma > \alpha$, so $\cos \gamma < \cos \alpha$ and $e < 1$; and for a hyperbola, we have $\gamma < \alpha$, so $\cos \gamma > \cos \alpha$ and $e > 1$.

The words "parabola," "ellipse," and "hyperbola" come from three Greek words meaning "a comparison," "a deficiency," and "an excess," referring to the fact that for the corresponding curves we have $e = 1$, $e < 1$, and $e > 1$. One should also compare these words with the words "parable," "ellipsis," and "hyperbole" in modern English.

15.6

(OPTIONAL) SECOND-DEGREE EQUATIONS. ROTATION OF AXES

The general equation of the second degree in x and y is

$$Ax^2 + Bxy + Cy^2 + Dx + Ey + F = 0, \tag{1}$$

where at least one of the coefficients A, B, C is different from zero. The latter requirement, of course, guarantees that the degree of the equation really is 2, rather than 1 or 0. In the preceding sections we have found that circles, parabolas, ellipses, and hyperbolas are all curves whose equations are special cases of (1). Thus, for example, the circle

$$(x - h)^2 + (y - k)^2 = r^2$$

can be obtained from (1) by taking

$$A = C = 1, \qquad B = 0, \qquad D = -2h, \qquad E = -2k,$$
$$F = h^2 + k^2 - r^2,$$

and the parabola

$$x^2 = 4py$$

by taking

$$A = 1, \qquad E = -4p, \qquad B = C = D = F = 0.$$

In addition to the conic sections mentioned here, we have also noted various "exceptional cases" that can arise as graphs of (1) from special choices of the coefficients. Thus, the graph of

$$x^2 + y^2 = 0$$

is a point, and the graph of

$$x^2 + y^2 + 1 = 0$$

is the empty set. Further, the graph of

$$x^2 = 0$$

is a single line, namely, the y-axis, and the graph of

$$x^2 - y^2 = 0, \qquad \text{or equivalently} \qquad (x + y)(x - y) = 0,$$

is a pair of lines, namely, $x + y = 0$ and $x - y = 0$. Our purpose in this section is to investigate the full range of possibilities of the curves represented by (1). Briefly, we shall find that the eight graphs we have just listed exhaust all possibilities:

The graph of every second-degree equation of the form (1) *is a circle, a parabola, an ellipse, a hyperbola, a point, the empty set, a single line, or a pair of lines.*

The main problem before us is posed by the so-called "mixed" term Bxy in (1), because when this term is present we have no idea how to identify the graph. No such terms have arisen in our previous work on the conic sections. The reason for this is that in every case we have been careful to choose the coordinate axes in a simple and natural position, so that at least one axis is parallel to an axis of symmetry of the curve under discussion. In order to see what can happen when a curve is placed in a skew position relative to the axes, let us find the equation of the hyperbola (see Fig. 15.30) with foci $F = (2, 2)$ and $F' = (-2, -2)$, where $PF' - PF = \pm 4$. We have

$$\sqrt{(x + 2)^2 + (y + 2)^2} - \sqrt{(x - 2)^2 + (y - 2)^2} = \pm 4,$$

and when we move the second radical to the right-hand side, square, solve for the radical that still remains, and square again, this reduces to

$$xy = 2. \qquad (2)$$

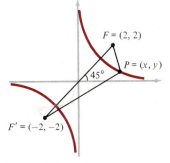

Figure 15.30

This is really a very simple equation, but nevertheless it does provide a special case of (1) in which the mixed term is present. The asymptotes of the hyperbola (2) are evidently the x- and y-axes, and its principal axis is the line $y = x$, which makes a 45° angle with the x-axis. It will become clear that a mixed term is present only when a curve is "tilted" in this way with respect to the coordinate axes, and also that this term can be removed by rotating the axes to "untilt" the curve. In the case of (2), it is easy to see by looking at the figure that this curve can be untilted by rotating the axes through a 45° angle in the counterclockwise direction.

To construct the machinery that is necessary for carrying out an arbitrary rotation of axes, we start with the xy-system and rotate these axes counterclockwise through an angle θ to obtain the $x'y'$-system, as shown in Fig. 15.31. A point P in the plane will then have two pairs of rectangular coordinates, (x, y) and (x', y'). To see how these coordinates are related, we observe from the figure that

$$x = OR = OQ - RQ = OQ - ST$$
$$= x' \cos \theta - y' \sin \theta$$

and

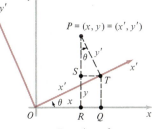

Figure 15.31   Rotation of axes.

$$y = RP = RS + SP = QT + SP$$
$$= x' \sin \theta + y' \cos \theta.$$

We write these equations together for convenient reference,

$$x = x' \cos \theta - y' \sin \theta,$$
$$y = x' \sin \theta + y' \cos \theta; \qquad (3)$$

they are called the *equations for rotation of axes*. For example, if $\theta = 45°$, then, since $\sin 45° = \cos 45° = \frac{1}{2}\sqrt{2} = 1/\sqrt{2}$, we have

$$x = \frac{x' - y'}{\sqrt{2}}, \qquad y = \frac{x' + y'}{\sqrt{2}}. \qquad (4)$$

And for another example, if $\theta = 30°$, then since $\sin 30° = \frac{1}{2}$ and $\cos 30° = \frac{1}{2}\sqrt{3}$, we have

$$x = \frac{\sqrt{3}x' - y'}{2}, \qquad y = \frac{x' + \sqrt{3}y'}{2}. \tag{5}$$

As a simple illustration of the use of these equations, we substitute (4) into (2) and obtain

$$\frac{x'^2 - y'^2}{2} = 2 \qquad \text{or} \qquad \frac{x'^2}{4} - \frac{y'^2}{4} = 1.$$

This is immediately recognizable as a rectangular hyperbola whose principal axis is the x'-axis. Of course, we already knew this from the way (2) was obtained. However, if we had started with (2) without knowing anything about the nature of its graph, then this procedure for removing the mixed term would have enabled us to identify the curve without difficulty.

In the case of equation (2), the $45°$ rotation represented by equations (4) worked. But how could we have known this in advance? Can we be sure that a suitable rotation will always remove the xy term if one is present? And if so, how do we find a suitable angle of rotation?

To answer these questions we return to the general second-degree equation (1),

$$Ax^2 + Bxy + Cy^2 + Dx + Ey + F = 0,$$

and we apply the general rotation (3) through an unspecified angle θ, which yields

$$A(x' \cos \theta - y' \sin \theta)^2 + B(x' \cos \theta - y' \sin \theta)(x' \sin \theta + y' \cos \theta)$$
$$+ C(x' \sin \theta + y' \cos \theta)^2 + D(x' \cos \theta - y' \sin \theta)$$
$$+ E(x' \sin \theta + y' \cos \theta) + F = 0.$$

When we collect coefficients for the various terms, we get a new equation of the same form,

$$A'x'^2 + B'x'y' + C'y'^2 + D'x' + E'y' + F' = 0, \tag{6}$$

with new coefficients related to the old ones by the following formulas:

$$A' = A \cos^2 \theta + B \sin \theta \cos \theta + C \sin^2 \theta,$$

$$B' = -2A \sin \theta \cos \theta + B(\cos^2 \theta - \sin^2 \theta) + 2C \sin \theta \cos \theta,$$

$$C' = A \sin^2 \theta - B \sin \theta \cos \theta + C \cos^2 \theta, \tag{7}$$

$$D' = D \cos \theta + E \sin \theta,$$

$$E' = -D \sin \theta + E \cos \theta,$$

$$F' = F.$$

We have written down all these formulas for future reference, but for the moment we are only interested in B'. If we start out with a second-degree equation (1) in which the mixed term is present, $B \neq 0$, then we can always find an angle θ of rotation such that the new mixed term is eliminated. To find a suitable angle θ, we simply put $B' = 0$ in (7) and solve for θ. To do this most easily, we use the double-angle formulas

$$\sin 2\theta = 2 \sin \theta \cos \theta$$

and

$$\cos 2\theta = \cos^2 \theta - \sin^2 \theta$$

to write

$$B' = B \cos 2\theta + (C - A) \sin 2\theta.$$

Then $B' = 0$ if we choose θ so that

$$\cot 2\theta = \frac{A - C}{B}. \tag{8}$$

Since we are assuming that $B \neq 0$, it is clear that this is always possible, and furthermore that θ can always be chosen in the first quadrant, $0 < \theta < \pi/2$.

Example 1   Determine the nature of the curve whose equation is

$$4x^2 + 2xy + 4y^2 = 15. \tag{9}$$

Solution   Here we have $A = 4$, $B = 2$, and $C = 4$. The mixed term will be removed by choosing θ according to (8), which in this case gives

$$\cot 2\theta = 0, \qquad 2\theta = 90°, \qquad \theta = 45°.$$

We therefore substitute equations (4) into (9) and obtain

$$5x'^2 + 3y'^2 = 15 \qquad \text{or} \qquad \frac{x'^2}{3} + \frac{y'^2}{5} = 1$$

after simplification. This is clearly an ellipse with its foci on the new y'-axis.

We observe that when $B \neq 0$ and $A = C$, a rotation through $45°$ is always appropriate.

Example 2   Determine the nature of the curve whose equation is

$$11x^2 + 10\sqrt{3}\, xy + y^2 - 32 = 0. \tag{10}$$

Solution   Here we have

$$\cot 2\theta = \frac{11 - 1}{10\sqrt{3}} = \frac{1}{\sqrt{3}}, \qquad 2\theta = 60°, \qquad \theta = 30°.$$

In this case we use equations (5), which transform (10) into

$$16x'^2 - 4y'^2 - 32 = 0 \qquad \text{or} \qquad \frac{x'^2}{2} - \frac{y'^2}{8} = 1$$

after simplification. This is a hyperbola with its principal axis along the new x'-axis.

We now return to our original problem of classifying all possible graphs of the second-degree equation (1). Since the axes can always be rotated to eliminate the mixed term, there is no loss of generality in assuming that this has been done. We are therefore confronted by equation (6) with $B' = 0$, and we drop the primes to simplify the notation,

$$Ax^2 + Cy^2 + Dx + Ey + F = 0. \tag{11}$$

Our experience in the preceding sections enables us to distinguish four cases and to be certain that there are no others. The graph of equation (11), with $B = 0$, is:

1 A circle if $A = C \neq 0$. In special cases the graph can be a single point or the empty set.
2 An ellipse if A and C are both positive or both negative and $A \neq C$. Again, in special cases the graph can be a single point or the empty set.
3 A hyperbola if A and C have opposite signs. In special cases the graph can be a pair of intersecting straight lines.
4 A parabola if either $A = 0$ or $C = 0$ (but not both). In special cases the graph can be one straight line, or two parallel lines, or the empty set.

In Chapter 16 we will meet many equations of the third degree, fourth degree, etc. The exhaustive list given here of possible graphs of second-degree equations is relatively simple and stands in sharp contrast to the wilderness of bizarre curves that awaits us in connection with these higher-degree equations.

PROBLEMS

In Problems 1 to 11, determine and carry out a suitable rotation of axes to eliminate the mixed term, find the new equation, and identify the curve.

1 $5x^2 - 6xy + 5y^2 = 8$.
2 $x^2 - 2xy + y^2 + x + y = \sqrt{2}$.
3 $2x^2 + 4\sqrt{3}xy - 2y^2 = 8$.
4 $11x^2 + 4\sqrt{3}xy + 7y^2 - 65 = 0$.
5 $x^2 + 2xy + y^2 + 8x - 8y = 0$.
6 $x^2 - 3xy + y^2 = 10$.
7 $3x^2 + 2xy + 3y^2 = 8$.
8 $x^2 + 2\sqrt{3}xy + 3y^2 + 2\sqrt{3}x - 2y = 0$.
9 $5x^2 + 4\sqrt{3}xy + 9y^2 = 33$.
10 $3x^2 + 4\sqrt{3}xy - y^2 = 30$.
*11 $6x^2 - 6xy + 14y^2 = 5$.
12 An ellipse has foci $(1, 0)$ and $(0, \sqrt{3})$ and passes through the point $(-1, 0)$. Use the focus definition to find its equation. Through what angle should the axes be rotated to eliminate the mixed term from this equation?
13 Use equations (7) to show that $B^2 - 4AC = B'^2 - 4A'C'$. For this reason, the number $B^2 - 4AC$ is said to be *invariant under rotations.*

14 The number $B^2 - 4AC$ is called the *discriminant* of equation (1). To understand why, rotate the axes to remove the mixed term and use Problem 13 to show that

$$B^2 - 4AC \begin{cases} <0 & \text{for circles and ellipses,} \\ =0 & \text{for parabolas,} \\ >0 & \text{for hyperbolas,} \end{cases}$$

where the various special cases are considered to belong to the appropriate categories.
15 Verify the statement in Problem 14 for Problems 1 to 11.
16 As a check on the equations for rotation of axes, show that the equation of a circle centered at the origin, $x^2 + y^2 = r^2$, is unchanged in form when the axes are rotated through an arbitrary angle θ.
17 If a rotation of axes through an angle θ is followed by a rotation through an angle ϕ, this obviously amounts to a rotation through the angle $\theta + \phi$. Use the formulas

$$x = x' \cos \theta - y' \sin \theta$$

and
$$y = x' \sin\theta + y' \cos\theta$$

$$x' = x'' \cos\phi - y'' \sin\phi$$
$$y' = x'' \sin\phi + y'' \cos\phi$$

to show that

$$x = x'' \cos(\theta + \phi) - y'' \sin(\theta + \phi)$$
$$y = x'' \sin(\theta + \phi) + y'' \cos(\theta + \phi).$$

18 Show that a 45° rotation of axes transforms the equation $x^4 + 6x^2y^2 + y^4 = 32$ into $x'^4 + y'^4 = 16$. Sketch the curve and both sets of axes.

ADDITIONAL PROBLEMS FOR CHAPTER 15

SECTION 15.2

1 The chord of a parabola through the focus and perpendicular to the axis is called the *latus rectum*. If (as usual) p is the distance between the vertex and the focus, find the length of the latus rectum.

2 Find the equation of the circle tangent to the directrix of the parabola $x^2 = 4py$ with the focus of the parabola as its center. What are the points of intersection of the parabola and the circle?

3 Show that every upward-opening parabola with focus at the origin has an equation of the form $x^2 = 4p(y + p)$, and every downward-opening parabola with focus at the origin has an equation of the form $x^2 = -4\bar{p}(y - \bar{p})$, where p and $\bar{p}$ are positive constants.

4 Show that every upward-opening parabola with focus at the origin intersects at right angles every downward-opening parabola with focus at the origin.

5 Let P be a point on the parabola $x^2 = 4py$ other than the vertex. If Q and R are the points at which the tangent and normal at P intersect the axis of the parabola, and if S is the foot of the perpendicular from P to this axis, then the segments QS and RS are called the *subtangent* and *subnormal.*

   (a) Show that the vertex V bisects the subtangent.

   (b) Show that the subnormal has constant length $2p$.

   (c) Show that P and R are the same distance from the focus F.

   (d) Show that P and Q are the same distance from the focus F, so that F bisects the segment QR.

   (e) If the tangent at P intersects the directrix at a point T, show that PFT is a right angle.

   (f) If the tangent at P intersects the tangent at V (the x-axis) at a point U, show that PUF is a right angle.

6 Show that the reflection property of parabolas follows easily (without calculation) from part (d) of Problem 5.

7 Show that the vertex is the point on a parabola that is closest to the focus.

SECTION 15.3

8 The reflection property of ellipses is an easy consequence (without calculation) of the following geometric property of ellipses. In Fig. 15.32, let the line T be tangent at P to the ellipse with foci F and F'. Let G and G' be the reflections in T of F and F', so that T is the perpendicular bisector of the segments FG and $F'G'$. Then, as suggested by the figure, the segments FG' and $F'G$ intersect at P. To prove this, let Q be any point on T different from P and verify that

   (a) $FP + PF' < FQ + QF'$;

   (b) $FP + PG' < FQ + QG'$;

   (c) P lies on the segment FG'.

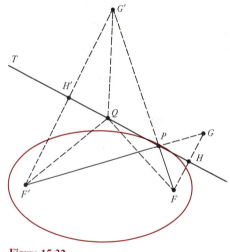

Figure 15.32

Similarly, P lies on the segment $F'G$, so it is the point of intersection of these segments. Finally, use this fact to infer that the angles FPH and $F'PH'$ are equal.

9 A barrel has the shape of a prolate spheroid with its ends cut off by planes through the foci. If the barrel is 4 ft high and the diameter of its top (and bottom) is 2 ft, find the volume.

10 A *latus rectum* of an ellipse is a chord through a focus and perpendicular to the major axis. If the equation of the ellipse is $x^2/a^2 + y^2/b^2 = 1$ with $a > b$, show that

(a) the length of a latus rectum is $2b^2/a$;

(b) the slope of the tangent line at the upper end of the latus rectum to the right of the y-axis is $-e$;

(c) the tangent line in part (b) intersects the corresponding directrix on the x-axis.

11 Show that every parabola of the form $y = Ax^2$ intersects every ellipse of the form $x^2 + 2y^2 = B$ at right angles.

12 Suppose that the tangent to an ellipse at a point P intersects a directrix at a point Q. If F is the corresponding focus, show that PFQ is a right angle.

13 Let P be a point on the ellipse $x^2/a^2 + y^2/b^2 = 1$ and let Q be the point where the tangent at P intersects the line $x = -a$. If A is the point $(a, 0)$, show that AP is parallel to OQ.

14 Show that the product of the distances from the foci of an ellipse to a tangent has the same value for all tangents. Hint: Use Additional Problem 21 in Chapter 1.

SECTION 15.4

15 Show that the product of the distances from the foci of a hyperbola to a tangent has the same value for all tangents.

16 Show that the product of the distances from a point P on a hyperbola to the asymptotes has the same value for all P's.

17 Just as for an ellipse, a *latus rectum* of a hyperbola $x^2/a^2 - y^2/b^2 = 1$ is a chord thorough a focus and perpendicular to the line on which the foci lie. Show that

(a) the length of a latus rectum is $2b^2/a$;

(b) the slope of the tangent line at the upper end of the latus rectum to the right of the origin is e;

(c) the tangent line in (b) intersects the corresponding directrix on the x-axis.

18 Suppose that the tangent to a hyperbola at a point P intersects the nearest directrix at a point Q. If F is the corresponding focus, show that PFQ is a right angle.

19 Show that the distance from either focus of the hyperbola $x^2/a^2 - y^2/b^2 = 1$ to either asymptote is b.

20 Sketch the graphs of the equations

$$\frac{x^2}{a^2} - \frac{y^2}{b^2} = \pm 1$$

on a single coordinate system. These two hyperbolas have the same asymptotes, and their four branches "enclose" a region that stretches out to infinity in four directions. Determine whether the area of this region is finite or infinite.

21 Let c be a given positive number.

(a) Show that for all positive values of h the ellipses

$$\frac{x^2}{c^2 + h} + \frac{y^2}{h} = 1$$

have the same foci $(\pm c, 0)$.

(b) Show that for all positive values of $k < c^2$ the hyperbolas

$$\frac{x^2}{c^2 - k} - \frac{y^2}{k} = 1$$

have the same foci as the ellipses in part (a).

(c) If $P_1 = (x_1, y_1)$ is a point of intersection of one of the ellipses in (a) with one of the hyperbolas in (b), show that the tangents to the two curves at this point are perpendicular.

22 Show that a line through a focus of a hyperbola and perpendicular to an asymptote intersects the asymptote on the corresponding directrix.

16

POLAR COORDINATES

As we know, a coordinate system in the plane allows us to associate an ordered pair of numbers with each point in the plane. This simple but powerful idea enables us to study many problems of geometry—especially the properties of curves—by the methods of algebra and calculus. Up to this stage of our work we have considered only the rectangular (or Cartesian) coordinate system, in which the emphasis is placed on the distances of a point from two perpendicular axes. However, it often happens that a curve appears to have a special affinity for the origin, like the path of a planet whose journey around its orbit is determined by the central attracting force of the sun. Such a curve is often best described as the path of a moving point whose position is specified by its direction from the origin and its distance out from the origin. This is exactly what polar coordinates do, as we now explain.

A point is located by means of its distance and direction from the origin, as shown in Fig. 16.1. Direction is specified by an angle θ (in radians), measured from the positive x-axis. This angle is understood to be described in the counterclockwise sense if θ is positive and in the clockwise sense if θ is negative, just as in trigonometry. Distance is given by the directed distance r, measured out from the origin along the terminal side of the angle θ. The two numbers r and θ, written in this order as an ordered pair (r, θ), are called *polar coordinates* of the point. The direction $\theta = 0$ (the positive x-axis) is called the *polar axis.*

Every point has many pairs of polar coordinates. For instance, the point P in Fig. 16.2 has polar coordinates $(3, \pi/4)$, but it also has polar coordinates $(3, \pi/4 + 2\pi)$, $(3, \pi/4 - 4\pi)$, etc. Any multiple of 2π added to or subtracted from the θ-coordinate of a point yields another angle with the same terminal side, and therefore another θ-coordinate of the same point.

The term "directed distance" is intended to suggest that we often meet situations in which r is negative. In this case it is understood that instead of moving out from the origin in the direction indicated by the terminal side of θ, we move *back through the origin* a distance $-r$ in the opposite direction. Thus, another pair of polar coordinates for the point P in Fig. 16.2 is $(-3, \pi/4 + \pi)$. In Fig. 16.3 we plot the two points $Q = (2, \pi/6)$ and $R = (-2, \pi/6)$.

16.1

THE POLAR COORDINATE SYSTEM

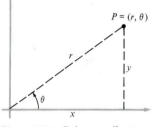

Figure 16.1   Polar coordinates.

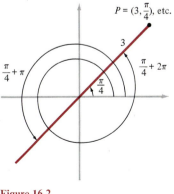

Figure 16.2

479

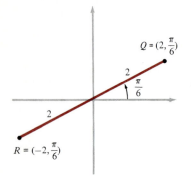

Figure 16.3

The value $r = 0$ specifies the origin, regardless of the value of θ. For instance, the pairs $(0, 0)$, $(0, \pi/2)$, $(0, -\pi/4)$ are all polar coordinates of the origin.

The fact that a point does not determine a unique pair of polar coordinates is a nuisance, but only a minor nuisance. Nevertheless, it *is* true that when any particular pair of polar coordinates is given, this pair determines the corresponding point without any ambiguity.

Even though it is incorrect to speak of *the* polar coordinates of a point because they are not unique, this error of usage is very common and is tolerated for the sake of euphony.

It is important to know the connection between rectangular and polar coordinates. Figure 16.1 shows at once that

$$x = r \cos \theta \quad \text{and} \quad y = r \sin \theta. \tag{1}$$

When r and θ are known, these equations tell us how to find x and y. We also have the equations

$$r^2 = x^2 + y^2 \quad \text{and} \quad \tan \theta = \frac{y}{x}, \tag{2}$$

which enable us to find r and θ when x and y are known. In using these equations, it is necessary to take a little care to make sure that the sign of r and the choice of θ are consistent with the quadrant in which the given point (x, y) lies.

Example 1 The rectangular coordinates of a point are $(-1, \sqrt{3})$. Find a pair of polar coordinates for this point.

Solution We have

$$r = \pm\sqrt{1 + 3} = \pm 2 \quad \text{and} \quad \tan \theta = -\sqrt{3}.$$

Since the point is in the second quadrant, we can use our knowledge of trigonometry to choose $r = 2$ and $\theta = 2\pi/3$, so one pair of polar coordinates for the point is $(2, 2\pi/3)$. Another acceptable pair with a negative value of r is $(-2, -\pi/3)$. Students should plot the point and have a clear visual understanding of each of these statements, as suggested by Fig. 16.4.

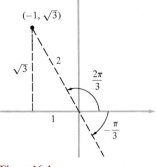

Figure 16.4

Just as in the case of rectangular coordinates, the *graph* of a polar equation

$$F(r, \theta) = 0 \tag{3}$$

is the set of all points $P = (r, \theta)$ whose polar coordinates satisfy the equation. Since the point P has many different pairs of coordinates, it is necessary to state explicitly that P lies on the graph if *any one* of its many different pairs of coordinates satisfies the equation.

Example 2 Show that the points $(1, \pi/2)$ and $(0, \pi/2)$ both lie on the graph of $r = \sin^2 \theta$.

Solution The point $(1, \pi/2)$ lies on the graph because the given coordinates satisfy the equation: $1 = \sin^2 \pi/2$. On the other hand, the point $(0, \pi/2)$ lies on the graph even though $0 \neq \sin^2 \pi/2$. The reason for this seemingly strange behavior is that $(0, 0)$ is also a pair of coordinates for the same point, and $0 = \sin^2 0$.

In most of the situations we will encounter, equation (3) can be solved for r and takes the form

$$r = f(\theta). \tag{4}$$

If the function $f(\theta)$ is reasonably simple, the graph is fairly easy to sketch. We merely choose a convenient sequence of values for θ, each determining its own direction from the origin, and compute the corresponding values of r that tell us how far out to go in each of these directions. We begin by discussing the simplest possible equations.

Example 3   The equation $\theta = \alpha$, where α is a constant, has as its graph the line through the origin that makes an angle α with the positive x-axis (Fig. 16.5).

Example 4   The equation $r = a$, where a is a positive constant, has as its graph the circle with center at the origin and radius a (Fig. 16.6).

Our next example is more complicated, and serves to introduce several important methods.

Example 5   The graph of $r = 2 \cos \theta$ is another circle, but this is not obvious. One way to try to get an idea of the shape of an unknown polar graph is to compute a short table of selected values and plot the corresponding points, as shown in Fig. 16.7.

A better procedure than computing values and plotting points is to sketch the graph as the path of a moving point, by direct analysis of the polar equation, as follows. When $\theta = 0$, $r = 2 \cos 0 = 2$. As θ increases through the first quadrant, from 0 to $\pi/2$, $2 \cos \theta$ decreases from 2 to 0, and we obtain the upper part of the curve shown in Fig. 16.7. As θ increases from $\pi/2$ to π, $2 \cos \theta$ decreases from 0 to -2, and the lower part of the curve is traced out. As θ increases from π to $3\pi/2$, the upper part of the curve is retraced, and as θ increases from $3\pi/2$ to 2π, the lower part is retraced.

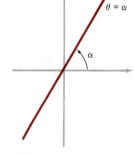

Figure 16.5

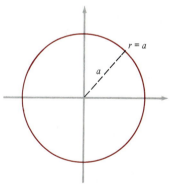

Figure 16.6

θ	r
0	2
$\pi/6$	$\sqrt{3}$
$\pi/4$	$\sqrt{2}$
$\pi/3$	1
$\pi/2$	0
$2\pi/3$	-1
$3\pi/4$	$-\sqrt{2}$
$5\pi/6$	$-\sqrt{3}$
π	-2

Figure 16.7

It is clear that the resulting graph is some kind of oval, perhaps even a circle. To verify that it really is a circle, we find and recognize the rectangular equation of the curve. To accomplish this, we multiply the given equation $r = 2 \cos \theta$ by r and use the change-of-variable equations (1) and (2) to write

$$r^2 = 2r \cos \theta,$$

$$x^2 + y^2 = 2x,$$

$$x^2 - 2x + y^2 = 0,$$

$$(x - 1)^2 + y^2 = 1.$$

This last equation tells us that the graph is a circle with center $(1, 0)$ and radius 1. It should be pointed out that multiplying the given equation by r introduces the origin as a point on the graph; however, since this point is already on the graph, nothing is changed.

The method illustrated here, sketching a polar graph by direct examination of the polar equation $r = f(\theta)$, will often be important in our future work. Briefly, the process is this: We imagine a radius swinging around the origin in the counterclockwise direction, with our curve being traced out by a point attached to this turning radius which is free to move toward the origin or away from it in accordance with the behavior of the function $f(\theta)$. In many of our examples and problems, $f(\theta)$ will be a simple expression involving the trigonometric functions $\sin \theta$ or $\cos \theta$. In these circumstances it will clearly be very useful to have a solid grasp of the way these functions vary as the radius makes one complete revolution, that is, as θ increases from 0 to $\pi/2$, then from $\pi/2$ to π, π to $3\pi/2$, and $3\pi/2$ to 2π.

PROBLEMS

1 Find the rectangular coordinates of the points with the given polar coordinates, and plot the points:
   (a) $(2, \pi/4)$; (b) $(4, -\pi/3)$;
   (c) $(0, -\pi)$; (d) $(-1, 7\pi/6)$;
   (e) $(2, -\pi/2)$; (f) $(4, 3\pi/4)$;
   (g) $(3, \pi)$; (h) $(-6, -\pi/4)$;
   (i) $(1, 0)$; (j) $(0, 1)$;
   (k) $(2, -5\pi/3)$; (l) $(13, \tan^{-1} \frac{12}{5})$;
   (m) $(-4, 11\pi/6)$; (n) $(3, -3\pi/2)$.
2 Find two pairs of polar coordinates, with r's having opposite signs, for the points with the following rectangular coordinates:
   (a) $(-2, 2)$; (b) $(4, 0)$;
   (c) $(2\sqrt{3}, 2)$; (d) $(2, 2\sqrt{3})$;
   (e) $(\sqrt{3}, 1)$; (f) $(0, 4)$;
   (g) $(-3, -3)$; (h) $(5, 5)$;
   (i) $(0, -2)$; (j) $(-\sqrt{3}, 1)$;
   (k) $(5, -12)$; (l) $(-3, 4)$;
   (m) $(-1, 0)$; (n) $(1, 2)$.
3 A regular pentagon is inscribed in the circle $r = 1$ with one vertex on the positive x-axis. Find the polar coordinates of all the vertices.

4 Show that the point $(3, 3\pi/4)$ lies on the curve $r = 3 \sin 2\theta$.
5 Show that the point $(3, 3\pi/2)$ lies on the curve $r^2 = 9 \sin \theta$.
6 Sketch each of the following curves and show that each is a circle by finding the equivalent rectangular equation:
   (a) $r = 6 \sin \theta$;
   (b) $r = -8 \cos \theta$;
   (c) $r = -4 \sin \theta$.
7 Sketch the curve $r = 4 (\sin \theta + \cos \theta)$, and identify it by finding the equivalent rectangular equation.
8 Show that the graph of $r = 2a \cos \theta + 2b \sin \theta$ is either a single point or a circle through the origin. Find the center and radius of the circle.
9 Sketch and identify each of the following graphs:
   (a) $r = 2 \csc \theta$; (b) $r = 4 \sec \theta$;
   (c) $r = -3 \csc \theta$; (d) $r = -2 \sec \theta$.

We continue our program of getting better acquainted with polar graphs. In this section we concentrate particularly on sketching polar equations $r = f(\theta)$ of the type mentioned earlier, where $f(\theta)$ involves $\sin \theta$ or $\cos \theta$ in some simple way.

We again emphasize the change in point of view that is necessary for sketching polar equations. With rectangular coordinates and $y = f(x)$, we are accustomed to the idea of a point x moving along the horizontal axis and y as the directed distance measured up or down to the corresponding point (x, y) in the plane. We think in terms of "left-right" and "up-down."

With polar coordinates and $r = f(\theta)$, however, we must think of the angle θ swinging around like the hand of a clock turning in the wrong direction. For each θ we measure out from the origin a directed distance $f(\theta)$, and our moving point is farther out or closer in according as $f(\theta)$ is larger or smaller. We must think in terms of "around and around" and "in and out."

Example 1 The curve $r = a(1 + \cos \theta)$ with $a > 0$ is called a *cardioid*. When $\theta = 0$, $\cos \theta = 1$ and $r = 2a$. As θ increases from 0 to $\pi/2$ and on to π, $\cos \theta$ decreases from 1 to 0 to -1, so r decreases steadily from $2a$ to a to 0. This is shown in the upper half of Fig. 16.8. As θ continues to increase through the third and fourth quadrants, we see that $\cos \theta$, and with it r, retraces its values in reverse order, reaching $\cos \theta = 1$ and $r = 2a$ at $\theta = 2\pi$. Since $\cos \theta$ is periodic with period 2π, values of θ less than 0 or greater than 2π give points already sketched. The complete cardioid shown in the figure is evidently symmetric about the x-axis. The strange name this curve bears is accounted for by its fancied resemblance to a heart.

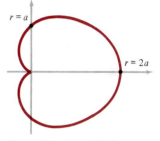

Figure 16.8   A cardioid.

When facing a polar equation, it is a natural temptation to try to return to familiar ground by converting immediately to rectangular coordinates. This is accomplished by using the transformation equations mentioned in Section 16.1,

$$r^2 = x^2 + y^2, \qquad \sin \theta = \frac{y}{r}, \qquad \cos \theta = \frac{x}{r}, \qquad \tan \theta = \frac{y}{x}.$$

In the case of the cardioid discussed in Example 1, its equation $r = a(1 + \cos \theta)$ becomes

$$r = a\left(1 + \frac{x}{r}\right), \qquad r^2 = a(r + x), \qquad x^2 + y^2 - ax = ar,$$

and finally,

$$(x^2 + y^2 - ax)^2 = a^2(x^2 + y^2).$$

This rectangular equation of the cardioid doesn't really tell us much. Clearly, it is better in this case to think exclusively in the language of polar coordinates. Nevertheless, there is a certain interest in seeing that the cardioid is a fourth-degree curve, in contrast to the second-degree curves discussed in Chapter 15.

Example 2 The curve $r = a(1 + 2 \cos \theta)$ with $a > 0$ is called a *limaçon* (French for "snail"). When $\theta = 0$, $r = 3a$. As θ increases, r decreases, be-

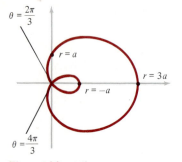

Figure 16.9 A limaçon.

coming 0 when $2 \cos \theta = -1$, that is, when $\theta = 2\pi/3$. As θ continues increasing to π, r continues to decrease through negative values from 0 to $-a$, and the point whose movement we are following traces the lower half of the inner loop shown in Fig. 16.9. Just as in Example 1, as θ continues to increase through the third and fourth quadrants, r retraces its values in reverse order; the inner loop is completed at $\theta = 4\pi/3$, and the outer loop is completed at $\theta = 2\pi$.

The curves in Examples 1 and 2 are both symmetric about the x-axis. We always have this kind of symmetry when r is a function only of $\cos \theta$, because of the identity $\cos(-\theta) = \cos \theta$. Similarly, if r is a function only of $\sin \theta$, then the curve is symmetric about the y-axis, because of the identity $\sin(\pi - \theta) = \sin \theta$.

We sometimes encounter curves whose equations have the form $r^2 = f(\theta)$. In this case, if θ is an angle for which $f(\theta) < 0$, then there is no corresponding point on the graph, because we must have $r^2 \geq 0$. But if θ is an angle for which $f(\theta) > 0$, then there are *two* corresponding points on the graph, with $r = \pm\sqrt{f(\theta)}$. These points are equally far from the origin in opposite directions, so the graph of $r^2 = f(\theta)$ is always symmetric with respect to the origin.

Example 3 The curve $r^2 = 2a^2 \cos 2\theta$ is called a *lemniscate*. For each θ there are two r's,

$$r = \pm\sqrt{2}a \sqrt{\cos 2\theta}. \tag{1}$$

As θ increases from 0 to $\pi/4$, 2θ increases from 0 to $\pi/2$ and $\cos 2\theta$ decreases from 1 to 0. Accordingly, the two r's in (1) simultaneously trace out the two parts of the curve shown on the left in Fig. 16.10. As θ continues to increase through the second half of the first quadrant and the first half of the second quadrant, 2θ varies through the second and third quadrants and $\cos 2\theta$ is negative, so there is no graph for these θ's. Through the second half of the second quadrant, $\cos 2\theta$ is positive again, and the two r's given by (1) simultaneously complete the two loops begun on the left in the figure. Further investigation reveals that no additional points are obtained, and the complete lemniscate is shown on the right. The name of this curve comes

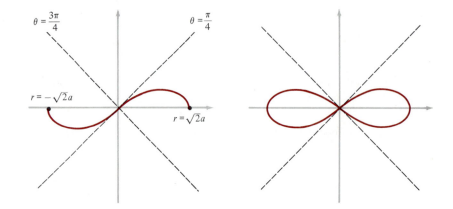

Figure 16.10 A lemniscate.

from the Latin word *lemniscus,* meaning a ribbon tied into a bow in the form of a figure eight.*

Example 4   The curve $r = a \sin 2\theta$ with $a > 0$ is called a *four-leaved rose,* for reasons that will become clear. To sketch it, we observe that as θ increases from 0 to $\pi/4$, 2θ increases from 0 to $\pi/2$ and r increases from 0 to a; and as θ increases from $\pi/4$ to $\pi/2$, 2θ increases from $\pi/2$ to π and r decreases from a to 0. This gives the leaf in the first quadrant (Fig. 16.11). Values of θ between $\pi/2$ and π (2θ between π and 2π) yield negative r's which trace out the leaf in the fourth quadrant; those between π and $3\pi/2$ (2θ between 2π and 3π) yield positive r's which trace out the leaf in the third quadrant; and those between $3\pi/2$ and 2π (2θ between 3π and 4π) produce negative r's and the leaf in the second quadrant.

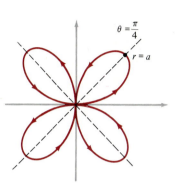

Figure 16.11   A four-leaved rose.

We sometimes need to find the points of intersection of two curves that are defined by polar equations. It is natural to try to do this by solving the equations simultaneously. Unfortunately, this may not give *all* the points of intersection. The reason for this is that a point can lie on each of two curves and yet not have a pair of polar coordinates that satisfies both equations simultaneously. An extreme example of this behavior is provided by the two equations

$$r = 1 + \cos^2 \theta \qquad \text{and} \qquad r = -1 - \cos^2 \theta,$$

whose graphs are identical. In this case there are no simultaneous solutions because all the first r's are positive and all the second r's are negative, and yet there are infinitely many points of intersection.

What can be done about finding intersections? The most sensible approach is to depend on drawing good enough graphs of both equations on a single figure to see whether there are any points of intersection. When there are, it is usually possible to find the polar coordinates of these points either by solving simultaneous equations or else by observing where the points are by direct inspection of the figure.

Remark   People who enjoy geometry in school usually take special pleasure in construction problems. As students will perhaps recall, the Greek mathematicians of antiquity learned how to perform a great variety of intricate constructions with only ruler and compass allowed as tools for drawing straight lines and circles: For instance, an angle can be bisected; a segment can be trisected; the perpendicular bisector of a segment can be drawn; regular polygons with n sides, where $n = 3, 4, 5, 6$, can be constructed; etc. All of these constructions and many more have been known since the time of Euclid and Archimedes, and the details form an important part of the study of plane geometry.†

---

* The lemniscate was introduced by James Bernoulli in 1694. It played a considerable role in some of the early work of Gauss (in 1797) and Abel (in 1826) on elliptic functions and ruler-and-compass constructions in geometry. See M. Rosen, "Abel's Theorem on the Lemniscate," *Amer. Math. Monthly,* 1981, pp. 387–395.

† See Chapters III, IV, and IX of H. Tietze, *Famous Problems of Mathematics* (Graylock Press, 1965).

The creation of geometric constructions with ruler and compass alone, when considered as an intellectual game played according to clearly understood rules, was certainly one of the most fascinating and enduring games ever invented. The complicated constructions that turn out to be possible must be seen to be believed. Nevertheless, after ingenious and persistent efforts extending over more than 2000 years, there were three classical Greek construction problems that still remained unsolved at the beginning of the nineteenth century. These problems were:

1   *to trisect an angle,* that is, to divide a given angle into three equal parts;
2   *to double a cube,* that is, to construct the edge of a cube with twice the volume of a given cube;
3   *to square a circle,* that is, to construct a square whose area equals that of a given circle.

In the course of the nineteenth century all three constructions were conclusively proved to be impossible under the stated conditions.

The traditional restriction to the use of ruler and compass alone seems to have originated with the ancient Greek philosophers, but the Greek mathematicians themselves did not hesitate to use other tools. In particular, they invented various bizarre curves for the specific purpose of solving one or another of the classical construction problems. Some of these curves are described in the problems that follow.

PROBLEMS

1   The following curves are also called *cardioids.* Sketch them, observing the way the position of the curve changes as the form of the equation changes.*
(a) $r = a(1 - \cos \theta)$.   (b) $r = a(1 + \sin \theta)$.
(c) $r = a(1 - \sin \theta)$.

2   All curves of the form $r = a \pm b \cos \theta$ or $r = a \pm b \sin \theta$ with $a, b > 0$ are called *limaçons.* If $a > b$, the graph is a single loop. If $a < b$, the graph consists of a smaller loop inside a larger one, as in Fig. 16.9. Sketch the following limaçons:
(a) $r = 3 + 2 \cos \theta$;   (b) $r = 1 + 2 \sin \theta$;
(c) $r = 1 - \sqrt{2} \cos \theta$;   (d) $r = 5 - 3 \sin \theta$.

3   Sketch the lemniscate $r^2 = 2a^2 \sin 2\theta$.

4   Sketch the graphs of the following polar equations:
(a) $r = 2a \cos \theta$;   (b) $r = 2a \sin \theta$;
(c) $r = 2 - \cos \theta$;   (d) $r = 2 + \cos \theta$;
(e) $r^2 = \cos \theta$;   (f) $r = 4 \sin^2 \theta$;
(g) $r = \cos 2\theta$;   (h) $r = 1 + \sin 2\theta$;
(i) $r = 2 + \sin 2\theta$;   (j) $r = \cos \frac{1}{2}\theta$;
(k) $r = \sin \frac{1}{2}\theta$;   (l) $r = 2 \sin^2 \frac{1}{2}\theta$;

(m) $r = 1 + 2 \sin \theta$;   (n) $r = \tan \theta$;
(o) $r = \cot \theta$;   (p) $r = \sin 3\theta$;
(q) $r = \cos 3\theta$.

5   Transform the given rectangular equation into an equivalent polar equation:
(a) $x = 5$;   (b) $y = -3$;
(c) $x^2 + y^2 = 9$;   (d) $x^2 - y^2 = 9$;
(e) $y = x^2$;   (f) $xy = 1$;

(g) $y^2 = x(x^2 - y^2)$;   (h) $y^2 = x^2 \left(\dfrac{2 + x}{2 - x} \right)$.

6   Transform the given polar equation into an equivalent rectangular equation:
(a) $r = 2$;   (b) $\theta = \pi/4$;
(c) $r \cos \theta = 3$;   (d) $r = 4 \sin \theta$;
(e) the limaçon of Example 2, $r = a(1 + 2 \cos \theta)$;
(f) the lemniscate of Example 3, $r^2 = 2a^2 \cos 2\theta$;
(g) the rose of Example 4, $r = a \sin 2\theta$;
(h) $r = \tan \theta$;   (i) $r^2 = \cos 4\theta$.

7   Let a be a positive number and consider the points $F = (a, 0)$ and $F' = (-a, 0)$. The lemniscate $r^2 = 2a^2 \cos 2\theta$ has the following simple geometric property: It is the set of all points P such that the product of the distances PF and PF' equals a^2. Prove this by first

---
* Unless the contrary is explicitly stated, it is customary to assume that the constant a that occurs in polar equations like these is a positive number.

finding the rectangular equation of the curve and then transforming this equation into its polar form $r^2 = 2a^2 \cos 2\theta$.

8 Use the formula $y = r \sin \theta$ to find the largest value of y on
(a) the cardioid $r = 2(1 + \cos \theta)$;
(b) the lemniscate $r^2 = 8 \cos 2\theta$.

9 Use the formula $x = r \cos \theta$ to find the polar coordinates of the points on the cardioid $r = 2(1 + \cos \theta)$ with the smallest x-coordinate. What is this smallest x-coordinate?

10 Find all points of intersection of each pair of curves:
(a) $r = 4 \cos \theta,\ r = 4\sqrt{3} \sin \theta$;
(b) $r = \sqrt{2} \sin \theta,\ r^2 = \cos 2\theta$;
(c) $r^2 = 4 \cos 2\theta,\ r^2 = 4 \sin 2\theta$;
(d) $r = 1 - \cos \theta,\ r = \cos \theta$;
(e) $r = a,\ r = 3a \sin \theta$;
(f) $r = a,\ r^2 = 2a^2 \sin 2\theta$.

11 A line segment of length $2a$ slides in such a way that one end is always on the x-axis and the other end is always on the y-axis. Find the polar equation of the locus of the point P in which a line from the origin perpendicular to the moving segment intersects the segment.

*12 Consider a circle of diameter $2a$ that is tangent to the y-axis at the origin (Fig. 16.12). Let OA be the diame-

ter that lies along the x-axis, AB a segment tangent to the circle at A, and C the point at which OB intersects the circle. If $P = (r, \theta)$ lies on OB in such a position that $OP = CB$, find the polar equation of the locus of P. This curve is called a *cissoid,* meaning "ivy-shaped"—or, more precisely, the *cissoid of Diocles* (Greek, second century B.C.).*

13 Find the rectangular equation of the cissoid in Problem 12.

14 Show that the line $x = 2a$ is an asymptote of the cissoid in Problem 12.

15 Let a and b be given positive numbers and consider the line whose rectangular equation is $x = a$ and whose polar equation is $r \cos \theta = a$ or $r = a \sec \theta$ (Fig. 16.13). The line OA in the figure intersects the line $x = a$ at the point A, and P is a distance b beyond A.

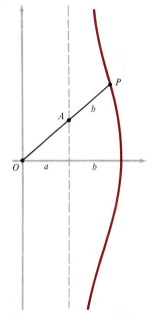

Figure 16.13   A conchoid.

The locus of P is called a *conchoid,* meaning "shell-shaped"—or, more precisely, the *conchoid of Nicomedes* (Greek, third century B.C.). The polar equation of the conchoid is clearly $r = a \sec \theta + b$. Find its rectangular equation.†

---

* In the Additional Problems we explain how the cissoid can be used to solve the problem of doubling a cube.

† In the Additional Problems we explain how the conchoid can be used to solve the problem of trisecting an angle.

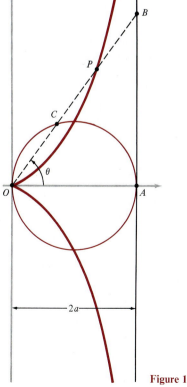

Figure 16.12   A cissoid.

16 If a rotation of the polar axis through a specified angle transforms one polar equation into another, then their graphs are clearly congruent. Show that a counterclockwise rotation of the polar axis through an angle α can be accomplished by replacing θ by $\theta + \alpha$. Use this method with suitable choices of α to show that

(a) the cardioids in Problem 1 are congruent to the cardioid $r = a(1 + \cos \theta)$ discussed in Example 1;

(b) the limaçon $r = a \pm b \sin \theta$ in Problem 2 is congruent to the limaçon $r = a \pm b \cos \theta$;

(c) the lemniscate $r^2 = 2a^2 \sin 2\theta$ in Problem 3 is congruent to the lemniscate $r^2 = 2a^2 \cos 2\theta$ discussed in Example 3.

16.3

POLAR EQUATIONS OF CIRCLES, CONICS, AND SPIRALS

We have already had considerable experience in transforming the rectangular equation of a given curve into an equivalent polar equation for the same curve. Our basic tools for this procedure are the transformation equations listed in Section 16.2.

Consider, for example, the circle (Fig. 16.14, left) with center $(a, 0)$ and radius a:

$$(x - a)^2 + y^2 = a^2 \qquad \text{or} \qquad x^2 + y^2 = 2ax. \tag{1}$$

Since $x^2 + y^2 = r^2$ and $x = r \cos \theta$, this equation becomes

$$r^2 = 2ar \cos \theta,$$

which is equivalent to

$$r = 2a \cos \theta \tag{2}$$

because the origin $r = 0$ lies on the graph of (2).

This example illustrates one way to find the polar equation of a curve, namely, transform its rectangular equation into polar coordinates. Another method that is better whenever it is feasible is to obtain the polar equation directly from some characteristic geometric property of the curve. In the case of the circle just discussed, we use the fact that the angle OPA in the figure on the right is a right angle. Since OPA is a right triangle with r the adjacent side to the acute angle θ, we clearly have

$$r = 2a \cos \theta,$$

which of course is the same equation previously obtained, but derived in a very different way.

We shall use this second and more natural method to find the polar equations of various curves in the following examples.

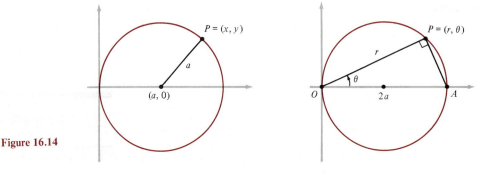

Figure 16.14

Example 1   Find the polar equation of the circle with radius a and center at the point C with polar coordinates (b, α), where b is assumed to be positive.

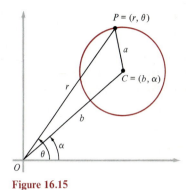

Solution   Let $P = (r, \theta)$ be any point on the circle, as shown in Fig. 16.15, and apply the law of cosines to the triangle OPC to obtain

$$a^2 = r^2 + b^2 - 2br \cos(\theta - \alpha).$$

This is the polar equation of the circle. For circles that pass through the origin we have $b = a$, and the equation can be written as

$$r = 2a \cos(\theta - \alpha). \tag{3}$$

In particular, when $\alpha = 0$, then (3) reduces to (2), and when $\alpha = \pi/2$, so that the center lies on the y-axis, then $\cos(\theta - \pi/2) = \sin\theta$, and (3) reduces to

Figure 16.15

$$r = 2a \sin\theta. \tag{4}$$

In this case the right triangle OPA in Fig. 16.16 provides a more direct geometric way of obtaining (4), since here r is the opposite side to the acute angle θ.

Example 2   Let F_1 and F_2 be the two points whose rectangular coordinates are $(a, 0)$ and $(-a, 0)$, as shown in Fig. 16.17. If b is a positive constant, find the polar equation of the locus of a point P that moves in such a way that the product of its distances from F_1 and F_2 is b^2.

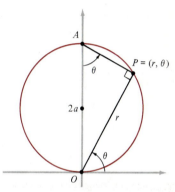

Solution   If $P = (r, \theta)$ is an arbitrary point on the curve, then the defining condition is

Figure 16.16

$$d_1 d_2 = b^2 \qquad \text{or} \qquad d_1^2 d_2^2 = b^4,$$

where $d_1 = PF_1$ and $d_2 = PF_2$. We apply the law of cosines twice, first to the triangle OPF_1,

$$d_1^2 = r^2 + a^2 - 2ar \cos\theta, \tag{5}$$

and then to the triangle OPF_2,

$$d_2^2 = r^2 + a^2 - 2ar \cos(\pi - \theta). \tag{6}$$

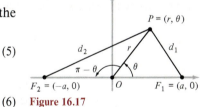

Figure 16.17

Since $\cos(\pi - \theta) = -\cos\theta$, we can write (6) as

$$d_2^2 = r^2 + a^2 + 2ar \cos\theta, \tag{7}$$

and by multiplying (5) and (7) we obtain

$$d_1^2 d_2^2 = (r^2 + a^2)^2 - (2ar \cos\theta)^2$$

or

$$b^4 = r^4 + a^4 + 2a^2r^2(1 - 2\cos^2\theta).$$

The trigonometric identity $2\cos^2\theta = 1 + \cos 2\theta$ permits us to write this equation as

$$b^4 = r^4 + a^4 - 2a^2r^2 \cos 2\theta. \tag{8}$$

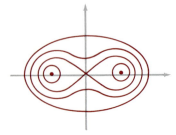

Figure 16.18   The ovals of Cassini.

In the special case $b = a$, the curve passes through the origin, and the equation takes the much simpler form

$$r^2 = 2a^2 \cos 2\theta. \tag{9}$$

We recognize this as the equation of the lemniscate discussed in Example 3 of Section 16.2. When $b > a$, the curve consists of a single loop, but when $b < a$ it breaks into two separate loops. The cases $b < a$ and $b = a$ are illustrated in Fig. 16.18, along with two cases of $b > a$. Collectively, these curves are called the *ovals of Cassini.**

Polar coordinates are particularly well suited to working with conic sections, as we see in the next example.

Example 3   Find the polar equation of the conic section with eccentricity e if the focus is at the origin and the corresponding directrix is the line $x = -p$ to the left of the origin.

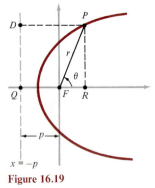

Figure 16.19

Solution   With the notation of Fig. 16.19, the focus-directrix-eccentricity characterization of the conic section is

$$\frac{PF}{PD} = e \qquad \text{or} \qquad PF = e \cdot PD. \tag{10}$$

We recall that the curve is an ellipse, a parabola, or a hyperbola according as $e < 1$, $e = 1$, or $e > 1$. By examining the figure, we see that $PF = r$ and

$$PD = QR = QF + FR$$
$$= p + r \cos \theta,$$

so (10) is

$$r = e(p + r \cos \theta).$$

This is easily solved for r, which gives

$$r = \frac{ep}{1 - e \cos \theta} \tag{11}$$

as the polar equation of our conic section.

We give two concrete illustrations of the ideas in Example 3.

Example 4   Find the polar equation of the conic with eccentricity $\frac{1}{3}$, focus at the origin, and directrix $x = -4$.

---

* The Italian astronomer Giovanni Domenico Cassini thought of these ovals in 1680 in connection with his efforts to understand the relative motions of the earth and the sun. He proposed them as alternatives to Kepler's ellipses before Newton settled the matter with his theory of the solar system in 1687. Cassini discovered several of the satellites of Saturn, and also the so-called "Cassini division" in Saturn's ring, thereby showing that this ring consists of more than one piece.

Solution   We merely substitute $e = \frac{1}{3}$ and $p = 4$ in equation (11), which yields

$$r = \frac{\frac{1}{3}(4)}{1 - \frac{1}{3}\cos\theta} = \frac{4}{3 - \cos\theta}.$$

The curve is an ellipse. Observe that the denominator here is never zero, so r is bounded for all θ's.

Example 5   Given the conic with equation

$$r = \frac{25}{4 - 5\cos\theta},$$

find the eccentricity, locate the directrix, and identify the curve.

Solution   We begin by dividing numerator and denominator by 4 to put the equation in the exact form of (11),

$$r = \frac{\frac{25}{4}}{1 - \frac{5}{4}\cos\theta}.$$

This tells us that $e = \frac{5}{4}$ and $ep = \frac{25}{4}$, so $p = 5$. The directrix is the line $x = -5$, and the curve is a hyperbola. Observe that the denominator here is zero when $\cos\theta = \frac{4}{5}$, so r becomes infinite near these directions.

In connection with Example 3, it is worth pointing out that if the directrix is the line $x = p$ to the right of the origin, as in Fig. 16.20, then $PD = p - r\cos\theta$. The equation $PF = e \cdot PD$ now has the form

$$r = e(p - r\cos\theta),$$

and instead of (11) we have

$$r = \frac{ep}{1 + e\cos\theta}.$$

Polar coordinates are very convenient for describing certain spirals.

Example 6   The *spiral of Archimedes* (Fig. 16.21) can be defined as the locus of a point P that starts at the origin and moves outward at a constant speed along a radius which, in turn, is rotating counterclockwise at a constant speed from its initial position along the polar axis, where both motions start at the same time.* Since both r and θ are proportional to the time t measured from the beginning of the motions, r is proportional to θ and the polar equation of the spiral is $r = a\theta$, where a is a positive constant. In the figure, it is assumed that θ starts at zero and increases into positive values, as implied

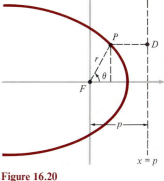

Figure 16.20

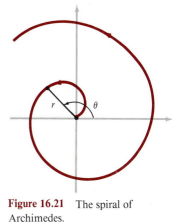

Figure 16.21   The spiral of Archimedes.

_____
* These are almost the same words which Archimedes himself uses to define his spiral. See his treatise "On Spirals" in *The Works of Archimedes,* T. L. Heath (ed.), Dover, n.d., especially p. 154.

by the definition. However, if we wish to allow θ to be negative, then there is another part to the spiral which we have deliberately not sketched for the sake of keeping the figure uncluttered.

Example 7 In the spiral discussed in Example 6, r is directly proportional to θ, $r = a\theta$. We now consider the case in which r is inversely proportional to θ,

$$r = \frac{a}{\theta} \qquad \text{or} \qquad r\theta = a, \tag{12}$$

where a is a positive constant. For positive values of θ, the graph is the curve shown in Fig. 16.22; it is called a *hyperbolic spiral* because of the resemblance of $r\theta = a$ to the equation $xy = a$, which represents a hyperbola in rectangular coordinates.

The essential features of the graph are easy to see by considering $r = a/\theta$. When $\theta = 0$, there is no r; when θ is small and positive, r is large and positive; and as θ increases to ∞, r decreases to 0. This tells us that a variable point P on the graph comes in from infinity and winds around the origin in the counterclockwise direction in an infinite number of shrinking coils as θ increases indefinitely. To understand the behavior of this curve for small positive θ's, we need to think about what happens to

$$y = r \sin \theta = a \frac{\sin \theta}{\theta}.$$

We know that

$$\lim_{\theta \to 0} \frac{\sin \theta}{\theta} = 1,$$

and therefore

$$\lim_{\theta \to 0} y = \lim_{\theta \to 0} a \frac{\sin \theta}{\theta} = a.$$

It follows that the line $y = a$ is an asymptote of the curve, as shown in the figure.

If θ is allowed to be negative, we get another part of the curve, which again we do not sketch in order to avoid cluttering up the figure. The nature of this other part is easily understood by observing that if r and θ are replaced by $-r$ and $-\theta$, then equation (12) is unaltered. This means that for every point (r, θ) on the curve, the point $(-r, -\theta)$, which is symmetrically located with respect to the y-axis, is also on the curve. Thus, the other part is a second mirror-image spiral that winds in to the origin in the clockwise direction as $\theta \to -\infty$.

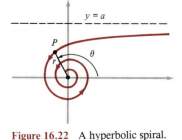

Figure 16.22   A hyperbolic spiral.

PROBLEMS

In Problems 1 to 6, find the polar equation of the circle determined by the stated conditions.

1 Center $(4, \pi/6)$, radius 3.

2 Center $(-3, \pi/3)$, radius 4.

3 Center $(5, 0)$, radius 5.

4 Center $(3, \pi/2)$, radius 3.

5 Center on the line $\theta = \pi/3$ and passing through $(6, \pi/2)$ and $(0, 0)$.

6 Center $(5, \pi/4)$ and passing through $(8, 0)$.

7 A line is drawn from the origin perpendicular to a tangent to the circle $r = 2a \cos \theta$. Find the equation of the locus of the point of intersection and sketch the curve.

8 Find the rectangular equation of the ovals of Cassini (Example 2) by direct use of the condition $d_1{}^2 d_2{}^2 = b^4$.

9 The largest and smallest values of r on the lemniscate (9) are clearly $r = \sqrt{2}a$ and $r = 0$. Find the largest and smallest values of r on the ovals of Cassini (8)
(a) in the one-loop case $b = 2a$;
(b) in the two-loop case $b = \frac{1}{2}a$.

10 The equation $r = 4/(3 - \cos \theta)$ in Example 4 represents an ellipse with one focus at the origin. Sketch the curve, find both of its directrices, and locate the center.

11 If a conic section with eccentricity e has focus at the origin and directrix $y = -p$ below the origin, show that its polar equation is $r = ep/(1 - e \sin \theta)$. What is the polar equation if the directrix is the line $y = p$ above the origin?

Find the eccentricity of each of the following conic sections (in Problems 12 to 15) and sketch the curve.

12 $r = \dfrac{6}{1 - \cos \theta}.$

13 $r = \dfrac{10}{2 - \cos \theta}.$

14 $r = \dfrac{4}{2 + 4 \cos \theta}.$

15 $r = \dfrac{18}{6 + \cos \theta}.$

16 One focus of a hyperbola with eccentricity $e = \frac{4}{3}$ is at the origin, and the corresponding directrix is the line $x = 7$ (or $r \cos \theta = 7$). Find the polar equation and the polar coordinates of the second focus and center, and sketch the curve.

17 When $e > 1$, the equation $r = ep/(1 - e \cos \theta)$ represents a hyperbola. Use this equation to determine the slopes of the asymptotes.

18 Transform the equation $r = ep/(1 - e \cos \theta)$ into rectangular coordinates. Use the facts established in Chapter 15 about the equations of conics in rectangular coordinates to show that the given equation repre-

sents a parabola if $e = 1$, an ellipse if $0 < e < 1$, and a hyperbola if $e > 1$.

19 If $e < 1$, use calculus to find the polar coordinates of the point on the ellipse $r = ep/(1 - e \cos \theta)$ that is
(a) farthest from the origin;
(b) closest to the origin.

20 Find the length of the major axis of the ellipse in Problem 19. Also find the polar coordinates of its center.

*21 In his attempts to trisect an angle, Hippias of Elis, a leading sophist of the time of Socrates, invented a new curve, as follows. Consider a square $OABC$ of side a located in the first quadrant of the xy-plane (Fig. 16.23). Let OA rotate clockwise about O at a constant speed to the position OC. In the same time, let AB move downward at a constant speed to the position OC. The *quadratrix APG* is the locus of the point P of intersection of the turning radius OD and the moving segment EF.*

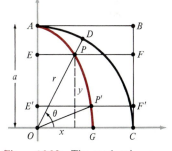

Figure 16.23 The quadratrix.

(a) Find the rectangular equation of the quadratrix [hint: $y/a = \theta/(\pi/2)$; why?].

(b) Find the polar equation of the quadratrix.

(c) Use part (b) to show that $OG = 2a/\pi$.

(d) Pappus of Alexandria (fourth century A.D.) proved geometrically that $ADC/OC = OC/OG$. Show that this is equivalent to the result stated in part (c).

(e) Verify the validity of the following procedure for using the quadratrix as a tool for trisecting an arbitrary acute angle θ: Construct the point E' that trisects OE, so that $OE' = \frac{1}{3}OE$; draw $E'F'$

---

* The point G is not defined as part of the quadratrix because OA reaches OC at the same moment AB reaches OC, so there is no point of intersection. However, G is the limiting position of the points on the quadratrix that approach the x-axis.

parallel to OC, and let P' be the point where this line intersects the quadratrix; draw OP' and conclude that $\angle COP' = \frac{1}{3}\theta$.* [Part (e) requires that the segment OE be trisected by the point E'. Figure 16.24 shows how this point can be produced by a Euclidean (ruler-and-compass) construction: Measure off any length b three times in any direction OQ, join R to E, and draw the line parallel to RE through the first point of division, intersecting OE at E'.]

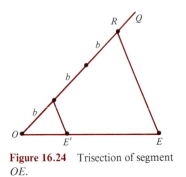

Figure 16.24   Trisection of segment OE.

that the area under the quadratrix $y = x \cot x$ from 0 to $x \le \pi/2$ is

$$x - \tfrac{1}{9}x^3 - \tfrac{1}{225}x^5 - \tfrac{2}{6615}x^7 - \cdots.$$

23   Verify the validity of the following method for trisecting an angle AOB by using the spiral of Archimedes, $r = a\theta$ (Fig. 16.25):

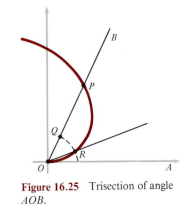

Figure 16.25   Trisection of angle AOB.

*22   In Problem 21, let $a = \pi/2$, so that the rectangular equation of the quadratrix is $x = y \cot y$. Interchange x and y so that the equation becomes $y = x \cot x$, and use division of power series to obtain Newton's result

(a) let OB intersect the spiral at P, and construct the point Q that trisects OP, so that $OQ = \frac{1}{3}OP$;
(b) construct the circle with center O and radius OQ, and let this circle intersect the spiral at R;
(c) draw OR and conclude that $\angle AOR = \frac{1}{3} \angle AOB$.†

---

* In the Additional Problems we show how the quadratrix can also be used to square a circle.

† In the Additional Problems we show how the spiral of Archimedes can also be used to square a circle.

16.4

ARC LENGTH AND TANGENT LINES

Consider a curve whose polar equation is $r = f(\theta)$, and let s denote arc length measured along the curve from a specified point in a specified direction (Fig. 16.26). By Section 7.5 we know that the differential element of arc length ds is given by the formula

$$ds^2 = dx^2 + dy^2.$$

But $x = r \cos \theta$ and $y = r \sin \theta$, and by differentiating with respect to θ by the product rule, we obtain

$$\frac{dx}{d\theta} = -r \sin \theta + \cos \theta \, \frac{dr}{d\theta} \qquad \text{and} \qquad \frac{dy}{d\theta} = r \cos \theta + \sin \theta \, \frac{dr}{d\theta},$$

or equivalently, in the notation of differentials,

$$dx = -r \sin \theta \, d\theta + \cos \theta \, dr \qquad \text{and} \qquad dy = r \cos \theta \, d\theta + \sin \theta \, dr. \qquad (1)$$

It follows from these formulas that

$ds^2 = dx^2 + dy^2$

$\quad = r^2 \sin^2 \theta \, d\theta^2 - 2r \sin \theta \cos \theta \, dr \, d\theta + \cos^2 \theta \, dr^2$

$\quad\quad\quad\quad\quad + r^2 \cos^2 \theta \, d\theta^2 + 2r \sin \theta \cos \theta \, dr \, d\theta + \sin^2 \theta \, dr^2$

$\quad = r^2 \, d\theta^2 + dr^2.$

Thus, we have

$$ds^2 = r^2 \, d\theta^2 + dr^2 \qquad (2)$$

or

$$ds = \sqrt{r^2 \, d\theta^2 + dr^2} = \sqrt{\left(r^2 + \frac{dr^2}{d\theta^2}\right) d\theta^2}$$

$$= \sqrt{r^2 + \left(\frac{dr}{d\theta}\right)^2} \, d\theta.$$

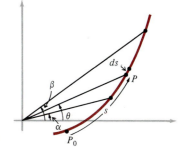

Figure 16.26

This formula enables us to compute arc lengths of polar curves by integration, as suggested by the figure:

$$\text{Arc length from } \theta = \alpha \text{ to } \theta = \beta \text{ equals } \int ds = \int_\alpha^\beta \sqrt{r^2 + \left(\frac{dr}{d\theta}\right)^2} \, d\theta.$$

Example 1 Find the total length of the cardioid $r = a(1 - \cos \theta)$.

Solution This curve is quite familiar to us and is shown in Fig. 16.29. From the equation of the curve, we have $dr = a \sin \theta \, d\theta$, so formula (2) gives

$$ds^2 = a^2(1 - \cos \theta)^2 \, d\theta^2 + a^2 \sin^2 \theta \, d\theta^2$$

$$= a^2[(1 - \cos \theta)^2 + \sin^2 \theta] \, d\theta^2$$

$$= 2a^2(1 - \cos \theta) \, d\theta^2.$$

Therefore

$$ds = \sqrt{2}a\sqrt{1 - \cos \theta} \, d\theta$$

$$= 2a|\sin \tfrac{1}{2}\theta| \, d\theta,$$

since $1 - \cos \theta = 2 \sin^2 \tfrac{1}{2}\theta$. We know that $\sin \tfrac{1}{2}\theta \geq 0$ for $0 \leq \theta \leq 2\pi$, so we can drop the absolute value signs and write

$$s = \int ds = \int_0^{2\pi} 2a \sin \frac{1}{2}\theta \, d\theta$$

$$= -4a \cos \frac{1}{2}\theta \bigg]_0^{2\pi} = 4a - (-4a) = 8a.$$

The symmetry of this curve about the horizontal axis tells us that we can also obtain the total length by integrating from 0 to π and multiplying by 2,

$$s = 2 \int_0^\pi 2a \sin \frac{1}{2}\theta \, d\theta = -8a \cos \frac{1}{2}\theta \bigg]_0^\pi$$

$$= 0 - (-8a) = 8a.$$

As a matter of routine, we should accustom ourselves to simplifying the calculation of integrals as much as possible by exploiting whatever symmetry is available.

The above formula for ds in polar coordinates can also be used to find areas of surfaces of revolution, as explained in Section 7.6.

Example 2 Find the area of the surface generated when the lemniscate $r^2 = 2a^2 \cos 2\theta$ is revolved about the x-axis.

Solution An element of arc length ds (Fig. 16.27) generates an element of surface area

$$dA = 2\pi y \, ds,$$

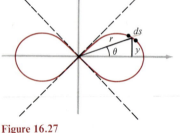

Figure 16.27

where

$$y = r \sin \theta \quad \text{and} \quad ds = \sqrt{r^2 \, d\theta^2 + dr^2},$$

so

$$dA = 2\pi r \sin \theta \sqrt{r^2 \, d\theta^2 + dr^2} = 2\pi \sin \theta \sqrt{r^4 \, d\theta^2 + r^2 \, dr^2}. \tag{3}$$

From the equation of the curve we have

$$r \, dr = -2a^2 \sin 2\theta \, d\theta,$$

so

$$r^4 \, d\theta^2 + r^2 \, dr^2 = (4a^4 \cos^2 2\theta + 4a^4 \sin^2 2\theta) \, d\theta^2$$
$$= 4a^4 \, d\theta^2$$

and (3) becomes

$$dA = 4\pi a^2 \sin \theta \, d\theta.$$

The total surface area is twice the area of the right half, which is generated as ds moves along the part of the lemniscate in the first quadrant, that is, as θ increases from 0 to $\pi/4$. We therefore have

$$A = 2\int_0^{\pi/4} 4\pi a^2 \sin \theta \, d\theta = -8\pi a^2 \cos \theta \Big]_0^{\pi/4}$$

$$= -8\pi a^2 \left(\frac{\sqrt{2}}{2} - 1\right) = 4\pi a^2(2 - \sqrt{2}).$$

When working with rectangular coordinates, we specify the direction of a curve $y = f(x)$ at a point by the angle α from the positive x-axis to the tangent line. However, in the case of a polar curve $r = f(\theta)$, it is easier to work with the angle ψ (psi) from the radius vector to the tangent line, as shown in Fig. 16.28. We see from the figure that $\alpha = \theta + \psi$, so $\psi = \alpha - \theta$; and since $\tan \alpha = dy/dx$ and $\tan \theta = y/x$, the subtraction formula for the tangent gives

$$\tan \psi = \tan(\alpha - \theta)$$

$$= \frac{\tan \alpha - \tan \theta}{1 + \tan \alpha \tan \theta}$$

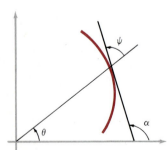

Figure 16.28

$$= \frac{dy/dx - y/x}{1 + (dy/dx) \cdot (y/x)}$$

$$= \frac{x \, dy - y \, dx}{x \, dx + y \, dy}. \tag{4}$$

The reason why ψ is a convenient angle to use with polar coordinates is that (4) can be put into a very simple form. First, the fact that $x^2 + y^2 = r^2$ tells us that $x \, dx + y \, dy = r \, dr$. Next, from (1) we obtain

$$x \, dy - y \, dx = r^2 \cos^2 \theta \, d\theta + r \sin \theta \cos \theta \, dr + r^2 \sin^2 \theta \, d\theta - r \sin \theta \cos \theta \, dr$$
$$= r^2 \, d\theta.$$

By substituting these expressions into (4), we find that

$$\tan \psi = \frac{r \, d\theta}{dr} = \frac{r}{dr/d\theta}. \tag{5}$$

This formula is the basic tool for working with tangent lines to polar curves.

Example 3 Find the angle ψ for the cardioid $r = a(1 - \cos \theta)$.

Solution This curve is shown in Fig. 16.29. The equation of the curve gives

$$\frac{dr}{d\theta} = a \sin \theta,$$

so

$$\tan \psi = \frac{r}{dr/d\theta} = \frac{a(1 - \cos \theta)}{a \sin \theta}$$

$$= \frac{2 \sin^2 \frac{1}{2}\theta}{2 \sin \frac{1}{2}\theta \cos \frac{1}{2}\theta}$$

$$= \tan \frac{1}{2}\theta.$$

We therefore have $\psi = \frac{1}{2}\theta$, and as θ increases from 0 to 2π, ψ increases from 0 to π, as indicated in the figure.

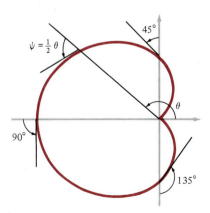

Figure 16.29

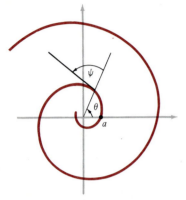

Figure 16.30   The equiangular spiral.

As another example of the use of the formula for tan ψ, we consider an interesting curve called the *exponential spiral.*

Example 4   Find the angle ψ for the curve $r = ae^{b\theta}$, where $a > 0$ and $b \neq 0$.

Solution   If $b > 0$, we see that r increases as θ increases, as shown in Fig. 16.30. Further, it is clear that $r \to \infty$ as $\theta \to \infty$ and $r \to 0$ as $\theta \to -\infty$. The distinctive feature of this curve is that ψ is constant, because

$$\tan \psi = \frac{r}{dr/d\theta} = \frac{ae^{b\theta}}{abe^{b\theta}} = \frac{1}{b}.$$

This enables us to find ψ in the form $\psi = \tan^{-1}(1/b)$. If $b < 0$, the curve spirals in to the origin instead of outward as θ increases. The curve $r = ae^{b\theta}$ is sometimes called the *equiangular spiral* because of the constancy of ψ.

The two main facts of this section, formulas (2) and (5), are easy to remember by using Fig. 16.31 as a mnemonic device. In this figure we have an arc of length ds joining two points with polar coordinates r, θ and $r + dr$, $\theta + d\theta$. The outer part of the figure is approximately a rectangle, and the "differential triangle" on the right is approximately a right triangle with hypotenuse ds and with $r\,d\theta$ and dr as the legs opposite and adjacent to the angle ψ. The formulas

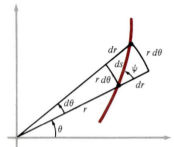

Figure 16.31

$$ds^2 = r^2\,d\theta^2 + dr^2$$

and

$$\tan \psi = \frac{r\,d\theta}{dr}$$

are now self-evident from this triangle, by the theorem of Pythagoras and the right triangle definition of the tangent. Needless to say, this way of reasoning is not a proof, but it is very useful nevertheless. It is also a good example of the true Leibnizian spirit in calculus, in the sense repeatedly explained in Chapter 7.

PROBLEMS

1   For the spiral of Archimedes $r = a\theta \ (\theta \geq 0)$, show that $\psi = 45°$ when $\theta = 1$ radian, and also that $\psi \to 90°$ as the spiral winds on around the origin in the counterclockwise direction. Sketch the curve and show the angle ψ for the direction $\theta = 1$ radian.

2   If two curves $r = f_1(\theta)$ and $r = f_2(\theta)$ intersect at a point other than the origin for a common value of θ, show by examining a figure that the angle γ between their

tangents can be found from the formula

$$\tan \gamma = \frac{\tan \psi_2 - \tan \psi_1}{1 + \tan \psi_2 \tan \psi_1}.$$

Under what circumstances will the two curves intersect orthogonally (at right angles)?

3   Sketch each of the following pairs of curves on a single figure and use the result of Problem 2 to show that they intersect orthogonally:
(a) $r = 2a \sin \theta$, $r = 2b \cos \theta$;

(b) $r = a(1 + \cos\theta)$, $r = b(1 - \cos\theta)$, except at the origin;

(c) $r = a/(1 - \cos\theta)$, $r = b/(1 + \cos\theta)$;

(d) $r = a/(1 - \cos\theta)$, $r = a(1 - \cos\theta)$;

(e) $r^2 = 2a^2 \cos 2\theta$, $r^2 = 2b^2 \sin 2\theta$, except at the origin.

4 Show that the spirals $r = \theta$ and $r = 1/\theta$ intersect orthogonally at $\theta = 1$.

5 Find the area of the surface generated by revolving the cardioid $r = a(1 - \cos\theta)$ about the x-axis.

6 The lemniscate $r^2 = 2a^2 \cos 2\theta$ is revolved about the y-axis. Find the area of the surface of revolution generated in this way.

7 Consider the tangent at a point P on the spiral $r = a\theta$ ($\theta \geq 0$), and let the line OT which is perpendicular to OP at the origin O meet this tangent at T (Fig. 16.32). Show that the segment OT equals the circular arc ASP with center O which is drawn from the polar axis to the point P.

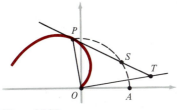

Figure 16.32

8 At what angle does the lemniscate $r^2 = 2a^2 \cos 2\theta$ intersect the circle $r = a$?

9 On the upper half of the parabola $r = a/(1 - \cos\theta)$, show that $\tan\psi = -\tan\frac{1}{2}\theta$, and conclude that $\psi = \pi - \frac{1}{2}\theta$. Show that this establishes the following reflection property: The tangent at any point on the parabola makes equal angles with the horizontal through that point and the line from the origin through the point.

10 Show that the part of the hyperbolic spiral $r\theta = 1$ from $\theta = \pi/2$ to $\theta = \infty$ has infinite length.

11 Find the surface area generated by revolving the circle $r = 2a \cos\theta$ about the line

(a) $\theta = 0$;

(b) $\theta = \pi/2$.

12 Find the length of one turn of the spiral $r = \theta$, from $\theta = 0$ to $\theta = 2\pi$.

13 If a curve $r = f(\theta)$ has the property that ψ is a constant $\neq \pi/2$, show that the curve must be the exponential spiral $r = ae^{b\theta}$.

14 Use integration to find the circumference of the circle $r = 2a \cos\theta$.

15 Show that if a point moves at constant speed along the exponential spiral $r = ae^{b\theta}$, then the radius r changes at a constant rate.

*16 If a curve $r = f(\theta)$ has the property that $\psi = \frac{1}{2}\theta$, show that the curve must be the cardioid $r = a(1 - \cos\theta)$.

17 Find the length of the exponential spiral $r = e^{-\theta}$ from $r = 1$ to the origin.

18 Sketch the exponential spiral $r = ae^{b\theta}$ for the case in which a is positive and b is negative, and show that the arc length from $\theta = 0$ to $\theta = \infty$ is equal to the length of the part of the tangent at $\theta = 0$ that is cut off by the x- and y-axes.

*19 Show that the length L of the right loop of the lemniscate $r^2 = 2a^2 \cos 2\theta$ can be expressed in the form

$$L = \sqrt{2}a \int_{-\pi/4}^{\pi/4} \frac{d\theta}{\sqrt{\cos 2\theta}}$$

$$= \sqrt{2}a \int_{-\pi/4}^{\pi/4} \frac{d\theta}{\sqrt{1 - 2\sin^2\theta}}.$$

Introduce the new variable $u = \tan\theta$ and show that

$$\sin^2\theta = \frac{u^2}{1 + u^2} \quad \text{and} \quad d\theta = \frac{du}{1 + u^2},$$

and that therefore

$$L = \sqrt{2}a \int_{-1}^{1} \frac{du}{\sqrt{1 - u^4}}.^*$$

---

* This is a special elliptic integral that played a large part in the investigations of Gauss mentioned in a footnote of Section 16.2. See also Problem 16 in Section 10.8.

16.5
AREAS IN POLAR COORDINATES

Our problem here is to find the area A of the region bounded by a polar curve $r = f(\theta)$ and two half-lines $\theta = \alpha$ and $\theta = \beta$, as shown in Fig. 16.33. Our approach is modeled on the "differential element of area" idea described in Section 7.1.

In working with areas in rectangular coordinates, we use thin rectangular strips and rely on the fact that the area of a rectangle equals length times

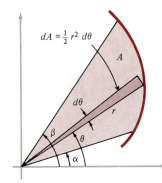

Figure 16.33

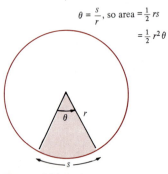

Figure 16.34

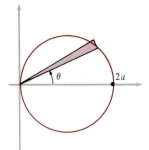

Figure 16.35

width. Here we need the fact, illustrated in Fig. 16.34, that the area of a sector of a circle of radius r and central angle θ (measured in radians) is $\frac{1}{2}r^2\theta$. In Fig. 16.33 our element of area dA is the area of the very thin sector with radius r and central angle $d\theta$, so

$$dA = \tfrac{1}{2}r^2\,d\theta. \qquad (1)$$

In the manner of Section 7.1, we think of the total area A as the result of adding up these elements of area dA as our thin sector sweeps across the region, that is, as θ increases from α to β:

$$A = \int dA = \int_\alpha^\beta \frac{1}{2} r^2\,d\theta. \qquad (2)$$

Again, the essence of the process of integration is that we calculate the whole of a quantity by cutting it up into a great many convenient small pieces and then adding up these pieces.

We shall give a more mathematically sophisticated approach to formula (2) in Remark 2. First, however, we illustrate its use in several examples. As students will observe in these examples, it is always essential in solving area problems to have a good idea of what the curve looks like, because the correct limits of integration will be determined from the figure.

Example 1   Use integration to find the area of the circle $r = 2a\cos\theta$.

Solution   The complete circle (Fig. 16.35) is swept out as θ increases from $-\pi/2$ to $\pi/2$. By symmetry we can integrate from 0 to $\pi/2$ and multiply by 2,

$$A = 2\int_0^{\pi/2} \frac{1}{2} r^2\,d\theta = 2\int_0^{\pi/2} \frac{1}{2}\cdot 4a^2\cos^2\theta\,d\theta$$

$$= 4a^2 \int_0^{\pi/2} \frac{1}{2}(1 + \cos 2\theta)\,d\theta$$

$$= 2a^2\left(\theta + \frac{1}{2}\sin 2\theta\right)\bigg]_0^{\pi/2} = \pi a^2.$$

Naturally, we expected this answer because our circle has radius a, but it is reassuring to obtain a familiar result by a new method.

Example 2   Find the total area enclosed by the lemniscate $r^2 = 2a^2\cos 2\theta$ (Fig. 16.36).

Solution   By symmetry, we calculate the area of the first quadrant part and multiply by 4:

$$A = 4\int_0^{\pi/4} \frac{1}{2} r^2\,d\theta = 4\int_0^{\pi/4} \frac{1}{2}\cdot 2a^2\cos 2\theta\,d\theta$$

$$= 4a^2 \int_0^{\pi/4} \cos 2\theta\,d\theta = 2a^2\sin 2\theta\bigg]_0^{\pi/4} = 2a^2.$$

This problem provides a good illustration of the value of exploiting symmetry; for if we carelessly integrate all the way around from 0 to 2π, forgetting that r^2 is sometimes positive and sometimes negative, then our final answer turns out to be 0, which is obviously wrong.

Example 3 Find the area inside the circle $r = 6a \cos \theta$ and outside the cardioid $r = 2a(1 + \cos \theta)$.

Solution By equating the r's and solving for θ, we see that the curves intersect in the first quadrant at $\theta = \pi/3$, as shown in Fig. 16.37. The indicated element of area is

$$dA = \tfrac{1}{2}(r_{\text{circle}})^2 \, d\theta - \tfrac{1}{2}(r_{\text{cardioid}})^2 \, d\theta$$

$$= \tfrac{1}{2}[(r_{\text{circle}})^2 - (r_{\text{cardioid}})^2] \, d\theta$$

$$= \tfrac{1}{2}[36a^2 \cos^2 \theta - 4a^2(1 + \cos \theta)^2] \, d\theta$$

$$= 2a^2(8 \cos^2 \theta - 1 - 2 \cos \theta) \, d\theta.$$

By symmetry, the area we seek is double the first quadrant area, so

$$A = 2 \int_0^{\pi/3} 2a^2(8 \cos^2 \theta - 1 - 2 \cos \theta) \, d\theta$$

$$= 4a^2 \int_0^{\pi/3} [4(1 + \cos 2\theta) - 1 - 2 \cos \theta] \, d\theta$$

$$= 4a^2 \left[3\theta + 2 \sin 2\theta - 2 \sin \theta \right]_0^{\pi/3} = 4\pi a^2.$$

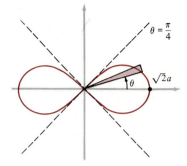

Figure 16.36

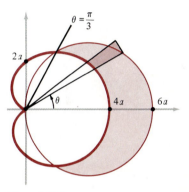

Figure 16.37

Remark 1 The ideas of this section have an important application to the astronomy of the solar system. Consider a point P moving along a polar curve $r = f(\theta)$. We can think of P as a planet moving along its orbit, with the sun at the origin. If A is the area swept out by the radius OP from a fixed direction α to a variable direction θ, as shown in Fig. 16.38, then we have

$$dA = \tfrac{1}{2}r^2 \, d\theta.$$

If both A and θ are thought of as functions of time t, then we see that

$$\frac{dA}{dt} = \frac{1}{2} r^2 \frac{d\theta}{dt}.$$

The derivative dA/dt is, of course, the rate of change of the area A. Kepler's second law of planetary motion states that a planet moves in such a way that the radius joining the planet to the sun sweeps out area at a constant rate. This means that dA/dt is constant, which in turn means that

$$r^2 \frac{d\theta}{dt} = \text{a constant} \tag{3}$$

for any given planet. Thus, for example, if a planet's orbit takes it in twice as close to the origin, then its angular velocity $d\theta/dt$ must increase by a factor of

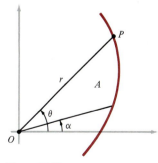

Figure 16.38

4. This fact has far-reaching implications which we shall examine more thoroughly in the last section of the next chapter.

Remark 2 We briefly reconsider formula (2) for the area A shown in Fig. 16.33. Our purpose is to remind students of the point of view developed in Section 6.4, namely, that a definite integral is defined to be a limit of approximating sums. As usual, we begin with a subdivision of the interval of integration $[\alpha, \beta]$:

$$\alpha = \theta_0 < \theta_1 < \cdots < \theta_n = \beta.$$

For each $k = 1, 2, \ldots, n$, let m_k and M_k be the minimum and maximum values of $f(\theta)$ on the kth subinterval $[\theta_{k-1}, \theta_k]$ of length $\Delta\theta_k = \theta_k - \theta_{k-1}$. Also, let ΔA_k be the area within the curve $r = f(\theta)$ corresponding to this subinterval. In Fig. 16.39 we show the area ΔA_k squeezed between the areas of the inscribed sector with radius $r = m_k$ and the circumscribed sector with radius $r = M_k$. We therefore have

$$\tfrac{1}{2}m_k^2\,\Delta\theta_k \le \Delta A_k \le \tfrac{1}{2}M_k^2\,\Delta\theta_k.$$

By adding these inequalities from $k = 1$ to $k = n$, we obtain

$$\sum_{k=1}^{n} \frac{1}{2}\,m_k^2\,\Delta\theta_k \le A \le \sum_{k=1}^{n} \frac{1}{2}\,M_k^2\,\Delta\theta_k,$$

because A is the sum of the ΔA_k's. We now vary the subdivision in the manner described in Section 6.4, so that $\max\Delta\theta_k \to 0$. Then each of these sums approaches the definite integral

$$\int_{\alpha}^{\beta} \frac{1}{2}f(\theta)^2\,d\theta$$

for any continuous function $r = f(\theta)$, and since the area A is squeezed between the sums, we legitimately conclude that

$$A = \int_{\alpha}^{\beta} \frac{1}{2}f(\theta)^2\,d\theta = \int_{\alpha}^{\beta} \frac{1}{2}r^2\,d\theta,$$

which is (2).

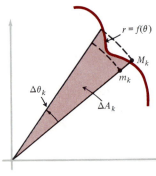

Figure 16.39

PROBLEMS

1 Find the area enclosed by the cardioid $r = a(1 + \cos\theta)$.

2 Find the area between the parabola $r = 8/(1 - \cos\theta)$ and the y-axis.

3 Show that the area inside the first turn of the spiral $r = a\theta$, that is, for $0 \le \theta \le 2\pi$, equals one-third the area of the circle that passes through the endpoint of this turn and has center at the origin.*

4 Find the total area inside the rose $r = \sin 2\theta$.

5 Find the total area inside the rose $r = \sin 3\theta$.

6 Find the area inside the smaller loop of the limaçon $r = 1 + 2\cos\theta$.

7 Find the area between the two loops of the limaçon $r = 1 + 2\cos\theta$.

8 Find the area between the circle $r = 2a\cos\theta$ and the line $y = x$
(a) by integration;
(b) by elementary geometry.

9 Find the area that lies inside both curves $r = a\cos\theta$ and $r = a(1 - \cos\theta)$.

---

* This statement and the result of Problem 7 in Section 16.4 are the main theorems proved by Archimedes in his treatise *On Spirals* (Propositions 20 and 24).

10 Find the area that lies inside both curves $r = a$ and $r^2 = 2a^2 \cos 2\theta$.

11 Use integration in polar coordinates to show that the area of the rectangle bounded by $x = 0$, $y = 0$, $x = a$, $y = b$ is ab.

12 Find the area common to the two circles $r = 2a \cos \theta$ and $r = 2b \sin \theta$.

13 Show that the area between the cissoid $r =$ $2a \, (\sec \theta - \cos \theta)$ and its asymptote $x = 2a$ (see Problem 12 in Section 16.2) is 3 times the area of the generating circle.

14 In equation (3), show that the value of the constant is $2A_e/T$, where A_e is the area of the elliptical orbit and T is the time required for the planet to go once around its orbit.

ADDITIONAL PROBLEMS FOR CHAPTER 16

SECTION 16.2

1 Transform the given rectangular equation into an equivalent polar equation:
(a) $y = 4x$;
(b) $4x^2 + 9y^2 = 36$;
(c) $x^2 + y^2 - 2x + 4y = 0$;
(d) $2x - 5y = 3$;
(e) $y^2 = 4x$;
(f) $x^2 + y^2 - 4y = \sqrt{x^2 + y^2}$;
(g) $x^3 + y^3 = 12xy$.

2 Transform the given polar equation into an equivalent rectangular equation:
(a) $r = -3$; (b) $\theta = 3\pi/4$;
(c) $r \sin \theta = -5$; (d) $r = 2 \sec \theta$;
(e) $r^2 = \sin 2\theta$; (f) $r = \cos 2\theta$;
(g) $r = \sin 3\theta$; (h) $r = \cos 3\theta$;
(i) $r^2 = \sin^2 \theta \tan \theta$.

3 Find all points of intersection of each pair of curves:
(a) $r \sin \theta = a$, $r \cos \theta = a$;
(b) $r = a(1 + \cos \theta)$, $r = a(1 - \sin \theta)$;
(c) $r = a \cos 2\theta$, $4r \cos \theta = \sqrt{3}a$;
(d) $r \sin \theta = 3$, $r = 6 \sin \theta$;
(e) $r = 1 + \cos \theta$, $r^2 = \frac{1}{2} \cos \theta$;
(f) $r = a$, $r^2 = 2a^2 \cos 2\theta$;
(g) $r = a(1 + \sin \theta)$, $r = 2a \cos \theta$;
(h) $r \cos \theta = 2$, $r \sin \theta + 2\sqrt{3} = 0$;
(i) $r = 2 \sin^2 \theta$, $r = -2$;
(j) $r \cos \theta = 1$, $r = 2 \cos \theta + 1$;
(k) $r = 1 + \cos \theta$, $r = 3 \cos \theta$;
(l) $r = a(1 + \cos \theta)$, $r = a(1 + \sin \theta)$;
(m) $r = a \sin 2\theta$, $r = a(1 - \cos 2\theta)$.

4 Diocles invented his cissoid to solve the classical Greek problem of doubling a cube, that is, constructing a second cube whose volume is twice that of a given cube. If OA in Problem 12 of Section 16.2 is the edge of the given cube, then the edge of the second cube must have length $\sqrt[3]{2} \, OA$. Verify that the following construction produces such a length. Let D be a point on the positive y-axis such that $OD = 2OA$, let AD intersect the cissoid at E, and extend OE to inter-

sect the asymptote $x = 2a$ at F. Then $(AF)^3 = 2(OA)^3$, so AF is the required length.* Hint: If $E = (x, y)$, show that $y^3 = 2x^3$.

5 Verify that the following construction, using the conchoid discussed in Problem 15 of Section 16.2, trisects the given angle BOP (Fig. 16.40). Draw any line $x = a$, and let A be the point where it intersects OP. Form the conchoid determined by the numbers a and b, where $b = 2OA$. If C is the point where the horizontal line through A intersects this conchoid, let D be the

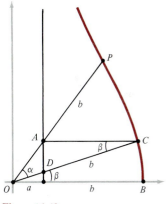

Figure 16.40

---

* Hippocrates of Chios (see Section 6.2) reduced the problem of doubling a cube of edge a to the problem of constructing two mean proportionals x and y between a and $2a$:

$$\frac{a}{x} = \frac{x}{y} = \frac{y}{2a}.$$

From these equations we have $x^2 = ay$ and $xy = 2a^2$, and eliminating y we find that $x^3 = 2a^3$. Thus x is the edge of a cube that has twice the volume of the first cube. This value is the x-coordinate of the point of intersection of the parabola $x^2 = ay$ and the hyperbola $xy = 2a^2$, so the problem of doubling a cube is solved if we allow ourselves to use these curves as tools. Historians of mathematics believe that the conic sections may have arisen in just this way.

point where OC intersects the line $x = a$, so that $DC = b = 2OA$. Let α be the angle POC and β the angle BOC, so that the given angle BOP is $\alpha + \beta$. Then by the law of sines we have

$$\frac{OA}{\sin \beta} = \frac{AC}{\sin \alpha},$$

so

$$\frac{OA}{\sin \beta} = \frac{2 \, OA \cos \beta}{\sin \alpha}$$

and $\sin \alpha = 2 \sin \beta \cos \beta = \sin 2\beta$. Therefore $\alpha = 2\beta$ and $\angle BOP = 3\beta$, so $\beta = \frac{1}{3} \angle BOP$ and the line OC trisects the given angle.

6   A general conchoid can be defined as follows. Let O be a fixed point and C a given curve. On the line OA from O to a point A on C, continue to a point P, where AP is a positive constant b. Then the locus of P is called the *conchoid of C with pole O and constant b.* If C is a straight line and O is any point not on C, we get the conchoid of Nicomedes (Problem 15 in Section 16.2) as a special case. Show that a conchoid of a circle $r = a \cos \theta$ with pole O at the origin and constant b is the limaçon $r = a \cos \theta + b$, so that conchoids include limaçons—and therefore cardioids—as special cases.*

SECTION 16.3

7   Verify the validity of the following procedure for squaring a circle of a radius a by using the quadratrix defined in Problem 21 of Section 16.3:
(a) By part (d) of the problem just mentioned,

$$\frac{OG}{OC} = \frac{OC}{ADC},$$

so a segment whose length is $\frac{1}{4}$ the circumference of our circle can be constructed as the third proportional to the segments OG and OC. (In an equation of the form $c/d = d/x$, x is called the *third proportional* to the given segments of lengths c and d, and x can be constructed by ruler and compass as indicated in Fig. 16.41.)

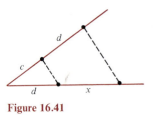

Figure 16.41

(b) By part (a), a segment can be constructed whose length b is $\frac{1}{2}$ the circumference of our circle, with ab the area of the circle. The side s of a square whose area is ab can now be constructed as the mean proportional to a and b. (In an equation of the form $a/s = s/b$, s is called the *mean proportional* to a and b, and can be constructed by ruler and compass as indicated in Fig. 16.42.)

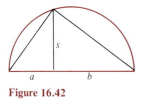

Figure 16.42

8   Verify the validity of the following procedure for squaring a circle of radius a by using the spiral of Archimedes:
(a) draw the circle with center O and radius a, and

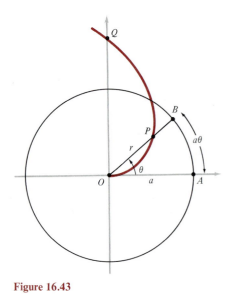

Figure 16.43

---

superimpose the spiral $r = a\theta$ with constant of proportionality equal to the radius of this circle (Fig. 16.43);

(b) since OP and the arc AB both have length $a\theta$ and are therefore equal, it follows that the radius OQ for $\theta = \pi/2$ is a segment whose length is $\frac{1}{4}$ the circumference of our circle;

(c) the squaring of the circle is now completed exactly as in part (b) of Problem 7.

SECTION 16.4

9 For the hyperbolic spiral $r\theta = a$ $(\theta > 0)$, show that $\psi = 135°$ when $\theta = 1$ radian, and also that $\psi \to 90°$ as the spiral winds on around the origin in the counterclockwise direction. Sketch the curve and show the angle ψ for the direction $\theta = 1$ radian.

10 If a point moves along a polar curve $r = f(\theta)$ at a constant speed, and is also moving away from the origin at a constant speed, show that the curve must be the exponential spiral $r = ae^{b\theta}$.

*11 Show that the arc length of one leaf of each of the following roses equals the total arc length of the corresponding ellipse (but do not try to evaluate the integrals involved, because it cannot be done):
(a) $r = 2 \sin 2\theta$, $x^2 + 4y^2 = 1$;
(b) $r = 6 \cos 3\theta$, $x^2 + 9y^2 = 9$.

*12 The distances r_1 and r_2 from the foci to any point on an ellipse satisfy the equation

$$r_1 + r_2 = \text{a constant.}$$

By differentiating both sides of this equation with respect to arc length s and interpreting the result in terms of differentials, show that the tangent at any point on the ellipse makes equal angles with the lines to the foci.

13 Establish the reflection property of parabolas by an adaptation of the argument used in Problem 12.

14 Consider the part of the lemniscate $r^2 = 2a^2 \cos 2\theta$ that lies in the first quadrant, and show that at any point on this curve the angle between the radial direction and the outward normal is 2θ.

15 Find the angle at which the circles $r = a$ and $r = 2a \cos \theta$ intersect.

16 Show that the length of an arc of the exponential spiral $r = ae^{b\theta}$ is proportional to the difference of the radii at its ends.

17 Find the length of the spiral $r = a\theta^2$ from $\theta = 0$ to $\theta = 2\pi$. Sketch the curve.

18 Find the length of the curve $r = a \sin^3 \frac{1}{3}\theta$ from $\theta = 0$ to $\theta = 3\pi/2$. What is the total length of this curve?

SECTION 16.5

*19 Suppose that a polar curve $r = r(\theta)$, $\alpha \le \theta \le \beta$, also has a representation as a rectangular curve $y = y(x)$, $a \le x \le b$, as shown in Fig. 16.44. (It is convenient

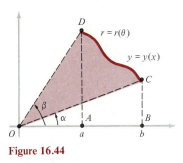

Figure 16.44

here to denote each function and its dependent variable by the same letter.) From the point of view of rectangular coordinates, the area of the region OCD is

$$A_{OCD} = A_{OAD} + A_{ABCD} - A_{OBC}$$

$$= \tfrac{1}{2}r^2(\beta) \sin \beta \cos \beta + \int_a^b y\, dx$$
$$- \tfrac{1}{2}r^2(\alpha) \sin \alpha \cos \alpha$$

$$= \tfrac{1}{4}\left[r^2(\theta) \sin 2\theta \right]_\alpha^\beta$$

$$+ \int_\beta^\alpha r(\theta) \sin \theta[r'(\theta) \cos \theta - r(\theta) \sin \theta]\, d\theta.$$

By integrating by parts at the right moment, show that this formula reduces to

$$A_{OCD} = \int_\alpha^\beta \frac{1}{2}r^2\, d\theta.$$

20 Find the area inside the circle $r = a$ and outside the cardioid $r = a(1 - \cos \theta)$.

21 Find the area outside the circle $r = 4 \cos \theta$ and inside the limaçon $r = 1 + 2 \cos \theta$.

22 Find the area outside the circle $r = a$ and inside the circle $r = 2a \cos \theta$.

23 Show that the area inside the first turn of the exponential spiral $r = ae^{b\theta}$ is $(r_2{}^2 - r_1{}^2)/4b$, where r_1 is the initial radius and r_2 is the terminal radius.

24 Find the area inside one loop of the curve $r^2 = a^2 \sin \theta$.

25 Find the total area outside the circle $r = a$ and inside the lemniscate $r^2 = 2a^2 \cos 2\theta$.

17

PARAMETRIC EQUATIONS. VECTORS IN THE PLANE

17.1

PARAMETRIC EQUATIONS OF CURVES

When we think of a curve as the path of a moving point, it is often more convenient to study the curve by using two simple equations for x and y in terms of a third independent variable t,

$$x = f(t) \quad \text{and} \quad y = g(t), \tag{1}$$

than by using a single more complicated equation of the form

$$F(x, y) = 0. \tag{2}$$

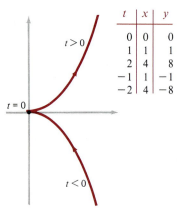

Figure 17.1

In physical problems we often consider a moving point, and t is understood to be the time measured from the moment at which the motion begins. The point P whose coordinates are x and y then traces out the curve as t traverses some definite interval, say $t_1 \leq t \leq t_2$. This provides not only a description of the path on which the point moves, but also information about the direction of its motion and its location on the path for various values of t, as suggested in Fig. 17.1. The third variable in terms of which x and y are expressed is called a *parameter* (from the Greek *para,* meaning together, and *meter,* meaning measure), and equations (1) are called parametric equations of the curve. If we want the rectangular equation of the curve in the form (2), we must eliminate the parameter from equations (1).

Example 1 Sketch the curve $x = t^2$, $y = t^3$ and find its rectangular equation.

Solution We can plot a few points by calculating x and y for several values of t, as indicated by the table in Fig. 17.2. A few calculations are worthwhile to give us something concrete to start with. However, it is more profitable to study how x and y vary as t varies, instead of merely plotting points. Here we see that as t increases from 0 to ∞, x and y both start at 0 and increase into positive values, but y increases faster than x. This means that for large t's the point $P = (x, y)$ moves away from the x-axis faster than from the y-axis, as shown. For negative t's x is still positive, but y is negative, so this part of the curve is a reflection about the x-axis of the upper part which we have just described. The general shape of the curve can be discovered from the behavior of the slope of the tangent dy/dx, which can easily be calculated as a function of t by dividing $dy = 3t^2 \, dt$ by $dx = 2t \, dt$:

t	x	y
0	0	0
1	1	1
2	4	8
-1	1	-1
-2	4	-8

Figure 17.2

506

$$\frac{dy}{dx} = \frac{3t^2\,dt}{2t\,dt} = \frac{3}{2}\,t \begin{cases} \to 0 & \text{as } t \to 0, \\ \to \infty & \text{as } t \to \infty, \\ \to -\infty & \text{as } t \to -\infty. \end{cases}$$

Finally, we notice from the parametric equations that the square of y equals the cube of x, so

$$y^2 = x^3 \qquad \text{or} \qquad y = x^{3/2}$$

is the rectangular equation of the curve.

As we remarked earlier, the use of parametric equations is very natural if we think of a curve as the path of a moving point whose position depends on the time t measured from some convenient initial moment.

Example 2 Let a projectile be fired from the origin at time $t = 0$ with an initial velocity of magnitude v_0 ft/s (or m/s) and direction given by the angle of elevation α (Fig. 17.3), and assume that the only force acting on the projectile is the force of gravity. Discuss the subsequent motion.

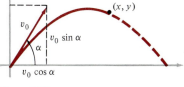

Solution We consider the x- and y-components of the acceleration separately. Since the force of gravity acts downward, we have

$$a_x = \frac{dv_x}{dt} = 0 \qquad \text{and} \qquad a_y = \frac{dv_y}{dt} = -g,$$

Figure 17.3

where g $= 32$ ft/s^2 (or 9.80 m/s^2) is the acceleration due to gravity. Therefore

$$v_x = c_1 \qquad \text{and} \qquad v_y = -gt + c_2$$

for certain constants c_1 and c_2. But when $t = 0$, we have $v_x = v_0 \cos \alpha$ and $v_y = v_0 \sin \alpha$, and consequently

$$v_x = \frac{dx}{dt} = v_0 \cos \alpha \qquad \text{and} \qquad v_y = \frac{dy}{dt} = -gt + v_0 \sin \alpha.$$

Another integration yields

$$x = (v_0 \cos \alpha)t + c_3 \qquad \text{and} \qquad y = -\tfrac{1}{2}gt^2 + (v_0 \sin \alpha)t + c_4.$$

But $x = y = 0$ when $t = 0$, so $c_3 = c_4 = 0$ and

$$x = (v_0 \cos \alpha)t, \qquad y = -\tfrac{1}{2}gt^2 + (v_0 \sin \alpha)t. \qquad (3)$$

These are parametric equations for the path of the projectile. We can use equations (3) to show that the projectile follows a parabolic path. To do this, we eliminate the parameter by solving the first equation for t and substituting in the second:

$$t = \frac{x}{v_0 \cos \alpha},$$

$$y = -\frac{1}{2}\,g \cdot \frac{x^2}{v_0^2 \cos^2 \alpha} + (v_0 \sin \alpha) \cdot \frac{x}{v_0 \cos \alpha}$$

$$= -\frac{g}{2v_0^2 \cos^2 \alpha}\,x^2 + (\tan \alpha)\,x.$$

The fact that y is a quadratic function of x shows that the point (x, y) moves on a parabola. Further properties of this motion are developed in Problem 13.

In any motion problem like that discussed in Example 2, it is natural to use the time t as the parameter. However, in problems that are more concerned with geometry than with physics, the most convenient parameter is likely to have some geometric significance. In Example 1, for instance, it is perfectly possible to think of the parameter t as a pure variable, without any connotation at all. On the other hand, in this example we have $y/x = t^3/t^2 = t$, so we can also think of t as the slope of the radial line from the origin to a variable point on the curve, and this certainly lends additional vividness to our conception of the way the curve is traced out as the parameter varies. Needless to say, there is nothing sacred about the letter t, and we are always free to use any letter we wish as a parameter.

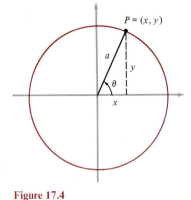

Figure 17.4

Example 3   Consider the circle shown in Fig. 17.4, with radius a and center at the origin. It is easy to see that

$$x = a \cos \theta, \qquad y = a \sin \theta \qquad (4)$$

are parametric equations for this circle, where θ is the indicated central angle. As θ varies from 0 to 2π, the point $P = (x, y)$ starts at $(a, 0)$ and moves once around the circle in the counterclockwise direction. If we didn't already know the rectangular equation of this circle, we could obtain it from the identity $\cos^2 \theta + \sin^2\theta = 1$, which yields

$$\frac{x^2}{a^2} + \frac{y^2}{a^2} = 1 \qquad \text{or} \qquad x^2 + y^2 = a^2. \qquad (5)$$

But students should notice that this equation is a static thing, and in passing from (4) to (5) we lose our sense of the circle as a curve traversed by a moving point.

Example 4   The ellipse

$$\frac{x^2}{a^2} + \frac{y^2}{b^2} = 1$$

shown in Fig. 17.5 can be parametrized as follows. Since

$$\left(\frac{x}{a}\right)^2 + \left(\frac{y}{b}\right)^2 = 1,$$

there exists an angle θ such that $\cos \theta = x/a$ and $\sin \theta = y/b$, so

$$x = a \cos \theta, \qquad y = b \sin \theta.$$

As θ varies from 0 to 2π, the point $P = (x, y)$ starts at $(a, 0)$ and moves once around the ellipse in the counterclockwise direction. Observe from the figure that θ is not the central angle of the point $P = (x, y)$; instead, it is the central angle of the points A and B on the two circles, one circumscribed about the ellipse and the other inscribed in the ellipse, and P is the intersection of the vertical line through A and the horizontal line through B.

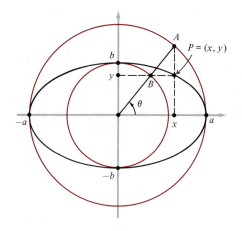

Figure 17.5

Example 5 The parabola $x^2 = 4py$ (Fig. 17.6) can be parametrized in many ways. One method is to use the slope of the tangent at (x, y) as parameter,

$$t = \frac{dy}{dx}.$$

Since

$$2x = 4p\,\frac{dy}{dx} \qquad \text{or} \qquad \frac{dy}{dx} = \frac{x}{2p},$$

the parametric equations in this case are

$$x = 2pt, \qquad y = pt^2.$$

Another method is to use as parameter the number

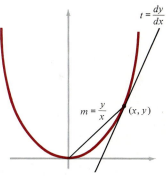

Figure 17.6

$$m = \frac{y}{x},$$

which is the slope of the radial line to (x, y). Here we have

$$y = mx \qquad \text{and} \qquad x^2 = 4py = 4pmx,$$

so

$$x = 4pm, \qquad y = 4pm^2$$

are the parametric equations. In each case the entire parabola is traced out as the parameter increases from $-\infty$ to ∞.

Our next example illustrates the fact that a parametric curve is often only a part of the corresponding rectangular curve.

Example 6 Sketch the curve $x = \cos^2 (\pi/2)t$, $y = \sin^2 (\pi/2)t$ and find its rectangular equation.

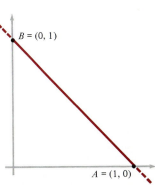

Figure 17.7

Solution Since $\cos^2 (\pi/2)t + \sin^2 (\pi/2)t = 1$, the point $P = (x, y)$ moves on the straight line $x + y = 1$ (Fig. 17.7). But neither x nor y can be negative, so the point is confined to that portion of this line that lies in the first quadrant. It is easy to see that the point is at $A = (1, 0)$ when $t = 0$; that it moves to

$B = (0, 1)$ as t increases from 0 to 1; that it moves back to A as t increases from 1 to 2; and so on.

Example 7 In Example 6 we discussed a parametric curve consisting of only part of a straight line. We now show that the parametric equations

$$x = t - 1, \qquad y = 2t + 3 \tag{6}$$

represent *all* of a straight line. By eliminating t from these equations, that is, by multiplying the first by 2 and subtracting the second, we obtain

$$2x - y = -5. \tag{7}$$

Thus all points (x, y) satisfying (6) also satisfy (7), which is the equation of a straight line. Conversely, given a point (x, y) satisfying (7), let $t = 1 + x$ [we obtain this by solving the first equation in (6) for t]. Then

$$x = t - 1,$$

and from (7) we have

$$y = 2x + 5 = 2(t - 1) + 5 = 2t + 3,$$

so the point (x, y) lies on the parametric curve (6), as we wished to show.

Our previous ways of representing curves, by rectangular coordinates and by polar coordinates, are easy to fit into our present system of parametric representation. Thus, if we have a curve $y = f(x)$, then we can write

$$y = f(x) \qquad \text{and} \qquad x = x,$$

so that x itself is used as the parameter. Also, a curve that is given in polar coordinates by the polar equation $r = F(\theta)$ can be viewed as a parametric curve with parameter θ. To see this, we use the transformation equations $x = r \cos \theta$ and $y = r \sin \theta$ to write

$$x = F(\theta) \cos \theta, \qquad y = F(\theta) \sin \theta.$$

For example, the spiral of Archimedes $r = a\theta$ becomes

$$x = a\theta \cos \theta, \qquad y = a\theta \sin \theta;$$

and the cardioid $r = a(1 + \cos \theta)$ can be expressed as

$$x = a(\cos \theta + \cos^2 \theta), \qquad y = a(\sin \theta + \sin \theta \cos \theta).$$

PROBLEMS

1 In each case, sketch the curve represented by the given parametric equations, describe the way the point (x, y) moves as t varies from large negative to large positive values, and find the rectangular equation:
(a) $x = 1 + t$, $y = 1 - t$;
(b) $x = -1 + 2t$, $y = 2 + 4t$.

2 If x and y are linear functions of t,

$$x = x_0 + at, \qquad y = y_0 + bt,$$

show that the graph is always a complete straight line unless both $a = 0$ and $b = 0$. Can every straight line be represented in this way?

3 Sketch the graph of $x = 1 - t^2$, $y = 2 + t^2$. Describe the way the point (x, y) moves as t varies from large negative to large positive values, and find the rectangular equation of the curve.

For each of the following pairs of parametric equations (in

Problems 4 to 9), sketch the curve and find its rectangular equation.

4 $x = 3\cos t$, $y = 2\sin t$.
5 $x = 1 + t^2$, $y = 3 - t$.
6 $x = \sin t$, $y = -3 + 2\cos t$.
7 $x = \sec t$, $y = \tan t$.
8 $x = t^3$, $y = 1 - t^2$.
9 $x = \sin t$, $y = \cos 2t$.
10 Sketch the graph represented by

$$x = t + \frac{1}{t}, \qquad y = t - \frac{1}{t}$$

and find its rectangular equation. Hint: Square and subtract.

11 Are the parametric curves

$$x = t + \frac{1}{t}, \qquad y = t - \frac{1}{t}$$

and

$$x = e^t + e^{-t}, \qquad y = e^t - e^{-t}$$

identical? Explain.

12 We know by Example 3 that the unit circle $x^2 + y^2 = 1$ (Fig. 17.8) can be parametrized by the equations

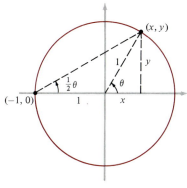

Figure 17.8

$$x = \cos\theta, \qquad y = \sin\theta, \qquad 0 \le \theta \le 2\pi.$$

A very different parametrization can be obtained by using as the parameter the tangent of the angle $\frac{1}{2}\theta$ shown in the figure,

$$t = \frac{y}{1 + x}.$$

(a) By first eliminating y from this equation and $x^2 + y^2 = 1$, show that the equations

$$x = \frac{1 - t^2}{1 + t^2}, \qquad y = \frac{2t}{1 + t^2}, \qquad -\infty < t < \infty,$$

parametrize the entire circle except for the point $(-1, 0)$.

(b) A point (x, y) in the plane such that both x and y are rational numbers is called a *rational point*. Show that for rational values of t the equations in (a) give all rational points on the unit circle, except the point $(-1, 0)$.*

13 Consider the motion of the projectile described in Example 2.

(a) Show that the maximum height of the projectile is

$$y_{\max} = \frac{v_0^2 \sin^2\alpha}{2g}.$$

(b) Show that the range R of the projectile, i.e., the distance from the origin to the point where the projectile reaches the x-axis on its descent, is given by the formula

$$R = \frac{v_0^2}{g}\sin 2\alpha.$$

(c) What angle of elevation α produces the maximum range?

(d) Show that doubling the magnitude of the initial velocity multiplies both the maximum height and the range by a factor of 4.

14 The *witch* is a bell-shaped curve that can be defined as follows. Consider the circle of radius a which is tangent to the x-axis at the origin (Fig. 17.9). The variable line OA through the origin intersects the line $y = 2a$ at the point A and the circle at the point B. The point P is the intersection of the vertical line through A and the horizontal line through B, and the witch is the locus of P as OA varies. Find parametric equations for this curve by using as parameter the angle θ from the posi-

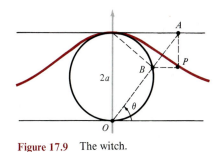

Figure 17.9 The witch.

---

* In general it is very difficult to find rational points on curves, and it is quite remarkable that we are able to do so for the case of the unit circle. For instance, it has been asserted that if n is any integer > 2, the only rational points on the curve $x^n + y^n = 1$ are those for which $x = 0$ or $y = 0$. This statement is called *Fermat's last theorem*. It has been proved for many particular values of n over the past 300 years, but in its full generality it remains to this day one of the most famous (and intractable) unsolved problems of mathematics.

tive *x*-axis to the line *OA.* Also find its rectangular equation.*

15 The *involute* of a circle is the curve traced out by the point at the end of a thread as the thread is held taut and unwound from a fixed spool, as shown in Fig. 17.10. If the center of the spool is placed at the origin and its radius is *a*, and if the thread begins to unwind at the point $A = (a, 0)$, find parametric equations for the involute by using the angle *AOT* shown in the figure as the parameter θ.

*16 The *folium of Descartes,* shown in Fig. 17.11, is the graph of the equation $x^3 + y^3 = 3axy.$†

(a) Introduce the parameter $t = y/x$ and find parametric equations for the curve.

(b) Use the equations found in (a) to show that the line $x + y + a = 0$ is an asymptote, by showing that $x + y \to -a$ as $t \to -1$.

(c) The folium is clearly symmetric about the line $y = x$, because interchanging *x* and *y* leaves the equation unaltered. Use this, together with the geometric meaning of *t* and the results of (a) and (b), to verify as much as possible of the general nature of the curve as suggested by the figure. In

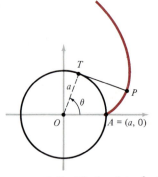

Figure 17.10   The involute of a circle.

particular, decide how various parts of the curve are traced out as *t* varies over various ranges of values.

---

* The witch is sometimes called the *witch of Agnesi* after the Italian mathematician Maria Agnesi (1718–1799), who referred to it in her book on calculus, published in 1748. This curve had previously been called the *versoria,* a Latin word meaning a guy rope attached to a sail, but she apparently confused this word with a different Latin word, *versiera,* meaning a witch, and the name has stuck.

† The word *folium* means leaf in Latin. This curve was originally used by Descartes as a challenge to Fermat to find its tangent line at an arbitrary point. Fermat succeeded immediately, much to Descartes's dismay.

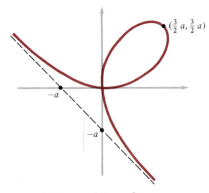

Figure 17.11   The folium of Descartes.

17.2

(OPTIONAL) THE CYCLOID AND OTHER SIMILAR CURVES

The *cycloid* is the curve traced out by a point on the circumference of a circle when the circle rolls along a straight line in its own plane, as shown in Fig. 17.12. We shall see that this curve has many remarkable geometric and physical properties.

The only convenient way of representing a cycloid is by means of parametric equations. We assume that the rolling circle has radius *a* and that it rolls along the *x*-axis, starting from a position in which the center of the circle is on the positive *y*-axis. The curve is the locus of the point *P* on the circle which is located at the origin *O* when the center *C* is on the *y*-axis. The angle θ in the figure is the angle through which the radius *CP* turns as the circle rolls to a new position. If *x* and *y* are the coordinates of *P*, then the rolling of the circle implies that $OB = $ arc $BP = a\theta$, so $x = OB - AB = OB - PQ = a\theta - a \sin \theta = a(\theta - \sin \theta)$. Also, $y = BC - QC = a - a \cos \theta = a(1 - \cos \theta)$.

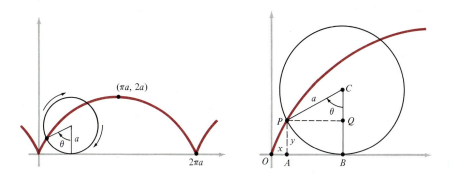

Figure 17.12 The cycloid.

The cycloid therefore has the parametric representation

$$x = a(\theta - \sin \theta), \qquad y = a(1 - \cos \theta). \tag{1}$$

It is clear from Fig. 17.12 that y is a function of x, but it is also clear from equations (1) that it is not possible to find a simple formula for this function. The cycloid is one of many curves for which the parametric equations are much simpler and easier to work with than the rectangular equation.

From equations (1) we have

$$y' = \frac{dy}{dx} = \frac{a \sin \theta \, d\theta}{a(1 - \cos \theta) \, d\theta} = \frac{\sin \theta}{1 - \cos \theta}$$

$$= \frac{2 \sin \tfrac{1}{2}\theta \cos \tfrac{1}{2}\theta}{2 \sin^2 \tfrac{1}{2}\theta} = \cot \tfrac{1}{2}\theta. \tag{2}$$

We observe that the derivative y' is not defined for $\theta = 0, \pm 2\pi, \pm 4\pi$, etc. These values of θ correspond to the points where the cycloid touches the x-axis; these points are called *cusps*. The tangent to the cycloid is vertical at the cusps.

In the following examples we establish the main geometric properties of the cycloid.

Example 1 Show that the area under one arch of the cycloid is three times the area of the rolling circle.

Solution One arch is traced out as the circle turns through one complete revolution. The usual area integral can therefore be written as follows, using the parameter θ as the variable of integration:

$$A = \int_0^{2\pi a} y \, dx = \int_0^{2\pi} y \frac{dx}{d\theta} \, d\theta = \int_0^{2\pi} a(1 - \cos \theta) a(1 - \cos \theta) \, d\theta$$

$$= a^2 \int_0^{2\pi} (1 - \cos \theta)^2 \, d\theta = a^2 \int_0^{2\pi} (1 - 2 \cos \theta + \cos^2 \theta) \, d\theta$$

$$= a^2 \int_0^{2\pi} (1 + \cos^2 \theta) \, d\theta = a^2 \int_0^{2\pi} d\theta + a^2 \int_0^{2\pi} \tfrac{1}{2}(1 + \cos 2\theta) \, d\theta = 3\pi a^2.$$

Example 2   Show that the length of one arch of the cycloid is four times the diameter of the rolling circle.

Solution   Since $dx = a(1 - \cos \theta)\, d\theta$ and $dy = a \sin \theta\, d\theta$, the element of arc length ds is given by

$$ds^2 = dx^2 + dy^2 = a^2[(1 - \cos \theta)^2 + \sin^2 \theta]\, d\theta^2$$

$$= 2a^2[1 - \cos \theta]\, d\theta^2 = 4a^2 \sin^2 \tfrac{1}{2}\theta\, d\theta^2,$$

so

$$ds = 2a \sin \tfrac{1}{2}\theta\, d\theta.$$

The length of one arch is therefore

$$L = \int ds = \int_0^{2\pi} 2a \sin \tfrac{1}{2}\theta\, d\theta = -4a \cos \tfrac{1}{2}\theta \Big]_0^{2\pi} = 8a.$$

Example 3   Show that the tangent to the cycloid at the point P in Fig. 17.12 passes through the top of the rolling circle.

Solution   The point at the top of the circle has coordinates $(a\theta, 2a)$. The slope of the tangent at P is given by (2). The equation of the tangent at P is therefore

$$y - a(1 - \cos \theta) = \frac{\sin \theta}{1 - \cos \theta}(x - a\theta + a \sin \theta).$$

We substitute $x = a\theta$ in this equation and solve for y, which gives

$$y = a(1 - \cos \theta) + \frac{\sin \theta}{1 - \cos \theta} \cdot a \sin \theta = \frac{a(1 - \cos \theta)^2 + a \sin^2 \theta}{1 - \cos \theta} = 2a.$$

This shows that the tangent at P does indeed pass through the point $(a\theta, 2a)$ at the top of the circle.

Galileo seems to have been the first to notice the cycloid and investigate its properties, in the early 1600s. He didn't actually discover any of these properties, but he gave the curve its name and recommended its study to his friends, including Mersenne in Paris. Mersenne informed Descartes and others about it, and in 1638 Descartes found a construction for the tangent which is equivalent to the property given in Example 3. In 1644 Galileo's disciple Torricelli (who invented the barometer) published his discovery of the area under one arch. The length of one arch was discovered in 1658 by the great English architect Christopher Wren.* The list of famous men who have worked on the cycloid will be continued, but first we consider some other related curves.

If a circle rolls on the *inside* of a fixed circle, the locus of a point on the rolling circle is called a *hypocycloid.* If a circle rolls on the *outside* of a fixed circle, the locus of a point on the rolling circle is called an *epicycloid.*†

---

* Wren was an astronomer and a mathematician — in fact, Savilian Professor of Astronomy at Oxford — before the Great Fire of London in 1666 gave him his opportunity to build St. Paul's Cathedral, as well as dozens of smaller churches throughout the city.

† The distinction between these words is easy to remember because the Greek prefix *hypo* means under or beneath, as in "hypodermic," and *epi* means on or above, as in "epicenter."

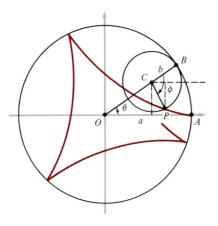

Figure 17.13   The hypocycloid.

We show how to represent a hypocycloid parametrically. Let the fixed circle have radius a and the rolling circle radius b, where $b < a$. Let the fixed circle have its center at the origin (Fig. 17.13), and let the smaller rolling circle start in a position internally tangent to the fixed circle at the point A on the positive x-axis. We consider the point P on the rolling circle that was initially at A. With θ and ϕ as shown in the figure, the rolling of the small circle implies that the arcs AB and BP are equal: $a\theta = b\phi$. We can then see that the coordinates of P are

$$x = (a - b) \cos \theta + b \cos (\phi - \theta),$$

$$y = (a - b) \sin \theta - b \sin (\phi - \theta).$$

But $\phi - \theta = [(a - b)/b]\theta$, so the parametric equations of the hypocycloid are

$$x = (a - b) \cos \theta + b \cos \frac{a - b}{b} \theta,$$

$$\tag{3}$$

$$y = (a - b) \sin \theta - b \sin \frac{a - b}{b} \theta.$$

The arc length along the fixed circle between successive cusps of the hypocycloid is $2\pi b$. If $2\pi a$ is an integral multiple of $2\pi b$, so that a/b is an integer n, then the hypocycloid has n cusps and the point P returns to A after the smaller circle rolls off its circumference n times on the fixed circle. We leave it to students to decide when P will return to A if a/b is a rational number but not an integer, for example if $a/b = \frac{3}{2}$. A discussion of the case in which a/b is irrational is beyond the scope of this book; it suffices to say that as the smaller circle rolls around and around indefinitely, the cusps of the resulting hypocycloid are evenly and densely distributed on the fixed circle.*

The parametric equations of a hypocycloid of four cusps can be written in a very simple form by using some trigonometric identities. If $a = 4b$, equations (3) become

---

* The curious reader will find additional information in Theorem 439 of G. H. Hardy and E. M. Wright, *Introduction to the Theory of Numbers* (Oxford, 1954); or in Theorem 6.3 of I. Niven, *Irrational Numbers* (Wiley, 1956).

$$x = 3b \cos \theta + b \cos 3\theta, \qquad y = 3b \sin \theta - b \sin 3\theta.$$

But

$$\cos 3\theta = \cos (2\theta + \theta) = \cos 2\theta \cos \theta - \sin 2\theta \sin \theta$$
$$= (2 \cos^2 \theta - 1) \cos \theta - 2 \sin^2 \theta \cos \theta$$
$$= [2 \cos^2 \theta - 1 - 2(1 - \cos^2 \theta)] \cos \theta$$
$$= 4 \cos^3 \theta - 3 \cos \theta,$$

and a similar calculation yields

$$\sin 3\theta = 3 \sin \theta - 4 \sin^3 \theta.$$

Our parametric equations therefore become

$$x = 4b \cos^3 \theta = a \cos^3 \theta, \qquad y = 4b \sin^3 \theta = a \sin^3 \theta. \qquad (4)$$

From these equations it is easy to obtain the corresponding rectangular equation,

$$x^{2/3} + y^{2/3} = a^{2/3}. \qquad (5)$$

Because of its appearance (Fig. 17.14), a hypocycloid of four cusps is often called an *astroid*.

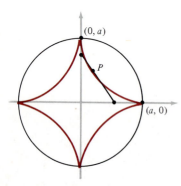

$(0, a)$

P

$(a, 0)$

Figure 17.14   The astroid.

Example 4   Consider the tangent to the astroid at a point P in the first quadrant. Show that the part of this tangent which is cut off by the coordinate axes has constant length, independent of the position of P.

Solution   By equations (4), the slope of the tangent is

$$y' = \frac{dy}{dx} = \frac{3a \sin^2 \theta \cos \theta \, d\theta}{-3a \cos^2 \theta \sin \theta \, d\theta} = -\tan \theta,$$

so the equation of the tangent is

$$y - a \sin^3 \theta = -\tan \theta \, (x - a \cos^3 \theta).$$

We find the x-intercept by putting $y = 0$ and solving for x,

$$x = a \cos^3 \theta + a \sin^2 \theta \cos \theta = a \cos \theta.$$

Similarly, the y-intercept is $y = a \sin \theta$. The length of the part of the tangent cut off by the axes is therefore

$$\sqrt{a^2 \cos^2 \theta + a^2 \sin^2 \theta} = a,$$

which is constant.

x

y

Figure 17.15

We now return to the cycloid discussed earlier, and reflect both it and the y-axis about the x-axis, as shown in Fig. 17.15. The parametric equations (1) are still valid, and the resulting curve has several interesting physical properties, which we now describe and analyze.

In 1696 John Bernoulli conceived and solved the now famous *brachistochrone problem*. He published the problem (but not the solution) as a challenge to other mathematicians of the time. The problem is this: Among all smooth curves in a vertical plane that join a given point P_0 to a given lower

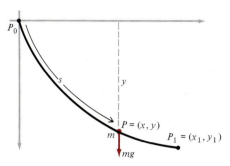

Figure 17.16

point P_1 not directly below it, find that particular curve along which a particle will slide down from P_0 to P_1 in the shortest possible time.* We can think of the particle as a bead of mass m sliding down an ideal frictionless wire, with the downward force of gravity mg as the only force acting on the bead.

If we assume that the points P_0 and P_1 lie at the origin and at (x_1, y_1) in the first quadrant, as shown in Fig. 17.16, then Bernoulli's problem can be stated in mathematical language as follows. The bead is released from rest at P_0, so its initial velocity and initial kinetic energy are zero. The work done by gravity in pulling it down from the origin to an arbitrary point $P = (x, y)$ is mgy. This must equal the increase in the kinetic energy of the bead as it slides down the wire to this point, so

$$\tfrac{1}{2} mv^2 = mgy,$$

and therefore

$$v = \frac{ds}{dt} = \sqrt{2gy}. \tag{6}$$

This can be written as

$$dt = \frac{ds}{\sqrt{2gy}} = \frac{\sqrt{dx^2 + dy^2}}{\sqrt{2gy}} = \frac{\sqrt{1 + (dy/dx)^2}\,dx}{\sqrt{2gy}}. \tag{7}$$

The total time T_1 required for the bead to slide down the wire from P_0 to P_1 will depend on the shape of the wire as specified by its equation $y = f(x)$; it is given by

$$T_1 = \int dt = \int_0^{x_1} \sqrt{\frac{1 + (y')^2}{2gy}}\,dx. \tag{8}$$

The brachistochrone problem therefore amounts to this: to find the particular curve $y = f(x)$ that passes through P_0 and P_1 and minimizes the value of the integral (8).

Since the straight line joining P_0 and P_1 is clearly the shortest path, we might guess that this line also yields the shortest time. However, a moment's consideration of the possibilities will make us more skeptical about this conjecture. There might be an advantage in having the bead slide down more steeply at first, thereby increasing its speed more quickly at the beginning of

_____
* The word *brachistochrone* comes from two Greek words meaning shortest time.

the motion; for with a faster start, it is reasonable to suppose that the bead might reach P_1 in a shorter time, even though it travels over a longer path. And this is the way it turns out: The brachistochrone curve is an arc of a cycloid through P_0 and P_1 with a cusp at the origin.

Leibniz and Newton, as well as John Bernoulli and his older brother James, solved the problem. John's solution, which is very ingenious but rather specialized in the methods it uses, is given in Appendix A.19. The cycloid was well known to all these men through the earlier work of the great Dutch scientist Huygens on pendulum clocks (see below). When John found that the cycloid is also the solution of his brachistochrone problem, he was astounded and delighted. He wrote: "With justice we admire Huygens because he first discovered that a heavy particle slides down to the bottom of a cycloid in the same time, no matter where it starts. But you [his readers] will be petrified with astonishment when I say that this very same cycloid, the tautochrone of Huygens, is also the brachistochrone we are seeking."*

Huygens was a profound student of the theory of the pendulum, and in fact was the inventor of the pendulum clock. He was very well aware of the theoretical flaw in such a clock, which is due to the fact that the period of oscillation of a pendulum is not strictly independent of the amplitude of the swing.† We can express this flaw in another way by saying that if a bead is released on a frictionless circular wire in a vertical plane, then the time the bead takes to slide down to the bottom will depend on the height of the starting point. Huygens wondered what would happen if the circular wire were replaced by one having the shape of an inverted cycloidal arch. But he did more than merely wonder, for he then went on to make the remarkable discovery referred to in the passage previously quoted, that for a wire of this shape the bead will slide down from any point to the bottom in exactly the same time, no matter where it is released (Fig. 17.17). This is the *tautochrone* ("equal time") *property* of the cycloid, and we now prove it by using the formulas given above.

If we write (8) in the equivalent form

$$T_1 = \int \sqrt{\frac{dx^2 + dy^2}{2gy}}$$

and substitute equations (1) into this, we obtain

$$T_1 = \int_0^{\theta_1} \sqrt{\frac{2a^2(1 - \cos\theta)}{2ag(1 - \cos\theta)}} \, d\theta = \theta_1 \sqrt{\frac{a}{g}}$$

as the time required for the bead to slide down a cycloidal wire from P_0 to P_1. The time needed for the bead to reach the bottom of this wire is the value of T_1 when $\theta_1 = \pi$, namely, $\pi\sqrt{a/g}$. Huygens' tautochrone property amounts to the statement that the bead will reach the bottom in exactly the same time if it starts at any intermediate point (x_0, y_0). To prove this, we replace (6) by

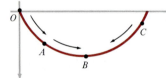

Figure 17.17   Beads released on the cycloidal wire at *O, A, C* will reach *B* in the same amount of time.

---

* For an English translation of Bernoulli's writings on this subject, see pp. 644–655 of D. E. Smith, *A Source Book in Mathematics* (Macmillan, 1929). Bernoulli's vivid, enthusiastic, personal style is in sharp contrast to the dead, gray, impersonal style of most of the writing in scientific journals nowadays.

† See the remark about the "circular error" in Example 3 in Section 9.6.

$$v = \frac{ds}{dt} = \sqrt{2g(y - y_0)}.$$

The total time required for the bead to slide down to the bottom is therefore

$$T = \int_{\theta_0}^{\pi} \sqrt{\frac{2a^2(1 - \cos\theta)}{2ag(\cos\theta_0 - \cos\theta)}} \, d\theta = \sqrt{\frac{a}{g}} \int_{\theta_0}^{\pi} \sqrt{\frac{1 - \cos\theta}{\cos\theta_0 - \cos\theta}} \, d\theta$$

$$= \sqrt{\frac{a}{g}} \int_{\theta_0}^{\pi} \frac{\sin\frac{1}{2}\theta \, d\theta}{\sqrt{\cos^2\frac{1}{2}\theta_0 - \cos^2\frac{1}{2}\theta}}, \qquad (9)$$

where the last step makes use of the trigonometric identities $2\sin^2\theta = 1 - \cos 2\theta$ and $\cos 2\theta = \cos^2\theta - \sin^2\theta = 2\cos^2\theta - 1$. If we now use the substitution

$$u = \frac{\cos\frac{1}{2}\theta}{\cos\frac{1}{2}\theta_0}, \qquad du = -\frac{1}{2}\frac{\sin\frac{1}{2}\theta \, d\theta}{\cos\frac{1}{2}\theta_0},$$

then the integral (9) becomes

$$T = -2\sqrt{\frac{a}{g}} \int_{1}^{0} \frac{du}{\sqrt{1 - u^2}} = 2\sqrt{\frac{a}{g}} \sin^{-1} u \bigg]_{0}^{1} = \pi\sqrt{\frac{a}{g}}.$$

This shows that T has the same value as before and is therefore independent of the starting point, and the argument is complete.

Once Huygens established the tautochrone property of the cycloid, a further problem presented itself: How could he arrange for a pendulum in a clock to move along a cycloidal, rather than a circular, path? Here he made a further beautiful discovery. If we suspend from the point P at the cusp between two equal inverted cycloidal semiarches a flexible pendulum whose length equals the length of one of the semiarches (Fig. 17.18), then the bob will draw up as it swings to the side in such a way that its path is another cycloid.*

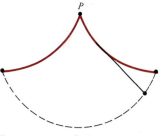

Figure 17.18 A flexible pendulum constrained by cycloidal jaws swings along another cycloid.

------

* A proof of this statement is given in Appendix A.20, but to understand this appendix, one must first understand the ideas discussed in Section 17.5.

PROBLEMS

1 Find the rectangular equation of the cycloid by eliminating θ from the parametric equations (1). Observe how hopeless it is to try to solve this for y as a simple function of x.

2 Show that for the cycloid (1) the second derivative is given by $y'' = dy'/dx = -a/y^2$. Observe that this fact implies that the cycloid is concave down between the cusps, as shown in Fig. 17.12.

3 Use the equation of the normal to the cycloid at P (in Fig. 17.12) to show that this normal passes through the point B at the bottom of the rolling circle. Also, obtain this conclusion from the result of Example 3 by using elementary geometry.

4 Assume that the circle in Fig. 17.12 rolls to the right along the x-axis at a constant speed, with the center C moving at v_0 units per second. (a) Find the rates of change of the coordinates x and y of the point P. (b) What is the greatest rate of increase of x, and where is P when this is attained? (c) What is the greatest rate of increase of y, and for what value of θ is this attained?

5 If a polygon $ABCD$ rolls (awkwardly) on a straight line $A'D'$, as shown in Fig. 17.19, then the point A will trace out in succession several arcs of circles with centers B', C', D'. The tangent to any such arc is evidently perpendicular to the line joining the point of tangency to the corresponding center. Therefore, if the rolling circle

that generates a cycloid is thought of as a polygon with an infinite number of sides, then the tangent to the cycloid at any point is the line perpendicular to the line joining the point of tangency to the bottom of the rolling circle. This is Descartes's method for finding the tangent at any point of a cycloid. Verify that it is correct.

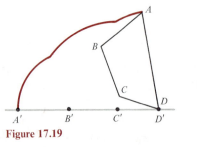

Figure 17.19

6 Find the area inside the astroid (5).
7 Find the total length of the astroid (5).
8 Find the area of the surface generated by revolving the astroid (5) about the x-axis.
9 Show that the hypocycloid of two cusps, with $a = 2b$, is simply the diameter of the fixed circle that lies along

the x-axis. In this case, if the center C of the rolling circle moves around with constant angular velocity ω, so that $d\theta/dt = \omega$, show that P moves back and forth on the x-axis with simple harmonic motion (Section 9.6) of period $2\pi/\omega$ and maximum speed $a\omega$.

10 If the astroid (4) is generated by the small circle rolling around counterclockwise with constant angular velocity ω, find the position of the point P in the first quadrant for which y is increasing most rapidly.

11 The hypocycloid of three cusps, with $a = 3b$, is called a *deltoid.* Sketch this curve, find its parametric equations, and find its total length.

12 Find parametric equations for the epicycloid generated by a circle of radius b rolling on the outside of a fixed circle of radius a. Use a figure similar to Fig. 17.13, where the fixed circle has its center at the origin and the point P is initially at $(a, 0)$.

13 Show that the equations in Problem 12 can be obtained from equations (3) in the text by replacing b by $-b$.

14 The epicycloid of two cusps, with $a = 2b$, is called a *nephroid* (meaning kidney-shaped). Sketch this curve, find its parametric equations, and calculate its total length.

17.3
VECTOR ALGEBRA. THE UNIT VECTORS i AND j

A physical quantity such as mass, temperature, or kinetic energy is completely determined by a single real number that specifies its magnitude. These are called *scalar quantities,* or simply *scalars.* In contrast to this, other entities called *vector quantities* or *vectors* possess both magnitude and direction. As examples we mention velocities, forces, and displacements.

Example 1 Let us briefly consider the case of velocity. When we discuss a point moving along a straight line, we can specify its position by means of a coordinate, which may be positive or negative, and the velocity of the moving point is the derivative of this coordinate with respect to time, that is, the rate of change of position. Direction is certainly important in such a discussion, but in this simple one-dimensional case all questions about direction are easy to handle by using positive and negative numbers.

However, to specify the velocity of a point moving along a curved path in the plane, it is essential to give both the speed of the point (the rate at which it traverses distance) and the direction of its motion. This combination of two ingredients is the *velocity vector,* or simply the *velocity,* of the moving point. It is natural to represent this vector (see Fig. 17.20) by an arrow or directed line segment **v** whose tail is placed at the current position of the point, whose length is the speed in some agreed system of measurement, and whose direction is the direction of motion.

Figure 17.20

Example 2 A force applied to an object is also a vector quantity, whose magnitude is the strength of the force and whose direction is the direction in which the force acts. For instance, the gravitational force **F** exerted by the earth on a circling artificial satellite (Fig. 17.21) is directed toward the center of the earth and its magnitude is proportional to $1/r^2$, where r is the distance from the satellite to the center of the earth.

Figure 17.21

From the mathematical point of view, we don't merely *represent* a vector by a directed line segment; we say that a vector *is* a directed line segment. This frees us to develop the algebra of vectors independently of any particular physical interpretation.

As we have indicated, vectors are often denoted in print by boldface type. A good substitute for this in the case of handwritten work is to use letters with arrows over them. Thus, **v** and $\vec{v}$ denote the same vector. Also, if a vector extends from a point P to a point Q, we can place an arrowhead at Q and denote the vector by $\overrightarrow{PQ}$ (Fig. 17.22). We then call P the *tail* or *initial point* and Q the *head* or *terminal point* of the vector. The vector $\overrightarrow{PQ}$ can be thought of as representing the *displacement* of a point along the line segment from P to Q, that is, the path taken by a point as it moves from P to Q. Such vectors describe the relative positions of points. The length or magnitude of a vector $\overrightarrow{PQ}$ is denoted by the symbol $|\overrightarrow{PQ}|$; this notation is used because the length of a vector is in many ways similar to the absolute value of a real number.

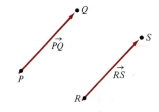
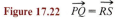

Figure 17.22 $\overrightarrow{PQ} = \overrightarrow{RS}$

Two vectors $\overrightarrow{PQ}$ and $\overrightarrow{RS}$ are said to be *equal,* and we write $\overrightarrow{PQ} = \overrightarrow{RS}$, if they have the same length and direction (Fig. 17.22). This definition of equality enables us to move a vector from one position to another without changing it, as long as its length and direction are unaltered. Thus, the vectors shown in Fig. 17.23 are all equal to each other; in other words, they are the same vector in different positions. The *position vector* of a point P in the coordinate plane is the vector $\overrightarrow{OP}$ from the origin O to the point P (Fig. 17.24). Such vectors describe the positions of points relative to the origin. As suggested in Fig. 17.23, any vector **A** can be placed with its tail at the origin, and thereby becomes the position vector of the point P that lies at its head.

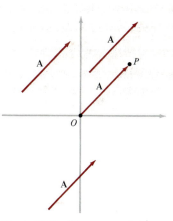

Figure 17.23 The same vector in different positions.

We shall discuss two algebraic operations on vectors. The first operation is that of adding two vectors to get another vector, and the second is that of multiplying a vector by a number to get another vector. In any discussion involving vectors, it is customary to refer to numbers as *scalars*, and this second operation is usually called *scalar multiplication*.

First, addition. Suppose a vector $\mathbf{A} = \overrightarrow{PQ}$ represents the displacement of a point along the line segment from P to Q. As shown in Fig. 17.25, when the displacement $\mathbf{A} = \overrightarrow{PQ}$ is followed by a displacement $\mathbf{B} = \overrightarrow{QR}$, the final result is equivalent to the single displacement $\overrightarrow{PR}$. It is therefore natural to think of $\overrightarrow{PR}$ as the sum of $\overrightarrow{PQ}$ and $\overrightarrow{QR}$, and to write

$$\overrightarrow{PR} = \overrightarrow{PQ} + \overrightarrow{QR}.$$

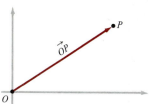

Figure 17.24 The position vector of **P**.

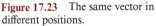

Figure 17.25 Addition.

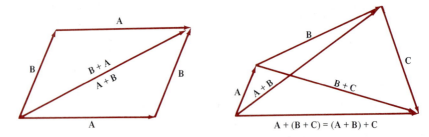

Figure 17.26   The commutative and associative laws.

This suggests the definition we adopt for vector addition: If **A** and **B** are any two vectors, we add them as shown in the figure, by placing the tail of **B** at the head of **A**; the vector from the tail of **A** to the head of **B** is then written **A** + **B** and called the *sum* of **A** and **B**. Figure 17.26 shows that addition is commutative and associative,

$$\mathbf{A} + \mathbf{B} = \mathbf{B} + \mathbf{A} \qquad \text{and} \qquad \mathbf{A} + (\mathbf{B} + \mathbf{C}) = (\mathbf{A} + \mathbf{B}) + \mathbf{C}.$$

The associative law enables us to omit parentheses, writing **A** + **B** + **C** for **A** + (**B** + **C**).

These ideas suggest another equivalent way to find the sum of two vectors **A** and **B**. If we place their tails together, as in Fig. 17.27, and form the parallelogram with **A** and **B** as adjacent sides, then **A** + **B** is the vector from the common tail to the opposite vertex. This shows that our definition of addition is well suited to working with forces in physics; for if **A** and **B** are interpreted as two forces acting at their common tail, then it is known from experiment that **A** + **B** is the *resultant force,* that is, the single force that produces the same effect as the two combined forces. This is called the *parallelogram rule,* for the addition of forces and also for vector addition.

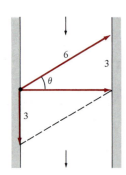

Figure 17.27

Example 3   Velocities are also combined by the parallelogram rule. For instance, a man in a canoe wishes to paddle across a river to the point on the other bank directly opposite to his starting point (Fig. 17.28). The river flows at 3 mi/h, and he can paddle at 6 mi/h. In what direction should he aim his canoe?

Solution   His actual velocity is the vector sum of the velocity of the water and his velocity relative to the water. For this sum to be perpendicular to the bank, he must aim his canoe upstream at an angle θ for which $\sin \theta = \frac{3}{6} = \frac{1}{2}$, so $\theta = 30°$.

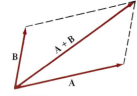

Figure 17.28

Now for scalar multiplication. If we add a vector **A** to itself, we obtain a vector in the same direction but twice as long, and it is natural to write this as **A** + **A** = 2**A**. By a natural extension, if c is any real number, then $c\mathbf{A}$ is defined to be the vector which is $|c|$ times as long as **A**, in the same direction as **A** if c is positive and in the opposite direction if c is negative (Fig. 17.29). A vector of zero length is denoted by **0** and called the *zero vector*; this vector has no direction. Evidently $1 \cdot \mathbf{A} = \mathbf{A}$ and $0 \cdot \mathbf{A} = \mathbf{0}$. The properties

$$c(d\mathbf{A}) = (cd)\mathbf{A},$$

$$(c + d)\mathbf{A} = c\mathbf{A} + d\mathbf{A},$$

$$c(\mathbf{A} + \mathbf{B}) = c\mathbf{A} + c\mathbf{B}$$

Figure 17.29   Scalar multiplication.

are valid and easy to establish, but we shall not pause to discuss them in detail. It is worth noting, however, that a proof of the last property for the case $c > 0$ is implicit in Fig. 17.30. Also, we agree that the factor c can be written on either side of the vector, $c\mathbf{A} = \mathbf{A}c$; we will not employ this clumsy usage very often, but it is occasionally convenient.

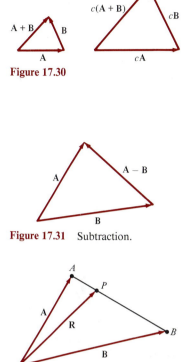

Figure 17.30

The vector $(-1) \cdot \mathbf{B}$ is written $-\mathbf{B}$; it is evidently a vector equal in length to **B** but having the opposite direction. Just as in elementary algebra, $\mathbf{A} + (-\mathbf{B})$ is written $\mathbf{A} - \mathbf{B}$. There is a simple geometric construction for $\mathbf{A} - \mathbf{B}$, resulting from the fact that $\mathbf{A} - \mathbf{B}$ is what must be added to **B** to give **A**: When **A** and **B** are placed so that their tails coincide, then $\mathbf{A} - \mathbf{B}$ is the vector from the head of **B** to the head of **A** (Fig. 17.31).

Since the laws governing addition and scalar multiplication of vectors are identical with those that we know from elementary algebra, we are justified in using the familiar rules of algebra to solve linear equations involving vectors. The following examples illustrate the efficiency of these procedures for solving certain types of geometric problems.

Figure 17.31   Subtraction.

Example 4   In Fig. 17.32, the ratio of the segment AP to the segment AB is t, where $0 < t < 1$. Express the vector **R** in terms of **A**, **B**, and t.

Solution   The vector $\overrightarrow{AB}$ is $\mathbf{B} - \mathbf{A}$, and the vector $\overrightarrow{AP}$ is $t(\mathbf{B} - \mathbf{A})$. Since $\mathbf{R} = \mathbf{A} + \overrightarrow{AP}$, we have

$$\mathbf{R} = \mathbf{A} + t(\mathbf{B} - \mathbf{A}) = (1 - t)\mathbf{A} + t\mathbf{B}.$$

In particular, if P is the midpoint of AB, so that $t = \frac{1}{2}$, then

$$\mathbf{R} = \tfrac{1}{2}\mathbf{A} + \tfrac{1}{2}\mathbf{B} = \tfrac{1}{2}(\mathbf{A} + \mathbf{B}).$$

Figure 17.32

Example 5   Use vector methods to show that the three medians of any triangle intersect at a point which is two-thirds of the way from each vertex to the midpoint of the opposite side.*

Solution   Let A, B, C be the vertices of a triangle (Fig. 17.33), and let **A**, **B**, **C** be the vectors from an outside point O to these vertices. If M is the midpoint of BC, then $\overrightarrow{OM} = \frac{1}{2}(\mathbf{B} + \mathbf{C})$, $\overrightarrow{AM} = \overrightarrow{OM} - \mathbf{A} = \frac{1}{2}(\mathbf{B} + \mathbf{C}) - \mathbf{A}$, and if P is the point two-thirds of the way from A to M, then we have

$$\overrightarrow{OP} = \mathbf{A} + \tfrac{2}{3}\overrightarrow{AM} = \mathbf{A} + \tfrac{2}{3}[\tfrac{1}{2}(\mathbf{B} + \mathbf{C}) - \mathbf{A}]$$

$$= \tfrac{1}{3}\mathbf{A} + \tfrac{1}{3}(\mathbf{B} + \mathbf{C}) = \tfrac{1}{3}(\mathbf{A} + \mathbf{B} + \mathbf{C}). \tag{1}$$

Similarly, if N is the midpoint of AC and Q is two-thirds of the way from B to N, then we see that

$$\overrightarrow{OQ} = \mathbf{B} + \tfrac{2}{3}\overrightarrow{BN} = \mathbf{B} + \tfrac{2}{3}(\overrightarrow{ON} - \mathbf{B})$$

$$= \mathbf{B} + \tfrac{2}{3}[\tfrac{1}{2}(\mathbf{A} + \mathbf{C}) - \mathbf{B}] = \tfrac{1}{3}(\mathbf{A} + \mathbf{B} + \mathbf{C}). \tag{2}$$

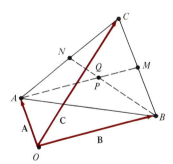

Figure 17.33

---

* Recall that a *median* of a triangle is a segment drawn from a vertex to the midpoint of the opposite side.

Comparison of (1) and (2) shows that the two points P and Q coincide, and in the same way we obtain the same point yet again if we go two-thirds of the way from C to the midpoint of AB. This completes the proof. We can also draw our conclusion more elegantly by observing that since (1) is symmetric in the three vectors $\mathbf{A}$, $\mathbf{B}$, $\mathbf{C}$, we clearly get the same point no matter which midpoint we start with.

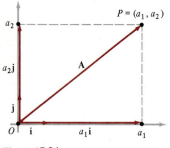

Figure 17.34

A vector of length 1 is called a *unit vector.* It is easy to see that if we divide any nonzero vector $\mathbf{A}$ by its own length, we obtain a unit vector $\mathbf{A}/|\mathbf{A}|$ in the same direction. This simple fact is surprisingly useful.

When we are working with vectors in the coordinate plane, it is often convenient to use the standard unit vectors $\mathbf{i}$ and $\mathbf{j}$; as shown in Fig. 17.34, $\mathbf{i}$ points in the direction of the positive x-axis and $\mathbf{j}$ points in the direction of the positive y-axis. We have seen that any vector $\mathbf{A}$ in the xy-plane can be placed with its tail at the origin, and in this way becomes the position vector $\overrightarrow{OP}$ of the point P at its head. If P has coordinates a_1 and a_2, then the vectors $a_1\mathbf{i}$ and $a_2\mathbf{j}$ run from the origin to the points a_1 and a_2 on the axes, and by the parallelogram rule we have

$$\mathbf{A} = a_1\mathbf{i} + a_2\mathbf{j}. \tag{3}$$

The number a_1 in (3) is called the *x-component* or $\mathbf{i}$-*component* of the vector $\mathbf{A}$, and a_2 is called its *y-component* or $\mathbf{j}$-*component*. These components are scalars, and should be distinguished from the vector components $a_1\mathbf{i}$ and $a_2\mathbf{j}$. By the Pythagorean theorem, we clearly have

$$|A| = \sqrt{a_1{}^2 + a_2{}^2}.$$

Formula (3) tells us that every vector in the plane is a *linear combination* of $\mathbf{i}$ and $\mathbf{j}$. The value of this formula is based on the fact that such linear combinations can be manipulated by the ordinary rules of algebra. Thus, if

$$\mathbf{A} = a_1\mathbf{i} + a_2\mathbf{j} \qquad \text{and} \qquad \mathbf{B} = b_1\mathbf{i} + b_2\mathbf{j},$$

then

$$\begin{aligned} \mathbf{A} + \mathbf{B} &= (a_1\mathbf{i} + a_2\mathbf{j}) + (b_1\mathbf{i} + b_2\mathbf{j}) \\ &= (a_1 + b_1)\mathbf{i} + (a_2 + b_2)\mathbf{j}. \end{aligned}$$

Also,

$$c\mathbf{A} = c(a_1\mathbf{i} + a_2\mathbf{j}) = (ca_1)\mathbf{i} + (ca_2)\mathbf{j}.$$

Example 6 If $\mathbf{A} = 3\mathbf{i} + 4\mathbf{j}$ and $\mathbf{B} = 2\mathbf{i} - 5\mathbf{j}$, find $|\mathbf{A}|$ and express $3\mathbf{A} - 4\mathbf{B}$ in terms of $\mathbf{i}$ and $\mathbf{j}$.

Solution Clearly, $|\mathbf{A}| = \sqrt{9 + 16} = 5$ and

$$\begin{aligned} 3\mathbf{A} - 4\mathbf{B} &= 3(3\mathbf{i} + 4\mathbf{j}) - 4(2\mathbf{i} - 5\mathbf{j}) \\ &= \mathbf{i} + 32\mathbf{j}. \end{aligned}$$

PROBLEMS

1 For each of the following pairs of vectors **A** and **B**, find |**A**|, $3\mathbf{A} - 5\mathbf{B}$, and $6\mathbf{A} + \mathbf{B}$:
  (a) $\mathbf{A} = \mathbf{i} - 3\mathbf{j}$, $\mathbf{B} = -2\mathbf{i} + 5\mathbf{j}$;
  (b) $\mathbf{A} = -7\mathbf{i} + 2\mathbf{j}$, $\mathbf{B} = 3\mathbf{i} + 2\mathbf{j}$;
  (c) $\mathbf{A} = -6\mathbf{j}$, $\mathbf{B} = -2\mathbf{i} + 3\mathbf{j}$;
  (d) $\mathbf{A} = 3\mathbf{i} + 5\mathbf{j}$, $\mathbf{B} = 2\mathbf{i} - 8\mathbf{j}$.

2 For each of the following pairs of points P and Q, find the vector $\overrightarrow{PQ}$ in terms of **i** and **j**:
  (a) $P = (-5, 0)$, $Q = (1, 3)$;
  (b) $P = (-1, -4)$, $Q = (-2, 3)$;
  (c) $P = (1, -5)$, $Q = (6, 4)$;
  (d) $P = (0, 2)$, $Q = (3, -5)$.

3 For any three points P, Q, R in the plane, we have
$$\overrightarrow{PQ} + \overrightarrow{QR} + \overrightarrow{RP} = \mathbf{0}.$$
Why? Verify this for the special case $P = (2, -4)$, $Q = (-3, 5)$, $R = (-4, 0)$ by expressing each vector in terms **i** and **j** and carrying out the addition.

4 Find a vector in the same direction as $6\mathbf{i} - 2\mathbf{j}$ that has (a) three times its length; (b) half its length.

5 For each of the following vectors **A**, find two unit vectors parallel to **A**:
  (a) $\mathbf{A} = 3\mathbf{i} - 4\mathbf{j}$;   (b) $\mathbf{A} = -5\mathbf{i} + 12\mathbf{j}$;
  (c) $\mathbf{A} = 5\mathbf{i} - 7\mathbf{j}$;   (d) $\mathbf{A} = 24\mathbf{i} - 7\mathbf{j}$.

6 Find a vector of length 3 which has (a) the same direction as $5\mathbf{i} - 2\mathbf{j}$; (b) the opposite direction to $4\mathbf{i} + 5\mathbf{j}$.

7 Find two vectors of length 26 and slope $\frac{5}{12}$.

8 Find a unit vector which, if its tail is placed at the point $(4, 4)$ on $x^2 = 4y$, is normal to the curve and points toward the positive y-axis.

9 If **A** is a nonzero vector, we know that $\mathbf{A}/|\mathbf{A}|$ is a unit vector with the same direction as **A**. Use this fact to write down a vector that bisects the angle between two nonzero vectors **A** and **B** whose tails coincide.

10 Three vectors are drawn from the vertices of a triangle to the midpoints of the opposite sides. Show that the sum of these vectors is zero.

11 Use vector methods to show that the diagonals of a parallelogram bisect each other.

12 Use vector methods to show that a line from a vertex of a parallelogram to the midpoint of a nonadjacent side trisects a diagonal.

13 Solve the canoe problem in Example 3 when the current and canoe speeds are 2 and $2\sqrt{2} \cong 2.8$ mi/h, respectively.

14 If the velocity of the wind is $\mathbf{v}_w$ and an airplane flies with velocity $\mathbf{v}_a$ relative to the air, then the velocity of the plane relative to the ground is
$$\mathbf{v}_g = \mathbf{v}_w + \mathbf{v}_a.$$
The vectors $\mathbf{v}_a$ and $\mathbf{v}_g$ are called the *apparent velocity* and the *true velocity*, respectively.
  (a) If the wind is blowing from the northeast at 60 mi/h and the pilot wishes to fly straight east at 600 mi/h, what should be the plane's apparent velocity?
  (b) Repeat part (a) if the pilot wishes to fly southeast at 600 mi/h.

17.4
DERIVATIVES OF VECTOR FUNCTIONS. VELOCITY AND ACCELERATION

In Section 17.3 we became acquainted with the *algebra* of vectors. In the rest of this chapter we shall be interested in problems of motion, and this requires us to work with the *calculus* of vectors. When vectors and calculus are allowed to interact with each other, the result is a mathematical discipline of great power and efficiency for studying multidimensional problems of geometry and physics. This vector calculus—usually called *vector analysis*—is one of the major topics of advanced courses in calculus. In this chapter we can only introduce the subject and discuss a few of the classic applications, culminating in a treatment of Kepler's laws of planetary motion and Newton's law of universal gravitation.

We begin by pointing out the connection between vectors and the parametric equations of curves discussed in the first two sections of this chapter.

Suppose a point $P = (x, y)$ moves along a curve in the xy-plane, and suppose further that we know its position at any time t (Fig. 17.35). This means that the coordinates x and y are known as functions of the scalar variable t, so that

$$x = x(t) \qquad \text{and} \qquad y = y(t).$$

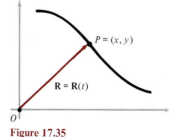

Figure 17.35

These are parametric equations for the path in terms of the time parameter t.* A more concise description of the motion is obtained by using the position vector of the moving point,

$$\mathbf{R} = \overrightarrow{OP} = x(t)\mathbf{i} + y(t)\mathbf{j}.$$

We can emphasize that $\mathbf{R}$ is a vector function of t by writing $\mathbf{R} = \mathbf{R}(t)$. Thus, the study of a pair of parametric equations is equivalent to the study of a single vector function, and, as we shall see, the latter is often much more effective at revealing the essence of what is going on.

Just as in ordinary calculus, $\mathbf{R}(t)$ is said to be *continuous* at $t = t_0$ if

$$\lim_{t \to t_0} \mathbf{R}(t) = \mathbf{R}(t_0), \tag{1}$$

which means that $|\mathbf{R}(t) - \mathbf{R}(t_0)|$ can be made as small as we please by taking t sufficiently close to t_0. It follows easily from (1) that

$$\mathbf{R}(t) = x(t)\mathbf{i} + y(t)\mathbf{j} \tag{2}$$

is continuous if and only if $x(t)$ and $y(t)$ are both continuous.

We define the derivative of the vector function $\mathbf{R}(t)$ exactly as might be expected. When t changes to $t + \Delta t$, the change in $\mathbf{R}$ is $\Delta\mathbf{R} = \mathbf{R}(t + \Delta t) - \mathbf{R}(t)$, and the *derivative* of $\mathbf{R}(t)$ with respect to t is defined as the limit

$$\frac{d\mathbf{R}}{dt} = \lim_{\Delta t \to 0} \frac{\Delta\mathbf{R}}{\Delta t}. \tag{3}$$

That is, we divide the vector $\Delta\mathbf{R}$ by Δt and then find the limit of the new vector $\Delta\mathbf{R}/\Delta t$ as $\Delta t \to 0$. This vector will approach a limit if and only if its head approaches a limiting position, and this happens if and only if each of its components approaches a limit. It is clear that in terms of components we have

$$\frac{\Delta\mathbf{R}}{\Delta t} = \frac{\mathbf{R}(t + \Delta t) - \mathbf{R}(t)}{\Delta t}$$

$$= \frac{x(t + \Delta t) - x(t)}{\Delta t}\mathbf{i} + \frac{y(t + \Delta t) - y(t)}{\Delta t}\mathbf{j}.$$

Thus, if the definition (3) is applied to (2), we see at once that $\mathbf{R}(t)$ is differentiable if and only if $x(t)$ and $y(t)$ are, and in this case

$$\frac{d\mathbf{R}}{dt} = \frac{dx}{dt}\mathbf{i} + \frac{dy}{dt}\mathbf{j}. \tag{4}$$

As in ordinary calculus, we often write $\mathbf{R}'(t)$ for $d\mathbf{R}/dt$ and $\mathbf{R}''(t)$ for $d^2\mathbf{R}/dt^2$.

Several familiar differentiation rules can now be extended to vector functions. One of the most important is this: If a vector function is multiplied by a scalar function, and if both can be differentiated, then their product can be differentiated according to the rule

* In Section 17.1 we wrote the parametric equations of a curve as $x = f(t)$, $y = g(t)$. However, it is more direct and convenient to use the same letter for the function as for the dependent variable, as we do here.

$$\frac{d}{dt}(u\mathbf{R}) = u\frac{d\mathbf{R}}{dt} + \mathbf{R}\frac{du}{dt}.$$

This is proved in just the same way as the product rule for two scalar functions. Also, the rule for the sum of two vector functions is just what we expect, the derivative of a constant vector function is the vector **0**, and the chain rule is valid.

It is important to understand the meaning of the derivative $d\mathbf{R}/dt$ *as a vector,* and not merely in terms of its components as given by (4). To do this, we follow the geometric meaning of the various steps expressed in the definition (3). First, the change Δt of the independent variable t carries the position vector from $\mathbf{R}(t)$ to $\mathbf{R}(t + \Delta t)$, as shown in Fig. 17.36. The vector $\Delta\mathbf{R} = \mathbf{R}(t + \Delta t) - \mathbf{R}(t)$ is directed along the chord from the head of $\mathbf{R}(t)$ to the head of $\mathbf{R}(t + \Delta t)$. Dividing $\Delta\mathbf{R}$ by the scalar Δt changes its length and produces another vector $\Delta\mathbf{R}/\Delta t$ parallel to $\Delta\mathbf{R}$. Since the limiting direction of the chord as $\Delta t \to 0$ is the direction of the tangent, the derivative $d\mathbf{R}/dt$ is tangent to the path at the head of $\mathbf{R}$. As we know, every vector can be thought of as having its tail at the origin, but in Fig. 17.36 we place the tail of $d\mathbf{R}/dt$ at the head of $\mathbf{R}$ in order better to visualize what is happening.

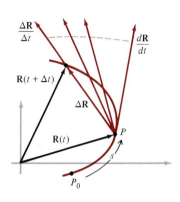

Figure 17.36

To interpret the length of the vector $d\mathbf{R}/dt$, let s be the length of the curve from a fixed point P_0 given by $t = t_0$ to the variable point P given by t, where $t \ge t_0$. By (4) we have

$$\left|\frac{d\mathbf{R}}{dt}\right| = \sqrt{\left(\frac{dx}{dt}\right)^2 + \left(\frac{dy}{dt}\right)^2} = \frac{\sqrt{dx^2 + dy^2}}{dt} = \frac{ds}{dt}. \tag{5}$$

Since t is time in the present discussion, the derivative ds/dt is the rate at which the moving point P traverses distance, that is, its speed.

These observations tell us that the vector $d\mathbf{R}/dt$ has as its direction and length the direction and speed of our moving point. It is therefore natural to adopt the following formal definitions. Just as in the case of one-dimensional motion, we define the *velocity* **v** of a moving point as the rate of change of its position,

$$\mathbf{v} = \frac{d\mathbf{R}}{dt},$$

and the *speed* v as the magnitude of the velocity,

$$v = |\mathbf{v}| = \left|\frac{d\mathbf{R}}{dt}\right|.$$

Example 1 If $\mathbf{R} = (4\cos 2t)\mathbf{i} + (3\sin 2t)\mathbf{j}$, find the path of the moving point, the velocity **v**, and the points on the path where the speed v is greatest and least.

Solution The curve has parametric equations $x = 4\cos 2t$, $y = 3\sin 2t$, so the path is the ellipse shown in Fig. 17.37,

$$\frac{x^2}{16} + \frac{y^2}{9} = 1.$$

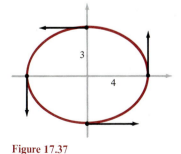

Figure 17.37

The point $P = (x, y)$ moves around this ellipse in the counterclockwise direction, as indicated by the arrows in the figure. The velocity is

$$\mathbf{v} = (-8 \sin 2t)\mathbf{i} + (6 \cos 2t)\mathbf{j}, \tag{6}$$

and the speed is

$$v = |\mathbf{v}| = (64 \sin^2 2t + 36 \cos^2 2t)^{1/2} = (28 \sin^2 2t + 36)^{1/2}.$$

It is clear from this formula for v that the smallest speed is 6, and this occurs when $\sin 2t = 0$, and by the given formula for $\mathbf{R}$, this happens when P is at either end of the major axis. The greatest speed is 8, and this occurs when $\sin 2t = 1$, so that $\cos 2t = 0$, that is, at either end of the minor axis.

Just as the velocity $\mathbf{v}$ of our moving point is the rate of change of its position, the *acceleration* $\mathbf{a}$ is the rate of change of its velocity,

$$\mathbf{a} = \frac{d\mathbf{v}}{dt} = \frac{d^2\mathbf{R}}{dt^2}.$$

Thus, our present concepts of velocity and acceleration are direct extensions of more limited versions of these concepts from our earlier studies of one-dimensional motion.

If the moving point P is the location of a moving physical object, and can therefore be thought of as a particle of mass m moving under the action of an applied force $\mathbf{F}$, then *Newton's second law of motion* states that

$$\mathbf{F} = m\mathbf{a}. \tag{7}$$

This vector form of Newton's law shows that the force and acceleration vectors both have the same direction. Since we visualize the force $\mathbf{F}$ as being applied to the particle, so that its tail is at P, it is customary also to place the tail of $\mathbf{a}$ at P, as shown in Fig. 17.38. Both $\mathbf{F}$ and $\mathbf{a}$ usually point toward the concave side of the curve, but in exceptional cases they may be tangent to the curve.

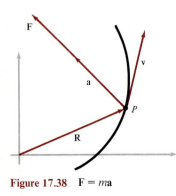

Figure 17.38 $\mathbf{F} = m\mathbf{a}$

Example 1 (cont.) To find the acceleration $\mathbf{a}$ of the motion given by $\mathbf{R} = (4 \cos 2t)\mathbf{i} + (3 \sin 2t)\mathbf{j}$, we have only to differentiate the velocity (6) with respect to t,

$$\mathbf{a} = \frac{d\mathbf{v}}{dt} = (-16 \cos 2t)\mathbf{i} + (-12 \sin 2t)\mathbf{j}.$$

Since this can be written in the form

$$\mathbf{a} = -4[(4 \cos 2t)\mathbf{i} + (3 \sin 2t)\mathbf{j}]$$
$$= -4\mathbf{R},$$

the acceleration vector is always directed toward the center of the elliptical path.

A simple but important situation is that in which a particle travels at constant speed around a circular path.

Example 2 *Uniform circular motion.* A particle of mass m moves counterclockwise around the circle $x^2 + y^2 = r^2$ with constant speed v. Find the acceleration of the particle and the force needed to produce this motion.

Solution By using the notation in Fig. 17.39, the path can be written as

$$\mathbf{R} = (r\cos\theta)\mathbf{i} + (r\sin\theta)\mathbf{j}, \tag{8}$$

with θ as the parameter. Since $s = r\theta$, we have

$$v = \frac{ds}{dt} = r\frac{d\theta}{dt},$$

and therefore $d\theta/dt = v/r$. This enables us to find the velocity and acceleration from (8) by using the chain rule:

$$\mathbf{v} = \frac{d\mathbf{R}}{dt} = \frac{d\mathbf{R}}{d\theta} \cdot \frac{d\theta}{dt}$$

$$= [(-r\sin\theta)\mathbf{i} + (r\cos\theta)\mathbf{j}] \cdot \frac{v}{r}$$

$$= v[(-\sin\theta)\mathbf{i} + (\cos\theta)\mathbf{j}];$$

and

$$\mathbf{a} = \frac{d\mathbf{v}}{dt} = \frac{d\mathbf{v}}{d\theta} \cdot \frac{d\theta}{dt}$$

$$= v[(-\cos\theta)\mathbf{i} + (-\sin\theta)\mathbf{j}] \cdot \frac{v}{r}$$

$$= -\frac{v^2}{r}[(\cos\theta)\mathbf{i} + (\sin\theta)\mathbf{j}].$$

By multiplying and dividing by r, it is easy to see that $\mathbf{a} = -(v^2/r^2)\mathbf{R}$. This tells us that the acceleration vector $\mathbf{a}$ points in toward the center of the circle and has magnitude

$$|\mathbf{a}| = \frac{v^2}{r^2}|\mathbf{R}| = \frac{v^2}{r}.$$

By Newton's law (7), the force $\mathbf{F}$ needed to produce this motion must point toward the center of the circle and have constant magnitude mv^2/r. Such a force is called a *centripetal force.*

It is obvious by now that the time t is a parameter of fundamental importance for studying the motion of a point P along a curved path. Another important parameter is the arc length s, measured along the curve from a fixed point P_0 to P, as shown in Fig. 17.40. We now consider $\mathbf{R}$ as a function of s and examine the meaning of the derivative $d\mathbf{R}/ds$. If P moves along the curve to Q when s changes to $s + \Delta s$, then

$$\frac{\Delta\mathbf{R}}{\Delta s} = \frac{\overrightarrow{PQ}}{\Delta s}$$

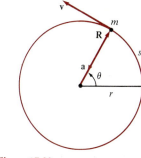

Figure 17.39

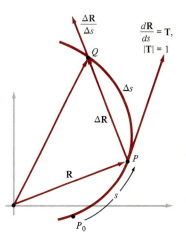

Figure 17.40

is a vector in the direction of the chord from P to Q whose length is

$$\frac{PQ}{\Delta s} = \frac{\text{chord}}{\text{arc}}.$$

When $\Delta s \to 0$, the direction of the chord approaches the direction of the tangent and the ratio of the chord to the arc approaches 1. Therefore the vector $\mathbf{T}$, which is defined by

$$\mathbf{T} = \frac{d\mathbf{R}}{ds} = \lim_{\Delta s \to 0} \frac{\Delta \mathbf{R}}{\Delta s},$$

is a vector of unit length which is tangent to the curve at P and points in the direction of increasing s. $\mathbf{T}$ is called the *unit tangent vector.*

To clarify our first use of $\mathbf{T}$, we recall that the formulas for the velocity $\mathbf{v}$ and acceleration $\mathbf{a}$ in terms of their components are

$$\mathbf{v} = \frac{dx}{dt}\mathbf{i} + \frac{dy}{dt}\mathbf{j} \quad \text{and} \quad \mathbf{a} = \frac{d^2x}{dt^2}\mathbf{i} + \frac{d^2y}{dt^2}\mathbf{j}. \tag{9}$$

These formulas are convenient for calculation, but they don't contribute much to our intuitive understanding of the nature of the vectors $\mathbf{v}$ and $\mathbf{a}$. However, the chain rule enables us to write the velocity $\mathbf{v}$ in the form

$$\mathbf{v} = \frac{d\mathbf{R}}{dt} = \frac{d\mathbf{R}}{ds}\frac{ds}{dt} = \mathbf{T}\frac{ds}{dt}. \tag{10}$$

For the purpose of conveying insight, this formula is much superior to the first of formulas (9), because in (10) the direction of $\mathbf{v}$ is given by $\mathbf{T}$ and its magnitude is given by ds/dt, and the meaning of each is visible at a glance. Our main aim in the next two sections is to obtain a corresponding formula for the acceleration $\mathbf{a}$.

Remark In most of our work we restrict ourselves to parametrized curves $\mathbf{R} = \mathbf{R}(t)$ that are *smooth,* in the sense that the derivative $\mathbf{R}'(t)$ is continuous and nonzero at every point. In principle, the continuity of $\mathbf{R}'(t)$ enables us to find s as a function of t from formula (5),

$$s = \int_{t_0}^{t} |\mathbf{R}'(t)|\, dt = \int_{t_0}^{t} \sqrt{\left(\frac{dx}{dt}\right)^2 + \left(\frac{dy}{dt}\right)^2}\, dt.$$

And since

$$\frac{ds}{dt} = |\mathbf{R}'(t)| > 0,$$

the function $s = s(t)$ is strictly increasing and therefore has an inverse function $t = t(s)$. This permits us to introduce s as a parameter for the curve,

$$\mathbf{R} = \mathbf{R}(t) = \mathbf{R}[t(s)].$$

However, in most cases it is difficult or impossible to carry out these calculations. The integral for s may be hard to evaluate; and even if $s = s(t)$ is known explicitly, it may be hard to find the inverse function $t = t(s)$. Fortunately these difficulties are not a serious obstacle, because there is seldom any real need to have $\mathbf{R}$ expressed explicitly as a function of s. It is the *idea* of using arc length as a parameter that is important for understanding motion along curves — as in the preceding paragraph — and not the actual act of doing so in specific problems.

PROBLEMS

1 Give a geometric description of the locus of the head of **R** if **R** = **A** + *t***B**, where neither **A** nor **B** is **0** and **B** is not parallel to **A**. Draw a sketch.

2 What is the locus of the head of **R** if **R** = at**i** + $b(1 - t)$**j**, where a and b are nonzero constants?

3 Show that the locus of the head of **R** = t**i** + $(mt + b)$**j** is the line $y = mx + b$.

4 What is the locus of the head of **R** = $(t + 1)$**i** + $(t^2 + 2t + 3)$**j**?

In Problems 5 to 9, **R** is the position of a moving point at time t. In each case compute the velocity, acceleration, and speed.

5 **R** = $(t^2 + 1)$**i** + $(t - 1)$**j**.

6 **R** = t^2**i** + t^3**j**.

7 **R** = t**i** + $(t^3 - 3t)$**j**.

8 **R** = $(\cos 2t)$**i** + $(\sin t)$**j**.

9 **R** = $(\tan t)$**i** + $(\sec t)$**j**.

10 If the position vector of a moving particle is **R** = $(a \cos kt)$**i** + $(b \sin kt)$**j**, where a, b, k are positive constants, then the particle moves on the ellipse $x^2/a^2 + y^2/b^2 = 1$. Show that **a** = $-k^2$**R** and describe the force **F** that produces such a motion.

11 If the acceleration of a moving particle is **a** = a**j**, where a is a constant, find **R** by two successive integrations with respect to t and show that the path is a parabola, a straight line, or a single point.

12 If a moving particle is acted on by no force, so that **a** = **0**, show that the particle moves with constant speed along a straight line. This is Newton's first law of motion.

17.5

CURVATURE AND THE UNIT NORMAL VECTOR

In Section 17.4 we expressed the velocity **v** of our moving point P in terms of the unit tangent vector **T** shown in Fig. 17.41, where **T** was obtained as the derivative of the position vector **R** with respect to arc length s,

$$\mathbf{T} = \frac{d\mathbf{R}}{ds}.$$

As our first step toward the general acceleration formula derived in Section 17.6, we must now analyze the derivative of **T** with respect to s, and this requires us to examine the purely geometric concept of the "curvature" of a curve.

If we consult our intuitive feelings about the notion of curvature, most of us will agree that a straight line does not curve at all; that is, it has zero curvature. Also, a circle has the same curvature at every point, and a small circle has greater curvature than a large one, as suggested in Fig. 17.42. In the case of a nonuniform curve like the one on the right, the curvature ought to

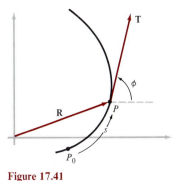

Figure 17.41

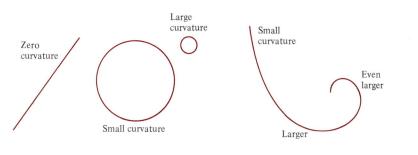

Zero curvature

Large curvature

Small curvature

Small curvature

Larger

Even larger

Figure 17.42 The meaning of curvature.

be smaller where the curve is relatively straight and larger where the curve bends more sharply.

These opinions are based on the idea that curvature at a point ought to measure how rapidly the direction of a curve is changing at that point with respect to distance along the curve. Since direction is specified by the angle ϕ from the x-axis to the tangent line (Fig. 17.41), we consider this angle as a function of the arc length s and define the *curvature k* to be the rate of change of ϕ with respect to s,

$$k = \frac{d\phi}{ds}. \tag{1}$$

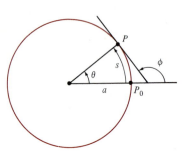

Figure 17.43

The curvature can be either positive or negative, and it may be zero in certain cases. Since $k > 0$ means that ϕ is increasing as s increases, it is clear that this means that the curve turns away to the left of the tangent as we move along the curve in the positive direction. Similarly, $k < 0$ means the curve turns away to the right of the tangent.

It is obvious from the definition (1) that the curvature of a straight line is zero, since ϕ does not change as we move along the line. In the case of a circle of radius a (Fig. 17.43), we have

$$\phi = \theta + \frac{\pi}{2},$$

and by using the fact that $\theta = s/a$, we easily see that the curvature is

$$k = \frac{d\phi}{ds} = \frac{d\theta}{ds} = \frac{1}{a}. \tag{2}$$

Since the curvature of a circle is clearly constant, we can also obtain the result (2) by observing that a complete revolution amounts to a change in direction of 2π radians over a curve of length $2\pi a$, so

$$k = \frac{d\phi}{ds} = \frac{2\pi}{2\pi a} = \frac{1}{a}.$$

We note in passing that formula (2) shows that smaller circles have larger curvatures, as indicated in Fig. 17.42.

Apart from these very simple cases, the actual calculation of the curvature is carried out by various rather complicated formulas, depending on how the curve is defined.

The simplest situation is that in which the curve is the graph of a function $y = f(x)$. Since $\tan \phi = dy/dx$, we have

$$\phi = \tan^{-1} \frac{dy}{dx} \quad \text{and} \quad d\phi = \frac{d^2y/dx^2}{1 + (dy/dx)^2} \, dx.$$

Also, the expression $ds = \sqrt{dx^2 + dy^2}$ for the differential of arc length gives

$$ds = \sqrt{1 + \left(\frac{dy}{dx}\right)^2} \, dx = \left[1 + \left(\frac{dy}{dx}\right)^2\right]^{1/2} dx. \tag{3}$$

On dividing $d\phi$ by ds, we see that in this case the curvature is given by the formula

$$k = \frac{d\phi}{ds} = \frac{d^2y/dx^2}{[1 + (dy/dx)^2]^{3/2}}. \tag{4}$$

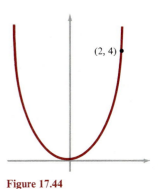

In Section 4.2 we used the sign of the second derivative d^2y/dx^2 to find out which direction the curve is bending, concave up or concave down. Formula (4) gives us this information and much more — it tells us precisely *how much* the curve is bending.

Example 1   Show that the curvature of the parabola $y = x^2$ is greatest at the vertex.

Solution   We are familiar enough with the general shape of parabolas (Fig. 17.44) to accept this statement without difficulty, because the curve visibly flattens out as $|x| \to \infty$. To verify it by calculation, we use the fact that $dy/dx = 2x$ and $d^2y/dx^2 = 2$ to write

$$k = \frac{d^2y/dx^2}{[1 + (dy/dx)^2]^{3/2}} = \frac{2}{(1 + 4x^2)^{3/2}}.$$

It is clear that this quantity has its greatest value when $x = 0$, which is at the vertex, and also that $k \to 0$ as $|x| \to \infty$. To illustrate how quickly the curve flattens out as we move away from the vertex, we notice that at the point $(2, 4)$ — which is fairly close to the vertex — we have

$$k = \frac{2}{(1 + 16)^{3/2}} < \frac{2}{16^{3/2}} = \frac{1}{32}.$$

Thus, at this point the parabola is flatter than a circle of radius 32, which is quite surprising.

Figure 17.44

If a curve is defined by parametric equations $x = x(t)$ and $y = y(t)$, then its curvature is computed from a slightly different formula. This time we start with

$$\phi = \tan^{-1} \frac{dy/dt}{dx/dt} \tag{5}$$

and

$$ds = \sqrt{\left(\frac{dx}{dt}\right)^2 + \left(\frac{dy}{dt}\right)^2}\, dt = \left[\left(\frac{dx}{dt}\right)^2 + \left(\frac{dy}{dt}\right)^2\right]^{1/2} dt. \tag{6}$$

The calculations leading to the curvature formula are a bit more complicated because of the quotient in (5), and the result is

$$k = \frac{d\phi}{ds} = \frac{(dx/dt)(d^2y/dt^2) - (dy/dt)(d^2x/dt^2)}{[(dx/dt)^2 + (dy/dt)^2]^{3/2}}$$

$$= \frac{x'y'' - y'x''}{[(x')^2 + (y')^2]^{3/2}}. \tag{7}$$

Students will notice that (7) includes (4) as a special case, when a curve $y = f(x)$ is thought of as a parametric curve $x = x$, $y = f(x)$, with x replacing t as the parameter.

There is a slight difficulty with signs that should be mentioned. By choosing the positive square root in both (3) and (6), we are assuming that the direction of increasing arc length s is the same as the direction in which the parameter increases. If this is not the case in applying (4) or (7) to a specific problem, then it is necessary to change the sign to get the actual curvature.

Example 2 Show that a circle of radius a has curvature $1/a$ by using the parametric equations $x = a \cos \theta$, $y = a \sin \theta$.

Solution We apply formula (7) with the understanding that primes denote derivatives with respect to θ. First we calculate

$$x' = -a \sin \theta, \qquad y' = a \cos \theta,$$
$$x'' = -a \cos \theta, \qquad y'' = -a \sin \theta.$$

Formula (7) now gives

$$k = \frac{a^2 \sin^2 \theta + a^2 \cos^2 \theta}{(a^2 \sin^2 \theta + a^2 \cos^2 \theta)^{3/2}} = \frac{1}{a},$$

as expected.

Now that we understand the concept of curvature, we are ready to deal quickly with the main problem of this section, which is to analyze the derivative of the unit tangent vector $\mathbf{T}$ with respect to s.

We begin by observing that in terms of the slope angle ϕ (Fig. 17.45) we have

$$\mathbf{T} = \mathbf{i} \cos \phi + \mathbf{j} \sin \phi,$$

so

$$\frac{d\mathbf{T}}{d\phi} = -\mathbf{i} \sin \phi + \mathbf{j} \cos \phi. \tag{8}$$

The derivative (8) is clearly a unit vector, because its length is

$$\left| \frac{d\mathbf{T}}{d\phi} \right| = \sqrt{\sin^2 \phi + \cos^2 \phi} = 1.$$

Also, it is perpendicular to $\mathbf{T}$, because its slope is

$$\frac{\cos \phi}{-\sin \phi} = -\frac{1}{\tan \phi},$$

which is the negative reciprocal of the slope of $\mathbf{T}$. In fact, the derivative (8) is the *unit normal vector* $\mathbf{N}$ shown in the figure,

$$\frac{d\mathbf{T}}{d\phi} = \mathbf{N}, \tag{9}$$

where $\mathbf{N}$ is obtained by rotating $\mathbf{T}$ through an angle $\pi/2$ in the counterclockwise direction. This is established by comparing (8) with

$$\mathbf{N} = \mathbf{i} \cos \left(\phi + \frac{\pi}{2} \right) + \mathbf{j} \sin \left(\phi + \frac{\pi}{2} \right) = -\mathbf{i} \sin \phi + \mathbf{j} \cos \phi.$$

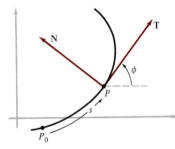

Figure 17.45

By using the chain rule together with (1) and (9), we now easily obtain the main result of this section,

$$\frac{d\mathbf{T}}{ds} = \frac{d\mathbf{T}}{d\phi}\frac{d\phi}{ds} = \mathbf{N}k.$$

It should be clear at this stage why it was necessary to discuss curvature in analyzing the meaning of $d\mathbf{T}/ds$: Since $\mathbf{T}$ has constant length, only its direction changes as s varies, and this is what brings us to the curvature. We also point out that regardless of whether k is positive or negative, $\mathbf{N}k$ always points toward the concave side of the curve.

Remark Let P be a point on a curve at which the curvature k is not zero, and draw the normal toward the concave side of the curve, as shown in Fig. 17.46. Every circle through P whose center lies on this normal will be tangent to the curve at P. That particular circle whose curvature is equal to $|k|$ is called the *circle of curvature.* Also, the center C of this circle is called the *center of curvature,* and its radius r is called the *radius of curvature.* We know from (2) that in the case of a circle, the radius is the reciprocal of the curvature, so the radius of curvature is given by the formula

$$r = \frac{1}{|k|} = \frac{[1 + (dy/dx)^2]^{3/2}}{|d^2y/dx^2|},$$

Figure 17.46 The circle of curvature.

if the curve is the graph of a function $y = f(x)$. A similar formula holds for a parametric curve. As P moves along the given curve, the locus of the corresponding center of curvature C is called the *evolute* of the given curve. In Appendix A.20 we discuss some remarkable applications of the theory of evolutes to cycloids.

PROBLEMS

1 Find the curvature of the given curve as a function of x or t:
(a) $y = \sqrt{x}$;
(b) $y = \ln \sec x$;
(c) $y = x + \dfrac{1}{x}$;
(d) $x = e^t \sin t, y = e^t \cos t$;
(e) $x = t^2, y = \ln t$.

2 Find the radius of curvature of the given curve at the given point:
(a) $x = t^2, y = t^3$ at $t = 2$;
(b) $x = e^t, y = e^{-t}$ at $t = 0$;
(c) $y = \dfrac{1}{x}$ at $(1, 1)$;
(d) $x = \tan t, y = \cot t$ at $t = \pi/4$.

3 In each case find the largest value (if any) of the curvature:

(a) $y = \sin x$;   (b) $y = \frac{1}{3}x^3$;
(c) $y = \ln x$.

4 Carry out the details of establishing the parametric curvature formula (7).

5 Find the curvature of the circle $x^2 + y^2 = a^2$ by applying formula (4) separately to $y = \sqrt{a^2 - x^2}$ and $y = -\sqrt{a^2 - x^2}$. What difficulty arises, and how can it be fixed?

6 For the curve $y = e^x$, find the radius of curvature and the equation of the circle of curvature at the point $(0, 1)$. Sketch the curve and this circle. Use the equation of the circle to calculate the values of dy/dx and d^2y/dx^2 at the point $(0, 1)$, and verify that these derivatives have the same values there as the corresponding derivatives of $y = e^x$.

7 At what point on the curve $y = e^x$ is the radius of curvature smallest? What is this smallest radius?

8 We know that if $y = f(x)$ is a straight line, then $k = 0$.

Show, conversely, that if $k = 0$, then $y = f(x)$ is a straight line.

9 Find the largest value of the radius of curvature on the first quadrant part of the hypocycloid of four cusps $x = a \cos^3 \theta$, $y = a \sin^3 \theta$. Where does the radius have this largest value?

10 By Problem 15 in Section 17.1, the equations

$$x = \cos \theta + \theta \sin \theta,$$

$$y = \sin \theta - \theta \cos \theta$$

represent the involute of a circle of radius 1. Find the curvature at any point.

11 Find the radius of curvature of the cycloid

$$x = a(\theta - \sin \theta),$$

$$y = a(1 - \cos \theta)$$

at any point.

12 (a) Sketch the ellipse $x = a \cos \theta$, $y = b \sin \theta$, where

$0 < b < a$, and find its curvature k at an arbitrary point.

(b) Without calculating $dk/d\theta$, use the formula in (a) to show that k has its largest values at the ends of the major axis and its smallest values at the ends of the minor axis. Show that these values are a/b^2 and b/a^2, respectively. Notice that if $b = a$, then we have a circle, and both of these formulas give $k = 1/a$, as they should.*

***13** Let a be a positive number and consider the curve $y = x^a$ for $x > 0$. Show that the curvature approaches a finite limit as $x \to 0$ if $a \leq \frac{1}{2}$ or $a = 1$ or $a \geq 2$, and only in these cases.

---

* For curves in general, a point where the curvature has a maximum or minimum value is called a *vertex*. By Problem 12, an ellipse has four vertices. An ellipse is a special case of an *oval*, which is a convex closed curve whose parametric equations $x = x(t)$, $y = y(t)$ have continuous second derivatives. The famous *four vertex theorem* of differential geometry states that every oval has at least four vertices.

17.6

TANGENTIAL AND NORMAL COMPONENTS OF ACCELERATION

Consider a moving particle whose position at time t is given by the parametric equations $x = x(t)$ and $y = y(t)$. The position vector of this particle is $\mathbf{R} = x\mathbf{i} + y\mathbf{j}$, and its velocity and acceleration are

$$\mathbf{v} = \frac{d\mathbf{R}}{dt} = \frac{dx}{dt}\mathbf{i} + \frac{dy}{dt}\mathbf{j}, \qquad \mathbf{a} = \frac{d\mathbf{v}}{dt} = \frac{d^2x}{dt^2}\mathbf{i} + \frac{d^2y}{dt^2}\mathbf{j}. \tag{1}$$

Unfortunately, the $\mathbf{i}$- and $\mathbf{j}$-components of these vectors have no physical meaning, because they depend on the coordinate system, and the choice of the coordinate system is arbitrary; it is not determined by the intrinsic nature of the motion itself. However, we saw in Section 17.4 that the velocity can also be written as

$$\mathbf{v} = \frac{d\mathbf{R}}{dt} = \frac{d\mathbf{R}}{ds}\frac{ds}{dt}$$

or

$$\mathbf{v} = \mathbf{T}\frac{ds}{dt}, \tag{2}$$

where $\mathbf{T}$ is the unit tangent vector (Fig. 17.47). This expression for the

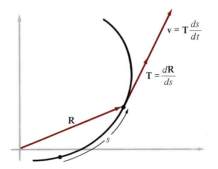

Figure 17.47

velocity *does* have physical meaning, because **T** gives the direction of the motion and ds/dt gives its magnitude, the speed.

To obtain a similar revealing expression for the acceleration, we differentiate (2) with respect to t,

$$\mathbf{a} = \frac{d\mathbf{v}}{dt} = \mathbf{T}\frac{d^2s}{dt^2} + \frac{ds}{dt}\frac{d\mathbf{T}}{dt}. \tag{3}$$

By Section 17.5 we know that

$$\frac{d\mathbf{T}}{dt} = \frac{d\mathbf{T}}{ds}\frac{ds}{dt} = \frac{d\mathbf{T}}{d\phi}\frac{d\phi}{ds}\frac{ds}{dt}$$

$$= \mathbf{N}k\frac{ds}{dt}, \tag{4}$$

where k is the curvature and **N** is the unit normal vector shown in Fig. 17.48. When (4) is substituted in (3) we get our fundamental result,

$$\mathbf{a} = \mathbf{T}\frac{d^2s}{dt^2} + \mathbf{N}k\left(\frac{ds}{dt}\right)^2. \tag{5}$$

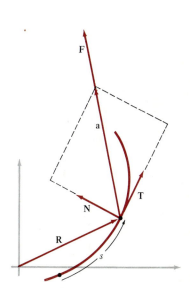

This is an important equation in mechanics. The vectors **T** and **N** serve as reference unit vectors much like **i** and **j**. They enable us to resolve the acceleration into two "natural" components, in the direction of the motion and normal to this direction, in contrast to the arbitrary components given by the second of equations (1). The *tangential component, d^2s/dt^2*, is simply the derivative of the speed $v = ds/dt$ of the particle along its path. The *normal component, $k(ds/dt)^2 = kv^2$*, has magnitude

$$|k|v^2 = \frac{v^2}{r}, \tag{6}$$

where r is the radius of curvature. It is clear from (5) that when $k \neq 0$ and the particle is actually moving, the acceleration is always directed toward the concave side of the curve.

Figure 17.48

Of course, the great importance of the acceleration lies in the fact that when a particle of mass m is acted on by a force **F**, it moves in accordance with Newton's second law of motion $\mathbf{F} = m\mathbf{a}$. The vectors **F** and **a** therefore have the same direction, as shown in the figure, and this fits with our intuitive understanding that when a force changes the direction of a moving particle, it pulls the particle away from the direction of the tangent toward the concave side of the path.

Since the curvature k is available whenever the curve is given in parametric form, the tangential and normal components of acceleration can be calculated from (5). However, it is often more efficient to use the following procedure. The acceleration vector is the same whether it is expressed in terms of **i**- and **j**-components or tangential and normal components, so

$$\mathbf{a} = a_x\mathbf{i} + a_y\mathbf{j} = a_t\mathbf{T} + a_n\mathbf{N},$$

where the a's have the obvious meanings. This tells us that

$$|\mathbf{a}|^2 = a_x^2 + a_y^2 = a_t^2 + a_n^2,$$

so

$$a_n = \sqrt{|\mathbf{a}|^2 - a_t^2} \tag{7}$$

where $a_t = d^2s/dt^2$.

Example 1 If a particle moves along the curve whose parametric equations are

$$x = \cos t + t \sin t, \qquad y = \sin t - t \cos t,$$

find the tangential and normal components of acceleration.

Solution The position vector is

$$\mathbf{R} = (\cos t + t \sin t)\mathbf{i} + (\sin t - t \cos t)\mathbf{j},$$

so

$$\begin{aligned}
\mathbf{v} &= \frac{d\mathbf{R}}{dt} \\
&= (-\sin t + t \cos t + \sin t)\mathbf{i} + (\cos t + t \sin t - \cos t)\mathbf{j} \\
&= (t \cos t)\mathbf{i} + (t \sin t)\mathbf{j}
\end{aligned}$$

and

$$\mathbf{a} = \frac{d\mathbf{v}}{dt} = (-t \sin t + \cos t)\mathbf{i} + (t \cos t + \sin t)\mathbf{j}.$$

The speed $v = ds/dt$ is given by

$$\frac{ds}{dt} = |\mathbf{v}| = \sqrt{(t \cos t)^2 + (t \sin t)^2} = t,$$

so the tangential component of acceleration is

$$a_t = \frac{d^2s}{dt^2} = \frac{d}{dt} t = 1.$$

The normal component of acceleration can be computed directly, by finding k and using $a_n = k(ds/dt)^2$. However, it is easier to use (7), which gives

$$\begin{aligned}
a_n &= \sqrt{|\mathbf{a}|^2 - a_t^2} \\
&= \sqrt{(-t \sin t + \cos t)^2 + (t \cos t + \sin t)^2 - 1} = t.
\end{aligned}$$

Example 2 If a particle of mass m moves around a circular path of radius r with constant speed $v = ds/dt$, then d^2s/dt^2 is zero. By equations (5) and (6), the acceleration is directed toward the center of the circle and has magnitude v^2/r. Further, the centripetal force acting on the particle has magnitude

$$F_1 = \frac{mv^2}{r}. \tag{8}$$

Thus, for an automobile going around a given unbanked curve, it takes four times as much normal force between the tires and the road to "hold the road" at 60 mi/h as at 30 mi/h, and the required force is doubled again if the radius is halved. These are the results about uniform circular motion that we obtained in Example 2 of Section 17.4.

Now suppose that our particle is an artificial satellite in a circular orbit around the earth, as shown in Fig. 17.49. If M is the mass of the earth, then Newton's law of gravitation tells us that the force of attraction which the earth exerts on the satellite has magnitude

$$F_2 = G\frac{Mm}{r^2}, \tag{9}$$

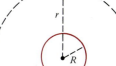

where r is the distance from the satellite to the center of the earth and G is a constant of proportionality called the constant of gravitation. We know that the weight of the satellite is the force which gravity exerts on it at the surface of the earth, and this is mg. Therefore, if R denotes the radius of the earth, then $F_2 = mg$ when $r = R$, so (9) tells us that

Figure 17.49

$$mg = G\frac{Mm}{R^2} \quad \text{or} \quad GM = gR^2.$$

This enables us to write (9) in the more convenient form

$$F_2 = \frac{gR^2 m}{r^2}. \tag{10}$$

For a satellite in stable circular orbit, the centripetal force is precisely equal to the gravitational force, so $F_1 = F_2$ and

$$v^2 = \frac{gR^2}{r}. \tag{11}$$

This formula gives the speed at which a satellite must move in order to maintain a circular orbit at a specified distance r from the center of the earth.

We make two observations about formula (11). First, if our satellite is moving in a circular orbit at a relatively low altitude above the surface of the earth, then $r \cong R$ and $v \cong \sqrt{gR}$. This orbital speed is approximately 5 mi/s, which should be compared with the escape speed of $\sqrt{2gR}$ or 7 mi/s that we calculated in Example 3 of Section 5.5.

Second, we consider a communications relay satellite that is placed in a circular orbit around the earth and has a period of revolution of $T = 24$ hours. This is a so-called *synchronous orbit,* in which the satellite moves with the turning earth and appears to hang motionless in the sky. If r is the radius of this orbit, then the orbital speed is $v = 2\pi r/T$, and when this is substituted in (11) we find that

$$r^3 = \frac{gR^2 T^2}{4\pi^2}.$$

With suitable adjustments applied to the values $g = 32$ ft/s², $R = 4000$ mi, and $T = 24$ h, we easily find that r is approximately 26,000 mi, which means that the satellite must be about 22,000 mi above the surface of the earth. Such satellites were first conceived in 1945 by the famous science fiction writer Arthur C. Clarke, and are the crucial links in our present-day worldwide television communications.

Of course, the ideas discussed in this example assume a circular path, which is approximately true for some satellites. We shall give a detailed treatment of elliptical orbits in Section 17.7.

PROBLEMS

1 In Example 1, find a_n the other way, by first calculating k and then using $a_n = k(ds/dt)^2$.

In Problems 2 to 7, find the velocity and acceleration vectors, then find the speed and the tangential and normal components of the acceleration.

2 $\mathbf{R} = (2t - 5)\mathbf{i} + (t^2 + 3)\mathbf{j}$.

3 $\mathbf{R} = a \cos \omega t\, \mathbf{i} + a \sin \omega t\, \mathbf{j}$, where a and ω are positive constants.

4 $\mathbf{R} = \cos t^2\, \mathbf{i} + \sin t^2\, \mathbf{j}$.

5 $\mathbf{R} = e^t \cos t\, \mathbf{i} + e^t \sin t\, \mathbf{j}$.

6 $\mathbf{R} = t \cos t\, \mathbf{i} + t \sin t\, \mathbf{j}$.

7 $\mathbf{R} = 2 \ln (t^2 + 1)\mathbf{i} + (2t - 4 \tan^{-1} t)\mathbf{j}$.

In Problems 8 and 9, find the normal component of the acceleration for the given values of t.

8 $\mathbf{R} = a \cos t\, \mathbf{i} + b \sin t\, \mathbf{j}; t = 0, \pi/2$.

9 $\mathbf{R} = t\, \mathbf{i} + \sin t\, \mathbf{j}; t = 0, \pi/2$.

10 Deduce from equation (5) that
  (a) the path of a moving particle will be a straight line if the normal component of acceleration is zero;
  (b) if the speed of a moving particle is constant, then the force is always directed along the normal;
  (c) if the force acting on a moving particle is always directed along the normal, then the speed is constant.

11 A road has the shape of the parabola $120y = x^2$. A truck is loaded in such a way that it will tip over if the normal component of its acceleration exceeds 30. What speeds will guarantee disaster for the truck as it swings around the vertex of the parabola?

12 If a particle moves along a path whose curvature k is never zero, how must the speed be adjusted if the normal component of the acceleration is to be held at a constant magnitude?

17.7
(OPTIONAL) KEPLER'S LAWS AND NEWTON'S LAW OF GRAVITATION

As we know, Isaac Newton conceived the basic ideas of calculus in the years 1665 and 1666 (at age 22 and 23) for the purpose of helping him to understand the movements of the planets against the background of the fixed stars. In order to appreciate what was involved in this achievement, we briefly recall the main stages in the development of astronomical thinking up to his time.

The ancient Greeks constructed an elaborate mathematical model to account for the complicated movements of the sun, moon, and planets as viewed from the earth. A combination of uniform circular motions was used to describe the motion of each body about the earth. It was very natural for them — as it is for all people — to adopt the geocentric point of view that the earth is fixed at the center of the universe and everything else moves around it. Also, they were influenced by the semimystical Pythagorean belief that nothing but motion at constant speed in a perfect circle is worthy of a celestial body.

In this Greek model, each planet P moves uniformly around a small circle (called an *epicycle*) with center C, and at the same time C moves uniformly around a larger circle centered at the earth E, as shown in Fig. 17.50. The radius of each circle and the angular speeds of P and C around the centers C and E are chosen to match the observed motion of the planet as closely as possible. This theory of epicycles was given its definitive form in Ptolemy's massive treatise *Almagest* in the second century A.D., and the theory itself is called the *Ptolemaic system*.

The next great step forward was taken by the Polish astronomer Copernicus. Shortly before his death in 1543, when he was presumably almost beyond the reach of a wrathful Church, he at last allowed the publication of his heretical book, *On the Revolution of the Celestial Spheres*. This work changed the Ptolemaic point of view by placing the sun, instead of the earth,

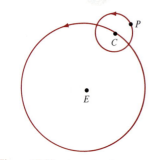

Figure 17.50   An epicycle.

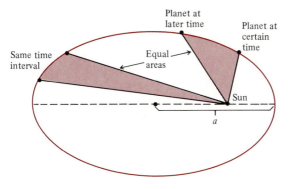

Figure 17.51 Kepler's second law.

at the center of each primary circle. Nevertheless, this *heliocentric system* was of much greater cultural than scientific importance. It enlarged the consciousness of many educated Europeans by giving them a better understanding of their place in the scheme of things, but it also kept the clumsy machinery of Ptolemy's circles whose centers move around on other circles.

It was Johannes Kepler (1571–1630) who finally eliminated this jumble of circles. Kepler was the assistant of the wealthy Danish astronomer Tycho Brahe, and when Brahe died in 1601, Kepler inherited the great masses of raw data they had accumulated on the positions of the planets at various times. Kepler worked incessantly on this material for 20 years, and at last succeeded in distilling from it his three beautifully simple laws of planetary motion, which were the climax of thousands of years of purely observational astronomy:

1   The orbit of each planet is an ellipse with the sun at one focus.
2   The line segment joining a planet to the sun sweeps out equal areas in equal times. See Fig. 17.51.
3   The *square* of the period of revolution of a planet is proportional to the *cube* of the semimajor axis of the planet's elliptical orbit. That is, if T is the time required for a planet to make one complete revolution about the sun and a is the semimajor axis shown in the figure, then the ratio T^2/a^3 is the same for all planets in the solar system.

From Kepler's point of view, these were empirical statements that fitted the data, and he had no idea of why they might be true or how they might be related to one another. In short, there was no theory to provide a context within which they could be understood.

Newton created such a theory. In the 1660s he discovered how to derive the inverse square law from Kepler's laws by mathematical reasoning, and also how to derive Kepler's laws from the inverse square law. We recall that *Newton's inverse square law of universal gravitation* states that any two particles of matter in the universe attract each other with a force directed along the line between them and of magnitude

$$G\frac{Mm}{r^2}, \tag{1}$$

where M and m are the masses of the particles, r is the distance between them,

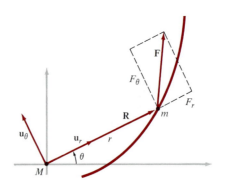

Figure 17.52

and G is a constant of nature called the gravitational constant. With this simple, clean, clear law as the unifying principle of his thinking, Newton published his theory of gravitation in 1687 in his *Principia Mathematica.* In this one book — perhaps the greatest of all scientific treatises — his success in using mathematical methods to explain the most diverse natural phenomena was so profound and far-reaching that he essentially created the sciences of physics and astronomy where only a handful of disconnected observations and simple inferences had existed before. These achievements launched the modern age of science and technology and radically altered the direction of human history.

We now derive Kepler's laws of planetary motion from Newton's law of gravitation, and to this end we discuss the motion of a small particle of mass m (a planet) under the attraction of a fixed large particle of mass M (the sun).

For problems involving a moving particle in which the force acting on it is always directed along the line from the particle to a fixed point, it is usually simplest to resolve the velocity, acceleration, and force into components along and perpendicular to this line. We therefore place the fixed particle M at the origin of a polar coordinate system (Fig. 17.52) and express the position vector of the moving particle m in the form

$$\mathbf{R} = r\,\mathbf{u}_r, \tag{2}$$

where $\mathbf{u}_r$ is the unit vector in the direction of $\mathbf{R}$. It is clear that

$$\mathbf{u}_r = \mathbf{i}\cos\theta + \mathbf{j}\sin\theta, \tag{3}$$

and also that the corresponding unit vector $\mathbf{u}_\theta$, perpendicular to $\mathbf{u}_r$, in the direction of increasing θ, is given by

$$\mathbf{u}_\theta = -\mathbf{i}\sin\theta + \mathbf{j}\cos\theta. \tag{4}$$

It is easy to see by componentwise differentiation that

$$\frac{d\mathbf{u}_r}{d\theta} = \mathbf{u}_\theta \quad \text{and} \quad \frac{d\mathbf{u}_\theta}{d\theta} = -\mathbf{u}_r. \tag{5}$$

Thus, differentiating $\mathbf{u}_r$ and $\mathbf{u}_\theta$ with respect to θ has the effect of rotating these vectors 90° in the counterclockwise direction. We shall need the derivatives of $\mathbf{u}_r$ and $\mathbf{u}_\theta$ with respect to the time t. By means of the chain rule we at once obtain the formulas

$$\frac{d\mathbf{u}_r}{dt} = \frac{d\mathbf{u}_r}{d\theta}\frac{d\theta}{dt} = \mathbf{u}_\theta\frac{d\theta}{dt} \quad \text{and} \quad \frac{d\mathbf{u}_\theta}{dt} = \frac{d\mathbf{u}_\theta}{d\theta}\frac{d\theta}{dt} = -\mathbf{u}_r\frac{d\theta}{dt}, \tag{6}$$

which are essential for computing the velocity and acceleration vectors $\mathbf{v}$ and $\mathbf{a}$.

Direct calculation from (2) now yields

$$\mathbf{v} = \frac{d\mathbf{R}}{dt} = r\frac{d\mathbf{u}_r}{dt} + \mathbf{u}_r\frac{dr}{dt} = r\frac{d\theta}{dt}\mathbf{u}_\theta + \frac{dr}{dt}\mathbf{u}_r, \tag{7}$$

and

$$\mathbf{a} = \frac{d\mathbf{v}}{dt} = \frac{dr}{dt}\frac{d\theta}{dt}\mathbf{u}_\theta + r\frac{d^2\theta}{dt^2}\mathbf{u}_\theta + r\frac{d\theta}{dt}\frac{d\mathbf{u}_\theta}{dt} + \frac{d^2r}{dt^2}\mathbf{u}_r + \frac{dr}{dt}\frac{d\mathbf{u}_r}{dt};$$

and by keeping formulas (6) in mind and rearranging, the latter equation can be written in the form

$$\mathbf{a} = \left(r\frac{d^2\theta}{dt^2} + 2\frac{dr}{dt}\frac{d\theta}{dt}\right)\mathbf{u}_\theta + \left[\frac{d^2r}{dt^2} - r\left(\frac{d\theta}{dt}\right)^2\right]\mathbf{u}_r. \tag{8}$$

If the force $\mathbf{F}$ acting on m is written as

$$\mathbf{F} = F_\theta\mathbf{u}_\theta + F_r\mathbf{u}_r, \tag{9}$$

then, from (8) and (9) and Newton's second law of motion $m\mathbf{a} = \mathbf{F}$, we get

$$m\left(r\frac{d^2\theta}{dt^2} + 2\frac{dr}{dt}\frac{d\theta}{dt}\right) = F_\theta \quad \text{and} \quad m\left[\frac{d^2r}{dt^2} - r\left(\frac{d\theta}{dt}\right)^2\right] = F_r. \tag{10}$$

These differential equations govern the motion of the particle m and are called the *equations of motion*; they are valid regardless of the nature of the force $\mathbf{F}$. Our next task is to extract the desired conclusions from these equations by making suitable assumptions about the direction and magnitude of $\mathbf{F}$.

CENTRAL FORCES AND KEPLER'S SECOND LAW

$\mathbf{F}$ is called a *central force* if it has no component perpendicular to $\mathbf{R}$, that is, if $F_\theta = 0$. Under this assumption the first of equations (10) becomes

$$r\frac{d^2\theta}{dt^2} + 2\frac{dr}{dt}\frac{d\theta}{dt} = 0.$$

On multiplying through by r, we obtain

$$r^2\frac{d^2\theta}{dt^2} + 2r\frac{dr}{dt}\frac{d\theta}{dt} = 0$$

or

$$\frac{d}{dt}\left(r^2\frac{d\theta}{dt}\right) = 0,$$

so

$$r^2\frac{d\theta}{dt} = h \tag{11}$$

for some constant h. We shall assume that h is positive, or equivalently that $d\theta/dt$ is positive, which evidently means that m is moving around the origin in a counterclockwise direction.

If $A = A(t)$ is the area swept out by $\mathbf{R}$ from some fixed position of reference, so that $dA = \frac{1}{2}r^2\,d\theta$, then (11) implies that

$$dA = \frac{1}{2}\left(r^2\frac{d\theta}{dt}\right)dt = \tfrac{1}{2}h\,dt.$$

On integrating this from t_1 to t_2, we get

$$A(t_2) - A(t_1) = \tfrac{1}{2}h(t_2 - t_1). \tag{12}$$

This yields Kepler's second law: The line segment joining the sun to a planet sweeps out equal areas in equal intervals of time.

CENTRAL GRAVITATIONAL FORCES AND KEPLER'S FIRST LAW

We now specialize even further, and assume that $\mathbf{F}$ is a central attractive force whose magnitude is given by the inverse square law (1), so that

$$F_r = -G\frac{Mm}{r^2}. \tag{13}$$

If we write (13) in the slightly simpler form

$$F_r = -\frac{km}{r^2}$$

where $k = GM$, then the second of equations (10) becomes

$$\frac{d^2r}{dt^2} - r\left(\frac{d\theta}{dt}\right)^2 = -\frac{k}{r^2}. \tag{14}$$

The next step in this line of thought is difficult to motivate, because it involves considerable technical ingenuity, but we will try. Our purpose is to use the differential equation (14) to obtain the equation of the orbit in the polar form $r = f(\theta)$, so we want to eliminate t from (14) and consider θ as the independent variable. Also, we want r to be the dependent variable, but if (11) is used to put (14) in the form

$$\frac{d^2r}{dt^2} - \frac{h^2}{r^3} = -\frac{k}{r^2}, \tag{15}$$

then the presence of powers of $1/r$ suggests that it might be temporarily convenient to introduce a new dependent variable $z = 1/r$.

To accomplish these various aims, we must first express d^2r/dt^2 in terms of $d^2z/d\theta^2$, by calculating

$$\frac{dr}{dt} = \frac{d}{dt}\left(\frac{1}{z}\right) = -\frac{1}{z^2}\frac{dz}{dt} = -\frac{1}{z^2}\frac{dz}{d\theta}\frac{d\theta}{dt}$$

$$= -\frac{1}{z^2}\frac{dz}{d\theta}\frac{h}{r^2} = -h\frac{dz}{d\theta}$$

and

$$\frac{d^2r}{dt^2} = -h\frac{d}{dt}\left(\frac{dz}{d\theta}\right) = -h\frac{d}{d\theta}\left(\frac{dz}{d\theta}\right)\frac{d\theta}{dt}$$

$$= -h\frac{d^2z}{d\theta^2}\frac{h}{r^2} = -h^2z^2\frac{d^2z}{d\theta^2}.$$

When the latter expression is inserted in (15), and r is replaced by $1/z$, we get

$$-h^2z^2\frac{d^2z}{d\theta^2} - h^2z^3 = -kz^2$$

or

$$\frac{d^2z}{d\theta^2} + z = \frac{k}{h^2}. \tag{16}$$

To solve this equation, we observe that, except for the constant term on the right, it is the differential equation of simple harmonic motion discussed in Section 9.6. To eliminate the constant term, we put

$$w = z - \frac{k}{h^2},$$

so that $d^2w/d\theta^2 = d^2z/d\theta^2$ and (16) becomes

$$\frac{d^2w}{d\theta^2} + w = 0.$$

As we know, the general solution of this familiar equation is

$$w = A\sin\theta + B\cos\theta,$$

so

$$z = A\sin\theta + B\cos\theta + \frac{k}{h^2}. \tag{17}$$

For the sake of simplicity, we now shift the direction of the polar axis in such a way that r is minimal (that is, m is closest to the origin) when $\theta = 0$. This means that z is to be maximal in this direction, so

$$\frac{dz}{d\theta} = 0 \qquad \text{and} \qquad \frac{d^2z}{d\theta^2} < 0$$

when $\theta = 0$. By calculating $dz/d\theta$ and $d^2z/d\theta^2$ from (17), we easily see that these conditions imply that $A = 0$ and $B > 0$. If we now replace z by $1/r$, then (17) can be written as

$$r = \frac{1}{k/h^2 + B\cos\theta} = \frac{h^2/k}{1 + (Bh^2/k)\cos\theta};$$

and if we put $e = Bh^2/k$, then our equation for the orbit becomes

$$r = \frac{h^2/k}{1 + e\cos\theta}, \tag{18}$$

where e is a positive constant.

We recall from Section 16.3 that (18) is the polar equation of a conic section with focus at the origin and vertical directrix to the right; and furthermore, that this conic section is an ellipse, a parabola, or a hyperbola according as $e < 1$, $e = 1$, or $e > 1$. Since the planets remain in the solar system and do not move infinitely far away from the sun, the ellipse is the only possibility. This yields Kepler's first law: The orbit of each planet is an ellipse with the sun at one focus.*

KEPLER'S THIRD LAW

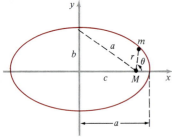

Figure 17.53

We now restrict ourselves to the case in which m has an elliptic orbit (Fig. 17.53) whose polar and rectangular equations are (18) and

$$\frac{x^2}{a^2} + \frac{y^2}{b^2} = 1.$$

We know that $e = c/a$ and $c^2 = a^2 - b^2$, so $e^2 = (a^2 - b^2)/a^2$ and

$$b^2 = a^2(1 - e^2). \tag{19}$$

In astronomy the semimajor axis a of the elliptical orbit is called the *mean distance*, because it is one-half the sum of the least and greatest values of r. These are the values of r corresponding to $\theta = 0$ and $\theta = \pi$ in (18), so by (18) and (19) we have

$$a = \frac{1}{2}\left(\frac{h^2/k}{1 + e} + \frac{h^2/k}{1 - e}\right) = \frac{h^2}{k(1 - e^2)} = \frac{h^2 a^2}{k b^2},$$

which yields

$$b^2 = \frac{h^2 a}{k}. \tag{20}$$

If T is the period of m (that is, the time required for one complete revolution in its orbit), then, since the area of the ellipse is πab, it follows from (12) that $\pi ab = \frac{1}{2}hT$, so $T = 2\pi ab/h$. By using (20), we now obtain

$$T^2 = \frac{4\pi^2 a^2 b^2}{h^2} = \left(\frac{4\pi^2}{k}\right)a^3. \tag{21}$$

Since the constant $k = GM$ depends on the central attracting mass M but not on m, (21) holds for all the planets in our solar system and we have Kepler's third law: The squares of the periods of revolution of the planets are proportional to the cubes of their mean distances.

As we explained in Section 15.3, the standard unit of distance among astronomers who work with the solar system is the *astronomical unit*. This is the mean distance from the earth to the sun, which is approximately

---

* In the discussion of equation (17) we have ignored the possibility that r might have a constant value and therefore not be minimal in any direction, so that z has a constant value and is not maximal in any direction. This happens when both $A = 0$ and $B = 0$, so that $z = k/h^2$ and $r = h^2/k$. Under these circumstances we have a circular orbit with radius h^2/k, and this can be included under equation (18) by allowing the possibility that $e = 0$. However, we saw in Section 17.6 that a circular orbit of given radius requires a certain precise orbital speed, which is infinitely unlikely for an actual planet and can be disregarded as a genuine possibility.

93,000,000 mi or 150,000,000 km. Equation (21) takes the more convenient form

$$T^2 = a^3 \qquad (22)$$

when time is measured in years and distance in astronomical units. The reason for this, of course, is that 1 year is by definition the period of revolution of the earth in its orbit, so that with these units of measurement $T = 1$ when $a = 1$.

We would like to point out that the mathematical theory discussed in this section is just the beginning of what Newton accomplished, and constitutes only a first approximation to the full story of planetary motion. For instance, we have assumed that only the sun and *one* planet are present. But actually, of course, all the other planets are present as well, and each exerts its own independent gravitational force on the planet under consideration. These additional influences introduce what are called "perturbations" into the idealized elliptical orbit derived here, and the main purpose of the science of celestial mechanics is to take all these complexities into account. One of the great events of nineteenth century astronomy arose in just this way, namely, the discovery of the planet Neptune by Adams and Leverrier, through their attempts to explain the relatively large deviations of Uranus from its Keplerian orbit.*

Also, we have assumed that the sun and planet under discussion are particles, that is, points at which mass is concentrated. In fact, of course, they are extended bodies with substantial dimensions. One of Newton's most remarkable achievements was to prove that the sun and planets behave like particles under the inverse square law of attraction. We will prove this statement ourselves in Chapter 20, where we study three-dimensional integrals.

Newton's enormous success revived and greatly intensified the almost-forgotten Greek belief that it is possible to understand the universe in a rational way. This new confidence in its own intellectual powers permanently altered humanity's perception of itself, and over the past 300 years almost every department of human life has felt its consequences.

---

* For the details of this dramatic story, see pp. 820–839 of *The World of Mathematics*, ed. James R. Newman, Simon and Schuster, 1956.

PROBLEMS

1 Newton himself did not know the value of the constant of gravitation G. This was determined by means of a classic experiment in 1789 by the English scientist Henry Cavendish. Once G is known, explain how equation (21), written in the form

$$T^2 = \left(\frac{4\pi^2}{GM} \right) a^3,$$

might be used to calculate the mass of the sun.

2 What is the period of revolution T (in years) of a planet whose mean distance from the sun is
(a) twice that of the earth?
(b) three times that of the earth?
(c) twenty-five times that of the earth?

*3 Kepler's first two laws, in the form of equations (11) and (18), imply that m is attracted toward the origin with a force whose magnitude is inversely proportional to the square of r. This was Newton's fundamental discovery, for it caused him to propound his law of gravitation and investigate its consequences. Prove this by assuming (11) and (18) and verifying the following statements:

(a) $F_\theta = 0$;

(b) $\dfrac{dr}{dt} = \dfrac{ke}{h}\sin\theta$;

(c) $\dfrac{d^2r}{dt^2} = \dfrac{ke\cos\theta}{r^2}$;

(d) $F_r = -\dfrac{mk}{r^2} = -G\dfrac{Mm}{r^2}$.

*4 Use formula (17) to show that the speed v of a planet at any point of its orbit is given by

$$v^2 = k\left(\frac{2}{r} - \frac{1}{a}\right).$$

5 Suppose that the earth explodes into fragments which fly off at the same speed in different directions into orbits of their own. Use Kepler's third law and the result of Problem 4 to show that all fragments that do not fall into the sun or escape from the solar system will reunite later at the same point.

ADDITIONAL PROBLEMS FOR CHAPTER 17

SECTION 17.1

1 Find the area of the loop of the folium of Descartes shown in Fig. 17.11. Hint: Use the polar equation of the folium and evaluate the area integral with the aid of the substitution $u = \tan\theta$.

2 Find parametric equations for the right loop of the lemniscate $r^2 = 2a^2\cos 2\theta$ by using the slope of the radial line $t = y/x$ as parameter. How can the left loop be represented?

SECTION 17.2

3 Consider the cycloid discussed in Section 17.2.
   (a) Find the volume of the solid generated by revolving the region under one arch about the x-axis.
   (b) Find the area of the surface generated by revolving one arch about the x-axis.

*4 Let a be fixed and consider the hypocycloid of n cusps, so that $a = nb$. Find the total length L_n of this curve, and also the limit approached by L_n as $n \to \infty$.

5 Find the length of one arch of the epicycloid generated by a circle of radius b rolling on the outside of a fixed circle of radius a.

*6 Let a be fixed and consider the epicycloid of n cusps, so that $a = nb$. Find the total length L_n of this curve, and also the limit approached by L_n as $n \to \infty$.

*7 Consider an ideal pendulum consisting of a particle of mass m at the end of a weightless string of length L (Fig. 17.54). If it is pulled aside through an angle α and released, show that its period of oscillation T can be expressed in the form

---

* Newton discovered this length, and obtained Wren's Theorem (Example 2 in Section 17.2) by letting $a \to \infty$. See Book I of the *Principia*, Prop. 48 and Cor. 2 to Prop. 52. It is interesting to try to bridge the gap between Newton's language and our own.

$$T = 4\sqrt{\frac{L}{g}}\int_0^1 \frac{du}{\sqrt{(1-u^2)(1-k^2u^2)}},$$

where $k = \sin\frac{1}{2}\alpha$ and $u = (1/k)\sin\frac{1}{2}\theta$. Hint: $x = L\sin\theta$ and $y = L\cos\theta$, and $\frac{1}{2}mv^2 = mg(L\cos\theta - L\cos\alpha)$. This integral is called a *complete elliptic integral of the first kind*, and it cannot be evaluated by means of elementary functions. When α is small, so that k^2 is very small, we have the approximation

$$T \cong 4\sqrt{\frac{L}{g}}\sin^{-1}u\Big]_0^1 = 2\pi\sqrt{\frac{L}{g}},$$

as in Example 3 in Section 9.6.

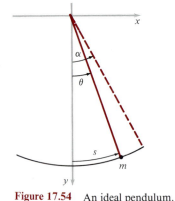

Figure 17.54 An ideal pendulum.

*8 Consider a wire bent into the shape of a cycloid with parametric equations $x = a(\theta - \sin\theta)$ and $y = a(1 - \cos\theta)$, and invert it as in Fig. 17.15. If a bead is released on the wire and slides without friction under the influence of gravity alone, show that its velocity v satisfies the equation

$$4av^2 = g(s_0^2 - s^2),$$

where s_0 and s are the arc lengths from the lowest point to the bead's initial position and its position at any later time, respectively. By differentiation obtain the equation

$$\frac{d^2s}{dt^2} + \frac{g}{4a}s = 0,$$

which shows that the bead moves in simple harmonic motion. Use the ideas of Section 9.6 to find s as a function of t, determine the period of the motion, and observe that this establishes in another way the tautochrone property of the cycloid proved in Section 17.2.

SECTION 17.3

9 Use vector methods to show that the line joining the midpoints of two sides of a triangle is parallel to the third side and half its length.

10 Generalize Problem 9 by using vector methods to show that the line joining the midpoints of the nonparallel sides of a trapezoid is parallel to the parallel sides and half the sum of their lengths.

SECTION 17.5

11 Locate the points on the curve $y = \frac{1}{4}x^4$ where the radius of curvature is smallest. What is this smallest radius?

12 Show that the radius of curvature at any point (x, y) on the hypocycloid of four cusps $x^{2/3} + y^{2/3} = a^{2/3}$ is three times the distance from the origin to the line which is tangent to the curve at (x, y).

SECTION 17.7

13 The Cavendish value for G is 6.7×10^{-8} cm$^3 \cdot$g$^{-1} \cdot$s^{-2} when mass is measured in grams, distance in centimeters, and time in seconds. In Example 2 in Section 17.6 we used the fact that $GM_e = gR^2$, where M_e is the mass of the earth. Calculate M_e (approximately) in grams by using the values $g = 980$ cm/s^2 and $R = 6.37 \times 10^8$ cm.

14 With the notation of Section 17.7, the inverse square law can be written as

$$\mathbf{F} = -\frac{km}{r^2}\mathbf{u}_r,$$

where $k = GM$, and since $m\mathbf{a} = \mathbf{F}$, we have

$$\frac{d\mathbf{v}}{dt} = -\frac{k}{r^2}\mathbf{u}_r.$$

Verify the following steps to obtain another derivation of Kepler's first law from the inverse square law:

(a) Use (6) and (11) to write

$$\frac{d\mathbf{v}}{dt} = -\frac{k}{r^2}\left(-\frac{1}{d\theta/dt}\frac{d\mathbf{u}_\theta}{dt}\right) = \frac{k}{h}\frac{d\mathbf{u}_\theta}{dt}.$$

(b) Integrate the equation in (a) to obtain

$$\mathbf{v} = \frac{k}{h}\mathbf{u}_\theta + \left(v_0 - \frac{k}{h}\right)\mathbf{j},$$

by assuming initial conditions in the following form (Fig. 17.55): At $t = 0$, m has its closest approach to the origin and crosses the polar axis at the point $\mathbf{R} = r_0\mathbf{i}$ with velocity $\mathbf{v} = v_0\mathbf{j}$.

(c) Equate $\mathbf{u}_\theta$-components of the equation in (b) to obtain

$$r\frac{d\theta}{dt} = \frac{k}{h} + \left(v_0 - \frac{k}{h}\right)\cos\theta,$$

and use (11) to write this in the form

$$\frac{h}{r} = \frac{k}{h} + \left(v_0 - \frac{k}{h}\right)\cos\theta.$$

(d) Solve the equation in (c) for r to obtain

$$r = \frac{h^2/k}{1 + (v_0h/k - 1)\cos\theta}.$$

(e) Use (7) and (11) to show that $h = r_0v_0$, and write the equation in (d) in the form

$$r = \frac{(r_0v_0)^2/k}{1 + (r_0v_0^2/k - 1)\cos\theta} = \frac{(r_0v_0)^2/k}{1 + e\cos\theta},$$

where $e = r_0v_0^2/k - 1$.

(f) Observe that the equation in (e) represents

A circle if $r_0v_0^2 = GM$;
An ellipse if $GM < r_0v_0^2 < 2GM$;
A parabola if $r_0v_0^2 = 2GM$;
A hyperbola if $r_0v_0^2 > 2GM$.

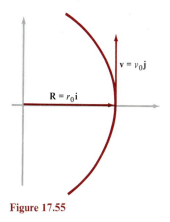

Figure 17.55

18

VECTORS IN THREE-DIMENSIONAL SPACE. SURFACES

18.1
COORDINATES AND VECTORS IN THREE-DIMENSIONAL SPACE

In the preceding seventeen chapters we have discussed many aspects of the calculus of functions of a *single* variable. The geometry of these functions is two-dimensional because the graph of a function of a single variable is a curve in the plane. Most of the remainder of this book is concerned with the calculus of functions of *several* (two or more) independent variables. The geometry of functions of two variables is three-dimensional, because in general the graph of such a function is a curved surface in space.

In this chapter we discuss the analytic geometry of three-dimensional space. Our treatment will emphasize vector algebra, partly because this approach provides a more direct and intuitive understanding of the equations of lines and planes, and partly because the concepts of dot and cross products as developed in the next two sections are indispensable in many other parts of mathematics and physics.

Rectangular coordinates in the plane can be generalized in a natural way to rectangular coordinates in space. The position of a point in space is described by giving its location relative to three mutually perpendicular *coordinate axes* passing through the *origin O*. We always draw the x-, y-, and z-axes as shown in Fig. 18.1, with equal units of length on all three axes and

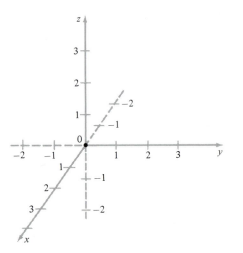

Figure 18.1 Coordinate axes.

550

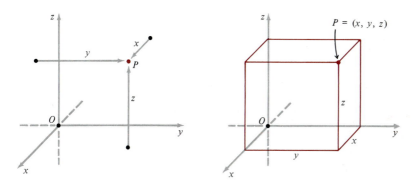

Figure 18.2   Locating a point by its rectangular coordinates.

with arrows indicating the positive directions. Each pair of axes determines a *coordinate plane*: the *x*-axis and *y*-axis determine the *xy*-plane, etc. The configuration of axes in this figure is called *right-handed,* because if the thumb of the right hand points in the direction of the positive *z*-axis, then the curl of the fingers gives the positive direction of rotation in the *xy*-plane, from the positive *x*-axis to the positive *y*-axis.

Since many people have trouble visualizing space figures from plane drawings, we point out that Fig. 18.1 can be thought of as part of a rectangular room drawn in perspective, with the origin *O* at the far left corner of the floor. The *xy*-plane is the floor, and has the normal appearance of the *xy*-plane if we look down on it from a point on the positive *z*-axis; the *yz*-plane is the back wall of the room, in the plane of the paper; and the *xz*-plane is the wall on the left side of the room.

A point *P* in space (see Fig. 18.2) is said to have *rectangular* (or *Cartesian*) *coordinates x, y, z* if

x is its signed distance from the *yz*-plane;

y is its signed distance from the *xz*-plane;

z is its signed distance from the *xy*-plane.

Just as in plane analytic geometry, we write $P = (x, y, z)$ and identify the point *P* with the ordered triple of its coordinates. On the right in the figure we attempt to strengthen the illusion of three dimensions by completing the box that has *O* and *P* as opposite vertices.*

The three coordinate planes divide all of space into eight cells called *octants.* The cell emphasized in Fig. 18.2, where *x*, *y*, and *z* are all positive numbers, is called the *first octant.* (No one bothers to number the other seven octants.)

Even before plunging into a general study of the equations of lines and planes in Section 18.4, we can notice a few obvious facts. The *xy*-plane is the set of all points $(x, y, 0)$; it consists precisely of those points in space whose *z*-coordinate is 0, so its equation is

$$z = 0.$$

---

* The technical term for the object shown on the right is "rectangular parallelepiped." We prefer the simpler word "box."

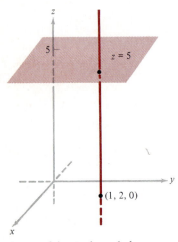

Figure 18.3   Horizontal plane.

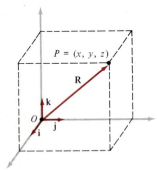

Figure 18.4   The position vector of P.

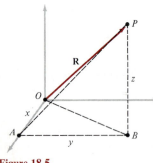

Figure 18.5

Similarly, the equation of the yz-plane is $x = 0$, and the equation of the xz-plane is $y = 0$.

The z-axis is the set of all points $(0, 0, z)$. It is therefore represented by the pair of equations

$$x = 0, \qquad y = 0. \tag{1}$$

These are the equations of the yz-plane and the xz-plane, respectively, so equations (1) taken together characterize the z-axis as the intersection of these two coordinate planes. Similarly, the equations of the x-axis are $y = 0$, $z = 0$; and the equations of the y-axis are $x = 0$, $z = 0$.

There is nothing special about the number 0 in these remarks. For instance, the equation of the horizontal plane 5 units above the xy-plane is $z = 5$; and the equations of the vertical line that passes through the point $(1, 2, 0)$ in the xy-plane are $x = 1$, $y = 2$. See Fig. 18.3.

Almost all of the ideas about vectors that were presented in Section 17.3 are valid in three-dimensional space and require no further discussion. This remark applies to the concept of a vector, to the definition of equality for vectors, and to the definitions of addition and scalar multiplication. In all this material there is no need at all to suppose that the vectors lie in a plane.

The only real difference is that a vector in space has three components rather than two. In computing with vectors in the plane, we used the unit vectors $\mathbf{i}$ and $\mathbf{j}$ in the positive x- and y-directions. In order to compute with vectors in three-dimensional space, we introduce a third unit vector $\mathbf{k}$ in the positive z-direction, as shown in Fig. 18.4. If $P = (x, y, z)$ is any point in space, the position vector $\mathbf{R} = \overrightarrow{OP}$ can be written in the form

$$\mathbf{R} = x\mathbf{i} + y\mathbf{j} + z\mathbf{k},$$

and the numbers x, y, and z are called its $\mathbf{i}$-, $\mathbf{j}$-, and $\mathbf{k}$-*components*.

The *length* of the vector $\mathbf{R}$ is given by the formula

$$|\mathbf{R}| = \sqrt{x^2 + y^2 + z^2}. \tag{2}$$

This can be proved by a double application of the theorem of Pythagoras, as illustrated in Fig. 18.5:

$$|\mathbf{R}|^2 = OP^2 = OB^2 + BP^2$$
$$= OA^2 + AB^2 + BP^2$$
$$= x^2 + y^2 + z^2.$$

If $P_1 = (x_1, y_1, z_1)$ and $P_2 = (x_2, y_2, z_2)$ are any two points in space (Fig. 18.6), the distance between them is the length of the vector $\overrightarrow{P_1P_2}$ from P_1 to P_2. Since

$$\overrightarrow{P_1P_2} = \mathbf{R}_2 - \mathbf{R}_1 = (x_2\mathbf{i} + y_2\mathbf{j} + z_2\mathbf{k}) - (x_1\mathbf{i} + y_1\mathbf{j} + z_1\mathbf{k})$$
$$= (x_2 - x_1)\mathbf{i} + (y_2 - y_1)\mathbf{j} + (z_2 - z_1)\mathbf{k},$$

we can use (2) to obtain

$$|\overrightarrow{P_1P_2}| = \sqrt{(x_2 - x_1)^2 + (y_2 - y_1)^2 + (z_2 - z_1)^2}. \tag{3}$$

This is the important *distance formula*; it has many uses.

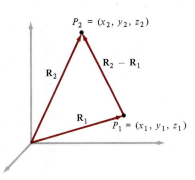

Figure 18.6

Since a sphere is the set of all points P at a given distance r from a given fixed point P_0 (Fig. 18.7), the equation of a sphere can be written as

$$|\overrightarrow{P_0P}| = r. \tag{4}$$

If $P = (x, y, z)$ and $P_0 = (x_0, y_0, z_0)$, then (3) enables us to write (4) in the equivalent form

$$(x - x_0)^2 + (y - y_0)^2 + (z - z_0)^2 = r^2. \tag{5}$$

This is the standard equation of the sphere with center $P_0 = (x_0, y_0, z_0)$ and radius r.

Example   If we complete the squares in the equation

$$x^2 + y^2 + z^2 + 4x - 2y - 6z + 8 = 0, \tag{6}$$

it becomes

$$(x + 2)^2 + (y - 1)^2 + (z - 3)^2 = 6.$$

By comparing this with (5), we see at once that (6) is the equation of the sphere with center $(-2, 1, 3)$ and radius $\sqrt{6}$.

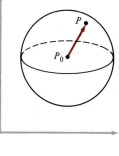

Figure 18.7   Sphere.

PROBLEMS

1   Sketch the box with the vertices $(1, -1, 0)$, $(1, 4, 0)$, $(-2, 4, 0), (-2, -1, 0), (1, -1, 5), (1, 4, 5), (-2, 4, 5)$, $(-2, -1, 5)$. Write down the equations of the faces and edges of the box that pass through the vertex $(1, 4, 5)$.

2   Sketch the box bounded by the planes $x = 1$, $x = 3$, $y = 0$, $y = 4$, $z = 1$, $z = 5$. Write down the vertices.

3   Sketch the tetrahedron whose base vertices are $(3, 4, 0)$, $(3, -4, 0)$, and $(-5, 4, 0)$ and whose fourth vertex is $(0, 0, 6)$. Use the fact that the volume of a tetrahedron is one-third the area of the base times the height to find the volume of this tetrahedron.

4   Sketch the straight lines whose equations are given:
(a) $x = 2$, $z = 3$;   (b) $y = 1$, $z = 4$;
(c) $x = -3$, $y = 1$.

5   Describe the graph of the equation
(a) $xy = 0$;   (b) $xyz = 0$.

6   Describe and sketch the locus of all points $P = (x, y, z)$ that satisfy the given pairs of simultaneous equations:
(a) $x^2 + y^2 = 4$, $z = 3$;
(b) $x = 4$, $z = 4y^2$;
(c) $y = x$, $x = 5$;
(d) $z = -x^2$, $y = 0$.

7   Find the point on the y-axis which is equidistant from $(2, 5, -3)$ and $(-3, 6, 1)$.

8   Find and simplify the equation of the locus of all points that are equidistant from $(7, 0, -4)$ and $(-3, 2, 2)$. Describe this locus in geometric language.

9   Write the equation of the sphere with radius 7 and center on the positive z-axis, if the sphere is tangent to the plane $z = 0$.

10 Find the equation of the sphere with center $(3, -2, 5)$ which is

(a) tangent to the xy-plane;

(b) tangent to the yz-plane;

(c) tangent to the xz-plane.

11 Identify the graph of each of the following equations, and if it is a sphere, give its center and radius:

(a) $x^2 + y^2 + z^2 + 2x - 6y - 10z + 26 = 0$;

(b) $x^2 + y^2 + z^2 - 10x + 2y - 6z + 35 = 0$;

(c) $2x^2 + 2y^2 + 2z^2 - 16x + 8y + 4z + 49 = 0$;

(d) $x^2 + y^2 + z^2 + 2x - 14y - 6z + 59 = 0$;

(e) $4x^2 + 4y^2 + 4z^2 - 16x + 24y + 51 = 0$.

12 A point P moves in such a way that it is always twice as far from $(3, 2, 0)$ as from $(3, 2, 6)$. Show that the locus of P is a sphere and find its center and radius.

13 If $P_1 = (x_1, y_1, z_1)$ and $P_2 = (x_2, y_2, z_2)$, use vectors to show that the coordinates of the midpoint of the segment P_1P_2 are

$$x = \tfrac{1}{2}(x_1 + x_2), \qquad y = \tfrac{1}{2}(y_1 + y_2), \qquad \text{and}$$

$$z = \tfrac{1}{2}(z_1 + z_2).$$

14 Find the equation of the sphere that has the two given points as ends of a diameter:

(a) $(6, 2, -1), (-2, 4, 3)$;   (b) $(0, 1, -7), (-6, 7, 3)$.

15 Show that the triangle with vertices $(4, 3, 6)$, $(-2, 0, 8)$, and $(1, 5, 0)$ is a right triangle. Find its area.

16 For each of the following pairs of points, find the vector from the first point to the second, and also the distance between the points:

(a) $(2, 0, -3), (5, 1, 2)$;   (b) $(-2, 1, 7), (1, -4, 2)$;

(c) $(8, 3, 6), (2, -2, 0)$;   (d) $(1, 5, -3), (9, 7, 1)$.

17 Find the vector from the origin O to the intersection of the medians of the triangle whose vertices are $A = (3, 2, 2)$, $B = (-1, 0, 4)$, and $C = (5, 3, -2)$.

18 If $\mathbf{A}, \mathbf{B}, \mathbf{C}$ are any three distinct vectors, their endpoints form a triangle. Find the position vector of the intersection of the medians of this triangle.

*19 If $\mathbf{A}, \mathbf{B}, \mathbf{C}, \mathbf{D}$ are any four distinct vectors, their endpoints form a tetrahedron. Show that the four lines joining each vertex to the intersection of the medians of the opposite face are concurrent, and find the position vector of their common point.

18.2
THE DOT PRODUCT OF TWO VECTORS

Up to this point in our work we have not defined the product of two vectors $\mathbf{A}$ and $\mathbf{B}$. There are two different ways of doing this, both of which have important uses in geometry and physics. Since there is no reason to choose one of these definitions in preference to the other, we keep both, using a dot for one definition and a cross for the other. The *dot product* (or *scalar product*) of $\mathbf{A}$ and $\mathbf{B}$ is denoted by $\mathbf{A} \cdot \mathbf{B}$ and is a number. The *cross product* (or *vector product*) is denoted by $\mathbf{A} \times \mathbf{B}$ and is a vector. These two kinds of multiplication are totally different. We discuss the first in this section and the second in Section 18.3.

The *dot product* $\mathbf{A} \cdot \mathbf{B}$ of two vectors $\mathbf{A}$ and $\mathbf{B}$ is defined to be the product of their lengths and the cosine of the angle between them. This definition can be written as

$$\mathbf{A} \cdot \mathbf{B} = |\mathbf{A}|\,|\mathbf{B}| \cos \theta, \tag{1}$$

where θ ($0 \le \theta \le \pi$) is the angle between $\mathbf{A}$ and $\mathbf{B}$ when they are placed so that their tails coincide (Fig. 18.8). It is clear from the definition that $\mathbf{A} \cdot \mathbf{B}$ is a *scalar* (or number), not a vector.

As Fig. 18.8 shows, the number $|\mathbf{B}| \cos \theta$ is the *scalar projection* of $\mathbf{B}$ on $\mathbf{A}$, denoted by $\text{proj}_{\mathbf{A}} \mathbf{B}$. Definition (1) can therefore be interpreted geometrically as follows:

$$\mathbf{A} \cdot \mathbf{B} = |\mathbf{A}|(|\mathbf{B}| \cos \theta) = |\mathbf{A}|\,\text{proj}_{\mathbf{A}} \mathbf{B}$$

$$= (\text{length of } \mathbf{A}) \times (\text{scalar projection of } \mathbf{B} \text{ on } \mathbf{A}).$$

By interchanging the roles of $\mathbf{A}$ and $\mathbf{B}$, we also have

$$\mathbf{A} \cdot \mathbf{B} = |\mathbf{B}|(|\mathbf{A}| \cos \theta) = |\mathbf{B}|\,\text{proj}_{\mathbf{B}} \mathbf{A}$$

$$= (\text{length of } \mathbf{B}) \times (\text{scalar projection of } \mathbf{A} \text{ on } \mathbf{B}).$$

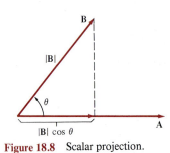

Figure 18.8   Scalar projection.

The *vector projection* of **B** on **A** is also indicated in the figure. Both types of projections are useful in applications.

It is easy to see from the definition (1) that the dot product has the properties

$$\mathbf{A} \cdot \mathbf{B} = \mathbf{B} \cdot \mathbf{A}, \qquad \text{the commutative law,} \tag{2}$$

and

$$(c\mathbf{A}) \cdot \mathbf{B} = c(\mathbf{A} \cdot \mathbf{B}) = \mathbf{A} \cdot (c\mathbf{B}). \tag{3}$$

It also has the property

$$\mathbf{A} \cdot (\mathbf{B} + \mathbf{C}) = \mathbf{A} \cdot \mathbf{B} + \mathbf{A} \cdot \mathbf{C}, \qquad \text{the distributive law,} \tag{4}$$

but this is not quite as evident as (2) and (3). To establish (4), we observe from Fig. 18.9 that

$$\begin{aligned}
\mathbf{A} \cdot (\mathbf{B} + \mathbf{C}) &= |\mathbf{A}|[\text{proj}_{\mathbf{A}}(\mathbf{B} + \mathbf{C})] \\
&= |\mathbf{A}|(\text{proj}_{\mathbf{A}} \mathbf{B} + \text{proj}_{\mathbf{A}} \mathbf{C}) \\
&= |\mathbf{A}|\text{proj}_{\mathbf{A}} \mathbf{B} + |\mathbf{A}|\text{proj}_{\mathbf{A}} \mathbf{C} \\
&= \mathbf{A} \cdot \mathbf{B} + \mathbf{A} \cdot \mathbf{C}.
\end{aligned}$$

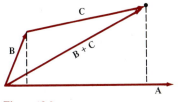

Figure 18.9

If we combine (4) with the commutative law (2), we also have

$$(\mathbf{A} + \mathbf{B}) \cdot \mathbf{C} = \mathbf{A} \cdot \mathbf{C} + \mathbf{B} \cdot \mathbf{C}. \tag{5}$$

Properties (4) and (5) permit us to multiply out sums of vectors by the ordinary procedures of elementary algebra, as in

$$(\mathbf{A} + \mathbf{B}) \cdot (\mathbf{C} + \mathbf{D}) = \mathbf{A} \cdot \mathbf{C} + \mathbf{A} \cdot \mathbf{D} + \mathbf{B} \cdot \mathbf{C} + \mathbf{B} \cdot \mathbf{D}.$$

Another simple consequence of the definition (1) is the fact that

$$\mathbf{A} \cdot \mathbf{A} = |\mathbf{A}|^2 \tag{6}$$

for any vector **A**.

Example 1   In the notation of Fig. 18.10, the cosine law of trigonometry states that

$$c^2 = a^2 + b^2 - 2ab \cos \theta.$$

This can be proved very easily by using property (6) to write

$$\begin{aligned}
c^2 = |\mathbf{C}|^2 = |\mathbf{A} - \mathbf{B}|^2 &= (\mathbf{A} - \mathbf{B}) \cdot (\mathbf{A} - \mathbf{B}) \\
&= \mathbf{A} \cdot \mathbf{A} + \mathbf{B} \cdot \mathbf{B} - 2\mathbf{A} \cdot \mathbf{B} \\
&= |\mathbf{A}|^2 + |\mathbf{B}|^2 - 2\mathbf{A} \cdot \mathbf{B} \\
&= a^2 + b^2 - 2ab \cos \theta.
\end{aligned}$$

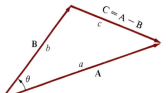

Figure 18.10

If we apply the definition (1) to the mutually perpendicular unit vectors **i**, **j**, and **k** introduced in Section 18.1, we obtain

$$\begin{aligned}
\mathbf{i} \cdot \mathbf{i} = \mathbf{j} \cdot \mathbf{j} = \mathbf{k} \cdot \mathbf{k} &= 1, \\
\mathbf{i} \cdot \mathbf{j} = \mathbf{i} \cdot \mathbf{k} = \mathbf{j} \cdot \mathbf{k} &= 0.
\end{aligned} \tag{7}$$

These facts enable us to find a convenient formula for computing the dot product of any two vectors given in $\mathbf{i}, \mathbf{j}, \mathbf{k}$ form,

$$\mathbf{A} = a_1\mathbf{i} + a_2\mathbf{j} + a_3\mathbf{k} \qquad \text{and} \qquad \mathbf{B} = b_1\mathbf{i} + b_2\mathbf{j} + b_3\mathbf{k}.$$

If we expand $\mathbf{A} \cdot \mathbf{B}$ by using (7) together with the general properties previously discussed, we get

$$\mathbf{A} \cdot \mathbf{B} = a_1b_1 + a_2b_2 + a_3b_3, \tag{8}$$

since six of the nine terms in the expansion vanish. Thus, to compute $\mathbf{A} \cdot \mathbf{B}$, we simply multiply their respective $\mathbf{i}$-, $\mathbf{j}$-, and $\mathbf{k}$-components, and add.

If $\mathbf{A}$ and $\mathbf{B}$ are nonzero vectors, the definition (1) can be written in the form

$$\cos \theta = \frac{\mathbf{A} \cdot \mathbf{B}}{|\mathbf{A}||\mathbf{B}|}. \tag{9}$$

This formula displays the main significance of the dot product in geometry: It provides a simple way to find the angle between two vectors and, in particular, to decide when two vectors are perpendicular. Indeed, if we agree that the zero vector is perpendicular to every vector, then by (9) we see at once that

$$\mathbf{A} \perp \mathbf{B} \qquad \text{if and only if} \qquad \mathbf{A} \cdot \mathbf{B} = 0.$$

Formula (8) makes it possible for us to use the dot product in these ways as a convenient computational tool.

Example 2   Find the cosine of the angle θ between the vectors $\mathbf{A} = \mathbf{i} + 2\mathbf{j} + 2\mathbf{k}$ and $\mathbf{B} = -3\mathbf{i} + 4\mathbf{j}$.

Solution   It is clear that

$$|\mathbf{A}| = \sqrt{1 + 4 + 4} = 3, \qquad |\mathbf{B}| = \sqrt{9 + 16} = 5, \qquad \mathbf{A} \cdot \mathbf{B} = -3 + 8 + 0 = 5.$$

Therefore by (9) we have

$$\cos \theta = \frac{\mathbf{A} \cdot \mathbf{B}}{|\mathbf{A}||\mathbf{B}|} = \frac{5}{3 \cdot 5} = \frac{1}{3}.$$

If we want the angle θ itself, we can use trigonometric tables (or a calculator) to find that $\theta \cong 70.5°$.

Example 3   Compute the cosine of the angle θ between $\mathbf{A}$ and $\mathbf{B}$ if $\mathbf{A} = \mathbf{i} - 2\mathbf{j} + 2\mathbf{k}$ and $\mathbf{B} = -\mathbf{i} + c\mathbf{k}$, and find a value of c for which $\mathbf{A} \perp \mathbf{B}$.

Solution   We have

$$|\mathbf{A}| = \sqrt{1 + 4 + 4} = 3, \qquad |\mathbf{B}| = \sqrt{1 + c^2}, \qquad \text{and} \qquad \mathbf{A} \cdot \mathbf{B} = -1 + 2c,$$

so

$$\cos \theta = \frac{\mathbf{A} \cdot \mathbf{B}}{|\mathbf{A}||\mathbf{B}|} = \frac{2c - 1}{3\sqrt{1 + c^2}}.$$

When $c = \frac{1}{2}$, this quantity has the value 0, and hence the vectors are perpendicular.

The simplest physical illustration of the use of the dot product is furnished by the concept of work. We recall that the work W done by a constant force F exerted along the line of motion in moving a particle through a distance d is given by $W = Fd$. But what if the force is a constant vector $\mathbf{F}$ pointing in some direction other than the line of motion from P to Q, as shown in Fig. 18.11? Only the vector component of $\mathbf{F}$ in the direction of the line of motion does work, so in this case we have

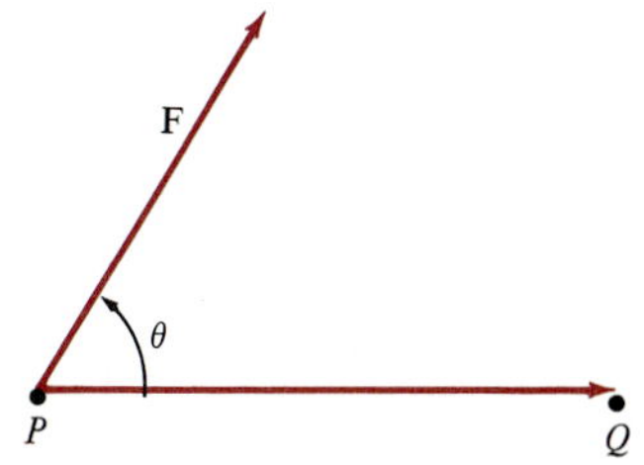

$$W = (|\mathbf{F}| \cos \theta)|\overrightarrow{PQ}|$$
$$= |\mathbf{F}||\overrightarrow{PQ}| \cos \theta = \mathbf{F} \cdot \overrightarrow{PQ},$$

that is,

$$W = \mathbf{F} \cdot \overrightarrow{PQ}. \tag{10}$$

Figure 18.11

In more advanced treatments of the physical uses of vectors, it is often necessary to calculate the work done by variable forces whose points of application move along curved paths, and formula (10) is the starting point for all such applications.

PROBLEMS

1 Show that the vectors

$$\mathbf{A} = \mathbf{i} + 3\mathbf{j} + 4\mathbf{k},$$
$$\mathbf{B} = 4\mathbf{i} + 4\mathbf{j} - 4\mathbf{k},$$

are perpendicular.

2 Show that

$$\mathbf{A} = \mathbf{i} - 2\mathbf{j} + \mathbf{k},$$
$$\mathbf{B} = \mathbf{i} - \mathbf{k},$$
$$\mathbf{C} = \mathbf{i} + \mathbf{j} + \mathbf{k},$$

are mutually perpendicular.

3 Find the angle between each of the given pairs of vectors:
   (a) $\mathbf{A} = \mathbf{i} + 2\mathbf{j} + \mathbf{k}$, $\mathbf{B} = -\mathbf{i} + \mathbf{j} + 2\mathbf{k}$;
   (b) $\mathbf{A} = \mathbf{i}$, $\mathbf{B} = \mathbf{i} + \mathbf{j}$;
   (c) $\mathbf{A} = 3\mathbf{i} + 4\mathbf{j}$, $\mathbf{B} = 4\mathbf{i} - 3\mathbf{j} + 9\mathbf{k}$.

4 Use dot products to show that the given three points are the vertices of a right triangle. Which is the vertex of the right angle?
   (a) $P = (1, 7, 3)$, $Q = (0, 7, -1)$, $R = (-1, 6, 2)$.
   (b) $P = (2, -5, -2)$, $Q = (-1, -2, 2)$, $R = (4, 1, -5)$.
   (c) $P = (2, 7, -2)$, $Q = (0, 4, -1)$, $R = (1, 4, 1)$.

5 Show that the vectors

$$\mathbf{A} = \mathbf{i} - 3\mathbf{j} - 5\mathbf{k},$$
$$\mathbf{B} = 2\mathbf{i} - \mathbf{j} + \mathbf{k},$$
$$\mathbf{C} = 3\mathbf{i} - 4\mathbf{j} - 4\mathbf{k},$$

form the sides of a right triangle if placed in the proper positions.

6 Find the angle θ between a diagonal of a cube and

   (a) an adjacent edge;
   (b) an adjacent diagonal of a face.

7 Let $\mathbf{A}$ be a nonzero vector, and suppose that $\mathbf{B}$ and $\mathbf{C}$ are two vectors such that $\mathbf{A} \cdot \mathbf{B} = \mathbf{A} \cdot \mathbf{C}$. Is it legitimate to cancel $\mathbf{A}$ from both sides of this equation and conclude that $\mathbf{B} = \mathbf{C}$? Explain.

8 Find a value of c for which the given vectors will be perpendicular:
   (a) $3\mathbf{i} - 2\mathbf{j} + 5\mathbf{k}$, $2\mathbf{i} + 4\mathbf{j} + c\mathbf{k}$;
   (b) $\mathbf{i} + \mathbf{j} + \mathbf{k}$, $\mathbf{i} + \mathbf{j} + c\mathbf{k}$.

9 If $a = |\mathbf{A}|$ and $b = |\mathbf{B}|$, show that the vector $b\mathbf{A} + a\mathbf{B}$ bisects the angle between $\mathbf{A}$ and $\mathbf{B}$.

10 With the notation of Problem 9, show that $b\mathbf{A} + a\mathbf{B}$ and $b\mathbf{A} - a\mathbf{B}$ are perpendicular.

11 Use the dot product to prove that an angle inscribed in a semicircle is a right angle. Hint: With the notation in Fig. 18.12, calculate $(\mathbf{A} + \mathbf{B}) \cdot (\mathbf{A} - \mathbf{B})$.

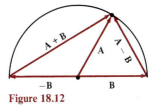

Figure 18.12

12 If $Q = (1, -1, 7)$, find the points $P = (0, c, c)$ on the line $z = y$ in the yz-plane such that the vector $\overrightarrow{OP}$ is perpendicular to the vector $\overrightarrow{PQ}$.

13 If $\mathbf{A} = \mathbf{i} + 2\mathbf{j} - 3\mathbf{k}$ and $\mathbf{B} = 4\mathbf{i} - 2\mathbf{k}$, find the vector component of $\mathbf{A}$ along $\mathbf{B}$. Solve this problem by finding a general formula for the vector component of **A** along **B** if **A** and **B** are any two vectors.

14 Use vector methods to show that the distance from a point (x_0, y_0) to a line $ax + by + c = 0$ (both in the xy-plane) is

$$\frac{|ax_0 + by_0 + c|}{\sqrt{a^2 + b^2}}.$$

15 Use property (6) to prove the *parallelogram law* of elementary geometry: The sum of the squares of the diagonals of a parallelogram equals the sum of the squares of the four sides. Hint: With the notation of Fig. 18.13, use (6) to expand $|\mathbf{A} + \mathbf{B}|^2 + |\mathbf{A} - \mathbf{B}|^2$.

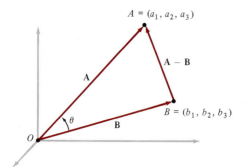

Figure 18.13   The parallelogram law.

16 For the triangle OAB in Fig. 18.14, the law of cosines states that

$$|\mathbf{A} - \mathbf{B}|^2 = |\mathbf{A}|^2 + |\mathbf{B}|^2 - 2|\mathbf{A}||\mathbf{B}| \cos \theta.$$

Give another proof of formula (8) by solving this equation for $|\mathbf{A}||\mathbf{B}| \cos \theta$ and simplifying the result.

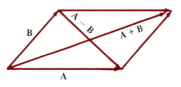

Figure 18.14

17 If a vector $\mathbf{V} = a\mathbf{i} + b\mathbf{j} + c\mathbf{k}$ makes angles α, β, and γ with the positive x-, y-, and z-axes (Fig. 18.15), then these angles are called the *direction angles* and $\cos \alpha$, $\cos \beta$, and $\cos \gamma$ are called the *direction cosines* of $\mathbf{V}$. Show that
(a) $(a\mathbf{i} + b\mathbf{j} + c\mathbf{k})/\sqrt{a^2 + b^2 + c^2}$ is a unit vector having the same direction as $\mathbf{V}$;

(b) $\cos \alpha = \dfrac{a}{\sqrt{a^2 + b^2 + c^2}}$,

$\cos \beta = \dfrac{b}{\sqrt{a^2 + b^2 + c^2}}$,

$\cos \gamma = \dfrac{c}{\sqrt{a^2 + b^2 + c^2}}$;

(c) $\cos^2 \alpha + \cos^2 \beta + \cos^2 \gamma = 1$.

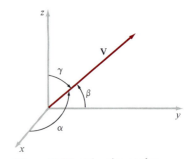

Figure 18.15   Direction angles.

18 How many lines through the origin make angles of $45°$ with both the positive x-axis and the positive y-axis?

19 How many lines through the origin make angles of $60°$ with both the positive x-axis and the positive y-axis? What angles do they make with the positive z-axis?

20 Find the work done by the force $\mathbf{F}$ when its point of application moves from P to Q:
(a) $\mathbf{F} = 2\mathbf{i} - 5\mathbf{j} + 3\mathbf{k}$, $P = (1, 2, -2)$, $Q = (3, -1, 1)$;
(b) $\mathbf{F} = 3\mathbf{i} + 2\mathbf{j} - 3\mathbf{k}$, $P = (-1, 2, 3)$, $Q = (1, 2, -1)$.

21 Find the work done by a force $\mathbf{F} = -c\mathbf{k}$ when its point of application moves from $P = (x_1, y_1, z_1)$ to $Q = (x_2, y_2, z_2)$.

22 Find the work done by a constant force $\mathbf{F}$ if its point of application moves around a closed polygonal path.

Many problems in geometry require us to find a vector that is perpendicular to each of two given vectors **A** and **B**. A routine way of doing this is provided by the *cross product* (or *vector product*) of **A** and **B**, denoted by **A** $\times$ **B**. This cross product is very different from the dot product **A·B**—for one thing, **A** $\times$ **B** is a vector, while **A·B** is a scalar. First we define this new product, then we describe its algebraic properties so that we can compute it with reasonable ease, and finally we illustrate some of its uses.

Consider two nonzero vectors **A** and **B**. Suppose that one of these vectors is translated, if necessary, so that their tails coincide, and let θ be the angle from **A** to **B** (*not* from **B** to **A**), with $0 \le \theta \le \pi$. If **A** and **B** are not parallel, so that $0 < \theta < \pi$, then these two vectors determine a plane, as shown in Fig. 18.16. We now choose the unit vector **n** which is normal (perpendicular) to this plane and whose direction is determined by the *right-hand thumb rule*. This means that if the right hand is placed so that the thumb is perpendicular to the plane of **A** and **B** and the fingers curl from **A** to **B** in the direction of the angle θ, then **n** points in the same direction as the thumb of this hand. This gives the direction of the vector **A** $\times$ **B** that we are defining. Not only do the vectors **A** and **B** determine the plane under consideration, but they also determine a parallelogram in this plane, of area $|\mathbf{A}||\mathbf{B}| \sin \theta$ (see Fig. 18.17). We take the area of this parallelogram as the magnitude of the vector **A** $\times$ **B**. With these preliminaries, we can now state the definition of the cross product of **A** and **B**, in this order, as follows:

$$\mathbf{A} \times \mathbf{B} = |\mathbf{A}||\mathbf{B}| \sin \theta \, \mathbf{n}. \tag{1}$$

Observe that if **A** or **B** is **0**, or if **A** and **B** are parallel, then they do not determine a plane, and hence the unit normal vector **n** is not defined. But in these cases $|\mathbf{A}| = 0$ or $|\mathbf{B}| = 0$, or $\sin \theta = 0$, so by (1) we have **A** $\times$ **B** = **0** and the determination of **n** is not necessary. If we agree that the zero vector is to be considered as parallel to every vector, then it is easy to see that

A is parallel to **B** if and only if **A** $\times$ **B** = **0**.

In particular, we have

$$\mathbf{A} \times \mathbf{A} = \mathbf{0}$$

for every **A**. If instead of **A** $\times$ **B** we consider **B** $\times$ **A**, then the direction of the angle θ is reversed, and we must flip the right hand over so that the thumb points in the opposite direction. This means that **n** is replaced by $-\mathbf{n}$, and therefore

$$\mathbf{B} \times \mathbf{A} = -\mathbf{A} \times \mathbf{B}. \tag{2}$$

This shows that the cross product is not commutative, and we must pay close attention to the order of the factors.

If we keep (2) in mind and apply the definition (1) to the unit vectors **i**, **j**, and **k** (Fig. 18.18), then we easily see that

$$\mathbf{i} \times \mathbf{j} = -\mathbf{j} \times \mathbf{i} = \mathbf{k},$$
$$\mathbf{j} \times \mathbf{k} = -\mathbf{k} \times \mathbf{j} = \mathbf{i}, \tag{3}$$
$$\mathbf{k} \times \mathbf{i} = -\mathbf{i} \times \mathbf{k} = \mathbf{j},$$

18.3

THE CROSS PRODUCT OF TWO VECTORS

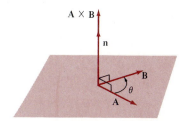

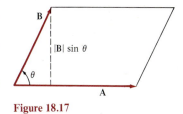

Figure 18.16

Figure 18.17

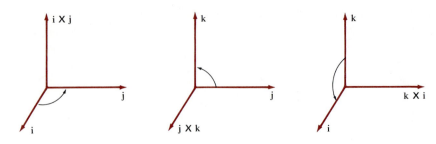

Figure 18.18

and also that

$$\mathbf{i} \times \mathbf{i} = \mathbf{j} \times \mathbf{j} = \mathbf{k} \times \mathbf{k} = \mathbf{0}.$$

For example, the right-hand thumb rule says that the direction of $\mathbf{i} \times \mathbf{j}$ is the same as the direction of $\mathbf{k}$. But the area of the parallelogram determined by $\mathbf{i}$ and $\mathbf{j}$ is 1, and since $\mathbf{k}$ itself has length 1, we have

$$\mathbf{i} \times \mathbf{j} = \mathbf{k}.$$

The products (3) are easy to remember by visualizing the figure. Another way to remember them is to arrange $\mathbf{i}$, $\mathbf{j}$, and $\mathbf{k}$ in cyclic order,

$$\mathbf{i} \rightarrow \mathbf{j} \rightarrow \mathbf{k},$$

and to observe that

(each unit vector) $\times$ (the next one) = (the third one).

Our next objective is to develop a convenient formula for calculating $\mathbf{A} \times \mathbf{B}$ in terms of the components of $\mathbf{A}$ and $\mathbf{B}$, where

$$\mathbf{A} = a_1\mathbf{i} + a_2\mathbf{j} + a_3\mathbf{k} \quad \text{and} \quad \mathbf{B} = b_1\mathbf{i} + b_2\mathbf{j} + b_3\mathbf{k}. \tag{4}$$

In order to multiply out the product

$$\mathbf{A} \times \mathbf{B} = (a_1\mathbf{i} + a_2\mathbf{j} + a_3\mathbf{k}) \times (b_1\mathbf{i} + b_2\mathbf{j} + b_3\mathbf{k}) = ?,$$

we need to know that the cross product possesses the following algebraic properties:

$$(c\mathbf{A}) \times \mathbf{B} = c(\mathbf{A} \times \mathbf{B}) = \mathbf{A} \times (c\mathbf{B}), \tag{5}$$

$$\mathbf{A} \times (\mathbf{B} + \mathbf{C}) = \mathbf{A} \times \mathbf{B} + \mathbf{A} \times \mathbf{C}, \tag{6}$$

$$(\mathbf{A} + \mathbf{B}) \times \mathbf{C} = \mathbf{A} \times \mathbf{C} + \mathbf{B} \times \mathbf{C}. \tag{7}$$

Property (5) is easily established directly from the definition (1). Property (7) follows from (6) by using (2),

$$(\mathbf{A} + \mathbf{B}) \times \mathbf{C} = -[\mathbf{C} \times (\mathbf{A} + \mathbf{B})]$$

$$= -(\mathbf{C} \times \mathbf{A} + \mathbf{C} \times \mathbf{B})$$

$$= -\mathbf{C} \times \mathbf{A} - \mathbf{C} \times \mathbf{B}$$

$$= \mathbf{A} \times \mathbf{C} + \mathbf{B} \times \mathbf{C}.$$

The real difficulty here is with the distributive law (6). There is no simple proof of this fact; and rather than hold up our progress by pausing to insert a complicated proof here, we simply take (6) for granted and continue on to our immediate objective. A proof of (6) is given in Remark 2 for the use of any students who may wish to examine it.

We continue with our task of multiplying out the cross product of the vectors (4). Remembering to pay close attention to the order of the factors, we have

$$\mathbf{A} \times \mathbf{B} = (a_1\mathbf{i} + a_2\mathbf{j} + a_3\mathbf{k}) \times (b_1\mathbf{i} + b_2\mathbf{j} + b_3\mathbf{k})$$
$$= a_1\mathbf{i} \times (b_1\mathbf{i} + b_2\mathbf{j} + b_3\mathbf{k}) + a_2\mathbf{j} \times (b_1\mathbf{i} + b_2\mathbf{j} + b_3\mathbf{k})$$
$$+ a_3\mathbf{k} \times (b_1\mathbf{i} + b_2\mathbf{j} + b_3\mathbf{k})$$

so

$$\mathbf{A} \times \mathbf{B} = a_1 b_1 \mathbf{i} \times \mathbf{i} + a_1 b_2 \mathbf{i} \times \mathbf{j} + a_1 b_3 \mathbf{i} \times \mathbf{k}$$
$$+ a_2 b_1 \mathbf{j} \times \mathbf{i} + a_2 b_2 \mathbf{j} \times \mathbf{j} + a_2 b_3 \mathbf{j} \times \mathbf{k}$$
$$+ a_3 b_1 \mathbf{k} \times \mathbf{i} + a_3 b_2 \mathbf{k} \times \mathbf{j} + a_3 b_3 \mathbf{k} \times \mathbf{k}.$$

By using (3), we now obtain the rather awkward formula

$$\mathbf{A} \times \mathbf{B} = \mathbf{i}(a_2 b_3 - a_3 b_2) - \mathbf{j}(a_1 b_3 - a_3 b_1) + \mathbf{k}(a_1 b_2 - a_2 b_1). \qquad (8)$$

(The slightly strange way of writing the signs here has a purpose that will become clear below.)

It is not necessary to memorize formula (8), because there is an equivalent version involving determinants that is easy to remember. We recall that a determinant of order 2 is defined by

$$\begin{vmatrix} a_1 & a_2 \\ b_1 & b_2 \end{vmatrix} = a_1 b_2 - a_2 b_1.$$

For example,

$$\begin{vmatrix} 3 & -2 \\ 4 & 5 \end{vmatrix} = 3 \cdot 5 - (-2) \cdot 4 = 23.$$

A determinant of order 3 can be defined in terms of determinants of order 2:

$$\begin{vmatrix} a_1 & a_2 & a_3 \\ b_1 & b_2 & b_3 \\ c_1 & c_2 & c_3 \end{vmatrix} = a_1 \begin{vmatrix} b_2 & b_3 \\ c_2 & c_3 \end{vmatrix} - a_2 \begin{vmatrix} b_1 & b_3 \\ c_1 & c_3 \end{vmatrix} + a_3 \begin{vmatrix} b_1 & b_2 \\ c_1 & c_2 \end{vmatrix}. \qquad (9)$$

Here we see that each number in the first row on the left is multiplied by the determinant of order 2 that remains when that number's row and column are deleted. We particularly notice the minus sign attached to the middle term on the right-hand side of formula (9).

Even though a determinant of order 3 can be expanded along any row or column, we use only expansions along the first row, as in (9). For example,

$$\begin{vmatrix} 3 & 2 & -1 \\ 4 & 3 & 3 \\ -2 & 7 & 1 \end{vmatrix} = 3 \begin{vmatrix} 3 & 3 \\ 7 & 1 \end{vmatrix} - 2 \begin{vmatrix} 4 & 3 \\ -2 & 1 \end{vmatrix} + (-1) \begin{vmatrix} 4 & 3 \\ -2 & 7 \end{vmatrix}$$
$$= 3(3 \cdot 1 - 3 \cdot 7) - 2[4 \cdot 1 - 3 \cdot (-2)] + (-1)[4 \cdot 7 - 3 \cdot (-2)]$$
$$= -54 - 20 - 34 = -108.$$

Formula (8) for the vector product of $\mathbf{A} = a_1\mathbf{i} + a_2\mathbf{j} + a_3\mathbf{k}$ and $\mathbf{B} = b_1\mathbf{i} + b_2\mathbf{j} + b_3\mathbf{k}$ is clearly equivalent to

$$\mathbf{A} \times \mathbf{B} = \mathbf{i}\begin{vmatrix} a_2 & a_3 \\ b_2 & b_3 \end{vmatrix} - \mathbf{j}\begin{vmatrix} a_1 & a_3 \\ b_1 & b_3 \end{vmatrix} + \mathbf{k}\begin{vmatrix} a_1 & a_2 \\ b_1 & b_2 \end{vmatrix}. \tag{10}$$

Motivated by (9), we now write (10) in the form

$$\mathbf{A} \times \mathbf{B} = \begin{vmatrix} \mathbf{i} & \mathbf{j} & \mathbf{k} \\ a_1 & a_2 & a_3 \\ b_1 & b_2 & b_3 \end{vmatrix}. \tag{11}$$

This is the concise and easily remembered formula for $\mathbf{A} \times \mathbf{B}$ that we have been seeking. The "symbolic determinant" here is to be evaluated by expanding along its first row, just as in equation (9). We emphasize that the components of the first vector $\mathbf{A}$ in $\mathbf{A} \times \mathbf{B}$ form the *second* row of the determinant in (11), and that the components of the *second* vector $\mathbf{B}$ form the *third* row of this determinant.*

Example 1   Calculate the cross product of $\mathbf{A} = 2\mathbf{i} - \mathbf{j} + 4\mathbf{k}$ and $\mathbf{B} = \mathbf{i} + 5\mathbf{j} - 3\mathbf{k}$.

Solution   By formula (11) we have

$$\mathbf{A} \times \mathbf{B} = \begin{vmatrix} \mathbf{i} & \mathbf{j} & \mathbf{k} \\ 2 & -1 & 4 \\ 1 & 5 & -3 \end{vmatrix} = \mathbf{i}\begin{vmatrix} -1 & 4 \\ 5 & -3 \end{vmatrix} - \mathbf{j}\begin{vmatrix} 2 & 4 \\ 1 & -3 \end{vmatrix} + \mathbf{k}\begin{vmatrix} 2 & -1 \\ 1 & 5 \end{vmatrix}$$

$$= -17\mathbf{i} + 10\mathbf{j} + 11\mathbf{k}.$$

As a routine check to help guard ourselves against computational errors, we observe that our answer is perpendicular to $\mathbf{A}$ because $-34 - 10 + 44 = 0$, and is perpendicular to $\mathbf{B}$ because $-17 + 50 - 33 = 0$.

Example 2   Find all unit vectors perpendicular to both of the vectors $\mathbf{A} = 2\mathbf{i} - \mathbf{j} + 3\mathbf{k}$ and $\mathbf{B} = -4\mathbf{i} + 3\mathbf{j} - 5\mathbf{k}$.

Solution   Since $\mathbf{A} \times \mathbf{B}$ is automatically perpendicular to both $\mathbf{A}$ and $\mathbf{B}$, we compute

$$\mathbf{A} \times \mathbf{B} = \begin{vmatrix} \mathbf{i} & \mathbf{j} & \mathbf{k} \\ 2 & -1 & 3 \\ -4 & 3 & -5 \end{vmatrix} = \mathbf{i}\begin{vmatrix} -1 & 3 \\ 3 & -5 \end{vmatrix} - \mathbf{j}\begin{vmatrix} 2 & 3 \\ -4 & -5 \end{vmatrix} + \mathbf{k}\begin{vmatrix} 2 & -1 \\ -4 & 3 \end{vmatrix}$$

$$= -4\mathbf{i} - 2\mathbf{j} + 2\mathbf{k}.$$

We next convert this into a unit vector in the same direction by dividing by its own length, which is $\sqrt{16 + 4 + 4} = \sqrt{24} = 2\sqrt{6}$:

---

* Some authors *define* $\mathbf{A} \times \mathbf{B}$ by formula (11). This approach has several disadvantages, one of which is that considerable effort is needed before the geometric nature of $\mathbf{A} \times \mathbf{B}$ (that is, its length and direction) can be understood. We prefer to define $\mathbf{A} \times \mathbf{B}$ directly, in terms of its length and direction, and to consider formula (11) as simply a convenient tool for making calculations. Definitions of vector operations that avoid dependence on explicit representations of vectors in terms of any particular coordinate system are called *invariant* or *coordinate-free*.

$$\frac{-4\mathbf{i} - 2\mathbf{j} + 2\mathbf{k}}{2\sqrt{6}} = \frac{-2\mathbf{i} - \mathbf{j} + \mathbf{k}}{\sqrt{6}}.$$

And finally, we introduce a plus-or-minus sign,

$$\pm \frac{-2\mathbf{i} - \mathbf{j} + \mathbf{k}}{\sqrt{6}},$$

because there are two possible directions.

Example 3 Find the area of the triangle whose vertices are $P = (2, -1, 3)$, $Q = (1, 2, 4)$, and $R = (3, 1, 1)$.

Solution Two sides of the triangle are represented by the vectors

$$\mathbf{A} = \overrightarrow{PQ} = (1 - 2)\mathbf{i} + (2 + 1)\mathbf{j} + (4 - 3)\mathbf{k} = -\mathbf{i} + 3\mathbf{j} + \mathbf{k},$$

$$\mathbf{B} = \overrightarrow{PR} = (3 - 2)\mathbf{i} + (1 + 1)\mathbf{j} + (1 - 3)\mathbf{k} = \mathbf{i} + 2\mathbf{j} - 2\mathbf{k}.$$

The vector

$$\mathbf{A} \times \mathbf{B} = \begin{vmatrix} \mathbf{i} & \mathbf{j} & \mathbf{k} \\ -1 & 3 & 1 \\ 1 & 2 & -2 \end{vmatrix} = -8\mathbf{i} - \mathbf{j} - 5\mathbf{k}$$

has magnitude $\sqrt{64 + 1 + 25} = \sqrt{90} = 3\sqrt{10}$, and this is equal to the area of the parallelogram with $\mathbf{A} = \overrightarrow{PQ}$ and $\mathbf{B} = \overrightarrow{PR}$ as adjacent sides. The area of the given triangle is clearly half the area of this parallelogram, and is therefore $\frac{3}{2}\sqrt{10}$.

Remark 1 The cross product arises quite naturally in many situations in physics. For example, if a force $\mathbf{F}$ is applied to a body at a point P (Fig. 18.19), and if $\mathbf{R}$ is the vector from a fixed origin O to P, then this force tends to rotate the body about an axis through O and perpendicular to the plane of $\mathbf{R}$ and $\mathbf{F}$. The *torque vector* $\mathbf{T}$ defined by

$$\mathbf{T} = \mathbf{R} \times \mathbf{F}$$

specifies the direction and magnitude of this rotational effect, since $|\mathbf{R}||\mathbf{F}| \sin \theta$ is the moment of the force about the axis, namely, the product of the length of the lever arm and the scalar component of $\mathbf{F}$ perpendicular to $\mathbf{R}$.

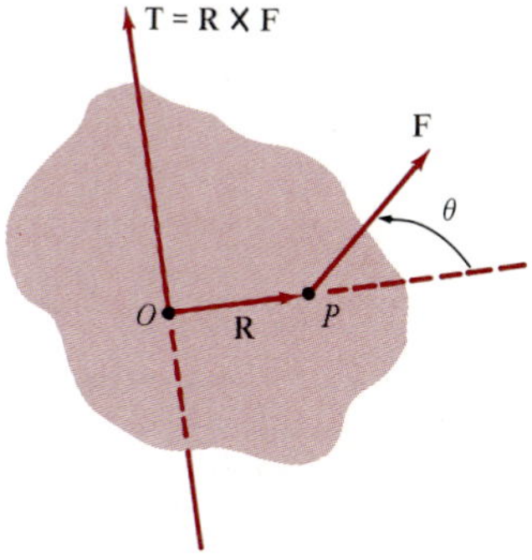

As another example, we mention the force $\mathbf{F}$ exerted on a moving charged particle by a magnetic field $\mathbf{B}$. It turns out that

$$\mathbf{F} = q\mathbf{V} \times \mathbf{B},$$

where $\mathbf{V}$ is the velocity of the charged particle and q is the magnitude of its charge. This is the primary fact that causes the aurora borealis or "northern lights," which are produced by blasts of charged particles from the sun streaming through the magnetic field of the earth. This basic principle of electromagnetism also underlies the design and operation of cyclotrons and television receivers.

Figure 18.19 Torque vector.

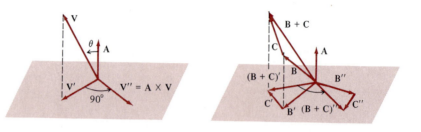

Figure 18.20   The distributive law.

Remark 2   We now return to the problem of establishing the distributive law (6). We prove (6) only for unit vectors **A**, because once this has been done, an application of (5) allows us to obtain (6) immediately for vectors **A** of arbitrary length.

With a unit vector **A** and an arbitrary vector **V**, **A** × **V** can be constructed by performing the following two operations, shown on the left in Fig. 18.20: First, project **V** on the plane perpendicular to **A** to obtain a vector **V'** of length $|V| \sin \theta$; then rotate **V'** in this plane through an angle of 90° in the positive direction to obtain **V''**, which is **A** × **V** since **A** is a unit vector. Each of these operations transforms a triangle into a triangle; so if we start with the three vectors **B**, **C**, and **B** + **C** shown on the right, the final three vectors **B''**, **C''**, and (**B** + **C**)'' still form a triangle, and therefore (**B** + **C**)'' = **B''** + **C''**. But this means that

$$\mathbf{A} \times (\mathbf{B} + \mathbf{C}) = \mathbf{A} \times \mathbf{B} + \mathbf{A} \times \mathbf{C},$$

and the argument is complete.

PROBLEMS

1   Calculate **A** × **B** and check the result by showing that it is perpendicular to both **A** and **B**:
   (a) **A** = 3**i** − 2**j** + 4**k**, **B** = 2**i** + **j** − 2**k**;
   (b) **A** = 2**i** + 2**j** − **k**, **B** = **i** + **j** + **k**;
   (c) **A** = 5**i** − 4**j** + 3**k**, **B** = −3**i** − 2**j** + **k**;
   (d) **A** = **i**, **B** = **i** + **j**.

2   Find a vector **N** perpendicular to the plane of the three points $P = (1, -1, 4)$, $Q = (2, 0, 1)$, $R = (0, 2, 3)$.

3   Find the area of the triangle PQR in Problem 2.

4   Find the distance from the origin to the plane in Problem 2 by finding the scalar projection of $\overrightarrow{OP}$ along the vector **N**.

5   If **A** · (**B** × **C**) = 0, what can be concluded about the configuration of **A**, **B**, and **C**?

6   Show that $|\mathbf{A} \times \mathbf{B}|^2 = |\mathbf{A}|^2 |\mathbf{B}|^2 - (\mathbf{A} \cdot \mathbf{B})^2$.

7   Show that the cross product is not associative by showing that

$$\mathbf{A} \times (\mathbf{B} \times \mathbf{C}) \neq (\mathbf{A} \times \mathbf{B}) \times \mathbf{C}$$

for the three vectors **A** = **i** + **j**, **B** = **j**, **C** = **k**.

8   Show that the cross product of each pair of the following vectors is parallel to the third: **i** − 2**j** + **k**, **i** + **j** + **k**, **i** − **k**. What does this tell us about the configuration of the vectors?

9   Show that if **A** is a nonzero vector and **A** × **B** = **A** × **C**, then **B** = **C** is not necessarily true.

10   If **A**, **B**, and **C** are mutually perpendicular, show that **A** × (**B** × **C**) = **0**.

11   Let P_1 and Q_1 be two points on a line L_1, and let P_2 and Q_2 be two points on a line L_2. If L_1 and L_2 are not parallel, then the perpendicular distance d between them is the absolute value of the scalar projection of $\overrightarrow{P_1 P_2}$ on a unit vector that is perpendicular to both lines. Why?
   (a) Show that

$$d = \left| \overrightarrow{P_1 P_2} \cdot \frac{\overrightarrow{P_1 Q_1} \times \overrightarrow{P_2 Q_2}}{|\overrightarrow{P_1 Q_1} \times \overrightarrow{P_2 Q_2}|} \right|.$$

   (b) Find d if L_1 is the line determined by $P_1 = (-1, 1, 1)$ and $Q_1 = (1, 0, 0)$, and L_2 is the line determined by $P_2 = (3, 1, 0)$ and $Q_2 = (4, 5, -1)$.

12 If $\mathbf{A} = 2\mathbf{i} - 3\mathbf{j} + \mathbf{k}$ is normal to one plane and $\mathbf{B} = -\mathbf{i} + 4\mathbf{j} - 2\mathbf{k}$ is normal to another plane, do the planes necessarily intersect? Give a reason for your answer. If they do intersect, find a vector parallel to their line of intersection.

Since all the machinery of vector *algebra* is now in place, it might be expected that we would next turn to the *calculus* of vector functions in three-dimensional space. This calculus is concerned with vector-valued functions of one or more scalar variables, and deals with limits, derivatives, velocity, acceleration, and curvature, just as in Chapter 17. These concepts are essential for working with spatial motions, vector fields, and the differential geometry of curves and surfaces in space. However, a careful study of these topics belongs to a later course in advanced calculus or vector analysis, and is not part of our purpose in this book.

Nevertheless, on a few occasions we will need to consider the position vector $\mathbf{R}(t) = x(t)\mathbf{i} + y(t)\mathbf{j} + z(t)\mathbf{k}$ of a point P that moves along a space curve, as shown in Fig. 18.21. The derivative of this function is defined in the obvious way,

$$\frac{d\mathbf{R}}{dt} = \lim_{\Delta t \to 0} \frac{\mathbf{R}(t + \Delta t) - \mathbf{R}(t)}{\Delta t},$$

and has all the properties we expect on the basis of our experience in Chapter 17. In particular, $d\mathbf{R}/dt$ is tangent to the path at the point P, and is the velocity of P if the parameter t is time, and the unit tangent vector if t is arc length.

With these brief remarks we put aside the calculus of vector functions and turn to the main subject of the rest of this chapter, namely, the analytic geometry of lines, planes, and curved surfaces in three-dimensional space. We shall find that the vector algebra discussed in the preceding sections is a very valuable tool for this work.

As we know, in plane analytic geometry a single first-degree equation,

$$ax + by + c = 0,$$

is the equation of a straight line (assuming that a and b are not both zero). However, we shall see that in the geometry of three dimensions such an equation represents a plane, and therefore it is not possible to represent a line in space by any single first-degree equation.

We begin with the study of lines. A line in space can be given geometrically in three ways: as the line through two points, as the intersection of two planes, or as the line through a point in a specified direction. The third way is the most important for us.

Suppose L is the line in space that passes through a given point $P_0 = (x_0, y_0, z_0)$ and is parallel to a given nonzero vector

$$\mathbf{V} = a\mathbf{i} + b\mathbf{j} + c\mathbf{k},$$

as shown in Fig. 18.22. Then another point $P = (x, y, z)$ lies on the line L if and only if the vector $\overrightarrow{P_0P}$ is parallel to the vector $\mathbf{V}$. That is, P lies on L if and only if $\overrightarrow{P_0P}$ is a scalar multiple of $\mathbf{V}$, so that

$$\overrightarrow{P_0P} = t\mathbf{V} \tag{1}$$

18.4
LINES AND PLANES

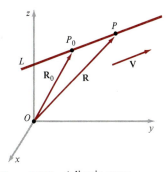

Figure 18.21 A space curve.

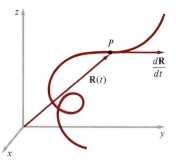

Figure 18.22 A line in space.

for some real number t. If $\mathbf{R}_0 = \overrightarrow{OP_0}$ and $\mathbf{R} = \overrightarrow{OP}$ are the position vectors of P_0 and P, then $\overrightarrow{P_0P} = \mathbf{R} - \mathbf{R}_0$ and (1) gives

$$\mathbf{R} = \mathbf{R}_0 + t\mathbf{V}, \tag{2}$$

which is the *vector equation* of L. As t varies from $-\infty$ to ∞, the point P traverses the entire infinite line L, moving in the direction of $\mathbf{V}$.

If we write (2) in the form

$$x\mathbf{i} + y\mathbf{j} + z\mathbf{k} = x_0\mathbf{i} + y_0\mathbf{j} + z_0\mathbf{k} + t(a\mathbf{i} + b\mathbf{j} + c\mathbf{k})$$

and equate the coefficients of $\mathbf{i}$, $\mathbf{j}$, and $\mathbf{k}$, we get the three scalar equations

$$\begin{aligned}
x &= x_0 + at, \\
y &= y_0 + bt, \\
z &= z_0 + ct.
\end{aligned} \tag{3}$$

These are the *parametric equations* of the line L through the point $P_0 = (x_0, y_0, z_0)$ and parallel to the vector $\mathbf{V} = a\mathbf{i} + b\mathbf{j} + c\mathbf{k}$. Observe that the parametric equations of a straight line are not unique. The numbers x_0, y_0, and z_0 can be replaced by the coordinates of any other point on L, and a, b, and c can be replaced by the components of any other nonzero vector parallel to L, and the resulting parametric equations will be completely equivalent to equations (3) in the sense that they describe the same line.

In order to obtain the Cartesian equations of the line, we eliminate the parameter from equations (3) by equating the three expressions obtained by solving for t. This gives

$$\frac{x - x_0}{a} = \frac{y - y_0}{b} = \frac{z - z_0}{c}. \tag{4}$$

These are called the *symmetric equations* of the line L. If any one of the constants a, b, c is zero in a denominator of (4), then the corresponding numerator must also be zero. This is easy to see from the parametric form (3), which shows, for example, that if

$$x = x_0 + at \quad \text{and} \quad a = 0,$$

then $x = x_0$. Thus, when one of the denominators in (4) vanishes, we interpret this as meaning that the corresponding numerator must also vanish. With this interpretation, equations (4) can always be used, even though division by zero is normally forbidden.

Example 1 A line L goes through the points $P_0 = (3, -2, 1)$ and $P_1 = (5, 1, 0)$. Find the parametric equations and the symmetric equations of L. Also find the points at which this line pierces the three coordinate planes.

Solution The line L is parallel to the vector $\overrightarrow{P_0P_1} = 2\mathbf{i} + 3\mathbf{j} - \mathbf{k}$, so by using P_0 as the known point on the line, equations (3) give the parametric equations

$$\begin{aligned}
x &= 3 + 2t, \\
y &= -2 + 3t, \\
z &= 1 - t.
\end{aligned}$$

By eliminating t, we obtain the symmetric equations

$$\frac{x-3}{2} = \frac{y+2}{3} = \frac{z-1}{-1}.$$

To find the point at which L pierces the xy-plane, we set $z = 0$ in the third parametric equation and see that $t = 1$. With this value of t, $x = 5$ and $y = 1$, so the point is $(5, 1, 0)$. Similarly, $x = 0$ implies that $t = -\frac{3}{2}$, so the point in the yz-plane is $(0, -\frac{13}{2}, \frac{5}{2})$; and $y = 0$ implies $t = \frac{2}{3}$, so the point in the xz-plane is $(\frac{13}{3}, 0, \frac{1}{3})$.

Now we turn to the study of planes. A plane can also be characterized in several ways: as the plane through three noncollinear points, as the plane through a line and a point not on the line, or as the plane through a point and perpendicular to a specified direction. Again, the third approach is the most convenient for us.

Consider the plane that passes through a given point $P_0 = (x_0, y_0, z_0)$ and is perpendicular to a given nonzero vector

$$\mathbf{N} = a\mathbf{i} + b\mathbf{j} + c\mathbf{k}, \tag{5}$$

as shown in Fig. 18.23. Another point $P = (x, y, z)$ lies on this plane if and only if the vector $\overrightarrow{P_0P}$ is perpendicular to the vector $\mathbf{N}$, which means that

$$\mathbf{N} \cdot \overrightarrow{P_0P} = 0. \tag{6}$$

If $\mathbf{R}_0 = \overrightarrow{OP_0}$ and $\mathbf{R} = \overrightarrow{OP}$ are the position vectors of P_0 and P, so that $\overrightarrow{P_0P} = \mathbf{R} - \mathbf{R}_0$, then (6) becomes

$$\mathbf{N} \cdot (\mathbf{R} - \mathbf{R}_0) = 0. \tag{7}$$

This is the *vector equation* of the plane under discussion.

Since $\mathbf{R} - \mathbf{R}_0 = (x - x_0)\mathbf{i} + (y - y_0)\mathbf{j} + (z - z_0)\mathbf{k}$, (7) can be written out in the scalar form

$$a(x - x_0) + b(y - y_0) + c(z - z_0) = 0. \tag{8}$$

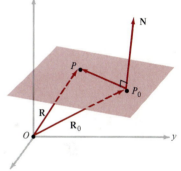

Figure 18.23 A plane in space.

This is the *Cartesian equation* of the plane through the point $P_0 = (x_0, y_0, z_0)$ with normal vector $\mathbf{N} = a\mathbf{i} + b\mathbf{j} + c\mathbf{k}$. For example, the equation of the plane through $P_0 = (5, -3, 1)$ with normal vector $\mathbf{N} = 4\mathbf{i} + 3\mathbf{j} - 2\mathbf{k}$ is

$$4(x - 5) + 3(y + 3) - 2(z - 1) = 0$$

or

$$4x + 3y - 2z = 9.$$

Observe that the coefficients of x, y, and z in the last equation are the components of the normal vector. This is always the case, for equation (8) can be written in the form

$$ax + by + cz = d, \tag{9}$$

where $d = ax_0 + by_0 + cz_0$; and the coefficients of x, y, and z in this equation are clearly the components of the normal vector (5). Conversely, every *linear equation* in x, y, and z of the form (9) represents a plane with normal vector $\mathbf{N} = a\mathbf{i} + b\mathbf{j} + c\mathbf{k}$ if the coefficients a, b, and c are not all zero. To see this, we notice that if (for instance) $a \neq 0$, then this permits us to choose y_0 and z_0

arbitrarily and solve the equation $ax_0 + by_0 + cz_0 = d$ for x_0. With these values, (9) can be written as

$$ax + by + cz = ax_0 + by_0 + cz_0$$

or

$$a(x - x_0) + b(y - y_0) + c(z - z_0) = 0,$$

and this is immediately recognizable as the equation of the plane through (x_0, y_0, z_0) with normal vector $\mathbf{N} = a\mathbf{i} + b\mathbf{j} + c\mathbf{k}$.

Example 2 Find an equation for the plane through the three points $P_0 = (3, 2, -1)$, $P_1 = (1, -1, 3)$, and $P_2 = (3, -2, 4)$.

Solution To use equation (8), we must find a vector $\mathbf{N}$ that is normal to the plane. This is easy to do by using the cross product. We compute

$$\mathbf{N} = \overrightarrow{P_0P_1} \times \overrightarrow{P_0P_2} = \begin{vmatrix} \mathbf{i} & \mathbf{j} & \mathbf{k} \\ -2 & -3 & 4 \\ 0 & -4 & 5 \end{vmatrix} = \mathbf{i} + 10\mathbf{j} + 8\mathbf{k}.$$

Since $\overrightarrow{P_0P_1}$ and $\overrightarrow{P_0P_2}$ lie in the plane, their cross product $\mathbf{N}$ is normal to the plane. Using equation (8) with P_0 as the given point, our plane has the equation

$$(x - 3) + 10(y - 2) + 8(z + 1) = 0$$

or

$$x + 10y + 8z = 15,$$

after simplification.

Example 3 Find the point at which the line

$$\frac{x - 2}{1} = \frac{y + 3}{2} = \frac{z - 4}{2}$$

pierces the plane $x + 2y + 2z = 22$.

Solution To find parametric equations for the line, we introduce t as the common ratio in the given symmetric equations,

$$\frac{x - 2}{1} = \frac{y + 3}{2} = \frac{z - 4}{2} = t,$$

which gives

$$x = 2 + t, \qquad y = -3 + 2t, \qquad z = 4 + 2t.$$

We want the value of t for which the variable point (x, y, z) on the line lies on the given plane. By substituting these equations into the equation of the plane, we obtain

$$(2 + t) + 2(-3 + 2t) + 2(4 + 2t) = 22,$$

so $t = 2$ at the point where the line pierces the plane. By substituting $t = 2$ back in the parametric equations of the line, we find that the desired point is $(4, 1, 8)$.

Example 4 Find the cosine of the angle between the two planes $x + 4y - 4z = 9$ and $x + 2y + 2z = -3$. Also, find parametric equations for the line of intersection of these planes.

Solution Clearly the angle θ between two planes is the angle between their normals (Fig. 18.24). By inspecting the equations of the given planes, we see at once that their normals are

$$\mathbf{N}_1 = \mathbf{i} + 4\mathbf{j} - 4\mathbf{k}, \qquad \mathbf{N}_2 = \mathbf{i} + 2\mathbf{j} + 2\mathbf{k}.$$

We therefore use the dot product to obtain

$$\cos\theta = \frac{\mathbf{N}_1 \cdot \mathbf{N}_2}{|\mathbf{N}_1||\mathbf{N}_2|} = \frac{1}{3\sqrt{33}}.$$

From this we can find the angle θ if we wish, by tables or otherwise.

To find parametric equations for the line of intersection, we need a vector $\mathbf{V}$ parallel to this line and a point on the line. We find $\mathbf{V}$ by computing the cross product of $\mathbf{N}_1$ and $\mathbf{N}_2$,

$$\mathbf{V} = \mathbf{N}_1 \times \mathbf{N}_2 = \begin{vmatrix} \mathbf{i} & \mathbf{j} & \mathbf{k} \\ 1 & 4 & -4 \\ 1 & 2 & 2 \end{vmatrix} = 16\mathbf{i} - 6\mathbf{j} - 2\mathbf{k}.$$

Since any vector parallel to the line will do, we divide by 2 and use the slightly simpler vector $8\mathbf{i} - 3\mathbf{j} - \mathbf{k}$. To find a point on the line, we can set $z = 0$ and solve the resulting system in the unknowns x and y,

$$x + 4y = 9,$$

$$x + 2y = -3.$$

This yields $x = -15$, $y = 6$. The desired point is therefore $(-15, 6, 0)$, and the parametric equations of the line are

$$x = -15 + 8t,$$

$$y = 6 - 3t,$$

$$z = -t.$$

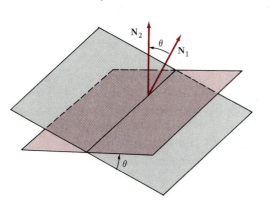

Figure 18.24

We repeat that there is nothing unique about these equations, for we could have found a point on the line in many other ways and there are many different vectors parallel to the line.

As we remarked at the beginning of this section, any two intersecting planes determine a straight line in space. The equations of the two planes are satisfied simultaneously only by points on the line of intersection. From this point of view, a pair of linear equations considered as a simultaneous system can be interpreted as representing a line, namely, the line of intersection of the two planes represented by the individual equations. (Of course, the planes must actually intersect, and not be parallel or identical.) Thus, in Example 4 the pair of simultaneous equations

$$x + 4y - 4z = 9,$$
$$x + 2y + 2z = -3,$$

represents the line discussed in that example. We also point out that the symmetric equations (4) are equivalent to the three simultaneous equations

$$b(x - x_0) - a(y - y_0) = 0,$$
$$c(x - x_0) - a(z - z_0) = 0,$$
$$c(y - y_0) - b(z - z_0) = 0.$$

These are the equations of three planes that intersect in the line L represented by (4). The first has normal vector $b\mathbf{i} - a\mathbf{j}$, which is parallel to the xy-plane, so the first plane is perpendicular to the xy-plane. Similarly, the second plane is perpendicular to the xz-plane and the third is perpendicular to the yz-plane. Any pair of these equations represents the line L, which is the intersection of the corresponding pair of planes.

PROBLEMS

1 Label each of the following statements as true or false:
(a) Two planes perpendicular to a line are parallel.
(b) Two lines perpendicular to a third line are parallel.
(c) Two planes parallel to a third plane are parallel.
(d) Two lines perpendicular to a plane are parallel.
(e) Two planes parallel to a line are parallel.
(f) Two lines parallel to a third line are parallel.
(g) Two planes perpendicular to a third plane are parallel.
(h) Two lines parallel to a plane are parallel.

2 What conclusion can be drawn about the lines

$$\frac{x - x_0}{a} = \frac{y - y_0}{b} = \frac{z - z_0}{c},$$

$$\frac{x - x_0}{A} = \frac{y - y_0}{B} = \frac{z - z_0}{C}$$

if $aA + bB + cC = 0$?

3 What conclusion can be drawn about the lines

$$\frac{x - x_0}{a} = \frac{y - y_0}{b} = \frac{z - z_0}{c},$$

$$\frac{x - x_1}{A} = \frac{y - y_1}{B} = \frac{z - z_1}{C}$$

if $a/A = b/B = c/C$?

4 Write symmetric equations for the line through the point $(3, 0, -2)$ and parallel to
(a) the vector $4\mathbf{i} - 3\mathbf{j} + 5\mathbf{k}$;
(b) the line $(x + 1)/7 = (y - 2)/2 = z/(-3)$;
(c) the x-axis.

5 Write parametric equations for the line through the point $(2, -1, -3)$ and parallel to
(a) the vector $\mathbf{i} + 4\mathbf{j} - 2\mathbf{k}$;
(b) the line $x/3 = (y + 7)/(-1) = (z - 3)/6$;
(c) the line $x = 2t - 3$, $y = 3 - 2t$, $z = 5t - 4$.

6 Write symmetric equations for the line through the points
(a) $(2, -1, 3)$ and $(5, 2, -2)$;
(b) $(7, 3, -1)$ and $(3, -1, 3)$.

7 Write parametric equations for the line through the points
(a) $(2, 0, 3)$ and $(-1, 3, 5)$;
(b) $(4, 2, -1)$ and $(0, 2, -1)$.

8 If L is the line through the points $(-6, 6, -4)$ and $(12, -6, 2)$, find the points where L pierces the coordinate planes.

9 Show that the lines

$$x = 1 + t, \qquad y = 2t, \qquad z = 1 + 3t$$

and

$$x = 3s, \qquad y = 2s, \qquad z = 2 + s$$

intersect, and find their point of intersection.

10 Find the distance between the lines

(a) $\dfrac{x-2}{-1} = \dfrac{y-3}{4} = \dfrac{z}{2}$

and

$\dfrac{x+1}{1} = \dfrac{y-2}{0} = \dfrac{z+1}{2}$;

(b) $x = 2t - 4, \qquad y = 4 - t, \qquad z = -2t - 1$

and $x = 4t - 5, \quad y = -3t + 5, \quad z = -5t + 5$.

11 Find the distance from the origin to the line

$$\frac{x-4}{3} = \frac{y-2}{4} = \frac{4-z}{5}.$$

12 (a) As a function of t, find the distance D from the point $P_0 = (1, 2, 3)$ to a variable point on the line

$$x = 3 + t, \qquad y = 2 + t, \qquad z = 1 + t.$$

(b) By differentiating, find the value of t that minimizes D, find the actual minimum distance, and find the corresponding point P_1 on the line.

(c) Verify that the vector $\overrightarrow{P_0 P_1}$ is perpendicular to the line.

13 Find the equation of the plane that contains the point $(1, 3, 1)$ and the line $x = t, y = t, z = t + 2$.

14 Find symmetric equations for the line of intersection of each of the following pairs of planes:
(a) $2x + y + z = 0$, $3x + 4y - z = 10$;
(b) $2x + 3y + 5z = 21$, $3x - 2y + z = 12$.

15 Use vector methods to show that the distance D from the point (x_0, y_0, z_0) to the plane $ax + by + cz + d = 0$ is given by the formula

$$D = \frac{|ax_0 + by_0 + cz_0 + d|}{\sqrt{a^2 + b^2 + c^2}}.$$

16 Show that the planes $ax + by + cz + d_1 = 0$ and $ax + by + cz + d_2 = 0$ are parallel, and that the distance between them is

$$\frac{|d_1 - d_2|}{\sqrt{a^2 + b^2 + c^2}}.$$

17 Find the distance between the planes $x - 2y + 4z = 1$ and $2x - 4y + 8z = -14$.

18 Find an equation for the plane that is parallel to the plane $2x - 5y + 3z = 7$ and passes through the point $(5, 2, 3)$.

19 Consider the sphere of radius 3 with its center at the origin. The plane tangent to this sphere at $(1, 2, 2)$ intersects the x-axis at a point P. Find the coordinates of P.

20 Find the value of the parameter t for which the planes

$$3x - 4y + 2z + 9 = 0,$$
$$3x + 4y - tz + 7 = 0,$$

are perpendicular.

21 Verify that the planes

$$2x + 3y + 4z - 1 = 0,$$
$$x - 2y + 3z - 4 = 0,$$

intersect in a line L. Find symmetric equations for L in two ways:
(a) by finding two points on L and going on from there;
(b) by first eliminating x from the given equations, and then y, to find two planes through L that are perpendicular to the yz-plane and the xz-plane, respectively, and then solving each of these equations for z and equating all the z's.

22 Show that the two sets of equations

$$\frac{x-4}{3} = \frac{y-6}{4} = \frac{z+9}{-12} \qquad \text{and}$$

$$\frac{x-1}{-6} = \frac{y-2}{-8} = \frac{z-3}{24}$$

represent the same straight line.

23 Let p_1 and p_2 be two planes that intersect in a line L and have equations

$$a_1 x + b_1 y + c_1 z + d_1 = 0,$$
$$a_2 x + b_2 y + c_2 z + d_2 = 0.$$

If k is a constant, show that

$$(a_1 x + b_1 y + c_1 z + d_1) + k(a_2 x + b_2 y + c_2 z + d_2) = 0$$

is the equation of a plane containing L. For various values of k, this equation represents every member of the family of all planes containing L, with one exception. What is this exception?

24 Find the equation of the plane that contains the intersection of the planes $2x + 3y - z = 1$ and $3x - y + 5z = 2$ and passes through $(1, 4, 1)$.

25 Find the equation of the plane that contains the intersection of the planes $x - 2y - 5z = 3$ and $5x + y - z = 1$ and is parallel to $4x + 3y + 4z + 7 = 0$.

26 Find the coordinates of the point P at which the line

$$\frac{x-1}{3} = \frac{y+3}{4} = \frac{z-3}{2}$$

pierces the plane $3x + 4y + 5z = 76$.

27 Show that the line

$$\frac{x+8}{9} = \frac{y-10}{-4} = \frac{z-9}{-6}$$

lies in the plane $2x - 3y + 5z = -1$.

28 Show that the line of intersection of the planes

$$x + y - z = 0 \quad \text{and} \quad x - y - 5z + 7 = 0$$

is parallel to the line

$$\frac{x+3}{3} = \frac{y-1}{-2} = \frac{z-5}{1}.$$

29 Find the cosine of the angle between the given planes:
(a) $2x - y + 2z = 3$, $3x + 2y - 6z = 7$;
(b) $5x - 3y + 2z = 3$, $x + 3y + 2z = -11$.

30 Show that the single equation

$$(2x + y - z - 3)^2 + (x + 2y - 3z + 5)^2 = 0$$

represents a straight line in space. (But this equation is of the second degree.)

18.5
CYLINDERS AND SURFACES OF REVOLUTION

We know that the graph of an equation $f(x, y) = 0$ is usually a curve in the xy-plane. In just the same way, the graph of an equation

$$F(x, y, z) = 0 \tag{1}$$

is usually a surface in xyz-space. The simplest surfaces are planes, and we saw in Section 18.4 that the equation of a plane is a linear equation that can be written in the form

$$ax + by + cz + d = 0;$$

that is, it contains only first-degree terms in the variables x, y, and z. In this section and the next we examine a few other simple surfaces containing terms of higher degree that often appear in multivariable calculus.

Cylinders are the next surfaces after planes in order of complexity. To understand what these surfaces are, we consider a plane curve C and a line L not parallel to the plane of C. By a *cylinder* we mean the geometric figure in space that is generated (or swept out) by a straight line moving parallel to L and passing through C (Fig. 18.25).* The moving line is called the *generator* of the cylinder. The cylinder can be thought of as consisting of infinitely many parallel lines, called *rulings,* corresponding to various positions of the generator. This is suggested in the figure.

For example, suppose that the given curve C is the curve

$$f(x, y) = 0 \tag{2}$$

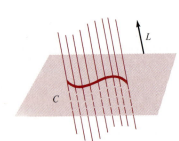

Figure 18.25 A general cylinder.

---

* This concept includes the familiar *right circular* cylinders of elementary geometry, for which the curve C is a circle and the line L is perpendicular to the plane of the circle. In geometry the adjectives are often omitted, because no other kinds of cylinders are considered. However, it should be noticed that when C is itself a straight line, the cylinder is a plane, so cylinders also include planes as special cases.

in the xy-plane, and let the generator be parallel to the z-axis, as shown in Fig. 18.26. *Then exactly the same equation* (2) *is the equation of the cylinder in three-dimensional space.* The reason for this is that the point $P = (x, y, z)$ lies on the cylinder if and only if the point $P_0 = (x, y, 0)$ lies on the curve C, and this happens if and only if $f(x, y) = 0$. The essential feature of (2) as the equation of the cylinder is that it is an equation of the form (1) from which the variable z is missing. To express this in another way, the fact that we are dealing with a cylinder whose rulings are parallel to the z-axis means that for a point $P = (x, y, z)$, the value of z has no bearing on whether P lies on the cylinder or not; and since only the variables x and y are relevant to this issue, only the variables x and y can be present in the equation of the cylinder — that is, z must be missing from this equation.

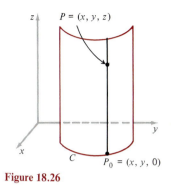

Figure 18.26

Example 1   Sketch the cylinder

$$\frac{x^2}{9} + \frac{y^2}{4} = 1.$$

Solution   This appears to be the equation of an ellipse in the xy-plane. However, it is stated that this is a cylinder, and since the variable z is missing from the equation, the rulings of this cylinder are parallel to the z-axis. In Fig. 18.27, the ellipse in the xy-plane is drawn first, then two vertical rulings, then a horizontal elliptical cross section above the xy-plane. In spite of the limitations of our figure (which we hope students will try to overcome by an active use of imagination), it should be remembered that all rulings on a cylinder extend to infinity in both directions. This surface is called an *elliptic cylinder*.

It is clear that this discussion can be carried through for a cylinder with rulings parallel to any coordinate axis. We therefore have the conclusion that *any equation in rectangular coordinates x, y, z with one variable missing represents a cylinder whose rulings are parallel to the axis corresponding to the missing variable.*

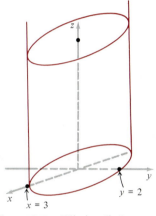

Figure 18.27   Elliptic cylinder.

Example 2   Sketch the cylinder $z = x^2$.

Solution   In the xz-plane, this is the equation of a parabola with vertex at the origin that opens in the positive z-direction. However, we know that we are dealing with a cylinder, and since the variable y is missing from the equation, the rulings of this cylinder are parallel to the y-axis. In Fig. 18.28 the parabola in the xz-plane is drawn first, then several rulings, and then a second parabolic cross section located to the right of the xz-plane. This surface can be described as a *parabolic cylinder*.

Another way to generate a surface by using a plane curve C is to revolve the curve (in space) about a line L in its plane. The resulting surface is called a *surface of revolution* with *axis L*. In Chapter 7 we became acquainted with surfaces of revolution by calculating their areas as an application of definite integrals. We now consider the equations of these surfaces.

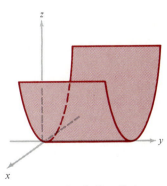

Figure 18.28   Parabolic cylinder.

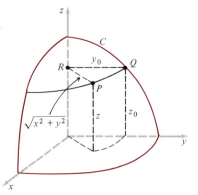

Figure 18.29

Suppose, for example, that the curve C lies in the yz-plane and has equation

$$f(y, z) = 0. \tag{3}$$

As this curve is revolved about the z-axis, a typical point $P = (x, y, z)$ on the resulting surface comes from a point Q on C, as shown in Fig. 18.29. Since Q lies on C, its coordinates (y_0, z_0) satisfy (3),

$$f(y_0, z_0) = 0. \tag{4}$$

But the relation of P to Q tells us that $z_0 = z$ and $y_0 = \sqrt{x^2 + y^2}$, and so (4) yields

$$f(\sqrt{x^2 + y^2}, z) = 0 \tag{5}$$

as the equation of the surface of revolution. Briefly, as Q swings out to the point P on the surface, the distances QR and PR to the z-axis are equal, and we get equation (5) by replacing y in (3) by $\sqrt{x^2 + y^2}$. Equation (5) assumes that $y \geq 0$ on C. If y is positive on some parts of C and negative on others, we must replace y in (3) by $\pm \sqrt{x^2 + y^2}$ to get

$$f(\pm \sqrt{x^2 + y^2}, z) = 0$$

as the equation of the complete surface. The awkward radical with its plus-or-minus sign can usually be eliminated by squaring.

Example 3 If the line $z = 3y$ in the yz-plane is revolved about the z-axis, the resulting surface of revolution is clearly a right circular cone of two nappes with vertex at the origin and axis the z-axis (Fig. 18.30). To get the equation of this cone, we replace y in the equation $z = 3y$ by $\pm\sqrt{x^2 + y^2}$ and then rationalize by squaring:

$$z = \pm 3\sqrt{x^2 + y^2}, \qquad z^2 = 9(x^2 + y^2).$$

If we had merely replaced y by $\sqrt{x^2 + y^2}$ to obtain

$$z = 3\sqrt{x^2 + y^2},$$

we would have had the equation of only the upper nappe of the cone, the part where $z \geq 0$.

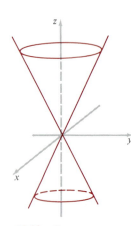

Figure 18.30 Cone.

In essentially the same way, we can obtain equations for surfaces of revolution with the x-axis or the y-axis as the axis of symmetry.

PROBLEMS

Sketch the cylinders whose equations are given in Problems 1 to 8. If a cylinder has an obvious name, state it.

1   $y = x^2$.   2 $y^2 + 4z^2 = 16$.
3   $x = \sin y$.   4 $xz = 4$.
5   $x + 3z = 6$.   6 $x^2 + z^2 = 9$.
7   $x = \tan y, -\pi/2 < y < \pi/2$.
8   $y = e^x$.
9   The rulings of a cylinder are parallel to the y-axis. Its intersection with the xz-plane is a circle with center $(0, 0, a)$ and radius a. Sketch the cylinder and find its equation.
10 The rulings of a cylinder are parallel to the x-axis. Its intersection with the yz-plane is a parabola with vertex at $(0, 0, 0)$ and focus at $(0, 0, -p)$. Sketch the cylinder and find its equation.
11 Find the equation of the surface of revolution generated by revolving the curve $z = e^{-y^2}$ about
    (a) the z-axis;   (b) the y-axis.
    Sketch both surfaces.
12 Find the equation of the surface of revolution generated by revolving the circle $(y - b)^2 + z^2 = a^2 \ (a < b)$ about
    (a) the z-axis;   (b) the y-axis.
    Sketch both surfaces.
13 In each of the following, write the equation for the surface of revolution generated by revolving the given curve about the indicated axis, and sketch the surface:
    (a) $y = z^2$, the y-axis;
    (b) $9x^2 + 4y^2 = 36$, the y-axis;
    (c) $z = 4 - x^2$, the z-axis;
    (d) $x = y^2$, the x-axis.

14 Any direction in space not parallel to the xy-plane can be specified by a vector of the form $\mathbf{V} = a\mathbf{i} + b\mathbf{j} + \mathbf{k}$ (why?). If a curve C in the xy-plane has the equation $f(x, y) = 0$, show that the equation of the cylinder generated by a moving line that is parallel to $\mathbf{V}$ and passes through C (Fig. 18.31) is

$$f(x - az, y - bz) = 0.$$

Hint: Write the symmetric equations of the line through a point $(x_0, y_0, 0)$ on C and parallel to $\mathbf{V}$.

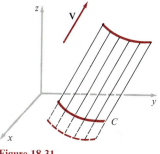

Figure 18.31

15 Find the equation of the cylinder generated by a line through the circle $x^2 + y^2 = 6x$ in the xy-plane that moves parallel to the vector $\mathbf{V} = 2\mathbf{i} + 3\mathbf{j} + \mathbf{k}$.
16 Find the equation of the cylinder generated by a line through the parabola $y = x^2$ in the xy-plane that moves parallel to the vector $\mathbf{V} = -2\mathbf{i} - 3\mathbf{j} + 5\mathbf{k}$.

In Section 15.6 we learned that the graph of a second-degree equation in the variables x and y is always a conic section—a parabola, an ellipse, a hyperbola, or perhaps some degenerate form of one of these curves, such as a point, the empty set, or a pair of straight lines.

In three-dimensional space the most general equation of the second degree is

$$Ax^2 + By^2 + Cz^2 + Dxy + Exz + Fyz + Gx + Hy + Iz + J = 0. \qquad (1)$$

We assume that not all of the coefficients $A, B, \ldots, F$ are zero, so that the degree of the equation is really 2 instead of 1 or 0. The graph of such an equation is called a *quadric surface*. We have already encountered several quadric surfaces, such as spheres and parabolic, elliptic, and hyperbolic cylinders, but there are a number of others as well. Indeed, if we set aside the familiar case of cylinders, then by suitable rotations and translations of the coordinate axes—which we do not discuss—it is possible to simplify any equation of the form (1) and thereby show that there are exactly six distinct kinds of nondegenerate quadric surfaces:

18.6
QUADRIC SURFACES

1   The ellipsoid.
2   The hyperboloid of one sheet.
3   The hyperboloid of two sheets.
4   The elliptic cone.
5   The elliptic paraboloid.
6   The hyperbolic paraboloid.

In the following we give an example of each type of surface in which the equation appears in as simple a form as possible.

Students should become familiar with these surfaces and their equations, and in particular should try to understand how the shape of each surface is related to the special features of its equation. For the purpose of visualizing and sketching a surface, it is often useful to examine its *sections,* which are the curves of intersection of the surface with planes

$$x = k, \qquad y = k, \qquad z = k$$

parallel to the coordinate planes. We point out explicitly that every second-degree section of every quadric surface is a conic section. Sections that are closed curves are usually the easiest to sketch, and therefore we look for elliptic sections and sketch these first. Symmetry considerations should also be kept in mind.

In the following examples, the numbers a, b, and c are all assumed to be positive. We comment informally rather than exhaustively on the surface considered in each example.

Example 1   The *ellipsoid*

$$\frac{x^2}{a^2} + \frac{y^2}{b^2} + \frac{z^2}{c^2} = 1 \tag{2}$$

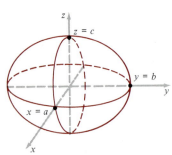

Figure 18.32   Ellipsoid.

is shown in Fig. 18.32. Since only even powers of x, y, and z occur in the equation, this surface is symmetric about each coordinate plane. The sections in the xz- and yz-planes are the ellipses

$$\frac{x^2}{a^2} + \frac{z^2}{c^2} = 1, \qquad \frac{y^2}{b^2} + \frac{z^2}{c^2} = 1$$

with a common vertical axis. The section in a horizontal plane $z = k$ is the ellipse

$$\frac{x^2}{a^2} + \frac{y^2}{b^2} = 1 - \frac{k^2}{c^2},$$

and this decreases in size as k varies from 0 to c or $-c$. The numbers a, b, and c are the intercepts on the coordinate axes, and are called the *semiaxes.* If two of the semiaxes are equal, the ellipsoid is called a *spheroid*—an *oblate* spheroid if it is flattened like a "flying saucer," and a *prolate* spheroid if it is elongated like a football. Of course, if $a = b = c$, then the ellipsoid is a sphere.

Example 2   The graph of the equation

$$\frac{x^2}{a^2} + \frac{y^2}{b^2} - \frac{z^2}{c^2} = 1 \tag{3}$$

is a *hyperboloid of one sheet* (Fig. 18.33). If we write the equation in the form

$$\frac{x^2}{a^2} + \frac{y^2}{b^2} = \frac{z^2}{c^2} + 1, \tag{4}$$

then we see that all its horizontal sections in planes $z = k$ are ellipses, and that these ellipses grow larger as their planes move up or down from the xy-plane, the smallest ellipse being the one in the xy-plane. The section of the surface in the yz-plane is the hyperbola

$$\frac{y^2}{b^2} - \frac{z^2}{c^2} = 1.$$

It is this hyperbola that binds together the horizontal elliptical sections into a smooth surface. The phrase "of one sheet" is used because this surface consists of one piece, in contrast to the hyperboloid discussed in the next example, which consists of two pieces. Observe that equation (3) is obtained from (2) by changing the sign of the third term on the left; we get the same kind of surface no matter which of these terms has its sign changed.

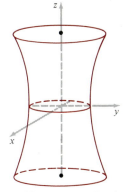

Figure 18.33 Hyperboloid of one sheet.

Example 3 The *hyperboloid of two sheets*

$$-\frac{x^2}{a^2} - \frac{y^2}{b^2} + \frac{z^2}{c^2} = 1 \tag{5}$$

is shown in Fig. 18.34. This equation is obtained from (2) by changing the signs of the first two terms on the left. (The reason for this choice is explained below.) If we write the equation in the form

$$\frac{x^2}{a^2} + \frac{y^2}{b^2} = \frac{z^2}{c^2} - 1, \tag{6}$$

then we see that all its horizontal sections in planes $z = k$ with $k \geq c$ or $k \leq -c$ are ellipses or single points, while sections in planes $z = k$ with $|k| < c$ are empty. The section in the yz-plane is the hyperbola

$$\frac{z^2}{c^2} - \frac{y^2}{b^2} = 1,$$

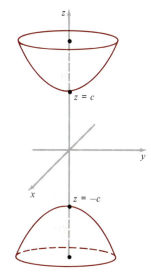

Figure 18.34 Hyperboloid of two sheets.

and it is this hyperbola that unifies the horizontal sections into a smooth surface — of "two sheets." Observe that (6) is identical with (4) except for the presence of the minus sign on the right, and it is this sign that makes all the difference between the surfaces in these two examples; for the right-hand side of (4) is positive for all z's, whereas the right-hand side of (6) is negative for $|z| < c$.

Example 4 The graph of the equation

$$\frac{x^2}{a^2} + \frac{y^2}{b^2} = \frac{z^2}{c^2} \tag{7}$$

is an *elliptic cone* (Fig. 18.35). This surface intersects the xz-plane and the yz-plane in the pairs of intersecting straight lines

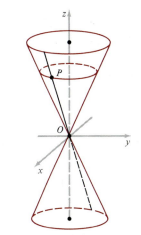

Figure 18.35 Elliptic cone.

$$z = \pm \frac{c}{a} x \qquad \text{and} \qquad z = \pm \frac{c}{b} y,$$

respectively. It intersects the xy-plane at the origin alone. All horizontal sections in planes $z = k$ with $k \neq 0$ are ellipses. (In Chapter 15 it was convenient to distinguish circles from ellipses; here we include circles among the ellipses.) It is clear from the form of (7) that if (x, y, z) is a point on the surface, then (tx, ty, tz) is also on the surface for any number t. This tells us that the entire surface can be thought of as generated by a moving line through the origin O and a variable point P on any horizontal elliptical section. When $a = b$, the cone is the familiar right circular cone.

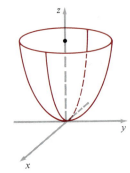

Example 5   The *elliptic paraboloid*

$$z = ax^2 + by^2 \tag{8}$$

is shown in Fig. 18.36. The vertical sections of this surface in the xz-plane and yz-plane are the parabolas

$$z = ax^2 \qquad \text{and} \qquad z = by^2,$$

respectively. The horizontal section in the plane $z = k$ is an ellipse if $k > 0$, the origin alone if $k = 0$, and empty if $k < 0$.

Figure 18.36   Elliptic paraboloid.

Example 6   In Fig. 18.37 we sketch the *hyperbolic paraboloid*

$$z = by^2 - ax^2. \tag{9}$$

The section in the yz-plane is the parabola $z = by^2$ opening upward, and that in the xz-plane is the parabola $z = -ax^2$ opening downward. In all planes $y = k$ parallel to the xz-plane, the sections are downward-opening parabolas that are identical with one another and can be thought of as hanging from their vertices at various points along the parabola $z = by^2$; this is emphasized in the way we have drawn the figure. Near the origin the surface rises in the y-direction and falls in the x-direction, and thus has the general shape of a saddle or a mountain pass. For this reason, the surface is often called a *saddle surface,* with the origin as the *saddle point.* It is clear from (9) that in the horizontal plane $z = k$, the section is a hyperbola with principal axis in the y-direction if $k > 0$, and a hyperbola with principal axis in the x-direction if $k < 0$; if $k = 0$, the section is a pair of intersecting straight lines through the origin.

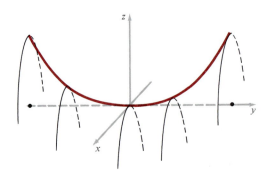

Figure 18.37   Hyperbolic paraboloid.

PROBLEMS

Sketch and identify the surfaces in Problems 1 to 14.

1 $2x^2 + y^2 + 4z^2 = 16.$ **2** $z^2 = 4(x^2 + y^2).$
3 $z = 4(x^2 + y^2).$ **4** $x^2 + z^2 - 4y^2 = 4.$
5 $y^2 - 4x^2 - 9z^2 = 36.$ **6** $z = 4 - 2x^2 - 3y^2.$
7 $z = x^2 - 2y^2.$ **8** $x^2 = y^2 + 4z^2.$
9 $x^2 - 4y^2 - z^2 = 4.$ **10** $x^2 + 9y^2 - 4z^2 = 36.$
11 $36x^2 + 4y^2 + 9z^2 = 36.$ **12** $y = 1 - x^2 - 2y^2.$
13 $z + 4x^2 = y^2.$
14 $x^2 + y^2 - z^2 - 2x - 4y + 1 = 0.$

15 Find the points at which the line

$$\frac{x-6}{3} = \frac{y+2}{-6} = \frac{z-2}{4}$$

pierces the ellipsoid

$$\frac{x^2}{81} + \frac{y^2}{36} + \frac{z^2}{9} = 1.$$

16 Show that the plane $2x - 2z - y = 10$ intersects the paraboloid

$$2z = \frac{x^2}{9} + \frac{y^2}{4}$$

at a single point, and find the point.

17 (a) Consider the ellipsoid

$$\frac{x^2}{a^2} + \frac{y^2}{b^2} + \frac{z^2}{c^2} = 1,$$

and find the area $A(k)$ of the elliptical section in the horizontal plane $z = k$. Hint: Recall that πAB is the area of an ellipse with semiaxes A and B.

(b) Use the formula found in (a) to calculate the volume of the ellipsoid by integration.

18 Consider the elliptic paraboloid $z = ax^2 + by^2$, and use integration to show that the volume of the segment cut off by the plane $z = k$ ($k > 0$) is half the area of its base times its height.

19 Show that the projection on the xy-plane of the curve of intersection of the surfaces $z = 1 - x^2$ and $z = x^2 + y^2$ is an ellipse. Hint: What does it mean geometrically to eliminate z from these equations?

20 Show that the projection on the yz-plane of the curve of intersection of the plane $x = 2y$ and the paraboloid $x = y^2 + z^2$ is a circle.

21 Show that the projection on the xy-plane of the intersection of the paraboloids $z = 3x^2 + 5y^2$ and $z = 8 - 5x^2 - 3y^2$ is a circle.

22 The two equations

$$x^2 + 3y^2 - z^2 + 3x = 0,$$

$$2x^2 + 6y^2 - 2z^2 - 4y = 3,$$

when taken together as a simultaneous system, define the space curve in which the corresponding surfaces

intersect. Show that this curve lies in a plane. Hint: Project onto a coordinate plane.

23 Use the methods of Section 15.6 to discover the nature of the graph of $z = xy$. Sketch the surface.

A *ruled surface* is a surface S with the property that for each point P on S there is a straight line through P that lies entirely on S. All cones and cylinders are ruled surfaces, while ellipsoids, hyperboloids of two sheets, and elliptic paraboloids obviously are not. It is very surprising that all hyperboloids of one sheet and all hyperbolic paraboloids are ruled surfaces.

24 Show that the hyperboloid of one sheet $x^2 + y^2 - z^2 = 1$ is a ruled surface, as follows:

(a) The section of the surface in the xy-plane is the circle C whose equation is $x^2 + y^2 = 1$. Let $P_0 = (x_0, y_0, 0)$ be a point on C, and show that the line L whose equations are

$$x = x_0 + y_0 t, \qquad y = y_0 - x_0 t, \qquad z = t,$$

passes through P_0 and lies entirely on the surface.

(b) If $P = (x, y, z)$ is an arbitrary point on the surface, show that the line L in part (a) passes through P for a suitable point $P_0 = (x_0, y_0, 0)$. Thus, as P_0 moves around C, the lines L cover the surface.*

25 Show that the hyperbolic paraboloid $z = y^2 - x^2$ is a ruled surface by showing that if $P_0 = (x_0, y_0, y_0^2 - x_0^2)$ is any point on the surface, then the line

$$x = x_0 + t, \qquad y = y_0 + t,$$

$$z = (y_0^2 - x_0^2) + 2(y_0 - x_0)t$$

passes through P_0 and lies entirely on the surface.†

The two families of straight lines constituting the doubly ruled surfaces discussed in Problems 24 and 25 are shown in Fig. 18.38.

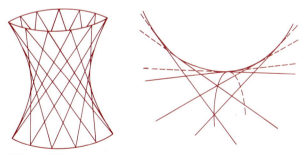

Figure 18.38 Doubly ruled surfaces.

---

* The family of lines $x = x_0 + y_0 t$, $y = y_0 - x_0 t$, $z = -t$ also covers the surface, and for this reason the hyperboloid of one sheet is often called a *doubly ruled* surface.

† The family of lines $x = x_0 + t$, $y = y_0 - t$, $z = (y_0^2 - x_0^2) - 2(y_0 + x_0)t$ also covers the surface, so the hyperbolic paraboloid is also doubly ruled.

18.7

CYLINDRICAL AND SPHERICAL COORDINATES

In plane analytic geometry we used a rectangular coordinate system for some types of problems and a polar coordinate system for others. We saw that there are many situations in which one system is more convenient than the other. The same is true for the study of geometry and calculus in space. We now describe two other three-dimensional coordinate systems, in addition to the now-familiar rectangular coordinate system, that are often useful for dealing with special kinds of problems.

Consider a point P in space whose rectangular coordinates are (x, y, z). The *cylindrical coordinates* of this point are obtained by replacing x and y with the corresponding polar coordinates r and θ, and allowing z to remain unchanged. That is, we place a z-axis on top of a polar coordinate system and describe the location of a point in space by the three coordinates $(r, \theta\, z)$. We will always assume that this cylindrical coordinate system is superimposed on a rectangular coordinate system in the manner shown in Fig. 18.39, so that the transformation equations connecting the two sets of coordinates of a given point are

$$x = r \cos \theta, \qquad y = r \sin \theta, \qquad z = z,$$

and

$$r^2 = x^2 + y^2, \qquad \tan \theta = \frac{y}{x}, \qquad z = z.$$

It is easy to see that the graph of the equation $r =$ a constant is a right circular cylinder whose axis is the z-axis; this is the reason for the term "cylindrical coordinates." Similarly, the graph of $\theta =$ a constant is a plane containing the z-axis, and the graph of $z =$ a constant is a horizontal plane.

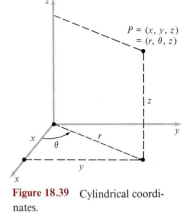

$P = (x, y, z)$
$= (r, \theta, z)$

Figure 18.39   Cylindrical coordinates.

Example 1   Find cylindrical coordinates for the points P_1 and P_2 whose rectangular coordinates are $(3, 3, 7)$ and $(2\sqrt{3}, 2, 5)$, respectively.

Solution   For P_1 we have $r = \sqrt{9 + 9} = 3\sqrt{2}$, $\tan \theta = 1$, $z = 7$, so a set of cylindrical coordinates is $(3\sqrt{2}, \pi/4, 7)$. For P_2 we have $r = \sqrt{12 + 4} = 4$, $\tan \theta = 1/\sqrt{3} = \frac{1}{3}\sqrt{3}$, $z = 5$, so a set of cylindrical coordinates is $(4, \pi/6, 5)$.

Example 2   Describe the surfaces

(a) $r + z = 3$, and

(b) $r(2 \cos \theta + 5 \sin \theta) + 3z = 0$.

Solution   (a) The intersection of the surface $r + z = 3$ with the yz-plane is the straight line $y + z = 3$, because $r = y$ in the yz-plane. But θ is missing from the given equation, so the desired surface is symmetric about the z-axis, and is therefore the cone generated by revolving the line $y + z = 3$ about the z-axis. More generally, it follows from our discussion of surfaces of revolution in Section 18.5 that if a curve $f(y, z) = 0$ is revolved about the z-axis, then the cylindrical equation of the resulting surface is $f(r, z) = 0$.

(b) Since $r \cos \theta = x$ and $r \sin \theta = y$, the given equation transforms into $2x + 5y + 3z = 0$, which is the plane through the origin with normal vector $\mathbf{N} = 2\mathbf{i} + 5\mathbf{j} + 3\mathbf{k}$.

Example 3 Find a cylindrical equation for (a) the spheroid $x^2 + y^2 + 2z^2 = 4$, and (b) the hyperbolic paraboloid $z = x^2 - y^2$.

Solution The equation in (a) transforms at once into $r^2 + 2z^2 = 4$. For (b), we have

$$z = x^2 - y^2$$
$$= r^2 \cos^2 \theta - r^2 \sin^2 \theta = r^2 (\cos^2 \theta - \sin^2 \theta)$$
$$= r^2 \cos 2\theta,$$

so $z = r^2 \cos 2\theta$ is the desired equation.

In physics, cylindrical coordinates are particularly convenient for studying situations in which there is axial symmetry, that is, symmetry about a line in space. As examples we mention two important classes of problems: those dealing with the flow of heat in solid cylindrical rods, and those concerned with the movements of a vibrating circular membrane—for instance, a drumhead.

Again consider a point P in space whose rectangular coordinates are (x, y, z). The *spherical coordinates* of P are the numbers (ρ, ϕ, θ) shown in Fig. 18.40. Here ρ (the Greek letter *rho*) is the distance from the origin O to P, so $\rho \geq 0$. The angle ϕ is the angle down from the positive z-axis to the radial line OP, and it is understood that ϕ is restricted to the interval $0 \leq \phi \leq \pi$. Finally, the angle θ has exactly the same meaning in spherical coordinates as it has in cylindrical coordinates; that is, θ is the angle from the positive x-axis to the line OP', where P' is the projection of P on the xy-plane. It is clear from the figure that $OP' = \rho \sin \phi$, and since $x = OP' \cos \theta$ and $y = OP' \sin \theta$, we have the transformation equations

$$x = \rho \sin \phi \cos \theta, \quad y = \rho \sin \phi \sin \theta, \quad z = \rho \cos \phi,$$

and

$$\rho^2 = x^2 + y^2 + z^2, \quad \tan \phi = \frac{\sqrt{x^2 + y^2}}{z}, \quad \tan \theta = \frac{y}{x}.$$

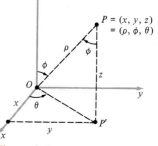

Figure 18.40 Spherical coordinates.

The term "spherical coordinates" is used because the graph of the equation $\rho = $ a constant is a sphere with center at the origin. The graph of $\phi = $ a constant α is the upper nappe of a cone with vertex at the origin and vertex angle α, if $0 < \alpha < \pi/2$. The graph of $\theta = $ a constant is a plane containing the z-axis, just as in cylindrical coordinates.

Example 4 Find an equation in spherical coordinates for the sphere $x^2 + y^2 + z^2 - 2az = 0$, where $a > 0$.

Solution Since $\rho^2 = x^2 + y^2 + z^2$ and $z = \rho \cos \phi$, the given equation can be written as

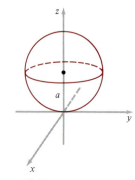

Figure 18.41

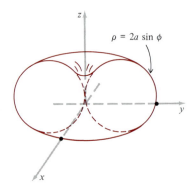

Figure 18.42

$$\rho^2 - 2a\rho \cos \phi = 0 \qquad \text{or} \qquad \rho(\rho - 2a \cos \phi) = 0.$$

The graph of this equation is the graph of $\rho = 0$ together with the graph of $\rho - 2a \cos \phi = 0$. But the graph of $\rho = 0$ (namely, the origin) is part of the graph of $\rho = 2a \cos \phi$, so the desired equation is

$$\rho = 2a \cos \phi.$$

This is the sphere of radius a that is tangent to the xy-plane at the origin, as shown in Fig. 18.41.

Example 5   What is the graph of the spherical equation $\rho = 2a \sin \phi$?

Solution   The variable θ is missing from this equation, so we have a surface of revolution about the z-axis. In the yz-plane the equation $\rho = 2a \sin \phi$ represents a circle of radius a, as shown in Fig. 18.42. Since the graph we are seeking is obtained by revolving this circle about the z-axis, this graph is a torus (doughnut) in which the hole has radius zero.

There are many physical uses of spherical coordinates, ranging from problems about heat conduction to problems in the theory of gravitation. We shall discuss some of these applications in Chapter 20.

PROBLEMS

1 Find a set of cylindrical coordinates for the point whose rectangular coordinates are
   (a) $(2, 2, -1)$;
   (b) $(1, -\sqrt{3}, 7)$;
   (c) $(3, \sqrt{3}, 2)$;
   (d) $(3, 6, 5)$.

2 Find the rectangular coordinates of the point with cylindrical coordinates
   (a) $(\sqrt{2}, \pi/4, -2)$;
   (b) $(\sqrt{3}, 5\pi/6, 11)$;
   (c) $(1, 1, 1)$;
   (d) $(2, \pi/3, \pi)$.

3 Find a set of spherical coordinates for the point whose rectangular coordinates are
   (a) $(1, 1, \sqrt{6})$;
   (b) $(1, -1, -\sqrt{6})$;
   (c) $(1, 1, \sqrt{2})$;
   (d) $(0, -1, \sqrt{3})$.

4 Find the rectangular coordinates of the point with spherical coordinates
   (a) $(3, \pi/2, \pi/2)$;
   (b) $(4, \pi/2, \pi)$;
   (c) $(4, \pi/3, \pi/3)$;
   (d) $(4, 2\pi/3, \pi/3)$.

In Problems 5 to 11, find a cylindrical equation for the surface whose rectangular equation is given. Sketch the surface.

5 $x^2 + y^2 + z^2 = 16$.   **6** $x^2 + y^2 = 6z$.

7 $x^2 + y^2 = z^2$.   **8** $x^2 - y^2 = 3$.

9 $x^2 + y^2 - 2y = 0$.   **10** $x^2 + y^2 - 4x = 0$.

11 $x^2 + y^2 = 9$.

12 Find a cylindrical equation for the surface whose rectangular equation is $z^2(x^2 - y^2) = 4xy$.

In Problems 13 to 18, find a spherical equation for the surface whose rectangular equation is given. Sketch the surface.

13 $x^2 + y^2 + z^2 = 16$.

14 $x^2 + y^2 + z^2 + 4z = 0$.

15 $x^2 + y^2 + z^2 - 6z = 0$.

16 $x^2 + y^2 = 9$.

17 $z = 4 - x^2 - y^2$.

18 $(x^2 + y^2 + z^2)^3 = (x^2 + y^2)^2$.

19

PARTIAL DERIVATIVES

19.1
FUNCTIONS OF SEVERAL VARIABLES

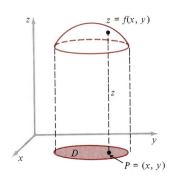

Figure 19.1 A surface in space.

Many of the functions that arise in mathematics and its applications involve two or more independent variables. We have already met functions of this kind in our study of solid analytic geometry. Thus, the equation $z = x^2 - y^2$ is the equation of a certain saddle surface, but it also defines z as a function of the two variables x and y, and the surface can be thought of as the graph of this function.

We usually denote an arbitrary function of the two variables x and y by writing $z = f(x, y)$, and we can visualize such a function by sketching—or imagining—its graph in xyz-space, as suggested in Fig. 19.1. In this figure, $P = (x, y)$ is a "suitable" point in the xy-plane—that is, a point in the domain D of the function—and z is the directed distance up or down to the corresponding point on the surface. This surface is thought of as lying "over" the domain D, even though part of it may actually be below the xy-plane.

By an obvious extension of the notation used here, $w = f(x, y, z, t, u, v)$ is a function of the six variables displayed in parentheses. For example, if the temperature T at a point P inside a solid iron sphere depends on the three rectangular coordinates x, y, and z of P, then we write $T = f(x, y, z)$; and if we also allow for the possibility that the temperature at a given point may vary with the time t, then T is a function of all four variables, $T = f(x, y, z, t)$.

In this chapter we shall see that the main themes of single-variable differential calculus—derivatives, rates of change, chain rule computations, maximum-minimum problems, and differential equations—can all be extended to functions of several variables. However, students should be prepared for the fact that there are striking differences between single-variable calculus and multivariable calculus. Since most of these differences already show up in functions of only two independent variables, we usually emphasize this case, and refer more briefly to functions of three or more variables. In the next chapter we turn to the integral calculus of functions of several variables.

DOMAIN

Just as in our previous work, the *domain* (or *domain of definition*) of a function $z = f(x, y)$ is the set of all points $P = (x, y)$ in the xy-plane for which there exists a corresponding z, and similarly for functions defined in xyz-space, $xyzt$-space, etc. Most of the functions we deal with are defined by formulas, and in these cases the domain is understood to be the largest set of points for which the formula makes sense. For example, the domain of

$$z = f(x, y) = \frac{1}{x - y}$$

is understood to be the set of all points (x, y) with $x \neq y$, that is, all points in the xy-plane that do not lie on the line $y = x$. The domain of

$$z = g(x, y) = \sqrt{9 - x^2 - y^2}$$

is the set of all points (x, y) for which $9 - x^2 - y^2 \geq 0$, that is, the circular disk $x^2 + y^2 \leq 9$ of radius 3 with center at the origin. And the domain of

$$w = h(x, y, z) = \frac{2x + 3y + 4z}{x^2 + y^2 + z^2}$$

is the set of all points (x, y, z) for which $x^2 + y^2 + z^2 \neq 0$, that is, all points of xyz-space except the origin.

In discussing a general function $z = f(x, y)$, we shall often require that this function be defined at a certain point P_0 and throughout some *neighborhood* of this point. This means that the domain of $f(x, y)$ must include not only P_0 itself, but also every point "sufficiently close" to P_0, that is, every point in some small circular disk centered on P_0. Similar remarks apply to functions defined in xyz-space, etc.

CONTINUITY

There are several places in this chapter where it will be necessary to mention continuity in order to state things correctly. This concept extends in a natural way from the one-variable case to functions $f(x, y)$, as follows.

A function $f(x, y)$ is said to be *continuous* at a point (x_0, y_0) in its domain if its value $f(x, y)$ can be made as close as we please to $f(x_0, y_0)$ by taking the point (x, y) close enough to (x_0, y_0), that is, if $|f(x, y) - f(x_0, y_0)|$ can be made as small as we please by making both $|x - x_0|$ and $|y - y_0|$ small enough. For example, $f(x, y) = xy$ is continuous at any point (x_0, y_0), because

$$|xy - x_0y_0| = |xy - xy_0 + xy_0 - x_0y_0|$$
$$= |x(y - y_0) + y_0(x - x_0)|$$
$$\leq |x||y - y_0| + |y_0||x - x_0|,$$

and it is easy to see that the quantity last written can be made as small as we please by making both $|x - x_0|$ and $|y - y_0|$ small enough.

On the other hand, the function defined by

$$f(x, y) = \begin{cases} \dfrac{xy}{x^2 + y^2} & \text{if } (x, y) \neq (0, 0), \\ 0 & \text{if } (x, y) = (0, 0), \end{cases} \tag{1}$$

is not continuous at the origin $(0, 0)$. For, if we let (x, y) approach $(0, 0)$ along a line $y = mx$ with $m \neq 0$, then

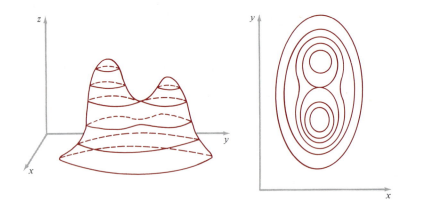

Figure 19.2   Level curves.

$$f(x, y) = \frac{mx^2}{x^2 + m^2x^2} = \frac{m}{1 + m^2}, \tag{2}$$

which is a nonzero constant, and these values cannot be made as close as we please to $f(0, 0) = 0$ by making (x, y) close enough to $(0, 0)$. To express this in another way, (2) shows that the values of the function approach different limiting values as the point (x, y) approaches the origin from different directions, and this is impossible if the function is continuous at the origin.

We shall not pursue the details of this topic any further, beyond making the rather loose statement that any finite combination of elementary functions is continuous at each point of its domain. Also, continuity is defined in essentially the same way for functions of three or more variables.

LEVEL CURVES

Many simple functions $z = f(x, y)$ have graphs that are much too difficult to sketch. Fortunately there is another way to understand and express the geometric nature of such a function.

The basic idea comes from the art of the mapmaker. In mapping terrain with valleys, hills, and mountains, it is common practice to draw curves joining points of constant elevation. When these curves are included on a map and properly labeled, the resulting topographical map enables an experienced user to obtain a clear mental picture of the contours of the land in three-dimensional space from this two-dimensional representation.

We can do the same thing to portray a function $z = f(x, y)$ of two variables. For any value c that $f(x, y)$ assumes, we can sketch the curve

$$f(x, y) = c$$

in the xy-plane, as shown in Fig. 19.2. Such a curve is called a *level curve*; it lies in the domain of the function, and on it $z = f(x, y)$ has the constant value c.

A collection of level curves is called a *contour map*; it can give a good idea of the shape of the graph, and is the next best thing to a three-dimensional sketch. For instance, the graph of $z = xy$ is difficult — though not impossible — to draw. However, a reasonably clear idea of the shape of this graph is given by the contour map shown in Fig. 19.3, which is easy to draw.

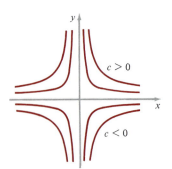

Figure 19.3

Each level curve $xy = c$ is a hyperbola in the first and third quadrants if $c > 0$, a hyperbola in the second and fourth quadrants if $c < 0$, and the two axes taken together if $c = 0$. We ascend the surface as we leave the origin going into the first and third quadrants, and descend it as we leave going into the second and fourth quadrants, and in this way we see that the origin is the saddle point of a saddle surface. Students should try to use this figure to visualize the shape of the surface as it appears in three-dimensional space, looking down on it from above.

LEVEL SURFACES

Drawing graphs for functions of two variables is often difficult, but drawing graphs for functions of three variables is always impossible. We would need a visible space of four dimensions to contain such a graph, and no such space is available.

However, the concept of level curves suggests a way to visualize the behavior of a function $w = f(x, y, z)$ of three variables: examine its *level surfaces*. These are the surfaces

$$f(x, y, z) = c \qquad (3)$$

for various values of the constant c. Of course, level surfaces can be hard to draw, but a knowledge of what they are can help us form a useful intuitive idea of the nature of the function. In Fig. 19.4 we present a schematic view of three adjacent level surfaces of the form (3) for three values of the constant c, where $c_1 < c_2 < c_3$. As a point $P = (x, y, z)$ moves along the lowest surface, the value of $w = f(x, y, z)$ is constantly equal to c_1; but as this point hops to the next surface above it, the value of the function increases to c_2; and so on.

We consider two simple examples. In the case of the function $w = x + 2y + 3z$, the level surfaces are easily seen to be the planes

$$x + 2y + 3z = c$$

with normal vector $\mathbf{N} = \mathbf{i} + 2\mathbf{j} + 3\mathbf{k}$; and for $w = \sqrt{x^2 + y^2 + z^2}$, the level surfaces are the concentric spheres

$$x^2 + y^2 + z^2 = c^2.$$

In applications, if the function $w = f(x, y, z)$ represents the temperature at the point $P = (x, y, z)$, then the level surfaces are called *isothermal surfaces*; if it represents potential, they are called *equipotential surfaces*.

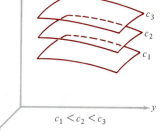

Figure 19.4 Level surfaces.

HIGHER DIMENSIONS

Level surfaces have a certain limited value, but in a sense they avoid the real question: How do we go about trying to obtain an intuitive understanding of the behavior of functions of three or more variables?

Briefly, what we do is work by analogy with the one- and two-variable cases. For example, there is nothing to prevent us from considering the set of all quadruples of numbers such as $(2, -3, 1, 4)$ as forming a perfectly legitimate four-dimensional space, with an origin $(0, 0, 0, 0)$, four coordinate axes, and a satisfactory concept of the distance from an arbitrary point (x, y, z, w) to the origin,

$$d = \sqrt{x^2 + y^2 + z^2 + w^2}.$$

We can now consider the graph of a function

$$w = f(x, y, z)$$

as forming a three-dimensional "surface" in this four-dimensional space, with the domain D of the function lying in the three-dimensional "coordinate plane" consisting of all points of the form $(x, y, z, 0)$.

In a similar way, if n is any positive integer we can think of the graph of a function of n variables as forming an n-dimensional "surface" in $(n + 1)$-dimensional space. It is true that for $n \geq 3$ we can no longer draw pictures, but we can still bolster our intuition by using geometric language, and we can still think geometrically, but in a looser way. However, as we move farther away from the kind of mathematics that we can study and understand by drawing pictures, it is necessary to give more attention to the algebraic and analytic aspects of what we are doing, in order to avoid being misled by words and analogies. Nevertheless, the words, analogies, and geometric intuition remain indispensable, for they suggest worthwhile things to think about and prevent us from feeling totally lost among abstractions.

PROBLEMS

In Problems 1 to 12, find the domain of the given function.

1   $f(x, y) = \dfrac{xy}{y - 2x}$.

2   $f(x, y) = \dfrac{1}{x} + \dfrac{1}{y}$.

3   $f(x, y) = \sqrt{xy}$.

4   $f(x, y) = \dfrac{1}{(e^x + e^y)^2}$.

5   $f(x, y) = \ln(y - 3x)$.

6   $f(x, y, z) = \sqrt{x^2 + y^2 + z^2}$.

7   $f(x, y, z) = \dfrac{1}{\sqrt{x^2 + y^2 + z^2}}$.

8   $f(x, y, z) = \dfrac{z}{4x^2 - y^2}$.

9   $f(x, y, z) = \sqrt{16 - x^2 - y^2 - z^2}$.

10   $f(x, y, z) = \dfrac{1}{xyz}$.

11   $f(x, y, z) = xy \ln z + 3 \tan \frac{1}{2}z$.

12   $f(x, y, z) = \ln(x^2 + y^2 + z^2 - 1)$.

13   Show that the function defined by

$$f(x, y) = \begin{cases} \dfrac{xy}{\sqrt{x^2 + y^2}} & \text{if } (x, y) \neq (0, 0), \\ 0 & \text{if } (x, y) = (0, 0), \end{cases}$$

is continuous at the origin. Hint: Use $x = r \cos \theta$ and $y = r \sin \theta$ to transform to polar coordinates.

In Problems 14 to 24, represent the given function by drawing a few level curves, and try to visualize the surface from the resulting contour map.

14   $z = x^2 + y^2$.

15   $z = x^2 + 2y^2$.

16   $z = x + y$.

17   $z = x - y$.

18   $z = 2x - y$.

19   $z = x^2 - y$.

20   $z = x^3 - y$.

21   $z = y/x$.

22   $z = y/x^2$.

23   $z = x^2 - y^2$.

24   $z = \sqrt{x^2 - y^2}$.

In each of the following problems, sketch a few level surfaces for the given function and use these to estimate the general direction in which the values of the function increase.

25   $w = \dfrac{x^2}{4} + \dfrac{y^2}{9} + \dfrac{z^2}{16}$.

26   $w = \dfrac{1}{x^2 + y^2 + z^2}$.

27   $w = 2x - 5y + 3z$.

28   $w = x^2 + y^2 - z^2$.

Suppose that $y = f(x)$ is a function of only one variable. We know that its derivative, defined by

$$\frac{dy}{dx} = \lim_{\Delta x \to 0} \frac{f(x + \Delta x) - f(x)}{\Delta x},$$

can be interpreted as the rate of change of y with respect to x. In the case of a function $z = f(x, y)$ of two variables, we shall need similar mathematical machinery for working with the rate at which z changes as both x and y vary. The key idea is to allow only one variable to change at a time, while holding the other fixed. For functions of more than two variables, we vary one of them while holding *all* the others fixed. Specifically, we differentiate with respect to only one variable at a time, regarding all the others as constants, and this gives us one derivative corresponding to each of the independent variables. These individual derivatives are the constituents from which we build the more complicated machinery that will be needed later.

To return to our function $z = f(x, y)$ of two variables, we first hold y fixed and let x vary. The rate of change of z with respect to x is denoted by $\partial z/\partial x$ and defined by

$$\frac{\partial z}{\partial x} = \lim_{\Delta x \to 0} \frac{f(x + \Delta x, y) - f(x, y)}{\Delta x}.$$

This limit (if it exists) is called the *partial derivative of z with respect to x,* and is read "partial z, partial x." The most commonly used notations for this partial derivative are

$$\frac{\partial z}{\partial x}, \qquad z_x, \qquad \frac{\partial f}{\partial x}, \qquad f_x, \qquad f_x(x, y),$$

and we shall use all of these from time to time in order to help students become accustomed to them. The symbol ∂ in the notation $\partial z/\partial x$ is called the "roundback d" or "curly d"; it is used to emphasize that there are other independent variables present during the process of differentiating with respect to x.

Similarly, if x is held fixed and y is allowed to vary, then the *partial derivative of z with respect to y* is defined by

$$\frac{\partial z}{\partial y} = \lim_{\Delta y \to 0} \frac{f(x, y + \Delta y) - f(x, y)}{\Delta y},$$

and the standard notations in this case are

$$\frac{\partial z}{\partial y}, \qquad z_y, \qquad \frac{\partial f}{\partial y}, \qquad f_y, \qquad f_y(x, y).$$

The actual calculation of partial derivatives for most functions is very easy: Treat every independent variable except the one we are interested in as if it were a constant, and apply the familiar rules.

Example 1 Calculate the partial derivatives $\partial f/\partial x$ and $\partial f/\partial y$ of the function $f(x, y) = x^3 - 3x^2y^3 + y^2$.

Solution   To find the partial of f with respect to x, we think of y as a constant and differentiate in the usual way,

$$\frac{\partial f}{\partial x} = 3x^2 - 6xy^3.$$

When we regard x as a constant and differentiate with respect to y, we obtain

$$\frac{\partial f}{\partial y} = -9x^2y^2 + 2y.$$

The notations $f_x(x, y)$ and $f_y(x, y)$ are useful for indicating the values of partial derivatives at specific points.

Example 2   (a) If $f(x, y) = xy^2 + x^3$, then

$$f_x(x, y) = y^2 + 3x^2, \qquad f_y(x, y) = 2xy,$$
$$f_x(2, 1) = 13, \qquad\qquad f_y(2, 1) = 4.$$

In the other notation, the numerical values given here by the simple and convenient symbols $f_x(2, 1)$ and $f_y(2, 1)$ would have to be written more clumsily as

$$\left(\frac{\partial f}{\partial x}\right)_{(2,1)} \quad\text{and}\quad \left(\frac{\partial f}{\partial y}\right)_{(2,1)}.$$

(b) If $g(x, y) = xe^{xy^2}$, then

$$g_x(x, y) = xy^2e^{xy^2} + e^{xy^2}, \qquad g_y(x,y) = 2x^2ye^{xy^2}.$$

(c) If $h(x, y) = \sin x^2 \cos 3y$, then

$$h_x(x, y) = 2x \cos x^2 \cos 3y, \qquad h_y(x, y) = -3 \sin x^2 \sin 3y.$$

These examples illustrate the fact that the partial derivatives of a function of x and y are themselves functions of x and y.

These ideas and notations apply just as easily to functions of any number of variables.

Example 3   If $w = f(x, y, z, u, v) = xy^2 + 2x^3 + xyz + zu + \tan uv$, then

$$\frac{\partial w}{\partial x} = y^2 + 6x^2 + yz, \qquad \frac{\partial w}{\partial y} = 2xy + xz, \qquad \frac{\partial w}{\partial z} = xy + u,$$

$$\frac{\partial w}{\partial u} = z + v \sec^2 uv, \qquad \frac{\partial w}{\partial v} = u \sec^2 uv.$$

In the one-variable case, we know that the derivative dy/dx can legitimately be thought of as a fraction, the quotient of the differentials dy and dx. The notation $\partial z/\partial x$ for the partial derivative $f_x(x, y)$ suggests that something similar might be done with ∂z and ∂x. However, it is not possible to treat partial derivatives as fractions. We give an example to emphasize this point.

Example 4   The *ideal gas law* states that for a given quantity of gas, the pressure p, volume V, and absolute temperature T are connected by the equation $pV = nRT$, where n is the number of moles of gas in the sample and R is a constant. Show that

$$\frac{\partial p}{\partial V}\frac{\partial V}{\partial T}\frac{\partial T}{\partial p} = -1.$$

Solution Since

$$p = \frac{nRT}{V}, \qquad V = \frac{nRT}{p}, \qquad T = \frac{pV}{nR},$$

we have

$$\frac{\partial p}{\partial V} = -\frac{nRT}{V^2}, \qquad \frac{\partial V}{\partial T} = \frac{nR}{p}, \qquad \frac{\partial T}{\partial p} = \frac{V}{nR}.$$

It follows that

$$\frac{\partial p}{\partial V}\frac{\partial V}{\partial T}\frac{\partial T}{\partial p} = \left(-\frac{nRT}{V^2}\right)\frac{nR}{p}\frac{V}{nR} = -\frac{nRT}{pV} = -1.$$

The fact that this result is -1 instead of $+1$ shows that we cannot treat the partial derivatives on the left as fractions.

When we are working with a function $z = f(x, y)$ of only two variables, the partial derivatives have the following simple geometric interpretation. The graph of this function is a surface, as shown in Fig. 19.5. Let (x_0, y_0) be a given point in the xy-plane, with (x_0, y_0, z_0) the corresponding point on the surface. To hold y fixed at the value y_0 means to intersect the surface with the plane $y = y_0$, and the intersection is the curve

$$z = f(x, y_0)$$

in that plane. The number

$$\left(\frac{\partial z}{\partial x}\right)_{(x_0,\, y_0)} = f_x(x_0, y_0)$$

is the slope of the tangent line to this curve at $x = x_0$. Thus, in the figure we have

$$\tan \alpha = \left(\frac{\partial z}{\partial x}\right)_{(x_0,\, y_0)} = f_x(x_0, y_0).$$

Similarly, the intersection of the surface with the plane $x = x_0$ is the curve

$$z = f(x_0, y),$$

and the other partial derivative is the slope of the tangent to this curve at $y = y_0$,

$$\tan \beta = \left(\frac{\partial z}{\partial y}\right)_{(x_0,\, y_0)} = f_y(x_0, y_0).$$

No such interpretation is available when there are more than two independent variables.

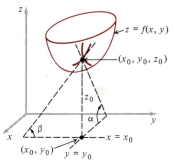

Figure 19.5

We remarked that for a function $z = f(x, y)$ of two variables, the partial derivatives f_x and f_y are also functions of two variables, and may themselves have partial derivatives. As we might expect, these *second-order partial derivatives* are denoted by several symbols. If we start with the first derivatives

$$\frac{\partial f}{\partial x} = f_x \quad \text{and} \quad \frac{\partial f}{\partial y} = f_y,$$

then the derivatives with respect to x are

$$\frac{\partial}{\partial x}\left(\frac{\partial f}{\partial x}\right) = \frac{\partial^2 f}{\partial x^2} = \frac{\partial}{\partial x} f_x = f_{xx}$$

and

$$\frac{\partial}{\partial x}\left(\frac{\partial f}{\partial y}\right) = \frac{\partial^2 f}{\partial x\, \partial y} = \frac{\partial}{\partial x} f_y = f_{yx};$$

and the derivatives with respect to y are

$$\frac{\partial}{\partial y}\left(\frac{\partial f}{\partial x}\right) = \frac{\partial^2 f}{\partial y\, \partial x} = \frac{\partial}{\partial y} f_x = f_{xy}$$

and

$$\frac{\partial}{\partial y}\left(\frac{\partial f}{\partial y}\right) = \frac{\partial^2 f}{\partial y^2} = \frac{\partial}{\partial y} f_y = f_{yy}.$$

This notation may seem a bit confusing at first, but it is actually quite reasonable. Observe that in f_{yx} we differentiate first with respect to the "inside" variable y, then with respect to the "outside" variable x. This is the natural order, since f_{yx} ought to mean $(f_y)_x$. Thus, in the symbols f_{yx} and f_{xy}, the subscript letters accumulate from left to right, because this is the order in which the differentiations are performed. For the same reason, in the symbols

$$\frac{\partial^2 f}{\partial x\, \partial y} \quad \text{and} \quad \frac{\partial^2 f}{\partial y\, \partial x},$$

it is natural for the letters indicating the variable of differentiation to accumulate from right to left: first y, then x in the first of these; and first x, then y in the second.

The *pure* second partial derivatives,

$$f_{xx} = \frac{\partial^2 f}{\partial x^2} \quad \text{and} \quad f_{yy} = \frac{\partial^2 f}{\partial y^2},$$

don't represent anything really new. Each is found by holding one variable constant and differentiating twice with respect to the other variable, and each gives the rate of change of the rate of change of f in the direction of one of the axes.

Example 5   If $f(x, y) = x^3 e^{5y} + y \sin 2x$, then

$$f_x = 3x^2 e^{5y} + 2y \cos 2x, \qquad f_y = 5x^3 e^{5y} + \sin 2x,$$
$$f_{xx} = 6x e^{5y} - 4y \sin 2x, \qquad f_{yy} = 25x^3 e^{5y}.$$

On the other hand, the *mixed* second partial derivatives,

$$f_{xy} = \frac{\partial^2 f}{\partial y\,\partial x} \quad \text{and} \quad f_{yx} = \frac{\partial^2 f}{\partial x\,\partial y},$$

represent new ideas. The mixed partial derivative f_{xy} gives the rate of change in the y-direction of the rate of change of f in the x-direction, and f_{yx} gives the rate of change in the x-direction of the rate of change of f in the y-direction. It is not at all clear how these two mixed partials are related to each other, if indeed they are related at all.

Example 5 (cont.) For the function being considered, $f(x, y) = x^3 e^{5y} + y \sin 2x$, we easily see that

$$f_x = 3x^2 e^{5y} + 2y \cos 2x, \qquad f_y = 5x^3 e^{5y} + \sin 2x,$$

$$f_{xy} = 15x^2 e^{5y} + 2 \cos 2x, \qquad f_{yx} = 15x^2 e^{5y} + 2 \cos 2x.$$

For the particular function considered in this example, we obviously have

$$f_{xy} = f_{yx}, \tag{1}$$

or equivalently,

$$\frac{\partial^2 f}{\partial y\,\partial x} = \frac{\partial^2 f}{\partial x\,\partial y},$$

so the order of differentiation seems to be unimportant — at least in this case. But this is not an accident, and (1) is true for almost all functions that normally arise in applications. More precisely, *if both f_{xy} and f_{yx} exist for all points near (x_0, y_0) and are continuous at (x_0, y_0), then*

$$f_{xy}(x_0, y_0) = f_{yx}(x_0, y_0).$$

A proof of this statement is given in Appendix C.15.

Partial derivatives of order greater than two, as well as higher-order derivatives of functions of more than two variables, are defined in the obvious way. For example, if $w = f(x, y, z)$, then

$$\frac{\partial^3 f}{\partial x\,\partial y\,\partial z} = \frac{\partial}{\partial x}\left(\frac{\partial^2 f}{\partial y\,\partial z}\right) = (f_{zy})_x = f_{zyx},$$

$$\frac{\partial^4 f}{\partial z\,\partial y\,\partial x^2} = \frac{\partial}{\partial z}\left(\frac{\partial^3 f}{\partial y\,\partial x^2}\right) = (f_{xxy})_z = f_{xxyz},$$

etc. In general, with suitable continuity, it is immaterial in what order a sequence of partial differentiations is carried out, for by (1) we can reverse the order of any two successive differentiations. For example, $f_{xxyz} = f_{xyxz} = f_{xyzx} = f_{yxzx} = f_{yzxx}.$

PROBLEMS

In Problems 1 to 14, find $\partial z/\partial x$ and $\partial z/\partial y$.

1 $z = 2x + 3y.$

2 $z = 5x^2y.$

3 $z = \dfrac{2y^2}{3x + 1}.$

4 $z = y \cos x.$

5 $z = x^2 \sin y.$

6 $z = \tan 3x + \cot 4y.$

7 $z = x \tan 2y + y \tan 3x.$

8 $z = \sin xy.$

9 $z = \cos(3x - y).$

10 $z = xye^{xy}.$

11 $z = e^x \sin y.$

12 $z = \tan^{-1} \dfrac{x}{y}.$

13 $z = e^y \ln x^2.$

14 $z = \ln(3x + y^2).$

In Problems 15 to 18, find the partial derivatives with respect to x, y, and z.

15 $w = x^2y^5z^7.$

16 $w = \sin^{-1} \dfrac{z}{xy}.$

17 $w = x \ln \dfrac{y}{z}.$

18 $w = e^{x^2 + y^3 + z^4}.$

19 Consider the surface $z = 2x^2 + y^2.$
  (a) The plane $y = 3$ intersects the surface in a curve. Find the equations of the tangent line to this curve at $x = 2.$
  (b) The plane $x = 2$ intersects the surface in a curve. Find the equations of the tangent line to this curve at $y = 3.$

20 Consider the surface $z = x^2/(y^2 - 3).$
  (a) The plane $y = 2$ intersects the surface in a curve. Find the equations of the tangent line to this curve at $x = 3.$
  (b) The plane $x = 3$ intersects the surface in a curve. Find the equations of the tangent line to this curve at $y = 2.$

21 Show that all of the following functions $z = f(x, y)$ satisfy the equation $x \dfrac{\partial z}{\partial x} + y \dfrac{\partial z}{\partial y} = 0$:
  (a) $z = \dfrac{x}{y}$;
  (b) $z = \dfrac{x}{x + y}$;
  (c) $z = \ln \dfrac{2y^2}{x^2}$;
  (d) $z = \dfrac{xy^2}{x^3 + y^3}.$

22 If $z = ye^{x/y}$, show that $xz_x + yz_y = z.$

23 If $z = x^5 - 2x^4y + 5x^2y^3$, show that

$$x \frac{\partial z}{\partial x} + y \frac{\partial z}{\partial y} = 5z.$$

In Problems 24 to 28, verify that $\partial^2 z/\partial x\, \partial y = \partial^2 z/\partial y\, \partial x.$

24 $z = \tan^{-1} \dfrac{x}{y}.$

25 $z = \ln(x + 5y).$

26 $z = e^{xy} \cos(y - 2x).$

27 $z = f(x)g(y).$

28 $z = x^3 \tan 2x \csc 3y^4 \sin^{-1} \sqrt{x^2 + 1}.$

29 Show that each of the following functions satisfies Laplace's equation $\partial^2 f/\partial x^2 + \partial^2 f/\partial y^2 = 0$:
  (a) $f(x, y) = \ln(x^2 + y^2)$;   (b) $f(x, y) = e^x \sin y$;
  (c) $f(x, y) = e^{-3x} \cos 3y$;   (d) $f(x, y) = \tan^{-1} \dfrac{y}{x}.$

30 Show that each of the following functions satisfies the wave equation $a^2\, \partial^2 f/\partial x^2 = \partial^2 f/\partial t^2$:
  (a) $f(x, t) = (x + at)^3$;   (b) $f(x, t) = (x - at)^5$;
  (c) $f(x, t) = \sin(x + at)$;   (d) $f(x, t) = e^{x - at}.$

31 Find a function $f(x, y)$ such that

$$\frac{\partial f}{\partial x} = 3y^2 - 2x \cos y \quad \text{and} \quad \frac{\partial f}{\partial y} = 6xy + x^2 \sin y + 2.$$

It is sometimes convenient to define functions by integrals of the form

$$F(x) = \int_a^b f(x, y)\, dy.$$

It is usually possible to calculate the derivative $F'(x)$ of such a function by "differentiating under the integral sign":

$$F'(x) = \frac{d}{dx} \int_a^b f(x, y)\, dy = \int_a^b \left[\frac{\partial}{\partial x} f(x, y) \right] dy.^*$$

32 Verify the formula just stated in the following cases:
  (a) $a = 0, b = 1, f(x, y) = x + y$;
  (b) $a = 0, b = 1, f(x, y) = x^3y^2 + x^2y^3$;
  (c) $a = 0, b = \pi, f(x, y) = \sin xy.$

---

* A proof is given in Appendix C.16.

The concept of a tangent plane to a surface corresponds to the concept of a tangent line to a curve. Geometrically, the tangent plane to a surface at a point is the plane that "best approximates" the surface near the point. It will be necessary for us to think rather carefully about what this means, because —as we shall see in Sections 19.5 and 19.6—weighty practical consequences depend on it.

Consider a surface $z = f(x, y)$, as shown in Fig. 19.6. As we pointed out in Section 19.2, the plane $y = y_0$ intersects this surface in a curve C_1 whose equation is

$$z = f(x, y_0),$$

and the plane $x = x_0$ intersects it in a curve C_2 whose equation is

$$z = f(x_0, y);$$

and the slopes of the tangent lines to these curves at the point $P_0 = (x_0, y_0, z_0)$ are the partial derivatives

$$f_x(x_0, y_0) \quad \text{and} \quad f_y(x_0, y_0). \tag{1}$$

19.3

THE TANGENT PLANE TO A SURFACE

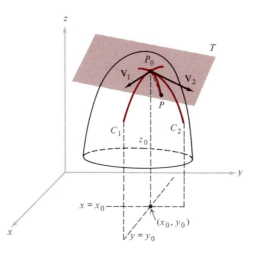

Figure 19.6   The tangent plane.

These two tangent lines determine a plane, and, as Fig. 19.6 suggests, if the surface is sufficiently smooth near P_0, then this plane will be tangent to the surface at P_0.

It is important to be quite clear about what we mean by a tangent plane, so we give a definition. In this context, where P_0 is a point on a surface $z = f(x, y)$, let T be a plane through P_0 and let P be any other point on the surface. If, as P approaches P_0 along the surface, the angle between the segment P_0P and the plane T approaches zero, then T is called the *tangent plane* to the surface at P_0.

It is easy to see that a surface need not have a tangent plane at a point P_0. A very simple example is provided by the half-cone $z = \sqrt{x^2 + y^2}$ shown in Fig. 19.7. It is clear that no plane is tangent to this surface at the origin. In this case the curves C_1 and C_2 have no tangent lines at the origin, and the partial derivatives (1) do not exist there. However, even when the curves C_1 and C_2 are smooth enough to have tangent lines at P_0, the surface may still not have

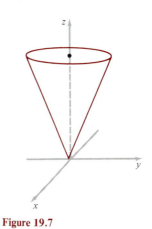

Figure 19.7

a tangent plane at P_0, because of nonsmooth behavior near P_0 in the regions between C_1 and C_2. In Section 19.4 we discuss a vital lemma to the effect that this cannot happen if the partial derivatives $f_x(x, y)$ and $f_y(x, y)$ exist at all points in some neighborhood of (x_0, y_0) and are continuous at (x_0, y_0) itself.

Meanwhile, we assume that the tangent plane exists at P_0, and we develop a method of finding its equation. Since the point $P_0 = (x_0, y_0, z_0)$ lies on this tangent plane, we know that the equation has the form

$$a(x - x_0) + b(y - y_0) + c(z - z_0) = 0, \tag{2}$$

where $\mathbf{N} = a\mathbf{i} + b\mathbf{j} + c\mathbf{k}$ is any normal vector. It remains to find $\mathbf{N}$, and to do this we use the cross product of two vectors $\mathbf{V}_1$ and $\mathbf{V}_2$ that are tangent to the curves C_1 and C_2 at P_0 (see Fig. 19.6). To find $\mathbf{V}_1$, we use the fact that along the tangent line to C_1, an increase of 1 unit in x produces a change $f_x(x_0, y_0)$ in z, while y does not change at all. Thus, the vector

$$\mathbf{V}_1 = \mathbf{i} + 0 \cdot \mathbf{j} + f_x(x_0, y_0)\mathbf{k}$$

is tangent to C_1 at P_0. Similarly, the vector

$$\mathbf{V}_2 = 0 \cdot \mathbf{i} + \mathbf{j} + f_y(x_0, y_0)\mathbf{k}$$

is tangent to C_2 at P_0. Since $\mathbf{V}_1$ and $\mathbf{V}_2$ lie in the tangent plane, we are now able to obtain our normal vector $\mathbf{N}$ by calculating

$$\mathbf{N} = \mathbf{V}_2 \times \mathbf{V}_1 = \begin{vmatrix} \mathbf{i} & \mathbf{j} & \mathbf{k} \\ 0 & 1 & f_y(x_0, y_0) \\ 1 & 0 & f_x(x_0, y_0) \end{vmatrix} = f_x(x_0, y_0)\mathbf{i} + f_y(x_0, y_0)\mathbf{j} - \mathbf{k}. \tag{3}$$

(The order of factors in this cross product is chosen only for convenience, to produce one minus sign in the result instead of two.) When the components of (3) are inserted in (2), we see that the desired equation is

$$f_x(x_0, y_0)(x - x_0) + f_y(x_0, y_0)(y - y_0) - (z - z_0) = 0,$$

or equivalently,

$$z - z_0 = f_x(x_0, y_0)(x - x_0) + f_y(x_0, y_0)(y - y_0). \tag{4}$$

Example 1 Find the tangent plane to the surface

$$z = f(x, y) = 2xy^3 - 5x^2$$

at the point $(3, 2, 3)$.

Solution The first step should be to check that this point actually lies on the given surface, and we assume that this has been done. Here we have $f_x = 2y^3 - 10x$ and $f_y = 6xy^2$, so $f_x(3, 2) = -14$ and $f_y(3, 2) = 72$. The equation of the tangent plane is therefore

$$z - 3 = -14(x - 3) + 72(y - 2).$$

Tangent planes to surfaces where z is not explicitly given as a function of x and y will be discussed in Section 19.5. However, we can get a preliminary idea of what to expect by applying our present method to simple cases.

Example 2 Find the tangent plane to the sphere

$$x^2 + y^2 + z^2 = 14 \tag{5}$$

at the point $(1, 2, 3)$.

Solution Even though this sphere is not a surface of the form $z = f(x, y)$, it can be thought of as a combination of two such surfaces, the upper and lower hemispheres. By solving (5) for z, we see that the upper hemisphere is given by

$$z = f(x, y) = \sqrt{14 - x^2 - y^2},$$

so

$$f_x = \frac{-x}{\sqrt{14 - x^2 - y^2}} \quad \text{and} \quad f_y = \frac{-y}{\sqrt{14 - x^2 - y^2}}.$$

These formulas give

$$f_x(1, 2) = -\tfrac{1}{3} \quad \text{and} \quad f_y(1, 2) = -\tfrac{2}{3},$$

so the equation of the tangent plane is

$$z - 3 = -\tfrac{1}{3}(x - 1) - \tfrac{2}{3}(y - 2),$$

or

$$x + 2y + 3z = 14.$$

In this example we solved equation (5) explicitly for z, then proceeded as before. An alternative method that is often easier is to assume that the given equation defines z implicitly as a function of x and y, and to find the partial derivatives by implicit differentiation. With this method we use equation (4) in the slightly different form

$$z - z_0 = \left(\frac{\partial z}{\partial x}\right)_{P_0} (x - x_0) + \left(\frac{\partial z}{\partial y}\right)_{P_0} (y - y_0), \tag{6}$$

where the coefficients are written this way because $\partial z/\partial x$ and $\partial z/\partial y$ need not depend only on x and y.

Example 3 To find the tangent plane of Example 2 by the method just suggested, we first hold y fixed and differentiate (5) implicitly with respect to x, which gives

$$2x + 2z \frac{\partial z}{\partial x} = 0,$$

so $\partial z/\partial x = -x/z$. Similarly, $\partial z/\partial y = -y/z$. At the point $P_0 = (1, 2, 3)$, these partial derivatives have the numerical values

$$\left(\frac{\partial z}{\partial x}\right)_{P_0} = -\frac{1}{3} \quad \text{and} \quad \left(\frac{\partial z}{\partial y}\right)_{P_0} = -\frac{2}{3},$$

so by (6) the tangent plane is

$$z - 3 = -\tfrac{1}{3}(x - 1) - \tfrac{2}{3}(y - 2),$$

just as before. Of course, this method is of particular value when the equation of the surface is difficult or impossible to solve for z.

PROBLEMS

In Problems 1 to 10, find an equation for the tangent plane to the given surface at the indicated point.

1 $z = (x^2 + y^2)^2$, (1, 2, 25).

2 $z = 4xy$, $(4, \frac{1}{4}, 4)$.

3 $z = \sin x + \sin 2y + \sin 3(x + y)$, (0, 0, 0).

4 $z = x^2 + xy + y^2 - 10y + 5$, (3, 2, 4).

5 $z = x^2 - 2y^2$, (3, 2, 1).

6 $z = \dfrac{2x + y}{x - 2y}$, (3, 1, 7).

7 $z = e^y \cos x$, (0, 0, 1).

8 $z = \tan^{-1} \dfrac{x}{y}$, $\left(4, 4, \dfrac{\pi}{4}\right)$.

9 $xy^2 + yz^2 + zx^2 = 25$, (1, 2, 3).

10 $z^3 + xyz = 33$, (1, 2, 3).

11 Let $P_0 = (x_0, y_0, z_0)$ with $z_0 > 0$ be a point on the sphere

$$x^2 + y^2 + z^2 = a^2.$$

Show that the tangent plane at this point is perpendicular to the radius vector to the point, in agreement with the definition given in geometry.

12 Use implicit differentiation to show that the equation of the tangent plane to the sphere

$$x^2 + y^2 + z^2 = a^2$$

at the point $P_0 = (x_0, y_0, z_0)$ is $x_0 x + y_0 y + z_0 z = a^2$.

13 Use implicit differentiation to find the equation of the tangent plane to the ellipsoid

$$\frac{x^2}{a^2} + \frac{y^2}{b^2} + \frac{z^2}{c^2} = 1$$

at the point $P_0 = (x_0, y_0, z_0)$.

14 Let a be a positive constant and consider the tangent plane to the surface $xyz = a$ at a point in the first octant. Show that the tetrahedron formed by this plane and the coordinate planes has constant volume, independent of the point of tangency. What is this volume?

15 The *angle between two surfaces* at a common point is the smallest positive angle between the normals to these surfaces at this point. Find the angle between $z = e^{xy} - 1$ and $z = \ln \sqrt{x^2 + y^2}$ at (0, 1, 0).

16 If $P_0 = (x_0, y_0, z_0)$ is a point on the curve of intersection of two surfaces $z = f(x, y)$ and $z = g(x, y)$, devise a method for finding a tangent vector to this curve at P_0. Apply this method to find a vector tangent to the curve of intersection of the cone $z^2 = 3x^2 + 4y^2$ and the plane $3x - 2y + z = 8$ at the point $P_0 = (2, 1, 4)$.

17 If a surface has an equation of the form $z = xf(x/y)$, show that all of its tangent planes have a common point. What is this point?

18 If $P_0 = (x_0, y_0, z_0)$ is a point on the cone $z^2 = a(x^2 + y^2)$ other than the vertex, show that the tangent plane at P_0 has $z_0 z = a(x_0 x + y_0 y)$ as its equation. Conclude that every such plane passes through the vertex. Show that the normal line at P_0 has

$$x = x_0 + ax_0 t, \qquad y = y_0 + ay_0 t, \qquad z = z_0 - z_0 t$$

as parametric equations.

19 On the cone in Problem 18, consider all points of fixed height h above the xy-plane and draw normal lines at these points. Show that the points where these lines intersect the xy-plane form a circle, and find the radius of this circle.

20 Let normal lines be drawn at all points on the surface $z = ax^2 + by^2$ which are at a given fixed height h above the xy-plane, and find the equation of the curve in which these lines intersect the xy-plane.

19.4

INCREMENTS AND DIFFERENTIALS. THE FUNDAMENTAL LEMMA

Most of calculus can be understood by using geometric intuition mixed with a little common sense, without getting bogged down in the underlying theory of the subject. In a few places, however, this theory is inescapable, because without it there is no way to grasp what is going on in the main developments of the subject itself. This is true for infinite series and the theory of convergence. It is also true for the topics of the next two sections—directional derivatives and the chain rule—which cannot be understood without a certain degree of attention to the theoretical issues that we now briefly discuss.

In order to see what these issues are, we begin by considering a function $y = f(x)$ of one variable that has a derivative at a point x_0. If Δx is an increment that carries x_0 to a nearby point $x_0 + \Delta x$ (see Fig. 19.8), we are interested in the corresponding increment in y,

$$\Delta y = f(x_0 + \Delta x) - f(x_0).$$

The definition of the derivative $f'(x_0)$ is

$$f'(x_0) = \lim_{\Delta x \to 0} \frac{\Delta y}{\Delta x}, \tag{1}$$

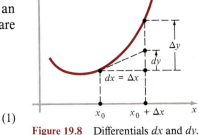

Figure 19.8 Differentials dx and dy.

and this can be written in the equivalent form

$$\frac{\Delta y}{\Delta x} = f'(x_0) + \epsilon,$$

where $\epsilon \to 0$ as $\Delta x \to 0$. Accordingly, with no further hypotheses than the assumed existence of the derivative (1), we can write the increment Δy in the form

$$\Delta y = f'(x_0)\,\Delta x + \epsilon\,\Delta x, \qquad \text{where } \epsilon \to 0 \text{ as } \Delta x \to 0. \tag{2}$$

The situation is entirely different for a function of two (or more) variables, as we now explain.

Consider a function $z = f(x, y)$ and let (x_0, y_0) be a point at which the partial derivatives $f_x(x_0, y_0)$ and $f_y(x_0, y_0)$ both exist. The increment in z produced by moving from (x_0, y_0) to a nearby point $(x_0 + \Delta x, y_0 + \Delta y)$ is

$$\Delta z = f(x_0 + \Delta x, y_0 + \Delta y) - f(x_0, y_0),$$

as shown in Fig. 19.9. In order to develop the tools we shall need in Sections 19.5 and 19.6, it will be necessary to express Δz in a form analogous to (2),

$$\Delta z = f_x(x_0, y_0)\,\Delta x + f_y(x_0, y_0)\,\Delta y + \epsilon_1\,\Delta x + \epsilon_2\,\Delta y, \tag{3}$$

where ϵ_1 and $\epsilon_2 \to 0$ as Δx and $\Delta y \to 0$. Unfortunately, in sharp contrast to the one-variable case, the mere existence of the partial derivatives f_x and f_y at

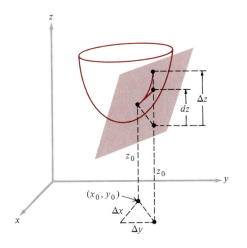

Figure 19.9 The differential dz.

(x_0, y_0) is not enough to guarantee the validity of (3). Sufficient conditions for this conclusion are given in the

---

Fundamental Lemma   *Suppose that a function $z = f(x, y)$ and its partial derivatives f_x and f_y are defined at a point (x_0, y_0), and also throughout some neighborhood of this point. Suppose further that f_x and f_y are continuous at (x_0, y_0). Then the increment Δz can be expressed in the form (3), where ϵ_1 and $\epsilon_2 \rightarrow 0$ as Δx and $\Delta y \rightarrow 0$.*

This statement is called a "lemma" for the usual reason: its significance lies not in itself, but rather in the use that can be made of it elsewhere. A proof is given in Appendix C.17.

We do not wish to dwell upon these matters, but nevertheless a few brief remarks are in order.

Remark 1   In the case of a function of one variable, (1) and (2) are equivalent, and if either condition holds it is customary to denote Δx by dx and to write $dy = f'(x_0)\, dx$, so that dy is the change in y along the tangent line. A function $z = f(x, y)$ for which $f_x(x_0, y_0)$ and $f_y(x_0, y_0)$ both exist is said to be *differentiable* at (x_0, y_0) if the conclusion of the lemma is valid—so that more is required than merely the existence of the partial derivatives. In this case— and *only* in this case!—we denote Δx and Δy by dx and dy, and we define the *differential dz* by*

$$dz = f_x(x_0, y_0)\, dx + f_y(x_0, y_0)\, dy.$$

Under these circumstances it can be proved that the surface $z = f(x, y)$ has a tangent plane at (x_0, y_0, z_0) and that dz is the change in z along this plane, as suggested in Fig. 19.9. The differential dz is usually written in the equivalent forms

$$dz = \frac{\partial z}{\partial x}\, dx + \frac{\partial z}{\partial y}\, dy \qquad \text{or} \qquad df = \frac{\partial f}{\partial x}\, dx + \frac{\partial f}{\partial y}\, dy.$$

Remark 2   A function $z = f(x, y)$ which is differentiable at a point is automatically continuous there. This follows at once from (3), which shows that $\Delta z \rightarrow 0$ if Δx and $\Delta y \rightarrow 0$. In the single-variable case, we know that if a function has a derivative at a point, then it is necessarily continuous there. However, this is not true for functions of more than one variable: the mere existence of the partial derivatives f_x and f_y at a point does not imply the continuity of $f(x, y)$ at that point. This is shown by the example of the bizarre function discussed in Section 19.1, for which $f_x(0, 0) = f_y(0, 0) = 0$ and yet the function is discontinuous at $(0, 0)$.

The concepts of a differentiable function and its differential, and also the Fundamental Lemma, can be extended in an obvious way to functions of any finite number of variables. This would involve much additional writing but no new ideas, and we shall not burden the reader with the details.

---

* Sometimes dz is called the *total differential.*

Let $f(x, y, z)$ be a function defined throughout some region of three-dimensional space, and let P be a point in this region. At what rate does f change as we move away from P in a specified direction? In the directions of the positive x-, y-, and z-axes, we know that the rates of change of f are given by the partial derivatives $\partial f/\partial x$, $\partial f/\partial y$, and $\partial f/\partial z$. But how do we calculate the rate of change of f if we move away from P in a direction that is not a coordinate direction? In analyzing this problem, we will encounter the very important concept of the gradient of a function.

Suppose that the point P under consideration has coordinates x, y, and z, so that $P = (x, y, z)$; let $\mathbf{R} = x\mathbf{i} + y\mathbf{j} + z\mathbf{k}$ be the position vector of P, and let the specified direction be given by a unit vector $\mathbf{u}$, as shown in Fig. 19.10. If we move away from P in this direction to a nearby point $Q = (x + \Delta x, y + \Delta y, z + \Delta z)$, then the function f will change by an amount Δf. If we now divide this change Δf by the distance $\Delta s = |\Delta\mathbf{R}|$ between P and Q, then the quotient $\Delta f/\Delta s$ is the average rate of change of f (with respect to distance) as we move from P to Q. For instance, if the value of f at P is the temperature at this point, then $\Delta f/\Delta s$ is the average rate of change of temperature along the segment PQ. The limiting value of $\Delta f/\Delta s$ as Q approaches P, namely,

$$\frac{df}{ds} = \lim_{\Delta s \to 0} \frac{\Delta f}{\Delta s},$$

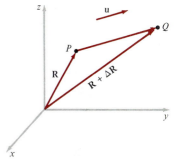

Figure 19.10

is called the *derivative of f at the point P in the direction* $\mathbf{u}$, or simply the *directional derivative* of f. In the case of the temperature function, df/ds represents the instantaneous rate of change of temperature with respect to distance — roughly speaking, how fast it is getting hotter — at the point P as we move away from P in the direction specified by $\mathbf{u}$.

This is all very well, but how do we actually calculate df/ds in a specific case? To discover how to do this, we assume that $f(x, y, z)$ has continuous partial derivatives with respect to x, y, and z. Indeed, to avoid the tedious repetition of hypotheses, we make this a blanket assumption for every function we discuss, unless we explicitly state otherwise. With this, the Fundamental Lemma enables us to write Δf in the form

$$\Delta f = \frac{\partial f}{\partial x} \Delta x + \frac{\partial f}{\partial y} \Delta y + \frac{\partial f}{\partial z} \Delta z + \epsilon_1 \Delta x + \epsilon_2 \Delta y + \epsilon_3 \Delta z, \tag{1}$$

where $\epsilon_1, \epsilon_2, \epsilon_3 \to 0$ as $\Delta x, \Delta y$, and $\Delta z \to 0$, that is, as $\Delta s \to 0$. Dividing (1) by Δs now gives

$$\frac{\Delta f}{\Delta s} = \frac{\partial f}{\partial x} \frac{\Delta x}{\Delta s} + \frac{\partial f}{\partial y} \frac{\Delta y}{\Delta s} + \frac{\partial f}{\partial z} \frac{\Delta z}{\Delta s} + \epsilon_1 \frac{\Delta x}{\Delta s} + \epsilon_2 \frac{\Delta y}{\Delta s} + \epsilon_3 \frac{\Delta z}{\Delta s}, \tag{2}$$

and by taking the limit as $\Delta s \to 0$, we see that the last three terms in (2) approach zero and we obtain the formula

$$\frac{df}{ds} = \frac{\partial f}{\partial x} \frac{dx}{ds} + \frac{\partial f}{\partial y} \frac{dy}{ds} + \frac{\partial f}{\partial z} \frac{dz}{ds}. \tag{3}$$

This formula should be recognized as a special kind of chain rule, in the sense that as we move along the line through P and parallel to $\mathbf{u}$, f is a function of x, y, and z, where x, y, and z are in turn functions of the arc length s, and (3) shows how to differentiate f with respect to s.

We observe that the first factor in each product on the right of (3) depends only on the function f and the coordinates of the point P at which the partial derivatives of f are evaluated, while the second factor in each product is independent of f and depends only on the direction in which df/ds is being calculated. These facts suggest that the right side of (3) ought to be thought of—and written—as the dot product of two vectors, as follows:

$$\frac{df}{ds} = \left(\frac{\partial f}{\partial x}\mathbf{i} + \frac{\partial f}{\partial y}\mathbf{j} + \frac{\partial f}{\partial z}\mathbf{k}\right) \cdot \left(\frac{dx}{ds}\mathbf{i} + \frac{dy}{ds}\mathbf{j} + \frac{dz}{ds}\mathbf{k}\right)$$

$$= \left(\frac{\partial f}{\partial x}\mathbf{i} + \frac{\partial f}{\partial y}\mathbf{j} + \frac{\partial f}{\partial z}\mathbf{k}\right) \cdot \frac{d\mathbf{R}}{ds}. \tag{4}$$

The first factor here is a vector called the *gradient* of f. It is denoted by the symbol grad f, so that by definition

$$\text{grad } f = \frac{\partial f}{\partial x}\mathbf{i} + \frac{\partial f}{\partial y}\mathbf{j} + \frac{\partial f}{\partial z}\mathbf{k}. \tag{5}$$

With this notation, (4) can be written as

$$\frac{df}{ds} = (\text{grad } f) \cdot \frac{d\mathbf{R}}{ds}. \tag{6}$$

But we know that $d\mathbf{R}/ds$ is a unit vector, and since it has the same direction as $\mathbf{u}$, it equals $\mathbf{u}$. Formula (6) is therefore equivalent to

$$\frac{df}{ds} = (\text{grad } f) \cdot \mathbf{u}. \tag{7}$$

This tells us how to calculate df/ds, because (5) is presumably simple to compute from the given function f, and then to evaluate at the given point P, and the dot product (7) of two known vectors is easy to find.

For a given function f and a given point P, grad f is a fixed vector which can be placed so that its tail lies at P. We also place the tail of $\mathbf{u}$ at P, as shown in Fig. 19.11. To understand the significance of grad f, we use the definition of the dot product and the fact that $\mathbf{u}$ is a unit vector to write (7) in the form

$$\frac{df}{ds} = |\text{grad } f| \cos \theta, \tag{8}$$

where θ is the angle between grad f and $\mathbf{u}$. Since the direction of $\mathbf{u}$ can be chosen to suit our convenience, (8) immediately yields the first fundamental property of the gradient:

Property 1   The directional derivative df/ds in any given direction is the scalar projection of grad f in that direction (see Fig. 19.11).

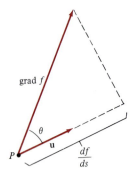

Figure 19.11   Directional derivative.

In this sense, the single vector grad f contains within itself the directional derivatives of f at P in all possible directions.

Next, if u is chosen to point in the same direction as grad f, so that $\theta = 0$ and $\cos \theta = 1$, then (8) shows that df/ds has its maximum value — that is, f increases most rapidly — in this direction. Also, this maximum value equals $|\text{grad } f|$. These remarks give the next two fundamental properties of the gradient:

Property 2 The vector grad f points in the direction in which f increases most rapidly.

Property 3 The length of the vector grad f is the maximum rate of increase of f.

As these remarks show, even though formulas (7) and (8) are equivalent, they play very different roles in our thinking, for we use (7) to calculate df/ds and (8) to understand the intuitive meaning of the vector grad f.

Example 1 If $f(x, y, z) = x^2 - y + z^2$, find the directional derivative df/ds at the point $(1, 2, 1)$ in the direction of the vector $4\mathbf{i} - 2\mathbf{j} + 4\mathbf{k}$.

Solution At the point $(1, 2, 1)$, we have grad $f = 2x\mathbf{i} - \mathbf{j} + 2z\mathbf{k} = 2\mathbf{i} - \mathbf{j} + 2\mathbf{k}$. We obtain a unit vector $\mathbf{u}$ in the desired direction by dividing the given vector by its own length,

$$\mathbf{u} = \frac{4\mathbf{i} - 2\mathbf{j} + 4\mathbf{k}}{\sqrt{16 + 4 + 16}} = \frac{2}{3}\mathbf{i} - \frac{1}{3}\mathbf{j} + \frac{2}{3}\mathbf{k}.$$

Formula (7) now gives

$$\frac{df}{ds} = (\text{grad } f) \cdot \mathbf{u}$$

$$= (2\mathbf{i} - \mathbf{j} + 2\mathbf{k}) \cdot (\tfrac{2}{3}\mathbf{i} - \tfrac{1}{3}\mathbf{j} + \tfrac{2}{3}\mathbf{k}) = 3.$$

Thus, the function f is increasing at the rate of 3 units per unit distance as we leave $(1, 2, 1)$ in the given direction.

Example 2 Let the temperature of the air at points in space be given by the function $f(x, y, z) = x^2 - y + z^2$. A mosquito located at $(1, 2, 1)$ wishes to get cool as soon as possible. In what direction should it fly?

Solution We saw in Example 1 that grad $f = 2\mathbf{i} - \mathbf{j} + 2\mathbf{k}$ at the point $(1, 2, 1)$. Since the direction of grad f is that in which the temperature increases most rapidly, the mosquito should fly in the opposite direction, that of $-\text{grad } f = -2\mathbf{i} + \mathbf{j} - 2\mathbf{k}$.

The fourth fundamental property of the gradient is useful in geometry. In order to explain what it is, we denote the point under consideration by $P_0 = (x_0, y_0, z_0)$ to emphasize that it is fixed in this discussion, and we let c_0

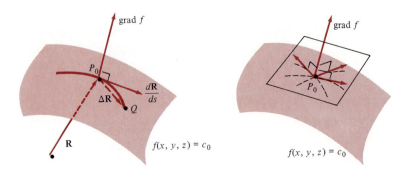

Figure 19.12   Gradient is normal to level surface.

be the value of our function f at the point P_0. Then the set of all points in space at which $f(x, y, z)$ has the same value c_0 constitutes, in general, a level surface through P_0 whose equation is $f(x, y, z) = c_0$. We wish to show that the vector grad f is normal (perpendicular) to this level surface at the point P_0, as suggested on the left in Fig. 19.12. To this end, we consider a curve that lies on the surface and passes through P_0. If we move to a nearby point Q on this curve and measure s along the curve, then $\Delta f = 0$ because f has the same value at all points on the surface, and therefore $df/ds = 0$ at P_0 in the direction of the tangent to the curve. Formula (6) remains valid and implies that

$$(\operatorname{grad} f) \cdot \frac{d\mathbf{R}}{ds} = 0, \tag{9}$$

where $d\mathbf{R}/ds$ is the unit tangent vector to the curve at P_0. The vanishing of the dot product in (9) tells us that grad f is perpendicular to this tangent vector. But the same reasoning applies to every curve on the surface that passes through P_0, so grad f is perpendicular to the tangent vectors to all these curves (Fig. 19.12, right). Since these tangent vectors determine the tangent plane at P_0, and being normal to the surface means being normal to this tangent plane, we have:

Property 4   The gradient of a function $f(x, y, z)$ at a point P_0 is normal to the level surface of f that passes through P_0.

In the context of this discussion, we point out that the equation of any surface can be written in the form $f(x, y, z) = c_0$, and can therefore be regarded as a level surface of the function $f(x, y, z)$. If $P_0 = (x_0, y_0, z_0)$ is a point on this surface, then Property 4 tells us that the vector

$$\mathbf{N} = \operatorname{grad} f = \left(\frac{\partial f}{\partial x}\right)_{P_0} \mathbf{i} + \left(\frac{\partial f}{\partial y}\right)_{P_0} \mathbf{j} + \left(\frac{\partial f}{\partial z}\right)_{P_0} \mathbf{k}$$

is normal to the tangent plane at P_0, so if $\mathbf{N} \neq \mathbf{0}$, the equation of this tangent plane is

$$\left(\frac{\partial f}{\partial x}\right)_{P_0} (x - x_0) + \left(\frac{\partial f}{\partial y}\right)_{P_0} (y - y_0) + \left(\frac{\partial f}{\partial z}\right)_{P_0} (z - z_0) = 0. \tag{10}$$

We observe that this equation includes equation (4) in Section 19.3 as a special case; for if the surface is given in the form $z = g(x, y)$, then this can be written as $g(x, y) - z = 0$, so the surface is a level surface of the function $f(x, y, z) = g(x, y) - z$, and this makes the coefficients in (10) equal to $g_x(x_0, y_0), g_y(x_0, y_0), -1$.

Example 3 Find the equation of the tangent plane to the surface $xy^2z^3 = 12$ at the point $(3, -2, 1)$.

Solution This surface is a level surface of the function $f(x, y, z) = xy^2z^3$. The vector grad f at the point $(3, -2, 1)$ is normal to the surface at this point. This vector is

$$\text{grad } f = y^2z^3\mathbf{i} + 2xyz^3\mathbf{j} + 3xy^2z^2\mathbf{k}$$

$$= 4\mathbf{i} - 12\mathbf{j} + 36\mathbf{k} = 4(\mathbf{i} - 3\mathbf{j} + 9\mathbf{k}).$$

Therefore the equation of the tangent plane is

$$(x - 3) - 3(y + 2) + 9(z - 1) = 0$$

or

$$x - 3y + 9z = 18.$$

Remark 1 The main uses of directional derivatives and gradients are found in the geometry and physics of three-dimensional space. However, these concepts can also be defined in two dimensions, and they have similar (but thinner) properties. Thus, a curve $f(x, y) = c_0$ can be thought of as a level curve of the function $z = f(x, y)$; and if the gradient of this function is defined by

$$\text{grad } f = \frac{\partial f}{\partial x}\mathbf{i} + \frac{\partial f}{\partial y}\mathbf{j},$$

then the value of this gradient at a point $P_0 = (x_0, y_0)$ on the curve is a vector that is normal to the curve.

Remark 2 The gradient of a function $f(x, y, z)$ can be written in "operational form" as

$$\text{grad } f = \left(\frac{\partial}{\partial x}\mathbf{i} + \frac{\partial}{\partial y}\mathbf{j} + \frac{\partial}{\partial z}\mathbf{k} \right) f.$$

The *del operator* preceding the function f is usually denoted by the symbol ∇ (an inverted delta, read "del"), so that

$$\nabla = \frac{\partial}{\partial x}\mathbf{i} + \frac{\partial}{\partial y}\mathbf{j} + \frac{\partial}{\partial z}\mathbf{k}.$$

This del operator is similar to, but more complicated than, the familiar differentiation operator d/dx. When del is applied to a function f, it produces a vector, namely, the vector grad f. In this notation, formulas (5), (6), and (7) become

$$\text{grad } f = \nabla f, \qquad \frac{df}{ds} = \nabla f \cdot \frac{d\mathbf{R}}{ds}, \qquad \text{and} \qquad \frac{df}{ds} = \nabla f \cdot \mathbf{u}.$$

In courses on advanced calculus and vector analysis, a much deeper study is made of the operator ∇ and its uses than we can carry out here.

PROBLEMS

1 Find the gradient of f at P if
(a) $f(x, y, z) = xy + xz + yz$, $P = (-1, 3, 5)$;
(b) $f(x, y, z) = e^{xy} \cos z$, $P = (0, 2, 0)$;
(c) $f(x, y, z) = \ln (x^2 + y^2 + z^2)$, $P = (1, 2, -2)$;
(d) $f(x, y, z) = xy/z$, $P = (2, -1, 5)$.

2 Find the directional derivative of f at P in the direction of the given vector:
(a) $f(x, y, z) = xy^2 + x^2z + yz$, $P = (1, 1, 2)$, $\mathbf{i} + 2\mathbf{j} - \mathbf{k}$;
(b) $f(x, y, z) = \ln (x^2 + y^2 + z^2)$, $P = (0, 0, 1)$, vector from P to $(2, 2, 0)$;
(c) $f(x, y, z) = x \sin y + y \sin z + z \sin x$, $P = (1, 0, 0)$, $2\sqrt{3}\mathbf{i} + 2\mathbf{j}$;
(d) $f(x, y, z) = xye^z + yze^x$, $P = (1, 0, 0)$, vector from P to $(2, 2, 1)$.

3 Find the maximum value of the directional derivative of f at P, and the direction in which it occurs:
(a) $f(x, y, z) = \sin xy + \cos yz$, $P = (-3, 0, 7)$;
(b) $f(x, y, z) = e^x \cos y + e^y \cos z + e^z \cos x$, $P = (0, 0, 0)$;
(c) $f(x, y, z) = 2xyz + y^2 + z^2$, $P = (2, 1, 1)$;
(d) $f(x, y, z) = e^{xyz}$, $P = (2, 1, 1)$.

4 In what direction should one travel, starting at the origin, to obtain the most rapid rate of decrease of the function

$$f(x, y, z) = (2 - x - y)^3 + (3x + 2y - z + 1)^2?$$

5 Find the unit vectors normal to the surface $xyz = 4$ at the point $(2, -2, -1)$.

6 If $f(x, y, z) = x^2 + 4y^2 - 8z$, find df/ds at $(4, 1, 0)$ (a) along the line $(x - 4)/2 = (y - 1)/1 = z/(-2)$ in the direction of decreasing x; (b) along the normal to the plane $3(x - 4) - (y - 1) + 2z = 0$ in the direction of increasing x; (c) in the direction in which f increases most rapidly.

7 Suppose that the temperature T at a point $P = (x, y, z)$ is given by $T = 2x^2 - y^2 + 4z^2$. Find the rate of change of T at the point $(1, -2, 1)$ in the direction of the vector $4\mathbf{i} - \mathbf{j} + 2\mathbf{k}$. In what direction does T increase most rapidly at this point? What is this maximum rate of increase?

8 Find the tangent plane and normal line to the hyperboloid $x^2 + y^2 - z^2 = 5$ at the point $(4, 5, 6)$.

9 Show that the tangent plane to the quadric surface $ax^2 + by^2 + cz^2 = d$ at the point (x_0, y_0, z_0) has $ax_0x + by_0y + cz_0z = d$ as its equation.

10 Show that the del operator has the following properties that demonstrate its close similarity to the differentiation operator d/dx:
(a) $\nabla(f + g) = \nabla f + \nabla g$;
(b) $\nabla(fg) = f\nabla g + g\nabla f$;
(c) $\nabla\left(\dfrac{f}{g}\right) = \dfrac{g\nabla f - f\nabla g}{g^2}$;
(d) $\nabla f^n = nf^{n-1}\nabla f$.

19.6
THE CHAIN RULE FOR PARTIAL DERIVATIVES

The single-variable chain rule for ordinary derivatives tells us how to differentiate composite functions. It says that if w is a function of x where x is in turn a function of a third variable t, say $w = f(x)$ where $x = g(t)$, then

$$\frac{dw}{dt} = \frac{dw}{dx}\frac{dx}{dt}. \tag{1}$$

We know from ample experience that this is an indispensable tool of calculus; it is used more frequently than any other differentiation rule.

The simplest multivariable chain rule involves a function $w = f(x, y)$ of two variables x and y, where x and y are each functions of another variable t, $x = g(t)$ and $y = h(t)$. Then w is a function of t,

$$w = f[g(t), h(t)] = F(t),$$

and we shall prove that the derivative of this composite function is given by the formula

$$\frac{dw}{dt} = \frac{\partial w}{\partial x}\frac{dx}{dt} + \frac{\partial w}{\partial y}\frac{dy}{dt}. \tag{2}$$

This is the *chain rule* for this situation.

The proof of (2) is easy. We begin by changing t to $t + \Delta t$, where $\Delta t \neq 0$. This increment in t produces increments Δx and Δy in x and y, which in turn produce an increment Δw in w. Since all the functions we discuss are assumed to have continuous partial derivatives, the Fundamental Lemma enables us to write Δw in the form

$$\Delta w = \frac{\partial w}{\partial x}\Delta x + \frac{\partial w}{\partial y}\Delta y + \epsilon_1\,\Delta x + \epsilon_2\,\Delta y, \tag{3}$$

where ϵ_1 and $\epsilon_2 \to 0$ as Δx and $\Delta y \to 0$. On dividing (3) by Δt, we obtain

$$\frac{\Delta w}{\Delta t} = \frac{\partial w}{\partial x}\frac{\Delta x}{\Delta t} + \frac{\partial w}{\partial y}\frac{\Delta y}{\Delta t} + \epsilon_1\frac{\Delta x}{\Delta t} + \epsilon_2\frac{\Delta y}{\Delta t}. \tag{4}$$

If we now form the limit as $\Delta t \to 0$, then Δx and Δy also $\to 0$, so ϵ_1 and $\epsilon_2 \to 0$, and (4) immediately yields (2).

Example 1 If $w = 3x^2 + 2xy - y^2$ where $x = \cos t$ and $y = \sin t$, find dw/dt.

Solution Formula (2) tells us that

$$\frac{dw}{dt} = (6x + 2y)(-\sin t) + (2x - 2y)\cos t.$$

By substituting $x = \cos t$ and $y = \sin t$, we can express this in terms of t alone,

$$\begin{aligned}
\frac{dw}{dt} &= (6\cos t + 2\sin t)(-\sin t) + (2\cos t - 2\sin t)(\cos t) \\
&= -6\sin t\cos t - 2\sin^2 t + 2\cos^2 t - 2\sin t\cos t \\
&= 2(\cos^2 t - \sin^2 t) - 8\sin t\cos t = 2\cos 2t - 4\sin 2t.
\end{aligned}$$

We can check this result by first substituting and then differentiating, which gives

$$w = 3\cos^2 t + 2\sin t\cos t - \sin^2 t$$

and

$$\begin{aligned}
\frac{dw}{dt} &= 6\cos t\,(-\sin t) + 2\sin t\,(-\sin t) + 2\cos^2 t - 2\sin t\cos t \\
&= 2(\cos^2 t - \sin^2 t) - 8\sin t\cos t = 2\cos 2t - 4\sin 2t,
\end{aligned}$$

as before.

In the situation of formula (2), it is convenient to call w the *dependent variable*, x and y the *intermediate variables*, and t the *independent variable*. We notice that the right side of (2) has two terms, one for each intermediate variable, and that each of these terms resembles the right side of the single-variable chain rule (1).

Formula (2) extends in an obvious way to any number of intermediate variables. For instance, if $w = f(x, y, z)$ where x, y, and z are each functions of t, then

$$\frac{dw}{dt} = \frac{\partial w}{\partial x}\frac{dx}{dt} + \frac{\partial w}{\partial y}\frac{dy}{dt} + \frac{\partial w}{\partial z}\frac{dz}{dt}. \tag{5}$$

The proof of this is essentially the same as the proof of (2), except that it uses the Fundamental Lemma for three variables instead of two.

Further, x, y, and z here need not be functions of only one independent variable, but can be functions of two or more variables. Thus, if x, y, and z are each functions of the variables t and u, then w is also a function of t and u, and its partial derivatives are given by

$$\frac{\partial w}{\partial t} = \frac{\partial w}{\partial x}\frac{\partial x}{\partial t} + \frac{\partial w}{\partial y}\frac{\partial y}{\partial t} + \frac{\partial w}{\partial z}\frac{\partial z}{\partial t} \tag{6}$$

and

$$\frac{\partial w}{\partial u} = \frac{\partial w}{\partial x}\frac{\partial x}{\partial u} + \frac{\partial w}{\partial y}\frac{\partial y}{\partial u} + \frac{\partial w}{\partial z}\frac{\partial z}{\partial u}. \tag{7}$$

We use roundback d's everywhere here because every function depends on more than one variable. It is necessary to be very clear about the meanings of the letters in formulas like these. For example, on the left-hand side of (6), w is considered a function of t and u, while on the right-hand side it is considered a function of x, y, and z. The proofs are the same as before, and all of these formulas—(2), (5), (6), (7), and their extensions to any number of intermediate and independent variables—are collectively called the *chain rule.*

In Section 19.4 we defined the differential dw of a function $w = f(x, y, z)$ by the formula

$$dw = \frac{\partial w}{\partial x}dx + \frac{\partial w}{\partial y}dy + \frac{\partial w}{\partial z}dz. \tag{8}$$

The chain rule (5) tells us that if x, y, z are themselves functions of a single independent variable t, then it is permissible to calculate dw/dt by formally dividing (8) by dt. Similarly, if x, y, z are functions of the independent variables t, u and we want to calculate $\partial w/\partial t$, then the chain rule (6) tells us that we can find $\partial w/\partial t$ by dividing (8) by dt and writing roundback d's in place of ordinary d's to show that there is another independent variable present which is being held fixed.

The individual terms on the right side of (8) are sometimes called the *partial differentials* of w with respect to x, y, z. From this point of view, the quantity dw defined by (8) deserves the name *total differential,* as we remarked in Section 19.4.

Example 2 A function of several variables is said to be *homogeneous of degree n* if multiplying each variable by t (where $t > 0$) has the same effect as multiplying the original function by t^n. Thus, $f(x, y)$ is homogeneous of degree n if

$$f(tx, ty) = t^n f(x, y). \tag{9}$$

For example, $f(x, y) = x^2 + 3xy$ is homogeneous of degree 2, because $f(tx, ty) = (tx)^2 + 3(tx)(ty) = t^2(x^2 + 3xy) = t^2 f(x, y)$. Similarly, $f(x, y) = (x + y)/(x - y)$ is homogeneous of degree 0, $f(x, y) = (xy - x^2 e^{x/y})/y$ is homogeneous of degree 1, and $f(x, y, z) = \sqrt{x^3 - 3xy^2 + 2z^3}$ is homogeneous of degree $\frac{3}{2}$. Most functions, for instance $f(x, y) = y^2 + x \sin y$, are not homogeneous at all.

There is a theorem of Euler about homogeneous functions that has several important applications: *If $f(x, y)$ is homogeneous of degree n, then*

$$x \frac{\partial f}{\partial x} + y \frac{\partial f}{\partial y} = nf(x, y). \tag{10}$$

To prove this, we hold x and y fixed and differentiate both sides of (9) with respect to t. We can clarify this process by writing $u = tx$ and $v = ty$, so that (9) becomes

$$f(u, v) = t^n f(x, y).$$

Then by using the chain rule to differentiate with respect to t, we obtain

$$\frac{\partial f}{\partial u} \frac{\partial u}{\partial t} + \frac{\partial f}{\partial v} \frac{\partial v}{\partial t} = nt^{n-1} f(x, y)$$

or

$$x \frac{\partial f}{\partial u} + y \frac{\partial f}{\partial v} = nt^{n-1} f(x, y),$$

and putting $t = 1$ yields (10). Similarly, if $f(x_1, x_2, \ldots, x_m)$ is homogeneous of degree n, then the same argument shows that

$$x_1 \frac{\partial f}{\partial x_1} + x_2 \frac{\partial f}{\partial x_2} + \cdots + x_m \frac{\partial f}{\partial x_m} = nf(x_1, x_2, \ldots, x_m).$$

Euler's theorem has some interesting consequences for economics. As an example, suppose that $f(x, y)$ is the production (measured in dollars) of x units of capital and y units of labor. If the amounts of capital and labor are doubled, then it is reasonable to expect that the resulting production will also double, that is, that $f(2x, 2y) = 2f(x, y)$. More generally, we expect that

$$f(tx, ty) = tf(x, y),$$

so that the production function is homogeneous of degree 1. [In economics, this property of $f(x, y)$ is called "constant returns to scale."] Euler's theorem now says that

$$f(x, y) = x \frac{\partial f}{\partial x} + y \frac{\partial f}{\partial y}. \tag{11}$$

The partial derivatives $\partial f/\partial x$ and $\partial f/\partial y$ are called the *marginal product of capital* and the *marginal product of labor,* respectively. In this language, (11) is a theorem of quantitative economics whose verbal statement is, "The total value of production equals the cost of capital plus the cost of labor if each is paid for at the rate of its marginal product." Under these circumstances there are no surplus earnings, and in the real world this is a Very Bad Thing.*

Example 3 Many applications of the chain rule involve calculating the effect on some equation or expression when new variables are introduced. As an illustration of a method that will be useful for solving the wave equation in Section 19.9, we now solve the partial differential equation

$$a\frac{\partial w}{\partial x} = \frac{\partial w}{\partial y}, \qquad a \neq 0. \tag{12}$$

That is, we find the most general function $w = f(x, y)$ that satisfies this equation. To do this, we introduce new independent variables u, v by writing

$$u = x + ay, \qquad v = x - ay. \tag{13}$$

We think of w as a function of u and v,

$$w = F(u, v),$$

and we find the uv-equation equivalent to (12) by using the chain rule to write

$$\frac{\partial w}{\partial x} = \frac{\partial w}{\partial u}\frac{\partial u}{\partial x} + \frac{\partial w}{\partial v}\frac{\partial v}{\partial x} = \frac{\partial w}{\partial u} + \frac{\partial w}{\partial v},$$

$$\frac{\partial w}{\partial y} = \frac{\partial w}{\partial u}\frac{\partial u}{\partial y} + \frac{\partial w}{\partial v}\frac{\partial v}{\partial y} = a\frac{\partial w}{\partial u} - a\frac{\partial w}{\partial v}.$$

By substituting in these expressions, we see that (12) transforms into the partial differential equation

$$2a\frac{\partial w}{\partial v} = 0 \qquad \text{or} \qquad \frac{\partial w}{\partial v} = 0.$$

This equation is very easy to solve, because it says that the function $w = F(u, v)$ is constant when u is held fixed and v is allowed to vary, and therefore is a function of u alone. This means that our desired solution of (12) is

$$w = g(u) = g(x + ay),$$

where $g(u)$ is a *completely arbitrary* (continuously differentiable) function of u. We apologize to students for introducing "out of the blue" the apparently unmotivated transformation equations (13). However, some of the developments of Section 19.9 will make this procedure seem fairly natural.

Example 4 Partial derivatives are the main mathematical tools used in thermodynamics. It is the universal practice in this science to avoid confusion by

---

* For further information on these matters, see pp. 81–84 of J. M. Henderson and R. E. Quandt, *Microeconomic Theory* (McGraw-Hill, 1971); or Chapter 12, "Homogeneous Functions and Euler's Theorem," in D. E. James and C. D. Throsby, *Quantitative Methods in Economics* (Wiley, 1973). For an application of Euler's theorem to advanced theoretical mechanics, see p. 382 of the present writer's text, *Differential Equations* (McGraw-Hill, 1972).

using subscripts on partial derivatives to specify the variable (or variables) held fixed in the differentiation. Thus, if $w = F(x, y)$ then $\partial w/\partial x$ would be denoted by

$$\left(\frac{\partial w}{\partial x}\right)_y.$$

This notation tells us that w is being thought of as a function of x and y, and that y is held fixed and x is the variable of differentiation. This usage may seem superfluous, but the following situation — which is quite common in thermodynamics — shows that it is not.

If $w = f(x, y)$ where y is a function $g(x, t)$ of x and another variable t, so that w is a composite function of x and t, we find its partial derivative with respect to x.

This is a typical chain rule situation with x and y the intermediate variables and x and t the independent variables:

$$w = f(x, y) \qquad \text{where} \qquad \begin{cases} x = x, \\ y = g(x, t). \end{cases}$$

The chain rule therefore gives

$$\frac{\partial w}{\partial x} = \frac{\partial w}{\partial x}\frac{\partial x}{\partial x} + \frac{\partial w}{\partial y}\frac{\partial y}{\partial x}, \tag{14}$$

so

$$\frac{\partial w}{\partial x} = \frac{\partial w}{\partial x} + \frac{\partial w}{\partial y}\frac{\partial y}{\partial x}. \tag{15}$$

Unfortunately this equation contains two partial derivatives of w with respect to x. By an effort of thought one can keep in mind that $\partial w/\partial x$ on the left of (15) is the derivative of the composite function, while $\partial w/\partial x$ on the right is the derivative of $w = f(x, y)$. Nevertheless, this ambiguous notation invites confusion and is contrary to the overall spirit of mathematical symbols, which are intended to make it easy to be correct with a minimum of thought. However, if we use the subscript notation of thermodynamics, then (14) can be written as

$$\left(\frac{\partial w}{\partial x}\right)_t = \left(\frac{\partial w}{\partial x}\right)_y\left(\frac{\partial x}{\partial x}\right)_t + \left(\frac{\partial w}{\partial y}\right)_x\left(\frac{\partial y}{\partial x}\right)_t.$$

Since $(\partial x/\partial x)_t = 1$, this becomes

$$\left(\frac{\partial w}{\partial x}\right)_t = \left(\frac{\partial w}{\partial x}\right)_y + \left(\frac{\partial w}{\partial y}\right)_x\left(\frac{\partial y}{\partial x}\right)_t, \tag{16}$$

which is somewhat clumsy but much less vulnerable to misunderstanding than (15).*

---

* Students whose main interest is physics may wish to read discussions of these matters in some of the standard treatises. See, for example, p. 19 of Enrico Fermi, *Thermodynamics* (Dover, 1956); p. 28 of Philip M. Morse, *Thermal Physics* (W. A. Benjamin, 1969); or pp. 30–33, 52–55 of F. W. Sears, *Thermodynamics* (Addison-Wesley, 1953).

PROBLEMS

In Problems 1 to 4, find dw/dt in two ways, (a) by using the chain rule and then expressing everything in terms of t, and (b) by first substituting and then differentiating.

1 $w = e^{x^2+y^2}$, $x = \cos t$, $y = \sin t$.

2 $w = xy + yz + zx$, $x = 3t^2$, $y = e^t$, $z = e^{-t}$.

3 $w = \dfrac{3xy}{x^2 - y^2}$, $x = t^2$, $y = 3t$.

4 $w = \ln(x^4 + 2x^2y + 3y^2)$, $x = t$, $y = 2t^2$.

In Problems 5 and 6, find $\partial w/\partial t$ and $\partial w/\partial u$ by the chain rule and check your answers by using a different method.

5 $w = x^2 + y^2$, $x = t^2 - u^2$, $y = 2tu$.

6 $w = \dfrac{x}{x^2 + y^2}$, $x = t\cos u$, $y = t\sin u$.

7   If f is any (continuously differentiable) function, show that $w = f(x^2 - y^2)$ is a solution of the partial differential equation

$$y\frac{\partial w}{\partial x} + x\frac{\partial w}{\partial y} = 0.$$

Hint: Write $w = f(u)$ where $u = x^2 - y^2$ and apply the chain rule with only one intermediate variable.

8   If a and b are constants and $w = f(ax + by)$, show that

$$b\frac{\partial w}{\partial x} = a\frac{\partial w}{\partial y}.$$

9   If $w = f(x^2 - y^2, y^2 - x^2)$, show that

$$y\frac{\partial w}{\partial x} + x\frac{\partial w}{\partial y} = 0.$$

10   If $w = f\left(\dfrac{y-x}{xy}, \dfrac{z-y}{yz}\right)$, show that

$$x^2\frac{\partial w}{\partial x} + y^2\frac{\partial w}{\partial y} + z^2\frac{\partial w}{\partial z} = 0.$$

11   The differential dw of a function $w = f(x, y, z)$ is defined by (8) only for the case in which x, y, z are independent variables. If x, y, z are not independent, but instead are functions of independent variables t, u, then dw must be defined by

$$dw = \frac{\partial w}{\partial t}\,dt + \frac{\partial w}{\partial u}\,du.$$

Show that the definition implies (8), so that (8) remains valid regardless of whether x, y, z are independent or not.

12   If u and v are both functions of x, y, z, show that
(a) $d(c) = 0$, c a constant;

(b) $d(cu) = c\,du$;
(c) $d(u + v) = du + dv$;
(d) $d(uv) = u\,dv + v\,du$;

(e) $d\left(\dfrac{u}{v}\right) = \dfrac{v\,du - u\,dv}{v^2}$;

(f) if $w = f(u)$, then $dw = f'(u)\,du$.

13   Verify Euler's theorem (10) for each of the following functions:
(a) $f(x, y) = xy^2 + x^2y - y^3$;
(b) $f(x, y) = e^{x/y}$;
(c) $f(x, y) = \sqrt{x^2 + y^2}$;

(d) $f(x, y) = \dfrac{\sqrt{x + y}}{x}$.

14   If $w = f(x, y)$ where $x = r\cos\theta$ and $y = r\sin\theta$, show that

$$\left(\frac{\partial w}{\partial x}\right)^2 + \left(\frac{\partial w}{\partial y}\right)^2 = \left(\frac{\partial w}{\partial r}\right)^2 + \frac{1}{r^2}\left(\frac{\partial w}{\partial \theta}\right)^2.$$

15   If α is a constant and $w = f(x, y)$ where $x = u\cos\alpha - v\sin\alpha$ and $y = u\sin\alpha + v\cos\alpha$, show that

$$\left(\frac{\partial w}{\partial x}\right)^2 + \left(\frac{\partial w}{\partial y}\right)^2 = \left(\frac{\partial w}{\partial u}\right)^2 + \left(\frac{\partial w}{\partial v}\right)^2.$$

16   If $w = f(x, y)$ where $x = e^u\cos v$ and $y = e^u\sin v$, show that

$$\left(\frac{\partial w}{\partial x}\right)^2 + \left(\frac{\partial w}{\partial y}\right)^2 = e^{-2u}\left[\left(\frac{\partial w}{\partial u}\right)^2 + \left(\frac{\partial w}{\partial v}\right)^2\right].$$

17   Obtain formula (16) by using differentials, as follows:
(a) Write

$$dw = \left(\frac{\partial w}{\partial x}\right)_y dx + \left(\frac{\partial w}{\partial y}\right)_x dy,$$

$$dw = \left(\frac{\partial w}{\partial x}\right)_t dx + \left(\frac{\partial w}{\partial t}\right)_x dt,$$

$$dy = \left(\frac{\partial y}{\partial x}\right)_t dx + \left(\frac{\partial y}{\partial t}\right)_x dt.$$

(b) Substitute dy from the third formula into the first, and compare the result with the second.

*18   Let the pressure, volume, and temperature of a given quantity of a certain gas be denoted (as usual) by p, V, T. These variables are not independent, but are connected by an equation of the general form

$$f(p, V, T) = 0,$$

which is called the *equation of state*. This equation

determines any one of the variables as a function of the other two.

(a) By calculating dp and dV, and eliminating dV, show that

$$\left(\frac{\partial p}{\partial V}\right)_T = \frac{1}{(\partial V/\partial p)_T},$$

$$\left(\frac{\partial p}{\partial V}\right)_T\left(\frac{\partial V}{\partial T}\right)_p + \left(\frac{\partial p}{\partial T}\right)_V = 0,$$

$$\left(\frac{\partial p}{\partial V}\right)_T\left(\frac{\partial V}{\partial T}\right)_p\left(\frac{\partial T}{\partial p}\right)_V = -1.$$

(b) If the internal energy E of the quantity of gas under discussion is a function of V and T, then, since T is a function of p and V, E is indirectly a function of p and V. Show that

$$\left(\frac{\partial E}{\partial V}\right)_p = \left(\frac{\partial E}{\partial V}\right)_T + \left(\frac{\partial E}{\partial T}\right)_V\left(\frac{\partial T}{\partial V}\right)_p.$$

It is sometimes necessary to work with functions of the form

$$w = F(u, v, x) = \int_u^v f(x, y)\, dy,$$

where $u = u(x)$ and $v = v(x)$ are functions of x. The chain rule yields

$$\frac{dw}{dx} = \frac{\partial w}{\partial u}\frac{du}{dx} + \frac{\partial w}{\partial v}\frac{dv}{dx} + \frac{\partial w}{\partial x},$$

and by applying the Fundamental Theorem of Calculus and differentiating under the integral sign (Problem 32 in Section 19.2), we obtain

$$\frac{d}{dx}\int_u^v f(x, y)\, dy = -f(x, u)\frac{du}{dx} + f(x, v)\frac{dv}{dx}$$
$$+ \int_u^v \left[\frac{\partial}{\partial x} f(x, y)\right] dy.$$

This is known as *Leibniz's formula.*

19 Verify Leibniz's formula in the following cases:
(a) $u = x$, $v = x^2$, $f(x, y) = x + y$;
(b) $u = x$, $v = x^2$, $f(x, y) = x^3 y^2 + x^2 y^3$;
(c) $u = x$, $v = x^2$, $f(x, y) = \ln y$.

In the case of functions of a single variable, one of the main applications of derivatives is to the study of maxima and minima. In Chapter 4 we developed various tests involving first and second derivatives, and we used these tests for graphing functions and attacking a wide variety of geometric and physical problems. Maximum and minimum problems for functions of two or more variables can be much more complicated. We confine ourselves here to an introduction to such problems, including a two-variable version of the second derivative test (Remark 3, Section 4.2).

Suppose that a function $z = f(x, y)$ has a maximum value at a point $P_0 = (x_0, y_0)$ in the interior of its domain. This means that $f(x, y)$ is defined, and also $f(x, y) \leq f(x_0, y_0)$, throughout some neighborhood of P_0, as shown on the left in Fig. 19.13.* If we hold y fixed at the value y_0, then $z = f(x, y_0)$ is a function of x alone, and since it has a maximum value at $x = x_0$, its derivative must be zero there, as in Chapter 4. That is, $\partial z/\partial x = 0$ at this point. In just the same way, $\partial z/\partial y = 0$ at this point. The equations

$$\frac{\partial z}{\partial x} = 0 \quad \text{and} \quad \frac{\partial z}{\partial y} = 0 \tag{1}$$

are therefore two equations in two unknowns that are satisfied at the maximum point (x_0, y_0). In many cases we can solve these equations simulta-

19.7

MAXIMUM AND MINIMUM PROBLEMS

---

* In this discussion we are considering only a so-called *relative* (or *local*) maximum, which takes into account only points near to P_0, but for simplicity we drop the adjective.

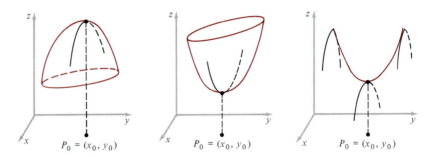

Figure 19.13

neously to find the point (x_0, y_0), and thus the actual maximum value of the function.

Exactly the same considerations apply to the minimum value shown in the center of the figure. However, when we try to locate maximum or minimum values of a function by solving equations (1), it is necessary to keep in mind that these equations are also satisfied at a saddle point like that shown on the right, where the function has a maximum in one direction and a minimum in some other direction. Equations (1) mean only that the tangent plane is horizontal, and it is then up to us to decide what significance this fact has.

By analogy with our earlier definition in Chapter 4 for functions of one variable, we call a point (x_0, y_0) where both partial derivatives are zero a *critical point* of $f(x, y)$.

Example 1 Find the dimensions of the rectangular box with open top and a fixed volume of 4 ft³ which has the smallest possible surface area.

Solution If x and y are the edges of the base, and z is the height, then the area is

$$A = xy + 2xz + 2yz.$$

Since $xyz = 4$, we have $z = 4/xy$, and the area to be minimized can be expressed as a function of the two variables x and y,

$$A = xy + \frac{8}{y} + \frac{8}{x}. \tag{2}$$

We seek a critical point of this function, that is, a point where

$$\frac{\partial A}{\partial x} = y - \frac{8}{x^2} = 0, \qquad \frac{\partial A}{\partial y} = x - \frac{8}{y^2} = 0.$$

To solve these equations simultaneously, we first write them as

$$x^2y = 8, \qquad xy^2 = 8.$$

Dividing gives $x/y = 1$, so $y = x$ and either equation becomes $x^3 = 8$. Therefore $x = y = 2$, and it follows from this that $z = 1$, so the box with minimum surface area has a square base and a height one-half the edge of the base.

In this example it is geometrically clear that the critical point $(2, 2)$ is actually a minimum point, and not a maximum or saddle point. However, in a more complicated situation we might find a critical point and yet be completely unable to state its nature, based on common-sense considerations alone. A useful tool for classifying critical points is provided by the *second derivative test*:

If $f(x, y)$ has continuous second partial derivatives in a neighborhood of a critical point (x_0, y_0), and if a number D (called the discriminant) is defined by

$$D = f_{xx}(x_0, y_0) f_{yy}(x_0, y_0) - [f_{xy}(x_0, y_0)]^2, \tag{3}$$

then (x_0, y_0) is

  (i)   *a maximum point if $D > 0$ and $f_{xx}(x_0, y_0) < 0$;*
 (ii)   *a minimum point if $D > 0$ and $f_{xx}(x_0, y_0) > 0$;*
(iii)   *a saddle point if $D < 0$.*

Further, if $D = 0$, then no conclusion can be drawn, and any of the behaviors described in (i) *to* (iii) *can occur.*

Since there are many erroneous proofs of this theorem abroad in the land, and a valid proof requires machinery that is not available to us, we refer interested students to more advanced books.*

Example 1 (cont.)   As an illustration of the use of the second derivative test, we apply it to verify that the critical point $(2, 2)$ found in Example 1 is a minimum point of the function (2). Here we have

$$A_{xx} = \frac{16}{x^3}, \qquad A_{yy} = \frac{16}{y^3}, \qquad A_{xy} = 1,$$

so the discriminant (3) has the value $D = 2 \cdot 2 - 1^2 = 3 > 0$. Since A_{xx} is also positive at the point $(2, 2)$, the test tells us that this critical point is indeed a minimum point, as claimed.

Example 2   Find the critical points of the function

$$z = 3x^2 + 2xy + y^2 + 10x + 2y + 1,$$

and use the second derivative test to classify them.

Solution   Here we have

$$\frac{\partial z}{\partial x} = 6x + 2y + 10 = 0 \qquad \text{and} \qquad \frac{\partial z}{\partial y} = 2x + 2y + 2 = 0,$$

so the system of equations we must solve is

$$3x + y = -5,$$
$$x + y = -1.$$

By simple manipulations we easily see that $x = -2$ and $y = 1$, so there is a single critical point $(-2, 1)$. At this point we have

---

* See, for example, pp. 157–159 of R. C. Buck, *Advanced Calculus* (McGraw-Hill, 1978).

$$D = z_{xx} z_{yy} - z_{xy}{}^2 = 6 \cdot 2 - 2^2 = 8 > 0,$$

and since $z_{xx} = 6 > 0$, the critical point is a minimum point.

Success in finding maximum and minimum points for a function $z = f(x, y)$ clearly depends on our ability to solve the two simultaneous equations $f_x = 0$ and $f_y = 0$. In Examples 1 and 2 these equations were very easy to solve. However, as students can readily imagine, there are many complicated situations that arise in which routine methods of solving simultaneous equations are quite useless. The only general advice we can give is to try to solve one of the equations for one of the unknowns in terms of the other, substitute this in the second equation, and try to solve the result. Apart from this, make good guesses and be ingenious—advice that is easier to give than to follow!

PROBLEMS

In Problems 1 to 8, find the critical points and classify them by means of the second derivative test.

1   $z = 5x^2 - 3xy + y^2 - 15x - y + 2$.
2   $z = 2x^2 + xy + 3y^2 + 10x - 9y + 11$.
3   $z = x^5 + y^4 - 5x - 32y - 3$.
4   $z = x^2 + y^3 - 6xy$.
5   $z = x^2 y + 3xy - 3x^2 - 4x + 2y$.
6   $z = 3xy^2 + y^2 - 3x - 6y + 7$.
7   $z = x^3 + y^3 + 3xy + 5$.
8   $z = xy(2x + 4y + 1)$.
9   For each of the following functions $z = f(x, y)$, show that f_x, f_y, and D are all 0 at the origin. Also show that at the origin (a) has a minimum, (b) has a maximum, and (c) has a saddle point.
    (a) $f(x, y) = x^4 + y^4$.
    (b) $f(x, y) = -x^4 - y^4$.
    (c) $f(x, y) = x^3 y^3$.
10   Show that a rectangular box with a top and fixed volume has the smallest surface area if it is a cube.
11   Show that a rectangular box with a top and fixed surface area has the largest volume if it is a cube.
12   If the equations $f(x) = 0$ and $f'(x) = 0$ have no common roots, show that any critical points of $z = yf(x) + g(x)$ must be saddle points.
13   If the sum of three numbers x, y, and z is 12, what must these numbers be for the product of x, y^2, and z^3 to be as large as possible?
*14   The function $z = (y - x^2)(y - 2x^2)$ has a saddle point at $(0, 0)$.
    (a) Verify that $(0, 0)$ is the only critical point.
    (b) Show that the second derivative test fails to establish that this critical point is a saddle point.
    (c) Show that this critical point is a saddle point by direct examination of the sign of the function near $(0, 0)$.

(d) The "erroneous proofs" of the second derivative test referred to in the text are based on the idea that a critical point $P_0 = (x_0, y_0)$ of $z = f(x, y)$ will necessarily be a minimum point for this surface if every vertical section of the surface through P_0 has P_0 as a minimum point. Show that this idea is false by examining the vertical section of $z = (y - x^2)(y - 2x^2)$ in the plane $y = mx$.
15   A rectangular box has three faces in the coordinate planes and one vertex $P = (x, y, z)$ in the first octant on the plane $ax + by + cz = 1$. Find the volume of the largest such box.
16   Solve Problem 15 if P lies on the ellipsoid $x^2/a^2 + y^2/b^2 + z^2/c^2 = 1$. Hint: Use implicit differentiation.
17   Solve Problem 15 if P lies on the paraboloid $z = 1 - x^2 - y^2$.
18   If $z = f(s, t)$ is the square of the distance between a variable point on the line

$$x = -2 + 4s, \qquad y = 3 + s, \qquad z = -1 + 5s$$

and a variable point on the line

$$x = -1 - 2t, \qquad y = 3t, \qquad z = 3 + t,$$

show that this function has one critical point which is a minimum. In this way find the distance between the lines.
19   Find the distance from the origin to the plane $x + 2y + 3z = 14$. Hint: Minimize $w = x^2 + y^2 + z^2$ by treating y and z as the independent variables.
20   The sides of an open rectangular box cost twice as much per square foot as the base. Find the relative dimensions of the largest box that can be made for a given cost.
21   Find the equation of the plane through $(2, 2, 1)$ that cuts off the smallest volume from the first octant.

22 If (x_1, y_1), (x_2, y_2), and (x_3, y_3) are the vertices of a triangle, find the point (x, y) such that the sum of the squares of its distances from the vertices is as small as possible.

23 If α, β, γ are the angles of a triangle, find the maximum value of $\sin \alpha + \sin \beta + \sin \gamma$.

***24** Among all triangles with given fixed perimeter, show that the equilateral triangle has the largest area. Hint: Let the perimeter be $2s$, use Heron's formula $A = \sqrt{s(s-a)(s-b)(s-c)}$, and maximize $\ln A$. (Can you solve this problem without calculation, by merely thinking about it?)

***25** Among all triangles inscribed in a given circle, show that the equilateral triangle has the greatest area. Hint: If α, β, γ are the central angles subtending the sides of the triangle, so that $\alpha + \beta + \gamma = 2\pi$, observe that the area of the triangle is a constant multiple of $\sin \alpha + \sin \beta - \sin (\alpha + \beta)$.

26 When an electric current of magnitude I flows through a wire of resistance R, the heat generated is proportional to $I^2 R$. Two terminals are connected by three wires of resistances R_1, R_2, R_3. A given current flowing between the terminals will divide in such a way as to minimize the total heat produced. Show that the currents I_1, I_2, I_3 in the three wires will satisfy the equations $I_1 R_1 = I_2 R_2 = I_3 R_3$.

***27** A pentagon consists of an isosceles triangle on top of a rectangle. If the perimeter P is fixed, find the dimensions of the rectangle and the height of the triangle that yield the maximum area.

***28** Show that the surface $z = (2x^2 + y^2)e^{1-x^2-y^2}$ looks like two mountain peaks joined by two ridges with a hollow depression between them.

29 A laboratory scientist performs an experiment n times and obtains n pairs of data,

$$(x_1, y_1), (x_2, y_2), \ldots, (x_n, y_n).$$

The theory underlying her experiment suggests that these points should lie on a straight line $y = mx + b$, but they do not because of experimental error. She then determines the line that gives the "best fit" for

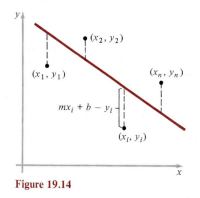

Figure 19.14

the data in the sense of the *method of least squares*: She chooses m and b to minimize the sum of the squares of the vertical deviations (Fig. 19.14),

$$S = S(m, b) = \sum_{i=1}^{n} (mx_i + b - y_i)^2.$$

Show that m and b are determined as the simultaneous solution of the equations

$$m \sum x_i^2 + b \sum x_i = \sum x_i y_i,$$
$$m \sum x_i + nb = \sum y_i.$$

30 Use the method of least squares explained in Problem 29 to find the line that best fits the data $(1, 1.7)$, $(2, 1.8)$, $(3, 2.3)$, $(4, 3.2)$.

***31** Use the second derivative test to verify that S in Problem 29 is actually minimized by the stated values of m and b. Hint: It is necessary to use the fact that $(\Sigma x_i)^2 < n \Sigma x_i^2$ unless the x_i's are all equal; as a start toward establishing this, show that the maximum value of $f(x, y, z) = x + y + z$ on the sphere $x^2 + y^2 + z^2 = a^2$ is $\sqrt{3}a$, and conclude that $(x + y + z)^2 \leq 3(x^2 + y^2 + z^2)$ for any three numbers x, y, z.

19.8

(OPTIONAL)
CONSTRAINED
MAXIMA
AND MINIMA.
LAGRANGE
MULTIPLIERS

In this section we explain the method of Lagrange multipliers by means of intuitive ideas that depend on the geometric meaning of gradients. This method is used for maximizing or minimizing functions of several variables subject to one or more constraints. It is an important tool in economics, differential geometry, and advanced theoretical mechanics.

We begin with the simplest case, that of two variables and one constraint.

In Section 19.7 we learned how to calculate maximum and minimum values of a function $z = f(x, y)$ of two independent variables x and y. However, in many problems x and y are not independent, but instead are connected by a *side condition* or *constraint* in the form of an equation

$$g(x, y) = 0. \tag{1}$$

In Chapter 4 we became thoroughly familiar with situations of this kind. The following is a simple illustration.

Example 1 Find the dimensions of the rectangle of maximum area that can be inscribed in a semicircle of radius a.

Solution It is clear from Fig. 19.15 that the problem is to maximize the function

$$A = 2xy \tag{2}$$

subject to the constraint

$$x^2 + y^2 = a^2. \tag{3}$$

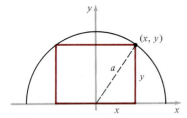

Figure 19.15

In Example 3 of Section 4.3 we solved this problem by using the constraint (3) to express A as a function of only one variable,

$$y = \sqrt{a^2 - x^2}, \qquad \text{so that} \qquad A = 2x\sqrt{a^2 - x^2}.$$

We then calculated dA/dx, set it equal to zero, solved the resulting equation, and so on. This example will be continued after some remarks and explanations.

The procedure we have just described works well enough for this problem, but as a general method it has two defects. First, in this particular case equation (3) is easy to solve for y, but in another problem the constraint (1) might be so complicated that it would be difficult or impossible to solve. The other defect lies in the fact that even though the variables x and y play equal roles in the problem, they are handled differently in the solution: We singled out one variable, x, to be the independent variable, and the other, y, to be the dependent variable. It is often more convenient, and certainly more elegant, to treat such problems in a symmetric form, in which no preference is given to any one of the variables over the others.*

We now return to the general problem of maximizing a function $f(x, y)$ subject to the constraint $g(x, y) = 0$. To understand what is going on, we

---

* The great physicist Einstein once said—probably in a fit of impatience with mathematicians and their ways—that "Elegance is for tailors," but he was wrong. For mathematicians and theoretical physicists alike, the aesthetic factor in their thinking is as indispensable as the senses of taste and smell are for a master chef.

sketch the graph of $g(x, y) = 0$ (Fig. 19.16) together with several level curves $f(x, y) = c$ of the function $f(x, y)$, noting the direction in which c increases. In the figure, for instance, we suppose that $c_1 < c_2 < c_3 < c_4$. To find the maximum value of $f(x, y)$ along the curve $g(x, y) = 0$, we look for the largest c for which $f(x, y) = c$ intersects $g(x, y) = 0$. At such an intersection point (P_0 in the figure) the two curves have the same tangent line, so they also have the same normal line. But the vectors

$$\text{grad } f = \frac{\partial f}{\partial x}\mathbf{i} + \frac{\partial f}{\partial y}\mathbf{j}$$

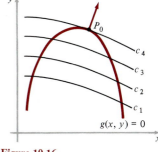

Figure 19.16

and

$$\text{grad } g = \frac{\partial g}{\partial x}\mathbf{i} + \frac{\partial g}{\partial y}\mathbf{j}$$

are normal to these curves, and are therefore parallel to each other at the point P_0. Hence one vector is a scalar multiple of the other at P_0, that is,

$$\text{grad } f = \lambda \text{ grad } g \tag{4}$$

for some number λ. (This argument assumes that grad $g \neq \mathbf{0}$ at P_0, so that the curve $g(x, y) = 0$ actually has a tangent at this point.)

The vector equation (4), together with $g(x, y) = 0$, yields the three scalar equations

$$\frac{\partial f}{\partial x} = \lambda \frac{\partial g}{\partial x}, \qquad \frac{\partial f}{\partial y} = \lambda \frac{\partial g}{\partial y}, \qquad g(x, y) = 0. \tag{5}$$

Accordingly, we have three equations that we can try to solve simultaneously for the three unknowns x, y, and λ. The points (x, y) that we find are the only possible locations for the maximum (or minimum) values of $f(x, y)$ with the constraint $g(x, y) = 0$. The corresponding values of λ may emerge from the process of solving (5), but they are usually not of much interest to us. The final step is to calculate the value of $f(x, y)$ at each of the solution points (x, y) in order to distinguish maximum values from minimum values.

The *method of Lagrange multipliers* is simply the following handy device for obtaining equations (5): Define a function $L(x, y, \lambda)$ of the three variables x, y, and λ by

$$L(x, y, \lambda) = f(x, y) - \lambda g(x, y), \tag{6}$$

and observe that equations (5) are equivalent, in the same order, to the equations

$$\frac{\partial L}{\partial x} = 0, \qquad \frac{\partial L}{\partial y} = 0, \qquad \frac{\partial L}{\partial \lambda} = 0. \tag{7}$$

The variable λ is called the *Lagrange multiplier*. Thus, to find the *constrained* maximum or minimum values of $f(x, y)$ with the constraint $g(x, y) = 0$, we look for the *unconstrained* (or *free*) maximum or minimum values of the function L defined by (6). We emphasize that this method has two major features that can be of practical value, and are often important for theoretical work: It does not disturb the symmetry of the problem by making an arbitrary choice of the independent variable, and it removes the constraint at the small expense of introducing λ as another variable.

Example 1 (cont.) To solve the inscribed rectangle problem by this new method, we first express the constraint (3) in the form $x^2 + y^2 - a^2 = 0$ and then write down the function

$$L = 2xy - \lambda(x^2 + y^2 - a^2).$$

The equations (7) are

$$\frac{\partial L}{\partial x} = 2y - 2\lambda x = 0, \tag{8}$$

$$\frac{\partial L}{\partial y} = 2x - 2\lambda y = 0, \tag{9}$$

$$\frac{\partial L}{\partial \lambda} = -(x^2 + y^2 - a^2) = 0. \tag{10}$$

Equations (8) and (9) yield $y = \lambda x$ and $x = \lambda y$, and substituting in (10) gives

$$\lambda^2(x^2 + y^2) = a^2.$$

But (10) tells us that $x^2 + y^2 = a^2$, so $\lambda^2 = 1$ and $\lambda = \pm 1$. The value $\lambda = -1$ would imply that $y = -x$, which is impossible because both x and y are positive numbers, so $\lambda = 1$ and $y = x$. This gives the shape of the largest inscribed rectangle, namely, twice as long as it is wide, because

$$\text{Length} = 2x = 2y = 2(\text{width}).$$

If we want the actual dimensions of this largest rectangle, we substitute $y = x$ into $x^2 + y^2 = a^2$ to find that $x = y = \frac{1}{2}\sqrt{2}a$, so the length $= 2x = \sqrt{2}a$ and the width $= y = \frac{1}{2}\sqrt{2}a$.

One of the merits of the method of Lagrange multipliers is that it extends very easily to situations with more variables or more constraints. For instance, to maximize $f(x, y, z)$ subject to the constraint $g(x, y, z) = 0$, the gradient

$$\text{grad } f = \frac{\partial f}{\partial x}\mathbf{i} + \frac{\partial f}{\partial y}\mathbf{j} + \frac{\partial f}{\partial z}\mathbf{k}$$

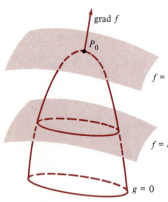

must be normal to the surface $g(x, y, z) = 0$ (Fig. 19.17), so grad f must be parallel to grad g, and again we have

$$\text{grad } f = \lambda \text{ grad } g.$$

The four equations

$$\frac{\partial f}{\partial x} = \lambda \frac{\partial g}{\partial x}, \qquad \frac{\partial f}{\partial y} = \lambda \frac{\partial g}{\partial y}, \qquad \frac{\partial f}{\partial z} = \lambda \frac{\partial g}{\partial z}, \qquad g(x, y, z) = 0,$$

in the four unknowns x, y, z, λ are again equivalent to the simpler equations

$$\frac{\partial L}{\partial x} = 0, \qquad \frac{\partial L}{\partial y} = 0, \qquad \frac{\partial L}{\partial z} = 0, \qquad \frac{\partial L}{\partial \lambda} = 0,$$

Figure 19.17

where $L = f(x, y, z) - \lambda g(x, y, z)$.

Similarly, suppose we want to maximize or minimize $f(x, y, z)$ subject to *two* constraints $g(x, y, z) = 0$ and $h(x, y, z) = 0$. Each constraint defines a surface, and in general these two surfaces have a curve of intersection. As before, a point P_0 where $f(x, y, z)$ has a maximum or minimum value on this curve is a point where a level surface of f is tangent to the curve, that is, a point where grad f is normal to the curve (Fig. 19.18). But the vectors grad g and grad h determine the normal plane to the curve at P_0, and since grad f lies in this plane, there must be scalars λ and μ (*two* Lagrange multipliers this time) with the property that

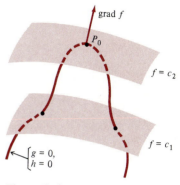

$$\text{grad } f = \lambda \text{ grad } g + \mu \text{ grad } h.$$

(This argument assumes that grad $g \neq \mathbf{0}$ and grad $h \neq \mathbf{0}$, and that these vectors are not parallel.) Just as before, this vector equation and the two constraint equations are easily seen to be equivalent to the following five equations in five unknowns:

Figure 19.18

$$\frac{\partial L}{\partial x} = 0, \qquad \frac{\partial L}{\partial y} = 0, \qquad \frac{\partial L}{\partial z} = 0, \qquad \frac{\partial L}{\partial \lambda} = 0, \qquad \frac{\partial L}{\partial \mu} = 0,$$

where $L = f(x, y, z) - \lambda g(x, y, z) - \mu h(x, y, z)$.

We illustrate these methods with two examples.

Example 2   Find the point on the plane $x + 2y + 3z = 6$ that is closest to the origin.

Solution   We want to minimize the distance $\sqrt{x^2 + y^2 + z^2}$ subject to the constraint $x + 2y + 3z - 6 = 0$. If the distance is a minimum, its square is a minimum, so we simplify the calculations a bit by minimizing $x^2 + y^2 + z^2$ with the same constraint. Let

$$L = x^2 + y^2 + z^2 - \lambda(x + 2y + 3z - 6).$$

Then the equations we must solve are

$$\frac{\partial L}{\partial x} = 2x - \lambda = 0,$$

$$\frac{\partial L}{\partial y} = 2y - 2\lambda = 0,$$

$$\frac{\partial L}{\partial z} = 2z - 3\lambda = 0,$$

$$\frac{\partial L}{\partial \lambda} = -(x + 2y + 3z - 6) = 0.$$

Substituting the values for x, y, and z from the first three equations into the fourth gives

$$\tfrac{1}{2}\lambda + 2\lambda + \tfrac{9}{2}\lambda = 6 \qquad \text{or} \qquad \tfrac{14}{2}\lambda = 6 \qquad \text{or} \qquad \lambda = \tfrac{6}{7}.$$

It now follows that $x = \tfrac{3}{7}$, $y = \tfrac{6}{7}$, and $z = \tfrac{9}{7}$, so the desired point is $(\tfrac{3}{7}, \tfrac{6}{7}, \tfrac{9}{7})$.

Example 3 Find the point on the line of intersection of the planes $x + y + z = 1$ and $3x + 2y + z = 6$ that is closest to the origin.

This time we want to minimize $x^2 + y^2 + z^2$ subject to the two constraints $x + y + z - 1 = 0$ and $3x + 2y + z - 6 = 0$. If we write

$$L = x^2 + y^2 + z^2 - \lambda(x + y + z - 1) - \mu(3x + 2y + z - 6),$$

then our equations are

$$\frac{\partial L}{\partial x} = 2x - \lambda - 3\mu = 0,$$

$$\frac{\partial L}{\partial y} = 2y - \lambda - 2\mu = 0,$$

$$\frac{\partial L}{\partial z} = 2z - \lambda - \mu = 0,$$

$$\frac{\partial L}{\partial \lambda} = -(x + y + z - 1) = 0,$$

$$\frac{\partial L}{\partial \mu} = -(3x + 2y + z - 6) = 0.$$

The first three equations give

$$x = \tfrac{1}{2}(\lambda + 3\mu), \qquad y = \tfrac{1}{2}(\lambda + 2\mu), \qquad z = \tfrac{1}{2}(\lambda + \mu).$$

When these expressions are substituted in the fourth and fifth equations and the results are simplified, we get

$$3\lambda + 6\mu = 2,$$

$$3\lambda + 7\mu = 6,$$

so $\mu = 4$ and $\lambda = -\tfrac{22}{3}$. These values give $x = \tfrac{7}{3}$, $y = \tfrac{1}{3}$, $z = -\tfrac{5}{3}$, and so the desired point is $(\tfrac{7}{3}, \tfrac{1}{3}, -\tfrac{5}{3})$.

Remark 1 In economics, Lagrange multipliers are used to analyze the problem of maximizing the total production of a manufacturing firm subject to the constraint of fixed available resources. For example, let

$$P = f(x, y) = Ax^\alpha y^\beta, \qquad \alpha + \beta = 1,$$

be the production (measured in dollars) resulting from x units of capital and y units of labor. Then this function—known to economists as the *Cobb-Douglas production function*—is homogeneous of degree 1 in the sense explained in Example 2 of Section 19.6. If the cost of each unit of capital is a dollars, and of each unit of labor is b dollars, and if a total of c dollars is available to cover the combined costs of capital and labor, then we want to maximize the production $P = f(x, y)$ subject to the constraint $ax + by = c$. In Problem 23 we ask students to show that production is maximized when $x = \alpha c/a$ and $y = \beta c/b$.*

---

* For further details, interested students can look up Cobb-Douglas production functions in the indexes of the books mentioned in the first footnote of Section 19.6.

Remark 2 At the beginning of this section we said that Lagrange multipliers also have applications to differential geometry and advanced theoretical mechanics. These applications are too complicated to describe here, but the details can be found on pp. 372–374 and 380 of the present writer's text, *Differential Equations* (McGraw-Hill, 1972).

PROBLEMS

Solve all of the following problems by Lagrange multipliers.

1 A rectangle with sides parallel to the axes is inscribed in the region bounded by the axes and the line $x + 2y = 2$. Find the maximum area of this rectangle.

2 Find the rectangle of maximum perimeter (with sides parallel to the axes) that can be inscribed in the ellipse $x^2 + 4y^2 = 4$.

3 Find the rectangle of maximum area (with sides parallel to the axes) that can be inscribed in the ellipse $x^2 + 4y^2 = 4$.

*4 On each of the following curves, find the points that are closest to the origin and those that are farthest from the origin:
   (a) $x^2 + xy + y^2 = 3$;   (b) $x^4 + 3xy + y^4 = 2$.

5 If a cylinder has fixed volume V_0, find the relation between the height h and the radius of the base r that minimizes the surface area.

6 Find the maximum and minimum values of $f(x, y) = 2x^2 + y + y^2$ on the circle $x^2 + y^2 = 1$.

7 Find the maximum and minimum values of $f(x, y) = x^2 - xy + y^2$ on the circle $x^2 + y^2 = 1$.

8 Find the ellipse $x^2/a^2 + y^2/b^2 = 1$ that passes through the point $(4, 1)$ and has the smallest area. Hint: The area of this ellipse is πab.

*9 Find the ellipsoid $x^2/a^2 + y^2/b^2 + z^2/c^2 = 1$ that passes through the point $(1, 2, 3)$ and has the smallest volume. Hint: The volume of this ellipsoid is $\frac{4}{3}\pi abc$.

10 Find the maximum value of $f(x, y, z) = 2x + 2y - z$ on the sphere $x^2 + y^2 + z^2 = 4$.

11 Find the minimum value of $f(x, y, z) = x^2 + 2y^2 + 3z^2$ on the plane $x - y - z = 1$.

12 Find the maximum value of $f(x, y, z) = x + y + z$ on the ellipsoid $x^2/a^2 + y^2/b^2 + z^2/c^2 = 1$.

13 Find the maximum volume of a rectangular box if the sum of the lengths of its edges is $12a$.

14 Find the maximum volume of a rectangular box if the sum of the areas of its faces is $6a^2$.

15 (a) Show that of all triangles inscribed in a given circle, the equilateral triangle has the maximum perimeter. Hint: If a is the radius of the circle and α, β, γ are the central angles subtending the three sides, what is the perimeter?

(b) In part (a), show that the inscribed equilateral triangle also has the maximum area.

16 Find the point on the line of intersection of the planes $x + 2y + z = 1$ and $-3x - y + 2z = 4$ that is closest to the origin.

17 (a) Find the point on the plane $ax + by + cz + d = 0$ that is closest to the origin, and use this information to write down a formula for the distance from the origin to the plane.

(b) Adapt the method used in part (a) to show that the distance from an arbitrary point (x_0, y_0, z_0) to the given plane is

$$\frac{|ax_0 + by_0 + cz_0 + d|}{\sqrt{a^2 + b^2 + c^2}}.$$

18 (a) Show that the triangle with the greatest area A for a given perimeter is equilateral. Hint: If x, y, z are the sides, then $A = \sqrt{s(s - x)(s - y)(s - z)}$, where $2s = x + y + z$.

(b) Show that the triangle with the smallest perimeter for a given area is equilateral.

19 If the sum of n positive numbers $x_1, x_2, \ldots, x_n$ has a fixed value s, show that their product $x_1 x_2 \cdots x_n$ has s^n/n^n as its maximum value, and conclude from this that the geometric mean of n positive numbers can never exceed their arithmetic mean:

$$\sqrt[n]{x_1 x_2 \cdots x_n} \le \frac{x_1 + x_2 + \cdots + x_n}{n}.$$

*20 (a) Find the maximum value of

$$\sum_{i=1}^{n} x_i y_i$$

with the constraints

$$\sum_{i=1}^{n} x_i^2 = 1 \quad \text{and} \quad \sum_{i=1}^{n} y_i^2 = 1.$$

(b) Use part (a) to prove that for any numbers $a_1, a_2, \ldots, a_n$ and $b_1, b_2, \ldots, b_n$,

$$\sum_{i=1}^{n} a_i b_i \le \left(\sum_{i=1}^{n} a_i^2\right)^{1/2} \left(\sum_{i=1}^{n} b_i^2\right)^{1/2}.$$

Hint: Put

$$x_i = \frac{a_i}{\left(\sum\limits_{i=1}^{n} a_i^2\right)^{1/2}} \qquad \text{and} \qquad y_i = \frac{b_i}{\left(\sum\limits_{i=1}^{n} b_i^2\right)^{1/2}}.$$

The inequality in (b) is an important fact in higher mathematics called the *Schwarz inequality*.

*21 Use the method of Problem 20 to establish *Hölder's inequality*: If $1/p + 1/q = 1$ and the a_i's and b_i's are nonnegative numbers, then

$$\sum_{i=1}^{n} a_i b_i \le \left(\sum_{i=1}^{n} a_i^p\right)^{1/p} \left(\sum_{i=1}^{n} b_i^q\right)^{1/q}.$$

Observe that when $p = q = 2$, Hölder's inequality reduces to the Schwarz inequality.

22 Refer back to Example 4 in Section 4.4, and the notation in Fig. 4.26, to obtain Snell's law of refraction by minimizing the total time of travel,

$$T = \frac{a}{v_a \cos \alpha} + \frac{b}{v_w \cos \beta},$$

subject to the constraint $a \tan \alpha + b \tan \beta = a$ constant.

23 Show that to maximize the Cobb-Douglas production function $P = f(x, y) = Ax^\alpha y^\beta$ ($\alpha + \beta = 1$) subject to the constraint of fixed total costs, $ax + by = c$, we must put $x = \alpha c/a$ and $y = \beta c/b$.

19.9

(OPTIONAL) LAPLACE'S EQUATION, THE HEAT EQUATION, AND THE WAVE EQUATION

A very large part of mathematical physics is concerned with three classic partial differential equations: *Laplace's equation,*

$$\frac{\partial^2 w}{\partial x^2} + \frac{\partial^2 w}{\partial y^2} + \frac{\partial^2 w}{\partial z^2} = 0; \tag{1}$$

the *heat equation,*

$$a^2 \left(\frac{\partial^2 w}{\partial x^2} + \frac{\partial^2 w}{\partial y^2} + \frac{\partial^2 w}{\partial z^2}\right) = \frac{\partial w}{\partial t}; \tag{2}$$

and the *wave equation,*

$$a^2 \left(\frac{\partial^2 w}{\partial x^2} + \frac{\partial^2 w}{\partial y^2} + \frac{\partial^2 w}{\partial z^2}\right) = \frac{\partial^2 w}{\partial t^2}. \tag{3}$$

As the notation indicates, in (2) and (3) the variable w is understood to be a function of the time t and the three space coordinates x, y, z of a point P, and in (1) w depends only on x, y, z and is independent of t. The quantity a is a constant. Each of our three equations also has simpler two-dimensional and one-dimensional versions, depending on whether two space coordinates are present, or only one. Thus,

$$\frac{\partial^2 w}{\partial x^2} + \frac{\partial^2 w}{\partial y^2} = 0 \tag{4}$$

is the two-dimensional Laplace equation, and

$$a^2 \frac{\partial^2 w}{\partial x^2} = \frac{\partial w}{\partial t} \qquad \text{and} \qquad a^2 \frac{\partial^2 w}{\partial x^2} = \frac{\partial^2 w}{\partial t^2}$$

are the one-dimensional heat equation and wave equation.

A full study of these equations can occupy years, because their physical meaning is extraordinarily rich, and also much concentrated thought is necessary to master the various branches of advanced mathematics that are needed to solve and interpret them. In this section we consider several aspects of these equations that do not require too much technical background.

LAPLACE'S EQUATION

If a number of particles of masses $m_1, m_2, \ldots, m_n$, attracting according to the inverse square law of gravitation, are placed at points $P_1, P_2, \ldots, P_n$, then the *potential* due to these particles at any point P (that is, the work done against their attractive forces in moving a unit mass from P to an infinite distance) is

$$w = \frac{Gm_1}{PP_1} + \frac{Gm_2}{PP_2} + \cdots + \frac{Gm_n}{PP_n}, \tag{5}$$

where G is the gravitational constant.* If the points $P, P_1, P_2, \ldots, P_n$ have rectangular coordinates (x, y, z), (x_1, y_1, z_1), $(x_2, y_2, z_2), \ldots$, (x_n, y_n, z_n), so that

$$PP_1 = \sqrt{(x - x_1)^2 + (y - y_1)^2 + (z - z_1)^2},$$

with similar expressions for the other distances, then it is quite easy to verify that the potential w satisfies Laplace's equation (1). This equation does not involve either the particular masses or the coordinates of the points at which they are located, so it is satisfied by the potential produced in empty space by an arbitrary discrete or continuous distribution of mass.

   The function w defined by (5) is called a *gravitational potential*. If we work instead with electrically charged particles of charges $q_1, q_2, \ldots, q_n$, then their *electrostatic potential* has the same form as (5) with the m's replaced by q's and G by Coulomb's constant, so it also satisfies Laplace's equation. In fact, this equation has such a wide variety of applications that its study is a branch of mathematics in its own right, known as *potential theory*.

THE HEAT EQUATION

When we study the flow of heat in thermally conducting bodies, we encounter an entirely different type of problem leading to a partial differential equation.

   In the interior of a body where heat is flowing from one region to another, the temperature generally varies from point to point at any one time, and from time to time at any one point. Thus, the temperature w is a function of the space coordinates x, y, z and the time t, say $w = f(x, y, z, t)$. The precise form of this function naturally depends on the shape of the body, the thermal characteristics of its material, the initial distribution of temperature, and the conditions maintained on the surface of the body. The French physicist-mathematician Fourier studied this problem in his classic treatise of 1822, *Théorie Analytique de la Chaleur (Analytic Theory of Heat)*. He used physical principles to show that the temperature function w must satisfy the heat equation (2).† We shall retrace his reasoning in a simple one-dimensional situation, and thereby derive the one-dimensional heat equation.

---

* See Example 2 in Section 7.8. In this example we show that if two particles of masses M and m are separated by a distance a, then the work done in separating them to an infinite distance is GMm/a.

† The same partial differential equation also describes a more general class of diffusion processes, and is sometimes called the *diffusion equation*.

The following are the physical principles that will be needed.

(a) Heat flows in the direction of decreasing temperature, that is, from hot regions to cold regions.

(b) The rate at which heat flows across an area is proportional to the area and to the rate of change of temperature with respect to distance in a direction perpendicular to the area. (This proportionality factor is denoted by k and called the *thermal conductivity* of the substance.)

(c) The quantity of heat gained or lost by a body when its temperature changes, that is, the change in its thermal energy, is proportional to the mass of the body and to the change of temperature. (This proportionality factor is denoted by c and called the *specific heat* of the substance.)

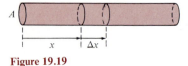

Figure 19.19

We now consider the flow of heat in a thin cylindrical rod of cross-sectional area A (Fig. 19.19) whose lateral surface is perfectly insulated so that no heat flows through it. This use of the word "thin" means that the temperature is assumed to be uniform on any cross section, and is therefore a function only of the time and the position of the cross section, say $w = f(x, t)$. We examine the rate of change of the heat contained in a thin slice of the rod between the positions x and $x + \Delta x$.

If ρ is the density of the rod, that is, its mass per unit volume, then the mass of the slice is

$$\Delta m = \rho A \, \Delta x.$$

Furthermore, if Δw is the temperature change at the point x in a small time interval Δt, then (c) tells us that the quantity of heat stored in the slice in this time interval is

$$\Delta H = c \, \Delta m \, \Delta w = c \rho A \, \Delta x \, \Delta w,$$

so the rate at which heat is being stored is approximately

$$\frac{\Delta H}{\Delta t} = c \rho A \, \Delta x \, \frac{\Delta w}{\Delta t}. \tag{6}$$

We assume that no heat is generated inside the slice—for instance, by chemical or electrical processes—so that the slice gains heat only by means of the flow of heat through its faces. By (b) the rate at which heat flows into the slice through the left face is

$$-kA \left. \frac{\partial w}{\partial x} \right|_x .$$

The negative sign here is chosen in accordance with (a), so that this quantity will be positive if $\partial w/\partial x$ is negative. Similarly, the rate at which heat flows into the slice through the right face is

$$kA \left. \frac{\partial w}{\partial x} \right|_{x+\Delta x},$$

so the total rate at which heat flows into the slice is

$$kA \left. \frac{\partial w}{\partial x} \right|_{x+\Delta x} - kA \left. \frac{\partial w}{\partial x} \right|_x . \tag{7}$$

If we equate the expressions (6) and (7), the result is

$$kA \left. \frac{\partial w}{\partial x} \right|_{x+\Delta x} - kA \left. \frac{\partial w}{\partial x} \right|_{x} = c\rho A \, \Delta x \, \frac{\Delta w}{\Delta t},$$

or

$$\frac{k}{c\rho} \left[\frac{\partial w/\partial x|_{x+\Delta x} - \partial w/\partial x|_{x}}{\Delta x} \right] = \frac{\Delta w}{\Delta t}.$$

Finally, by letting Δx and $\Delta t \to 0$ we obtain the desired equation,

$$a^2 \frac{\partial^2 w}{\partial x^2} = \frac{\partial w}{\partial t},$$

where $a^2 = k/c\rho$. This is the physical reasoning that leads to the one-dimensional heat equation. The three-dimensional equation (2) can be derived in essentially the same way.

THE WAVE EQUATION

All phenomena of wave propagation, for example, of light or sound or radio waves, are governed by the wave equation (3). We shall consider the simple case of a one-dimensional wave described by the one-dimensional wave equation

$$a^2 \frac{\partial^2 w}{\partial x^2} = \frac{\partial^2 w}{\partial t^2}. \tag{8}$$

Such a wave involves some property $w = f(x, t)$, such as the position of a particle, the intensity of an electric field, or the pressure in a column of air, that depends not only on the position x but also on the time t.

In order to understand the connection between waves and equation (8), we consider a function $w = F(x - at)$. At $t = 0$, it defines the curve $w = F(x)$, and at any later time $t = t_1$, it defines the curve $w = F(x - at_1)$. It is easy to see that these curves are identical except that the latter is translated to the right through a distance at_1, and therefore with velocity

$$v = \frac{at_1}{t_1} = a.$$

This shows that the function $w = F(x - at)$ represents a traveling wave that moves to the right with velocity a, as suggested in Fig. 19.20. If we assume that $w = F(u)$ has a second derivative, then by the chain rule applied to $w = F(u)$ where $u = x - at$, we have

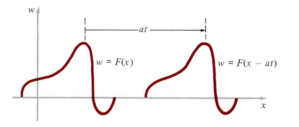

Figure 19.20   A traveling wave.

$$\frac{\partial w}{\partial x} = F'(u), \qquad \frac{\partial w}{\partial t} = F'(u) \cdot (-a) = -aF'(u),$$

$$\frac{\partial^2 w}{\partial x^2} = F''(u), \qquad \frac{\partial^2 w}{\partial t^2} = -aF''(u) \cdot (-a) = a^2 F''(u).$$

It is clear from this that $w = F(x - at)$ satisfies the one-dimensional wave equation (8).

Similarly, the function $w = G(x + at)$ represents a traveling wave that moves to the left with velocity a, and it is equally easy to show that this function is a solution of (8). By the linearity of differentiation, it follows that the sum

$$w = F(x - at) + G(x + at) \tag{9}$$

is also a solution. In fact, it can be shown (see Problem 8) that if F and G are *arbitrary* twice-differentiable functions, then (9) is the *general* solution of (8), in the sense that every solution of (8) has the form (9). It is fairly clear that the function (9) represents the most general one-dimensional wave, and this result confirms it.

PROBLEMS

1 (a) Verify that the function

$$w = \frac{1}{\sqrt{(x - x_1)^2 + (y - y_1)^2 + (z - z_1)^2}}$$

satisfies Laplace's equation (1).

(b) A differential equation is called *linear* if the sum of two solutions is a solution, and any constant times a solution is a solution. Verify that Laplace's equation (1) is linear, and conclude from part (a) that the potential (5) is a solution.

2 (a) Determine whether or not the function

$$w = \frac{1}{\sqrt{(x - x_1)^2 + (y - y_1)^2}}$$

is a solution of Laplace's equation (4) in two dimensions.

(b) Show that the function $w = \ln [(x - x_1)^2 + (y - y_1)^2]$ is a solution of Laplace's equation (4) in two dimensions.

3 Verify that each of the following functions satisfies Laplace's equation (1):

(a) $w = x^2 + 2y^2 - 3z^2$;
(b) $w = x^2 - y^2 + 57z$;
(c) $w = 4z^3 - 6(x^2 + y^2)z$;
(d) $w = e^{-2x} \sin 2y + 3z$;
(e) $w = e^{3x} e^{4y} \cos 5z$;
(f) $w = e^{13x} \sin 12y \cos 5z$.

*4 If $w = f(x, y)$ is transformed into $w = F(r, \theta)$ by the equations $x = r \cos \theta$, $y = r \sin \theta$ (these are the transformation equations from rectangular to polar coordinates), show that the two-dimensional Laplace equation

$$\frac{\partial^2 w}{\partial x^2} + \frac{\partial^2 w}{\partial y^2} = 0$$

becomes

$$\frac{\partial^2 w}{\partial r^2} + \frac{1}{r}\frac{\partial w}{\partial r} + \frac{1}{r^2}\frac{\partial^2 w}{\partial \theta^2} = 0.$$

5 Use Problem 4 to show that each of the functions $w_1 = r^n \sin n\theta$ and $w_2 = r^n \cos n\theta$ satisfies Laplace's equation in two dimensions.

6 Suppose that a solution $w = f(x, t)$ of the one-dimensional heat equation

$$a^2 \frac{\partial^2 w}{\partial x^2} = \frac{\partial w}{\partial t}$$

has the form $f(x, t) = g(x)h(t)$, that is, is the product of a function of x and a function of t.

(a) Show that $a^2 g''(x)h(t) = g(x)h'(t)$.
(b) Part (a) implies that

$$\frac{g''(x)}{g(x)} = \frac{h'(t)}{a^2 h(t)},$$

where the left side is a function of x alone and the right side is a function of t alone. Deduce that there is a constant λ such that $g''/g = \lambda$ and $h'/(a^2h) = \lambda$.

(c) Assume that the constant λ in part (b) is negative, and can therefore be written in the form $\lambda = -k^2$ for some positive number k. Show that $g(x) = c_1 \sin kx + c_2 \cos kx$ is a solution of the equation $g''/g = -k^2$ for every choice of the constants c_1 and c_2.

(d) Show that $h(t) = ce^{-a^2k^2t}$ is a solution of the equation $h'/(a^2h) = -k^2$ for every choice of the constant c. Thus all of the functions

$$w = f(x, t) = ce^{-a^2k^2t}(c_1 \sin kx + c_2 \cos kx)$$

are solutions of the heat equation, and any sum of such solutions is a solution.

7 A steady-state solution of the heat equation (2) is one that does not depend on t, and in this case the heat equation reduces to Laplace's equation (1). Solve the one-dimensional Laplace equation.

*8 Use the chain rule to show that under the change of variables specified by $u = x - at, v = x + at$, the equation

$$a^2 \frac{\partial^2 w}{\partial x^2} = \frac{\partial^2 w}{\partial t^2} \qquad \text{becomes} \qquad \frac{\partial^2 w}{\partial u\, \partial v} = 0.$$

Hint: See Example 3 in Section 19.6. Use this result to show that $w = F(x - at) + G(x + at)$ is the most general solution of the one-dimensional wave equation.

19.10

(OPTIONAL) IMPLICIT FUNCTIONS

In Section 3.4 we stated that when we are given an equation

$$F(x, y) = 0, \tag{1}$$

there usually exists at least one function

$$y = f(x) \tag{2}$$

that "solves" (1), in the sense that (2) reduces (1) to an identity in x. With the idea in mind that y in (1) stands for this function of x, we then differentiated the identity (1) with respect to x and went on to solve the resulting equation for dy/dx, calling the process "implicit differentiation." For instance, if we have the equation

$$x^2y^5 - 2xy + 1 = 0, \tag{3}$$

then by differentiating with respect to x we obtain

$$x^2 \cdot 5y^4 \frac{dy}{dx} + 2xy^5 - 2x \frac{dy}{dx} - 2y = 0, \tag{4}$$

so

$$\frac{dy}{dx} = \frac{2y - 2xy^5}{5x^2y^4 - 2x}. \tag{5}$$

Most students feel slightly uncomfortable about implicit differentiation, and with good reason. For one thing, in this particular case we have no idea whether (3) actually defines y as a function of x or not; and if it doesn't, then the subsequent calculation leading to (5) has no meaning at all. Also, the procedure itself is a bit clumsy, because it requires us to keep in mind the different roles played by the variables x and y. We are now in a position to clarify the meaning of this process, and also to give a precise statement of the conditions under which an equation of the form (1) defines a differentiable function (2).

We broaden the discussion slightly, and instead of (1) consider an equation of the form

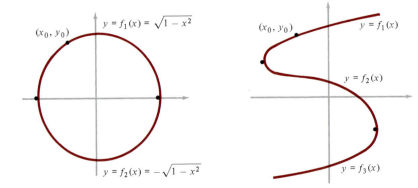

Figure 19.21

$$F(x, y) = c,\tag{6}$$

whose graph is a level curve of the function $z = F(x, y)$. For example, the graph of

$$x^2 + y^2 = 1\tag{7}$$

is a circle about the origin (Fig. 19.21, left), and this is a level curve of the function $F(x, y) = x^2 + y^2$. Generally, as in this case, the graph of (6) will be some sort of curve that is not the graph of a single function. However, even though the entire graph of (7) is not the graph of a single function, it is clear that every point (x_0, y_0) on this graph with $y_0 \neq 0$ lies on a *portion* of the graph that *is* the graph of a function — indeed, of a differentiable function. Specifically, if $y_0 > 0$ then (x_0, y_0) lies on the graph of the function

$$y = f_1(x) = \sqrt{1 - x^2},\tag{8}$$

and if $y_0 < 0$ then (x_0, y_0) lies on the graph of the function

$$y = f_2(x) = -\sqrt{1 - x^2}.\tag{9}$$

Similarly, the graph of (6) might consist of the graphs of two or more differentiable functions $y = f(x)$, as suggested on the right in the figure.

We next point out that the function $z = F(x, y)$ has the constant value c along the graph of any such function $y = f(x)$,

$$z = F[x, f(x)] = c.$$

As usual, we assume that $F(x, y)$ has continuous partial derivatives, so it is permissible to write

$$\frac{dz}{dx} = \frac{\partial F}{\partial x} + \frac{\partial F}{\partial y}\frac{dy}{dx} = 0,$$

or equivalently,

$$\frac{dz}{dx} = F_x(x, y) + F_y(x, y)\frac{dy}{dx} = 0.\tag{10}$$

The middle term here is just the chain rule evaluation of dz/dx when $z = F(x, y)$ and y is a function of x, and the result is zero because z is constant as a function of x. If $F_y(x, y) \neq 0$, equation (10) can be solved for dy/dx,

$$\frac{dy}{dx} = -\frac{F_x(x, y)}{F_y(x, y)}. \tag{11}$$

In the language of Section 3.4, any differentiable function $y = f(x)$ with the property that

$$F[x, f(x)] = c$$

is an *implicit function* defined by (6), and (11) provides a general formula for the derivative of such a function.

If we apply formula (11) to equation (7), where $F(x, y) = x^2 + y^2$, we obtain

$$F_x = 2x \quad \text{and} \quad F_y = 2y, \quad \text{so} \quad \frac{dy}{dx} = -\frac{F_x}{F_y} = -\frac{x}{y}, \quad y \neq 0. \tag{12}$$

In this case we know from (8) and (9) that (7) actually determines two implicit functions $y = f(x)$, so the calculations (12) are legitimate and apply to either function as long as we avoid points where $y = 0$. However, suppose that instead of (7) we have one of the equations

$$x^2 + y^2 = -1 \quad \text{or} \quad x^2 + y^2 = 0. \tag{13}$$

By acting blindly without thinking, we can write down the calculations (12) just as before and "find" dy/dx. The difficulty with this is obvious: Since the graph of (13) is either empty or consists of a single point, no implicit function $y = f(x)$ exists, and these calculations would be nothing more than a kind of mathematical doubletalk, which seems to be saying something but really says nothing at all.

In order to avoid committing such nonsense, it is necessary to have definite knowledge that implicit functions exist. This is the purpose of the

---

Implicit Function Theorem *Let $F(x, y)$ have continuous partial derivatives throughout some neighborhood of a point (x_0, y_0), and assume that $F(x_0, y_0) = c$ and $F_y(x_0, y_0) \neq 0$. Then there is an interval I about x_0 with the property that there exists exactly one differentiable function $y = f(x)$ defined on I such that $y_0 = f(x_0)$ and*

$$F[x, f(x)] = c.$$

Further, the derivative of this function is given by the formula

$$\frac{dy}{dx} = -\frac{F_x}{F_y},$$

and is therefore continuous.

It should be understood that this theorem is a purely theoretical statement to the effect that the specified implicit function $y = f(x)$ does in fact exist, and it has no bearing on the issue of whether a simple formula can be found for this function. A proof is given in Appendix C.18.

Example 1 We consider once more the equation mentioned earlier,

$$F(x, y) = x^2 y^5 - 2xy + 1 = 0. \tag{3}$$

It is clear that the point $(1, 1)$ lies on the graph, so the graph is not empty. Since $F_x = 2xy^5 - 2y$ and $F_y = 5x^2y^4 - 2x$, our theorem guarantees that equation (3) determines an implicit function $y = f(x)$ about any point of the graph where $F_y = 5x^2y^4 - 2x \neq 0$, for instance the point $(1, 1)$. It is instructive to write down equation (10) for this case,

$$(2xy^5 - 2y) + (5x^2y^4 - 2x)\frac{dy}{dx} = 0, \tag{14}$$

and to compare the result with (4), where implicit differentiation is carried out by the old method. Equation (14) evidently yields

$$\frac{dy}{dx} = \frac{2y - 2xy^5}{5x^2y^4 - 2x},$$

just as before.

The simplicity of our present method is even more clearly visible when there are three variables in the given equation.

Thus, suppose an equation $F(x, y, z) = c$ defines a certain implicit function $z = f(x, y)$, and let us find $\partial z/\partial x$ in terms of the function $F(x, y, z)$. The equations

$$w = F(x, y, z),$$

$$x = x, \qquad y = y, \qquad z = f(x, y),$$

give w as a composite function of x and y. Also,

$$w = F[x, y, f(x, y)] = c,$$

so if we differentiate this with respect to x, the chain rule yields

$$\frac{\partial w}{\partial x} = \frac{\partial F}{\partial x}\frac{\partial x}{\partial x} + \frac{\partial F}{\partial y}\frac{\partial y}{\partial x} + \frac{\partial F}{\partial z}\frac{\partial z}{\partial x} = \frac{\partial F}{\partial x} + \frac{\partial F}{\partial z}\frac{\partial z}{\partial x} = 0.$$

We therefore obtain

$$\frac{\partial z}{\partial x} = -\frac{\partial F/\partial x}{\partial F/\partial z}, \tag{15}$$

and this formula is valid wherever $\partial F/\partial z \neq 0$. In just the same way, we also have

$$\frac{\partial z}{\partial y} = -\frac{\partial F/\partial y}{\partial F/\partial z}. \tag{16}$$

As students have surely guessed, there is also an Implicit Function Theorem that covers this situation. Briefly, it says that if $\partial F/\partial z \neq 0$ at a point (x_0, y_0, z_0) on a surface $F(x, y, z) = c$, then in a neighborhood of this point the surface defines a unique implicit function $z = f(x, y)$ such that $z_0 = f(x_0, y_0)$, and furthermore the partial derivatives of this function are given by (15) and (16).

Example 2 It is easy to verify that the point $(1, 2, -1)$ lies on the graph of the equation

$$x^2z + yz^5 + 2xy^3 = 13, \tag{17}$$

so this graph is not empty. If the equation defines an implicit function $z = f(x, y)$ in a neighborhood of this point, then we can calculate $\partial z/\partial x$ by implicit differentiation in the old way. This means we differentiate (17) implicitly with respect to x, thinking of y as a constant, which gives

$$x^2 \frac{\partial z}{\partial x} + 2xz + y \cdot 5z^4 \frac{\partial z}{\partial x} + 2y^3 = 0,$$

so

$$\frac{\partial z}{\partial x} = -\frac{2xz + 2y^3}{x^2 + 5yz^4}.$$

This procedure is unsatisfactory because we don't know in the beginning whether any such function $z = f(x, y)$ actually exists—after all, (17) is a fifth-degree equation in z—and also because in the implicit differentiation each of the three variables has to be treated in a different way, and it is quite easy to lose track of what is going on. Our present ideas provide a much better method. We have

$$F(x, y, z) = x^2z + yz^5 + 2xy^3,$$

so

$$\frac{\partial F}{\partial x} = 2xz + 2y^3 \quad \text{and} \quad \frac{\partial F}{\partial z} = x^2 + 5yz^4.$$

It is easy to see that $\partial F/\partial z = 11 \neq 0$ at $(1, 2, -1)$, so the Implicit Function Theorem guarantees that $z = f(x, y)$ exists. Also, by (15) we have

$$\frac{\partial z}{\partial x} = -\frac{\partial F/\partial x}{\partial F/\partial z} = -\frac{2xz + 2y^3}{x^2 + 5yz^4},$$

which avoids the messy implicit differentiation.

Remark The two-variable version of the Implicit Function Theorem enables us to complete a long-standing piece of unfinished business. In the earlier chapters of this book we gave quite a bit of attention to the important problem of finding the inverse function of a given function $g(y) = x$, in other words, the problem of solving the equation

$$F(x, y) = g(y) - x = 0 \tag{18}$$

for the variable y. Specifically, this is the way the familiar functions $y = \ln x$, $y = \sin^{-1} x$, and $y = \tan^{-1} x$ were defined. Each of these inverse functions was discussed earlier in an *ad hoc* but perfectly legitimate way. We are now in a position to draw the general inference that when $g(y)$ has a continuous derivative and $\partial F/\partial y = g'(y) \neq 0$, then (18) can indeed be solved for y, $y = f(x)$, and also that this function has a continuous derivative given by

$$\frac{dy}{dx} = -\frac{\partial F/\partial x}{\partial F/\partial y} = -\frac{-1}{g'(y)} = \frac{1}{dx/dy}.$$

This completes the line of thought that was briefly described in Remark 2 of Section 9.5.

PROBLEMS

In Problems 1 to 6, use formula (11) to compute dy/dx.

1 $y^2 - 3x^2 - 1 = 0$.

2 $x^6 + 2y^4 = 1$.

3 $x \sin y = x + y$.

4 $\sin y + \tan y = x^2 + x^3$.

5 $e^{xy} = 2xy^2$.

6 $e^x \sin y = e^y \cos x$.

In Problems 7 to 10, use formulas (15) and (16) to compute $\partial z/\partial x$ and $\partial z/\partial y$.

7 $\ln z = z + 2y - 3x$.

8 $\tan^{-1} x + \tan^{-1} y + \tan^{-1} z = 9$.

9 $z = xy \sin xz$.

10 $\sin xy + \sin yz + \sin xz = 1$.

11   Use formulas (15) and (16) to find the largest value of z on the ellipsoid $2x^2 + 3y^2 + z^2 + yz - xz = 1$.

12   The folium of Descartes (Problem 16 in Section 17.1) has $x^3 + y^3 = 3axy$ as its equation. Use formula (11) to find the highest point on the loop.

13   If $F(x, y)$ has continuous second partial derivatives and the equation $F(x, y) = c$ defines $y = f(x)$ as a twice-differentiable function, show that if $F_y \neq 0$,

$$\frac{d^2y}{dx^2} = - \frac{F_{xx}F_y{}^2 - 2F_{xy}F_xF_y + F_{yy}F_x{}^2}{F_y{}^3}.$$

14   Compute d^2y/dx^2 by using Problem 13 if
(a) $x^4y^5 = 1$;   (b) $e^y = x + y$.

20

MULTIPLE INTEGRALS

A continuous function $f(x, y)$ of two variables can be integrated over a plane region R in much the same way that a continuous function of one variable can be integrated over an interval. The result is a number called the *double integral* of $f(x, y)$ over R and denoted by

$$\iint_R f(x, y)\, dA \quad \text{or} \quad \iint_R f(x, y)\, dx\, dy.$$

A different but closely related concept is that of an *iterated* (or *repeated*) *integral.* We discuss iterated integrals in this section, and in the next section return to the topic of double integrals and explain what they are and how they are related to iterated integrals.

In Section 7.3 we discussed the "method of moving slices" for finding volumes. Thus, if $A(x)$ is the area of the section cut from a solid by a plane perpendicular to the x-axis at a distance x from the origin, then the formula

$$V = \int_a^b A(x)\, dx \tag{1}$$

gives the volume of the solid between the planes $x = a$ and $x = b$. The essence of this formula lies in the idea that

$$dV = A(x)\, dx$$

is the volume of a thin slice of the solid of thickness dx. The total volume (1) is then found by adding together (or integrating) these elements of volume as our typical slice sweeps through the complete solid, that is, as x increases from a to b.

However, if the section itself has curved boundaries—as happens in many cases—then the determination of $A(x)$ also requires integration. For instance, the section shown in Fig. 20.1 extends from the xy-plane $z = 0$ up to the curved surface $z = f(x, y)$. By considering x to be arbitrary but momentarily fixed between a and b, we see that the area of this section is

$$A(x) = \int_{y_1(x)}^{y_2(x)} f(x, y)\, dy, \tag{2}$$

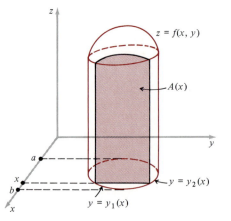

Figure 20.1

where $y = y_1(x)$ and $y = y_2(x)$ are the equations of the curves that bound the base on the left and right. To find the total volume V, we now insert (2) in (1) and obtain the *iterated integral*

$$V = \int_a^b \left[\int_{y_1(x)}^{y_2(x)} f(x, y)\, dy \right] dx. \tag{3}$$

Students should notice particularly that in (3) we first integrate $f(x, y)$ with respect to y, holding x fixed. The limits of integration depend on this fixed but arbitrary value of x, and so does the resulting value of the inner integral. This inner integral is precisely the function $A(x)$ given by (2), which we then integrate with respect to x from a to b to obtain the iterated integral (3). To summarize, we start with a positive function $f(x, y)$ of two variables; we first "integrate y out," which gives a function of x alone; and then we "integrate x out," which gives a number—the volume of the solid.

On the other hand, in some cases it may be more convenient to cut the solid by a plane perpendicular to the y-axis and to form the iterated integral in the other order, first integrating x and then y,

$$V = \int_c^d \left[\int_{x_1(y)}^{x_2(y)} f(x, y)\, dx \right] dy. \tag{4}$$

These two possible orders of integration are suggested in Fig. 20.2, represent-

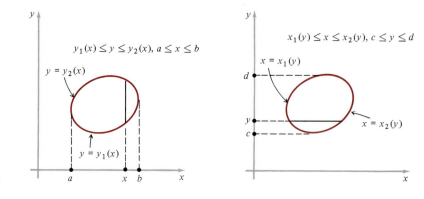

Figure 20.2

ing the base of the solid, with (3) shown on the left and (4) on the right. The iterated integrals (3) and (4) are usually written without brackets, as

$$\int_a^b \int_{y_1(x)}^{y_2(x)} f(x, y)\, dy\, dx \qquad \text{and} \qquad \int_c^d \int_{x_1(y)}^{x_2(y)} f(x, y)\, dx\, dy;$$

however, we can always retain the brackets for additional clarity if we wish to do so. The order in which the integrations are carried out (first with respect to y and then with respect to x, or the reverse) is determined by the order in which the differentials dx and dy are written in these iterated integrals: *We always work from the inside out.*

Example 1 Use an iterated integral to find the volume of the tetrahedron bounded by the coordinate planes and the plane $x + y + z = 1$.

Solution The section in the plane $x =$ a constant is the triangle shown in Fig. 20.3, with base extending from $y = 0$ to the line $y = 1 - x$. Its area is

$$A(x) = \int_0^{1-x} z\, dy = \int_0^{1-x} (1 - x - y)\, dy.$$

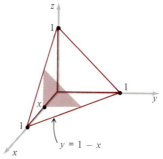

Figure 20.3

We now find the desired volume by integrating this from $x = 0$ to $x = 1$,

$$V = \int_0^1 \int_0^{1-x} (1 - x - y)\, dy\, dx = \int_0^1 \left[y - xy - \frac{1}{2} y^2 \right]_0^{1-x} dx$$

$$= \int_0^1 \left(\frac{1}{2} - x + \frac{1}{2} x^2 \right) dx = \frac{1}{6}.$$

The correctness of this result can be verified by elementary geometry, from the fact that the volume of any tetrahedron is one-third the area of the base times the height.

Example 2 Determine the region in the xy-plane over which the iterated integral

$$\int_{-1}^2 \int_{x^2}^4 f(x, y)\, dy\, dx$$

extends.

Solution In the inner integral, with x fixed between -1 and 2, y varies from the curve $y = x^2$ up to the line $y = 4$ (see Fig. 20.4). In the second integration x increases from -1 to 2. The region is that shown in the figure, and is bounded by the curve $y = x^2$ and the lines $y = 4$ and $x = -1$. Students should notice particularly how we determine what the region is by examining the limits of integration.

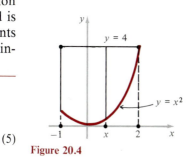

Figure 20.4

Example 3 The iterated integral

$$\int_0^1 \int_{x^2}^x 2y\, dy\, dx \qquad\qquad (5)$$

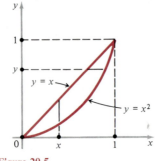

Figure 20.5

extends over a certain region in the xy-plane. Write an equivalent integral with the order of integration reversed, and evaluate both integrals.

Solution We see that the given integral extends over the region shown in Fig. 20.5, between the curves $y = x^2$ and $y = x$, where $0 \leq x \leq 1$. With the order of integration reversed, y is first held fixed between $y = 0$ and $y = 1$, and x increases from $x = y$ to $x = y^{1/2}$. The required integral is therefore

$$\int_0^1 \int_y^{y^{1/2}} 2y \, dx \, dy = \int_0^1 [2xy]_y^{y^{1/2}} \, dy = \int_0^1 (2y^{3/2} - 2y^2) \, dy = \frac{2}{15}.$$

The given integral (5) has the same value,

$$\int_0^1 \int_{x^2}^x 2y \, dy \, dx = \int_0^1 [y^2]_{x^2}^x \, dx = \int_0^1 (x^2 - x^4) \, dx = \frac{2}{15},$$

because both iterated integrals give the volume of a certain solid, and this volume must be the same regardless of how it is calculated. In computational problems of this kind, we are naturally free to use any methods of integration we wish from our past experience — trigonometric substitution, integration by parts, etc.

PROBLEMS

Determine the regions over which the iterated integrals in Problems 1 and 2 extend.

1 $\displaystyle\int_0^1 \int_0^y f(x, y) \, dx \, dy.$

2 $\displaystyle\int_0^4 \int_0^{\sqrt{x}} f(x, y) \, dy \, dx.$

Evaluate each of the iterated integrals in Problems 3 to 14. Also sketch the region R over which the integral extends.

3 $\displaystyle\int_0^1 \int_{x^2}^x (2x + 2y) \, dy \, dx.$

4 $\displaystyle\int_0^1 \int_0^1 xy^2 \, dy \, dx.$

5 $\displaystyle\int_0^4 \int_0^y 3\sqrt{y^2 + 9} \, dx \, dy.$

6 $\displaystyle\int_1^2 \int_{y^2}^{y^3} dx \, dy.$

7 $\displaystyle\int_0^{\pi/2} \int_0^{\cos x} y \, dy \, dx.$

8 $\displaystyle\int_1^{e^3} \int_0^{1/y} e^{xy} \, dx \, dy.$

9 $\displaystyle\int_1^3 \int_0^{\ln y} ye^x \, dx \, dy.$

10 $\displaystyle\int_0^1 \int_{-\sqrt{1-y^2}}^{\sqrt{1-y^2}} y \, dx \, dy.$

11 $\displaystyle\int_0^\pi \int_0^x x \cos y \, dy \, dx.$

12 $\displaystyle\int_0^\pi \int_0^{\sin x} y^2 \, dy \, dx.$

13 $\displaystyle\int_1^2 \int_x^{2x} \frac{dy \, dx}{(x + y)^2}.$

14 $\displaystyle\int_0^\pi \int_0^{\pi - y} \sin (x + y) \, dx \, dy.$

In Problems 15 to 18, write an equivalent iterated integral with the order of integration reversed.

15 $\displaystyle\int_0^1 \int_y^1 f(x, y) \, dx \, dy.$

16 $\displaystyle\int_0^1 \int_0^{\sqrt{2-2x^2}} f(x, y) \, dy \, dx.$

17 $\displaystyle\int_1^2 \int_{e^y}^{e^2} f(x, y) \, dx \, dy.$

18 $\displaystyle\int_{-2}^2 \int_{1-\sqrt{2-x}}^{\frac{1}{4}x} f(x, y) \, dy \, dx.$

In Problems 19 to 24, write an equivalent iterated integral with the order of integration reversed, and evaluate both integrals.

19 $\displaystyle\int_0^1 \int_{\sqrt{y}}^1 2x^3 \, dx \, dy.$

20 $\displaystyle\int_0^2 \int_0^{4-x^2} 2xy \, dy \, dx.$

21 $\displaystyle\int_0^2 \int_0^1 (5 - 2x - y) \, dy \, dx.$

22 $\displaystyle\int_1^{e^3} \int_{\ln y}^3 dx \, dy.$

23 $\displaystyle\int_{-5}^4 \int_{2-\sqrt{4-y}}^{\frac{1}{3}(y+2)} dx \, dy.$

24 $\displaystyle\int_0^{\sqrt{2}} \int_{-\sqrt{4-2x^2}}^{\sqrt{4-2x^2}} x \, dy \, dx.$

In Problems 25 to 28, use iterated integrals to find the volumes of the given regions of space. Sketch each region.

25 The region in the first octant bounded by the coordinate planes and the plane

$$\frac{x}{a} + \frac{y}{b} + \frac{z}{c} = 1,$$

where a, b, c are positive numbers.

26 The region in the first octant bounded by the plane $x + y = 1$ and the cylinder $z = 1 - x^2$.

27 The region in the first octant bounded by the plane $y = x$ and the cylinder $z = 4 - y^2$.

28 The region in the first octant bounded by the surface $z = 4 - x - y^2$.

*29 Use any method to find the volume of the region bounded by the surface $x^{2/3} + y^{2/3} + z^{2/3} = a^{2/3}$.

20.2
DOUBLE INTEGRALS AND ITERATED INTEGRALS

The double integral of a function of two variables is the two-dimensional analog of the definite integral of a function of one variable. It is convenient here to call this latter type of integral a *single integral,* in contrast to the term *double integral.*

As we know, the value of the single integral $\int_a^b f(x)\,dx$ is determined by the function $f(x)$ and the interval $[a, b]$. In the case of a double integral, the interval $[a, b]$ is replaced by a region R in the xy-plane, and the double integral of $f(x, y)$ over R is denoted by the symbol

$$\iint\limits_R f(x, y)\,dA. \tag{1}$$

The reason for the dA notation will be explained below.

We recall that in Section 6.4 a single integral was defined as the limit of certain sums. We now define the double integral (1) in much the same way.

Consider a continuous function $f(x, y)$ defined on a region R in the xy-plane. It is necessary to assume that R is *bounded,* in the sense that it can be enclosed in a sufficiently large rectangle and doesn't go off to infinity in any direction; otherwise, just as in the case of a single integral where a or b is infinite, the double integral will be *improper.*

We begin by covering R with a network of lines parallel to the axes, as shown in Fig. 20.6, where the distances between consecutive parallel lines are permitted to be equal or unequal. These lines divide the plane into many small rectangles. Some rectangles will lie entirely or partly outside of R, and these we ignore. Other rectangles will lie entirely inside R, and if there are n of these altogether — we assume there is at least one — then we number them in

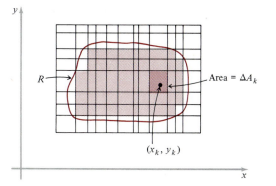

Figure 20.6

any order from 1 to n, denoting by ΔA_k the area of the kth rectangle. We now choose an arbitrary point (x_k, y_k) in the kth rectangle and form the sum

$$\sum_{k=1}^{n} f(x_k, y_k)\, \Delta A_k. \tag{2}$$

Finally, suppose that many more parallel lines are added to produce a network that divides the given rectangles into even smaller rectangles, and consider the sum (2) corresponding to this finer partition of the plane. If these sums approach a unique limit as n becomes infinite and the maximum diagonal of the rectangles (that is, the longest diagonal of any of the rectangles) approaches zero — independent of the choice of dividing lines and the points (x_k, y_k) in the rectangles — then the double integral (1) is defined to be this limit:

$$\iint\limits_{R} f(x, y)\, dA = \lim \sum_{k=1}^{n} f(x_k, y_k)\, \Delta A_k. \tag{3}$$

So far, it appears that the definition (3) differs very little from the corresponding definition of a single integral. However, there are certain technical difficulties in two dimensions that do not arise in one dimension. For one thing, plane regions can be much more complicated than intervals $[a, b]$. Nevertheless, the existence of double integrals can be rigorously proved under assumptions that are general enough for all practical purposes. In particular, it is enough to assume that the regions we consider contain their boundaries and that these boundaries consist of a finite number of smooth curves.

We shall not attempt a careful theoretical treatment of double integrals. This is a difficult subject, and is best left to courses in advanced calculus.* Instead, we prefer to emphasize the intuitive meaning of double integrals, and to concentrate our attention on their geometric and physical applications.

As an illustration of this point of view, suppose that $z = f(x, y)$ is the equation of a surface in xyz-space that lies above the region R, so that $f(x, y) > 0$ in R, as shown in Fig. 20.7. Then $f(x_k, y_k)\, \Delta A_k$ is approximately the volume (height times area of base) of the thin column in the figure; the sum (2) is the sum of many such volumes and therefore approximates the total volume of the solid under the surface; and the limit (3), which is the double integral

$$\iint\limits_{R} f(x, y)\, dA, \tag{1}$$

gives the exact volume of this solid.†

---

* Even at this level, one needs an advanced calculus course of the traditional kind. For example, see Philip Franklin, *A Treatise on Advanced Calculus* (Wiley, 1940); or Angus E. Taylor, *Advanced Calculus* (Ginn, 1955).

† The double integral (1) is actually the volume of a *region* in three-dimensional space, but it seems to be more natural to speak of the volume of a *solid*.

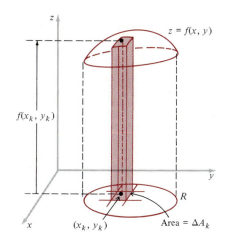

Figure 20.7

It is clear that if $f(x, y)$ has a constant value, say $f(x, y) = c$, then

$$\iint\limits_{R} f(x, y)\, dA = cA,$$

where A is the area of the region R. In particular, if $f(x, y) = 1$ we have

$$\iint\limits_{R} dA = A.$$

We also point out that in the definition (3) there is no requirement that $f(x, y)$ must be positive. If $f(x, y)$ takes both positive and negative values, then the double integral represents an *algebraic volume* instead of a geometric volume; that is, the volume between the surface $z = f(x, y)$ and the xy-plane counts positively when $f(x, y) > 0$ and negatively when $f(x, y) < 0$.

Since the area of a rectangle with sides parallel to the axes can be written as $\Delta A = \Delta x\, \Delta y$, it is reasonable to use

$$\iint\limits_{R} f(x, y)\, dx\, dy \qquad (4)$$

as an alternative notation for the double integral (1). In this form the double integral resembles an iterated integral, and in fact, as we next explain, when the region R has a certain simple shape the double integral (1) is always equal to a suitably chosen iterated integral. This equality often misleads students into thinking that double integrals are essentially the same as iterated integrals, but they are not. We shall say more below about the distinction between these two types of integrals.

A region R is called *vertically simple* if it can be described by inequalities of the form

$$a \le x \le b, \qquad y_1(x) \le y \le y_2(x), \qquad (5)$$

where $y = y_1(x)$ and $y = y_2(x)$ are continuous functions on $[a, b]$. A region of this kind is shown in Fig. 20.8. Similarly, a region R is called *horizontally simple* if it can be described by inequalities of the form

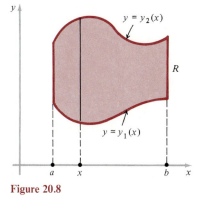

Figure 20.8

$$c \leq y \leq d, \qquad x_1(y) \leq x \leq x_2(y), \tag{6}$$

where $x = x_1(y)$ and $x = x_2(y)$ are continuous functions on $[c, d]$. The region in Fig. 20.9 has this property.

The following are the basic facts about the use of iterated integrals to compute double integrals: *if R is the vertically simple region given by* (5), *then*

$$\iint_R f(x, y) \, dA = \int_a^b \int_{y_1(x)}^{y_2(x)} f(x, y) \, dy \, dx; \tag{7}$$

and if R is the horizontally simple region given by (6), *then*

$$\iint_R f(x, y) \, dA = \int_c^d \int_{x_1(y)}^{x_2(y)} f(x, y) \, dx \, dy. \tag{8}$$

In addition to their obvious practical value for the computation of double integrals, these equations also serve to clarify the conceptual distinction between double integrals and iterated integrals. A double integral is a number associated with a function $f(x, y)$ and a region R, and this number exists and has a meaning independently of any particular method of computing it. On the other hand, an iterated integral is a double integral *plus a built-in computational procedure.* Thus, every iterated integral is a double integral, but not vice versa.

Example 1   Compute the double integral $\iint_R 2xy \, dA$ in two different ways, where R is the region bounded by the parabola $x = y^2$ and the straight line $y = x$.

Solution   *It is essential to always sketch the region R of integration before trying to evaluate a double integral.* In this case the region is shown in Fig. 20.10. It is clear that R is vertically simple with $a = 0$, $b = 1$, $y_1(x) = x$, $y_2(x) = x^{1/2}$, so by (7)

$$\iint_R 2xy \, dA = \int_0^1 \int_x^{x^{1/2}} 2xy \, dy \, dx = \int_0^1 [xy^2]_x^{x^{1/2}} \, dx$$

$$= \int_0^1 (x^2 - x^3) \, dx = \frac{1}{3} - \frac{1}{4} = \frac{1}{12}.$$

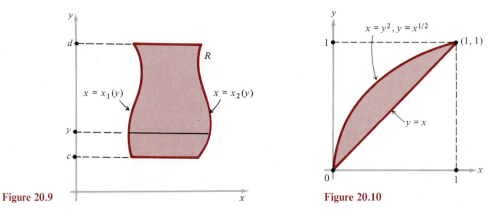

Figure 20.9   **Figure 20.10**

The region R is also horizontally simple with $c = 0$, $d = 1$, $x_1(y) = y^2$, $x_2(y) = y$, so by (8)

$$\iint_R 2xy \, dA = \int_0^1 \int_{y^2}^y 2xy \, dx \, dy = \int_0^1 [x^2 y]_{y^2}^y \, dy$$

$$= \int_0^1 (y^3 - y^5) \, dy = \frac{1}{4} - \frac{1}{6} = \frac{1}{12}.$$

Example 2   Compute $\iint_R (1 + 2x) \, dA$, where R is the region bounded by $x = y^2$ and $x - y = 2$.

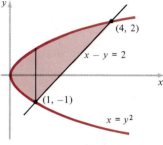

Solution   This region is shown in Fig. 20.11. In order to integrate first with respect to y and then with respect to x, we would need to compute two separate integrals, one to the left of the line $x = 1$ and the other to the right, because the limits of the y-integration are different in these two parts of the region:

$$\iint_R (1 + 2x) \, dA = \int_0^1 \int_{-\sqrt{x}}^{\sqrt{x}} (1 + 2x) \, dy \, dx + \int_1^4 \int_{x-2}^{\sqrt{x}} (1 + 2x) \, dy \, dx.$$

Figure 20.11

The other order is easier, and yields

$$\iint_R (1 + 2x) \, dA = \int_{-1}^2 \int_{y^2}^{y+2} (1 + 2x) \, dx \, dy = \int_{-1}^2 [x + x^2]_{y^2}^{y+2} \, dy$$

$$= \int_{-1}^2 (6 + 5y - y^4) \, dy = \frac{189}{10}.$$

Example 2 shows that even when the region R is both vertically and horizontally simple, it may be easier to integrate in one order than in the other, and we naturally prefer to do things in the easiest way. Sometimes the choice of the order of integration is determined by the nature of the integrand $f(x, y)$, for it may be difficult — or even impossible — to compute an integral in one order, but easy to do so *if the order of integration is reversed*.

Example 3   Compute

$$\int_0^1 \int_{2y}^2 4e^{x^2} \, dx \, dy.$$

Solution   We cannot integrate in this order because $\int e^{x^2} \, dx$ is not an elementary function. We therefore try the other order. This requires us to sketch the region R by examining the limits on the given iterated integral. R is shown in Fig. 20.12, and in the other order the underlying double integral has the value

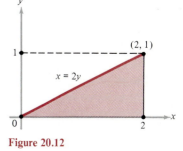

$$\iint_R 4e^{x^2} \, dA = \int_0^2 \int_0^{\frac{1}{2}x} 4e^{x^2} \, dy \, dx = \int_0^2 \left[4ye^{x^2} \right]_0^{\frac{1}{2}x} \, dx$$

$$= \int_0^2 2xe^{x^2} \, dx = e^{x^2} \Big]_0^2 = e^4 - 1.$$

Figure 20.12

PROBLEMS

In Problems 1 to 6, use double integrals to find the areas of the regions bounded by the given curves and lines.

1 The parabola $x = y^2$ and the line $y = x - 2$.

2 The parabola $y = x - x^2$ and the line $x + y = 0$.

3 The axes and the line $2x + y = 2a$ ($a > 0$).

4 The y-axis, the line $y = 3x$, and the line $y = 6$.

5 The x-axis, the curve $y = e^{-x}$, and the lines $x = 0$, $x = a$ ($a > 0$).

6 The parabolas $y = x^2$ and $y = 2x - x^2$.

In Problems 7 to 10, find the volumes above the xy-plane bounded by the given surfaces.

7 The paraboloid $z = x^2 + y^2$ and the planes $x = \pm 1$, $y = \pm 1$.

8 The cylinder $x^2 + y^2 = 1$ and the plane $x + y + z = 2$.

9 The cylinder $y = 4 - x^2$ and the planes $y = 3x$, $z = x + 4$.

10 The cylinder $x^2 + y^2 = a^2$ and the paraboloid $az = x^2 + y^2$.

11 Find the volume of the solid bounded by the coordinate planes, the planes $x = 2$ and $y = 5$, and the surface $2z = xy$.

12 Find the volume of the solid in the first octant bounded by the cylinder $4y = x^2$ and the planes $x = 0$, $z = 0$, $y = 4$, and $x - y + 2z = 2$.

In Problems 13 to 16, set up a double integral whose value is the stated volume, express this double integral in two ways as an iterated integral, and evaluate one of these.

13 The volume under the plane $z = 2y$ and above the first-quadrant region bounded by $y = 0$, $x = 2$, $x^2 + y^2 = 16$.

14 The volume under the plane $z = x + y$ and above the first-quadrant region inside the ellipse $9x^2 + 4y^2 = 36$.

15 The volume under the cylinder $x = z^2$ and above the region in the xy-plane bounded by $x = 0$ and $y^2 + 9x = 9$.

16 The volume in the first octant bounded by the cylinder $z = 4 - y^2$ and the planes $x = 0$, $y = 0$, $z = 0$, $3x + 4y = 12$.

17 Calculate the value of $\iint_R x \, dA$ if R is the first-quadrant part of the ring between the circles $x^2 + y^2 = a^2$ and $x^2 + y^2 = b^2$, where $a < b$. Do this two ways, corresponding to the two possible orders of integration.

18 Compute the double integral in Example 2 in the other order, requiring two separate iterated integrals.

20.3
PHYSICAL APPLICATIONS OF DOUBLE INTEGRALS

We have seen that the double integral

$$\iint_R f(x, y) \, dA \tag{1}$$

gives the volume of a certain solid if $f(x, y) \geq 0$. This integral has many other useful interpretations that arise by making special choices of the function $f(x, y)$. Before we discuss these, it will be convenient to return to the way of thinking about integration that was described and extensively illustrated in Chapter 7.

The limit-of-sums definition of (1) that was given in Section 20.2 is necessary from the point of view of logic and mathematical legitimacy. However, for working with applications it is better to think of the volume given by (1) as composed of infinitely many infinitely thin columns, as suggested in Fig. 20.13a. A typical column stands on an infinitely small rectangular *element of area dA* whose sides are dx and dy, so that

$$dA = dx \, dy = dy \, dx. \tag{2}$$

The height of this column is $f(x, y)$, so its volume is

$$dV = f(x, y) \, dA.$$

The total volume V is now obtained by adding together — or integrating — all of these infinitely small elements of volume,

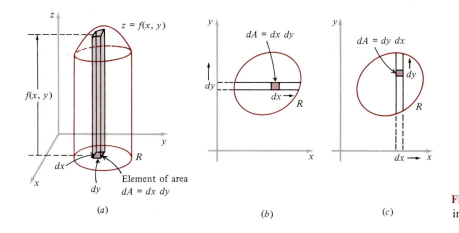

Figure 20.13 Two orders of integration.

$$V = \iint dV = \iint\limits_{R} f(x, y)\, dA. \tag{3}$$

We understand here that the complete double integral (1) is produced by allowing dA to sweep in any manner over the whole of the region R. In parts (b) and (c) of Fig. 20.13 we indicate the two ways of calculating (3) as an iterated integral: in (b), we first allow dA to move across R along a thin horizontal strip, corresponding to integrating first x and then y; and in (c), we first allow dA to move across R along a thin vertical strip, integrating first y and then x. As suggested by formula (2), the double integral (3) can be written in either of the forms

$$\iint\limits_{R} f(x, y)\, dx\, dy \qquad \text{or} \qquad \iint\limits_{R} f(x, y)\, dy\, dx,$$

depending on which iterated integral we wish to consider; and to apply these ideas to a particular problem, all that remains is to insert suitable limits of integration and carry out the calculations.

This description of the intuitive meaning of the double integral (1) expresses the essence of the Leibniz approach to integration: to find the whole of a quantity, imagine it to be judiciously divided into a great many small pieces, and then add these pieces together. This is the unifying theme of the following applications, and also of many further developments in the rest of this chapter. And here again, as so often before, the superb Leibniz notation almost does our thinking for us.

In Chapter 11 we discussed the concepts of moment, center of mass, and moment of inertia for a thin plate of homogeneous material that occupies a given region R in the xy-plane. The word "homogeneous" meant that the density δ of the material ($=$ mass per unit area) was assumed to be constant, that is, to have the same value at every point $P = (x, y)$ in R. We are now in a position to allow δ to be a function of x and y, $\delta = \delta(x, y)$, so that thin plates of varying density can be brought within the scope of our methods.

I.   MASS

If $\delta = \delta(x, y)$ is the density of our thin plate, then $\delta(x, y)\, dA$ is the mass contained in the element of area dA, and the total mass of the plate is

$$M = \iint_R \delta(x, y)\, dA. \tag{4}$$

II.   MOMENT

The moment of the element of mass $\delta(x, y)\, dA$ with respect to the x-axis is the mass multiplied by the "lever arm" y, namely, $y\delta(x, y)\, dA$, and the total moment of the plate with respect to the x-axis is

$$M_x = \iint_R y\delta(x, y)\, dA. \tag{5}$$

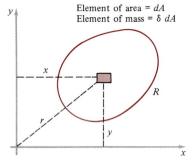

Element of area = dA
Element of mass = $\delta\, dA$

Figure 20.14

See Fig. 20.14. Similarly, the total moment with respect to the y-axis is

$$M_y = \iint_R x\delta(x, y)\, dA. \tag{6}$$

III.   CENTER OF MASS

This is the point $(\bar{x}, \bar{y})$ whose coordinates are defined by

$$\bar{x} = \frac{M_y}{M} = \frac{\iint_R x\delta(x, y)\, dA}{\iint_R \delta(x, y)\, dA} \tag{7}$$

and

$$\bar{y} = \frac{M_x}{M} = \frac{\iint_R y\delta(x, y)\, dA}{\iint_R \delta(x, y)\, dA}. \tag{8}$$

Physically, this is the point at which the total mass of the plate could be concentrated without changing its moment with respect to either axis. When the density δ is constant, so that the mass of the plate is uniformly distributed, then the δ's can be removed from the integrals in (7) and (8) and canceled away. In this case the center of mass becomes the geometric center of the region R, and for this reason is usually called the *centroid*.

IV.   MOMENT OF INERTIA

When the square of the lever arm distance is used instead of its first power [as in (5) and (6)], we get the moment of inertia of the plate about the corresponding axis. Thus, the moment of inertia I_x about the x-axis is defined by

$$I_x = \iint_R y^2\delta(x, y)\, dA. \tag{9}$$

Similarly, the moment of inertia I_y about the y-axis is

$$I_y = \iint\limits_R x^2\delta(x, y)\, dA. \tag{10}$$

Also of interest is the moment of inertia of the plate about the z-axis. This is often called the *polar moment of inertia,* and is defined by

$$I_z = \iint\limits_R r^2\delta(x, y)\, dA, \tag{11}$$

where $r^2 = x^2 + y^2$. As we explained in Section 11.4, the moment of inertia of a body about an axis is its capacity to resist angular acceleration about that axis; this quantity plays the same role in rotational motion as mass does in linear motion.

Students should explicitly notice that in each of the formulas (4), (5), (6), (9), (10), (11) we obtain the total quantity under discussion by adding together — or integrating — the "infinitesimal" parts of it associated with the element of area dA, as dA sweeps over the region R.

Example   A thin plate of material of variable density occupies the square R whose vertices are $(0, 0), (a, 0), (a, a), (0, a)$. The density at a point $P = (x, y)$ is the product of the distances from P to the axes, $\delta = xy$. Find the mass of the plate, its center of mass, and its moment of inertia about the x-axis.

Solution   A sketch of the situation is shown in Fig. 20.15. We have

$$M = \iint\limits_R \delta\, dA = \int_0^a \int_0^a xy\, dy\, dx = \int_0^a \left[\frac{1}{2}xy^2\right]_0^a dx$$

$$= \frac{1}{2} a^2 \int_0^a x\, dx = \frac{1}{4} a^4.$$

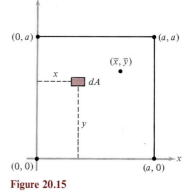

Figure 20.15

The x-coordinate of the center of mass is

$$\bar{x} = \frac{M_y}{M} = \frac{4}{a^4} \iint\limits_R x\delta\, dA = \frac{4}{a^4} \int_0^a \int_0^a x^2 y\, dy\, dx$$

$$= \frac{4}{a^4} \int_0^a \left[\frac{1}{2}x^2y^2\right]_0^a dx = \frac{2}{a^2}\int_0^a x^2\, dx = \frac{2}{3}a,$$

and by symmetry we have $\bar{x} = \bar{y} = \frac{2}{3}a$. The desired moment of inertia is

$$I_x = \iint\limits_R y^2\delta\, dA = \int_0^a \int_0^a xy^3\, dy\, dx = \int_0^a \left[\frac{1}{4}xy^4\right]_0^a dx$$

$$= \frac{1}{4} a^4 \int_0^a x\, dx = \frac{1}{8} a^6.$$

It is customary to express the moment of inertia of a body about an axis in terms of its total mass M, which in this case gives

$$I_x = \tfrac{1}{2}Ma^2.$$

Remark 1 We emphasize that the symbols dA, dx, and dy in formula (2) do *not* designate differentials in the sense discussed in Section 19.4. Instead, they are merely notational aids that enable us to write down appropriate double integrals directly, without repeatedly going back to the complicated limit-of-sums definitions of these integrals.

Remark 2 A surprising application of our present ideas is given in Appendix A.21, where Euler's formula

$$\sum_{n=1}^{\infty} \frac{1}{n^2} = \frac{\pi^2}{6}$$

is obtained by evaluating a certain double integral.

PROBLEMS

In Problems 1 to 8, find the total mass M and the center of mass $(\bar{x}, \bar{y})$ of the thin plate of material that lies in the given region R and has the given density δ. Use symmetry wherever possible to simplify calculations.

1   R is the square with vertices $(0, 0)$, $(a, 0)$, (a, a), $(0, a)$; $\delta = x + y$.
2   R is the first-quadrant region bounded by the axes and the circle $x^2 + y^2 = 1$; $\delta = xy$.
3   R is the region bounded by the parabola $x = y^2$ and the line $x = 4$; $\delta = x$.
4   R is the region bounded by the axes and the line $x + y = a$; $\delta = x^2 + y^2$.
5   R is the region bounded by $x = 0$ and the right half of the circle $x^2 + y^2 = a^2$; $\delta = x$.
6   R is the region bounded by the parabola $y = x^2$ and the line $y = x$; $\delta = \sqrt{x}$.
7   R is the region between $y = \sin x$ and the x-axis from $x = 0$ to $x = \pi$; $\delta = x$.

8   R is the region bounded by the parabola $y = x^2$ and the line $y = x + 2$; $\delta = x^2$.
9   Find the moment of inertia I_x for the square plate considered in the text if the density δ is constant.
10 Show that $I_z = I_x + I_y$. Use this and the result of Problem 9 to find the moment of inertia of a uniform (constant density) cube of edge a and mass M about one of its edges.
11 If the density δ is constant, find the moment of inertia I_x of the thin triangular plate bounded by the line $x + y = a$ and the axes $x = 0$, $y = 0$.
12 Solve Problem 11 for the triangular plate bounded by the lines $x + y = a$, $x = a$, $y = a$.
13 Solve Problem 11 if the density is $\delta = xy$.
14 Find the polar moment of inertia I_z of the circular plate bounded by $x^2 + y^2 = a^2$ if the density δ is constant.

20.4
DOUBLE INTEGRALS IN POLAR COORDINATES

It is often more convenient to describe the boundaries of a region by using polar coordinates r, θ than by using rectangular coordinates x, y. In these circumstances we can usually save ourselves a lot of work by expressing a double integral

$$\iint_R f(x, y) \, dA \tag{1}$$

in terms of polar coordinates. The integrand is easy to transform by using the equations $x = r \cos \theta$, $y = r \sin \theta$ to write $f(x, y)$ as a function of r and θ,

$$f(x, y) = f(r \cos \theta, r \sin \theta).$$

For example, if $f(x, y) = x^2 + y^2$, this becomes $(r \cos \theta)^2 + (r \sin \theta)^2 = r^2 (\cos^2 \theta + \sin^2 \theta) = r^2$. But what do we do about the element of area dA?

The answer to this question is suggested by Fig. 20.16. We recall that the element of area in rectangular coordinates,

$$dA = dx\,dy,$$

is intended to remind us of the small rectangles with sides parallel to the axes that were used to define the double integral (1) in Section 20.2. In working with polar coordinates it is natural to subdivide the plane in another way, by a series of circles with centers at the origin and a series of rays emanating from the origin. These circles and rays form many small cells that resemble rectangles, as shown by the shaded part of the figure. The double integral (1) can now be given an equivalent definition by means of a limit-of-sums process that uses these small "polar rectangles." However, we omit the details and use Fig. 20.16 only to suggest the line of thought we should follow, as we now explain.

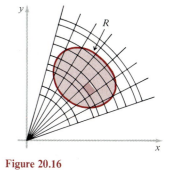

Figure 20.16

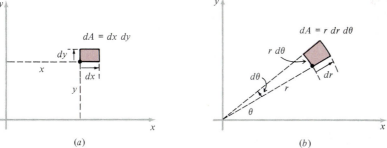

(a)   (b)

Figure 20.17   The polar element of area.

The element of area $dA = dx\,dy$ in rectangular coordinates is the area of the small rectangle swept out by an increase dx in x and an increase dy in y (see Fig. 20.17a). Figure 20.16 suggests the approach to be used with polar coordinates: If r increases to $r + dr$ and θ increases to $\theta + d\theta$ (Fig. 20.17b), then a small polar rectangle is swept out whose sides are dr, the change in r, and $r\,d\theta$.* The area of the small polar rectangle is therefore approximately

$$dA = (dr)(r\,d\theta) = r\,dr\,d\theta. \tag{2}$$

This is the basic formula of this section. It gives the element of area in polar coordinates, and it enables us to write the double integral (1) in polar form, as

$$\iint\limits_{R} f(x, y)\,dA = \iint\limits_{R} f(r\cos\theta, r\sin\theta)\,r\,dr\,d\theta. \tag{3}$$

Many of the regions R we deal with are *radially simple,* in the sense that they can be described by inequalities of the form

$$\alpha \le \theta \le \beta, \qquad r_1(\theta) \le r \le r_2(\theta).$$

Figure 20.18 shows a region of this kind, and also suggests how the figure can be used to write the double integral (3) as an iterated integral,

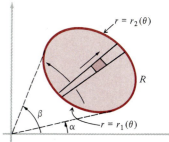

Figure 20.18

---

* The second side of this polar rectangle is a short arc of a circle of radius r that is cut off by a central angle $d\theta$, and its length s is given by the formula $s = r \cdot d\theta$, *because the angle is measured in radians.*

$$\iint\limits_{R} f(x, y)\, dA = \int_{\alpha}^{\beta} \int_{r_1(\theta)}^{r_2(\theta)} f(r\cos\theta, r\sin\theta)\, r\, dr\, d\theta.$$

Here we integrate first r and then θ, working from the inside out as always. We visualize the element of area dA as first moving out across R along the indicated radial strip, from the inner curve $r = r_1(\theta)$ to the outer curve $r = r_2(\theta)$. The resulting strip is then rotated from $\theta = \alpha$ to $\theta = \beta$ in order to sweep over all of R. Iterated integrals can also be set up in the other order, but these are seldom used.

Example 1 Find the area of the region R enclosed by the cardioid $r = a(1 + \cos\theta)$.

Solution This cardioid is shown in Fig. 20.19, and we find the area by integrating the element of area $dA = r\, dr\, d\theta$ over the region,

$$A = \iint\limits_{R} dA = \iint\limits_{R} r\, dr\, d\theta.$$

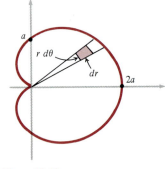

Figure 20.19

For fixed θ, we allow r to increase from $r = 0$ to $r = a(1 + \cos\theta)$. As usual, we exploit all available symmetry, so we next allow θ to increase from 0 to π and obtain the total area by multiplying by 2,

$$A = 2 \int_0^\pi \int_0^{a(1+\cos\theta)} r\, dr\, d\theta = 2 \int_0^\pi \left[\frac{1}{2} r^2\right]_0^{a(1+\cos\theta)} d\theta$$

$$= 2 \int_0^\pi \frac{1}{2} a^2 (1 + \cos\theta)^2\, d\theta = a^2 \int_0^\pi (1 + 2\cos\theta + \cos^2\theta)\, d\theta$$

$$= a^2 \int_0^\pi \left(1 + 2\cos\theta + \frac{1}{2}[1 + \cos 2\theta]\right) d\theta$$

$$= a^2 \left[\theta + 2\sin\theta + \frac{1}{2}\theta + \frac{1}{4}\sin 2\theta\right]_0^\pi = \frac{3}{2}\pi a^2.$$

This problem can also be solved by the method of Section 16.5, which would have started with the third integral in our calculation. However, our present method has much greater flexibility. It allows us, for example, to find the centroid of the region R by thinking of it as a thin plate of material of constant density $\delta = 1$. It is clear by symmetry that $\bar{y} = 0$, and we find $\bar{x}$ by writing

$$\bar{x} = \frac{M_y}{M} = \frac{2}{3\pi a^2} \iint\limits_{R} x\, dA.$$

We ask students to complete the details of this calculation in Problem 25.

Example 2 Derive the formula for the volume of a sphere of radius a by our present methods.

Solution If the sphere has center at the origin, its equation is $x^2 + y^2 + z^2 = a^2$ or $r^2 + z^2 = a^2$, and the equation of the upper hemisphere is $z =$

$\sqrt{a^2 - r^2}$. By symmetry, we calculate the volume in the first octant (Fig. 20.20) and multiply by 8. The region R over which we integrate is defined by $0 \le \theta \le \pi/2$ and $0 \le r \le a$, so

$$V = 8 \iint\limits_R z \, dA = 8 \int_0^{\pi/2} \int_0^a \sqrt{a^2 - r^2} \, r \, dr \, d\theta$$

$$= 8 \int_0^{\pi/2} \int_0^a -\frac{1}{2}(a^2 - r^2)^{1/2}(-2r \, dr) \, d\theta$$

$$= -4 \int_0^{\pi/2} \left[\frac{2}{3}(a^2 - r^2)^{3/2}\right]_0^a d\theta$$

$$= -4 \int_0^{\pi/2} \left(-\frac{2}{3}a^3\right) d\theta = \frac{4}{3}\pi a^3.$$

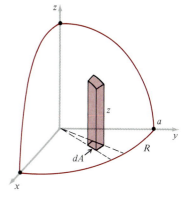

Figure 20.20

Students should notice particularly how the presence of the r in the inner integral makes this calculation work out smoothly.

Example 3   The improper integral

$$\int_0^\infty e^{-x^2} \, dx$$

is important in the theory of probability and elsewhere. We shall find its value by a clever device that depends on an improper double integral in polar coordinates. Write

$$I = \int_0^\infty e^{-y^2} \, dy.$$

Since it doesn't matter what letter we use for the variable of integration, we have

$$I^2 = \left(\int_0^\infty e^{-x^2} \, dx\right)\left(\int_0^\infty e^{-y^2} \, dy\right).$$

By moving the first factor past the second integral sign, this can be written in the form

$$I^2 = \int_0^\infty \left(\int_0^\infty e^{-x^2} \, dx\right) e^{-y^2} \, dy = \int_0^\infty \left(\int_0^\infty e^{-x^2} e^{-y^2} \, dx\right) dy$$

$$= \int_0^\infty \int_0^\infty e^{-(x^2+y^2)} \, dx \, dy.$$

This double integral is extended over the entire first quadrant of the xy-plane. In polar coordinates it becomes

$$I^2 = \int_0^{\pi/2} \int_0^\infty e^{-r^2} r \, dr \, d\theta = \int_0^{\pi/2} \left[-\frac{1}{2} e^{-r^2}\right]_0^\infty d\theta = \int_0^{\pi/2} \frac{1}{2} \, d\theta = \frac{\pi}{4},$$

so $I = \frac{1}{2}\sqrt{\pi}$ or

$$\int_0^\infty e^{-x^2} \, dx = \frac{1}{2}\sqrt{\pi}. \tag{4}$$

This formula is especially remarkable because it is known that the indefinite integral

$$\int e^{-x^2}\, dx$$

is impossible to express as an elementary function.*

---

* There is a famous story about the nineteenth century Scottish physicist Lord Kelvin. "Do you know what a mathematician is?" Kelvin once asked a class. He stepped to the blackboard and wrote

$$\int_{-\infty}^{\infty} e^{-x^2}\, dx = \sqrt{\pi},$$

which is clearly equivalent to (4). "A mathematician," he continued, "is one to whom *that* is as obvious as twice two makes four is to you." As a matter of fact, this formula is *not* obvious, either to the present writer or to any of the many mathematicians he has known. The conclusion seems to be that Kelvin was both showing off and trying to put down his class in a rather mean-spirited way.

PROBLEMS

In Problems 1 to 13, use double integrals in polar coordinates to find the areas of the indicated regions.

1 The circle $r = a$.

2 The circle $r = 2a \cos \theta$.

3 The region common to the circles $r = a$ and $r = 2a \cos \theta$.

4 One loop of $r = a \cos 2\theta$.

5 The right loop of the lemniscate $r^2 = 2a^2 \cos 2\theta$.

6 The region inside the curve $r = 2 + \sin 3\theta$.

7 The region inside the lemniscate $r^2 = 2a^2 \cos 2\theta$ and outside the circle $r = a$.

8 The region inside $r = \tan \theta$ and between $\theta = 0$ and $\theta = \pi/4$.

9 The region inside the cardioid $r = a(1 + \cos \theta)$ and outside the circle $r = a$.

10 The region inside the circle $r = a$ and outside the cardioid $r = a(1 + \cos \theta)$.

11 The region inside the cardioid $r = 2a(1 + \cos \theta)$ and outside the circle $r = 3a$.

*12 The region between $r = \pi/4$ and $r = \pi/2$, between $r = \theta$ and $r = \frac{1}{2}\theta$ ($\theta \geq 0$).

13 The region inside the cardioid $r = 1 + \cos \theta$ and to the right of the line $x = \frac{3}{4}$.

*14 If R is the region bounded by the lines $y = x$, $y = 0$, $x = 1$, evaluate the double integral

$$\iint_R \frac{dx\, dy}{(1 + x^2 + y^2)^{3/2}}$$

by changing to polar coordinates.

15 Evaluate the integral

$$\int_0^{2a} \int_0^{\sqrt{2ax - x^2}} (x^2 + y^2)\, dy\, dx$$

by changing to polar coordinates

In Problems 16 to 22, write the given integral in the form

$$\int_\alpha^\beta \int_{r_1(\theta)}^{r_2(\theta)} z\, r\, dr\, d\theta.$$

16 $\displaystyle\int_0^2 \int_0^{\sqrt{4 - x^2}} z\, dy\, dx.$

17 $\displaystyle\int_0^3 \int_{-\sqrt{9 - x^2}}^{\sqrt{9 - x^2}} z\, dy\, dx.$

18 $\displaystyle\int_{-1}^0 \int_{-\sqrt{1 - y^2}}^{\sqrt{1 - y^2}} z\, dx\, dy.$

19 $\displaystyle\int_0^1 \int_{x^2}^x z\, dy\, dx.$

20 $\displaystyle\int_0^4 \int_0^{\sqrt{4 - (x - 2)^2}} z\, dy\, dx.$

21 $\displaystyle\int_0^{\sqrt{2}/2} \int_y^{\sqrt{1 - y^2}} z\, dx\, dy.$

22 $\displaystyle\int_0^2 \int_0^{\sqrt{2y - y^2}} z\, dx\, dy.$

23 A cylindrical hole of radius b is drilled through the center of a sphere of radius a.

    (a) Find the volume of the hole. Notice that this formula gives the volume of the sphere when $b = a$.

    (b) Find the volume of the ring-shaped solid that remains. Express this volume in terms of the height h of the ring. Notice the remarkable fact that this volume depends only on h, and not on either the radius a of the sphere or the radius b of the hole.

*24 Find the centroid of the region enclosed by the loop of $r = a \cos 2\theta$ that lies in the first and fourth quadrants.

25 Find the centroid of the region enclosed by the cardioid $r = a(1 + \cos \theta)$.

*26 Find the centroid of the region enclosed by the right loop of the lemniscate $r^2 = 2a^2 \cos 2\theta$.

27 Find the centroid of the semicircular disk $x^2 + y^2 \leq a^2$, $y \geq 0$.

28 Find the volume of the solid cone $0 \leq z \leq h(a - r)/a$.

29 Find the volume of the solid under the cone $z = 2a - r$ whose base is bounded by the cardioid $r = a(1 + \cos\theta)$.

30 Find the volume cut out of the sphere $x^2 + y^2 + z^2 = 4a^2$ by the cylinder $x^2 + y^2 = 2ax$.

31 For the solid bounded by the xy-plane, the cylinder $x^2 + y^2 = a^2$, and the paraboloid $z = b(x^2 + y^2)$ with $b > 0$, find (a) the volume, (b) the centroid.

32 Find the polar moment of inertia I_z of the circular plate bounded by $r = a$ if the density δ is constant. (Compare this very easy calculation with the work needed to solve the same problem using rectangular coordinates, in Problem 14 in Section 20.3.)

33 Find the polar moment of inertia I_z of a thin plate of constant density δ that has the shape of the circle $r = 2a \cos\theta$.

34 Solve Problem 33 for a plate that has the shape of the cardioid $r = a(1 + \cos\cdot\theta)$.

35 A thin plate is bounded by the circle $r = a$ and has density $\delta = a^2/(a^2 + r^2)$. Find its mass M and polar moment of inertia I_z.

36 The center of a circle of radius $2a$ lies on a circle of radius a. Find the centroid of the region between the two circles.

37 A thin plate of constant density δ has the shape of a circular sector of radius a and central angle 2α. Find the moment of inertia about the bisector of the angle.

38 Find the centroid of the circular sector described in Problem 37. Obtain the result of Problem 27 as a special case of this.

*39 Use the fact that $\int_0^\infty e^{-x^2}\,dx = \frac{1}{2}\sqrt{\pi}$ to show that

(a) $\displaystyle\int_0^\infty e^{-2x^2}\,dx = \frac{1}{4}\sqrt{2\pi}$;

(b) $\displaystyle\int_0^\infty e^{-3x^2}\,dx = \frac{1}{6}\sqrt{3\pi}$;

(c) $\displaystyle\int_0^\infty e^{-4x^2}\,dx = \frac{1}{4}\sqrt{\pi}$;

(d) $\displaystyle\int_0^{\pi/2} \frac{e^{-\tan^2 x}}{\cos^2 x}\,dx = \frac{1}{2}\sqrt{\pi}$;

(e) $\displaystyle\int_0^\infty \frac{e^{-x}}{\sqrt{x}}\,dx = \sqrt{\pi}$;

(f) $\displaystyle\int_0^\infty x^2 e^{-x^2}\,dx = \frac{1}{4}\sqrt{\pi}$;

(g) $\displaystyle\int_0^\infty \sqrt{x}\,e^{-x}\,dx = \frac{1}{2}\sqrt{\pi}$;

(h) $\displaystyle\int_0^1 \frac{dx}{\sqrt{-\ln x}} = \sqrt{\pi}$;

(i) $\displaystyle\int_0^1 \sqrt{-\ln x}\,dx = \frac{1}{2}\sqrt{\pi}$.

40 Use the method of Example 3 to evaluate
$$\int_0^\infty \int_0^\infty \frac{dx\,dy}{(1 + x^2 + y^2)^2}.$$

41 There is a slight difficulty with the calculation of I^2 in Example 3, because we have not discussed improper double integrals. In this problem we outline a somewhat less cavalier approach to formula (4). In Fig. 20.21 we show a quadrant of a circle of radius a, which is inside a square of side a, which in turn is inside a quadrant of a circle of radius $\sqrt{2}a$. Denote these regions by R_1, R_2, R_3.

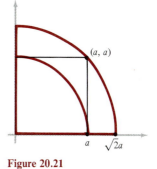

Figure 20.21

(a) Show that
$$\iint_{R_1} e^{-r^2}\,dA = \frac{\pi}{4}\left(1 - e^{-a^2}\right)$$
and
$$\iint_{R_3} e^{-r^2}\,dA = \frac{\pi}{4}\left(1 - e^{-2a^2}\right).$$

(b) Show that
$$\iint_{R_2} e^{-r^2}\,dA = \int_0^a \int_0^a e^{-(x^2+y^2)}\,dx\,dy$$
$$= \left(\int_0^a e^{-x^2}\,dx\right)^2.$$

(c) Use (a) and (b) to show that
$$\frac{\pi}{4}\left(1 - e^{-a^2}\right) < \left(\int_0^a e^{-x^2}\,dx\right)^2 < \frac{\pi}{4}\left(1 - e^{-2a^2}\right).$$

(d) Use (c) to conclude that
$$\int_0^\infty e^{-x^2}\,dx = \frac{1}{2}\sqrt{\pi}.$$

20.5
TRIPLE INTEGRALS

The definition of a triple integral follows the same pattern of ideas that was used to define a double integral in Section 20.2. We shall therefore confine ourselves to a very brief explanation.

A triple integral involves a function $f(x, y, z)$ defined on a three-dimensional region R. We divide R into many small rectangular boxes (and parts of boxes) by planes parallel to the coordinate planes, and we denote the volume of the kth box that lies wholly inside R by ΔV_k. Next, we evaluate the function at a point (x_k, y_k, z_k) in the kth box and form the product $f(x_k, y_k, z_k) \Delta V_k$. Finally, we form the sum of these products over all the boxes that lie inside R,

$$\sum_{k=1}^{n} f(x_k, y_k, z_k) \, \Delta V_k.$$

The triple integral of $f(x, y, z)$ over R is now defined to be the limit of these sums as n becomes infinite and the maximum diagonal of the boxes (that is, the longest diagonal of any of the boxes) approaches zero,

$$\iiint_R f(x, y, z) \, dV = \lim \sum_{k=1}^{n} f(x_k, y_k, z_k) \, \Delta V_k. \tag{1}$$

Sometimes we use the alternative notation

$$\iiint_R f(x, y, z) \, dx \, dy \, dz, \tag{2}$$

with no implication intended about the order of integration. This arises from the fact that since the volume of a box with faces parallel to the coordinate planes can be written as $\Delta V = \Delta x \, \Delta y \, \Delta z$, we have the element of volume formula

$$dV = dx \, dy \, dz. \tag{3}$$

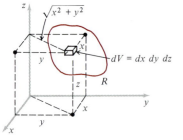

Figure 20.22

Figure 20.22 suggests the way the triple integral can be formed directly from the function $f(x, y, z)$ and the element of volume dV, in the manner explained in the previous two sections: that is, we multiply dV by $f(x, y, z)$ and integrate (or add together) the quantities $f(x, y, z) \, dV$ as the element of volume dV sweeps over the entire region R. As before, this way of thinking is merely a convenient abbreviation of the complex limit-of-sums process that constitutes the actual definition of the triple integral.

The main theoretical fact is that the triple integral (1) [or (2)] exists if $f(x, y, z)$ is continuous and the boundary of R is reasonably well behaved. We shall not pursue this issue any further. And the main practical fact is that triple integrals can often be calculated as iterated integrals.

Before we discuss iterated triple integrals, we quickly extend the ideas of Section 20.3 to the present context. First, if the region R is thought of as a solid body of variable density $\delta = \delta(x, y, z)$ [=mass per unit volume], then δdV is the element of mass—that is, the mass contained in the element of volume—and the total mass is

$$M = \iiint_R \delta dV.$$

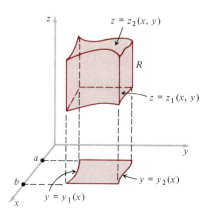

Figure 20.23

Similar considerations lead to formulas for the moments with respect to the various coordinate planes, denoted by M_{yz}, M_{xz}, and M_{xy}; and also to formulas for the moments of inertia about the various axes, denoted by I_x, I_y, and I_z. These formulas (see Fig. 20.22) are

$$M_{yz} = \iiint_R x\, \delta dV, \qquad M_{xz} = \iiint_R y\, \delta dV, \qquad M_{xy} = \iiint_R z\, \delta dV;$$

and

$$I_x = \iiint_R (y^2 + z^2)\, \delta dV, \qquad I_y = \iiint_R (x^2 + z^2)\, \delta dV,$$

$$I_z = \iiint_R (x^2 + y^2)\, \delta dV.$$

Also, the equations

$$\bar{x} = \frac{M_{yz}}{M}, \qquad \bar{y} = \frac{M_{xz}}{M}, \qquad \bar{z} = \frac{M_{xy}}{M},$$

define the center of mass of the body, or the centroid if δ is constant.

Just as we did with double integrals, we usually evaluate triple integrals by iteration. For example, if R is described by inequalities of the form

$$a \le x \le b, \qquad y_1(x) \le y \le y_2(x), \qquad z_1(x, y) \le z \le z_2(x, y),$$

as shown in Fig. 20.23, then

$$\iiint_R f(x, y, z)\, dV = \int_a^b \left[\int_{y_1(x)}^{y_2(x)} \left(\int_{z_1(x,y)}^{z_2(x,y)} f(x, y, z)\, dz \right) dy \right] dx.$$

We usually omit the parentheses and brackets, and write this in the form

$$\int_a^b \int_{y_1(x)}^{y_2(x)} \int_{z_1(x,y)}^{z_2(x,y)} f(x, y, z)\, dz\, dy\, dx.$$

As always, we integrate from the inside out, here integrating first with respect to z, then with respect to y, and finally with respect to x. Other orders of

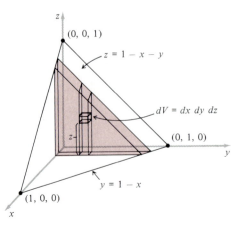

Figure 20.24

integration are often possible, and the order we choose in any specific problem is determined by a little foresight and our preference for easy calculations over hard ones.

Example 1 Find the centroid of the tetrahedron bounded by the coordinate planes and the plane $x + y + z = 1$.

Solution We can treat the tetrahedron (Fig. 20.24) as a solid of density $\delta = 1$, so that mass equals volume. By geometry the volume of the tetrahedron is $V = \frac{1}{6}$, and $\bar{z}$ is defined by

$$\bar{z} = \frac{1}{V} \iiint_R z \, dV.$$

If we integrate first z, then y, then x, this means that we must write

$$\bar{z} = \frac{1}{V} \iiint_R z \, dz \, dy \, dx,$$

with suitable limits of integration inserted. To find the z-limits we use the indicated equation of the slanting plane and imagine that the element of volume shown in the figure—like an elevator car in an elevator shaft—moves up from $z = 0$ to $z = 1 - x - y$. Next, the resulting column generates a slice by moving across the solid from left to right, from $y = 0$ to $y = 1 - x$. And finally, the slice moves through the solid from back to front, from $x = 0$ to $x = 1$. Thus,

$$\bar{z} = \frac{1}{\frac{1}{6}} \int_0^1 \int_0^{1-x} \int_0^{1-x-y} z \, dz \, dy \, dx = 6 \int_0^1 \int_0^{1-x} \left[\frac{1}{2} z^2 \right]_0^{1-x-y} dy \, dx$$

$$= 3 \int_0^1 \int_0^{1-x} (1 - x - y)^2 \, dy \, dx = 3 \int_0^1 \left[-\frac{1}{3} (1 - x - y)^3 \right]_0^{1-x} dx$$

$$= \int_0^1 (1 - x)^3 \, dx = -\frac{1}{4} (1 - x)^4 \Big]_0^1 = \frac{1}{4}.$$

By the symmetry of the situation we see that the centroid is the point $(\frac{1}{4}, \frac{1}{4}, \frac{1}{4})$.

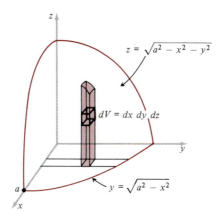

$$z = \sqrt{a^2 - x^2 - y^2}$$

$$dV = dx\ dy\ dz$$

$$y = \sqrt{a^2 - x^2}$$

Figure 20.25

We could also have found $\bar{z}$ by integrating in any other order, for instance, first x, then y, then z,

$$\bar{z} = \frac{1}{\frac{1}{6}} \int_0^1 \int_0^{1-z} \int_0^{1-y-z} z\ dx\ dy\ dz,$$

where the limits of integration are determined as they are above, that is, by examining the figure. Students should verify that this integral gives the same result as before.

Example 2   Use a triple integral to find the volume of the sphere $x^2 + y^2 + z^2 = a^2$.

Solution   The total volume is 8 times the volume in the first octant, so by integrating in the order z, y, x (see Fig. 20.25) we have

$$V = 8 \int_0^a \int_0^{\sqrt{a^2-x^2}} \int_0^{\sqrt{a^2-x^2-y^2}} dz\ dy\ dx$$

$$= 8 \int_0^a \int_0^{\sqrt{a^2-x^2}} \sqrt{a^2 - x^2 - y^2}\ dy\ dx. \qquad (4)$$

To calculate the inner integral here, we use the method of trigonometric substitution with $y = A \sin\theta$, $dy = A \cos\theta\ d\theta$ to obtain the auxiliary formula

$$\int_0^A \sqrt{A^2 - y^2}\ dy = A^2 \int_0^{\pi/2} \cos^2\theta\ d\theta = \frac{1}{2} A^2 \int_0^{\pi/2} (1 + \cos 2\theta)\ d\theta$$

$$= \frac{1}{2} A^2 \left[\theta + \frac{1}{2} \sin 2\theta \right]_0^{\pi/2} = \frac{1}{4} \pi A^2.$$

With $A = \sqrt{a^2 - x^2}$, this enables us to write (4) as

$$V = 8 \int_0^a \frac{1}{4} \pi(a^2 - x^2)\ dx = 2\pi \left[a^2 x - \frac{1}{3} x^3 \right]_0^a = \frac{4}{3} \pi a^3,$$

and the calculation is complete. Of course, we are thoroughly familiar with this result, which we have already obtained by a number of different methods. Our purpose here is to provide another illustration of the technique of triple integration.

PROBLEMS

In Problems 1 to 10, evaluate the given iterated integral.

1 $\displaystyle\int_0^1 \int_0^{x^2} \int_0^{xy^3} 18x^3y^2z \, dz \, dy \, dx.$

2 $\displaystyle\int_0^1 \int_{y^2}^1 \int_0^{1-x} x \, dz \, dx \, dy.$

3 $\displaystyle\int_0^a \int_0^b \int_0^c \sin\frac{\pi x}{a} \, dz \, dy \, dx.$

4 $\displaystyle\int_0^1 \int_0^{1-y} \int_0^{x^2+y^2} y \, dz \, dx \, dy.$

5 $\displaystyle\int_0^2 \int_0^{\pi} \int_0^{\ln 4} x^3 \cos\frac{y}{2} e^z \, dz \, dy \, dx.$

6 $\displaystyle\int_0^1 \int_0^{1-x} \int_0^{2-x} xyz \, dz \, dy \, dx.$

7 $\displaystyle\int_0^2 \int_0^{\sqrt{4-z^2}} \int_{y^2+z^2-4}^{4-y^2-z^2} dx \, dy \, dz.$

8 $\displaystyle\int_0^1 \int_0^{\sqrt{3}z} \int_0^{\sqrt{3(y^2+z^2)}} xyz\sqrt{x^2+y^2+z^2} \, dx \, dy \, dz.$

9 $\displaystyle\int_0^2 \int_0^{\sqrt{4-y^2}} \int_0^{4-x^2-y^2} y \, dz \, dx \, dy.$

10 $\displaystyle\int_0^1 \int_1^{2y} \int_0^x (x+2z) \, dz \, dx \, dy.$

11   Change the order of integration by putting suitable limits on the right side:

$$\int_0^a \int_0^x \int_0^y f(x, y, z) \, dz \, dy \, dx$$

$$= \iiint f(x, y, z) \, dx \, dy \, dz.$$

12   Evaluate both integrals in Problem 11 if $f(x, y, z) = 1$.

13   Evaluate both integrals in Problem 11 if $f(x, y, z) = x$.

14   Evaluate both integrals in Problem 11 if $f(x, y, z) = yz$.

15   Change the order of integration by putting suitable limits on the right side:

$$\int_{-1}^1 \int_{-\sqrt{1-x^2}}^{\sqrt{1-x^2}} \int_0^{1-x^2-y^2} f(x, y, z) \, dz \, dy \, dx$$

$$= \iiint f(x, y, z) \, dx \, dy \, dz.$$

16   Same directions as Problem 15:

$$\int_0^6 \int_0^{6-x} \int_0^{6-x-y} f(x, y, z) \, dz \, dy \, dx$$

$$= \iiint f(x, y, z) \, dx \, dy \, dz.$$

In Problems 17 to 24, use triple integration to find the volumes of the given regions.

17   The region in the first octant bounded by the cylinder $x = 4 - y^2$ and the planes $y = z$, $x = 0$, $z = 0$.

18   The region above the xy-plane bounded by the surfaces $z^2 = 16y$, $z^2 = y$, $y = x$, $y = 4$, and $x = 0$.

19   The region bounded by the paraboloids $z = x^2 + 9y^2$ and $z = 18 - x^2 - 9y^2$.

20   The region bounded by the paraboloids $z = 8 - x^2 - y^2$ and $z = x^2 + 3y^2$.

21   The region bounded by the ellipsoid

$$\frac{x^2}{a^2} + \frac{y^2}{b^2} + \frac{z^2}{c^2} = 1.$$

22   The region bounded by the cylinder $z = 4 - y^2$ and the paraboloid $z = x^2 + 3y^2$.

23   The tetrahedron bounded by the coordinate planes and the plane

$$\frac{x}{a} + \frac{y}{b} + \frac{z}{c} = 1,$$

where a, b, c are positive numbers.

24   The region bounded by the cylinder $x^2 + y^2 = 4x$, the xy-plane, and the paraboloid $4z = x^2 + y^2$.

25   The density of a cube is proportional to the square of the distance from one corner. Show that the mass is what it would be if the density were constant and equal to the original density at another corner adjacent to the first.

26   If the density $\delta = xy$, find the moment with respect to the xy-plane of the part of the sphere $x^2 + y^2 + z^2 \leq a^2$ that lies in the first octant.

27   The cube bounded by the coordinate planes and the planes $x = a$, $y = a$, $z = a$ has density $\delta = cz$ where c

is a constant. Find its moment of inertia I_z about the z-axis.

*28 Show that

$$\int_0^a \int_0^b \int_0^c \cos(x+y+z)\, dz\, dy\, dx$$

$$= 8 \sin\frac{a}{2} \sin\frac{b}{2} \sin\frac{c}{2} \cos\frac{a+b+c}{2}.$$

*29 Show that the four-dimensional "sphere" $x^2 + y^2 + z^2 + u^2 = a^2$ has volume

$$V = 16 \int_0^a \int_0^{\sqrt{a^2-x^2}} \int_0^{\sqrt{a^2-x^2-y^2}} \int_0^{\sqrt{a^2-x^2-y^2-z^2}} du\, dz\, dy\, dx$$

$$= \frac{1}{2}\pi^2 a^4.$$

Hint: Notice that the inner triple integral is the volume of the first octant of a three-dimensional sphere of radius $\sqrt{a^2 - x^2}$.

*30 Use the result of Problem 29 to find the volume of the five-dimensional "sphere" $x^2 + y^2 + z^2 + u^2 + v^2 = a^2$.

If a solid has axial symmetry — that is, symmetry about a line in space — it is often convenient to place its axis of symmetry on the z-axis and use cylindrical coordinates r, θ, z (Fig. 20.26) for the calculation of triple integrals. Instead of the element of volume in rectangular coordinates,

$$dV = dx\, dy\, dz,$$

we use the element of volume in cylindrical coordinates,

$$dV = r\, dr\, d\theta\, dz. \qquad (1)$$

It is easy to understand this formula by starting at a point (r, θ, z) and giving the coordinates small increments dr, $d\theta$, dz. These increments sweep out a small cell in space which is approximately a rectangular box with edges $r\, d\theta$, dr, and dz, as shown in Fig. 20.27, and dV as given by (1) is simply the product of these edges. Triple integrals now have the form

$$\iiint_R f(x, y, z)\, dV = \iiint_R f(r\cos\theta, r\sin\theta, z)\, r\, dr\, d\theta\, dz.$$

We can often calculate such an integral by writing it as an iterated integral, in the manner illustrated in the following examples.

20.6
CYLINDRICAL COORDINATES

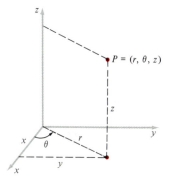

Figure 20.26 Cylindrical coordinates.

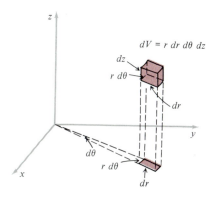

Figure 20.27 The cylindrical element of volume.

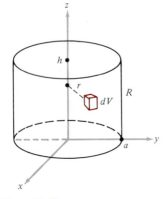

Figure 20.28

Example 1 Use a triple integral in cylindrical coordinates to find the moment of inertia of a uniform solid cylinder of height h, base radius a, and mass M about its axis.

Solution Place the cylinder in the position shown in Fig. 20.28. The word "uniform" in this context means that the density δ is constant. The mass contained in dV is δdV, and the moment of inertia of this mass about the z-axis is $r^2 \, \delta dV$. The total moment of inertia of the cylinder about its axis is therefore

$$\iiint\limits_R r^2 \, \delta dV = \iiint\limits_R r^2 \, \delta \, r \, dr \, d\theta \, dz$$

$$= \delta \int_0^{2\pi} \int_0^a \int_0^h r^3 \, dz \, dr \, d\theta$$

$$= \delta h \int_0^{2\pi} \int_0^a r^3 \, dr \, d\theta = \delta h \cdot \frac{1}{4} a^4 \int_0^{2\pi} d\theta$$

$$= \delta \cdot \frac{1}{2} \pi a^4 h = \frac{1}{2} M a^2,$$

since $M = \delta \cdot \pi a^2 h$. The fact that the limits on these integrals are all constants is a consequence of the fact that cylindrical coordinates are perfectly suited to this problem.

Example 2 Use a triple integral in cylindrical coordinates to find the volume of the sphere $x^2 + y^2 + z^2 = a^2$.

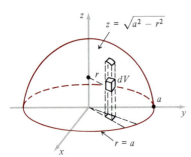

Figure 20.29

Solution The cylindrical equation of the sphere is $r^2 + z^2 = a^2$, so the equation of the upper hemisphere is $z = \sqrt{a^2 - r^2}$. We multiply the volume above the xy-plane by 2, and find this volume by integrating in the order z, r, θ, as suggested in Fig. 20.29:

$$V = 2 \int_0^{2\pi} \int_0^a \int_0^{\sqrt{a^2-r^2}} r \, dz \, dr \, d\theta = 2 \int_0^{2\pi} \int_0^a r\sqrt{a^2 - r^2} \, dr \, d\theta$$

$$= 2 \int_0^{2\pi} \left[-\frac{1}{3} (a^2 - r^2)^{3/2} \right]_0^a d\theta = 2 \int_0^{2\pi} \frac{1}{3} a^3 \, d\theta = \frac{4}{3} \pi a^3.$$

Of course, we obtain the same result as in Example 2 in Section 20.5, but the calculation is much easier here because cylindrical coordinates are better than rectangular coordinates for working with spheres.

Example 3 Find the moment of inertia of a uniform solid sphere of radius a and mass M about a diameter.

Solution We may assume that our present sphere occupies the region bounded by the sphere $r^2 + z^2 = a^2$ in Example 2. If the constant density is denoted by δ, then the moment of inertia about the z-axis is

$$I_z = 2 \int_0^{2\pi} \int_0^a \int_0^{\sqrt{a^2-r^2}} r^2 \cdot \delta r \, dz \, dr \, d\theta = 2\delta \int_0^{2\pi} \int_0^a r^3 \sqrt{a^2 - r^2} \, dr \, d\theta$$

$$= \delta \cdot 4\pi \int_0^a r^3 \sqrt{a^2 - r^2} \, dr.$$

(In the last step here we integrated out of the indicated order for the purpose of disposing of the simple θ-integral so that we could concentrate our attention on the harder r-integral. Students will become accustomed to this type of short cut.) To evaluate this integral we use the substitution $r = a \sin \phi$, $dr = a \cos \phi \, d\phi$ to write

$$\int r^3 \sqrt{a^2 - r^2} \, dr = a^5 \int \sin^3 \phi \cos^2 \phi \, d\phi$$

$$= a^5 \int (\cos^2 \phi - \cos^4 \phi) \sin \phi \, d\phi$$

$$= a^5 \left(\frac{1}{5} \cos^5 \phi - \frac{1}{3} \cos^3 \phi \right).$$

This gives

$$I_z = \delta \cdot 4\pi a^5 \left[\frac{1}{5} \cos^5 \phi - \frac{1}{3} \cos^3 \phi \right]_0^{\pi/2} = \delta \cdot \frac{8}{15} \pi a^5 = \frac{2}{5} Ma^2,$$

since $M = \delta \cdot \frac{4}{3}\pi a^3$.

PROBLEMS

Use cylindrical coordinates to solve the following problems.

1   Find the volume of the solid bounded above by the paraboloid $z = 1 - x^2 - y^2$ and below by the xy-plane.

2   Find the mass of the solid in Problem 1 if the density is
   (a) proportional to the distance from the xy-plane, $\delta = cz$;
   (b) proportional to the distance from the z-axis, $\delta = cr$;
   (c) proportional to the square of the distance from the origin, $\delta = c(r^2 + z^2)$.

3   A uniform solid cone of height h and base radius a rests on the xy-plane with its vertex on the positive z-axis. Find its center of mass.

4   If the mass of the cone in the preceding problem is M, find its moment of inertia I_z about the z-axis
   (a) by integrating first with respect to z;
   (b) by integrating first with respect to r.

5   A cylindrical hole of radius b is bored through the center of a uniform solid sphere of radius a. If the density is denoted by δ, find the mass of the ring-shaped solid that remains, and also its moment of inertia about the axis of the hole. Notice that this result generalizes the result of Example 3.

6   A wedge is cut from a uniform solid cylinder of radius a by a plane tangent to the base and inclined at a 45° angle to the base. Find its moment of inertia about the axis of the cylinder.

7   A uniform solid cone has height h, radius of base a, and mass M. Find its moment of inertia about an axis through the vertex and parallel to the base. Hint: Let the cone have its vertex at the origin and its axis on the z-axis, and find I_x.

8   A uniform solid cone has height h, radius of base a, and mass M. Find its moment of inertia about a diameter of the base. Hint: Let the cone have its base in the xy-plane and its axis on the z-axis, and find I_x.

9   A uniform solid hemisphere is bounded above by the sphere $x^2 + y^2 + z^2 = a^2$ and below by the xy-plane. Find its center of mass. (The result of this problem is a theorem of Archimedes.)

10   Find the mass of a cylindrical solid of height h and base radius a if the density at a point is proportional to the distance from the axis of the cylinder.

11 A cylindrical hole of radius a is bored through the center of a solid sphere of radius $2a$. Find the volume of the hole.

12 Find the volume of the region bounded above by the plane $z = 2x$ and below by the paraboloid $z = x^2 + y^2$.

13 Find the volume of the solid bounded above by the plane $z = x$ and below by the paraboloid $z = x^2 + y^2$.

14 Find the mass of the solid in Problem 13 if the density at each point is proportional to the square of the distance from the z-axis.

15 Find the volume of the region bounded above by the plane $z = x + y$, below by the xy-plane, and on the sides by the cylinder $x^2 + y^2 = a^2$ and the planes $x = a$, $y = a$.

*16 Find the volume of the region bounded above by the paraboloid $z = x^2 + y^2$, below by the xy-plane, and on the side by the hyperboloid

$$x^2 + y^2 = 1 + \frac{z^2}{4}.$$

17 Find the volume of the region bounded above and below by the sphere $x^2 + y^2 + z^2 = 4a^2$ and on the side by the cylinder $(x - a)^2 + y^2 = a^2$.

*18 If the region in Problem 17 is filled with matter of constant density $\delta = 1$, find the moment of inertia of this solid about the z-axis.

19 Find the moment of inertia of a uniform solid cylinder of radius a and mass M about a generator. Hint: Place the cylinder so that a generator lies on the z-axis.

20 Find the volume of the region bounded above by the sphere $x^2 + y^2 + z^2 = 2a^2$ and below by the paraboloid $az = x^2 + y^2$.

*21 Find the moment of inertia of a uniform solid sphere of radius a and mass M about a tangent line. Hint: Place the sphere with its center at the origin and let the tangent line be the line of intersection of the planes $x = a$, $y = 0$.

22 Find the volume of the region inside the cylinder $r = a \sin \theta$ which is bounded above by the sphere $x^2 + y^2 + z^2 = a^2$ and below by the upper half of the ellipsoid $x^2/a^2 + y^2/a^2 + z^2/b^2 = 1$ where $b < a$.

23 Find the volume of the region bounded above by the sphere $x^2 + y^2 + z^2 = a^2$ and below by the cone $z = r \cot \alpha$. Use this result to find the volume of a hemisphere of radius a.

*24 Find the volume of the spherical segment of height h which is cut from a sphere of radius a by a plane at a distance $a - h$ from the center.

20.7
SPHERICAL COORDINATES. GRAVITATIONAL ATTRACTION

Just as cylindrical coordinates help us deal with problems involving symmetry about a line, spherical coordinates are designed to fit situations with symmetry about a point, as in the case of a solid sphere whose density is proportional to the distance from its center. We became acquainted with the spherical coordinates ρ, ϕ, θ (see Fig. 20.30) in Section 18.7. We now put them to use in the calculation of certain triple integrals.

In order to express a triple integral

$$\iiint_R f(x, y, z)\, dV$$

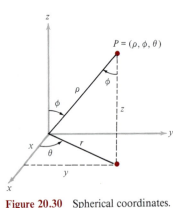

Figure 20.30 Spherical coordinates.

in spherical coordinates, we need to be able to write x, y, z as functions of ρ, ϕ, θ. This is easy to do by simply looking at Fig. 20.30:

$$z = \rho \cos \phi,$$

$$r = \rho \sin \phi,$$

$$x = \rho \sin \phi \cos \theta,$$

$$y = \rho \sin \phi \sin \theta.$$

We must now find a formula for the element of volume dV in terms of ρ, ϕ, θ. To do this we start at a point $P = (\rho, \phi, \theta)$ and give small increments $d\rho$, $d\phi$, $d\theta$ to its spherical coordinates. As we see in Fig. 20.31, the displacement of P in the ρ-direction has length $d\rho$, that in the ϕ-direction has length $\rho\, d\phi$, and that in the θ-direction has length $\rho \sin \phi\, d\theta$. These three lengths are the

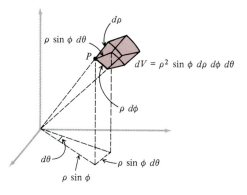

Figure 20.31 The spherical element of volume.

edges of the "spherical box" shown in the figure, so the volume of this box is $(d\rho)\,(\rho\,d\phi)\,(\rho\sin\phi\,d\theta)$ and we have

$$dV = \rho^2 \sin\phi\, d\rho\, d\phi\, d\theta.$$

To calculate a triple integral in spherical coordinates we therefore write

$$\iiint\limits_R f(x, y, z)\, dV = \iiint\limits_R f(\rho\sin\phi\cos\theta, \rho\sin\phi\,\sin\theta, \rho\cos\phi)\, \rho^2 \sin\phi\, d\rho\, d\phi\, d\theta.$$

In any particular problem we try to express this as an iterated integral in such a way that dV sweeps over the region R in a convenient manner. In most cases the nature of the region R will suggest an appropriate order of integration, together with corresponding limits of integration.

Example 1 Use a triple integral in spherical coordinates to find the volume of the sphere $x^2 + y^2 + z^2 = a^2$.

Solution The equation of this sphere in spherical coordinates is $\rho = a$. We calculate the integral

$$V = \iiint\limits_R dV = \iiint\limits_R \rho^2 \sin\phi\, d\rho\ d\phi\ d\theta$$

by integrating in the order ρ, ϕ, θ. The first integration, as ρ increases from 0 to a, adds the elements of volume dV to give the volume of the "spike" shown in Fig. 20.32; the second, as ϕ increases from 0 to π, adds the volumes of these spikes to give the volume of the wedge in the figure; and the third, as θ increases from 0 to 2π, adds the volumes of these wedges around the z-axis to give the volume of the entire sphere. The actual calculation is

$$V = \int_0^{2\pi} \int_0^{\pi} \int_0^{a} \rho^2 \sin\phi\, d\rho\ d\phi\ d\theta$$

$$= \left[\int_0^a \rho^2\, d\rho\right]\left[\int_0^{\pi} \sin\phi\, d\phi\right]\left[\int_0^{2\pi} d\theta\right]$$

$$= \frac{1}{3} a^3 \cdot 2 \cdot 2\pi = \frac{4}{3}\pi a^3,$$

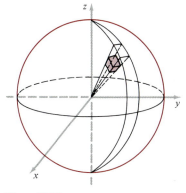

Figure 20.32

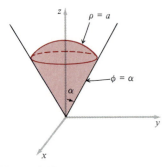

Figure 20.33

as expected. This problem is perfectly suited to spherical coordinates, as we see from the simplicity of this calculation compared with those given in the corresponding examples in Sections 20.5 and 20.6.

Example 2 Find the centroid of the region bounded by the sphere $\rho = a$ and the cone $\phi = \alpha$.

Solution This region (Fig. 20.33) is shaped like a filled ice cream cone. Its volume is

$$V = \int_0^{2\pi} \int_0^{\alpha} \int_0^{a} \rho^2 \sin \phi \, d\rho \, d\phi \, d\theta$$

$$= \frac{1}{3} a^3 \cdot 2\pi \int_0^{\alpha} \sin \phi \, d\phi$$

$$= \frac{2}{3} \pi a^3 (1 - \cos \alpha).$$

As a check, this gives $\frac{4}{3}\pi a^3$ as the volume of the sphere when $\alpha = \pi$. Now for the centroid. It is clear by symmetry that $\bar{x} = \bar{y} = 0$. To find $\bar{z}$, we must first find the moment of the region with respect to the xy-plane,

$$M_{xy} = \iiint_R z \, dV = \int_0^{2\pi} \int_0^{\alpha} \int_0^{a} (\rho \cos \phi) \, \rho^2 \sin \phi \, d\rho \, d\phi \, d\theta$$

$$= \frac{1}{2} \pi a^4 \int_0^{\alpha} \sin \phi \cos \phi \, d\phi$$

$$= \frac{1}{4} \pi a^4 \sin^2 \alpha.$$

Finally, we have

$$\bar{z} = \frac{M_{xy}}{V} = \frac{3}{2\pi a^3 (1 - \cos \alpha)} \cdot \frac{1}{4} \pi a^4 \sin^2 \alpha = \frac{3}{8} a(1 + \cos \alpha).$$

When $\alpha = \pi/2$ this specializes to $\bar{z} = \frac{3}{8}a$, which is the result of Problem 9 in Section 20.6.

In our next example we discuss an idea with important implications for several branches of physical science.

Example 3 *The gravitational attraction of a thin spherical shell.* Suppose that matter of total mass M is uniformly distributed on the surface of a sphere of radius a centered at the origin (Fig. 20.34). Show that the gravitational force **F** exerted by this thin spherical shell on a particle of mass m located at a point $(0, 0, b)$, with $b > a$, is exactly what it would be if all the mass of the shell were concentrated at its center. That is, show that

$$|\mathbf{F}| = G \frac{Mm}{b^2}, \tag{1}$$

where G is the constant of gravitation.

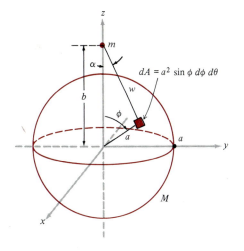

Figure 20.34

Solution By symmetry it is clear that the vector $\mathbf{F}$ is directed downward, so that $\mathbf{F} = F_z\mathbf{k}$ where F_z is negative. By Fig. 20.31 the element of area on the surface of the sphere is

$$dA = a^2 \sin \phi \, d\phi \, d\theta; \tag{2}$$

and since the mass per unit area on the surface is $M/4\pi a^2$, the mass contained in dA is

$$dM = \frac{M}{4\pi} \sin \phi \, d\phi \, d\theta.$$

Newton's law of gravitation states that the magnitude of the force this element of mass exerts on m (see Fig. 20.34) is

$$G \frac{dM \cdot m}{w^2} = \frac{GMm}{4\pi w^2} \sin \phi \, d\phi \, d\theta,$$

with downward component

$$G \frac{dM \cdot m}{w^2} \cos \alpha = \frac{GMm}{4\pi w^2} \cos \alpha \sin \phi \, d\phi \, d\theta.$$

We now find the magnitude of the total force the shell exerts on m by integrating this expression over the surface of the sphere,

$$|\mathbf{F}| = \int_0^{2\pi} \int_0^{\pi} \frac{GMm}{4\pi w^2} \cos \alpha \sin \phi \, d\phi \, d\theta$$

$$= \frac{GMm}{2} \int_0^{\pi} \frac{1}{w^2} \cos \alpha \sin \phi \, d\phi. \tag{3}$$

To calculate this integral we change the variable of integration from ϕ to w and integrate from $w = b - a$ to $w = b + a$ (see the figure). The reason for this strategy will become clear as we proceed. To accomplish the necessary transformation of the integral in (3), we first use the law of cosines to write

$$w^2 = a^2 + b^2 - 2ab \cos \phi, \tag{4}$$

so

$$2w \, dw = 2ab \sin \phi \, d\phi$$

or

$$\sin \phi \, d\phi = \frac{w \, dw}{ab}. \tag{5}$$

To write $\cos \alpha$ as a function of w, we use the fact that

$$w \cos \alpha + a \cos \phi = b$$

or

$$\cos \alpha = \frac{b - a \cos \phi}{w}.$$

With the aid of (4), this becomes

$$\cos \alpha = \frac{b - [(a^2 + b^2 - w^2)/2b]}{w} = \frac{b^2 - a^2 + w^2}{2bw}. \tag{6}$$

When (5) and (6) are substituted in (3), we obtain

$$|\mathbf{F}| = \frac{GMm}{2} \int_{b-a}^{b+a} \frac{1}{w^2} \left(\frac{b^2 - a^2 + w^2}{2bw} \right) \frac{w \, dw}{ab}$$

$$= \frac{GMm}{4ab^2} \int_{b-a}^{b+a} \left(\frac{b^2 - a^2}{w^2} + 1 \right) dw. \tag{7}$$

The value of the integral here is

$$\left[-\frac{(b^2 - a^2)}{w} + w \right]_{b-a}^{b+a} = [-(b-a) + (b+a) + (b+a) - (b-a)]$$

$$= 4a,$$

so (7) becomes

$$|\mathbf{F}| = \frac{GMm}{4ab^2} \cdot 4a = G \frac{Mm}{b^2},$$

and the proof of (1) is complete.

The conclusion reached in this example implies one of Newton's greatest theorems in mathematical astronomy: *Under the inverse square law of gravitation, a uniform solid sphere attracts an outside particle as if its mass were concentrated at its center;* for such a sphere can be thought of as if it were composed of a great many concentric thin spherical shells, like the layers of an onion, and each shell attracts in this way. Indeed, our discussion proves even more, namely, that the same statement holds for a solid sphere of variable density, provided that the density depends only on the distance from the center. Newton's theorem shows that in computing the mutual gravitational attraction of various bodies in the solar system, like the sun, the earth, and the moon, it is legitimate to replace these huge bodies by equal point masses — that is, particles — located at their centers. It is believed by some historians of science that Newton delayed the publication of his theory of the solar system for 20 years until he was able to prove this theorem.

PROBLEMS

Use spherical coordinates to solve the following problems.

1   If the region in Example 2 is filled with matter of constant density δ, find the moment of inertia of the resulting solid about the z-axis. Use this result to show that the moment of inertia of a uniform solid sphere of radius a and mass M about a diameter is $\frac{2}{5}Ma^2$.

2   In Example 2, $\bar{z} \to \frac{3}{4}a$ as $\alpha \to 0$. Explain this, in view of the fact that the region approaches a line segment as $\alpha \to 0$ and the centroid of a line segment is its midpoint.

3   Find the volume of the torus $\rho = 2a \sin \phi$ (see Fig. 18.42).

4   If $0 < b < a$ and $0 < \alpha < \pi$, find the volume of the region bounded by the concentric spheres $\rho = b, \rho = a$ and the cone $\phi = \alpha$.

5   Find the centroid of the hemispherical shell $0 < b \le \rho \le a, z \ge 0$.

6   Find the moment of inertia about the z-axis of the shell in Problem 5 if it is a solid of constant density δ.

7   A wedge is cut from a solid sphere of radius a by two planes that intersect on a diameter. If α is the angle between the planes, find the volume of the wedge.

8   Find the mass of a solid sphere of radius a if the density at each point equals the distance from the surface.

*9   Use a triple integral (in spherical coordinates) to verify that the volume of a cone of height h and base radius r is $\frac{1}{3}\pi r^2 h$.

10   If the density of a solid sphere of radius a is proportional to the distance from the center, $\delta = c\rho$, show that its mass is $c\pi a^4$.

11   Let n be a nonnegative constant, and consider a solid sphere of radius a centered at the origin whose density is proportional to the nth power of the distance from the center, $\delta = c\rho^n$.
   (a)   Find the moment of inertia of this sphere about the z-axis.
   (b)   Show how the result obtained in (a) yields the conclusion that the moment of inertia, about a diameter, of a uniform solid sphere of radius a and mass M is $\frac{2}{5}Ma^2$.

12   In Problem 11, allow the exponent n to be negative and determine what restriction must be placed on n if the mass of the sphere is to be finite. Hint: Find the mass between concentric spheres $\rho = b$ and $\rho = a$ with $0 < b < a$, and then let $b \to 0$.

13   Sketch the region bounded by the surface $\rho = a(1 - \cos \phi)$, and find its volume.

14   Find the mass of a solid sphere of radius a centered at the origin if the density at a point P equals the product of the distances from P to the origin and to the z-axis.

15   Consider a solid sphere of radius a centered at the origin with variable density $\delta = \delta(\rho, \phi, \theta)$.

(a)   Set up an iterated integral for the mass M with the integrations in the order θ, ϕ, ρ.
(b)   Simplify the integral in (a) as much as possible for the special case in which the density is a function of ρ alone, say $\delta(\rho, \phi, \theta) = f(\rho)$.
(c)   Show that the formula in (b) can be obtained directly by using thin spherical shells, without any use of iterated integrals.

16   Apply formula (2) to find the area of the polar cap on a sphere of radius a which is defined by $0 \le \phi \le \alpha$, and use this result to find the total surface area of the sphere.

17   In Example 3, assume that the particle m lies inside the spherical shell, so that $b < a$, and show that in this case the integral in (7) has the value zero. This proves the remarkable fact that a uniform thin spherical shell of matter exerts no gravitational force whatever on bodies located inside its cavity. Further, the same conclusion is also true for any nonthin spherical shell in which matter of variable density fills the space between two concentric spheres, provided that the density depends only on the distance from their common center.

18   Assume that the earth is spherical and of constant density, and imagine that a small tunnel is bored through the center. Neglecting the effect of this removal of matter, show that the gravitational attraction of the earth on a particle in the tunnel is *directly* proportional to the distance from the particle to the center of the earth. Is this necessarily true if the density is variable but depends only on the distance from the center?

19   Assume that the region discussed in Example 2 is filled with matter of constant density δ, and find the gravitational attraction it exerts on a particle of mass m placed at the origin.

20   If the rounded top is cut off the solid in Problem 19, leaving a cone of height $h = a \cos \alpha$, what now is the gravitational attraction exerted on a particle of mass m placed at the origin?

21   Assume that matter of constant density δ (= mass per unit area) is spread over the entire xy-plane and that a particle of mass m is located at the point $(0, 0, b)$ on the z-axis. Show that the gravitational attraction exerted by the planar mass on the particle is given by the following improper integral in polar coordinates,

$$\iint\limits_{R} \frac{Gm\delta b}{(r^2 + b^2)^{3/2}} \, r \, dr \, d\theta,$$

where R is the entire xy-plane. Evaluate this integral by computing it over a circle of radius a centered at the origin and then letting $a \to \infty$. Notice the remarkable fact that the value of this integral does not depend on b, so that the attractive force of the infinite plane on the particle is independent of the distance from the plane.

20.8

AREAS OF CURVED SURFACES

In Section 7.6 we discussed the problem of finding the area of a surface of revolution. We now consider the area problem for more general curved surfaces, specifically, those that have equations of the form

$$z = f(x, y),$$

where both partial derivatives $f_x(x, y)$ and $f_y(x, y)$ are continuous functions.

The method we describe rests on the simple fact that if two planes intersect at an angle γ (see Fig. 20.35), then all areas in one plane are multiplied by $\cos \gamma$ when projected on the other,

$$A = S \cos \gamma.$$

This is clearly true for the area of a rectangle with one side parallel to the line of intersection of the planes, and it follows for other regions by a limiting process. In just the same way, we project an element of surface area dS down from the given curved surface $z = f(x, y)$ onto an element of area dA in the xy-plane, as shown in Fig. 20.36. Here we have

$$dA = dS \cos \gamma,$$

where γ is the angle between the vertical line in the figure and the upward-pointing normal to the surface. This equation yields

$$dS = \frac{dA}{\cos \gamma},$$

so the total area of the curved surface is given by the formula

$$S = \iint dS = \iint_R \frac{dA}{\cos \gamma}, \qquad (1)$$

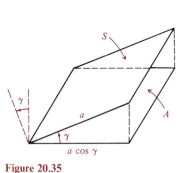

Figure 20.35

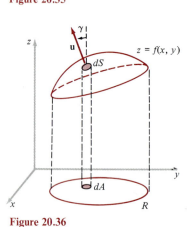

Figure 20.36

where R is the region in the xy-plane that lies under the part of the surface $z = f(x, y)$ whose area we wish to find. The element of area dA in Fig. 20.36 is drawn without any special shape, because the double integral (1) is sometimes used with rectangular coordinates and sometimes with polar coordinates.

In order to make (1) into a practical tool for actual calculations, we need a formula for $\cos \gamma$. We find this formula from the fact that the vector $f_x \mathbf{i} + f_y \mathbf{j} - \mathbf{k}$ is normal to the surface, as we saw in Section 19.3. This particular normal vector points downward, because its $\mathbf{k}$-component is negative. If we reverse the direction and divide by the length, then we see that the vector

$$\mathbf{u} = \frac{-f_x \mathbf{i} - f_y \mathbf{j} + \mathbf{k}}{\sqrt{f_x^2 + f_y^2 + 1}}$$

is the upward-pointing unit normal, and therefore $\cos \gamma$ is its $\mathbf{k}$-component,

$$\cos \gamma = \frac{1}{\sqrt{f_x^2 + f_y^2 + 1}}.$$

This enables us to write (1) in the form

$$S = \iint_R \sqrt{f_x^2 + f_y^2 + 1} \; dA, \qquad (2)$$

which is the basic formula of this section.

Example 1   Find the area of the upper half of the sphere $x^2 + y^2 + z^2 = a^2$ (Fig. 20.37).

Solution   The upper hemisphere is represented by the equation $z = \sqrt{a^2 - x^2 - y^2}$, so we have

$$f_x = \frac{-x}{\sqrt{a^2 - x^2 - y^2}},$$

with a similar formula for f_y. The integrand in (2) is therefore

$$\left[\frac{x^2}{a^2 - x^2 - y^2} + \frac{y^2}{a^2 - x^2 - y^2} + 1 \right]^{1/2} = \frac{a}{\sqrt{a^2 - x^2 - y^2}},$$

so the area of the hemisphere is

$$S = \iint\limits_{R} \frac{a}{\sqrt{a^2 - x^2 - y^2}} \, dA, \tag{3}$$

where R is the region in the xy-plane bounded by the circle $x^2 + y^2 = a^2$. [It is worth noticing that in this particular case the figure tells us directly that $\cos \gamma = z/a$, so the integrand in (1) is

$$\frac{1}{\cos \gamma} = \frac{a}{z} = \frac{a}{\sqrt{a^2 - x^2 - y^2}}$$

and the integral (3) can be written down at once, without calculation.] We now evaluate the integral (3) by introducing polar coordinates,

$$S = a \int_0^{2\pi} \int_0^a \frac{r \, dr \, d\theta}{\sqrt{a^2 - r^2}} = a \int_0^{2\pi} \left[-\sqrt{a^2 - r^2} \right]_0^a d\theta$$

$$= a^2 \int_0^{2\pi} d\theta = 2\pi a^2.$$

This result is in agreement with Archimedes' formula from elementary geometry, which states that the surface area of a sphere of radius a is $4\pi a^2$.

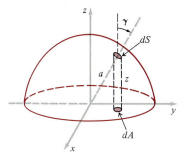

Figure 20.37

Example 2   Find the area of the part of the paraboloid $z = x^2 + y^2$ that lies inside the sphere $x^2 + y^2 + z^2 = 6$.

Solution   The boundary of the base region R is the projection on the xy-plane of the curve of intersection of the two surfaces. See Fig. 20.38. This is most easily determined by writing the surfaces in cylindrical coordinates, $z = r^2$ and $r^2 + z^2 = 6$. When z is eliminated, we find that the boundary of R is the circle $r^2 = 2$ or $r = \sqrt{2}$. In this case we have $f(x, y) = x^2 + y^2$, so $f_x = 2x$ and $f_y = 2y$, and therefore the desired surface area is

$$S = \iint\limits_{R} \sqrt{4x^2 + 4y^2 + 1} \, dA.$$

Again we carry out the calculation by using polar coordinates, which gives

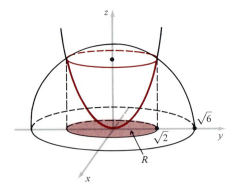

Figure 20.38

$$S = \int_0^{2\pi} \int_0^{\sqrt{2}} \sqrt{4r^2 + 1}\, r\, dr\, d\theta$$

$$= \int_0^{2\pi} \left[\frac{1}{12} (4r^2 + 1)^{3/2} \right]_0^{\sqrt{2}} d\theta$$

$$= 2\pi \cdot \frac{1}{12} (27 - 1) = \frac{13}{3}\, \pi.$$

Formulas (1) and (2) are the standard formulas of calculus for actually finding the areas of specific curved surfaces. They work well, and we hope they seem reasonable to students. Nevertheless, the *theory* of surface area is very difficult, in particular, the problem of giving a fully satisfactory definition of the concept itself. This problem has occupied the attention of mathematicians for almost a hundred years, and research on these matters continues to this day. Anyone who wishes to understand the nature of the difficulty should study the classic example of H. A. Schwarz (1890), which jolted the mathematical world of the time out of its complacency. Schwarz's example is a simple and familiar curved surface whose area can be computed in several equally reasonable ways to yield wildly different results.*

---

* This example is described in many places. See, for instance, p. 204 of D. V. Widder, *Advanced Calculus,* 2d ed. (Prentice-Hall, 1961).

PROBLEMS

Solve Problems 1 to 6 by using the ideas of this section but without integration.

1   Find the area of the triangle cut from the plane $x + 2y + 3z = 6$ by the coordinate planes.

2   Find the area above the xy-plane cut from the cone $z^2 = x^2 + y^2$ by the cylinder $x^2 + y^2 = 2ax$.

3   Find the area cut from the plane $x + y + z = 7$ by the cylinder $x^2 + y^2 = a^2$.

4   Find the area cut from the plane $z = by$ by the cylinder $x^2 + y^2 = a^2$.

5   Find the area of the part of the cone $z^2 = x^2 + y^2$ that lies between the xy-plane and the plane $2z + y = 3$. Hint: What is the area of an ellipse?

6   In Problem 5, find the area of the ellipse in which the plane intersects the cone.

7   Find the area of the part of the sphere $x^2 + y^2 + z^2 = a^2$ that lies above the xy-plane and inside the cylinder $x^2 + y^2 = ax$.

8   In Problem 7, find the area of the part of the cylinder above the xy-plane that lies inside the sphere. Hint: Find $\int h\, ds$, where h is the height of the cylinder and ds is the element of arc length in the xy-plane.

9 Find the area cut from the paraboloid $z = x^2 + y^2$ by the plane $z = 1$.

10 Find the area of the part of the surface $z^2 = 2xy$ that lies above the xy-plane and is bounded by the planes $x = 0$, $x = 2$ and $y = 0$, $y = 1$.

11 Find the area cut from the saddle surface $az = x^2 - y^2$ by the cylinder $x^2 + y^2 = a^2$.

12 Find the area of the part of the sphere $x^2 + y^2 + z^2 = 2a^2$ that lies inside the upper half of the cone $z^2 = x^2 + y^2$.

13 If R is any region in the xy-plane, show that the area of the part of the paraboloid $z = ax^2 + by^2$ that lies above R is equal to the area of the part of the saddle surface $z = ax^2 - by^2$ that lies above (or below) R. Show that this statement is also true for the pairs of surfaces $z = x^2 + y^2$, $z = 2xy$ and $z = \ln (x^2 + y^2)$, $z = 2 \tan^{-1} x/y$.

14 Find the area of the part of the paraboloid $z = 4 - x^2 - y^2$ that lies above the xy-plane.

*15 Find the area of the part of the cylinder $z = 1 - x^3$ that is cut out by the planes $y = 0$, $z = 0$, and $y = ax$ where $a > 0$.

16 Find the area of the part of the cylinder $x^2 + y^2 = a^2$ that is cut out by the cylinder $y^2/a^2 + z^2/b^2 = 1$.

17 Find the area of the part of the cylinder $x^2 + z^2 = a^2$ that lies in the first octant and between the planes $y = 3x$ and $y = 5x$.

18 Find the area of the part of the cylinder $y^2 + z^2 = a^2$ that lies inside the cylinder $x^2 + y^2 = a^2$.

19 The cylinder $r^2 = 4 \cos 2\theta$ intersects the xy-plane in a lemniscate. Find the area of the part of the paraboloid $4z = x^2 + y^2$ that lies inside this cylinder.

*20 In Section 7.6 we used the formula

$$S = \int_a^b 2\pi y \sqrt{1 + \left(\frac{dy}{dx}\right)^2} \, dx$$

to find the area of the surface of revolution obtained when the curve $y = f(x)$ is revolved about the x-axis. Show that our new method is consistent with the old, by deriving this formula from (2). Hint: The equation of the surface of revolution is $y^2 + z^2 = f(x)^2$.

*21 A *spherical triangle* is the figure on the surface of a sphere which is bounded by arcs of three great circles (Fig. 20.39). If a is the radius of the sphere, then *Legendre's formula** for the area of such a triangle is

$$S = a^2(\alpha + \beta + \gamma - \pi),$$

where α, β, γ are the angles between the sides. (The quantity $\alpha + \beta + \gamma - \pi$ is called the *spherical excess* of the triangle.) Thus, on a given sphere the area of a triangle depends only on its angles. Prove Legendre's formula by the following steps.

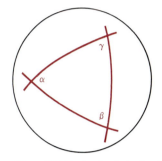

Figure 20.39   A spherical triangle.

(a) In the special triangle cut from the sphere $x^2 + y^2 + z^2 = a^2$ by the planes $y = 0$, $y = x \tan \alpha$, $z = x \tan \mu$, show that the angles between the sides are α, $\pi/2$, and β, where $\cos \beta = \sin \alpha \sin \mu$. Hint: Use vectors.

(b) Show that the projections on the xy-plane of the sides of the right triangle in (a) are $\theta = 0$, $\theta = \alpha$, and $r = a/\sqrt{1 + \tan^2 \mu \cos^2 \theta}$, and that the area of this triangle is

$$S = a^2 \int_0^\alpha \left[1 - \frac{\sin \mu \cos \theta}{\sqrt{1 - \sin^2 \mu \sin^2 \theta}} \right] d\theta.$$

(c) Carry out the integration in (b) and thereby show that the area of the right triangle is

$$a^2 \left(\alpha + \beta - \frac{\pi}{2} \right).$$

(d) Complete the proof of Legendre's formula by dividing an arbitrary triangle into two right triangles and using (c).

Apply Legendre's formula to show that the area of the complete sphere is $4\pi a^2$. Hint: Divide the surface into convenient triangles.

---

* A. M. Legendre (1752–1833) was an able French mathematician who had the bad luck to see most of his life's work rendered obsolete by the discoveries of younger and more brilliant men. In spite of this, he retained his amiable and generous disposition.

20.9

(OPTIONAL) CHANGE OF VARIABLES IN MULTIPLE INTEGRALS. JACOBIANS

Our basic tools for integrating in polar, cylindrical, and spherical coordinates are the formulas

$$dA = r \, dr \, d\theta, \qquad dV = r \, dr \, d\theta \, dz, \qquad \text{and} \qquad dV = \rho^2 \sin \phi \, d\rho \, d\phi \, d\theta, \quad (1)$$

for the elements of area and volume in these three coordinate systems. However, the justifications we gave in earlier sections of this chapter were purely intuitive and geometric. Our purpose in this brief final section is to describe a broader theoretical setting within which these formulas can be understood as merely different aspects of a single idea.

The problem that we now consider is the following: What happens to a multiple integral

$$\iint_R \cdots \int f(x, y, \ldots) \, dx \, dy \cdots$$

if we change the variables from $x, y, \ldots$ to $u, v, \ldots$?

We know the answer to this question in the case of a single variable: If $f(x)$ is continuous and the function $x = x(u)$ has a continuous derivative, then

$$\int_a^b f(x) \, dx = \int_c^d f[x(u)] \frac{dx}{du} \, du, \tag{2}$$

where $a = x(c)$ and $b = x(d)$. As an example of the use of this formula, we point out that the trigonometric substitution $x = \sin \theta$, $dx = \cos \theta \, d\theta$ enables us to write

$$\int_0^1 \sqrt{1 - x^2} \, dx = \int_0^{\pi/2} \cos \theta \cdot \cos \theta \, d\theta = \int_0^{\pi/2} \frac{1}{2} (1 + \cos 2\theta) \, d\theta = \frac{\pi}{4}.$$

Students should observe particularly that the change of variable in this calculation is accompanied by a corresponding change of the interval of integration.

Our only similar experience in the two-variable case is with changing double integrals from rectangular to polar coordinates by using the transformation equations

$$x = r \cos \theta, \qquad y = r \sin \theta. \tag{3}$$

Up to this stage we have interpreted these equations as expressing the rectangular coordinates of a given point in terms of its polar coordinates. However, they can also be interpreted as defining a *transformation* or *mapping* that carries points (r, θ) in the $r\theta$-plane over to points (x, y) in the xy-plane. That is, if a point (r, θ) is given, then equations (3) determine the corresponding point (x, y), as suggested in Fig. 20.40. Further, in order to make this correspondence one-to-one, it is customary to restrict the point (r, θ) to lie in the part of the $r\theta$-plane specified by the inequalities $0 \le r, 0 \le \theta < 2\pi$.

From this point of view, the formula for changing a double integral into polar coordinates [formula (3) in Section 20.4] can be written as

$$\iint_R f(x, y) \, dx \, dy = \iint_S f(r \cos \theta, r \sin \theta) \, r \, dr \, d\theta. \tag{4}$$

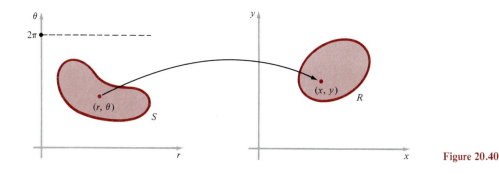

Figure 20.40

Thus, we are allowed to substitute $x = r \cos \theta$ and $y = r \sin \theta$ in the integral on the left, but we must then replace $dx \, dy$ by $r \, dr \, d\theta$ and R by the corresponding region S in the $r\theta$-plane. In our previous work we made no mention of the region S, but instead — and equivalently — changed the limits of integration on iterated integrals to describe the same region R in terms of polar coordinates.

Formula (4) is a special case of a very general formula for changing variables in double integrals. The detailed proof is beyond the scope of this book, but at least we can state the result. First we need a definition. Consider a pair of functions of two variables,

$$x = x(u, v), \qquad y = y(u, v), \tag{5}$$

and assume that they have continuous partial derivatives. The *Jacobian* of these functions is the determinant defined by

$$\frac{\partial(x, y)}{\partial(u, v)} = \begin{vmatrix} \dfrac{\partial x}{\partial u} & \dfrac{\partial x}{\partial v} \\[2mm] \dfrac{\partial y}{\partial u} & \dfrac{\partial y}{\partial v} \end{vmatrix} = \frac{\partial x}{\partial u} \frac{\partial y}{\partial v} - \frac{\partial x}{\partial v} \frac{\partial y}{\partial u}.^{*} \tag{6}$$

This is often called a "functional determinant," because it is a function of the variables u and v. As an example, we see that the Jacobian of the polar coordinates transformation (3) is

$$\frac{\partial(x, y)}{\partial(r, \theta)} = \begin{vmatrix} \dfrac{\partial x}{\partial r} & \dfrac{\partial x}{\partial \theta} \\[2mm] \dfrac{\partial y}{\partial r} & \dfrac{\partial y}{\partial \theta} \end{vmatrix} = \begin{vmatrix} \cos \theta & -r \sin \theta \\ \sin \theta & r \cos \theta \end{vmatrix} = r \cos^2 \theta + r \sin^2 \theta = r.$$

The general change of variables formula for double integrals can now be stated as follows: *If (5) is a one-to-one transformation of a region S in the uv-plane onto a region R in the xy-plane, and if the Jacobian (6) is positive, then*

---

* Determinants of this form were first discussed by the German mathematician C. G. J. Jacobi (1804–1851). He did important work in the theory of elliptic functions, and applied his discoveries in astonishing ways to the theory of numbers. He also created a new and fruitful approach to theoretical dynamics. The Hamilton-Jacobi equations are part of the standard equipment of every student of mathematical physics.

$$\iint\limits_{R} f(x, y)\, dx\, dy = \iint\limits_{S} f[x(u, v), y(u, v)]\, \frac{\partial(x, y)}{\partial(u, v)}\, du\, dv. \qquad (7)$$

Since r is the Jacobian of the polar coordinates transformation (3), it is clear that (4) is a special case of (7). Further, we can think of (7) as a two-dimensional extension of (2), with the derivative dx/du being replaced by the Jacobian $\partial(x, y)/\partial(u, v)$.

Formula (7) in turn can be extended to triple integrals. First, we define the *Jacobian* of the transformation

$$\begin{cases} x = x(u, v, w), \\ y = y(u, v, w), \\ z = z(u, v, w), \end{cases} \quad \text{by} \quad \frac{\partial(x, y, z)}{\partial(u, v, w)} = \begin{vmatrix} \dfrac{\partial x}{\partial u} & \dfrac{\partial x}{\partial v} & \dfrac{\partial x}{\partial w} \\[2mm] \dfrac{\partial y}{\partial u} & \dfrac{\partial y}{\partial v} & \dfrac{\partial y}{\partial w} \\[2mm] \dfrac{\partial z}{\partial u} & \dfrac{\partial z}{\partial v} & \dfrac{\partial z}{\partial w} \end{vmatrix}.$$

Then, under similar assumptions, we have the following extension of (7):

$$\iiint\limits_{R} f(x, y, z)\, dx\, dy\, dz = \iiint\limits_{S} F(u, v, w)\, \frac{\partial(x, y, z)}{\partial(u, v, w)}\, du\, dv\, dw, \qquad (8)$$

where $F(u, v, w) = f[x(u, v, w), y(u, v, w), z(u, v, w)]$. The main thing to notice here is that

$$dx\, dy\, dz \quad \text{is replaced by} \quad \frac{\partial(x, y, z)}{\partial(u, v, w)}\, du\, dv\, dw.$$

Two important special cases of (8) are those of *cylindrical coordinates,*

$$\iiint\limits_{R} f(x, y, z)\, dx\, dy\, dz = \iiint\limits_{S} F(r, \theta, z)\, r\, dr\, d\theta\, dz,$$

where $F(r, \theta, z) = f(r \cos \theta, r \sin \theta, z)$; and *spherical coordinates,*

$$\iiint\limits_{R} f(x, y, z)\, dx\, dy\, dz = \iiint\limits_{S} F(\rho, \phi, \theta)\, \rho^2 \sin \phi\, d\rho\, d\phi\, d\theta,$$

where $F(\rho, \phi, \theta) = f(\rho \sin \phi \cos \theta, \rho \sin \phi \sin \theta, \rho \cos \phi)$. We leave it to the student to verify the spherical coordinates formula by using the transformation equations

$$\begin{cases} x = \rho \sin \phi \cos \theta, \\ y = \rho \sin \phi \sin \theta, \\ z = \rho \cos \phi, \end{cases}$$

to calculate the Jacobian

$$\frac{\partial(x, y, z)}{\partial(\rho, \phi, \theta)} = \begin{vmatrix} \sin \phi \cos \theta & \rho \cos \phi \cos \theta & -\rho \sin \phi \sin \theta \\ \sin \phi \sin \theta & \rho \cos \phi \sin \theta & \rho \sin \phi \cos \theta \\ \cos \phi & -\rho \sin \phi & 0 \end{vmatrix} = \rho^2 \sin \phi.$$

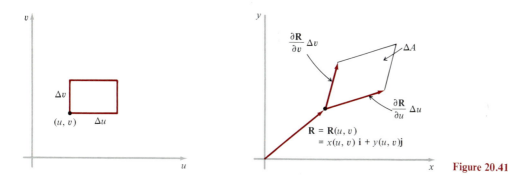

Figure 20.41

It is in this way that we can understand a little more fully what lies behind formulas (1).

One question remains, and for the sake of simplicity we state it only for the two-variable case: What is the underlying reason for the presence of the Jacobian on the right side of formula (7)? We now give a very brief intuitive explanation of this by means of vectors. In the uv-plane the equations $u = a$ constant and $v = a$ constant determine a network of straight lines parallel to the axes, whereas in the xy-plane these equations determine a network of intersecting curves. A small rectangle in the uv-plane with sides Δu and Δv corresponds to a small parallelogram in the xy-plane (see Fig. 20.41) with sides that can be written in vector form as

$$\frac{\partial \mathbf{R}}{\partial u} \Delta u = \left(\frac{\partial x}{\partial u} \mathbf{i} + \frac{\partial y}{\partial u} \mathbf{j} \right) \Delta u$$

and

$$\frac{\partial \mathbf{R}}{\partial v} \Delta v = \left(\frac{\partial x}{\partial v} \mathbf{i} + \frac{\partial y}{\partial v} \mathbf{j} \right) \Delta v,$$

approximately. In calculating the integral on the left side of (7) as a limit of sums, it is natural to abandon the usual rectangular cells and instead use these small parallelograms. If we denote by ΔA the area of the parallelogram in the figure, then ΔA equals the magnitude of the cross product of the two vectors given above. Since this cross product is

$$\begin{vmatrix} \mathbf{i} & \mathbf{j} & \mathbf{k} \\ \dfrac{\partial x}{\partial u} & \dfrac{\partial y}{\partial u} & 0 \\ \dfrac{\partial x}{\partial v} & \dfrac{\partial y}{\partial v} & 0 \end{vmatrix} \Delta u \, \Delta v = \left[\frac{\partial(x, y)}{\partial(u, v)} \Delta u \, \Delta v \right] \mathbf{k},$$

we have

$$\Delta A = \frac{\partial(x, y)}{\partial(u, v)} \Delta u \, \Delta v. \tag{9}$$

This shows that the Jacobian plays the role of a local magnification factor for areas. Further, these remarks constitute a sketch of a proof of (7), because all that remains to establish (7) is to form the integral on the left side as a limit of sums and make use of (9).

21

LINE INTEGRALS AND GREEN'S THEOREM

21.1

LINE INTEGRALS IN THE PLANE

This chapter brings together into a unified package several topics in multivariable calculus that are important for physical science and engineering, as well as for mathematics itself. The main focus of our work is the concept of a line integral, which provides yet another way (in addition to double and triple integrals) of extending ordinary integration to higher dimensions. Line integrals are used, for example, to compute the work done by a variable force in moving a particle along a curved path from one point to another. In their origin and applications, these integrals are therefore associated with mathematical physics as much as they are with mathematics. The main result of this chapter (Green's Theorem) uses partial derivatives to establish a connecting link between line integrals and double integrals, and this in turn enables us to distinguish those vector fields that have potential energy functions from those that do not. Here again, as so often in our earlier work, mathematics and physics constitute a seamless fabric in which neither ingredient has much meaning or value without the other.

Throughout this chapter we assume that the functions under discussion have all the continuity and differentiability properties that are needed in any given situation.

Our first problem is to formulate a satisfactory concept of work. If we push a particle along a straight path with a constant force $\mathbf{F}$ (constant in both direction and magnitude), then we know that the work done by this force is the product of the component of $\mathbf{F}$ in the direction of motion and the distance the particle moves. It is convenient to use the dot product to write this in the form

$$W = \mathbf{F} \cdot \Delta \mathbf{R}, \tag{1}$$

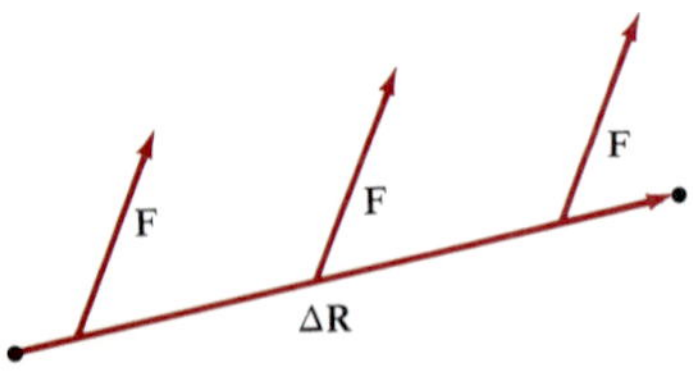

where $\Delta \mathbf{R}$ is the vector from the initial position of the particle to its final position (Fig. 21.1). Now suppose that the force $\mathbf{F}$ is not constant, but instead is a vector function that varies from point to point throughout a certain region of the plane, say

$$\mathbf{F} = \mathbf{F}(x, y) = M(x, y)\mathbf{i} + N(x, y)\mathbf{j}. \tag{2}$$

Figure 21.1

676

Suppose also that this variable force pushes a particle along a smooth curve C in the plane (Fig. 21.2), where C has parametric equations

$$x = x(t) \qquad \text{and} \qquad y = y(t), \qquad t_1 \le t \le t_2. \tag{3}$$

What is the work done by this force as the point of application moves along the curve from the initial point A to the final point B?

Before answering this question, we remark that the vector-valued function (2) is usually called a *force field*. More generally, a *vector field* in the plane is any vector-valued function that associates a vector with each point (x, y) in a certain plane region R. In this context a function whose values are numbers (scalars) is called a *scalar field*. For example, the function $f(x, y) = x^2 y^3$ is a scalar field defined on the entire xy-plane. Every scalar field $f(x, y)$ gives rise to the corresponding vector field

$$\nabla f(x, y) = \operatorname{grad} f(x, y) = \frac{\partial f}{\partial x}\mathbf{i} + \frac{\partial f}{\partial y}\mathbf{j}.$$

(Remember that the symbol ∇f is pronounced "del f.") This is called the *gradient field* of f; its intuitive meaning was described in Section 19.5. For the function just mentioned, we have $\nabla f = 2xy^3\mathbf{i} + 3x^2 y^2\mathbf{j}$. Some vector fields are gradient fields, but most are not. We shall see in the next section that those vector fields that are also gradient fields are of special importance.

We now return to the problem of calculating the work done by the variable force

$$\mathbf{F} = M(x, y)\mathbf{i} + N(x, y)\mathbf{j} \tag{2}$$

along the smooth curve C. This leads to a new kind of integral called a line integral and denoted by

$$\int_C \mathbf{F} \cdot d\mathbf{R} \qquad \text{or} \qquad \int_C M(x, y)\,dx + N(x, y)\,dy.$$

We begin the definition by approximating the curve by a polygonal path as shown in Fig. 21.3. That is, choose points $P_0 = A, P_1, P_2, \ldots, P_{n-1}$, $P_n = B$ along C in this order, let $\mathbf{R}_k$ be the position vector of P_k, and define the n incremental vectors shown in the figure by $\Delta\mathbf{R}_k = \mathbf{R}_{k+1} - \mathbf{R}_k$, where $k = 0, 1, \ldots, n-1$. If we now denote by $\mathbf{F}_k$ the value of the vector function $\mathbf{F}$ at P_k and form the sum

$$\sum_{k=0}^{n-1} \mathbf{F}_k \cdot \Delta\mathbf{R}_k, \tag{4}$$

then the *line integral of* $\mathbf{F}$ *along* C is defined to be the limit of sums of this form, and we write

$$\int_C \mathbf{F} \cdot d\mathbf{R} = \lim \sum_{k=0}^{n-1} \mathbf{F}_k \cdot \Delta\mathbf{R}_k. \tag{5}$$

In this limit the polygonal paths are understood to approximate the curve C more and more closely, in the sense that the number of points of division

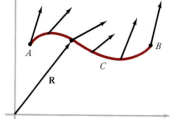

Figure 21.2

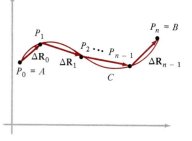

Figure 21.3

increases and the maximum length of the incremental vectors approaches zero.*

The idea behind the definition (5) is that $\mathbf{F}$ (being continuous) is almost constant along the short path segment $\Delta\mathbf{R}_k$, so by formula (1) we see that $\mathbf{F}_k \cdot \Delta\mathbf{R}_k$ is approximately the work done by $\mathbf{F}$ along the corresponding part of the curve, and therefore the sum (4) is approximately the work done by $\mathbf{F}$ along the entire curve C, with the limit (5) giving the exact value of this work.

A quick intuitive way of constructing the line integral (5) is illustrated in Fig. 21.4. If $d\mathbf{R}$ is the element of displacement along C, then the corresponding element of work done by $\mathbf{F}$ is $dW = \mathbf{F} \cdot d\mathbf{R}$. The total work is now obtained by integrating (or adding together) these elements of work along the entire curve C,

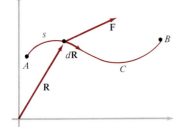

Figure 21.4

$$W = \int dW = \int_C \mathbf{F} \cdot d\mathbf{R}. \tag{6}$$

For additional insight into the meaning of this formula, we think of the position vector $\mathbf{R}$ as a function of the arc length s measured from the initial point A. Since we know that $d\mathbf{R}/ds$ is the unit tangent vector $\mathbf{T}$ (Section 17.4), we can write

$$\int_C \mathbf{F} \cdot d\mathbf{R} = \int_C \mathbf{F} \cdot \frac{d\mathbf{R}}{ds}\, ds = \int_C \mathbf{F} \cdot \mathbf{T}\, ds. \tag{7}$$

The line integral (6) can therefore be thought of as the integral of the tangential component of $\mathbf{F}$ along the curve C. It can be seen from (7) that line integrals include ordinary integrals as special cases; for if the curve C lies along the x-axis between $x = a$ and $x = b$, and if $\mathbf{F} = f(x)\mathbf{i}$, then (7) reduces to $\int_a^b f(x)\, dx$.

If the variable vector $\mathbf{F}$ is given by $\mathbf{F} = M(x, y)\mathbf{i} + N(x, y)\mathbf{j}$, then since $\mathbf{R} = x\mathbf{i} + y\mathbf{j}$ and $d\mathbf{R} = dx\mathbf{i} + dy\mathbf{j}$, the formula for computing the dot product of two vectors yields

$$\mathbf{F} \cdot d\mathbf{R} = M(x, y)\, dx + N(x, y)\, dy.$$

The line integral (6) can therefore be written in the form

$$\int_C \mathbf{F} \cdot d\mathbf{R} = \int_C M(x, y)\, dx + N(x, y)\, dy.$$

The parametric representation $x = x(t)$ and $y = y(t)$, $t_1 \leq t \leq t_2$, for the curve C allows us to express everything here in terms of t,

$$\int_C \mathbf{F} \cdot d\mathbf{R} = \int_C M(x, y)\, dx + N(x, y)\, dy$$

$$= \int_{t_1}^{t_2} \left[M(x, y)\frac{dx}{dt} + N(x, y)\frac{dy}{dt} \right] dt.$$

This is an ordinary single integral with t as the variable of integration, and it can be evaluated in the usual way.

---

* The term *line integral* for the limit (5) is perhaps unfortunate, because the curve C need not be a straight line segment. *Curve integral* would be more appropriate, but the terminology is well established and cannot be changed now.

So much for generalities. We now turn our attention to getting some practice in the actual calculation of line integrals.

Example 1 Evaluate the line integral

$$I = \int_C x^2 y \, dx + (x - 2y) \, dy,$$

where C is the segment of the parabola $y = x^2$ from $(0, 0)$ to $(1, 1)$ (see Fig. 21.5).

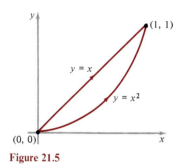

Figure 21.5

Solution We may parametrize the curve in any way that seems convenient. (It is not difficult to show that the value of the line integral does not depend on what parametric equations are used, provided that the orientation or direction is kept the same.) The simplest parametric representation of this curve is

$$x = t, \qquad y = t^2, \qquad \text{where } 0 \le t \le 1.$$

Here we have $dx = dt$ and $dy = 2t \, dt$, so the line integral is

$$I = \int_0^1 t^2 \cdot t^2 \, dt + (t - 2t^2) 2t \, dt$$

$$= \int_0^1 [t^4 + 2t^2 - 4t^3] \, dt = \left[\frac{1}{5} t^5 + \frac{2}{3} t^3 - t^4\right]_0^1 = -\frac{2}{15}.$$

To illustrate the fact that the value of the line integral is independent of the choice of parameter, let us use the representation

$$x = \sin t, \qquad y = \sin^2 t, \qquad \text{where } 0 \le t \le \pi/2.$$

This time we have $dx = \cos t \, dt$ and $dy = 2 \sin t \cos t \, dt$, so

$$I = \int_0^{\pi/2} \sin^2 t \cdot \sin^2 t \cdot \cos t \, dt + (\sin t - 2 \sin^2 t) 2 \sin t \cos t \, dt$$

$$= \int_0^{\pi/2} [\sin^4 t + 2 \sin^2 t - 4 \sin^3 t] \cos t \, dt$$

$$= \left[\frac{1}{5} \sin^5 t + \frac{2}{3} \sin^3 t - \sin^4 t\right]_0^{\pi/2} = -\frac{2}{15},$$

as before.

Every curve C that we use with line integrals is understood to have a direction, from its initial point to its final point. Even though the value of a line integral does not depend on the parameter, it *does* depend on the direction. If $-C$ denotes the same curve traversed in the opposite direction, then we have

$$\int_{-C} \mathbf{F} \cdot d\mathbf{R} = -\int_C \mathbf{F} \cdot d\mathbf{R},$$

or equivalently,

$$\int_{-C} M\,dx + N\,dy = -\int_{C} M\,dx + N\,dy.$$

That is, integrating in the opposite direction changes the sign of the integral. This can be seen at once from Fig. 21.3 and the definition (5), because the directions of all the incremental vectors $\Delta\mathbf{R}_k$ are reversed.

Example 2   Evaluate the line integral

$$I = \int_{C} x^2 y\,dx + (x - 2y)\,dy,$$

where C is the straight line segment $y = x$ from $(0, 0)$ to $(1, 1)$.

Solution   This is the same integrand as in Example 1, and the initial and final points of the curve are the same, but the curve itself is different (see Fig. 21.5). Using x as the parameter, so that the parametric equations are $x = x$ and $y = x$, we have $dx = dx$ and $dy = dx$, so

$$I = \int_{0}^{1} x^2 \cdot x\,dx + (x - 2x)\,dx$$

$$= \int_{0}^{1} [x^3 - x]\,dx = \left[\frac{1}{4}x^4 - \frac{1}{2}x^2\right]_{0}^{1} = -\frac{1}{4},$$

which we observe is different from the value $-\frac{2}{15}$ obtained along the parabolic path.

The integral in this example can be written as

$$\int_{C} \mathbf{F} \cdot d\mathbf{R}, \qquad \text{where } \mathbf{F} = x^2 y\mathbf{i} + (x - 2y)\mathbf{j}.$$

If $\mathbf{F}$ is thought of as a force field, then the work done by $\mathbf{F}$ in moving a particle from $(0, 0)$ to $(1, 1)$ is different for the two curves in Examples 1 and 2. This illustrates the fact that in general the line integral of a given vector field from one given point to another depends on the choice of the curve, and has different values for different curves.

If a curve C consists of a finite number of smooth curves joined at corners, then we say that C is a *piecewise smooth curve,* or a *path.* The value of a line integral along C is then defined as the sum of its values along the smooth pieces of C. This is illustrated in the first part of our next example.

Example 3   Evaluate the line integral

$$\int_{C} y\,dx + (x + 2y)\,dy$$

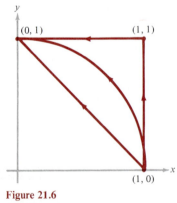

Figure 21.6

from $(1, 0)$ to $(0, 1)$, where C is (a) the broken line from $(1, 0)$ to $(1, 1)$ to $(0, 1)$; (b) the arc of the circle $x = \cos t$, $y = \sin t$; (c) the straight line segment $y = 1 - x$. See Fig. 21.6.

Solution (a) Along the segment from (1, 0) to (1, 1) we have $x = 1$ and $dx = 0$; and along the segment from (1, 1) to (0, 1) we have $y = 1$ and $dy = 0$. Since the complete line integral is the sum of the line integrals along each of the segments, we have

$$\int_C y\, dx + (x + 2y)\, dy = \int_0^1 (1 + 2y)\, dy + \int_1^0 dx$$

$$= \left[y + y^2 \right]_0^1 + \left[x \right]_1^0 = 1.$$

(b) Here we have $x = \cos t$ and $y = \sin t$ for $0 \le t \le \pi/2$, so $dx = -\sin t\, dt$ and $dy = \cos t\, dt$, and therefore

$$\int_C y\, dx + (x + 2y)\, dy = \int_0^{\pi/2} -\sin^2 t\, dt + (\cos t + 2 \sin t)\cos t\, dt$$

$$= \int_0^{\pi/2} (\cos^2 t - \sin^2 t + 2 \sin t \cos t)\, dt$$

$$= \int_0^{\pi/2} (\cos 2t + 2 \sin t \cos t)\, dt$$

$$= \left[\frac{1}{2} \sin 2t + \sin^2 t \right]_0^{\pi/2} = 1.$$

(c) To integrate along the segment $y = 1 - x$ we can use x as the parameter, so that $dy = -dx$. Since x varies from 1 to 0 along this path, the integral is

$$\int_C y\, dx + (x + 2y)\, dy = \int_1^0 (1 - x)\, dx + [x + 2(1 - x)](-dx)$$

$$= \int_1^0 (-1)\, dx = 1.$$

In this example all three line integrals have the same value, and we might suspect that perhaps with this integrand we get the same value for *any* path from (1, 0) to (0, 1). This is indeed true, as we shall see in Section 21.2, where we investigate the underlying reasons why some line integrals from one point to another have values that are independent of the path of integration.

It will often be necessary to consider situations in which the path of integration C is a *closed* curve, which means that the final point B is the same as the initial point A (Fig. 21.7). For the sake of emphasis, in this case a line integral is usually written with a small circle on the integral sign, as in

$$\oint_C \mathbf{F} \cdot d\mathbf{R} \qquad \text{or} \qquad \oint_C M\, dx + N\, dy.$$

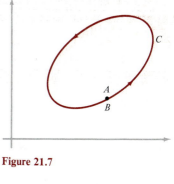

Figure 21.7

Example 4 Calculate $\oint_C \mathbf{F} \cdot d\mathbf{R}$, where $\mathbf{F} = y\mathbf{i} + 2x\mathbf{j}$ and C is the circle $x^2 + y^2 = 1$ described counterclockwise from $A = (1, 0)$ back to the same point (Fig. 21.8).

Solution A simple parametric representation is $x = \cos t$ and $y = \sin t$, where the counterclockwise orientation means that t increases from 0 to 2π.

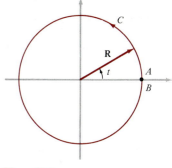

Figure 21.8

Since $\mathbf{R} = x\mathbf{i} + y\mathbf{j} = \cos t \, \mathbf{i} + \sin t \, \mathbf{j}$, we have

$$d\mathbf{R} = (-\sin t \, \mathbf{i} + \cos t \, \mathbf{j}) \, dt,$$

and therefore

$$\mathbf{F} \cdot d\mathbf{R} = (\sin t \, \mathbf{i} + 2 \cos t \, \mathbf{j}) \cdot (-\sin t \, \mathbf{i} + \cos t \, \mathbf{j}) \, dt$$
$$= (2 \cos^2 t - \sin^2 t) \, dt$$
$$= \left(\frac{1}{2} + \frac{3}{2} \cos 2t \right) dt,$$

by the half-angle formulas. It now follows that

$$\oint_C \mathbf{F} \cdot d\mathbf{R} = \int_0^{2\pi} \left(\frac{1}{2} + \frac{3}{2} \cos 2t \right) dt$$
$$= \left[\frac{1}{2} t + \frac{3}{4} \sin 2t \right]_0^{2\pi} = \pi.$$

PROBLEMS

1 Evaluate the line integral $I = \int_C xy^2 \, dx - (x + y) \, dy$, where C is
(a) the straight line segment from $(0, 0)$ to $(1, 2)$;
(b) the parabolic path $y = 2x^2$ from $(0, 0)$ to $(1, 2)$;
(c) the broken line from $(0, 0)$ to $(1, 0)$ to $(1, 2)$.
Sketch all paths.

2 Evaluate the line integral $I = \int_C x^2 y \, dx - xy^2 \, dy$, where C is the broken line joining the points $(0, 0)$, $(1, 1)$, $(2, 1)$ in this order.

3 Evaluate $\int_C dx/y + dy/x$, where C is the part of the hyperbola $xy = 4$ from $(1, 4)$ to $(4, 1)$.

4 Show that (a) $\int_C (x^2 - 2y) \, dx = -\frac{2}{3}$, (b) $\int_C 2xy^2 \, dy = \frac{1}{2}$, and (c) $\int_C (x^2 - 2y) \, dx + 2xy^2 \, dy = -\frac{1}{6}$, if C is the straight line segment $y = x$, $0 \le x \le 1$.

5 Show that $\int_C (x^2 + 3xy) \, dx + (3x^2 - 2y^2) \, dy = -\frac{71}{12}$, if C is the segment of the parabola $x = t$, $y = t^2$ from $t = 1$ to $t = 2$.

6 Evaluate $\int_C (dx + dy)/(x^2 + y^2)$, where C is the upper half of the circle $x^2 + y^2 = a^2$ from $(a, 0)$ to $(-a, 0)$.

7 Find the values of the line integral $\int_C (x - y) \, dx + \sqrt{x} \, dy$ along the following paths C from $(0, 0)$ to $(1, 1)$:
(a) $x = t$, $y = t$; (b) $x = t$, $y = t^2$; (c) $x = t^2$, $y = t$; (d) $x = t$, $y = t^3$. Sketch all paths.

8 Show that $\int_C (x^2 + y^2) \, dx = -\frac{2}{3}$, if C is the broken line from $(0, 0)$ to $(1, 1)$ to $(0, 1)$.

9 Evaluate $\int_C x \, dx + x^2 \, dy$ from $(-1, 0)$ to $(1, 0)$
(a) along the x-axis;
(b) along the semicircle $y = \sqrt{1 - x^2}$;
(c) along the broken line from $(-1, 0)$ to $(0, 1)$ to $(1, 1)$ to $(1, 0)$.
Sketch all paths.

10 Evaluate $\oint_C (3x + 4y) \, dx + (2x + 3y^2) \, dy$, where C is the circle $x^2 + y^2 = 4$ traversed counterclockwise from $(2, 0)$.

11 Find the values of the line integral $\int_C 2xy \, dx + (x^2 + y^2) \, dy$ along the following paths C from $(0, 0)$ to $(1, 1)$:
(a) $y = x$; (b) $y = x^2$; (c) $x = y^2$; (d) $y = x^3$; (e) $x = y^3$; (f) the broken line from $(0, 0)$ to $(1, 0)$ to $(1, 1)$. Sketch all paths.

12 Find the values of the line integral $\int_C x^2 \, dx + y^2 \, dy$ along the following paths C from $(0, 1)$ to $(1, 0)$: (a) the circular arc $x = \cos t$, $y = \sin t$; (b) the straight line segment; (c) the segment of the parabola $y = 1 - x^2$.

13 If $\mathbf{F} = (y\mathbf{i} - x\mathbf{j})/(x^2 + y^2)$, find $\int_C \mathbf{F} \cdot d\mathbf{R}$ from $(-1, 0)$ to $(1, 0)$
(a) along the semicircle $y = \sqrt{1 - x^2}$;
(b) along the broken line from $(-1, 0)$ to $(0, 1)$ to $(1, 1)$ to $(1, 0)$.

14 Compute $\oint_C \mathbf{F} \cdot d\mathbf{R}$ if $\mathbf{F} = (x + y)\mathbf{i} + (y^2 - x)\mathbf{j}$, where C is the closed curve that begins at $(1, 0)$, proceeds along the upper half of the unit circle to $(-1, 0)$, and returns to $(1, 0)$ along the x-axis.

15 Evaluate $\int_C y\sqrt{y} \, dx + x\sqrt{y} \, dy$, where C is the part of the curve $x^2 = y^3$ from $(1, 1)$ to $(8, 4)$.

16 If $\mathbf{F} = (2x + y)\mathbf{i} + (3x - 2y)\mathbf{j}$, evaluate $\int_C \mathbf{F} \cdot d\mathbf{R}$ along
(a) the straight line from $(0, 0)$ to $(1, 1)$;
(b) the parabola $y = x^2$ from $(0, 0)$ to $(1, 1)$;
(c) $y = \sin (\pi/2) x$ from $(0, 0)$ to $(1, 1)$;
(d) $x = y^n$ ($n > 0$) from $(0, 0)$ to $(1, 1)$.

17 Show that $\oint_C (-y \, dx + x \, dy)/(x^2 + y^2) = 2\pi$, where C is the circle $x^2 + y^2 = a^2$ traversed counterclockwise from $(a, 0)$.

18 If $\mathbf{F} = xy\mathbf{i} + (y^2 + 1)\mathbf{j}$, calculate $\int_C \mathbf{F} \cdot d\mathbf{R}$, where C is
  (a) the line segment from $(0, 0)$ to $(1, 1)$;
  (b) the broken line from $(0, 0)$ to $(1, 0)$ to $(1, 1)$;
  (c) the parabola $x = y^2$ from $(0, 0)$ to $(1, 1)$.

19 Find the values of $\int_C y \, dx + x \, dy$ along the following paths C from $(-a, 0)$ to $(a, 0)$:
  (a) the upper half of the circle $x^2 + y^2 = a^2$;
  (b) the broken line from $(-a, 0)$ to $(-a, a)$ to (a, a) to $(a, 0)$;
  (c) the straight line segment joining the points.

20 Evaluate $\int_C xy^2 \, dx + x^3 y \, dy$, where C is the broken line consisting of the segments from $(-1, -1)$ to $(2, -1)$ and from $(2, -1)$ to $(2, 4)$.

21 A particle is moved around a square path from $(0, 0)$ to $(1, 0)$ to $(1, 1)$ to $(0, 1)$ to $(0, 0)$ under the action of the force field $\mathbf{F} = (2x + y)\mathbf{i} + (x + 4y)\mathbf{j}$. Find the work done.

22 Calculate $\oint_C 2xy \, dx + (x^2 + y^2) \, dy$, where C is the boundary of the semicircular region $x^2 + y^2 \leq 1$, $y \geq 0$ described counterclockwise.

21.2
INDEPENDENCE OF PATH. CONSERVATIVE FIELDS

In Example 3 of Section 21.1 we calculated the line integral

$$\int_C \mathbf{F} \cdot d\mathbf{R} \tag{1}$$

of the vector field

$$\mathbf{F}(x, y) = y\mathbf{i} + (x + 2y)\mathbf{j} \tag{2}$$

along each of the three different paths from $(1, 0)$ to $(0, 1)$ shown in Fig. 21.9, and we obtained the same value 1 for the integral along all of these paths. This was not an accident. The underlying reason for this result is the fact that the vector field (2) is the *gradient of a scalar field,* namely, of the function

$$f(x, y) = xy + y^2, \tag{3}$$

because clearly

$$\nabla f = \frac{\partial f}{\partial x} \mathbf{i} + \frac{\partial f}{\partial y} \mathbf{j} = y\mathbf{i} + (x + 2y)\mathbf{j} = \mathbf{F}.$$

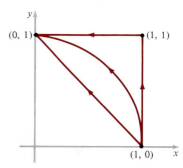

Figure 21.9

To understand the significance of this statement, recall from Section 19.5 that in multivariable calculus the gradient plays a similar role to that of the derivative in single-variable calculus. The Fundamental Theorem of (single-variable) Calculus can be expressed in the form

$$\int_a^b f'(x) \, dx = f(b) - f(a),$$

where $f(x)$ is a function of a single variable. The corresponding result here is

$$\int_C \nabla f \cdot d\mathbf{R} = f(B) - f(A), \tag{4}$$

where $f(x, y)$ is a function of two variables (a scalar field) and A and B are the initial and final points of the path C. For example, since the vector field (2) is the gradient of the scalar field (3), that is, $\mathbf{F} = \nabla f$, formula (4) tells us that the value of the line integral (1) along any of the paths C shown in Fig. 21.9 is

$$\int_C \mathbf{F} \cdot d\mathbf{R} = \int_C \nabla f \cdot d\mathbf{R} = f(0, 1) - f(1, 0) = 1 - 0 = 1,$$

without calculation.

Formula (4) is called the *Fundamental Theorem of Calculus for Line Integrals.* We can state this theorem more precisely as follows:

If a vector field $\mathbf{F}$ is the gradient of some scalar field f in a region R, so that $\mathbf{F} = \nabla f$ in R, and if C is any piecewise smooth curve in R with initial and final points A and B, then

$$\int_C \mathbf{F} \cdot d\mathbf{R} = f(B) - f(A). \tag{5}$$

To prove this, suppose that C is smooth with parametric equations $x = x(t)$ and $y = y(t)$, $a \le t \le b$. Then

$$\int_C \mathbf{F} \cdot d\mathbf{R} = \int_C \nabla f \cdot d\mathbf{R} = \int_a^b \left[\nabla f \cdot \frac{d\mathbf{R}}{dt} \right] dt$$

$$= \int_a^b \left[\frac{\partial f}{\partial x} \frac{dx}{dt} + \frac{\partial f}{\partial y} \frac{dy}{dt} \right] dt$$

$$= \int_a^b \frac{d}{dt} f[x(t), y(t)] \, dt$$

$$= f[x(b), y(b)] - f[x(a), y(a)]$$

$$= f(B) - f(A).$$

The crucial steps here depend on the multivariable chain rule (Section 19.6) and the one-variable Fundamental Theorem of Calculus. The argument for piecewise smooth curves now follows at once by applying (5) to each smooth piece separately, adding, and canceling the function values at the corners.

This theorem has several layers of meaning. We begin by illustrating its usefulness for evaluating line integrals.

Example 1 Compute the line integral of the vector field $\mathbf{F} = y \cos xy \, \mathbf{i} + x \cos xy \, \mathbf{j}$ along the parabolic path $y = x^2$ from $(0, 0)$ to $(1, 1)$.

Solution For this path it is natural to use x as the parameter, where x varies from 0 to 1. Since $dy = 2x \, dx$, we have

$$\mathbf{F} \cdot d\mathbf{R} = y \cos xy \, dx + x \cos xy \, dy$$

$$= x^2 \cos x^3 \, dx + x \cos x^3 \, 2x \, dx$$

$$= 3x^2 \cos x^3 \, dx,$$

so

$$\int_C \mathbf{F} \cdot d\mathbf{R} = \int_0^1 3x^2 \cos x^3 \, dx = \sin x^3 \Big]_0^1 = \sin 1 - \sin 0 = \sin 1.$$

This straightforward calculation of the line integral is easy to carry out, but a much easier method is now available. The first step is to notice that the vector field $\mathbf{F}$ is the gradient of the scalar field $f(x, y) = \sin xy$. (Students should verify this.) With this fact in hand, all that remains is to apply formula (5):

$$\int_C \mathbf{F} \cdot d\mathbf{R} = \sin xy \Big]_{(0,0)}^{(1,1)} = \sin 1 - \sin 0 = \sin 1.$$

The great advantage of this method is that no attention at all needs to be paid to the actual path of integration from the first point to the second.

As this example shows, the Fundamental Theorem can sometimes be used in the practical task of evaluating line integrals. Nevertheless, its main importance is theoretical. First, we point out that the right side of (5) depends only on the points A and B, and not at all on the path C that joins them. The line integral on the left side of (5) therefore has the same value for all paths C from A to B. This can be expressed by saying that *the line integral of a gradient field is independent of the path.* Next, it is clear from formula (5) that if C is a closed path, so that the final point B is the same as the initial point A, then $f(B) - f(A) = 0$ and therefore the line integral is zero. That is, *if **F** is a gradient field, then*

$$\oint_C \mathbf{F} \cdot d\mathbf{R} = 0$$

for every closed path C.

These arguments show that

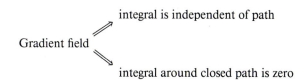

(The symbol $\Rightarrow$ means "implies.") Actually, these three properties are equivalent, in the sense that each implies the other two.

To begin the demonstration of equivalence, suppose that the line integral of the vector field $\mathbf{F}$ is independent of the path. We shall prove that the integral of $\mathbf{F}$ around a closed path is zero. To see why this is so, we examine Fig. 21.10, in which two points A and B are chosen on the closed path C. These points divide C into paths C_1 from A to B and C_2 from B to A. Since C_1 and $-C_2$ are both paths from A to B, the assumption of independence of path implies that

$$\int_{C_1} \mathbf{F} \cdot d\mathbf{R} = \int_{-C_2} \mathbf{F} \cdot d\mathbf{R} = -\int_{C_2} \mathbf{F} \cdot d\mathbf{R}.$$

It follows from this that

$$\oint_C \mathbf{F} \cdot d\mathbf{R} = \int_{C_1} \mathbf{F} \cdot d\mathbf{R} + \int_{C_2} \mathbf{F} \cdot d\mathbf{R} = 0,$$

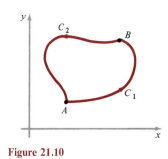

Figure 21.10

as asserted. Conversely, if we assume that the integral around every closed path is zero, then we can easily reverse this argument to show that the integral from A to B is independent of the path.

To complete the proof of the equivalence of the three properties, it suffices to show that if $\mathbf{F}$ is a vector field whose line integral is independent of path, then $\mathbf{F} = \nabla f$ for some scalar field f. To do this, we choose a fixed point (x_0, y_0) in the region under discussion and let (x, y) be an arbitrary point in this region. Given any path C from (x_0, y_0) to (x, y) [we assume there is such a path], we *define* the function $f(x, y)$ by means of the formula

$$f(x, y) = \int_C \mathbf{F} \cdot d\mathbf{R} = \int_{(x_0, y_0)}^{(x, y)} \mathbf{F} \cdot d\mathbf{R}.$$

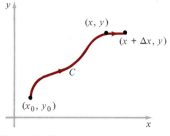

Figure 21.11

See Fig. 21.11. Because of the hypothesis of independence of path, the value of this integral depends only on the point (x, y) and not on the path C, and therefore provides an unambiguous definition for $f(x, y)$. To verify that $\nabla f = \mathbf{F}$, we suppose that the vector field $\mathbf{F}$ has the usual form, $\mathbf{F} = M(x, y)\mathbf{i} + N(x, y)\mathbf{j}$, so that

$$f(x, y) = \int_{(x_0, y_0)}^{(x, y)} M\, dx + N\, dy.$$

To show that $\partial f/\partial x = M$, we hold y fixed and move along the straight path from (x, y) to $(x + \Delta x, y)$, as shown in the figure. Since $dy = 0$ on this short path increment, we have

$$f(x + \Delta x, y) - f(x, y) = \int_x^{x + \Delta x} M\, dx,$$

so

$$\frac{\partial f}{\partial x} = \lim_{\Delta x \to 0} \frac{f(x + \Delta x, y) - f(x, y)}{\Delta x} = \lim_{\Delta x \to 0} \frac{1}{\Delta x} \int_x^{x + \Delta x} M\, dx = M,$$

by the Fundamental Theorem of Calculus. Similarly, $\partial f/\partial y = N$, so $\nabla f = \mathbf{F}$ and the argument is complete.

    As we suggested earlier, the main significance of these ideas lies in their applications to physics. In order to understand what is involved, let us suppose that $\mathbf{F}$ is a force field and that a particle of mass m is moved by this force along a curved path C from a point A to a point B. Let the path be parametrized by the time t, with parametric equations $x = x(t)$ and $y = y(t)$, $t_1 \leq t \leq t_2$. Then the work done by $\mathbf{F}$ in moving the particle along this path is

$$W = \int_C \mathbf{F} \cdot d\mathbf{R} = \int_{t_1}^{t_2} \left[\mathbf{F} \cdot \frac{d\mathbf{R}}{dt} \right] dt. \tag{6}$$

According to Newton's second law of motion we have

$$\mathbf{F} = m \frac{d\mathbf{v}}{dt},$$

where $\mathbf{v} = d\mathbf{R}/dt$ is the velocity. If v denotes the speed, so that $v = |\mathbf{v}|$, then we can write the integrand in (6) as

$$\mathbf{F} \cdot \frac{d\mathbf{R}}{dt} = m \frac{d\mathbf{v}}{dt} \cdot \mathbf{v} = \frac{1}{2} m \frac{d}{dt} (\mathbf{v} \cdot \mathbf{v}) = \frac{1}{2} m \frac{d}{dt} (v^2).*$$

Therefore (6) becomes

$$W = \frac{1}{2} m \int_{t_1}^{t_2} \frac{d}{dt} (v^2)\, dt = \frac{1}{2} mv^2 \Big]_{t_1}^{t_2} = \frac{1}{2} mv_B^2 - \frac{1}{2} mv_A^2, \tag{7}$$

---

* In this calculation we use the product rule for the derivative of the dot product of two vector functions of t. This is easy to prove from formula (8) in Section 18.2.

where v_A and v_B are the initial and final speeds, that is, the speeds at the points A and B. Since $\frac{1}{2}mv^2$ is the kinetic energy of the particle, (7) says that *the work done equals the change in kinetic energy.* (A similar discussion for the case of linear motion is given in Section 7.8.)

We continue this line of thought to its natural conclusion. The force field **F** is called *conservative* if it is the gradient of a scalar field. For reasons that will appear in a moment, it is customary in this context to introduce a minus sign and write $\mathbf{F} = -\nabla V$, so that V increases most rapidly in the direction opposite to **F**. The function $V(x, y)$ is then called the *potential energy.* This function is just the negative of what we have been denoting by f. It exists if and only if **F** is a gradient field, and when it does, the Fundamental Theorem (5) tells us that

$$W = \int_C \mathbf{F} \cdot d\mathbf{R} = -\int_C \nabla V \cdot dR = -[V(B) - V(A)] = V(A) - V(B), \qquad (8)$$

where A and B are the initial and final points of the arbitrary path C. If we now equate (7) and (8), we get

$$V(A) - V(B) = \frac{1}{2}mv_B^2 - \frac{1}{2}mv_A^2$$

or

$$\frac{1}{2}mv_A^2 + V(A) = \frac{1}{2}mv_B^2 + V(B). \qquad (9)$$

(We now see that the minus sign is introduced into the definition of potential energy in order to make the signs here come out right.) Equation (9) says that the sum of the kinetic energy and the potential energy is the same at the initial point as it is at the final point. Since these points are arbitrary, *the total energy is constant.* This is the *law of conservation of energy,* which is one of the basic principles of classical physics. This law is true in any conservative force field, such as the earth's gravitational field or the electric field produced by any distribution of electric charge.

Our work has demonstrated that a force field is conservative if and only if it satisfies any one of the following equivalent conditions:

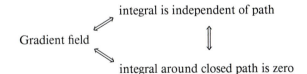

The importance of these fields justifies turning our attention to the practical problem of determining whether a given force field is or is not conservative. Since any vector field can be thought of as a force field, these remarks apply to vector fields in general.

Example 2 Show that the vector field $\mathbf{F} = xy\mathbf{i} + xy^2\mathbf{j}$ is not conservative.

Solution One way of doing this is to show that $\int_C \mathbf{F} \cdot d\mathbf{R}$ *does* depend on the path. We choose two convenient points, say $(0, 0)$ and $(1, 1)$, and integrate from the first to the second along any two convenient different paths, as

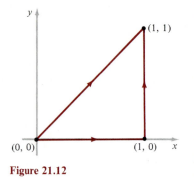

Figure 21.12

shown in Fig. 21.12. First, along the line $y = x$ we have

$$\int_C \mathbf{F} \cdot d\mathbf{R} = \int_C xy \, dx + xy^2 \, dy = \int_0^1 (x^2 + x^3) \, dx = \frac{1}{3} + \frac{1}{4} = \frac{7}{12}.$$

On the other hand, along the broken line from $(0, 0)$ to $(1, 0)$ to $(1, 1)$ we have

$$\int_C \mathbf{F} \cdot d\mathbf{R} = \int_0^1 0 \cdot dx + \int_0^1 y^2 \, dy = \frac{1}{3}.$$

Since the values of these integrals are not equal, the field is not conservative. In probing for unequal line integrals in this way, we are perfectly free to choose paths that make the calculations easy. (Of course, if these two line integrals had turned out to be equal, this would not have precluded unequal results for two other line integrals, so nothing would have been proved one way or the other.)

Another method is to assume that the field *is* conservative, so that $\mathbf{F} = \nabla f$ for some function $f(x, y)$, and to deduce a contradiction from this assumption. The assumption means that there exists a function f such that $\partial f / \partial x = xy$ and $\partial f / \partial y = xy^2$. But this is impossible, because the mixed partial derivatives would then be

$$\frac{\partial^2 f}{\partial y \, \partial x} = x \qquad \text{and} \qquad \frac{\partial^2 f}{\partial x \, \partial y} = y^2,$$

which are obviously not equal, whereas the theory of partial derivatives tells us that these derivatives must be equal. This contradiction tells us that no f exists, so $\mathbf{F}$ is not conservative.

The reasoning used in the second method of this example depends on the equality of mixed partial derivatives,

$$\frac{\partial^2 f}{\partial y \, \partial x} = \frac{\partial^2 f}{\partial x \, \partial y}, \tag{10}$$

which is valid in any region where both derivatives are continuous (Section 19.2). This reasoning can be extended as follows: if

$$\mathbf{F} = M(x, y)\mathbf{i} + N(x, y)\mathbf{j} \tag{11}$$

is a conservative vector field, so that an f exists with the property that $\nabla f = \mathbf{F}$ or

$$\frac{\partial f}{\partial x} = M \qquad \text{and} \qquad \frac{\partial f}{\partial y} = N,$$

then by (10) we know that

$$\frac{\partial M}{\partial y} = \frac{\partial N}{\partial x}. \tag{12}$$

Condition (12) is therefore necessary for a vector field to be conservative, and we have seen how this fact can be used. But is it also sufficient? That is, does (12) guarantee that (11) is conservative? We investigate this question in Section 21.3.

PROBLEMS

In Problems 1 to 4, use both methods of Example 2 to show that the vector field is not conservative.

1 $\mathbf{F} = y\mathbf{i} - x\mathbf{j}$.

2 $\mathbf{F} = x(y-1)\mathbf{i} + x\mathbf{j}$.

3 $\mathbf{F} = x^3y\mathbf{i} + xy^2\mathbf{j}$.

4 $\mathbf{F} = \dfrac{y\mathbf{i} + y\mathbf{j}}{x^2 + y^2}$.

In Problems 5 to 8, show that the given line integrals are not independent of path by integrating along two different paths from $(0, 0)$ to $(1, 1)$.

5 $\displaystyle\int_C 2xy\,dx + (y^2 - x^2)\,dy$.

6 $\displaystyle\int_C 2xy\,dx + (y - x^2)\,dy$.

7 $\displaystyle\int_C (x^2 - y^3)\,dx + 3xy^2\,dy$.

8 $\displaystyle\int_C (x - y)\,dx + (x + y)\,dy$.

9 Show that

$$\int_{(-2,1)}^{(1,4)} 2xy\,dx + x^2\,dy$$

is independent of the path, and evaluate the integral by
(a) using formula (5);
(b) integrating along any convenient path.

10 Show that

$$\int_{(-1,0)}^{(1,\pi)} \sin y\,dx + x\cos y\,dy$$

is independent of the path, and evaluate the integral by
(a) using formula (5);
(b) integrating along any convenient path.

In Problems 11 to 16, show that the integral is independent of the path and use any method to evaluate it.

11 $\displaystyle\int_{(-2,-1)}^{(1,5)} 2y\,dx + 2x\,dy$.

12 $\displaystyle\int_{(0,0)}^{(4,5)} y^2 e^x\,dx + 2ye^x\,dy$.

13 $\displaystyle\int_{(0,0)}^{(\pi/2,1)} e^y\cos x\,dx + e^y\sin x\,dy$.

14 $\displaystyle\int_{(-1,1)}^{(2,3)} 3x^2y^2\,dx + 2x^3y\,dy$.

15 $\displaystyle\int_{(-2,1)}^{(4,1)} 2xy\,dx + (x^2 + y^2)\,dy$.

16 $\displaystyle\int_{(0,0)}^{(1,1)} (x + y)\,dx + x\,dy$.

17 Suppose that a particle of mass m moves in the xy-plane under the influence of the constant gravitational force $\mathbf{F} = -mg\mathbf{j}$. If the particle moves from (x_1, y_1) to (x_2, y_2) along a path C, show that the work done by $\mathbf{F}$ is

$$W = mg(y_1 - y_2),$$

regardless of the path.

As we said at the beginning of this chapter, Green's Theorem establishes an important link between line integrals and double integrals. Our purpose in this section is to reveal the nature of this link.

Consider a vector field

$$\mathbf{F} = M(x, y)\mathbf{i} + N(x, y)\mathbf{j} \tag{1}$$

defined on a certain region in the xy-plane. We now take up the question of whether the condition

$$\frac{\partial M}{\partial y} = \frac{\partial N}{\partial x} \tag{2}$$

is sufficient to guarantee that $\mathbf{F}$ is conservative, that is, that $\mathbf{F}$ is the gradient of some scalar field f. In the light of what we have learned in Section 21.2, this is equivalent to asking whether condition (2) implies that the integral of $\mathbf{F}$ around every closed path is zero. We shall use our investigation of this

21.3

GREEN'S THEOREM

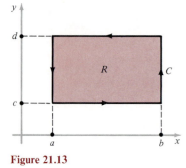

Figure 21.13

question as a means of discovering Green's Theorem, which we will then prove and apply in various ways.

The simplest type of closed path C is a rectangular path like the one shown in Fig. 21.13. We shall calculate the integral of $\mathbf{F}$ around this path and see what is needed to make its value zero. Integrating counterclockwise as shown, and beginning with the path segment on the lower edge of the rectangular region R, we have

$$\oint_C \mathbf{F} \cdot d\mathbf{R} = \oint_C M(x, y)\, dx + N(x, y)\, dy$$

$$= \int_a^b M(x, c)\, dx + \int_c^d N(b, y)\, dy + \int_b^a M(x, d)\, dx + \int_d^c N(a, y)\, dy$$

$$= \int_c^d [N(b, y) - N(a, y)]\, dy - \int_a^b [M(x, d) - M(x, c)]\, dx. \tag{3}$$

We next make an ingenious application of the Fundamental Theorem of Calculus to write these two integrands as

$$N(b, y) - N(a, y) = N(x, y)\Big]_{x=a}^{x=b} = \int_a^b \frac{\partial N}{\partial x}\, dx$$

and

$$M(x, d) - M(x, c) = M(x, y)\Big]_{y=c}^{y=d} = \int_c^d \frac{\partial M}{\partial y}\, dy.$$

This enables us to write (3) in the form

$$\oint_C \mathbf{F} \cdot d\mathbf{R} = \oint_C M\, dx + N\, dy$$

$$= \int_c^d \int_a^b \frac{\partial N}{\partial x}\, dx\, dy - \int_a^b \int_c^d \frac{\partial M}{\partial y}\, dy\, dx.$$

These iterated integrals can be written as double integrals over the region R enclosed by C, so we have

$$\oint_C \mathbf{F} \cdot d\mathbf{R} = \oint_C M\, dx + N\, dy$$

$$= \iint_R \frac{\partial N}{\partial x}\, dA - \iint_R \frac{\partial M}{\partial y}\, dA = \iint_R \left[\frac{\partial N}{\partial x} - \frac{\partial M}{\partial y} \right] dA. \tag{4}$$

Now we can see what is happening. Condition (2) implies that this double integral is zero, so $\oint_C \mathbf{F} \cdot d\mathbf{R} = 0$. It is tempting to infer from this that condition (2) implies $\mathbf{F}$ is conservative. However, this inference requires that $\oint_C \mathbf{F} \cdot d\mathbf{R} = 0$ for *every* closed path C, and we have demonstrated this only for rectangular paths like the one in Fig. 21.13.

If we pluck out the essence of this argument, we see that it lies in equation (4), which we can write in the form

$$\oint_C M\, dx + N\, dy = \iint_R \left[\frac{\partial N}{\partial x} - \frac{\partial M}{\partial y} \right] dA. \tag{5}$$

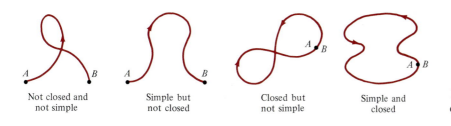

Figure 21.14 Various types of curves.

Not closed and
not simple

Simple but
not closed

Closed but
not simple

Simple and
closed

This statement, that a line integral around a closed curve equals a certain double integral over the region inside the curve, is called *Green's Theorem,* after the English mathematical physicist George Green.*

Strictly speaking, Green's Theorem is not merely equation (5), but rather a fairly careful statement of conditions under which (5) is valid. To state such conditions, it is necessary to introduce the concept of a simple closed curve. We already know that a closed curve is one for which the final point B is the same as the initial point A. A plane curve is said to be *simple* if it does not intersect itself anywhere between its endpoints (Fig. 21.14). Unless the contrary is explicitly stated, we assume that simple closed curves are *positively* oriented, which means that they are traversed in such a way that their interiors are always on the left, as shown on the right in the figure.

Green's Theorem can now be stated as follows:

If C is a piecewise smooth, simple closed curve that bounds a region R, and if M(x, y) and N(x, y) are continuous and have continuous partial derivatives along C and throughout R, then

$$\oint_C M \, dx + N \, dy = \iint_R \left[\frac{\partial N}{\partial x} - \frac{\partial M}{\partial y} \right] dA. \tag{5}$$

We have proved (5) only for rectangular regions R of the kind shown in Fig. 21.13. We now give a similar argument for the case in which R is both vertically simple and horizontally simple, in the sense described in Section 20.2. Then we shall indicate how to extend the theorem to more general regions.

Since R is assumed to be vertically simple (Fig. 21.15), its boundary C can be thought of as consisting of a lower curve $y = y_1(x)$ and an upper curve $y = y_2(x)$, possibly separated by vertical segments on the sides. The integral $\int M(x, y) \, dx$ over any part of C that consists of vertical segments is zero, since $dx = 0$ on such a segment. We therefore have

$$\oint_C M(x, y) \, dx = \int_a^b M[x, y_1(x)] \, dx + \int_b^a M[x, y_2(x)] \, dx, \tag{6}$$

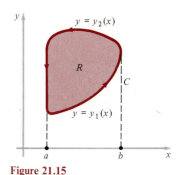

Figure 21.15

---

* Green (1793–1841) was obliged to leave school at an early age to work in his father's bakery, and consequently had little formal education. By assiduous study in his spare time, he taught himself mathematics and physics from library books, particularly Laplace's *Mécanique Céleste.* In 1828 he published locally at his own expense his most important work, *Essay on the Application of Mathematical Analysis to the Theories of Electricity and Magnetism.* Although Green's Theorem (in an equivalent form) appeared in this pamphlet, little notice was taken until the pamphlet was republished in 1846, five years after his death, and thereby came to the attention of other scientists who had the knowledge to appreciate its merits.

where the lower curve is traced from left to right and the upper curve from right to left. By the Fundamental Theorem of Calculus, (6) can be written as

$$\oint_C M\,dx = \int_a^b \{M[x, y_1(x)] - M[x, y_2(x)]\}\,dx$$

$$= \int_a^b \left[-M(x, y)\right]_{y=y_1(x)}^{y=y_2(x)}\,dx$$

$$= \int_a^b \int_{y_1(x)}^{y_2(x)} -\frac{\partial M}{\partial y}\,dy\,dx = \iint_R -\frac{\partial M}{\partial y}\,dA. \tag{7}$$

But R is also assumed to be horizontally simple, and a similar argument, which we ask students to give in Problem 22, shows that

$$\oint_C N\,dy = \iint_R \frac{\partial N}{\partial x}\,dA. \tag{8}$$

We now obtain Green's Theorem (5) for the region R by adding (7) and (8).

A complete and rigorous proof of Green's Theorem is beyond the scope of this book. Nevertheless, it is quite easy to extend the argument to cover any region R that can be subdivided into a finite number of regions R_1, $R_2, \ldots, R_n$ that are both vertically and horizontally simple. The validity of Green's Theorem for R then follows from its validity for each of the regions $R_1, R_2, \ldots, R_n$.

For example, the region R in Fig. 21.16 can be subdivided into the regions R_1 and R_2 by introducing the indicated cut, which becomes part of the boundary of R_1 when traced from right to left (C_3), and part of the boundary of R_2 when traced from left to right (C_4). By applying Green's Theorem separately to R_1 and R_2, we get

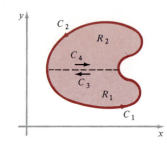

Figure 21.16

$$\oint_{C_1+C_3} M\,dx + N\,dy = \iint_{R_1} \left[\frac{\partial N}{\partial x} - \frac{\partial M}{\partial y}\right]\,dA$$

and

$$\oint_{C_2+C_4} M\,dx + N\,dy = \iint_{R_2} \left[\frac{\partial N}{\partial x} - \frac{\partial M}{\partial y}\right]\,dA.$$

If we add these two equations the result is

$$\oint_{C_1+C_2} M\,dx + N\,dy = \iint_R \left[\frac{\partial N}{\partial x} - \frac{\partial M}{\partial y}\right]\,dA,$$

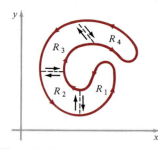

Figure 21.17

which is Green's Theorem for the region R. This occurs because the two line integrals along C_3 and C_4 cancel each other, since C_3 and C_4 are the same curve traced in opposite directions. Similarly, Green's Theorem can be extended to the region in Fig. 21.17 by subdividing it into the four simpler regions shown in the figure.

Example 1   Evaluate the line integral

$$I = \oint_C (3x - y)\, dx + (x + 5y)\, dy$$

around the unit circle $x = \cos t,\ y = \sin t,\ 0 \le t \le 2\pi$.

Solution   The straightforward calculation of this integral gives

$$I = \int_0^{2\pi} [(3\cos t - \sin t)(-\sin t) + (\cos t + 5\sin t)(\cos t)]\, dt$$

$$= \int_0^{2\pi} [2\sin t \cos t + 1]\, dt = \left[\sin^2 t + t\right]_0^{2\pi} = 2\pi.$$

This is easy enough, but Green's Theorem makes it even easier. Since $M = 3x - y$ and $N = x + 5y$, we have

$$\frac{\partial M}{\partial y} = -1 \quad \text{and} \quad \frac{\partial N}{\partial x} = 1,$$

so

$$I = \iint_R [1 - (-1)]\, dA$$

$$= 2 \iint_R dA = 2(\text{area of circle}) = 2\pi.$$

Example 2   Evaluate the line integral

$$I = \oint_C (2y + \sqrt{1 + x^5})\, dx + (5x - e^{y^2})\, dy$$

around the circle $x^2 + y^2 = 4$.

Solution   The actual calculation of this integral looks like a very forbidding task, but Green's Theorem provides another way. Since $M = 2y + \sqrt{1 + x^5}$ and $N = 5x - e^{y^2}$,

$$\frac{\partial M}{\partial y} = 2 \quad \text{and} \quad \frac{\partial N}{\partial x} = 5.$$

Therefore

$$I = \iint_R (5 - 2)\, dA = 3 \iint_R dA = 3(\text{area of circle}) = 3(4\pi) = 12\pi,$$

since R is a circular disk of radius 2.

Example 3   If R is any region to which Green's Theorem is applicable, show that the area A of R is given by the formula

$$A = \frac{1}{2} \oint_C -y\, dx + x\, dy. \tag{9}$$

Solution Since $M = -y$ and $N = x$, and therefore

$$\frac{\partial M}{\partial y} = -1 \quad \text{and} \quad \frac{\partial N}{\partial x} = 1,$$

Green's Theorem yields

$$\oint_C -y\, dx + x\, dy = \iint_R [1 - (-1)]\, dA = 2 \iint_R dA = 2A,$$

as stated.

Example 4 Use formula (9) to find the area bounded by the ellipse $x^2/a^2 + y^2/b^2 = 1$.

Solution We can parametrize the ellipse by $x = a \cos t$, $y = b \sin t$, where $0 \le t \le 2\pi$. Then formula (9) yields

$$A = \frac{1}{2} \int_0^{2\pi} [(-b \sin t)(-a \sin t) + (a \cos t)(b \cos t)]\, dt$$

$$= \frac{1}{2} \int_0^{2\pi} ab\, dt = \pi ab.$$

Our original problem in this section was to determine whether the condition

$$\frac{\partial M}{\partial y} = \frac{\partial N}{\partial x} \tag{2}$$

is sufficient to guarantee that the vector field

$$F = M(x, y)\mathbf{i} + N(x, y)\mathbf{j} \tag{1}$$

is conservative. Green's Theorem provides the solution. For if C is any simple closed path in the domain of $\mathbf{F}$, and if the region enclosed by C is also in the domain, then Green's Theorem tells us that

$$\oint_C \mathbf{F} \cdot d\mathbf{R} = \oint_C M\, dx + N\, dy = \iint_R \left(\frac{\partial N}{\partial x} - \frac{\partial M}{\partial y} \right) dA.$$

By using this equation we see that if $\partial M/\partial y = \partial N/\partial x$ then the double integral is zero, and therefore the line integral is zero. If the line integral is zero around every simple closed path, then it is also zero around every closed path, and this proves that $\mathbf{F}$ is conservative. We emphasize that for this reasoning to work, the region enclosed by C must lie entirely in the domain of $\mathbf{F}$. A convenient way to guarantee this is to require that the domain of $\mathbf{F}$ must be *simply connected,* which means that the inside of every simple closed path in the domain also lies in the domain. Roughly speaking, the domain of $\mathbf{F}$ is not allowed to have any holes. In Fig. 21.18 we show regions with one, two, and three holes, respectively; the points inside the inner curves do not belong to the regions R, so these regions are not simply connected. Our

Figure 21.18 Regions not simply connected.

overall conclusion can be stated as follows:

If the domain of definition of the vector field $\mathbf{F} = M(x, y)\mathbf{i} + N(x, y)\mathbf{j}$ *is simply connected, then* $\mathbf{F}$ *is conservative if and only if the condition* $\partial M/\partial y = \partial N/\partial x$ *is satisfied.*

One final question: If a given vector field $\mathbf{F}$ is known to be conservative, so that $\mathbf{F} = \nabla f$ for some function $f(x, y)$, how do we find f? Such a function is called a *potential function,* or simply a *potential,* for $\mathbf{F}$.* One way is by inspection, but this only works in simple cases. A more systematic method is illustrated in the following example. As the student will see, it amounts to integrating the equations

$$\frac{\partial f}{\partial x} = M, \qquad \frac{\partial f}{\partial y} = N,$$

and condition (2) guarantees that this can be done.

Example 5 Find a potential f for the vector field

$$\mathbf{F} = (y^2 + 1)\mathbf{i} + 2xy\mathbf{j}.$$

Solution Here we have $M = y^2 + 1$ and $N = 2xy$. It is easy to verify that $\partial M/\partial y = \partial N/\partial x$, and therefore f exists and our only problem is to find it. We know that

$$\frac{\partial f}{\partial x} = y^2 + 1 \qquad \text{and} \qquad \frac{\partial f}{\partial y} = 2xy. \tag{10}$$

In computing $\partial f/\partial x$, we differentiate with respect to x while holding y constant, so by integrating the first of equations (10) with respect to x, we obtain $f = xy^2 + x + g(y)$, where $g(y)$ is a function of y that is yet to be determined. By differentiating with respect to y, we see that $\partial f/\partial y = 2xy + g'(y)$, and by comparing this with the second of equations (10) we conclude that $g'(y) = 0$. It follows that $g(y)$ is a constant C that can be chosen arbitrarily, and therefore the potential we are seeking is $f(x, y) = xy^2 + x + C$. It is easy to check this result by verifying that $\nabla f = \mathbf{F}$.

---

* Recall that for physical reasons the *potential energy* associated with a force field $\mathbf{F}$ is any scalar function V (if one exists) such that $\mathbf{F} = -\nabla V$. The concepts of potential and potential energy are closely related but not identical.

PROBLEMS

In Problems 1 to 4, evaluate the line integrals directly, and also by Green's Theorem.

1 $\oint_C (xy - y^2)\, dx + xy^2\, dy$, where C is the simple closed path formed by $y = 0$, $x = 1$, $y = x$.

2 $\oint_C x\, dx + xy^2\, dy$, where C is the simple closed path formed by $y = x^2$ and $y = x$.

3 $\oint_C 1/y\, dx + 1/x\, dy$, where C is the simple closed path formed by $y = x$, $y = 4$, $x = 1$.

4 $\oint_C y^2\, dx + x^2\, dy$, where C is the simple closed path formed by $y = 0$, $x = 1$, $y = 1$, $x = 0$.

In Problems 5 to 12, use Green's Theorem to compute the given line integrals.

5 $\oint_C xy\, dx + (x + y)\, dy$, where C is the closed path (obviously simple) formed by $y = 0$, $x = 0$, $y = 1$, $x = -1$.

6 $\oint_C -xy/(1 + x)\, dx + \ln\,(1 + x)\, dy$, where C is the closed path formed by $y = 0$, $x + 2y = 4$, $x = 0$.

7 $\oint_C -x^2 y/(1 + x^2)\, dx + \tan^{-1} x\, dy$, where C is the closed path formed by $y = 0$, $x = 1$, $y = 1$, $x = 0$.

8 $\oint_C x\, dx + xy\, dy$, where C is the closed path formed by $y = 0$, $x^2 + y^2 = 1$ $(x, y \geq 0)$, $x = 0$.

9 $\oint_C (e^{x^3} + y^2)\, dx + (x + \sqrt{1 + y^7})\, dy$, where C is the closed path formed by $y = 0$, $x = 1$, $y = x$.

10 $\oint_C -y^3\, dx + x^3\, dy$, where C is the closed path formed by $y = x^3$ and $y = x$.

11 $\oint_C (-y^2 + \tan^{-1} x)\, dx + \ln y\, dy$, where C is the closed path formed by $y = x^2$ and $x = y^2$.

12 $\oint_C (x^2 - y)\, dx + x\, dy$, where C is the circle $x^2 + y^2 = 9$.

In Problems 13 to 20, use formula (9) to find the area bounded by the given curves.

13 $y = 3x$ and $y^2 = 9x$.

14 $y = 0$, $x + y = a\ (a > 0)$, $x = 0$.

15   The x-axis and one arch of the cycloid $x = a(\theta - \sin\theta)$, $y = a(1 - \cos\theta)$.

16 $y = x^2$ and $x = y^3$.

17 $x = a\cos^3\theta$ and $y = a\sin^3\theta$, $0 \leq \theta \leq 2\pi$ (an astroid or hypocycloid of four cusps).

18 $y = x^2$ and $x = y^2$.

19   The x-axis and one arch of $y = \sin x$.

20 $9y = x$, $xy = 1$, $y = x$.

21   The loop of the folium of Descartes (with Cartesian equation $x^3 + y^3 = 3axy$) is shown in Fig. 17.11. In Problem 16 of Section 17.1 we asked students to introduce the parameter $t = y/x$ and obtain the parametric equations

$$x = \frac{3at}{1 + t^3}, \qquad y = \frac{3at^2}{1 + t^3}.$$

Use formula (9) to find the area of the loop. Hint: The part of the loop below the line $y = x$ is traced out as t increases from 0 to 1.

22   Give the details of the argument establishing formula (8) for the case in which R is horizontally simple.

In Problems 23 to 28, verify that the given vector field is conservative and find a potential for it.

23 $\mathbf{F} = y^3\mathbf{i} + 3xy^2\mathbf{j}$.

24 $\mathbf{F} = e^y\cos x\,\mathbf{i} + e^y\sin x\,\mathbf{j}$.

25 $\mathbf{F} = (ye^{xy} - 2x)\mathbf{i} + (xe^{xy} + 2y)\mathbf{j}$.

26 $\mathbf{F} = y\cos xy\,\mathbf{i} + x\cos xy\,\mathbf{j}$.

27 $\mathbf{F} = (\sin y - y\sin x)\mathbf{i} + (x\cos y + \cos x)\mathbf{j}$.

28 $\mathbf{F} = x\mathbf{i} + y\mathbf{j}$.

29   Let C_1, C_2, and C_3 be the simple closed curves shown in Fig. 21.19, and let R be the region inside C_1 and outside C_2 and C_3. Assume that $M(x, y)$ and $N(x, y)$ are continuous and have continuous partial derivatives in R and along all the curves. Show that Green's Theorem

$$\oint_C M\, dx + N\, dy = \iint_R \left[\frac{\partial N}{\partial x} - \frac{\partial M}{\partial y} \right] dA$$

remains valid in this case, provided that C is understood to be the *total* boundary of R, consisting of C_1, C_2, and C_3 positively oriented as shown in the figure. (The curves C_2 and C_3 are oriented clockwise, but nevertheless the orientation is positive because they are traversed in such a way that the region R remains on the left.)

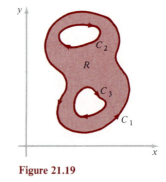

Figure 21.19

30   Can Green's Theorem be used to evaluate the line integral

$$\oint_C \frac{-y}{x^2 + y^2}\, dx + \frac{x}{x^2 + y^2}\, dy,$$

(a) where C is the circle $x^2 + y^2 = 1$?

(b) where C is the triangle with vertices $(1, 0)$, $(1, 2)$, $(2, 2)$?

31 If C_1 is the circle $x^2 + y^2 = 1$ and C_2 is an arbitrary simple closed path containing C_1, as shown in Fig. 21.20, use the idea of Problem 29 to show that

$$\oint_{C_2} \frac{-y}{x^2 + y^2} \, dx + \frac{x}{x^2 + y^2} \, dy$$
$$= \oint_{C_1} \frac{-y}{x^2 + y^2} \, dx + \frac{x}{x^2 + y^2} \, dy,$$

and evaluate the integral on the left by calculating the integral on the right.

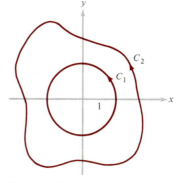

Figure 21.20

This brings us to the end of the present book. But every end is also a beginning, and so it is here.

For many students who have reached this stage, this is enough as far as mathematics is concerned, and they will turn aside to their own preferred disciplines. Most serious students of physical science need to know a good deal more about mathematics than this book contains. And serious students of mathematics itself also need to know something about totally different kinds of mathematics, for example, number theory and abstract algebra.

To all of these groups we wish to convey an impression of the nature of what lies ahead. This is the purpose of Fig. 21.21.

Our work in this chapter has been concerned with vector analysis in the plane. All of this material extends to three-dimensional space, with Green's

21.4
WHAT NEXT?

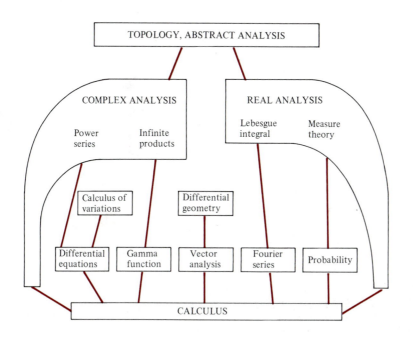

Figure 21.21 A schematic picture of mathematical analysis.

Theorem extending to Stokes' Theorem.* These concepts and tools have rich applications to mathematical physics. They also lead to the Gaussian differential geometry of curved surfaces, and on from there to Riemannian geometry.

One of the main recurring topics of this book has been differential equations. At a higher level this topic broadens into one of the major branches of classical mathematics, with profound applications to science and many far-reaching implications for mathematics itself.

We shall not attempt to describe any of the other items shown in the figure. It suffices to say that these branches of mathematics are interrelated in many ways that we have not tried to indicate, and also that any student who pushes on in these directions will encounter wonders beyond imagining.

---

* For the convenience of those teachers who wish to pursue these themes a little farther, brief expositions of the divergence theorem and Stokes' Theorem (with homework problems) are given in Appendixes A.22 and A.23.

APPENDIX A

A VARIETY OF ADDITIONAL TOPICS

It is difficult to give an idea of the vast scope of modern mathematics. The word "scope" is not the best; I have in mind an expanse swarming with beautiful details, not the uniform expanse of a bare plain, but a region of a beautiful country, first seen from a distance, but worthy of being surveyed from one end to the other and studied even in its smallest details: its valleys, streams, rocks, woods and flowers.
Arthur Cayley

A.1

MORE ABOUT NUMBERS: IRRATIONALS, PERFECT NUMBERS, AND MERSENNE PRIMES

We referred several times in Section 1.2 to the fact that $\sqrt{2}$ is irrational. The proof which is traditionally ascribed to Pythagoras is one of the earliest intellectual productions of Western civilization that still retains its vigor and interest. It deserves to be included here, both for its own sake and as an introduction to problems of irrationality.

We begin by recalling that the *even* numbers are the integers $0, \pm 2, \pm 4, \ldots$ that can be written in the form $2n$ for some integer n, and the *odd* numbers are the integers $\pm 1, \pm 3, \pm 5, \ldots$ that can be written in the form $2n + 1$. It is easy to see that the square of an even number is even, since $(2n)^2 = 4n^2 = 2(2n^2)$, and the square of an odd number is odd, since $(2n + 1)^2 = 4n^2 + 4n + 1 = 2(2n^2 + 2n) + 1$. After these preliminaries, we are now ready to prove that $\sqrt{2}$ is not rational. Let us suppose that it is—contrary to what we want to establish—so that $\sqrt{2} = a/b$ for certain positive integers a and b. We may specify that a and b have no common factors > 1; for if they have, we can cancel them away until none remain. It will soon be clear that the possibility of making this specification without loss of generality is crucial for the proof. Now, the equation $\sqrt{2} = a/b$ implies that $2 = a^2/b^2$, so $a^2 = 2b^2$ and a^2 is even. This implies that a is also even; for if it were odd, then a^2 would be odd, which it is not. Since a is even, it has the form $a = 2c$ for some integer c, and therefore $4c^2 = 2b^2$ or $b^2 = 2c^2$, so b^2 is even. As before, this implies that b is even. But since a and b are both even, they have 2 as a common factor. This contradicts our earlier specification and shows that it cannot be true that $\sqrt{2}$ is rational. We are therefore driven to the desired conclusion, that $\sqrt{2}$ is irrational.

It is often quite difficult to determine whether a given specific number is rational or not. For instance, the fact that π is irrational was not discovered until 1761. We shall prove this later with the aid of some rather complicated reasoning depending on the calculus of the trigonometric functions. Unfortunately, no really simple proof is known even to this day.

The Pythagorean argument given here for $\sqrt{2}$ is essentially an argument from elementary number theory, for it depends only on relatively simple properties of positive integers. There are many interesting kinds of positive integers, with a great variety of remarkable properties that have fascinated curious people through the ages. We mention the prime numbers $2, 3, 5, 7, 11, 13, 17, 19, 23, \ldots$, the squares $1, 4, 9, 16, 25, \ldots$, and the perfect numbers $6, 28, \ldots$.*

The primes are the multiplicative building blocks of the positive integers, in the sense that every integer > 1 either is itself prime or is expressible as a product of primes. To see this, notice that if $n > 1$ is not a prime, then $n = ab$, where a and b are $< n$; if either a or b is not a prime, it can be factored similarly; and continuing in this way until all factors are not further factorable proves that n can be written as a product of primes. For example, $198 = 2 \cdot 99 = 2 \cdot 3 \cdot 33 = 2 \cdot 3 \cdot 3 \cdot 11$. A fundamental theorem of number theory (called the *unique factorization theorem*) states that this factorization is always unique except for the order of the factors. In particular, a prime factorization of 198 can never involve 5 as a factor, and the factor 2 can never appear more than once.

The unique factorization theorem is deeper than it looks, but to most people it is obviously true. The following facts are much more surprising, and therefore have greater appeal to the imagination.

(a)   The *four squares theorem*: Every positive integer can be expressed as the sum of not more than four squares.

(b)   The *two squares theorem*: Every prime number of the form $4n + 1$ can be expressed as the sum of two squares in one and only one way.

(c)   To formulate our next statement, we consider the geometric progression whose first term is 1 and whose ratio is any number $r \neq 1$:

$$1, r, r^2, \ldots, r^n, \ldots$$

We recall from elementary algebra that the sum of the first n terms of this progression is given by the formula†

$$1 + r + r^2 + \cdots + r^{n-1} = \frac{1 - r^n}{1 - r}. \tag{1}$$

In particular, if $r = 2$, this formula yields

$$1 + 2 + 2^2 + \cdots + 2^{n-1} = 2^n - 1.$$

---

* We remind the reader that a *prime number* is an integer $p > 1$ that has no positive factors (or divisors) except 1 and p; equivalently, it cannot be written in the form $p = ab$, where a and b are both positive integers $< p$.

A *perfect number* is a positive integer like $6 = 1 + 2 + 3$ that equals the sum of its positive divisors other than itself. Note also that $28 = 1 + 2 + 4 + 7 + 14$. The next perfect numbers after 6 and 28 are 496, 8128, and 33,550,336.

† To prove (1), denote the sum on the left by s,

$$s = 1 + r + r^2 + \cdots + r^{n-1},$$

multiply by r,

$$rs = r + r^2 + r^3 + \cdots + r^n,$$

and subtract, carrying out all possible cancellations, to obtain

$$s - rs = 1 - r^n \quad \text{or} \quad s(1 - r) = 1 - r^n.$$

Since $r \neq 1$, formula (1) follows at once from the last equation.

The *theorem of even perfect numbers* asserts the following: If the sum $2^n - 1$ is prime, then the product $2^{n-1}(2^n - 1)$ of the last term and the sum is an even perfect number; and conversely, every even perfect number is of this form where $2^n - 1$ is prime.

The first part of the theorem in (c) was proved by Euclid about 300 B.C., and the second by Euler in the mid-eighteenth century. These two statements and their proofs constitute a gem of classical number theory with ramifications that continued to attract the attention of serious mathematicians and computer technologists as late as 1979. The details are brief enough for us to present them here.

Let us first notice that for $n = 1, 2, 3, 4, 5, 6, 7$ the corresponding values of $2^n - 1$ are $1, 3, 7, 15, 31, 63, 127$. The only primes in this list are $3, 7, 31, 127$. According to the theorem, the first four even perfect numbers are therefore $2 \cdot 3 = 6$, $4 \cdot 7 = 28$, $16 \cdot 31 = 496$, $64 \cdot 127 = 8128$. No odd perfect numbers are known, and the question of whether any exist is one of the oldest unsolved problems of mathematics.*

To prove the theorem, we shall need some tools. First, a piece of standard notation: If a is a positive integer, the sum of all the divisors of a (including 1 and a itself) is denoted by the symbol $\sigma(a)$, read "sigma of a." For example, $\sigma(1) = 1$, $\sigma(2) = 1 + 2 = 3$, $\sigma(3) = 1 + 3 = 4$, $\sigma(4) = 1 + 2 + 4 = 7$, $\sigma(5) = 1 + 5 = 6$, $\sigma(6) = 1 + 2 + 3 + 6 = 12$. Since a perfect number is one that equals the sum of its divisors other than itself, perfect numbers are precisely those for which $\sigma(a) = 2a$. The only other tool we need is the following fact.

Lemma   *If a and b are positive integers whose greatest common divisor is 1, then* $\sigma(ab) = \sigma(a)\sigma(b)$.

Proof   Since a and b have no common factor greater than 1, any divisor d of ab is expressible in the form

$$d = a_i b_j$$

in one and only one way, where a_i is a divisor of a and b_j is a divisor of b. We denote the divisors of a and b by

$$1, a_1, a_2, \ldots, a \quad \text{and} \quad 1, b_1, b_2, \ldots, b,$$

so that their sums are

$$\sigma(a) = 1 + a_1 + a_2 + \cdots + a$$

and

$$\sigma(b) = 1 + b_1 + b_2 + \cdots + b.$$

Now let us consider all divisors $d = a_i b_j$ of ab with the same a_i. Their sum is

$$a_i \cdot 1 + a_i b_1 + a_i b_2 + \cdots + a_i b = a_i(1 + b_1 + b_2 + \cdots + b)$$
$$= a_i \sigma(b).$$

Finally, by adding these numbers for all possible a_i's, we obtain the sum of all the divisors of ab:

$$\sigma(ab) = 1 \cdot \sigma(b) + a_1 \sigma(b) + a_2 \sigma(b) + \cdots + a\sigma(b)$$
$$= (1 + a_1 + a_2 + \cdots + a)\sigma(b) = \sigma(a)\sigma(b),$$

and the proof is complete.

---

* It is known, however, that there is no odd perfect number containing less than 100 digits.

As a last preliminary remark, we point out that formula (1) enables us to calculate the value of $\sigma(p^{n-1})$ whenever p is a prime number. Since the divisors of p^{n-1} are $1, p, p^2, \ldots, p^{n-1}$, we have

$$\sigma(p^{n-1}) = 1 + p + p^2 + \cdots + p^{n-1} = \frac{1 - p^n}{1 - p}$$

$$= \frac{p^n - 1}{p - 1}.$$

In particular, when $p = 2$, this gives

$$\sigma(2^{n-1}) = 2^n - 1.$$

We are now ready to prove the theorem of even perfect numbers, which we divide into two separate statements for the sake of clarity.

Theorem 1 (Euclid) *If n is a positive integer for which $2^n - 1$ is prime, then $a = 2^{n-1}(2^n - 1)$ is an even perfect number.*

Proof Since $2^n - 1$ is prime, n must be at least 2 and a is even. We prove that a is perfect by showing that $\sigma(a) = 2a$. First, $2^n - 1$ is odd, so 2^{n-1} and $2^n - 1$ have no common factor > 1. The lemma therefore tells us that

$$\sigma[2^{n-1}(2^n - 1)] = \sigma(2^{n-1})\sigma(2^n - 1).$$

Next, since $2^n - 1$ is prime, its only divisors are 1 and itself, so

$$\sigma(2^n - 1) = 1 + (2^n - 1) = 2^n.$$

Finally,

$$\sigma(a) = \sigma[2^{n-1}(2^n - 1)]$$

$$= \sigma(2^{n-1})\sigma(2^n - 1)$$

$$= (2^n - 1)2^n$$

$$= 2[2^{n-1}(2^n - 1)] = 2a,$$

so a is perfect.

Theorem 2 (Euler) *If a is an even perfect number, then $a = 2^{n-1}(2^n - 1)$ for some positive integer n such that $2^n - 1$ is prime.*

Proof Factor the highest possible power of 2 out of a, and in this way write a in the form

$$a = m2^{n-1},$$

where n is at least 2 and m is odd. We shall prove that $m = 2^n - 1$ and that $2^n - 1$ is prime. Since a is perfect and therefore $\sigma(a) = 2a$, we have

$$m2^n = 2a = \sigma(a) = \sigma(m2^{n-1}) = \sigma(m)\sigma(2^{n-1})$$

$$= \sigma(m)(2^n - 1),$$

so

$$\sigma(m) = \frac{m2^n}{2^n - 1}.$$

But $\sigma(m)$ is an integer, so $2^n - 1$ divides $m2^n$; and since $2^n - 1$ and 2^n have no common factor > 1, $2^n - 1$ divides m. We see from this that $m/(2^n - 1)$ is a divisor of m, and this divisor is less than m, since $2^n - 1$ is at least 3. The equation

$$\sigma(m) = \frac{m2^n}{2^n - 1} = m + \frac{m}{2^n - 1}$$

therefore exhibits $\sigma(m)$ as the sum of m and one other divisor of m. This implies that m has two and only two divisors, so it must be prime. Also, it must be true that

$$\frac{m}{2^n - 1} = 1,$$

so $m = 2^n - 1$ and $2^n - 1$ is prime, which completes the proof.

The ideas discussed here raise the natural question: What numbers of the form $2^n - 1$ are prime? The factorization formula

$$a^n - 1 = (a - 1)(a^{n-1} + a^{n-2} + a^{n-3} + \cdots + 1)$$

shows that $2^n - 1$ cannot be prime if n is not; for example,

$$2^6 - 1 = (2^2)^3 - 1 = (2^2 - 1)[(2^2)^2 + 2^2 + 1] = 3 \cdot 21.$$

We may therefore confine our attention to the case in which the exponent n is assumed to be a prime p, and our question becomes: What numbers of the form $2^p - 1$ are prime? Such primes are called *Mersenne primes,* after Father Mersenne, a French scientist-mathematician-priest of the seventeenth century. Until 1952 only 12 were known, those corresponding to

$$p = 2, 3, 5, 7, 13, 17, 19, 31, 61, 89, 107, 127.$$

The primality of $2^{127} - 1$, a number of 39 digits, was established in 1876.[*] Beginning in 1952, 15 more have been found with the aid of electronic computers, those corresponding to

$$p = 521, 607, 1279, 2203, 2281, 3217, 4253, 4423, 9689, 9941,$$
$$11213, 19937, 21701, 23209, 44497,$$

the last of all in 1979.[†] The largest currently known prime number is therefore

$$2^{44497} - 1.$$

This has been computed in decimal form and is a number with 13,395 digits. It is so enormous that it exceeds by far the total number of grains of sand that could be packed solidly into the entire visible universe.

---

[*] The English mathematician G. H. Hardy remarked, "We may be able to recognize directly that 5, or even 17, is prime, but nobody can convince himself that

$$2^{127} - 1$$

is prime except by studying a proof. No one has ever had an imagination so vivid and comprehensive as that."

[†] See D. Slowinski and H. Nelson, "Searching for the 27th Mersenne Prime," *Journal of Recreational Mathematics,* **11**(4): 258–261 (1978–1979). This search continues among those who like to play with supercomputers, and by the time the present book is published, additional Mersenne primes will probably be known.

PROBLEMS

1 Prove that $\sqrt{3}$ is irrational. Hint: Modify the method of the text and use the fact that every integer is of the form $3n$, $3n + 1$, or $3n + 2$.

2 Prove that $\sqrt{5}$ and $\sqrt{6}$ are irrational by the method of Problem 1. Why doesn't this method work for $\sqrt{4}$?

3 Prove that $\sqrt[3]{2}$ and $\sqrt[3]{3}$ are irrational.

4 Prove that $\sqrt{2} + \sqrt{3}$ is irrational. Hint: Use the fact that $\sqrt{6}$ is irrational.

5 Prove that $\sqrt{2} + \sqrt{3} - \sqrt{6}$ is irrational.

6 Use the unique factorization theorem to prove that if a positive integer m is not the nth power of another integer, then $\sqrt[n]{m}$ is irrational. Notice that this result includes Problems 1, 2, and 3 as special cases.

7 Prove that $\log_{10} 2$ is irrational. For what positive integers n is $\log_{10} n$ irrational?

8 Prove that if a real number x_0 is a root of a polynomial equation

$$x^n + c_{n-1}x^{n-1} + c_{n-2}x^{n-2} + \cdots + c_1 x + c_0 = 0$$

with integral coefficients, then x_0 is either an irrational number or an integer; and if it is an integer, show that it must be a factor of c_0.

9 Use Problem 8 to show that $\sqrt{2} + \sqrt{3}$, $\sqrt{2} + \sqrt[3]{2}$, and $\sqrt{3} + \sqrt[3]{2}$ are irrational.

10 (a) Verify the four squares theorem for all integers from 1 to 50.

(b) Every prime except 2 (that is, every odd prime) is of the form $4n + 1$ or $4n + 3$. For the integers from 1 to 50, verify that every prime of the first type is expressible as the sum of two squares, and that no prime of the second type is.

A.2

ARCHIMEDES' QUADRATURE OF THE PARABOLA

The title of this section is a slightly altered version of the title of a treatise by Archimedes (*On the Quadrature of the Parabola*). The word *quadrature* is an old-fashioned term meaning the act or process of finding areas. In Section 6.2 we briefly described the first stages of Archimedes' quadrature of a segment of a parabola. Our purpose here is to complete the argument.

For the convenience of the reader, we repeat the figure under discussion (see Fig. A.1). The crux of the proof is the fact that the combined area of triangles ACD and BCE is one-fourth the area of triangle ABC, and similarly at each succeeding stage of the process. We shall prove this fact, but first let us see how it enables us to accomplish the quadrature of the parabolic segment. The stated relation between the newly added area and the previous area tells us first that

$$ACD + BCE = \frac{1}{4} ABC;$$

second that

$$\text{The combined area of the next four triangles} = \frac{1}{4} ACD + \frac{1}{4} BCE$$

$$= \frac{1}{4} (ACD + BCE)$$

$$= \frac{1}{4^2} ABC;$$

and so on indefinitely. Thus, by the nature of the exhaustion process, it is clear that

$$\text{The area of the parabolic segment} = ABC + \frac{1}{4} ABC + \frac{1}{4^2} ABC + \cdots$$

$$= ABC \left(1 + \frac{1}{4} + \frac{1}{4^2} + \cdots \right).$$

But from elementary algebra we know the sum of a geometric series,

$$1 + r + r^2 + \cdots = \frac{1}{1 - r},$$

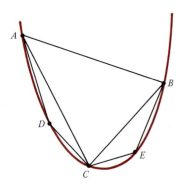

Figure A.1

where $-1 < r < 1.$* By putting $r = \frac{1}{4}$ we obtain

$$1 + \frac{1}{4} + \frac{1}{4^2} + \cdots = \frac{1}{1 - \frac{1}{4}} = \frac{1}{\frac{3}{4}} = \frac{4}{3},$$

and this brings us to our conclusion:

$$\text{The area of the parabolic segment} = \frac{4}{3} ABC,$$

as Archimedes discovered and proved.

We now establish the geometric fact stated above. Archimedes did this by purely geometric reasoning, but we shall use analytic geometry instead. Let us suppose that the parabola is the graph of $y = ax^2$, and that A and B have coordinates $A = (x_0, ax_0^2)$ and $B = (x_2, ax_2^2)$, as shown in Fig. A.2. If x_1 is the x-coordinate of C, then by using the formula $dy/dx = 2ax$ we see that the slope of the tangent at C is $2ax_1$. Since the tangent at C is parallel to AB, we have

$$2ax_1 = \frac{ax_0^2 - ax_2^2}{x_0 - x_2} \qquad \text{or} \qquad x_1 = \frac{1}{2}(x_0 + x_2).$$

This tells us that the vertical line through C bisects the chord AB at a point P. It clearly suffices to prove that

$$BCE = \frac{1}{4} BCP. \qquad (1)$$

To do this, we begin by completing the parallelogram $CPBQ$. By the same reasoning as before, the vertical line through E bisects the chord BC at a point G, and therefore also bisects the segment BP at H. If we can show that

$$EG = \frac{1}{2} GH, \qquad (2)$$

then this will imply that

$$BEG = \frac{1}{2} BGH \qquad \text{and} \qquad CEG = \frac{1}{2} BGH,$$

_____

* This is an immediate consequence of formula (1) in Section A.1.

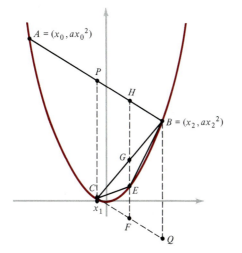

Figure A.2

so that

$$BCE = BGH;$$

and since clearly $BGH = \frac{1}{4}BCP$, this will prove (1). To establish (2), it suffices to show that $FE = \frac{1}{4}FH$, and we do this by showing that $FE = \frac{1}{4}QB$. The calculations are straightforward:

$$FE = a[\tfrac{1}{2}(x_1 + x_2)]^2 - [ax_1^2 + 2ax_1 \cdot \tfrac{1}{2}(x_2 - x_1)]$$
$$= \tfrac{1}{4}a[(x_1 + x_2)^2 - 4x_1^2 - 4x_1(x_2 - x_1)]$$
$$= \tfrac{1}{4}a(x_1^2 - 2x_1x_2 + x_2^2) = \tfrac{1}{4}a(x_1 - x_2)^2;$$

and

$$QB = ax_2^2 - [ax_1^2 + 2ax_1(x_2 - x_1)]$$
$$= a(x_1^2 - 2x_1x_2 + x_2^2) = a(x_1 - x_2)^2.$$

A.3

THE LUNES OF HIPPOCRATES

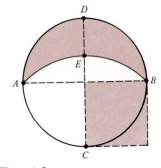

Figure A.3

According to one tradition, Hippocrates of Chios (ca. 430 B.C.) — not to be confused with his better-known contemporary, the physician Hippocrates of Cos — was originally a merchant whose goods were stolen by pirates.* He then went to Athens, where he lived for many years, studied mathematics, and compiled a book on the elements of geometry that strongly influenced Euclid more than a century later.

We recall Hippocrates' discovery as stated in Section 6.2: The lune (crescent-shaped region) in Fig. A.3 bounded by the circular arcs *ADB* and *AEB* (the latter having *C* as its center) has an area exactly equal to the area of the shaded square whose side is the radius of the circle. (Hippocrates also found the areas of two other kinds of lunes, but we do not discuss these here.)

This very surprising theorem seems to be the earliest precise determination of the area of a region bounded by curves. Its proof is simple but ingenious, and depends on the last of the following three geometric facts, each of which implies the next: (a) the areas of two circles are to each other as the squares of the radii (Fig. A.4); (b) sectors of two circles with equal central angles are to each other as the squares of the radii (Fig. A.5); (c) segments of two circles with equal central angles are to each other as the squares of the radii (Fig. A.6). We shall need (c) in the special case of right angles at the center.

---

* Aristotle, who rarely missed a chance to express his scorn for mathematicians, gives a more demeaning account of Hippocrates' misfortune. "It is well known," he wrote with relish, "that people brilliant in one particular field may be quite foolish in most other things. Thus Hippocrates, though skilled in geometry, was so stupid and spineless that he let a tax collector of Byzantium cheat him out of a fortune." This, from the man who asserted that heavier bodies fall to the ground more rapidly and that men have more teeth than women.

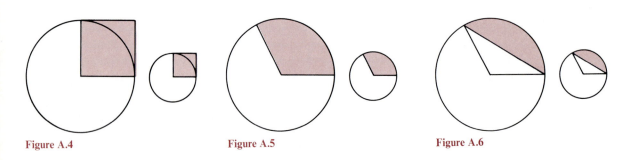

Figure A.4   Figure A.5   Figure A.6

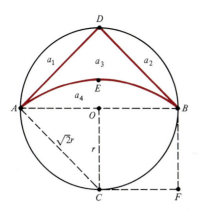

The proof of Hippocrates' theorem now proceeds as follows. Redraw the lune as shown in Fig. A.7. The chords joining D with A and B are tangent to the arc AEB and divide the lune into three regions with areas a_1, a_2, a_3. If the radius of the smaller circle is denoted by r, then the Pythagorean theorem tells us that the radius of the larger circle is $\sqrt{2}r$. It is easy to see that a_1 and a_2 are equal segments of the smaller circle and that a_4 is a segment of the larger circle, all with right angles at the center. We now use statement (c) to infer that

$$\frac{a_1}{a_4} = \frac{r^2}{(\sqrt{2}r)^2} = \frac{1}{2}.$$

This yields

$$a_1 = \frac{1}{2}\,a_4 \quad \text{and} \quad a_2 = \frac{1}{2}\,a_4,$$

so

$$a_1 + a_2 = a_4.$$

It now follows that

$$\text{Area of lune} = a_1 + a_2 + a_3$$
$$= a_4 + a_3$$
$$= \text{area of triangle } ABD$$
$$= r^2 = \text{area of square } OBFC,$$

and the argument is complete.

Hippocrates was a contemporary of Pericles, the great political and cultural leader of Athens during its Golden Age. But nothing Pericles achieved has the enduring quality of this beautiful geometric discovery; even the Parthenon, whose design and construction he supervised, is crumbling away. The reasoning of Hippocrates is a paragon of mathematical proof, untouched by time: In a few elegant steps it converts something easy to understand but difficult to believe into something impossible to doubt.

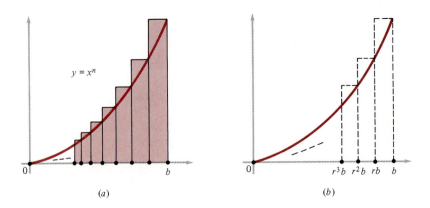

Figure A.8   (a)   (b)

A.4

FERMAT'S CALCULATION OF

$$\int_0^b x^n\,dx$$

FOR POSITIVE RATIONAL n

We shall need the following formulas from Appendixes A.1 and A.2: If $0 < r < 1$, then

$$1 + r + r^2 + \cdots + r^n = \frac{1 - r^{n+1}}{1 - r} \quad \text{and} \quad 1 + r + r^2 + \cdots = \frac{1}{1 - r}. \quad (1)$$

Fermat used upper sums similar to those of Section 6.4 but based on a division of the interval $[0, b]$ into an *infinite* number of unequal subintervals, as suggested in Fig. A.8a. We start with a fixed positive number $r < 1$ (but close to 1) and produce the points of division by moving downward from b by repeatedly multiplying by r, as shown in Fig. A.8b. The sum of the areas of all the upper rectangles, starting from the right, is an infinite sum depending on r,

$$S_r = b^n(b - rb) + (rb)^n(rb - r^2b) + (r^2b)^n(r^2b - r^3b) + \cdots$$

$$= b^{n+1}(1 - r) + b^{n+1}r^{n+1}(1 - r) + b^{n+1}r^{2n+2}(1 - r) + \cdots$$

$$= b^{n+1}(1 - r)[1 + r^{n+1} + (r^{n+1})^2 + \cdots]$$

$$= \frac{b^{n+1}(1 - r)}{1 - r^{n+1}} = \frac{b^{n+1}}{(1 - r^{n+1})/(1 - r)}$$

$$= \frac{b^{n+1}}{1 + r + r^2 + \cdots + r^n}. \quad (2)$$

Here we used first the second formula of (1) and then the first formula, the latter requiring that n be a positive integer. If we now let $r \to 1$, we see that each of the $n + 1$ terms in the denominator of the last expression also approaches 1, so we have our result:

$$\lim_{r \to 1} S_r = \frac{b^{n+1}}{n + 1}, \quad (3)$$

or equivalently,

$$\int_0^b x^n\,dx = \frac{b^{n+1}}{n + 1} \quad (4)$$

for every positive integer n.

The only part of this argument in which n must be a positive integer is the last step in arriving at (2). If we assume only that n is a positive rational number p/q, then we can get around this difficulty by making the substitution $s = r^{1/q}$ and calculating as follows:

$$\frac{1-r}{1-r^{n+1}} = \frac{1-s^q}{1-(s^q)^{p/q+1}} = \frac{1-s^q}{1-s^{p+q}}$$

$$= \frac{(1-s^q)/(1-s)}{(1-s^{p+q})/(1-s)} = \frac{1+s+s^2+\cdots+s^{q-1}}{1+s+s^2+\cdots+s^{p+q-1}}.$$

Now, as $r \to 1$, we also have $s \to 1$, and the expression last written tells us that

$$\frac{1-r}{1-r^{n+1}} \to \frac{q}{p+q} = \frac{1}{p/q+1} = \frac{1}{n+1},$$

so (3) and (4) remain valid for any positive rational exponent n.

Archimedes' discovery of the formula for the volume of a sphere was one of the greatest mathematical achievements of all time. The formula itself was of obvious importance, but even more so was his method of discovering it; for this method amounted to the earliest appearance of the basic idea of integral calculus.

He proved this formula in his treatise *On the Sphere and Cylinder,* by means of a long and rigorous argument of classic perfection. Unfortunately, however, this argument was of the kind that compels belief but provides little insight. It was like a great work of architecture whose architect has removed all the scaffolding, burnt the plans, and concealed his private thoughts from which the overall concept emerged. Mathematicians have always been aware — from his formal treatises — of what Archimedes discovered. However, his method of making discoveries remained a mystery until the year 1906, when the Danish scholar Heiberg uncovered a lost manuscript dealing with exactly this question.*

In this manuscript Archimedes described to his friend Eratosthenes how he "investigated some problems in mathematics by means of mechanics."† The most wonderful of these investigations was his discovery of the volume of a sphere. To understand this work, it is necessary to know a little about the level of knowledge from which he started.

As Archimedes states, it was Democritus two centuries earlier who discovered that the volume of a cone is one-third the volume of a cylinder with the same height and the same base. Nothing is definitely known about Democritus' method, but it is believed that he succeeded by considering first a three-sided pyramid (tetrahedron), then an arbitrary pyramid, and finally a cone as the limit of inscribed pyramids.‡

Also, the Greeks knew a little analytic geometry, but without our notation. They were acquainted with the idea that a locus in a plane can be studied by considering the distances from a moving point to two perpendicular lines; and if the sum of the squares of these distances is constant, then they knew that the locus is a circle. In our notation, this condition amounts to the equation $x^2 + y^2 = a^2$.

Further, Archimedes himself virtually created Greek mechanics. As everyone knows, he discovered the law of floating bodies. He also discovered the principle of the lever and many facts about centers of gravity.

We are now ready to follow Archimedes in his search for the volume of a sphere. He considered the sphere to be generated by revolving a circle about its diameter. In modern notation, we start with the circle

$$x^2 + y^2 = 2ax, \tag{1}$$

A.5

HOW ARCHIMEDES DISCOVERED INTEGRATION

---

* See the *Method* in *The Works of Archimedes,* T. L. Heath (ed.), Dover (no date).

† *Method,* p. 13.

‡ See Chapter 1 of the present writer's booklet, *Precalculus Mathematics in a Nutshell,* William Kaufmann, Inc., 1981.

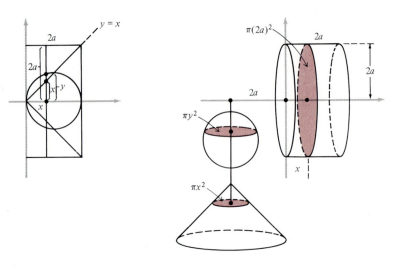

Figure A.9   Archimedes' balancing argument.

which has radius a and is tangent to the y-axis at the origin. This circle is shown on the left in Fig. A.9, which is almost identical with Archimedes' original figure. Equation (1) contains the term y^2, and since πy^2 is the area of the variable cross section of the sphere x units to the right of the origin, it is natural to multiply through by π and write (1) in the form

$$\pi x^2 + \pi y^2 = \pi 2ax. \tag{2}$$

This leads us to interpret πx^2 as the area of the variable cross section of the cone generated by revolving the line $y = x$ about the x-axis. This in turn suggests that we seek a similar interpretation for the term $\pi 2ax$ on the right side of (2). If we persist in this search, we might perhaps think of multiplying by $2a$ and thus rewriting (2) as

$$2a(\pi x^2 + \pi y^2) = x\pi (2a)^2. \tag{3}$$

The motivation for this change clearly lies in the fact that $\pi (2a)^2$ is the area of the cross section of the cylinder with the same height and base as the cone.

We therefore have on the left in Fig. A.9 three circular disks of areas πy^2, πx^2, and $\pi (2a)^2$ which are the intersections of a single plane with three solids of revolution. This plane is perpendicular to the x-axis at a distance x units to the right of the origin, and the solids are the sphere, the cone, and the cylinder, as indicated in the figure.

On the left-hand side of equation (3), the sum of the first two areas is multiplied by $2a$, and on the right-hand side, the third area is multiplied by x. This observation led Archimedes to the following great idea, as shown on the right in Fig. A.9. He left the disk with radius $2a$ where it is, in a vertical position x units to the right of the origin; and he moved the disks with radii y and x to a point $2a$ units to the left of the origin, where he hung them horizontally with their centers under this point, suspended by a weightless string. The purpose of this maneuver can be understood only if we think of the x-axis as a lever and the origin as its fulcrum or balancing point. It can now be seen that equation (3) deals with moments. (A *moment* is the product of the suspended weight and the length of the lever arm.) From this point of view, equation (3) states that the combined moments of the two disks on the left equals the moment of the single disk on the right, and so, by Archimedes' own principle of the lever, this lever is in equilibrium.

We now carry out the final step of the reasoning. As x increases from 0 to $2a$, the three cross sections sweep through their respective solids and fill these solids. Since the three cross sections are in equilibrium throughout this process, the solids themselves are also in equilibrium. Let V denote the volume of the sphere, which was unknown until Archimedes finished this calculation. If we use Democritus' formula for the volume of the cone, and also the volume of the cylinder and the obvious location of its center of gravity, then the equilibrium of the solids in the positions shown in the figure yields

$$2a[\tfrac{1}{3}\pi(2a)^2(2a) + V] = a\pi(2a)^2(2a). \tag{4}$$

It is now easy to solve (4) for V and obtain

$$V = \tfrac{4}{3}\pi a^3.$$

The ideas discussed here were created by one who has been described — with good reason — as "the greatest genius of the ancient world." But these ideas are, after all, only a beginning. The crux of the reasoning lies in the transition from (3) to (4), from the moving cross sections to the complete solids. With the advantage of historical perspective, we can recognize this transition as the essence of integration, which we know to be a process of great scope and diversity, with innumerable applications in science and mathematics. Archimedes himself had an inkling of the potential value of his ideas: "I am persuaded that this method will be of no little service to mathematics. For I foresee that once it is understood and established, it will be used to discover other theorems which have not yet occurred to me, by other mathematicians, now living or yet unborn."*

Consider a particle of mass m that starts from rest at the origin on the x-axis and moves in the positive direction under the influence of a constant force F. If we write Newton's second law of motion

$$F = ma \tag{1}$$

in the form

$$a = \frac{1}{m} F,$$

then we can think of the force F as producing the constant acceleration a. If this acceleration continues long enough, then the velocity v of the particle increases beyond all bounds, and in particular beyond c, the velocity of light. But according to Einstein this cannot happen; nothing can travel faster than light.

The way out of this impasse is to realize that Newton's law is actually somewhat more general than (1); it states that when a force F acts on a body of mass m, it produces *momentum* ($= mv$) at a rate equal to the force:

$$F = \frac{d}{dt}(mv). \tag{2}$$

This equation reduces to (1) when the mass m is constant. But according to Einstein, the mass is not constant. It increases as the velocity v increases, and is determined as a function of v by the formula

$$m = \frac{m_0}{\sqrt{1 - v^2/c^2}}, \tag{3}$$

where m_0 is the so-called *rest mass*. When this expression for m is inserted in (2), we obtain *Einstein's law of motion*,

$$F = m_0 \frac{d}{dt}\left(\frac{v}{\sqrt{1 - v^2/c^2}}\right). \tag{4}$$

A.6a

A SIMPLE APPROACH
TO $E = Mc^2$

---

* *Method*, p. 14.

It will be convenient to carry out the differentiation in (4), and in this way introduce the acceleration $a = dv/dt$. We have

$$\frac{d}{dt}\left(\frac{v}{\sqrt{1 - v^2/c^2}}\right) = \frac{d}{dv}\left(\frac{v}{\sqrt{1 - v^2/c^2}}\right)\frac{dv}{dt}$$

$$= a\left[\frac{\sqrt{1 - v^2/c^2} - v(\tfrac{1}{2})(1 - v^2/c^2)^{-1/2}(-2v/c^2)}{1 - v^2/c^2}\right]$$

$$= a\left[\frac{c^2(1 - v^2/c^2) + v^2}{c^2(1 - v^2/c^2)^{3/2}}\right] = \frac{a}{(1 - v^2/c^2)^{3/2}}.$$

This enables us to write (4) in the form

$$F = \frac{m_0 a}{(1 - v^2/c^2)^{3/2}}, \tag{5}$$

which shows how close Einstein's law is to Newton's law (1) when v is much less than c. However, when v is near the velocity of light, as in most phenomena of atomic physics, then the two laws differ considerably, and all the experimental evidence supports Einstein's version.

We now return to our original problem of the particle starting from rest at the origin on the x-axis, with the slight change that the force F is now assumed only to be positive, so that the acceleration a in (5) is also positive and the velocity is increasing. Our purpose is to show that if the energy of the particle at any stage of the process is understood to be the work done on it by F, then this energy E is related to the increase in the mass, which is $M = m - m_0$, by Einstein's famous equation $E = Mc^2$. If we begin by writing

$$a = \frac{dv}{dt} = \frac{dv}{dx}\frac{dx}{dt} = v\frac{dv}{dx},$$

then (5) yields

$$E = \int_0^x F\,dx = m_0 \int_0^x \frac{a}{(1 - v^2/c^2)^{3/2}}\,dx$$

$$= m_0 \int_0^v \frac{v\,dv}{(1 - v^2/c^2)^{3/2}}$$

$$= m_0\left(-\frac{c^2}{2}\right)\int_0^v \left(1 - \frac{v^2}{c^2}\right)^{-3/2}\left(-\frac{2v\,dv}{c^2}\right)$$

$$= m_0 c^2 \left(1 - \frac{v^2}{c^2}\right)^{-1/2}\Bigg]_0^v$$

$$= m_0 c^2\left(\frac{1}{\sqrt{1 - v^2/c^2}} - 1\right)$$

$$= c^2\left(\frac{m_0}{\sqrt{1 - v^2/c^2}} - m_0\right) = c^2(m - m_0) = Mc^2.$$

Needless to say, this is not a proof of Einstein's equation in all cases; it merely shows that this equation connects the increase in mass arising from (3) with the increase in energy associated with the greater velocity.

However, it is necessary to add that the crux of Einstein's equation is the much deeper fact that the rest mass m_0 also has energy associated with it, in the amount of $E = m_0 c^2$. This energy can be thought of as the "energy of being" of the particle, in the sense that mass possesses energy just by virtue of existing. The point of view of modern physics is even more direct than this: Matter *is* energy, in a highly concentrated and localized form. It should also be understood that the constant c^2 is so enormous that a small amount of mass is equivalent to a very large amount of energy. Thus, if the mass of a drop of water could be completely converted into energy in a controlled and useful way, then the resulting energy would be enough to lift several heavy trucks as far as the moon. This is the source of the energy that fuels the sun, in the so-called thermonuclear reactions which physicists are even now seeking to tame to our service.

At the beginning of Appendix A.6a we pointed out that Newton's second law of motion for a force F acting on a body of mass m moving with velocity v can be stated as

$$F = \frac{d}{dt}(mv),\qquad(1)$$

A.6b

ROCKET PROPULSION IN OUTER SPACE

and also that this assumes the more familiar form $F = ma$ when m is constant. In particular, if the body is moving under the action of no external force at all, so that $F = 0$, then (1) tells us that the momentum mv is constant.

As an illustration we consider a rocket of mass m, velocity v, and constant exhaust velocity c, and we assume that this rocket is moving in a straight line in outer space, where no external forces are present. The mass m consists of the structural mass of the rocket plus the mass of the fuel it carries, so m decreases as the fuel is burned. The exhaust gases are ejected at high speed from the tail of the rocket, and this propels it forward just as escaping air propels a toy balloon. We shall find the equation of motion.

Suppose that at time t the mass of the rocket (including fuel) is m and it is moving with velocity v, as shown in Fig. A.10, while at time $t + \Delta t$ the mass is $m + \Delta m$ and the velocity is $v + \Delta v$. The mass of fuel burned in this time is $-\Delta m$ (Δm is clearly negative), and the exhaust products, which are therefore of mass $-\Delta m$, are expelled backward at velocity c relative to the rocket, so that this material has actual velocity $v - c$. The fact that the total momentum of the system is constant means that

$$mv = (m + \Delta m)(v + \Delta v) + (-\Delta m)(v - c).$$

This yields

$$mv = mv + m\,\Delta v + (\Delta m)v + \Delta m\,\Delta v - (\Delta m)v + \Delta m\,c,$$

which reduces to

$$m\,\Delta v = -\Delta m(c + \Delta v).\qquad(2)$$

After division by Δt, (2) becomes

$$m\frac{\Delta v}{\Delta t} = -\frac{\Delta m}{\Delta t}(c + \Delta v),$$

and by letting $\Delta t \to 0$ we obtain

$$m\frac{dv}{dt} = -c\frac{dm}{dt}.\qquad(3)$$

This is the basic equation of rocket propulsion in outer space.

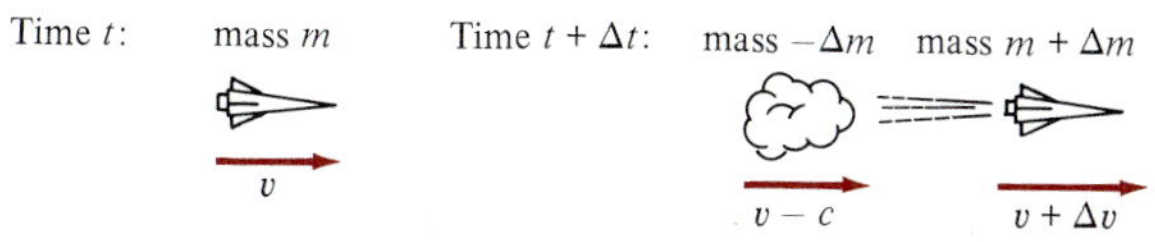

Figure A.10   Rocket acceleration by exhaust expulsion.

To illustrate the qualitative conclusions that can be drawn from (3), we use the fact that dv/dt is the acceleration a and write the equation as

$$a = \frac{1}{m}\left[c\left(-\frac{dm}{dt}\right)\right]. \tag{4}$$

Since dm/dt is negative, we see that a is positive, and this means that the velocity is increasing, as we expect. The quantity in brackets here — the product of the exhaust velocity and the rate at which fuel is consumed — is called the *thrust* of the rocket engine. It is clear from (4) that a large acceleration requires a rocket to have a large thrust, and a large thrust is obtained by designing an engine with a large exhaust velocity and a high rate of fuel consumption. Also, if the thrust is constant, then (4) tells us that the acceleration increases as m decreases, that is, as the fuel is burned.

In addition to qualitative inferences of the kind just mentioned, it is also possible to obtain quantitative information from (3).* If we write the equation in the form

$$dv = -c\,\frac{dm}{m}$$

and integrate from 0 to t, we get

$$v(t) = v(0) - c \ln \frac{m(t)}{m(0)}$$

$$= v(0) + c \ln \frac{m(0)}{m(t)}. \tag{5}$$

As an illustration of the way (5) can be used, suppose the initial mass of the rocket is one-tenth structure and nine-tenths fuel. If the exhaust velocity is $c = 2$ mi/s and the rocket starts from rest, then its speed at burnout is

$$v = 2 \ln 10 \cong 4.6 \text{ mi/s.}$$

A.7

A PROOF OF VIETA'S FORMULA

To prove Vieta's formula

$$\frac{2}{\pi} = \frac{\sqrt{2}}{2} \cdot \frac{\sqrt{2 + \sqrt{2}}}{2} \cdot \frac{\sqrt{2 + \sqrt{2 + \sqrt{2}}}}{2} \cdots, \tag{1}$$

we need the limit

$$\lim_{\theta \to 0} \frac{\sin \theta}{\theta} = 1, \tag{2}$$

the double-angle formula for the sine in the form

$$\sin \theta = 2 \sin \frac{\theta}{2} \cos \frac{\theta}{2}, \tag{3}$$

and the half-angle formula for the cosine in the form

$$\cos \frac{\theta}{2} = \frac{1}{2}\sqrt{2 + 2 \cos \theta}. \tag{4}$$

---

* The following material requires the reader to understand the meaning of the formula

$$\int \frac{dm}{m} = \ln m,$$

as explained in Chapter 8.

By repeatedly applying (3) we obtain

$$1 = \sin \frac{\pi}{2} = 2 \sin \frac{\pi}{4} \cos \frac{\pi}{4}$$

$$= 2^2 \sin \frac{\pi}{8} \cos \frac{\pi}{4} \cos \frac{\pi}{8}$$

$$= 2^3 \sin \frac{\pi}{16} \cos \frac{\pi}{4} \cos \frac{\pi}{8} \cos \frac{\pi}{16}$$

$$= \cdots = 2^{n-1} \sin \frac{\pi}{2^n} \cos \frac{\pi}{4} \cos \frac{\pi}{8} \cdots \cos \frac{\pi}{2^n}.$$

With the aid of (4), this can be written as

$$\frac{2}{\pi} \cdot \frac{\pi/2^n}{\sin \pi/2^n} = \frac{1}{2^{n-1} \sin \pi/2^n}$$

$$= \frac{\sqrt{2}}{2} \cdot \frac{\sqrt{2 + \sqrt{2}}}{2} \cdot \frac{\sqrt{2 + \sqrt{2 + \sqrt{2}}}}{2} \cdots \frac{\sqrt{2 + \sqrt{2 + \cdots}}}{2}, \qquad (5)$$

where the last factor contains $n - 1$ nested root signs. If we now let $n \to \infty$ and use (2), (5) yields Vieta's formula (1).*

A.8

AN ELEMENTARY PROOF OF LEIBNIZ'S FORMULA
$$\frac{\pi}{4} = 1 - \frac{1}{3} + \frac{1}{5} - \frac{1}{7} + \cdots$$

The formula in question is

$$\frac{\pi}{4} = 1 - \frac{1}{3} + \frac{1}{5} - \frac{1}{7} + \cdots, \qquad (1)$$

which was discussed in a loose, nonrigorous way in Section 9.5.

We begin the proof with the formula

$$\frac{1}{1 + t^2} = 1 - t^2 + t^4 - t^6 + t^8 - \cdots + t^{4n} - \frac{t^{4n+2}}{1 + t^2}, \qquad (2)$$

which is valid for all t and can be checked by multiplying through by $1 + t^2$. Now consider a number x such that $0 \le x \le 1$. Integrating (2) over the interval $0 \le t \le x$ gives

$$\tan^{-1} x = \int_0^x \frac{dt}{1 + t^2} = x - \frac{x^3}{3} + \frac{x^5}{5} - \frac{x^7}{7} + \cdots + \frac{x^{4n+1}}{4n+1} - R_n(x), \qquad (3)$$

where

$$R_n(x) = \int_0^x \frac{t^{4n+2}}{1 + t^2} \, dt.$$

---

* In the present book, Vieta's formula is only an isolated jewel in the early history of mathematics. However, it can be made the first step of a fascinating but demanding journey leading upward to some of the peaks of classical mathematical physics (the kinetic theory of gases, the second law of thermodynamics, etc.). See the little book by M. Kac, *Statistical Independence in Probability, Analysis and Number Theory* (John Wiley and Sons, 1959). As for François Vieta (1540–1603) himself, he was a Frenchman trained in the law who rose to become a royal privy councillor under Henry IV and cultivated mathematics as a hobby. He contributed to the early development of analytic trigonometry and algebra, in particular by his systematic use of letters to represent constants and unknowns.

It is clear that $1 \le 1 + t^2$, so

$$0 \le R_n(x) \le \int_0^x t^{4n+2}\, dt$$

or

$$0 \le R_n(x) \le \frac{x^{4n+3}}{4n+3}.$$

For the x's under discussion, this shows that

$$0 \le R_n(x) \le \frac{1}{4n+3},$$

so $R_n(x) \to 0$ as $n \to \infty$. This allows us to use (3) to deduce that the formula

$$\tan^{-1} x = x - \frac{x^3}{3} + \frac{x^5}{5} - \frac{x^7}{7} + \cdots \tag{4}$$

is valid for $0 \le x \le 1$. [The right-hand side of (4) is by definition the limit as $n \to \infty$ of the right-hand side of (3) if $R_n(x) \to 0$, which we have just shown to be true for the stated x's.] To obtain Leibniz's formula (1), we now — legally! — set $x = 1$ in (4).

A.9

THE CATENARY, OR CURVE OF A HANGING CHAIN

As a specific example of the use of the methods of integration discussed in Section 10.4, we solve the classical problem of determining the exact shape of the curve assumed by a flexible chain of uniform density which is suspended between two points and hangs under its own weight. This curve is called a *catenary*, from the Latin word for chain, *catena*.

Let the y-axis pass through the lowest point of the chain (Fig. A.11), let s be the arc length from this point to a variable point (x, y), and let w_0 be the linear density (weight per unit length) of the chain. We obtain the differential equation of the catenary from the fact that the part of the chain between the lowest point and (x, y) is in static equilibrium under the action of three forces: the tension T_0 at the lowest point; the variable tension T at (x, y), which acts in the direction of the tangent because of the flexibility of the chain; and a downward force $w_0 s$ equal to the weight of the chain between these two points.

Equating the horizontal component of T to T_0 and the vertical component of T to the weight of the chain gives

$$T \cos \theta = T_0 \qquad \text{and} \qquad T \sin \theta = w_0 s,$$

and by dividing we eliminate T and get $\tan \theta = w_0 s / T_0$ or

$$\frac{dy}{dx} = as, \qquad \text{where} \qquad a = \frac{w_0}{T_0}.$$

We next eliminate the variable s by differentiating with respect to x,

$$\frac{d^2y}{dx^2} = a\frac{ds}{dx} = a\sqrt{1 + \left(\frac{dy}{dx}\right)^2}. \tag{1}$$

This is the differential equation of the catenary.

We now solve equation (1) by two successive integrations. This process is facilitated by introducing the auxiliary variable $p = dy/dx$, so that (1) becomes

$$\frac{dp}{dx} = a\sqrt{1 + p^2}.$$

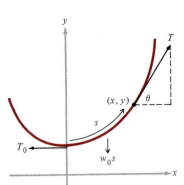

Figure A.11

On separating variables and integrating, we get

$$\int \frac{dp}{\sqrt{1 + p^2}} = \int a\, dx. \tag{2}$$

To calculate the integral on the left, we make the trigonometric substitution $p = \tan \phi$, so that $dp = \sec^2 \phi\, d\phi$ and $\sqrt{1 + p^2} = \sec \phi$. Then

$$\int \frac{dp}{\sqrt{1 + p^2}} = \int \frac{\sec^2 \phi\, d\phi}{\sec \phi} = \int \sec \phi\, d\phi$$
$$= \ln (\sec \phi + \tan \phi) = \ln (\sqrt{1 + p^2} + p),$$

so (2) becomes

$$\ln (\sqrt{1 + p^2} + p) = ax + c_1.$$

Since $p = 0$ when $x = 0$, we see that $c_1 = 0$, so

$$\ln (\sqrt{1 + p^2} + p) = ax.$$

It is easy to solve this equation for p, which yields

$$\frac{dy}{dx} = p = \frac{1}{2} (e^{ax} - e^{-ax}),$$

and by integrating we obtain

$$y = \frac{1}{2a} (e^{ax} + e^{-ax}) + c_2.$$

If we now place the origin of the coordinate system in Fig. A.11 at just the right level so that $y = 1/a$ when $x = 0$, then $c_2 = 0$ and our equation takes its final form,

$$y = \frac{1}{2a} (e^{ax} + e^{-ax}). \tag{3}$$

Equation (3) reveals the precise mathematical nature of the catenary and can be used as the basis for further investigations of its properties.*

The problem of finding the true shape of the catenary was proposed by James Bernoulli in 1690. Galileo had speculated long before that the curve was a parabola, but Huygens had shown in 1646 (at the age of 17), largely by physical reasoning, that this is not correct, without, however, shedding any light on what the shape might be. Bernoulli's challenge produced quick results, for in 1691 Leibniz, Huygens (now aged 62), and James's brother John all published independent solutions of the problem. John Bernoulli was exceedingly pleased that he had been successful in solving the problem, while his brother James, who proposed it, had failed. The taste of victory was still sweet 27 years later, as we see from this passage in a letter John wrote in 1718:

> The efforts of my brother were without success. For my part, I was more fortunate, for I found the skill (I say it without boasting; why should I conceal the truth?) to solve it in full. . . . It is true that it cost me study that robbed me of rest for an

---

* The hyperbolic cosine defined in Section 9.7 enables us to write the function (3) in the form

$$y = \frac{1}{a} \cosh ax.$$

This fact is sometimes thought to justify a detailed study of the hyperbolic functions, but the present writer remains skeptical.

entire night. It was a great achievement for those days and for the slight age and experience I then had. The next morning, filled with joy, I ran to my brother, who was still struggling miserably with this Gordian knot without getting anywhere, always thinking like Galileo that the catenary was a parabola. Stop! Stop! I say to him, don't torture yourself any more trying to prove the identity of the catenary with the parabola, since it is entirely false.

However, James evened the score by proving in the same year of 1691 that of all possible shapes a chain hanging between two fixed points might have, the catenary has the lowest center of gravity, and therefore the smallest potential energy. This was a very significant discovery, because it was the first hint of the profound idea that in some mysterious way the actual configurations of nature are those that minimize potential energy.

A.10
WALLIS'S PRODUCT

As an application of integration by parts in Section 10.7, we obtained the following reduction formula:

$$\int \sin^n x \, dx = -\frac{1}{n} \sin^{n-1} x \cos x + \frac{n-1}{n} \int \sin^{n-2} x \, dx. \tag{1}$$

This formula leads in an elementary but ingenious way to a very remarkable expression for the number $\pi/2$ as an infinite product,

$$\frac{\pi}{2} = \frac{2}{1} \cdot \frac{2}{3} \cdot \frac{4}{3} \cdot \frac{4}{5} \cdot \frac{6}{5} \cdot \frac{6}{7} \cdots \frac{2n}{2n-1} \cdot \frac{2n}{2n+1} \cdots. \tag{2}$$

This expression was discovered by the English mathematician John Wallis in 1656, and is called *Wallis's product.* Apart from its intrinsic interest, formula (2) underlies other important developments in both pure and applied mathematics, so we prove it here.

If we define I_n by

$$I_n = \int_0^{\pi/2} \sin^n x \, dx,$$

then (1) tells us that

$$I_n = \frac{n-1}{n} I_{n-2}. \tag{3}$$

It is clear that

$$I_0 = \int_0^{\pi/2} dx = \frac{\pi}{2} \quad \text{and} \quad I_1 = \int_0^{\pi/2} \sin x \, dx = 1.$$

We now distinguish the cases of even and odd subscripts, and use (3) to calculate I_{2n} and I_{2n+1}, as follows:

$$I_{2n} = \frac{2n-1}{2n} I_{2n-2} = \frac{2n-1}{2n} \cdot \frac{2n-3}{2n-2} I_{2n-4}$$

$$= \cdots = \frac{2n-1}{2n} \cdot \frac{2n-3}{2n-2} \cdot \frac{2n-5}{2n-4} \cdots \frac{1}{2} I_0$$

$$= \frac{1}{2} \cdot \frac{3}{4} \cdot \frac{5}{6} \cdots \frac{2n-1}{2n} \cdot \frac{\pi}{2}; \tag{4}$$

and

$$I_{2n+1} = \frac{2n}{2n+1} I_{2n-1} = \frac{2n}{2n+1} \cdot \frac{2n-2}{2n-1} I_{2n-3}$$

$$= \cdots = \frac{2n}{2n+1} \cdot \frac{2n-2}{2n-1} \cdot \frac{2n-4}{2n-3} \cdots \frac{2}{3} I_1$$

$$= \frac{2}{3} \cdot \frac{4}{5} \cdot \frac{6}{7} \cdots \frac{2n}{2n+1}. \tag{5}$$

As the next link in the chain of this reasoning, we need the fact that the ratio of these two quantities approaches 1 as $n \to \infty$,

$$\frac{I_{2n}}{I_{2n+1}} \to 1. \tag{6}$$

To establish this, we begin by noticing that on the interval $0 \le x \le \pi/2$ we have $0 \le \sin x \le 1$, and therefore

$$0 \le \sin^{2n+2} x \le \sin^{2n+1} x \le \sin^{2n} x.$$

This implies that

$$0 < \int_0^{\pi/2} \sin^{2n+2} x \, dx \le \int_0^{\pi/2} \sin^{2n+1} x \, dx \le \int_0^{\pi/2} \sin^{2n} x \, dx,$$

or equivalently,

$$0 < I_{2n+2} \le I_{2n+1} \le I_{2n}. \tag{7}$$

If we divide through by I_{2n} and use the fact that by (3) we have

$$\frac{I_{2n+2}}{I_{2n}} = \frac{2n+1}{2n+2},$$

then (7) yields

$$\frac{2n+1}{2n+2} \le \frac{I_{2n+1}}{I_{2n}} \le 1.$$

This implies that

$$\frac{I_{2n+1}}{I_{2n}} \to 1 \qquad \text{as} \qquad n \to \infty,$$

and this is equivalent to (6).

The final steps of the argument are as follows. On dividing (5) by (4), we obtain

$$\frac{I_{2n+1}}{I_{2n}} = \frac{2}{1} \cdot \frac{2}{3} \cdot \frac{4}{3} \cdot \frac{4}{5} \cdot \frac{6}{5} \cdot \frac{6}{7} \cdots \frac{2n}{2n-1} \cdot \frac{2n}{2n+1} \cdot \frac{2}{\pi},$$

so

$$\frac{\pi}{2} = \frac{2}{1} \cdot \frac{2}{3} \cdot \frac{4}{3} \cdot \frac{4}{5} \cdot \frac{6}{5} \cdot \frac{6}{7} \cdots \frac{2n}{2n-1} \cdot \frac{2n}{2n+1} \left(\frac{I_{2n}}{I_{2n+1}} \right).$$

On forming the limit as $n \to \infty$ and using (6), we obtain

$$\frac{\pi}{2} = \lim_{n \to \infty} \frac{2}{1} \cdot \frac{2}{3} \cdot \frac{4}{3} \cdot \frac{4}{5} \cdot \frac{6}{5} \cdot \frac{6}{7} \cdots \frac{2n}{2n-1} \cdot \frac{2n}{2n+1},$$

and this is what (2) means.

We also remark that Wallis's product (2) is equivalent to the formula

$$\left(1 - \frac{1}{2^2}\right)\left(1 - \frac{1}{4^2}\right)\left(1 - \frac{1}{6^2}\right) \cdots = \frac{2}{\pi}. \tag{8}$$

This is easy to see if we write each number in parentheses on the left in factored form. This gives

$$\left(1 - \frac{1}{2}\right)\left(1 + \frac{1}{2}\right)\left(1 - \frac{1}{4}\right)\left(1 + \frac{1}{4}\right)\left(1 - \frac{1}{6}\right)\left(1 + \frac{1}{6}\right) \cdots = \frac{2}{\pi}$$

or

$$\frac{1}{2} \cdot \frac{3}{2} \cdot \frac{3}{4} \cdot \frac{5}{4} \cdot \frac{5}{6} \cdot \frac{7}{6} \cdots = \frac{2}{\pi},$$

which is clearly equivalent to (2). Formula (8) will reappear in Appendix A.12 as a special case of another even more wonderful formula.*

A.11

HOW LEIBNIZ DISCOVERED HIS FORMULA

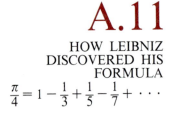

$$\frac{\pi}{4} = 1 - \frac{1}{3} + \frac{1}{5} - \frac{1}{7} + \cdots$$

The area of the quarter-circle of radius 1 shown in Fig. A.12 is obviously $\pi/4$. We follow Leibniz and calculate this area in a different way. The part that we actually calculate is the area A of the circular segment cut off by the chord OT, because the remainder of the quarter-circle is clearly an isosceles right triangle of area $\frac{1}{2}$.

We obtain the stated area A by integrating the sliverlike elements of area OPQ, where the arc PQ is considered to be so small that it is virtually straight. We think of OPQ as a triangle whose base is the segment PQ of length ds and whose height is the perpendicular distance OR from the vertex O to the base PQ extended. The two similar right triangles in the figure tell us that

$$\frac{ds}{dx} = \frac{OS}{OR} \qquad \text{or} \qquad OR\ ds = OS\ dx,$$

so the area dA of OPQ is

$$dA = \tfrac{1}{2} OR\ ds = \tfrac{1}{2} OS\ dx = \tfrac{1}{2} y\ dx,$$

---

* Wallis was Savilian Professor of Geometry at Oxford for 54 years, from 1649 until his death in 1703 at the age of 87, and played an important part in forming the climate of thought in which Newton flourished. He introduced negative and fractional exponents as well as the now-standard symbol ∞ for infinity, and was the first to treat conic sections as plane curves of the second degree. His infinite product stimulated his friend Lord Brouncker (first president of the Royal Society) to discover the astonishing formula

$$\frac{4}{\pi} = 1 + \cfrac{1^2}{2 + \cfrac{3^2}{2 + \cfrac{5^2}{2 + \cfrac{7^2}{2 + \cfrac{9^2}{2 + \cdots}}}}},$$

from which the theory of continued fractions later arose. (No one knows how Brouncker made this discovery, but a proof based on the work of Euler in the next century is given in the chapter on Brouncker in J. L. Coolidge's *The Mathematics of Great Amateurs,* Oxford University Press, 1949.) Among the activities of Wallis's later years was a lively quarrel with the famous philosopher Hobbes, who was under the impression that he had succeeded in squaring the circle and published his erroneous proof. Wallis promptly refuted it, but Hobbes was both arrogant and too ignorant to understand the refutation, and defended himself with a barrage of additional errors, as if a question about the validity of a mathematical proof could be settled by rhetoric and invective.

where y denotes the length of the segment OS. The element of area dA sweeps across the circular segment in question as x increases from 0 to 1, so

$$A = \int dA = \frac{1}{2} \int_0^1 y \, dx;$$

and integrating by parts in order to reverse the roles of x and y gives

$$A = \frac{1}{2} xy \Big]_0^1 - \frac{1}{2} \int_0^1 x \, dy = \frac{1}{2} - \frac{1}{2} \int_0^1 x \, dy, \tag{1}$$

where the limits on the two integrals are understood to be $y = 0$ and $y = 1$. To continue the calculation, we observe that since

$$y = \tan \tfrac{1}{2}\phi \qquad \text{and} \qquad x = 1 - \cos \phi = 2 \sin^2 \tfrac{1}{2}\phi,$$

the trigonometric identity

$$\tan^2 \frac{1}{2} \phi = \frac{\sin^2 \tfrac{1}{2}\phi}{\cos^2 \tfrac{1}{2}\phi} = \sin^2 \frac{1}{2} \phi \sec^2 \frac{1}{2} \phi = \sin^2 \frac{1}{2} \phi \left(1 + \tan^2 \frac{1}{2} \phi\right)$$

yields

$$\frac{x}{2} = \frac{y^2}{1 + y^2}.$$

The version of the geometric series given in formula (13) in Section 9.5 enables us to write this as

$$\frac{x}{2} = y^2(1 - y^2 + y^4 - y^6 + \cdots) = y^2 - y^4 + y^6 - y^8 + \cdots,$$

so (1) becomes

$$\begin{aligned}
A &= \frac{1}{2} - \int_0^1 (y^2 - y^4 + y^6 - y^8 + \cdots) \, dy \\
&= \frac{1}{2} - \left[\frac{1}{3} y^3 - \frac{1}{5} y^5 + \frac{1}{7} y^7 - \frac{1}{9} y^9 + \cdots\right]_0^1 \\
&= \frac{1}{2} - \left(\frac{1}{3} - \frac{1}{5} + \frac{1}{7} - \frac{1}{9} + \cdots\right) \\
&= \frac{1}{2} - \frac{1}{3} + \frac{1}{5} - \frac{1}{7} + \frac{1}{9} - \cdots.
\end{aligned}$$

When $\tfrac{1}{2}$ is added to this to account for the area of the isosceles right triangle, and the result is equated to the known area $\pi/4$ of the quarter-circle, we have Leibniz's formula

$$\frac{\pi}{4} = 1 - \frac{1}{3} + \frac{1}{5} - \frac{1}{7} + \cdots.$$

Is it any wonder that he took great pleasure and pride in this discovery for the rest of his life?

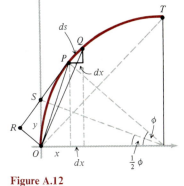

Figure A.12

A.12

EULER'S DISCOVERY OF
THE FORMULA $\sum_{1}^{\infty} \dfrac{1}{n^2} = \dfrac{\pi^2}{6}$

In Section 13.2 we encountered Euler's formula for the sum of the reciprocals of the squares,

$$1 + \frac{1}{4} + \frac{1}{9} + \frac{1}{16} + \cdots = \frac{\pi^2}{6}. \tag{1}$$

Our purpose in this appendix is to understand the heuristic reasoning that led Euler to this wonderful discovery.

We begin with some simple algebra. If a and b are $\neq 0$, then it is clear that these numbers are the roots of the equation

$$\left(1 - \frac{x}{a}\right)\left(1 - \frac{x}{b}\right) = 0. \tag{2}$$

This equation can also be written in the form

$$1 - \left(\frac{1}{a} + \frac{1}{b}\right)x + \frac{1}{ab}x^2 = 0, \tag{3}$$

in which it is evident that the negative of the coefficient of x is the sum of the reciprocals of the roots. If we replace x by x^2, and a and b by a^2 and b^2, then (2) and (3) become

$$\left(1 - \frac{x^2}{a^2}\right)\left(1 - \frac{x^2}{b^2}\right) = 0 \tag{4}$$

and

$$1 - \left(\frac{1}{a^2} + \frac{1}{b^2}\right)x^2 + \frac{1}{a^2b^2}x^4 = 0. \tag{5}$$

The roots of (4) are plainly $\pm a$ and $\pm b$, and (5) is the same equation in polynomial form, from which we see that the negative of the coefficient of x^2 is the sum of the reciprocals of the squares of the positive roots. This pattern persists as we move to equations of higher degree, for

$$\left(1 - \frac{x^2}{a^2}\right)\left(1 - \frac{x^2}{b^2}\right)\left(1 - \frac{x^2}{c^2}\right) = 0$$

(whose roots are obviously $\pm a$, $\pm b$, and $\pm c$) can be written as

$$1 - \left(\frac{1}{a^2} + \frac{1}{b^2} + \frac{1}{c^2}\right)x^2 + \left(\frac{1}{a^2b^2} + \frac{1}{a^2c^2} + \frac{1}{b^2c^2}\right)x^4 - \frac{1}{a^2b^2c^2}x^6 = 0,$$

and so on.

Let us now consider the transcendental equation

$$\sin x = 0$$

or

$$x - \frac{x^3}{3!} + \frac{x^5}{5!} - \frac{x^7}{7!} + \cdots = 0.$$

This can be thought of as "a polynomial equation of infinite degree" with an infinite number of roots $0, \pm\pi, \pm 2\pi, \pm 3\pi, \ldots$. The root 0 can be removed by dividing by x, which gives the equation

$$\frac{\sin x}{x} = 0$$

or

$$1 - \frac{x^2}{3!} + \frac{x^4}{5!} - \frac{x^6}{7!} + \cdots = 0,$$

with roots $\pm\pi, \pm 2\pi, \pm 3\pi, \ldots$. In the light of our knowledge of the roots of this equation, the situation described in the previous paragraph suggests that the infinite series

$$\frac{\sin x}{x} = 1 - \frac{x^2}{3!} + \frac{x^4}{5!} - \frac{x^6}{6!} + \cdots$$

can be written as an "infinite product,"

$$\frac{\sin x}{x} = \left(1 - \frac{x^2}{\pi^2}\right)\left(1 - \frac{x^2}{4\pi^2}\right)\left(1 - \frac{x^2}{9\pi^2}\right) \cdots . \tag{6}$$

Further, our analogy also suggests that

$$\frac{1}{\pi^2} + \frac{1}{4\pi^2} + \frac{1}{9\pi^2} + \cdots = \frac{1}{3!},$$

from which Euler's formula (1) follows at once. As an additional observation, it is interesting to note that if we put $x = \pi/2$ in (6), we find that

$$\frac{2}{\pi} = \left(1 - \frac{1}{2^2}\right)\left(1 - \frac{1}{4^2}\right)\left(1 - \frac{1}{6^2}\right) \cdots$$

$$= \left(\frac{1}{2} \cdot \frac{3}{2}\right)\left(\frac{3}{4} \cdot \frac{5}{4}\right)\left(\frac{5}{6} \cdot \frac{7}{6}\right) \cdots ,$$

which is equivalent to

$$\frac{\pi}{2} = \left(\frac{2}{1} \cdot \frac{2}{3}\right)\left(\frac{4}{3} \cdot \frac{4}{5}\right)\left(\frac{6}{5} \cdot \frac{6}{7}\right) \cdots .$$

This is *Wallis's product,* which was rigorously proved in Appendix A.10.

These daring speculations are characteristic of Euler's unique genius, but we hope that no students will suppose that they carry the force of rigorous proof. It will be seen that the crux of the matter is the question of the meaning and validity of (6), which is known as *Euler's infinite product for the sine.* The shortcomings of this discussion invite the construction of a general theory of infinite products, within which formulas like (6) can take their place as firmly established facts. This aim is achieved in more advanced fields of mathematics.

A rigorous elementary proof of Euler's formula (1) which is based on totally different ideas is given in the next appendix.

The proof given here is due to D. P. Giesy, and appeared in the *Mathematics Magazine,* vol. 45 (1972), pp. 148–149. Other elementary proofs (and some not so elementary) can be tracked down through Giesy's references.

We begin by defining a function $f_n(x)$ by

$$f_n(x) = \tfrac{1}{2} + \cos x + \cos 2x + \cdots + \cos nx. \tag{1}$$

We will need the closed formula for this function given by

$$f_n(x) = \frac{\sin [(2n + 1)x/2]}{2 \sin (x/2)}, \tag{2}$$

A.13

A RIGOROUS PROOF OF EULER'S FORMULA
$$\sum_{1}^{\infty} \frac{1}{n^2} = \frac{\pi^2}{6}$$

where x is not an integral multiple of 2π. To prove this, we use the trigonometric identity $2 \cos \theta \sin \phi = \sin (\theta + \phi) - \sin (\theta - \phi)$ to write the following identities (after the first):

$$2 \cdot \frac{1}{2} \sin \frac{1}{2} x = \sin \frac{1}{2} x,$$

$$2 \cos x \sin \frac{1}{2} x = \sin \frac{3}{2} x - \sin \frac{1}{2} x,$$

$$2 \cos 2x \sin \frac{1}{2} x = \sin \frac{5}{2} x - \sin \frac{3}{2} x,$$

$$\cdots$$

$$2 \cos nx \sin \frac{1}{2} x = \sin \frac{(2n+1)x}{2} - \sin \frac{(2n-1)x}{2}.$$

Formula (2) is now easy to obtain by adding these identities and carrying out the obvious cancellations. We now use (1) to define a number E_n, and also to calculate it, as follows:

$$E_n = \int_0^\pi x f_n(x)\, dx = \frac{\pi^2}{4} + \sum_{k=1}^n \left[\frac{(-1)^k - 1}{k^2} \right]. \tag{3}$$

Here each term of the integral after the first is integrated by parts to obtain the stated result. Since the even terms in the sum on the right are zero, (3) can be written in the form

$$\frac{1}{2} E_{2n-1} = \frac{\pi^2}{8} - \sum_{k=1}^n \frac{1}{(2k-1)^2}. \tag{4}$$

Our next purpose is to show that $\lim E_{2n-1} = 0$, because this will establish the formula

$$\sum_{k=1}^\infty \frac{1}{(2k-1)^2} = \frac{\pi^2}{8}, \tag{5}$$

which will be used to prove our final result. We accomplish this as follows. If we use formula (2) and define $g(x)$ by

$$g(x) = \frac{d}{dx} \left[\frac{x/2}{\sin (x/2)} \right],$$

then an integration by parts yields

$$E_{2n-1} = \frac{1}{4n-1} \left[2 + 2 \int_0^\pi g(x) \cos \frac{(4n-1)x}{2}\, dx \right], \tag{6}$$

where we make use in the calculation of the familiar fact that

$$\lim_{x \to 0} \frac{x/2}{\sin (x/2)} = 1.$$

Our desired conclusion that $\lim E_{2n-1} = 0$ will follow from (6) if we can show that $g(x)$ is bounded on the interval of integration. But $g(x)$ is increasing and is therefore bounded on this interval by $g(\pi) = \frac{1}{2}$, and this establishes (5).

To complete our proof of Euler's formula, we divide the positive integers into the evens and odds and use (5) to write

$$\sum_1^\infty \frac{1}{k^2} = \sum_1^\infty \frac{1}{(2k)^2} + \sum_1^\infty \frac{1}{(2k-1)^2} = \frac{1}{4} \sum_1^\infty \frac{1}{k^2} + \frac{\pi^2}{8}.$$

This yields

$$\frac{3}{4} \sum_1^\infty \frac{1}{k^2} = \frac{\pi^2}{8},$$

so

$$\sum_1^\infty \frac{1}{k^2} = \frac{4}{3} \cdot \frac{\pi^2}{8} = \frac{\pi^2}{6}.$$

A.14
THE SEQUENCE OF PRIMES

Number theory is mainly concerned with properties of the familiar positive integers 1, 2, 3, ... The notion of a positive integer is perhaps the simplest and clearest of all mathematical concepts, and yet, as we shall see, it is easy to ask elementary questions about these numbers which are incapable of being answered by the deepest resources of modern mathematics. This striking mixture of simplicity and profundity is part of the enduring attraction of the subject.

In Appendix A.1 we considered several interesting topics in the theory of numbers whose treatment did not depend on complicated mathematical machinery. We continue here with some further topics of this kind, and we also extend the range of our inquiry to include a few ideas that cannot be understood without some knowledge of calculus. We begin by reminding students of several concepts that were briefly discussed in the earlier appendix.

It is obvious that every positive integer is divisible by 1 and by itself. If an integer $p > 1$ has no positive divisors except 1 and p, it is called a *prime number*, or simply a *prime*; otherwise, it is said to be *composite*. The first few primes are easily seen to be

$$2, 3, 5, 7, 11, 13, 17, 19, 23, 29, 31, 37, 41, 43, \ldots .$$

It is a matter of common experience that every positive integer > 1 either is a prime or can be split into prime factors. Thus, for example, $84 = 2 \cdot 42 = 2 \cdot 2 \cdot 21 = 2 \cdot 2 \cdot 3 \cdot 7$ and $630 = 2 \cdot 315 = 2 \cdot 3 \cdot 105 = 2 \cdot 3 \cdot 3 \cdot 35 = 2 \cdot 3 \cdot 3 \cdot 5 \cdot 7$. We also feel that we ought to get the same prime factors no matter what method of factorization is used. These remarks are the content of the *unique factorization theorem*—also called the *fundamental theorem of arithmetic*—which we state formally as follows.

Theorem 1 *Every positive integer > 1 either is a prime or can be expressed as a product of primes, and this expression is unique except for the order of the prime factors.*

Proof We ask students to consider this in Problem 1.

This statement seems at first sight to be so obviously true that most people are inclined to take it for granted and accept it without proof. Nevertheless, it is far from trivial, and it takes on greater significance when one encounters systems of "integers" and "primes" for which it is false (see Problem 2). The questions that arise in this connection lead into the branch of modern mathematics known as *algebraic number theory*.

As we move out along the sequence of positive integers, we notice that the primes seem to occur less and less frequently. This is quite reasonable; a large number is more likely to be composite than a small one, since it lies beyond more numbers that might qualify as its factors. It is even conceivable that the primes might come to an end, and that all sufficiently large numbers are composite. Euclid's proof that this is not the case has been a model of mathematical elegance for more than two thousand years.

Theorem 2 (Euclid's Theorem)   *There are infinitely many primes.*

Proof It suffices to show that if p is any given prime, then there exists a prime $> p$. Let 2, 3, 5, . . . , p be the complete list of primes up to p. If we form the number $N = (2 \cdot 3 \cdot 5 \cdots p) + 1$, then it is clear that $N > p$ and also that N is not divisible by any of the primes 2, 3, 5, . . . , p. However, we know that N either is itself a prime or is divisible by some prime $q < N$, and in the latter case the preceding remark implies that $q > p$. Thus, in each case there exists a prime $> p$, and this is what we set out to prove.

Since every prime > 2 is odd, Euclid's theorem is equivalent to the assertion that the arithmetic progression

$$1, 3, 5, \ldots, 2n + 1, \ldots$$

of all odd positive integers contains infinitely many primes. It is therefore natural to wonder about primes in other arithmetic progressions. For example, it is clear that every odd prime lies in one of the two progressions

(a)   1, 5, 9, 13, 17, . . . , $4n + 1$, . . . ;
(b)   3, 7, 11, 15, 19, . . . , $4n + 3$, ...

We know that both progressions together contain infinitely many primes, but it is still possible that one of them might contain only a finite number. We can dispose of this possibility for progression (b) by a slight refinement of Euclid's argument.

Theorem 3   *The arithmetic progression* 3, 7, 11, . . . , $4n + 3$, . . . *contains infinitely many primes.*

Proof It is clear that the general term of our progression can also be written as $4n - 1$. Just as in Theorem 2, we show that if p is any given prime of this form, then there necessarily exists a larger prime of this form. Let 3, 7, 11, . . . , p be the complete list of primes in progression (b) up to p, and form the number $N = 4(3 \cdot 7 \cdot 11 \cdots p) - 1$. It is clear that $N > p$ and also that N is not divisible by 2 or any of the primes 3, 7, 11, . . . , p. If N is prime, then it is itself a prime of the form $4n - 1$ which is $> p$. Suppose that N is not prime. Then by the previous remark it is a product of odd primes $< N$ which cannot include any of the primes 3, 7, 11, . . . , p. We have noted that every odd prime is of the form $4n + 1$ or $4n - 1$, and since

$$(4m + 1)(4n + 1) = 4(4mn + m + n) + 1,$$

it is clear that any product of numbers of the form $4n + 1$ is again of this form. These facts imply that in our present situation, that is, where N is a product of odd primes which cannot include any of the primes 3, 7, 11, . . . , p, it must be true that at least one of its prime factors is of the form $4n - 1$ and therefore $> p$. We conclude that in each case there exists a prime in progression (b) which is $> p$, and this completes the proof.

It is also true that progression (a) contains infinitely many primes. However, the idea used in the proof of Theorem 3 breaks down in this case and must be replaced by another (see Problem 3). A similar situation arises with the two progressions whose terms are of the form $6n + 1$ and $6n + 5$, for together they clearly contain all primes except 2 and 3, and reasonably elementary methods suffice to show that there are infinitely many primes in each of them.

Let us now look for primes in a general arithmetic progression

$$a, a + b, a + 2b, \ldots, a + nb, \ldots,$$

where a and b are given positive integers. It is easy to see that none are present (except perhaps a itself) if a and b have a common factor > 1. If we exclude this case, then it is natural to conjecture that the progression will contain infinitely many primes. This generalizes our previous statements about primes in special arithmetic progressions, and is the content of a famous theorem proved by the German mathematician Dirichlet in 1837.

Theorem 4 (Dirichlet's Theorem) *If a and b are positive integers with no common factor > 1, then the arithmetic progression $a, a + b, a + 2b, \ldots,$ $a + nb, \ldots$ contains infinitely many primes.*

The methods of proof that work in the special cases we have discussed are unable to cope with the general arithmetic progression of this theorem. Dirichlet's proof used ideas and techniques from advanced analysis, and opened up new lines of thought in the theory of numbers which have been richly productive down to the present day. (*Analysis* is the standard term for that part of mathematics consisting of calculus and other subjects that depend on calculus more or less directly, such as differential equations, advanced calculus, etc. See Fig. 21.21.)

Many of the most interesting and important facts about primes were discovered by a combination of observation and experiment. In this kind of investigation it is useful to have available a list of all primes up to some prescribed limit N. An obvious method for constructing such a list is to write down in order all the integers from 2 to N and then systematically eliminate the composite numbers. Thus, since 2 is the first prime and every proper multiple of 2 is composite, we strike out 4, 6, 8, etc. The next remaining number is 3, which is prime because it is not a multiple of the only prime smaller than itself, namely 2. Since the proper multiples of 3 are composite, we strike out all of these numbers not already removed as multiples of 2. The next survivor is 5, which is prime because it is not a multiple of 2 or 3, so we strike out all proper multiples of 5 not already removed in the previous steps. And so on. We note that a composite number n must obviously have a prime factor $\leq \sqrt{n}$. This shows that the process is complete when we have eliminated the proper multiples of all primes $\leq \sqrt{N}$.

The procedure described here is called the *sieve of Eratosthenes*, after its discoverer.* The result of applying it to the case $N = 100$ is given in the following table. It

---

* The Greek scientist Eratosthenes (276–194 B.C.) was custodian of the famous Library at Alexandria. He wrote on astronomy, geography, chronology, ethics, mathematics, and other subjects. He is remembered mainly for his prime number sieve and for being the first to measure accurately (within 50 miles!) the circumference of the earth. Being informed that at noon on the summer solstice the sun illuminated the bottom of a well at Aswan (i.e., was directly overhead), he measured the angle between the zenith and the sun at noon on the summer solstice at Alexandria and also the distance between Alexandria and Aswan, which lie on roughly the same meridian. A simple calculation then gave the circumference of the earth. He was also a friend of Archimedes (287–212 B.C.)—the greatest intellect of antiquity—and was the recipient of a famous letter (called the *Method*) in which Archimedes revealed his method of making mathematical discoveries. In his old age Eratosthenes went blind, and is said to have committed suicide by starving himself to death.

should be noted that since $\sqrt{100} = 10$, the table is complete by the time all proper multiples of 7 have been stricken out. For the sake of emphasis, the primes are printed in color.

$$
\begin{array}{cccccccccc}
2 & 3 & 4 & 5 & 6 & 7 & 8 & 9 & 10 & 11 & 12 & 13 & 14 & 15 & 16 & 17 & 18 & 19 & 20 \\
21 & 22 & 23 & 24 & 25 & 26 & 27 & 28 & 29 & 30 & 31 & 32 & 33 & 34 & 35 & 36 & 37 & 38 & 39 & 40 \\
41 & 42 & 43 & 44 & 45 & 46 & 47 & 48 & 49 & 50 & 51 & 52 & 53 & 54 & 55 & 56 & 57 & 58 & 59 & 60 \\
61 & 62 & 63 & 64 & 65 & 66 & 67 & 68 & 69 & 70 & 71 & 72 & 73 & 74 & 75 & 76 & 77 & 78 & 79 & 80 \\
81 & 82 & 83 & 84 & 85 & 86 & 87 & 88 & 89 & 90 & 91 & 92 & 93 & 94 & 95 & 96 & 97 & 98 & 99 & 100
\end{array}
$$

Complete tables of primes up to more than 10,000,000 have been compiled by refinements of this process.* These tables provide the investigator with an enormous mass of raw data which can be used to formulate and test hypotheses, almost as if the study of prime numbers were a laboratory science. For example, an inspection of our short table shows that there are several chains of 5 consecutive composite numbers, and one of 7. Can the length of such a chain be made as large as we please? The answer to this question is easily seen to be Yes, for if n is a large positive integer, then $n! + 2, n! + 3, n! + 4, \ldots, n! + n$ is a chain of $n - 1$ consecutive composite numbers. On the other hand, the primes tend to cluster together here and there. In our table there are 8 pairs of *twin primes,* that is, primes like 3 and 5 which are separated by a single even number. Are there infinitely many such pairs? The longest tables suggest that there are, but no one knows for sure. In 1921 the Norwegian mathematician Viggo Brun was able to generalize the sieve of Eratosthenes to show that the sum of the reciprocals of the twin primes,

$$
\frac{1}{3} + \frac{1}{5} + \frac{1}{7} + \frac{1}{11} + \frac{1}{13} + \frac{1}{17} + \frac{1}{19} + \frac{1}{29} + \frac{1}{31} + \cdots,
$$

is either finite or convergent. This result is to be contrasted with the fact (which is proved in Appendix A.17) that the sum of the reciprocals of all the primes diverges.

We have seen that the primes are very irregularly distributed among all the positive integers. The problem of discovering the law that governs their occurrence—and of understanding the reasons for it—is one that has challenged human curiosity for hundreds of years.†

Many attempts have been made to find simple formulas for the nth prime p_n and for the exact number of primes among the first n positive integers. All such efforts have failed, and real progress was achieved only when mathematicians started instead to look for information about the *average* distribution of the primes among the positive integers. It is customary to denote by $\pi(x)$ the number of primes $\leq$ a positive number x. Thus, $\pi(1) = 0$, $\pi(2) = 1$, $\pi(3) = 2$, $\pi(4) = 2$, and $\pi(p_n) = n$ for every positive integer n. In his early youth the great Gauss studied this function by means of tables of primes, with the aim of finding a simple function that approximates $\pi(x)$ with a small relative error for large x. More precisely, he sought a function $f(x)$ with the property that

$$
\lim_{x \to \infty} \frac{f(x) - \pi(x)}{\pi(x)} = \lim_{x \to \infty} \left(\frac{f(x)}{\pi(x)} - 1 \right) = 0,
$$

---

* The standard reference is D. N. Lehmer, *List of Prime Numbers from 1 to 10,006,721,* Carnegie Institution of Washington Publication 165, 1914.

† In 1751 Euler expressed his own bafflement as follows: "Mathematicians have tried in vain to this day to discover some order in the sequence of prime numbers, and we have reason to believe that it is a mystery into which the human mind will never penetrate." Fortunately, Euler was wrong in this pessimistic forecast.

that is, such that

$$\lim_{x \to \infty} \frac{\pi(x)}{f(x)} = 1.$$

On the basis of his observations he conjectured (in 1792, at the age of 14 or 15) that both

$$\frac{x}{\ln x} \quad \text{and} \quad \text{li}(x) = \int_2^x \frac{dt}{\ln t}$$

are good approximations. The function li(x) is known as the *logarithmic integral*. The accompanying table shows how well these functions succeed.

x	$\pi(x)$	$x/\ln x$	$\text{li}(x)$
1,000	168	145	178
10,000	1,229	1,086	1,246
100,000	9,592	8,686	9,630
1,000,000	78,498	72,382	78,628
10,000,000	664,579	620,421	664,918

Even as an adult, Gauss was unable to prove his conjectures. The first solidly established results in this direction were attained around 1850 by the Russian mathematician Chebyshev, who showed that the inequalities

$$\frac{7}{8} < \frac{\pi(x)}{x/\ln x} < \frac{9}{8}$$

are true for all sufficiently large x. He also proved that if the limit

$$\lim_{x \to \infty} \frac{\pi(x)}{x/\ln x}$$

exists, then its value must be 1. The next step—a very long one—was taken by Riemann in 1859, in a brief paper of only 9 pages which is famous for its wealth of profound ideas. Riemann, however, merely sketched his proofs, and omitted some of them altogether, so his work was inconclusive in several respects. The end of this part of the story came in 1896 when Hadamard and de la Vallée Poussin, working independently of each other but building on Riemann's ideas, established the existence of this limit and thereby completed the proof of the *prime number theorem*:

$$\lim_{x \to \infty} \frac{\pi(x)}{x/\ln x} = 1. \tag{1}$$

This relatively simple law is one of the most remarkable facts in the whole of mathematics. If we write it in the form

$$\lim_{x \to \infty} \frac{\pi(x)/x}{1/\ln x} = 1, \tag{2}$$

then it admits the following interesting interpretation in terms of probability. If n is a positive integer, then the ratio $\pi(n)/n$ is the proportion of primes among the integers 1, 2, . . . , n, or equivalently, the probability that one of these integers chosen at random will be prime. We can think of (2) as asserting that this probability is approximately $1/\ln n$ for large n.

It follows quite easily from the prime number theorem that the nth prime is approximately $n \ln n$, in the sense that

$$\lim_{n\to\infty} \frac{p_n}{n \ln n} = 1. \tag{3}$$

To prove this, we use the fact that $\pi(p_n) = n$ and infer from (1) that

$$\lim_{n\to\infty} \frac{n}{p_n/\ln p_n} = 1 \quad \text{or} \quad \lim_{n\to\infty} \frac{p_n}{n \ln p_n} = 1. \tag{4}$$

If we now take the logarithm of (4) and use the continuity of the logarithm in the form $\ln \lim = \lim \ln$, we get

$$\lim_{n\to\infty} (\ln p_n - \ln n - \ln \ln p_n) = 0$$

or

$$\lim_{n\to\infty} \ln p_n \left[1 - \frac{\ln n}{\ln p_n} - \frac{\ln \ln p_n}{\ln p_n} \right] = 0.$$

This implies that the bracketed expression must approach 0; and since the third term in it also approaches 0 [recall that $(\ln n)/n \to 0$], we must have

$$\lim_{n\to\infty} \frac{\ln n}{\ln p_n} = 1. \tag{5}$$

With the aid of (5), (4) now yields

$$\lim_{n\to\infty} \frac{p_n}{n \ln n} = \lim_{n\to\infty} \frac{p_n}{n \ln p_n} \cdot \frac{\ln p_n}{\ln n} = 1,$$

which concludes the proof of (3).

It is also interesting to see that the prime number theorem is equivalent to the statement that

$$\lim_{x\to\infty} \frac{\pi(x)}{\text{li}(x)} = 1. \tag{6}$$

To prove this, it suffices to show that

$$\lim_{x\to\infty} \frac{\text{li}(x)}{x/\ln x} = 1; \tag{7}$$

for if this is so, then

$$\lim_{x\to\infty} \frac{\pi(x)}{x/\ln x} = \lim_{x\to\infty} \frac{\pi(x)}{\text{li}(x)} \cdot \frac{\text{li}(x)}{x/\ln x} = \lim_{x\to\infty} \frac{\pi(x)}{\text{li}(x)}.$$

We establish (7) as follows. On integrating $\text{li}(x)$ by parts, we get

$$\text{li}(x) = \int_2^x \frac{dt}{\ln t} = \frac{x}{\ln x} - \frac{2}{\ln 2} + \int_2^x \frac{dt}{(\ln t)^2}. \tag{8}$$

Since $1/(\ln t)^2$ is positive and decreasing for $t > 1$, if $x \geq 4$ we have

$$0 < \int_2^x \frac{dt}{(\ln t)^2} = \int_2^{\sqrt{x}} \frac{dt}{(\ln t)^2} + \int_{\sqrt{x}}^x \frac{dt}{(\ln t)^2}$$

$$< \frac{\sqrt{x} - 2}{(\ln 2)^2} + \frac{x - \sqrt{x}}{(\ln \sqrt{x})^2}$$

$$< \frac{\sqrt{x}}{(\ln 2)^2} + \frac{4x}{(\ln x)^2}.$$

This yields

$$0 < \frac{\int_2^x dt/(\ln t)^2}{x/\ln x} < \frac{\ln x}{\sqrt{x}(\ln 2)^2} + \frac{4}{\ln x},$$

so

$$\lim_{x \to \infty} \frac{\int_2^x dt/(\ln t)^2}{x/\ln x} = 0. \qquad (9)$$

If (8) is divided by $x/\ln x$, then (7) follows at once from (9), and the proof is complete. This result shows that both of Gauss's youthful conjectures were vindicated when the prime number theorem was finally proved.

The prime number theorem, as a statement involving the logarithm function and a limit, is obviously related to analysis. This is very surprising in view of the fact that the primes are discrete objects that have no apparent connection with the continuous functions and limit processes that are the essence of analysis. Nevertheless, almost all significant work on both Dirichlet's theorem and the prime number theorem has depended on the advanced analytical machinery of infinite series, complex variable theory, Fourier transforms, and the like. Accordingly, this part of mathematics has come to be known as *analytic number theory.**

---

* For additional information on the topics discussed above, see H. M. Edwards, *Riemann's Zeta Function,* Academic Press, 1974, pp. 1–6; and T. M. Apostol, *Introduction to Analytic Number Theory,* Springer-Verlag, 1976, pp. 1–12. For some extremely interesting plausibility discussions that convey an intuitive feeling for the meaning of the prime number theorem, see David Hawkins, "Mathematical Sieves," *Scientific American,* December 1958; and R. Courant and H. Robbins, *What Is Mathematics?* (Oxford University Press, 1941), pp. 482–486.

PROBLEMS

1 (a) If n is an integer > 1 which is not a prime, show that it can be expressed as a product of primes. Hint: There exist integers a and b, both of which are > 1 and $< n$, such that $n = ab$.

(b) If $p_1 p_2 \cdots p_m = q_1 q_2 \cdots q_n$, where the p's and q's are primes such that $p_1 \le p_2 \le \cdots \le p_m$ and $q_1 \le q_2 \le \cdots \le q_n$, show that $m = n$ and $p_1 = q_1$, $p_2 = q_2, \ldots, p_n = q_n$. Hint: Assume that a prime which divides a product of positive integers necessarily divides one of the factors (this fact is called *Euclid's lemma*).

2 (This problem is intended to suggest that the unique factorization theorem may not be as "obvious" as it looks.) The arithmetic progression 1, 4, 7, 10, . . . , $3n + 1$, . . . is a system of numbers which—like the positive integers—is closed under multiplication, in the sense that the product of any two numbers in the progression is again in the progression. Let a number $p > 1$ in this progression be called *pseudo-prime* if its only factorization into factors which are both in the progression is $p = 1 \cdot p$.

(a) Show that every number > 1 in the progression either is a pseudo-prime or can be expressed as a product of pseudo-primes.

(b) List all pseudo-primes ≤ 100.

(c) Find a number in the progression that can be expressed as a product of pseudo-primes in two different ways.

3 (a) It is asserted in the text that the arithmetic progression 1, 5, 9, 13, 17, . . . , $4n + 1$, . . . contains infinitely many primes. Try to prove this by imitating the proof of Theorem 3, that is, by listing all the

primes in the progression up to some given prime p (5, 13, 17, . . . , p) and considering the number $N = 4(5 \cdot 13 \cdot 17 \cdots p) + 1$. At what point does this attempted proof break down?

(b) It is known (and was first proved by Euler in 1749) that any odd prime factor of a number of the form $a^2 + 1$ is necessarily of the form $4n + 1$. Use this to prove the assertion in (a) by considering the number

$$M = (2 \cdot 5 \cdot 13 \cdot 17 \cdots p)^2 + 1$$
$$= 4(5 \cdot 13 \cdot 17 \cdots p)^2 + 1.$$

4 Prove that the arithmetic progression 5, 11, 17, . . . , $6n + 5$, . . . contains infinitely many primes. Hint: The general term of this progression can be written $6n - 1$.

5 Prove equation (7) by using L'Hospital's rule.

A.15

MORE ABOUT IRRATIONAL NUMBERS. π IS IRRATIONAL

Readers who have never thought about the matter before may wonder why we care about irrational numbers. In order to understand this, let us assume for a moment that the only numbers we have are the rationals—which, after all, are the only numbers ever used in making scientific measurements. Under these circumstances the symbol $\sqrt{2}$ has no meaning, since there is no rational number whose square is 2 (see Appendix A.1). One consequence of this is that the circle $x^2 + y^2 = 4$ and the straight line $y = x$ through its center do not intersect; that is, in spite of appearances, there is no point that lies on both, because both curves are discontinuous in the sense of having many missing points, and each threads its way through a gap in the other. This suggests that the system of rational numbers is an inadequate tool for representing the continuous objects of geometry and the continuous motions of physics. In addition, without irrational numbers most sequences and series would not converge and most integrals would not exist; and since it is also true that e and π would be meaningless (we prove below that π is irrational), the enormous and intricate structure of mathematical analysis would collapse into a heap of rubbish so insignificant as to be hardly worth sweeping up. As a practical matter, it is clear that if the irrationals did not exist, it would be necessary to invent them. It was the ancient Greeks who discovered that irrational numbers are indispensable in geometry, and this was one of their more important contributions to civilization.

In Section 14.3 we proved that e is irrational by assuming the contrary and constructing a number a which was then shown to be a positive integer < 1—an obvious impossibility. This strategy is also the key to the proofs of the following two theorems, but the details are somewhat more complicated.

We shall need a few properties of the function $f(x)$ defined by

$$f(x) = \frac{x^n(1-x)^n}{n!} = \frac{1}{n!} \sum_{k=n}^{2n} c_k x^k, \tag{1}$$

where the c_k's are certain integers and n is a positive integer to be specified later. First, it is clear that if $0 < x < 1$, then we have

$$0 < f(x) < \frac{1}{n!}. \tag{2}$$

Next, $f(0) = 0$ and $f^{(m)}(0) = 0$ if $m < n$ or $m > 2n$; also, if $n \le m \le 2n$, then

$$f^{(m)}(0) = \frac{m!}{n!} c_m,$$

and this number is an integer. Thus, $f(x)$ and all its derivatives have integral values at $x = 0$. Since $f(1 - x) = f(x)$, the same is true at $x = 1$.

Theorem 1 e^r is irrational for every rational number $r \ne 0$.

Proof If $r = p/q$ and e^r is rational, then so is $(e^r)^q = e^p$. Also, if e^{-p} is rational, so is e^p. It therefore suffices to prove that e^p is irrational for every positive integer p.

Assume that $e^p = a/b$ for certain positive integers a and b. We define $f(x)$ by (1) and $F(x)$ by

$$F(x) = p^{2n}f(x) - p^{2n-1}f'(x) + p^{2n-2}f''(x) - \cdots - pf^{(2n-1)}(x) + f^{(2n)}(x), \quad (3)$$

and we observe that $F(0)$ and $F(1)$ are integers. Next,

$$\frac{d}{dx}[e^{px}F(x)] = e^{px}[F'(x) + pF(x)] = p^{2n+1}e^{px}f(x), \quad (4)$$

where the last equality is obtained from a detailed examination of $F'(x) + pF(x)$ based on (3). Equation (4) shows that

$$b\int_0^1 p^{2n+1}e^{px}f(x)\, dx = b\left[e^{px}F(x) \right]_0^1 = aF(1) - bF(0),$$

which is an integer. However, (2) implies that

$$0 < b\int_0^1 p^{2n+1}e^{px}f(x)\, dx < \frac{bp^{2n+1}e^p}{n!} = bpe^p\,\frac{(p^2)^n}{n!};$$

and since the expression on the right $\to 0$ as $n \to \infty$ (by Example 3 in Section 14.2), it follows that the integer $aF(1) - bF(0)$ has the property that

$$0 < aF(1) - bF(0) < 1$$

if n is large enough. Since there is no positive integer < 1, this contradiction completes the proof.

If we say that a point (x, y) in the plane is a *rational point* whenever both x and y are rational numbers, then this theorem asserts that the curve $y = e^x$ traverses the plane in such a way that it misses all rational points except $(0, 1)$. An equivalent statement is that $y = \ln x$ misses all rational points except $(1, 0)$, so $\ln 2$, $\ln 3$, . . . are all irrational. It can also be proved that $y = \sin x$ misses all rational points except $(0, 0)$, and that $y = \cos x$ misses all rational points except $(0, 1)$.* Each of these theorems implies that π is irrational, since $\sin \pi = 0$ and $\cos \pi = -1$. However, we prefer to prove the irrationality of π by the following more direct argument.

Theorem 2 π *is irrational.*

Proof It is clearly sufficient to prove that π^2 is irrational, so we assume the contrary, that $\pi^2 = a/b$ for certain positive integers a and b. We again define $f(x)$ by (1), but this time we put

$$F(x) = b^n[\pi^{2n}f(x) - \pi^{2n-2}f''(x) + \pi^{2n-4}f^{(4)}(x) - \cdots + (-1)^n f^{(2n)}(x)], \quad (5)$$

and again we observe that $F(0)$ and $F(1)$ are integers. A calculation based on (5) shows that

$$\frac{d}{dx}[F'(x)\sin \pi x - \pi F(x)\cos \pi x] = [F''(x) + \pi^2 F(x)]\sin \pi x$$

$$= b^n\pi^{2n+2}f(x)\sin \pi x = \pi^2 a^n f(x)\sin \pi x,$$

---

* The details can be found in Chapter II of I. Niven's excellent book, *Irrational Numbers*, Wiley, 1956.

so

$$\int_0^1 \pi a^n f(x) \sin \pi x \, dx = \left[\frac{F'(x) \sin \pi x}{\pi} - F(x) \cos \pi x \right]_0^1$$
$$= F(1) + F(0),$$

which is an integer. But (2) implies that

$$0 < \int_0^1 \pi a^n f(x) \sin \pi x \, dx < \frac{\pi a^n}{n!} < 1$$

if n is large enough; and this contradiction—that $F(1) + F(0)$ is a positive integer < 1—concludes the proof.

The underlying method of proof in Theorems 1 and 2 was devised by the French mathematician Hermite in 1873, but the details of the latter argument were first published by Niven in 1947.

A.16
ALGEBRAIC AND TRANSCENDENTAL NUMBERS. *e* IS TRANSCENDENTAL

In Appendix A.15 we considered the classification of real numbers into the rationals and the irrationals. We now discuss a similar but much deeper and more substantive distinction, between algebraic and transcendental numbers.

A real number is said to be *algebraic* if it satisfies a polynomial equation of the form

$$a_n x^n + a_{n-1} x^{n-1} + \cdots + a_1 x + a_0 = 0 \tag{1}$$

with integral coefficients, where $a_n \neq 0$; if it satisfies no such equation, it is called *transcendental*.* For example, $\sqrt{2}$ and $\sqrt[3]{2} + \sqrt{2}$ are algebraic because they are roots of $x^2 - 2 = 0$ and $x^6 - 4x^3 + 2 = 0$. Any rational number p/q is algebraic, since it satisfies the first-degree equation $qx - p = 0$. The algebraic numbers can therefore be viewed as a natural generalization of the rational numbers. If an algebraic number satisfies (1) but no such equation of lower degree, it is said to be of *degree n*. Thus, the rationals are of degree 1, and $\sqrt{2}$ is of degree 2 because it is irrational and satisfies a second-degree equation.

As their name suggests, algebraic numbers are of great interest and importance for many reasons stemming from both algebra and number theory. We shall briefly explain how they also arise in geometry, and then conclude with some classical applications of analysis to this subject.

GEOMETRIC CONSTRUCTIONS

The ancient Greek mathematicians were fond of geometric problems requiring the construction of figures with specified properties. They often succeeded in solving their problems, but sometimes they failed, most notably in squaring a circle. By this they meant starting with a given circle and constructing a square of equal area. In this context the word "construct" has a special meaning, for the only instruments allowed in the game are an unmarked ruler and a compass.

Any ruler-and-compass construction starts with certain initial data, consisting of a finite collection of points, lines, and circles in a plane. The construction then proceeds to generate new points, lines, and circles by means of some combination of the following operations: With the ruler we can draw the line determined by two given points; with the compass we can draw the circle whose center is a given point and whose radius is the distance between two given points; and we can find the points of

---

* Complex numbers are classified in exactly the same way, but in most of this appendix we shall be concerned with real numbers alone.

intersection (if any) of given lines and circles. When it is understood in this way, the problem of squaring a circle reduces to that of starting with a segment of length 1 (we can take the radius of the given circle as the unit of measurement) and constructing a segment of length $\sqrt{\pi}$. This problem remained unsolved for more than 2000 years, and was settled only in the nineteenth century by the German mathematician Lindemann, who proved it to be unsolvable.

If we start with a single segment of length 1, then it is quite easy to show that the length of any constructible segment is an algebraic number whose degree is a power of 2.* In particular, if a real number is not algebraic at all, then it is certainly not constructible in our present sense. This is the way Lindemann proved that a circle cannot be squared: He demonstrated (in 1882) that π is transcendental and therefore not constructible, and from this it follows that $\sqrt{\pi}$ is also not constructible.† Lindemann's proof is rather difficult, so we now turn our attention to the simpler but still nontrivial problem of showing that transcendental numbers actually exist.

LIOUVILLE'S THEOREM

The first numbers that were known to be transcendental were exhibited by the French mathematician Liouville in 1851. He produced his examples with the aid of the following theorem, which in a loose interpretation says that an irrational algebraic number cannot be approximated very closely by rational numbers p/q unless the denominators q are allowed to be very large.

Theorem 1 *If α is a real algebraic number of degree $n > 1$, then there exists a constant $c > 0$ such that*

$$\left| \alpha - \frac{p}{q} \right| \geq \frac{c}{q^n}$$

for all rational numbers p/q with $q > 0$.

Proof The assumption is that α is a real root of a polynomial equation

$$f(x) = a_n x^n + a_{n-1} x^{n-1} + \cdots + a_1 x + a_0 = 0,$$

where the coefficients are integers, $n > 1$, and $a_n \neq 0$; and also that α does not satisfy any such equation of lower degree (and in particular is not rational). Let M be any upper bound for $|f'(x)|$ on the interval $\alpha - 1 \leq x \leq \alpha + 1$. If c is defined to be the smaller of the numbers 1 and $1/M$, we shall prove that c has the required property.

First, if $|\alpha - p/q| \geq 1$, then it is obvious that

$$\left| \alpha - \frac{p}{q} \right| \geq c \geq \frac{c}{q^n},$$

since q is a positive integer.

---

* A very clear elementary discussion can be found in Chapter III of R. Courant and H. Robbins, *What Is Mathematics?*, Oxford Press, 1941. For somewhat deeper treatments based more firmly on the theory of field extensions in modern algebra, see Burton W. Jones, *An Introduction to Modern Algebra*, Macmillan, 1975; I. T. Adamson, *Introduction to Field Theory*, Oliver and Boyd, 1964; or Seth Warner, *Modern Algebra*, Prentice-Hall, 1965.

† Ferdinand Lindemann (1852–1939) was a pupil of Weierstrass and taught at the University of Munich from 1893 until his retirement. There is a bust of him at this institution, and beneath his engraved name is the letter π, framed by a circle and a square in memory of his one great scientific achievement (one like this is enough!). He also tried fruitlessly for many years to prove Fermat's last theorem.

If $|\alpha - p/q| < 1$, we argue as follows. We begin by observing that p/q cannot be a root of $f(x) = 0$. For if it were, we could factor out $x - p/q$ from $f(x)$ and write the equation as $(x - p/q)g(x) = 0$, where $g(x)$ is a polynomial of degree $n - 1$ with integral coefficients; and α (being irrational) would then satisfy the equation $g(x) = 0$, contrary to the hypothesis. We therefore have $f(p/q) \neq 0$, so

$$\left| f\left(\frac{p}{q}\right) \right| = \frac{|a_n p^n + a_{n-1} p^{n-1} q + \cdots + a_0 q^n|}{q^n} \geq \frac{1}{q^n}.$$

Next, the mean value theorem implies that

$$\left| f\left(\frac{p}{q}\right) \right| = \left| f\left(\frac{p}{q}\right) - f(\alpha) \right| = \left| \frac{p}{q} - \alpha \right| |f'(a)|$$

for some number a that lies between p/q and α, and hence is in the interval on which $|f'(x)|$ is bounded by M. This yields

$$\frac{1}{q^n} \leq \left| f\left(\frac{p}{q}\right) \right| \leq \left| \alpha - \frac{p}{q} \right| M,$$

so

$$\left| \alpha - \frac{p}{q} \right| \geq \frac{1}{M} \cdot \frac{1}{q^n} \geq \frac{c}{q^n},$$

and the proof is complete.

As a direct consequence of this theorem, we have the following: If α is an irrational number with the property that for each $n > 1$ there exist rationals p/q making

$$q^n \left| \alpha - \frac{p}{q} \right|$$

as small as we please, then α cannot be algebraic and therefore must be transcendental. This fact enables us to produce many specific examples of transcendental numbers, for instance

$$\alpha = \frac{1}{10^{1!}} + \frac{1}{10^{2!}} + \frac{1}{10^{3!}} + \cdots + \frac{1}{10^{m!}} + \cdots$$

$$= 0.110001000000000000000000100 \ldots .$$

To see this, we first note that α is irrational because this decimal is nonrepeating. Next, if an integer $n > 1$ is given and for any $m > n$ we approximate α by the mth partial sum p/q of the series, then $q = 10^{m!}$ and

$$\left| \alpha - \frac{p}{q} \right| = \frac{1}{10^{(m+1)!}} + \frac{1}{10^{(m+2)!}} + \cdots < \frac{2}{10^{(m+1)!}}.$$

It now follows that

$$q^n \left| \alpha - \frac{p}{q} \right| = (10^{m!})^n \left| \alpha - \frac{p}{q} \right|$$

$$< 2 \cdot \frac{(10^{m!})^n}{10^{(m+1)!}} = 2 \left(\frac{10^n}{10^{m+1}} \right)^{m!} \to 0$$

as $m \to \infty$, so α is transcendental. It is clear that we can produce many other examples of this type by considering the sums of the series

$$\frac{a_1}{10^{1!}} + \frac{a_2}{10^{2!}} + \cdots + \frac{a_m}{10^{m!}} + \cdots ,$$

where the numerators are any of the digits 1, 2, . . . , 9. These irrational numbers are all transcendental, and the essential reason is that each is the limit of a very rapidly convergent sequence of rationals.

This approach makes it possible to exhibit many specimens of transcendental numbers, but it is much harder to prove the transcendence of particular numbers that occur more or less naturally in mathematics. The real beginning here was the work of Hermite, who showed in 1873 that *e* is transcendental (see below); Lindemann's 1882 proof for π was a fairly direct extension of Hermite's ideas. But subsequent progress was slow. In his famous Paris address of 1900, the great German mathematician David Hilbert proposed a list of 23 unsolved problems which he considered to be the outstanding challenges to the mathematicians of the future. The seventh of these (in part) was to prove that e^π and $2^{\sqrt{2}}$ are transcendental. Gelfond succeeded for e^π in 1929, and Kuzmin for $2^{\sqrt{2}}$ in 1930. This line of thought reached its climax in 1934–1935, when these facts took their place as special cases of a profound theorem of Gelfond and Schneider, which we state without proof as follows.[*]

Theorem 2 *If α and β are real or complex algebraic numbers such that α is neither 0 nor 1 and β is not a rational real number, then α^β is transcendental.*

The transcendence of $2^{\sqrt{2}}$ is an obvious consequence of this theorem, and for students who know something about complex numbers, that of e^π follows from the fact that e^π is one of the values of $e^{-2i \ln i} = i^{-2i}$, since the number last written has the required form. However, in spite of these advances, there are still many unsolved problems, and nothing is known about the nature of any of the following numbers: $e + \pi, e\pi, \pi^e, 2^e, 2^\pi$.

THE TRANSCENDENCE OF *e*

The argument we present is Hilbert's simplified version of Hermite's original proof.

We begin by assuming that *e* is algebraic of degree *n*, which means that it satisfies an equation of the form

$$a_n e^n + \cdots + a_2 e^2 + a_1 e + a_0 = 0, \tag{2}$$

where the coefficients are integers and $a_n, a_0 \neq 0$. The strategy of the proof is to reach a contradiction by closely approximating the various powers of *e* by rationals:

$$e^n = \frac{M_n + \epsilon_n}{M}, \qquad \ldots, \qquad e^2 = \frac{M_2 + \epsilon_2}{M}, \qquad e = \frac{M_1 + \epsilon_1}{M}, \tag{3}$$

where $M, M_n, \ldots, M_2, M_1$ are integers and the error terms $\epsilon_n/M, \ldots, \epsilon_2/M, \epsilon_1/M$ are very small. If we substitute (3) in (2) and multiply by M, then the result is

$$[a_n M_n + \cdots + a_1 M_1 + a_0 M] + [a_n \epsilon_n + \cdots + a_1 \epsilon_1] = 0. \tag{4}$$

The first expression in brackets is an integer, and we will choose the M's so that it is not zero. At the same time we will choose the ϵ's to be so small that the second expression in brackets is < 1 in absolute value,

$$|a_n \epsilon_n + \cdots + a_1 \epsilon_1| < 1. \tag{5}$$

---

[*] The main references are the book by Niven mentioned in Appendix A.15; C. L. Siegel, *Transcendental Numbers,* Princeton, 1949; and A. O. Gelfond, *Transcendental and Algebraic Numbers,* Dover, 1960. See also Hilbert's address, "Mathematical Problems," *Bull. Amer. Math. Soc.,* vol. 8(1902), pp. 437–479. The circumstances surrounding Hilbert's address, which perhaps was the second most important scientific lecture ever given, are interestingly described in C. Reid's biography, *Hilbert,* Springer-Verlag, 1970, pp. 69–84.

This contradiction (a nonzero integer plus a number of absolute value < 1 cannot be 0) will then complete the proof.

Students will surely agree that this is a simple and reasonable plan. But the remarkable feature of Hermite's proof is the extremely ingenious way in which he defines the M's and ϵ's by means of an integral whose structure is precisely adapted to its purpose. Hermite's integral defines M:

$$M = \int_0^\infty \frac{x^{p-1}[(x-1)(x-2) \cdots (x-n)]^p e^{-x}}{(p-1)!} \, dx, \tag{6}$$

where n is the degree of (2) and p is a prime to be specified later. We shall see that M is an integer for every choice of p, and it will be essential to know that p can be taken as large as we please (there are infinitely many primes!). It is clear from (3) that for each $k = 1, 2, \ldots, n$ we must have $M_k + \epsilon_k = e^k M$; and this relation will be satisfied if we define M_k and ϵ_k by breaking the interval of integration of $e^k M$ at the point k:

$$M_k = e^k \int_k^\infty \frac{x^{p-1}[(x-1) \cdots (x-n)]^p e^{-x}}{(p-1)!} \, dx, \tag{7}$$

$$\epsilon_k = e^k \int_0^k \frac{x^{p-1}[(x-1) \cdots (x-n)]^p e^{-x}}{(p-1)!} \, dx. \tag{8}$$

We now proceed to the detailed demonstration that for a suitable choice of p, these M's and ϵ's have the desired properties.

We will need the following formula from the elementary theory of the gamma function: For any positive integer m,

$$\int_0^\infty x^m e^{-x} \, dx = m!. \tag{9}$$

The proof is easy, and begins with an integration by parts (with $u = x^m$, $dv = e^{-x} \, dx$):

$$\int_0^\infty x^m e^{-x} \, dx = \left[-x^m e^{-x} \right]_0^\infty + m \int_0^\infty x^{m-1} e^{-x} \, dx$$

$$= m \int_0^\infty x^{m-1} e^{-x} \, dx.$$

The integral on the right is the same as the one on the left except that the exponent of x has been reduced by 1, so continuing in the same way yields

$$\int_0^\infty x^m e^{-x} \, dx = m \int_0^\infty x^{m-1} e^{-x} \, dx = m(m-1) \int_0^\infty x^{m-2} e^{-x} \, dx$$

$$= \cdots = m(m-1)(m-2) \cdots 2 \cdot 1 \int_0^\infty e^{-x} \, dx = m!,$$

since the value of the last integral is 1.

We now turn to the evaluation of the integral (6). If the expression $[(x-1)(x-2) \cdots (x-n)]$ is multiplied out, the result is a polynomial

$$x^n + \cdots \pm n!$$

with integral coefficients; raising this to the pth power yields

$$[(x-1)(x-2) \cdots (x-n)]^p = x^{np} + \cdots \pm (n!)^p,$$

where again the coefficients are integers. This enables us to write (6) in the form

$$M = \frac{\pm (n!)^p}{(p-1)!} \int_0^\infty x^{p-1} e^{-x} \, dx + \sum_{i=1}^{np} \frac{b_i}{(p-1)!} \int_0^\infty x^{p-1+i} e^{-x} \, dx,$$

where the b_i's are integers, and an application of (9) gives

$$M = \pm(n!)^p + \sum_{i=1}^{np} b_i \frac{(p-1+i)!}{(p-1)!},$$

which is an integer. If we now restrict ourselves to primes $p > n$, then the first term here is an integer not divisible by p. However,

$$b_i \frac{(p-1+i)!}{(p-1)!} = b_i p(p+1) \cdots (p-1+i)$$

is divisible by p, so M is plainly an integer which is not divisible by p. We may add the further condition that $p > |a_0|$; and since $a_0 \neq 0$, it follows that the term $a_0 M$ in (4) is not divisible by p (if a prime divides a product, it must divide one of the factors).

We next consider the integral (7) defining M_k:

$$M_k = \int_k^\infty \frac{x^{p-1}[(x-1) \cdots (x-n)]^p e^{-(x-k)}}{(p-1)!} \, dx.$$

If a new variable $y = x - k$ is introduced, this becomes

$$M_k = \int_0^\infty \frac{(y+k)^{p-1}[(y+k-1) \cdots y \cdots (y+k-n)]^p e^{-y}}{(p-1)!} \, dy.$$

Now the expression in brackets contains y in the kth place, so its pth power is a polynomial in y with terms running from y^p to y^{np}. This tells us that

$$(y+k)^{p-1}[(y+k-1) \cdots y \cdots (y+k-n)]^p$$

is a polynomial with integral coefficients whose terms run from y^p to y^{p-1+np}, so

$$M_k = \sum_{i=1}^{np} \frac{c_i}{(p-1)!} \int_0^\infty y^{p-1+i} e^{-y} \, dy = \sum_{i=1}^{np} c_i \frac{(p-1+i)!}{(p-1)!},$$

where the c_i's are integers. By the same reasoning as before, each M_k is an integer divisible by p, so the first bracketed expression in (4) is an integer not divisible by p. We conclude that for the primes p under consideration, the number

$$a_n M_n + \cdots + a_1 M_1 + a_0 M$$

is a nonzero integer, for if it were zero it would be divisible by p.

All that remains is to show that (5) is true if p is sufficiently large; and since n is fixed, it is enough to prove that each $|\epsilon_k|$ can be made as small as we please by taking p to be large enough. To establish this, we note that

$$|\epsilon_k| \leq e^k \int_0^k \frac{x^{p-1}|(x-1) \cdots (x-n)|^p e^{-x}}{(p-1)!} \, dx$$

$$\leq e^n \int_0^n \frac{x^{p-1}|(x-1) \cdots (x-n)|^p e^{-x}}{(p-1)!} \, dx$$

$$\leq \frac{e^n n^{p-1}}{(p-1)!} \int_0^n |(x-1) \cdots (x-n)|^p \, dx.$$

If B is an upper bound for $|(x-1) \cdots (x-n)|$ on the interval $0 \leq x \leq n$, then it follows that

$$|\epsilon_k| \leq \frac{e^n n^{p-1}}{(p-1)!} B^p n = e^n \frac{(nB)^p}{(p-1)!};$$

and since the expression on the right $\to 0$ as $p \to \infty$, the proof is complete.

A.17
THE SERIES $\Sigma 1/p_n$ OF THE RECIPROCALS OF THE PRIMES

In Appendix A.14 we gave Euclid's proof of the fact that there exist infinitely many prime numbers. About 2000 years after the time of Euclid, in 1737, Euler discovered two fundamentally different new proofs, and the methods he used laid the foundations of a new branch of mathematics that is now called *analytic number theory*.

In order to understand Euler's ideas, we begin by recalling that the harmonic series

$$1 + \frac{1}{2} + \frac{1}{3} + \cdots + \frac{1}{n} + \cdots$$

diverges. On the other hand, we know that for any exponent $s > 1$, the series

$$1 + \frac{1}{2^s} + \frac{1}{3^s} + \cdots + \frac{1}{n^s} + \cdots$$

converges, and the so-called *zeta function* is defined to be the sum of this series,

$$\zeta(s) = \sum_{n=1}^{\infty} \frac{1}{n^s} = 1 + \frac{1}{2^s} + \frac{1}{3^s} + \cdots ,$$

considered as a function of the variable s.*

Euler's basic discovery was a remarkable identity connecting the zeta function with the prime numbers,

$$\zeta(s) = \prod_p \frac{1}{1 - 1/p^s}, \tag{1}$$

where the expression on the right denotes the product of the numbers $1/(1 - p^{-s})$ for all primes $p = 2, 3, 5, 7, 11, \ldots$, that is, where

$$\prod_p \frac{1}{1 - 1/p^s} = \frac{1}{1 - 1/2^s} \cdot \frac{1}{1 - 1/3^s} \cdot \frac{1}{1 - 1/5^s} \cdot \frac{1}{1 - 1/7^s} \cdots .$$

To see how the identity (1) arises, we recall that the geometric series $1/(1 - x) = 1 + x + x^2 + \cdots$ is valid for $|x| < 1$, so for each prime p we have

$$\frac{1}{1 - 1/p^s} = 1 + \frac{1}{p^s} + \frac{1}{p^{2s}} + \frac{1}{p^{3s}} + \cdots .$$

Without stopping to justify the process, we now multiply these series together for all primes p, remembering that each integer $n > 1$ is uniquely expressible as a product of powers of different primes. This yields

$$\prod_p \frac{1}{1 - 1/p^s} = \prod_p \left(1 + \frac{1}{p^s} + \frac{1}{p^{2s}} + \frac{1}{p^{3s}} + \cdots \right)$$

$$= 1 + \frac{1}{2^s} + \frac{1}{3^s} + \cdots + \frac{1}{n^s} + \cdots$$

$$= \sum_{n=1}^{\infty} \frac{1}{n^s} = \zeta(s),$$

which is the identity (1).

One of Euler's arguments is based on (1) and goes this way. We begin by observing that if there were only a finite number of primes, then the product on the right-hand side of (1) would be an ordinary finite product and would clearly have a finite value for every $s > 0$, even for $s = 1$. However, the value of the left-hand side of (1) for $s = 1$ is the harmonic series,

---

* We denote the independent variable by s (instead of p, as in Chapter 14) in order to retain the notation that is customary in the theory of numbers.

$$\zeta(1) = 1 + \frac{1}{2} + \frac{1}{3} + \cdots ,$$

which diverges to infinity. This argument by contradiction, which can be made into a rigorous proof, shows that there must be an infinite number of primes. Euler's second argument rests on his discovery that *the series of the reciprocals of the primes diverges,*

$$\Sigma \frac{1}{p_n} = \frac{1}{2} + \frac{1}{3} + \frac{1}{5} + \frac{1}{7} + \frac{1}{11} + \cdots = \infty; \qquad (2)$$

for if there were only a finite number of primes, it is obvious that this series couldn't possibly diverge.

The proof of (2) that we give here starts with the geometric series

$$\frac{1}{1 - \frac{1}{2}} = 1 + \frac{1}{2} + \frac{1}{2^2} + \cdots ,$$

$$\frac{1}{1 - \frac{1}{3}} = 1 + \frac{1}{3} + \frac{1}{3^2} + \cdots ,$$

$$\frac{1}{1 - \frac{1}{5}} = 1 + \frac{1}{5} + \frac{1}{5^2} + \cdots ,$$

$$\cdots$$

$$\frac{1}{1 - 1/p_n} = 1 + \frac{1}{p_n} + \frac{1}{p_n{}^2} + \cdots .$$

If we multiply these series together by forming a new series whose terms are all possible products of one term selected from each of the series on the right, then this new series converges in any order to the product of the numbers on the left.* Since every integer greater than 1 is uniquely expressible as a product of powers of different primes, the product of these series is the series of the reciprocals of all positive integers whose prime factors are $\leq p_n$. In particular, all positive integers $\leq p_n$ have this property, so

$$\frac{1}{1 - \frac{1}{2}} \cdot \frac{1}{1 - \frac{1}{3}} \cdots \frac{1}{1 - 1/p_n} \geq \sum_{k=1}^{p_n} \frac{1}{k}$$

$$> \int_1^{p_n+1} \frac{dx}{x} = \ln (p_n + 1) > \ln p_n.$$

(It is in the transition here from the sum to the integral that we use the ideas of Section 14.5.) It follows that

$$\left(1 - \frac{1}{2}\right)\left(1 - \frac{1}{3}\right) \cdots \left(1 - \frac{1}{p_n}\right) < \frac{1}{\ln p_n},$$

and taking logarithms of both sides yields

$$\sum_{k=1}^{n} \ln \left(1 - \frac{1}{p_k}\right) < -\ln \ln p_n. \qquad (3)$$

We next show that

$$-\frac{2}{p_k} < \ln \left(1 - \frac{1}{p_k}\right), \qquad (4)$$

---

* This statement follows from one of the theorems proved in Appendix C.11.

for when this is applied to (3), we get

$$-2 \sum_{k=1}^{n} \frac{1}{p_k} < -\ln \ln p_n$$

or

$$\sum_{k=1}^{n} \frac{1}{p_k} > \frac{1}{2} \ln \ln p_n,$$

and our conclusion that $\Sigma 1/p_n$ diverges will follow from the fact that $\ln \ln p_n \to \infty$. To establish (4) and complete the argument, it suffices to observe that the line $y = 2x$ lies below the curve $y = \ln(1 + x)$ on the interval $-\frac{1}{2} \le x < 0$ and that every prime is ≥ 2.*

A.18

THE BERNOULLI NUMBERS AND SOME WONDERFUL DISCOVERIES OF EULER

In this appendix we derive several formulas discovered by Euler that rank among the most elegant truths in the whole of mathematics. We use the word "derive" instead of "prove" because some of our arguments are rather formal and require more advanced ideas than we can provide here to become fully rigorous in the sense demanded by modern concepts of mathematical proof. However, the mere fact that we are not able here to seal every crack in the reasoning seems a flimsy excuse for denying students an opportunity to glimpse some of the wonders that can be found in this part of calculus. For those who wish to dig deeper, full proofs are given in the treatise by K. Knopp mentioned in Section 14.11.

THE BERNOULLI NUMBERS

Since

$$\frac{e^x - 1}{x} = 1 + \frac{x}{2!} + \frac{x^2}{3!} + \cdots$$

for $x \ne 0$, and this power series has the value 1 at $x = 0$, the reciprocal function $x/(e^x - 1)$ has a power series expansion valid in some neighborhood of the origin if the value of this function is defined to be 1 at $x = 0$. We write this series in the form

$$\frac{x}{e^x - 1} = \sum_{n=0}^{\infty} \frac{B_n}{n!} x^n = B_0 + B_1 x + \frac{B_2}{2!} x^2 + \cdots. \tag{1}$$

The numbers B_n defined in this way are called the *Bernoulli numbers,* and it is clear that $B_0 = 1$. It is easy to see that

$$\frac{x}{e^x - 1} = \frac{x}{2}\left(\frac{e^x + 1}{e^x - 1} - 1\right) = -\frac{x}{2} + \frac{x}{2} \cdot \frac{e^x + 1}{e^x - 1}. \tag{2}$$

A routine check shows that the second term on the right is an even function, so $B_1 = -\frac{1}{2}$ and $B_n = 0$ if n is odd and > 1. If we write (1) in the form

$$\left(\frac{B_0}{0!} + \frac{B_1}{1!} x + \frac{B_2}{2!} x^2 + \cdots\right)\left(\frac{1}{1!} + \frac{x}{2!} + \frac{x^2}{3!} + \cdots\right) = 1,$$

then it is clear that the coefficient of x^{n-1} in the product on the left equals zero if $n > 1$. By the rule for multiplying power series, this yields

---

* For other proofs of (2), see I. Niven, *Amer. Math. Monthly,* 1971, pp. 272–273; and C. V. Eynden, *Amer. Math. Monthly,* 1980, pp. 394–397.

$$\frac{B_0}{0!} \cdot \frac{1}{n!} + \frac{B_1}{1!} \cdot \frac{1}{(n-1)!} + \frac{B_2}{2!} \cdot \frac{1}{(n-2)!} + \cdots + \frac{B_{n-1}}{(n-1)!} \cdot \frac{1}{1!} = 0,$$

and by multiplying through by $n!$ we obtain

$$\frac{n!}{0!n!} B_0 + \frac{n!}{1!(n-1)!} B_1 + \frac{n!}{2!(n-2)!} B_2 + \cdots + \frac{n!}{(n-1)!1!} B_{n-1} = 0. \qquad (3)$$

This equation can also be written more briefly as

$$\binom{n}{0} B_0 + \binom{n}{1} B_1 + \binom{n}{2} B_2 + \cdots + \binom{n}{n-1} B_{n-1} = 0$$

or

$$\sum_{k=0}^{n-1} \binom{n}{k} B_k = 0,$$

where $\binom{n}{k}$ is the binomial coefficient $n!/[k!(n-k)!]$. By taking $n = 3, 5, 7, 9,$ $11, \ldots$ in (3) and doing a little arithmetic, we easily find that

$$B_2 = \tfrac{1}{6}, \qquad B_4 = -\tfrac{1}{30}, \qquad B_6 = \tfrac{1}{42}, \qquad B_8 = -\tfrac{1}{30}, \qquad B_{10} = \tfrac{5}{66}, \ldots.$$

These calculations can be continued recursively as far as we please, and so all the Bernoulli numbers can be considered as known, even though considerable labor may be required to make any particular one of them visibly present. We also point out that it is obvious from (3) and the mode of calculation that every B_n is rational.

THE POWER SERIES FOR THE TANGENT

We now begin to explore the uses of these numbers.

In equation (2) we move the term $-x/2$ to the left and use the fact that

$$\frac{x}{2} \cdot \frac{e^x + 1}{e^x - 1} = \frac{x}{2} \cdot \frac{e^{x/2} + e^{-x/2}}{e^{x/2} - e^{-x/2}}$$

to obtain

$$\frac{x}{2} \cdot \frac{e^{x/2} + e^{-x/2}}{e^{x/2} - e^{-x/2}} = \sum_{n=0}^{\infty} \frac{B_{2n}}{(2n)!} x^{2n}. \qquad (4)$$

On the left-hand side of this, we now replace x by $2ix$, which yields

$$\frac{2ix}{2} \cdot \frac{e^{ix} + e^{-ix}}{e^{ix} - e^{-ix}} = x \frac{(e^{ix} + e^{-ix})/2}{(e^{ix} - e^{-ix})/2i} = x \cot x,$$

by the formulas for $\sin x$ and $\cos x$ that were derived in Section 14.12. Making the same substitution on the right-hand side of (4) gives

$$\sum_{n=0}^{\infty} \frac{B_{2n}}{(2n)!} (2ix)^{2n} = \sum_{n=0}^{\infty} (-1)^n \frac{2^{2n} B_{2n}}{(2n)!} x^{2n},$$

so

$$x \cot x = \sum_{n=0}^{\infty} (-1)^n \frac{2^{2n} B_{2n}}{(2n)!} x^{2n}. \qquad (5)$$

The trigonometric identity $\tan x = \cot x - 2 \cot 2x$ now enables us to use (5) to write

$$\tan x = \sum_{n=0}^{\infty} (-1)^n \frac{2^{2n} B_{2n}}{(2n)!} x^{2n-1} - 2 \sum_{n=0}^{\infty} (-1)^n \frac{2^{2n} B_{2n}}{(2n)!} (2x)^{2n-1}$$

$$= \sum_{n=0}^{\infty} (-1)^n \frac{2^{2n} B_{2n}}{(2n)!} x^{2n-1} - \sum_{n=0}^{\infty} (-1)^n \frac{2^{2n} B_{2n}}{(2n)!} 2^{2n} x^{2n-1}$$

$$= \sum_{n=0}^{\infty} (-1)^n \frac{2^{2n} B_{2n}}{(2n)!} (1 - 2^{2n}) x^{2n-1},$$

so

$$\tan x = \sum_{n=1}^{\infty} (-1)^{n+1} \frac{2^{2n}(2^{2n} - 1) B_{2n}}{(2n)!} x^{2n-1}.$$

This is the full power series for $\tan x$ that was encountered several times in truncated form in Section 14.11. Based on our knowledge of the Bernoulli numbers, the first few terms of this series are easy to calculate explicitly,

$$\tan x = x + \tfrac{1}{3}x^3 + \tfrac{2}{15}x^5 + \tfrac{17}{315}x^7 + \tfrac{62}{2835}x^9 + \cdots.$$

THE PARTIAL FRACTIONS EXPANSION OF THE COTANGENT

By using entirely different methods, Euler discovered another remarkable expansion of the cotangent: If x is not an integer, then

$$\pi \cot \pi x = \frac{1}{x} + 2x \sum_{n=1}^{\infty} \frac{1}{x^2 - n^2}. \tag{6}$$

We will examine this formula from two very different points of view, and give two derivations.

First, it is quite easy to see that (6) is analogous to the expansion of a rational function in partial fractions. For instance, if we consider the rational function $(2x + 1)/(x^2 - 3x + 2)$ and notice that the denominator has zeros 1 and 2 and can therefore be factored into $(x - 1)(x - 2)$, then this leads to the expansion

$$\frac{2x + 1}{x^2 - 3x + 2} = \frac{2x + 1}{(x - 1)(x - 2)} = \frac{c_1}{x - 1} + \frac{c_2}{x - 2}$$

for certain constants c_1 and c_2. The constant c_1 can now be determined by multiplying through by $x - 1$ and allowing x to approach 1, and similarly for c_2. Formally, (6) can be obtained in much the same way by noticing that $\cot \pi x = \cos \pi x / \sin \pi x$ has a denominator with zeros $0, \pm 1, \pm 2, \ldots$, and should therefore be expressible in the form

$$\cot \pi x = \frac{a}{x} + \sum_{n=1}^{\infty} \left(\frac{b_n}{x - n} + \frac{c_n}{x + n} \right). \tag{7}$$

From this, the constants a, b_n, and c_n can be found by the procedure suggested (they are all equal to $1/\pi$), and (7) can then be rearranged to yield (6). For reasons that will now be obvious, it is customary to refer to (6) as the *partial fractions expansion of the cotangent*. The main gap in this suggestive but rather tentative derivation is of course the fact that we have no prior guarantee that an expansion of the form (7) is possible.

Another way of approaching (6) is to begin with the infinite product (6) in Appendix A.12:

$$\frac{\sin x}{x} = \left(1 - \frac{x^2}{\pi^2} \right)\left(1 - \frac{x^2}{4\pi^2} \right)\left(1 - \frac{x^2}{9\pi^2} \right) \cdots \left(1 - \frac{x^2}{n^2\pi^2} \right) \cdots.$$

If we take the logarithm of both sides to obtain

$$\ln \frac{\sin x}{x} = \sum_{n=1}^{\infty} \ln\left(1 - \frac{x^2}{n^2\pi^2}\right),$$

and then differentiate, the result is easily seen to be

$$\cot x - \frac{1}{x} = \sum_{n=1}^{\infty} \frac{-2x}{n^2\pi^2 - x^2}$$

or

$$\cot x = \frac{1}{x} + 2x \sum_{n=1}^{\infty} \frac{1}{x^2 - n^2\pi^2};$$

and replacing x by πx and then multiplying through by πx yields

$$\pi x \cot \pi x = 1 + 2x^2 \sum_{n=1}^{\infty} \frac{1}{x^2 - n^2}, \tag{8}$$

which is equivalent to (6).

EULER'S FORMULA FOR $\Sigma 1/n^{2k}$

We now obtain a major payoff from (5) and (8) by replacing x by πx in (5) and equating the two expressions for $\pi x \cot \pi x$,

$$1 + \sum_{n=1}^{\infty} \frac{-2x^2}{n^2 - x^2} = 1 + \sum_{k=1}^{\infty} (-1)^k \frac{2^{2k}B_{2k}}{(2k)!}(\pi x)^{2k}, \tag{9}$$

where we use k as the index of summation on the right for reasons that will appear in a moment. Each term of the series on the left is easy to expand in a geometric series,

$$\frac{-2x^2}{n^2 - x^2} = -2\frac{x^2/n^2}{1 - x^2/n^2} = -2\sum_{k=1}^{\infty}\left(\frac{x^2}{n^2}\right)^k = -2\sum_{k=1}^{\infty}\frac{x^{2k}}{n^{2k}},$$

so (9) can be written as

$$1 + \sum_{n=1}^{\infty}\left(-2\sum_{k=1}^{\infty}\frac{x^{2k}}{n^{2k}}\right) = 1 + \sum_{k=1}^{\infty}(-1)^k\frac{2^{2k}B_{2k}}{(2k)!}\pi^{2k}x^{2k}.$$

We now interchange the order of summation on the left and obtain

$$1 + \sum_{k=1}^{\infty}\left(-2\sum_{n=1}^{\infty}\frac{1}{n^{2k}}\right)x^{2k} = 1 + \sum_{k=1}^{\infty}(-1)^k\frac{2^{2k}B_{2k}}{(2k)!}\pi^{2k}x^{2k},$$

and equating the coefficients of x^{2k} yields

$$\sum_{n=1}^{\infty}\frac{1}{n^{2k}} = (-1)^{k-1}\frac{2^{2k}B_{2k}}{2(2k)!}\pi^{2k}$$

for each positive integer k. In particular, for $k = 1, 2, 3$ we get

$$\sum_{n=1}^{\infty}\frac{1}{n^2} = \frac{\pi^2}{6}, \qquad \sum_{n=1}^{\infty}\frac{1}{n^4} = \frac{\pi^4}{90}, \qquad \sum_{n=1}^{\infty}\frac{1}{n^6} = \frac{\pi^6}{945}.$$

It is very remarkable that for almost 250 years there has been no progress whatever toward finding the exact sum of any one of the series

$$\sum_{n=1}^{\infty}\frac{1}{n^3}, \qquad \sum_{n=1}^{\infty}\frac{1}{n^5}, \qquad \sum_{n=1}^{\infty}\frac{1}{n^7}, \quad \ldots.$$

Perhaps a second Euler is needed for this breakthrough, but none is in sight.

A.19

BERNOULLI'S SOLUTION OF THE BRACHISTOCHRONE PROBLEM

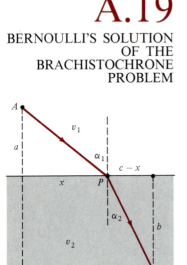

Figure A.13   The refraction of light.

As explained in Section 17.2, we begin with a point P_0 and a lower point P_1, and we seek the shape of the curved wire joining these points down which a bead will slide without friction in the shortest possible time.

We start by considering an apparently unrelated problem in optics. Figure A.13 illustrates a situation in which a ray of light travels from A to P with constant velocity v_1, and then, entering a denser medium, travels from P to B with a smaller velocity v_2. In terms of the notation in the figure, the total time T required for the journey is given by

$$T = \frac{\sqrt{a^2 + x^2}}{v_1} + \frac{\sqrt{b^2 + (c - x)^2}}{v_2}.$$

If we assume that this ray of light is able to select its path from A to B in such a way as to minimize T, then $dT/dx = 0$, and with a little work we see that the minimizing path is characterized by the equation

$$\frac{\sin \alpha_1}{v_1} = \frac{\sin \alpha_2}{v_2}.$$

This is *Snell's law of refraction.** The assumption that light travels from one point to another along the path requiring the shortest time is called *Fermat's principle of least time.* This principle not only provides a rational basis for Snell's law — which is an experimental fact — but also can be applied to find the path of a ray of light through a medium of variable density, where in general light will travel along curves instead of straight lines. In Fig. A.14a we have a stratified optical medium. In the individual layers the velocity of light is constant, but the velocity decreases from each layer to the one below it. As the descending ray of light passes from layer to layer, it is refracted more and more toward the vertical, and when Snell's law is applied to the boundaries between the layers, we obtain

$$\frac{\sin \alpha_1}{v_1} = \frac{\sin \alpha_2}{v_2} = \frac{\sin \alpha_3}{v_3} = \frac{\sin \alpha_4}{v_4}.$$

If we next allow these layers to grow thinner and more numerous, then in the limit the velocity of light decreases continuously as the ray descends, and we conclude that

$$\frac{\sin \alpha}{v} = \text{a constant.}$$

---

* See Example 4 in Section 4.4.

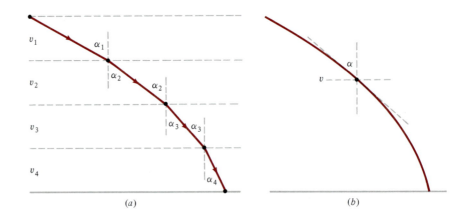

Figure A.14   Refraction in other optical media.

(a)

(b)

This situation is indicated in Fig. A.14b; it is approximately what happens to a ray of sunlight falling on the earth as it slows in descending through atmosphere of increasing density.

Returning now to the brachistochrone problem, we introduce a coordinate system as in Fig. A.15 and assume that the bead (like the ray of light) is capable of selecting the path down which it will slide from P_0 to P_1 in the shortest possible time. The argument given above yields

$$\frac{\sin \alpha}{v} = \text{a constant.} \tag{1}$$

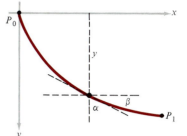

Figure A.15

If the bead has mass m, so that mg is the downward force that gravity exerts on it, then the fact that the work done by gravity in pulling the bead down the wire equals the increase in the kinetic energy of the bead tells us that $mgy = \frac{1}{2}mv^2$. This gives

$$v = \sqrt{2gy}. \tag{2}$$

From the geometry of the situation we also have

$$\sin \alpha = \cos \beta = \frac{1}{\sec \beta} = \frac{1}{\sqrt{1 + \tan^2 \beta}} = \frac{1}{\sqrt{1 + (y')^2}}. \tag{3}$$

On combining equations (1), (2), and (3)—obtained from optics, mechanics, and calculus—we get

$$y[1 + (y')^2] = c \tag{4}$$

as the differential equation of the brachistochrone.

We now complete our discussion, and discover what curve the brachistochrone actually is, by solving equation (4). When y' is replaced by dy/dx and the variables are separated, (4) becomes

$$dx = \sqrt{\frac{y}{c - y}}\, dy,$$

so

$$x = \int \sqrt{\frac{y}{c - y}}\, dy.$$

We evaluate this integral by starting with the algebraic substitution $u^2 = y/(c - y)$, so that

$$y = \frac{cu^2}{1 + u^2} \quad \text{and} \quad dy = \frac{2cu}{(1 + u^2)^2}\, du.$$

Then

$$x = \int \frac{2cu^2}{(1 + u^2)^2}\, du,$$

and the trigonometric substitution $u = \tan \phi$, $du = \sec^2 \phi\, d\phi$ enables us to write this as

$$x = \int \frac{2c \tan^2 \phi \sec^2 \phi}{(1 + \tan^2 \phi)^2}\, d\phi$$

$$= 2c \int \frac{\tan^2 \phi}{\sec^2 \phi}\, d\phi = 2c \int \sin^2 \phi\, d\phi$$

$$= c \int (1 - \cos 2\phi)\, d\phi = \frac{1}{2}c(2\phi - \sin 2\phi).$$

The constant of integration here is zero because $y = 0$ when $\phi = 0$, and since P_0 is at the origin, we also want to have $x = 0$ when $\phi = 0$. The formula for y gives

$$y = \frac{c \tan^2 \phi}{\sec^2 \phi} = c \sin^2 \phi = \frac{1}{2} c(1 - \cos 2\phi).$$

We now simplify our equations by writing $a = \frac{1}{2}c$ and $\theta = 2\phi$, which yields

$$x = a(\theta - \sin \theta), \qquad y = a(1 - \cos \theta).$$

These are the standard parametric equations of the cycloid with a cusp at the origin. We note that there is a single value of a that makes the first inverted arch of this cycloid pass through the point P_1 in Fig. A.15; for if a is allowed to increase from 0 to ∞, then the arch inflates, sweeps over the first quadrant of the plane, and clearly passes through P_1 for a single suitably chosen value of a.

A.20
EVOLUTES AND INVOLUTES

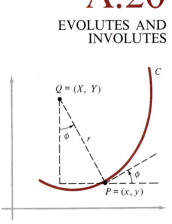

Figure A.16   The center of curvature.

Our main purpose in this appendix is to establish the property of cycloids stated in the last paragraph of Section 17.2. We shall do this by describing as briefly as possible the natural context for this property, within which it appears as a routine example.

Let P be a point moving along a given curve C, and assume that the curvature k is never zero on the part of C we consider, so that the radius of curvature $r = 1/|k|$ exists at every point. Draw the normal to C at P toward the concave side of the curve, and let Q be the point on this normal whose distance from P is r, as shown in Fig. A.16. The point Q is called the *center of curvature* of C corresponding to the point P. As P moves along C, the corresponding center of curvature Q generates a locus called the *evolute* of C. The evolute of a circle is a single point, its center. In general, however, the evolute of a curve is another curve.

We find the center of curvature Q as follows. Suppose C is the graph of a function $y = f(x)$, so that

$$k = \frac{y''}{(1 + y'^2)^{3/2}} \qquad \text{and} \qquad r = \frac{1}{|k|} = \frac{(1 + y'^2)^{3/2}}{|y''|}.$$

If P is (x, y), let the corresponding point Q be denoted by (X, Y). It is clear from Fig. A.16 that

$$x - X = \pm r \sin \phi, \qquad Y - y = \pm r \cos \phi, \tag{1}$$

where we use the plus or minus signs according as the curve is concave up or down. By recalling the meaning of the sign of the curvature, we see that (1) can be written as

$$x - X = \frac{1}{k} \sin \phi, \qquad Y - y = \frac{1}{k} \cos \phi. \tag{2}$$

Since

$$\sin \phi = \frac{dy}{ds} = \frac{y'}{\sqrt{1 + y'^2}} \qquad \text{and} \qquad \cos \phi = \frac{dx}{ds} = \frac{1}{\sqrt{1 + y'^2}},$$

equations (2) become

$$X = x - \frac{y'(1 + y'^2)}{y''}, \qquad Y = y + \frac{1 + y'^2}{y''}. \tag{3}$$

These equations give the coordinates of Q, and are therefore parametric equations for the evolute with the variable x used as the parameter. If the given curve C is defined by parametric equations $x = x(t)$ and $y = y(t)$, then we have

$$k = \frac{x'y'' - y'x''}{(x'^2 + y'^2)^{3/2}}, \qquad \sin \phi = \frac{y'}{\sqrt{x'^2 + y'^2}}, \qquad \cos \phi = \frac{x'}{\sqrt{x'^2 + y'^2}},$$

where now the primes denote derivatives with respect to t. In this case equations (3) are replaced by

$$X = x - \frac{y'(x'^2 + y'^2)}{x'y'' - y'x''}, \qquad Y = y + \frac{x'(x'^2 + y'^2)}{x'y'' - y'x''}, \tag{4}$$

which are parametric equations for the evolute with t as the parameter.

Example 1 Find the evolute of the parabola $y = x^2$.

Solution Since $y' = 2x$ and $y'' = 2$, equations (3) yield

$$X = -4x^3, \qquad Y = \frac{6x^2 + 1}{2}$$

after simplification. It is easy to eliminate the parameter x and put the equation of the evolute in the form

$$27X^2 = 16(Y - \tfrac{1}{2})^3.$$

The appearance of this curve in relation to the given parabola is shown in Fig. A.17.

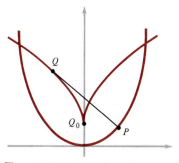

There are two important facts about the relation between a curve and its evolute that are easy to visualize by examining Fig. A.17. First, the line PQ is tangent to the evolute at Q. And second, if x is allowed to increase, so that P moves up the right side of the parabola, then the arc length Q_0Q along the evolute increases at exactly the same rate as the length of the tangent PQ.

These statements can be verified quite easily for the parabola and its evolute by using the formulas given in Example 1. However, instead of pausing to deal with this special case, we shall prove the statements for a general curve, as follows.

It is convenient to simplify the calculations by assuming that $k > 0$ at each point of the given curve C; the case $k < 0$ can be treated in a similar way. This assumption enables us to write equations (2) as

$$X = x - r \sin \phi, \qquad Y = y + r \cos \phi. \tag{5}$$

Since $k = d\phi/ds$ and $r = 1/k = ds/d\phi$, we have

$$r \sin \phi = \frac{ds}{d\phi} \frac{dy}{ds} = \frac{dy}{d\phi}, \qquad r \cos \phi = \frac{ds}{d\phi} \frac{dx}{ds} = \frac{dx}{d\phi}. \tag{6}$$

If we now differentiate equations (5) with respect to ϕ and use (6), we obtain

$$\frac{dX}{d\phi} = \frac{dx}{d\phi} - r \cos \phi - \frac{dr}{d\phi} \sin \phi = -\frac{dr}{d\phi} \sin \phi$$

and

$$\frac{dY}{d\phi} = \frac{dy}{d\phi} - r \sin \phi + \frac{dr}{d\phi} \cos \phi = \frac{dr}{d\phi} \cos \phi, \tag{7}$$

so

$$\frac{dY}{dX} = -\frac{\cos \phi}{\sin \phi} = -\frac{1}{\tan \phi}. \tag{8}$$

The slopes of the given curve C and its evolute E (Fig. A.18) are $\tan \phi$ and dY/dX. We can therefore conclude from equation (8) that *the normal PQ to the curve at P is*

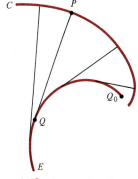

Figure A.17 The evolute of a parabola.

Figure A.18 A general evolute.

tangent to the evolute at the center of curvature Q. In getting from (7) to (8), it is evidently necessary to assume that $dr/d\phi \neq 0$ on C. That this condition is not superfluous is shown by the example of a circle, whose evolute is a single point.

To establish the second statement, let S be the length of the evolute from a fixed point Q_0 to the variable point Q corresponding to P. Then $dS^2 = dX^2 + dY^2$, and by (7) we have

$$\left(\frac{dS}{d\phi}\right)^2 = \left(\frac{dX}{d\phi}\right)^2 + \left(\frac{dY}{d\phi}\right)^2 = \left(\frac{dr}{d\phi}\right)^2. \tag{9}$$

If the direction of increasing S is chosen so that S and r both increase together, then (9) tells us that

$$\frac{dS}{d\phi} = \frac{dr}{d\phi},$$

and by integrating we obtain

$$S = r + \text{a constant.} \tag{10}$$

It follows from (10) that *the length of an arc of the evolute between any two points Q_1 and Q_2 is equal to the difference between the radii of curvature at the corresponding points P_1 and P_2.*

These results about the geometric relation between a curve C and its evolute E can be expressed in another way. Let a flexible inextensible string be wrapped around the evolute. If the string is held taut and unwound from the evolute, and if in addition the endpoint P of this string starts on the original curve C, then as we unwind the string, the point P traces out the curve C. This explains the name evolute, for it comes from the Latin *evolvere*, to unwind. Further, the original curve C is called an *involute* of the evolute E.

Example 2   Find the evolute of the cycloid $x = a(\theta - \sin \theta)$, $y = a(1 - \cos \theta)$.

Solution   We calculate

$$x' = a(1 - \cos \theta), \qquad y' = a \sin \theta,$$
$$x'' = a \sin \theta, \qquad y'' = a \cos \theta,$$

and apply formulas (4). After simplification the result is

$$X = a(\theta + \sin \theta),$$
$$Y = -a(1 - \cos \theta).$$

These are easily seen to be parametric equations for another cycloid congruent to the original one, but displaced $2a$ units down and πa units to the right, as shown in Fig. A.19. Since the lower cycloid is the evolute of the upper, the upper is an involute of the lower in the sense of the string property just discussed. This fact provides a complete justification for Huygens' construction of the cycloidal pendulum, described at the end of Section 17.2.

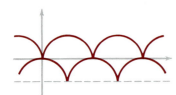

Figure A.19   The evolute of a cycloid.

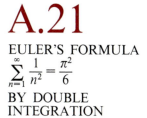

A.21

EULER'S FORMULA
$$\sum_{n=1}^{\infty} \frac{1}{n^2} = \frac{\pi^2}{6}$$
BY DOUBLE
INTEGRATION

The geometric series $1/(1 - r) = 1 + r + r^2 + \cdots$ enables us to write

$$\int_0^1 \int_0^1 \frac{dx\,dy}{1 - xy} = \int_0^1 \int_0^1 (1 + xy + x^2y^2 + \cdots)\,dx\,dy$$

$$= \int_0^1 \left(x + \frac{1}{2}x^2y + \frac{1}{3}x^3y^2 + \cdots\right)\Big]_0^1 dy$$

$$= \int_0^1 \left(1 + \frac{y}{2} + \frac{y^2}{3} + \cdots\right) dy$$

$$= \left(y + \frac{y^2}{2^2} + \frac{y^3}{3^2} + \cdots\right)\Big]_0^1 = 1 + \frac{1}{2^2} + \frac{1}{3^2} + \cdots.$$

The sum of Euler's series $\Sigma 1/n^2$ is therefore the value of the double integral

$$I = \int_0^1 \int_0^1 \frac{dx\,dy}{1 - xy}.$$

We evaluate this integral—and thereby determine the sum of the series—by means of a rotation of the coordinate system through the angle $\theta = \pi/4$.

If we rotate the xy-system into the uv-system through an arbitrary angle θ, as shown in Fig. A.20, then the transformation equations are

$$x = u \cos \theta - v \sin \theta,$$

$$y = u \sin \theta + v \cos \theta.$$

When $\theta = \pi/4$ these equations become

$$x = \tfrac{1}{2}\sqrt{2}(u - v),$$

$$y = \tfrac{1}{2}\sqrt{2}(u + v),$$

so we have

$$xy = \frac{1}{2}(u^2 - v^2) \quad \text{and} \quad 1 - xy = \frac{2 - u^2 + v^2}{2}.$$

By inspecting Fig. A.21, we see that the integral I can be written in the form

$$I = 4 \int_0^{\sqrt{2}/2} \int_0^u \frac{dv\,du}{2 - u^2 + v^2} + 4 \int_{\sqrt{2}/2}^{\sqrt{2}} \int_0^{\sqrt{2}-u} \frac{dv\,du}{2 - u^2 + v^2}.$$

If we denote the integrals on the right by I_1 and I_2, then

$$I_1 = 4 \int_0^{\sqrt{2}/2} \left[\int_0^u \frac{dv}{2 - u^2 + v^2}\right] du$$

$$= 4 \int_0^{\sqrt{2}/2} \left[\frac{1}{\sqrt{2 - u^2}} \tan^{-1}\left(\frac{v}{\sqrt{2 - u^2}}\right)\right]_0^u du$$

$$= 4 \int_0^{\sqrt{2}/2} \frac{1}{\sqrt{2 - u^2}} \tan^{-1}\left(\frac{u}{\sqrt{2 - u^2}}\right) du.$$

To continue the calculation, we use the substitution

$$u = \sqrt{2} \sin \theta, \quad \sqrt{2 - u^2} = \sqrt{2} \cos \theta, \quad du = \sqrt{2} \cos \theta\,d\theta,$$

$$\tan^{-1}\left(\frac{u}{\sqrt{2 - u^2}}\right) = \tan^{-1}\left(\frac{\sqrt{2} \sin \theta}{\sqrt{2} \cos \theta}\right) = \theta.$$

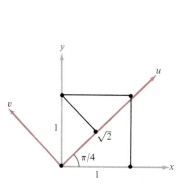

Figure A.20

Figure A.21

Then

$$I_1 = 4 \int_0^{\pi/6} \frac{1}{\sqrt{2} \cos \theta} \cdot \theta \cdot \sqrt{2} \cos \theta \, d\theta = 2\theta^2 \Big]_0^{\pi/6} = \frac{\pi^2}{18}.$$

To calculate I_2, we write

$$I_2 = 4 \int_{\sqrt{2}/2}^{\sqrt{2}} \left[\int_0^{\sqrt{2}-u} \frac{dv}{2 - u^2 + v^2} \right] du$$

$$= 4 \int_{\sqrt{2}/2}^{\sqrt{2}} \left[\frac{1}{\sqrt{2-u^2}} \tan^{-1} \left(\frac{v}{\sqrt{2-u^2}} \right) \right]_0^{\sqrt{2}-u} du$$

$$= 4 \int_{\sqrt{2}/2}^{\sqrt{2}} \frac{1}{\sqrt{2-u^2}} \tan^{-1} \left(\frac{\sqrt{2}-u}{\sqrt{2-u^2}} \right) du.$$

To continue the calculation, we use the same substitution as before, with the additional fact that

$$\tan^{-1} \left(\frac{\sqrt{2}-u}{\sqrt{2-u^2}} \right) = \tan^{-1} \left(\frac{\sqrt{2} - \sqrt{2} \sin \theta}{\sqrt{2} \cos \theta} \right) = \tan^{-1} \left(\frac{1 - \sin \theta}{\cos \theta} \right)$$

$$= \tan^{-1} \left(\frac{\cos \theta}{1 + \sin \theta} \right) = \tan^{-1} \left(\frac{\sin (\pi/2 - \theta)}{1 + \cos (\pi/2 - \theta)} \right)$$

$$= \tan^{-1} \left(\frac{2 \sin \frac{1}{2}(\pi/2 - \theta) \cos \frac{1}{2}(\pi/2 - \theta)}{2 \cos^2 \frac{1}{2}(\pi/2 - \theta)} \right) = \frac{1}{2} \left(\frac{\pi}{2} - \theta \right).$$

This enables us to write

$$I_2 = 4 \int_{\pi/6}^{\pi/2} \frac{1}{\sqrt{2} \cos \theta} \left(\frac{\pi}{4} - \frac{1}{2} \theta \right) \sqrt{2} \cos \theta \, d\theta = 4 \left[\frac{\pi}{4} \theta - \frac{1}{4} \theta^2 \right]_{\pi/6}^{\pi/2}$$

$$= 4 \left[\left(\frac{\pi^2}{8} - \frac{\pi^2}{16} \right) - \left(\frac{\pi^2}{24} - \frac{\pi^2}{144} \right) \right] = \frac{\pi^2}{9}.$$

We complete the calculation by putting these results together,

$$\sum_{n=1}^{\infty} \frac{1}{n^2} = I = I_1 + I_2 = \frac{\pi^2}{18} + \frac{\pi^2}{9} = \frac{\pi^2}{6}.$$

It is interesting to observe that

$$\int_0^1 \int_0^1 \int_0^1 \frac{dx \, dy \, dz}{1 - xyz} = \sum_{n=1}^{\infty} \frac{1}{n^3},$$

so that any person who can evaluate this triple integral will thereby discover the sum of the series on the right—which has remained an unsolved problem since Euler first raised the question in 1736.

In this section and the next, we extend the ideas of Chapter 21 to three dimensions and give a brief intuitive introduction to the two fundamental integral theorems of vector analysis. These theorems are roughly similar to each other, for both make assertions of the following kind:

The integral of a certain function over the boundary of a region is equal to the integral of a related function over the region itself.

It is possible to spend considerable time analyzing such purely mathematical issues as what is meant by a region and its boundary, but in this short sketch we shall proceed informally and concentrate instead on the physical meaning of what we are doing.

The concept of *gradient,* as we presented it in Chapter 19, applies only to scalar fields, that is, functions whose values are numbers. The gradient of a scalar field $f(x, y, z)$ is a vector field that represents the rate of change of f, because at any point its component in a given direction is the directional derivative of f in that direction. Our purpose here is to consider the more complicated problem of describing the rate of change of a *vector* field. There are two fundamental tools for measuring the rate of change of a vector field: the *divergence* and the *curl.*

We recall that the gradient of a scalar field $f(x, y, z)$ is defined by

$$\nabla f = \frac{\partial f}{\partial x}\,\mathbf{i} + \frac{\partial f}{\partial y}\,\mathbf{j} + \frac{\partial f}{\partial z}\,\mathbf{k},$$

where the symbol ∇ ("del") represents the vector differential operator

$$\nabla = \frac{\partial}{\partial x}\,\mathbf{i} + \frac{\partial}{\partial y}\,\mathbf{j} + \frac{\partial}{\partial z}\,\mathbf{k}.$$

If $\mathbf{F} = L\mathbf{i} + M\mathbf{j} + N\mathbf{k}$ is a given vector field, we can apply ∇ to $\mathbf{F}$ in two ways, by using the dot and cross products. We interpret the dot product of ∇ and $\mathbf{F}$ to mean

$$\nabla \cdot \mathbf{F} = \left(\frac{\partial}{\partial x}\,\mathbf{i} + \frac{\partial}{\partial y}\,\mathbf{j} + \frac{\partial}{\partial z}\,\mathbf{k}\right) \cdot (L\mathbf{i} + M\mathbf{j} + N\mathbf{k})$$

$$= \frac{\partial L}{\partial x} + \frac{\partial M}{\partial y} + \frac{\partial N}{\partial z}.$$

This scalar quantity is called the *divergence* of $\mathbf{F}$ and is often denoted by div $\mathbf{F}$, so that

$$\text{div } \mathbf{F} = \nabla \cdot \mathbf{F} = \frac{\partial L}{\partial x} + \frac{\partial M}{\partial y} + \frac{\partial N}{\partial z}. \tag{1}$$

The cross product of ∇ and $\mathbf{F}$ is interpreted to mean

$$\nabla \times \mathbf{F} = \begin{vmatrix} \mathbf{i} & \mathbf{j} & \mathbf{k} \\ \dfrac{\partial}{\partial x} & \dfrac{\partial}{\partial y} & \dfrac{\partial}{\partial z} \\ L & M & N \end{vmatrix} = \left(\frac{\partial N}{\partial y} - \frac{\partial M}{\partial z}\right)\mathbf{i} + \left(\frac{\partial L}{\partial z} - \frac{\partial N}{\partial x}\right)\mathbf{j} + \left(\frac{\partial M}{\partial x} - \frac{\partial L}{\partial y}\right)\mathbf{k}.$$

This vector quantity is called the *curl* of $\mathbf{F}$ and is often denoted by curl $\mathbf{F}$, so that

$$\text{curl } \mathbf{F} = \nabla \times \mathbf{F}. \tag{2}$$

Example 1 Compute the divergence and curl of the vector field $\mathbf{F} = 2x^2y\mathbf{i} + 3xz^3\mathbf{j} + xy^2z^2\mathbf{k}$.

A.22

SURFACE INTEGRALS AND THE DIVERGENCE THEOREM

Solution By using formulas (1) and (2) we at once obtain

$$\operatorname{div} \mathbf{F} = \nabla \cdot \mathbf{F} = \frac{\partial}{\partial x}(2x^2y) + \frac{\partial}{\partial y}(3xz^3) + \frac{\partial}{\partial z}(xy^2z^2)$$

$$= 4xy + 2xy^2z$$

and

$$\operatorname{curl} \mathbf{F} = \nabla \times \mathbf{F} = \begin{vmatrix} \mathbf{i} & \mathbf{j} & \mathbf{k} \\ \dfrac{\partial}{\partial x} & \dfrac{\partial}{\partial y} & \dfrac{\partial}{\partial z} \\ 2x^2y & 3xz^3 & xy^2z^2 \end{vmatrix}$$

$$= (2xyz^2 - 9xz^2)\mathbf{i} + (-y^2z^2)\mathbf{j} + (3z^3 - 2x^2)\mathbf{k}.$$

There is clearly no difficulty about performing calculations of this kind. The real questions are, What do they mean and what is their value? Our purpose in the rest of this section is to explore the meaning of the divergence, and to do this we need the concept of flux.

THE MEANING OF THE DIVERGENCE

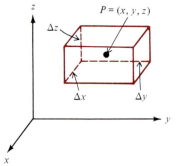

Figure A.22

We shall use an example from hydrodynamics to motivate the ideas. Suppose that a stream of fluid (gas or liquid) is flowing through a region of space. At a given point (x, y, z), let its density be the scalar function $\delta = \delta(x, y, z)$ and its velocity the vector function $\mathbf{v} = \mathbf{v}(x, y, z)$, and consider the vector field $\mathbf{F} = \delta\mathbf{v}$. Now consider a small flat patch of surface inside the fluid, with area ΔA and unit normal vector $\mathbf{n}$, as shown in Fig. A.22. If we think of this patch as a piece of screen or netting, so that the fluid can move through it without hindrance, we wish to find an expression for the amount of fluid that flows through the patch per unit time. It is clear from the figure that the fluid passing through the patch in a small time interval Δt forms a small tube of approximate volume $(\mathbf{v}\,\Delta t)\cdot\mathbf{n}\,\Delta A$, and the approximate mass of the fluid in this tube is $\delta(\mathbf{v}\,\Delta t)\cdot\mathbf{n}\,\Delta A$.* The approximate mass of fluid crossing the area ΔA per unit time is therefore $\delta\mathbf{v}\cdot\mathbf{n}\,\Delta A$ or $\mathbf{F}\cdot\mathbf{n}\,\Delta A$. This is called the *flux* of the vector field $\mathbf{F}$ through the area ΔA.

We now put forward an alternative definition for the divergence of $\mathbf{F}$ and then show that this new definition agrees with the one given above in formula (1). The purpose of this maneuver is to arrive at a way of thinking about the divergence that conveys an intuitive understanding of what it means.

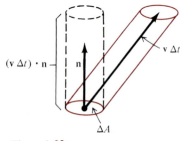

Figure A.23

Consider a point $P = (x, y, z)$ at the center of a small rectangular box with edges Δx, Δy, Δz, as shown in Fig. A.23. We compute the total flux of the vector field $\mathbf{F}$ outward through the six faces of this box (i.e., on each face we choose $\mathbf{n}$ to be the outward unit normal). We then divide this total flux by the volume $\Delta V = \Delta x\,\Delta y\,\Delta z$ of the box, and form the limit of this flux per unit volume as the dimensions of the box approach zero. This is our new definition for the divergence of $\mathbf{F}$ at the point $P = (x, y, z)$:

$$\operatorname{div} \mathbf{F} = \lim_{\Delta V \to 0} \frac{1}{\Delta V}(\text{flux of } \mathbf{F} \text{ out through the faces}). \tag{3}$$

---

* We assume in this discussion that all functions are continuous, so when ΔA and Δt are very small, the vector $\mathbf{v}$ changes very little in direction or magnitude from one point of ΔA to another, and the density δ changes very little from one point of the tube to another.

Physically, this represents the mass of fluid that emerges from a small element of volume containing the point P, per unit time per unit volume.

To show that this definition agrees with formula (1), we carry out a rough calculation of the limit (3), where $\mathbf{F} = L\mathbf{i} + M\mathbf{j} + N\mathbf{k}$. On the front face of the box in Fig. A.23 we see that the outward unit normal is $\mathbf{i}$, so the flux out through this face is approximately $L(x + \frac{1}{2}\Delta x, y, z)\,\Delta y\,\Delta z$. Since the outward unit normal on the back face is $-\mathbf{i}$, the flux out through this face is approximately $-L(x - \frac{1}{2}\Delta x, y, z)\,\Delta y\,\Delta z$, and therefore the combined flux out through the front and back faces is approximately

$$[L(x + \tfrac{1}{2}\Delta x, y\ z) - L(x - \tfrac{1}{2}\Delta x, y, z)]\,\Delta y\,\Delta z.$$

Similarly, the faces in the y-direction and z-direction contribute flux of approximate amounts

$$[M(x, y + \tfrac{1}{2}\Delta y, z) - M(x, y - \tfrac{1}{2}\Delta y, z)]\,\Delta x\,\Delta z$$

and

$$[N(x, y, z + \tfrac{1}{2}\Delta z) - N(x, y, z - \tfrac{1}{2}\Delta z)]\,\Delta x\,\Delta y.$$

We next divide the sum of these three quantities — the total flux out through all the faces of the box — by $\Delta V = \Delta x\,\Delta y\,\Delta z$ to obtain

$$\frac{L(x + \frac{1}{2}\Delta x, y, z) - L(x - \frac{1}{2}\Delta x, y, z)}{\Delta x} + \frac{M(x, y + \frac{1}{2}\Delta y, z) - M(x, y - \frac{1}{2}\Delta y, z)}{\Delta y}$$

$$+ \frac{N(x, y, z + \frac{1}{2}\Delta z) - N(x, y, z - \frac{1}{2}\Delta z)}{\Delta z}.$$

Finally, if we take the limit of this expression as $\Delta x, \Delta y, \Delta z \to 0$, then formula (3) yields the earlier definition (1), as stated,*

$$\text{div } \mathbf{F} = \frac{\partial L}{\partial x} + \frac{\partial M}{\partial y} + \frac{\partial N}{\partial z}.$$

This result permits us to consider (3) as the basic definition of the divergence and (1) as merely a formula for computing it in rectangular coordinates.

SURFACE INTEGRALS

Let S be a smooth surface and $f(x, y, z)$ a continuous function defined on S. The *surface integral* of f over S is denoted by

$$\iint\limits_{S} f(x, y, z)\, dA, \tag{4}$$

and is defined as a limit of sums in the following way. We begin by subdividing the surface into n small pieces with areas $\Delta A_1, \Delta A_2, \ldots, \Delta A_n$. We next choose a point (x_i, y_i, z_i) on the ith piece, find the value $f(x_i, y_i, z_i)$ of the function at this point, multiply this value by the area ΔA_i to obtain the product $f(x_i, y_i, z_i)\,\Delta A_i$, and form the sum of these products,

$$\sum_{i=1}^{n} f(x_i, y_i, z_i)\,\Delta A_i. \tag{5}$$

---

* Here we use a slightly different way of defining the derivative of a function. See Additional Problem 9 in Chapter 2.

Finally, we let n tend to infinity in such a way that the largest diameter of the pieces approaches zero; that is, we carry out a sequence of subdivisions of the surface S into smaller and smaller pieces, each time constructing a sum of the form (5). If these sums approach a limiting value, independent of the way the subdivisions are formed and the way the points (x_i, y_i, z_i) are chosen, then this limit is the definition of the surface integral (4):

$$\iint\limits_S f(x, y, z)\, dA = \lim_{n \to \infty} \sum_{i=1}^{n} f(x_i, y_i, z_i)\, \Delta A_i.$$

It may be encouraging to students to know that only rarely do we actually evaluate a surface integral. It is the *concept* of these integrals that is important, because they provide a convenient language for expressing certain basic ideas of mathematics and physics.

To see what a surface integral can represent, we return to our example from hydrodynamics. Consider a fluid flowing through a certain region of space, and let $\delta = \delta(x, y, z)$ and $\mathbf{v} = \mathbf{v}(x, y, z)$ be its density and velocity, as before. Suppose that S is a smooth surface lying inside the region, and think of S as a curved piece of screen or netting that permits the fluid to pass through it without any resistance (Fig. A.24). As we saw in our previous discussion, the mass of fluid crossing a surface element of area dA and unit normal $\mathbf{n}$ per unit time, is $\delta \mathbf{v} \cdot \mathbf{n}\, dA$ or $\mathbf{F} \cdot \mathbf{n}\, dA$, where $\mathbf{F} = \delta \mathbf{v}$. Accordingly, the surface integral

$$\iint\limits_S \mathbf{F} \cdot \mathbf{n}\, dA \tag{6}$$

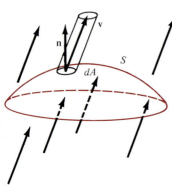

Figure A.24

gives the rate of flow of the fluid through the entire surface S in terms of mass per unit time. This is called the *flux* of $\mathbf{F}$ through S.

More generally, if $\mathbf{F}$ is any vector field whatever, the surface integral (6) is still called the *flux* of $\mathbf{F}$ through the surface S. The physical meaning of this integral clearly depends on the nature of the physical quantity represented by $\mathbf{F}$. A variety of interpretations and applications arise by letting $\mathbf{F}$ be a vector field related to heat flow, or gravitation, or electricity, or magnetism. Hydrodynamics is only one of many subjects in which these concepts are useful.

We restricted ourselves to a smooth surface in the above discussion in order to guarantee that the unit normal vector $\mathbf{n}$ will be a continuous function of the position of its tail, and this in turn is necessary in order to guarantee that the integrand $\mathbf{F} \cdot \mathbf{n}$ in (6) is a continuous scalar function. A surface S is called *piecewise smooth* if it consists of a finite number of smooth pieces. The surfaces we work with are understood to be piecewise smooth, and the value of an integral of the form (6) over such a surface is defined to be the sum of its values over the smooth pieces.

THE DIVERGENCE THEOREM

Surface integrals like (6) take on special importance when they are extended over closed surfaces. A surface S is said to be *closed* if it is the boundary of a bounded region of space. As examples we mention the surfaces of a sphere, a cube, a cylinder, and a tetrahedron.

The *divergence theorem* (also called *Gauss's Theorem*) states that

The flux of a vector field $\mathbf{F}$ out through a closed surface S equals the integral of the divergence of $\mathbf{F}$ over the region R bounded by S,

$$\iint_S \mathbf{F} \cdot \mathbf{n}\, dA = \iiint_R \operatorname{div} \mathbf{F}\, dV. \qquad (7)$$

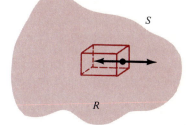

Figure A.25

This is a rather crude statement, without any of the hypotheses or carefully formulated restrictions that characterize most respectable mathematical theorems. We shall provide an equally crude "proof"—which, however, has the great merit of showing at a glance why the theorem is true.

First, we use planes parallel to the coordinate planes to subdivide the region R into a great many small rectangular boxes of the kind shown in Fig. A.25 (we ignore the incomplete boxes that do not lie wholly inside R). For the box in the figure, with volume ΔV, definition (3) tells us that the outward flux of $\mathbf{F}$ over the faces is given by the approximate formula

$$\text{Flux of } \mathbf{F} \text{ over faces} \cong (\operatorname{div} \mathbf{F})\, \Delta V. \qquad (8)$$

We now observe that the outward flux of $\mathbf{F}$ through the surface S is approximately equal to the total flux over all the faces of all the boxes, since for two adjacent boxes the outward flux from one through their common face precisely cancels the outward flux from the other through the same face, leaving only the flux through all the exterior faces. In view of (8), this tells us that

$$\iint_S \mathbf{F} \cdot \mathbf{n}\, dA \cong \sum (\operatorname{div} \mathbf{F})\, \Delta V.$$

Finally, by using the fact that the sum on the right is an approximating sum for the triple integral of the divergence of $\mathbf{F}$ over R, we obtain (7) by taking smaller and smaller subdivisions of R.

Example 2 Make a direct calculation of the flux of the vector field $\mathbf{F} = x\mathbf{i} + y\mathbf{j} + z\mathbf{k}$ out through the surface of the cylinder whose lateral surface is $x^2 + y^2 = a^2$ and whose bottom and top are $z = 0$ and $z = b$. Also find this flux by applying the divergence theorem.

Solution On the lateral surface L we have $\mathbf{n} = (x\mathbf{i} + y\mathbf{j})/a$, so the flux over L is

$$\iint_L \mathbf{F} \cdot \mathbf{n}\, dA = \iint_L \frac{x^2 + y^2}{a}\, dA = \iint_L a\, dA = a(2\pi ab) = 2\pi a^2 b.$$

On the top T we have $\mathbf{n} = \mathbf{k}$, so on T, $\mathbf{F} \cdot \mathbf{n} = z = b$ and the flux is

$$\iint_T \mathbf{F} \cdot \mathbf{n}\, dA = \iint_T b\, dA = b(\pi a^2) = \pi a^2 b.$$

On the bottom B we have $\mathbf{n} = -\mathbf{k}$, so on B, $\mathbf{F} \cdot \mathbf{n} = -z = 0$ and the flux is

$$\iint_B \mathbf{F} \cdot \mathbf{n}\, dA = \iint_B 0\, dA = 0.$$

Accordingly, the flux over the whole surface is $2\pi a^2 b + \pi a^2 b + 0 = 3\pi a^2 b$. To find this flux by applying the divergence theorem, we have only to notice that

$$\operatorname{div} \mathbf{F} = \frac{\partial}{\partial x}(x) + \frac{\partial}{\partial y}(y) + \frac{\partial}{\partial z}(z) = 3,$$

and therefore

$$\iint_S \mathbf{F} \cdot \mathbf{n} \, dA = \iiint_R \text{div } \mathbf{F} \, dV = \iiint_R 3 \, dV = 3(\text{volume}) = 3\pi a^2 b.$$

The divergence theorem is a profound theorem of mathematical analysis, with a wealth of important applications to many of the physical sciences. The cursory sketch of these ideas that we have given here—together with a similar sketch of Stokes' Theorem in the next section—is perhaps as far as an introductory calculus course should go in this direction. Students who wish to learn more are encouraged to continue and take advanced courses (vector analysis, potential theory, mathematical physics, etc.) in which these themes are fully developed.

PROBLEMS

1 Find the divergence of the vector field $\mathbf{F}$ if
(a) $\mathbf{F} = (y - z)\mathbf{i} + (z - x)\mathbf{j} + (x - y)\mathbf{k}$;
(b) $\mathbf{F} = (2z^2 - \sin e^y)\mathbf{i} + xy\mathbf{j} - xz\mathbf{k}$;
(c) $\mathbf{F} = xy\mathbf{i} + xz^3\mathbf{j} + (2z - yz)\mathbf{k}$;
(d) $\mathbf{F} = e^x \sin y\mathbf{i} + e^x \cos y\mathbf{j} + e^z \sin x\mathbf{k}$;
(e) $\mathbf{F} = \dfrac{x}{r}\mathbf{i} + \dfrac{y}{r}\mathbf{j} + \dfrac{z}{r}\mathbf{k}$, where $r = \sqrt{x^2 + y^2 + z^2}$.

In Problems 2 to 6, use the divergence theorem to find the flux of the given vector field over the given surface S.

2 $\mathbf{F} = x\mathbf{i} - y\mathbf{j} + z\mathbf{k}$; S is the surface of the cylinder in Example 2.

3 $\mathbf{F} = x\mathbf{i} + y\mathbf{j} + z\mathbf{k}$; S is the surface of the ellipsoid $x^2/a^2 + y^2/b^2 + z^2/c^2 = 1$.

4 $\mathbf{F} = x\mathbf{i} + y\mathbf{j} + z\mathbf{k}$; S is the surface of the tetrahedron formed by the plane $x/a + y/b + z/c = 1$ $(a, b, c > 0)$ and the coordinate planes.

5 $\mathbf{F} = x^3\mathbf{i} + y^3\mathbf{j} + z^3\mathbf{k}$; S is the surface of the sphere $x^2 + y^2 + z^2 = a^2$.

6 $\mathbf{F} = yz\mathbf{i} + xz\mathbf{j} + xy\mathbf{k}$; S is the surface of the cone $z^2 = m^2(x^2 + y^2)$, $0 \le z \le h$.

7 If $\mathbf{R}$ is the position vector $x\mathbf{i} + y\mathbf{j} + z\mathbf{k}$ and $r = \sqrt{x^2 + y^2 + z^2}$ is its length, find the divergence of the central force field $\mathbf{F} = f(r)(\mathbf{R}/r)$, where $f(r)$ is an arbitrary differentiable function.

8 Find the flux of the vector field defined in Problem 7 over the sphere $x^2 + y^2 + z^2 = a^2$ if
(a) $f(r) = r$;   (b) $f(r) = \dfrac{1}{r^2}$.

9 If n is a positive number and $f(r) = 1/r^n$ in Problem 7, show that the divergence of the force field $\mathbf{F}$ is zero if $n = 2$, and only in this case.

10 Verify the divergence theorem for the vector field $\mathbf{F} = 2z\mathbf{i} + (x - y)\mathbf{j} + (2xy + z)\mathbf{k}$ and the rectangular box whose faces are $x = 0$, $x = 1$, $y = 0$, $y = 2$, $z = 0$, $z = 3$.

11 Use the divergence theorem to find the flux of $\mathbf{F}$ over the surface of the box in Problem 10 if
(a) $\mathbf{F} = x^2\mathbf{i} + y^2\mathbf{j} + z^2\mathbf{k}$;
(b) $\mathbf{F} = xz\mathbf{i} + xy\mathbf{j} + yz\mathbf{k}$.

A.23
STOKES' THEOREM

Stokes' Theorem is an extension of Green's Theorem to three dimensions, involving curved surfaces and their boundaries rather than plane regions and their boundaries. Sir George Stokes (1819–1903) was a very eminent British mathematical physicist. He introduced the theorem known by his name in an examination question for students at Cambridge University in 1854. A fairly full account of Stokes' personality and scientific work can be found in G. E. Hutchinson, *The Enchanted Voyage,* Yale University Press, 1962. We shall state the theorem after a few preliminaries that will help us understand its meaning.

Suppose that $\mathbf{F} = L\mathbf{i} + M\mathbf{j} + N\mathbf{k}$ is a vector field defined in a certain region of space. It will be convenient in this section to think of $\mathbf{F}$ as the velocity field of a flowing fluid. Suppose also that C is a curve that lies in the region and is specified by certain parametric equations. The line integral of $\mathbf{F}$ along C, denoted by

$$\int_C \mathbf{F} \cdot d\mathbf{R} \qquad \text{or} \qquad \int_C L\,dx + M\,dy + N\,dz,$$

is defined and calculated in just the same way as in two dimensions, and requires no further explanation. If C is a closed curve, as shown in Fig. A.26, the line integral is usually written as

$$\oint_C \mathbf{F} \cdot d\mathbf{R}.$$

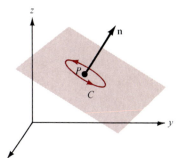

Figure A.26

This integral measures the tendency of the fluid to circulate or swirl around C, and is called the *circulation* of $\mathbf{F}$ around C.

Now suppose that C is a small simple closed curve that lies in a plane with unit normal vector $\mathbf{n}$, where the direction of $\mathbf{n}$ is related to the direction of C by the right-hand thumb rule, and let P be a point inside C (Fig. A.26). If ΔA is the area of the region enclosed by C, then

$$\frac{1}{\Delta A} \oint_C \mathbf{F} \cdot d\mathbf{R}$$

can be thought of as the circulation of $\mathbf{F}$ per unit area around P, and the limit

$$\lim_{\Delta A \to 0} \frac{1}{\Delta A} \oint_C \mathbf{F} \cdot d\mathbf{R}$$

is called the *circulation density* of $\mathbf{F}$ at P around $\mathbf{n}$. The point of these remarks is that this concept is closely related to the curl of the vector field $\mathbf{F}$, which was defined in Appendix A.22 by

$$\text{curl } \mathbf{F} = \nabla \times \mathbf{F} = \begin{vmatrix} \mathbf{i} & \mathbf{j} & \mathbf{k} \\ \dfrac{\partial}{\partial x} & \dfrac{\partial}{\partial y} & \dfrac{\partial}{\partial z} \\ L & M & N \end{vmatrix}. \tag{1}$$

In fact, it can be shown that the curl of $\mathbf{F}$ has the property that at any point its component in a given direction $\mathbf{n}$ is precisely the circulation density of $\mathbf{F}$ around $\mathbf{n}$,

$$(\text{curl } \mathbf{F}) \cdot \mathbf{n} = \lim_{\Delta A \to 0} \frac{1}{\Delta A} \oint_C \mathbf{F} \cdot d\mathbf{R}. \tag{2}$$

The proof of (2) is fairly complicated and will not be given here; it can be found in any good book on vector analysis.

We can visualize the meaning of (2) in a concrete way if we imagine a small paddle wheel placed in the flowing fluid at the point P with its axis pointing in the direction of $\mathbf{n}$ (Fig. A.27). The circulation of the fluid around $\mathbf{n}$ will cause the paddle wheel to turn, and the speed at which it spins will be proportional to the circulation density. The paddle wheel will spin fastest when it points in the direction in which the circulation density is largest, and (2) tells us that this happens when $\mathbf{n}$ points in the same direction as curl $\mathbf{F}$. We conclude that at each point of space the vector curl $\mathbf{F}$ has the direction in which the circulation density is largest, with magnitude equal to this largest circulation density. The paddle wheel we have described here can therefore be thought of as an imaginary instrument for sensing the direction and magnitude of the curl.

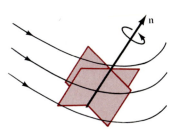

Figure A.27

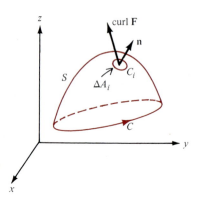

Figure A.28

We are now ready for the main theorem of this section. *Stokes' Theorem* asserts the following (see Fig. A.28):

If S is a surface in space with boundary curve C, then the circulation of a vector field **F** *around C is equal to the integral over S of the normal component of the curl of* **F**,

$$\oint_C \mathbf{F} \cdot d\mathbf{R} = \iint_S (\text{curl } \mathbf{F}) \cdot \mathbf{n} \, dA. \qquad (3)$$

Just as in the case of the divergence theorem in Appendix A.22, we choose to keep this statement as simple as possible and not complicate it with the hypotheses and restrictions that would be needed to convert it into a genuine mathematical theorem. For instance, it is necessary to assume that S is a two-sided (orientable) surface with the direction of the unit normal vector **n** related to the direction of C by the right-hand thumb rule, as shown in the figure. Also, of course, **n** must be continuous, **F** must be continuous, L, M, and N must have continuous partial derivatives, and so on.

This rough, intuitive version of Stokes' Theorem has a rough, intuitive "proof" based on equation (2). First, we subdivide the surface S into a large number of small patches with areas ΔA_i and boundary curves C_i. By applying (2) to the ith patch we obtain the approximate equation

$$\oint_{C_i} \mathbf{F} \cdot d\mathbf{R} \cong (\text{curl } \mathbf{F}) \cdot \mathbf{n} \, \Delta A_i.$$

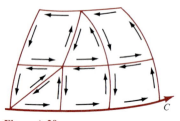

Figure A.29

If we add the left sides of these equations for all curves C_i, then the line integrals over all interior common boundaries cancel, being calculated once in each direction, leaving only the line integral around the exterior boundary C (see Fig. A.29). This gives

$$\oint_C \mathbf{F} \cdot d\mathbf{R} \cong \sum (\text{curl } \mathbf{F}) \cdot \mathbf{n} \, \Delta A_i, \qquad (4)$$

and by taking smaller and smaller subdivisions we obtain (3), since the sums on the right side of (4) are approximating sums for the surface integral on the right side of (3).

Example 1 If the surface S is a region R lying flat in the xy-plane, then $\mathbf{n} = \mathbf{k}$ and by (1) we see that

$$(\text{curl } \mathbf{F}) \cdot \mathbf{n} = \frac{\partial M}{\partial x} - \frac{\partial L}{\partial y},$$

so (3) reduces to

$$\oint_C L \, dx + M \, dy = \iint_R \left(\frac{\partial M}{\partial x} - \frac{\partial L}{\partial y} \right) dA.$$

This is Green's Theorem (Section 21.3), which is thus a special case of Stokes' Theorem.

Example 2 Evaluate the line integral

$$I = \oint_C y^3 z^2 \, dx + 3xy^2 z^2 \, dy + 2xy^3 z \, dz$$

around the closed curve C whose vector equation is $\mathbf{R} = a \sin t\mathbf{i} + b \cos t\mathbf{j} + c \cos t\mathbf{k}$, $0 \le t \le 2\pi$, where $abc \ne 0$.

Solution This integral is the circulation around C of the vector field $\mathbf{F} = y^3z^2\mathbf{i} + 3xy^2z^2\mathbf{j} + 2xy^3z\mathbf{k}$. An easy calculation shows that curl $\mathbf{F} = \nabla \times \mathbf{F} = \mathbf{0}$. If S is any surface whose boundary is C, then the right side of (3) has the value 0 in this case, and therefore Stokes' Theorem tells us that $I = 0$.

Also, for this particular $\mathbf{F}$ and C it is not too difficult to verify Stokes' Theorem by calculating I directly. This gives

$$I = \int_0^{2\pi} [(b^3 \cos^3 t)(c^2 \cos^2 t)(a \cos t)$$

$$+ 3(a \sin t)(b^2 \cos^2 t)(c^2 \cos^2 t)(-b \sin t)$$

$$+ 2(a \sin t)(b^3 \cos^3 t)(c \cos t)(-c \sin t)]\, dt$$

$$= ab^3c^2 \int_0^{2\pi} [\cos^6 t - 5 \cos^4 t \sin^2 t]\, dt = ab^3c^2 \sin t \cos^5 t \Big]_0^{2\pi} = 0.$$

In Section 21.2 we proved that three properties of vector fields in the plane are equivalent to one another. Stokes' Theorem makes it possible to extend these ideas in a natural way to three-dimensional space. Specifically, if $\mathbf{F}$ is a vector field defined in a simply connected region of space, then any one of the following four properties implies the remaining three:*

(a) $\oint_C \mathbf{F} \cdot d\mathbf{R} = 0$ for every simple closed curve C.
(b) $\int_C \mathbf{F} \cdot d\mathbf{R}$ is independent of the path.
(c) $\mathbf{F}$ is a gradient field, i.e., $\mathbf{F} = \nabla f$ for some scalar field f.
(d) curl $\mathbf{F} = \mathbf{0}$.

The equivalence of (a), (b), and (c) is established in just the same way as in two dimensions; the fact that (c) implies (d) is a straightforward calculation; and Stokes' Theorem enables us to show very easily that (d) implies (a). A vector field with any one of these properties is said to be *conservative* or *irrotational* [because of (d)].

We have seen that the divergence theorem relates an integral over a closed surface to a corresponding volume integral over the region of space enclosed by the surface, and Stokes' Theorem relates an integral around a closed curve to a corresponding surface integral over any surface bounded by the curve. As we suggested at the beginning of Appendix A.22, these statements are very similar and are presumably somehow connected with each other. It turns out that both are special cases of a powerful theorem of modern analysis called the *generalized Stokes Theorem.* Students who wish to understand these relationships must study the theory of differential forms.

---

* A region in three-dimensional space is said to be *simply connected* if every simple closed curve in the region can be shrunk continuously to a point without leaving the region.

PROBLEMS

1 Show that property (c) implies property (d), that is, show that curl $\nabla f = \mathbf{0}$.

2 If S is a closed surface that lies in a region of space in which a vector field $\mathbf{F}$ is defined, show that

$$\iint_S (\text{curl } \mathbf{F}) \cdot \mathbf{n} \, dA = 0.$$

3 If the vector field $\mathbf{F} = -y\mathbf{i} + x\mathbf{j} + 0\mathbf{k}$ is the velocity field of a flowing fluid, sketch enough of this field in the xy-plane (that is, sketch enough of the velocity vectors attached to various points) to understand the nature of the motion. Then calculate curl $\mathbf{F}$, let C be the circle $\mathbf{R} = r\cos\theta\mathbf{i} + r\sin\theta\mathbf{j} + 0\mathbf{k}$ ($0 \le \theta \le 2\pi$) in the xy-plane, and verify the formula

$$(\text{curl } \mathbf{F}) \cdot \mathbf{n} = \frac{1}{A} \oint_C \mathbf{F} \cdot d\mathbf{R}$$

for this circle and its interior in the xy-plane.

4 Repeat Problem 3 for the vector fields
(a) $\mathbf{F} = ax\mathbf{j}$, where a is a positive constant;
(b) $\mathbf{F} = f(r)\mathbf{R}$, where r is the length of the position vector $\mathbf{R} = x\mathbf{i} + y\mathbf{j} + z\mathbf{k}$ and $f(r)$ is an arbitrary differentiable function.

In Problems 5 to 11, apply Stokes' Theorem to find $\oint_C \mathbf{F} \cdot d\mathbf{R}$ for the given $\mathbf{F}$ and the given C. In each case let C be oriented counterclockwise as seen from above.

5 $\mathbf{F} = y(x - z)\mathbf{i} + (2x^2 + z^2)\mathbf{j} + y^3 \cos xz \; \mathbf{k}$; C is the boundary of the square $0 \le x \le 2$, $0 \le y \le 2$, $z = 5$.

6 $\mathbf{F} = (z - y)\mathbf{i} + y\mathbf{j} + x\mathbf{k}$; C is the intersection of the top half of the sphere $x^2 + y^2 + z^2 = 4$ with the cylinder $r = 2\cos\theta$.

7 $\mathbf{F} = y\mathbf{i} + (x + y)\mathbf{j} + (x + y + z)\mathbf{k}$; C is the ellipse in which the plane $z = x$ intersects the cylinder $x^2 + y^2 = 1$.

8 $\mathbf{F} = (y - x)\mathbf{i} + (x - z)\mathbf{j} + (x - y)\mathbf{k}$; C is the boundary of the triangular part of the plane $x + y + 2z = 2$ that lies in the first octant.

9 $\mathbf{F} = (3y + z)\mathbf{i} + (\sin y - 3x)\mathbf{j} + (e^z + x)\mathbf{k}$; C is the circle $x^2 + y^2 = 1$, $z = 5$.

10 $\mathbf{F} = 2z\mathbf{i} + 6x\mathbf{j} - 3y\mathbf{k}$; C is the ellipse in which the plane $z = y + 1$ intersects the cylinder $x^2 + y^2 = 1$.

11 $\mathbf{F} = e^{x^2}\mathbf{i} + (x + z)\sin y^3\mathbf{j} + (y^2 - x^2 + 2yz)\mathbf{k}$; C is the boundary of the triangular part of the plane $x + y + z = 3$ that lies in the first octant.

In Problems 12 to 15, verify Stokes' Theorem for the given $\mathbf{F}$, S, and C.

12 $\mathbf{F} = (z - y)\mathbf{i} + (x + z)\mathbf{j} - (x + y)\mathbf{k}$, S is the part of the paraboloid $z = 9 - x^2 - y^2$ that lies above the xy-plane, and C is its boundary circle $x^2 + y^2 = 9$ in the xy-plane, oriented counterclockwise as seen from above.

13 $\mathbf{F} = xy\mathbf{i} + yz\mathbf{j} + zx\mathbf{k}$, S is the part of the plane $x + y + z = 1$ that lies in the first octant, and C is its boundary oriented counterclockwise as seen from above.

14 $\mathbf{F} = y\mathbf{i} - x\mathbf{j}$, S is the top half of the sphere $x^2 + y^2 + z^2 = 4$, and C is its boundary circle $x^2 + y^2 = 4$ in the xy-plane, oriented counterclockwise as seen from above.

15 $\mathbf{F} = (x + y)\mathbf{i} + (y + z)\mathbf{j} + (z + x)\mathbf{k}$, S is the elliptical disk $x^2/a^2 + y^2/b^2 \le 1$, $z = 0$, and C is its boundary oriented counterclockwise as seen from above.

16 Let S be the top half of the ellipsoid $x^2 + y^2 + z^2/9 = 1$, oriented so that $\mathbf{n}$ is directed upward. If $\mathbf{F} = x^3\mathbf{i} + y^4\mathbf{j} + z^3 \sin xy \; \mathbf{k}$, evaluate

$$\iint_S (\text{curl } \mathbf{F}) \cdot \mathbf{n} \, dA$$

by replacing S by a simpler surface with the same boundary.

17 Repeat Problem 16 if $\mathbf{F} = xz^2\mathbf{i} + x^3\mathbf{j} + \cos xz\mathbf{k}$ and S is the top half of the ellipsoid $x^2 + y^2 + 4z^2 = 1$ with $\mathbf{n}$ directed upward.

BIOGRAPHICAL NOTES

Biographical history, as taught in our public schools, is still largely a history of boneheads: ridiculous kings and queens, paranoid political leaders, compulsive voyagers, ignorant generals—the flotsam and jetsam of historical currents. The men who radically altered history, the great creative scientists and mathematicians, are seldom mentioned if at all.

Martin Gardner

I will not go so far as to say that to construct a history of thought without profound study of the mathematical ideas of successive epochs is like omitting Hamlet from the play which is named after him. That would be claiming too much. But it is certainly analogous to cutting out the part of Ophelia. This simile is singularly exact. For Ophelia is quite essential to the play, she is very charming—and a little mad. Let us grant that the pursuit of mathematics is a divine madness of the human spirit, a refuge from the goading urgency of contingent happenings.

A. N. Whitehead

763

AN OUTLINE OF THE HISTORY OF CALCULUS

THE ANCIENTS

Pythagoras (c. 580–500 B.C.)	Pythagorean theorem about right triangles; irrationality of $\sqrt{2}$
Euclid (c. 300 B.C.)	Organized most of the mathematics known at his time; Euclid's theorems on perfect numbers and the infinity of primes
Archimedes (c. 287–212 B.C.)	Determined tangents, areas, and volumes, essentially by calculus; found volume and surface of a sphere; centers of gravity; spiral of Archimedes; calculated π
Pappus (fourth century A.D.)	Centers of gravity linked to solids and surfaces of revolution

THE FORERUNNERS

Descartes (1596–1650)	Putative discoverer of analytic geometry; introduced some good notations
Mersenne (1588–1648)	Lubricated the flow of ideas; cycloid; Mersenne primes
Fermat (1601–1665)	Actual discoverer of analytic geometry; calculated and used derivatives and integrals; founded modern number theory; probability
Pascal (1623–1662)	Mathematical induction; binomial coefficients; cycloid; Pascal's theorem in geometry; probability; influenced Leibniz
Huygens (1629–1695)	Catenary; cycloid; circular motion; Leibniz' mathematics teacher (what a pupil! what a teacher!)

THE EARLY MODERNS

Newton (1642–1727)	Invented his own version of calculus; discovered Fundamental Theorem; used infinite series; virtually created astronomy and physics as mathematical sciences
Leibniz (1646–1716)	Invented his own better version of calculus; discovered Fundamental Theorem; invented many good notations; teacher of the Bernoulli brothers
The Bernoullis (James 1654–1705, John 1667–1748)	Learned calculus from Leibniz, and developed and applied it extensively; infinite series; John is teacher of Euler
Euler (1707–1783)	Organized calculus and developed it very extensively; codified analytic geometry and trigonometry; introduced symbols $e, \pi, i, f(x), \sin x, \cos x$; infinite series and products; calculus of variations; number theory; topology; mathematical physics; etc.
Lagrange (1736–1813)	Calculus of variations; analytical mechanics
Laplace (1749–1827)	Laplace's equation; celestial mechanics; analytic probability
Fourier (1768–1830)	Fourier series; the heat equation

THE MATURE MODERNS

Gauss (1777–1855)	Initiated rigorous analysis with convergence proofs for infinite series; number theory; complex numbers in analysis, algebra, and number theory; differential geometry; non-Euclidean geometry; etc.
Cauchy (1789–1857)	Careful treatment of limits, continuity, derivatives, integrals, series; complex analysis
Abel (1802–1829)	Binomial series; fifth-degree equation; integral calculus; elliptic functions
Dirichlet (1805–1859)	Convergence of Fourier series; modern definition of function; analytic number theory
Liouville (1809–1882)	Integrals of elementary functions; transcendental numbers
Hermite (1822–1901)	Transcendence of e; Hermitian matrices; elliptic functions
Riemann (1826–1866)	Riemann integral; Riemann rearrangement theorem; Riemannian geometry; Riemann zeta function; complex analysis

PYTHAGORAS (c. 580–500 B.C.)

. . . three fifths of him genius and two fifths sheer fudge.
J. R. Lowell

Western civilization is like a great river flowing through time, nourished and strengthened by many rich tributaries from other cultures. Let us project our imagination backward along its course a few thousand years, to its headwaters in ancient Greece. There, near the primary source of the stream, stands the hazy, half-mythical figure of Pythagoras. At the present time most people think of him as a mathematician, but to his contemporaries he was many things—a teacher of wisdom, a religious prophet, a saint, a magician, a charlatan, a political agitator, depending on the point of view. His fanatical followers spread his ideas throughout the Greek world, which ignored some and absorbed others. In terms of his enduring influence on mathematics, science, and philosophy in European civilization, Pythagoras was as important as anyone who has ever lived.

Mathematics begins with him, in the sense that he first conceived it as an organized system of thought held together by deductive proof. He was even the first to use the word *mathematike* to mean mathematics. Before him there was only *mathemata,* which meant knowledge or learning in general.

Science begins with him, in the sense that he performed the first deliberate scientific experiment and was the first person to conceive the supremely daring conjecture that the world is an ordered, understandable whole. He was the first to apply the word *kosmos*—which previously meant order or harmony—to this whole.

Western philosophy begins with him, in the sense that his ideas about the nature of reality crystallized two centuries later into the core of Plato's metaphysical system, and subsequent philosophic thought in the West has often been described as a series of footnotes to Plato. Also, he seems to have originated the very word philosophy (*philosophia*, love of wisdom), in rejecting *sophia* (wisdom) as too pretentious to describe his own attempts at understanding.

Any one of these great beginnings would be a lofty achievement to claim for any man. Is it not too much to believe that all three are linked to a single individual? Let us see how they arose.

First, however, what can be said about his life? He was a contemporary of Confucius, Buddha, and Zoroaster. Like these other great figures from the childhood of the race, Pythagoras is known to us only through legends and traditions recorded hundreds of years after his death.

In broad outline, these traditions agree. He was born on the island of Samos, off the western coast of Asia Minor. He was studious as a youth (what sage was not?), and then traveled for about 30 years in Egypt, Babylonia, Phoenicia, Syria, and perhaps even Persia and India. In the course of his journeys he acquired a little astronomy and primitive empirical mathematics, and a good deal of ridiculous nonsense in the form of Oriental mysticism — the so-called "immemorial wisdom of the East." On returning at last to Samos, he disliked what he found there — an efficient but unsympathetic tyrant — and at the age of about 50 migrated to the Greek colony of Crotona in southern Italy.

Here his public life began. He established himself as a teacher and founded the famous Pythagorean school, which was a semisecret association of several hundred disciples with some claim to the honor of being the world's first university. In the beginning, this school seems to have been as much a religious brotherhood aiming at the moral reformation of society as it was a focus of intellectual activity. However, society does not always welcome moral reformation, and outsiders came to regard the Pythagoreans as an offensively puritanical political party. Eventually, their increasing political activities aroused the ire of the citizens to such an extent that they were violently stamped out and their buildings sacked and burned. Pythagoras himself fled to the nearby colony of Metapontum, where he died at an advanced age. The other surviving Pythagoreans, though scattered throughout the Mediterranean world, kept the faith and carried on as an active philosophical school for more than a century.

What was this faith? Its starting point was Pythagoras' theory of the soul as an objective entity, which no doubt evolved from his experiences in Egypt and Asia. He believed in the doctrine of metempsychosis, or the transmigration of the individual soul at death from one body into another, either human or animal.* Each soul continues this process of reincarnation indefinitely, moving up or down to higher or lower animals according to merit or demerit. The only way to escape from this "wheel of birth" and attain unity with the Divine is through purification, both of the body and of the mind. These ideas, though fantastic to modern minds, were widespread in antiquity, and played formative roles in many of the world's religions.

The Pythagorean community was bound together by vows of loyalty to one another and obedience to the Master, and purification was sought in a variety of ways. They practiced a communistic sharing of material things. They dressed plainly, behaved modestly, and did not laugh or swear. It was forbidden to eat beans or meat. The prohibition of beans was probably an echo of some primitive taboo, and vegetarianism was a natural precaution against the abomination of eating an ancestor. Also, the drinking of water instead of wine was recommended — advice of doubtful wisdom in southern Italy today.

It appears that Pythagoras himself surpassed all his students in the perfection of his life according to these standards. His moral and intellectual authority were so great

---

* "Do not hit him," he is said to have ordered a man who was beating a dog, "for in this dog lives the soul of a friend of mine; I recognize him by his voice."

that the phrase *autos epha*—"he himself has said it"—became their formula for a final decision on any issue. Also, it was customary to attribute all ideas and discoveries to the Master, which makes it almost impossible for us to distinguish his own achievements from those that may have originated among his disciples.

As suggested above, the Pythagoreans sought purification of the body through austerity, abstinence, and moderation. This was common then, and is common now in many lands of the East. The uniqueness of Pythagoras lies in his plan for attaining purification of the mind: through the active study of mathematics and science. This is diametrically opposed to the passive "meditation" urged by most mystical cults, which an unsympathetic observer might describe as little more than presiding over a vacuum. This plan of Pythagoras is the source of his germinal influence on Western civilization, and accounts in part for the distinctive character of that civilization as it has developed through the past 2500 years.

The course of study required by Pythagoras consisted of four subjects: geometry, arithmetic, music, and astronomy. In the Middle Ages this group of subjects came to be known as the *quadrivium* (or "fourfold way"), and it was then enlarged by adding the *trivium* of grammar, rhetoric, and logic. These were the seven liberal arts that came to be regarded as essential parts of the education of any cultured person.

Greek mathematics is certainly one of the half-dozen supreme intellectual achievements of human history. Pythagoras started it all, not in the practical sense of the Babylonian clerks and Egyptian surveyors, but for its own sake, as a mental discipline which is capable of lifting the mind to higher levels of order and clarity. Before him, there were only a few isolated rules of thumb in geometry, arrived at empirically and devoid of any suggestion that there might be logical relations among them. Pythagoras seems to have created the pattern of definitions, axioms, theorems, and proofs, according to which the intricate structure of geometry is produced from a small number of explicitly stated assumptions by the action of rigorous deductive reasoning. It appears that the very idea of mathematical proof is due to him. Tradition tells us that Pythagoras himself discovered many theorems, most notably, the fact that the sum of the angles in any triangle equals two right angles, and the famous Pythagorean theorem about the square of the hypotenuse of a right triangle. According to one source, his joy on discovering this magnificent theorem was so great that he sacrificed an ox in thanksgiving, but this is an unlikely story, since such an action would have been a shocking violation of Pythagorean beliefs. The brotherhood also knew many properties of parallel lines and similar triangles, and arranged all this material into a coherent logical system roughly equivalent to the first two books of Euclid's *Elements* (c. 300 B.C.). That is to say, starting almost from the beginning, they discovered for themselves about as much geometry as a student learns today in the first half of a high school course.

The Pythagoreans were also entranced by arithmetic—not in the sense of useful computational skills, but rather as the abstract theory of numbers. They seem to have been the first to classify numbers into even and odd, prime and factorable, etc. Their favorites were the figurate numbers, which arise by arranging dots or points in regular geometric patterns. We mention the triangular numbers 1, 3, 6, 10, . . . , which are the numbers of dots in the following triangular arrays:

These are evidently numbers of the form $1 + 2 + 3 + \cdots + n$. There are also the square numbers 1, 4, 9, 16, . . . :

As indicated, each square number can be obtained from its predecessor by adding an L-shaped border called a *gnomon*, meaning a carpenter's square. The Pythagoreans established many interesting facts about figurate numbers by merely inspecting pictures. For example, since the successive gnomons are the successive odd numbers, it is immediately clear from the square arrays that the sum of the first n odd numbers equals n^2:

$$1 + 3 + 5 + \cdots + (2n - 1) = n^2.$$

In much the same way, the formula

$$1 + 2 + 3 + \cdots + n = \tfrac{1}{2}n(n + 1)$$

for the nth triangular number can be made to appear visibly true by writing it in the form

$$2 + 4 + 6 + \cdots + 2n = n(n + 1),$$

for the left side of this is the sum of the first n even numbers, and the equality is seen at once when this sum is expressed in the form of a rectangular array with n dots along one side and $n + 1$ along the other, as follows:

The tremendous idea that mathematics is the key to the correct interpretation of nature originated with the Pythagoreans, and probably with Pythagoras himself. The discovery that suggested this idea arose from a simple experiment with music. Pythagoras stretched a lyre string between two pegs on a board. When this taut string was plucked, it emitted a certain note. He found that if the string is stopped at its midpoint by a movable wedge inserted between the string and the board, so that the vibrating part is reduced to $\frac{1}{2}$ its original length, then it emits a note one octave above the first; if it is reduced to $\frac{2}{3}$ its original length, it emits a note which is a "fifth" above the first; and if it is reduced to $\frac{3}{4}$ its original length, it emits a note which is a "fourth" above the first. The octave, the fifth, and the fourth were already well-known melodic concepts. The Pythagoreans were deeply impressed by this remarkable relation between the simple fractions $\frac{1}{2}$, $\frac{2}{3}$, and $\frac{3}{4}$ and musical intervals whose recognized significance was based on purely aesthetic considerations. Further, in what seemed to them a natural next step, they held that every body moving in space produces a sound whose pitch is proportional to its speed. It followed from this that the planets moving at different speeds in their various orbits around the earth produce a celestial harmony, which they called the "music of the spheres." As an additional contribution to astronomy, Pythagoras also asserted that the earth is spherical—probably for the simple reason that the sphere is the most beautiful solid body—and he was apparently the first person to do so.

The law of musical intervals described here was the first quantitative fact ever discovered about the natural world. Along with its "philosophically obvious" extension to the planets, this led the Pythagoreans to the conviction that numbers—which to them meant positive whole numbers and fractions—are the essence of all that is intelligible in the universe.* "Everything is Number" became their motto, apparently in the sense that the only basic and timeless aspects of any object or idea lie in the numerical attributes it possesses.

Almost at once, this doctrine tripped over geometry and fell flat on its face. Because everything is number — meaning rational number, since no others existed for them — it was evident that the length of any line segment must be a rational multiple of the length of any other line segment. Unfortunately this is false, as they soon discovered, for the Pythagorean theorem guarantees that a square of side 1 has diagonal of length $\sqrt{2}$, and, according to tradition, Pythagoras himself proved that there is no rational number whose square is 2. This fatal discovery confronted the brotherhood with two alternatives, one unthinkable and the other intolerable: Either the diagonal of a square of side 1 has no length, or it is not true that everything is number. To them the crumbling of their simple generalization reducing the universe to rational numbers shattered the necessary foundations of thought, and one legend states that they even went so far as to drown a renegade Pythagorean who revealed their unspeakable secret to the outside world. To us, however, it represents the discovery of irrational numbers, which was one of the finest achievements of early Greek mathematics. It is often seen in the history of ideas that one generation's disaster is an opportunity for the next.

In spite of this setback, Pythagoras and his followers kept their faith in Number as they conceived it. If Number contradicts reality, so much the worse for reality. In the ecstasy of their enthusiasm they abandoned all further interest in learning about the world by combining observation, experiment, and thought, and instead sought their own "higher reality" by plunging joyously into the quagmire of number mysticism.

Like the tenets of any religion, the numerological beliefs of the Pythagoreans are difficult to make plausible to the uninitiated. The central concept of their system seems to have been the sacred *tetractys,* consisting of the numbers 1, 2, 3, 4, whose sum is the holy number 10 — holy because 1 is the point, 2 is the line, 3 is the surface, and 4 is the solid, and therefore $1 + 2 + 3 + 4 = 10$ is everything, the number of the universe. It was doubtless a great day for them when they learned that the fractions $\frac{1}{2}$, $\frac{2}{3}$, $\frac{3}{4}$, which are the successive ratios of the numbers 1, 2, 3, 4, are closely linked to musical harmony; and it should be satisfying to us to realize that our own decimal system has a more rational foundation than the accidental fact that human beings have 10 fingers. Their next basic article of faith lies much deeper, so deep indeed that modern people can scarcely hope to comprehend it: Odd numbers (except 1) are male, and even numbers are female. In addition to such general principles as these, they believed that each number has its own individual significance and personality. Thus, 1 is the generator of all numbers, the omnipotent One; 2 is diversity, the first female number; $3 = 1 + 2$ is the first male number, being composed of unity and diversity; $4 = 2 + 2 = 2 \cdot 2$ is the number of justice, being evenly balanced; $5 = 3 + 2$ is the number of marriage, since it is the union of the first male and female numbers; $6 = 1 + 2 + 3$ is perfect, because it is the sum of its proper divisors, and these are unity, diversity, and the sacred trinity, whose meaning expanded considerably in early Christian numerology.† And so on, and so on.‡

---

* As Aristotle wrote in his *Metaphysics* (Book I, Chapter 5, c. 330 B.C.), "The so-called Pythagoreans, who were the first to take up mathematics, not only advanced this subject, but, saturated with it, they fancied that the principles of mathematics were the principles of all things."

† As St. Augustine said in *The City of God* (c. 420 A.D.): "Six is a number perfect in itself, and not because God created the world in six days; rather the contrary is true; God created the world in six days because this number is perfect." In a very different mood, this remarkable man also uttered the following unforgettable prayer as he approached the end of his licentious youth and looked forward to a monastic old age: "O Lord, give me chastity, but not quite yet."

‡ A more richly developed theology of numbers, confirmed by many numerological calculations, is given in E. T. Bell, *The Magic of Numbers,* McGraw-Hill, 1946.

The importance for us of this chaotic mass of fanciful mumbo-jumbo is that it passed into the mind of Plato (428–348 B.C.), and emerged in an altered form as part of a powerful torrent of belief that swept almost unabated through the early Christian era, the Middle Ages, and the Renaissance, and is still a potent influence in our own time.

Plato, of course, is one of the titans of world literature. His half-dozen greatest dialogues have held their place in the affection and respect of mankind mainly because of their dramatic and poetic qualities and the personality of their chief character, Socrates. The Socratic element in Plato's thought is primarily concerned with human affairs—with morality, politics, and the problem of how to live a good life. In addition to his love and admiration for Socrates, Plato was enthralled by mathematics, especially in its role as a body of knowledge that appears to be independent of the evidence of the senses. In his middle years he spent considerable time in southern Italy and came into personal contact with the Pythagorean communities there, whose philosophy was mathematical but whose driving forces were religious and mystical.* This decisive experience imparted a Pythagorean tinge to much of the rest of Plato's thought, which can be characterized as a mixture of gold nuggets and gravel, with the gravel gilded by his exalted literary style. Unfortunately, gold-plated gravel is still basically gravel.†

The quintessence of Plato's Pythagoreanism is found in his mystical doctrine of Ideas or Forms. This doctrine asserts a view of reality consisting of two worlds: first, the everyday world perceived by our senses, the world of change, appearance, and imperfect knowledge; and second, the world of Ideas perceived by the reason, the world of permanence, reality, and true knowledge. Thus Justice is an Idea imperfectly reflected in human efforts to be just, and Two is an Idea participated in by every pair of material objects. Each of Plato's Ideas was for him an objective reality lying outside of space and time; they can be approached by thought but are not created by it; they are the timeless and perfect patterns of Being, whose blurred and shadowy copies constitute the deceptive phenomena of the world around us.

These stable absolutes imagined by Plato appeal to all people battered by change and hungry for permanence. Aristotle tried to dilute them with a little common sense; but like oil and water, Platonism and common sense do not mix easily, and his efforts failed.‡ This Pythagorean apotheosis of abstract concepts is now called "Platonic realism." It has had a long and controversial history in Western thought, and was still alive and well in the early twentieth century.§

We hope these remarks have clarified the assertion made at the beginning that

---

* See F. M. Cornford, *Before and After Socrates,* Cambridge University Press, 1932, especially Chapter III.

† The following are a few of the gamiest items in Plato's numerology: his Geometrical Number, $60^4 = 12,960,000$, "which has control over the good and evil of births" (*Republic* 546); his apparent rejection of the well-known length of the year—approximately $365\frac{1}{4}$ days—in favor of $364\frac{1}{2}$ days, because this is 729 or 9^3 days and nights, and 9^3 is "the interval by which the tyrant is parted from the king" (*Republic* 588); his number 5040 ($= 1 \cdot 2 \cdot 3 \cdot 4 \cdot 5 \cdot 6 \cdot 7$), which he concludes is the exact number of citizens suitable for his ideal city (*Laws* 738, 741, 747, 771, 878); and the whole of the *Timaeus,* in which the structure of the universe and the nature of life are given thoroughgoing Pythagorean explanations in terms of triangles. For more details on the mysterious Geometrical Number, see T. L. Heath, *A History of Greek Mathematics,* Oxford University Press, 1921, pp. 305–307.

‡ "The Forms [Ideas] we can dispense with, for they are mere sound without sense."—*Posterior Analytics* 83a. See also *Metaphysics* 990b, 991a, 1079a, 1079b, 1090a, etc. Compared with Plato, Aristotle suffered the serious disadvantage of having about as much charm as an old shoe.

§ Consider the statement of the English astronomer Sir Arthur Eddington (1935): "I believe that all the laws of nature that are usually classed as fundamental can be foreseen wholly from epistomological considerations." Also that of the English mathematician G. H. Hardy (1940): "I believe that mathematical reality lies outside us, that our function is to discover or observe it, and that the theorems which we prove, and which we describe grandiloquently as our 'creations,' are simply the notes of our observations."

Pythagoras was much more than merely an ancient mathematician: He is entitled to be recognized as the founder of mathematics, science, and philosophy in European civilization. He was also the first to open up the enduring gulf of incomprehension between the scientific spirit, which hopes that the universe is ultimately understandable, and the mystical spirit, which hopes — perhaps unconsciously — that it is not.*

EUCLID (c. 300 B.C.)

Euclid's Elements *is certainly one of the greatest books ever written.*

Bertrand Russell

It is also one of the dullest, and by any educational standards whatever, it ought to have been a student's and teacher's nightmare throughout the twenty-three centuries of its existence.

The *Elements* purports to begin at the very beginning of geometry, with nothing required of the reader in the way of previous knowledge or experience. Yet it offers no preliminary explanations, and nowhere does it provide illuminating remarks of any kind. It has no direct scientific content and doesn't even hint at a single application. It makes no attempt to place its subject in any mathematical or historical context, and nowhere is the name of any person mentioned. Its stony impersonality stuns the mind. The Bible also starts impersonally enough — "In the beginning God created the heaven and the earth" — but even this compresses the greatest possible action into the fewest possible words, and after several sentences living creatures make their appearance. The *Elements* begins with a definition — "A point is that which has no part" — and marches with inhuman, undeviating monotony through 13 Books and 465 Propositions, none of which are discussed or motivated in any way. It conveys the impression of simply existing, like a rock, indifferent to human concerns. Such are the outward qualities of a book that has gone through more than a thousand editions since the invention of printing and has often been described as having had an influence on the human mind greater than that of any other work except the Bible.

In view of its tone and style, the most surprising thing about the *Elements* is that it seems to have had an author. Who was this Euclid, whose name was almost synonymous with geometry until the twentieth century? Only three facts are known about him, and two slender anecdotes.

The facts are these: He was younger than Plato (b. 428 B.C.), he was older than Archimedes (b. 287 B.C.), and he taught in Alexandria. When Alexander the Great died in 323 B.C., his African empire was inherited by Ptolemy, his favorite Macedon-

---

* The view expressed here, that in the core of his thinking Plato was a deep-dyed mystic, is rejected with such indignation by most modern philosophers that perhaps it requires some independent support. For example, in his *Mysticism and Logic,* Bertrand Russell writes: "In Plato, the same twofold impulse [toward mysticism and science] exists, though the mystic impulse is distinctly the stronger of the two, and secures ultimate victory whenever the conflict is sharp." Two other eminent contemporary philosophers who have no difficulty in seeing Plato for what he was — a very great writer but unfortunately a mystic and would-be tyrant — are Karl Popper (*The Open Society and its Enemies,* 1945) and Gilbert Ryle (review of Popper's book in *Mind,* 1947). And in a letter to John Adams (July 5, 1814), Thomas Jefferson — at the age of 71 — writes: "I amused myself with reading seriously Plato's republic [in the original Greek, of course]. I am wrong however in calling it amusement, for it was the heaviest task-work I ever went through. I had occasionally before taken up some of his other works, but scarcely ever had the patience to go through a whole dialogue. While wading thro' the whimsies, the puerilities, and unintelligible jargon of this work, I laid it down often to ask myself how it could have been that the world should have so long consented to give reputation to such nonsense as this? . . . But fashion and authority apart, and bringing Plato to the test of reason, take from him his sophisms, futilities, and incomprehensibilities, and what remains?" In other letters Jefferson expressed himself on this subject with much less restraint.

ian general, who ruled as king from 305 to 285 B.C. It is surmised that Ptolemy brought Euclid from Athens to Alexandria to join the staff of the great center of Hellenistic learning—known as the Museum, with its famous Library—that he founded there.*

And the anecdotes are these:

Ptolemy once asked Euclid if there was any shorter way to a knowledge of geometry than that of the *Elements,* and he replied that there is no royal road to geometry.

Someone who had begun to read geometry with Euclid, when he had learned the first proposition, asked him, "What shall I get by learning these things?" Euclid called his slave and said, "Give this person a penny, since he must make a profit out of what he learns."

The reliability of these smug little stories can be judged from the fact that their authors (Proclus and Stobaeus) lived in the fifth century A.D., more than 700 years after the time of Euclid.

In addition to its systematic account of elementary geometry, the *Elements* also contains all that was known at the time about the theory of numbers. Euclid's role as the author was mainly that of an organizer and arranger of the scattered discoveries of his predecessors. It is possible that he contributed some of the ideas and proofs himself, and in the absence of evidence to the contrary, several important theorems are traditionally credited to him.

Book I of the *Elements* begins with 23 definitions (point, straight line, circle, etc.), 5 postulates, and 5 axioms or "common notions." Among the Greek philosophers, axioms were thought of as general truths common to all fields of study ("The whole is greater than the part"), while postulates were considered to be assumptions that have meaning only in the specific subject under discussion ("It is possible to draw a straight line from any point to any other point"). This distinction has been dropped in modern mathematics, and the words *axiom* and *postulate* are now used interchangeably. Broadly speaking, Books I to VI deal with plane geometry, Books VII to IX with number theory, Book X with irrationals, and Books XI to XIII with solid geometry. The 47th proposition in Book I (usually designated as I 47) is the Pythagorean theorem.† The following are a few other individual items of particular interest: VII 1 and VII 2 give the Euclidean algorithm, a process for finding the greatest common divisor of two positive integers; VII 30 is Euclid's lemma, which asserts that a prime number that divides the product of two positive integers necessarily divides one of the factors; IX 20 is Euclid's theorem on the infinity of primes (Theorem 2 in our Appendix A.14); IX 36 is Euclid's theorem on perfect numbers (Theorem 1 in Appendix A.1); and XII 10 gives the volume of a cone.

Students will recall from their study of geometry that a regular polygon with n sides (also known as a regular n-gon) has all of its n sides equal and all of its n angles equal. Figure B.1 shows a regular 3-gon, 4-gon, 5-gon, and 6-gon, which of course are

---

* For more information about the origin and history of the Alexandrian Museum and Library, see two articles by David E. H. Jones in *Smithsonian Magazine,* December 1971 and January 1972.

† This theorem was the stimulus for the English philosopher Thomas Hobbes's first encounter with geometry at the age of 40. He happened to be in a friend's library and saw a copy of the *Elements* lying open at I 47. As Aubrey tells it in his *Brief Lives:* "He read the proposition. 'By G--,' sayd he, (he would now and then sweare, by way of emphasis), 'this is impossible!' So he reads the demonstration of it, which referred him back to such a proposition; which proposition he read. That referred him back to another, which he also read. *Et sic dienceps* [and so in succession] back to the first, so that at last he was demonstratively convinced of that trueth. This made him in love with geometry." We add Bertrand Russell's remark about this incident: "No one can doubt that this was for him a voluptuous moment, unsullied by the thought of the utility of geometry in measuring fields."

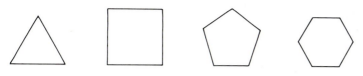

Figure B.1   Regular polygons.

usually called an equilateral triangle, a square, a regular pentagon, and a regular hexagon. Book IV of the *Elements* gives the classical constructions, using only a straightedge and compass, of the regular polygons with 3, 4, 5, 6, and 15 sides. These constructions were known to the Pythagoreans long before Euclid's time, and it was Plato and his pupils who insisted that the so-called Euclidean tools—a straightedge and compass—are the only ones "philosophically suitable" for use in geometry.* By means of angle bisections, it is easy to construct from a given regular polygon with n sides another regular polygon with $2n$ sides. The Greeks were therefore able to construct regular n-gons where n has the following values:

$$3, 6, 12, 24, \ldots ,$$

$$4, 8, 16, 32, \ldots ,$$

$$5, 10, 20, 40, \ldots ,$$

$$15, 30, 60, 120, \ldots .$$

The next obvious step was to seek Euclidean constructions for regular polygons with 7, 9, 11, 13, . . . sides. Many tried, but all such efforts failed, and there the matter rested for about 2100 years, until March 30, 1796.

On that day one of the greatest geniuses of recorded history, a young German named Carl Friedrich Gauss, proved the constructibility of the regular polygon with 17 sides. Gauss was 18 years old at the time, and his discovery pleased him so much that he decided to adopt mathematics as a career in preference to philology. He continued his investigations and quickly solved the constructibility problem completely. He proved by rather recondite methods involving algebra and number theory that a regular n-gon is constructible if and only if n is the product of a power of 2 (including $2^0 = 1$) and distinct prime numbers of the form $p_k = 2^{2^k} + 1$. In particular, when $k = 0, 1, 2, 3$, we see that each of the corresponding numbers $p_k = 3, 5, 17, 257$ is prime, so regular polygons with these numbers of sides are constructible. The prime number 7 is not of the stated form, so the regular polygon with 7 sides is not constructible.†

Book XIII of the *Elements* is devoted to the construction of the regular polyhedra, which are less familiar to most people than the regular polygons. A *polyhedron* is simply a solid whose surface consists of a number of polygonal faces; it is said to be *regular* if its faces are congruent regular polygons and if the solid angles at all its vertices are congruent. There are clearly an infinite number of regular polygons, but there are only five regular polyhedra. They are named for the numbers of faces they possess (Fig. B.2): the tetrahedron (4 triangular faces), the hexahedron or cube (6 square faces), the octahedron (8 triangular faces), the dodecahedron (12 pentagonal faces), and the icosahedron (20 triangular faces). Plato and his followers studied these

---

* Plato's Academy, which was the second genuine university in the Western world (after the school of Pythagoras) and which lasted over a thousand years, placed great importance on the study of mathematics. It is supposed to have had a sign over its entrance saying, "Let no one ignorant of geometry enter here." In all probability, Euclid was a member of the Academy before moving to Alexandria. The name "Academy" comes from the place where it was located, in a grove named after a certain hero, Hecademus (Diogenes Laertius, Loeb Edition, I, p. 283).

† Additional details on Euclidean constructions and constructibility are given very clearly and briefly in Chapter 3 of R. Courant and H. Robbins, *What Is Mathematics?,* Oxford University Press, 1941; and also in Chapter IX of H. Tietze, *Famous Problems of Mathematics,* Graylock Press, 1965. Gauss's prodigious creative life continued for another 60 years, and he is now recognized as the greatest of all mathematicians.

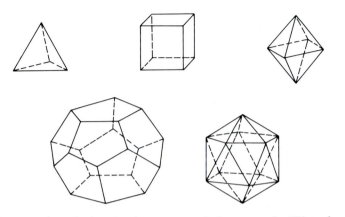

Figure B.2 Regular solids.

polyhedra so persistently that they have come to be known as the "Platonic solids." In his fantastic dialogue *Timaeus,* he associates the tetrahedron, octahedron, cube, and icosahedron with the four classical "elements" of fire, air, earth, and water (in this order), while in some mystical sense he makes the dodecahedron symbolize the entire universe. The first three regular polyhedra actually occur in nature as crystals, and the last two as the skeletons of microscopic sea animals called radiolarians.* However, it is the symmetry and beauty of these figures, and not their applications, that have fascinated people through the centuries. The construction of the regular polyhedra provides a superb climax to Euclid's geometry, and some have conjectured that this was the primary purpose for which the *Elements* was assembled — to glorify the Platonic solids.

It is so easy to prove that there are only five regular polyhedra that we venture to give the argument here. Let m be the number of sides of each regular polygonal face, and n the number of polygons meeting at each vertex. The size (in degrees) of each angle in each face is $180 - (360/m)$. Also, the sum of the angles at each vertex of the polyhedron is less than 360 degrees, so

$$n\left(180 - \frac{360}{m}\right) < 360$$

or

$$n\left(1 - \frac{2}{m}\right) < 2,$$

which is easily shown to be equivalent to

$$(m - 2)(n - 2) < 4.$$

But m and n are both greater than 2. Therefore if $m = 3$, n can only be 3, 4, or 5; if $m = 4$, n can only be 3; and if $m = 5$, n can only be 3, which gives five possibilities in all.

To mathematicians, some of the theorems in the *Elements* are important, some are interesting, and a few are both. However, the source of the immense influence of this book on all subsequent thought lies elsewhere, not so much in the exposition of particular facts as in the methodology of it all. It is clear that one of Euclid's main aims was to give a connected logical development of geometry in such a way that every theorem is rigorously deduced from the "self-evident truths" which are explicitly stated at the beginning. This pattern of thought was conceived by Pythagoras, but it was Euclid who worked it out in such stupefying detail that for more than 2000 years no one was capable of doubting that his success had been complete and final. It is true that from time to time critics pointed out definitions that do not define and

---

* See the illustrations on p. 75 of H. Weyl, *Symmetry,* Princeton University Press, 1952.

proofs that do not prove.* Nevertheless, these flaws were considered relatively minor and easy to mend. All thinking people continued to believe that the Euclidean system of geometry was *true,* in the sense that it described correctly the geometry of the actual world in which we live, and *necessary,* in the sense that it could be deduced by unassailable reasoning from axioms whose self-evident character is apparent to all.

This happy state of affairs in geometry led to the hope that in a similar way the remotest truths of science and society could be discovered and proved by simply pointing out those things that are self-evident and then reasoning from these foundations. No more attractive or tenacious idea has ever appeared in the intellectual history of the Western world. The prestige of geometry was so great, especially in the seventeenth and eighteenth centuries, that True Knowledge in any field almost required the Euclidean deductive form as a seal of legitimacy. The more disorderly branches of knowledge, which evaded this pattern, were considered to be somehow less respectable, a stage or two beneath the aristocratic disciplines.

Thus Spinoza's *Ethics,* whose subjects are God and the human passions, consists of definitions, axioms, and propositions which he attempts to support by means of proofs in the Euclidean manner.† Kant taught that the axioms of Euclidean geometry are imposed on our minds *a priori,* and are therefore necessary modes of perceiving space; and he built his entire system of philosophy on this foundation. Newton's *Principia,* with its empirical content centered on the laws of motion and the astronomy of the solar system, is wholly dominated by the Euclidean scheme of definitions, axioms, lemmas, propositions, corollaries, and proofs, with a liberal sprinkling of Q.E.D.'s. The seventeenth century doctrine of natural rights proclaimed by Locke was an attempt to deduce the laws of politics and government from axioms of a Euclidean type.‡ Even the American Declaration of Independence, in saying "We hold these truths to be self-evident," was seeking clarity and credibility by emulating the Euclidean model.

Unfortunately, self-evident truths are much scarcer now than they used to be. Since the advent of general relativity and cosmology, it has been known that Euclidean geometry is not an adequate mathematical framework for the universe at large, and in this sense is no longer "true." Since the advent of non-Euclidean geometries, it has been known that Euclid's axioms are not self-evident at all: On the contrary, they can be replaced by others which contradict them and have just as good a claim to acceptance from the point of view of logic. Axioms in government and human behavior are now recognized as hopes and expressions of preference rather than immutable truths.

In spite of these lost illusions, the axiomatic method first elaborated by Euclid is still widely used in the more abstract parts of higher mathematics as a convenient way of clearly delineating the mathematical system to be investigated. It is no exaggeration to say that modern abstract mathematics could hardly exist without this method.

All in all, for more than 2000 years the intellectual architecture of the *Elements* has rivaled the Parthenon as a symbol of the Greek genius. Both have deteriorated somewhat in recent centuries, but perhaps the book has sustained less damage than the building.

---

* Recall the definition of a point quoted earlier. Also: "A *line* is breadthless length"; "A *straight line* is a line which lies evenly with the points on itself"; "A *unit* is that by virtue of which each of the things that exist is called one"; "A *number* is a multitude composed of units." The defects in the proofs often consist of the use of additional assumptions that are not explicitly recognized.

† "I shall consider human actions and desires just as if I were studying lines, planes, and solid bodies."—*Ethics,* Part III, Introduction.

‡ "To understand political power right, and derive it from its original, we must consider what state all men are naturally in, and that is, a state of perfect freedom to order their actions, and dispose of their possessions and persons, as they think fit."—*Second Treatise of Government,* Section 4.

ARCHIMEDES (c. 287–212 B.C.)

There was more imagination in the head of Archimedes than in that of Homer.

Voltaire

Archimedes will be remembered when Aeschylus is forgotten, because languages die and mathematical ideas do not.

G. H. Hardy

Archimedes, who was certainly the greatest mathematician, physicist, and inventor of the ancient world, was one of the supreme intellects of Western civilization. Another genius of comparable power and creativity did not appear until Isaac Newton in the seventeenth century.

Archimedes was born in the Greek city of Syracuse on the island of Sicily. He was on intimate terms with the royal family, and was probably a relative of King Hieron II. In his youth he studied in the great intellectual center of Alexandria. It was perhaps during this period that he met his friend Eratosthenes, later director of the Alexandrian Library, to whom he communicated many of his discoveries. On his return to his native city, he settled down and devoted the rest of his life to mathematical research. At the age of 75 he was killed by a Roman soldier when Syracuse was conquered by the army of Marcellus during the Second Punic War.

Archimedes was famous throughout the Greek world during his lifetime, and has been a legendary figure ever since, not because of his profound mathematical discoveries, but rather because of his vivid and memorable achievements, his many ingenious inventions, and the manner of his death. Few solid facts are known, but traditional accounts of his activities are found in the writings of numerous Roman, Greek, Byzantine, and Arabic authors down through the centuries. He put the stamp of his personality on the world, and the world has not forgotten.*

Perhaps the most famous of the traditional stories concerns the occasion when he was asked by King Hieron to determine whether a newly made crown was made of pure gold, as specified, or whether the goldsmith had cheated him by substituting silver for some of the gold. Archimedes was perplexed until one day he noticed the overflow of water as he stepped into a public bath. He suddenly realized that since gold is denser than silver, a given weight of gold occupies a smaller volume than an equal weight of silver, and will therefore displace less water than an equal weight of silver. He was so overjoyed with his discovery that he forgot he was naked and ran home through the streets of the city without his clothes, shouting "Eureka, eureka!" which means "I have found it, I have found it!" He quickly determined that Hieron's new crown displaced more water than an equal weight of gold, thereby convicting the goldsmith of fraud. This story is often associated with his discovery of the basic law of hydrostatics, which states that a floating body displaces its own weight of liquid. From this beginning he created the science of hydrostatics, and proved many theorems about equilibrium positions of floating bodies of various shapes.† Further, one of his best-known inventions was a spiral-shaped water pump called the "screw of Archimedes." This device is still used along the Nile for raising water from the river to the adjoining fields.‡

---

* For information about the sources, see the introductory chapters (with references) of E. J. Dijksterhuis, *Archimedes,* Humanities Press, 1957; or T. L. Heath, *The Works of Archimedes,* Dover, n.d. The most detailed account is that given by Plutarch in his *Life of Marcellus.*

† See the treatise *On Floating Bodies, Works,* pp. 252–300.

‡ The present writer has actually seen this device being used on the banks of the Nile.

In mechanics, he discovered the principle of the lever, originated the concept of center of gravity, and found the centers of gravity of many plane and solid figures.* According to one writer, it was his study of levers that led him to utter his famous saying, "Give me a place to stand on, and I can move the earth." Plutarch gives another version:

Archimedes one day asserted to King Hieron, whose friend and kinsman he was, that with a given force he could move any given weight whatever; and he further claimed that if he were given another earth, he could cross over to it and move this one. When Hieron, full of wonder, begged him to give a demonstration of some great weight moved by a small force, Archimedes caused one of the king's galleys to be drawn up on shore by many men and great labor; and loading her with many passengers and much freight, he placed himself at a distance, and without visible effort, only moving with his hand the end of a machine, which consisted of a variety of ropes and pulleys, he drew the ship to him as smoothly and safely as if she were moving through the water.

Hieron was so astounded at this feat that he declared, "From this day forth Archimedes is to be believed in everything he says."

Archimedes' greatest fame in antiquity came from the many stories of the engines of war he devised to defend Syracuse against the army and navy of the Roman general Marcellus. Plutarch devotes several vivid pages of description to the attacks of the Romans and the shattering effect of Archimedes' defensive machines. There were catapults of adjustable range for hurling enormous stones, huge movable beams for projecting suddenly over the walls and dropping heavy weights on enemy galleys that approached too close, and giant grappling cranes that caught hold of ships by the prow, lifted them up, and plunged them to the bottom of the sea. There were even burning mirrors that set the enemy ships on fire at a distance.† As Plutarch writes:

The Romans, being infinitely distressed by an invisible enemy, began to think they were fighting against the gods. Marcellus escaped unhurt, and deriding his own engineers said, "We must give up fighting with this geometrical Briareus [a hundred-armed mythological monster], who sitting on the shore and acting as if it were only a game, plays pitch-and-toss with our ships, and striking us in a moment with such a multitude of bolts, outdoes even the hundred-handed giants of mythology." At last the Romans were so terrified, that if they saw only a rope or a stick put over the walls, they cried out that Archimedes was leveling some engine at them, and turned their backs and fled.

Thus Marcellus abandoned his intention of assaulting the city and put his hopes on a siege. This siege of Syracuse lasted three years and ended in 212 B.C. with the fall of the city.

According to all accounts Archimedes died in a manner consistent with his life, absorbed in mathematical contemplation. In the general confusion and slaughter that followed the fall of the city, he was found concentrating on some diagrams he had drawn in the sand, and was killed by a marauding soldier who did not know who he was. In one version of the story he said to the intruder, who came too close, "Do not disturb my circles," whereupon the enraged soldier ran a sword through his body. Marcellus was greatly saddened by this, since he had given strict orders to his men to spare the house and person of Archimedes. He mourned the death of his formidable antagonist, befriended his surviving relatives, and saw to it that he had an honorable

---

* See the treatise *On the Equilibrium of Planes, Works,* pp. 189–220.

† For the details of a modern experiment by the Greek navy showing that this use of solar power in warfare is quite feasible, see *Newsweek,* Nov. 26, 1973, p. 64.

burial. The eminent modern philosopher A. N. Whitehead found a larger meaning in this event than the death of a single man:

> The death of Archimedes at the hands of a Roman soldier is symbolic of a world change of the first magnitude. The Romans were a great race, but they were cursed by the sterility which waits upon practicality. They were not dreamers enough to arrive at new points of view, which could give more fundamental control over the forces of nature. No Roman lost his life because he was absorbed in the contemplation of a mathematical diagram.

Archimedes is said to have asked his friends to place on his tombstone a representation of a cylinder circumscribing a sphere, and in memory of his greatest mathematical achievement to inscribe the proportion ($\frac{3}{2}$) which the containing solid bears to the contained. This was done at the order of Marcellus. The Roman orator Cicero, when he was quaestor in Sicily in 75 B.C., searched out this monument, discovered it neglected and overgrown with brambles, and had it cleaned and restored out of respect for the great mathematician.*

Cicero also saw and described an invention of Archimedes that made such a deep impression on the ancient world that it is mentioned by many classical authors. This device was apparently a small planetarium, an open revolving sphere of bronze and glass with internal machinery driven by water, in which during one revolution the sun, moon, and five planets moved in the same way relative to the sphere of fixed stars as they do in the sky in one day, and in which one could also observe the phases and eclipses of the moon. Closed spheres like modern terrestrial globes that revolved uniformly and imitated the daily motion of the fixed stars had long been known, but that Archimedes was able to represent by one mechanism the independent and very different motions of the sun, moon, and planets, together with the revolution of the fixed stars, seemed to his contemporaries to be evidence of superhuman abilities. Cicero, having seen this planetarium himself, writes:

> When Gallus set the sphere in motion, one could actually see the moon rise above the earth's horizon after the sun, just as occurs in the sky every day; and then one saw how the sun disappeared and how the moon entered the shadow of the earth with the sun on the opposite side.†

This hydraulic mechanism was captured by the Romans as part of their booty in the sack of Syracuse, and it was evidently treasured by them for more than a hundred years afterward as one of the wonders of the world.

Archimedes was clearly a very ingenious and successful inventor, but Plutarch says that his inventions were only "the diversions of geometry at play." In a famous passage he purports to tell us about Archimedes' attitude toward practical life in general and his own inventions in particular:

> Archimedes possessed such a lofty spirit, so profound a soul, and such a wealth of scientific knowledge, that although his inventions had won him renown for superhuman sagacity, he would not deign to leave behind him any written work on these subjects; but regarding as ignoble and vulgar the construction of instruments and every art concerned with use and profit, he devoted his whole ambition and effort to those studies whose beauty and subtlety have no relation to the practical needs of life.

---

* See Cicero's *Tusculan Disputations,* Loeb Classical Library, p. 491. The Romans were so uninterested in mathematics that Cicero's act of respect in cleaning up Archimedes' grave was perhaps the most memorable contribution of any Roman to the history of mathematics.

† See Cicero's *De Re Publica,* Loeb Classical Library, p. 43.

Though eloquent, the truth of what Plutarch says here is more than doubtful, because Archimedes is known to have written a treatise that is now lost (*On Sphere-making*) which probably dealt with the detailed techniques required for the construction of his planetarium. Plutarch was thoroughly infected, as Archimedes certainly was not, by the Platonic contempt for scientific instruments and measurement that was one of the many foolish legacies of the philosopher Plato to his worshipful posterity.

Nevertheless, it is very clear that in pure mathematics Archimedes was able to satisfy to the full the deepest desires of his nature. Plutarch is more convincing when he tells us that few people have ever lived who were so preoccupied with mathematics as he was:

> We are not, therefore, to reject as incredible, what is commonly told of him, that being perpetually charmed by his familiar Siren, that is, by his geometry, he neglected to eat and drink and took no care of his person; that he was often carried by force to the baths, and when there he would trace geometrical figures in the ashes of the fire, and with his finger draw lines upon his body when it was anointed with oil, being in a state of great ecstasy and divinely possessed by his science.

What, precisely, *were* his achievements in mathematics? Most of his wonderful writings still survive, and even on cursory inspection are obvious works of genius. Virtually all of the subject matter of his nine treatises is completely original, and consists of entirely new discoveries of his own. Though he treated a wide range of subjects, including plane and solid geometry, arithmetic, astronomy, hydrostatics, and mechanics, he was no compiler of earlier discoveries like Euclid, no mere writer of textbooks. His aim was always to provide some new contribution to knowledge. As to the overall impression conveyed by his works, Heath says:

> The treatises are, without exception, monuments of mathematical exposition; the gradual revelation of the plan of attack, the masterly ordering of the propositions, the stern elimination of everything not immediately relevant to the purpose, the finish of the whole, are so impressive in their perfection as to create a feeling akin to awe in the mind of the reader. As Plutarch said [with understandable exaggeration], "It is not possible to find in geometry more difficult and troublesome questions or proofs set out in simpler and clearer propositions." There is at the same time a certain mystery veiling the way in which he arrived at his results. For it is clear that they were not *discovered* by the steps which lead up to them in the finished treatises.*

Thus, from one point of view his writings present an aspect of austere architectural perfection. On the other hand, in most of the mathematical treatises (though not in those concerned with physics) there is a personal preface in which he addresses his friends, explains his purposes, and generally sets the stage for the intellectual drama to come. Compared with Euclid, his writings are throbbing with life.

The range and importance of the mathematical work of Archimedes can best be understood from a brief account of his six geometrical treatises, three dealing with plane geometry, two with solid geometry, and one with his method of making discoveries.

1. Quadrature of the Parabola This treatise of 24 propositions contains two proofs of his theorem that the area of a segment of a parabola, that is, the region cut from a parabola by any transverse line, is $\frac{4}{3}$ the area of the triangle with the same base and height. A description of this theorem is given in our Section 6.2, with full details provided in Appendix A.2. In his last two propositions Archimedes sums the infinite

---

* T. L. Heath, *A History of Greek Mathematics,* Oxford University Press, 1921, vol. II, p. 20.

geometric series $1 + \frac{1}{4} + (\frac{1}{4})^2 + \cdots$ in such a way as to demonstrate that he is fully aware of the subtlety of the concept of limit. This did not become clear to other mathematicians until the early nineteenth century.

2. On Spirals   The subject of this treatise of 28 propositions is the curve now known as the *spiral of Archimedes*. He defines it as follows:

> If a straight line of which one extremity remains fixed be made to revolve at a uniform rate in a plane until it returns to the position from which it started, and if, at the same time as the straight line revolves, a point moves at a uniform rate along the straight line, starting from the fixed extremity, the point will describe a spiral in the plane.

His main achievements were to determine the tangent at any point (Prop. 20) and to find the area enclosed by the first turn (Prop. 24), the latter being $\frac{1}{3}$ the area of the circle whose radius is the distance the moving point travels along the moving line. This spiral and these properties are treated in our Sections 16.3 (Example 6), 16.4 (Problem 7), and 16.5 (Problem 3). Later mathematicians used the spiral as an auxiliary curve for trisecting an angle (Section 16.3, Problem 23) and squaring a circle (Chapter 16, Additional Problem 8). Also, it has been conjectured that he discovered how to determine the tangent at a point by methods verging on those of differential calculus.*

3. Measurement of a Circle   In this short work of three propositions he demonstrates with full rigor, as no one had done before him, and as we do more loosely in Section 6.2, that the area of a circle is equal to that of a triangle with base equal to its circumference and height equal to its radius, $A = \frac{1}{2}cr$; and since $c = 2\pi r$ by the definition of π, we have the familiar formula $A = \pi r^2$. He also establishes the inequalities

$$3\tfrac{10}{71} < \pi < 3\tfrac{1}{7},$$

by an elaborate calculation of the perimeters of regular polygons of 96 sides inscribed in and circumscribed about a given circle.

4. On the Sphere and Cylinder   This is the profoundest of the treatises, for it contains rigorous proofs of his great discoveries of the volume and surface area of a sphere (Props. 33 and 34), as well as much else besides. As to how he made these discoveries, see item 6.

5. On Conoids and Spheroids   By these terms Archimedes means solids of revolution generated by revolving parabolas, hyperbolas, and ellipses about their axes. He calculates the volumes of segments of these solids, and incidentally proves and uses the formulas

$$1 + 2 + \cdots + n = \frac{n(n+1)}{2}$$

and

$$1^2 + 2^2 + \cdots + n^2 = \frac{n(n+1)(2n+1)}{6}$$

for the sums of the first n integers and squares (see pp. 162 and 105–109 of the

---

* See the Appendix to vol. II of Heath's *History*.

Works). Also, he proves the formula πab for the area of an ellipse with semiaxes a and b (Prop. 4).

6. Method   This, the most interesting of all the treatises, takes the form of a letter to Eratosthenes in which Archimedes explains his method of making discoveries in geometry and illustrates his ideas with 15 propositions. This work was accidentally discovered on a palimpsest parchment in Constantinople in 1906, after having been lost for nearly a thousand years. As Heath says:

> The *Method,* so happily recovered, is of the greatest interest for the following reason. Nothing is more characteristic of the classical works of the great geometers of Greece, or more tantalizing, than the absence of any indication of the steps by which they worked their way to the discovery of their great theorems. As they have come down to us, these theorems are finished masterpieces which leave no traces of any rough-hewn stage, no hint of the method by which they were evolved. We cannot but suppose that the Greeks had some method or methods of analysis hardly less powerful than those of modern analysis; yet, in general, they seem to have taken pains to clear away all traces of the machinery used and all the litter, so to speak, resulting from tentative efforts, before they permitted themselves to publish, in sequence carefully thought out, and with definitive and rigorously scientific proofs, the results obtained. A partial exception is now furnished by the *Method*; for here we have a sort of lifting of the veil, a glimpse of the interior of Archimedes' workshop.

In one of Archimedes' illustrations of his method, he shows us how he discovered his favorite theorem about the volume of a sphere. The details of his ideas are given in our Appendix A.5; and, as we point out there, his way of thinking is essentially equivalent to the basic process of integral calculus. He then goes on to tell us how he was led by this to discover the surface area of a sphere, by thinking of a sphere as if it were a cone wrapped around its vertex:

> From this theorem, to the effect that a sphere is four times as great as the cone with a great circle of the sphere as base and height equal to the radius of the sphere, I conceived the notion that the surface of any sphere is four times as great as a great circle in it; for, judging from the fact that any circle is equal to a triangle with base equal to the circumference and height equal to the radius of the circle, I apprehended that, in like manner, a sphere is equal to a cone with base equal to the surface of the sphere and height equal to the radius.*

Among other discoveries included in this treatise are two standard examples (or problems) often used in modern calculus textbooks, about the location of the center of gravity of a solid hemisphere (Prop. 6), and the volume common to two equal cylinders whose axes intersect at right angles (Prop. 15, also preface).

In addition to these six geometrical treatises, and the two dealing with physics, there is one more that ought to be mentioned. This is concerned with arithmetic and astronomy and is called *The Sand-Reckoner.* In it he constructs a system of notation for designating very large numbers, a system that enables him (without our decimal system or exponent notation, which came two thousand years later) to express numbers as large as N^{NN} where N is 10^8. He then applies his ideas to find an upper limit for the number of grains of sand that would fill a sphere whose radius is the distance from the sun to what Aristarchus called "the sphere of the fixed stars," and this turns out to be a mere 10^{63}. It is here, among interesting related comments about astronomy, that

---

* This idea is worked out in a remark at the end of our Section 7.6.

we learn that Aristarchus had propounded the Copernican theory of the solar system a few decades earlier.

Finally, if we consider what all other men accomplished in mathematics and physics, on every continent and in every civilization, from the beginning of time down to the seventeenth century in Western Europe, the achievement of Archimedes outweighs it all. He was a great civilization all by himself.

PAPPUS (fourth century A.D.)

Pappus of Alexandria was an able, enthusiastic, and elegant mathematician who had many good ideas. However, he had the bad luck to be born when the great age of Greek mathematics—which had lasted roughly 900 years, from the time of Thales and Pythagoras—was drawing its last breath.

His main work, the *Mathematical Collection,* is a combined encyclopedia, commentary, and guidebook for Greek geometry as it existed in his time, enriched by many new theorems, extensions and new proofs of old ones, and valuable historical remarks. Unfortunately, the *Collection* turned out to be the requiem of Greek mathematics instead of a breath of new life, for after Pappus, mathematics shriveled and almost disappeared, and had to wait 1300 years for a rebirth in the early seventeenth century.

He is best known for his beautiful geometric theorems connecting centers of gravity with solids and surfaces of revolution (Section 11.3). The first of these theorems states that the volume generated by the complete revolution of a region bounded by a closed plane curve that lies entirely on one side of the axis of revolution equals the product of the area of the region and the distance traveled by the center of gravity around its circular path. Pappus was rightly proud of the generality of his theorems, for, as he says, "they include any number of theorems of all sorts about curves, surfaces, and solids, all of which are proved at once by one demonstration."[*]

He gave the first statement and proof of the focus-directrix-eccentricity characterization of the three conic sections (Section 15.5). Since he was scrupulous about crediting the souces of his material, and no source is given for this, it is reasonable to infer that it was his own discovery.

He gave the following interesting extension of the theorem of Pythagoras (see Fig. B.3): Let *ABC* be any triangle and *ACDE* and *BCFG* any parallelograms constructed externally on the sides *AC* and *BC*; if *DE* and *FG* intersect at *H*, and *AJ* and *BI* are equal and parallel to *HC*, then the area of the parallelogram *ABIJ* equals the sum of the areas of the parallelograms *ACDE* and *BCFG* (proof: *ACDE* = *ACHR* = *ATUJ* and *BCFG* = *BCHS* = *BIUT*). It is not difficult to see that this statement really does yield the theorem of Pythagoras as a special case, when the angle *C* is a right angle and the constructed parallelograms are squares.

Finally, we mention the important result of projective geometry known as Pappus' Theorem: If the vertices of a hexagon lie alternately on a pair of intersecting lines (Fig. B.4), then the three points of intersection of the opposite sides of the hexagon are collinear. (The "opposite" sides can be recognized from the numbered vertices of the schematic hexagon shown in the figure.) The full significance of this classic theorem was at last revealed only in 1899, by the great German mathematician David Hilbert, as part of his program to clarify the foundations of geometry.[†]

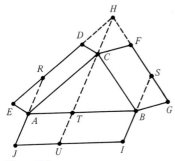

Figure B.3

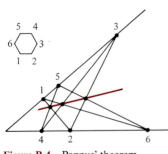

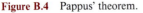

Figure B.4 Pappus' theorem.

---

[*] T. L. Heath, *A History of Greek Mathematics,* Oxford University Press, 1921, vol. II, p. 403.

[†] For mathematicians: Hilbert discovered that the validity of Pappus' Theorem in a Desarguesian projective plane is equivalent to the commutativity of the field of coefficients. See pp. 82–86 of A. Seidenberg, *Lectures in Projective Geometry,* Van Nostrand, 1962; or E. Artin, *Geometric Algebra,* Interscience, 1957.

DESCARTES (1596–1650)

Descartes commanded the future from his study more than Napoleon from his throne.

Oliver Wendell Holmes, Jr.

Modern philosophy was born in the year 1637, in a short book by Descartes entitled *Discourse on Method.* In this work he rejected the sterile scholasticism prevailing at the time and set himself the task of rebuilding knowledge from the ground up, on a foundation of reason and science instead of authority and faith. He provided the fresh points of view needed for the vigorous development of the Scientific Revolution, whose influence has been the dominant fact of modern history. Further, in an appendix to the *Discourse* on his ideas about geometry, he foreshadowed the new forms of mathematics — analytic geometry and calculus — without which this Revolution would have died in infancy.

The seventeenth century, like the twentieth, was an age of religious hatred, oppression, and plundering warfare, presided over, for the most part, by squalid rulers with the political ethics of cattle rustlers and the personal morality of gigolos. Nevertheless, history is not exclusively what Gibbon called it, a record of "the crimes, follies, and misfortunes of mankind." Intellectually, the century in which Descartes lived was one of the greatest periods in the history of civilization. It began with Galileo and Kepler; it ended with Newton and Leibniz; and it nurtured an array of remarkable men so gifted and diverse in their talents and achievements that it can only be compared with the golden age of ancient Greece. It is often called the Century of Genius.

René Descartes came from a family of the lesser nobility near Tours, in central France. His mother died soon after his birth, and left him some property, which later made it possible for him to enjoy a leisurely life of travel and study. When the boy was 8, his father sent him to the nearby Jesuit school at La Flèche, an excellent institution for the education of young gentlemen which Henry IV had recently established in one of his favorite palaces. Here Descartes was thoroughly trained in literature and the classical languages, rhetoric, philosophy, theology, science, and mathematics. He was treated with kindly consideration by the Jesuit fathers, and in view of his frail constitution and meditative disposition was allowed to lie in bed in the mornings as late as he pleased, long after the other boys were in their classes. He maintained this habit to the end of his life, and liked to say that many of his best thoughts came to him in those tranquil hours of the late morning. There is even a story that he conceived the basic idea of analytic geometry while lying in bed and watching a fly crawling on the ceiling of his room, by noticing that the path of the fly could be described if only one knew a relation connecting its distances from two adjacent walls.

The young Descartes was a born skeptic, and as he matured he began to suspect that the so-called humanistic learning he was absorbing at La Flèche was almost barren of human significance, with little power to enrich or improve human life. As a polite and circumspect youth, he kept most of his doubts to himself. Nevertheless, he saw more and more clearly that the principles of philosophy and theology taught by the Jesuits were often little more than baseless superstitions cloaked in a few tattered rags of scholastic logic. He was passionately interested in science, but the science he was offered was the worthless physics of Aristotle adjusted to the empty doctrines of Saint Thomas Aquinas. Only mathematics escaped his contempt, "because of the certainty of its proofs and the clarity of its reasoning"; and even here, he was "astonished that foundations so firm and solid should have nothing loftier built upon them." It must be remembered that mathematics at that time consisted of classical geometry and a few primitive fragments of elementary algebra, with almost all of its applications to science still beyond the horizon.

Like many intelligent young people in every century, Descartes left school filled with disgust at the arid emptiness of his studies. His state of mind is best expressed in his own words:

> This is why, as soon as my age permitted me to leave the control of my teachers, I completely gave up the study of letters. And resolving to seek no other knowledge than that which I could find in myself, or else in the great book of the world, I employed the rest of my youth in travel, in seeing courts and armies, in frequenting men of diverse temperaments and conditions . . . and above all in trying to learn from what I saw, so that I might derive some profit from my experience.*

Naturally, he first went to Paris, where he gambled, wenched, and generally played the dandy. These dissipations quickly palled, and he astonished his raffish friends by enlisting in the Dutch army as an unpaid gentleman volunteer. The army was inactive at that time, and in his enforced leisure he was once again attracted to the study of mathematics. A year later, in 1619, he transferred to the army of the Duke of Bavaria; and while staying in winter quarters in a small town on the Danube, he experienced an "illumination" that for him was comparable with the great mystical revelations in the lives of the saints.

In this frigid village where he was a stranger to everyone, Descartes shut himself up for the winter in a well-heated room and plunged into solitary study and meditation. He thought over the knowledge he had acquired in the various sciences and despairingly noted its confusion and uncertainty. What was needed was a totally fresh start, a new beginning that would sweep away all those systems of thought and belief that had become polluted over the centuries with half-truths, wishful thinking, and false reasoning. Only in mathematics had he found the certainty he desired, and the task that gradually formed in his mind was that of extending this certainty to all other fields of knowledge. But how could such a monumental project be accomplished? On November 10, 1619—a famous day in the history of philosophy—in a state of exhaustion and feverish excitement, he found his method and felt that he had glimpsed "the foundations of a marvelous science." Much more than merely showing him the way in an isolated problem, or even clarifying the principles of a particular science, his illumination revealed to him the essential unity of all the sciences—indeed, of all knowledge. And his method of displaying the various disciplines as branches of his single "marvelous science" would be that of mathematics:

> Those long chains of simple reasoning which geometers use to arrive at their most difficult conclusions made me believe that all things which are the objects of human knowledge are similarly interdependent; and that if we will only abstain from assuming something to be true which is not, and always follow the necessary order in deducing one thing from another, there is nothing so remote that we cannot reach it, nor so hidden that we cannot discover it.†

On that fateful day he made two decisions that shaped the future course of his life. First, he decided that he must systematically doubt everything he knew or thought he knew about all the sciences, and search for self-evident and certain foundations on which the edifice of knowledge could be rebuilt with confidence. Second, since a great work of art is always the product of a single master artist, he decided that he must carry out the entire project himself.

He left the army and traveled extensively in Germany, Switzerland, Italy, and Holland. He worked on mathematical problems, studied glaciers and avalanches,

---

* *Discourse,* Part I.

† *Discourse,* Part II.

computed the heights of mountains, and continued to harbor his great secret ambition for a total reform of human knowledge. In one of his private notebooks he wrote the following fragment of self-analysis: "As an actor, ready to appear on the stage, dons a mask to hide his timidity, so I go forward in a mask preparing to mount the stage of the world, which up to now I have known only as a spectator." Returning to Paris in 1625, he spent most of the next three years in the company of men with similar mathematical and scientific interests. In particular, he renewed his acquaintance with the Franciscan friar Marin Mersenne, a former schoolmate at La Flèche. Mersenne knew everyone worth knowing, and was to become and remain his closest friend and most faithful admirer.

Late in 1628 Descartes began to realize that he would never "mount the stage of the world" unless he started climbing, and soon. In order to find the peaceful leisure necessary for thinking and writing, he left the noisy bustle of Paris and went to Holland, where he lived for the next 21 years. He preferred the solitary life and greater intellectual freedom available in that country, and he also hoped to avoid the annoyance of oafish visitors who came to see him in the morning and rousted him out of bed. In order to further ensure his privacy, he changed his address an incredible 24 times during that period and carefully kept it a secret from all but his closest friends. The motto he adopted was more appropriate for a fugitive than a philosopher: *Bene vixit qui bene latuit*—"He has lived well who has hidden well." Though solitary, he was far from isolated, for he set aside one day a week to attend to his voluminous correspondence with learned men all over Europe. His chief correspondent was Mersenne, who acted as his link with Parisian intellectual circles for all manner of scientific news, philosophical questions, and mathematical problems. He read comparatively little because he trusted experiments more than books; and more than either, he trusted his own mind. But he didn't believe in overdoing meditation, either. His custom, he said, was "never to spend more than a few hours a day in thoughts which occupy the imagination, or more than a few hours a year in those which occupy the understanding, and to give all the remaining time to the relaxation of the senses and the repose of the mind." But perhaps this was partly bluff, since gentleman intellectuals in those days weren't supposed to work very hard; if they violated this code, as Descartes certainly did, then decency required a modicum of dissembling.

Descartes's first important book, *Rules for the Direction of the Mind,* was probably written during his first year in Holland. It contains the fullest description of his method for clear and correct thinking, but he left it unfinished, and it was not published until 1701. A few years later he decided to make public three of his shorter scientific treatises, accompanied by an explanatory preface. The result was his *Discourse on Method* (1637), with its appendixes entitled *Dioptric, Meteors,* and *Geometry,* which he intended to be convincing illustrations of the power of his method. These appendixes contained little of permanent interest or value, and have deservedly sunk into oblivion. However, the *Discourse* itself remains a landmark of philosophy which is also a literary classic. It is unique in both form and content — a mixture of philosophy and science, a manifesto and a prospectus, an intellectual autobiography of great charm in which the adventures of reason are narrated as vividly as Homer's account of the wanderings of Odysseus. As a natural outgrowth of his continuing study of science, Descartes developed an interest in the theory of knowledge. What do I know? he asked himself as a youth. And later, How do I know? What does it mean to know? In 1641 he published his ideas on this subject in his *Meditations,* a work that has been described as combining literary excellence and philosophical genius on a level unmatched by any thinker since Plato. We have already mentioned his decision in 1619 to doubt everything he thought he knew about the various sciences. The philosophic doubt of his middle years, as expressed in both the *Discourse* and the *Meditations,* was far more searching. It omitted nothing from its withering blast, not even his conviction of his own existence.

A brief summary will perhaps convey the flavor of his ideas. At least once in his life, Descartes said, a man who seeks truth must summon the courage to doubt everything. Most of our beliefs — about the physical world, religion, society, and ourselves — enmesh us in childhood before our defenses are up; and by the time we reach the age of critical judgment, their bonds have become so comfortable and familiar that we are scarcely aware of their presence. We acquire these beliefs from inadequate senses, credulous parents, unreliable teachers, and self-serving institutions. We must examine all our convictions, he said, in the search for those that resist our most strenuous efforts to doubt them, for only these can provide a solid and certain foundation on which to rebuild the temple of knowledge. This systematic search for bedrock, by digging down through the mud and sand of the mind, led him at last to his ultimate verity: *Cogito ergo sum* — "I think, therefore I am" — perhaps the most famous sentence in the history of philosophy. It was not an original thought, for St. Augustine expressed the same idea in almost the same words more than a thousand years earlier: "Who doubts that he lives? For if he doubts, he lives."* However, this perception was only incidental for Augustine, while Descartes made it the source of his entire system of thought. His epistomology and metaphysics arose from his effort to reconstruct the external world from the primal fact of his own thinking; and for better or worse, Western philosophy has been preoccupied ever since with the problem of whether this "external world" really exists except as an idea.†

The constructive part of his system is much less interesting and important than the destructive part just described. For example, his eloquent proof of the existence of God reduces in the end to the flimsiest reasoning one can imagine: "It is not possible that I could have in myself the idea of God, if God did not truly exist."‡ Arguing next that his "clear and distinct ideas" are necessarily true, since God is perfect and therefore cannot stoop to deception, he proceeds to "prove" the existence of the external world, the distinctness of mind and body, and so on. In effect, he throws his old ideas out the front door with resounding fanfare, and then, after a decent interval and with suitable ceremonies, quietly lets them in again at the back door.§ In spite of their flaws, the books of Descartes were widely read, and their skeptical parts gave a powerful push to the simplest, most obvious, yet most potent of all revolutionary principles: Base your beliefs on evidence, and give them only the degree of support that the evidence justifies. This unpopular principle has smouldered feebly ever since, bursting into flame now and then in a few individuals to touch off the explosions of scientific knowledge that have set apart the last three centuries of European civilization from all other periods of human development. Whatever one thinks of the ideas of Descartes, it is undeniable that a great part of his influence was due to his extraordinary skill as a writer. "When writing about transcendental issues," he said, "be transcendentally clear," and in this he usually followed his own advice. Descartes was one of the great masters of the art of language, and for a thinker who wishes to be remembered, this is often better than having important original ideas.

In his own eyes Descartes was mainly a scientist and mathematician, and only incidentally a philosopher. What was the nature of his scientific activity? In 1633 he

---

* *The Trinity,* Book X, Chapter 10.

† The modern Chinese philosopher Lin Yutang, educated in a different tradition yet profoundly familiar with Western thought, made the following acid comment on this somewhat absurd preoccupation: "How on earth did Descartes, who could not on *prima facie* evidence accept his existence as real, believe that his thinking was? That was the beginning of the dark ages of European philosophy."

‡ *Meditations,* Part III.

§ "The human brain is a complex organ with the wonderful power of enabling man to find reasons for continuing to believe whatever it is that he wants to believe." — Voltaire.

completed an ambitious treatise entitled *The World,* in which he undertook to explain "the nature of light, and of the sun and stars which emit it; of the heavens which transmit it; of the planets, the comets, and the earth which reflect it; of all terrestrial bodies which are colored by it; and of Man its spectator." The condemnation of Galileo by the Inquisition caused him to abandon all thought of publishing this work, and it has survived only in fragments. In addition to physics and astronomy, his scientific interests included meteorology, optics, embryology, anatomy, physiology, psychology, geology, and even medicine and nutrition, which he studied in the hope of prolonging his life. In meteorology he gave fanciful explanations of thunder and lightning, thunder being the sound made when a higher cloud falls on a lower one; nevertheless, his explanation of the rainbow, in terms of refraction and reflection within water droplets in the atmosphere, was quite correct.* His attempt to account for the colors of the rainbow was not successful, and this achievement was left for Newton. He was the first to publish the sine law of refraction, but he stated it incorrectly and "proved" his incorrect version by an *ad hoc* argument that suggested to many later scientists (including Newton and Huygens) that he did not discover the law himself, but learned of it from Snell, who is now generally given credit for the discovery.† In other fields, he dissected a fetus and described its anatomy; removed the back of an ox's eye to examine the image formed by an object placed in front of the eye; welcomed Harvey's discovery of the circulation of the blood, but engaged in an unsuccessful dispute with him over the action of the heart; and dissected the heads of various animals in an effort to locate the sources of memory and imagination. His doctrine that the body is a machine had considerable influence on the later history of physiology and psychology. According to him, animals are nothing but machines and a person is a machine distinguished by the possession of a soul, which probably resides in the pineal gland at the base of the brain. These mechanistic views also permeated his physics and astronomy. He rejected action at a distance and assumed that all physical influences—such as gravity, light, and magnetism—are transmitted mechanically by the pressures of adjacent particles. The entire universe, he said, is filled with these particles, moving ceaselessly in vortices; the earth, for example, moves in its orbit because it is swept around the vortex of the sun like a twig in a whirlpool. These pictorial fancies enjoyed a brief vogue, but were soon destroyed by the rise of Newtonian mathematical physics. In his science Descartes searched for the meaning of everything that exists and the cause of everything that happens, but little or nothing remains of his work. He was crippled as a scientist by philosophic preconceptions, unrestrained speculation, and excessive ambition—diseases to which philosophers are particularly vulnerable—and as a result his ideas were almost entirely erroneous. As Newton said, probably with Descartes in mind:

> To explain all nature is too difficult a task for any one man or even for any one age. 'Tis much better to do a little with certainty, and leave the rest for others that come after you, than to explain all things.

We come at last to the difficult question of Descartes's mathematics—difficult because one of the commonest items of second-hand gossip among historians of science is that he invented analytic geometry, and yet he did nothing of the kind.‡ Any qualified person who examines Descartes's treatise on geometry will soon convince himself that this work contains nothing about perpendicular axes, or the "Car-

---

* He probably did not know that this phenomenon had been correctly explained more than 300 years earlier. See A. I. Sabra, *Theories of Light from Descartes to Newton,* Oldbourne, 1967, p. 62.

† Ibid., pp. 99–105.

‡ It is said that history repeats itself, and historians repeat each other.

tesian" coordinates of a point, or equations of lines or circles, or any material at all that bears a recognizable relation to analytic geometry as this subject has been understood for the past 300 years. We find familiar notational conventions appearing here for the first time, such as the use of exponents and the custom of denoting constants and variables by the letters *a, b, c* and *x, y, z*, respectively; we find geometry and algebra, and algebra used as a language for discussing geometry; but we do not find analytic geometry, or for that matter any content whatever that justifies Descartes's mathematical reputation. His *Geometry* was little read then and is less read now, and deservedly so, for the entire work is a grotesque betrayal of what he earlier called "the transparency and unsurpassable clarity which are proper to a rightly ordered mathematics." It appears from the many deliberate obscurities and condescending remarks — so foreign to his usual way of writing — that he wrote it more to boast than to explain, and somehow he managed to cow most of his contemporaries into believing against the evidence that he had accomplished something worthwhile.*

By 1649 Descartes's books and ideas had spread to all corners of Europe, and Queen Christina of Sweden invited him to come to Stockholm to adorn her court and act as her private tutor in philosophy. Stockholm was a cold and disagreeable city, and at first he was not interested in this opportunity "to live in the land of bears among rocks and ice"; but at last he yielded, and, putting aside his fears for his habits and independence, he left Holland on a special warship which the Queen had sent to fetch him. This headstrong and strangely masculine young woman of 19 was one of history's most remarkable characters. A passionate huntress, an expert horsewoman capable of staying in the saddle all day without tiring, indifferent to women's dress and rarely combing her hair more than once a week, fluent in five languages, enthusiastic about the study of literature and philosophy, and filled with ambitions for turning Stockholm into "the Athens of the North" — she captured poor Descartes as a spider does a fly. Needing little rest herself and impervious to cold and discomfort, she set the bleak hour of 5 a.m. for her philosophy lessons. With his lifelong routine shattered, the unhappy man was forced to stumble out of his warm bed in the dark and make his way to the palace through the bitterest winter Stockholm had known in years. Worse yet, Christina took advantage of his numbed wits; when he tried to convince her that animals are only machines, she objected that she had never yet seen her watch give birth to baby watches (his reply — if he was able to think of one — is not recorded). Exhausted, weakened, and full of despair over his humiliating predicament, Descartes caught a chill and died of pneumonia only 4 months after his arrival in Sweden.

Is there anything left of the work of Descartes which still has meaning for the modern world? Very little, if we count only specific doctrines or discoveries in philosophy, science, or mathematics. However, he holds a secure place in the canonical succession of the high priests of thought by virtue of the rational temper of his mind and his vision of the unity of knowledge. He struck the gong, and Western civilization has vibrated ever since with the Cartesian spirit of skepticism and inquiry that he made common currency among educated people.

---

* Nevertheless, he was certainly a clever mathematician, as is shown by his trisection of an angle using a straightedge, compass, and fixed parabola; see H. Tietze, *Famous Problems of Mathematics,* Graylock Press, 1965, p. 53.

MERSENNE (1588–1648)

There is more in Mersenne than in all the universities together.
Thomas Hobbes

Scientists in the twentieth century are linked together into a worldwide communications network by hundreds of professional organizations whose meetings and journals stimulate a constant interchange of ideas and discoveries. In the early seventeenth century none of these organizations existed. Worse yet, most of the universities were hollow shells of medieval ritual and scholastic rigidity that were incapable of welcoming new knowledge. The intense intellectual ferment of the time found its main outlets in informal discussion groups and private correspondence. The circle of friends of Marin Mersenne was by far the most important of these groups.

Mersenne was a Franciscan friar who lived in a monastery in Paris near the Place Royale, but in middle life his interests shifted away from theology toward philosophy, science, and mathematics. His friends included almost everyone in Western Europe who was active in these fields — Pascal, Roberval, Desargues, Gassendi, and others in Paris; Descartes in Holland; Galileo, Torricelli, and Cavalieri in Italy; Fermat in Toulouse; Hobbes during his frequent long visits in Paris; and many more. The Mersenne circle constituted an "invisible college" in which most of the important intellectual activity of the period took place. Those members who lived in Paris met regularly in his rooms with the approval and support of Cardinal Richelieu, who was famous for his patronage of the sciences. These meetings continued for many years after Mersenne's death, and formed the nucleus of the French Academy of Sciences when it was chartered in 1666. Mersenne acted not only as the informal chairman of the group, but also as its corresponding secretary. He transmitted letters and manuscripts from one member to another according to their interests; and his own enormous correspondence, particularly with Descartes and Fermat, lubricated the flow of ideas so effectively that telling him about an interesting discovery amounted to publishing it throughout the whole of Europe.

In science, Mersenne's work on sound was of such fundamental importance that he is sometimes called "the father of acoustics." His 1636 treatise entitled *Harmonie Universelle* contains accounts of many ingenious experiments and the conclusions he drew from them. He laid out long hemp cords and brass wires, some more than 100 ft in length, and stretched them taut between two posts by means of weights. He found that their vibratory movements when plucked could easily be followed by the eye, and he timed them by using his own pulse. By varying the lengths and tensions, he discovered the following basic principles, which are now known as Mersenne's laws: The frequency of vibration of a stretched string is (1) inversely proportional to the length if the tension is constant, (2) directly proportional to the square root of the tension if the length is constant, and (3) inversely proportional to the square root of the mass per unit length (linear density) for different strings of the same length and tension. He next shortened a stretched brass wire until its sound became audible and tuned it to the pitch of one of his organ pipes. By applying his laws he found that the frequency of this note was 150 vibrations per second, and also that the frequency of its octave was 300 vibrations per second. This was the first determination of the frequency of a specific musical note, and was a remarkable scientific achievement. He was also the first to determine the speed of sound in air, using a seven-syllable shout ("Benedicam Dominum!") that took a second to utter and immediately echoed back from a measured distance of 519 ft. From this he concluded that sound travels 1038 ft/s, which is quite close to the currently accepted figure of about 1087 ft/s in dry air at 32°F.

Among his mathematical interests were the cycloid and perfect numbers. A cycloid is the archlike curve traced out by a point on the rim of a rolling wheel. Mersenne probably first heard of this beautiful curve from Galileo, and he suggested it to many of his friends as a worthy object for investigation. Over the next two centuries the cycloid turned out to have many remarkable geometric and physical properties, and was studied by Roberval, Torricelli, Pascal, Huygens, John Bernoulli, Leibniz, Newton, Euler, and Abel, among others.

Perfect numbers—those, like $6 = 1 + 2 + 3$, which equal the sum of their proper divisors—have fascinated mathematicians since the time of Pythagoras. Mersenne was aware of Euclid's proof that if $2^n - 1$ is prime, then $2^{n-1}(2^n - 1)$ is perfect (see our Appendix A.1). He also knew that $2^n - 1$ cannot be prime if n is not, so he was led to the problem of determining those prime numbers p for which $2^p - 1$ is also prime (the latter are now called *Mersenne primes*). Unfortunately, there are some primes p for which $2^p - 1$ is prime and others for which it is not, so this problem is far from simple. In 1644 Mersenne stated that among the 55 primes $p \leq 257$ the only ones for which $2^p - 1$ is also prime are $p = 2, 3, 5, 7, 13, 17, 19, 31, 67, 127, 257$. He did not give any of the evidence that led him to make this statement, but it is now known that he made five errors: 67 and 257 do not belong in the list, and 61, 89, and 107 do belong there. The factorability of $2^{67} - 1$ was discovered in 1903 by the American mathematician F. N. Cole, and was announced by him in a dramatic presentation to a meeting of the American Mathematical Society. When called upon for his lecture, Cole walked to the blackboard, silently calculated $2^{67} - 1$, and, still without saying a word, moved over to a clear space on the board and multiplied out

$$193{,}707{,}721 \quad \text{and} \quad 761{,}838{,}257{,}287.$$

The results were visibly the same, the audience applauded enthusiastically, and Cole returned to his seat after having delivered the only totally wordless lecture in recorded history. He later told a friend that this factorization had cost him "three years of Sundays."[*] In 1931 it was proved by D. H. Lehmer that $2^{257} - 1$ is factorable, but the proof was theoretical and did not exhibit any specific factors. Additional details about the current status of this subject are given in Appendix A.1.

FERMAT (1601–1665)

. . . a master of masters.
E. T. Bell

Pierre de Fermat was perhaps the greatest mathematician of the seventeenth century, but his influence was limited by his lack of interest in publishing his discoveries, which are known mainly from letters to friends and marginal notes in his copy of the *Arithmetica* of Diophantus.[†] By profession he was a lawyer and a member of the provincial supreme court in Toulouse, in southwestern France. However, his hobby and private passion was mathematics, and his casual creativity was one of the wonders of the age to the few who knew about it.

His letters suggest that he was a shy and retiring man, courteous and affable, but slightly remote. His outward life was as quiet and orderly as one would expect of a

---

[*] E. T. Bell, *Mathematics, Queen and Servant of Science,* McGraw-Hill, New York, 1951, p. 228.

[†] These marginal notes were reproduced in a new edition of this third century work on number theory that was published by Fermat's son in 1670.

provincial judge with a sense of responsibility toward his work. Fortunately, this work was not too demanding, and left ample leisure for the extraordinary inner life that flourished by lamplight in the silence of his study at night. He was a lover of classical learning, and his own mathematical ideas grew in part out of his intimate familiarity with the works of Archimedes, Apollonius, Diophantus, and Pappus. Though he was a genius of the first magnitude, he seems to have thought of himself as at best a rather clever fellow with a few good ideas, and not at all in the same class with the masters of Greek antiquity.

Father Mersenne in Paris heard about some of Fermat's researches from a mutual friend, and wrote to him in 1636 inviting him to share his discoveries with the Parisian mathematicians. If Fermat was surprised to receive this letter, Mersenne was even more surprised at the reply, and at the cascade of letters that followed over the years, to him and also to other members of his circle. Fermat's letters were packed with ideas and discoveries, and were sometimes accompanied by short expository essays in which he briefly described a few of his methods. These essays were handwritten in Latin and were excitedly passed from one person to another in the Mersenne group. To the mathematicians in Paris, who never met him personally, he sometimes seemed to be a looming, faceless shadow dominating all their efforts, a mysterious magician buried in the country who invariably solved the problems they proposed and in return proposed problems they could not solve — and then genially furnished the solutions on request. He enjoyed challenges himself, and naively took it for granted that his correspondents did too. For instance, Mersenne once wrote to him asking whether the very large number 100,895,598,169 is prime or not. Such questions often take years to answer, but Fermat replied without hesitation that this number is the product of 112,303 and 898,423, and that each of these factors is prime — and to this day no one knows how he did it. The unfortunate Descartes locked horns with him several times, on issues that he considered crucial both to his reputation as a mathematician and to the success of his philosophy. As an outsider Fermat knew nothing about Descartes's monumental egotism and touchy disposition, and with calm courtesy demolished him on each occasion. Wonder, exasperation, and chagrin were apparently common emotions among those who came into contact with Fermat's mind.

He invented analytic geometry in 1629 and described his ideas in a short work entitled *Introduction to Plane and Solid Loci,* which circulated in manuscript form from early 1637 on but was not published during his lifetime.* The credit for this achievement has usually been given to Descartes on the basis of his *Geometry,* which was published late in 1637 as an appendix to his famous *Discourse on Method.* However, nothing that we would recognize as analytic geometry can be found in Descartes's essay, except perhaps the idea of using algebra as a language for discussing geometric problems. Fermat had the same idea, but did something important with it: He introduced perpendicular axes and found the general equations of straight lines and circles and the simplest equations of parabolas, ellipses, and hyperbolas; and he further showed in a fairly complete and systematic way that every first- or second-degree equation can be reduced to one of these types. None of this is in Descartes's essay; but to give him his due, he did introduce several notational conventions that are still with us — which gives his work a modern appearance — while Fermat used an older and now archaic algebraic symbolism. The result is that superficially Descartes's essay looks as if it might be analytic geometry, but isn't; while Fermat's doesn't look it, but is. Descartes certainly knew some analytic geometry by the late 1630s; but since he had possession of the original manuscript of the *Introduction*

---

* Translations are given in D. E. Smith, *A Source Book in Mathematics,* McGraw-Hill, 1929, pp. 389–396; and in D. J. Struik, *A Source Book in Mathematics, 1200–1800,* Harvard, 1969, pp. 143–150.

several months before the publication of his own *Geometry,* it may be surmised that much of what he knew he learned from Fermat.

The invention of calculus is usually credited to Newton and Leibniz, whose ideas and methods were not published until about 20 years after Fermat's death. However, if differential calculus is considered to be the mathematics of finding maxima and minima of functions and drawing tangents to curves, then Fermat was the true creator of this subject as early as 1629, more than a decade before either Newton or Leibniz was born.* With his usual honesty in such matters, Newton stated—in a letter that was discovered only in 1934—that his own early ideas about calculus came directly "from Fermat's way of drawing tangents."†

So few curves were known before Fermat's time that no one had felt any need to improve upon the old and comparatively useless idea that a tangent is a line that touches a curve at one and only one point. However, with the aid of his new analytic geometry, Fermat was able not only to find the equations of familiar classical curves, but also to construct a multitude of new curves by simply writing down various equations and considering the corresponding graphs. This great increase in the variety of curves that were available for study aroused his interest in what came to be called "the problem of tangents."

What Newton acknowledged in the remark quoted above is that Fermat was the first to arrive at the modern concept of the tangent line to a given curve at a given point P (see Fig. B.5). In essence, he took a second nearby point Q on the curve, drew the secant line PQ, and considered the tangent at P to be the limiting position of the secant as Q slides along the curve toward P. Even more important, this qualitative idea served him as a stepping-stone to quantitative methods for calculating the exact slope of the tangent.

Fermat's methods were of such critical significance for the future of mathematics and science that we pause briefly to consider how they arose.

While sketching the graphs of certain polynomial functions $y = f(x)$, he hit upon a very ingenious idea for locating points at which such a function assumes a maximum or minimum value. He compared the value $f(x)$ at a point x with the value $f(x + E)$ at a nearby point $x + E$ (see Fig. B.6). For most x's the difference between these values, $f(x + E) - f(x)$, is not small compared with E, but he noticed that at the top or bottom of a curve this difference is much smaller than E and diminishes faster than E does. This idea gave him the approximate equation

$$\frac{f(x + E) - f(x)}{E} \cong 0,$$

which becomes more and more nearly correct as the interval E is taken smaller and smaller. With this in mind, he next put $E = 0$ to obtain the equation

$$\left[\frac{f(x + E) - f(x)}{E}\right]_{E=0} = 0.$$

According to Fermat, this equation is exactly correct at the maximum and minimum points on the curve, and solving it yields the values of x that correspond to these points. The legitimacy of this procedure was a subject of acute controversy for many years. However, students of calculus will recognize that Fermat's method amounts to calculating the derivative

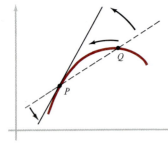

Figure B.5

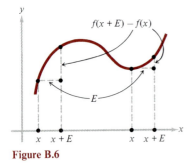

$f(x + E) - f(x)$

E

Figure B.6

---

* Fermat wrote several accounts of his methods, but as usual he made no effort to publish them. The earliest of these was the very short essay given on pp. 223–224 of Struik's *Source Book*; this was circulating in Paris in 1636, and according to Fermat's own statement was then about 7 years old.

† See L. T. More, *Isaac Newton,* Scribner's, 1934, p. 185.

$$f'(x) = \lim_{E \to 0} \frac{f(x + E) - f(x)}{E}$$

and setting this equal to zero, which is just what we do in calculus today, except that we customarily use the symbol Δx in place of his E.

In one of the first tests of his procedure, he gave the following proof of Euclid's theorem that the largest rectangle with a given perimeter is a square. If B is half the perimeter and one side is x, then $B - x$ is the adjacent side, and the area is $f(x) = x(B - x)$. To maximize this area by the process described above, compute

$$f(x + E) - f(x) = (x + E)[B - (x + E)] - x(B - x)$$
$$= EB - 2Ex - E^2,$$
$$\frac{f(x + E) - f(x)}{E} = B - 2x - E,$$

and

$$\left[\frac{f(x + E) - f(x)}{E}\right]_{E=0} = B - 2x.$$

Fermat's equation is therefore $B - 2x = 0$, so $x = \frac{1}{2}B$, $B - x = \frac{1}{2}B$, and the largest rectangle is a square. When he reached this conclusion, he remarked with justifiable pride, "We can hardly expect to find a more general method." He also found the shape of the largest cylinder that can be inscribed in a given sphere (ratio of height to diameter of base $= \frac{1}{2}\sqrt{2}$) and solved many similar problems that are familiar in calculus courses today.

Fermat's most memorable application of his method of maxima and minima was his analysis of the refraction of light. The qualitative phenomenon had of course been known for a very long time: That when a ray of light passes from a less dense medium into a denser medium — for instance, from air into water — it is refracted toward the perpendicular (see Fig. B.7). The quantitative description of refraction was apparently discovered experimentally by the Dutch scientist Snell in 1621. He found that when the direction of the incident ray is altered, the ratio of the sines of the two indicated angles remains constant,

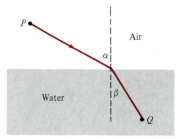

Air

Water

Figure B.7

$$\frac{\sin \alpha}{\sin \beta} = \text{a constant,}$$

but he had no idea why. This sine law was first published by Descartes in 1637 (without any mention of Snell), and he purported to prove it in a form equivalent to

$$\frac{\sin \alpha}{\sin \beta} = \frac{v_w}{v_a},$$

where v_a and v_w are the velocities of light in air and in water. Descartes based his argument on a fanciful model and on the metaphysically inspired opinion that light travels faster in a denser medium. Fermat rejected both the opinion ("shocking to common sense") and the argument ("demonstrations which do not force belief cannot bear this name"). After many years of passive skepticism, he actively confronted the problem in 1657 and proved the correct law himself,

$$\frac{\sin \alpha}{\sin \beta} = \frac{v_a}{v_w}.$$

The foundation of his reasoning was the hypothesis that the actual path along which the ray of light travels from P to Q is that which minimizes the total time of travel —

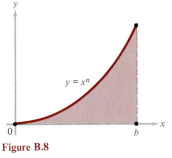

Figure B.8

now known as *Fermat's principle of least time.** This principle of least time led to the calculus of variations created by Euler and Lagrange in the next century, and on from this discipline to Hamilton's principle of least action, which has been one of the most important unifying ideas in modern physical science.

Fermat's method of finding tangents developed out of his approach to problems of maxima and minima, and was the occasion of yet another clash with Descartes. When the famous philosopher was informed of Fermat's method by Mersenne, he attacked its generality, challenged Fermat to find the tangent to the curve $x^3 + y^3 = 3axy$, and foolishly predicted that he would fail. Descartes was unable to cope with this problem himself, and was intensely irritated when Fermat solved it easily.†

These successes in the early stages of differential calculus were matched by comparable achievements in integral calculus. We mention only one: his calculation of the area under the curve $y = x^n$ from $x = 0$ to $x = b$ for any positive integer n (see Fig. B.8). In modern notation, this amounts to the evaluation of the integral

$$\int_0^b x^n \, dx = \frac{b^{n+1}}{n+1}.$$

The Italian mathematician Cavalieri had proved this formula by increasingly laborious methods for $n = 1, 2, \ldots, 9$, but bogged down at $n = 10$. Fermat devised a beautiful new approach that worked with equal ease for all n's.‡

In the light of all these accomplishments, it may reasonably be asked why Newton and Leibniz are commonly regarded as the inventors of calculus, and not Fermat. The answer is that Fermat's activities came a little too early, before the essential features of the subject had fully emerged. He had pregnant ideas and solved many individual calculus problems; but he did not isolate the explicit calculation of derivatives as a formal process, he had no notion of indefinite integrals, he apparently never noticed the Fundamental Theorem of Calculus that binds together the two parts of the subject, and he didn't even begin to develop the rich structure of computational machinery on which the more advanced applications depend. Newton and Leibniz did all these things, and thereby transformed a collection of ingenious devices into a problem-solving tool of great power and efficiency.

The mind of Fermat had as many facets as a well-cut diamond and threw off flashes of light in surprising directions. A minor but significant chapter in his intellectual life began when Blaise Pascal, the precocious dilettante of mathematics and physics, wrote to him in 1654 with some questions about certain gambling games that are played with dice. In the ensuing correspondence over the next several months, they jointly developed the basic concepts of the theory of probability.§ This was the effective beginning of a subject whose influence is now felt in almost every corner of modern life, ranging from such practical fields as insurance and industrial quality control to the esoteric disciplines of genetics, quantum mechanics, and the kinetic theory of gases. However, neither man carried his ideas very far. Pascal was soon caught up in the paroxysms of piety that blighted the remainder of his short life, and Fermat dropped the subject because he had other, more compelling mathematical interests.

The many remarkable achievements sketched here—in analytic geometry, calculus, optics, and the theory of probability—would have sufficed to place Fermat

---

* A full discussion, including the details of Fermat's proof, can be found in Chapter V of A. I. Sabra, *Theories of Light from Descartes to Newton,* Oldbourne, 1967. A proof by the methods of calculus is given in our Chapter 4.

† The curve $x^3 + y^3 = 3axy$ is now called the *folium of Descartes.*

‡ The details are given in Appendix A.4.

§ See Smith, *Source Book,* pp. 546–565.

among the outstanding mathematicians of the seventeenth century if he had done nothing else. But to him these activities were all of minor importance compared with the consuming passion of his life, the theory of numbers. It was here that his genius shone most brilliantly, for his insight into the properties of the familiar but mysterious positive integers has perhaps never been equaled. He was the sole and undisputed founder of the modern era in this subject, without any rivals and with few followers until the time of Euler and Lagrange in the next century. Pascal, who called him *le premier homme du monde*—"the foremost man in the world"—wrote to him and said: "Look elsewhere for someone who can follow you in your researches about numbers. For my part, I confess that they are far beyond me, and I am competent only to admire them."

The attractions of number theory are felt by many but are not easy to explain, being mainly aesthetic in nature. On the one hand, the positive whole numbers 1, 2, 3, . . . are perhaps the simplest and most transparent conceptions of the human mind; and on the other, many of their most easily understood properties have roots that strike so deep as to be almost beyond the reach of human ingenuity. A large part of the lure of the subject lies in the fact that its smooth and apparently simple surface conceals depths of the utmost profundity. In order to convey something of the flavor of Fermat's work in this field, we briefly describe several of his most characteristic and influential discoveries. It should be remembered that most of the truths he uncovered are known only because he wrote about them to his friends or jotted them down in the margins of his copy of Diophantus.* Unfortunately, many of his proofs went unrecorded and were lost forever when he died.

(1) The following is known as *Fermat's theorem*: If p is a prime and n is a positive integer not divisible by p, then p divides $n^{p-1} - 1$. For instance, if $p = 5$ and $n = 4$, then $n^{p-1} - 1 = 4^4 - 1 = 255$, which is divisible by 5; and if $p = 3$ and $n = 8$, then $n^{p-1} - 1 = 8^2 - 1 = 63$, which is divisible by 3. This theorem is of fundamental importance in both number theory and modern algebra.† Fermat stated it in a letter in 1640, and the first published proof was given by Euler in 1736.

(2) Our second example is his profound and beautiful theorem on polygonal numbers. In his own words, as written in the margin of his copy of Diophantus:

> Every positive integer is triangular or the sum of 2 or 3 triangular numbers; a square or the sum of 2, 3 or 4 squares; a pentagonal number or the sum of 2, 3, 4 or 5 pentagonal numbers; and so on to infinity, whether it is a question of hexagonal, heptagonal, or any polygonal numbers.

To understand this statement, recall that triangular numbers are the numbers 1, 3, 6, 10, . . . that can be obtained by building up triangular arrays of dots,

that the squares 1, 4, 9, 16 . . . arise from square arrays of dots,

---

* A few were proposed as challenges to certain English mathematicians whom he hoped (in vain) to interest in his ideas. For instance, it is clear that $x = 5$ and $y = 3$ is a positive integer solution of $x^2 + 2 = y^3$, and he asked for a proof that this is the *only* such solution. As E. T. Bell remarked, "It requires more innate intellectual capacity to dispose of this apparently childish thing than it does to grasp the theory of relativity." See J. V. Uspensky and M. A. Heaslet, *Elementary Number Theory,* McGraw-Hill, 1939, p. 398.

† See Chapter VI of G. H. Hardy and E. M. Wright, *An Introduction to the Theory of Numbers*, Oxford, 1938.

and similarly with pentagonal numbers and the rest. After the statement quoted, Fermat continued as follows: "I cannot give the proof here, for it depends on many abstruse mysteries of numbers; but I intend to devote an entire book to this subject, and to present in this part of number theory astonishing advances beyond previously known boundaries." It will not come as a surprise to anyone that this book was never written, though no one doubts that he could have done so. Euler struggled off and on for nearly 40 years to find a proof of the part of Fermat's theorem relating to squares, but he didn't quite reach his goal.* Lagrange at last succeeded in 1772, with an argument based heavily on Euler's ideas. Gauss established the part about triangular numbers in 1796, and Cauchy proved the complete theorem in 1815.

(3) A far more famous marginal note—as familiar to mathematicians as the activities of Napoleon are to historians—occurs next to a passage in Diophantus dealing with positive integer solutions of the equation $x^2 + y^2 = z^2$. It is easy to see that $3^2 + 4^2 = 5^2$ and $5^2 + 12^2 = 13^2$, so the triples 3, 4, 5 and 5, 12, 13 are obvious solutions. There are infinitely many such triples. They have been completely known since the time of Euclid, and were discussed by Diophantus.† Fermat's note in its entirety reads as follows:

> In contrast to this, it is impossible to separate a cube into two cubes, a fourth power into two fourth powers, or, generally, any power above the second into two powers of the same degree. I have discovered a truly wonderful proof which this margin is too narrow to contain.

This simple statement is now known as *Fermat's last theorem*: In modern notation, the equation $x^n + y^n = z^n$ has no positive integer solutions whatever for any exponent $n > 2$. Generations of mathematicians have cursed the narrowness of that margin, for in spite of intense efforts by some of the most penetrating minds in the world for more than 300 years, no proof has ever been found by anyone else.‡ In another place, Fermat himself left a sketch of a proof for the case $n = 4$. Euler published a proof for $n = 4$ (1747) and also for the more difficult case $n = 3$ (1770).§ Gauss, Legendre, Dirichlet, and others settled the cases $n = 5$ and $n = 7$, and at the present time the theorem is known to be true for all exponents $n \leq 125,000$.¶ No one doubts its truth for all n, but its interest and unique reputation lie in the resistance it offers to complete and rigorous proof. Instant immortality awaits anyone who can find such a proof, but those considering this as a research project for their next free weekend should remember what David Hilbert, perhaps the greatest mathematician of the twentieth century, said when asked why he did not try: "Before beginning I would have to put in three years of intensive study, and I haven't that much time to

---

* He did manage to prove Fermat's two squares theorem (every prime of the form $4n + 1$ is expressible as the sum of two squares in one and only one way) after a mere 7 years of effort.
† The general solution of this problem is given in H. Rademacher and O. Toeplitz, *The Enjoyment of Mathematics,* Princeton, 1957, Chapter 14. See also Chapter XIII of H. Tietze, *Famous Problems of Mathematics,* Graylock, 1965.
‡ Not even the Devil. See the charming short story "The Devil and Simon Flagg," by Arthur Porges, in Clifton Fadiman's anthology, *Fantasia Mathematica,* Simon and Schuster, 1958.
§ Struik, *Source Book,* pp. 36–40.
¶ The current status of the subject is discussed in H. M. Edwards, "Fermat's Last Theorem," *Scientific American,* October 1978. See also H. S. Vandiver's article *Fermat's Last Theorem* in the *Encyclopaedia Britannica* (any recent edition before 1974).

waste on a probable failure." Some experts believe that Fermat deceived himself in thinking he had a proof. However, he was a man of complete integrity and a number theorist of unsurpassed ability. It should also be remembered that he has never been caught in a mistake; with this single exception — which has not been refuted — others have succeeded in proving every theorem of which he definitely stated that he had a proof. Fermat's last theorem remains to his day the most celebrated of the enigmatic legacies which he left to his baffled posterity.*

PASCAL (1623–1662)

Pascal stamped his works with the passionate conviction of a man in love with the absolute.

Jean Orcibal

A modern office worker, setting off in the morning, may glance at a wrist watch, inspect the barometer, buy a newspaper at the corner store and receive change from the cash register, and board a bus for the trip downtown to the business district. What has all this to do with a French mathematician who was mixed up in musty theological disputes when Louis XIV was still a teenager? Pascal invented that wrist watch, originated that barometer, invented that calculating machine, and was the first to think of a bus system and organize a public transportation company.†

Blaise Pascal was one of the most gifted and tragic figures in the whole history of Western thought. A child prodigy, he was even more prodigious as a man. And yet his life was twisted and stunted by mystic visions and religious neuroses, and of all the great things he had it in him to do, none progressed beyond memorable beginnings.

Pascal was born in Clermont-Ferrand, in the Auvergne region of central France. His mother died when he was only 3 years old, so that he and his two sisters were brought up by their father Étienne, a man of strong character and broad learning. In 1631 the family moved to Paris for the sake of the children's development. Blaise never attended ordinary schools, but instead was taught exclusively by his father.

Étienne Pascal became a member of Mersenne's weekly discussion group, which was sponsored by Cardinal Richelieu and later developed into the French Academy. Its purpose was to encourage interest in scientific matters, especially mathematics and physics. It included the philosopher Descartes (when he happened to be in town), the mathematicians Roberval and Desargues, the Englishman Hobbes, who was present in the winter of 1636–1637, and several others. From the age of 12 or 13, Pascal often participated in these gatherings, where he listened avidly and sometimes entered into disputes himself. There is an old saying: "Genius is like a fire; a single burning log will smoulder or go out, while a heap of logs piled loosely together will flame fiercely." And so it was with the young Pascal and his father's eminent friends.

When he was 16 he published his famous *Essai sur les coniques* ("Essay on Conic Sections"). Descartes was jealous of its success, and refused to believe that it was produced by a mere boy. This brief work contains what is still the most important theorem of projective geometry, known as *Pascal's theorem*: If a hexagon is inscribed in a conic section (Fig. B.9), then the three points of intersection of its opposite sides always lie on a straight line.

At 17, while watching his father's arithmetical drudgery over tax assessments, he conceived the possibility of a calculating machine, by 18 or 19 he had completed the

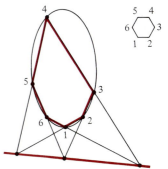

Figure B.9 Pascal's theorem.

---

* We feel obliged to mention that there does exist a book about Fermat and his mathematical work; however, this book was dissected and destroyed in a famous review by the eminent French mathematician André Weil (*Bull. Amer. Math. Soc.*, vol. 79 (1973), pp. 1138–1149).

† See the excellent biography by Ernest Mortimer, *Blaise Pascal,* Harper, 1959, p. 12.

first working model, and in the next few years he manufactured over 50 machines which he offered for sale. He hoped that his project would make him rich, but it never did. The manufacturing costs were too high, and the increasing use of logarithms reduced the demand.

When he was 23 he heard a sketchy report of Torricelli's experiment involving a 3-ft glass tube closed at one end and filled with mercury. If the open end is closed with a thumb and immersed in a bowl of mercury, and then the thumb is removed, the mercury in the tube sinks to a level about 30 in above the level of the mercury in the bowl. Pascal repeated this experiment with many variations and great care, and gave a complete and correct explanation of the results. His explanation agreed with Torricelli's, which he had not heard: namely, that the empty space in the top of the tube is a vacuum, and the mercury is held up in the tube by the weight of the ocean of air that presses down on the surface of the mercury in the bowl. Also, as a byproduct of his work, Pascal invented the syringe. The conclusions established by these investigations stirred up a storm of controversy with the scholastic philosophers, who tried vainly to maintain the Aristotelian doctrine that "nature abhors a vacuum."

A year or two later Pascal conceived the famous Puy-de-Dôme experiment, in which a Torricelli tube was carried up a mountain in a single day, so that the fall of the mercury level in the tube with increasing altitude could actually be observed. Further, among the notes found after his death was a series of observations of the variations of the Torricelli tube according to changes in the weather. He writes, "This knowledge can be very useful to farmers, travellers, etc., to learn the present state of the weather, and that which is to follow immediately, but not to know that which is to come in three weeks." Thus was the barometer born.

During this period he also studied hydrostatics, and discovered what is now called *Pascal's Principle* for the transmission of pressure through an enclosed fluid. This led him to the idea of the hydraulic press, which he described very clearly even though technical difficulties prevented him from making a successful working model.

There is a French saying that "Many people know the whole history of human thought without ever having had one." It is clearly true that most educated people in any period think other people's thoughts and little else. However, Pascal was trained by his father from infancy in the art of original thought, which is a rare and precious thing. He wanted to know the reason for everything, and reasons consisting of mere words threw his mind into a turmoil of frustration. In his scientific work he was strongly committed to the experimental method, with its emphasis on empirical facts combined with logical thinking, and with its total disregard for the appeals to authority that constituted most of the act of reasoning for the scholastic philosophers. However, he was also convinced that evidence and thought are not enough in the domain of religion: In this part of human experience, faith is necessary in order to arrive at the truth.

Religion played an important part in Pascal's life from the time of his "first conversion" in 1646, when he felt a strong compulsion to turn away from the world toward God. This impulse soon weakened, and he became absorbed again by his scientific interests and his fashionable friends. It was near the end of this period, in 1654, that in correspondence with Fermat he assisted in formulating some of the ideas that led to the mathematical theory of probability. At this time he also wrote his *Traité du triangle arithmétique* ("Treatise on the Arithmetical Triangle"), in which he studied a triangular arrangement of the binomial coefficients and discovered and proved many of their properties.* It is in this work that he gives what seems to be the first satisfactory statement of the principle of proof by mathematical induction.†

---

* See pp. 21–26 of D. J. Struik (ed.), *A Source Book in Mathematics, 1200–1800*, Harvard University Press, 1969.

† See also G. Polya, *Mathematical Discovery*, Wiley, 1962, vol. 1, pp. 70–75, especially p. 74.

In the years 1653 and 1654 Pascal worked harder and harder on science and mathematics, as a desperate diversion from the increasing emptiness within. Late in 1654 he had the decisive experience of his life, an overwhelming mystic vision that caused him to turn away from his worldly, free-thinking friends and immerse himself permanently in religious contemplation. He returned to mathematics only once again, in 1658. While suffering from a severe toothache, he began to think about some problems concerning the cycloid, the archlike curve traced out by a point on the rim of a rolling wheel. His tooth suddenly stopped aching, and he took this as a sign of divine approval. Over the next few months he worked feverishly on this topic and solved a number of problems in a series of small treatises. The main problems were to find the area and length of one arch, but unfortunately for him these had already been solved by Torricelli and Christopher Wren. Nevertheless, these works had a consequence of very great importance, for about 15 years later they suggested to Leibniz an idea that was crucial for his own invention of differential and integral calculus. Leibniz later wrote that "a great light" burst upon him when he read a particular passage, and he wondered how Pascal could have missed the idea.*

It was toward the end of his brief life that he wrote the works that earned him a place among the very greatest figures of French literature. His *Provincial Letters* is a series of polemical pamphlets against the Jesuits. They took the form of "Letters Written to a Provincial by One of His Friends," and were signed with a fictitious name. In these letters we encounter for the first time the variety, brevity, and tautness of style that distinguish the best modern French prose. As Voltaire said, "The first work of genius in prose that we find is the collected edition of the *Provincial Letters.* All kinds of eloquence are displayed here. It is in this work that our language takes its final form." At the end of Letter XVI, Pascal makes his memorable apology for the length of the letter he has just written, his reason being that "I don't have time enough to make it shorter."

The *Provincial Letters* has only a limited historical interest for us today, but Pascal's *Pensées* ("Thoughts") will presumably endure as long as the French language lasts. His purpose was to write a monumental and irresistible defense of the Christian religion against the unbelievers. However, during the last years of his life he was often weak, semidelirious, and racked with the pain of his illnesses, which after his death were determined to be a brain tumor and cancer of the stomach. He was incapable of connected work, and took to scribbling down on any bit of paper that came to hand the thoughts that flashed into his mind. Thus we have about a thousand scraps of paper containing ideas and fragments of ideas for his intended great work — phrases, single sentences, sometimes several whole paragraphs together.† The specific subjects are very diverse, but in general the theme is the grandeur and misery of man. Here are a few of Pascal's *pensées:*

The heart has its reasons that the reason does not know. (277)

The eternal silence of these infinite spaces terrifies me. (206)

I have discovered that all the misfortune of men comes from one single thing, not knowing how to remain quiet in a room. (139)

Men never do evil so completely and cheerfully as when they do it from religious conviction. (894)

What a chimera is man! What a novelty! What a monster, what a chaos, what a

---

* See Struik, *Source Book,* pp. 239–241.

† About 300 years ago these scraps of paper were glued helter-skelter (upside-down, sideways, at all angles) into a large album that is now protected as one of the most precious cultural treasures of France in the Bibliothèque Nationale in Paris. The present writer has had the privilege of spending a couple of hours examining this album, looking for familiar passages and occasionally finding them.

contradiction, what a prodigy! Judge of all things, feeble worm of the earth, depository of truth, a sink of uncertainty and error, the glory and shame of the universe. (434)

Man is only a reed, the feeblest thing in nature, but he is a thinking reed. There is no need for the whole universe to arm itself to crush him; one vapor, one drop of water, will suffice to kill him. But when the universe crushes him, man is still nobler than that which kills him, because he knows what kills him and he understands the advantage the universe has over him. The universe knows nothing of this. (347)

As his biographer says, "In the general estimation of his own countrymen Pascal occupies not only a high place but the top place. 'He is to France,' writes Professor Chevalier, 'what Plato is to Greece, Dante to Italy, Cervantes and S. Theresa to Spain, Shakespeare to England.'"*

The achievements of Pascal's short life were remarkable enough. However, if he had not died at the early age of 39, if he had not been ill of body and mind during those last years, if he had not deliberately rejected mathematics and science for a handful of dust, there is little doubt that he could have discovered calculus—and probably much else besides—years before Newton and Leibniz. He was surely the greatest "might-have-been" in the history of mathematics.

HUYGENS (1629–1695)

The extraordinary astronomer-mathematician-physicist Christiaan Huygens was undoubtedly Holland's greatest scientist, and he deserves to be much more widely known among people who are interested in the great thinkers of our past.

Even apart from its scientific immortals like Huygens and his friend Leeuwenhoek, the inventor of the microscope, Holland in the mid-seventeenth century was a rich garden of civilization. Its worldwide trading empire provided peace and comfort for its people, its serene light inspired great artists like Rembrandt and Vermeer, and its religious liberty provided safe haven for philosophers and free-thinkers like Spinoza and Descartes. Also, as an outward sign of intense intellectual ferment, the Dutch cities were teeming with publishers and books. In the whole world at that time there were not more than ten or a dozen cities where books were printed on any substantial scale. England had only two centers of the publishing trade, London and Oxford; France also had only two, Paris and Lyons; but in Holland there were five— Amsterdam, Rotterdam, Leiden, The Hague, and Utrecht—all printing books in Greek, Latin, English, French, German, Italian, and Hebrew as well as in Dutch. In Amsterdam alone there were more than four hundred printers or booksellers. A touchstone like this is an almost infallible guide to the quality of a society.†

As we see, Holland was probably the most civilized nation in Europe at that time, and Constantijn Huygens, the father of Christiaan, was surely its most civilized citizen. He was a statesman and diplomat who spent most of his life in the service of the Princes of Orange; a scholar who knew seven languages; a poet, playwright, musician, composer, and amateur scientist; friend of Francis Bacon, Descartes, and Mersenne; friend and translator of the English poet John Donne; knighted by James I of England; friend and patron of Rembrandt, whom he persuaded to move from Leiden to Amsterdam; and head of one of the great families of his country. Descartes described his reaction after first meeting Constantijn: "I could not believe that a single mind could occupy itself so well with so many things." Eminent thinkers and travelers from other nations were often guests at the Huygens home in The Hague. As

---

* Ernest Mortimer, *Pascal*, p. 183.
† See p. 88 of P. Hazard, *The European Mind, 1680–1715,* Yale University Press, 1953.

a young man growing up in such an environment, it was almost inevitable that Christiaan Huygens would become skilled in languages, art, music, science, and mathematics. "The world is my country," he said, "and science is my religion."*

Huygens made his first notable discovery in 1646, at the age of 17. Galileo had maintained that a flexible chain hangs in the shape of a parabola, but Huygens proved that this is not correct. His father informed Mersenne, who responded with gratifying enthusiasm. In 1691 Huygens returned to this problem and determined the true shape of this interesting curve.†

In 1655 he developed an improved method of grinding and polishing lenses for telescopes, and a flood of new knowledge quickly followed. He discovered that Saturn is surrounded by rings that nowhere touch the planet.‡ He discovered Titan, the moon of Saturn that is now known to be the largest moon in the solar system. He was the first to notice a surface feature on Mars, and by observing the movement of this feature as the planet rotates, he was the first to determine that the Martian day is approximately 24 hours long, very nearly the same as our own. In 1656–1657 he invented the pendulum clock, which sprang from the need for a more precise way of measuring time in astronomical observations.

In 1657 he published a little booklet that was the first formal treatise on the theory of probability, *De rationiis in ludo aleae* ("On Reasoning in Games of Dice"). He had visited Paris and heard about the 1654 correspondence between Fermat and Pascal on this subject; and since neither of these men seemed inclined to write up their ideas, he sought his own answers. Among other things, he introduced the important concept of "mathematical expectation."

In 1666 Huygens moved to Paris at the urging of Colbert, the great minister of Louis XIV who was chiefly responsible for much of the economic and political power of France over the next several centuries. He became one of the first salaried members of Colbert's newly created French Academy of Sciences, and made his home in Paris for the next 15 years.

In 1663 he was elected a fellow of the Royal Society of London, and in 1669 he presented that organization with the first clear and correct statement of the laws of impact for elastic bodies. His laws refuted Descartes's erroneous laws of impact as set forth in his *Principia Philosophiae* (1644).§ One of Huygens's laws states that in the mutual impact of two bodies, the sum of the products of the masses and the squares of their velocities is the same before and after impact. This appears to be the earliest version of the principle of conservation of energy.¶

In 1673 Huygens published his greatest work, his treatise *Horologium oscillatorium* ("The Pendulum Clock"). Here he dug deeply into the theory of the pendulum clock that he had invented 16 years earlier, and uncovered many valuable nuggets of mathematics and physics. He had long been acutely aware of the so-called circular

---

* See p. 251 of G. N. Clark, *The Seventeenth Century*, Oxford University Press, 1931.

† Leibniz and John Bernoulli also found the equation of this curve in the same year, independently of Huygens and each other, and Leibniz named it the *catenary*. See Appendix A.9.

‡ Galileo had earlier seen these rings through his own more primitive telescope, but he had no idea what they were. To him they appeared to be two strange protuberances attached to the planet like ears.

§ For example, one of Descartes's laws states that if a small ball collides with a large ball at rest, then the small ball will rebound while the large ball remains immobile, which is obviously false. Some devastating criticisms of Descartes's attitudes toward science and scientific experiment are given at the end of Chapter 6 of H. Butterfield, *The Origins of Modern Science, 1300–1800*, G. Bell & Sons, 1957.

¶ Some of his reasoning is described on pp. 16–19 of C. Lanczos, *Albert Einstein and the Cosmic World Order*, Interscience, 1965. On p. 19 Lanczos remarks, "He [Huygens] employs here in his proof for the first time that 'principle of relativity' which in Einstein's hands attained such fundamental importance."

error inherent in such clocks, namely, the fact that the period of oscillation is not determined strictly by the length of the pendulum alone, but also depends on the magnitude of the swing. To express this differently, if a frictionless ball is placed on the side of a smooth hemispherical bowl and released, the time it takes to reach the lowest point will be almost, but not quite, independent of the height from which it starts. It happened that various properties of the cycloid were widely discussed in Western Europe in the late 1650s, and it occurred to Huygens to wonder what would happen if the hemispherical bowl were replaced by one whose vertical cross section is an inverted arch of a cycloid. He was overjoyed to discover that in this case the ball will reach the lowest point in exactly the same time no matter where it is released on the side of the bowl. This is the *tautochrone* ("same time") *property* of the cycloid, and is the main theorem in the second part of his treatise.* In the third part he introduced the concepts of the evolute and involute of a plane curve and determined the evolutes of a parabola and a cycloid.† And in the last part he applied his mathematical discoveries to formulating the theory of a cycloidal pendulum clock, in which the pendulum bob is compelled to move along a cycloidal path instead of a circular path, and in which the period of oscillation is therefore exactly the same regardless of the magnitude of the swing, thereby eliminating the circular error. Huygens actually built several of these cycloidal clocks, but because of construction difficulties they turned out to be impractical as a way of obtaining greater accuracy. At the end of his treatise he gave a number of theorems about circular motion, proving, among other things, that for a body moving around a circular path with constant speed, the centripetal force is directly proportional to the square of the speed and inversely proportional to the radius of the path.‡ Newton greatly respected Huygens, and used many of these discoveries in his own work a few years later.

Early in 1673 Huygens had some memorable conversations with Leibniz that had very far-reaching consequences. Leibniz was then 26 years old, a young diplomat on a mission to Paris for his employer in Germany, and largely ignorant of contemporary mathematics.

> Huygens came to like the studious and intelligent young German more and more, gave him a copy of the *Horologium* as a present and talked to him about this latest work of his, the fruit of ten years of study, of the deep theoretical research to which he had been led in connection with the problem of pendular motion, and how eventually everything went back to Archimedes' methods for centers of gravity. Leibniz listened intently; at the close he felt he had to say something, but what he brought up was clumsy to a degree; surely a straight line drawn through the centroid of a plane (convex) area will always bisect the area, will it not? This was nearly too much: if it had been one of his mathematical rivals like Gregory or Newton then Huygens would probably never have condoned such a remark, but what this innocent young German had to say one could not really take amiss; good-humoredly, Huygens corrected his error and advised him to seek out further details from the relevant works of Pascal, etc. . . . Leibniz very readily and willingly took refuge in science. He procured the books named by Huygens and a few more from the Royal Library, made excerpt after excerpt and went really deeply into mathematics. As he learned, his personality rapidly matured, digesting what he read and systematically penetrating its essence; he was concerned to acquire not facility in calculation or a mere catalog of results, but basic insights and methods, and what he took in inspired continually in turn a surge of creative activity within him-

---

* We prove this at the end of Section 17.2, where the property is stated in terms of a bead sliding down a frictionless wire.

† These results are established in Appendix A.20. See also pp. 263–269 of D. J. Struik, *A Source Book in Mathematics, 1200–1800,* Harvard University Press, 1969.

‡ See formula (8) in Section 17.6.

self. . . . It was, to begin with, a diversion for a mind deprived of its customary field of action, but soon became a unified passion for knowledge.*

Even though Huygens discovered many fine things in mathematics, his most important discovery was certainly the mind of Leibniz.

In the late 1670s Huygens began to feel an atmosphere of increasing intolerance for Protestants, and in 1681 he decided to leave Paris and move back to his home in The Hague. The next years were spent partly on microscopy in loose association with his friend Leeuwenhoek, and partly working on his wave theory of light.

His originality as a protozoologist was entirely unknown until a few years ago. As a modern expert says:

> Christiaan Huygens never himself published any serious contributions to protozoology: and the records of his own observations, which were made in an attempt to repeat Leeuwenhoek's experiments, remained in manuscript and unknown until only a few years ago. Consequently, his private work had no influence whatsoever upon the progress of protozoology. Had it been published in his lifetime, it would have assured him a place in the very forefront of the founders of the science.†

Among other things, he explained how microorganisms develop in water previously sterilized by boiling. He suggested that these creatures are small enough to float through the air and reproduce when they fall into the water, a speculation that was proved correct by Louis Pasteur two centuries later.

In 1690 he published his *Traité de la lumière* ("Treatise on Light"), in which he propounded his wave theory and used this as a basis for deducing geometrically the laws of reflection and refraction, and explained the phenomenon of double refraction in Iceland Crystal.‡

Huygens's last work, and his most popular, was his posthumously published *Cosmotheoros*, in which he summed up man's knowledge of the universe at that time and frankly and freely speculated about the nature of possible inhabitants of other planets.§ He declined to allow this book to be published during his lifetime because he had no wish to be attacked for his unorthodox religious ideas. As he said to his sister-in-law, "If people knew my opinions and sentiments on religion, they would tear me apart." Toward the end of this book we find what is perhaps his most brilliant contribution to astronomy, the first reasonable estimate of the distance to a fixed star. He compared the remembered brightness of the star Sirius from the previous night with the observed brightness of a tiny piece of the sun as seen through a very small hole; and by calculating the fraction of the sun's diameter visible through the hole, he concluded that Sirius is 27,664 times as far away as the sun. This result was slightly in error, because Sirius turned out to have greater intrinsic brightness than the sun, but Huygens had the right idea, and his estimate was the best that was available for more than a century.

The eminent modern philosopher Alfred North Whitehead has written:

---

* See pp. 47–48 of J. E. Hofmann, *Leibniz in Paris, 1672–1676,* Cambridge University Press, 1974. An account of these conversations in Leibniz's own words is given on p. 215 of J. M. Child, *The Early Mathematical Manuscripts of Leibniz,* Open Court, 1920.

† See pp. 163–164 of C. Dobell, *Antony van Leeuwenhock and his "Little Animals,"* Dover, 1960.

‡ This treatise has been translated into English by S. P. Thompson, University of Chicago Press, 1945.

§ A charming English translation was published in 1698 under the title *The Celestial Worlds Discover'd.* This was reprinted in 1968 by Frank Cass & Co.

A brief, and sufficiently accurate, description of the intellectual life of the European races during the succeeding two centuries and a quarter up to our own times [that is, from about 1700 to the early 20th century] is that they have been living upon the accumulated capital of ideas provided for them by the genius of the seventeenth century. The men of this epoch inherited a ferment of ideas attendant upon the historical revolt of the sixteenth century, and they bequeathed formed systems of thought touching every aspect of human life. It is the one century which consistently, and throughout the whole range of human activities, provided intellectual genius adequate for the greatness of its occasions.*

Christiaan Huygens was one of the brightest stars in that galaxy of brilliant men whose light still shines undiminished over the world of our own century.

NEWTON (1642–1727)

Nature to him was an open book, whose letters he could read without effort.

Albert Einstein

Most people are acquainted in some degree with the name and reputation of Isaac Newton, for his universal fame as the discoverer of the law of gravitation has continued undiminished over the two and a half centuries since his death. It is less well known, however, that in the immense sweep of his vast achievements he virtually created modern physical science, and in consequence has had a deeper influence on the direction of civilized life than the rise and fall of nations. Those in a position to judge have been unanimous in considering him one of the very few supreme intellects that the human race has produced.

Newton was born to a farm family in the village of Woolsthorpe in northern England. Little is known of his early years, and his undergraduate life at Cambridge seems to have been outwardly undistinguished. In 1665 an outbreak of the plague caused the universities to close, and Newton returned to his home in the country, where he remained until 1667. There, in 2 years of rustic solitude—from age 22 to 24—his creative genius burst forth in a flood of discoveries unmatched in the history of human thought: the binomial series for negative and fractional exponents; differential and integral calculus; universal gravitation as the key to the mechanism of the solar system; and the resolution of sunlight into the visual spectrum by means of a prism, with its implications for understanding the colors of the rainbow and the nature of light in general. In his old age he reminisced as follows about this miraculous period of his youth: "In those days I was in the prime of my age for invention and minded Mathematicks and Philosophy [i.e., science] more than at any time since."†

Newton was always an inward and secretive man, and for the most part kept his monumental discoveries to himself. He had no itch to publish, and most of his great works had to be dragged out of him by the cajolery and persistence of his friends. Nevertheless, his unique ability was so evident to his teacher, Isaac Barrow, that in 1669 Barrow resigned his professorship in favor of his pupil (an unheard-of event in academic life), and Newton settled down at Cambridge for the next 27 years. His mathematical discoveries were never really published in connected form; they be-

---

* See pp. 57–58 of Whitehead's *Science and the Modern World,* Macmillan, 1946.

† The full text of this autobiographical statement (probably written sometime in the period 1714–1720) is given on pp. 291–292 of I. Bernard Cohen, *Introduction to Newton's 'Principia,'* Harvard University Press, 1971. The present writer owns a photograph of the original document.

came known in a limited way almost by accident, through conversations and replies to questions put to him in correspondence. He seems to have regarded his mathematics mainly as a fruitful tool for the study of scientific problems, and of comparatively little interest in itself. Meanwhile, Leibniz in Germany had also invented calculus independently; and by his active correspondence with the Bernoullis and the later work of Euler, leadership in the new analysis passed to the Continent, where it remained for 200 years.*

Not much is known about Newton's life at Cambridge in the early years of his professorship, but it is certain that optics and the construction of telescopes were among his main interests. He experimented with many techniques for grinding lenses (using tools which he made himself), and about 1670 built the first reflecting telescope, the earliest ancestor of the great instruments in use today at Mount Palomar and throughout the world. The pertinence and simplicity of his prismatic analysis of sunlight have always marked this early work as one of the timeless classics of experimental science. But this was only the beginning, for he went further and further in penetrating the mysteries of light, and all his efforts in this direction continued to display experimental genius of the highest order. He published some of his discoveries, but they were greeted with such contentious stupidity by the leading scientists of the day that he retired back into his shell with a strengthened resolve to work thereafter for his own satisfaction alone. Twenty years later he unburdened himself to Leibniz in the following words: "As for the phenomena of colours . . . I conceive myself to have discovered the surest explanation, but I refrain from publishing books for fear that disputes and controversies may be raised against me by ignoramuses."†

In the late 1670s Newton lapsed into one of his periodic fits of distaste for science, and directed his energies into other channels. As yet he had published nothing about dynamics or gravity, and the many discoveries he had already made in these areas lay unheeded in his desk. At last, however, under the skillful prodding of the astronomer Edmund Halley (of Halley's Comet), he turned his mind once again to these problems and began to write his greatest work, the *Principia*.‡

It all seems to have started in 1684 with three men in deep conversation in a London inn—Halley, and his friends Christopher Wren and Robert Hooke. By thinking about Kepler's third law of planetary motion, Halley had come to the conclusion that the attractive gravitational force holding the planets in their orbits was probably inversely proportional to the square of the distance from the sun.§ However, he was unable to do anything more with the idea than formulate it as a

---

* It is interesting to read Newton's correspondence with Leibniz (via Oldenburg) in 1676 and 1677 (see *The Correspondence of Isaac Newton,* Cambridge University Press, 1959–1976, 6 volumes so far). In Items 165, 172, 188, and 209, Newton discusses his binomial series but conceals in anagrams his ideas about calculus and differential equations, while Leibniz freely reveals his own version of calculus. Item 190 is also of considerable interest, for in it Newton records what is probably the earliest statement and proof of the Fundamental Theorem of Calculus.

† *Correspondence,* Item 427.

‡ The full title is *Philosophiae Naturalis Principia Mathematica (Mathematical Principles of Natural Philosophy).*

§ At that time this was quite easy to prove under the simplifying assumption—which contradicts Kepler's other two laws—that each planet moves with constant speed v in a circular orbit of radius r. [Proof: In 1673 Huygens had shown, in effect, that the acceleration a of such a planet is given by $a = v^2/r$. If T is the periodic time, then

$$a = \frac{(2\pi r/T)^2}{r} = \frac{4\pi^2}{r^2} \cdot \frac{r^3}{T^2}.$$

By Kepler's third law, T^2 is proportional to r^3, so r^3/T^2 is constant, and a is therefore inversely proportional to r^2. If we now suppose that the attractive force F is proportional to the acceleration, then it follows that F is also inversely proportional to r^2.]

conjecture. As he later wrote (in 1686):

> I met with Sir Christopher Wren and Mr. Hooke, and falling in discourse about it, Mr. Hooke affirmed that upon that principle all the Laws of the celestiall motions were to be demonstrated, and that he himself had done it. I declared the ill success of my attempts; and Sir Christopher, to encourage the Inquiry, said that he would give Mr. Hooke or me two months' time to bring him a convincing demonstration therof, and besides the honour, he of us that did it, should have from him a present of a book of 40 shillings. Mr. Hooke then said that he had it, but that he would conceale it for some time, that others triing and failing, might know how to value it, when he should make it publick; however I remember Sir Christopher was little satisfied that he could do it, and tho Mr. Hooke then promised to show it him, I do not yet find that in that particular he has been as good as his word.*

It seems clear that Halley and Wren considered Hooke's assertions to be merely empty boasts. A few months later Halley found an opportunity to visit Newton in Cambridge, and put the question to him: "What would be the curve described by the planets on the supposition that gravity diminishes as the square of the distance?" Newton answered immediately, "An ellipse." Struck with joy and amazement, Halley asked him how he knew that. "Why," said Newton, "I have calculated it." Not guessed, or surmised, or conjectured, but *calculated.* Halley wanted to see the calculations at once, but Newton was unable to find the papers. It is interesting to speculate on Halley's emotions when he realized that the age-old problem of how the solar system works had at last been solved—but that the solver hadn't bothered to tell anybody and had even lost his notes. Newton promised to write out the theorems and proofs again and send them to Halley, which he did. In the course of fulfilling his promise he rekindled his own interest in the subject, and went on, and greatly broadened the scope of his researches.†

In his scientific efforts Newton somewhat resembled a live volcano, with long periods of quiescence punctuated from time to time by massive eruptions of almost superhuman activity. The *Principia* was written in 18 incredible months of total concentration, and when it was published in 1687 it was immediately recognized as one of the supreme achievements of the human mind. It is still universally considered to be the greatest contribution to science ever made by one man. In it he laid down the basic principles of theoretical mechanics and fluid dynamics; gave the first mathematical treatment of wave motion; deduced Kepler's laws from the inverse square law of gravitation, and explained the orbits of comets; calculated the masses of the earth, the sun, and the planets with satellites; accounted for the flattened shape of the earth, and used this to explain the precession of the equinoxes; and founded the theory of tides. These are only a few of the splendors of this prodigious work.‡ The *Principia* has always been a difficult book to read, for the style has an inhuman quality of icy remoteness, which perhaps is appropriate to the grandeur of the theme. Also, the densely packed mathematics consists almost entirely of classical geometry, which was little cultivated then and is less so now.§ In his dynamics and celestial mechanics, Newton achieved the victory for which Copernicus, Kepler, and Galileo

---

* *Correspondence,* Item 289.

† For additional details and the sources of our information about these events, see Cohen, op. cit., pp. 47–54.

‡ A valuable outline of the contents of the *Principia* is given in Chapter VI of W. W. Rouse Ball, *An Essay on Newton's Principia* (first published in 1893; reprinted in 1972 by Johnson Reprint Corp.)

§ The nineteenth century British philosopher Whewell has a vivid remark about this: "Nobody since Newton has been able to use geometrical methods to the same extent for the like purposes; and as we read the *Principia* we feel as when we are in an ancient armoury where the weapons are of gigantic size; and as we look at them we marvel what manner of man he was who could use as a weapon what we can scarcely lift as a burden."

had prepared the way. This victory was so complete that the work of the greatest scientists in these fields over the next two centuries amounted to little more than footnotes to his colossal synthesis. It is also worth remembering in this context that the science of spectroscopy, which more than any other has been responsible for extending astronomical knowledge beyond the solar system to the universe at large, had its origin in Newton's spectral analysis of sunlight.

After the mighty surge of genius that went into the creation of the *Principia,* Newton again turned away from science. However, in a famous letter to Bentley in 1692, he offered the first solid speculations on how the universe of stars might have developed out of a primordial featureless cloud of cosmic dust:

> It seems to me, that if the matter of our Sun and Planets and all the matter in the Universe was evenly scattered throughout all the heavens, and every particle had an innate gravity towards all the rest . . . some of it would convene into one mass and some into another, so as to make an infinite number of great masses scattered at great distances from one to another throughout all that infinite space. And thus might the Sun and Fixt stars be formed, supposing the matter were of a lucid nature.*

This was the beginning of scientific cosmology, and later led, through the ideas of Thomas Wright, Kant, Herschel, and their successors, to the elaborate and convincing theory of the nature and origin of the universe provided by late twentieth century astronomy.

In 1693 Newton suffered a severe mental illness accompanied by delusions, deep melancholy, and fears of persecution. He complained that he could not sleep, and said that he lacked his "former consistency of mind." He lashed out with wild accusations in shocking letters to his friends Samuel Pepys and John Locke. Pepys was informed that their friendship was over and that Newton would see him no more; Locke was charged with trying to entangle him with women and with being a "Hobbist" (a follower of Hobbes, i.e., an atheist and materialist).† Both men feared for Newton's sanity. They responded with careful concern and wise humanity, and the crisis passed.

In 1696 Newton left Cambridge for London to become Warden (and soon Master) of the Mint, and during the remainder of his long life he entered a little into society and even began to enjoy his unique position at the pinnacle of scientific fame. These changes in his interests and surroundings did not reflect any decrease in his unrivaled intellectual powers. For example, late one afternoon, at the end of a hard day at the Mint, he learned of a now-famous problem that the Swiss scientist John Bernoulli had posed as a challenge "to the most acute mathematicians of the entire world." The problem can be stated as follows: Suppose two nails are driven at random into a wall, and let the upper nail be connected to the lower by a wire in the shape of a smooth curve. What is the shape of the wire down which a bead will slide (without friction) under the influence of gravity so as to pass from the upper nail to the lower in the least possible time? This is Bernoulli's *brachistochrone* ("shortest time") *problem.* Newton recognized it at once as a challenge to himself from the Continental mathematicians; and in spite of being out of the habit of scientific thought, he summoned his resources and solved it that evening before going to bed. His solution was published anonymously, and when Bernoulli saw it, he wryly remarked, "I recognize the lion by his claw."

Of much greater significance for science was the publication of his *Opticks* in 1704. In this book he drew together and extended his early work on light and color. As an

---

* *Correspondence,* Item 398.

† *Correspondence,* Items 420, 421, and 426.

appendix he added his famous Queries, or speculations on areas of science that lay beyond his grasp in the future. In part the Queries relate to his lifelong preoccupation with chemistry (or alchemy, as it was then called). He formed many tentative but exceedingly careful conclusions — always founded on experiment — about the probable nature of matter; and though the testing of his speculations about atoms (and even nuclei) had to await the refined experimental work of the late nineteenth and early twentieth centuries, he has been proven absolutely correct in the main outlines of his ideas.* So, in this field of science too, in the prodigious reach and accuracy of his scientific imagination, he passed far beyond not only his contemporaries but also many generations of his successors. In addition, we quote two astonishing remarks from Queries 1 and 30, respectively: "Do not Bodies act upon Light at a distance, and by their action bend its Rays?" and "Are not gross Bodies and Light convertible into one another?" It seems as clear as words can be that Newton is here conjecturing the gravitational bending of light and the equivalence of mass and energy, which are prime consequences of the theory of relativity. The former phenomenon was first observed during the total solar eclipse of May 1919, and the latter is now known to underlie the energy generated by the sun and the stars. On other occasions as well he seems to have known, in some mysterious intuitive way, far more than he was ever willing or able to justify, as in this cryptic sentence in a letter to a friend: "It's plain to me by the fountain I draw it from, though I will not undertake to prove it to others."† Whatever the nature of this "fountain" may have been, it undoubtedly depended on his extraordinary powers of concentration. When asked how he made his discoveries, he said, "I keep the subject constantly before me and wait till the first dawnings open little by little into the full light." This sounds simple enough, but everyone with experience in science or mathematics knows how very difficult it is to hold a problem continuously in mind for more than a few seconds or a few minutes. One's attention flags; the problem repeatedly slips away and repeatedly has to be dragged back by an effort of will. From the accounts of witnesses, Newton seems to have been capable of almost effortless sustained concentration on his problems for hours and days and weeks, with even the need for occasional food and sleep scarcely interrupting the steady squeezing grip of his mind.

In 1695 Newton received a letter from his Oxford mathematical friend John Wallis, containing news that cast a cloud over the rest of his life. Writing about Newton's early mathematical discoveries, Wallis warned him that in Holland "your Notions" are known as "Leibniz's *Calculus Differentialis,*" and he urged Newton to take steps to protect his reputation.‡ At that time the relations between Newton and Leibniz were still cordial and mutually respectful. However, Wallis's letters soon curdled the atmosphere, and initiated the most prolonged, bitter, and damaging of all scientific quarrels: the famous (or infamous) Newton-Leibniz priority controversy over the invention of calculus.

It is now well established that each man developed his own form of calculus independently of the other, that Newton was first by 8 or 10 years but did not publish his ideas, and that Leibniz's papers of 1684 and 1686 were the earliest publications on the subject. However, what are now perceived as simple facts were not nearly so clear at the time. There were ominous minor rumblings for years after Wallis's letters, as the storm gathered:

> What began as mild innuendoes rapidly escalated into blunt charges of plagiarism on both sides. Egged on by followers anxious to win a reputation under his aus-

---

* See S. I. Vavilov, "Newton and the Atomic Theory," in *Newton Tercentenary Celebrations,* Cambridge University Press, 1947.

† *Correspondence,* Item 193.

‡ *Correspondence,* Items 498 and 503.

pices, Newton allowed himself to be drawn into the centre of the fray; and, once his temper was aroused by accusations of dishonesty, his anger was beyond constraint. Leibniz's conduct of the controversy was not pleasant, and yet it paled beside that of Newton. Although he never appeared in public, Newton wrote most of the pieces that appeared in his defense, publishing them under the names of his young men, who never demurred. As president of the Royal Society, he appointed an "impartial" committee to investigate the issue, secretly wrote the report officially published by the society [in 1712], and reviewed it anonymously in the *Philosophical Transactions.* Even Leibniz's death could not allay Newton's wrath, and he continued to pursue the enemy beyond the grave. The battle with Leibniz, the irrepressible need to efface the charge of dishonesty, dominated the final 25 years of Newton's life. Almost any paper on any subject from those years is apt to be interrupted by a furious paragraph against the German philosopher, as he honed the instruments of his fury ever more keenly.*

All this was bad enough, but the disastrous effect of the controversy on British science and mathematics was much more serious. It became a matter of patriotic loyalty for the British to use Newton's geometrical methods and clumsy calculus notations, and to look down their noses at the upstart work being done on the Continent. However, Leibniz's analytical methods proved to be far more fruitful and effective, and it was his followers who were the moving spirits in the richest period of development in mathematical history. What has been called "the Great Sulk" continued; for the British, the work of the Bernoullis, Euler, Lagrange, Laplace, Gauss, and Riemann remained a closed book; and British mathematics sank into a coma of impotence and irrelevancy that lasted through most of the eighteenth and nineteenth centuries.

Newton has often been thought of and described as the ultimate rationalist, the embodiment of the Age of Reason. His conventional image is that of a worthy but dull absent-minded professor in a foolish powdered wig. But nothing could be further from the truth. This is not the place to discuss or attempt to analyze his psychotic flaming rages; or his monstrous vengeful hatreds that were unquenched by the death of his enemies and continued at full strength to the end of his own life; or the 58 sins he listed in the private confession he wrote in 1662; or his secretiveness and shrinking insecurity; or his peculiar relations with women, especially with his mother, who he thought had abandoned him at the age of 3. And what are we to make of the bushels of unpublished manuscripts (millions of words and thousands of hours of thought!) that reflect his secret lifelong studies of ancient chronology, early Christian doctrine, and the prophecies of Daniel and St. John? Newton's desire to know had little in common with the smug rationalism of the eighteenth century; on the contrary, it was a form of desperate self-preservation against the dark forces that he felt pressing in around him. As an original thinker in science and mathematics he was a stupendous genius whose impact on the world can be seen by everyone; but as a man he was so strange in every way that normal people can scarcely begin to understand him.† It is perhaps most accurate to think of him in medieval terms—as a consecrated, solitary, intuitive mystic for whom science and mathematics were means of reading the riddle of the universe.

---

* Richard S. Westfall, in the *Encyclopaedia Britannica.*

† The best effort is Frank E. Manuel's excellent book, *A Portrait of Isaac Newton,* Harvard University Press, 1968.

LEIBNIZ (1646–1716)

It would be difficult to name a man more remarkable for the greatness and universality of his intellectual powers than Leibniz.

John Stuart Mill

The ideas of calculus were "in the air" in the 1650s and 1660s. The ingenious area calculations and tangent constructions of Cavalieri, Fermat, Pascal, Barrow, and others were so suggestive that the final discovery of calculus as an autonomous discipline was almost inevitable within a very few years. The last steps of putting it all together were taken by two men of great genius working independently of each other: by Isaac Newton in what he called "the two plague years of 1665 and 1666," and also by Gottfried Wilhelm Leibniz during his sojourn in Paris from 1672 to 1676.

Leibniz is probably better known to most people as a philosopher than as a mathematician. The history of philosophy has long recognized him as one of its greatest system builders, and also as the producer of most of the ammunition with which Kant later attacked Hume. But this too was only a small fraction of his total thought. He made memorable creative contributions across the entire spectrum of intellectual life, from mathematics and logic through the various sciences to history, law, diplomacy, politics, philology, metaphysics, and theology. No other thinker except Aristotle has rivaled him in the range and variety of his abilities and achievements. Leibniz lived in a period when it was still possible — as his own astounding career demonstrated — for a very highly intelligent and hard-working scholar to absorb all the knowledge of his time. To Oswald Spengler he was "without doubt the greatest intellect in Western philosophy"; and Admiral Mahan — perhaps the most influential historian of modern times — called him "one of the world's great men." What manner of man was he, and how did he live and what did he think?

Leibniz was born in 1646 at Leipzig, where his father was professor of moral philosophy at the university. He was sent to a good school, but after his father's death in 1652 he seems to have acted for the most part as his own teacher, leading a self-propelled intellectual life even as a small child. The German books that were available to him were quickly read through. He began teaching himself Latin at the age of 8, and soon mastered it sufficiently to read it with ease and compose acceptable Latin verse; he started the study of Greek a few years later. He had acquired a love of history from his father, and he spent most of his childhood eagerly devouring the large library of choice books that his father had collected, including Herodotus, Xenophon, Homer, Plato, Aristotle, Cicero, Quintilian, Seneca, Pliny, Polybius, and many others. By his early teens Leibniz had thus become familiar with a broad range of classical literature, and was well embarked on the omnivorous reading that was to be his custom throughout his life.*

At this stage of his mental development, classical studies no longer satisfied him. He turned his attention to logic, zealously reading the scholastic philosophers and attempting already to reform the doctrines of Aristotle. In a letter written in 1696 he recalled this period of his life as follows:

> As soon as I began to learn logic, I was fascinated by the classification and order which I perceived in its principles. I soon observed, as much as a boy of 13 could,

---

* His willingness to read almost anything led Fontenelle to remark of him that he bestowed the honor of reading them on a great mass of bad books. However, as Leibniz himself said, "When a new book reaches me, I search for what I can learn, not for what I can criticize in it." For more on his reading habits, see p. 237 of L. E. Loemker (ed. and trans.), *Leibniz: Philosophical Papers and Letters,* University of Chicago Press, 1956.

that there must be something great to the subject. My strongest pleasure lay in the categories, which seemed to me to call the roll of all the things in the world.*

He had an insatiable appetite for discovering the meaning and purpose of everything around him. The thoughts of youth stoke secret fires, and few such fires could have burned as intensely as his.†

At the age of 15, Leibniz entered the University of Leipzig as a law student. During the first two years he studied mainly philosophy and mathematics as far as Euclid. At that time the University was firmly congealed in the sterile Aristotelian tradition and did nothing to encourage science. It was by his own efforts that he became acquainted with those thinkers who had already launched the modern age in science and philosophy: Francis Bacon, Kepler, Galileo, and Descartes. As with most men of really great intellect, Leibniz's formal education was only a minor eddy in the torrent of thought and study and learning that was the essence of his life.

The next 3 years were devoted to legal studies, and in 1666 he applied for the degree of doctor of law with the aim of seeking appointment to a judicial position. This application was refused, ostensibly on the grounds of his youth but probably because of the small-minded jealousy of the faculty, and he left Leipzig in disgust. At Altdorf, the university town of the free imperial city of Nuremberg, his brilliant dissertation *De casibus perplexis in jure* ("On Perplexing Cases in the Law") procured him the doctor's degree at once and the immediate offer of a professorship in the University. He declined this offer, having, as he said, "very different things in view." He called the universities "monkish," and charged that they possessed learning but little common sense and were preoccupied with empty trivialities. His purpose was to enter public rather than academic life. It is remarkable how few of the major philosophers have been professors in universities. Leibniz spent the next year in Nuremberg, which was then a center of the secret mystical order of the Rosicrucians, and he made himself so familiar with the ideas and writings of the alchemists—much as Newton was doing at Cambridge—that he was elected secretary of the local Rosicrucian society.

At the age of 20, Leibniz had not only earned his doctorate in law, but had also published several highly original essays on logic and jurisprudence. His *Dissertatio de arte combinatoria* ("Dissertation on the Art of Combinations") initiated his lifelong project of reducing all knowledge and reasoning to what he called a "universal characteristic." By this he meant a precise system of notation—a symbolic mathematical language analogous to algebra—in which the symbols themselves and their rules of combination would automatically analyze all concepts into their ultimate constituents in such a way as to provide the means for obtaining knowledge of the essential nature of all things.‡ Needless to say, this grandiose project was not realized in his lifetime or thereafter, but its spirit continued to propel and guide his thinking, and later led to the invention of his differential and integral calculus and his first tentative steps toward the creation of symbolic logic. There was also his *Nova methodus docendae discendaeque jurisprudentiae* ("A New Method for Teaching and Learning Jurisprudence"), which he wrote during the rest stops of his journey from Leipzig to Altdorf. This essay is remarkable for containing the first clear recognition

---

* Loemker, op. cit., p. 756.

† It is worth noting that Leibniz's IQ has been estimated by experts as 180 at the very least and probably much higher, "close to the maximum for the human race." See pp. 155 and 702–705 of Lewis M. Terman (ed.), *Genetic Studies of Genius:* Vol. II, *The Early Mental Traits of Three Hundred Geniuses,* Stanford University Press, 1926.

‡ For Leibniz's own explanation of what he had in mind, see Loemker, op. cit., pp. 339–346; or pp. 12–25 of Philip P. Wiener (ed.), *Leibniz Selections,* Scribner's, 1951.

of the importance of the historical approach to law.* It also had the practical effect of securing him a position (in 1667) as legal advisor to the Prince Elector of Mainz.†

Leibniz remained at Mainz for 5 years, at first as an assistant in recodifying the laws and later as a trusted councilor and diplomat serving the political goals of the Elector. The most important of these goals was survival, for at that time the swollen arrogance of Louis XIV was like a boil on the face of Europe, and his armies were threatening the Low Countries and the small German states along the Rhine. Leibniz conceived a plan to divert Louis from Germany by persuading him to conquer Egypt and build a colonial empire in North Africa, thereby satisfying his imperial ambitions at little cost to his European neighbors. A detailed memorandum was sent to the French government; and in March of 1672, at the invitation of the French foreign minister, the young diplomat traveled to Paris to present his proposals to the King. Unfortunately, however, Louis had an irrational hatred for the Dutch, and declared war on them a few weeks later. Leibniz never met the King, and his plan for a French conquest of Egypt disappeared from practical politics until the time of Napoleon, who revived it in 1798.‡ The important thing for Leibniz himself was the visit to the great city of Paris, where he spent most of the next four years. This experience was crucial for his intellectual development, for he mastered the French language, became personally acquainted with the leaders in science and philosophy, and immersed himself in the mainstream of European thought.

The world Leibniz entered in 1672 — in France, England, and Holland — was bubbling with the ferment of new ideas and teeming with men of genius. It was a garden of intellectual civilization, in comparison with which his former world of Leipzig, Nuremberg, and Mainz was little better than dull barbarism. Philosophy and theology were in a state of upheaval, and the champions of both the new and the old ways of regarding man and God — Hobbes, Spinoza, Locke, Arnauld, Malebranche, Bossuet — were well known in all learned circles. Leibniz met Arnauld and Malebranche in Paris, visited Spinoza in Holland, and corresponded with the others. The sciences were enjoying a period of unprecedented growth, and Leibniz kept himself informed of all the latest discoveries and made a number of contributions of his own. The Dutch physicist Huygens — creator of the wave theory of light and inventor of the pendulum clock — became Leibniz's friend and mathematical mentor during his years in Paris. Another of his friends was the Danish astronomer Roemer, who in 1675 first calculated the speed of light from the observations of the moons of Jupiter that he made at the Paris observatory. And in 1676 Leibniz entered into a mathematical correspondence with Newton that had fateful consequences for the future development of European science. Other eminent men who were active during the second half of the seventeenth century were Boyle, Hooke, von Guericke, and Halley in chemistry, physics, and astronomy; Leeuwenhoek, Malpighi, and Swammerdam in biology; and Wallis, Wren, Roberval, Tschirnhaus, and the Bernoullis in mathematics — and Leibniz knew and corresponded with almost all of them. Western

---

* See the chapter on Leibniz in H. Cairns, *Legal Philosophy from Plato to Hegel,* Johns Hopkins University Press, 1949.

† At that time Mainz was more than just a city on the Rhine; it was one of the most powerful member states of the Holy Roman Empire, that strange agglomeration that Voltaire characterized as "neither holy, nor Roman, nor an empire." The ruler of Mainz was one of the seven regional princes empowered to elect the Emperor.

‡ It was a great misfortune for France that Louis ignored Leibniz's proposals; for if they had been adopted and pursued vigorously, it would probably have been France instead of England that captured India and the mastery of the seas, and the subsequent history of Europe would have been very different. As it was, the King's folly "ruined the prosperity of France, and was felt in its consequences from generation to generation afterward." See pp. 106–107 and 141–143 of A. T. Mahan, *The Influence of Sea Power upon History: 1660–1783,* Little, Brown, 12th ed., 1944; first published in 1890.

Europe was drunk with the wine of reason, and Leibniz enthusiastically joined the party when he moved to Paris at the age of 26.

In January of 1673 Leibniz crossed the Channel to England on a diplomatic mission for the Elector of Mainz. In London he quickly became acquainted with Henry Oldenburg, the German-born first secretary of the Royal Society, and also with others of its members, including Boyle and Hooke.* He had invented a calculating machine for performing more complicated operations than the earlier machine of Pascal—multiplying and dividing as well as adding and subtracting. He exhibited a rough model of his invention to the Royal Society, and was elected a member of the Society shortly after his return to Paris in March.†

When Leibniz first arrived in Paris in 1672, he had little knowledge of mathematics beyond the simplest parts of Euclid and some fragmentary ideas from Cavalieri. He quickly became aware that in that age to be ignorant of mathematics was to be negligible in the eyes of most educated men, and he began his mathematical studies with the aim of establishing his credibility as a serious thinker.‡ However, once started, he was irresistibly drawn to the subject. When he returned to Paris from London, he spent more and more of his time on higher geometry, under the general guidance of Huygens, and began the series of investigations that led over the next few years to his invention of the differential and integral calculus. In 1673 he made one of his most remarkable discoveries, the infinite series expansion

$$\frac{\pi}{4} = 1 - \frac{1}{3} + \frac{1}{5} - \frac{1}{7} + \frac{1}{9} - \cdots .$$

This beautiful formula reveals a striking relation between the mysterious number π and the familiar sequence of all the odd numbers.§

During this entire period Leibniz read, wrote, and thought continually, pursuing ideas with a strength and intensity known to ordinary people only in their pursuit of wealth and power. His motto at the time was, "With every lost hour a part of life perishes." In 1673 the Elector of Mainz died, and Leibniz tentatively entered the service of the learned John Frederick, Duke of Brunswick-Lüneburg, with whom he had already been corresponding for several years, as curator of the ducal library at Hanover. However, he was held by the magnetic attraction of Paris and continued to hope that he could find some way of staying on there indefinitely.¶ Nothing turned up, and late in 1676, at the Duke's insistence, he left Paris and traveled slowly to Hanover by way of London and Holland. At The Hague he had several long conversations with Spinoza, who permitted him to read and copy passages from his unpub-

---

* Oldenburg knew almost everyone worth knowing in Western Europe at the time, from Milton and Cromwell to Newton, Leibniz, Spinoza, Leeuwenhoek, and many others. He was a major crossroads in the intellectual life of the period, and deserves a full-scale scholarly biography. See the 9 volumes of *The Correspondence of Henry Oldenburg,* ed. and trans. by A. Rupert Hall and Marie Boas Hall, University of Wisconsin Press, 1965–1973. The introductions to these volumes provide a good running biography of Oldenburg.

† For descriptions of several stages in the development of this machine, see pp. 23, 79, and 126 of J. E. Hofmann, *Leibniz in Paris: 1672–1676,* Cambridge University Press, 1974. A picture, together with Leibniz's own explanation, can be found in pp. 173–181 of D. E. Smith, *A Source Book in Mathematics,* McGraw-Hill, 1929. See also pp. 7–9 of H. H. Goldstine, *The Computer from Pascal to von Neumann,* Princeton University Press, 1972.

‡ Loemker, op. cit., pp. 400–401.

§ A modern expert on infinite series (K. Knopp) has said: "It is as if, by this expansion, the veil which hung over that strange number had been drawn aside." Leibniz found his formula—of which he was justifiably proud all his life—by a very ingenious calculation of the area of one quarter of a circle of radius 1. We show how he did it in Appendix A.11. See also Items 123, 126, 130, and 134 in *The Correspondence of Isaac Newton,* Cambridge University Press, 1959–1976, 6 volumes so far.

¶ Hofmann, op. cit., pp. 46–47 and 160–163.

lished *Ethics*.* He also visited Antony van Leeuwenhoek, discoverer of the tiny forms of life that can be seen only through a microscope. The miniature universes that Leeuwenhoek was able to find in every drop of pond water made a deep impression on Leibniz, and years later contributed to the metaphysical system in which he portrayed the whole world as consisting of tiny, invisible centers of awareness called monads.†

Leibniz reached Hanover by the end of November, and through the remaining 40 years of his life served three successive dukes as librarian, family historian, and informal minister in charge of scientific and cultural affairs. Though he lived in the atmosphere of the petty local politics of a small German principality, his activities and outlook were always constructive and cosmopolitan. He supervised the mint, and suggested various improvements in the coinage and the economic theory behind it. He reorganized the Harz silver mines, the basis of the currency, and acted as an engineer in designing windmill-powered pumps intended to protect the mines against the seeping water that threatened them. He was both an engineer and a landscape architect in planning the fountains for the great formal garden of the summer palace at Herrenhausen. He wrote many pamphlets and position papers to support various rights and claims of his patrons. He also wrote a masque that was performed at court by the nobility; and his memorial poem on the occasion of John Frederick's death in 1679 contained a description of the recently discovered element phosphorus that was considered one of the finest passages in modern Latin poetry.

In the midst of all this miscellaneous activity, Leibniz's main continuing responsibilities were as librarian and historian. His ideas about the purposes, organization, and administration of scholarly libraries were so far-sighted that he has been called "the greatest librarian of his age."‡ His historical work began with an assignment to compile a genealogy of the Brunswick family for use as a weapon in the dynastic political struggles of the day. The necessary research meant that Leibniz had to travel, which has been one of the main benefits of the profession of history since the time of Herodotus. He spent three years (1687–1690) examining the archives and private libraries of Southern Germany and Italy, and at last was able to prove the ancestral connection between the ducal houses of Brunswick and Este. This achievement was influential in winning for Hanover (Brunswick-Lüneburg) the status of an electorate of the Empire (in 1692). Leibniz's collection of historical documents enabled him not only to undertake an extensive history of the House of Brunswick *(Annales Brunsvicenses),* but also to publish two important volumes of source material for a code of international law *(Codex juris gentium diplomaticus,* 1693 and 1700). Over the years his history expanded into an exhaustive study of the German Empire in the Middle Ages, and was later used by Gibbon.

But all this was only the froth on the surface of Leibniz's life, the visible career of the courtier and public official. Beneath was a churning sea of private intellectual activity, on so vast and varied a scale as to be almost beyond belief. We briefly summarize his main interests and achievements (apart from mathematics) under four headings: (1) logic, (2) theology, (3) metaphysics, (4) science.

(1) He was the first to perceive that the laws of thought are essentially algebraic in nature, and by this insight and his subsequent efforts he founded symbolic logic. He imagined a distant future when philosophical discussions would be carried on by

---

* Little is known of these tantalizing conversations between the two greatest metaphysical thinkers of the time; see pp. 37–39 of F. Pollock, *Spinoza: His Life and Philosophy,* 2d ed., Duckworth, 1899. A reconstruction—vivid and dramatic, but mostly imaginary—is attempted in pp. 281–292 of R. Kayser, *Spinoza: Portrait of a Spiritual Hero,* Greenwood Press, 1968.

† Loemker, op. cit., p. 1056 ("The Monadology," 66–69).

‡ By Sir Frank C. Francis, Director and Principal Librarian of the British Museum, 1959–1968.

means of logical symbolism and would reach conclusions as certain as those of mathematics. He expressed his vision as follows:

> If controversies were to arise, there would be no more need of disputation between two philosophers than between two accountants. For it would suffice to take their pencils in their hands, to sit down to their slates, and to say to each other (with a friend as witness if they liked): Let us calculate.*

Philosophy has not yet reached this stage, and perhaps it never will, but much of what Leibniz foresaw can be recognized in the computerized decision-making processes of modern business, government, and military strategy. During the 1670s and 1680s he made considerable progress on his project to deal with logic by algebraic methods.† In modern terminology, he stated the main formal properties of logical addition, multiplication, and negation; he considered the empty set and set inclusion; and he pointed out the similarity between certain properties of set inclusion and the relation of implication for propositions. Though unfortunately most of this work was not published until two centuries later, it was the historical source of the symbolic logic (Boolean algebra) developed by George Boole in the nineteenth century and carried forward by Whitehead and Russell in the early part of the twentieth century. There is evidently little exaggeration in the judgment, "Leibniz deserves to be ranked among the greatest of all logicians."‡

(2) Leibniz had been a close student of Protestant theology ever since his early teens, and his interest in such matters was greatly stimulated by his contact with Catholic clergymen and ritual at the court of Mainz. He observed that Catholic and Protestant doctrines differ only in minor ways, and he began to dream of reuniting the divided creeds of Christianity into a monolithic Christendom. In 1686 he wrote his *Systema theologicum* as a statement of basic belief on which he hoped all Christians could agree, but it was only in the late twentieth century that such ecumenical ideas began to find a favorable climate. His main abstract theological purpose was to establish logical proofs for the existence of God, and he put the four standard arguments—one of which he invented himself—into their final form.§ However, this project also had little influence, since most people find it difficult to feel affection or reverence for a deity whose main role is to fill a gap in a metaphysical jigsaw puzzle. The idea of proving religious creeds as if they were theorems of geometry has never been widely popular, for in our moments of clarity we recognize that such creeds have always been matters of culture, custom, and preference, not logical truth or falsity. He also tried to give a rational explanation for the presence of evil in the world, and this led him to write a long, dull book (*Essais de Théodicée,* 1710) that brought him European fame for his doctrine that this is "the best of all possible worlds."¶ This unfortunate production stimulated Voltaire to satirize his ideas through the character of Doctor Pangloss in his most famous work, *Candide* (1758).**

---

* See p. 170 of Bertrand Russell, *A Critical Exposition of the Philosophy of Leibniz,* 2d ed., George Allen and Unwin, 1937; first published in 1900.

† See pp. 123–132 of D. J. Struik, *A Source Book in Mathematics, 1200–1800,* Harvard University Press, 1969.

‡ See pp. 320–345 of W. and M. Kneale, *The Development of Logic,* Oxford University Press, 1962.

§ See pp. 585–589 of Bertrand Russell, *A History of Western Philosophy,* Simon and Schuster, 1945.

¶ Because God, being all-wise, must know all possible worlds, being all-powerful, must be able to create whatever kind of world he chooses, and being all-good, must choose the best.

** Leibniz was also the object of Voltaire's laughter in the wittiest of his satirical romances, *Micromégas* (1752), in which a visitor from Sirius comes to the earth and amuses himself by listening to the arguments of philosophers.

(3) Leibniz lived at a time when the passion for metaphysics was deep and strong, when it was still believed possible to understand the world purely by thought. He struggled for years to penetrate the mysteries of nature and God by the use of reason alone — by constructing a great metaphysical system that explains all things by the *a priori* method of deducing necessary consequences from a few self-evident principles. His main starting point was his Principle of Sufficient Reason: "Nothing happens without a reason, there is no effect without a cause." From this slender beginning he purported to demonstrate by irresistible logic the chief tenets of his metaphysical system.* However, by equally irresistible logic Spinoza had earlier established the main features of his own totally different metaphysical system. And so with most of the other eminent philosophers of the seventeenth and eighteenth centuries — each thought he was looking through a window at the great outside world of reality, but instead looked into a mirror and saw only his own face. As Spinoza said in another context, "What St. Paul tells us about God tells us more about St. Paul than it does about God." The growth of empiricism and the rise of science over the past three centuries have made it almost impossible to take seriously the extravagant pretensions of the *a priori* philosopher, who sits in his study and spins a web of words, fanciful imaginings, and empty speculations out of the material of his own consciousness. Faith in reason alone is alien to us, and we believe that only careful observation and experiment can reveal anything of substance about the actual universe. We no longer study philosophy for the old reason, the hope of learning the truth about the nature of things, but rather, for the fascination of learning what people have thought, and if possible why they have thought it.†

(4) Leibniz was torn between the contradictory claims of science and philosophy as ways of knowing reality, but he leaned in the direction of science. In 1691 he wrote as follows to his friend Huygens: "I prefer a Leeuwenhoek who tells me what he sees to a Cartesian who tells me what he thinks. It is, however, necessary to add reasoning to observation." He often urged the pursuit of real knowledge — chemistry, physics, geology, botany, zoology, anatomy, history, and geography — in contrast to the learned nonsense of the academics. He was fascinated by every aspect of the developing sciences and technology, and his own contributions to these fields alone would have filled several distinguished lifetimes. In 1693 he published his ideas about the beginnings of the earth, in a paper in the *Acta Eruditorum.*‡ He explained his geological theories more fully in his remarkable treatise *Protogaea,* which unfortunately was not published until 1749, long after his death. The earth, he believed, was originally an incandescent globe; it slowly cooled, contracted, and formed a crust; and as it cooled, the surrounding vapor condensed into oceans, which gradually became salty by dissolving salts in the crust.§ He was the first to distinguish igneous from sedimentary rocks. He also gave a good explanation for fossils, suggested that the different kinds of fossils found in different layers of the crust might be clues to the earth's

---

* This system is mostly concerned with the nature, activities, and interrelations of the invisible percipient points called monads, whose existence Leibniz deduced and which he believed to be the ultimate constituents of reality and the causes of all phenomena.

† For readers who wish to sample Leibniz's metaphysics at the source, we suggest three brief essays given in Loemker, op. cit., pp. 346–350, 411–417 and 1033–1043. A useful general commentary is provided by L. Couturat in his article "On Leibniz's Metaphysics," pp. 19–45 in H. G. Frankfurt (ed.), *Leibniz: A Collection of Critical Essays,* Doubleday, 1972. In addition, we recommend the article on Metaphysics by Gilbert Ryle in the *Encyclopaedia Britannica*; here the reader is treated to the enthralling spectacle of an eminent Professor of Metaphysical Philosophy at Oxford University suavely demonstrating that metaphysical philosophy does not exist.

‡ This periodical was the most influential European journal of the time in science and mathematics. It was founded by Leibniz in 1682, and he was its editor-in-chief for many years.

§ See p. 352 of A. Wolf, *A History of Science, Technology and Philosophy in the 16th and 17th Centuries,* George Allen and Unwin, 1935.

history, gave the earliest reasonably satisfactory definition of the concept of species, and — foreshadowing the evolutionists of the eighteenth and nineteenth centuries — thought it likely that species have undergone many drastic changes through the long history of the earth. All this, in an age in which even the most intelligent and well-educated people considered *Genesis* to be the final authority in such matters. No new idea or discovery escaped his attention, and he had a hand in many of the notable achievements of the day: Papin's steam engine, for which he suggested a self-regulating device; the discovery of European porcelain by his friend Tschirnhaus at Meissen; the use of microscopes in biological research; the use of vital statistics in coping with problems of public health, on which he wrote several articles; and the principle of the aneroid barometer, which he was the first to propose. It was at his urging that a continuous series of observations of barometric pressures and weather conditions was carried out at Kiel from 1679 to 1714, with the purpose of testing the value of the barometer in forecasting the weather. The science of linguistics originated in his efforts to construct a comparative system of linguistic genealogy for the main languages of Europe and Asia.* Also, it was he who destroyed the prevailing belief that Hebrew was the primordial language of the human race.† His *New Essays on Human Understanding* (written in French in 1704 but not published until 1765) was one of the most influential books in the history of psychology, for in it he introduced for the first time the concept of subconscious or unconscious mental processes, and thereby permanently altered the development of psychology in ways that are well known to everyone in the twentieth century.‡ In physics, his contributions to the emerging concept of kinetic energy were so significant that "he stood beside Newton as one of the creators of modern dynamics."§ In fact, the very word "dynamics" is due to Leibniz, in its French form "dynamique." He also solved the problem of the catenary curve, gave the first analysis of the tension in the interior fibers of a loaded beam, and studied many other physical problems with the aid of his own differential and integral calculus.¶ Leibniz was obviously a scientific thinker of rare talent and vision, and there was so much of real value to do and learn at that time that it seems a pity he didn't spend more of his time and energy on these projects and less on the fantasies of theology and metaphysics.

Aristotle said that by their nature all men desire to know. In the case of Leibniz this desire was intensified into an overwhelming passion. The great modern philosopher A. N. Whitehead once remarked, "There is a book to be written, and its title should be *The Mind of Leibniz.*" But who could write such a book and do justice to the subject? The prodigious variety of his interests was an essential part of his genius, but he paid a price, for he scattered himself in so many directions that he left mostly fragments behind him. In a letter written in 1695 he expressed his occasional despair in these words:

> How extremely distracted I am cannot be described. I dig up various things from the archives, examine ancient documents, and collect unpublished manuscripts. From these I strive to throw light on the history of Brunswick. I receive and send letters in great numbers. I have, indeed, so much that is new in mathematics, so

---

* This work was published in 1710 in the first volume of *Miscellania Berolinensia,* the official journal of the Berlin Academy of Sciences. This volume contained 58 papers on science and mathematics, of which 12 were by Leibniz himself. He had founded the Berlin Academy in 1700 with the support of the Queen of Prussia, whom he had tutored when she was a child.

† See pp. 9–10 of H. Pedersen, *Linguistic Science in the 19th Century,* Harvard University Press, 1931.

‡ See Wolf, op. cit., pp. 579–581.

§ See pp. 283–322 of Richard S. Westfall, *Force in Newton's Physics,* American Elsevier, 1971.

¶ See many index references in C. Truesdell, *Essays in the History of Mechanics,* Springer Verlag, 1968.

many thoughts in philosophy, so many other literary observations which I do not wish to have perish, that I am often bewildered as to where to begin.*

With Leibniz the mind and the hand went together so closely that thinking and writing were almost a single act, and he coped with the flood of his thoughts by writing them down. Since he rarely discarded written material, he accumulated over the years an immense chaotic mass of papers as big as a haystack, which was hastily packed away in crates soon after his death and stored in the Royal Library at Hanover. Only a small fraction of this mountain of material was published in his lifetime. Partial excavations have been made by various scholars, but most of it remains unpublished to this day.† There are rough notes and early drafts and almost-finished projects; innumerable essays and memoranda; book-length manuscripts; dialogues with himself; and letters — more than 15,000 letters written by Leibniz in correspondence with 1063 different people.

All his life Leibniz collected correspondents on topics in science and philosophy as other people collect stamps or works of art. The letters exchanged with a particular person often constitute an absorbing intellectual drama. Thus we have, among others, the famous correspondence of 1715 – 1716 between Leibniz and Dr. Samuel Clarke, Newton's disciple and spokesman, which quickly heated up into a pitched battle over the validity of Newton's ideas about absolute space and absolute time.‡ Leibniz believed that these ideas were devoid of meaning, and that space and time are purely relative concepts; and to Newton's intense annoyance, he claimed to have proved his contention.§ Whatever we may think of Leibniz's arguments today, the fact remains that modern science has discovered serious defects in the Newtonian concepts of the framework of the physical universe, and these defects were remedied only by the discoveries of Einstein in the early twentieth century.

On looking back at Leibniz through the haze of 300 years of history, we seem to see a multitude of men — philosopher, mathematician, scientist, logician, diplomat, lawyer, historian, etc.— instead of a single individual. But what were his qualities as a living, breathing human being?

Like many of his great contemporaries he never married, and we know little of his personal life. At the age of 50, according to Fontenelle, he proposed to a certain lady, but "the lady asked for time to consider the matter, so Leibniz had a chance to think again, and he withdrew his offer." He had an astonishing capacity for rapid and sustained work, often spending days on end at his desk, except for occasional hurried meals or brief naps; even when traveling in the jolting, uncomfortable carriages of the day, he used his time to work on mathematical problems; and to the end of his life he preserved the indomitable energy without which all his ambitions and plans and projects would have come to nothing. He is described as a man of moderate habits in everything but work, quick of temper but easily appeased, very self-assured, and tolerant of differences of opinion though confident of the correctness of his own opinions. He enjoyed social life of all kinds, and was firmly convinced that there was something interesting to be learned from everyone he met. According to his secre-

---

* Loemker, op. cit., p. 21.

† In 1900 the Berlin Academy began planning a complete critical edition of Leibniz's works in 40 volumes. Only about a dozen of these volumes have so far been published, and it seems unlikely that this edition will be finished until sometime in the twenty-first century.

‡ In the *Principia* Newton wrote: "Absolute space, in its own nature, without regard to anything external, remains always similar and immovable. . . . Absolute, true, and mathematical time, of itself, and from its own nature, flows equably and without regard to anything external."

§ See H. G. Alexander (ed.), *The Leibniz-Clarke Correspondence,* Manchester University Press, 1956, especially paragraphs 5 and 6 of Leibniz's third letter. For a further commentary of great interest and value, see Koyré's article "Leibniz and Newton," in H. G. Frankfurt, op. cit., pp. 239–279.

tary, he spoke well of everybody and made the best of everything. On the less amiable side, he is said to have been fond of money to the point of avarice, and he had a reputation as a pennypincher.

When Queen Anne died in August 1714, Leibniz's master, the Elector George Louis of Hanover, succeeded her as George I, the first German King of England. Leibniz was in Vienna at the time and returned to Hanover as quickly as he could, but George and his entourage had already departed. Leibniz hoped to join them in London as court historian and councilor of state, but the new King refused to consider it and ordered him to remain at Hanover and finish his history of the House of Brunswick. It appears that George disliked Leibniz and had no interest in any of his ideas or projects except those that might add luster to his own family background.* Leibniz's last years were made difficult by neglect and the agonies of gout and kidney stones, but he struggled on with his work. His death in November 1716 was ignored by the court, both in London and Hanover. Only his secretary followed him to his grave, and an eyewitness of his funeral wrote that "he was buried more like a robber than what he really was, the ornament of his country." But the books are now balanced, for he is recognized today as one of the few universal geniuses in human history, and the first in the line of those eminent Germans who were also towering figures of world culture: Leibniz, Bach, Goethe, Beethoven, Gauss, Einstein.

It remains to add a few words of a more detailed nature about Leibniz's greatest creation, his differential and integral calculus, for this incomparable tool of thought is the means by which his genius continues to make itself felt on a day-by-day basis in every civilized country of the modern world.

Leibniz published many sketchy papers on his calculus beginning in 1684, and we shall say more about these below. However, the development of his ideas and the sequence of his discoveries can be followed in full detail through the hundreds of pages of his private notes made from 1673 on.†

It all seems to have started in a memorable conversation Leibniz had with Huygens in the spring of 1673, which he referred to repeatedly in later years.‡ As a result of this encounter, and on Huygens's advice, he began an intensive study of some of the mathematical writings of Pascal and others. In particular, it was on reading Pascal's brief paper *Traité des sinus du quart de cercle* ("Treatise on the Sines of a Quadrant of a Circle") that Leibniz later reported that "a great light" burst upon him. He suddenly realized that the tangent to (or slope of) a given curve can be found by forming the ratio of the *differences* in the ordinates and abscissas of two neighboring points on the curve as these differences become infinitely small (see Fig. B.10). He also saw that the area under the curve is the *sum* of the infinitely thin rectangles making up this area. Most important of all, he observed that the two processes of differencing and summing—in our terminology, differentiating and integrating—are inverses of each other, and are linked together by means of the infinitesimal or differential triangle (dx, dy, ds) shown in the figure.

It was in a famous manuscript dated Oct. 29, 1675, that he first introduced the modern integral sign, a long letter S suggesting the first letter of the Latin word *summa* (sum). He was doing integrations by forming sums of Cavalieri's indivisibles, and he abbreviated "omnes lineae"—all lines—to "omn 1." He then remarked, "It

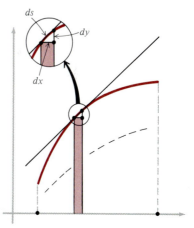

Figure B.10

---

* This dislike was perhaps natural, for George I was ridiculous and repulsive both as a man and as a monarch, and inferior men in positions of power often dislike superior men who happen to be their subordinates. Jonathan Swift called him "the ruling Yahoo," and a later British prime minister (Winston Churchill, in his role as historian) described him as "a humdrum German martinet with dull brains and coarse tastes."

† These notes were found in the Hanover Library in the middle of the nineteenth century. See J. M. Child (ed.), *The Early Mathematical Manuscripts of Leibniz,* Open Court, 1920.

‡ Hofmann, op. cit., pp. 47–48.

will be useful to write $\int$ for omn, thus $\int l$ for omn l, that is, the sum of those l's." On the same day he introduced the differential symbol d, and soon he was writing dx, dy, and dy/dx as we do today, as well as integrals like $\int y\, dy$ and $\int y\, dx$. All this time he was formulating and solving new problems and learning to use the machinery he was developing. Throughout his notes for these years he carried on a fascinating conversation with himself, filled with warnings, reminders, and efforts to cheer himself up, as well as occasional expressions of triumph.

Leibniz's first published account of his differential calculus was in a seven-page paper in the *Acta Eruditorum* of 1684.* The meaning of the differentials dx and dy is far from clear, and in fact he never did clarify this issue to anyone's satisfaction. He states the formulas $d(xy) = x\, dy + y\, dx$, $d(x/y) = (y\, dx - x\, dy)/y^2$, and $d(x^n) = nx^{n-1}\, dx$, but he makes no attempt to explain or justify them. As we know from his notes and letters, he thought of dx and dy as infinitely small increments, or "infinitesimals," and he derived these formulas by discarding infinitesimals of higher order, but none of this is in the paper.† However, he does give the condition $dy = 0$ for maxima and minima and $ddy = 0$ for points of inflection, and he makes several geometric applications. This paper was followed in 1686 by a second,‡ in which he casually introduces his integral symbol $\int$ with what amounts to no explanation at all and claims that $\int$ and d "are each other's converse." These early papers appear to be hastily and carelessly written, and are so unclear that they are barely intelligible. Even the Bernoulli brothers, who somehow understood Leibniz's intentions and realized that something profound was being born, spoke of these papers as presenting "an enigma rather than an explanation."

This early work of Leibniz, though obscure and fragmentary, was a fertile seed of great potential. It aroused no interest in Germany or England, but James and John Bernoulli of Basle found it exciting and richly suggestive. They eagerly absorbed Leibniz's ideas and methods and contributed many of their own; and before the end of the century these three men, in constant correspondence and stimulating one another like athletes in a race, had discovered much of the content of our modern college calculus courses. In fact, between 1695 and 1700 every monthly issue of the *Acta Eruditorum* contained at least one paper—and often several—by Leibniz or the Bernoulli brothers in which they treated, using notation almost identical with that used today, a great variety of problems in differential and integral calculus, differential equations, infinite series, and even the calculus of variations. In this headlong rush to exploit the wealth of applications of the new analysis, there was little interest in pausing to indulge in leisurely examinations of the basic ideas. This uncritical spirit prevailed throughout the eighteenth century, and it was not until the early decades of the nineteenth century that serious attention was given to the logical foundations of the subject.

In addition to the actual content of his work, Leibniz was also one of the greatest inventors of mathematical symbols. Few people have understood so well that a really good notation smooths the way and is almost capable of doing our thinking for us. He wrote about this to his friend Tschirnhaus as follows:

In symbols one observes an advantage in discovery which is greatest when they

---

* A translation is given in Struik, op. cit., pp. 271–280.

† Thus, for example, $d(xy) = (x + dx)(y + dy) - xy = x\, dy + y\, dx + dx\, dy$, and since dx and dy are infinitely small, the product $dx\, dy$ is infinitely infinitely small and—according to Leibniz—can be dropped, giving the correct formula $d(xy) = x\, dy + y\, dx$.

‡ Struik, op. cit., pp. 281–282.

express the exact nature of a thing briefly and, as it were, picture it; then indeed the labor of thought is wonderfully diminished.*

His flexible and suggestive calculus notations $dx, dy, dy/dx,$ and $\int y\, dx$ are perfect illustrations of this remark and are still in standard use, as are the English versions of his descriptive phrases "calculus differentialis" and "calculus integralis."† It was mainly through his influence that the symbol $=$ is universally used for equality, and he advocated the dot ($\cdot$) instead of the cross ($\times$) for multiplication.‡ His colon for division ($x:y$ for x/y) and his symbols for geometric similarity and congruence ($\sim$ and $\simeq$) are still widely used. He introduced the terms "constant," "variable," "parameter," and "transcendental" (in the sense of "nonalgebraic"), as well as "abscissa" and "ordinate," which together he called "coordinates." Also, it was he who first used the word "function" with essentially its modern meaning.

Leibniz is sometimes criticized for not producing any great work that can be pointed to and admired, like Newton's *Principia.* But he did produce such a work, even though it was not a book. The line of descent for all the greatest mathematicians of modern times begins with him — not with Newton — and extends in unbroken succession down to the twentieth century. He was the intellectual father of the Bernoullis; John Bernoulli was Euler's teacher; Euler adopted Lagrange as his scientific protégé; then came Gauss, Riemann, and the rest — all direct intellectual descendants of Leibniz. He had predecessors, of course, as every great thinker does. But apart from this, he was the true founder of modern European mathematics.

THE BERNOULLI BROTHERS

With justice we admire Huygens because he first discovered that a heavy particle falls down along a cycloid in the same time no matter from what point on the cycloid it begins its motion. But you will be petrified with astonishment when I say that precisely this cycloid, the tautochrone *of Huygens, is our required* brachistochrone.

John Bernoulli

Most people are aware that Johann Sebastian Bach was one of the greatest composers of all time. However, it is less well known that his prolific family was so consistently talented in this direction that several dozen Bachs were eminent musicians from the sixteenth to the nineteenth centuries. In fact, there were parts of Germany where the very word *bach* meant a musician. What the Bach clan was to music, the Bernoullis were to mathematics and science. In three generations this remarkable Swiss family

---

* See F. Cajori, *A History of Mathematical Notations,* Open Court, 1929, vol. II, p. 184. On pp. 180–196 and 201–205 Cajori gives a comprehensive discussion of Leibniz's use of mathematical symbols. See also the long quotation from A. N. Whitehead on pp. 332–333 ("By relieving the brain of all unnecessary work, a good notation sets it free to concentrate on more advanced problems, and in effect increases the mental power of the race. . . . Civilization advances by extending the number of important operations which we can perform without thinking about them.").

† Leibniz first suggested "calculus summatorius," but in 1696 he and John Bernoulli agreed on "calculus integralis."

‡ The equality sign was first introduced by the Englishman Robert Recorde in 1557, as follows: "To avoide the tediouse repetition of these woordes: is equalle to: I will sette as I doe often in woorke use, a paire of paralleles, or Gemowe [twin] lines of one lengthe, thus: ====, bicause noe .2. thynges, can be moare equalle."

produced eight mathematicians—two of them outstanding—who in turn had a swarm of descendants who distinguished themselves in many fields.* These two were the brothers James (1654–1705) and John (1667–1748), who played indispensable roles in the development of modern European mathematics.

At the insistence of their merchant father, James studied theology and John studied medicine. However, they found their true vocation when Leibniz's early papers of 1684 and 1686 were published in the *Acta Eruditorum.* They taught themselves the new calculus, entered into extensive correspondence with Leibniz, and became his most important students and disciples. James was professor of mathematics at Basel from 1687 until his death. John became a professor at Groningen in Holland in 1695, and on James's death succeeded his brother in the chair at Basel, where he flourished for another 43 years.

James was interested in infinite series, and among other things established the divergence of the sum of the reciprocals of the positive integers,

$$1 + \frac{1}{2} + \frac{1}{3} + \frac{1}{4} + \cdots,$$

and the convergence of the sum of the reciprocals of the squares,

$$1 + \frac{1}{4} + \frac{1}{9} + \frac{1}{16} + \cdots.$$

He often speculated about the sum of the latter series, but the question wasn't settled until 1736, when Euler discovered that this sum is $\pi^2/6$. James invented polar coordinates, studied many special curves (including the catenary, the tractrix, the lemniscate, and the exponential spiral), and introduced the Bernoulli numbers that appear in the power series expansion of the function tan x. It was he (in 1690) who first used the word "integral." In his book *Ars Conjectandi* (published posthumously in 1713) he formulated the basic principle in the theory of probability known as *Bernoulli's theorem* or the *law of large numbers*: If the probability of a certain event is p, and if n independent trials are made with k successes, then $k/n \to p$ as $n \to \infty$. At first sight this statement may seem to be a triviality, but beneath its surface lies a tangled thicket of philosophical (and mathematical) problems that have been a source of controversy down to the present day.

Like his older brother, John was fascinated by the almost magical power of Leibniz's calculus. He quickly mastered it and applied it to many problems in geometry, differential equations, and mechanics. Many of the ideas of Leibniz and the Bernoulli brothers were given wide circulation in 1696 in the first calculus textbook, *Analyse des infiniment petits,* by G. F. A. de l'Hospital (1661–1701). This man, a French nobleman and good amateur mathematician, openly acknowledged his debt to his teachers: "I have made free use of their discoveries, so that I frankly return to them whatever they please to claim as their own." L'Hospital's book is best known for its rule about indeterminate forms of the type 0/0. After l'Hospital's death John Bernoulli stated that much of the content of the book, and in particular this rule, was his own property. This minor mystery was cleared up in 1955 by the publication of the correspondence between the two men. L'Hospital was interested in mathematics but lacked confidence in his ability to learn calculus by himself. John was willing to tutor him, and in return for a yearly allowance agreed to sell him some of his own discoveries. The arrangement was discussed in a letter from l'Hospital of March 17, 1694, and the rule for 0/0 is contained in a letter from Bernoulli of July 22, 1694.†

The Bernoulli brothers sometimes worked on the same problems, which was unfortunate in view of their suspicious natures and surly dispositions. On occasion

---

* See Francis Galton, *Hereditary Genius,* Macmillan, 1892, pp. 195–196.
† See D. J. Struik's article in *The Mathematics Teacher,* vol. 56, 1963, pp. 257–260.

the friction between them flared up into a bitter and abusive public feud, as it did over the brachistochrone problem. In 1696 John proposed the problem (which is stated in our note on Newton) as a challenge to the mathematicians of Europe. It aroused great interest, and was solved by Newton and Leibniz as well as by the two Bernoullis. John's solution was more elegant, while James's—though rather clumsy and laborious—was more general. This situation started an acrimonious quarrel that dragged on for several years and was often conducted in rough language more suited to a street brawl than a scientific discussion.

After the deaths of his brother and Leibniz, John Bernoulli became the acknowledged leader of the Continental mathematicians in their battle against the English. He continued to produce good mathematical ideas and biting invective for many years, and was a major force in the ultimate triumph of Leibniz's calculus over the fluxions of Newton. Perhaps his greatest contribution was his student, the prodigious Euler, whose incredible flood of discoveries dominated mathematics through most of the eighteenth century.

EULER (1707–1783)

Read Euler: he is our master in everything.
            Pierre Simon de Laplace

Leonhard Euler was Switzerland's foremost scientist and one of the three greatest mathematicians of modern times (the other two being Gauss and Riemann).

He was perhaps the most prolific author of all time in any field. From 1727 to 1783 his writings poured out in a seemingly endless flood, constantly adding knowledge to every known branch of pure and applied mathematics, and also to many that were not known until he created them. He averaged about 800 printed pages a year throughout his long life, and yet he almost always had something worthwhile to say and never seems long-winded. The publication of his complete works was started in 1911, and the end is not in sight. This edition was planned to include 887 titles in 72 volumes, but since that time extensive new deposits of previously unknown manuscripts have been unearthed, and it is now estimated that more than 100 large volumes will be required for completion of the project. Euler evidently wrote mathematics with the ease and fluency of a skilled speaker discoursing on subjects with which he is intimately familiar. His writings are models of relaxed clarity. He never condensed, and he reveled in the rich abundance of his ideas and the vast scope of his interests. The French physicist Arago, in speaking of Euler's incomparable mathematical facility, remarked that "He calculated without apparent effort, as men breathe, or as eagles sustain themselves in the wind." He suffered total blindness during the last 17 years of his life, but with the aid of his powerful memory and fertile imagination, and with helpers to write his books and scientific papers from dictation, he actually increased his already prodigious output of work.

Euler was a native of Basel and a student of John Bernoulli at the University, but he soon outstripped his teacher. His working life was spent as a member of the Academies of Science at Berlin and St. Petersburg, and most of his papers were published in the journals of these organizations. His business was mathematical research, and he knew his business. He was also a man of broad culture, well versed in the classical languages and literatures (he knew the *Aeneid* by heart), many modern languages, physiology, medicine, botany, geography, and the entire body of physical science as it was known in his time. However, he had little talent for metaphysics or disputation, and came out second best in many good-natured verbal encounters with Voltaire at the court of Frederick the Great. His personal life was as placid and uneventful as is possible for a man with 13 children.

Though he was not himself a teacher, Euler has had a deeper influence on the teaching of mathematics than any other individual. This came about chiefly through his three great treatises: *Introductio in Analysin Infinitorum* (1748); *Institutiones Calculi Differentialis* (1755); and *Institutiones Calculi Integralis* (1768–1794). There is considerable truth in the old saying that all elementary and advanced calculus textbooks since 1748 are essentially copies of Euler or copies of copies of Euler.* These works summed up and codified the discoveries of his predecessors, and are full of Euler's own ideas. He extended and perfected plane and solid analytic geometry, introduced the analytic approach to trigonometry, and was responsible for the modern treatment of the functions $\ln x$ ($= \log_e x$) and e^x. He created a consistent theory of logarithms of negative and imaginary numbers, and discovered that $\ln x$ has an infinite number of values. It was through his work that the symbols e, π, and i ($= \sqrt{-1}$) became common currency for all mathematicians, and it was he who linked them together in the astonishing relation $e^{\pi i} = -1$. This is merely a special case (put $\theta = \pi$) of his famous formula $e^{i\theta} = \cos \theta + i \sin \theta$, which connects the exponential and trigonometric functions and is absolutely indispensable in higher analysis.† Among his other contributions to standard mathematical notation were $\sin x$, $\cos x$, the use of $f(x)$ for an unspecified function, and the use of Σ for summation.‡ Good notations are important, but the ideas behind them are what really count, and in this respect, Euler's fertility was almost beyond belief. He preferred concrete special problems to the general theories in vogue today, and his unique insight into the connections between apparently unrelated formulas blazed many trails into new areas of mathematics which he left for his successors to cultivate.

He was the first and greatest master of infinite series, infinite products, and continued fractions, and his works are crammed with striking discoveries in these fields. James Bernoulli (John's older brother) found the sums of several infinite series, but he was not able to find the sum of the reciprocals of the squares, $1 + \frac{1}{4} + \frac{1}{9} + \frac{1}{16} + \cdots$. He wrote, "If someone should succeed in finding this sum, and will tell me about it, I shall be much obliged to him." In 1736, long after James's death, Euler made the wonderful discovery that

$$1 + \frac{1}{4} + \frac{1}{9} + \frac{1}{16} + \cdots = \frac{\pi^2}{6}.$$

He also found the sums of the reciprocals of the fourth and sixth powers,

$$1 + \frac{1}{2^4} + \frac{1}{3^4} + \cdots = 1 + \frac{1}{16} + \frac{1}{81} + \cdots = \frac{\pi^4}{90}$$

and

$$1 + \frac{1}{2^6} + \frac{1}{3^6} + \cdots = 1 + \frac{1}{64} + \frac{1}{729} + \cdots = \frac{\pi^6}{945}.$$

---

* See C. B. Boyer, "The Foremost Textbook of Modern Times," *Amer. Math. Monthly,* vol. 58, 1951, pp. 223–226.

† An even more astonishing consequence of his formula is the fact that an imaginary power of an imaginary number can be real, in particular $i^i = e^{-\pi/2}$; for if we put $\theta = \pi/2$, we obtain $e^{\pi i/2} = i$, so

$$i^i = (e^{\pi i/2})^i = e^{\pi i^2/2} = e^{-\pi/2}.$$

Euler further showed that i^i has infinitely many values, of which this calculation produces only one.

‡ See F. Cajori, *A History of Mathematical Notations,* Open Court, 1929.

When John heard about these feats, he wrote, "If only my brother were alive now."* Few would believe that these formulas are related — as they are — to Wallis's infinite product (1656),

$$\frac{\pi}{2} = \frac{2}{1} \cdot \frac{2}{3} \cdot \frac{4}{3} \cdot \frac{4}{5} \cdot \frac{6}{5} \cdot \frac{6}{7} \cdots .$$

Euler was the first to explain this in a satisfactory way, in terms of his infinite product expansion of the sine,

$$\frac{\sin x}{x} = \left(1 - \frac{x^2}{\pi^2}\right)\left(1 - \frac{x^2}{4\pi^2}\right)\left(1 - \frac{x^2}{9\pi^2}\right) \cdots .$$

Wallis's product is also related to Brouncker's remarkable continued fraction,

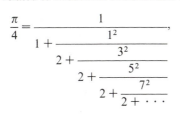

$$\frac{\pi}{4} = \cfrac{1}{1 + \cfrac{1^2}{2 + \cfrac{3^2}{2 + \cfrac{5^2}{2 + \cfrac{7^2}{2 + \cdots}}}}},$$

which became understandable only in the context of Euler's extensive researches in this field.†

His work in all departments of analysis strongly influenced the further development of this subject through the next two centuries. He contributed many important ideas to differential equations, including substantial parts of the theory of second-order linear equations and the method of solution by power series. He gave the first systematic discussion of the calculus of variations, which he founded on his basic differential equation for a minimizing curve. He introduced the number now known as *Euler's constant,*

$$\gamma = \lim_{n \to \infty} \left(1 + \frac{1}{2} + \frac{1}{3} + \cdots + \frac{1}{n} - \ln n\right) = 0.5772 \ldots ,$$

which is the most important special number in mathematics after π and e. He discovered the integral defining the gamma function,

$$\Gamma(x) = \int_0^\infty t^{x-1} e^{-t} \, dt,$$

which is often the first of the so-called higher transcendental functions that students meet beyond the level of calculus, and he developed many of its applications and special properties.‡ He also worked with Fourier series, encountered the Bessel functions in his study of the vibrations of a stretched circular membrane, and applied Laplace transforms to solve differential equations — all before Fourier, Bessel, and Laplace were born. Even though Euler died about 200 years ago, he lives everywhere in analysis.

E. T. Bell, the well-known historian of mathematics, observed that "One of the most remarkable features of Euler's universal genius was its equal strength in both of

---

* The world is still waiting — more than 200 years later — for someone to discover the sum of the reciprocals of the cubes.

† The ideas described in this paragraph are explained more fully in Appendix A.18.

‡ A few of the simpler properties of the gamma function are discussed on pp. 234–237 of G. F. Simmons, *Differential Equations,* McGraw-Hill, 1972.

the main currents of mathematics, the continuous and the discrete." In the realm of the discrete, he was one of the originators of number theory and made many far-reaching contributions to this subject throughout his life. In addition, the origins of topology — one of the dominant forces in modern mathematics — lie in his solution of the Königsberg bridge problem and his formula $V - E + F = 2$ connecting the numbers of vertices, edges, and faces of a simple polyhedron. In the following paragraphs, we briefly describe his activities in these fields.

In number theory, Euler drew much of his inspiration from the challenging marginal notes left by Fermat in his copy of the works of Diophantus, and some of his achievements are mentioned in our account of Fermat. He also initiated the theory of partitions, a little-known branch of number theory that turned out much later to have applications in statistical mechanics and the kinetic theory of gases. A typical problem of this subject is to determine the number $p(n)$ of ways in which a given positive integer n can be expressed as a sum of positive integers, and if possible to discover some properties of this function. For example, 4 can be partitioned into $4 = 3 + 1 = 2 + 2 = 2 + 1 + 1 = 1 + 1 + 1 + 1$, so $p(4) = 5$, and similarly $p(5) = 7$ and $p(6) = 11$. It is clear that $p(n)$ increases very rapidly with n, so rapidly, in fact, that*

$$p(200) = 3,972,999,029,388.$$

Euler began his investigations by noticing (only geniuses notice such things) that $p(n)$ is the coefficient of x^n when the function $[(1 - x)(1 - x^2)(1 - x^3) \cdots]^{-1}$ is expanded in a power series:

$$\frac{1}{(1 - x)(1 - x^2)(1 - x^3) \cdots} = 1 + p(1)x + p(2)x^2 + p(3)x^3 + \cdots.$$

By building on this foundation, he derived many other remarkable identities related to a variety of problems about partitions.†

The Königsberg bridge problem originated as a pastime of Sunday strollers in the town of Königsberg (now Kaliningrad) in what was formerly East Prussia. There were seven bridges across the river that flows through the town (see Fig. B.11). The residents used to enjoy walking from one bank to the islands and then to the other bank and back again, and the conviction was widely held that it is impossible to do this by crossing all seven bridges without crossing any bridge more than once. Euler analyzed the problem by examining the schematic diagram given on the right in the figure, in which the land areas are represented by points and the bridges by lines connecting these points. The points are called vertices, and a vertex is said to be odd or even according as the number of lines leading to it is odd or even. In modern terminology, the entire configuration is called a *graph,* and a path through the graph that traverses every line but no line more than once is called an *Euler path.* An Euler path need not end at the vertex where it began, but if it does, it is called an *Euler*

---

* This evaluation required a month's work by a skilled computer in 1918. His motive was to check an approximate formula for $p(n)$, namely

$$p(n) \cong \frac{1}{4n\sqrt{3}} e^{\pi\sqrt{2n/3}}$$

(the error was extremely small).

† See Chapter XIX of G. H. Hardy and E. M. Wright, *An Introduction to the Theory of Numbers,* Oxford, 1938; or Chapters 12–14 of G. E. Andrews, *Number Theory,* W. B. Saunders, 1971. These treatments are "elementary" in the technical sense that they do not use the high-powered machinery of advanced analysis, but nevertheless they are far from simple. For students who wish to experience some of Euler's most interesting work in number theory at first hand, and in a context not requiring much previous knowledge, we recommend Chapter VI of G. Polya's fine book, *Induction and Analogy in Mathematics,* Princeton, 1954.

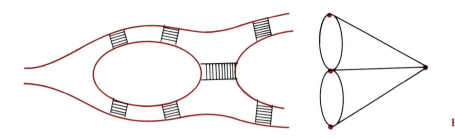

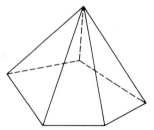

Figure B.11   The Königsberg bridges.

circuit. By the use of combinatorial reasoning, Euler arrived at the following theorems about any such graph: (1) there are an even number of odd vertices; (2) if there are no odd vertices, there is an Euler circuit starting at any point; (3) if there are two odd vertices, there is no Euler circuit, but there is an Euler path starting at one odd vertex and ending at the other; (4) if there are more than two odd vertices, there are no Euler paths.* The graph of the Königsberg bridges has four odd vertices, and therefore, by the last theorem, has no Euler paths.† The branch of mathematics that has developed from these ideas is known as *graph theory*; it has applications to chemical bonding, economics, psychosociology, the properties of networks of roads and railroads, and other subjects.

In our note on Euclid we remarked that a polyhedron is a solid whose surface consists of a number of polygonal faces, and Fig. B.2 of that note displays the five regular polyhedra. The Greeks studied these figures assiduously, but it remained for Euler to discover the simplest of their common properties: If V, E, and F denote the numbers of vertices, edges, and faces of any one of them, then in every case we have

$$V - E + F = 2.$$

This fact is known as *Euler's formula for polyhedra*, and it is easy to verify from the data summarized in the following table.

	V	E	F
Tetrahedron	4	6	4
Cube	8	12	6
Octahedron	6	12	8
Dodecahedron	20	30	12
Icosahedron	12	30	20

This formula is also valid for any irregular polyhedron as long as it is *simple*—which means that it has no "holes" in it, so that its surface can be deformed continuously into the surface of a sphere. Figure B.12 shows two simple irregular polyhedra for which $V - E + F = 6 - 10 + 6 = 2$ and $V - E + F = 6 - 9 + 5 = 2$. However, Euler's formula must be extended to

$$V - E + F = 2 - 2p$$

in the case of a polyhedron with p holes (a simple polyhedron is one for which $p = 0$). Figure B.13 illustrates the cases $p = 1$ and $p = 2$; here we have $V - E + F = 16 - 32 + 16 = 0$ when $p = 1$, and $V - E + F = 24 - 44 + 18 = -2$ when $p = 2$. The

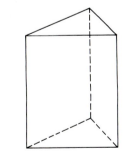

Figure B.12

---

* Euler's original paper of 1736 is interesting to read and easy to understand; it can be found on pp. 573–580 of J. R. Newman (ed.), *The World of Mathematics,* Simon and Schuster, 1956.

† It is easy to see—without appealing to any theorems—that this graph contains no Euler circuit, for if there were such a circuit, it would have to enter each vertex as many times as it leaves it, and therefore every vertex would have to be even. Similar reasoning shows also that if there were an Euler path that is not a circuit, there would have to be two odd vertices.

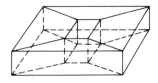 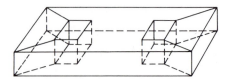

Figure B.13

significance of these ideas can best be understood by imagining a polyhedron to be a hollow figure with a surface made of thin rubber, and inflating it until it becomes smooth. We no longer have flat faces and straight edges, but instead a map on the surface consisting of curved regions, their boundaries, and points where boundaries meet. The number $V - E + F$ has the same value for all maps on our surface, and is called the *Euler characteristic* of this surface. The number p is called the *genus* of the surface. These two numbers, and the relation between them given by the equation $V - E + F = 2 - 2p$, are evidently unchanged when the surface is continuously deformed by stretching or bending. Intrinsic geometric properties of this kind — which have little connection with the type of geometry concerned with lengths, angles, and areas — are called *topological.* The serious study of such topological properties has greatly increased during the past century, and has furnished valuable insights to many branches of mathematics and science.*

The distinction between pure and applied mathematics did not exist in Euler's day, and for him the entire physical universe was a convenient object whose diverse phenomena offered scope for his methods of analysis. The foundations of classical mechanics had been laid down by Newton, but Euler was the principal architect. In his treatise of 1736 he was the first to explicitly introduce the concept of a mass-point or particle, and he was also the first to study the acceleration of a particle moving along any curve and to use the notion of a vector in connection with velocity and acceleration. His continued successes in mathematical physics were so numerous, and his influence was so pervasive, that most of his discoveries are not credited to him at all and are taken for granted by physicists as part of the natural order of things. However, we do have Euler's equations of motion for the rotation of a rigid body, Euler's hydrodynamic equation for the flow of an ideal incompressible fluid, Euler's law for the bending of elastic beams, and Euler's critical load in the theory of the buckling of columns. On several occasions the thread of his scientific thought led him to ideas his contemporaries were not ready to assimilate. For example, he foresaw the phenomenon of radiation pressure, which is crucial for the modern theory of the stability of stars, more than a century before Maxwell rediscovered it in his own work on electromagnetism.

Euler was the Shakespeare of mathematics — universal, richly detailed, and inexhaustible.†

---

* Proofs of Euler's formula and its extension are given on pp. 236–240 and 256–259 of R. Courant and H. Robbins, *What Is Mathematics?*, Oxford, 1941. See also G. Polya, op. cit., pp. 35–43.

† For further information, see C. Truesdell, "Leonhard Euler, Supreme Geometer (1707–1783)," in *Studies in Eighteenth-Century Culture,* Case Western Reserve University Press, 1972. Also, the November 1983 issue of *Mathematics Magazine* is wholly devoted to Euler and his work.

LAGRANGE (1736–1813)

*The "generalized coordinates" of our mechanics of today
were conceived and installed by Lagrange, and this was an
achievement of unmatchable magnitude.*

Salomon Bochner

Joseph Louis Lagrange detested geometry, but made outstanding discoveries in the calculus of variations and analytical mechanics. He also contributed to number theory and algebra, and fed the stream of thought that later nourished Gauss and Abel. His mathematical career can be viewed as a natural extension of the work of his older and greater contemporary, Euler, which in many respects he carried forward and refined.

Lagrange was born in Turin of mixed French-Italian ancestry. As a boy, his tastes were more classical than scientific, but his interest in mathematics was kindled while he was still in school by reading a paper by Edmund Halley on the uses of algebra in optics. He then began a course of independent study, and progressed so rapidly that at the age of 19 he was appointed professor at the Royal Artillery School in Turin.*

Lagrange's contributions to the calculus of variations were among his earliest and most important works. In 1755 he communicated to Euler his method of multipliers for solving isoperimetric problems. These problems had baffled Euler for years, since they lay beyond the reach of his own semigeometrical techniques. Euler was immediately able to answer many questions he had long contemplated, but he replied to Lagrange with admirable kindness and generosity, and withheld his own work from publication "so as not to deprive you of any part of the glory which is your due." Lagrange continued working for a number of years on his analytic version of the calculus of variations, and both he and Euler applied it to many new types of problems, especially in mechanics.

In 1766, when Euler left Berlin for St. Petersburg, he suggested to Frederick the Great that Lagrange be invited to take his place. Lagrange accepted and lived in Berlin for 20 years until Frederick's death in 1786. During this period he worked extensively in algebra and number theory and wrote his masterpiece, the treatise *Mécanique Analytique* (1788), in which he unified general mechanics and made of it, as Hamilton later said, "a kind of scientific poem." Among the enduring legacies of this work are Lagrange's equations of motion, generalized coordinates, and the concept of potential energy.

Men of science found the atmosphere of the Prussian court rather uncongenial after the death of Frederick, so Lagrange accepted an invitation from Louis XVI to move to Paris, where he was given apartments in the Louvre. Lagrange was extremely modest and undogmatic for a man of his great gifts; and though he was a friend of aristocrats—and indeed an aristocrat himself—he was respected and held in affection by all parties throughout the turmoil of the French Revolution. His most important work during these years was his leading part in establishing the metric system of weights and measures. In mathematics, he tried to provide a satisfactory foundation for the basic processes of analysis, but these efforts were largely abortive. Toward the end of his life, Lagrange felt that mathematics had reached a dead end, and that chemistry, physics, biology, and other sciences would attract the ablest minds of the future. His pessimism might have been relieved if he had been able to foresee the coming of Gauss and his successors, who made the nineteenth century the richest in the long history of mathematics.

---

* See George Sarton's valuable essay, "Lagrange's Personality," *Proc. Am. Phil. Soc.,* vol. 88, 1944, pp. 457–496.

LAPLACE (1749–1827)

*Laplace is the great example of the wisdom of directing all of
one's efforts to a single central objective worthy of the best
that a man has in him.*

E. T. Bell

Pierre Simon de Laplace was a French mathematician and theoretical astronomer
who was so famous in his own time that he was known as the Newton of France. His
main interests throughout his life were celestial mechanics, the theory of probability,
and personal advancement.

At the age of 24 he was already deeply engaged in the detailed application of
Newton's law of gravitation to the solar system as a whole, in which the planets and
their satellites are not governed by the sun alone, but interact with one another in a
bewildering variety of ways. Even Newton had been of the opinion that divine
intervention would occasionally be needed to prevent this complex mechanism from
degenerating into chaos. Laplace decided to seek reassurance elsewhere, and succeeded in proving that the ideal solar system of mathematics is a stable dynamical
system that will endure unchanged for all time. This achievement was only one of the
long series of triumphs recorded in his monumental treatise *Mécanique Céleste*
(published in five volumes from 1799 to 1825), which summed up the work on
gravitation of several generations of illustrious mathematicians. Unfortunately for
his later reputation, he omitted all reference to the discoveries of his predecessors and
contemporaries, and left it to be inferred that the ideas were entirely his own. Many
anecdotes are associated with this work. One of the best known describes the occasion
on which Napoleon tried to get a rise out of Laplace by protesting that he had written
a huge book on the system of the world without once mentioning God as the author
of the universe. Laplace is supposed to have replied, "Sire, I had no need of that
hypothesis." The principal legacy of the *Mécanique Céleste* to later generations lay in
Laplace's wholesale development of potential theory, with its far-reaching implications for a dozen different branches of physical science ranging from gravitation and
fluid mechanics to electromagnetism and atomic physics. Even though he lifted the
idea of the potential from Lagrange without acknowledgment, he exploited it so
extensively that ever since his time the fundamental differential equation of potential
theory has been known as Laplace's equation.

His other masterpiece was the treatise *Théorie Analytique des Probabilités* (1812),
in which he incorporated his own discoveries in probability from the preceding 40
years. Again he failed to acknowledge the many ideas of others he mixed in with his
own; but even discounting this, his book is generally agreed to be the greatest contribution to this part of mathematics by any one person. In the introduction he says, "At
bottom, the theory of probability is only common sense reduced to calculation." This
may be so, but the following 700 pages of intricate analysis—in which he freely used
Laplace transforms, generating functions, and many other highly nontrivial tools—
has been said by some to surpass in complexity even the *Mécanique Céleste.*

After the French Revolution, Laplace's political talents and greed for position
came to full flower. His compatriots speak ironically of his "suppleness" and "versatility" as a politician. What this really means is that each time there was a change of
regime (and there were many), Laplace smoothly adapted himself by changing his
principles—back and forth between fervent republicanism and fawning royalism—
and each time he emerged with a better job and grander titles. He has been aptly
compared with the apocryphal Vicar of Bray in English literature, who was twice a
Catholic and twice a Protestant. The Vicar is said to have replied as follows to the

charge of being a turncoat: "Not so, neither, for if I changed my religion, I am sure I kept true to my principle, which is to live and die the Vicar of Bray."

To balance his faults, Laplace was always generous in giving assistance and encouragement to younger scientists. From time to time he helped forward in their careers such men as the chemist Gay-Lussac, the traveler and naturalist Humboldt, the physicist Poisson, and—appropriately—the young Cauchy, who was destined to become one of the chief architects of nineteenth century mathematics.

FOURIER (1768–1830)

The profound study of nature is the most fruitful source of mathematical discoveries.

Joseph Fourier

Jean Baptiste Joseph Fourier, an excellent mathematical physicist, was a friend of Napoleon (so far as such people have friends) and accompanied his master to Egypt in 1798. On his return he became prefect of the district of Isère in southeastern France, and in this capacity built the first real road from Grenoble to Turin. He also befriended the boy Champollion, who later deciphered the Rosetta Stone.

During these years he worked on the theory of the conduction of heat, and in 1822 published his famous *Théorie Analytique de la Chaleur,* in which he made extensive use of the series that now bear his name. These series were of profound significance in connection with the evolution of the concept of a function. The general attitude at that time was to call $f(x)$ a function if it could be represented by a single expression like a polynomial, a finite combination of elementary functions, a power series $\sum_{n=0}^{\infty} a_n x^n$, or a trigonometric series of the form

$$\sum_{n=0}^{\infty} (a_n \cos nx + b_n \sin nx).$$

If the graph of $f(x)$ were "arbitrary" — for example, a polygonal line with a number of corners and even a few gaps — then $f(x)$ would not have been accepted as a genuine function. Fourier claimed that "arbitrary" graphs can be represented by trigonometric series and should therefore be treated as legitimate functions, and it came as a shock to many that he turned out to be right. It was a long time before these issues were completely clarified, and it was no accident that the definition of a function that is now almost universally used was first formulated by Dirichlet in 1837 in a research paper on the theory of Fourier series. Also, the classical definition of the definite integral due to Riemann was first given in his fundamental paper of 1854 on the subject of Fourier series. Indeed, many of the most important mathematical discoveries of the nineteenth century are directly linked to the theory of Fourier series, and the applications of this subject to mathematical physics have been scarcely less profound.

Fourier himself is one of the lucky few whose name has become rooted in all civilized languages as an adjective that is well known to physical scientists and mathematicians in every part of the world.

GAUSS (1777–1855)

The name of Gauss is linked to almost everything that the mathematics of our century [the nineteenth] has brought forth in the way of original scientific ideas.

L. Kronecker

Carl Friedrich Gauss was the greatest of all mathematicians and perhaps the most richly gifted genius of whom there is any record. This gigantic figure, towering at the beginning of the nineteenth century, separates the modern era in mathematics from all that went before. His visionary insight and originality, the extraordinary range and depth of his achievements, his repeated demonstrations of almost superhuman power and tenacity—all these qualities combined in a single individual present an enigma as baffling to us as it was to his contemporaries.

Gauss was born in the city of Brunswick in northern Germany. His exceptional skill with numbers was clear at a very early age, and in later life he joked that he knew how to count before he could talk. It is said that Goethe wrote and directed little plays for a puppet theater when he was 6 and that Mozart composed his first childish minuets when he was 5, but Gauss corrected an error in his father's payroll accounts at the age of 3.* His father was a gardener and bricklayer without either the means or the inclination to help develop the talents of his son. Fortunately, however, Gauss's remarkable abilities in mental computation attracted the interest of several influential men in the community, and eventually brought him to the attention of the Duke of Brunswick. The Duke was impressed with the boy and undertook to support his further education, first at the Caroline College in Brunswick (1792–1795) and later at the University of Göttingen (1795–1798).

At the Caroline College, Gauss completed his mastery of the classical languages and explored the works of Newton, Euler, and Lagrange. Early in this period— perhaps at the age of 14 or 15—he discovered the prime number theorem, which was finally proved in 1896 after great efforts by many mathematicians (see our note on Riemann). He also invented the method of least squares for minimizing the errors inherent in observational data, and conceived the Gaussian (or normal) law of distribution in the theory of probability.

At the university, Gauss was attracted by philology but repelled by the mathematics courses, and for a time the direction of his future was uncertain. However, at the age of 18 he made a wonderful geometric discovery that caused him to decide in favor of mathematics and gave him great pleasure to the end of his life. The ancient Greeks had known ruler-and-compass constructions for regular polygons of 3, 4, 5, and 15 sides, and for all others obtainable from these by bisecting angles. But this was all, and there the matter rested for 2000 years, until Gauss solved the problem completely. He proved that a regular polygon with n sides is constructible if and only if n is the product of a power of 2 and distinct prime numbers of the form $p_k = 2^{2^k} + 1$. In particular, when $k = 0, 1, 2, 3$, we see that each of the corresponding numbers $p_k = 3$, 5, 17, 257 is prime, so regular polygons with these numbers of sides are constructible.†

During these years Gauss was almost overwhelmed by the torrent of ideas which flooded his mind. He began the brief notes of his scientific diary in an effort to record

---

* See W. Sartorius von Waltershausen, *Gauss zum Gedächtniss.* These personal recollections appeared in 1856, and a translation by Helen W. Gauss (the mathematician's great-granddaughter) was privately printed in Colorado Springs in 1966.

† Details of some of these constructions are given in H. Tietze, *Famous Problems of Mathematics,* Chapter IX, Graylock, Baltimore, 1965.

his discoveries, since there were far too many to work out in detail at that time. The first entry, dated March 30, 1796, states the constructibility of the regular polygon with 17 sides, but even earlier than this he was penetrating deeply into several unexplored continents in the theory of numbers. In 1795 he discovered the law of quadratic reciprocity, and as he later wrote, "For a whole year this theorem tormented me and absorbed my greatest efforts, until at last I found a proof."* At that time Gauss was unaware that the theorem had already been imperfectly stated without proof by Euler, and correctly stated with an incorrect proof by Legendre. It is the core of the central part of his famous treatise *Disquisitiones Arithmeticae,* which was published in 1801 although completed in 1798.† Apart from a few fragmentary results of earlier mathematicians, this great work was wholly original. It is usually considered to mark the true beginning of modern number theory, to which it is related in much the same way as Newton's *Principia* is to physics and astronomy. In the introductory pages Gauss develops his method of congruences for the study of divisibility problems and gives the first proof of the fundamental theorem of arithmetic (also called the unique factorization theorem), which asserts that every integer $n > 1$ can be expressed uniquely as a product of primes. The central part is devoted mainly to quadratic congruences, forms, and residues. The last section presents his complete theory of the cyclotomic (circle-dividing) equation, with its applications to the constructibility of regular polygons. The entire work was a gargantuan feast of pure mathematics, which his successors were able to digest only slowly and with difficulty.

In his *Disquisitiones* Gauss also created the modern rigorous approach to mathematics. He had become thoroughly impatient with the loose writing and sloppy proofs of his predecessors, and resolved that his own works would be beyond criticism in this respect. As he wrote to a friend, "I mean the word proof not in the sense of the lawyers, who set two half proofs equal to a whole one, but in the sense of the mathematician, where $\frac{1}{2}$ proof $= 0$ and it is demanded for proof that every doubt becomes impossible." The *Disquisitiones* was composed in this spirit and in Gauss's mature style, which is terse, rigorous, devoid of motivation, and in many places so carefully polished that it is almost unintelligible. In another letter he said, "You know that I write slowly. This is chiefly because I am never satisfied until I have said as much as possible in a few words, and writing briefly takes far more time than writing at length." One of the effects of this habit is that his publications concealed almost as much as they revealed, for he worked very hard at removing every trace of the train of thought that led him to his discoveries. Abel remarked, "He is like the fox, who effaces his tracks in the sand with his tail." Gauss replied to such criticisms by saying that no self-respecting architect leaves the scaffolding in place after completing his building. Nevertheless, the difficulty of reading his works greatly hindered the diffusion of his ideas.

Gauss's doctoral dissertation (1799) was another milestone in the history of mathematics. After several abortive attempts by earlier mathematicians—d'Alembert, Euler, Lagrange, Laplace—the fundamental theorem of algebra was here given its first satisfactory proof. This theorem asserts the existence of a real or complex root for any polynominal equation with real or complex coefficients. Gauss's success inaugurated the age of existence proofs, which ever since have played an important part in pure mathematics. Furthermore, in this first proof (he gave four altogether), Gauss

---

* See D. E. Smith, *A Source Book in Mathematics,* McGraw-Hill, New York, 1929, pp. 112–118. This selection includes a statement of the theorem and the fifth of eight proofs that Gauss found over a period of many years. There are probably more than 50 known today.

† There is a translation by Arthur A. Clarke (Yale University Press, New Haven, CT, 1966).

appears as the earliest mathematician to use complex numbers and the geometry of the complex plane with complete confidence.*

The next period of Gauss's life was heavily weighted toward applied mathematics, and with a few exceptions the great wealth of ideas in his diary and notebooks lay in suspended animation.

In the last decades of the eighteenth century, many astronomers were searching for a new planet between the orbits of Mars and Jupiter, where Bode's law (1772) suggested that there ought to be one. The first and largest of the numerous minor planets known as asteroids was discovered in that region in 1801, and was named Ceres. This discovery ironically coincided with an astonishing publication by the philosopher Hegel, who jeered at astronomers for ignoring philosophy: This science (he said) could have saved them from wasting their efforts by demonstrating that no new planet could possibly exist.† Hegel continued his career in a similar vein, and later rose to even greater heights of clumsy obfuscation. Unfortunately the tiny new planet was difficult to see under the best of circumstances, and it was soon lost in the light of the sky near the sun. The sparse observational data made it difficult to calculate the orbit with sufficient accuracy to locate Ceres again after it had moved away from the sun. The astronomers of Europe attempted this task without success for many months. Finally, Gauss was attracted by the challenge, and with the aid of his method of least squares and his unparalleled skill at numerical computation he determined the orbit and told the astronomers where to look with their telescopes, and there it was. He had succeeded in rediscovering Ceres after all the experts had failed.

This achievement brought him fame, an increase in his pension from the Duke, and in 1807 an appointment as professor of astronomy and first director of the new observatory at Göttingen. He carried out his duties with his customary thoroughness, but, as it turned out, he disliked administrative chores, committee meetings, and all the tedious red tape involved in the business of being a professor. He also had little enthusiasm for teaching, which he regarded as a waste of his time and as essentially useless (for different reasons) for both talented and untalented students. However, when teaching was unavoidable, he apparently did it superbly. One of his students was the eminent algebraist Richard Dedekind, for whom Gauss's lectures after the passage of 50 years remained "unforgettable in memory as among the finest which I have ever heard."‡ Gauss had many opportunities to leave Göttingen, but he refused all offers and remained there for the rest of his life, living quietly and simply, traveling rarely, and working with immense energy on a wide variety of problems in mathematics and its applications. Apart from science and his family—he had two wives and six children, two of whom emigrated to America—his main interests were history and world literature, international politics, and public finance. He owned a large library of about 6000 volumes in many languages, including Greek, Latin, English, French, Russian, Danish, and of course German. His acuteness in handling his own financial affairs is shown by the fact that although he started with virtually nothing, he left an estate over a hundred times as great as his average annual income during the last half of his life.

---

* The idea of this proof is very clearly explained by F. Klein, *Elementary Mathematics from an Advanced Standpoint,* Dover, New York, 1945, pp. 101–104.

† See the last few pages of "De Orbitis Planetarum," vol. 1 of Georg Wilhelm Hegel, *Sämtliche Werke,* Frommann Verlag, Stuttgart, 1965.

‡ Dedekind's detailed recollections of this course are given in G. Waldo Dunnington, *Carl Friedrich Gauss: Titan of Science,* Hafner, New York, 1955, pp. 259–261. This book is useful mainly for its many quotations, its bibliography of Gauss's publications, and its list of the courses he offered (but often did not teach) from 1808 to 1854.

In the first two decades of the nineteenth century Gauss produced a steady stream of works on astronomical subjects, of which the most important was the treatise *Theoria Motus Corporum Coelestium* (1809). This remained the bible of planetary astronomers for over a century. Its methods for dealing with perturbations later led to the discovery of Neptune. Gauss thought of astronomy as his profession and pure mathematics as his recreation, and from time to time he published a few of the fruits of his private research. His great work on the hypergeometric series (1812) belongs to this period. This was a typical Gaussian effort, packed with new ideas in analysis that have kept mathematicians busy ever since.

Around 1820 he was asked by the government of Hanover to supervise a geodetic survey of the kingdom, and various aspects of this task — including extensive field work and many tedious triangulations — occupied him for a number of years. It is natural to suppose that a mind like his would have been wasted on such an assignment, but the great ideas of science are born in many strange ways. These apparently unrewarding labors resulted in one of his deepest and most far-reaching contributions to pure mathematics, without which Einstein's general theory of relativity would have been quite impossible.

Gauss's geodetic work was concerned with the precise measurement of large triangles on the earth's surface. This provided the stimulus that led him to the ideas of his paper *Disquisitiones generales circa superficies curvas* (1827), in which he founded the intrinsic differential geometry of general curved surfaces.* In this work he introduced curvilinear coordinates u and v on a surface; he obtained the fundamental quadratic differential form $ds^2 = E\,du^2 + 2F\,du\,dv + G\,dv^2$ for the element of arc length ds, which makes it possible to determine geodesic curves; and he formulated the concepts of Gaussian curvature and integral curvature.† His main specific results were the famous *theorema egregium*, which states that the Gaussian curvature depends only on E, F, and G, and is therefore invariant under bending; and the Gauss-Bonnet theorem on integral curvature for the case of a geodesic triangle, which in its general form is the central fact of modern differential geometry in the large. Apart from his detailed discoveries, the crux of Gauss's insight lies in the word *intrinsic*, for he showed how to study the geometry of a surface by operating only on the surface itself and paying no attention to the surrounding space in which it lies. To make this more concrete, let us imagine an intelligent two-dimensional creature who inhabits a surface but has no awareness of a third dimension or of anything not on the surface. If this creature is capable of moving about, measuring distances along the surface, and determining the shortest path (geodesic) from one point to another, then he is also capable of measuring the Gaussian curvature at any point and of creating a rich geometry on the surface — and this geometry will be Euclidean (flat) if and only if the Gaussian curvature is everywhere zero. When these conceptions are generalized to more than two dimensions, they open the door to Riemannian geometry, tensor analysis, and the ideas of Einstein.

Another great work of this period was his 1831 paper on biquadratic residues. Here he extended some of his early discoveries in number theory with the aid of a new method, his purely algebraic approach to complex numbers. He defined these numbers as ordered pairs of real numbers with suitable definitions for the algebraic operations, and in so doing laid to rest the confusion that still surrounded the subject and prepared the way for the later algebra and geometry of n-dimensional spaces. But this was only incidental to his main purpose, which was to broaden the ideas of

---

* A translation by A. Hiltebeitel and J. Morehead was published under the title *General Investigations of Curved Surfaces* by the Raven Press, Hewlett, New York, in 1965.

† These ideas are explained in nontechnical language in C. Lanczos, *Albert Einstein and the Cosmic World Order*, Interscience-Wiley, New York, 1965, chap. 4.

number theory into the complex domain. He defined complex integers (now called Gaussian integers) as complex numbers $a + ib$ with a and b ordinary integers; he introduced a new concept of prime numbers, in which 3 remains prime but $5 = (1 + 2i)(1 - 2i)$ does not; and he proved the unique factorization theorem for these integers and primes. The ideas of this paper inaugurated algebraic number theory, which has grown steadily from that day to this.*

From the 1830s on, Gauss was increasingly occupied with physics, and he enriched every branch of the subject he touched. In the theory of surface tension, he developed the fundamental idea of conservation of energy and solved the earliest problem in the calculus of variations involving a double integral with variable limits. In optics, he introduced the concept of the focal length of a system of lenses and invented the Gauss wide-angle lens (which is relatively free of chromatic aberration) for telescope and camera objectives. He virtually created the science of geomagnetism, and in collaboration with his friend and colleague Wilhelm Weber he built and operated an iron-free magnetic observatory, founded the Magnetic Union for collecting and publishing observations from many places in the world, and invented the electromagnetic telegraph and the bifilar magnetometer. There are many references to his work in James Clerk Maxwell's famous *Treatise on Electricity and Magnetism* (1873). In his preface, Maxwell says that Gauss "brought his powerful intellect to bear on the theory of magnetism and on the methods of observing it, and he not only added greatly to our knowledge of the theory of attractions, but reconstructed the whole of magnetic science as regards the instruments used, the methods of observation, and the calculation of results, so that his memoirs on Terrestrial Magnetism may be taken as models of physical research by all those who are engaged in the measurement of any of the forces in nature." In 1839 Gauss published his fundamental paper on the general theory of inverse square forces, which established potential theory as a coherent branch of mathematics.† As usual, he had been thinking about these matters for many years, and among his discoveries were the divergence theorem (also called Gauss's theorem) of modern vector analysis, the basic mean value theorem for harmonic functions, and the very powerful statement which later became known as "Dirichlet's principle" and was finally proved by Hilbert in 1899.

We have discussed the published portion of Gauss's total achievement, but the unpublished and private part was almost equally impressive. Much of this came to light only after his death, when a great quantity of material from his notebooks and scientific correspondence was carefully analyzed and included in his collected works. His scientific diary has already been mentioned. This little booklet of 19 pages, one of the most precious documents in the history of mathematics, was unknown until 1898, when it was found among family papers in the possession of one of Gauss's grandsons. It extends from 1796 to 1814 and consists of 146 very concise statements of the results of his investigations, which often occupied him for weeks or months.‡ All of this material makes it abundantly clear that the ideas Gauss conceived and worked out in considerable detail, but kept to himself, would have made him the greatest mathematician of his time if he had published them and done nothing else.

For example, the theory of functions of a complex variable was one of the major accomplishments of nineteenth century mathematics, and the central facts of this discipline are Cauchy's integral theorem (1827) and the Taylor and Laurent expansions of an analytic function (1831, 1843). In a letter written to his friend Bessel in

---

* See E. T. Bell, "Gauss and the Early Development of Algebraic Numbers," *National Math. Mag.*, vol. 18, pp. 188–204, 219–233, 1944.

† George Green's "Essay on the Application of Mathematical Analysis to the Theories of Electricity and Magnetism" (1828) was neglected and almost completely unknown until it was reprinted in 1846.

‡ See Gauss's *Werke*, vol. X, pp. 483–574, 1917.

1811, Gauss explicitly states Cauchy's theorem and then remarks, "This is a very beautiful theorem whose fairly simple proof I will give on a suitable occasion. It is connected with other beautiful truths which are concerned with series expansions."* Thus, many years in advance of those officially credited with these important discoveries, he knew Cauchy's theorem and probably knew both series expansions. However, for some reason the "suitable occasion" for publication did not arise. A possible explanation for this is suggested by his comments in a letter to Wolfgang Bolyai, a close friend from his university years with whom he maintained a lifelong correspondence. "It is not knowledge but the act of learning, not possession but the act of getting there, which grants the greatest enjoyment. When I have clarified and exhausted a subject, then I turn away from it in order to go into darkness again." His was the temperament of an explorer who is reluctant to take the time to write an account of his last expedition when he could be starting another. As it was, Gauss wrote a great deal, but to have published every fundamental discovery he made in a form satisfactory to himself would have required several long lifetimes.

Another prime example is non-Euclidean geometry, which has been compared with the Copernican revolution in astronomy for its impact on the minds of civilized people. From the time of Euclid to the boyhood of Gauss, the postulates of Euclidean geometry were universally regarded as necessities of thought. Yet there was a flaw in the Euclidean structure that had long been a focus of attention: the so-called parallel postulate, stating that through a point not on a line there exists a single line parallel to the given line. This postulate was thought not to be independent of the others, and many had tried without success to prove it as a theorem. We now know that Gauss joined in these efforts at the age of 15, and that he also failed. But he failed with a difference, for he soon came to the shattering conclusion—which had escaped all his predecessors—that the Euclidean form of geometry is not the only one possible. He worked intermittently on these ideas for many years, and by 1820 he was in full possession of the main theorems of non-Euclidean geometry (the name is due to him).† But he did not reveal his conclusions, and in 1829 and 1832 Lobachevsky and Johann Bolyai (son of Wolfgang) published their own independent work on the subject. One reason for Gauss's silence in this case is quite simple. The intellectual climate of the time in Germany was totally dominated by the philosophy of Kant, and one of the basic tenets of his system was the idea that Euclidean geometry is the only possible way of thinking about space. Gauss knew that this idea was totally false and that the Kantian system was a structure built on sand. However, he valued his privacy and quiet life, and held his peace in order to avoid wasting his time on disputes with the philosophers. In 1829 he wrote as follows to Bessel: "I shall probably not put my very extensive investigations on this subject [the foundations of geometry] into publishable form for a long time, perhaps not in my lifetime, for I dread the shrieks we would hear from the Boeotians if I were to express myself fully on this matter."‡

The same thing happened again in the theory of elliptic functions, a very rich field of analysis that was launched primarily by Abel in 1827 and also by Jacobi in 1828–1829. Gauss had published nothing on this subject, and claimed nothing, so the mathematical world was filled with astonishment when it gradually became known that he had found many of the results of Abel and Jacobi before these men were born. Abel was spared this devastating knowledge by his early death in 1829, at the age of 26, but Jacobi was compelled to swallow his disappointment and go on

---

* *Werke,* vol. VIII, p. 91, 1900.

† Everything he is known to have written about the foundations of geometry was published in his *Werke,* vol. VIII, pp. 159–268, 1900.

‡ *Werke,* vol. VIII, p. 200. The Boeotians were a dull-witted tribe of the ancient Greeks.

with his work. The facts became known partly through Jacobi himself. His attention was caught by a cryptic passage in the *Disquisitiones* (Article 335), whose meaning can only be understood if one knows something about elliptic functions. He visited Gauss on several occasions to verify his suspicions and tell him about his own most recent discoveries, and each time Gauss pulled 30-year-old manuscripts out of his desk and showed Jacobi what Jacobi had just shown him. The depth of Jacobi's chagrin can readily be imagined. At this point in his life Gauss was indifferent to fame and was actually pleased to be relieved of the burden of preparing the treatise on the subject which he had long planned. After a week's visit with Gauss in 1840, Jacobi wrote to his brother, "Mathematics would be in a very different position if practical astronomy had not diverted this colossal genius from his glorious career."

Such was Gauss, the supreme mathematician. He surpassed the levels of achievement possible for ordinary men of genius in so many ways that one sometimes has the eerie feeling that he belonged to a higher species.

CAUCHY (1789 – 1857)

His [Cauchy's] scientific production was enormous. For long periods he appeared before the Academy once a week to present a new paper, so that the Academy, largely on his account, was obliged to introduce a rule restricting the number of articles a member could request to be published in a year.

Oystein Ore

Augustin Louis Cauchy was one of the most influential French mathematicians of the nineteenth century. He began his career as a military engineer, but when his health broke down in 1813 he followed his natural inclination and devoted himself wholly to mathematics.

In mathematical productivity Cauchy was surpassed only by Euler, and his collected works fill 27 fat volumes. He made substantial contributions to number theory and determinants; is considered to be the originator of the theory of finite groups; and did extensive work in astronomy, mechanics, optics, and the theory of elasticity.

His greatest achievements, however, lay in the field of analysis. Together with his contemporaries Gauss and Abel, he was a pioneer in the rigorous treatment of limits, continuous functions, derivatives, integrals, and infinite series. Several of the basic tests for the convergence of series are associated with his name. He also provided the first existence proof for solutions of differential equations, gave the first proof of the convergence of a Taylor series (using his form of the remainder), and was the first to feel the need for a careful study of the convergence behavior of Fourier series. However, his most important work was in the theory of functions of a complex variable, which in essence he created and which has continued to be one of the dominant branches of both pure and applied mathematics. In this field, Cauchy's integral theorem and Cauchy's integral formula are fundamental tools without which modern analysis could hardly exist.

Unfortunately, his personality did not harmonize with the fruitful power of his mind. He was an arrogant royalist in politics and a self-righteous, preaching, pious believer in religion—all this in an age of republican skepticism—and most of his fellow scientists disliked him and considered him a smug hypocrite. It might be fairer to put first things first and describe him as a great mathematician who happened also to be a sincere but narrow-minded bigot.

ABEL (1802–1829)

Abel has left mathematicians enough to keep them busy for 500 years.

Hermite

Niels Henrik Abel was one of the foremost mathematicians of the nineteenth century and probably the greatest genius produced by the Scandinavian countries. Along with his older contemporaries Gauss and Cauchy, Abel was one of the pioneers in the development of modern mathematics, which is characterized by its insistence on rigorous proof. His career was a poignant blend of good-humored optimism under the strains of poverty and neglect, modest satisfaction in the many towering achievements of his brief maturity, and patient resignation in the face of an early death.

Abel was one of six children of a poor Norwegian country minister. His great abilities were recognized and encouraged by one of his teachers when he was only 16, and soon he was reading and digesting the works of Newton, Euler, and Lagrange. As a comment on this experience, he inserted the following remark in one of his later mathematical notebooks: "It appears to me that if one wants to make progress in mathematics, one should study the masters and not the pupils." When Abel was only 18, his father died and left the family destitute. They subsisted with the aid of friends and neighbors, and somehow the boy, helped by contributions from several professors, managed to enter the University of Oslo in 1821. His earliest researches were published in 1823, and included his solution of the classic tautochrone problem by means of an integral equation that is now known by his name. This was the first solution of an equation of this kind, and it foreshadowed the extensive development of integral equations in the late nineteenth and early twentieth centuries. He also proved that the general fifth-degree equation $ax^5 + bx^4 + cx^3 + dx^2 + ex + f = 0$ cannot be solved in terms of radicals, as is possible for equations of lower degree, and thus disposed of a problem that had baffled mathematicians for 300 years. He published his proof in a small pamphlet at his own expense.

In his scientific development Abel soon outgrew Norway, and longed to visit France and Germany. With the backing of his friends and professors, he applied to the government, and after the usual red tape and delays, he received a fellowship for a mathematical grand tour of the Continent. He spent most of his first year abroad in Berlin. Here he had the great good fortune to make the acquaintance of August Leopold Crelle, an enthusiastic mathematical amateur who became his close friend, advisor, and protector. In turn, Abel inspired Crelle to launch his famous *Journal für die Reine und Angewandte Mathematik,* which was the world's first periodical devoted wholly to mathematical research. The first three volumes contained 22 contributions by Abel.

Abel's early mathematical training had been exclusively in the older formal tradition of the eighteenth century, as typified by Euler. In Berlin he came under the influence of the new school of thought led by Gauss and Cauchy, which emphasized rigorous deduction as opposed to formal calculation. Except for Gauss's great work on the hypergeometric series, there were hardly any proofs in analysis that would be accepted as valid today. As Abel expressed it in a letter to a friend: "If you disregard the very simplest cases, there is in all of mathematics not a single infinite series whose sum has been rigorously determined. In other words, the most important parts of mathematics stand without a foundation." In this period he wrote his classic study of the binomial series, in which he founded the general theory of convergence and gave the first satisfactory proof of the validity of this series expansion.

Abel had sent to Gauss in Göttingen his pamphlet on the fifth-degree equation, hoping that it would serve as a kind of scientific passport. However, for some reason Gauss put it aside without looking at it, for it was found uncut among his papers after his death 30 years later. Unfortunately for both men, Abel felt that he had been snubbed, and decided to go on to Paris without visiting Gauss.

In Paris he met Cauchy, Legendre, Dirichlet, and others, but these meetings were perfunctory and he was not recognized for what he was. He had already published a number of important articles in Crelle's *Journal,* but the French were hardly aware yet of the existence of this new periodical, and Abel was much too shy to speak of his own work to people he scarcely knew. Soon after his arrival he finished his great *Mémoire sur une Propriété Générale d'une Classe Très Étendue des Fonctions Transcendantes,* which he regarded as his masterpiece. This work contains the discovery about integrals of algebraic functions now known as Abel's theorem, and is the foundation for the later theory of Abelian integrals, Abelian functions, and much of algebraic geometry. Decades later, Hermite is said to have remarked of this *Mémoire,* "Abel has left mathematicians enough to keep them busy for 500 years." Jacobi described Abel's theorem as the greatest discovery in integral calculus of the nineteenth century. Abel submitted his manuscript to the French Academy. He hoped that it would bring him to the notice of the French mathematicians, but he waited in vain until his purse was empty and he was forced to return to Berlin. What happened was this: The manuscript was given to Cauchy and Legendre for examination; Cauchy took it home, mislaid it, and forgot all about it; and it was not published until 1841, when again the manuscript was lost before the proof sheets were read. The original finally turned up in Florence in 1952.* In Berlin, Abel finished his first revolutionary article on elliptic functions, a subject he had been working on for several years, and then went back to Norway, deeply in debt.

He had expected on his return to be appointed to a professorship at the University, but once again his hopes were dashed. He lived by tutoring, and for a brief time held a substitute teaching position. During this period he worked incessantly, mainly on the theory of the elliptic functions that he had discovered as the inverses of elliptic integrals. This theory quickly took its place as one of the major fields of nineteenth century analysis, with many applications to number theory, mathematical physics, and algebraic geometry. Meanwhile, Abel's fame had spread to all the mathematical centers of Europe and he stood among the elite of the world's mathematicians, but isolated in Norway he was unaware of it. By early 1829 the tuberculosis he had contracted on his journey had progressed to the point where he was unable to work, and in the spring of that year he died, at the age of 26. As an ironic postscript, shortly after his death Crelle wrote that his efforts had been successful, and that Abel would be appointed to the chair of mathematics in Berlin.

Crelle eulogized Abel in his *Journal* as follows: "All of Abel's works carry the imprint of an ingenuity and force of thought which is amazing. One may say that he was able to penetrate all obstacles down to the very foundations of the problem, with a force of thought which appeared irresistible. . . . He distinguished himself equally by the purity and nobility of his character and by a rare modesty which made his person cherished to the same unusual degree as was his genius."

Mathematicians, however, have their own ways of remembering their great men, and so we speak of Abel's integral equation, Abelian integrals and functions, Abelian groups, Abel's series, Abel's partial summation formula, Abel's limit theorem in the theory of power series, and Abel summability. Few have had their names linked to so many concepts and theorems in modern mathematics, and what he might have accomplished in a normal lifetime is beyond conjecture.

---

* For the details of this astonishing story, see the fine book by O. Ore, *Niels Henrik Abel: Mathematician Extraordinary,* University of Minnesota Press, 1957.

DIRICHLET (1805–1859)

The story was told that young Dirichlet had as a constant companion on all his travels, like a devout man with his prayer book, an old, worn copy of the Disquisitiones Arithmeticae *of Gauss.*

Heinrich Tietze

Peter Gustav Lejeune Dirichlet was a German mathematician who made many contributions of lasting value to analysis and number theory. As a young man he was drawn to Paris by the reputations of Cauchy, Fourier, and Legendre, but he was most deeply influenced by his encounter and lifelong contact with Gauss's *Disquisitiones Arithmeticae* (1801). This prodigious but cryptic work contained many of the great master's far-reaching discoveries in number theory, but it was understood by very few mathematicians at that time. As Kummer later said, "Dirichlet was not satisfied to study Gauss's *Disquisitiones* once or several times, but continued throughout his life to keep in close touch with the wealth of deep mathematical thoughts which it contains by perusing it again and again. For this reason the book was never put on the shelf but had an abiding place on the table at which he worked. Dirichlet was the first one who not only fully understood this work, but also made it accessible to others." In later life Dirichlet became a friend and disciple of Gauss, and also a friend and advisor of Riemann, whom he helped in a small way with his doctoral dissertation. In 1855, after lecturing at Berlin for many years, he succeeded Gauss in the professorship at Göttingen.

One of Dirichlet's earliest achievements was a milestone in analysis: In 1829 he gave the first satisfactory proof that certain specific types of functions are actually the sums of their Fourier series. Previous work in this field had consisted wholly of the uncritical manipulation of formulas; Dirichlet transformed the subject into genuine mathematics in the modern sense. As a byproduct of this research, he also contributed greatly to the correct understanding of the nature of a function, and gave the definition which is now most often used, namely, that y is a function of x when to each value of x in a given interval there corresponds a unique value of y. He added that it does not matter whether y depends on x according to some "formula" or "law" or "mathematical operation," and he emphasized this by giving the example of the function of x which has the value 1 for all rational x's and the value 0 for all irrational x's.

Perhaps his greatest works were two long memoirs of 1837 and 1839 in which he made very remarkable applications of analysis to the theory of numbers. It was in the first of these that he proved his wonderful theorem that there are an infinite number of primes in any arithmetic progression of the form $a + nb$, where a and b are positive integers with no common factor. His discoveries about absolutely convergent series also appeared in 1837. His convergence test, discussed in Appendix C.12, was published posthumously in his *Vorlesungen über Zahlentheorie* (1863). These lectures went through many editions and had a very wide influence.

He was also interested in mathematical physics, and formulated the so-called Dirichlet principle of potential theory, which asserts the existence of harmonic functions (functions that satisfy Laplace's equation) with prescribed boundary values. Riemann—who gave the principle its name—used it with great effect in some of his profoundest researches. Hilbert gave a rigorous proof of Dirichlet's principle in the early twentieth century.

LIOUVILLE (1809–1882)

I would rather discover one cause than be King of Persia.
Democritus

Joseph Liouville was a highly respected professor at the Collège de France in Paris and the founder and editor of the *Journal des Mathématiques Pures et Appliquées,* a famous periodical that played an important role in French mathematical life through the latter part of the nineteenth century. For some reason, however, his own remarkable achievements as a creative mathematician have not received the appreciation they deserve. The fact that his collected works have never been published is an unfortunate and rather surprising oversight.

He was the first to solve a boundary value problem by solving an equivalent integral equation, a method developed by Fredholm and Hilbert in the early 1900s into one of the major fields of modern analysis. His ingenious theory of fractional differentiation answered the long-standing question of what reasonable meaning can be assigned to the symbol $d^n y/dx^n$ when n is not a positive integer. He discovered the fundamental result in complex analysis now known as *Liouville's theorem*—that a bounded entire function is necessarily a constant—and used it as the basis for his own theory of elliptic functions. There is also a well-known Liouville theorem in Hamiltonian mechanics, which states that volume integrals are time-invariant in phase space. His theory of integrals of elementary functions was perhaps the most original of all his achievements, for in it he proved that such integrals as

$$\int e^{-x^2}\, dx, \qquad \int \frac{e^x}{x}\, dx, \qquad \int \frac{\sin x}{x}\, dx, \qquad \int \frac{dx}{\ln x},$$

as well as the elliptic integrals of the first and second kinds, cannot be expressed in terms of a finite number of elementary functions.

The fascinating and difficult theory of transcendental numbers is another important branch of mathematics that originated in Liouville's work. The irrationality of π and e (that is, the fact that these numbers are not roots of any linear equation $ax + b = 0$ whose coefficients are integers) had been proved in the eighteenth century by Lambert and Euler. In 1844 Liouville showed that e is also not a root of any quadratic equation with integral coefficients. This led him to conjecture that e is *transcendental,* which means that it does not satisfy any polynomial equation

$$a_n x^n + a_{n-1} x^{n-1} + \cdots + a_1 x + a_0 = 0$$

with integral coefficients. His efforts to prove this failed, but his ideas contributed to Hermite's success in 1873 and then to Lindemann's 1882 proof that π is also transcendental. Lindemann's result showed at last that the age-old problem of squaring the circle by a ruler-and-compass construction is impossible. One of the great mathematical achievements of modern times was Gelfond's 1929 proof that e^π is transcendental, but nothing is yet known about the nature of any of the numbers $\pi + e$, πe, or π^e. Liouville also discovered a sufficient condition for transcendence and used it in 1844 to produce the first examples of real numbers that are provably transcendental. One of these is

$$\sum_{n=1}^{\infty} \frac{1}{10^{n!}} = \frac{1}{10^1} + \frac{1}{10^2} + \frac{1}{10^6} + \cdots = 0.11000100 \ldots .$$

His methods here have also led to extensive further work in the twentieth century.

What he accomplished was certainly better than being King of Persia, or being any king or political leader whatsoever. He was a thinker whose work will live as long as people care about beautiful ideas.

HERMITE (1822–1901)

Talk with M. Hermite. He never evokes a concrete image,
yet you soon perceive that the most abstract entities are to
him like living creatures.

Henri Poincaré

Charles Hermite, one of the most eminent French mathematicians of the nine-teenth century, was particularly distinguished for the clean elegance and high artistic quality of his work. As a student, he courted disaster by neglecting his routine assigned work to study the classic masters of mathematics; and though he nearly failed his examinations, he became a first-rate creative mathematician himself while still in his early twenties. In 1870 he was appointed to a professor-ship at the Sorbonne, where he trained a whole generation of well-known French mathematicians, including Picard, Borel, and Poincaré.

The character of his mind is suggested by the remark of Poincaré quoted above. He disliked geometry, but was strongly attracted to number theory and analysis, and his favorite subject was elliptic functions, where these two fields touch in many remarkable ways. Earlier in the century the Norwegian genius Abel had proved that the general equation of the fifth degree cannot be solved by functions involving only rational operations and root extractions. One of Hermite's most surprising achievements (in 1858) was to show that this equation can be solved by elliptic functions.

His 1873 proof of the transcendence of e was another high point of his career. If he had been willing to dig even deeper into this vein, he could probably have disposed of π as well, but apparently he'd had enough of a good thing. As he wrote to a friend, "I shall risk nothing on an attempt to prove the transcendence of the number π. If others undertake this enterprise, no one will be happier than I at their success, but believe me, my dear friend, it will not fail to cost them some efforts." As it turned out, Lindemann's proof 9 years later rested on extending Hermite's method by using complex integrals to show that no equation of the form

$$a_n e^{b_n} + \cdots + a_2 e^{b_2} + a_1 e^{b_1} + a_0 = 0$$

can be true if the b's are distinct nonzero algebraic numbers and the a's are algebraic numbers not all of which are zero. The transcendence of π now follows from Euler's equation $e^{\pi i} + 1 = 0$, since if π is algebraic, then πi is also.

Several of his purely mathematical discoveries have had unexpected applications many years later to mathematical physics. For example, the Hermitian forms and matrices which he invented in connection with certain problems of number theory turned out to be crucial for Heisenberg's 1925 formulation of quantum mechanics, and Hermite polynomials and Hermite functions are useful in solving Schrödinger's wave equation. The reason is not clear, but it seems to be true that mathematicians do some of their most valuable practical work when thinking about problems that appear to have nothing whatever to do with physical reality.*

---

* On this theme, see the article by E. P. Wigner, "The Unreasonable Effectiveness of Mathe-matics in the Natural Sciences," *Communications on Pure and Applied Mathematics,* vol. 13, pp. 1–14, 1960.

RIEMANN (1826–1866)

. . . an extraordinary mathematician.
S. Bochner

No great mind of the past has exerted a deeper influence on the mathematics of the twentieth century than Bernhard Riemann, the son of a poor country minister in northern Germany. He studied the works of Euler and Legendre while he was still in secondary school, and it is said that he mastered Legendre's treatise on the theory of numbers in less than a week. But he was shy and modest, with little awareness of his own extraordinary abilities, so at the age of 19 he went to the University of Göttingen with the aim of pleasing his father by studying theology and becoming a minister himself. Fortunately, this worthy purpose soon stuck in his throat, and with his father's willing permission he switched to mathematics.

The presence of the legendary Gauss automatically made Göttingen the center of the mathematical world. But Gauss was remote and unapproachable—particularly to beginning students—and after only a year Riemann left this unsatisfying environment and went to the University of Berlin. There he attracted the friendly interest of Dirichlet and Jacobi, and learned a great deal from both men. Two years later he returned to Göttingen, where he obtained his doctor's degree in 1851. During the next 8 years he endured debilitating poverty and created his greatest works. In 1854 he was appointed Privatdozent (unpaid lecturer), which at that time was the necessary first step on the academic ladder. Gauss died in 1855, and Dirichlet was called to Göttingen as his successor. Dirichlet helped Riemann in every way he could, first with a small salary (about one-tenth of that paid to a full professor) and then with a promotion to an assistant professorship. In 1859 he also died, and Riemann was appointed as a full professor to replace him. Riemann's years of poverty were over, but his health was broken. At the age of 39 he died of tuberculosis in Italy, on the last of several trips he undertook in order to escape the cold, wet climate of northern Germany. Riemann had a short life and published comparatively little, but his works permanently altered the course of mathematics in analysis, geometry, and number theory.*

His first published paper was his celebrated dissertation of 1851 on the general theory of functions of a complex variable.† Riemann's fundamental aim here was to free the concept of an analytic function from any dependence on explicit expressions such as power series, and to concentrate instead on general principles and geometric ideas. He founded his theory on what are now called the Cauchy-Riemann equations, created the ingenious device of Riemann surfaces for clarifying the nature of multiple-valued functions, and was led to the Riemann mapping theorem. Gauss was rarely enthusiastic about the mathematical achievements of his contemporaries, but in his official report to the faculty he warmly praised Riemann's work: "The dissertation submitted by Herr Riemann offers convincing evidence of the author's thorough and penetrating investigations in those parts of the subject treated in the dissertation, of a creative, active, truly mathematical mind, and of a gloriously fertile originality."

Riemann later applied these ideas to the study of hypergeometric and Abelian functions. In his work on Abelian functions he relied on a remarkable combination of geometric reasoning and physical insight, the latter in the form of Dirichlet's principle from potential theory. He used Riemann surfaces to build a bridge between

---

* His *Gesammelte Mathematische Werke* (reprinted by Dover in 1953) occupy only a single volume, of which two-thirds consists of posthumously published material. Of the nine papers Riemann published himself, only five deal with pure mathematics.

† "Grundlagen für eine allgemeine Theorie der Functionen einer veränderlichen complexen Grösse," in *Werke,* pp. 3–43.

analysis and geometry which made it possible to give geometric expression to the deepest analytic properties of functions. His powerful intuition often enabled him to discover such properties—for instance, his version of the Riemann-Roch theorem —by simply thinking about possible configurations of closed surfaces and performing imaginary physical experiments on these surfaces. Riemann's geometric methods in complex analysis constituted the true beginning of topology, a rich field of geometry concerned with those properties of figures that are unchanged by continuous deformations.

In 1854 he was required to submit a probationary essay in order to be admitted to the position of Privatdozent, and his response was another pregnant work whose influence is indelibly stamped on the mathematics of our own time.* The problem he set himself was to analyze Dirichlet's conditions (1829) for the representability of a function by its Fourier series. One of these conditions was that the function must be integrable. But what does this mean? Dirichlet had used Cauchy's definition of integrability, which applies only to functions that are continuous or have at most a finite number of points of discontinuity. Certain functions that arise in number theory suggested to Riemann that this definition should be broadened. He developed the concept of the Riemann integral as it now appears in most textbooks on calculus, established necessary and sufficient conditions for the existence of such an integral, and generalized Dirichlet's criteria for the validity of Fourier expansions. Cantor's famous theory of sets was directly inspired by a problem raised in this paper, and these ideas led in turn to the concept of the Lebesgue integral and even more general types of integration. Riemann's pioneering investigations were therefore the first steps in another new branch of mathematics, the theory of functions of a real variable.

The Riemann rearrangement theorem in the theory of infinite series was an incidental result in the paper just described. He was familiar with Dirichlet's example showing that the sum of a conditionally convergent series can be changed by altering the order of its terms:

$$1 - \frac{1}{2} + \frac{1}{3} - \frac{1}{4} + \frac{1}{5} - \frac{1}{6} + \frac{1}{7} - \frac{1}{8} + \cdots = \ln 2, \tag{1}$$

$$1 + \frac{1}{3} - \frac{1}{2} + \frac{1}{5} + \frac{1}{7} - \frac{1}{4} + \cdots = \frac{3}{2} \ln 2. \tag{2}$$

It is apparent that these two series have different sums but the same terms; for in (2) the first two positive terms in (1) are followed by the first negative term, then the next two positive terms are followed by the second negative term, and so on. Riemann proved that it is possible to rearrange the terms of any conditionally convergent series in such a manner that the new series will converge to an arbitrary preassigned sum or diverge to ∞ or $-\infty$.

In addition to his probationary essay, Riemann was also required to present a trial lecture to the faculty before he could be appointed to his unpaid lectureship. It was the custom for the candidate to offer three titles, and the head of his department usually accepted the first. However, Riemann rashly listed as his third topic the foundations of geometry, a profound subject on which he was unprepared but which Gauss had been turning over in his mind for 60 years. Naturally, Gauss was curious to see how this particular candidate's "gloriously fertile originality" would cope with such a challenge, and to Riemann's dismay he designated this as the subject of the lecture. Riemann quickly tore himself away from his other interests at the time—"my investigations of the connection between electricity, magnetism, light, and gravitation"—and wrote his lecture in the next two months. The result was one

---

* "Ueber die Darstellbarkeit einer Function durch eine trigonometrische Reihe," in *Werke*, pp. 227–264.

of the great classical masterpieces of mathematics, and probably the most important scientific lecture ever given.* It is recorded that even Gauss was surprised and enthusiastic.

Riemann's lecture presented in nontechnical language a vast generalization of all known geometries, both Euclidean and non-Euclidean. This field is now called Riemannian geometry, and apart from its great importance in pure mathematics, it turned out 60 years later to be exactly the right framework for Einstein's general theory of relativity. Like most of the great ideas of science, Riemannian geometry is quite easy to understand if we set aside the technical details and concentrate on its essential features. Let us recall the intrinsic differential geometry of curved surfaces which Gauss had discovered 25 years earlier. If a surface imbedded in three-dimensional space is defined parametrically by three functions $x = x(u, v)$, $y = y(u, v)$, and $z = z(u, v)$, then u and v can be interpreted as the coordinates of points on the surface. The distance ds along the surface between two nearby points (u, v) and $(u + du, v + dv)$ is given by Gauss's quadratic differential form

$$ds^2 = E\, du^2 + 2F\, du\, dv + G\, dv^2,$$

where E, F, and G are certain functions of u and v. This differential form makes it possible to calculate the lengths of curves on the surface, to find the geodesic (or shortest) curves, and to compute the Gaussian curvature of the surface at any point — all in total disregard of the surrounding space. Riemann generalized this by discarding the idea of a surrounding Euclidean space and introducing the concept of a continuous n-dimensional manifold of points $(x_1, x_2, \ldots, x_n)$. He then imposed an arbitrarily given distance (or metric) ds between nearby points

$$(x_1, x_2, \ldots, x_n) \quad \text{and} \quad (x_1 + dx_1, x_2 + dx_2, \ldots, x_n + dx_n)$$

by means of a quadratic differential form

$$ds^2 = \sum_{i,j=1}^{n} g_{ij}\, dx_i\, dx_j, \tag{3}$$

where the g_{ij}'s are suitable functions of $x_1, x_2, \ldots, x_n$ and different systems of g_{ij}'s define different Riemannian geometries on the manifold under discussion. His next steps were to examine the idea of curvature for these Riemannian manifolds and to investigate the special case of constant curvature. All of this depends on massive computational machinery, which Riemann mercifully omitted from his lecture but included in a posthumous paper on heat conduction. In that paper he explicitly introduced the Riemann curvature tensor, which reduces to the Gaussian curvature when $n = 2$ and whose vanishing he showed to be necessary and sufficient for the given quadratic metric to be equivalent to a Euclidean metric. From this point of view, the curvature tensor measures the deviation of the Riemannian geometry defined by formula (3) from Euclidean geometry. Einstein has summarized these ideas in a single statement: "Riemann's geometry of an n-dimensional space bears the same relation to Euclidean geometry of an n-dimensional space as the general geometry of curved surfaces bears to the geometry of the plane."

The physical significance of geodesics appears in its simplest form as the following consequence of Hamilton's principle in the calculus of variations: If a particle is constrained to move on a curved surface, and if no force acts on it, then it glides along a geodesic. A direct extension of this idea is the heart of the general theory of relativity, which is essentially a theory of gravitation. Einstein conceived the geometry of

---

* "Ueber die Hypothesen, welche der Geometrie zu Grunde liegen," in *Werke,* pp. 272–286. There is a translation in D. E. Smith, *A Source Book in Mathematics,* McGraw-Hill, New York, 1929.

space as a Riemannian geometry in which the curvature and geodesics are determined by the distribution of matter; in this curved space, planets move in their orbits around the sun by simply coasting along geodesics instead of being pulled into curved paths by a mysterious force of gravity whose nature no one has ever really understood.

In 1859 Riemann published his only work on the theory of numbers, a brief but exceedingly profound paper of less than 10 pages devoted to the prime number theorem.* This mighty effort started tidal waves in several branches of pure mathematics, and its influence will probably still be felt a thousand years from now. His starting point was a remarkable identity discovered by Euler over a century earlier: If s is a real number greater than 1, then

$$\sum_{n=1}^{\infty} \frac{1}{n^s} = \prod_p \frac{1}{1 - (1/p^s)}, \tag{4}$$

where the expression on the right denotes the product of the numbers $(1 - p^{-s})^{-1}$ for all primes p. To understand how this identity arises, we note that $1/(1 - x) = 1 + x + x^2 + \cdots$ for $|x| < 1$, so for each p we have

$$\frac{1}{1 - (1/p^s)} = 1 + \frac{1}{p^s} + \frac{1}{p^{2s}} + \cdots.$$

On multiplying these series for all primes p and recalling that each integer $n > 1$ is uniquely expressible as a product of powers of different primes, we see that

$$\prod_p \frac{1}{1 - (1/p^s)} = \prod_p \left(1 + \frac{1}{p^s} + \frac{1}{p^{2s}} + \cdots\right)$$

$$= 1 + \frac{1}{2^s} + \frac{1}{3^s} + \cdots + \frac{1}{n^s} + \cdots$$

$$= \sum_{n=1}^{\infty} \frac{1}{n^s},$$

which is the identity (4). The sum of the series on the left of (4) is evidently a function of the real variable $s > 1$, and the identity establishes a connection between the behavior of this function and properties of the primes. Euler himself exploited this connection in several ways, but Riemann perceived that access to the deeper features of the distribution of primes can be gained only by allowing s to be a complex variable. He denoted the resulting function by $\zeta(s)$, and it has since been known as the Riemann zeta function:

$$\zeta(s) = 1 + \frac{1}{2^s} + \frac{1}{3^s} + \cdots, \qquad s = \sigma + it.$$

In his paper he proved several important properties of this function, and in a sovereign way simply stated a number of others without proof. During the century since his death, many of the finest mathematicians in the world have exerted their strongest efforts and created rich new branches of analysis in attempts to prove these statements. The first success was achieved in 1893 by J. Hadamard, and with one exception every statement has since been settled in the sense Riemann expected.† This exception is the famous Riemann hypothesis: that all the zeros of $\zeta(s)$ in the strip

---

* "Ueber die Anzahl der Primzahlen unter einer gegebenen Grösse," in *Werke*, pp. 145–153. See the statement of the prime number theorem in Appendix A.14.

† Hadamard's work led him to his 1896 proof of the prime number theorem. See E. C. Titchmarsh, *The Theory of the Riemann Zeta Function*, Oxford University Press, London, 1951, Chapter 3. This treatise has a bibliography of 326 items.

$0 \le \sigma \le 1$ lie on the central line $\sigma = \frac{1}{2}$. It stands today as the most important unsolved problem of mathematics, and is probably the most difficult problem that the human mind has ever conceived. In a fragmentary note found among his posthumous papers, Riemann wrote that these theorems "follow from an expression for the function $\zeta(s)$ which I have not yet simplified enough to publish."[*] Writing about this fragment in 1944, Hadamard remarked with justified exasperation, "We still have not the slightest idea of what the expression could be."[†] He adds the further comment: "In general, Riemann's intuition is highly geometrical; but this is not the case for his memoir on prime numbers, the one in which that intuition is the most powerful and mysterious."

---

[*] *Werke,* p. 154.
[†] *The Psychology of Invention in the Mathematical Field,* Dover, New York, 1954, p. 118.

APPENDIX C

THE THEORY OF CALCULUS

When a student begins seriously to study mathematics, he believes he knows what a fraction is, what continuity is, and what the area of a curved surface is; he considers as evident, for example, that a continuous function cannot change its sign without vanishing. If, without any preparation, you say to him: No, that is not at all evident, I must demonstrate it to you; and if the demonstration rests on premises which do not appear to him more evident than the conclusion, what would this unfortunate student think? He will think that the science of mathematics is only an arbitrary accumulation of useless subtleties; either he will be disgusted with it, or he will amuse himself with it as a game and arrive at a state of mind similar to that of the Greek sophists.

Henri Poincaré

What is time? If no one asks me, I know; if I wish to explain to him who asks, I do not know.

St. Augustine

Certitude is not the test of certainty. We have been cocksure of many things that were not so.

O. W. Holmes, Jr.

C.1
THE REAL NUMBER SYSTEM

When considered for its own sake, and apart from any uses it may have, the real number system appears as an intricate intellectual structure whose endless complexities are of interest mainly to mathematicians. However, from the practical point of view, it is the foundation on which virtually all other branches of mathematics rest, and as such, it underlies every quantitative aspect of civilized life.

Most of us learn in school how to use real numbers for counting, measurement, and solving algebraic problems. Nevertheless, no matter how much skill of this kind we develop, few of us ever confront the question of just what the real numbers *are.* Our purpose here is to answer this question as briefly and clearly as possible. In doing so, we will also provide an adequate basis for the capsule discussions of the theory of calculus that are given in the following sections.

There are several ways to introduce the real number system. We adopt the most efficient of these—the axiomatic approach—in which we start with the real numbers themselves as given undefined objects possessing certain simple properties that we use as axioms. This means we assume there exists a set R of objects, called real numbers, that satisfy the ten axioms listed in the following pages. All further properties of real numbers, regardless of how profound they may be, are ultimately provable as logical consequences of these. The axioms fall into three natural groups. Those of the first group are stated in terms of the two operations $+$ and $\cdot$, addition and multiplication, which can be applied to any pair x and y of real numbers to produce their sum $x + y$ and their product $x \cdot y$ (also denoted more simply by xy).

Algebra Axioms

1. Commutative laws: $x + y = y + x$, $xy = yx$.
2. Associative laws: $x + (y + z) = (x + y) + z$, $x(yz) = (xy)z$.
3. Distributive law: $x(y + z) = xy + xz$.
4. Existence of identity elements: There exist two distinct real numbers, denoted by 0 and 1, such that $0 + x = x + 0 = x$ and $1 \cdot x = x \cdot 1 = x$ for every x.
5. Existence of negatives: For each x there exists a unique y such that $x + y = y + x = 0$.
6. Existence of reciprocals: For each $x \neq 0$ there exists a unique z such that $xz = zx = 1$.

The number y in (5) is customarily denoted by $-x$, and z in (6) by $1/x$ or x^{-1}. Subtraction and division can now be defined by $x - y = x + (-y)$ and $x/y = x(1/y)$. All the usual laws of elementary algebra can be deduced from these axioms and these definitions. We illustrate this process by giving three very brief proofs.

Example 1 (i) $x + y = x + z$ implies $y = z$ (the cancellation law of addition). *Proof* Since $x + y = x + z$, $(-x) + (x + y) = (-x) + (x + z)$; by (2), $[(-x) + x] + y = [(-x) + x] + z$; by (5), $0 + y = 0 + z$; and by (4), $y = z$.

(ii) $x \cdot 0 = 0$. *Proof* (4) gives $0 + 1 = 1$, so $x(0 + 1) = x \cdot 1$; by (3), $x \cdot 0 + x \cdot 1 = x \cdot 1$; by (4), $x \cdot 0 + x = x = 0 + x$; by (1), $x + x \cdot 0 = x + 0$; and by (i), $x \cdot 0 = 0$.

(iii) $(-1)(-1) = 1$. *Proof* (5) gives $1 + (-1) = 0$, so on multiplying by -1 and using (3), (4), and (ii) we obtain $(-1) + (-1)(-1) = 0$; and adding 1 to both sides of this yields $(-1)(-1) = 1$ after careful reduction.

The next group of axioms enables us to establish an order relation in the real number system. It is convenient to introduce this relation indirectly, by basing it on a concept of positiveness. This means we assume there exists in R a special subset P, called the set of positive numbers, that satisfies the three axioms listed below. The statement that a number x is in the set P is symbolized by writing $0 < x$, or equivalently $x > 0$.

Order Axioms

7. For each x, one and only one of the following possibilities is true: $x = 0$, $x > 0$, $-x > 0$.
8. If x and y are positive, so is $x + y$.
9. If x and y are positive, so is xy.

We now introduce the familiar order relations $<$ and $>$ as follows: $x < y$ is defined to mean $y - x > 0$, and $x > y$ is equivalent to $y < x$. As usual, $x \leq y$ means that $x < y$ or $x = y$, and $x \geq y$ is equivalent to $y \leq x$. All the customary rules for working with inequalities can be proved as theorems on the basis of these axioms and definitions.

Example 2   It is quite easy to show that for any real numbers x and y, one and only one of these possibilities is true: $x = y$, $x < y$, $x > y$ (proof: apply (7) to the number $y - x$). We next consider the proofs of the following familiar facts:

If $x < y$ and $y < z$, then $x < z$.

If $x > 0$ and $y < z$, then $xy < xz$.

If $x < 0$ and $y < z$, then $xy > xz$.

If $x < y$, then $x + z < y + z$ for any z.

The definitions allow us to express these statements in equivalent forms that are more convenient from the point of view of providing proofs:

If $y - x > 0$ and $z - y > 0$, then $z - x > 0$.

If $x > 0$ and $z - y > 0$, then $xz - xy > 0$.

If $-x > 0$ and $z - y > 0$, then $xy - xz > 0$.

If $y - x > 0$, then $(y + z) - (x + z) > 0$ for any z.

The first of these assertions is an obvious consequence of (8), the second and third follow directly from (9), and the fourth is trivial, since $(y + z) - (x + z) = y - x$.

The program of carefully deducing all the algebraic and order properties of R from axioms (1) to (9) is rather long and boring, and no useful purpose would be served by pursuing this aspect of the matter any further. It is quite enough for students to understand that this program can be carried out, and we omit the details.

The nine axioms given above do not fully determine the real number system. This is most easily seen by noticing that the set Q of all rational numbers is a number system different from R that also satisfies all nine axioms. Of course, the difference between Q and R is simply that Q lacks the irrationals, which any workable number system ought to have. One more axiom is needed to guarantee that R is free from this defect, or equivalently that the real number system has no "gaps" or "holes."

Two preliminary definitions are necessary before our final axiom can be stated. Both refer to an arbitrary set S of real numbers. A real number b is called an *upper bound* for S if $x \leq b$ for every x in S. Further, a real number b_0 is called a *least upper bound* for S if (i) b_0 is an upper bound for S, and (ii) $b_0 \leq b$ for every upper bound b of S. A set has many upper bounds if it has one, but it can have only one least upper bound. The proof is easy: If b_0 and b_1 are both least upper bounds for S, then $b_0 \leq b_1$ (since b_0 is a least upper bound and b_1 is an upper bound) and $b_1 \leq b_0$ (since b_1 is a least upper bound and b_0 is an upper bound), so $b_0 = b_1$. This argument permits us to speak of *the* least upper bound of S. These concepts can be visualized in the usual way, as suggested by Fig. C.1.

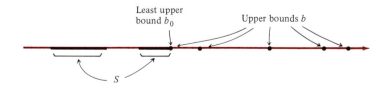

Figure C.1

Example 3   The set of all positive integers has no upper bound. If S is the closed interval $0 \leq x \leq 1$, then the numbers 1, 2, 3.74, and 513 (among others) are all upper bounds of S, and 1 is its least upper bound. The same statements are true if S is the open interval $0 < x < 1$. In the first case, the least upper bound 1 belongs to the set S, but in the second case it does not. The set S consisting of all numbers in the sequence

$$\frac{1}{2}, \frac{2}{3}, \frac{3}{4}, \ldots, \frac{n}{n+1}, \ldots \text{ also has 1 as its least upper bound.}$$

The following is the final axiom for the real number system R.

Least Upper Bound Axiom

10   Every nonempty set of real numbers that has an upper bound also has a least upper bound.

This axiom guarantees that the real number system has the property of "completeness" or "continuity" that is absolutely essential for the development of calculus. The best way to grasp the significance of this axiom is to observe that it is not true for the set Q of rational numbers: If S is taken to be the set of positive rationals r such that $r^2 < 2$, then S has an upper bound in Q but does not have a least upper bound in Q (the least upper bound of S in R is $\sqrt{2}$, but this number is not in Q).

Remark 1   We have implied, but not actually stated, that the 10 axioms given here completely characterize the real number system R. The meaning of this statement can be clarified by formulating our ideas on a more abstract level, as follows. In modern algebra a set of objects that satisfies axioms (1) to (6) is called a *field*. There are many different fields, some finite and others infinite. The simplest consists of the two elements 0 and 1 alone, with addition and multiplication defined by

$$0 + 0 = 0, \qquad 0 + 1 = 1 + 0 = 1, \qquad 1 + 1 = 0,$$
$$0 \cdot 0 = 0, \qquad 0 \cdot 1 = 1 \cdot 0 = 0, \qquad 1 \cdot 1 = 1.$$

A field that satisfies the additional axioms (7) to (9) is called an *ordered field*. Both Q and R are ordered fields, but there are also a number of others. It can be proved that an ordered field must have infinitely many distinct elements, so some fields — including the two-element field just mentioned — cannot be ordered. We use axiom (10) to narrow our scope still further, and an ordered field that satisfies this axiom is called a *complete ordered field*. It can be proved that any two complete ordered fields are abstractly identical in a very precise sense, so there is really only one, namely R.* It is therefore possible to define a real number very simply, as an element in a complete ordered field. However, it is clear that no such definition can be considered satisfactory without a good deal of preliminary explanation and proof.

Remark 2   There may be a few exceedingly skeptical readers who find themselves thinking thoughts like these: "What this writer says sounds reasonable enough, *provided the real number system R exists in the first place.* But how do we know that it does? After all, this number system is not a physical object that can be seen and touched, but a creation of the mind — like a unicorn — and perhaps we deceive ourselves by supposing that it exists."

There are two ways to answer this objection. One is to give a concrete definition of

---

* For a fuller discussion, with proofs (or sketches of proofs), see pp. 1–8 in the first volume of E. Hille's *Analytic Function Theory,* Ginn and Co., 1959.

R as the set of all infinite decimals, with the usual agreement that such decimals as 0.25000 . . . and 0.24999 . . . are to be considered equal. Addition, multiplication, and the set of positive numbers must now be given satisfactory definitions, and in this scheme of things our axioms (1) to (10) become theorems whose proofs lean heavily on these definitions. This program is surprisingly difficult to carry out.*

A second approach is to use the much more basic positive integers as a given supply of building materials for the explicit step-by-step construction of the real number system — first the integers, then the rationals, and finally the reals. This time the axioms (1) to (10) appear as theorems that can be deduced from assumed properties of the positive integers.†

We do not encourage students to investigate these matters any further, for there is no part of mathematics more tedious and unrewarding than the detailed construction of the real number system by either of these methods.

We begin by recalling the definition of the limit of a function from Section 2.5. Consider a function $f(x)$ that is defined for values of x arbitrarily close to a point a on the x-axis but not necessarily at a itself. Another way to express this requirement is to say that there are x's in the domain of the function that satisfy the inequalities $0 < |x - a| < \delta$ for every positive number δ. Under these circumstances, the statement that

$$\lim_{x \to a} f(x) = L$$

is defined to mean the following: For each positive number ϵ there exists a positive number δ with the property that

$$|f(x) - L| < \epsilon$$

for every x in the domain of the function that satisfies the inequalities

$$0 < |x - a| < \delta.$$

In the hope of clarifying the meaning of this definition, we examine the way it is used in a simple special case. It is obvious by inspection that

$$\lim_{x \to 1} (3x - 1) = 2. \tag{1}$$

However, to prove this by using the definition, we must start with an arbitrary positive number ϵ and find a $\delta > 0$ that "works" for this ϵ, in the sense that

$$0 < |x - 1| < \delta \qquad \text{implies} \qquad |(3x - 1) - 2| < \epsilon. \tag{2}$$

But the last inequality here is the same as $|3x - 3| < \epsilon$ or $-\epsilon < 3x - 3 < \epsilon$; and after division by 3 this becomes $-\frac{1}{3}\epsilon < x - 1 < \frac{1}{3}\epsilon$. This suggests that $\delta = \frac{1}{3}\epsilon$ might work. To show that it does, we observe that if $0 < |x - 1| < \frac{1}{3}\epsilon$ then $-\frac{1}{3}\epsilon < x - 1 < \frac{1}{3}\epsilon$, which implies $-\epsilon < 3x - 3 < \epsilon$ or $|(3x - 1) - 2| < \epsilon$. Thus, for any $\epsilon > 0$ the number $\delta = \frac{1}{3}\epsilon$ actually has the property stated in (2). The requirement of the definition is therefore satisfied, and (1) is proved.

It is natural to object to this procedure, and to feel that carefully proving a transparent statement like (1) is empty mumbo-jumbo and a waste of time. However, the point is this: (1) is obviously true and doesn't really need a proof, but many important

C.2

THEOREMS ABOUT LIMITS

---

* See Chapter 1 of J. F. Ritt's *Theory of Functions,* King's Crown Press, 1947.

† The classical source for this construction is E. Landau, *Foundations of Analysis,* Chelsea, 1951.

THE THEORY OF CALCULUS

limits are far from obvious and cannot be dealt with by simple inspection. For instance, it is no exaggeration to say that large parts of advanced mathematics would disappear like a puff of smoke without the ideas and methods that depend on the vital limits

$$\lim_{x \to 0} \frac{\sin x}{x} = 1 \qquad \text{and} \qquad \lim_{x \to 0} (1 + x)^{1/x} = e.$$

(The fundamental constant denoted by e is officially introduced in Chapter 8; its approximate value is 2.71828.) We clearly need powerful tools to cope with limits like these, not vague ideas and fuzzy concepts. We have proved (1) not for its own sake, but in order to illustrate the use of the definition of the limit of a function. This definition is not intended to be merely a passive description in the sense of most dictionary definitions. On the contrary, it is a sharp-edged instrument of proof that is capable of being manipulated effectively in complex and subtle arguments where sloppy thinking brings nothing but confusion. We have two purposes in the theorems and proofs given below: first, to establish the results themselves, and thereby provide a solid logical foundation for all of our work that depends on limits of functions; and second, to further illustrate the use of the definition in the machinery of formal proofs.

Our first theorem states a fact that most people take for granted without fully realizing it, namely, that a function $f(x)$ cannot approach two different limits as x approaches a.

Theorem 1 *If $\lim_{x \to a} f(x) = L_1$ and $\lim_{x \to a} f(x) = L_2$, then $L_1 = L_2$.*

Proof Our method of proof is to show that the assumption $L_1 \neq L_2$ leads to the absurd conclusion $|L_1 - L_2| < |L_1 - L_2|$. We therefore assume that $L_1 \neq L_2$, so that $|L_1 - L_2|$ is positive, and we let ϵ be the positive number $\frac{1}{2}|L_1 - L_2|$. By the first hypothesis there exists a number $\delta_1 > 0$ such that

$$0 < |x - a| < \delta_1 \qquad \text{implies} \qquad |f(x) - L_1| < \epsilon,$$

and by the second hypothesis there exists a number $\delta_2 > 0$ such that

$$0 < |x - a| < \delta_2 \qquad \text{implies} \qquad |f(x) - L_2| < \epsilon.$$

Define δ to be the smaller of the numbers δ_1 and δ_2. Then $0 < |x - a| < \delta$ implies both

$$|f(x) - L_1| < \epsilon \qquad \text{and} \qquad |f(x) - L_2| < \epsilon,$$

and therefore

$$|L_1 - L_2| = |[L_1 - f(x)] + [f(x) - L_2]|$$
$$\leq |L_1 - f(x)| + |f(x) - L_2|$$
$$< \epsilon + \epsilon = 2\epsilon = |L_1 - L_2|.$$

This contradiction—that the number $|L_1 - L_2|$ is less than itself—shows that it cannot be true that $|L_1 - L_2|$ is positive, so $L_1 = L_2$.

Theorem 2 *If $f(x) = x$, then $\lim_{x \to a} f(x) = a$; that is,*

$$\lim_{x \to a} x = a.$$

Proof Let $\epsilon > 0$ be given, and choose $\delta = \epsilon$. Then $0 < |x - a| < \delta = \epsilon$ implies that $|f(x) - a| < \epsilon$, since $f(x) = x$.

Theorem 3 *If $f(x) = c$, where c is a constant, then $\lim_{x \to a} f(x) = c$; that is,*

$$\lim_{x \to a} c = c.$$

Proof Since $|f(x) - c| = |c - c| = 0$ for all x, any $\delta > 0$ will do, because $|f(x) - c|$ will be $< \epsilon$ for any given $\epsilon > 0$ and all x.

Theorem 4 *If $\lim_{x \to a} f(x) = L$ and $\lim_{x \to a} g(x) = M$, then*
(i) $\lim_{x \to a} [f(x) + g(x)] = L + M$;
(ii) $\lim_{x \to a} [f(x) - g(x)] = L - M$; and
(iii) $\lim_{x \to a} f(x)g(x) = LM$.

Proof For (i), let $\epsilon > 0$ be given, let $\delta_1 > 0$ be a number such that

$$0 < |x - a| < \delta_1 \quad \text{implies} \quad |f(x) - L| < \tfrac{1}{2}\epsilon,$$

and let $\delta_2 > 0$ be a number such that

$$0 < |x - a| < \delta_2 \quad \text{implies} \quad |g(x) - M| < \tfrac{1}{2}\epsilon.$$

Define δ to be the smaller of the numbers δ_1 and δ_2. Then $0 < |x - a| < \delta$ implies

$$|[f(x) + g(x)] - (L + M)| = |[f(x) - L] + [g(x) - M]|$$
$$\leq |f(x) - L| + |g(x) - M|$$
$$< \tfrac{1}{2}\epsilon + \tfrac{1}{2}\epsilon = \epsilon,$$

and this proves (i).

The argument for (ii) is almost identical with that just given and will be omitted.

In proving (iii), we wish to make the difference $f(x)g(x) - LM$ depend on the differences $f(x) - L$ and $g(x) - M$. This can be accomplished by subtracting and adding $f(x)M$, as follows:

$$|f(x)g(x) - LM| = |[f(x)g(x) - f(x)M] + [f(x)M - LM]|$$
$$\leq |f(x)g(x) - f(x)M| + |f(x)M - LM|$$
$$= |f(x)||g(x) - M| + |M||f(x) - L|$$
$$\leq |f(x)||g(x) - M| + (|M| + 1)|f(x) - L|.$$

Let $\epsilon > 0$ be given. We know that there exist positive numbers $\delta_1, \delta_2, \delta_3$ such that

$$0 < |x - a| < \delta_1 \quad \text{implies} \quad |f(x) - L| < 1, \text{ which in turn implies } |f(x)| < |L| + 1;$$

$$0 < |x - a| < \delta_2 \quad \text{implies} \quad |g(x) - M| < \frac{1}{2}\epsilon\left(\frac{1}{|L| + 1}\right);$$

$$0 < |x - a| < \delta_3 \quad \text{implies} \quad |f(x) - L| < \frac{1}{2}\epsilon\left(\frac{1}{|M| + 1}\right).$$

Define δ to be the smallest of the numbers $\delta_1, \delta_2, \delta_3$. Then $0 < |x - a| < \delta$ implies

$$|f(x)g(x) - LM| < \tfrac{1}{2}\epsilon + \tfrac{1}{2}\epsilon = \epsilon,$$

and the proof of (iii) is complete.

Theorem 5   *If* $\lim_{x \to a} f(x) = L$ *and* $\lim_{x \to a} g(x) = M$ *where* $M \neq 0$, *then*

$$\lim_{x \to a} \frac{f(x)}{g(x)} = \frac{L}{M}.$$

Proof By Theorem 4 [part (iii)] and the fact that

$$\frac{f(x)}{g(x)} = f(x) \cdot \frac{1}{g(x)},$$

it suffices to prove that

$$\lim_{x \to a} \frac{1}{g(x)} = \frac{1}{M}.$$

We begin with the fact that if $g(x) \neq 0$, then

$$\left| \frac{1}{g(x)} - \frac{1}{M} \right| = \frac{|g(x) - M|}{|M| \, |g(x)|}.$$

Choose $\delta_1 > 0$ so that

$$0 < |x - a| < \delta_1 \qquad \text{implies} \qquad |g(x) - M| < \tfrac{1}{2}|M|.$$

For these x's we have

$$|g(x)| > \frac{1}{2}|M| \qquad \text{or} \qquad \frac{1}{|g(x)|} < \frac{2}{|M|},$$

and therefore

$$\left| \frac{1}{g(x)} - \frac{1}{M} \right| < \frac{2}{|M|^2} |g(x) - M|.$$

Let $\epsilon > 0$ be given and choose $\delta_2 > 0$ so that

$$0 < |x - a| < \delta_2 \qquad \text{implies} \qquad |g(x) - M| < \frac{|M|^2}{2} \epsilon.$$

We now define δ to be the smaller of the numbers δ_1 and δ_2 and observe that

$$0 < |x - a| < \delta \qquad \text{implies} \qquad \left| \frac{1}{g(x)} - \frac{1}{M} \right| < \frac{2}{|M|^2} \cdot \frac{|M|^2}{2} \epsilon = \epsilon,$$

and this concludes the argument.

Theorem 6   *If there exists a positive number p with the property that*

$$g(x) \leq f(x) \leq h(x)$$

for all x that satisfy the inequalities $0 < |x - a| < p$, and if $\lim_{x \to a} g(x) = L$ and $\lim_{x \to a} h(x) = L$, then

$$\lim_{x \to a} f(x) = L.$$

Proof This statement is sometimes called the "squeeze theorem," because it says that a function squeezed between two functions approaching the same limit L must also approach L (see Fig. C.2). For the proof, let $\epsilon > 0$ be given, and choose positive numbers δ_1 and δ_2 so that

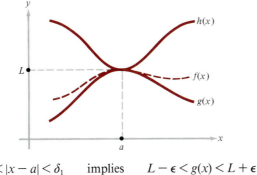

Figure C.2

$$0 < |x - a| < \delta_1 \quad \text{implies} \quad L - \epsilon < g(x) < L + \epsilon$$

and

$$0 < |x - a| < \delta_2 \quad \text{implies} \quad L - \epsilon < h(x) < L + \epsilon.$$

Define δ to be the smallest of the numbers p, δ_1, δ_2. Then $0 < |x - a| < \delta$ implies

$$L - \epsilon < g(x) \leq f(x) \leq h(x) < L + \epsilon,$$

so $|f(x) - L| < \epsilon$ and the proof is complete.

We continue with the proofs of a few simple facts about continuous functions that follow almost immediately from these theorems about limits. First, however, let us recall that a function $f(x)$ is said to be *continuous at a point a* if

$$\lim_{x \to a} f(x) = f(a).$$

It is sometimes convenient to use the epsilon-delta version of this statement: for each $\epsilon > 0$ there exists a $\delta > 0$ with the property that

$$|f(x) - f(a)| < \epsilon$$

for every x in the domain of the function that satisfies the inequality

$$|x - a| < \delta.$$

Theorem 7   *If $f(x)$ and $g(x)$ are continuous at a point a, then $f(x) + g(x)$, $f(x) - g(x)$, and $f(x)g(x)$ are also continuous at a. Further, $f(x)/g(x)$ is continuous at a if $g(a) \neq 0$.*

Proof   We prove only the statement about $f(x) + g(x)$, the other arguments being similar. Since $f(x)$ and $g(x)$ are continuous at a, we have

$$\lim_{x \to a} f(x) = f(a) \quad \text{and} \quad \lim_{x \to a} g(x) = g(a).$$

Part (i) of Theorem 4 now guarantees that

$$\lim_{x \to a} [f(x) + g(x)] = f(a) + g(a),$$

and this proves that $f(x) + g(x)$ is continuous at a.

Theorem 8   *The functions $f(x) = x$ and $g(x) = c$, where c is a constant, are continuous for all values of x.*

Proof   These statements follow at once from Theorems 2 and 3.

Theorem 9 *Any polynomial*

$$P(x) = a_n x^n + a_{n-1} x^{n-1} + \cdots + a_1 x + a_0 \tag{3}$$

is continuous for all values of x.

Proof By Theorem 8 and the multiplication part of Theorem 7, each of the following functions is continuous for all values of x: x, $x^2 = x \cdot x$, $x^3 = x \cdot x^2$, . . . , x^k for any positive integer k, and cx^k where c is any constant. Since the constant term a_0 is continuous, this tells us that each term of (3) is continuous for all values of x, and we obtain the conclusion by repeated application of the addition part of Theorem 7.

Theorem 10 *Any rational function*

$$R(x) = \frac{P(x)}{Q(x)},$$

where $P(x)$ and $Q(x)$ are polynomials, is continuous for all values of x for which $Q(x) \neq 0$.

Proof This is an immediate consequence of Theorem 9 and the division part of Theorem 7.

We conclude this section by proving that "a continuous function of a continuous function is continuous."

Theorem 11 *If $g(x)$ is continuous at a and $f(x)$ is continuous at $g(a)$, then the composite function $f(g(x))$ is continuous at a.*

Proof Let $\epsilon > 0$ be given. Since $f(x)$ is continuous at $g(a)$, we know that there exists a $\delta_1 > 0$ such that

$$|f(g(x)) - f(g(a))| < \epsilon \tag{4}$$

if

$$|g(x) - g(a)| < \delta_1. \tag{5}$$

But $g(x)$ is continuous at a, so there exists a $\delta > 0$ such that $|x - a| < \delta$ implies $|g(x) - g(a)| < \delta_1$. We therefore see that $|x - a| < \delta$ implies (5), which in turn implies (4), and this is all that is needed to complete the proof.

C.3
SOME DEEPER PROPERTIES OF CONTINUOUS FUNCTIONS

We recall that a *closed interval* $[a, b]$ on the x-axis is an interval which includes its endpoints a and b. A function is said to be *continuous on a closed interval* if it is defined and continuous at each point of the interval. Functions of this kind have several important properties that we now discuss and prove.

Theorem 1 (Boundedness Theorem) *Let $f(x)$ be a function continuous on a closed interval $[a, b]$. Then $f(x)$ is bounded on $[a, b]$, that is, there exists a number C with the property that $|f(x)| \leq C$ for all x in $[a, b]$.*

A good way to study a theorem like this critically is to see what happens if the hypotheses are weakened or removed. In Theorem 1 there are two main hypotheses: (1) the interval $[a, b]$ is closed; and (2) the function $f(x)$ is continuous at each point of the interval. We show by examples that if either hypothesis is weakened, then the conclusion of the theorem can be false.

Example 1 The function $f(x) = 1/x$ is clearly continuous on the closed interval $[1, 2]$, so according to Theorem 1, $f(x)$ should be bounded on this interval. Indeed, a bound C is easy to find:

$$|f(x)| \le 1 \text{ for all } x \text{ in } [1, 2].$$

Further (see Fig. C.3), $f(x)$ is continuous on the closed interval $[1/n, 2]$ for any positive integer n, and in this case the number n is a bound:

$$|f(x)| \le n \text{ for all } x \text{ in } [1/n, 2].$$

On the other hand, $f(x)$ is also continuous on the nonclosed interval $(0, 2]$, but $f(x)$ is not bounded on this interval. For, no matter how large a value of C we take, there are points in the interval for which $f(x) > C$; specifically, if $0 < x < 1/C$, then $f(x) = 1/x > C$. This shows that the hypothesis requiring that the interval $[a, b]$ be closed is necessary.

We now extend the definition of $f(x)$ to include the point $x = 0$, by putting

$$f(x) = \begin{cases} 1/x & \text{if } 0 < x \le 2, \\ 0 & \text{if } x = 0. \end{cases}$$

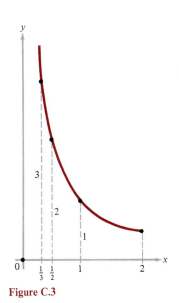

Figure C.3

This function is defined on the entire closed interval $[0, 2]$, and it is unbounded on this interval for the same reason. This time the conclusion of Theorem 1 is false because the function $f(x)$ is not continuous at each point of the closed interval: it is discontinuous at the single point $x = 0$.

These remarks show that the hypotheses of Theorem 1 cannot be weakened, and the following proof demonstrates that with both hypotheses in place the conclusion of the theorem is inescapable.

Proof of Theorem 1* Our proof uses the fact that a nonempty set of real numbers with an upper bound necessarily has a least upper bound (see Appendix C.1). Let S be the set of all points c in $[a, b]$ with the property that $f(x)$ is bounded on $[a, c]$. It is clear that S is nonempty and has b as an upper bound, and therefore has a least upper bound which we denote by c_0. We claim that $c_0 = b$. To establish this, suppose that $c_0 < b$. Since $f(x)$ is continuous at $x = c_0$, it is easy to see that $f(x)$ is bounded on $[c_0 - \epsilon, c_0 + \epsilon]$ for some $\epsilon > 0$. Since $f(x)$ is also bounded on $[a, c_0 - \epsilon]$, it is clearly bounded on $[a, c_0 + \epsilon]$. This contradicts the fact that c_0 is the least upper bound of S, so $c_0 = b$. This tells us that $f(x)$ is bounded on $[a, c]$ for every $c < b$. One more step is needed to finish the proof. Since $f(x)$ is continuous at $x = b$, it is bounded on some closed interval $[b - \epsilon, b]$. By what we just proved, $f(x)$ is also bounded on $[a, b - \epsilon]$, so it is bounded on all of $[a, b]$.

If a function $f(x)$ is bounded on $[a, b]$, then its range—the set of all its values—has an upper bound and a lower bound. If M and m are the least upper bound and greatest lower bound of the range, then

$$m \le f(x) \le M \text{ for all } x \text{ in } [a, b].$$

For bounded functions in general, the numbers M and m need not belong to the range. However, our next theorem asserts that if $f(x)$ is continuous, then both numbers M and m are actually assumed as values of the function.

---

* Some of the details of the proofs in this appendix are left for students to fill in.

Theorem 2 (Extreme Value Theorem) *Let $f(x)$ be a function continuous on a closed interval $[a, b]$. Then $f(x)$ assumes a maximum value M and a minimum value m, that is, there exist points x_1 and x_2 in $[a, b]$ such that*

$$f(x_1) \leq f(x) \leq f(x_2)$$

for all x in $[a, b]$.

This statement is intuitively clear if we think of a continuous function on a closed interval as one whose graph consists of a single continuous piece, without any gaps or holes; for as we move along the curve from the left endpoint $(a, f(a))$ to the right endpoint $(b, f(b))$, we feel compelled to believe that there must be a high point on the curve where $f(x)$ has its maximum value and a low point where $f(x)$ has its minimum value. This is true, but the situation is again very delicate, because if either hypothesis is weakened—even slightly—then the conclusion of the theorem can be false.

Example 2 Consider the function $f(x)$ defined by $f(x) = x$ on the nonclosed interval $[0, 1)$, and also the function $g(x)$ defined by

$$g(x) = \begin{cases} x & \text{if } 0 \leq x < 1, \\ 0 & \text{if } 1 \leq x \leq 2 \end{cases}$$

on the closed interval $[0, 2]$. Both functions are shown in Fig. C.4. The function $f(x)$ does not assume a maximum value even though it is continuous on the interval $[0, 1)$, because this interval is not closed; and the function $g(x)$ does not assume a maximum value even though the interval $[0, 2]$ is closed, because $g(x)$ is discontinuous at the single point $x = 1$. In each case the values of the function get close to the number 1 (which is the least upper bound M of the range) as $x \to 1$ from the left, but there is no point where the function actually *has* the value 1.

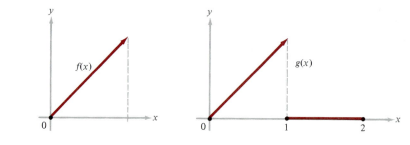

Figure C.4

Proof of Theorem 2 We prove the statement about assuming a maximum value. By Theorem 1, $f(x)$ is bounded on $[a, b]$, so the range has an upper bound and therefore a least upper bound M. We must show that there exists a point x_2 in $[a, b]$ such that $f(x_2) = M$. Suppose there is no such point, that is, suppose that $f(x) < M$ for all x in $[a, b]$. Then $M - f(x)$ is positive on $[a, b]$, the function

$$g(x) = \frac{1}{M - f(x)}$$

is continuous on $[a, b]$, and Theorem 1 implies that this function is bounded. This means that there exists a number C such that

$$\frac{1}{M - f(x)} \leq C$$

for all x in $[a, b]$, so

$$\frac{1}{C} \le M - f(x) \qquad \text{or} \qquad f(x) \le M - \frac{1}{C}.$$

This contradicts the fact that M is the least upper bound of the set of all $f(x)$'s, and we are thereby forced to the desired conclusion: there exists at least one point x_2 in $[a, b]$ for which $f(x_2) = M$. The statement that $f(x)$ assumes a minimum value at some point x_1 is proved similarly.

The Extreme Value Theorem says that a function continuous on a closed interval actually takes on a maximum value and a minimum value. There is a companion to this theorem which states that such a function also takes on every value between its maximum and minimum values. Thus, a function continuous on a closed interval has a range which is itself a closed interval. To put it another way, such a function does not skip any values. We begin with a preliminary theorem that has many applications of its own (see Section 4.6).

Theorem 3 *Let $f(x)$ be a function continuous on a closed interval $[a, b]$. If $f(a)$ and $f(b)$ have opposite signs, that is, if*

$$f(a) < 0 < f(b) \qquad \text{or} \qquad f(a) > 0 > f(b),$$

then there exists a point c between a and b such that $f(c) = 0$.

This says — in effect — that the graph of a function continuous on a closed interval cannot get from one side of the x-axis to the other side without actually crossing this axis at a definite point (Fig. C.5, left). However, this conclusion can be false if the function fails to be continuous even at a single point. This is shown (Fig. C.5, right) by the function $f(x)$ defined on the interval $[1, 3]$ by

$$f(x) = \begin{cases} -1 & \text{if } 1 \le x < 2, \\ 1 & \text{if } 2 \le x \le 3. \end{cases}$$

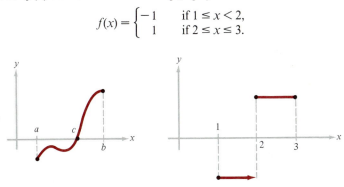

Figure C.5

Proof of Theorem 3   Suppose first that $f(a) < 0 < f(b)$. Since $f(a) < 0$ and $f(x)$ is continuous at $x = a$, there exists a number d in the open interval (a, b) such that $f(x)$ is negative on $[a, d)$. Let c be the least upper bound of the set of all such d's, and observe that $f(x)$ is negative for all $x < c$. It cannot be true that $f(c) > 0$, for by continuity this would imply that $f(x)$ is positive on some interval $(c - \epsilon, c]$, contrary to what we have just observed. Also, it cannot be true that $f(c) < 0$, for by continuity this would imply that $f(x)$ is negative on some interval $[a, c + \epsilon)$, contrary to the definition of c. We conclude that $f(c) = 0$. The argument for the other case is similar.

Theorem 4 (The Intermediate Value Theorem)   *Let $f(x)$ be a function continuous on a closed interval $[a, b]$. If M and m are the maximum and minimum values of* *f(x)* on [*a*, *b*], and if *C* is any number between *M* and *m*, so that $m < C < M$, then there exists a point *c* in [*a*, *b*] such that $f(c) = C$.

Proof   The function $g(x) = f(x) - C$ is also continuous on [*a*, *b*]. If x_1 and x_2 are points in [*a*, *b*] at which $f(x_1) = m$ and $f(x_2) = M$, then $g(x)$ is negative at x_1 and positive at x_2:

$$g(x_1) = f(x_1) - C = m - C < 0$$

and

$$g(x_2) = f(x_2) - C = M - C > 0.$$

By Theorem 3, there exists a point *c* between x_1 and x_2 (and therefore in [*a*, *b*]) such that $g(c) = 0$. But this means $f(c) - C = 0$ or $f(c) = C$.

As another consequence of Theorem 3, we have

Theorem 5   *Let f(x) be a function continuous on the closed unit interval* [0, 1] *which has the further property that its values also lie in this interval (Fig. C.6). Then there exists at least one point c in* [0, 1] *such that f(c) = c.*

Proof   The function $g(x) = f(x) - x$ is continuous on [0, 1] and has the property that $g(0) = f(0) - 0 = f(0) \geq 0$ and $g(1) = f(1) - 1 \leq 0$. By Theorem 3, there exists a point *c* in [0, 1] such that $g(c) = f(c) - c = 0$, so $f(c) = c$.

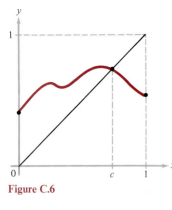

Figure C.6

A function $f(x)$ with the properties assumed here is often called a *continuous mapping of the interval* [0, 1] *into itself,* and the point *c* is called a *fixed point* of this mapping. Theorem 5 is a special case of a famous and far-reaching theorem of modern mathematics called *Brouwer's fixed point theorem,* which asserts that continuous mappings of certain very general spaces into themselves always have fixed points.

C.4

THE MEAN VALUE THEOREM

This theorem is one of the most useful facts in the theoretical part of calculus. In geometric language, it is easy to state and intuitively plausible. It asserts that between any two points *P* and *Q* on the graph of a differentiable function there must exist at least one point where the tangent line is parallel to the chord joining *P* and *Q*, as shown in Fig. C.7. For the curve in the figure there are two such points. There may be many or there may be only one, but the theorem guarantees that there must always be at least one such point. By using the notation in the figure, we can express the statement of the theorem analytically by saying that there exists at least one number *c* between *a* and *b* ($a < c < b$) with the property that

$$f'(c) = \frac{f(b) - f(a)}{b - a}.$$

The significance of the mean value theorem lies not in itself but in its consequences, for it provides a convenient way of getting a grip on many theoretical facts of practical importance. This will become clear in Theorems 3 and 4, and also in later sections of this appendix.

A rigorous proof of the mean value theorem is usually developed in the following way. We begin by establishing the special case of the theorem in which the points *P* and *Q* both lie on the *x*-axis:

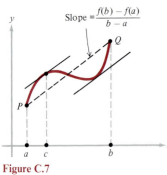

Figure C.7

Theorem 1 (Rolle's Theorem) *If a function $f(x)$ is continuous on the closed interval $a \leq x \leq b$ and differentiable in the open interval $a < x < b$, and if $f(a) = f(b) = 0$, then there exists at least one number c between a and b with the property that $f'(c) = 0$.*

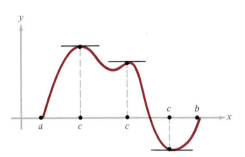

Figure C.8

This theorem says that if a differentiable curve touches the x-axis at two points, then there must be at least one point on the curve between these points at which the tangent is horizontal (Fig. C.8). Equivalently, the zeros of a differentiable function are always separated by zeros of its derivative.

Economists have a maxim, "There is no such thing as a free lunch." For us — in the realm of pure mathematics — this means we cannot get something for nothing; or in other words, strong conclusions require strong hypotheses. The conclusion of Rolle's Theorem depends heavily on its hypotheses, and the following examples show that these hypotheses cannot be weakened without destroying the conclusion.

Example 1 The function

$$f(x) = \begin{cases} x, & 0 \leq x \leq 1, \\ 2 - x, & 1 \leq x \leq 2 \end{cases}$$

(see Fig. C.9) is zero at $x = 0$ and $x = 2$, and is continuous on the closed interval $0 \leq x \leq 2$. It is differentiable in the open interval $0 < x < 2$, except at the single point $x = 1$, where the derivative does not exist. The derivative $f'(x)$ is clearly not zero at any point in the interval, and this failure of the conclusion of Rolle's Theorem arises from the fact that the function fails to be differentiable at a single crucial point.

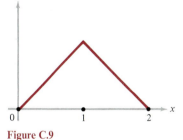

Figure C.9

Example 2 The function

$$f(x) = \begin{cases} x, & 0 \leq x < 1, \\ 0, & x = 1 \end{cases}$$

(see Fig. C.10) is zero at $x = 0$ and $x = 1$, and is differentiable in the open interval $0 < x < 1$. It is continuous on the closed interval $0 \leq x \leq 1$, except at the single point $x = 1$. The derivative $f'(x)$ is not zero at any point in the interval, and in this case the failure of the conclusion of Rolle's Theorem arises from the discontinuity of the function at a single point.

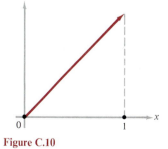

Figure C.10

Proof of Theorem 1 By Theorem 2 in Section C.3, our continuity hypothesis implies that $f(x)$ assumes a maximum value M and a minimum value m on $[a, b]$. The fact that $f(x)$ is zero at the endpoints a and b tells us that $m \leq 0 \leq M$. If $f(x)$ is zero at every point of $[a, b]$, then clearly $f'(c) = 0$ for every c in (a, b), and in this trivial case the conclusion is true. We may therefore suppose that the function assumes nonzero values, so that either $M > 0$ or $m < 0$ (or perhaps both). We first consider the case in which $M > 0$. If c is a point at which $f(c) = M$, then $a < c < b$ because the function is zero at the endpoints a and b. Since $f(x)$ is differentiable in the open interval $a < x < b$, the derivative

$$f'(c) = \lim_{x \to c} \frac{f(x) - f(c)}{x - c} \tag{1}$$

exists.* It is part of the meaning of (1) that this limit must exist and have the same value when x approaches c from the left and from the right. If x approaches c from the left, we have

$$x - c < 0 \quad \text{and} \quad f(x) - f(c) \le 0,$$

where the second inequality follows from the fact that $f(c) = M$ is a maximum value. This implies that

$$f'(c) = \lim_{x \to c-} \frac{f(x) - f(c)}{x - c} \ge 0. \tag{2}$$

Similarly, if x approaches c from the right, we have

$$x - c > 0 \quad \text{and} \quad f(x) - f(c) \le 0,$$

so

$$f'(c) = \lim_{x \to c+} \frac{f(x) - f(c)}{x - c} \le 0. \tag{3}$$

We conclude from (2) and (3) that $f'(c) = 0$, as asserted. If $M = 0$, then $m < 0$, and this case can be treated by a similar argument.

Our main theorem can now be stated as follows (see Fig. C.7).

Theorem 2 (Mean Value Theorem) *If a function $f(x)$ is continuous on the closed interval $a \le x \le b$ and differentiable in the open interval $a < x < b$, then there exists at least one number c between a and b with the property that*

$$f'(c) = \frac{f(b) - f(a)}{b - a}. \tag{4}$$

Proof It is easy to see that the equation of the chord joining P and Q in Fig. C.7 is

$$y = f(a) + \left[\frac{f(b) - f(a)}{b - a} \right] (x - a).$$

The function

$$F(x) = f(x) - f(a) - \left[\frac{f(b) - f(a)}{b - a} \right] (x - a) \tag{5}$$

is therefore the vertical distance from the chord up to the graph of $y = f(x)$. It is easy to see that the function (5) satisfies the hypotheses of Theorem 1, so there exists a point c between a and b with the property that $F'(c) = 0$. But this is equivalent to

---

* Equation (1) is clearly an equivalent way of writing

$$f'(c) = \lim_{\Delta x \to 0} \frac{f(c + \Delta x) - f(c)}{\Delta x}.$$

$$f'(c) - \frac{f(b) - f(a)}{b - a} = 0,$$

which in turn is equivalent to (4), so the proof is complete.

We now consider some of the applications of this theorem.

It is clear that the derivative of a constant function is zero. Is the converse true? That is, if the derivative of a function is zero on an interval, is the function necessarily constant on that interval? At the beginning of Section 5.3 we encountered an important piece of reasoning about indefinite integrals in which this converse was needed, and we took it for granted. We are now in a position to prove it by using the Mean Value Theorem.

Theorem 3   *If a function $f(x)$ is continuous on a closed interval I, and if $f'(x)$ exists and is zero in the interior of I, then $f(x)$ is constant on I.*

Proof   To say that $f(x)$ is constant on I means that it has only a single value there. To prove that this is the case, suppose it has two different values, say $f(a) \neq f(b)$ for $a < b$ in I. Then the Mean Value Theorem implies that for some c between a and b we have

$$f'(c) = \frac{f(b) - f(a)}{b - a} \neq 0.$$

But this cannot be true, since $f'(x) = 0$ at all points in the interior of I. This contradiction shows that $f(x)$ cannot have different values in I, and is therefore constant on I, as we wished to prove.

At the beginning of Chapter 4 we based our work on curve-sketching on the "intuitively obvious" fact that a function is increasing or decreasing according as its derivative is positive or negative. The Mean Value Theorem makes it possible to give a rigorous proof of this.

Theorem 4   *Let $f(x)$ be a function continuous on a closed interval I and differentiable in the interior of I. If $f'(x) > 0$ in the interior of I, then $f(x)$ is increasing on I. Similarly, if $f'(x) < 0$ in the interior of I, then $f(x)$ is decreasing on I.*

Proof   We shall prove only the first statement, in which we assume that $f'(x) > 0$ in the interior of I. For any two points $a < b$ in I, the Mean Value Theorem tells us that

$$f'(c) = \frac{f(b) - f(a)}{b - a}$$

for some c between a and b. But $f'(c) > 0$, so the fraction on the right side of this equation is positive. Since $b - a$ is positive, it follows that $f(b) - f(a)$ is also positive, so $f(a) < f(b)$ and consequently $f(x)$ is increasing on I.

Finally, we use Rolle's Theorem to prove a technical extension of the Mean Value Theorem that is needed for establishing L'Hospital's rule in Chapter 12.

Theorem 5 (Generalized Mean Value Theorem)   *Let $f(x)$ and $g(x)$ be continuous on the closed interval $a \leq x \leq b$ and differentiable in the open interval $a < x < b$, and assume further that $g'(x) \neq 0$ for $a < x < b$. Then there exists at least one number c between a and b with the property that*

$$\frac{f'(c)}{g'(c)} = \frac{f(b) - f(a)}{g(b) - g(a)}. \tag{6}$$

Proof   We begin by noticing that if $g(a) = g(b)$, then by Rolle's Theorem $g'(x)$ vanishes at some point between a and b, contrary to hypothesis. Therefore $g(a) \neq g(b)$, and the right side of (6) makes sense. To prove the theorem, consider the function

$$F(x) = [f(b) - f(a)][g(x) - g(a)] - [f(x) - f(a)][g(b) - g(a)].$$

It is easy to see that this function satisfies the hypotheses of Rolle's Theorem, so there exists a point c between a and b with the property that $F'(c) = 0$. But this is equivalent to

$$[f(b) - f(a)]g'(c) - f'(c)[g(b) - g(a)] = 0,$$

which is equivalent to (6).

Students should notice that this theorem reduces to Theorem 2 if $g(x) = x$.

C.5

THE INTEGRABILITY OF CONTINUOUS FUNCTIONS

In Section 6.4 the definite integral of a function over an interval was defined by means of a complicated passage to the limit, as follows.

We start with an arbitrary bounded function $f(x)$ defined on a closed interval $[a, b]$. We subdivide this interval into n equal or unequal subintervals by inserting $n - 1$ points of division $x_1, x_2, \ldots, x_{n-1}$, so that

$$a = x_0 < x_1 < x_2 < \cdots < x_{n-1} < x_n = b. \tag{1}$$

These points are said to constitute a *partition* P of $[a, b]$ into the subintervals

$$[x_0, x_1], [x_1, x_2], \ldots, [x_{n-1}, x_n].$$

If $\Delta x_k = x_k - x_{k-1}$ is the length of the kth subinterval, then the length of the longest subinterval is called the *norm* of the partition and is denoted by the symbol $\|P\|$,

$$\|P\| = \max \{\Delta x_1, \Delta x_2, \ldots, \Delta x_n\}.$$

In each of the subintervals $[x_{k-1}, x_k]$ we choose an arbitrary point x_k^*. We now multiply the value of the function $f(x)$ at the point x_k^* by the length Δx_k of the corresponding subinterval and form the sum of these products as the subscript k varies from 1 to n,

$$\sum_{k=1}^{n} f(x_k^*) \, \Delta x_k. \tag{2}$$

For each positive integer n we consider all possible partitions (1) and all possible choices of the points x_k^*, and therefore all possible values of the sum (2). If there exists a number I such that the sum (2) approaches I as $n \to \infty$ and $\|P\| \to 0$, regardless of how the partitions P are formed and the points x_k^* are chosen, then we call this number I the *definite integral* (or briefly the *integral*) of $f(x)$ on $[a, b]$ and denote it by the symbol

$$I = \int_a^b f(x) \, dx.$$

Under these circumstances the function $f(x)$ is said to be *integrable* on $[a, b]$. It is customary to express these ideas by writing

$$\int_a^b f(x) \, dx = \lim_{\|P\| \to 0} \sum_{k=1}^{n} f(x_k^*) \, \Delta x_k, \tag{3}$$

where there is no need to specify that $n \to \infty$ because this is implied by the stronger condition $\|P\| \to 0$.

As we said at the beginning, the limit operation in (3) is quite complicated and bears only a superficial resemblance to such straightforward limits as

$$\lim_{x \to 2} (x^2 + 1) = 5 \qquad \text{and} \qquad \lim_{n \to \infty} \left(2 + \frac{1}{n} \right) = 2.$$

In each of these cases we consider the behavior of a certain function in terms of the behavior of an independent variable, but (3) does not lend itself to this way of thinking. We could try to use $\|P\|$ as an independent variable, and describe the limit in terms of the idea expressed by the symbol $\|P\| \to 0$. But this is difficult, because the sum (2) is not a single-valued function of the quantity $\|P\|$; to a given value of $\|P\|$ there correspond an infinite number of different partitions P and an infinite number of ways of choosing the points x_k^*, and therefore an infinite number of values of the sum (2).

The complexity of the limit operation in (3) is a considerable inconvenience when it comes to giving rigorous proofs of theorems. The cumbersome notation required for such proofs forces the reasoning itself to be awkward and clumsy. For this reason, it is customary in modern treatments of the theory of integration to define the definite integral in a very different way, one which avoids appealing to any kind of passage to the limit. We now describe this more convenient approach and use it to prove our main theorem.

We therefore ignore our previous definition and begin all over again at the beginning, with an arbitrary bounded function $f(x)$ defined on a closed interval $[a, b]$. Since $f(x)$ is bounded, it has a greatest lower bound m and a least upper bound M. If P is any given partition of $[a, b]$, we denote by m_k and M_k the greatest lower bound and least upper bound of $f(x)$ on the kth subinterval $[x_{k-1}, x_k]$. (If $f(x)$ were assumed to be continuous on $[a, b]$, then by Theorem 2 in Appendix C.3 the m's and M's would be minimum values and maximum values of the function. But we are not assuming continuity at this stage, so we must work instead with greatest lower bounds and least upper bounds.) We now form the *lower sum*

$$s_P = \sum_{k=1}^{n} m_k \, \Delta x_k$$

and the *upper sum*

$$S_P = \sum_{k=1}^{n} M_k \, \Delta x_k.$$

It is obvious that $s_P \le S_P$. Further, we have the important

---

Lemma *Every lower sum is less than or equal to every upper sum; that is, if P_1 and P_2 are any two partitions of $[a, b]$, then $s_{P_1} \le S_{P_2}$.*

Proof It is easy to see that if a single point is added to a partition, then the lower sum is unchanged or increases and the upper sum is unchanged or decreases; and the same is true if any finite number of points are added to produce a refinement of the given partition. We now apply this fact to the new partition P_3 which is formed from the points of P_1 and P_2 taken together. Since P_3 is clearly a refinement of both P_1 and P_2, it follows that

$$s_{P_1} \le s_{P_3} \le S_{P_3} \le S_{P_2},$$

which completes the argument.

Among other things, this lemma tells us that every upper sum is an upper bound for the set of all lower sums, and that every lower sum is a lower bound for the set of all upper sums. We can therefore form the least upper bound of all possible lower sums, which is called the *lower integral* and denoted by

$$\underline{I} = \int_a^b f(x)\,dx.$$

Similarly, the greatest lower bound of all upper sums is called the *upper integral* and denoted by

$$\bar{I} = \int_a^{\bar{b}} f(x)\,dx.$$

At this point we make a further application of the lemma to conclude that

$$\underline{I} \le \bar{I}.$$

Accordingly, every bounded function defined on a closed interval has a lower integral and an upper integral, and these two integrals are defined without making any appeal to the concept of a limit. If the lower and upper integrals coincide, then we call their common value the *integral* of $f(x)$ on $[a, b]$ and denote it by the usual symbol,

$$I = \int_a^b f(x)\,dx;$$

and in this case the function $f(x)$ is said to be *integrable* on $[a, b]$. On the other hand, it is quite possible to have $\underline{I} < \bar{I}$, in which case $f(x)$ is *not* integrable. The function described in Remark 4 of Section 6.4 provides a good example of this recalcitrant behavior.

We now come to our main theorem, which guarantees that most of the functions we meet in practice are integrable. First, a bit of new terminology that will be useful in the proof. If $f(x)$ is a bounded function defined on an interval $[a, b]$, and if m and M are its greatest lower bound and least upper bound on this interval, then the difference $M - m$ is called the *oscillation* of $f(x)$ on $[a, b]$.

Theorem *If a function $f(x)$ is continuous on a closed interval $[a, b]$, then it is integrable on $[a, b]$.*

Proof Consider a partition P of $[a, b]$ into subintervals $[x_{k-1}, x_k]$, and form the lower and upper sums

$$s_P = \sum_{k=1}^n m_k\,\Delta x_k \qquad \text{and} \qquad S_P = \sum_{k=1}^n M_k\,\Delta x_k.$$

The difference between these sums is

$$S_P - s_P = \sum_{k=1}^n (M_k - m_k)\,\Delta x_k, \tag{4}$$

where $M_k - m_k$ is the oscillation of $f(x)$ on the kth subinterval $[x_{k-1}, x_k]$. If we can show that the difference (4) can be made as small as we please by choosing a suitable partition P, then this will clearly be enough to prove the theorem. We accomplish this in the following way. Let ϵ be a given small positive number. If it can be shown that there exists a partition P such that the oscillation of the function is less than $\epsilon/(b - a)$ on every subinterval, that is,

$$M_k - m_k < \frac{\epsilon}{b - a} \qquad \text{for } k = 1, 2, \ldots, n,$$

then it will follow that

$$S_P - s_P = \sum_{k=1}^{n} (M_k - m_k)\,\Delta x_k < \frac{\epsilon}{b-a} \sum_{k=1}^{n} \Delta x_k = \frac{\epsilon}{b-a}(b-a) = \epsilon.$$

Since ϵ can be made as small as we please, this will complete the proof.

We must therefore prove the existence of a partition P with the required property. If we simplify the notation by writing $\epsilon_1 = \epsilon/(b-a)$, so that ϵ_1 is perceived as merely another positive number that can be made as small as we please, then this property of the partition P can be stated as follows: The oscillation of the continuous function $f(x)$ on every subinterval of the partition must be less than ϵ_1.*

We give an indirect proof, that is, we assume that for at least one number $\epsilon_1 > 0$ no partition of the desired type exists, and we show that this assumption leads to a contradiction. Let c be the midpoint of $[a, b]$. Then no partition of the desired type exists for at least one of the two subintervals $[a, c]$ and $[c, b]$, for if each of these subintervals has such a partition, then the full interval $[a, b]$ also has. Let $[a_1, b_1]$ be that half of $[a, b]$ with no such partition; and if both halves have no such partition, let $[a_1, b_1]$ be the left half, $[a, c]$. Now bisect $[a_1, b_1]$, and in the same way produce one of its halves, say $[a_2, b_2]$, with no such partition; and continue the process indefinitely. We observe that the oscillation of $f(x)$ on the nth subinterval $[a_n, b_n]$ is at least ϵ_1, and also that the length of this subinterval is $(b-a)/2^n$. Let a_0 be the least upper bound of the set of left endpoints $a_1, a_2, a_3, \ldots$ of this nested sequence of subintervals. Then a_0 certainly lies in the interval $[a, b]$; and by the continuity of $f(x)$ at a_0, there exists an interval $(a_0 - \delta, a_0 + \delta)$ in which the oscillation of $f(x)$ is less than ϵ_1. However, if n is large enough, the interval $[a_n, b_n]$ lies wholly within the interval $(a_0 - \delta, a_0 + \delta)$, and therefore the oscillation of $f(x)$ on $[a_n, b_n]$ must also be less than ϵ_1, contradicting our previous inference that the oscillation of $f(x)$ on $[a_n, b_n]$ is at least ϵ_1. This contradiction finally concludes the proof of the theorem.

If students wonder whether a discontinuous function can be integrable, the answer is Yes. The function whose graph is shown in Fig. C.11 provides an example of this assertion. It is defined on the closed interval $[0, 1]$, and its values are

$$\frac{1}{2} \text{ for } 0 \leq x < \frac{1}{2},$$

$$\frac{3}{4} \text{ for } \frac{1}{2} \leq x < \frac{3}{4},$$

$$\frac{7}{8} \text{ for } \frac{3}{4} \leq x < \frac{7}{8},$$

$$\cdots$$

$$1 \text{ for } x = 1.$$

This function has an infinite number of points of discontinuity, but it also has the property of being *nondecreasing,* in the sense that $x_1 < x_2$ implies $f(x_1) \leq f(x_2)$, and any such function is integrable on any closed interval $[a, b]$. Students are invited to prove this for themselves by noticing that in this case the difference (4) can be written as

$$S_P - s_P = \sum_{k=1}^{n} (M_k - m_k)\,\Delta x_k$$

$$\leq \|P\| \sum_{k=1}^{n} (M_k - m_k) = \|P\|[f(b) - f(a)].$$

---

* This fact about a continuous function defined on a closed interval is usually referred to in the literature as the *theorem on uniform continuity.*

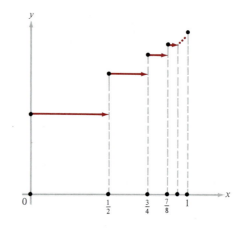

Figure C.11

The set of all integrable functions can be characterized in a simple and absolutely precise way, but we do not pursue this matter any further here.

C.6

ANOTHER PROOF OF THE FUNDAMENTAL THEOREM OF CALCULUS

The proof given here uses the Mean Value Theorem established in Appendix C.4, and assumes that students understand the concepts developed in Appendix C.5.

To set the stage for the argument, we consider a function $f(x)$ that is continuous on a closed interval $[a, b]$. If $F(x)$ is any function such that $F'(x) = f(x)$, we must prove that

$$\int_a^b f(x)\, dx = F(b) - F(a). \tag{1}$$

We accomplish this by showing that the number on the right of (1) lies between the lower sum and the upper sum associated with an arbitrary partition

$$a = x_0 < x_1 < x_2 < \cdots < x_{n-1} < x_n = b \tag{2}$$

of the interval $[a, b]$.

The reasoning is as follows. The function $F(x)$ satisfies the hypotheses of the Mean Value Theorem on each subinterval of the partition (2). This theorem therefore guarantees the existence of points $x_1^*, x_2^*, \ldots, x_n^*$ in these subintervals such that

$$F(x_1) - F(a) = F'(x_1^*)(x_1 - a) = f(x_1^*)\, \Delta x_1,$$
$$F(x_2) - F(x_1) = F'(x_2^*)(x_2 - x_1) = f(x_2^*)\, \Delta x_2,$$
$$\cdots$$
$$F(b) - F(x_{n-1}) = F'(x_n^*)(b - x_{n-1}) = f(x_n^*)\, \Delta x_n.$$

If we add these equations and take advantage of the cancellations on the left, we get

$$F(b) - F(a) = \sum_{k=1}^n f(x_k^*)\, \Delta x_k. \tag{3}$$

The right-hand side of (3) clearly lies between the lower sum and the upper sum associated with the partition (2), so the proof is complete.

In this discussion we begin by defining e to be the limit

$$e = \lim_{n \to \infty} \left(1 + \frac{1}{n}\right)^n. \qquad (1)$$

We then carefully extend this formula step by step until we reach the more general conclusion that

$$e = \lim_{h \to 0} (1 + h)^{1/h}, \qquad (2)$$

where h is allowed to approach 0 in any manner whatever, through rational or irrational, positive or negative, values.

Our first task is to prove the existence of the limit (1), and thereby to legitimize this definition of e. By the binomial theorem the quantity

$$x_n = \left(1 + \frac{1}{n}\right)^n$$

can be expressed as the following sum of $n + 1$ terms,

$$1 + n \cdot \frac{1}{n} + \frac{n(n-1)}{1 \cdot 2} \cdot \frac{1}{n^2} + \frac{n(n-1)(n-2)}{1 \cdot 2 \cdot 3} \cdot \frac{1}{n^3} + \cdots + \frac{1}{n^n}$$

$$= 1 + 1 + \frac{1}{1 \cdot 2}\left(1 - \frac{1}{n}\right) + \frac{1}{1 \cdot 2 \cdot 3}\left(1 - \frac{1}{n}\right)\left(1 - \frac{2}{n}\right) + \cdots + \frac{1}{n^n}. \qquad (3)$$

As n increases, the number of terms in this sum increases, and also each term after the second increases. This shows that

$$x_1 < x_2 < x_3 < \cdots < x_n < x_{n+1} < \cdots. \qquad (4)$$

Also, the expansion (3) tells us that

$$x_n < 1 + 1 + \frac{1}{1 \cdot 2} + \frac{1}{1 \cdot 2 \cdot 3} + \cdots + \frac{1}{1 \cdot 2 \cdot 3 \cdots n}$$

$$< 1 + 1 + \left(\frac{1}{2} + \frac{1}{2^2} + \cdots + \frac{1}{2^{n-1}}\right) < 1 + 1 + 1 = 3, \qquad (5)$$

since the expression in parentheses is part of the familiar geometric series

$$\frac{1}{2} + \frac{1}{2^2} + \cdots + \frac{1}{2^{n-1}} + \cdots = 1.$$

By (4) and (5) the x_n's steadily increase but always remain < 3, so they necessarily approach a limiting value. In the present context this limiting value is e by definition. This argument proves (1), and also we clearly have

$$e = \lim_{n \to \infty} \left(1 + \frac{1}{n + 1}\right)^{n+1},$$

which will be needed below.

We next consider the limit (2) for the special case in which h is required to approach 0 through positive values. When $h < 1$, there exists a unique positive integer n such that

$$n \leq \frac{1}{h} < n + 1.$$

This implies that

$$\left(1 + \frac{1}{n+1}\right)^n < (1 + h)^{1/h} < \left(1 + \frac{1}{n}\right)^{n+1},$$

which in turn can be written as

$$\frac{[1 + 1/(n+1)]^{n+1}}{1 + 1/(n+1)} < (1 + h)^{1/h} < \left(1 + \frac{1}{n}\right)^n\left(1 + \frac{1}{n}\right). \tag{6}$$

As $h \to 0$, $n \to \infty$ and the first and third terms of the inequality (6) approach e. Since $(1 + h)^{1/h}$ is caught between the two, it must have the same limit, and therefore (2) is proved for the case in which $h \to 0$ through positive values.

We conclude our analysis by establishing (2) for the case in which h approaches 0 through negative values. If we put $h = -k$, then

$$(1 + h)^{1/h} = (1 - k)^{-1/k} = \left(\frac{1}{1-k}\right)^{1/k}$$

$$= \left(1 + \frac{k}{1-k}\right)^{1/k} = \left(1 + \frac{k}{1-k}\right)^{(1-k)/k}\left(1 + \frac{k}{1-k}\right) \to e \cdot 1 = e,$$

by the result of the previous paragraph.

C.8
THE VALIDITY OF INTEGRATION BY INVERSE SUBSTITUTION

In making the direct substitutions discussed in Section 10.2, our procedure was to put $u = g(x)$ where $g(x)$ was part of the integrand. For this method to work, we had to have $du = g'(x)\, dx$ as another part of the integrand, and this meant that altogether the integrand had to have a rather special form.

A much more natural way to change the variable in an integral $\int f(x)\, dx$ is to introduce a new variable u by writing $x = h(u)$ and $dx = h'(u)\, du$, where $h(u)$ is some function that is suggested by the form of the integral. This means that if we translate the given integral from the x-notation to the u-notation by writing

$$\int f(x)\, dx = \int f[h(u)]h'(u)\, du = \int g(u)\, du, \tag{1}$$

where $g(u) = f[h(u)]h'(u)$, then we hope the integral on the right will be easy to calculate. In fact, if

$$\int g(u)\, du = G(u), \tag{2}$$

then we expect to have

$$\int f(x)\, dx = G[k(x)], \tag{3}$$

where $u = k(x)$ is the inverse function of $x = h(u)$.* This process is called *inverse substitution*. It is a very useful method, *if* we can find $G(u)$ and *if* we know the inverse function $u = k(x)$. These remarks constitute a general description of what is going on in the method of trigonometric substitutions, and also in some of the methods suggested in Section 10.8.

---

* That is, $u = k(x)$ is the result of solving $x = h(u)$ for u in terms of x. The concept of an inverse function is discussed in Remark 2 in Section 9.5.

We can prove the validity of inverse substitution as follows. The point is this: In direct substitution as discussed in Section 10.2 we used the integral transformation (1) in the other direction, to calculate $\int g(u)\,du$. We showed that if

$$\int f(x)\,dx = F(x),$$

then

$$\int g(u)\,du = F[h(u)].$$

Thus, in the present context, where we also have (2), it follows that

$$F[h(u)] = G(u) + c$$

for some constant c. But this says that

$$F(x) = G[k(x)] + c,$$

and therefore $G[k(x)]$ is an integral of $f(x)$, as claimed in (3).

Even more, we can use the same method for dealing with definite integrals if the limits of integration are correctly changed; that is,

$$\int_a^b f(x)\,dx = \int_c^d g(u)\,du,$$

where $c = k(a)$ and $d = k(b)$. This can be established very easily by thinking of it in the other direction, as

$$\int_c^d g(u)\,du = \int_a^b f(x)\,dx,$$

where $a = h(c)$ and $b = h(d)$, because this second version was proved in Section 10.2.

Our purpose here is to establish the validity of the partial fractions decomposition as stated in a piecemeal manner in Section 10.6. We are considering a rational function $P(x)/Q(x)$, and we assume that $Q(x)$ is a polynomial of degree n that is completely factored into real linear and quadratic factors of various multiplicities. In the beginning we do not assume that $P(x)/Q(x)$ is proper. This enables us to understand more clearly the significance of this assumption when it becomes necessary to make it. Our basic tool is the following lemma about the removal of a linear factor from the denominator.

C.9
PROOF OF THE PARTIAL FRACTIONS THEOREM

Lemma   *Let $x - r$ be a linear factor of $Q(x)$ of multiplicity 1, so that $Q(x) = (x - r)Q_1(x)$ with $Q_1(r) \neq 0$. Then $P(x)/Q(x)$ can be written in the form*

$$\frac{P(x)}{Q(x)} = \frac{P(x)}{(x - r)Q_1(x)} = \frac{A}{x - r} + \frac{P_1(x)}{Q_1(x)}, \qquad (1)$$

where A is a constant and $P_1(x)$ is a polynomial such that $P_1(x)/Q_1(x)$ is a proper rational function whenever $P(x)/Q(x)$ is. The constant A can be calculated from either of the formulas

$$A = \frac{P(r)}{Q_1(r)} = \frac{P(r)}{Q'(r)}. \qquad (2)$$

Proof We must find a suitable A and $P_1(x)$, and we do this by letting (1) suggest what their definitions ought to be. With these definitions we then show that (1) is valid.

By combining the fractions on the right-hand side of (1), we see that A and $P_1(x)$ must be chosen so that the numerators are identical,

$$P(x) = AQ_1(x) + (x - r)P_1(x). \qquad (3)$$

Since this is to be an identity, it must hold in particular for $x = r$. This gives $P(r) = AQ_1(r) + 0$, so we put

$$A = \frac{P(r)}{Q_1(r)}. \qquad (4)$$

This is a legitimate definition because $Q_1(r) \neq 0$. Since

$$Q(x) = (x - r)Q_1(x) \qquad \text{and} \qquad Q'(x) = (x - r)Q_1'(x) + Q_1(x),$$

we see that $Q'(r) = Q_1(r)$, and this establishes the second formula for A stated in (2). Using the formula for A given by (4), we now solve (3) for $P_1(x)$,

$$P_1(x) = \frac{P(x) - AQ_1(x)}{x - r} = \frac{P(x) - [P(r)/Q_1(r)]Q_1(x)}{x - r}$$

$$= \frac{1}{Q_1(r)} \frac{P(x)Q_1(r) - P(r)Q_1(x)}{x - r}. \qquad (5)$$

We adopt this as our definition of $P_1(x)$. It may appear that this function is not a polynomial. However, the numerator of this fraction is clearly a polynomial that has the value 0 for $x = r$, so by the factor theorem of algebra it has $x - r$ as a factor. The common factor $x - r$ can now be canceled from the numerator and denominator, and we conclude that $P_1(x)$ is indeed a polynomial. We now show that (1) is valid when A and $P_1(x)$ are defined as they are above:

$$\frac{A}{x - r} + \frac{P_1(x)}{Q_1(x)} = \frac{AQ_1(x) + (x - r)P_1(x)}{(x - r)Q_1(x)}$$

$$= \frac{[P(r)/Q_1(r)]Q_1(x) + [1/Q_1(r)][P(x)Q_1(r) - P(r)Q_1(x)]}{(x - r)Q_1(x)}$$

$$= \frac{P(x)}{(x - r)Q_1(x)}.$$

Finally, the statement that $P_1(x)/Q_1(x)$ is proper whenever $P(x)/Q(x)$ is proper follows from (3) by using the fact that the degree of $Q_1(x)$ is $n - 1$; for if the degree of $P_1(x)$ is $\geq n - 1$, then (3) shows that the degree of $P(x)$ is $\geq n$.

This lemma enables us to do everything we wish with respect to splitting off partial fractions generated by linear factors of $Q(x)$. At this stage we specifically assume that $P(x)/Q(x)$ is proper, so that each time the lemma is applied the residual rational function $P_1(x)/Q_1(x)$ will also be proper.

We first observe that if $Q(x)$ can be factored completely into distinct linear factors, so that

$$Q(x) = (x - r_1)(x - r_2) \cdots (x - r_n),$$

then

$$\frac{P(x)}{Q(x)} = \frac{A_1}{x - r_1} + \frac{A_2}{x - r_2} + \cdots + \frac{A_n}{x - r_n},$$

for we can remove the factors from the denominator one at a time in accordance with the lemma. At the last step the residual denominator is $x - r_n$, and since the numerator is necessarily of lower degree, this numerator must be a constant.

Suppose next that $x - r$ is a linear factor of $Q(x)$ of multiplicity m, so that $Q(x) = (x - r)^m Q_1(x)$ with $Q_1(r) \neq 0$. To cope with this situation we apply the lemma repeatedly in a slightly different way. First, by (1) we have

$$\frac{P(x)}{(x - r)Q_1(x)} = \frac{B_m}{x - r} + \frac{P_1(x)}{Q_1(x)}.$$

Dividing through by $x - r$ and applying (1) again yields

$$\frac{P(x)}{(x - r)^2 Q_1(x)} = \frac{B_m}{(x - r)^2} + \frac{P_1(x)}{(x - r)Q_1(x)} = \frac{B_m}{(x - r)^2} + \frac{B_{m-1}}{x - r} + \frac{P_2(x)}{Q_1(x)}.$$

By continuing in this way, we find in the end that

$$\frac{P(x)}{Q(x)} = \frac{P(x)}{(x - r)^m Q_1(x)} = \frac{B_m}{(x - r)^m} + \cdots + \frac{B_1}{x - r} + \frac{P_m(x)}{Q_1(x)}$$

$$= \frac{P_m(x)}{Q_1(x)} + \frac{B_1}{x - r} + \cdots + \frac{B_m}{(x - r)^m}.$$

In this manner we strip off all the linear factors from the denominator of our proper rational function $P(x)/Q(x)$ and generate the corresponding partial fractions as described in Section 10.6.

The rest of the proof requires an acquaintance with complex numbers, because the imaginary zeros of a real polynomial come in conjugate pairs and this fact plays an essential role in the argument. Before we begin, it is necessary to observe that our fundamental lemma works in just the same way if the number r happens to be imaginary.

Now suppose that $x^2 + bx + c$ is a quadratic factor of $Q(x)$ of multiplicity 1 which is irreducible in the sense that $b^2 - 4c < 0$, so that the roots r_1 and r_2 of the equation $x^2 + bx + c = 0$ are conjugate complex numbers.* Then

$$Q(x) = (x^2 + bx + c)Q_2(x) = (x - r_1)(x - r_2)Q_2(x),$$

and by two applications of our lemma we can find constants A_1 and A_2 and a polynomial $P_2(x)$ such that

$$\frac{P(x)}{Q(x)} = \frac{P(x)}{(x - r_1)(x - r_2)Q_2(x)} = \frac{A_1}{x - r_1} + \frac{A_2}{x - r_2} + \frac{P_2(x)}{Q_2(x)}.$$

By using (2) we see that

$$A_1 = \frac{P(r_1)}{Q'(r_1)} \quad \text{and} \quad A_2 = \frac{P(r_2)}{Q'(r_2)},$$

and these formulas imply that A_1 and A_2 are also conjugate complex numbers. By combining the corresponding partial fractions, we can now write

$$\frac{P(x)}{Q(x)} = \frac{(A_1 + A_2)x - (A_1 r_2 + A_2 r_1)}{(x - r_1)(x - r_2)} + \frac{P_2(x)}{Q_2(x)} = \frac{Ax + B}{x^2 + bx + c} + \frac{P_2(x)}{Q_2(x)},$$

where the numbers $A = A_1 + A_2$ and $B = -(A_1 r_2 + A_2 r_1)$ are real because r_1, r_2 and A_1, A_2 are conjugate pairs of complex numbers. Also, we know from the last expression that $P_2(x)$ is a real polynomial. If the factor $x^2 + bx + c$ occurs with multiplicity $m > 1$, then we simply remove it over and over in the way used above with repeated linear factors. This produces exactly the partial fractions decomposition described in Section 10.6.

---

* If $x^2 + bx + c$ were not irreducible, it would already have been factored into real linear factors in the "complete" factorization of $Q(x)$ previously mentioned.

When these procedures have been applied to each of the real linear and quadratic factors of $Q(x)$, and all the corresponding partial fractions have been stripped away, then there is nothing left of $Q(x)$, the decomposition is complete, and the partial fractions theorem is fully proved.

Up to this point we have said nothing about uniqueness, but it is worth remarking that a proper rational function can be decomposed into partial fractions in only one way. This will follow at once from our overall discussion if we can show in the lemma that the expansion (1) is unique. But this is easy; for if we assume two forms for this expansion,

$$\frac{A}{x-r} + \frac{P_1(x)}{Q_1(x)} = \frac{B}{x-r} + \frac{P_2(x)}{Q_1(x)},$$

then we have

$$AQ_1(x) + (x-r)P_1(x) = BQ_1(x) + (x-r)P_2(x).$$

By letting $x \to r$, we see from this that $B = A$, so A in (1) is unique; and this implies that $P_1(x)$ in (1) is also unique.

C.10
THE EXTENDED RATIO TESTS OF RAABE AND GAUSS

The convergence tests we discuss in this appendix are more delicate than the ratio test and enable us to arrive at a definite conclusion for many series with the property that $a_{n+1}/a_n \to 1$ from below. We begin with the following general theorem of Kummer.*

Theorem 1   *Assume that $a_n > 0$, $b_n > 0$, and $\Sigma 1/b_n$ diverges. If*

$$\lim \left(b_n - \frac{a_{n+1}}{a_n} \cdot b_{n+1} \right) = L,$$

then Σa_n converges if $L > 0$ and diverges if $L < 0$.

Proof   If $L > 0$, then

$$b_n - \frac{a_{n+1}}{a_n} \cdot b_{n+1} \geq h > 0$$

for all $n \geq$ some n_0, so

$$a_n b_n - a_{n+1} b_{n+1} \geq h a_n > 0 \tag{1}$$

for these n's. This shows that $\{a_n b_n\}$ is a decreasing sequence of positive numbers for $n \geq n_0$, so $K = \lim a_n b_n$ exists. It is now clear that $\Sigma_{n=n_0}^{\infty}(a_n b_n - a_{n+1} b_{n+1})$ is a convergent telescopic series (with sum $a_{n_0} b_{n_0} - K$), so by (1) and the comparison test we conclude that $\Sigma h a_n$ converges, and therefore Σa_n also converges.

Next, if $L < 0$ we have

$$a_n b_n - a_{n+1} b_{n+1} \leq 0$$

---

* The German mathematician Ernst Eduard Kummer (1810–1893) is remembered mainly for his work on the arithmetic of algebraic number fields, by means of which he proved Fermat's last theorem for many prime exponents. He also contributed to geometry (the entity known as "Kummer's surface" was much later found by Eddington to be related to Dirac's theory of the electron) and extended Gauss's work on hypergeometric series. He was a good-humored and rather easygoing man with a ready (and sometimes racy) wit. He taught at Breslau until 1855, when the death of Gauss dislocated the mathematical map of Europe. Dirichlet succeeded Gauss at Göttingen, and Kummer replaced Dirichlet at Berlin.

for all $n \geq$ some n_0, so $\{a_n b_n\}$ is an increasing sequence of positive numbers for these n's. It follows that

$$a_n b_n \geq a_{n_0} b_{n_0} \qquad \text{or} \qquad a_n \geq (a_{n_0} b_{n_0}) \cdot \frac{1}{b_n}$$

for $n \geq n_0$, so Σa_n diverges because $\Sigma 1/b_n$ diverges.

Students will observe that if we take $b_n = 1$ in Kummer's theorem, we obtain the ratio test. As another application we deduce *Raabe's test.**

Theorem 2   *If $a_n > 0$ and*

$$\frac{a_{n+1}}{a_n} = 1 - \frac{A}{n} + \frac{A_n}{n} \tag{2}$$

where $A_n \to 0$, then Σa_n converges if $A > 1$ and diverges if $A < 1$.

Proof   Take $b_n = n$ in Kummer's theorem. Then

$$\lim \left(b_n - \frac{a_{n+1}}{a_n} \cdot b_{n+1} \right) = \lim \left[n - \left(1 - \frac{A}{n} + \frac{A_n}{n} \right)(n+1) \right]$$

$$= \lim \left[-1 + \frac{A(n+1)}{n} - \frac{A_n(n+1)}{n} \right]$$

$$= A - 1,$$

and by Kummer's theorem it follows that Σa_n converges if $A > 1$ and diverges if $A < 1$.

For practical purposes, it is worth noting that Raabe's test can be formulated more conveniently as follows: If $a_n > 0$ and

$$\lim n \left[1 - \frac{a_{n+1}}{a_n} \right] = A, \tag{3}$$

then Σa_n converges if $A > 1$ and diverges if $A < 1$. To prove this, it suffices to express (3) in the equivalent form

$$n \left[1 - \frac{a_{n+1}}{a_n} \right] = A - A_n \tag{4}$$

where $A_n \to 0$, since (4) is merely another way of writing (2).

Example 1   If we apply the ratio test to the series

$$\frac{1}{2} + \frac{1 \cdot 3}{2 \cdot 4} + \frac{1 \cdot 3 \cdot 5}{2 \cdot 4 \cdot 6} + \cdots + \frac{1 \cdot 3 \cdot 5 \cdots (2n-1)}{2 \cdot 4 \cdot 6 \cdots (2n)} + \cdots , \tag{5}$$

---

* Joseph Ludwig Raabe (1801–1859) was born of poor parents in Galicia and studied in Vienna. When cholera swept that city in 1831 he moved to Zürich. In 1833 the Austrian embassy in Bern demanded that the government in Zürich return Raabe to Austria, because he had broken Austrian law by taking a position with the University of Zürich. This ludicrous demand was very sensibly ignored, and Raabe spent the rest of his life in various posts at the University. He was a man of unusual modesty and was considered a very gifted teacher. He is now known only for the convergence test discussed here, but he also worked on the summation of series, on systems of linear differential equations, and on the problem of the motion of the center of gravity of the planets.

then we find that

$$\frac{a_{n+1}}{a_n} = \frac{1 \cdot 3 \cdot 5 \ \cdots \ (2n-1)(2n+1)}{2 \cdot 4 \ \cdots \ (2n)(2n+2)} \cdot \frac{2 \cdot 4 \ \cdots \ (2n)}{1 \cdot 3 \ \cdots \ (2n-1)}$$

$$= \frac{2n+1}{2n+2} \rightarrow 1 \text{ from below,}$$

so the test fails. However,

$$n\left[1 - \frac{a_{n+1}}{a_n}\right] = n\left(1 - \frac{2n+1}{2n+2}\right) = \frac{n}{2n+2} \rightarrow \frac{1}{2},$$

so (5) diverges by Raabe's test.

Example 2   Now consider the related series in which each term is squared,

$$\left[\frac{1}{2}\right]^2 + \left[\frac{1 \cdot 3}{2 \cdot 4}\right]^2 + \left[\frac{1 \cdot 3 \cdot 5}{2 \cdot 4 \cdot 6}\right]^2 + \ \cdots \ + \left[\frac{1 \cdot 3 \cdot 5 \ \cdots \ (2n-1)}{2 \cdot 4 \cdot 6 \ \cdots \ (2n)}\right]^2 + \ \cdots \ . \quad (6)$$

Here we see that

$$\frac{a_{n+1}}{a_n} = \frac{(2n+1)^2}{(2n+2)^2} \rightarrow 1 \text{ from below,}$$

so the ratio test fails again. Furthermore, we also have

$$n\left[1 - \frac{a_{n+1}}{a_n}\right] = n\left[1 - \frac{4n^2 + 4n + 1}{4n^2 + 8n + 4}\right] = \frac{4n^2 + 3n}{4n^2 + 8n + 4} \rightarrow 1,$$

so even Raabe's test fails in this case.

When $A = 1$ in Raabe's test, we turn to *Gauss's test.*

Theorem 3   *If* $a_n > 0$ *and*

$$\frac{a_{n+1}}{a_n} = 1 - \frac{A}{n} + \frac{A_n}{n^{1+c}}$$

where $c > 0$ *and* A_n *is bounded as* $n \rightarrow \infty$, *then* Σa_n *converges if* $A > 1$ *and diverges if* $A \leq 1$.

Proof   If $A \neq 1$, the statement follows from Raabe's test, since $A_n/n^c \rightarrow 0$. We therefore confine our attention to the case $A = 1$. Take $b_n = n \ln n$ in Kummer's theorem. Then

$$\lim\left[b_n - \frac{a_{n+1}}{a_n} \cdot b_{n+1}\right] = \lim\left[n \ln n - \left(1 - \frac{1}{n} + \frac{A_n}{n^{1+c}}\right)(n+1)\ln(n+1)\right]$$

$$= \lim\left[n \ln n - \frac{(n^2-1)}{n}\ln(n+1) - \frac{(n+1)}{n} \cdot \frac{A_n \ln(n+1)}{n^c}\right]$$

$$= \lim\left[n \ln\left(\frac{n}{n+1}\right) + \frac{\ln(n+1)}{n} - \frac{(n+1)}{n} \cdot \frac{A_n \ln(n+1)}{n^c}\right]$$

$$= -1 + 0 - 0 = -1,$$

and the divergence of the series in this case is a consequence of Kummer's theorem.

Gauss actually expressed his test in a specialized form adapted to series in which a_{n+1}/a_n is a quotient of two polynomials having the same term of highest degree. This version is also known as *Gauss's test*.

Theorem 4 *If $a_n > 0$ and*

$$\frac{a_{n+1}}{a_n} = \frac{n^k + \alpha n^{k-1} + \cdots}{n^k + \beta n^{k-1} + \cdots},\tag{7}$$

then Σa_n converges if $\beta - \alpha > 1$ and diverges if $\beta - \alpha \leq 1$.

Proof If the quotient on the right of (7) is worked out by long division, we get

$$\frac{a_{n+1}}{a_n} = 1 - \frac{\beta - \alpha}{n} + \frac{A_n}{n^2},$$

where A_n is a quotient of the form

$$\frac{\gamma n^{k-2} + \cdots}{n^{k-2} + \cdots}$$

and is therefore clearly bounded as $n \to \infty$. The statement now follows from Theorem 3 with $c = 1$.

Example 3 For a simple application we consider the series (6), for which Raabe's test failed. Here we have

$$\frac{a_{n+1}}{a_n} = \frac{4n^2 + 4n + 1}{4n^2 + 8n + 4} = \frac{n^2 + n + \frac{1}{4}}{n^2 + 2n + 1},$$

so $\beta - \alpha = 2 - 1 = 1$ and the series diverges by Gauss's test.

Example 4 Gauss's original purpose in devising his test was to study the important *hypergeometric series*

$$1 + \sum_{n=1}^{\infty} \frac{a(a+1) \cdots (a+n-1)b(b+1) \cdots (b+n-1)}{n!c(c+1) \cdots (c+n-1)} x^n \tag{8}$$

when $x = 1$:

$$1 + \frac{a \cdot b}{1 \cdot c} + \frac{a(a+1)b(b+1)}{1 \cdot 2c(c+1)} + \cdots$$

$$+ \frac{a(a+1) \cdots (a+n-1)b(b+1) \cdots (b+n-1)}{n!c(c+1) \cdots (c+n-1)} + \cdots. \tag{9}$$

Here we assume that none of the constants a, b, c is zero or a negative integer. This condition on a and b keeps the series from terminating, while that on c avoids division by zero. The ratio

$$\frac{a_{n+1}}{a_n} = \frac{(a+n)(b+n)}{(n+1)(c+n)} = \frac{n^2 + (a+b)n + ab}{n^2 + (c+1)n + c}$$

is positive for all sufficiently large n, so the terms of (9) ultimately have the same sign. Any such series can be treated by Gauss's test (or the ratio test, or Raabe's test); and since in this case $\beta - \alpha = (c+1) - (a+b)$, we see that (9) converges if $c > a + b$ and diverges if $c \leq a + b$.

The hypergeometric series (8) is extremely interesting and versatile, and is capable of representing virtually every function that occurs in elementary analysis.* Here we confine ourselves to remarking that when $a = 1$ and $b = c$ it reduces to the ordinary geometric series—hence its name.

---

* See Problem 1 on p. 178 of the present writer's book, *Differential Equations,* McGraw-Hill, 1972.

PROBLEMS

1 Show that

$$\sum_{n=1}^{\infty} \left[\frac{1 \cdot 3 \cdot 5 \, \cdots \, (2n-1)}{2 \cdot 4 \cdot 6 \, \cdots \, (2n)} \right]^3$$

converges. More generally, let k be an arbitrary positive integer and show that

$$\sum_{n=1}^{\infty} \left[\frac{1 \cdot 3 \cdot 5 \, \cdots \, (2n-1)}{2 \cdot 4 \cdot 6 \, \cdots \, (2n)} \right]^k$$

converges if $k > 2$ and diverges if $k \le 2$.

2 Determine the convergence behavior of the following series:

(a) $\displaystyle\sum_{n=1}^{\infty} \frac{1 \cdot 3 \cdot 5 \, \cdots \, (2n-1)}{4 \cdot 6 \cdot 8 \, \cdots \, (2n+2)}$;

(b) $\displaystyle\sum_{n=1}^{\infty} \frac{2 \cdot 7 \cdot 12 \, \cdots \, (5n-3)}{6 \cdot 11 \cdot 16 \, \cdots \, (5n+1)}$;

(c) $\displaystyle\sum_{n=1}^{\infty} \frac{2 \cdot 4 \cdot 6 \, \cdots \, (2n)}{3 \cdot 5 \cdot 7 \, \cdots \, (2n+1)}$;

(d) $\displaystyle\sum_{n=1}^{\infty} \frac{2 \cdot 4 \cdot 6 \, \cdots \, (2n)}{5 \cdot 7 \cdot 9 \, \cdots \, (2n+3)}$;

(e) $\displaystyle\sum_{n=1}^{\infty} \frac{1 \cdot 3 \cdot 5 \, \cdots \, (2n-1)}{2 \cdot 4 \cdot 6 \, \cdots \, (2n)} \cdot \frac{1}{n}$.

3 Find the positive integers k for which the following series converge:

(a) $\displaystyle\sum_{n=1}^{\infty} \left[\frac{2 \cdot 4 \cdot 6 \, \cdots \, (2n)}{3 \cdot 5 \cdot 7 \, \cdots \, (2n+1)} \right]^k$;

(b) $\displaystyle\sum_{n=1}^{\infty} \left[\frac{2 \cdot 4 \cdot 6 \, \cdots \, (2n)}{5 \cdot 7 \cdot 9 \, \cdots \, (2n+3)} \right]^k$;

(c) $\displaystyle\sum_{n=1}^{\infty} \frac{1 \cdot 3 \cdot 5 \, \cdots \, (2n-1)}{2 \cdot 4 \cdot 6 \, \cdots \, (2n)} \cdot \frac{1}{n^k}$.

4 Determine the values of a and b (where neither is zero or a negative integer) for which the following series converge:

(a) $\dfrac{a}{b} + \dfrac{a(a+1)}{b(b+1)} + \cdots$

$\qquad + \dfrac{a(a+1) \, \cdots \, (a+n-1)}{b(b+1) \, \cdots \, (b+n-1)} + \cdots$;

(b) $\dfrac{a}{b} + \dfrac{a(a+2)}{b(b+2)} + \cdots$

$\qquad + \dfrac{a(a+2) \, \cdots \, (a+2n-2)}{b(b+2) \, \cdots \, (b+2n-2)} + \cdots$.

C.11

ABSOLUTE VS. CONDITIONAL CONVERGENCE

We begin with two simple examples:

$$\sum_{n=1}^{\infty} (-1)^{n+1} \frac{1}{n} = 1 - \frac{1}{2} + \frac{1}{3} - \frac{1}{4} + \cdots \tag{1}$$

and

$$\sum_{n=1}^{\infty} (-1)^{n+1} \frac{1}{n^2} = 1 - \frac{1}{2^2} + \frac{1}{3^2} - \frac{1}{4^2} + \cdots . \tag{2}$$

As we saw in Section 14.7, there is an important distinction between these series. By the alternating series test, each is convergent as it stands. However, if we change the signs of all the negative terms—that is, if we replace each term by its absolute value—then the series become

$$\sum_{n=1}^{\infty} \frac{1}{n} = 1 + \frac{1}{2} + \frac{1}{3} + \frac{1}{4} + \cdots$$

and

$$\sum_{n=1}^{\infty} \frac{1}{n^2} = 1 + \frac{1}{2^2} + \frac{1}{3^2} + \frac{1}{4^2} + \cdots;$$

and the first of these altered series now diverges, while the second still converges.

It was this phenomenon that led us in Section 14.7 to make the following definition: A series Σa_n is said to be *absolutely convergent* if $\Sigma |a_n|$ converges. Thus, (2) is absolutely convergent but (1) is not. The careful reader will notice that this definition in itself says nothing about the convergence of Σa_n. However, we proved in Section 14.7 that absolute convergence does indeed imply ordinary convergence.

The series (1) shows that the converse of this theorem is false, that is, convergence does not imply absolute convergence. Absolute convergence is therefore a stronger property than ordinary convergence, and we shall see that absolutely convergent series have several important properties which they do not share with convergent series in general. This brings us to another definition that was given but not pursued very far in Section 14.7: A series that is convergent but not absolutely convergent is said to be *conditionally convergent*. Our present purpose is to establish some of the general properties of absolutely convergent series, and also to emphasize the sharp contrast between these series and those which are only conditionally convergent. For instance, we shall see that rearranging the terms of an absolutely convergent series has no effect on its behavior or its sum, but that rearranging a conditionally convergent series can have a drastic effect.

It is convenient to begin this program by considering an arbitrary series Σa_n and defining p_n and q_n by

$$p_n = \frac{|a_n| + a_n}{2} \quad \text{and} \quad q_n = \frac{|a_n| - a_n}{2}. \tag{3}$$

It is clear that $p_n = a_n$ and $q_n = 0$ if $a_n \geq 0$, and that $p_n = 0$ and $q_n = -a_n$ if $a_n < 0$. Accordingly, if the given series is a mixture of positive and negative terms, then we can think of Σp_n as consisting of the positive terms of Σa_n, and of Σq_n as consisting of the negatives of its negative terms. This is not quite correct because many p_n's and q_n's can be zero, but it does provide a point of view which is useful for understanding the following theorems.

Theorem 1   *Consider a series Σa_n and define p_n and q_n by (3). If Σa_n converges conditionally, then Σp_n and Σq_n both diverge; and if Σa_n converges absolutely, then Σp_n and Σq_n both converge and the sums of these series are related by the equation $\Sigma a_n = \Sigma p_n - \Sigma q_n$.*

Proof   It is clear from (3) that $a_n = p_n - q_n$ and $|a_n| = p_n + q_n$. Our basic tools are these equations and the fact that convergent series can be added or subtracted term by term.

To prove the first statement, we assume that Σa_n converges and $\Sigma |a_n|$ diverges. If Σq_n converges, then the equation $p_n = a_n + q_n$ tells us that Σp_n must also converge. Similarly, if Σp_n converges, the equation $q_n = p_n - a_n$ tells us that Σq_n also converges. Thus, if either Σp_n or Σq_n converges, both must converge; and in this case the equation $|a_n| = p_n + q_n$ implies that $\Sigma |a_n|$ converges—contrary to the assumption. This proves that the conditional convergence of Σa_n implies that Σp_n and Σq_n both diverge. To establish the second statement, we assume that $\Sigma |a_n|$ converges. We know that Σa_n also converges, so equations (3) imply that Σp_n and Σq_n both converge. It follows from this that

$$\sum p_n - \sum q_n = \sum (p_n - q_n) = \sum a_n,$$

and the proof is complete.

The first part of this theorem is illustrated by the conditionally convergent series (1), in which Σp_n and Σq_n are the divergent series

$$1 + 0 + \frac{1}{3} + 0 + \frac{1}{5} + \cdots \quad \text{and} \quad 0 + \frac{1}{2} + 0 + \frac{1}{4} + \cdots .$$

In the case of the absolutely convergent series (2), Σp_n and Σq_n are

$$1 + 0 + \frac{1}{3^2} + 0 + \frac{1}{5^2} + \cdots \quad \text{and} \quad 0 + \frac{1}{2^2} + 0 + \frac{1}{4^2} + \cdots ,$$

both of which are convergent. Briefly, Theorem 1 tells us that the convergence of an absolutely convergent series is due to the smallness of its terms, while that of a conditionally convergent series is due not only to the smallness of its terms but also to cancellations between its positive and negative terms.

In the last paragraph of Section 14.4 we proved a theorem about rearranging a convergent series of nonnegative terms. We now extend this theorem to the case of absolutely convergent series.

Theorem 2   *If Σa_n is an absolutely convergent series with sum s, and if the a_n's are rearranged in any way to form a new series Σb_n, then this new series is also absolutely convergent with sum s.*

Proof   The series $\Sigma |a_n|$ is convergent and consists of nonnegative terms. Since the b_n's are just the a_n's in a different order, it follows from the theorem just mentioned that $\Sigma |b_n|$ also converges, and therefore Σb_n is absolutely convergent. If $\Sigma b_n = t$, then Theorem 1 allows us to write

$$s = \sum a_n = \sum p_n - \sum q_n \tag{4}$$

and

$$t = \sum b_n = \sum P_n - \sum Q_n, \tag{5}$$

where each of the constituent series on the right is convergent and consists of non-negative terms. But the P_n's and Q_n's are simply the p_n's and q_n's in a different order, so by another application of the theorem in Section 14.4 we have $\Sigma P_n = \Sigma p_n$ and $\Sigma Q_n = \Sigma q_n$. The fact that $t = s$ now follows at once from (4) and (5).

This theorem was proved in 1837 by Dirichlet, who discovered the phenomenon discussed in Problem 10 of Section 14.3 and was the first to understand the significance of absolutely convergent series.

In striking contrast to the behavior of absolutely convergent series as stated in Theorem 2, we find that the sum of a conditionally convergent series depends in an essential way on the order of its terms, and that the value of this sum can be changed at will by a suitable rearrangement of these terms. This fact was discovered and proved in 1854 by the great German mathematician Riemann, and is known as *Riemann's rearrangement theorem*. It can be formulated as follows.

Theorem 3   *Let Σa_n be a conditionally convergent series. Then its terms can be rearranged to yield a convergent series whose sum is an arbitrary preassigned number, or a series that diverges to ∞, or a series that diverges to $-\infty$.*

Proof The idea of the proof is surprisingly simple. We begin by using Theorem 1 to form the two divergent series of nonnegative terms Σp_n and Σq_n.

To establish the first statement, let s be any number and construct a rearrangement of the given series as follows. Start by writing down p's in order until the partial sum

$$p_1 + p_2 + \cdots + p_{n_1}$$

is first $\geq s$; next, continue with $-q$'s until the total partial sum

$$p_1 + p_2 + \cdots + p_{n_1} - q_1 - q_2 - \cdots - q_{m_1}$$

is first $\leq s$; then continue with p's until the total partial sum

$$p_1 + \cdots + p_{n_1} - q_1 - \cdots - q_{m_1} + p_{n_1+1} + \cdots + p_{n_2}$$

is first $\geq s$; and so on. The possibility of each of these steps is guaranteed by the divergence of Σp_n and Σq_n; and the resulting rearrangement of Σa_n converges to s because $p_n \to 0$ and $q_n \to 0$.

In order to make the rearrangement diverge to ∞, it suffices to write down enough p's to yield

$$p_1 + p_2 + \cdots + p_{n_1} \geq 1,$$

then to insert $-q_1$, then to continue with p's until

$$p_1 + \cdots + p_{n_1} - q_1 + p_{n_1+1} + \cdots + p_{n_2} \geq 2,$$

then to insert $-q_2$, and so on. We can produce divergence to $-\infty$ by a similar construction.

One of the principal applications of Theorem 2 relates to the multiplication of series. In this connection it is notationally convenient to index the terms of the series we consider by $n = 0, 1, 2, \ldots$. If we multiply two series

$$a_0 + a_1 + \cdots + a_n + \cdots \qquad \text{and} \qquad b_0 + b_1 + \cdots + b_n + \cdots \qquad (6)$$

by forming all possible products $a_i b_j$ (as in the case of finite sums), then we obtain the following doubly infinite array:

$$
\begin{array}{cccccc}
a_0 b_0 & a_0 b_1 & a_0 b_2 & a_0 b_3 & \cdots \\
a_1 b_0 & a_1 b_1 & a_1 b_2 & a_1 b_3 & \cdots \\
a_2 b_0 & a_2 b_1 & a_2 b_2 & a_2 b_3 & \cdots \\
a_3 b_0 & a_3 b_1 & a_3 b_2 & a_3 b_3 & \cdots \\
& & \cdots
\end{array}
\qquad (7)
$$

There are many ways of arranging these products into a single infinite series, of which two are important for us. The first is to group them by diagonals, as indicated by the arrows:

$$a_0 b_0 + (a_0 b_1 + a_1 b_0) + (a_0 b_2 + a_1 b_1 + a_2 b_0) + \cdots . \qquad (8)$$

This series can be defined as $\Sigma_{n=0}^{\infty} c_n$, where

$$c_n = a_0 b_n + a_1 b_{n-1} + \cdots + a_n b_0.$$

It is called the *product* (or sometimes the *Cauchy product*) of the two series (6), and is particularly useful in working with power series.

A second method of arranging (7) into a series is by squares, as suggested by the broken lines:

$$a_0b_0 + (a_0b_1 + a_1b_1 + a_1b_0) + (a_0b_2 + a_1b_2 + a_2b_2 + a_2b_1 + a_2b_0) + \cdots . \quad (9)$$

The advantage of this arrangement is that the nth partial sum s_n of (9) is given by

$$s_n = (a_0 + a_1 + \cdots + a_n)(b_0 + b_1 + \cdots + b_n), \quad (10)$$

and this is useful in proving a preliminary fact about the multiplication of series.

Theorem 4 *If the two series* (6) *have nonnegative terms and converge to s and t, then their product* (8) *converges to st.*

Proof It is clear from (10) that (9) converges to st. Now denote the series (8) and (9) *without parentheses* by (8′) and (9′). The series (9′) of nonnegative terms still converges to st; for if m is an integer such that $n^2 \le m \le (n+1)^2$, then the mth partial sum of (9′) lies between s_{n-1} and s_n, and both of these converge to st. By Theorem 2, the terms of (9′) can be rearranged to yield (8′) without changing the sum st; and when parentheses are suitably inserted, we see that (8) converges to st.

The force of this result can best be appreciated by observing that the product of the convergent series

$$\sum_{n=0}^{\infty} \frac{(-1)^n}{\sqrt{n+1}} = \frac{1}{\sqrt{1}} - \frac{1}{\sqrt{2}} + \frac{1}{\sqrt{3}} - \cdots \quad (11)$$

with itself does not converge at all. We ask students to convince themselves of this in Problem 9.

We now extend Theorem 4 to the case of absolutely convergent series.

Theorem 5 *If the series* $\sum_{n=0}^{\infty} a_n$ *and* $\sum_{n=0}^{\infty} b_n$ *are absolutely convergent, with sums s and t, then their product*

$$\sum_{n=0}^{\infty} (a_0b_n + a_1b_{n-1} + \cdots + a_nb_0) = a_0b_0 + (a_0b_1 + a_1b_0)$$

$$+ (a_0b_2 + a_1b_1 + a_2b_0) + \cdots$$

$$+ (a_0b_n + a_1b_{n-1} + \cdots + a_nb_0) + \cdots \quad (12)$$

is also absolutely convergent, with sum st.

Proof The series $\sum_{n=0}^{\infty} |a_n|$ and $\sum_{n=0}^{\infty} |b_n|$ are convergent and have nonnegative terms, so by the proof of Theorem 4 we see that the series

$$|a_0||b_0| + |a_0||b_1| + |a_1||b_0|$$

$$+ |a_0||b_2| + |a_1||b_1| + |a_2||b_0| + \cdots$$

$$+ |a_0||b_n| + |a_1||b_{n-1}| + \cdots + |a_n||b_0| + \cdots$$

$$= |a_0b_0| + |a_0b_1| + |a_1b_0|$$

$$+ |a_0b_2| + |a_1b_1| + |a_2b_0| + \cdots$$

$$+ |a_0b_n| + |a_1b_{n-1}| + \cdots + |a_nb_0| + \cdots \quad (13)$$

converges, and therefore

$$a_0b_0 + a_0b_1 + a_1b_0 + \cdots + a_0b_n + \cdots + a_nb_0 + \cdots \quad (14)$$

is absolutely convergent. It follows from Theorem 2 that the sum of (14) will not change if we rearrange its terms and write it as

$$a_0 b_0 + a_0 b_1 + a_1 b_1 + a_1 b_0$$
$$+ a_0 b_2 + a_1 b_2 + a_2 b_2 + a_2 b_1 + a_2 b_0 + \cdots . \qquad (15)$$

We now observe that the sum of the first $(n+1)^2$ terms of (15) is $(a_0 + a_1 + \cdots + a_n)(b_0 + b_1 + \cdots + b_n)$, so it is clear that (15), and with it (14), converges to st. Since (12) is obtained by suitably inserting parentheses in (14), we see that (12) also converges to st. All that remains is to show that (12) converges absolutely; but this follows by the comparison test from the inequality

$$|a_0 b_n + a_1 b_{n-1} + \cdots + a_n b_0| \le |a_0 b_n| + |a_1 b_{n-1}| + \cdots + |a_n b_0|$$

and the fact that the series

$$|a_0 b_0| + (|a_0 b_1| + |a_1 b_0|) + \cdots$$
$$+ (|a_0 b_n| + |a_1 b_{n-1}| + \cdots + |a_n b_0|) + \cdots ,$$

obtained from (13) by inserting parentheses, is convergent.

This theorem shows that the absolute convergence of both of the given series is a sufficient condition for the convergence of their product to st. It is an interesting fact that this conclusion can also be obtained from the weaker hypothesis that only one of the two series is absolutely convergent (the example following Theorem 4 shows that the product of two conditionally convergent series need not converge at all!). A proof of this result is outlined in Problem 10.

PROBLEMS

*1 Use the formula $1 + \frac{1}{2} + \cdots + 1/n = \ln n + \gamma + o(1)$ to establish the validity of the stated sums of the following rearrangements of the series $1 - \frac{1}{2} + \frac{1}{3} - \frac{1}{4} + \cdots = \ln 2$:
   (a) $1 + \frac{1}{3} + \frac{1}{5} - \frac{1}{2} - \frac{1}{4} - \frac{1}{6} + \cdots = \ln 2$;
   (b) $1 + \frac{1}{3} + \frac{1}{5} - \frac{1}{2} - \frac{1}{4} + \cdots = \frac{1}{2} \ln 6$;
   (c) $1 - \frac{1}{2} - \frac{1}{4} + \frac{1}{3} - \frac{1}{6} - \frac{1}{8} + \cdots = \frac{1}{2} \ln 2$;
   (d) $1 - \frac{1}{2} - \frac{1}{4} - \frac{1}{6} - \frac{1}{8} + \frac{1}{3} - \frac{1}{10} - \frac{1}{12} - \frac{1}{14} - \frac{1}{16} + \cdots = 0$.

2 A series Σa_n is sometimes said to be *unconditionally convergent* if it converges and every rearrangement converges to the same sum. Show that Σa_n converges unconditionally if and only if it converges absolutely.

3 If Σa_n is absolutely convergent, prove that Σa_n^2 converges.

4 If $a_1 + a_2 + a_3 + a_4 + \cdots$ is absolutely convergent, prove that $a_1 + a_2 + a_3 + a_4 + \cdots = (a_1 + a_3 + \cdots) + (a_2 + a_4 + \cdots)$. Is this necessarily true for any convergent series?

5 Let Σa_n be a convergent series with sum s. If Σb_n is a rearrangement in which no term of the first series is moved more than n_0 places from its original position, prove that Σb_n still converges to s.

6 What is the sum of

$$1 + \frac{1}{3} - \frac{1}{2} + \frac{1}{5} - \frac{1}{4} + \frac{1}{7} - \frac{1}{6} + \cdots ?$$

*7 Prove that the series

$$1 - \frac{1}{2} + \frac{1}{3} - \frac{1}{4} + \cdots = \ln 2$$

converges to $\ln 2\sqrt{p/q}$ if its terms are rearranged by writing the first p positive terms, then the first q negative terms, then the next p positive terms, then the next q negative terms, etc.

*8 We know that the harmonic series $1 + \frac{1}{2} + \frac{1}{3} + \frac{1}{4} + \cdots$ diverges but the alternating harmonic series $1 - \frac{1}{2} + \frac{1}{3} - \frac{1}{4} + \cdots$ converges.
   (a) Show that $1 + \frac{1}{2} - \frac{1}{3} + \frac{1}{4} + \frac{1}{5} - \frac{1}{6} + + - \cdots$ diverges.
   (b) If the signs are changed in the harmonic series in such a way that p positive signs are followed by q negative signs, then p positive signs by q negative signs, etc., show that the resulting series diverges if $p \ne q$ and converges if $p = q$.

9 Show that the product of the series (11) with itself diverges.

*10 If $\Sigma_{n=0}^{\infty} a_n$ and $\Sigma_{n=0}^{\infty} b_n$ converge to s and t, and if $\Sigma_{n=0}^{\infty} a_n$ converges absolutely, prove that their product

$$\sum_{n=0}^{\infty} (a_0 b_n + a_1 b_{n-1} + \cdots + a_n b_0)$$

converges to st (this result is known as *Mertens' theorem**). Hint: Put $s_n = a_0 + \cdots + a_n$, $t_n = b_0 + \cdots + b_n$, and $\alpha_n = t_n - t$; show that the nth partial

sum of the product can be written as

$$s_n t + a_0 \alpha_n + a_1 \alpha_{n-1} + \cdots + a_n \alpha_0;$$

and prove that if β_n is defined by

$$\beta_n = a_0 \alpha_n + a_1 \alpha_{n-1} + \cdots + a_n \alpha_0$$
$$= \alpha_0 a_n + \alpha_1 a_{n-1} + \cdots + \alpha_n a_0,$$

then $\beta_n \to 0$.

---

* Franz Mertens (1840–1927) was born in Poland, studied at Berlin under Kummer and Kronecker, and taught at Cracow, Graz, and Vienna. He retained extraordinary vigor of mind and body to an advanced age, and wrote the last of his more than 100 research papers at the age of 86. His main interest was analytic number theory, where he was a master in the use of elementary methods to simplify difficult proofs. He discovered the theorem given here in connection with his success (after Euler had failed) in proving the convergence of the series $\Sigma (\pm 1/p_n)$, where the p_n's are the primes and the signs are $+$ or $-$ according as p_n is of the form $4n + 1$ or $4n + 3$.

C.12
DIRICHLET'S TEST

With one exception, all of our convergence tests in Chapter 14 are tests that apply only to series of positive (or nonnegative) terms, that is, they are tests for absolute convergence. This exception is the alternating series test, which says that the series

$$b_1 - b_2 + b_3 - b_4 + \cdots \tag{1}$$

converges if the b_n's form a decreasing sequence of positive numbers and $b_n \to 0$. We can think of (1) as generated by multiplying the terms of the series $1 - 1 + 1 - 1 + \cdots$ by the terms of the sequence $b_1, b_2, b_3, b_4, \ldots$. From this point of view it is natural (and profitable) to generalize by considering a series

$$a_1 + a_2 + \cdots + a_n + \cdots \tag{2}$$

and sequence

$$b_1, b_2, \ldots, b_n, \ldots, \tag{3}$$

and the problem is to find conditions under which the series

$$a_1 b_1 + a_2 b_2 + \cdots + a_n b_n + \cdots \tag{4}$$

converges. It is obvious that if (2) is absolutely convergent and (3) is bounded, then (4) is also absolutely convergent. Our purpose here, however, is to obtain criteria for the convergence of (4) that are not merely tests for absolute convergence.

To accomplish this, we need *Abel's partial summation formula*: If $s_n = a_1 + a_2 + \cdots + a_n$, then

$$a_1 b_1 + a_2 b_2 + \cdots + a_n b_n = s_1(b_1 - b_2) + s_2(b_2 - b_3) + \cdots$$
$$+ s_{n-1}(b_{n-1} - b_n) + s_n b_n. \tag{5}$$

The proof is easy. Since $a_1 = s_1$ and $a_n = s_n - s_{n-1}$ for $n > 1$, we have

$$a_1 b_1 = s_1 b_1$$
$$a_2 b_2 = s_2 b_2 - s_1 b_2$$
$$a_3 b_3 = s_3 b_3 - s_2 b_3$$
$$\cdots$$
$$a_{n-1} b_{n-1} = s_{n-1} b_{n-1} - s_{n-2} b_{n-1}$$
$$a_n b_n = s_n b_n - s_{n-1} b_n.$$

On adding these equations and suitably grouping the terms on the right, we obtain (5). This result enables us to establish *Dirichlet's test*.

Theorem 1 *If the series* (2) *has bounded partial sums, and if* (3) *is a decreasing sequence of positive numbers such that* $b_n \to 0$, *then the series* (4) *converges.*

Proof If we put $S_n = a_1 b_1 + a_2 b_2 + \cdots + a_n b_n$, then (5) tells us that

$$S_n = T_n + s_n b_n, \tag{6}$$

where

$$T_n = s_1(b_1 - b_2) + s_2(b_2 - b_3) + \cdots + s_{n-1}(b_{n-1} - b_n).$$

We must prove that $\lim S_n$ exists, and we do this by showing that $\lim T_n$ and $\lim s_n b_n$ both exist. Our first assumption says that there is a constant M such that $|s_n| \le M$ for every n, so $|s_n b_n| \le M b_n$; and since $b_n \to 0$, we conclude that $s_n b_n \to 0$. Next, T_n is the $(n-1)$st partial sum of the series

$$s_1(b_1 - b_2) + s_2(b_2 - b_3) + \cdots, \tag{7}$$

so $\lim T_n$ will certainly exist if (7) converges. To establish this, it suffices to show that (7) is absolutely convergent. We now use the assumption that the b_n's are positive and decreasing, which yields

$$|s_1(b_1 - b_2)| + |s_2(b_2 - b_3)| + \cdots + |s_{n-1}(b_{n-1} - b_n)|$$
$$\le M(b_1 - b_2) + M(b_2 - b_3) + \cdots + M(b_{n-1} - b_n)$$
$$= M(b_1 - b_n) \le M b_1.$$

This implies that (7) is absolutely convergent, and the proof is complete.

In order to make effective use of Dirichlet's test, we must be acquainted with a few series having bounded partial sums. Naturally, all convergent series have this property, but many that are divergent also have it. Perhaps the simplest is $1 - 1 + 1 - 1 + \cdots$; and from this we see at once that the alternating series test is an immediate consequence of Theorem 1.

Dirichlet's test is particularly useful in demonstrating the convergence of certain trigonometric series, and here we encounter some further series that do not converge but nevertheless have bounded partial sums.

As an example, we show that if x is any number $\ne k\pi$, where $k = 0, \pm 1, \pm 2, \ldots$, then the series

$$\cos x + \frac{\cos 3x}{3} + \frac{\cos 5x}{5} + \cdots + \frac{\cos (2n-1)x}{2n-1} + \cdots \tag{8}$$

converges. Here the b_n's can be taken as $1, \frac{1}{3}, \frac{1}{5}, \ldots$, so it suffices to prove that the partial sums of

$$\cos x + \cos 3x + \cos 5x + \cdots + \cos (2n-1)x + \cdots \tag{9}$$

are bounded. To do this, we use the trigonometric identity

$$2 \cos a \sin b = \sin (a + b) - \sin (a - b).$$

By putting $b = x$ and $a = x, 3x, 5x, \ldots, (2n-1)x$, we obtain

$$2 \cos x \sin x = \sin 2x - 0$$
$$2 \cos 3x \sin x = \sin 4x - \sin 2x$$
$$2 \cos 5x \sin x = \sin 6x - \sin 4x$$
$$\cdots$$
$$2 \cos (2n-1)x \sin x = \sin 2nx - \sin (2n-2)x;$$

and adding yields

$$2 \sin x[\cos x + \cos 3x + \cdots + \cos (2n-1)x] = \sin 2nx$$

or

$$\cos x + \cos 3x + \cdots + \cos (2n-1)x = \frac{\sin 2nx}{2 \sin x}.$$

But $|\sin 2nx| \leq 1$, so

$$|\cos x + \cos 3x + \cdots + \cos (2n-1)x| \leq \frac{1}{2|\sin x|}. \qquad (10)$$

This proves that the partial sums of (9) are bounded, so (8) converges. It should now be apparent why we assumed that $x \neq k\pi$: We must have $\sin x \neq 0$ in (10). Actually, of course, it is obvious that (8) diverges if $x = k\pi$.

PROBLEMS

1 Let p be a positive constant. It is clear from the discussion in the text that

$$\cos x + \frac{\cos 3x}{3^p} + \frac{\cos 5x}{5^p} + \cdots$$

converges for every $x \neq k\pi$; and when $x = k\pi$, the series converges if $p > 1$ and diverges if $p \leq 1$. Use the identity

$$2 \sin a \sin b = \cos (a-b) - \cos (a+b)$$

to show that

$$\sin x + \frac{\sin 3x}{3^p} + \frac{\sin 5x}{5^p} + \cdots$$

converges for all x.

2 Prove the identities

$$2 \sin \frac{x}{2} (\sin x + \sin 2x + \cdots + \sin nx)$$

$$= \cos \frac{x}{2} - \cos \frac{(2n+1)x}{2}$$

and

$$2 \sin \frac{x}{2} (\cos x + \cos 2x + \cdots + \cos nx)$$

$$= \sin \frac{(2n+1)x}{2} - \sin \frac{x}{2}.$$

Let $b_1, b_2, \ldots, b_n, \ldots$ be a decreasing sequence of positive numbers such that $b_n \to 0$, and use these identities to show that
(a) $\sum_{n=1}^{\infty} b_n \sin nx$ converges for all x;
(b) $\sum_{n=1}^{\infty} b_n \cos nx$ converges for all $x \neq 2k\pi$.

3 Investigate the convergence behavior of

$$\frac{\cos 3x}{2} + \frac{\cos 6x}{3} + \frac{\cos 9x}{4} + \cdots.$$

4 Show that the following series converge:

$$1 + \frac{1}{2} - \frac{1}{3} - \frac{1}{4} + \frac{1}{5} + \frac{1}{6} - \frac{1}{7} - \frac{1}{8} + \cdots,$$

$$\frac{1}{\ln 2} + \frac{1}{\ln 3} - \frac{1}{\ln 4} - \frac{1}{\ln 5} + \frac{1}{\ln 6} + \frac{1}{\ln 7} - \cdots,$$

and

$$\frac{3}{1} - \frac{4}{2} + \frac{3}{3} - \frac{2}{4} + \frac{3}{5} - \frac{4}{6} + \frac{3}{7} - \frac{2}{8} + \cdots.$$

5 A series of the form $\sum_{n=1}^{\infty} a_n/n^x$ is called a *Dirichlet series*. If it converges for $x = x_0$, show that it also converges for every $x > x_0$.

6 Prove *Abel's test*: If the series (2) converges and (3) is a bounded increasing or decreasing sequence, then the series (4) also converges. (Note the relation of this statement to Dirichlet's test—more is assumed about the series, but less about the sequence. It is almost certain that Abel knew Dirichlet's test.)

7 Use Abel's test to show that the following series converge:
(a) $\cos 1 + \frac{1}{3} \cos \frac{1}{2} - \frac{1}{2} \cos \frac{1}{3} + \frac{1}{5} \cos \frac{1}{4} + \frac{1}{7} \cos \frac{1}{5} - \frac{1}{4} \cos \frac{1}{6} + \cdots$;

(b) $2 + \frac{3}{2^2} - \frac{4}{3^2} - \frac{5}{4^2} + \frac{6}{5^2} + \frac{7}{6^2} - \cdots$.

8 If $\sum a_n$ converges, so do $\sum a_n/n$, $\sum a_n/\ln n$, $\sum a_n \cos 1/n$, $\sum a_n \sin 1/n$, $\sum(1 + 1/n)a_n$, $\sum(1 + 1/n)^n a_n$, and $\sum \sqrt[n]{n}\, a_n$. Why?

Consider a power series $\Sigma a_n x^n$ with positive radius of convergence R, and let $f(x)$ be its sum. Our purpose here is to prove that $f(x)$ is continuous and differentiable on $(-R, R)$, and also that its derivative and integral can legitimately be calculated by differentiating and integrating the series term by term.

Let $S_n(x)$ be the nth partial sum of the series, so that

$$S_n(x) = a_0 + a_1 x + a_2 x^2 + \cdots + a_n x^n.$$

We write

$$f(x) = S_n(x) + R_n(x)$$

and call $R_n(x)$ the *remainder*. Evidently,

$$R_n(x) = a_{n+1} x^{n+1} + a_{n+2} x^{n+2} + \cdots.$$

For each x in the interval of convergence, we know that $R_n(x) \to 0$ as $n \to \infty$; that is, for any given $\epsilon > 0$ we have

$$|R_n(x)| < \epsilon \text{ for sufficiently large } n, \; n \geq n_0. \tag{1}$$

We emphasize here that this is true for each x individually, and is merely an equivalent way of expressing the fact that $\Sigma a_n x^n$ converges to $f(x)$. However, much more can be said: Throughout any given closed interval inside the interval of convergence, say $|x| \leq |x_1| < R$, (1) holds for all x simultaneously. Since $R_n(x) = f(x) - S_n(x)$, we can express this in another way by saying that throughout the given closed interval, $S_n(x)$ can be made to approximate $f(x)$ as closely as we please by taking n large enough.

To prove this, we observe that for every x in the given closed interval $|x| \leq |x_1| < R$, we have

$$|R_n(x)| = |a_{n+1} x^{n+1} + a_{n+2} x^{n+2} + \cdots|$$

$$\leq |a_{n+1} x^{n+1}| + |a_{n+2} x^{n+2}| + \cdots$$

$$\leq |a_{n+1} x_1^{n+1}| + |a_{n+2} x_1^{n+2}| + \cdots.$$

The argument is completed by using the fact that the last-written sum can be made $< \epsilon$ by taking n large enough, $n \geq n_0$, because of the absolute convergence of $\Sigma a_n x_1^n$. The point is, that the same n_0 works for all x's in the given closed interval. The conclusion proved here, that $R_n(x)$ can be made small *independently of x in the given closed interval,* is expressed by saying that the series is *uniformly convergent* in this interval.

CONTINUITY OF THE SUM

We will prove that $f(x)$ is continuous at each interior point x_0 of the interval of convergence. Consider a closed subinterval $|x| \leq |x_1| < R$ containing x_0 in its interior. If $\epsilon > 0$ is given, then by uniform convergence we can find an n with the property that $|R_n(x)| < \epsilon$ for all x's in the subinterval. Since the polynomial $S_n(x)$ is continuous at x_0, we can find a $\delta > 0$ so small that $|x - x_0| < \delta$ implies x lies in the subinterval and $|S_n(x) - S_n(x_0)| < \epsilon$. By putting these conditions together we find that $|x - x_0| < \delta$ implies

$$|f(x) - f(x_0)| = |[S_n(x) + R_n(x)] - [S_n(x_0) + R_n(x_0)]|$$

$$= |[S_n(x) - S_n(x_0)] + R_n(x) - R_n(x_0)|$$

$$\leq |S_n(x) - S_n(x_0)| + |R_n(x)| + |R_n(x_0)|$$

$$< \epsilon + \epsilon + \epsilon = 3\epsilon.$$

Since 3ϵ can be taken as small as we please, this proves that $f(x)$ is continuous at x_0.

C.13

UNIFORM
CONVERGENCE FOR
POWER SERIES

INTEGRATING TERM BY TERM

We have just proved that

$$f(x) = \sum a_n x^n \tag{2}$$

is continuous on $(-R, R)$. We can therefore integrate this function between limits a and b that lie inside the interval,

$$\int_a^b f(x)\, dx = \int_a^b \left(\sum a_n x^n \right) dx. \tag{3}$$

Our purpose here is to show that the right side of this can legitimately be integrated term by term,

$$\int_a^b \left(\sum a_n x^n \right) dx = \sum \int_a^b a_n x^n\, dx.$$

In words, the integral of the sum equals the sum of the integrals. An equivalent statement is that (3) can be written as

$$\int_a^b f(x)\, dx = \sum \int_a^b a_n x^n\, dx. \tag{4}$$

To prove this, we begin by observing that since $S_n(x)$ is a polynomial, and therefore continuous everywhere, all three of the functions in

$$f(x) = S_n(x) + R_n(x)$$

are continuous on $(-R, R)$. This allows us to write

$$\int_a^b f(x)\, dx = \int_a^b S_n(x)\, dx + \int_a^b R_n(x)\, dx. \tag{5}$$

Since any polynomial can be integrated term by term, the first integral on the right of (5) is

$$\int_a^b S_n(x)\, dx = \int_a^b (a_0 + a_1 x + a_2 x^2 + \cdots + a_n x^n)\, dx$$

$$= \int_a^b a_0\, dx + \int_a^b a_1 x\, dx + \int_a^b a_2 x^2\, dx + \cdots + \int_a^b a_n x^n\, dx.$$

To prove (4), it therefore suffices to show that

$$\int_a^b R_n(x)\, dx \to 0 \qquad \text{as} \qquad n \to \infty.$$

For this we use uniform convergence, which tells us that if $\epsilon > 0$ is given and $|x| \le |x_1| < R$ is a closed subinterval of $(-R, R)$ that contains both a and b, then $|R_n(x)| < \epsilon$ for all x in the subinterval if n is large enough. All that remains is to write

$$\left| \int_a^b R_n(x)\, dx \right| \le \int_a^b |R_n(x)|\, dx < \epsilon |b - a|,$$

and to notice that this can be made as small as we wish.

As a special case of (4) we take the limits 0 and x instead of a and b, and obtain

$$\int_0^x f(t)\, dt = \sum \frac{1}{n+1} a_n x^{n+1}$$

$$= a_0 x + \frac{1}{2} a_1 x^2 + \frac{1}{3} a_2 x^3 + \cdots + \frac{1}{n+1} a_n x^{n+1} + \cdots, \tag{6}$$

where the "dummy variable" t is used in the integral for the reason explained in Section 6.7.

DIFFERENTIATING TERM BY TERM

We now prove that the function $f(x)$ in (2) is not only continuous but also differentiable, and that its derivative can be calculated by differentiating (2) term by term,

$$f'(x) = \sum na_n x^{n-1}. \tag{7}$$

To do this, we begin by recalling from Section 14.7 that the series on the right of (7) converges on $(-R, R)$. If we denote its sum by $g(x)$, so that

$$g(x) = \sum na_n x^{n-1} = a_1 + 2a_2 x + 3a_3 x^2 + \cdots + na_n x^{n-1} + \cdots,$$

then (6) tells us that

$$\int_0^x g(t)\, dt = a_1 x + a_2 x^2 + a_3 x^3 + \cdots$$

$$= f(x) - a_0.$$

Since the left side of this has a derivative, so does the right side, and by differentiating we obtain

$$f'(x) = g(x) = \sum na_n x^{n-1},$$

as required.

C.14

THE DIVISION OF POWER SERIES

In order to justify the division of power series as described in Section 14.11, it suffices to justify dividing a power series into 1. To see this, we have only to notice that

$$\frac{\Sigma a_n x^n}{\Sigma b_n x^n} = (\Sigma a_n x^n) \cdot \left(\frac{1}{\Sigma b_n x^n} \right);$$

for this tells us that if we can expand $1/(\Sigma b_n x^n)$ in a power series with positive radius of convergence, then we can achieve our purpose by multiplying this series by $\Sigma a_n x^n$. It is clearly necessary to assume that $b_0 \neq 0$ (why?). We may assume that $b_0 = 1$ without any loss of generality, because if b_0 has any other nonzero value, we simply factor it out, leaving the leading coefficient equal to 1:

$$\frac{1}{b_0 + b_1 x + b_2 x^2 + \cdots} = \frac{1}{b_0} \cdot \frac{1}{1 + (b_1/b_0)x + (b_2/b_0)x^2 + \cdots}.$$

In view of these remarks, we direct our efforts at proving the following statement:

If $\Sigma b_n x^n$ has $b_0 = 1$ and positive radius of convergence R, then $1/(\Sigma b_n x^n)$ can be expanded in a power series $\Sigma c_n x^n$ which also has positive radius of convergence.

We begin by determining the c_n's. The condition $1/(\Sigma b_n x^n) = \Sigma c_n x^n$ means that $(\Sigma b_n x^n)(\Sigma c_n x^n) = 1$, so

$$b_0 c_0 + (b_0 c_1 + b_1 c_0)x + (b_0 c_2 + b_1 c_1 + b_2 c_0)x^2$$
$$+ \cdots + (b_0 c_n + b_1 c_{n-1} + \cdots + b_n c_0)x^n + \cdots = 1,$$

and therefore

$$b_0 c_0 = 1, \qquad b_0 c_1 + b_1 c_0 = 0, \qquad b_0 c_2 + b_1 c_1 + b_2 c_0 = 0,$$

$$\ldots, \qquad b_0 c_n + b_1 c_{n-1} + \cdots + b_n c_0 = 0, \ldots.$$

Since $b_0 = 1$, these equations determine the c_n's recursively:

$$c_0 = 1, \qquad c_1 = -b_1 c_0, \qquad c_2 = -b_1 c_1 - b_2 c_0, \ldots,$$

$$c_n = -b_1 c_{n-1} - b_2 c_{n-2} - \cdots - b_n c_0, \ldots.$$

All that remains is to prove that the power series $\Sigma c_n x^n$ with these coefficients has positive radius of convergence, and for this it suffices to show that the series converges for at least one nonzero x. This we now do.

Let r be any number such that $0 < r < R$, so that $\Sigma b_n r^n$ converges. Then there exists a constant $K \geq 1$ with the property that $|b_n r^n| \leq K$ or $|b_n| \leq K/r^n$ for all n. It now follows that

$$|c_0| = 1 \leq K,$$

$$|c_1| = |b_1 c_0| = |b_1| \leq \frac{K}{r},$$

$$|c_2| \leq |b_1 c_1| + |b_2 c_0| \leq \frac{K}{r} \cdot \frac{K}{r} + \frac{K}{r^2} \cdot K = \frac{2K^2}{r^2},$$

$$|c_3| \leq |b_1 c_2| + |b_2 c_1| + |b_3 c_0| \leq \frac{K}{r} \cdot \frac{2K^2}{r^2} + \frac{K}{r^2} \cdot \frac{K}{r} + \frac{K}{r^3} \cdot K$$

$$\leq (2 + 1 + 1)\frac{K^3}{r^3} = \frac{4K^3}{r^3} = \frac{2^2 K^3}{r^3},$$

since $K^2 \leq K^3$. In general,

$$|c_n| \leq |b_1 c_{n-1}| + |b_2 c_{n-2}| + \cdots + |b_n c_0|$$

$$\leq \frac{K}{r} \cdot \frac{2^{n-2}K^{n-1}}{r^{n-1}} + \frac{K}{r^2} \cdot \frac{2^{n-3}K^{n-2}}{r^{n-2}} + \cdots + \frac{K}{r^n} \cdot K$$

$$\leq (2^{n-2} + 2^{n-3} + \cdots + 1 + 1)\frac{K^n}{r^n} = \frac{2^{n-1}K^n}{r^n} \leq \frac{2^n K^n}{r^n}.$$

We therefore have $|c_n x^n| \leq |(2K/r)x|^n$, so the series $\Sigma c_n x^n$ is absolutely convergent—and therefore convergent—for any x that satisfies the condition $|x| < r/2K$. This shows that $\Sigma c_n x^n$ has nonzero radius of convergence, and the argument is complete.

C.15

THE EQUALITY OF MIXED PARTIAL DERIVATIVES

We shall prove that

$$f_{xy}(x_0, y_0) = f_{yx}(x_0, y_0), \tag{1}$$

under the assumption that both mixed partials f_{xy} and f_{yx} exist at all points near (x_0, y_0) and are continuous at (x_0, y_0).

Let Δx and Δy be so small that f_{xy} and f_{yx} exist throughout the rectangle with vertices (x_0, y_0), $(x_0 + \Delta x, y_0)$, $(x_0, y_0 + \Delta y)$, $(x_0 + \Delta x, y_0 + \Delta y)$ (see Fig. C.12). We carry out the proof by applying the Mean Value Theorem (Appendix C.4) in various ways to the expression

$$D = f(x_0 + \Delta x, y_0 + \Delta y) - f(x_0 + \Delta x, y_0) - f(x_0, y_0 + \Delta y) + f(x_0, y_0). \tag{2}$$

We begin by considering the function

$$F(x) = f(x, y_0 + \Delta y) - f(x, y_0),$$

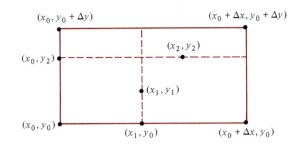

Figure C.12

where Δy is held fixed. The expression (2) can be written in the form

$$D = [f(x_0 + \Delta x, y_0 + \Delta y) - f(x_0 + \Delta x, y_0)] - [f(x_0, y_0 + \Delta y) - f(x_0, y_0)]$$
$$= F(x_0 + \Delta x) - F(x_0),$$

so by applying the Mean Value Theorem we obtain

$$D = \Delta x\, F'(x_1)$$
$$= \Delta x[f_x(x_1, y_0 + \Delta y) - f_x(x_1, y_0)],$$

where x_1 is some number between x_0 and $x_0 + \Delta x$. Since $f_x(x_1, y)$ is differentiable as a function of y, we can apply the Mean Value Theorem again to obtain

$$D = \Delta x\, \Delta y\, f_{xy}(x_1, y_1), \tag{3}$$

where y_1 lies between y_0 and $y_0 + \Delta y$.

We now start all over again with the function

$$G(y) = f(x_0 + \Delta x, y) - f(x_0, y),$$

where Δx is held fixed. The expression (2) can be written in the form

$$D = [f(x_0 + \Delta x, y_0 + \Delta y) - f(x_0, y_0 + \Delta y)] - [f(x_0 + \Delta x, y_0) - f(x_0, y_0)]$$
$$= G(y_0 + \Delta y) - G(y_0).$$

Just as before, by two applications of the Mean Value Theorem we find that

$$D = \Delta y\, G'(y_2) = \Delta y[f_y(x_0 + \Delta x, y_2) - f_y(x_0, y_2)]$$
$$= \Delta y\, \Delta x\, f_{yx}(x_2, y_2), \tag{4}$$

where y_2 lies between y_0 and $y_0 + \Delta y$ and x_2 lies between x_0 and $x_0 + \Delta x$.

Finally, by equating (3) and (4) we see that

$$f_{xy}(x_1, y_1) = f_{yx}(x_2, y_2). \tag{5}$$

Now let Δx and Δy approach zero, so that the rectangle shrinks toward the point (x_0, y_0). As this happens, it is clear that the points (x_1, y_1) and (x_2, y_2), which lie inside the rectangle, approach (x_0, y_0), and we obtain our conclusion (1) from (5) and the continuity of f_{xy} and f_{yx} at (x_0, y_0).

C.16

DIFFERENTIATION UNDER THE INTEGRAL SIGN

In Problems 32 in Section 19.2 and 19 in Section 19.6, we used the formula

$$\frac{d}{dx} \int_a^b f(x, y) \, dy = \int_a^b f_x(x, y) \, dy. \tag{1}$$

Our purpose here is to prove this under the assumption that $f(x, y)$ and its partial derivative $f_x(x, y)$ are both continuous functions on the closed rectangle $x_0 \le x \le x_1$, $a \le y \le b$.

It is convenient to write

$$F(x) = \int_a^b f(x, y) \, dy.$$

If x and $x + \Delta x$ both lie in the interval $x_0 \le x \le x_1$, then

$$F(x + \Delta x) - F(x) = \int_a^b f(x + \Delta x, y) \, dy - \int_a^b f(x, y) \, dy$$

$$= \int_a^b [f(x + \Delta x, y) - f(x, y)] \, dy.$$

Next, the Mean Value Theorem enables us to write this integrand in the form

$$f(x + \Delta x, y) - f(x, y) = \Delta x \, f_x(\bar{x}, y),$$

where $\bar{x}$ lies between x and $x + \Delta x$. Further, since $f_x(x, y)$ is assumed to be continuous on the closed rectangle, it can be shown that the absolute value of the difference

$$f_x(\bar{x}, y) - f_x(x, y)$$

is less than a positive number ϵ which is independent of x and y and approaches zero with Δx.* By putting these ingredients together, we obtain

$$\left| \frac{F(x + \Delta x) - F(x)}{\Delta x} - \int_a^b f_x(x, y) \, dy \right| = \left| \int_a^b [f_x(\bar{x}, y) - f_x(x, y)] \, dy \right|$$

$$\le \int_a^b |f_x(\bar{x}, y) - f_x(x, y)| \, dy$$

$$< \int_a^b \epsilon \, dy = \epsilon(b - a).$$

If we now let Δx approach zero, then ϵ also approaches zero, and we have

$$\lim_{\Delta x \to 0} \frac{F(x + \Delta x) - F(x)}{\Delta x} = \int_a^b f_x(x, y) \, dy,$$

which is (1).

C.17

A PROOF OF THE FUNDAMENTAL LEMMA

This lemma is stated and discussed in Section 19.4, and is the linchpin that holds together the main tools of Chapter 19. It has to do with a function $z = f(x, y)$ whose partial derivatives f_x and f_y exist at (x_0, y_0) and at all nearby points, and are continuous at the point (x_0, y_0) itself. The assertion of the lemma is that under these conditions the increment

$$\Delta z = f(x_0 + \Delta x, y_0 + \Delta y) - f(x_0, y_0)$$

---

* This is the two-dimensional analog of the property of uniform continuity that we discussed and proved in Appendix C.5.

can be expressed in the form

$$\Delta z = f_x(x_0, y_0)\, \Delta x + f_y(x_0, y_0)\, \Delta y + \epsilon_1\, \Delta x + \epsilon_2\, \Delta y, \tag{1}$$

where ϵ_1 and ϵ_2 are quantities that $\to 0$ as Δx and $\Delta y \to 0$.

To prove this statement, we analyze the change Δz in two steps, as shown in Fig. C.13, first changing x alone and moving from (x_0, y_0) to $(x_0 + \Delta x, y_0)$, and then changing y alone and moving from $(x_0 + \Delta x, y_0)$ to $(x_0 + \Delta x, y_0 + \Delta y)$. We denote the first change in z by $\Delta_1 z$, so that

$$\Delta_1 z = f(x_0 + \Delta x, y_0) - f(x_0, y_0).$$

By the Mean Value Theorem we can write this as

$$\Delta_1 z = \Delta x\, f_x(x_1, y_0), \tag{2}$$

where x_1 is between x_0 and $x_0 + \Delta x$. Similarly, if we denote the second part of the change in z by $\Delta_2 z$, so that

$$\Delta_2 z = f(x_0 + \Delta x, y_0 + \Delta y) - f(x_0 + \Delta x, y_0),$$

then

$$\Delta_2 z = \Delta y\, f_y(x_0 + \Delta x, y_1), \tag{3}$$

where y_1 is between y_0 and $y_0 + \Delta y$. As Δx and $\Delta y \to 0$, $x_1 \to x_0$ and $y_1 \to y_0$. By the assumed continuity of f_x and f_y at (x_0, y_0), we can therefore write

$$f_x(x_1, y_0) = f_x(x_0, y_0) + \epsilon_1$$

and

$$f_y(x_0 + \Delta x, y_1) = f_y(x_0, y_0) + \epsilon_2,$$

where ϵ_1 and $\epsilon_2 \to 0$ as Δx and $\Delta y \to 0$. This permits us to write (2) and (3) as

$$\Delta_1 z = f_x(x_0, y_0)\, \Delta x + \epsilon_1\, \Delta x \tag{4}$$

and

$$\Delta_2 z = f_y(x_0, y_0)\, \Delta y + \epsilon_2\, \Delta y. \tag{5}$$

Since $\Delta z = \Delta_1 z + \Delta_2 z$, we now complete the proof by adding (4) and (5) to obtain (1).

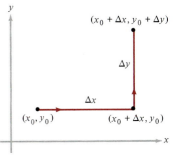

Figure C.13

C.18

A PROOF OF THE IMPLICIT FUNCTION THEOREM

We stated and discussed the Implicit Function Theorem in Section 19.10, and our purpose here is to provide a proof.

To restate the situation, we assume that $F(x, y)$ has continuous partial derivatives throughout some neighborhood of a point $P_0 = (x_0, y_0)$, and also that $F(x_0, y_0) = c$ and $F_y(x_0, y_0) \neq 0$. We shall prove that there exists a rectangle centered on P_0 within which the graph of $F(x, y) = c$ is the graph of a single differentiable function $y = f(x)$, and also that the derivative of this function is given by the formula

$$\frac{dy}{dx} = -\frac{F_x}{F_y}. \tag{1}$$

First, we know from the Fundamental Lemma (Section 19.4) that since F has continuous partial derivatives in the neighborhood mentioned above, F itself is continuous in this neighborhood.

Let us suppose, for definiteness, that $F_y > 0$ at P_0. (A similar proof can be constructed if $F_y < 0$ at P_0.) We observe that if $F_y > 0$ along a vertical segment, then F is

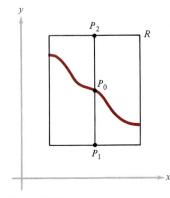

Figure C.14

an increasing function of y along that segment. It follows that no value of F (such as $F = c$, in which we are interested) can be taken on more than once on such a segment.

We begin by constructing a rectangle R_0 centered on P_0 within and on the boundary of which F and F_y are continuous and $F_y > 0$. This is possible because of the continuity of the functions. Along any vertical segment across R_0 the function F is an increasing function of y. By the Intermediate Value Theorem (Appendix C.3) we have $F = c$ on this segment if and only if $f < c$ at the lower end and $F > c$ at the upper end. Consider, for example, the vertical segment $P_1 P_2$ through P_0. Since $F = c$ at P_0, we have $F < c$ at P_1, and by continuity we have $F < c$ in some neighborhood of P_1 on the lower edge of R_0. Similarly, we have $F > c$ in some neighborhood of P_2 on the upper edge of R_0. A vertical segment close enough to $P_1 P_2$ will have its ends in both neighborhoods and will therefore intersect the graph of $F = c$ in exactly one point.

It is evident from this that we can shrink the base of R_0, if necessary, to form a new rectangle R centered on P_0 such that $F < c$ on its lower edge and $F > c$ on its upper edge. See Fig. C.14. Inside this rectangle there is one y that corresponds to each x in such a way that $F(x, y) = c$, and this defines our function $y = f(x)$. It is clear that this function is continuous, because the height of R_0 can be taken as small as we please to begin with.

To establish differentiability and formula (1), we consider a change Δx and the corresponding change $\Delta y = f(x_0 + \Delta x) - f(x_0)$. Since the new point $(x_0 + \Delta x, y_0 + \Delta y)$ is still on the graph of $F(x, y) = c$, we have

$$F(x_0, y_0) = c \quad \text{and} \quad F(x_0 + \Delta x, y_0 + \Delta y) = c, \quad \text{so} \quad \Delta F = 0.$$

But the Fundamental Lemma gives

$$\Delta F = F_x \, \Delta x + F_y \, \Delta y + \epsilon_1 \, \Delta x + \epsilon_2 \, \Delta y,$$

where $\epsilon_1, \epsilon_2 \to 0$ as Δx and $\Delta y \to 0$, and therefore

$$0 = F_x + F_y \frac{\Delta y}{\Delta x} + \epsilon_1 + \epsilon_2 \frac{\Delta y}{\Delta x}$$

or

$$\frac{\Delta y}{\Delta x} = -\frac{F_x + \epsilon_1}{F_y + \epsilon_2}. \tag{2}$$

We know that $y = f(x)$ is continuous. If $\Delta x \to 0$, it therefore follows that $\Delta y \to 0$, so $\epsilon_1, \epsilon_2 \to 0$ and (2) implies that $y = f(x)$ is differentiable with derivative given by (1). Finally, this proof of differentiability applies in just the same way to any other point on the graph inside the rectangle R.

The binomial theorem is a general formula for the expanded n-factor product

$$(a + b)^n = (a + b)(a + b) \cdots (a + b). \qquad (1)$$

If we compute a few cases by repeated laborious multiplication, we find that

$$(a + b)^1 = a + b,$$

$$(a + b)^2 = a^2 + 2ab + b^2,$$

$$(a + b)^3 = a^3 + 3a^2b + 3ab^2 + b^3,$$

$$(a + b)^4 = a^4 + 4a^3b + 6a^2b^2 + 4ab^3 + b^4,$$

$$(a + b)^5 = a^5 + 5a^4b + 10a^3b^2 + 10a^2b^3 + 5ab^4 + b^5.$$

It is clear that the expansion of (1) begins with a^n and ends with b^n, and also that the intermediate terms involve steadily decreasing powers of a and steadily increasing powers of b so that the sum of the two exponents is exactly n in each term. What is not so clear is the way the coefficients are calculated. To anticipate our final result, the expansion in question (the *binomial theorem*) is

$$(a + b)^n = a^n + na^{n-1}b + \frac{n(n-1)}{2} a^{n-2}b^2$$

$$+ \frac{n(n-1)(n-2)}{2 \cdot 3} a^{n-3}b^3 + \cdots$$

$$+ \frac{n(n-1)(n-2) \cdots (n-k+1)}{1 \cdot 2 \cdot 3 \cdots k} a^{n-k}b^k$$

$$+ \cdots + b^n. \qquad (2)$$

Our purpose is to understand the reasons behind the form of these coefficients. The best way to do this is to take a short detour through the closely related topics of permutations and combinations.

Before starting this detour, we remind students that if n is a positive integer, then the product of all the positive integers up to n is denoted by $n!$, called "n factorial":

$$n! = 1 \cdot 2 \cdot 3 \cdots n.$$

Thus, $1! = 1$, $2! = 1 \cdot 2 = 2$, $3! = 1 \cdot 2 \cdot 3 = 6$, $4! = 1 \cdot 2 \cdot 3 \cdot 4 = 24$, etc. For reasons that will appear below, we define 0! to be 1. These numbers increase very rapidly, as we see by doing a little arithmetic:

$$5! = 120, \quad 6! = 720, \quad 7! = 5040, \quad 8! = 40{,}320,$$
$$9! = 362{,}880, \quad 10! = 3{,}628{,}800.$$

Further, with the aid of mathematical tables we learn that

$$20! \cong 2.433 \times 10^{18} \quad \text{and} \quad 40! \cong 8.159 \times 10^{47}.$$

Any product of consecutive positive integers can easily be written in terms of factorials. For example,

$$6 \cdot 7 \cdot 8 \cdot 9 \cdot 10 = \frac{1 \cdot 2 \cdot 3 \cdot 4 \cdot 5 \cdot 6 \cdot 7 \cdot 8 \cdot 9 \cdot 10}{1 \cdot 2 \cdot 3 \cdot 4 \cdot 5} = \frac{10!}{5!}.$$

In general, if $k < n$, then

$$(k + 1)(k + 2) \cdots n = \frac{n!}{k!}.$$

PERMUTATIONS

We now discuss certain methods of counting that are useful in many applications of mathematics.

The reasoning on which our work is based can be illustrated by a simple example. Consider a journey from a city A through a city B to a city C. Suppose it is possible to go from A to B by 3 different routes and from B to C by 5 different routes. Then the total number of different routes from A through B to C is $3 \cdot 5 = 15$; for we can go from A to B in any one of 3 ways, and for each of these ways there are 5 ways of going on from B to C.

The basic principle here is this: If two successive independent decisions are to be made, and if there are c_1 choices for the first and c_2 choices for the second, then the total number of ways of making these two decisions is the product $c_1 c_2$. It is clear that the same principle is valid for any number of successive independent decisions.

The following is our main application of this idea. Given n distinct objects, in how many ways can we arrange them in order, that is, with a first, a second, a third, and so on? The answer is easy. There are n choices for the first object. After the first object is chosen, there are $n - 1$ choices for the second, then $n - 2$ choices for the third, etc. By the basic principle stated above, the total number of orderings is therefore

$$n(n - 1)(n - 2) \cdots 2 \cdot 1 = n!.$$

Each ordering of a set of objects is called a *permutation* of those objects. We have reached the following conclusion:

The number of permutations of n objects is $n!$.

Example 1 (a) There are $5! = 120$ ways of arranging 5 books on a shelf. (b) There are $9! = 362{,}880$ possible batting orders for the 9 players on a baseball team. (c) There are $52! \cong 8.066 \times 10^{67}$ ways of shuffling a deck of 52 cards.

We next consider a slight generalization. Suppose again that we have n distinct objects. This time we ask how many ways k of them can be chosen in order. Each such ordering is called a *permutation of n objects taken k at a time,* and the total number of these permutations is denoted by $P(n, k)$. There are evidently n choices for the first,

$n - 1$ choices for the second, $n - 2$ choices for the third, and $n - (k - 1) = n - k + 1$ choices for the kth. The total number of these permutations is therefore

$$P(n, k) = n(n - 1)(n - 2) \cdots (n - k + 1).$$

If we write this number in terms of factorials, then our conclusion can be formulated as follows:

The number of permutations of n objects taken k at a time is

$$P(n, k) = n(n - 1)(n - 2) \cdots (n - k + 1) = \frac{n!}{(n - k)!}.$$

Example 2   (a) If we have 7 books and only 3 spaces on a bookshelf, then the number of ways of filling these spaces with the available books (counting the order of the books) is

$$P(7, 3) = \frac{7!}{(7 - 3)!} = \frac{7!}{4!} = 7 \cdot 6 \cdot 5 = 210.$$

(b) The number of ways (counting the order of the cards) in which a 5-card poker hand can be dealt from a deck of 52 cards is

$$P(52, 5) = \frac{52!}{(52 - 5)!} = \frac{52!}{47!} = 52 \cdot 51 \cdot 50 \cdot 49 \cdot 48 = 311{,}875{,}200.$$

Of course, the order of the cards in a poker hand is immaterial to the value of the hand, so the number of distinct poker hands is a considerably smaller number. We take account of this below, in our discussion of combinations.

COMBINATIONS

A set of k objects chosen from a given set of n objects, without regard to the order in which they are arranged, is called a *combination of n objects taken k at a time.* The total number of such combinations is sometimes denoted by $C(n, k)$, but more frequently by $\binom{n}{k}$. For reasons to be explained, the numbers $\binom{n}{k}$ are called *binomial coefficients.*

Permutations and combinations are related in a simple way. Each permutation of n objects taken k at a time consists of a choice of k objects (a combination) followed by an ordering of these k objects. But there are $\binom{n}{k}$ ways to choose k objects, and then $k!$ ways to arrange them in order, so

$$P(n, k) = \binom{n}{k} \cdot k! \qquad \text{or} \qquad \binom{n}{k} = \frac{P(n, k)}{k!}.$$

Our formula for $P(n, k)$ now yields the following conclusion:

The number of combinations of n objects taken k at a time is

$$\binom{n}{k} = \frac{n!}{k!(n - k)!}.$$

The binomial coefficients have many properties, of which we mention only a few:

$$\binom{n}{0} = \binom{n}{n} = \frac{n!}{0!n!} = 1, \qquad \binom{n}{1} = \binom{n}{n - 1} = \frac{n!}{1!(n - 1)!} = n,$$

and

$$\binom{n}{k} = \binom{n}{n-k}.$$

The last fact can be established easily from the formula, or, more directly, by simply observing that a choice of k objects from a set of n objects is equivalent to a choice of the $n-k$ objects that are left behind.

Example 3 (a) The number of committees of 3 people that can be chosen from a group of 8 people is

$$\binom{8}{3} = \frac{8!}{3!5!} = \frac{8\cdot7\cdot6}{2\cdot3} = 56.$$

(b) A certain governmental commission is to consist of 2 economists and 3 engineers. If 6 economists and 5 engineers are candidates for the appointments, how many different commissions are possible? From the 6 economists, 2 can be chosen in $\binom{6}{2}$ ways; and from the 5 engineers, 3 can be chosen in $\binom{5}{3}$ ways. The number of possible commissions is therefore

$$\binom{6}{2}\binom{5}{3} = \frac{6!}{2!4!}\cdot\frac{5!}{3!2!} = \frac{6\cdot5}{2}\cdot\frac{5\cdot4}{2} = 150.$$

(c) The number of different 5-card poker hands that can be dealt from a deck of 52 cards is

$$\binom{52}{5} = \frac{52!}{5!47!} = \frac{52\cdot51\cdot50\cdot49\cdot48}{2\cdot3\cdot4\cdot5} = 2{,}598{,}960.$$

THE BINOMIAL THEOREM

To establish the binomial theorem (2), all that is necessary is to look at (1) and observe that each term of the expansion can be thought of as the product of n letters, one taken from each factor of the product

$$(a+b)(a+b)\ \cdots\ (a+b), \qquad n \text{ factors.}$$

Thus, a product $a^{n-k}b^k$ is obtained by choosing k b's and the rest a's. The number of ways this can be done is $\binom{n}{k}$. The coefficient of $a^{n-k}b^k$ on the right side of (2) is therefore $\binom{n}{k}$, and the proof is complete.

PROBLEMS

1 Write in terms of factorials
   (a) $5\cdot6\cdot7\cdot8\cdot9$;   (b) $22\cdot21\cdot20\cdot19\cdot18\cdot17$;
   (c) $\dfrac{52\cdot51\cdot50\cdot49\cdot48}{1\cdot2\cdot3\cdot4\cdot5}$.

2 Compute
   (a) $\dfrac{8!}{5!}$;   (b) $\dfrac{11!}{8!}$;   (c) $\dfrac{15!}{3!12!}$;

(d) $\dfrac{25!}{4!21!}$;   (e) $P(22, 2)$;   (f) $P(7, 5)$.

3 If 6 horses run in a race, how many different orders of finishing are there? How many possibilities are there for the first 3 places (win, place, and show)?

4 A club has 10 members. In how many ways can a president, a vice president, and a secretary be chosen?

5 How many batting orders are possible for a baseball team if the 4 best hitters are the first 4 to bat?

6 How many batting orders are possible for a baseball team if the fielders are the first 3 to bat and the pitcher bats last?

7 How many 10-digit numbers can be formed from all 10 digits 0, 1, 2, 3, 4, 5, 6, 7, 8, 9 if 0 is not allowed as the first digit?

8 How many 5-digit numbers can be formed from the 10 digits if the first digit cannot be 0 and no repetitions are allowed? If repetitions are allowed?

9 How many license plates can be made using 7 symbols, of which the first 3 are different letters of the alphabet and the last 4 are digits of which the first cannot be 0?

10 How many ways can 3 history books and 4 physics books be put on a shelf if books on the same subject must be kept together? If the history books must be kept together but the physics books need not be?

11 How many different signals can be made from 5 different flags if each signal consists of 5 flags placed one above the other on a flagpole? If each signal consists of 3 flags? If each signal consists of one or more flags?

12 In how many ways can 6 people be seated in a row of 6 chairs?

13 In how many ways can 6 people be seated in a row of 8 chairs?

14 In how many ways can 6 people be seated in a row of 6 chairs
(a) if a certain 2 insist on sitting next to each other?
(b) if a certain 2 refuse to sit next to each other?

15 In how many ways can 3 men and 3 women be seated in a row of 6 chairs if the men and women alternate?

16 Compute

(a) $\dbinom{100}{2}$;   (b) $\dbinom{15}{3}$;   (c) $\dbinom{10}{5}$.

17 A contractor employs 10 workers. In how many ways can he choose 4 of them to do a certain job?

18 In how many ways can 12 jurors be selected from a panel of 20 eligible citizens?

19 At a meeting, 28 people all shake hands with one another. How many handshakes are there?

20 In an examination a student has a choice of any 10 out of 12 questions. In how many ways can she choose her questions?

21 From a committee of 10, in how many ways can one choose a subcommittee of 4 and another subcommittee of 3, with no overlap?

22 How many committees of 4 can be chosen from a group of 12 men? How many of these will include a specific man? How many will exclude this man?

23 In how many ways can a committee of 5 men and 4 women be chosen from a club of 10 men and 7 women?

24 In how many ways can we select a committee of 5 from a group of 11
(a) if 2 of the group insist on serving together or not at all?
(b) if 2 of the group refuse to serve together?

25 In how many different orders can we shelve sets of 5 books, each set consisting of 3 history books and 2 chemistry books, if the books are to be chosen from a set of 9 history books and 7 chemistry books?

26 Three bags contain 8 black, 6 white, and 10 red marbles, respectively. In how many ways can we choose 6 black, 4 white, and 7 red marbles?

27 From a deck of 52 cards, how many 5-card poker hands are flushes (all cards of the same suit)? How many are full houses (3 cards of one kind together with a pair of another kind)?

28 How many lines are determined by 12 points in a plane if no 3 of the points are collinear?

29 How many triangles are determined by 13 points in a plane if no 3 of the points are collinear?

30 How many lines are determined by m points in a plane if k of them (where $k < m$) lie on the same line and, except for these, no 3 of the points are collinear?

31 How many planes are determined by 9 points in space if no 4 of the points are coplanar?

32 How many rectangles are formed by 5 vertical lines intersecting 8 horizontal lines?

33 How many diagonals can be drawn in a regular polygon with n sides? (A diagonal is a segment joining 2 nonadjacent vertices.)

34 Use the binomial theorem to expand each of the following:
(a) $(2x - y)^7$;   (b) $(3a + 2b)^6$;
(c) $(x^2y - 3z^3)^5$.

35 Find the term of
(a) $(x - 4)^{15}$ that involves x^{11};
(b) $(2x - 3)^9$ that involves x^4;
(c) $(a^2 - b^3)^{14}$ that involves a^{10}.

D.2

MATHEMATICAL INDUCTION

Discoveries in mathematics are sometimes made by carefully examining empirical evidence. As an illustration, let us try to find a formula for the sum of the first n odd numbers, where n is any positive integer. We compute:

$$
\begin{aligned}
\text{for } n = 1, \quad & 1 = 1 & = 1^2, \\
\text{for } n = 2, \quad & 1 + 3 = 4 & = 2^2, \\
\text{for } n = 3, \quad & 1 + 3 + 5 = 9 & = 3^2, \\
\text{for } n = 4, \quad & 1 + 3 + 5 + 7 = 16 & = 4^2, \\
\text{for } n = 5, \quad & 1 + 3 + 5 + 7 + 9 = 25 & = 5^2.
\end{aligned}
$$

The pattern emerging from this evidence seems to suggest that the value of the sum always equals the square of the number of terms in the sum. Since $2n - 1$ is the nth odd number, we can formulate this conjecture as follows:

$$1 + 3 + 5 + \cdots + (2n - 1) = n^2 \tag{1}$$

for every positive integer n.

The evidence for (1) is suggestive but far from conclusive. If we continue to test our conjecture for $n = 6, 7, 8$, and so on, and if it continues to hold up for these additional values of n, then this will certainly increase our confidence that (1) is probably true for every positive integer n. However, verifications of this kind can never constitute a proof, no matter how far they may be carried. If we verify (1) for all values of n up to $n = 1000$, then the logical possibility still remains that (1) might fail to be true for $n = 1001$.* There is an infinite chasm between "probably true" and "absolutely certain." What is needed is a logical argument proving that (1) is *always* true, for *all* values of n, beyond any doubt whatsoever. This is what the method of proof by mathematical induction accomplishes. We explain this method of reasoning by showing how it works in the case of formula (1), and then we state it as a formal principle.

Example 1 To prove (1) by mathematical induction, we begin by observing that this formula is true for $n = 1$, because it reduces to $1 = 1^2$. (We already knew this.) We next prove that if k is a value of n for which (1) is true, then (1) is necessarily also true for the next integer, $n = k + 1$. Thus, we assume that (1) is true for $n = k$,

$$1 + 3 + 5 + \cdots + (2k - 1) = k^2. \tag{2}$$

With the aid of this hypothesis we try to prove that (1) is also true for $n = k + 1$,

$$1 + 3 + 5 + \cdots + (2k - 1) + (2k + 1) = (k + 1)^2. \tag{3}$$

(The next to the last term on the left here is displayed for the sake of clarity in our next step.) By using (2) we see that the left-hand side of (3) can be written as

$$
\begin{aligned}
1 + 3 + 5 + \cdots + (2k - 1) + (2k + 1) &= k^2 + (2k + 1) \\
&= (k + 1)^2,
\end{aligned}
$$

so (3) is true if (2) is true. But this is enough to guarantee that (1) is actually true for all n. To see this, suppose we wish to assure ourselves that (1) is true for some specific value of n, say $n = 37$. The reasoning is as follows: We know by actual computation

---

* As a trivial illustration of this point, the equation

$$n^2 - 1 = (n + 1)(n - 1) + [(n - 1)(n - 2) \cdots (n - 1000)]$$

is clearly true for the first thousand values of n, because the bracketed expression is zero, and yet it is false for $n = 1001, 1002, \ldots$.

that (1) is true for $n = 1$; since it is true for $n = 1$, the argument just given tells us that it is also true for $n = 2$; since it is true for $n = 2$, it must be true for $n = 3$; and so on, up to $n = 37$ (or any other value of n).

Our main principle is little more than a distillation of the essence of this example.

Principle of Mathematical Induction *Let $S(n)$ be a proposition depending on a positive integer n.* Suppose that each of the following conditions is known to be satisfied.*

I *$S(1)$ is true.*
II *If $S(n)$ is assumed to be true for an integer $n = k$, then it is necessarily true for the next integer, $n = k + 1$.*

Under these circumstances it follows that $S(n)$ is true for every positive integer n.

Briefly, if we write down the propositions $S(n)$ in order,

$$S(1), S(2), S(3), \ldots ,$$

then the process of verification is started by I, and II is a link from each to the next guaranteeing that the process continues without end.

The idea of induction can be illustrated in many nonmathematical ways. For instance, imagine a row of dominoes standing on end. Suppose they are spaced in such a way that if any one of them falls, then it will knock over the next one. Suppose further that we actually knock over the first domino. In this situation we know that all the dominoes will fall. Our knowledge is based on two facts, which are closely analogous to I and II:

(i) The first domino *does* fall, because we knock it over.
(ii) *If* any domino falls, *then* it will knock over the next one.

We must be careful with the meaning of (ii); it does not state that any domino actually does fall, only that each domino is related to the next one in a certain way.

We continue with two additional examples of the method, in which we establish two formulas that are needed in Chapter 6. These are formulas for the sum of the first n positive integers, and for the sum of the first n squares:

$$1 + 2 + 3 + \cdots + n = \frac{n(n + 1)}{2}, \tag{4}$$

and

$$1^2 + 2^2 + 3^2 + \cdots + n^2 = \frac{n(n + 1)(2n + 1)}{6}. \tag{5}$$

Several of the remarks and problems that follow are concerned with the natural question of how such formulas can be discovered and understood. For the moment, however, we confine our attention to proving them by the method of mathematical induction.

---

* This means that for each specific value of n, $S(n)$ is a statement that is either true or false, without any ambiguity.

Example 2 To prove (4) by induction, we start by verifying I in this case, that is, we note that (4) is obviously true for $n = 1$:

$$1 = \frac{1 \cdot 2}{2}.$$

To verify II, we begin by assuming (4) for $n = k$,

$$1 + 2 + 3 + \cdots + k = \frac{k(k+1)}{2}, \tag{6}$$

and we hope to prove (4) for $n = k + 1$,

$$1 + 2 + 3 + \cdots + k + (k+1) = \frac{(k+1)(k+2)}{2}. \tag{7}$$

By using (6) we can write the left-hand side of (7) as

$$1 + 2 + 3 + \cdots + k + (k+1) = \frac{k(k+1)}{2} + (k+1)$$

$$= (k+1)\left(\frac{k}{2} + 1\right)$$

$$= \frac{(k+1)(k+2)}{2}.$$

Condition II is therefore satisfied, so by induction (4) is valid for all positive integers n.

Example 3 The proof of (5) is just as easy. To verify I, put $n = 1$:

$$1^2 = \frac{1 \cdot 2 \cdot 3}{6}.$$

To verify II, we must assume

$$1^2 + 2^2 + 3^2 + \cdots + k^2 = \frac{k(k+1)(2k+1)}{6}$$

and use this to prove

$$1^2 + 2^2 + 3^2 + \cdots + k^2 + (k+1)^2 = \frac{(k+1)(k+2)(2k+3)}{6}.$$

The details are routine,

$$1^2 + 2^2 + 3^2 + \cdots + k^2 + (k+1)^2 = \frac{k(k+1)(2k+1)}{6} + (k+1)^2$$

$$= (k+1)\left[\frac{k(2k+1)}{6} + (k+1)\right]$$

$$= (k+1)\left[\frac{2k^2 + 7k + 6}{6}\right]$$

$$= \frac{(k+1)(k+2)(2k+3)}{6},$$

so by induction the proof of (5) is complete.

Remark 1   Mathematical induction is a venerable method of proof that every student of mathematics ought to understand. Our purpose here has been to explain this method, and also to illustrate its use by proving two formulas [(4) and (5)] that are necessary for other parts of our work. However, much remains to be said.

Proofs by induction produce belief without insight, and are therefore fundamentally unsatisfying. It is important to know that a mathematical theorem *is* true, but it is often more important to understand *why* it is true. There are other proofs of formulas (1), (4), and (5) which convey much more insight into these formulas, and which also suggest how they might have been discovered in the first place. We begin with (4).

If we denote the sum of the integers from 1 to n by S, so that

$$S = 1 + 2 + \cdots + (n - 1) + n,$$

then it might occur to us to write this sum in the reverse order, as

$$S = n + (n - 1) + \cdots + 2 + 1.$$

If we now notice that the two first terms on the right add up to $n + 1$, and also the two second terms, and so on, then it is natural to add these two equations together to get

$$2S = n(n + 1) \qquad \text{or} \qquad S = \frac{n(n + 1)}{2}.$$

These ideas serve to discover formula (4) and also to prove it, simultaneously.

We next turn to formula (1), which again we discover and prove at a single stroke. Consider the sum of the first n odd numbers,

$$1 + 3 + 5 + \cdots + (2n - 1).$$

We notice certain obvious gaps in this sum, where the even numbers ought to be. If we fill in these gaps, and at the same time compensate for this filling in, then, using (4), we easily obtain

$$1 + 3 + 5 + \cdots + (2n - 1) = (1 + 2 + 3 + \cdots + 2n) - (2 + 4 + \cdots + 2n)$$
$$= (1 + 2 + 3 + \cdots + 2n) - 2(1 + 2 + \cdots + n)$$
$$= \frac{2n(2n + 1)}{2} - 2 \cdot \frac{n(n + 1)}{2}$$
$$= 2n^2 + n - n^2 - n = n^2,$$

which is (1).

We have discovered and proved the formula for the sum of the first n positive integers,

$$1 + 2 + 3 + \cdots + n = \frac{n(n + 1)}{2}. \tag{4}$$

It is somewhat more difficult to discover (5), that is, a formula for the sum of the first n squares,

$$1^2 + 2^2 + 3^2 + \cdots + n^2.$$

We know the answer by Example 3, but let us disregard this for a moment and try to think how we might discover it. It is natural to consider the two sums together:

n	1	2	3	4	5	6	$\cdots$
$1 + 2 + \cdots + n$	1	3	6	10	15	21	$\cdots$
$1^2 + 2^2 + \cdots + n^2$	1	5	14	30	55	91	$\cdots$

How are these sums related? It might occur to us to consider their ratio:

n	1	2	3	4	5	6	$\cdots$
$\dfrac{1+2+\cdots+n}{1^2+2^2+\cdots+n^2}$	1	$\dfrac{3}{5}$	$\dfrac{3}{7}$	$\dfrac{1}{3}$	$\dfrac{3}{11}$	$\dfrac{3}{13}$	$\cdots$

If we write these ratios in the form

$$\frac{3}{3} \quad \frac{3}{5} \quad \frac{3}{7} \quad \frac{3}{9} \quad \frac{3}{11} \quad \frac{3}{13} \quad \cdots \, ,$$

then it is difficult to miss the pattern that emerges. It seems clear that

$$\frac{1+2+\cdots+n}{1^2+2^2+\cdots+n^2} = \frac{3}{2n+1},$$

and by using (4) we easily find that

$$1^2 + 2^2 + \cdots + n^2 = \frac{n(n+1)(2n+1)}{6}. \tag{5}$$

This is certainly not a proof of (5). Nevertheless, it presents us with a plausible conjecture which we can then try to prove by induction, as we have done in Example 3.

Remark 2   There is another very ingenious way of discovering (5) that also constitutes a proof. It begins with the expansion

$$(k+1)^3 = k^3 + 3k^2 + 3k + 1,$$

expressed in the more convenient form

$$(k+1)^3 - k^3 = 3k^2 + 3k + 1.$$

If we write down this identity for $k = 1, 2, \ldots, n$ and add, then by taking advantage of wholesale cancellations we find that

$$2^3 - 1^3 = 3 \cdot 1^2 + 3 \cdot 1 + 1$$
$$3^3 - 2^3 = 3 \cdot 2^2 + 3 \cdot 2 + 1$$
$$\cdots$$
$$\underline{(n+1)^3 - n^3 = 3 \cdot n^2 + 3 \cdot n + 1}$$
$$(n+1)^3 - 1^3 = 3(1^2 + 2^2 + \cdots + n^2) + 3(1 + 2 + \cdots + n) + n$$

This enables us to obtain a formula for the sum of the squares in terms of our known formula (4) for the sum $1 + 2 + \cdots + n$:

$$1^2 + 2^2 + \cdots + n^2 = \tfrac{1}{3}[n^3 + 3n^2 + 3n - \tfrac{3}{2}n(n+1) - n]$$

$$= \tfrac{1}{6}(2n^3 + 6n^2 + 6n - 3n^2 - 3n - 2n)$$

$$= \frac{n}{6}(2n^2 + 3n + 1)$$

$$= \frac{n(n+1)(2n+1)}{6}.$$

The idea of this proof is due to the great French writer-scientist-mathematician-theologian Blaise Pascal. It can be extended quite easily to yield the sum of the first n cubes,

$$1^3 + 2^3 + \cdots + n^3 = \left[\frac{n(n+1)}{2}\right]^2, \tag{8}$$

the sum of the first n fourth powers, and so on indefinitely.

Remark 3   There is an exceedingly beautiful geometric proof of (8) that was known to the Arab mathematicians almost a thousand years ago. This proof depends on the square shown in Fig. D.1, which is constructed as follows. Beginning at the point O, we lay off successive segments of lengths 1, 2, 3, etc., and finally one of length n extending up to the point A. We do the same on a line OB perpendicular to OA, so that

$$OA = OB = 1 + 2 + \cdots + n$$

$$= \frac{n(n+1)}{2}.$$

The area of the square is therefore

$$S = \left[\frac{n(n+1)}{2}\right]^2. \tag{9}$$

However, the square is the sum of n L-shaped regions, as indicated:

$$S = L_1 + L_2 + \cdots + L_n.$$

What is the area of L_n? This region can be split into two rectangles, as shown in the figure, so that

$$L_n = n\left[\frac{n(n+1)}{2}\right] + n\left[\frac{(n-1)n}{2}\right]$$

$$= \frac{1}{2}\,n^2[(n+1)+(n-1)] = n^3.$$

Accordingly,

$$S = L_1 + L_2 + \cdots + L_n$$

$$= 1^3 + 2^3 + \cdots + n^3, \tag{10}$$

and on comparing (9) and (10) we have (8). There are also geometrically flavored proofs of (1) and (4). These are given in our note on Pythagoras in Appendix B, and we do not repeat them here.

Remark 4 Mathematical induction as a method of demonstrative proof originated in the work of Pascal on the binomial coefficients. The interested reader will find this work described and quoted in vol. 1 of G. Polya's remarkable book, *Mathematical Discovery* (Wiley, 1962), pp. 73–75.

PROBLEMS

1 Use (1) and (4) to find a formula for each of the following:
(a) $2 + 4 + 6 + \cdots + 2n$;
(b) $(n + 1) + (n + 2) + (n + 3) + \cdots + 3n$;
(c) $1 + 3 + 5 + \cdots + (4n - 1)$;
(d) $(2n + 1) + (2n + 3) + (2n + 5) + \cdots + (4n - 1)$;
(e) $3 + 8 + 13 + \cdots + (5n - 2)$.

2 Discover a formula for $1^2 + 3^2 + 5^2 + \cdots + (2n - 1)^2$ by using its relation to the sum $1^2 + 2^2 + 3^2 + \cdots + (2n)^2$.

3 Prove each of the following by induction:

(a) $\dfrac{1}{1 \cdot 2} + \dfrac{1}{2 \cdot 3} + \dfrac{1}{3 \cdot 4} + \cdots + \dfrac{1}{n(n + 1)} = \dfrac{n}{n + 1}$;

(b) $1 \cdot 2 + 2 \cdot 3 + 3 \cdot 4 + \cdots + n(n + 1) =$
$$\frac{n(n + 1)(n + 2)}{3};$$

(c) $\dfrac{1}{1 \cdot 3} + \dfrac{1}{3 \cdot 5} + \dfrac{1}{5 \cdot 7} + \cdots + \dfrac{1}{(2n - 1)(2n + 1)} =$
$$\frac{n}{2n + 1};$$

(d) $1 \cdot 3 + 3 \cdot 5 + 5 \cdot 7 + \cdots + (2n - 1)(2n + 1) =$
$$\frac{n(4n^2 + 6n - 1)}{3}.$$

Prove (a) without using mathematical induction, by means of the algebraic identity
$$\frac{1}{n(n + 1)} = \frac{1}{n} - \frac{1}{n + 1};$$
also, devise a similar method of proving (c).

4 Prove each of the following by induction:

(a) $1 + \dfrac{1}{2} + \dfrac{1}{4} + \cdots + \dfrac{1}{2^n} = 2 - \dfrac{1}{2^n}$;

(b) $1 + r + r^2 + \cdots + r^n = \dfrac{1 - r^{n+1}}{1 - r}$ $(r \neq 1)$;

(c) $\dfrac{1}{2} + \dfrac{2}{2^2} + \dfrac{3}{2^3} + \cdots + \dfrac{n}{2^n} = 2 - \dfrac{(n + 2)}{2^n}$;

(d) $r + 2r^2 + 3r^3 + \cdots + nr^n =$
$$\frac{r - (n + 1)r^{n+1} + nr^{n+2}}{(1 - r)^2} \quad (r \neq 1);$$

(e) $\dfrac{1}{n} + \dfrac{1}{n + 1} + \dfrac{1}{n + 2} + \cdots + \dfrac{1}{2n - 1} =$
$$1 - \frac{1}{2} + \frac{1}{3} - \frac{1}{4} + \cdots + \frac{1}{2n - 1};$$

(f) $\dfrac{1}{1 + x} + \dfrac{2}{1 + x^2} + \dfrac{4}{1 + x^4} + \cdots + \dfrac{2^n}{1 + x^{2^n}} =$
$$\frac{1}{x - 1} + \frac{2^{n+1}}{1 - x^{2^{n+1}}} \quad (x \neq \pm 1);$$

(g) $1^3 + 2^3 + 3^3 + \cdots + n^3 = \left[\dfrac{1}{2} n(n + 1)\right]^2$;

(h) $1^4 + 2^4 + 3^4 + \cdots + n^4 =$
$$\frac{1}{30} n(n + 1)(6n^3 + 9n^2 + n - 1).$$

5 Use the method of Remark 2 to discover and prove the formulas in parts (g) and (h) of Problem 4.

6 In each of the following, guess the general law suggested by the given facts and prove it by induction:
(a) $1 = 1$,
$1 - 4 = -(1 + 2)$,
$1 - 4 + 9 = 1 + 2 + 3$,
$1 - 4 + 9 - 16 = -(1 + 2 + 3 + 4)$;
(b) $1 - \frac{1}{2} = \frac{1}{2}$,
$(1 - \frac{1}{2})(1 - \frac{1}{3}) = \frac{1}{3}$,
$(1 - \frac{1}{2})(1 - \frac{1}{3})(1 - \frac{1}{4}) = \frac{1}{4}$,
$(1 - \frac{1}{2})(1 - \frac{1}{3})(1 - \frac{1}{4})(1 - \frac{1}{5}) = \frac{1}{5}$.

7 Guess the formulas that simplify the following products, and prove them by induction:

(a) $\left(1 - \dfrac{1}{4}\right)\left(1 - \dfrac{1}{9}\right)\left(1 - \dfrac{1}{16}\right) \cdots \left(1 - \dfrac{1}{n^2}\right)$;

(b) $(1 - x)(1 + x)(1 + x^2)(1 + x^4) \cdots (1 + x^{2^n})$.

8 Let $S(n)$ be the following statement:

$$1 + 2 + 3 + \cdots + n = \frac{(n - 1)(n + 2)}{2}.$$

(a) Prove that if $S(n)$ is true for $n = k$, then it is also true for $n = k + 1$.

(b) Criticize the assertion, "By induction we therefore know that $S(n)$ is true for all positive integers n."

APPENDIX E

NUMERICAL TABLES

Table 1. Trigonometric functions

ANGLE		SINE	COSINE	TANGENT	ANGLE		SINE	COSINE	TANGENT
DEGREE	RADIAN				DEGREE	RADIAN			
0°	0.000	0.000	1.000	0.000					
1°	0.017	0.017	1.000	0.017	46°	0.803	0.719	0.695	1.036
2°	0.035	0.035	0.999	0.035	47°	0.820	0.731	0.682	1.072
3°	0.052	0.052	0.999	0.052	48°	0.838	0.743	0.669	1.111
4°	0.070	0.070	0.998	0.070	49°	0.855	0.755	0.656	1.150
5°	0.087	0.087	0.996	0.087	50°	0.873	0.766	0.643	1.192
6°	0.105	0.105	0.995	0.105	51°	0.890	0.777	0.629	1.235
7°	0.122	0.122	0.993	0.123	52°	0.908	0.788	0.616	1.280
8°	0.140	0.139	0.990	0.141	53°	0.925	0.799	0.602	1.327
9°	0.157	0.156	0.988	0.158	54°	0.942	0.809	0.588	1.376
10°	0.175	0.174	0.985	0.176	55°	0.960	0.819	0.574	1.428
11°	0.192	0.191	0.982	0.194	56°	0.977	0.829	0.559	1.483
12°	0.209	0.208	0.978	0.213	57°	0.995	0.839	0.545	1.540
13°	0.227	0.225	0.974	0.231	58°	1.012	0.848	0.530	1.600
14°	0.244	0.242	0.970	0.249	59°	1.030	0.857	0.515	1.664
15°	0.262	0.259	0.966	0.268	60°	1.047	0.866	0.500	1.732
16°	0.279	0.276	0.961	0.287	61°	1.065	0.875	0.485	1.804
17°	0.297	0.292	0.956	0.306	62°	1.082	0.883	0.469	1.881
18°	0.314	0.309	0.951	0.325	63°	1.100	0.891	0.454	1.963
19°	0.332	0.326	0.946	0.344	64°	1.117	0.899	0.438	2.050
20°	0.349	0.342	0.940	0.364	65°	1.134	0.906	0.423	2.145
21°	0.367	0.358	0.934	0.384	66°	1.152	0.914	0.407	2.246
22°	0.384	0.375	0.927	0.404	67°	1.169	0.921	0.391	2.356
23°	0.401	0.391	0.921	0.424	68°	1.187	0.927	0.375	2.475
24°	0.419	0.407	0.914	0.445	69°	1.204	0.934	0.358	2.605
25°	0.436	0.423	0.906	0.466	70°	1.222	0.940	0.342	2.748
26°	0.454	0.438	0.899	0.488	71°	1.239	0.946	0.326	2.904
27°	0.471	0.454	0.891	0.510	72°	1.257	0.951	0.309	3.078
28°	0.489	0.469	0.883	0.532	73°	1.274	0.956	0.292	3.271
29°	0.506	0.485	0.875	0.554	74°	1.292	0.961	0.276	3.487
30°	0.524	0.500	0.866	0.577	75°	1.309	0.966	0.259	3.732
31°	0.541	0.515	0.857	0.601	76°	1.326	0.970	0.242	4.011
32°	0.559	0.530	0.848	0.625	77°	1.344	0.974	0.225	4.332
33°	0.576	0.545	0.839	0.649	78°	1.361	0.978	0.208	4.705
34°	0.593	0.559	0.829	0.675	79°	1.379	0.982	0.191	5.145
35°	0.611	0.574	0.819	0.700	80°	1.396	0.985	0.174	5.671
36°	0.628	0.588	0.809	0.727	81°	1.414	0.988	0.156	6.314
37°	0.646	0.602	0.799	0.754	82°	1.431	0.990	0.139	7.115
38°	0.663	0.616	0.788	0.781	83°	1.449	0.993	0.122	8.144
39°	0.681	0.629	0.777	0.810	84°	1.466	0.995	0.105	9.514
40°	0.698	0.643	0.766	0.839	85°	1.484	0.996	0.087	11.43
41°	0.716	0.656	0.755	0.869	86°	1.501	0.998	0.070	14.30
42°	0.733	0.669	0.743	0.900	87°	1.518	0.999	0.052	19.08
43°	0.750	0.682	0.731	0.933	88°	1.536	0.999	0.035	28.64
44°	0.768	0.695	0.719	0.966	89°	1.553	1.000	0.017	57.29
45°	0.785	0.707	0.707	1.000	90°	1.571	1.000	0.000	

Table 2. **Exponential functions**

x	e^x	e^{-x}	x	e^x	e^{-x}
0.00	1.0000	1.0000	2.5	12.182	0.0821
0.05	1.0513	0.9512	2.6	13.464	0.0743
0.10	1.1052	0.9048	2.7	14.880	0.0672
0.15	1.1618	0.8607	2.8	16.445	0.0608
0.20	1.2214	0.8187	2.9	18.174	0.0550
0.25	1.2840	0.7788	3.0	20.086	0.0498
0.30	1.3499	0.7408	3.1	22.198	0.0450
0.35	1.4191	0.7047	3.2	24.533	0.0408
0.40	1.4918	0.6703	3.3	27.113	0.0369
0.45	1.5683	0.6376	3.4	29.964	0.0334
0.50	1.6487	0.6065	3.5	33.115	0.0302
0.55	1.7333	0.5769	3.6	36.598	0.0273
0.60	1.8221	0.5488	3.7	40.447	0.0247
0.65	1.9155	0.5220	3.8	44.701	0.0224
0.70	2.0138	0.4966	3.9	49.402	0.0202
0.75	2.1170	0.4724	4.0	54.598	0.0183
0.80	2.2255	0.4493	4.1	60.340	0.0166
0.85	2.3396	0.4274	4.2	66.686	0.0150
0.90	2.4596	0.4066	4.3	73.700	0.0136
0.95	2.5857	0.3867	4.4	81.451	0.0123
1.0	2.7183	0.3679	4.5	90.017	0.0111
1.1	3.0042	0.3329	4.6	99.484	0.0101
1.2	3.3201	0.3012	4.7	109.95	0.0091
1.3	3.6693	0.2725	4.8	121.51	0.0082
1.4	4.0552	0.2466	4.9	134.29	0.0074
1.5	4.4817	0.2231	5	148.41	0.0067
1.6	4.9530	0.2019	6	403.43	0.0025
1.7	5.4739	0.1827	7	1096.6	0.0009
1.8	6.0496	0.1653	8	2981.0	0.0003
1.9	6.6859	0.1496	9	8103.1	0.0001
2.0	7.3891	0.1353	10	22026	0.00005
2.1	8.1662	0.1225			
2.2	9.0250	0.1108			
2.3	9.9742	0.1003			
2.4	11.023	0.0907			

Table 3. Natural logarithms (ln $x = \log_e x$)

This table contains logarithms of numbers from 1 to 10 to the base e. To obtain the natural logarithms of other numbers use the formulas:

$$\ln(10^r x) = \ln x + \ln 10^r \qquad \ln\left(\frac{x}{10^r}\right) = \ln x - \ln 10^r$$

$\ln 10 \ = 2.302585 \qquad \ln 10^2 = 4.605170 \qquad \ln 10^3 = 6.907755$
$\ln 10^4 = 9.210340 \qquad \ln 10^5 = 11.512925 \qquad \ln 10^6 = 13.815511$

x	0	1	2	3	4	5	6	7	8	9
1.0	0.0 0000	0995	1980	2956	3922	4879	5827	6766	7696	8618
1.1	0.0 9531	*0436	*1333	*2222	*3103	*3976	*4842	*5700	*6551	*7395
1.2	0.1 8232	9062	9885	*0701	*1511	*2314	*3111	*3902	*4686	*5464
1.3	0.2 6236	7003	7763	8518	9267	*0010	*0748	*1481	*2208	*2930
1.4	0.3 3647	4359	5066	5767	6464	7156	7844	8526	9204	9878
1.5	0.4 0547	1211	1871	2527	3178	3825	4469	5108	5742	6373
1.6	0.4 7000	7623	8243	8858	9470	*0078	*0682	*1282	*1879	*2473
1.7	0.5 3063	3649	4232	4812	5389	5962	6531	7098	7661	8222
1.8	0.5 8779	9333	9884	*0432	*0977	*1519	*2078	*2594	*3127	*3658
1.9	0.6 4185	4710	5233	5752	6269	6783	7294	7803	8310	8813
2.0	0.6 9315	9813	*0310	*0804	*1295	*1784	*2271	*2755	*3237	*3716
2.1	0.7 4194	4669	5142	5612	6081	6547	7011	7473	7932	8390
2.2	0.7 8846	9299	9751	*0200	*0648	*1093	*1536	*1978	*2418	*2855
2.3	0.8 3291	3725	4157	4587	5015	5442	5866	6289	6710	7129
2.4	0.8 7547	7963	8377	8789	9200	9609	*0016	*0422	*0826	*1228
2.5	0.9 1629	2028	2426	2822	3216	3609	4001	4391	4779	5166
2.6	0.9 5551	5935	6317	6698	7078	7456	7833	8208	8582	8954
2.7	0.9 9325	9695	*0063	*0430	*0796	*1160	*1523	*1885	*2245	*2604
2.8	1.0 2962	3318	3674	4028	4380	4732	5082	5431	5779	6126
2.9	1.0 6471	6815	7158	7500	7841	8181	8519	8856	9192	9527
3.0	1.0 9861	*0194	*0526	*0856	*1186	*1514	*1841	*2168	*2493	*2817
3.1	1.1 3140	3462	3783	4103	4422	4740	5057	5373	5688	6002
3.2	1.1 6315	6627	6938	7248	7557	7865	8173	8479	8784	9089
3.3	1.1 9392	9695	9996	*0297	*0597	*0896	*1194	*1491	*1788	*2083
3.4	1.2 2378	2671	2964	3256	3547	3837	4127	4415	4703	4990
3.5	1.2 5276	5562	5846	6130	6413	6695	6976	7257	7536	7815
3.6	1.2 8093	8371	8647	8923	9198	9473	9746	*0019	*0291	*0563
3.7	1.3 0833	1103	1372	1641	1909	2176	2442	2708	2972	3237
3.8	1.3 3500	3763	4025	4286	4547	4807	5067	5325	5584	5841
3.9	1.3 6098	6354	6609	6864	7118	7372	7624	7877	8128	8379
4.0	1.3 8629	8879	9128	9377	9624	9872	*0118	*0364	*0610	*0854
4.1	1.4 1099	1342	1585	1828	2070	2311	2552	2792	3031	3270
4.2	1.4 3508	3746	3984	4220	4456	4692	4927	5161	5395	5629
4.3	1.4 5862	6094	6326	6557	6787	7018	7247	7476	7705	7933
4.4	1.4 8160	8387	8614	8840	9065	9290	9515	9739	9962	*0185
4.5	1.5 0408	0630	0851	1072	1293	1513	1732	1951	2170	2388
4.6	1.5 2606	2823	3039	3256	3471	3687	3902	4116	4330	4543
4.7	1.5 4756	4969	5181	5393	5604	5814	6025	6235	6444	6653
4.8	1.5 6862	7070	7277	7485	7691	7898	8104	8309	8515	8719
4.9	1.5 8924	9127	9331	9534	9737	9939	*0141	*0342	*0543	*0744
5.0	1.6 0944	1144	1343	1542	1741	1939	2137	2334	2531	2728
x	0	1	2	3	4	5	6	7	8	9

Note: The * indicates that the first two digits are those at the beginning of the next row.

Table 3. Natural logarithms (ln $x = \log_e x$) (*Cont.*)

x	0	1	2	3	4	5	6	7	8	9
5.0	1.6 0944	1144	1343	1542	1741	1939	2137	2334	2531	2728
5.1	1.6 2924	3120	3315	3511	3705	3900	4094	4287	4481	4673
5.2	1.6 4866	5058	5250	5441	5632	5823	6013	6203	6393	6582
5.3	1.6 6771	6959	7147	7335	7523	7710	7896	8083	8269	8455
5.4	1.6 8640	8825	9010	9194	9378	9562	9745	9928	*0111	*0293
5.5	1.7 0475	0656	0838	1019	1199	1380	1560	1740	1919	2098
5.6	1.7 2277	2455	2633	2811	2988	3166	3342	3519	3695	3871
5.7	1.7 4047	4222	4397	4572	4746	4920	5094	5267	5440	5613
5.8	1.7 5786	5958	6130	6302	6473	6644	6815	6985	7156	7326
5.9	1.7 7495	7665	7843	8002	8171	8339	8507	8675	8842	9009
6.0	1.7 9176	9342	9509	9675	9840	*0006	*0171	*0336	*0500	*0665
6.1	1.8 0829	0993	1156	1319	1482	1645	1808	1970	2132	2294
6.2	1.8 2455	2616	2777	2938	3098	3258	3418	3578	3737	3896
6.3	1.8 4055	4214	4372	4530	4688	4845	5003	5160	5317	5473
6.4	1.8 5630	5786	5942	6097	6253	6408	6563	6718	6872	7026
6.5	1.8 7180	7334	7487	7641	7794	7947	8099	8251	8403	8555
6.6	1.8 8707	8858	9010	9160	9311	9462	9612	9762	9912	*0061
6.7	1.9 0211	0360	0509	0658	0806	0954	1102	1250	1398	1545
6.8	1.9 1692	1839	1986	2132	2279	2425	2571	2716	2862	3007
6.9	1.9 3152	3297	3442	3586	3730	3874	4018	4162	4305	4448
7.0	1.9 4591	4734	4876	5019	5161	5303	5445	5586	5727	5869
7.1	1.9 6009	6150	6291	6431	6571	6711	6851	6991	7130	7269
7.2	1.9 7408	7547	7685	7824	7962	8100	8238	8376	8513	8650
7.3	1.9 8787	8924	9061	9198	9334	9470	9606	9742	9877	*0013
7.4	2.0 0148	0283	0418	0553	0687	0821	0956	1089	1223	1357
7.5	2.0 1490	1624	1757	1890	2022	2155	2287	2419	2551	2683
7.6	2.0 2815	2946	3078	3209	3340	3471	3601	3732	3862	3992
7.7	2.0 4122	4252	4381	4511	4640	4769	4898	5027	5156	5284
7.8	2.0 5412	5540	5668	5796	5924	6051	6179	6306	6433	6560
7.9	2.0 6686	6813	6939	7065	7191	7317	7443	7568	7694	7819
8.0	2.0 7944	8069	8194	8318	8443	8567	8691	8815	8939	9063
8.1	2.0 9186	9310	9433	9556	9679	9802	9924	*0047	*0169	*0291
8.2	2.1 0413	0535	0657	0779	0900	1021	1142	1263	1384	1505
8.3	2.1 1626	1746	1866	1986	2106	2226	2346	2465	2585	2704
8.4	2.1 2823	2942	3061	3180	3298	3417	3535	3653	3771	3889
8.5	2.1 4007	4124	4242	4359	4476	4593	4710	4827	4943	5060
8.6	2.1 5176	5292	5409	5524	5640	5756	5871	5987	6102	6217
8.7	2.1 6332	6447	6562	6677	6791	6905	7020	7134	7248	7361
8.8	2.1 7475	7589	7702	7816	7929	8042	8155	8267	8380	8493
8.9	2.1 8605	8717	8830	8942	9054	9165	9277	9389	9500	9611
9.0	2.1 9722	9834	9944	*0055	*0166	*0276	*0387	*0497	*0607	*0717
9.1	2.2 0827	0937	1047	1157	1266	1375	1485	1594	1703	1812
9.2	2.2 1920	2029	2138	2246	2354	2462	2570	2678	2786	2894
9.3	2.2 3001	3109	3216	3324	3431	3538	3645	3751	3858	3965
9.4	2.2 4071	4177	4284	4390	4496	4601	4707	4813	4918	5024
9.5	2.2 5129	5234	5339	5444	5549	5654	5759	5863	5968	6072
9.6	2.2 6176	6280	6384	6488	6592	6696	6799	6903	7006	7109
9.7	2.2 7213	7316	7419	7521	7624	7727	7829	7932	8034	8136
9.8	2.2 8238	8340	8442	8544	8646	8747	8849	8950	9051	9152
9.9	2.2 9253	9354	9455	9556	9657	9757	9858	9958	*0058	*0158
10.0	2.3 0259	0358	0458	0558	0658	0757	0857	0956	1055	1154
x	0	1	2	3	4	5	6	7	8	9

Table 4. Common logarithms ($\log_{10} x$)

x	0	1	2	3	4	5	6	7	8	9
10	0000	0043	0086	0128	0170	0212	0253	0294	0334	0374
11	0414	0453	0492	0531	0569	0607	0645	0682	0719	0755
12	0792	0828	0864	0899	0934	0969	1004	1038	1072	1106
13	1139	1173	1206	1239	1271	1303	1335	1367	1399	1430
14	1461	1492	1523	1553	1584	1614	1644	1673	1703	1732
15	1761	1790	1818	1847	1875	1903	1931	1959	1987	2014
16	2041	2068	2095	2122	2148	2175	2201	2227	2253	2279
17	2304	2330	2355	2380	2405	2430	2455	2480	2504	2529
18	2553	2577	2601	2625	2648	2672	2695	2718	2742	2765
19	2788	2810	2833	2856	2878	2900	2923	2945	2967	2989
20	3010	3032	3054	3075	3096	3118	3139	3160	3181	3201
21	3222	3243	3263	3284	3304	3324	3345	3365	3385	3404
22	3424	3444	3464	3483	3502	3522	3541	3560	3579	3598
23	3617	3636	3655	3674	3692	3711	3729	3747	3766	3784
24	3802	3820	3838	3856	3874	3892	3909	3927	3945	3962
25	3979	3997	4014	4031	4048	4065	4082	4099	4116	4133
26	4150	4166	4183	4200	4216	4232	4249	4265	4281	4298
27	4314	4330	4346	4362	4378	4393	4409	4425	4440	4456
28	4472	4487	4502	4518	4533	4548	4564	4579	4594	4609
29	4624	4639	4654	4669	4683	4698	4713	4728	4742	4757
30	4771	4786	4800	4814	4829	4843	4857	4871	4886	4900
31	4914	4928	4942	4955	4969	4983	4997	5011	5024	5038
32	5051	5065	5079	5092	5105	5119	5132	5145	5159	5172
33	5185	5198	5211	5224	5237	5250	5263	5276	5289	5302
34	5315	5328	5340	5353	5366	5378	5391	5403	5416	5428
35	5441	5453	5465	5478	5490	5502	5514	5527	5539	5551
36	5563	5575	5587	5599	5611	5623	5635	5647	5658	5670
37	5682	5694	5705	5717	5729	5740	5752	5763	5775	5786
38	5798	5809	5821	5832	5843	5855	5866	5877	5888	5899
39	5911	5922	5933	5944	5955	5966	5977	5988	5999	6010
40	6021	6031	6042	6053	6064	6075	6085	6096	6107	6117
41	6128	6138	6149	6160	6170	6180	6191	6201	6212	6222
42	6232	6243	6253	6263	6274	6284	6294	6304	6314	6325
43	6335	6345	6355	6365	6375	6385	6395	6405	6415	6425
44	6435	6444	6454	6464	6474	6484	6493	6503	6513	6522
45	6532	6542	6551	6561	6571	6580	6590	6599	6609	6618
46	6628	6637	6646	6656	6665	6675	6684	6693	6702	6712
47	6721	6730	6739	6749	6758	6767	6776	6785	6794	6803
48	6812	6821	6830	6839	6848	6857	6866	6875	6884	6893
49	6902	6911	6920	6928	6937	6946	6955	6964	6972	6981
50	6990	6998	7007	7016	7024	7033	7042	7050	7059	7067
51	7076	7084	7093	7101	7110	7118	7126	7135	7143	7152
52	7160	7168	7177	7185	7193	7202	7210	7218	7226	7235
53	7243	7251	7259	7267	7275	7284	7292	7300	7308	7316
54	7324	7332	7340	7348	7356	7364	7372	7380	7388	7396
55	7404	7412	7419	7427	7435	7443	7451	7459	7466	7474
56	7482	7490	7497	7505	7513	7520	7528	7536	7543	7551
57	7559	7566	7574	7582	7589	7597	7604	7612	7619	7627
58	7634	7642	7649	7657	7664	7672	7679	7686	7694	7701
59	7709	7716	7723	7731	7738	7745	7752	7760	7767	7774

Table 4. Common logarithms ($\log_{10} x$) (*Cont.*)

x	0	1	2	3	4	5	6	7	8	9
60	7782	7789	7796	7803	7810	7818	7825	7832	7839	7846
61	7853	7860	7868	7875	7882	7889	7896	7903	7910	7917
62	7924	7931	7938	7945	7952	7959	7966	7973	7980	7987
63	7993	8000	8007	8014	8021	8028	8035	8041	8048	8055
64	8062	8069	8075	8082	8089	8096	8102	8109	8116	8122
65	8129	8136	8142	8149	8156	8162	8169	8176	8182	8189
66	8195	8202	8209	8215	8222	8228	8235	8241	8248	8254
67	8261	8267	8274	8280	8287	8293	8299	8306	8312	8319
68	8325	8331	8338	8344	8351	8357	8363	8370	8376	8382
69	8388	8395	8401	8407	8414	8420	8426	8432	8439	8445
70	8451	8457	8463	8470	8476	8482	8488	8494	8500	8506
71	8513	8519	8525	8531	8537	8543	8549	8555	8561	8567
72	8573	8579	8585	8591	8597	8603	8609	8615	8621	8627
73	8633	8639	8645	8651	8657	8663	8669	8675	8681	8686
74	8692	8698	8704	8710	8716	8722	8727	8733	8739	8745
75	8751	8756	8762	8768	8774	8779	8785	8791	8797	8802
76	8808	8814	8820	8825	8831	8837	8842	8848	8854	8859
77	8865	8871	8876	8882	8887	8893	8899	8904	8910	8915
78	8921	8927	8932	8938	8943	8949	8954	8960	8965	8971
79	8976	8982	8987	8993	8998	9004	9009	9015	9020	9025
80	9031	9036	9042	9047	9053	9058	9063	9069	9074	9079
81	9085	9090	9096	9101	9106	9112	9117	9122	9128	9133
82	9138	9143	9149	9154	9159	9165	9170	9175	9180	9186
83	9191	9196	9201	9206	9212	9217	9222	9227	9232	9238
84	9243	9248	9253	9258	9263	9269	9274	9279	9284	9289
85	9294	9299	9304	9309	9315	9320	9325	9330	9335	9340
86	9345	9350	9355	9360	9365	9370	9375	9380	9385	9390
87	9395	9400	9405	9410	9415	9420	9425	9430	9435	9440
88	9445	9450	9455	9460	9465	9469	9474	9479	9484	9489
89	9494	9499	9504	9509	9513	9518	9523	9528	9533	9538
90	9542	9547	9552	9557	9562	9566	9571	9576	9581	9586
91	9590	9595	9600	9605	9609	9614	9619	9624	9628	9633
92	9638	9643	9647	9652	9657	9661	9666	9671	9675	9680
93	9685	9689	9694	9699	9703	9708	9713	9717	9722	9727
94	9731	9736	9741	9745	9750	9754	9759	9763	9768	9773
95	9777	9782	9786	9791	9795	9800	9805	9809	9814	9818
96	9823	9827	9832	9836	9841	9845	9850	9854	9859	9863
97	9868	9872	9877	9881	9886	9890	9894	9899	9903	9908
98	9912	9917	9921	9926	9930	9934	9939	9943	9948	9952
99	9956	9961	9965	9969	9974	9978	9983	9987	9991	9996

Note: Decimal points are omitted in this table; the entries

	0	1	2
10	0000	0043	0086

mean that $\log_{10}(1.00) = 0.0000$, $\log_{10}(1.01) = 0.0043$, and $\log_{10}(1.02) = 0.0086$ (to four-decimal-place accuracy).

Table 5. **Powers and roots**

x	x^2	$\sqrt{x}$	x^3	$\sqrt[3]{x}$	x	x^2	$\sqrt{x}$	x^3	$\sqrt[3]{x}$
1	1	1.000	1	1.000	51	2,601	7.141	132,651	3.708
2	4	1.414	8	1.260	52	2,704	7.211	140,608	3.733
3	9	1.732	27	1.442	53	2,809	7.280	148,877	3.756
4	16	2.000	64	1.587	54	2,916	7.348	157,464	3.780
5	25	2.236	125	1.710	55	3,025	7.416	166,375	3.803
6	36	2.449	216	1.817	56	3,136	7.483	175,616	3.826
7	49	2.646	343	1.913	57	3,249	7.550	185,193	3.849
8	64	2.828	512	2.000	58	3,364	7.616	195,112	3.871
9	81	3.000	729	2.080	59	3,481	7.681	205,379	3.893
10	100	3.162	1,000	2.154	60	3,600	7.746	216,000	3.915
11	121	3.317	1,331	2.224	61	3,721	7.810	226,981	3.936
12	144	3.464	1,728	2.289	62	3,844	7.874	238,328	3.958
13	169	3.606	2,197	2.351	63	3,969	7.937	250,047	3.979
14	196	3.742	2,744	2.410	64	4,096	8.000	262,144	4.000
15	225	3.873	3,375	2.466	65	4,225	8.062	274,625	4.021
16	256	4.000	4,096	2.520	66	4,356	8.124	287,496	4.041
17	289	4.123	4,913	2.571	67	4,489	8.185	300,763	4.062
18	324	4.243	5,832	2.621	68	4,624	8.246	314,432	4.082
19	361	4.359	6,859	2.668	69	4,761	8.307	328,509	4.102
20	400	4.472	8,000	2.714	70	4,900	8.367	343,000	4.121
21	441	4.583	9,261	2.759	71	5,041	8.426	357,911	4.141
22	484	4.690	10,648	2.802	72	5,184	8.485	373,248	4.160
23	529	4.796	12,167	2.844	73	5,329	8.544	389,017	4.179
24	576	4.899	13,824	2.884	74	5,476	8.602	405,224	4.198
25	625	5.000	15,625	2.924	75	5,625	8.660	421,875	4.217
26	676	5.099	17,576	2.962	76	5,776	8.718	438,976	4.236
27	729	5.196	19,683	3.000	77	5,929	8.775	456,533	4.254
28	784	5.292	21,952	3.037	78	6,084	8.832	474,552	4.273
29	841	5.385	24,389	3.072	79	6,241	8.888	493,039	4.291
30	900	5.477	27,000	3.107	80	6,400	8.944	512,000	4.309
31	961	5.568	29,791	3.141	81	6,561	9.000	531,441	4.327
32	1,024	5.657	32,768	3.175	82	6,724	9.055	551,368	4.344
33	1,089	5.745	35,937	3.208	83	6,889	9.110	571,787	4.362
34	1,156	5.831	39,304	3.240	84	7,056	9.165	592,704	4.380
35	1,225	5.916	42,875	3.271	85	7,225	9.220	614,125	4.397
36	1,296	6.000	46,656	3.302	86	7,396	9.274	636,056	4.414
37	1,369	6.083	50,653	3.332	87	7,569	9.327	658,503	4.431
38	1,444	6.164	54,872	3.362	88	7,744	9.381	681,472	4.448
39	1,521	6.245	59,319	3.391	89	7,921	9.434	704,969	4.465
40	1,600	6.325	64,000	3.420	90	8,100	9.487	729,000	4.481
41	1,681	6.403	68,921	3.448	91	8,281	9.539	753,571	4.498
42	1,764	6.481	74,088	3.476	92	8,464	9.592	778,688	4.514
43	1,849	6.557	79,507	3.503	93	8,649	9.644	804,357	4.531
44	1,936	6.633	85,184	3.530	94	8,836	9.695	830,584	4.547
45	2,025	6.708	91,125	3.557	95	9,025	9.747	857,375	4.563
46	2,116	6.782	97,336	3.583	96	9,216	9.798	884,736	4.579
47	2,209	6.856	103,823	3.609	97	9,409	9.849	912,673	4.595
48	2,304	6.928	110,592	3.634	98	9,604	9.899	941,192	4.610
49	2,401	7.000	117,649	3.659	99	9,801	9.950	970,299	4.626
50	2,500	7.071	125,000	3.684	100	10,000	10.000	1,000,000	4.642

Table 6. Factorials

n	$n!$	n	$n!$	n	$n!$
0	1.00000 00000 E00	20	2.43290 20082 E18	35	1.03331 47966 E40
1	1.00000 00000 E00	21	5.10909 42172 E19	36	3.71993 32679 E41
2	2.00000 00000 E00	22	1.12400 07278 E21	37	1.37637 53091 E43
3	6.00000 00000 E00	23	2.58520 16739 E22	38	5.23022 61747 E44
4	2.40000 00000 E01	24	6.20448 40173 E23	39	2.03978 82081 E46
5	1.20000 00000 E02	25	1.55112 10043 E25	40	8.15915 28325 E47
6	7.20000 00000 E02	26	4.03291 46113 E26	41	3.34525 26613 E49
7	5.04000 00000 E03	27	1.08888 69450 E28	42	1.40500 61178 E51
8	4.03200 00000 E04	28	3.04888 34461 E29	43	6.04152 63063 E52
9	3.62880 00000 E05	29	8.84176 19937 E30	44	2.65827 15748 E54
10	3.62880 00000 E06	30	2.65252 85981 E32	45	1.19622 22087 E56
11	3.99168 00000 E07	31	8.22283 86542 E33	46	5.50262 21598 E57
12	4.79001 60000 E08	32	2.63130 83693 E35	47	2.58623 24151 E59
13	6.22702 08000 E09	33	8.68331 76188 E36	48	1.24139 15593 E61
14	8.71782 91200 E10	34	2.95232 79904 E38	49	6.08281 86403 E62
15	1.30767 43680 E12			50	3.04140 93202 E64
16	2.09227 89888 E13				
17	3.55687 42810 E14				
18	6.40237 37057 E15				
19	1.21645 10041 E17				

Note: Values are given in scientific notation with the exponent denoted by E; for example, 2.65252 85981 E32 denotes $2.6525285981 \times 10^{32}$.

ANSWERS

CHAPTER 1

Section 1.2, p. 6

1. (a) $5, -5$; (b) $-1, -7$; (c) $6, -2$; (d) $\frac{1}{2}$; (e) $3, \frac{1}{3}$; (f) $\pm 3, \pm 1$; (g) $[-2, 8]$.
3. (a) $-2 \le x \le 2$; (b) $x \le -3$, $x \ge 3$; (c) $x < \frac{4}{3}$; (d) $x < -3, x > 4$.
7. $\frac{1}{3}(2a + b), \frac{1}{3}(a + 2b)$.

Section 1.3, p. 10

5. Center $(-2, \frac{3}{2})$, radius $\frac{1}{2}\sqrt{113}$.
7. $(-1, -1)$.
9. Symmetric with respect to the straight line through the origin that bisects the first and third quadrants.
11. $\frac{1}{2}\sqrt{2}h$.

Section 1.4, p. 14

1. (a) $-\frac{2}{7}$; (b) $\frac{8}{3}$; (c) $\frac{1}{6}$; (d) -1; (e) 0; (f) 10.
3. (a) Yes; (b) no; (c) no; (d) yes.
5. (a) $y = -4x + 5$;
(b) $3x + 7y = 2$;
(c) $2x - 3y = 12$;
(d) $y = -4$;
(e) $x = 1$; (f) $x + 3y + 2 = 0$;
(g) $x + 2y = 11$;
(h) $3y - 2x = 17$;
(i) $x + 2y = 9$; (j) $x + y = 1$.

7. (a) $\dfrac{x}{-3} + \dfrac{y}{-5} = 1$;

(b) $\dfrac{x}{-8} + \dfrac{y}{3} = 1$;

(c) $\dfrac{x}{1} + \dfrac{y}{6} = 1$;

(d) $\dfrac{x}{\frac{9}{2}} + \dfrac{y}{-3} = 1$.

9. $(\frac{19}{5}, -\frac{11}{10})$.
11. $F = \frac{9}{5}C + 32$ or $C = \frac{5}{9}(F - 32)$.

Section 1.5, p. 20

1. (a) $(x - 4)^2 + (y - 6)^2 = 9$;
(b) $(x + 3)^2 + (y - 7)^2 = 5$;
(c) $(x + 5)^2 + (y + 9)^2 = 49$;
(d) $(x - 1)^2 + (y + 6)^2 = 2$;
(e) $(x - a)^2 + y^2 = a^2$ or $x^2 + y^2 = 2ax$;
(f) $x^2 + (y - a)^2 = a^2$ or $x^2 + y^2 = 2ay$.
3. (a) Circle, center $(2, 2)$ and radius $2\sqrt{2}$; (b) point $(9, 7)$; (c) circle, center $(-4, -5)$ and radius 1; (d) circle, center $(-\frac{3}{2}, 4)$ and radius 3; (e) empty; (f) point $(\frac{1}{2}\sqrt{2}, -\frac{1}{2}\sqrt{2})$; (g) circle, center $(8, -3)$ and radius 11.
5. Distinct real roots, $b^2 - 4ac > 0$; equal real roots, $b^2 - 4ac = 0$; no real roots, $b^2 - 4ac < 0$.
7. $y = \pm 2\sqrt{2}x + 4$.
9. (a) $y^2 = -12x$; (b) $x^2 = 4y$; (c) $y^2 = 8x$; (d) $3x^2 = -4y$; (e) $y^2 + 12x + 12 = 0$; (f) $x^2 - 6x - 8y + 17 = 0$.
11. 20 ft.

Section 1.6, p. 24

1. $f(1) = 0, f(2) = 2, f(3) = 10$, $f(0) = -2, f(-1) = -10$, $f(-2) = -30$.
5. (a) $x \ge 0$; (b) $x \le 0$; (c) all x; (d) $x \le -2, x \ge 2$; (e) all x except 2, -2; (f) all x; (g) $x \le -2, x \ge 1$; (h) $x < -2, x > 1$; (i) $-3 \le x \le 1$; (j) $x \le 0, x > 2$.
7. $f(0) = 0, f(1)$ does not exist, $f(2) = 2, f(3) = \frac{3}{2}, f(f(3)) = 3$. In the last part, it is tacitly understood

that x is restricted to those values for which $f(f(x))$ exists: that is, $x \ne 1$.
9. $f(0) = 1, f(1)$ does not exist, $f(2) = -1, f(f(2)) = \frac{1}{2}$, $f(f(f(2))) = 2$.
11. $f(x_1)f(x_2) = f(x_1 + x_2)$.
13. No; it is true if and only if $ad + b = bc + d$.
15. (a) $a = 4, b = -5, c = 3$.

Section 1.7, p. 27

1. (a) No; (b) $y = 1 - 3x^2$; (c) $y = \dfrac{x + 1}{x - 1}$; (d) no.
3. $A = \frac{1}{4}\sqrt{3}x^2$.
5. $V = x^3, A = 6x^2, d = \sqrt{3}x$.
7. $A = x^2 + \dfrac{1}{4\pi}(L - 4x)^2$.
9. $V = \dfrac{1}{6\sqrt{\pi}}A^{3/2}$.
11. $C = \frac{2}{3}S$.
13. $V = \dfrac{\pi H}{R}(Rr^2 - r^3)$.

Section 1.8, p. 34

1. (a)

919

(b)

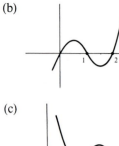

(c)

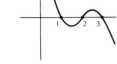

(d)

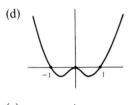

(e)

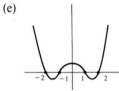

3. (a)

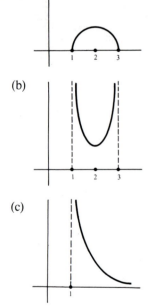

(b)

(c)

(d)

(e)

(f)

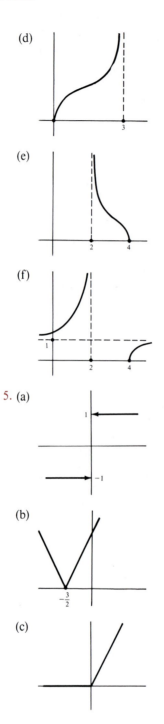

5. (a)

(b)

(c)

(d)

(e)

(f)

(g)

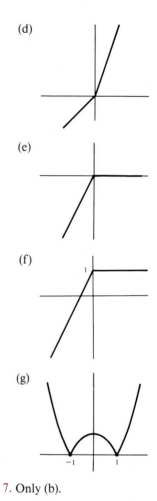

7. Only (b).

Additional Problems, p. 35
9. No, to both questions.
15. $(y_1 - y_2)x + (x_2 - x_1)y = x_2y_1 - x_1y_2$.

19. (a) $\left(b, \dfrac{ab - b^2}{c} \right)$;

(b) $\left(\dfrac{a}{2}, \dfrac{b^2 + c^2 - ab}{2c} \right)$;

(c) $\left(\dfrac{a + b}{3}, \dfrac{c}{3} \right)$.

23. (a) $x - 7y + 5 = 0$,
$7x + y - 15 = 0$; (b) $x = (1 \pm \sqrt{2})y$.
25. $|b| \le 2\sqrt{10}$.
27. (a) $(x - \frac{4}{3}a)^2 + y^2 = \frac{4}{9}a^2$;
(b) $(x^2 + y^2)^2 = 2a^2(x^2 - y^2)$.
31. $7x + y = 10$ and $x - y + 2 = 0$.
33. $y = -2x + 2$; $(0, 2)$ and $(\frac{4}{3}, \frac{2}{3})$.
35. $x^2 + y^2 - 2xy - 4x - 4y + 4 = 0$.

37. The line is $x = 2pm$.
41. No.
43. $g(x) = x^3$.
45. $V = \frac{1}{2}Ar - \pi r^3$.
47. $V = \frac{2}{3}\pi a\left(\dfrac{r^4}{r^2 - a^2}\right)$.
49. $\alpha = \dfrac{d}{ad - bc}$, $\beta = \dfrac{-b}{ad - bc}$,

$\gamma = \dfrac{-c}{ad - bc}$, $\delta = \dfrac{a}{ad - bc}$.

51. The following are some further hints: for (a), $f(x) = f(x \cdot 1) = f(x)\cdot f(1)$; for (e), if $x < f(x)$ for some real x, and r is a rational number such that $x < r < f(x)$, then $f(x) < f(r) = r < f(x)$, which is impossible.
53. $(x - 1)(x - 2) \cdots (x - n)$; $x^n + 1$; x^n.
55. (a) Odd; (b) even; (c) even; (d) odd; (e) neither; (f) odd; (g) neither; (h) neither.
57. (a) Even; (b) even; (c) odd.

59. $y = 275(x - 1)(x - 2) - \sqrt{3}(x - 1)(x - 3) + \dfrac{\pi}{2}(x - 2)(x - 3)$.

61. (a)

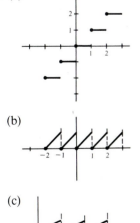

(b)

(c)

(d)

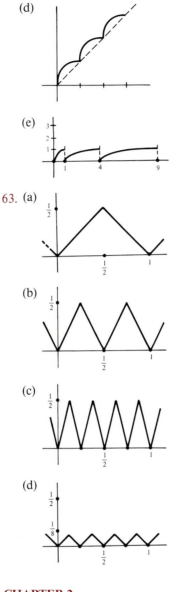

(e)

63. (a)

(b)

(c)

(d)

CHAPTER 2

Section 2.2, p. 45
1. (a) $4x + y + 4 = 0$; (b) $8x - y = 16$; (c) $8x - y = 16$.
5. (a) $2x_0 - 4$; (b) $2x_0 - 2$; (c) $4x_0$; (d) $2x_0$.
7. $8x + y + 7 = 0$.

Section 2.3, p. 50
1. $2ax + b$.
3. $6x^2 - 6x + 6$.
5. $1 + \dfrac{1}{x^2}$.

7. $\dfrac{1}{(x + 1)^2}$.
9. $\dfrac{-3}{x^4}$.
11. $\dfrac{1}{\sqrt{2x}}$.
13. (b) Area = 2.

Section 2.4, p. 54
1. $v = 6t - 12$; (a) $t = 2$, (b) $t > 2$.
3. $v = 4t + 28$; (a) $t = -7$, (b) $t > -7$.
5. $v = 14t$; (a) $t = 0$, (b) $t > 0$.
7. $v = 8t - 24$; (a) $t = 3$, (b) $t > 3$.
9. 10 seconds.
11. (a) 3200 gal/min; (b) 2400 gal/min.
13. dr/dt decreases as r increases.

Section 2.5, p. 59
1. 15.
3. -5.
5. 3.
7. -3.
9. 4.
11. 5.
13. 0.
15. $\frac{4}{3}$.
17. (a) 6; (b) 4; (c) -2; (d) 0; (e) does not exist; (f) $\frac{1}{4}$.
19. (a) None; (b) 1, -1; (c) 1; (d) all $x < 0$; (e) all $x \leq 0$; (f) none; (g) 3, -4; (h) none. (Remember that a function is automatically discontinuous at every point not in its domain; thus, $1/x$ is discontinuous at $x = 0$ even though it is a continuous function.)

Additional Problems, p. 60
1. $b = -6$.
5. (b) Drop the perpendicular from P to a point A on the axis of the parabola. Draw the circle whose center is the vertex V and which passes through A. Let B be the second point at which this circle intersects the axis, and draw the line PB. This line will be tangent to the parabola at P.
7. (a) $x = 0$; (b) $x = \pm 2$; (c) $x = \frac{3}{2}$; (d) differentiable at all points.
13. $m = 2a$, $b = -a^2$.
15. When $t = \frac{3}{4}$; 8 ft/s.

19. Does not exist.
21. -5.
23. Does not exist.
25. Does not exist.
27. 2.
29. 2.
31. -3.
33. $\frac{1}{7}$.
35. -5.
37. $\frac{1}{2}$.
39. 4.
41. $3a/2$.
43. 1.
45. 0.
47. Does not exist.
49. Does not exist.
51. 3.
53. 0.
55. 0.
57. 1.
59. $\lim\limits_{x\to 0+} f(x)$, $\lim\limits_{x\to 0-} f(x)$, and $\lim\limits_{x\to 0} f(x)$ do not exist.
61. Because there are rationals as close as we please to every irrational, and irrationals as close as we please to every rational.

CHAPTER 3

Section 3.1, p. 66

1. (a) $54x^8$; (b) 0; (c) $-60x^3$;
(d) $1500x^{99}(x^{400} + 1)$; (e) $2x - 6$;
(f) $x^4 + x^3 + x^2 + x + 1$; (g) $4x^3 + 3x^2 + 2x + 1$; (h) $5x^4 - 40x^3 + 120x^2 - 160x + 80$; (i) $12x(x^{10} + x^4 - x - 1)$; (j) $18x^2 - 6x + 4$.
3. (a) x^3; (b) $\frac{4}{3}x^3$; (c) $x^3 + x^2 - 5x$.
5. $(1, -4)$ and $(-2, 50)$.
7. $(4, 2)$.
9. $a = -8$, $b = 18$, $c = 4$.
11. $y = 2x - 1$.

Section 3.2, p. 70

1. (a) $2x$; (b) $18x^2 - 36x + 18$; (c) $15x^4 + 57x^2 + 6$; (d) $5x^4$.
3. (a) $-4/x^2$;
(b) $(-8 - 6x^3 - 2x^5)/x^5$;
(c) $(2x^4 - 2)/x^3$.
5. (a) tangent $2x + 3y = 8$, normal $3x - 2y + 1 = 0$; (b) tangent $4x + 5y = 13$, normal $5x - 4y = 6$;

(c) $3x - y + 4 = 0$; (d) $2y = x + 2$.
7. Area $= 1$.
9. $(0, 2)$, $(\pm 1, 1)$.

Section 3.3, p. 75

1. (a) $4(x^5 - 3x)^3 \cdot (5x^4 - 3)$;
(b) $500(x^2 - 2)^{499} \cdot 2x$;
(c) $6(x + x^2 - 2x^5)^5$ $\cdot (1 + 2x - 10x^4)$;
(d) $3/(1 - 3x)^2$; (e) $4x/(12 - x^2)^3$;
(f) $4[1 - (3x - 2)^3]^3$ $\cdot (-3)(3x - 2)^2 \cdot 3$.
3. (a) $\dfrac{-2(2t - 1)^2(t^2 - 2t - 9)}{(t^2 + 3)^3}$;
(b) $\dfrac{-4}{(2t - 3)^3}$; (c) $\dfrac{72}{(5 - 4t)^4}$;
(d) $\dfrac{4t(30 - t^2)}{(t^2 - 6)^3}$.
5. (a) $3u^2 \dfrac{du}{dx}$; (b) $4(2u - 1) \dfrac{du}{dx}$;
(c) $4u(u^2 - 2) \dfrac{du}{dx}$.

Section 3.4, p. 80

1. (a) $-\dfrac{3x^2}{4y^2}$; (b) $\dfrac{y^2 - 2xy + 2x}{x^2 - 2xy - 4y}$;
(c) $\dfrac{1}{1 - 7y^6}$; (d) $\dfrac{3y - 4x^3y^3}{3x^4y^2 - 3x}$;
(e) $\dfrac{3x^2 - 4y}{3y^2 + 4x}$; (f) $-\dfrac{y^2}{x^2}$; (g) $-\sqrt{\dfrac{y}{x}}$.
3. (a) $\frac{4}{5}x^{-1/5}$; (b) $\frac{5}{6}x^{-1/6}$; (c) $-\frac{3}{4}x^{-7/4}$;
(d) $-\frac{7}{11}x^{-18/11}$; (e) $\frac{6}{5}x^{-3/5}$;
(f) $\dfrac{(1 + x^{2/3})^{1/2}}{x^{1/3}}$;
(g) $\dfrac{3(x^3 - 16)}{4x^3} \sqrt[4]{\dfrac{x^2}{x^3 + 8}}$;
(h) $\dfrac{1}{4\sqrt{1 + x}\,\sqrt{1 + \sqrt{1 + x}}}$.

Section 3.5, p. 83

1. (a) 8, 0, 0, 0; (b) $16x - 11$, 16, 0, 0; (c) $24x^2 + 14x - 1$, $48x + 14$, 48, 0; (d) $4x^3 - 39x^2 + 10x + 3$, $12x^2 - 78x + 10$, $24x - 78$, 24; (e) $\frac{5}{2}x^{3/2}$, $\frac{15}{4}x^{1/2}$, $\frac{15}{8}x^{-1/2}$, $-\frac{15}{16}x^{-3/2}$.
3. (a) $n!(1 - x)^{-(n+1)}$;
(b) $(-1)^n n! 3^n(1 + 3x)^{-(n+1)}$;
(c) $(-1)^{n+1} n!(1 + x)^{-(n+1)}$.
5. $-\dfrac{(n - 1)a^n x^{n-2}}{y^{2n-1}}$.
7. (a) $t = \frac{1}{2}$, $s = 0$, $v = 12$; (b) $t = 4$, $s = 32$, $v = 6$; (c) $t = 1$, $s = 6$,

$v = -3$.
9. $3, \frac{1}{3}$.

Additional Problems, p. 84

1. $(-1, 10)$ and $(3, -22)$.
3. $(1, 2)$ and $(-1, -2)$; the smallest slope $= 1$, at $(0, 0)$.
5. Slope $= 4x^3 - 4x$; $x = 0, \pm 1$; $-1 < x < 0$, $x > 1$.
7. $a = 1$, $b = 1$, $c = 0$.
9. $a = 1$, $b = 0$, $c = -1$.
13. $a = 1$, $b = -2$, $c = 2$, $d = -1$.
17. $(6, 9)$, $(-2, 1)$, $(-4, 4)$.
19. (a) $\dfrac{-4x}{(x^2 - 1)^2}$; (b) $\dfrac{-4(x + 1)}{(x - 1)^3}$;
(c) $\dfrac{x(4 - x^3)}{(x^3 + 2)^2}$; (d) $\dfrac{-2x^2 - 6x - 11}{(x^2 + x - 4)^2}$;
(e) $\dfrac{x^2(3 - x^2)}{(1 - x^2)^2}$; (f) $\dfrac{-2}{(1 + x)^2}$;
(g) $\dfrac{18x^4 - 24x^3 - 9}{(x - 1)^2}$;
(h) $\dfrac{-10(x + 3)}{(x - 2)^3}$.
23. (a) $(x + 2)(x + 3) + (x + 1)(x + 3) + (x + 1)(x + 2)$; (b) $(x^3 + 3x^2)(x^4 + 4)(2x + 2) + (x^2 + 2x)(x^4 + 4)(3x^2 + 6x) + (x^2 + 2x)(x^3 + 3x^2)(4x^3)$.
25. $(0, 10\sqrt{5})$, $(\pm 3, \sqrt{5})$.
27. $(2, -2)$ and $(-10, \frac{2}{3})$.
29. (a) $-6(1 + 2x)^2(4 - 5x)^5$ $\cdot (15x + 1)$;
(b) $10x(x^2 + 1)^9(x^2 - 1)^{14}(5x^2 + 1)$;
(c) $\dfrac{-2x(2x^2 - 19)}{(16 + x^2)^4}$;
(d) $-3x^6(3 - 2x)^2(4x - 9)$ $\cdot (32x^2 - 96x + 63)$.
31. (a) $y = (x^4 + 1)^3$;
(b) $y = 2(x^6 + 1)^6$.
33. (a) $-\dfrac{2x^3 + y^3}{3xy^2 + 4y^3}$; (b) $\dfrac{y + 2x^2}{x(1 - x)}$;
(c) $\dfrac{-4x}{y(x^2 - 2)^2}$; (d) $-\dfrac{y^5}{x^5}$;
(e) $\dfrac{\sqrt{y} - y}{x + 4\sqrt{xy}}$.
35. (a) $\frac{1}{2}\sqrt{x}(5x - 3)$;
(b) $\frac{8}{3}x(x^2 + 2)^{-5/9}$;
(c) $\dfrac{2 + 5x^{3/2}}{6(x + x^{5/2})^{2/3}}$; (d) $\dfrac{2x - x^3}{(1 - x^2)^{3/2}}$;
(e) $\dfrac{x - 1}{2x\sqrt{x}}$; (f) $\dfrac{x}{(2x^2 - 1)^{3/4}}$;

(g) $\dfrac{2x}{(x^2+1)^2}\sqrt{\dfrac{x^2+1}{x^2-1}}$;

(h) $-\dfrac{1}{\sqrt{2-x}\,\sqrt{2+\sqrt{2-x}}}$.

39. (a) $-2(1+3x)^{-5/3}$;

(b) $-\dfrac{x+4}{4(x+1)^{5/2}}$; (c) $-\tfrac{4}{25}x^{-6/5}$;

(d) $\tfrac{35}{4}x^{3/2}$; (e) $-\tfrac{1}{4}x^{-3/2}+\tfrac{3}{4}x^{-5/2}$;

(f) $20(x^2+1)(x^2+4)^{1/2}$.

CHAPTER 4

Section 4.1, p. 91

1.
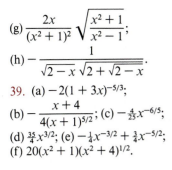

3.
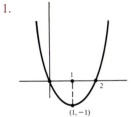

5.
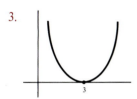

7.
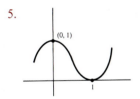

9.
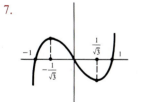

11.
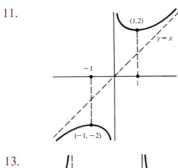

13.
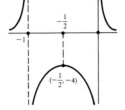

15.
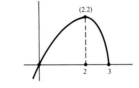

19.
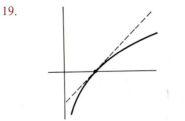

21. (a)

(b)
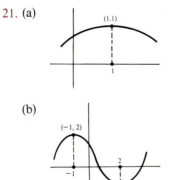

(c)
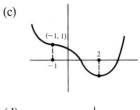

(d)
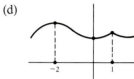

Section 4.2, p. 94

1.
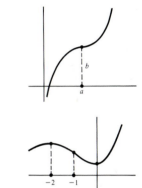

3.

5.

7.
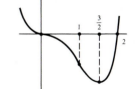

9.
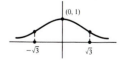

11.

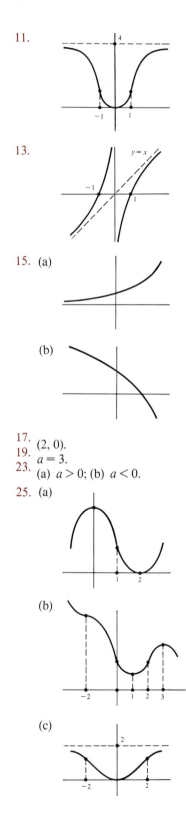

13.

15. (a)

(b)

17. (2, 0).
19. $a = 3$.
23. (a) $a > 0$; (b) $a < 0$.

25. (a)

(b)

(c)

(d)

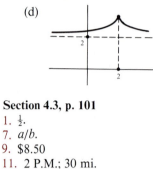

Section 4.3, p. 101
1. $\frac{1}{2}$.
7. a/b.
9. $8.50
11. 2 P.M.; 30 mi.
13. 4, 4.
15. 108.
17. 4 by 8 in.
19. 1.
21. 24 in.
23. $\sqrt{3}$.
25. $\frac{3}{2}a$.
27. 1.
29. $(a^{2/3} + b^{2/3})^{3/2}$ (convince yourself that this and Problem 28 are essentially the same).

Section 4.4, p. 108
1. 1.
3. $\frac{2}{3}R$.
5. 4 by 4 in.
7. $\frac{1}{2}$.
9. (1, 1).
11. (a) $\frac{3}{2}$ mi; (b) 1 h and 44 min; (c) 8 min longer.
13. $15/\sqrt{10}$ mi/h.
19. 1.
21. $a = 2$.
23. $x = \frac{1}{2}\sqrt{2}\,a$.
25. (a) 0; (b) 1.
27. (2, 4).

Section 4.5, p. 112
1. (a) 120π ft²/s; (b) 240π ft²/s.
3. $\dfrac{2}{\pi}$ ft/min.
5. 4 ft/s at each of the stated moments.
7. 3 ft/s.
9. $4\frac{1}{3}$ ft/s.
13. 52 mi/h.
15. $\frac{1}{3}$ lb/in² per min.
17. (a) $\dfrac{1}{\pi}$ in/min; (b) $\dfrac{1}{2\sqrt[3]{2}\pi}$ in/min.
19. $\dfrac{1}{5\pi}$ in/s.

Section 4.6, p. 115
3. 0.62.
5. 2.15.
7. 1.31 ft.
11. 0.92 and 2.86.

Section 4.7, p. 120
1. $x_0 = \sqrt{a/c}$.
3. (a) 11,250 at a price of $187.50; (b) absorb $12.50, pass on $12.50.
7. 30.

Additional Problems, p. 121
1.

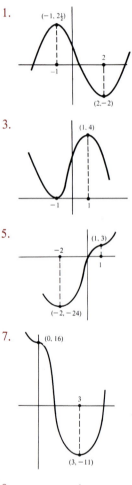

3.

5.

7.

9.

11.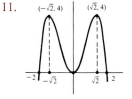

13.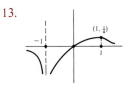

15.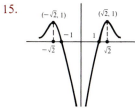

17.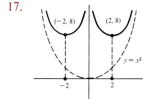

19.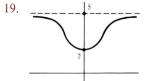

21.

23.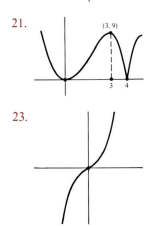

25.

27.

29.

31.

33.

35.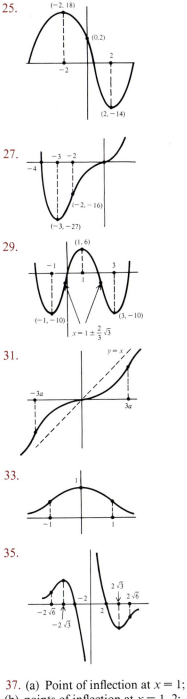

37. (a) Point of inflection at $x = 1$;
(b) points of inflection at $x = 1, 2$;
(c) points of inflection at $x = -2, 0, 1, 2, 3$.

39. $a = -3$.

41. $\frac{1}{3}\sqrt{3}$.

43. $x = 21$, $y = 35$.

45. $18 = 16 + 2$.

47. 5, 5, 5.

51. $\frac{1}{3}\sqrt{3}\,a$, $\frac{2}{3}b$.

59. 2 in.

63. 4000 knives at a price of $18 apiece.

65. 20 days.

67. (a) 120 ft; (b) 312 ft.

69. $\frac{1}{4}$.

71. $\sqrt{2}$.

73. $\pi/4$.

75. 4.

77. $4/\pi$.

79. A square with side $\frac{1}{6}(a + b - \sqrt{a^2 - ab + b^2})$.

81. $\sqrt{2}$.

83. 3 by 6 by 12 in.

85. $a - \dfrac{bs}{\sqrt{r^2 - s^2}}$ meters, if this number is positive.

87. $x^2 + y^2 = 32$.

89. (3, 3).

91. (5, 0) and $(-5, 0)$.

99. (a) 12 ft/s; (b) 3 ft/s.

101. $1/\pi$ ft/min.

103. At least 9 ft.

107. $\dfrac{dy}{dt} = \dfrac{ax}{\sqrt{x^2 + r^2}}$ in/s.

109. 0.32 lb/min.

111. 144π m^3/min.

115. Decreasing 1 in^2/min.

117. When $t = \dfrac{R_0\sqrt{b} - r_0\sqrt{a}}{a\sqrt{a} - b\sqrt{b}}$.

119. (a) 3.32; (b) 1.90; (c) 2.09.

123. 1.86 in.

127. Every 4 days.

CHAPTER 5

Section 5.2, p. 135

1. $(63x^8 - 15x^4)\,dx$.

3. $\dfrac{(2x - 3x^3)\,dx}{\sqrt{1 - x^2}}$.

5. $\dfrac{(2 - x)\,dx}{\sqrt{4x - x^2}}$.

7. $(2x^{-1/3} + 2x^{-4/5} - 17)\,dx$.

9. $\dfrac{(15x^2 + 8x)\,dx}{2\sqrt{3x + 2}}$.

11. $\dfrac{dy}{dx} = \dfrac{3x}{8y^2}$, $\dfrac{dx}{dy} = \dfrac{8y^2}{3x}$.

13. $\dfrac{dy}{dx} = \dfrac{2xy - x^2}{y^2 - x^2}, \dfrac{dx}{dy} = \dfrac{y^2 - x^2}{2xy - x^2}.$

15. $\dfrac{dy}{dx} =$
$\dfrac{-15(3u^2 - 2u + 1)x^2(x^3 + 2)^4}{(u^2 - u)^2}.$

17. $\dfrac{dy}{dx} = \dfrac{18x^2u^{5/2}}{(4y^3 + 3)(2u^{3/2} - 1)}.$

19. $\Delta V = 4\pi r^2\,\Delta r + 4\pi r\,\Delta r^2 + \frac{4}{3}\pi\,\Delta r^3$ and $dV = 4\pi r^2\,dr = 4\pi r^2\,\Delta r.$ The differential dV can be thought of as the area of the inner sphere times the thickness of the spherical shell.

Section 5.3, p. 141

1. $\frac{1}{2}x^2 + x + c.$
3. $\frac{1}{3}x^3 + \frac{1}{4}x^4 + \frac{1}{5}x^5 + c.$
5. $2\sqrt{x} + c.$
7. $\frac{4}{7}x^{7/4} + c.$
9. $\frac{3}{2}x^{2/3} + c.$
11. $\frac{2}{3}x^{3/2} - 2\sqrt{x} + c.$
13. $6\sqrt{x} + \frac{4}{3}x^{3/2} + c.$
15. $\frac{1}{9}(3x^2 + 1)^{3/2} + c.$
17. $-\frac{5}{72}(1 - 4x^3)^{6/5} + c.$
19. $\dfrac{3}{20(2 - x^{5/3})^4} + c.$
21. $\sqrt{1 + 4x + 3x^2} + c.$
23. $\frac{1}{5}x^{15} + c.$
25. $y = x^3 + 2.$

Section 5.4, p. 144

1. $y = 2x^3 + 2x^2 - 5x + c.$
3. $y = 6x^4 + 6x^3 - 4x^2 + 3x + c.$
5. $3y^2 - 4y^{3/2} = 3x^2 + 4x^{3/2} + c.$
7. $y = -\dfrac{1}{x} + \dfrac{1}{2}x^2 + c.$
9. $y = \dfrac{1}{5 - x^2}.$
11. $y = \sqrt{\dfrac{2x^2 - 31}{33 - 2x^2}}.$
13. $3\sqrt{y} = x\sqrt{x} - 3.$

Section 5.5, p. 151

1. 8 s; velocity $= -128$ ft/s and speed $= 128$ ft/s.
3. $v = -32t + 128;\ s = -16t^2 + 128t.$
5. $10\sqrt{10}$ s.
7. 40 ft/s.
9. 96 ft/s.
11. $v_0^2/64$ ft; 96 ft/s.

Additional Problems, p. 152

1. $\frac{3}{5}x^5 - \frac{7}{4}x^4 + 10x + c.$
3. $\frac{1}{3}x^3 - \frac{3}{2}x^2 + x - 4\sqrt{x} + c.$
5. $\frac{1}{4}x^4 + \frac{2}{3}x^3 + \frac{1}{2}x^2 + c.$
7. $17x^3 - 27x^4 + c.$
9. $6x - \frac{2}{3}x^{3/2} - \frac{1}{2}x^2 + c.$
11. $-\frac{2}{9}(2 - 3x)^{3/2} + c.$
13. $\frac{1}{825}(5x + 2)^{165} + c.$
15. $5\sqrt{1 + x^2} + c.$
17. $\frac{1}{3}\sqrt{2x^3 - 1} + c.$
19. $\frac{3}{4}(x^2 - 2x + 3)^{2/3} + c.$
21. $-\frac{3}{2}\sqrt[3]{2 - x^2} + c.$
23. $-\dfrac{1}{3}\left(1 + \dfrac{1}{x}\right)^3 + c.$
25. $\frac{3}{7}(x + 1)^{7/3} + c.$
27. $\frac{3}{7}(1 + x)^{7/3} - \frac{3}{4}(1 + x)^{4/3} + c.$
29. $\frac{2}{11}(x^3 + x + 32)^{11/2} + c.$
31. $\frac{1}{28}(x^3 - 1)^{4/3}(4x^3 + 3) + c.$
33. (a) $y = \sqrt{\dfrac{7x^2 - 3}{3x^2 + 13}};$

(b) $3\sqrt{y - 4} = (x - 1)^{3/2} - 2.$
35. $x^2 - y^2 = c.$
39. (a) 25 s, 1200 ft/s;
(b) $25\sqrt{2} \cong 35$ s, $800\sqrt{2} \cong 1120$ ft/s.
41. 30 m/s.
43. (a) 44 ft; (b) 680 ft.
45. About 2 mi.

CHAPTER 6

Section 6.3, p. 159

1. (a) 55; (b) 62; (c) 206.
5. (a) $\dfrac{(n - 1)n}{2};$

(b) $\dfrac{(n - 1)n(2n - 1)}{6};$

(c) $\left[\dfrac{(n - 1)n}{2}\right]^2.$

Section 6.6, p. 172

1. 9.
3. (a) $\frac{32}{3}$; (b) 16; (c) 12.
5. (a) $\frac{35}{3}$; (b) 4; (c) $\frac{21}{2}$; (d) $\frac{62}{3}$; (e) $\frac{26}{3}$; (f) 2; (g) 33; (h) 18; (i) $\frac{26}{3}$; (j) 2; (k) $\frac{4}{9}$.
7. (a) $\frac{2}{3}$; (b) 4; (c) 1; (d) $\frac{14}{3}$; (e) $\frac{13}{3}$; (f) 9; (g) $5/48a^2$; (h) $(\sqrt{5} - 1)b$; (i) $\frac{1}{6}$; (j) $\frac{9}{2}$; (k) $a^4/4$; (l) $\frac{1023}{10}$; (m) $b^2/6$; (n) $\frac{2}{15}$; (o) $\frac{1}{30}$; (p) $\frac{29}{6}$.

Section 6.7, p. 176

1. (a) $\frac{31}{6}$; (b) $\frac{22}{3}$; (c) 19; (d) $\frac{13}{2}$.
3. $\frac{128}{5}$.
13. $\pi a^2/2.$

Additional Problems, p. 176

7. (a) $\frac{8}{3}$; (b) $\frac{38}{3}$; (c) 12; (d) $5\sqrt{5}/3$; (e) $\frac{9}{20}$.
9. (a) $4(2\sqrt{2} - 1)$; (b) $\frac{8}{3}$; (c) 3; (d) $\frac{35}{4}$.
11. (a) $\dfrac{4x^3}{1 + x^4}$; (b) $\dfrac{2x}{1 + x^2}$;

(c) $\dfrac{3x^2}{\sqrt{3x^3 + 7}}$; (d) $\dfrac{5x^9}{\sqrt{1 + x^{10}}}.$

CHAPTER 7

Section 7.2, p. 181

1. $\frac{4}{3}$.
3. $\frac{8}{3}$.
5. $\frac{32}{3}$.
7. $\frac{64}{15}\sqrt{2}$.
9. 4.
11. $\frac{64}{3}$.
13. (a) $\frac{1}{12}$; (b) $\frac{4}{3}$.
15. $\frac{27}{4}$.
17. $2(\sqrt{b} - 1) \to \infty$ as $b \to \infty$.

Section 7.3, p. 184

1. (a) 8π; (b) $16\pi/15$; (c) $3\pi/5$; (d) $2\pi/3$; (e) $8\pi/3$; (f) $16\pi a^3/105$.
5. 4π.
7. $\frac{4}{3}a^3$.
9. $\frac{16}{3}a^3$.
11. $2\pi^2 a^2 b$.
13. $\frac{1}{6}\sqrt{3}a^3$.
15. $9\sqrt{2}/2$.
17. $\frac{16}{3}a^3$.

Section 7.4, p. 187

3. $128\pi/5$.
5. $486\pi/5$.
7. $2\pi(b - a)$.
9. $\dfrac{8\pi}{3}(2 - \sqrt{2})$.
11. (a) 8π; (b) $256\pi/15$.
13. $\pi h^3/6$.
15. $8\pi/9$.

Section 7.5, p. 191

1. $\frac{8}{27}(10\sqrt{10} - 1)$.
3. $\frac{53}{6}$.
5. $\frac{118}{9}$.
7. 12.

Section 7.6, p. 195

1. $253\pi/20$.
3. 12π.
5. $\dfrac{\pi}{27}(10\sqrt{10} - 1)$.

7. $\frac{8\pi}{3}(2\sqrt{2}-1)p^2$.

9. $\frac{12}{5}\pi a^2$.

Section 7.7, p. 198
1. 150 tons.
3. 125 tons.
5. $\frac{5}{6}$ ton.
7. $\frac{7}{6}$ ton.
9. 300π lb.
11. $10\frac{5}{16}$ tons; 11 ft.
13. $300\sqrt{2}w$ or approximately 13.2 tons.

Section 7.8, p. 203
1. 64 ft-lb.
3. 6000 ft-lb.
5. 550 ft-lb.
7. 50,000 ft-lb.
11. (b) 250 ft-lb.
13. 5 ft-lb.
15. $GMm/2a$.
17. $mgRh/(R+h)$.
19. $240\,\pi w$ ft-lb.
21. $\frac{4}{3}\pi a^3 w(h+a)$.
23. $m_1 = \frac{1}{4}m_2$.

Additional Problems, p. 205
1. $\frac{1}{6}$.
3. $\frac{128}{15}$.
5. 36.
7. $\frac{125}{6}$.
9. 18.
11. $\frac{8}{3}$.
13. 64.
15. $\frac{37}{6}$.
17. $\frac{136}{3}$.
19. (a) $56\pi/15$; (b) $56\pi/15$; (c) $32\pi/3$; (d) $48\pi/5$; (e) $\pi a^3/15$.
21. (a) $512\pi/15$; (b) $128\pi/3$.
23. $\frac{8}{3}a^3$.
25. $\frac{1}{3}\pi(b^5 - a^5)$.
27. (a) $2\pi a^3$; (b) $\frac{8}{3}\pi a^3$; (c) $\frac{16}{15}\pi a^3$.
29. $a^2 h$.
31. $\frac{4}{3}a^2 h$.
33. $128\pi/3$.
35. $11\pi/15$.
37. $135\pi/2$.
39. 2π.
43. $\frac{14}{3}$.
45. $\frac{14}{3}$.
47. $\frac{17}{6}$.
49. $\frac{495}{8}$.
51. $AB = \frac{1}{32}$.

53. 90π.
55. 168π.
57. $\pi a^2/2$.
59. 7.5 tons.
61. $\frac{1}{3}wBH^2$.
63. $\frac{14}{3}w$ or approximately 292 lb.
65. 36 ft.
67. 36 ft-lb.
71. $3 \cdot 2^8$ ft-lb.
73. $\frac{5}{2}p_1 V_1\left[\left(\frac{V_1}{V_2}\right)^{2/5} - 1\right]$.
75. aB.
79. 1575π ft-lb.

CHAPTER 8

Section 8.2, p. 211
1. (a) $\log_4 16 = 2$; (b) $\log_3 81 = 4$; (c) $\log_{81} 9 = 0.5$; (d) $\log_{32} 16 = \frac{4}{5}$.
3. (a) 4; (b) 6; (c) -4; (d) $\frac{2}{3}$.
5. (a) $a = 32$; (b) $a = \frac{1}{16}$; (c) $a = 6$; (d) $a = 49$.
9. (a) 7;
(b) acidic pH < 7, basic pH > 7.

Section 8.3, p. 216
1. $\frac{1}{2}(e^x - e^{-x})$.
3. $(x^2 + 2x)e^x$.
5. $e^{e^x}e^x$.
7. xe^{ax}.
9. $4x^2 e^{2x}$.
11. $\frac{1}{3}e^{3x} + c$.
13. $5e^{x/5} + c$.
15. $2e^{x^3} + c$.
17. (a) Max. pt. $(0, 1)$, no min. pt., pts. of infl. $(\pm\frac{1}{2}\sqrt{2}, 1/\sqrt{e})$; (b) no max. pt., min. pt. $(-3, -3/e)$, pt. of infl. $(-6, -6/e^2)$.
19. $\frac{1}{2}(e^b - e^{-b})$.
23. Area $= 1 - e^{-b} \to 1$ as $b \to \infty$.
25. (a) e; (b) e; (c) e; (d) e^2; (e) $\sqrt{e}$.

Section 8.4, p. 222
1. (a) 2; (b) 3; (c) $1/x$; (d) $1/x$; (e) $-x$; (f) $1/x$; (g) x; (h) $3x$; (i) 0; (j) $\frac{4}{3}$; (k) $\frac{4}{3}$; (l) 0; (m) $x^3 y^2$; (n) 8; (o) $2e^3$; (p) $x^2 e^x$.
3. (a) $\frac{y(1+2x)}{x(3y-1)}$; (b) $\frac{y(1+xy)}{x(1-xy)}$.
5. (a) $\frac{1}{3}\ln(3x+1) + c$;
(b) $\frac{1}{6}\ln(3x^2+2) + c$; (c) $\frac{3}{2}x^2 + 2\ln x + c$; (d) $x + \ln x + c$;
(e) $x - \ln(x+1) + c$; (f) $\frac{1}{2}\ln(x^2+1) + c$; (g) $-\frac{1}{4}\ln(3-2x^2) + c$;

(h) $\ln x(x-1) + c$; (i) $\frac{1}{2}(\ln x)^2 + c$; (j) $\ln(\ln x) + c$; (k) $2\ln(\sqrt{x}+1) + c$; (l) $\ln(e^x + e^{-x}) + c$.
9. No max., no pt. of infl., min. $(3, 9 - 18\ln 3) \cong (3, -11)$.
11. $\pi\ln 4$.
15. min. $(1/e, -1/e)$.
17. $1/\sqrt{e}$.
19. (a) $y\left(1 + \frac{2x}{x^2-1} - \frac{3}{6x-2}\right)$;
(b) $\frac{y}{5}\left(\frac{2x}{x^2+3} - \frac{1}{x+5}\right)$.
21. (a) $x^{x^x}x^x\left[\frac{1}{x} + (\ln x)(1 + \ln x)\right]$;
(b) $\sqrt[x]{x}\left[\frac{1 - \ln x}{x^2}\right]$. Max. $= \sqrt[e]{e}$.

Section 8.5, p. 228
1. (a) 2520; (b) 13.8.
3. 20.1
5. 95.8 percent.
7. When $t = 6$. Can you solve this problem without calculation, by merely thinking about it?
9. $x = 10(\frac{4}{5})^t$.
11. 2 more hours.
13. (a) About 3330 years (1380 B.C.); (b) about 3850 years (1900 B.C.); (c) about 10,510 years; (d) about 7010 years.
15. $x \to A$ if $A < B$; $x \to B$ if $A > B$.

Section 8.6, p. 234
1. $\frac{1}{2}N_1$.
3. $x = \frac{x_0 x_1}{x_0 + (x_1 - x_0)e^{-cx_1 t}}$.
5. $s = \frac{v_0}{c}(1 - e^{-ct})$.
7. When $v < 1$, the resisting force in the second case becomes very small.
9. About 53.4 lb.

Additional Problems, p. 234
1. $-xe^{\sqrt{1-x^2}}/\sqrt{1-x^2}$.
3. $(2x-2)e^{x^2-2x+1}$.
5. $\frac{e^{\sqrt{x}}}{2\sqrt{x}} + \frac{1}{2}\sqrt{e^x}$.
7. $-\frac{1}{3}e^{-3x} + c$.
9. $-e^{1/x} + c$.
11. $2\sqrt{e^x + 1} + c$.
13. $(1/a, e)$.

17. $\frac{\pi}{2}(e^6 - 1)$.

23. (a) $\frac{y(3x+1)}{x(2y^2-1)}$; (b) $\frac{y(2x^2+1)}{x(1-3y)}$.

25. (a) $\frac{1}{2}\ln(1+2x) + c$;
(b) $-\frac{1}{3}\ln(1-3x) + c$; (c) $\frac{1}{4}\ln 2$;
(d) $\frac{1}{2}\ln 10$; (e) $\ln 3$; (f) $2\sqrt{\ln x} + c$;
(g) $\frac{1}{2}\ln 7$; (h) $-\frac{1}{2}\ln(1-x^2) + c$;
(i) $\frac{1}{2}(\ln 3)^2$; (j) $\frac{1}{3}\ln(3x^2 - 3x +$
$7) + c$; (k) $\ln(e^x + 1) + c$; (l) $\ln(x +$
$1)(x+2) + c$; (m) $\frac{1}{3}(\ln x)^3 + c$;
(n) $\frac{1}{4}(\ln x)^2 + c$; (o) $\frac{1}{2}[\ln(\ln x)]^2 + c$;
(p) $-\frac{1}{2}(\ln x)^2 + c$.

29. $\frac{3}{2a} + \frac{a}{4}\ln 2; a = \sqrt{\dfrac{6}{\ln 2}}$.

31. (a) $(\ln 10)10^x$; (b) $(\ln 3)3^x$;
(c) $(\ln \pi)\pi^x$; (d) $(3\ln 7)7^{3x}$;
(e) $(\ln 6)(2x - 2)6^{x^2 - 2x}$;
(f) $\left(\dfrac{\ln 5}{2\sqrt{x}}\right)5^{\sqrt{x}}$.

33. Max. at $x = \dfrac{2}{\ln 5}$; pts. of infl. at
$x = \dfrac{2 \pm \sqrt{2}}{\ln 5}$.

35. (a) $(\ln x)^x\left[\dfrac{1}{\ln x} + \ln(\ln x)\right]$;
(b) $(2\ln x)x^{\ln x - 1}$;
(c) $\dfrac{(\ln x)^{\ln x}}{x}[1 + \ln(\ln x)]$;
(d) $\dfrac{x^{\sqrt{x}}}{\sqrt{x}}(1 + \frac{1}{2}\ln x)$;
(e) $\dfrac{x^{\sqrt[3]{x}}}{x^{2/3}}(1 + \frac{1}{3}\ln x)$.

37. In the year 3524, approximately.
39. In 4 more hours.
41. 73.12°F.
47. 17 more days.

49. $v^2 = \dfrac{g}{c}(1 - e^{-2cs}); v \to \sqrt{\dfrac{g}{c}}$ as
$s \to \infty$.

51. When $t = 4.86$ min; when
$t = 21.50$ min.
53. About 1.39 h.

CHAPTER 9

Section 9.1, p. 246
1. (a) $\pi/12$; (b) $7\pi/12$; (c) $2\pi/3$;
(d) $5\pi/12$; (e) $5\pi/6$; (f) $3\pi/4$;
(g) $5\pi/4$; (h) $7\pi/6$; (i) $7\pi/2$; (j) 5π.
3. $\theta = 2$ radians.
5. $A = 25\cot\frac{1}{2}\theta$.

7. $H = L\tan\theta$.
11. $\sin 3\theta = 3\sin\theta - 4\sin^3\theta$,
$\cos 3\theta = 4\cos^3\theta - 3\cos\theta$.
13. $\sin 4\theta =$
$(4\sin\theta - 8\sin^3\theta)\cos\theta$.
15. (a) $\frac{1}{4}(\sqrt{6} - \sqrt{2})$; (b) $\frac{1}{2}\sqrt{2 - \sqrt{3}}$.
21. For a proof by geometry, use the
fact that the vertex opposite the
fixed side must lie on a circle of
which this side is a chord.
23. $\sqrt[3]{2}$ ft.

Section 9.2, p. 252
1. $3\cos(3x - 2)$.
3. $48\cos 16x$.
5. $2x\cos x^2$.
7. $15(\cos 3x - \sin 5x)$.
9. $x\cos x + \sin x$.
11. $\frac{3}{2}\sin 12x$.
13. $\cos^5 x$.
15. $3e^{2x}\cos 3x + 2e^{2x}\sin 3x$.
17. $-\tan x$.
21. 45°.
23. 5.
31. $\pi/3$, $\tan^{-1} 3$.
33. 1.
35. 1.
37. $\frac{1}{2}$.
39. 1.
41. 1.
43. -1.

Section 9.3, p. 256
1. $-\frac{1}{5}\cos 5x + c$.
3. $\frac{1}{9}\cos(1 - 9x) + c$.
5. $\sin^2 x + c$ or $-\cos^2 x + c$ or
$-\frac{1}{2}\cos 2x + c$.
7. $\frac{1}{8}\sin^4 2x + c$.
9. $\frac{1}{4}\sin^8\frac{1}{2}x + c$.
11. $-2\cos\sqrt{x} + c$.
13. $\frac{1}{2}\sin(\sin 2x) + c$.
15. $\dfrac{3}{2}\sec\left(\dfrac{2x-1}{3}\right) + c$.
17. $-\ln(\cos x) + c$.
19. $\sin(x^2 + x) + c$.
21. $\frac{2}{3}$.
23. $\sqrt{2} - 1$.
25. $\frac{2}{3}$.
27. 3.
29. $\frac{1}{4}\pi^2$.

Section 9.4, p. 259
1. $8x\sec^2 4x^2$.
3. $2\tan(\sin x)\cdot\sec^2(\sin x)\cdot\cos x$.

5. 0.
7. $24\csc(-6x)\cot(-6x)$.
9. $-\sqrt{\csc 2x}\cot 2x$.
11. $\sec^2 x\, e^{\tan x}$.
13. $-\frac{1}{6}\cot 6x + c$.
15. $-\frac{1}{2}\cot 2x + c$.
17. $\frac{1}{5}\tan^5 x + c$.
19. $-\frac{1}{7}\csc 7x + c$.
21. $\frac{1}{2}$.
23. $\frac{1}{2}(\pi - 2)$.
25. 2.
27. $4\sqrt{3}$; no.
29. 3.2π mi/s.

Section 9.5, p. 265
1. $\frac{1}{2}\sqrt{3}, -\frac{1}{3}\sqrt{3}, -\sqrt{3}, \frac{2}{3}\sqrt{3}, -2$.
3. (a) π; (b) $\pi/2$; (c) 0.123; (d) 0.8;
(e) 0.96; (f) $\pi/7$; (g) $\pi/6$; (h) $\pi/4$.
5. $1/(25 + x^2)$.
7. $\dfrac{1}{\sqrt{x}(x+1)}$.
9. $\sin^{-1} x$.
11. $(\sin^{-1} x)^2$.
13. $\dfrac{4}{5 + 3\cos x}$.
15. $\pi/6$.
17. $\frac{1}{2}\sin^{-1} 2x + c$.
19. $\pi/8$.
21. $\frac{1}{2}\sin^{-1}\frac{2}{3}x + c$.
23. $\frac{1}{6}\tan^{-1}\frac{3}{2}x + c$.
25. $-\pi/12$.
27. $\pi/4$.
29. (a) $\sin^{-1}\frac{3}{5}$; (b) $\frac{1}{4}$ rad/s.
31. The formula is invalid, because
the integrand $1/\sqrt{1-x^2}$ is discontinuous at the point $x = 1$ in the
interval of integration.

Section 9.6, p. 270
1. (a) $x = 5\sqrt{2}\sin\left(t - \dfrac{\pi}{4}\right)$,
$A = 5\sqrt{2}, T = 2\pi$;
(b) $x = 2\sin\left(3t + \dfrac{2\pi}{3}\right), A = 2$,
$T = 2\pi/3$; (c) $x = \sqrt{2}\sin\left(t + \dfrac{\pi}{4}\right)$,
$A = \sqrt{2}, T = 2\pi$;
(d) $x = 4\sin\left(2t - \dfrac{\pi}{6}\right), A = 4$,
$T = \pi$.
3. $A = \sqrt{145}/2$ in; $T = \pi/4$.

5. $T = 2\pi\sqrt{R/g} \cong 89$ min.

7. About 39 in.

Additional Problems, p. 271

1. $-9\cos(1-9x)$.

3. $-2\sin x \cos x = -\sin 2x$.

5. $-10\sin 5x \cos 5x = -5\sin 10x$.

7. $-6\sin 6x$.

9. $-x^2 \sin x + 2x\cos x$.

11. $x\cos x$.

13. $(\sin x)[\sin(\cos x)]$.

15. $-(\cos x)[\sin(\sin x)]$.

17. $\cos x$.

25. 0.

27. 1.

29. 2.

31. $\frac{2}{3}$.

33. 2.

35. $\pi/4$.

37. $\frac{1}{3}\sin 3x + c$.

39. $-2\sin(1 - \frac{1}{2}x) + c$.

41. $\frac{1}{18}\sin^6 3x + c$.

43. $-\frac{2}{3}\cos 3x + \frac{1}{9}\cos^3 3x + c$.

45. $\frac{1}{3}\sin x^3 + c$.

47. $\frac{1}{2}\cos(\cos 2x) + c$.

49. $-\frac{1}{4}\csc 4x + c$.

51. $\dfrac{1}{2(3 + 2\cos x)} + c$.

53. $-\frac{2}{5}\sqrt{7 - \sin 5x} + c$.

55. $\frac{1}{7}$.

57. $\frac{1}{2}$.

59. $2\sqrt{2}$.

61. 21π.

67. $12\sec^2 3x$.

69. $\dfrac{-\csc^2 2x}{\sqrt{\cot 2x}}$.

71. $4\sec^2 x \tan x$.

73. $-10\cot 5x \csc^2 5x$.

75. $\tan\dfrac{1}{x} - \dfrac{1}{x}\sec^2\dfrac{1}{x}$.

77. $\dfrac{\sqrt{\sec\sqrt{x}} + \tan\sqrt{x}}{4\sqrt{x}}$.

79. $\sec^2 x \sec^2(\tan x)$.

81. $-3\csc\frac{1}{3}x + c$.

83. $-\frac{1}{3}\cot 3x + c$.

85. $-\frac{1}{4}\csc^4 x + c$.

87. $-\frac{1}{4}\cot^4 x + c$.

89. $4\pi/3$.

91. 300 km/h.

93. (a) $-\pi/3$; (b) $\pi/3$; (c) $-\pi$;
(d) 0.7; (e) 0.7; (f) -1; (g) $\pi/3$.

95. $\dfrac{1}{\sqrt{25 - x^2}}$.

97. $\dfrac{x^4}{1 + x^{10}}$.

99. $\dfrac{1}{x\sqrt{x^2 - 1}}$.

101. $\dfrac{1 + x}{1 + x^2}$.

103. $\dfrac{\sqrt{x^2 - 1}}{x}$.

105. $\pi/2$.

107. $\dfrac{1}{\sqrt{5}}\tan^{-1}\sqrt{5}x + c$.

109. $\frac{1}{2}\sin^{-1}\frac{2}{3}x + c$.

111. $\frac{1}{4}\tan^{-1}x^4 + c$.

113. 36 ft from the point on the road closest to the billboard.

115. $\frac{18}{250}$ rad/s.

117. $T = 2\pi/\sqrt{2w} \cong 0.56$ s.

119. $A = 5, f = 1/\pi$.

CHAPTER 10

Add a constant of integration to the answer for each indefinite integral in this chapter.

Section 10.2, p. 282

1. $-\frac{1}{3}(3 - 2x)^{3/2}$.

3. $\frac{1}{2}\ln[1 + (\ln x)^2]$.

5. $-\frac{1}{2}\cos 2x$.

7. $\frac{1}{3}\ln[\sin(3x - 1)]$.

9. $\frac{1}{3}(x^2 + 1)^{3/2}$.

11. $\frac{1}{5}e^{5x}$.

13. $-\frac{1}{3}\cot(3x + 2)$.

15. 2.

17. $2\sqrt{1 - \cos x}$.

19. $e^{\tan^{-1}x}$.

21. $\frac{1}{5}\sec 5x$.

23. $\frac{1}{2}(\ln x)^2$.

25. $\ln 2$.

27. $\sin^{-1}e^x$.

29. $\frac{1}{3}\sin^3 x$.

31. $\ln(1 + e^x)$.

33. $-\frac{1}{3}\ln(\cos 3x)$.

35. $4\sqrt{x^2 + 1}$.

37. $\tan^{-1}e^x$.

39. $\frac{1}{7}(e^x + 1)^7$.

41. $\frac{1}{5}\tan 5x$.

43. $-\frac{1}{2}\csc 2x$.

45. $4(\sqrt{2} - 1)$.

47. $\frac{2}{3}$.

49. (a) $n = 3, \frac{1}{4}e^{x^4}$; (b) $n = 2$, $\frac{1}{3}\sin x^3$; (c) $n = -1, \frac{1}{2}(\ln x)^2$; (d) $n = -\frac{1}{2}, 2\tan\sqrt{x}$.

Section 10.3, p. 286

1. $\frac{1}{2}x - \frac{1}{4}\sin 2x$.

3. $\frac{5}{16}x + \frac{1}{4}\sin 2x + \frac{3}{64}\sin 4x - \frac{1}{48}\sin^3 2x$.

5. $-\frac{1}{3}\cos^3 x + \frac{1}{5}\cos^5 x$.

7. $\sin x - \frac{1}{3}\sin^3 x$.

9. $\frac{2}{3}\sin^{3/2} x - \frac{2}{7}\sin^{7/2} x$.

11. $\frac{1}{8}x - \frac{1}{96}\sin 12x$.

13. $\frac{4}{3}$.

15. $\frac{1}{7}\sec^7 x - \frac{2}{5}\sec^5 x + \frac{1}{3}\sec^3 x$.

17. $-\cot x - x$.

19. $-\frac{1}{4}\cot 4x$.

21. $-\frac{1}{2}\cot 2x - \frac{1}{2}\csc 2x$.

23. $\frac{1}{3}\sin 3x$.

25. (a) $\tan x - x, \frac{1}{3}\tan^3 x - \tan x + x, \frac{1}{5}\tan^5 x - \frac{1}{3}\tan^3 x + \tan x - x$; (b) $\frac{1}{2}\tan^2 x + \ln(\cos x), \frac{1}{4}\tan^4 x - \frac{1}{2}\tan^2 x - \ln(\cos x), \frac{1}{6}\tan^6 x - \frac{1}{4}\tan^4 x + \frac{1}{2}\tan^2 x + \ln(\cos x)$.

27. (a) $\pi^2/2$; (b) π; (c) $(4\pi - \pi^2)/8$; (d) $3\pi^2/16$.

29. $\frac{1}{2}[\sec x \tan x + \ln(\sec x + \tan x)]$.

Section 10.4, p. 290

1. $-\sin^{-1}\dfrac{x}{a} - \dfrac{\sqrt{a^2 - x^2}}{x}$.

3. $\dfrac{1}{2a^3}\tan^{-1}\dfrac{x}{a} + \dfrac{x}{2a^2(a^2 + x^2)}$.

5. $-\frac{1}{3}\sqrt{9 - x^2}(x^2 + 18)$.

7. $\dfrac{1}{a}\ln\left(\dfrac{x}{a + \sqrt{a^2 + x^2}}\right)$.

9. $\ln(x + \sqrt{x^2 - a^2})$.

11. $\frac{1}{2}x\sqrt{a^2 + x^2} + \frac{1}{2}a^2\ln(x + \sqrt{a^2 + x^2})$.

13. $\dfrac{1}{2a}\ln\dfrac{a + x}{a - x}$.

15. $\sqrt{a^2 + x^2} - a\ln\left(\dfrac{a + \sqrt{a^2 + x^2}}{x}\right)$.

17. $\ln(x + \sqrt{x^2 - a^2}) - \dfrac{\sqrt{x^2 - a^2}}{x}$.

19. $\dfrac{1}{8}\left[a^4\sin^{-1}\dfrac{x}{a} + \sqrt{a^2 - x^2}(2x^3 - a^2 x)\right]$.

21. $-\sqrt{4-x^2}$.

23. $\dfrac{1}{a}\tan^{-1}\dfrac{x}{a}$.

25. $-\tfrac{1}{3}(9-x^2)^{3/2}$.

27. $\sqrt{9+x^2}$.

31. $2\pi^2 ba^2$.

33. $3-\sqrt{2}+\ln(1+\tfrac{1}{2}\sqrt{2})$.

Section 10.5, p. 293

1. $\sin^{-1}(x-1)$.

3. $\tan^{-1}(x+2)$.

5. $-\sqrt{2x-x^2}+2\sin^{-1}(x-1)$.

7. $\dfrac{27}{2}\sin^{-1}\left(\dfrac{x-3}{3}\right)$
$-6\sqrt{6x-x^2}-\dfrac{1}{2}(x-3)\sqrt{6x-x^2}$.

9. $\dfrac{1}{2}\ln(x^2+2x+5)$
$+3\tan^{-1}\left(\dfrac{x+1}{2}\right)$.

11. $\ln(x-1+\sqrt{x^2-2x-8})$.

13. $\tfrac{1}{2}\ln(2x+1+\sqrt{4x^2+4x+17})$.

15. $-\dfrac{x-1}{4\sqrt{x^2-2x-3}}$.

Section 10.6, p. 299

1. (a) $x+1+\dfrac{1}{x-1}$,
$\dfrac{1}{2}x^2+x+\ln(x-1)$;

(b) $\dfrac{1}{3}x^2-\dfrac{2}{9}x+\dfrac{4}{27}-\dfrac{8/27}{3x+2}$,
$\dfrac{1}{9}x^3-\dfrac{1}{9}x^2+\dfrac{4}{27}x$
$-\dfrac{8}{81}\ln(3x+2)$;

(c) $x-\dfrac{x}{x^2+1}$,
$\dfrac{1}{2}x^2-\dfrac{1}{2}\ln(x^2+1)$;

(d) $1+\dfrac{1}{x+2}$, $x+\ln(x+2)$;

(e) $1-\dfrac{2}{x^2+1}$, $x-2\tan^{-1}x$.

3. $3\ln(x-3)+4\ln(x+2)$.

5. $5\ln(x-7)-3\ln x$.

7. $2\ln x-4\ln(x+8)$
$+3\ln(x-3)$.

9. $3\ln x+2\ln(x+13)$
$-\ln(x-3)$.

11. $-\ln(x+1)-\dfrac{2}{x+1}-3\ln x$.

13. $2\ln x+\tfrac{1}{2}\ln(x^2+2x+2)$
$-6\tan^{-1}(x+1)$.

Section 10.7, p. 305

1. $\tfrac{1}{2}x^2\ln x-\tfrac{1}{4}x^2$.

3. $\tfrac{1}{2}x^2\tan^{-1}x-\tfrac{1}{2}x+\tfrac{1}{2}\tan^{-1}x$.

5. $\tfrac{1}{2}e^x(\sin x-\cos x)$.

7. $\tfrac{1}{2}x\sqrt{1-x^2}+\tfrac{1}{2}\sin^{-1}x$.

9. $\tfrac{1}{2}x^2\sin^{-1}x-\tfrac{1}{4}\sin^{-1}x$
$+\tfrac{1}{4}x\sqrt{1-x^2}$.

11. $\tfrac{1}{3}x\sin(3x-2)+\tfrac{1}{9}\cos(3x-2)$.

13. $x\tan x+\ln(\cos x)$.

15. $x\ln(a^2+x^2)-2x$
$+2a\tan^{-1}\dfrac{x}{a}$.

17. $\tfrac{1}{2}(\ln x)^2$.

19. $\pi(\pi-2)$.

23. (b) $x(\ln x)^5-5x(\ln x)^4$
$+20x(\ln x)^3-60x(\ln x)^2$
$+120x\ln x-120x$.

25. $2\pi[\sqrt{2}+\ln(\sqrt{2}+1)]$.

Section 10.8, p. 308

5. $2\sqrt{x}-2\ln(1+\sqrt{x})$.

7. $2\sqrt{x}-3\sqrt[3]{x}+6\sqrt[6]{x}$
$-6\ln(\sqrt[6]{x}+1)$.

9. $2\sqrt{x}-2\tan^{-1}\sqrt{x}$.

11. $\tfrac{4}{3}x^{3/4}-4\sqrt[4]{x}+4\tan^{-1}\sqrt[4]{x}$.

13. $2\sqrt{x+2}-2\tan^{-1}\sqrt{x+2}$.

Section 10.9, p. 313

1. (a) 0.643; (b) 0.656.

3. 2.2845.

5. 0.881.

7. 3.14156.

Additional Problems, p. 314

1. $\tfrac{2}{9}(3x+5)^{3/2}$.

3. $\ln(1+3x^2)$.

5. $-\tfrac{1}{5}\sin(1-5x)$.

7. $2\sec\sqrt{x}$.

9. $\tan^{-1}x^2$.

11. $\tfrac{1}{4}\ln(\sin 4x)$.

13. $-1/\ln x$.

15. $\ln(\tan x)$.

17. $-\dfrac{2}{3}\cos\left(\dfrac{3x-5}{2}\right)$.

19. $-2\csc x^3$.

21. $\tan^{-1}(\ln x)$.

23. $-\dfrac{1}{3(3x+5)}$.

25. $-\tfrac{1}{2}\ln(3-2x)$.

27. $\tfrac{1}{3}\sin(1+x^3)$.

29. $-\tfrac{1}{2}\cot(x^2+1)$.

31. $\tan^{-1}(\sin x)$.

33. $\tfrac{1}{2}\ln(\sin 2x)$.

35. $\tfrac{1}{2}(\tan^{-1}x)^2$.

37. $\tfrac{1}{2}\ln(2x+1)$.

39. $3e^{x/3}$.

41. $\tan(\sin x)$.

43. $\tfrac{1}{5}\sin^{-1}5x$.

45. $\tan^{-1}(\sec x)$.

47. $\tfrac{1}{3}(\ln x)^3$.

49. $-\ln(1+\cos x)$.

51. $-\tfrac{1}{3}e^{-3x}$.

53. $-\cos(\ln x)$.

55. $\csc\dfrac{1}{x}$.

57. $\tfrac{1}{2}\tan^{-1}e^{2x}$.

59. $\tfrac{1}{6}(2+x^4)^{3/2}$.

61. $\ln(e^x+x)$.

63. $-4/\sqrt{e^x}$.

65. $-\cot x$.

67. $-\tfrac{1}{2}\ln(\cos x^2)$.

69. $\tfrac{1}{2}\ln(1+x^2)$.

71. $\tfrac{1}{6}e^{3x^2-2}$.

73. $\tan x+\sec x$.

75. $\tfrac{2}{3}(1+x^{5/3})^{3/2}$.

77. $e^{\tan x}$.

79. $-\tfrac{1}{6}(1+\cos x)^5$.

81. $\sin(\tan x)$.

83. $\pi/6$.

85. $\tfrac{1}{4}$.

87. $\tfrac{196}{3}$.

89. $\tfrac{1}{2}x-\tfrac{1}{20}\sin 10x$.

91. $\tfrac{1}{2}x+\tfrac{1}{28}\sin 14x$.

93. $-\tfrac{1}{3}\cos^3 x+\tfrac{2}{5}\cos^5 x-\tfrac{1}{7}\cos^7 x$.

95. $\tfrac{1}{4}\sin 4x-\tfrac{1}{12}\sin^3 4x$.

97. $\csc x-\tfrac{1}{3}\csc^3 x$.

99. $\tfrac{5}{8}\sin^{8/5}x$.

101. $\tfrac{1}{5}\tan^5 x+\tfrac{2}{3}\tan^3 x+\tan x$.

103. $\tfrac{1}{9}\sec^9 x-\tfrac{1}{7}\sec^7 x$.

105. $-\tfrac{1}{4}\cot^4 x+\tfrac{1}{2}\cot^2 x$
$+\ln(\sin x)$.

107. $\tfrac{1}{3}\tan 3x-\tfrac{1}{3}\cot 3x$
$-\tfrac{2}{3}\ln(\csc 6x+\cot 6x)$.

109. $\dfrac{3}{2}\sin^{-1}\dfrac{x}{\sqrt{3}}+\dfrac{1}{2}x\sqrt{3-x^2}$.

111. $x-a\tan^{-1}\dfrac{x}{a}$.

113. $-\tfrac{1}{15}(a^2-x^2)^{3/2}(3x^2+2a^2)$.

115. $\ln(x+\sqrt{a^2+x^2})-\dfrac{\sqrt{a^2+x^2}}{x}$.

117. $-\dfrac{1}{3a^4x^3}\sqrt{a^2-x^2}\,(2x^2+a^2)$.

119. $\dfrac{1}{2a}\tan^{-1}\dfrac{x}{a}-\dfrac{x}{2(a^2+x^2)}$.

121. $\dfrac{\sqrt{x^2-9}}{9x}$.

123. $\dfrac{x}{\sqrt{1-9x^2}}$.

125. $-\dfrac{1}{3}\ln\left(\dfrac{3+\sqrt{9+4x^2}}{2x}\right)$.

127. $\dfrac{x}{\sqrt{a^2-x^2}}-\sin^{-1}\dfrac{x}{a}$.

129. $\ln(x+\sqrt{a^2+x^2})-\dfrac{x}{\sqrt{a^2+x^2}}$.

131. $\dfrac{1}{2}x\sqrt{x^2-a^2}$
$\quad +\dfrac{1}{2}a^2\ln(x+\sqrt{x^2-a^2})$.

133. $\sin^{-1}\left(\dfrac{x+4}{9}\right)$.

135. $\dfrac{1}{10}\sqrt{2}\tan^{-1}\left(\dfrac{x+1}{\sqrt{2}}\right)$.

137. $\dfrac{1}{\sqrt{3}}\sin^{-1}\left(\dfrac{3x-1}{\sqrt{7}}\right)$.

139. $\dfrac{3}{2}\sin^{-1}(x-1)-2\sqrt{2x-x^2}$
$\quad -\dfrac{1}{2}(x-1)\sqrt{2x-x^2}$.

141. $\dfrac{1}{6}\tan^{-1}\left(\dfrac{x-1}{2}\right)$.

143. $-\dfrac{1}{2}\sin^{-1}\left(\dfrac{2}{x-1}\right)$.

145. $3\sqrt{x^2+4x+8}$
$\quad +\ln(x+2+\sqrt{x^2+x+8})$.

147. $\dfrac{5x-3}{4\sqrt{x^2+2x-3}}$.

149. $19\ln(x-4)-3\ln(x+3)$.

151. $3\ln(2x+1)-5\ln(2x-1)$.

153. $5\ln x+\ln(x+4)$
$\quad -3\ln(x-3)$.

155. $-2\ln x+3\ln(x+3)$
$\quad -3\ln(x-3)$.

157. $2\ln x+\dfrac{1}{x}-\dfrac{3}{2x^2}$
$\quad -5\ln(x+1)$.

159. $-\ln x+\ln(x^2+4x+8)$
$\quad -\dfrac{5}{2}\tan^{-1}\left(\dfrac{x+2}{2}\right)$.

161. $\dfrac{1}{3}x^3\tan^{-1}x-\dfrac{1}{6}x^2$
$\quad +\dfrac{1}{6}\ln(1+x^2)$.

163. $\dfrac{1}{2}x\,[\cos(\ln x)+\sin(\ln x)]$.

165. $x^3\sin x+3x^2\cos x-$
$\quad 6x\sin x-6\cos x$.

167. $-\dfrac{\ln x}{x+1}+\ln x-\ln(x+1)$.

169. $\dfrac{1}{3}\sqrt{1+x^2}(x^2-2)$.

171. $\dfrac{e^{ax}}{a^2+b^2}\,(a\sin bx-b\cos bx)$.

173. $-xe^{-x}-e^{-x}$.

175. $-\dfrac{1}{2}x^3e^{-2x}-\dfrac{3}{4}x^2e^{-2x}$
$\quad -\dfrac{3}{4}xe^{-2x}-\dfrac{3}{8}e^{-2x}$.

177. 2π.

179. $\dfrac{16}{105}a^{7/2}$.

183. (b) $\dfrac{x^6(\ln x)^3}{6}-\dfrac{x^6(\ln x)^2}{12}$
$\quad +\dfrac{x^6\ln x}{36}-\dfrac{x^6}{216}$.

185. (b) $\dfrac{1}{2}[\sec x\tan x$
$\quad +\ln(\sec x+\tan x)]$.

CHAPTER 11

Section 11.2, p. 325

1. $(\frac{3}{2},\frac{6}{5})$.

3. $\left(0,\dfrac{4}{3\pi}a\right)$.

5. $(0,\frac{3}{5}a)$.

7. $(\frac{32}{7},\frac{4}{3})$.

9. $\left(\dfrac{2}{3(4-\pi)}a,\dfrac{2}{3(4-\pi)}a\right)$.

13. $\left(0,\dfrac{2}{\pi}a\right)$.

Section 11.3, p. 326

1. (a) $\left(0,\dfrac{4}{3\pi}a\right)$; (b) $\left(0,\dfrac{2}{\pi}a\right)$.

3. (a) $\frac{12}{5}\pi a^2$; (b) $6\sqrt{2}\pi a^2$.

5. $\frac{9}{2}\pi a^3$; $6\sqrt{3}\pi a^2$.

7. (a) $\pi r^2 h$; (b) $\frac{1}{3}\pi r^2 h$.

Section 11.4, p. 330

1. $\frac{1}{3}Ma^2$.

3. $\frac{1}{6}Mh^2$.

5. $\frac{5}{4}Ma^2$.

7. $\frac{1}{2}Ma^2$.

9. $\frac{3}{10}Ma^2$.

11. (a) $\frac{1}{2}\sqrt{2}a\cong 0.707a$;
(b) $\frac{1}{10}\sqrt{30}a\cong 0.548a$;
(c) $\frac{1}{3}\sqrt{10}a\cong 0.632a$.

Additional Problems, p. 330

3. (a) $\left(\dfrac{1}{2},\dfrac{2}{5}\right)$; (b) $\left(0,\dfrac{4}{5}\right)$;

(c) $\left(1,\dfrac{2}{5}\right)$; (d) $\left(\dfrac{5}{9},\dfrac{5}{27}\right)$;

(e) $\left(0,\dfrac{10}{7}\right)$; (f) $\left(\dfrac{16}{15},\dfrac{64}{21}\right)$;

(g) $\left(\dfrac{1}{e-1},0\right)$.

5. $8\pi abc$; $8\pi(a+b)c$.

7. $\frac{1}{6}Ma^2$.

CHAPTER 12

Section 12.2, p. 337

1. 3.

3. $\frac{1}{14}$.

5. $\frac{1}{6}$.

7. $-\frac{1}{9}$.

9. $\frac{1}{2}$.

11. -6.

13. 3.

15. 4.

17. $-\frac{1}{2}$.

19. $1/\pi$.

21. 16.

23. $\frac{1}{4}$.

25. 6.

27. $f(\theta)=\frac{1}{2}a^2(\sin\theta-\sin\theta\cos\theta)$,
$g(\theta)=\frac{1}{2}a^2(\theta-\sin\theta\cos\theta)$;
limit $=\frac{3}{4}$.

Section 12.3, p. 342

1. -3.

3. 1.

5. 3.

7. 0.

9. 2.

11. 1.

13. 0.

15. 0.

17. 0.

19. 2.

21. $\frac{1}{2}$.

23. 1.

25. 1.

27. 1.

29. 1.

31. 1.

33. e^a.

35. 1.

37. 1.

39. $1/\sqrt{e}$.

41. e^p.

Section 12.4, p. 348

1. $1/(2e^6)$.

3. $\frac{3}{2}$.

5. $1-\cos 1$.

7. 1.

9. 0.

11. $\ln\sqrt{3}$.

13. $\sqrt{2}(\ln 4 - 4)$.

15. 1.

17. Converges if $p < 1$, diverges if $p \geq 1$.

19. (a) $\pi/5$; (b) π.

Additional Problems, p. 348

1. $\frac{5}{2}$.

3. 44.

5. 12.

7. $\frac{1}{6}$.

9. $-\frac{2}{75}$.

11. ∞.

13. 6.

15. 3.

17. $\frac{1}{2}$.

19. ∞.

21. 0.

23. $-\frac{1}{9}$.

25. 0.

27. $\frac{1}{24}$.

29. $\frac{1}{8}$.

31. 9.

33. 3.

35. $\frac{1}{3}$.

37. $\frac{10}{9}$.

39. $\frac{1}{3}$.

41. $\frac{1}{16}$.

43. 0. No; instead, it emphasizes the logical point that L'Hospital's rule makes a definite statement only when the limit on the right exists.

49. 0.

51. 0.

53. 0.

55. 0.

57. 0.

59. 0.

61. p.

63. $-\frac{1}{3}$.

65. 0.

67. 0.

69. $\frac{1}{6}$.

71. 1.

73. 1.

75. 1.

77. 1.

79. 1.

81. 1.

83. $-\infty$.

85. 1.

87. 1.

89. 1.

91. 1.

93. e^2.

95. e^4.

97. e^3.

99. 1.

101. $1/(3e^6)$.

103. 1.

105. $\frac{1}{2}$.

107. $\pi/4$.

109. $\pi/8$.

111. $\frac{1}{3}$.

113. 2.

115. Diverges.

117. Diverges.

119. 3.

CHAPTER 13

Section 13.2, p. 362

5. (a) C; (b) D; (c) D; (d) C; (e) D; (f) C; (g) D; (h) D.

9. 40 mi.

Section 13.3, p. 368

1. (a) $-4 < x < -2$, $s = -1/(x^2 + 6x + 8)$; (b) $x > -1$ or $x < -2$, $s = (2x + 3)/(2x + 2)$; (c) $|x - k\pi| < \pi/6$, $k = 0, \pm 1$, $\pm 2, \ldots, s = 1/(1 - 2 \sin x)$; (d) $|x| < \frac{3}{4}$, $s = 1/(3 - 4x)$; (e) $|x| > \frac{3}{4}$, $s = 1/(3 - 4x)$; (f) $-\frac{5}{4} < x < \frac{3}{4}$, $s = 1/(3 - 4x)$.

5. 2.

Section 13.4, p. 373

5. $x^3 = 0 + 0 \cdot x + 0 \cdot x^2 + 1 \cdot x^3 + 0 \cdot x^4 + \cdots$.

Additional Problems, p. 374

1. (a) $\frac{3}{2}$; (b) $\frac{27}{2}$; (c) $\sin \theta/(2 - \sin \theta)$; (d) $1/(1 + x^2)$.

3. (a) $\displaystyle\sum_{n=1}^{\infty} (-1)^{n+1} \frac{n}{2n + 1}$.

17. (a) 1; (b) $2e - 3$; (c) $e + 1$.

CHAPTER 14

Section 14.2, p. 383

1. (a) D; (b) C, 0; (c) C, 0; (d) C, 0; (e) D; (f) C, $\frac{1}{2}$; (g) C, 0; (h) C, $\frac{1}{2}$; (i) C, 0; (j) D; (k) C, 0; (l) C, 0; (m) D; (n) C, 0; (o) C, π; (p) C, $-\frac{1}{4}$.

5. (a) $a/2$; (b) $4a^3$.

9. A decreasing sequence of positive numbers converges.

Section 14.3, p. 392

3. (a) $|x| < 1$, $ax/(1 - x^2)$; (b) $|x| > 1$, $1/(x - 1)$; (c) $|1 + x| > 1$, $1 + x$ (also, if $x = 0$ the sum is 0); (d) $e^{-1} < x < e$, $\ln x/(1 - \ln x)$.

5. $x < 0$.

Section 14.4, p. 396

1. (a) D; (b) C; (c) C; (d) C; (e) C; (f) D; (g) C; (h) C.

3. D.

5. D.

7. D.

9. D.

11. D.

13. C.

15. C.

17. D.

19. C.

21. C if $p > 1$, D if $p \leq 1$.

23. C.

25. C if $p > 1$, D if $p \leq 1$.

29. ln 2.

Section 14.5, p. 402

1. C.

3. D.

5. C.

7. C.

Section 14.6, p. 406

1. C.

3. D.

5. C.

7. C.

9. D.

11. C.

13. D.

17. D.

19. C.

21. D.

23. C.

Section 14.7, p. 411

1. CC.

3. D.

5. AC.

7. AC.

9. D.

11. CC.

13. CC.

15. AC.

17. AC.

19. AC.

21. CC.

23. D.

25. CC.

27. (a) F; (b) T; (c) F; (d) F; (e) T; (f) F.

Section 14.8, p. 417

1. $(-4, 4)$.

3. $R = 0$.

5. $[-1, 1)$.

7. $[-1, 1]$.

9. $R = 0$.

11. $(-\sqrt{3}, \sqrt{3})$.

13. $[-1, 1]$.

15. $(-\frac{1}{2}, \frac{1}{2}]$.

17. $[-1, 1)$.

19. $[-1, 1]$.

21. $(2, 6)$.

23. $R = \infty$.

25. $R = 0$.

27. $(0, 2e)$.

29. (a) $R = 1$; (b) $R = \infty$.

Section 14.9, p. 423

1. (a) $\Sigma (-1)^{n+1} n x^{n-1}, |x| < 1$;

(b) $\Sigma (-1)^n \dfrac{(n+2)(n+1)}{2} x^n$,

$|x| < 1$.

3. (a) $\dfrac{1}{2} \ln \dfrac{1+x}{1-x}$;

(b) $f(x) = (e^x - 1)/x$ if $x \neq 0$,

$f(0) = 1$; (c) $\dfrac{x}{(1-x)^2}$; (d) $\dfrac{x}{(1-x^2)^2}$.

Section 14.10, p. 429

15. (a) $x - \dfrac{x^3}{3 \cdot 3!} + \dfrac{x^5}{5 \cdot 5!} - \cdots$;

(b) $x + \frac{1}{8}x^4 - \frac{1}{56}x^7 + \cdots$;

(c) $x - \frac{1}{10}x^5 + \frac{1}{24}x^9 - \cdots$.

Section 14.11, p. 436

1. $x + x^2 + \frac{1}{3}x^3 - \frac{1}{30}x^5$
$\qquad - \frac{1}{90}x^6 + \cdots$.

31. $x + \frac{1}{3}x^3 + \frac{2}{15}x^5 + \frac{17}{315}x^7 + \cdots$;
$\qquad R = \pi/2$.

35. (a) $\frac{1}{2}$; (b) $\frac{1}{6}$.

37. 272.

Additional Problems, p. 440

1. (a) 0; (b) $\frac{1}{2}$; (c) 0; (d) 1.

5. (a) $\frac{1}{4}$; (b) $\frac{1}{16}$; (c) $\frac{1}{256}$.

7. $x_n = \dfrac{A^n - B^n}{\sqrt{5}}$, where A and B are the positive and negative roots of $x^2 - x - 1 = 0$.

11. $\dfrac{1 + \sqrt{1 + 4a}}{2}$.

23. $|x| > \sqrt{2}$.

25. $s_n = \dfrac{x(1 - x^n)}{(1 - x)^2} - \dfrac{nx^{n+1}}{1 - x}$.

31. (a) $-\ln 2$; (b) 1.

35. (a) C; (b) D; (c) C; (d) D; (e) C; (f) D; (g) C; (h) D; (i) C; (j) D; (k) C; (l) C; (m) C; (n) C; (o) C; (p) D; (q) C; (r) C; (s) C; (t) D; (u) C; (v) C.

51. C.

53. Inconclusive.

55. D.

57. C.

59. D.

67. (a) ∞; (b) ∞; (c) e.

71. $(-1, 1)$.

73. (a) $\displaystyle\int_0^x \dfrac{\tan^{-1} t}{t} \, dt$;

(b) $(1 + x) \ln (1 + x) - x$;

(c) $\dfrac{1 + x}{(1 - x)^3}$; (d) $-\frac{1}{4} \ln (1 - x^4)$;

(e) $\dfrac{x + 4x^2 + x^3}{(1 - x)^4}$; (f) $\dfrac{4 - 3x}{(1 - x)^2}$.

77. $f_2(x) = 2x/(1 - x)^3$.

81. (a) $-\frac{1}{3}$; (b) 0; (c) $-\frac{1}{12}$.

CHAPTER 15

Section 15.2, p. 453

1. (a) circle, center $(1, 3)$ and radius 5; (b) empty set; (c) point $(5, -1)$; (d) circle, center $(8, -6)$ and radius 2; (e) point $(-3, 7)$; (f) empty set.

3. (a) $(-2, -1)$, $(-2, 0)$, $y = -2$; (b) $(3, 1)$, $(5, 1)$, $x = 1$; (c) $(-2, 5)$, $(-2, 1)$, $y = 9$; (d) $(-2, 1)$, $(-5, 1)$, $x = 1$; (e) $(-1, 2)$, $(-1, \frac{9}{4})$, $y = \frac{7}{4}$.

5. $b^2 y = 4hx(b - x)$.

Section 15.3, p. 460

1. (a) $x^2/25 + y^2/21 = 1$; (b) $x^2/36 + y^2/52 = 1$; (c) $4x^2/9 + y^2/4 = 1$; (d) $x^2/16 + y^2/7 = 1$; (e) $x^2/27 + y^2/36 = 1$; (f) $24x^2/2500 + y^2/100 = 1$.

3. (a) $(0, 0)$, $(0, \pm 5)$, $(0, \pm 4)$, $e = \frac{4}{5}$; (b) $(0, 0)$, $(\pm 2, 0)$, $(\pm \sqrt{3}, 0)$, $e = \sqrt{3}/2$; (c) $(-2, 1)$, $(-2, 1 \pm \sqrt{2})$, $(-2, 2)$ and $(-2, 0)$, $e = \sqrt{2}/2$; (d) $(1, 0)$, $(2, 0)$ and $(0, 0)$, $(1 \pm \frac{1}{2}\sqrt{3}, 0)$, $e = \sqrt{3}/2$; (e) $(2, -1)$, $(5, -1)$ and $(-1, -1)$, $(2 \pm \sqrt{5}, -1)$,

$e = \sqrt{5}/3$; (f) $(0, 2)$, $(\pm 2\sqrt{2}, 2)$, $(\pm 2, 2)$, $e = \sqrt{2}/2$.

7. (a) $\frac{4}{3}\pi a b^2$; (b) $\frac{4}{3}\pi a^2 b$.

13. $\left(\pm \dfrac{a}{d} \sqrt{d^2 - b^2}, \dfrac{b^2}{d} \right)$.

17. $\frac{5}{39}$.

21. $y = mx \pm \sqrt{b^2 + a^2 m^2}$.

Section 15.4, p. 469

1. $(\pm 2, 0)$, $(\pm\sqrt{13}, 0)$, $2y = \pm 3x$, $e = \sqrt{13}/2$, $x = \pm 4/\sqrt{13}$.

3. $(0, \pm 2)$, $(0, \pm\sqrt{13})$, $3y = \pm 2x$, $e = \sqrt{13}/2$, $y = \pm 4/\sqrt{13}$.

5. $(0, \pm 2)$, $(0, \pm 2\sqrt{5})$, $2y = \pm x$, $e = \sqrt{5}$, $y = \pm 2/\sqrt{5}$.

7. $(0, \pm 1)$, $(0, \pm\sqrt{2})$, $y = \pm x$, $e = \sqrt{2}$, $\sqrt{2}y = \pm x$.

9. $y^2/9 - x^2/16 = 1$.

11. $x^2/9 - y^2/36 = 1$.

13. $x^2/36 - y^2/28 = 1$.

15. $x^2/36 - y^2/45 = 1$.

17. Hyperbola with center $(1, -2)$ and horizontal principal axis.

19. Two straight lines $5(y + 1) = \pm 6(x + 2)$.

31. $\left(\pm \dfrac{a}{d} \sqrt{b^2 + d^2}, -\dfrac{b^2}{d} \right)$.

Section 15.6, p. 476

1. $\theta = 45°$, $x'^2/4 + y'^2 = 1$, ellipse.

3. $\theta = 30°$, $y'^2/2 - x'^2/2 = 1$, hyperbola.

5. $\theta = 45°$, $x'^2 = 4\sqrt{2}y'$, parabola.

7. $\theta = 45°$, $x'^2/2 + y'^2/4 = 1$, ellipse.

9. $\theta = 60°$, $x'^2/3 + y'^2/11 = 1$, ellipse.

11. $\theta = \sin^{-1} 1/\sqrt{10}$, $x'^2 + 3y'^2 = 1$, ellipse.

Additional Problems, p. 477

1. $4p$.

9. $\frac{4}{3}\pi(2 + \sqrt{17})$ ft.³.

CHAPTER 16

Section 16.1, p. 482

1. (a) $(\sqrt{2}, \sqrt{2})$; (b) $(2, -2\sqrt{3})$; (c) $(0, 0)$; (d) $(\frac{1}{2}\sqrt{3}, \frac{1}{2})$; (e) $(0, -2)$; (f) $(-2\sqrt{2}, 2\sqrt{2})$; (g) $(-3, 0)$; (h) $(-3\sqrt{2}, 3\sqrt{2})$; (i) $(1, 0)$; (j) $(0, 0)$; (k) $(1, \sqrt{3})$; (l) $(5, 12)$; (m) $(-2\sqrt{3}, 2)$; (n) $(0, 3)$.

3. $(1, 0)$, $(1, 2\pi/5)$, $(1, 4\pi/5)$, $(1, 6\pi/5)$, $(1, 8\pi/5)$.

7. $(x - 2)^2 + (y - 2)^2 = 8$; circle

with center (2, 2) and radius $2\sqrt{2}$.
9. (a) line $y = 2$; (b) line $x = 4$; (c) line $y = -3$; (d) line $x = -2$.

Section 16.2, p. 486

5. (a) $r = 5 \sec \theta$; (b) $r = -3 \csc \theta$; (c) $r = 3$; (d) $r^2 = 9 \sec 2\theta$; (e) $r = \tan \theta \sec \theta$; (f) $r^2 = 2 \csc 2\theta$; (g) $r = \sin^2\theta/(\cos \theta \cos 2\theta)$; (h) $r = (2 \cos \theta)/(\tan^2 \theta - 1)$.
9. $(1, 2\pi/3)$ and $(1, 4\pi/3)$; $-\frac{1}{2}$.
11. $r = a \sin 2\theta$.
13. $x^3 = y^2(2a - x)$.
15. $(x - a)^2(x^2 + y^2) = b^2x^2$.

Section 16.3, p. 493

1. $9 = r^2 + 16 - 8r \cos (\theta - \pi/6)$.
3. $r = 10 \cos \theta$.
5. $r = 4\sqrt{3} \cos (\theta - \pi/3)$.
7. $r = a(1 + \cos \theta)$.
9. (a) $\sqrt{5}a, \sqrt{3}a$; (b) $\frac{1}{2}\sqrt{5}a, \frac{1}{2}\sqrt{3}a$.
11. $r = ep/(1 + e \sin \theta)$.
13. $e = \frac{1}{2}$.
15. $e = \frac{1}{6}$.
17. $\pm\sqrt{e^2 - 1}$.
19. (a) $\left(\dfrac{ep}{1 - e}, 0\right)$; (b) $\left(\dfrac{ep}{1 + e}, \pi\right)$.
21. (a) $x = y \cot \dfrac{\pi y}{2a}$;

(b) $r = \dfrac{2a}{\pi} \theta \csc \theta$.

Section 16.4, p. 498

5. $\frac{32}{5}\pi a^2$.
11. (a) $4\pi a^2$; (b) $4\pi^2 a^2$.
17. $\sqrt{2}$.

Section 16.5, p. 502

1. $\frac{3}{2}\pi a^2$.
5. $\pi/4$.
7. $\pi + 3\sqrt{3}$.
9. $a^2\left(\dfrac{7\pi}{12} - \sqrt{3}\right)$.

Additional Problems, p. 503

1. (a) $\tan \theta = 4$; (b) $r^2 = 36/(4 + 5 \sin^2 \theta)$; (c) $r = 2 \cos \theta - 4 \sin \theta$; (d) $r = 3/(2 \cos \theta - 5 \sin \theta)$; (e) $r = 4\cot \theta \csc \theta$; (f) $r = 1 + 4 \sin \theta$; (g) $r = 6 \sin 2\theta/(\sin^3 \theta + \cos^3 \theta)$.
3. (a) $(\sqrt{2}a, \pi/4)$; (b) the origin and $\left(\dfrac{2 - \sqrt{2}}{2} a, 3\pi/4\right)$,

$\left(\dfrac{2 + \sqrt{2}}{2} a, -\pi/4\right)$; (c) $(a/2, \pm \pi/6)$; (d) $(3\sqrt{2}, \pi/4), (3\sqrt{2}, 3\pi/4)$; (e) the origin and $(\frac{1}{2}, 2\pi/3), (\frac{1}{2}, 4\pi/3)$; (f) $(\pm a, \pi/6), (\pm a, -\pi/6)$; (g) the origin and $(8a/5, \sin^{-1} \frac{3}{5})$; (h) $(4, -\pi/3)$; (i) $(\pm 2, \pi/2)$; (j) $(2, \pm\pi/3), (-1, \pi)$; (k) the origin and $(\frac{3}{2}, \pm\pi/3)$; (l) the origin and $\left(\dfrac{2 + \sqrt{2}}{2} a, \pi/4\right)$,

$\left(\dfrac{2 - \sqrt{2}}{2} a, 5\pi/4\right)$; (m) the origin and $(\pm a, \pi/4), (\pm a, 3\pi/4)$.
15. Larger angle $= 2\pi/3$.
17. $\frac{1}{3}a[(4 + 4\pi^2)^{3/2} - 8]$.
21. $\sqrt{3} - \frac{1}{3}\pi$.
25. $\frac{1}{3}a^2(3\sqrt{3} - \pi)$.

CHAPTER 17

Section 17.1, p. 510

1. (a) $x + y = 2$; (b) $2x - y = -4$.
3. $x + y = 3$.
5. $x - 1 = (y - 3)^2$.
7. $x^2 - y^2 = 1$.
9. $y = 1 - 2x^2$.
11. No; the second is part of the first.
13. (c) $45°$.
15. $x = a \cos \theta + a\theta \sin \theta$, $y = a \sin \theta - a\theta \cos \theta$.

Section 17.2, p. 519

1. $a \sin^{-1} \dfrac{\sqrt{2ay - y^2}}{a} = \sqrt{2ay - y^2} + x$.
7. $6a$.
11. $x = 2b \cos \theta + b \cos 2\theta$, $y = 2b \sin \theta - b \sin 2\theta$; $\frac{16}{3}a$.

Section 17.3, p. 525

1. (a) $\sqrt{10}, 13\mathbf{i} - 34\mathbf{j}, 4\mathbf{i} - 13\mathbf{j}$;
(b) $\sqrt{53}, -36\mathbf{i} - 4\mathbf{j}, -39\mathbf{i} + 14\mathbf{j}$;
(c) $6, 10\mathbf{i} - 33\mathbf{j}, -2\mathbf{i} - 33\mathbf{j}$; (d) $\sqrt{34}, -\mathbf{i} + 55\mathbf{j}, 20\mathbf{i} + 22\mathbf{j}$.
5. (a) $\pm\dfrac{(3\mathbf{i} - 4\mathbf{j})}{5}$;

(b) $\pm\dfrac{(-5\mathbf{i} + 12\mathbf{j})}{13}$; (c) $\pm\dfrac{(5\mathbf{i} - 7\mathbf{j})}{\sqrt{74}}$;

(d) $\pm\dfrac{(24\mathbf{i} - 7\mathbf{j})}{25}$.
7. $\pm2(12\mathbf{i} + 5\mathbf{j})$.
9. $\dfrac{\mathbf{A}}{|\mathbf{A}|} + \dfrac{\mathbf{B}}{|\mathbf{B}|}$.
13. $\theta = 45°$.

Section 17.4, p. 531

1. The line through the head of $\mathbf{A}$ which is parallel to $\mathbf{B}$.
5. $2t \mathbf{i} + \mathbf{j}, 2\mathbf{i}, \sqrt{4t^2 + 1}$.
7. $\mathbf{i} + (3t^2 - 3)\mathbf{j}, 6t \mathbf{j}$, $\sqrt{1 + 9(t^2 - 1)^2}$.
9. $\sec^2 t \, \mathbf{i} + \sec t \tan t \, \mathbf{j}, 2 \sec^2 t \cdot \tan t \, \mathbf{i} + (\sec^3 t + \sec t \tan^2 t)\mathbf{j}$, $|\sec t|\sqrt{2 \sec^2 t - 1}$.
11. $\mathbf{R} = \frac{1}{2}at^2 \, \mathbf{j} + \mathbf{v}_0 t + \mathbf{R}_0$.

Section 17.5, p. 535

1. (a) $-2/(1 + 4x)^{3/2}$; (b) $\cos x$; (c) $2x^3/(2x^4 - 2x^2 + 1)^{3/2}$; (d) $-\frac{1}{4}\sqrt{2}e^{-t}$; (e) $-4t^2/(4t^4 + 1)^{3/2}$.
3. (a) 1; (b) $\frac{1}{18} 5^{5/4}6^{1/2}$; (c) none.
5. If (as usual) s increases on the circle in the counterclockwise direction, then k as calculated from (4) has the wrong sign on the upper half-circle, because s increases in the direction of decreasing x. Change the sign of this result to get $1/a$ on both halves of the circle.
7. $(-\frac{1}{2} \ln 2, \frac{1}{2}\sqrt{2}), \frac{2}{3}\sqrt{3}$.
9. $\frac{3}{2}a$ at $\theta = \pi/4$.
11. $4a \sin \frac{1}{2}\theta$.

Section 17.6, p. 540

3. $(-a\omega \sin \omega t)\mathbf{i} + (a\omega \cos \omega t)\mathbf{j}$, $(-a\omega^2 \cos \omega t)\mathbf{i} + (-a\omega^2 \sin \omega t)\mathbf{j}$, $a\omega, 0, a\omega^2$.
5. $(-e^t \sin t + e^t \cos t)\mathbf{i} + (e^t \cos t + e^t \sin t)\mathbf{j}, (-2e^t \sin t)\mathbf{i} + (2e^t \cos t)\mathbf{j}, \sqrt{2} e^t, \sqrt{2} e^t, \sqrt{2} e^t$.
7. $\dfrac{4t}{t^2 + 1}\mathbf{i} + \dfrac{2t^2 - 2}{t^2 + 1}\mathbf{j}$, $\dfrac{4(1 - t^2)}{(t^2 + 1)^2}\mathbf{i} + \dfrac{8t}{(t^2 + 1)^2}\mathbf{j}, 2, 0, 4/(t^2 + 1)$.
9. $a_n = 0$ when $t = 0$, $a_n = -1$ when $t = \pi/2$.
11. $v > 30\sqrt{2}$.

Section 17.7, p. 547

1. Since $M = \left(\dfrac{4\pi^2}{G}\right)\left(\dfrac{a^3}{T^2}\right)$, determine the ratio a^3/T^2 for any particular planet (for instance, the earth) and proceed with the arithmetic.

Additional Problems, p. 548

1. $3a^2/2$.
3. (a) $5\pi^2a^3$; (b) $\frac{64}{3}\pi a^2$.

5. $8b + \dfrac{8b^2}{a}$.

11. Smallest radius $= 9/(7\sqrt[6]{28})$, at $t = \pm \sqrt[6]{\frac{2}{7}}$.

13. Approximately 5.94×10^{27} grams.

CHAPTER 18

Section 18.1, p. 553
1. Faces: $x = 1$, $y = 4$, $z = 5$. Edges: $x = 1$, $y = 4$; $y = 4$, $z = 5$; $x = 1$, $z = 5$.
3. 64.
5. (a) The yz-plane and the xz-plane taken together; (b) all three coordinate planes taken together.
7. $(0, 4, 0)$.
9. $x^2 + y^2 + (z - 7)^2 = 49$ or $x^2 + y^2 + z^2 = 14z$.
11. (a) The sphere with center $(-1, 3, 5)$ and radius 3; (b) the point $(5, -1, 3)$; (c) the empty set; (d) the point $(-1, 7, 3)$; (e) the sphere with center $(2, -3, 0)$ and radius $\frac{1}{2}$.
15. $\frac{49}{2}$.
17. $\frac{1}{3}(7\mathbf{i} + 5\mathbf{j} + 4\mathbf{k})$.
19. $\frac{1}{4}(\mathbf{A} + \mathbf{B} + \mathbf{C} + \mathbf{D})$.

Section 18.2, p. 557
3. (a) $60°$; (b) $45°$; (c) $90°$.
7. No.
13. $2\mathbf{i} - \mathbf{k}$.
19. Two; $45°$ and $135°$.
21. $c(z_1 - z_2)$.

Section 18.3, p. 564
1. (a) $14\mathbf{j} + 7\mathbf{k}$; (b) $3\mathbf{i} - 3\mathbf{j}$; (c) $2\mathbf{i} - 14\mathbf{j} - 22\mathbf{k}$; (d) $\mathbf{k}$.
3. $2\sqrt{6}$.
5. Assuming that their tails coincide, all three vectors lie in a plane.
11. (b) $11/\sqrt{107}$.

Section 18.4, p. 570
1. (a) T; (b) F; (c) T; (d) T; (e) F; (f) T; (g) F; (h) F.
3. They are parallel.
5. (a) $x = 2 + t$, $y = -1 + 4t$, $z = -3 - 2t$; (b) $x = 2 + 3t$, $y = -1 - t$, $z = -3 + 6t$;

(c) $x = 2 + 2t$, $y = -1 - 2t$, $z = -3 + 5t$.
7. (a) $x = 2 + 3t$, $y = -3t$, $z = 3 - 2t$; (b) $x = 4 + 4t$, $y = 2$, $z = -1$.
9. $(\frac{3}{2}, 1, \frac{5}{2})$.
11. 6.
13. $2x - y - z + 2 = 0$.
17. $8/\sqrt{21}$.
19. $(9, 0, 0)$.
21. $\dfrac{x - 2}{17} = \dfrac{y + 1}{-2} = \dfrac{z}{-7}$.
23. The second plane.
25. $4x + 3y + 4z + 2 = 0$.
29. (a) $\frac{8}{21}$; (b) 0.

Section 18.5, p. 575
1. Parabolic cylinder.
5. Plane.
9. $x^2 + (z - a)^2 = a^2$.
11. (a) $z = e^{-(x^2+y^2)}$; (b) $x^2 + z^2 = e^{-2y^2}$.
13. (a) $y = x^2 + z^2$; (b) $9(x^2 + z^2) + 4y^2 = 36$; (c) $z = 4 - x^2 - y^2$; (d) $x = y^2 + z^2$.
15. $(x - 2z)^2 + (y - 3z)^2 - 6(x - 2z) = 0$.

Section 18.6, p. 579
1. Ellipsoid.
3. Circular paraboloid.
5. Hyperboloid of two sheets.
7. Hyperbolic paraboloid.
9. Hyperboloid of two sheets.
11. Ellipsoid.
13. Hyperbolic paraboloid.
15. $(6, -2, 2)$, $(3, 4, -2)$.
17. (a) $A(k) = \pi ab \left(1 - \dfrac{k^2}{c^2} \right)$; (b) $\frac{4}{3}\pi abc$.
23. Hyperbolic paraboloid.

Section 18.7, p. 582
1. (a) $(2\sqrt{2}, \pi/4, -1)$; (b) $(2, -\pi/3, 7)$; (c) $(2\sqrt{3}, \pi/6, 2)$; (d) $(3\sqrt{5}, \tan^{-1} 2, 5)$.
3. (a) $(2\sqrt{2}, \pi/6, \pi/4)$; (b) $(2\sqrt{2}, 5\pi/6, -\pi/4)$; (c) $(2, \pi/4, \pi/4)$; (d) $(2, \pi/6, -\pi/2)$.
5. $r^2 + z^2 = 16$.
7. $r^2 = z^2$.
9. $r = 2 \sin \theta$.
11. $r = 3$.
13. $\rho = 4$.

15. $\rho = 6 \cos \phi$.
17. $\rho^2 \sin^2 \phi + \rho \cos \phi = 4$.

CHAPTER 19

Section 19.1, p. 588
1. The entire plane except the line $y = 2x$.
3. The first and third quadrants, including the axes.
5. The part of the plane above the line $y = 3x$.
7. All of xyz-space except the origin.
9. The solid sphere $x^2 + y^2 + z^2 \le 16$.
11. All of xyz-space for which $z > 0$ except the planes $z = \pi$, 3π, $\ldots$.
25. Away from the origin.
27. In the positive x, negative y, positive z direction.

Section 19.2, p. 594
1. 2, 3.
3. $-6y^2/(3x + 1)^2$, $4y/(3x + 1)$.
5. $2x \sin y$, $x^2 \cos y$.
7. $\tan 2y + 3y \sec^2 3x$, $2x \sec^2 2y + \tan 3x$.
9. $-3 \sin (3x - y)$, $\sin (3x - y)$.
11. $e^x \sin y$, $e^x \cos y$.
13. $2e^y/x$, $e^y \ln x^2$.
15. $2xy^5z^7$, $5x^2y^4z^7$, $7x^2y^5z^6$.
17. $\ln \dfrac{y}{z}$, $\dfrac{x}{y}$, $-\dfrac{x}{z}$.
19. (a) $y = 3$, $z = 8x + 1$; (b) $x = 2$, $z = 6y - 1$.
31. $f(x, y) = 3xy^2 - x^2 \cos y + 2y$.

Section 19.3, p. 598
1. $z - 25 = 20(x - 1) + 40(y - 2)$.
3. $z = 4x + 5y$.
5. $z - 1 = 6(x - 3) - 8(y - 2)$.
7. $z - 1 = y$.
9. $10x + 13y + 13z = 75$.
13. $x_0 x/a^2 + y_0 y/b^2 + z_0 z/c^2 = 1$.
15. $60°$.
17. The origin.
19. $h(1 + a)/\sqrt{a}$.

Section 19.5, p. 606
1. (a) $8\mathbf{i} + 4\mathbf{j} + 2\mathbf{k}$; (b) $2\mathbf{i}$; (c) $\frac{2}{3}(\mathbf{i} + 2\mathbf{j} - 2\mathbf{k})$; (d) $\frac{1}{3}(-\mathbf{i} + 2\mathbf{j} + \frac{2}{3}\mathbf{k})$.
3. (a) $3, -\mathbf{j}$; (b) $\sqrt{3}$, $\mathbf{i} + \mathbf{j} + \mathbf{k}$; (c) $2\sqrt{19}$, $\mathbf{i} + 3\mathbf{j} + 3\mathbf{k}$;

(d) $3e^2$, $\mathbf{i} + 2\mathbf{j} + 2\mathbf{k}$.

5. $\pm(\mathbf{i} - \mathbf{j} - 2\mathbf{k})/\sqrt{6}$.

7. $28/\sqrt{21}$; $\mathbf{i} + \mathbf{j} + 2\mathbf{k}$; $4\sqrt{6}$.

Section 19.6, p. 612

1. 0.

3. $-9(t^2 + 9)/(t^2 - 9)^2$.

5. $4t^3 + 4tu^2$, $4u^3 + 4t^2u$.

Section 19.7, p. 616

1. $(3, 5)$, a minimum.

3. $(1, 2)$, a minimum; $(-1, 2)$, a saddle point.

5. $(-1, -2)$ and $(-2, 8)$, both saddle points.

7. $(0, 0)$, a saddle point; $(-1, -1)$, a maximum.

13. 2, 4, 6.

15. $1/(27abc)$.

17. $\frac{1}{8}$.

19. $\sqrt{14}$.

21. $x/6 + y/6 + z/3 = 1$.

23. $3\sqrt{3}/2$.

27. Base of rectangle $= (2 - \sqrt{3})P$; height of rectangle $= \frac{1}{6}(3 - \sqrt{3})P$; height of triangle $= \frac{1}{6}(2\sqrt{3} - 3)P$.

Section 19.8, p. 623

1. $\frac{1}{2}$.

3. Corner in first quadrant is $(\sqrt{2}, \frac{1}{2}\sqrt{2})$.

5. $2r = h$.

7. $\frac{3}{2}$, $\frac{1}{2}$.

9. $x^2/3 + y^2/12 + z^2/27 = 1$.

11. $\frac{6}{11}$.

13. a^3.

17. (a) $|d|/\sqrt{a^2 + b^2 + c^2}$.

Section 19.9, p. 628

7. $w = c_1 x + c_2$.

Section 19.10, p. 634

1. $3x/y$.

3. $(1 - \sin y)/(x \cos y - 1)$.

5. $(ye^{xy} - 2y^2)/(4xy - xe^{xy})$.

7. $3z/(z - 1)$, $2z/(1 - z)$.

9. $(xyz \cos xz + y \sin xz)/(1 - x^2 y \cos xz)$, $(x \sin xz)/(1 - x^2 y \cos xz)$.

11. $\sqrt{\frac{24}{19}}$.

CHAPTER 20

Section 20.1, p. 638

1. The triangle bounded by $x = 0$, $y = 1$, $y = x$.

3. $\frac{3}{10}$.

5. 98.

7. $\pi/8$.

9. $\frac{14}{3}$.

11. π.

13. $\frac{1}{6} \ln 2$.

15. $\displaystyle\int_0^1 \int_0^x f(x, y)\, dy\, dx$.

17. $\displaystyle\int_e^{e^2} \int_1^{\ln x} f(x, y)\, dy\, dx$.

19. $\displaystyle\int_0^1 \int_0^{x^2} 2x^3\, dy\, dx = \frac{1}{3}$.

21. $\displaystyle\int_0^1 \int_0^2 (5 - 2x - y)\, dx\, dy = 5$.

23. $\displaystyle\int_{-1}^2 \int_{3x-2}^{4x-x^2} dy\, dx = \frac{9}{2}$.

25. $\frac{1}{6} abc$.

27. 4.

29. $\frac{4}{35}\pi a^3$.

Section 20.2, p. 644

1. $\frac{9}{2}$.

3. a^2.

5. $1 - e^{-a}$.

7. $\frac{8}{3}$.

9. $\frac{625}{12}$.

11. $\frac{25}{2}$.

13. $\frac{40}{3}$.

15. $\frac{3}{4}\pi$.

17. $\frac{1}{3}(b^3 - a^3)$.

Section 20.3, p. 648

1. $M = a^3$; $\bar{x} = \bar{y} = \frac{7}{12}a$.

3. $M = \frac{128}{5}$; $\bar{x} = \frac{20}{7}$, $\bar{y} = 0$.

5. $M = \frac{2}{3}a^3$; $\bar{x} = \frac{3}{16}\pi a$, $\bar{y} = 0$.

7. $M = \pi$; $\bar{x} = (\pi^2 - 4)/\pi$, $\bar{y} = \pi/8$.

9. $\frac{1}{3}Ma^2$.

11. $\frac{1}{6}Ma^2$.

13. $\frac{1}{3}Ma^2$.

Section 20.4, p. 652

1. πa^2.

3. $\frac{1}{6}(4\pi - 3\sqrt{3})a^2$.

5. a^2.

7. $\frac{1}{3}(3\sqrt{3} - \pi)a^2$.

9. $\frac{1}{4}(8 + \pi)a^2$.

11. $\frac{1}{2}(9\sqrt{3} - 2\pi)a^2$.

13. $\frac{1}{16}(9\sqrt{3} + 8\pi)$.

15. $\frac{3}{4}\pi a^4$.

17. $\displaystyle\int_{-\pi/2}^{\pi/2} \int_0^3 z\, r\, dr\, d\theta$.

19. $\displaystyle\int_0^{\pi/4} \int_0^{\tan\theta \sec\theta} z\, r\, dr\, d\theta$.

21. $\displaystyle\int_0^{\pi/4} \int_0^1 z\, r\, dr\, d\theta$.

23. (a) $\frac{4}{3}\pi[a^3 - (a^2 - b^2)^{3/2}]$;

(b) $\frac{4}{3}\pi(a^2 - b^2)^{3/2} = \frac{1}{6}\pi h^3$.

25. $\bar{x} = \frac{5}{6}a$, $\bar{y} = 0$.

27. $\bar{x} = 0$, $\bar{y} = \dfrac{4}{3\pi}a$.

29. $\frac{4}{3}\pi a^3$.

31. $\frac{1}{2}\pi a^4 b$; $\bar{x} = \bar{y} = 0$, $\bar{z} = \frac{1}{3}a^2 b$.

33. $\frac{2}{3}Ma^2$.

35. $Ma^2 \left(\dfrac{1 - \ln 2}{\ln 2} \right)$.

37. $\frac{1}{4}Ma^2 \left(1 - \dfrac{\sin 2\alpha}{2\alpha} \right)$.

Section 20.5, p. 658

1. $\frac{1}{24}$.

3. $2abc/\pi$.

5. 24.

7. 4π.

9. $\frac{64}{15}$.

11. $\displaystyle\int_0^a \int_z^a \int_y^a f(x, y, z)\, dx\, dy\, dz$.

13. $a^4/8$.

15. $\displaystyle\int_0^1 \int_{-\sqrt{1-z}}^{\sqrt{1-z}} \int_{-\sqrt{1-y^2-z}}^{\sqrt{1-y^2-z}} f(x, y, z)\, dx\, dy\, dz$.

17. 4.

19. 27π.

21. $\frac{4}{3}\pi abc$.

23. $\frac{1}{6} abc$.

27. $\frac{2}{3}Ma^2$.

Section 20.6, p. 661

1. $\pi/2$.

3. $\bar{x} = \bar{y} = 0$, $\bar{z} = \frac{1}{4}h$.

5. $\frac{1}{5}M(2a^2 + 3b^2)$.

7. $\frac{3}{20}M(a^2 + 4h^2)$.

9. $\bar{x} = \bar{y} = 0$, $\bar{z} = \frac{3}{8}a$.

11. $\frac{4}{3}(8 - 3\sqrt{3})\pi a^3$.

13. $\pi/32$.

15. $a^3/3$.

17. $\frac{16}{9}a^3(3\pi - 4)$.

19. $\frac{3}{5}Ma^2$.

21. $\frac{7}{5}Ma^2$.

23. $\frac{2}{3}\pi a^3(1 - \cos\alpha)$.

Section 20.7, p. 667

1. $\delta \cdot \frac{2}{3}\pi a^5(\frac{2}{3} - \cos\alpha + \frac{1}{3}\cos^3\alpha) = \frac{2}{3}Ma^2[\frac{2}{3} - \frac{1}{3}\cos\alpha\,(1 + \cos\alpha)]$.

3. $2\pi^2a^3$.

5. $\dfrac{3}{8}\left(\dfrac{a^4-b^4}{a^3-b^3}\right)$.

7. $\frac{2}{3}\alpha a^3$.

11. $\dfrac{c8\pi a^{n+5}}{3(n+5)}$.

13. $\frac{8}{3}\pi a^3$.

15. (a) $M=\displaystyle\int_0^a\int_0^\pi\int_0^{2\pi}\delta(\rho,\phi,\theta)$ $\rho^2\sin\phi\,d\theta\,d\phi\,d\rho$;

(b) $M=4\pi\displaystyle\int_0^a\rho^2f(\rho)\,d\rho$.

19. $\pi Gm\delta a\sin^2\alpha$.

21. $2\pi Gm\delta$.

Section 20.8, p. 670

1. $3\sqrt{14}$.
3. $\pi a^2\sqrt{3}$.
5. $2\pi\sqrt{6}$.
7. $a^2(\pi-2)$.
9. $\frac{1}{6}\pi(5\sqrt{5}-1)$.
11. $\frac{1}{6}\pi a^2(5\sqrt{5}-1)$.
15. $\frac{1}{12}a[3\sqrt{10}+\ln(3+\sqrt{10})]$.
17. $2a^2$.
19. $\frac{4}{9}(20-3\pi)$.

CHAPTER 21

Section 21.1, p. 682

1. (a) -2; (b) $-\frac{8}{3}$; (c) -4.
3. 0.
7. (a) $\frac{2}{3}$; (b) $\frac{29}{30}$; (c) $\frac{1}{3}$; (d) $\frac{31}{28}$.
9. (a) 0; (b) 0; (c) $-\frac{2}{3}$.
11. $\frac{4}{3}$ for all paths.
13. π for both paths.
15. $\frac{105}{2}$.
19. 0 along all paths.
21. 0.

Section 21.2, p. 689

9. 0.
11. 6.
13. e.
15. 12.

Section 21.3, p. 696

1. $\frac{1}{12}$.
3. $2\ln 4-\frac{15}{4}$.
5. $\frac{3}{2}$.
7. 1.
9. $\frac{1}{6}$.
11. $\frac{3}{10}$.
13. $\frac{1}{2}$.
15. $3\pi a^2$.

17. $\frac{3}{8}\pi a^2$.
19. 2.
21. $\frac{3}{2}a^2$.
23. xy^3.
25. $e^{xy}-x^2+y^2$.
27. $x\sin y+y\cos x$.
31. 2π.

APPENDIX A

Section 1, p. 704

2. Because $(4n+2)^2$ is a multiple of 4.
6. If $\sqrt[n]{m}=a/b$ where a and b are positive integers with no common factor >1, then $a^n=mb^n$ and every prime factor of b must be a prime factor of a (why?), so $b=1$.
7. All n's except 1, 10, 10^2, 10^3, $\ldots$.
8. Assume that x is a rational number a/b (with $b>0$) in lowest terms, and show that $b=1$ as in Problem 6.

Section 14, p. 731

2. (c) $100=4\cdot25=10\cdot10$.
3. (a) If N is not prime, then since $(4m-1)(4n-1)=4(4mn-m-n)+1$, there is no guarantee that N has a prime factor of the form $4n+1$.

Section 22, p. 758

1. (a) 0; (b) 0; (c) 2; (d) $e^z\sin x$; (e) $2/r$.
2. πa^2b.
3. $4\pi abc$.
4. $\frac{1}{2}abc$.
5. $\frac{12}{5}\pi a^5$.
6. 0.
7. $2\dfrac{f(r)}{r}+f'(r)$.
8. (a) $4\pi a^3$; (b) 4π.
11. (a) 36; (b) 18.

Section 23, p. 762

5. 32.
6. π.
7. $-\pi$.
8. -1.
9. -6π.
10. 4π.
11. 27.
12. 18π.

13. $-1/2$.
14. -8π.
15. $-\pi ab$.
16. 0.
17. $\frac{3}{4}\pi$.

APPENDIX C

Section 10, p. 880

2. (a) C; (b) D; (c) D; (d) C; (e) C.
3. (a) $k\geq3$; (b) all k; (c) all k.
4. (a) $b-a>1$; (b) $b-a>2$.

Section 11, p. 885

4. No.
6. $\ln 2$.

Section 12, p. 888

3. C if $x\neq\frac{2}{3}k\pi$, D if $x=\frac{2}{3}k\pi$.
8. By Abel's test.

APPENDIX D

Section 1, p. 901

1. (a) $\dfrac{9!}{4!}$; (b) $\dfrac{22!}{16!}$; (c) $\dfrac{52!}{5!47!}$.
2. (a) 336; (b) 990; (c) 455; (d) $12{,}650$; (e) 462; (f) 2520.
3. 720; 120.
4. 720.
5. 2880.
6. 720.
7. $9\cdot9!=3{,}265{,}920$.
8. $9\cdot P(9,4)=27{,}216$; $9\cdot10^4=90{,}000$.
9. $140{,}400{,}000$.
10. $2(3!4!)=288$; $5(3!4!)=720$.
11. 120; 60; 325.
12. 720.
13. $20{,}160$.
14. $2\cdot5\cdot4!=240$; $[P(6,2)-10]\cdot4!=480$.
15. $2(3!3!)=72$.
16. (a) 4950; (b) 455; (c) 252.
17. 210.
18. $125{,}970$.
19. 378.
20. 66.
21. 4200.
22. 495; 165; 330.
23. 8820.
24. (a) 210; (b) 378.
25. $211{,}680$.
26. $50{,}400$.

27. $4\binom{13}{5} = 5148$;

$\left[13\binom{4}{3}\right]\left[12\binom{4}{2}\right] = 3744$.

28. 66.

29. 286.

30. $\binom{m-k}{2} + k(m-k) + 1 =$

$\frac{1}{2}m(m-1) - \frac{1}{2}k(k-1) + 1$.

31. 84.

32. 280.

33. $\binom{n}{2} - n$.

34. (a) $128x^7 - 448x^6y + 672x^5y^2 - 560x^4y^3 + 280x^3y^4 -$

$84x^2y^5 + 14xy^6 - y^7$; (b) $729a^6 + 2916a^5b + 4860a^4b^2 + 4320a^3b^3 + 2160a^2b^4 + 576ab^5 + 64b^6$;

(c) $x^{10}y^5 - 15x^8y^4z^3 + 90x^6y^3z^6 - 270x^4y^2z^9 + 405x^2yz^{12} - 243z^{15}$.

35. (a) $349,440x^{11}$;

(b) $-489,888x^4$; (c) $-2002a^{10}b^{27}$.

Section 2, p. 908

1. (a) $n(n+1)$; (b) $n(4n+1)$;

(c) $4n^2$; (d) $3n^2$; (e) $\frac{1}{2}n(5n+1)$.

2. $\frac{1}{3}n(4n^2 - 1)$.

4. (f) This induction starts at $n = 0$. Notice that if condition I is replaced by "$S(n_0)$ is true," and if II is still satisfied as it stands, then in just the same way a slightly general-

ized principle of mathematical induction guarantees that $S(n)$ is true for all $n \geq n_0$.

6. (a) $1^2 - 2^2 + 3^2 - \cdots + (-1)^{n+1}n^2 = (-1)^{n+1}(1 + 2 +$

$\cdots + n)$; (b) $\left(1 - \frac{1}{2}\right)\left(1 - \frac{1}{3}\right)$

$\cdots \left(1 - \frac{1}{n}\right) = \frac{1}{n}$, $n \geq 2$.

7. (a) $\frac{n+1}{2n}$, $n \geq 2$; (b) $1 - x^{2^{n+1}}$,

$n \geq 0$.

8. (b) $S(1)$ is not true, so the method of induction is not applicable. The assertion is unfounded.

INDEX

Abel, N. H., 839–840, 399*n*., 485*n*., 765, 837, 838, 843
  on Gauss's mathematical writing, 833
  partial summation formula, 886
  series, 399*n*.
  test, 888
Abscissa, 8
Absolute convergence, 881
Absolute value, 4
Absolutely convergent series, 409
Absorption, Lambert's law of, 229
Acceleration, 54, 145, 528
  angular, 327
  due to gravity, 146, 148
Adams, J. C., 547
Adamson, I. T., 735*n*.
Addition formulas, 243
Adiabatic gas law, 126, 204, 230
Age of rocks, 227
Agnesi, M., 512*n*.
Air resistance, 232
Alexander, H. G., 818*n*.
Algebra, fundamental theorem of, 439
Algebraic area, 173
Algebraic function, 26
Algebraic number, 734
Algebraic number theory, 725
Algebraic volume, 641
Alternating series, 407
Alternating series test, 407
Amplitude, 266
Analysis, 727
Analytic geometry, 15
Analytic number theory, 731, 740
Andrews, G. E., 826*n*.
Angular acceleration, 327
Angular velocity, 327
Antibiotic, 237
Antiderivative, 136
Apollonius, 449
Apostol, T. M., 731*n*.

Aquinas, empty doctrines of, 783
Arago, F., 823
Arc, 188
Archimedes, 776–782, 154, 156, 166, 449, 454, 502*n*., 661, 704, 727*n*., 764
  law of the lever, 318
  principle in hydrostatics, 204, 271
  spiral of, 491, 504, 780
  on surface area of sphere, 195, 669
  on volume of sphere, 184, 709–711
Area, 155
  algebraic, 173
  of circle, 155
  of curved surface, 668
  geometric, 173
  of parabolic segment, 156
Areas, problem of, 39
Aristarchus, 781
Aristotle, 769*n*., 770, 798, 810, 811
  on Hippocrates' misfortune, 706*n*.
  on man's desire to know, 817
  on Plato's Forms, 770*n*.
  worthless physics of, 783
Arithmetic mean, 7
Artin, E., 782*n*.
Astroid (*see* Hypocycloid, of four cusps)
Asymptote, 30
Aubrey, John, 772*n*.
Augustine, Saint, 769*n*., 786, 849
Average velocity, 51
Axis of parabola, 18, 452

Ball, W. W. Rouse, 806*n*.
Barrow, Isaac, 171, 804
Beckmann, P., 368
Bell, E. T., 769*n*., 795*n*., 836*n*.
  on Euler, 825
  on Fermat, 790
  on Laplace, 830

Bentley, Richard, 807
Bernoulli, James, 821–823, 134, 485*n*., 518, 717, 718, 764, 805, 820, 824
Bernoulli, John, 821–823, 134, 335, 424*n*., 516, 518, 764, 801*n*., 805, 807, 820
  on Euler's feats, 825
  solution of brachistochrone problem, 746–748
  on solving the catenary problem, 717–718
Bernoulli numbers, 742
Bessel, F. W., 836, 837
Bessel equation, 356
Bessel function, 356, 418, 423
Binomial coefficients, 899
Binomial series, 373, 429, 446
Binomial theorem, 63, 82, 353, 897
Black holes, 151
Bochner, S., 829, 844
Boeotians, shrieks of, 837
Bolyai, Johann, 837
Bolyai, Wolfgang, 837
Boole, George, 815
Boundedness theorem, 858
Boyer, C. B., 824*n*.
Boyle's law, 113, 204
Brachistochrone problem, 516–518, 807
  solution of, 746–748
Bracket symbol, 169
Brahe, Tycho, 541
Bray, Vicar of, 830
Bridge problem, Königsberg, 826
Brouncker, Lord, 720*n*.
Brouncker's continued fraction, 825
Brouwer's fixed point theorem, 862
Brown, P. L., 460*n*.
Brun, Viggo, 728
Buck, R.C., 615*n*.
Buffon's needle problem, 254
Butterfield, H., 801*n*.

Cairns, H., 812n.
Cajori, F., 821n., 824n.
Calculus:
    differential, 154
    fundamental theorem of, 40, 154,
        169, 171, 173n., 178, 683, 870
        for line integrals, 684
    infinitesimal, 134
    integral, 154
Calder, N., 460n.
Cardioid, 483, 486
Cartesian coordinates, 15
Cartesian plane, 15
Cassegrain telescope, 468
Cassini, G. D., 490n.
    ovals of, 490
Catenary, 290, 716
Cauchy, A. L., 838, 398n., 406, 765,
    796, 839, 840
Cauchy condensation test, 444
Cauchy integral test, 398n.
Cauchy product of two series, 883
Cavalieri, B., 166, 794
Cavendish, Henry, 547, 549
Cayley, Arthur, 699
Center:
    of curvature, 535, 748
    of gravity, 319
    of mass, 319–320, 322, 646, 655
    of population, 320
Central force, 543
Centripetal force, 529
Centroid, 322, 646, 655
Century of Genius, 783
Chain rule, 71, 74, 607–608
Champollion, J., 831
Chebyshev, P., 309, 729
Child, J. M., 803n., 819n.
Christina, queen of Sweden, 788
Churchill, Winston, 819n.
Cicero, Marcus Tullius, 778
Circle, 16
    center of, 451
    of curvature, 535
    equation of, 451
    radius of, 451
Circulation, 759
Circulation density, 759
Cissoid, 487, 503
Clark, G. N., 801n.
Clarke, A. A., 833n.
Clarke, Arthur C., 459n., 539
Clarke, Samuel, 818
Closed curve, 681
Cobb-Douglas production function, 622
Coefficient of friction, 272
Cohen, I. Bernard, 804n., 806n.
Colbert, J. B., 801
Cole, F. N., 790

Combination, 899
Common logarithm, 211
Comparison test, 393
    for improper integrals, 348
Completing the square, 17, 291
Complex numbers, 438
Composite function, 71
Composite number, 725
Concave down, 92
Concave up, 92
Conchoid, 487, 503, 504
Conditional convergence, 410, 881
Cone, 448
    elliptic, 577
Conic section, 448
Conservation of energy, law of, 203,
    687
Conservative field, 687, 761
Constant of integration, 137
Constraint, 618
Construction problems, 485
Continuity, uniform, 869n., 894n.
Continuous function, 58, 585, 857
Continuous vector, 526
Continuously compounded interest, 215
Contour map, 586
Convergence:
    absolute, 881
    conditional, 410, 881
    interval of, 364, 415, 417
    of series, 359, 385
    unconditional, 885
    uniform, 419
        for power series, 889
Convergent sequence, 379
Coolidge, J. L., 720n.
Cooling, Newton's law of, 229
Coordinate plane, 8
Coordinate system, rectangular, 7
Coordinates:
    Cartesian, 15
    cylindrical, 580, 659
    polar, 479
    rectangular, 551
    spherical, 581, 662
Copernicus, Nicolas, 540
Cornford, F. M., 770n.
Corridor problem, 102, 259
Cosine, 106
    series, 370, 376
    Taylor series for, 426
Cosines, law of, 246, 555, 558
Coulomb's law, 204
Counterexample, 7
Courant, R., 439n., 731n., 735n., 773n.,
    828n.
Couturat, L., 816n.
Crelle, A. L., 839, 840
Critical point, 88, 614

Critical value, 88
Cross product, 559
Curl, 753
Curvature, 532
    center of, 535, 748
    circle of, 535
    radius of, 535
Curve:
    closed, 681
    level, 586
    piecewise smooth, 680
    simple, 691
    smooth, 191, 530
Curvilinear motion, 145
Cycloid, 512, 790
    evolute of, 750
Cylinder, 572
Cylindrical coordinates, 580, 659

Decay:
    exponential, 226
    radioactive, 225
Decimal, 392
    infinite, 353
    repeating, 360
Declaration of Independence, 775
Decomposition into partial fractions,
    294
Decreasing function, 87
Dedekind, R., 834
Definite integral, 156, 162, 866, 868
Del operator, 605, 677, 753
Delta notation, 43
Deltoid, 520
Demand, elasticity of, 120
Demand curve, 117
Demand function, 117
Democritus, 195, 709, 842
Dependent variable, 22
Derivative, 46
    directional, 601
    partial, 589
        second-order, 592
    second, 80
    of vector, 526
Descartes, René, 783–788, 15, 514, 520,
    764, 789, 791, 797
    dismay of, 512n., 794
    folium of, 512, 548, 634, 696, 794n.
    on Constantijn Huygens, 800
    refuted by Huygens, 801
    on Snell's law, 793
Determinants, 561
Differentiable function, 46, 600
Differential, 129, 600
    partial, 608
    total, 600n., 608
Differential calculus, 154

Differential equations, 142
  general solution of, 142
  infinite series in, 354–356
  order of, 142
  particular solution of, 142
Differential triangle, 189
Differentiation, 62
  implicit, 76, 629
  logarithmic, 223
Diffusion equation, 625n.
Dijksterhuis, E. J., 776n.
Diogenes Laertius, 773n.
Diophantus, 790, 795, 796, 826
Direction angles, 558
Direction cosines, 558
Directional derivative, 601
Directrix, 18
  of ellipse, 457
  of hyperbola, 465
  of parabola, 18, 451
Dirichlet, P. G. L., 841, 765, 796, 831,
  844
Dirichlet's rearrangement theorem, 882
Dirichlet's test, 886
Dirichlet's theorem on primes, 727
Disk method, 183
Distance formula, 9, 552
Distance from point to line, 36, 125
Divergence:
  of series, 359
  of vector field, 753, 754
Divergence theorem, 756
Dobell, C., 803n.
Domain, 22, 584
Dot product, 554
Double-angle formulas, 243
Double integral, 639
Doubly ruled surface, 579n.
Dummy variable, 174
Dunnington, G. W., 834n.

e, 208, 212–214, 220
  definition of, 871
  irrationality of, 391
  series for, 389
  transcendence of, 737
e^x, power series for, 355, 372–373, 420,
  425
Eccentricity:
  of ellipse, 455
  of hyperbola, 464
Eddington, Arthur, 770n., 876n.
Edwards, H. M., 308n., 731n., 796n.
Einstein, Albert, 146n.
  on elegance, 618n.
  equation, 712
  law of motion, 711
  on Newton, 804

Einstein, Albert (Cont.):
  on Riemannian geometry, 846
Elasticity of demand, 120
Electric circuit, 229
Element:
  of arc length, 189
  of area, 179
  of force, 197
  of volume, 181
  of work, 199
Elementary function, 276
Eliot, T. S., vi
Ellipse, 448, 449, 454
  area of, 176, 460
  center of, 455
  directrix of, 457
  eccentricity of, 455
  equation of, 456
  focal radius of, 457
  focus of, 455
  latus rectum of, 477
  major axis of, 455
  minor axis of, 455
  semimajor axis of, 455
  semiminor axis of, 455
  string property of, 455
  vertex of, 455
Ellipsoid, 576
Elliptic cone, 577
Elliptic integral, 306n., 461, 548
  of the first kind, 309
Elliptic paraboloid, 578
Energy:
  kinetic, 201, 275, 329
  law of conservation of, 203, 687
  potential, 202–203, 275, 687, 695n.
  total, 203
Epicycloid, 514
Equals sign, 821n.
Equation(s):
  differential (see Differential
    equations)
  diffusion, 625n.
  general linear, 13
  heat, 624, 625
  intercept, 15
  Laplace's, 594, 624, 625
  of line, 11
  parametric, 506
    of line, 566
  of plane, 567
  point-slope, 12
  slope-intercept, 13
  of a sphere, 553
  of state, 612
  wave, 594, 624, 627
Equiangular spiral, 498
Eratosthenes, 709, 727n., 776, 781
  sieve of, 727

Escape velocity, 150, 153
Euclid, 771–775, 35, 706, 764, 779, 796
  lemma, 731
  theorem: on perfect numbers, 701,
    702, 790
    on primes, 726, 740
Euler, L., 823–828, 134, 740, 764, 795,
  796, 805, 833, 838, 886n.
  characteristic, 828
  constant, 400, 402, 445, 825
  discovery of e, 221
  elegant truths of, 742
  expansion of cotangent, 744
  formula, 437, 439, 722, 723, 751
    for $\Sigma\, 1/n^2$, 373, 648, 822
    for $\Sigma\, 1/n^{2k}$, 745
    for polyhedra, 827
  generosity to Lagrange, 829
  identity, 740, 847
  infinite product for sine, 723, 825
  irrationality of e, 391
  quoted: on fabric of world, 107
    on mystery of primes, 728n.
  theorem: on homogeneous functions,
    609
    on perfect numbers, 701, 702
Even function, 38, 434, 435
Even perfect numbers, theorem of, 701
Evolute, 535, 748
  of the cycloid, 750
Exponential decay, 226
Exponential function:
  general, 208, 209
  the, 213
Exponential growth, 216
Exponential spiral, 498
Exponents, fractional, 77
Extreme value theorem, 860
Eynden, C. V., 742

Factorial, 355
Factorial notation, 81
Fadiman, Clifton, 796n.
Fechner-Weber law, 237
Fermat, P., 790–797, 15n., 40, 101n.,
  107, 154, 166, 171, 708, 764, 789,
  801, 826
  concept of tangent, 41
  and Descartes, 512n., 791
  last theorem, 511n., 796, 876n.
  and Pascal, 794–795
  principle of least time, 107, 746
Fermi, Enrico, 611n.
Fibonacci, 440n.
Fibonacci sequence, 440
Field (algebraic), 852
  ordered, 852
    complete, 852

Field (in vector analysis):
  conservative, 687, 761
  force, 677
  gradient, 677
  irrotational, 761
  scalar, 677
  vector, 677
First law of motion, Newton's, 146*n.*
Fixed point, 862
Fixed point theorem, Brouwer's, 862
Flux, 754, 756
Focus, 18
  of ellipse, 455
  of hyperbola, 462
  of parabola, 18, 451
Folium of Descartes, 512, 548, 634, 696, 794*n.*
Fontenelle, B., 810*n.*, 818
Force:
  central, 543
  centripetal, 529
  of gravity, 146
Force field, 677
Four squares theorem, 700
Four vertex theorem, 536*n.*
Fourier, J., 831, 625, 764
Fractional exponents, 77
Francis, Frank C., 814*n.*
Frankfurt, H. G., 816*n.*
Franklin, P., 640*n.*
Frequency, 267
Function, 22, 25
  algebraic, 26
  Bessel, 356, 418, 423
  composite, 71
  continuous, 58, 585, 857
  decreasing, 87
  differentiable, 46, 600
  domain of, 584
  elementary, 276
  even and odd, 38, 434–435
  exponential, 213
  gamma, 347, 825
  general exponential, 208, 209
  general logarithm, 208, 210
  homogeneous, 608
    Euler's theorem on, 609
  hyperbolic, 271
  implicit, 76, 631
  increasing, 87
  integrable, 163, 866, 868
  inverse, 263, 633
  limit of, 55
  linear, 24
  periodic, 241
  quadratic, 24
  rational, 25–26, 293
    improper, 293
    proper, 293

Function *(Cont.)*:
  transcendental, 26
  trigonometric, 240–241
    inverse, 260–261
  zero of, 29
  zeta, 444, 740
Fundamental lemma, 600
Fundamental theorem of algebra, 439
Fundamental theorem of calculus, 40, 154, 169, 171, 173*n.*, 178, 683, 870
  for line integrals, 684

Galileo, 1, 449, 514, 717, 801
Galton, Francis, 822*n.*
Gamma function, 347, 825
Gardner, Martin, 763
Gas law:
  adiabatic, 126, 204, 230
  ideal, 590
Gauss, C. F., 832–838, 485*n.*, 499*n.*, 765, 796, 823, 839
  and Abel, 840
  and Dirichlet, 841
  and Jacobi, 837–838
  on lawyers, 833
  on prime number theorem, 728, 731
  on regular polygons, 773, 832
  and Riemann, 844–846
Gauss's test, 878
Gauss's theorem on divergence, 756
Gauss, Helen W., 832*n.*
Gelfond, A.O., 737, 842
General solution, 142, 252, 269, 272
Generalized mean value theorem, 335*n.*, 865
Geometric area, 173
Geometric mean, 7
Geometric progression, 357
Geometric series, 347*n.*, 357, 386
Geometric volume, 641
Geometry, analytic, 15
George I, king of England, 819
Gibbon, Edward, 783, 814
Giesy, D. P., 723
Goldstine, H. H., 813*n.*
Gradient, 602, 753
Gradient field, 677
Graph, 16, 22
Graph theory, 827
Gravitation, Newton's law of, 149, 151, 200, 541, 665
Gravitational attraction of spherical shell, 664
Gravity:
  acceleration due to, 146, 148
  center of, 319
  force of, 146
"Great Sulk," 809

Greek navy, 777*n.*
Green, George, 691, 836*n.*
Green's theorem, 691
Growth:
  exponential, 216
  population, 224
    inhibited, 230–232
Growth curve, inhibited or sigmoid, 232

Hadamard, J., 729, 847, 848
Half-angle formulas, 243, 284
Half-life, 226
Hall, A. R., 813*n.*
Hall, M. B., 813*n.*
Halley, Edmund, 460, 805, 806, 829
Halley's Comet, 460, 468
Halmos, P. R., 363*n.*
Hamilton, W., 107, 829
Hardy, G. H., 515*n.*, 703*n.*, 770*n.*, 776, 795*n.*, 826*n.*
Harmonic series, 359, 387
Harvey, William, 787
Hawkins, David, 731*n.*
Hazard, P., 800*n.*
Heaslet, M. A., 795*n.*
Heat equation, 624, 625
Heath, T. L., 491*n.*, 709*n.*, 770*n.*, 776*n.*, 779*n.*, 780*n.*, 781, 782*n.*
Hegel, G. W., 834
Heiberg, J. L., 709
Heilbroner, R. L., 118*n.*
Heisenberg, W., 843
Henderson, J. M., 610*n.*
Hermite, C., 843, 734, 737, 765, 839, 840, 842
Heron of Alexandria, 106
Heron's formula, 122
Hieron, king of Syracuse, 776–777
Hilbert, David, 737, 782, 796, 836, 841
Hille, E., 852*n.*
Hiltebeitel, A., 835*n.*
Hippocrates of Chios, 157, 503*n.*, 706
  lune of, 157, 706–707
Hobbes, Thomas, 720*n.*, 772*n.*, 789, 797, 807
Hofmann, J. E., 803*n.*, 813*n.*, 819*n.*
Hölder's inequality, 624
Holmes, Oliver Wendell, Jr., 783, 849
Homogeneous function, 608
  Euler's theorem on, 609
Hooke, Robert, 805, 806
Hooke's law, 199
Hume, David, 810
Hutchinson, G. Evelyn, 758
Huygens, Christiaan, 800–804, 518, 717, 750, 764, 787, 812, 813, 816, 819
  and Leibniz, 802–803, 812, 813, 819
  and tautochrone, 518–519

Huygens, Constantijn, 800
Hydrostatics, 196
Hyperbola, 448, 450, 462
    asymptote of, 464
    branch of, 463
    center of, 463
    conjugate axis of, 463
    directrix of, 465
    eccentricity of, 464
    focus of, 462
    latus rectum of, 478
    principal axis of, 463
    rectangular, 466
    reflection property of, 467
    vertex of, 463
Hyperbolic functions, 271
Hyperbolic paraboloid, 578
Hyperbolic spiral, 492
Hyperboloid:
    of one sheet, 577
    of two sheets, 577
Hypergeometric series, 418, 879
Hypocycloid, 514
    of four cusps, 85, 191, 192, 195, 515, 696

i, 437
I-beam, 330
Ideal gas law, 590
Ideal lot problem, 119
Implicit differentiation, 76, 629
Implicit function, 76, 631
Implicit function theorem, 631, 632
Improper integral, 343, 345
    comparison test for, 348
    convergent, 344, 346, 397
    divergent, 344, 346, 397
Inclination, 12
Increasing function, 87
Increment, 43
Indefinite integral, 137, 156
Independent variable, 22
Indeterminate form, 332, 338, 339
Index of refraction, 107
Induction, mathematical, 440n., 798, 903
Inequalities, 3
Inequality:
    Hölder's, 624
    Schwarz's, 125, 624
Inertia:
    moment of, 328, 646, 655
    polar moment of, 647
Infinite decimal, 353
Infinite discontinuity, 31
Infinite series (see Series)
Infinitesimal, 133
Infinitesimal calculus, 134

Infinity, 5
Inflection, point of, 92
Inhibited growth curve, 232
Inhibited population growth, 230
Initial condition, 143
Instantaneous velocity, 52
Integers, 2
    positive, 2
Integrable function, 163, 866, 868
Integral, 137
    definite, 156, 162, 866, 868
    double, 639
    elliptic, 306n., 461, 548
        of the first kind, 309
    improper, 343, 345
        comparison test for, 348
        convergent, 344, 346, 397
        divergent, 344, 346, 397
    indefinite, 137, 156
    iterated, 636
    line, 677
        fundamental theorem of calculus
            for, 684
    Riemann, 164, 831
    triple, 654
Integral calculus, 154
Integral sign, 137, 163
Integral test, 398
Integrand, 137, 163
Integration, 137
    constant of, 137
    Leibniz approach to, 179, 321, 645
    limits of, 163
    by parts, 300
    by substitution, 140
    variable of, 141, 163
Intercept equation, 15
Interest, continuously compounded, 215
Intermediate value theorem, 861
Interval:
    closed, 4
    of convergence, 364, 415, 417
    open, 4
Inverse function, 263, 633
Inverse problem of tangents, 128
Inverse substitution, 872
Inverse tangent series, 367
Inverse trigonometric functions, 260–261
Involute, 750
    of circle, 512
Irrational numbers, 2, 732
Irrotational field, 761
Iterated integral, 636

Jacobi, C. G. J., 673n., 837, 838, 844
Jacobian, 673, 674
James, D. E., 610n.

Jefferson, Thomas, on Plato's Republic, 771n.
Jones, Burton W., 735n.
Jones, David E. H., 772n.

Kac, M., 715n.
Kant, Immanuel, 775, 810, 837
Kayser, R., 814n.
Kelvin, Lord (W. Thomson), 652n.
Kepler, Johannes, 449, 459, 490n., 541
    first law, 546
    laws, 806
    problem of, 124
    second law, 501, 544
    third law, 546
Keynes, J. M., 118
Kinetic energy, 201, 275, 329
Klein, F., 834n.
Kneale, M., 815n.
Kneale, W., 815n.
Knopp, K., 434n., 742, 813n.
Königsberg bridge problem, 826
Koyré, A., 818n.
Kronecker, L., 832
Kummer, E. E., 876n.
Kummer's theorem, 876
Kuzmin, R., 737

Lagrange, J. L., 829, 134, 764, 796
Lagrange multiplier, 619
Lambert's law of absorption, 229
Lanczos, C., 801n., 835n.
Landau, E., 853n.
Laplace, P. S., 830–831, 691n., 764, 823
Laplace transform, 347
Laplace's equation, 594, 624, 625
Law:
    of conservation of energy, 203, 687
    of cosines, 246, 555, 558
    of gravitation, Newton's, 149, 151, 200, 541, 665
    ideal gas, 590
    Kepler's, 806
        first, 546
        second, 501, 544
        third, 546
    of mass action, 229
    of motion: Einstein's, 711
        Newton's second, 81, 146, 201, 327, 528, 711
    of reflection, 105–106
Lawyers, 833
Least squares, method of, 617
Least time, Fermat's principle of, 107, 746

Least upper bound, 851
Least upper bound axiom, 852
Leeuwenhoek, Antony van, 800, 803, 814
Legendre, A. M., 671*n.*, 796, 833, 844
Legendre's formula, 671
Legendre polynomials, 447
Lehmer, D. N., 728*n.*, 790
Leibniz, G. W., 810–821, 40, 128, 137, 154, 167, 171, 424*n.*, 518, 717, 764, 799, 801*n.*, 805, 822
  alternating series test, 407
  approach to calculus, 179, 321, 498, 645
  conversations with Huygens, 802–803
  formula: for differentiating integral, 594, 613
    for π/4, 264–265, 304, 366, 715–716, 720–721
  notation, 72, 130, 163, 189, 264
  notation for derivatives, 49
  quarrel with Newton, 424*n.*, 808–809
  series, 264–265, 304, 366, 715–716, 720–721
  use of differentials, 131–135
Leibnizian myth, 133
Lemniscate, 36, 484
Length of vector, 552
Level curve, 586
Level surface, 587
Leverrier, U. J., 547
Lewis, J. H. (*see* Montaigne's cube)
L'Hospital, G. F. A. de, 822
L'Hospital's rule, 335
Libby, W., 227, 228*n.*
Limaçon, 483, 486
Limit:
  of a function, 55
  of a sequence, 378
Limit comparison test, 394
Limit-of-sums process, 178
Limits of integration, 163
Lin Yutang, 786*n.*
Lindemann, F., 735*n.*, 737, 842, 843
Line:
  equation of, 11
  parametric equations of, 566
  symmetric equations of, 566
Line integral, 677
  fundamental theorem of calculus for, 684
Linear equation, 13
Linear function, 24
Liouville, J., 306, 765, 842
Liouville's theorem, 735
Little-oh notation, 400
Lobachevsky, N. I., 290, 837
Locke, John, 775, 807
Locus, 20

Loemker, L. E., 810*n.*, 811*n.*, 813*n.*, 814*n.*, 816*n.*, 818*n.*
Logarithm, 210
  common, 211
  natural, 217
Logarithm function, general, 208, 210
Logarithm series, 366
Logarithmic differentiation, 223
Long division, 294, 353, 432
Lost illusions, 775
Louis XIV, boil on the face of Europe, 812
Lowell, J. R., 765
Lower sum, 161
Lune of Hippocrates, 157, 706–707

Machin, J., 368
Maclaurin's series, 428
Mahan, Admiral A. T., 810, 812*n.*
Manuel, Frank E., 809*n.*
Marcellus, 776–778
Marginal cost, 54, 116
Mass, 146, 646, 654
  center of, 319–320, 322, 646, 655
Mass action, law of, 229
Mathematical induction, 440*n.*, 798, 903
Maximum value, 88
  at endpoints, cusps, and corners, 90
Maxwell, James Clerk, 828, 836
Mead, D. G., 306*n.*
Mean:
  arithmetic, 7
  geometric, 7
Mean value theorem, 189*n.*, 333, 864
  generalized, 335*n.*, 865
Median, 10
Mersenne, Marin, 789–790, 514, 703, 764, 785, 791, 794, 801
Mersenne primes, 703, 790
Mertens, F., 886*n.*
Method:
  of exhaustion, 154
  of least squares, 617
  of moving slices, 183
  of partial fractions, 293
  of substitution, 279
Midpoint formulas, 9
Mill, John Stuart, 810
Millikan, R. A., 230*n.*
Minimum value, 88
  at endpoints, cusps, and corners, 90
Mixing, 233
Moment, 318–319, 321, 646, 655, 710
  of inertia, 328, 646, 655
    polar, 647
Momentum, 711
Montaigne's cube (*see* Lewis, J. H.)
More, L. T., 792*n.*

Morehead, J., 835*n.*
Morgan, J. P., quoted, 356
Morse, P. M., 611*n.*
Mortimer, Ernest, 797*n.*, 800*n.*
Motion:
  curvilinear, 145
  Einstein's law of, 711
  Newton's second law of, 81, 146, 201, 327, 528, 711
  rectilinear, 145
  simple harmonic, 266
  uniform circular, 529

Napoleon, 830, 831
Natural logarithm, 217
Needle problem, Buffon's, 254
Neighborhood, 585
Nelson, H., 703*n.*
Nephroid, 520
Neumann, J. von, 363*n.*
Newman, J. R., 547*n.*, 827*n.*
Newton, Isaac, 804–809, 40, 50, 128, 154, 167, 171, 442*n.*, 449, 459, 490*n.*, 518, 540, 542, 547, 548*n.*, 720*n.*, 764, 775, 776, 787, 802, 810–812, 817, 830
  area under quadratrix, 494
  and John Bernoulli, 807
  and Fermat, 792
  first law of motion, 146*n.*
  and Halley, 805–806
  law of cooling, 229
  law of gravitation, 149, 151, 200, 541, 665
  method of approximate solution, 113–116
  quarrel with Leibniz, 424*n.*, 808–809
  second law of motion, 81, 146, 201, 327, 528, 711
  on space and time, 818*n.*
  theorem on attraction of spheres, 666
Newtonian mechanics, 145
Niven, I., 515*n.*, 733*n.*, 734, 737*n.*, 742*n.*
Normal, 67
Notation, factorial, 81
*n*th term test, 360, 388
Number:
  algebraic, 734
  Bernoulli, 742
  complex, 438
  composite, 725
  irrational, 2, 732
  perfect, 700*n.*, 790
    Euclid's theorem on, 702
    Euler's theorem on, 702
  prime, 383, 700*n.*, 725
  rational, 2

Number *(Cont.)*:
   real, 2, 852
   transcendental, 734, 842
Number theory:
   algebraic, 725
   analytic, 731, 740

Oblate spheroid, 184, 461, 576
Odd function, 38, 434, 435
Oldenburg, Henry, 805, 813
Orbit, synchronous, 539
Orbital speed, 539
Orcibal, Jean, 797
Ordinate, 8
Ore, Oystein, 838, 840*n.*
Oscillation, 868
Ovals of Cassini, 490

p-series, 395, 398
Paint paradox, 348
Pappus of Alexandria, 325, 493, 764,
   782
Pappus' theorems, 325–326
Parabola, 18, 448, 450, 451
   axis of, 18, 452
   directrix of, 18, 451
   equation of, 452
   focus of, 18, 451
   latus rectum of, 477
   reflection property of, 67, 453
   vertex of, 18, 452
Paraboloid:
   elliptic, 578
   hyperbolic, 578
Parallel axis theorem, 331
Parallelogram law, 558
Parallelogram rule, 522
Parameter, 506
Parametric equations, 506
   of line, 566
Parthenon, 775
Partial derivative, 589
   second-order, 592
Partial differential, 608
Partial fractions:
   decomposition into, 294
   method of, 293
   theorem, 297*n.*, 873
Partial sums, 357
   sequence of, 359
Particle, 145
Particular solution, 142
Partition, 866
Parts, integration by, 300
Pascal, Blaise, 797–800, 154, 171, 504*n.*,
   764, 801, 813, 819, 907, 908
   and Fermat, 794–795

Pascal, Étienne, 504*n.*, 797
Pasteur, Louis, 803
Path, 680
Pedersen, H., 817*n.*
Pendulum, 269
   period of, 275
Pepys, Samuel, 807
Perfect numbers, 700*n.*, 790
   Euclid's theorem on, 702
   Euler's theorem on, 702
Pericles, 707
Period, 241, 266
Periodic functions, 241
Permutation, 898
pH, 212
Pi (π):
   computation of, 368
   definition of, 155*n.*
   experimental method of calculating,
     256
   irrationality of, 733
   Leibniz's formula for $\pi/4$, 264–265,
     304, 366, 715–716, 720–721
   Wallis's product for $\pi/2$, 304, 441,
     718, 723, 825
Piecewise smooth curve, 680
Plane, equation of, 567
Planetary motion, Kepler's second law
   of, 501
Plato, 765, 770, 771*n.*, 779, 785
Plato's Academy, 773*n.*
Platonic solids, 774
Plutarch, 776*n.*, 777–779
Poincaré, Henri, 843, 849
Point of inflection, 92
Point-slope equation, 12
Poisson's gas equation, 230
Polar coordinates, 479
Polar moment of inertia, 647
Pollock, F., 814*n.*
Polya, G., 798*n.*, 826*n.*, 908
Polygon, regular, 772
Polyhedra:
   Euler's formula for, 827
   regular, 773
Polynomial, 25
   zeros of, 38
Popper, Karl, 771*n.*
Population growth, 224
   inhibited, 230–232
Porges, A., 796*n.*
Potential, 200, 625, 695
Potential energy, 202–203, 275, 687,
   695*n.*
Power rule, 73
Power series, 365, 369, 412, 417
   for cos x, 426
   for e^x, 355, 372–373, 420, 425
   interval of convergence, 364, 415, 417

Power series *(Cont.)*:
   properties, 418
   radius of convergence, 415, 417
   for sin x, 426
   for tan x, 744
   uniform convergence, 889
Powers of belief, 145
Pressure, 196
Prey-predator equations, Volterra's, 237
Prime number, 383, 700*n.*, 725
Prime number theorem, 729
Primes:
   Euclid's theorem on, 726, 740
   Mersenne, 703, 790
   reciprocals of, 741
   twin, 728
Probability, 254
Problem:
   of areas, 39
   of tangents, 39, 128
Proclus, 772
Product:
   cross, 559
   dot, 554
   of numerical series, 883
   of power series, 431
Product rule, 67, 84
Projection:
   scalar, 554
   vector, 555
Prolate spheroid, 184, 461, 576
Properties of power series, 418
Pseudosphere, 290
Psychophysics, 237
Ptolemy, 540, 772
Pythagoras, 765–771, 540, 699, 764, 774
Pythagorean theorem, 8, 35, 782

Quadratic formula, 20
Quadratic function, 24
Quadratrix, 493, 504
Quadrature, 704
Quadric surface, 575
Quandt, R. E., 610*n.*
Quotient rule, 68

Raabe, J. L., 877*n.*
Raabe's test, 877
Rademacher, H., 796*n.*
Radian, 239
Radioactive decay, 225
Radiocarbon dating, 227
Radius, 451
   of convergence, 415, 417
   of curvature, 535
   of gyration, 330
Range, 22

Rate of change, 53, 109
Ratio test, 403
Rational function, 25–26, 293
   improper, 293
   proper, 293
Rational numbers, 2
Rationalizing substitution, 308
Real line, 3
Real number, 2, 852
Reciprocals of primes, 741
Recorde, Robert, 821n.
Rectangular coordinate system, 7
Rectangular coordinates, 551
Rectilinear motion, 145
Reduction formula, 287, 298, 302, 317
Reflection, law of, 105–106
Reflection property:
   of ellipse, 459
   of hyperbola, 467
   of parabola, 67, 453
Refraction:
   index of, 107
   Snell's law of, 106, 746
Region, simple, 641, 649
Regular polygon, 772
Regular polyhedra, 773
Reid, C., 737n.
Rembrandt, 800
Repeated integral (see Iterated integral)
Repeating decimal, 360
Rest mass, 711
Restoring force, 267
Revenue, 118
Richelieu, Cardinal, 789, 797
Richter magnitude, 211
Riemann, B., 844–848, 729, 765, 823
   and complex analysis, 844–845
   and geometry, 845–846
   integral, 164, 831
   and number theory, 847–848
   rearrangement theorem, 882
Ritt, J. F., 306n., 853n.
Robbins, H., 439n., 731n., 735n., 773n., 828n.
Roberval, G., 504n.
Robinson, A., 135n.
Rocket, 237
Rocket propulsion, 713
Roemer, O., 812
Rolle's theorem, 863
Root test, 405
Rose, four-leaved, 485
Rosen, M., 485n.
Rotation of axes, 473
Ruled surface, 579
Ruler-and-compass constructions, 734, 773, 832, 842

Rumor, spread of, 234
Russell, Bertrand, 771, 772n., 815
Ryle, Gilbert, 771n., 816n.

Sabra, A. I., 787n., 794n.
Saddle surface, 578
Santayana, G., vi
Sarton, G., 829n.
Scalar, 520
Scalar field, 677
Scalar projection, 554
Schneider, T., 737
Schrödinger, E., 843
Schwarz, H. A., 670
Schwarz's inequality, 125, 624
Sears, F. W., 611n.
Second derivative, 80
Second derivative test, 94, 615
Second law of motion, Newton's, 81, 146, 201, 327, 528, 711
Second-order reaction, 229
Seidenberg, A., 782n.
Self-evident truths, 775
Separation of variables, 142
Sequence, 377
   bounded, 378
   convergent, 379
   decreasing, 384n.
   increasing, 380
   limit of, 378
   lower bound of, 378
   of partial sums, 359
   unbounded, 378
   upper bound of, 378
Series:
   Abel's, 399n.
   absolutely convergent, 409
   alternating, 407
   binomial, 373, 429, 446
   conditionally convergent, 410, 881
   convergent, 359, 385
   cosine, 370, 376
   divergent, 359, 385
   for e, 389
   geometric, 347n., 357, 386
   harmonic, 359, 387
   hypergeometric, 418, 879
   infinite, 352
   interval of convergence of, 364
   inverse tangent, 367
   Leibniz's, 264–265, 304, 366, 715–716, 720–721
   ln 2, 401
   logarithm, 366
   Maclaurin's, 428
   p-, 395, 398

Series (Cont.):
   partial sums of, 385
   power (see Power series)
   product of, 431
   sine, 369, 376, 426
   sum of, 359, 385
   tangent, 433, 436, 437
   Taylor, 424
      for even and odd functions, 435
   telescopic, 387, 392
Shell method, 185–186
Siegel, C. L., 737n.
Sieve of Eratosthenes, 727
Sigma notation, 157, 352
Sigmoid growth curve, 232
Simmons, G. F., xvin., 347n., 610n., 623, 709n., 776n., 799n., 804n., 825n., 880n.
Simple curve, 691
Simple harmonic motion, 266
Simple region, 641, 649
Simply connected domain, 694
Simply connected region in space, 761n.
Simpson's rule, 311
Sine, 106
   Euler's infinite product for, 723, 825
   series, 369, 376, 426
Sirius, distance to, 803
Skepticism, 72n., 786
Slope, 11
Slope-intercept equation, 13
Slowinski, D., 703n.
Smith, D. E., 518n., 791n., 794n., 813n., 833n., 846n.
Smooth curve, 191, 530
Snell, W., 793
Snell's law of refraction, 106, 746
Solid of revolution, 181
Speed, 52, 145, 527
   orbital, 539
Spengler, Oswald, 810
Sphere:
   equation of, 553
   five-dimensional, 659
   four-dimensional, 659
Spherical coordinates, 581, 662
Spherical shell, gravitational attraction of, 664
Spherical triangle, 671
Spheroid:
   oblate, 184, 461, 576
   prolate, 184, 461, 576
Spinoza, Baruch, 775, 800, 813, 816

Spiral:
  of Archimedes, 491, 504, 780
  equiangular, 498
  exponential, 498
  hyperbolic, 492
Square, completing the, 17, 291
Squeeze theorem, 856
Stirling, J., 442*n.*
Stirling's formula, 442
Stobaeus, 772
Stokes, George, 758
Stokes' theorem, 760
String property of ellipse, 455
Struik, D. J., 791*n.*, 792*n.*, 796*n.*,
  798*n.*, 799*n.*, 802*n.*, 815*n.*, 820*n.*,
  822*n.*
Substitution:
  inverse, 872
  method of, 279
  rationalizing, 308
  trigonometric, 287
Subtraction formulas, 243
Sum:
  lower, 161
  of series, 359
  upper, 162
Surface:
  area of, 668
  doubly ruled, 579*n.*
  integral, 755
  level, 587
  quadric, 575
  of revolution, 192, 573
  ruled, 579
  saddle, 578
Surface integral, 755
Swift, Jonathan, 819*n.*
Symmetric equations of line, 566
Synchronous orbit, 539

Tan *x*, power series for, 744
Tangent:
  inverse problem of, 128
  problem of, 39, 128
  series, 433, 436, 437
Tangent line, 40
Tangent plane, 595
Tautochrone property, 518
Taylor, A. E., 640*n.*
Taylor, B., 424*n.*
  coefficient, 424
  formula, 425
  series, 424
    for even and odd functions,
      435
Telescopic series, 387, 392

Terman, Lewis M., 811*n.*
Terminal velocity, 233
Test:
  Abel's, 888
  alternating series, 407
  Cauchy: condensation, 444
    integral, 398*n.*
  comparison, 393
  Dirichlet's, 886
  Gauss's, 878
  integral, 398
  limit comparison, 394
  *n*th term, 360, 388
  Raabe's, 877
  ratio, 403
  root, 405
  second derivative, 94, 615
Theorem:
  binomial, 63, 82, 353, 897
  Pythagorean, 8, 35, 782
Thompson, S. P., 803*n.*
Throsby, C. D., 610*n.*
Thrust, 714
Tietze, H., 485*n.*, 773*n.*, 788*n.*, 796*n.*,
  832*n.*, 841
Titchmarsh, E. C., 439, 847*n.*
Toeplitz, O., 796*n.*
Topological properties, 828
Torque, 327, 563
Torricelli, Evangelista, 514, 798, 799
Torus, 185, 206, 291, 326
Total differential, 600*n.*, 608
Total energy, 203
Tractrix, 289
Transcendence of *e*, 737
Transcendental function, 26
Transcendental number, 734, 842
Trapezoidal rule, 310
Trigonometric functions, 240–241
  inverse, 260–261
Trigonometric substitution, 287
Triple integral, 654
Truesdell, C., 817*n.*, 828*n.*
Tunnel through earth, 268, 271
Twin primes, 728
Two squares theorem, 700

Ultimate Certainty, 91
Unconditional convergence, 885
Uniform circular motion, 529
Uniform continuity, 869*n.*, 894*n.*
Uniform convergence, 419
  for power series, 889
Unique factorization theorem, 700,
  725
Upper sum, 162

Uspensky, J. V., 795*n.*

Vallée Poussin, C. de la, 729
Value:
  critical, 88
  maximum, 88
  minimum, 88
Vandiver, H. S., 796*n.*
Variable:
  dependent, 22
  dummy, 174
  independent, 22
  of integration, 141, 163
Vavilov, S. I., 808*n.*
Vector, 520
  continuous, 526
  derivative of, 526
  length of, 552
Vector analysis, 525
Vector field, 677
Vector projection, 555
Velocity, 52, 145, 520, 527
  angular, 327
  average, 51
  escape, 150, 153
  instantaneous, 52
  terminal, 233
Vermeer, 800
Vertex of parabola, 18, 452
Very Bad Thing, 610
Vicar of Bray, 830
Vieta, F., 715*n.*
Vieta's formula, 251, 714
Vilenkin, N. Ya., 191*n.*
Voltaire, 776, 786*n.*, 799, 815, 823
Volterra's prey-predator equations, 237
Volume, 640
  algebraic, 641
  geometric, 641

Wallis, John, 718, 720*n.*, 808
Wallis's product, 304, 441, 718, 723, 825
Waltershausen, W. Sartorius von, 832*n.*
Warner, S., 735*n.*
Washer method, 184
Watson, Fletcher G., 459*n.*
Wave equation, 594, 624, 627
Weber, Wilhelm, 836
Weight, 146, 199
Weil, André, 797*n.*
Weil, Simone, vi
Westfall, Richard S., 809*n.*, 817*n.*
Weyl, H., 774*n.*
Whewell, W., 806*n.*

Whispering gallery, 459
Whitehead, A. N., 79, 763, 778, 804n.,
    815, 817, 821n.
Widder, D. V., 670n.
Wiener, N., vi
Wiener, Philip P., 811n.
Wigner, E. P., 843n.
Witch, 511, 512n.
Wolf, A., 816n., 817n.

Work, 198, 557, 676, 678
Wren, Christopher, 514, 548n., 799,
    805, 806
Wright, E. M., 515n., 795n.,
    826n.
Wundt, W., 237

xy-plane, 8

Yahoo, ruling (see George I)

Zeno, 374
Zeros:
    of a function, 29
    of a polynomial, 38
Zeta function, 444, 740
Zeuner, F. E., 227n.

BASIC DIFFERENTIATION FORMULAS

Product Rule $\quad \dfrac{d}{dx}(uv) = u\dfrac{dv}{dx} + v\dfrac{du}{dx}$

Quotient Rule $\quad \dfrac{d}{dx}\left(\dfrac{u}{v}\right) = \dfrac{v\dfrac{du}{dx} - u\dfrac{dv}{dx}}{v^2}$

Chain Rule $\quad \dfrac{dy}{dx} = \dfrac{dy}{du}\dfrac{du}{dx}$

$\dfrac{d}{dx}x^n = nx^{n-1}$ $\qquad\qquad\qquad$ $\dfrac{d}{dx}u^n = nu^{n-1}\dfrac{du}{dx}$

$\dfrac{d}{dx}e^x = e^x$ $\qquad\qquad\qquad$ $\dfrac{d}{dx}e^u = e^u\dfrac{du}{dx}$

$\dfrac{d}{dx}\ln x = \dfrac{1}{x}$ $\qquad\qquad\qquad$ $\dfrac{d}{dx}\ln u = \dfrac{1}{u}\dfrac{du}{dx}$

$\dfrac{d}{dx}\sin x = \cos x$ $\qquad\qquad\qquad$ $\dfrac{d}{dx}\sin u = \cos u\dfrac{du}{dx}$

$\dfrac{d}{dx}\cos x = -\sin x$ $\qquad\qquad\qquad$ $\dfrac{d}{dx}\cos u = -\sin u\dfrac{du}{dx}$

$\dfrac{d}{dx}\tan x = \sec^2 x$ $\qquad\qquad\qquad$ $\dfrac{d}{dx}\tan u = \sec^2 u\dfrac{du}{dx}$

$\dfrac{d}{dx}\cot x = -\csc^2 x$ $\qquad\qquad\qquad$ $\dfrac{d}{dx}\cot u = -\csc^2 u\dfrac{du}{dx}$

$\dfrac{d}{dx}\sec x = \sec x \tan x$ $\qquad\qquad$ $\dfrac{d}{dx}\sec u = \sec u \tan u\dfrac{du}{dx}$

$\dfrac{d}{dx}\csc x = -\csc x \cot x$ $\qquad\qquad$ $\dfrac{d}{dx}\csc u = -\csc u \cot u\dfrac{du}{dx}$

$\dfrac{d}{dx}\sin^{-1} x = \dfrac{1}{\sqrt{1-x^2}}$ $\qquad\qquad$ $\dfrac{d}{dx}\sin^{-1} u = \dfrac{1}{\sqrt{1-u^2}}\dfrac{du}{dx}$

$\dfrac{d}{dx}\tan^{-1} x = \dfrac{1}{1+x^2}$ $\qquad\qquad$ $\dfrac{d}{dx}\tan^{-1} u = \dfrac{1}{1+u^2}\dfrac{du}{dx}$

SERIES

$$\sin x = x - \frac{x^3}{3!} + \frac{x^5}{5!} - \cdots, \qquad \cos x = 1 - \frac{x^2}{2!} + \frac{x^4}{4!} - \cdots$$

$$e^x = 1 + x + \frac{x^2}{2!} + \frac{x^3}{3!} + \cdots$$